EARTH SCIENCE

The Earth, the Atmosphere, and Space

SECOND EDITION

EARTH SCIENCE

The Earth, the Atmosphere, and Space

SECOND EDITION

Stephen Marshak

UNIVERSITY OF ILLINOIS AT URBANA-CHAMPAIGN

Robert Rauber

UNIVERSITY OF ILLINOIS AT URBANA-CHAMPAIGN

W. W. NORTON & COMPANY

Independent Publishers Since 1923

W. W. Norton & Company has been independent since its founding in 1923, when William Warder Norton and Mary D. Herter Norton first published lectures delivered at the People's Institute, the adult education division of New York City's Cooper Union. The firm soon expanded its program beyond the Institute, publishing books by celebrated academics from America and abroad. By midcentury, the two major pillars of Norton's publishing program—trade books and college texts—were firmly established. In the 1950s, the Norton family transferred control of the company to its employees, and today—with a staff of five hundred and hundreds of trade, college, and professional titles published each year—W. W. Norton & Company stands as the largest and oldest publishing house owned wholly by its employees.

Editor: Jake Schindel	**Marketing Manager, Geology:** Katie Sweeney
Senior Project Editor: Thom Foley	**Design Director:** Rubina Yeh
Production Manager: Ashley Horna	**Designer (interior and cover):** Anne DeMarinis
Associate Editor: Rachel Goodman	**Director of College Permissions:** Megan Schindel
Copy Editor: Norma Sims-Roche	**Photo Editor:** Stephanie Romeo
Managing Editor, College: Marian Johnson	**Photo Researcher:** Jane Miller
Managing Editor, College digital media: Kim Yi	**Composition:** MPS North America LLC
Digital Media Editor: Robert Bellinger	**MPS Project Manager:** Jackie Strohl
Associate Media Editor: Arielle Holstein	**Illustrations for the First Edition:** Stan Maddock and Joanne Brummett
Media Project Editor: Marcus van Harpen	**Illustrations for the Second Edition:** Troutt Visual Services
Editorial Assistant, Media: Jasvir Singh	**Manufacturing:** Transcontinental Printing

Permission to use copyrighted material is included in the backmatter of this book.

978-0-393-68066-9

W. W. Norton & Company, Inc., 500 Fifth Avenue, New York, NY 10110
wwnorton.com

W. W. Norton & Company Ltd., 15 Carlisle Street, London W1D 3BS
1 2 3 4 5 6 7 8 9 0

BRIEF CONTENTS

CONTENTS

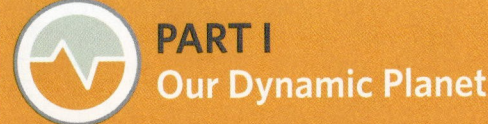

PART I
Our Dynamic Planet

We begin our introduction to geology, the study of the Earth, with a discussion of how the Earth, and the elements that comprise it, formed. We then introduce the theory of Plate Tectonics—the grand, unifying idea that provides a foundation for understanding how and why geologic processes operate. We next examine the materials that make up the solid Earth—minerals (the building blocks of rock), then igneous, sedimentary, and metamorphic rocks. This background allows us to discuss the Earth's mountain building and earthquakes; the Earth's history; and its energy and mineral resources.

Chapter 5

A SURFACE VENEER
Sedimentary Rocks 144

Chapter 6

A PROCESS OF CHANGE
Metamorphism and the Rock Cycle 170

PART II
Ever-Changing Landscapes

Picture the jagged peaks of a mountain range, the grassy plains of a continental interior, the sand of a desert, or the glassy surface of a pond. Clearly, our planet hosts an amazing variety of landscapes. In Part II, we explore the characteristics of these landscapes and the processes that produce them, such as erosion and deposition, the hydrologic cycle, downslope movement, and the impact of humanity. We then conclude with a focus on the Earth's harshest environments, deserts and glacial landscapes.

PART III
Restless Seas

Most of our planet's surface lies hidden beneath a vast global ocean of saltwater, in some places, deeper than mountains are high. Part III focuses on this watery realm. We begin with oceanography, the study of seawater (its composition, and its currents, tides, waves, and life). Then we turn our attention to marine geology (the study of ocean basins and the seafloor). We conclude with a focus on coasts, the boundary between land and sea, and how their landforms take shape and evolve.

Chapter 15

OCEAN WATERS
The Blue of the Blue Marble 452

PART IV
A Blanket of Gas: The Earth's Atmosphere

A thin blanket of gas, the atmosphere, surrounds our planet. Without it, there could be no life on Earth. Part IV explores the properties and behavior of the atmosphere, by describing its composition and character, and its variance with elevation and through time. Next, we turn our attention to clouds, precipitation, wind, and solar radiation, properties that define weather conditions. We also study large weather systems: monsoons, hurricanes, cyclones, and thunderstorms. And we conclude by studying the Earth's climate and why temperature and precipitation vary across the globe and through Earth's history.

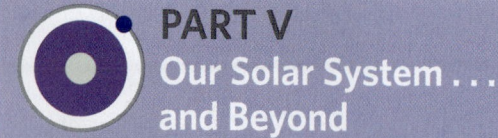

PART V
Our Solar System . . . and Beyond

In Part V, we look into the darkness of deep space. We begin by reviewing the history of astronomy and learn the reasons why planets orbit the Sun, why the Moon has phases, and why eclipses occur. Next, we explore our Solar System, visiting each of the planets, their moons, and other objects that orbit our Sun. We scrutinize the Sun's interior to see what it's made of and how it generates energy, and we learn how stars evolve over time. Finally, we probe deep space, inspecting galaxies, supermassive black holes, and quasars. We conclude with a look at models which address the formation and fate of the Universe.

Chapter 23
THE SUN, THE STARS, AND DEEP SPACE 728

PREFACE

Narrative Themes

Why do earthquakes, volcanoes, floods, and landslides happen? What causes mountains to rise? Does climate change through time? Where's the record of evolution? When did the Earth form and by what process? How can we find valuable metals and where do we drill to find oil? What drives violent storms? Is the size of the Universe constant? Could there be life on other planets or moons? Why are there oceans? The study of Earth science addresses these important questions and many more.

Earth Science, Second Edition, provides an introduction to the study of our planet and to its context in space. It addresses the essence of four different disciplines: geology, oceanography, atmospheric science (including meteorology), and astronomy. As such, this book will help you understand what's beneath your feet, over the horizon, up in the sky, and far beyond the highest clouds.

Because of the diversity of topics covered in *Earth Science,* it's important to keep in mind a set of narrative themes while reading this book. These themes, listed below, serve as the building blocks from which you can construct your personal understanding of our amazing planet and of the Universe around it.

1. Our planet's land, water, atmosphere, and living inhabitants are dynamically interconnected, and materials constantly cycle among various living and nonliving reservoirs on, above, and within the planet. Researchers refer to this complex network of interconnecting features and phenomena as the *Earth System.*

2. The Earth is a planet, formed like many other planets from dust and gas that was once floating in space. But, in contrast to other planets, the Earth is a dynamic place on which new geologic features continue to form and old ones continue to be destroyed. A combination of special conditions, such as the distance from the Sun, chemical composition, the appearance of liquid oceans, and the presence of life, makes our planet truly unique in our Solar System. Quite literally, there's no place like home! While there may be exoplanets (planets orbiting other stars) that resemble the Earth, we'll very likely never interact with them.

3. The Universe, Solar System, and the Earth are all very old. Researchers conclude that the Universe began in the Big Bang about 13.8 billion years ago, that our Solar System originated around 4.57 billion years ago, and that the birth of the Earth took place 4.56–4.54 billion years ago. During this deep time, our planet's surface, subsurface, and atmosphere have changed, and life has evolved.

4. Unlike all other planets or moons in the Solar System, the Earth's outer shell, the lithosphere, consists of about 20 plates. The *theory of plate tectonics* states that these plates slowly move relative to one another so that the map of our planet continuously changes. Plate interactions cause earthquakes and volcanoes, build mountains, shift oceans, provide gases that make up the atmosphere, and affect the distribution of life on Earth.

5. Internal processes (driven by the Earth's internal heat—and manifested by plate motions) and external processes (driven by heat from the Sun) interact at the Earth's surface to produce complex landscapes.

6. Knowledge of the Earth, its oceans, and its atmosphere can help society to understand dangerous natural hazards such as earthquakes, tsunamis, volcanoes, landslides, storms, and floods. In some cases, this knowledge can help to reduce the injury and devastation that these hazards can cause.

7. Energy and mineral resources come from the Earth and are formed by geologic phenomena. Geologic studies can help locate these resources and mitigate the consequences of their use.

8. Physical features of the Earth are linked to life processes, and vice versa. Therefore, the history of life links intimately to the history of the physical Earth. The careful study of rocks can provide a rich record of this history.

9. The water in the oceans constantly moves and moderates the climate by transporting heat around the world. And where the ocean interacts with the land, fascinating coastal landscapes develop.

10. Our planet's atmosphere has changed dramatically over the course of the Earth's history. If you went back a billion years in time, the atmosphere would be unbreathable and you would suffocate. In modern times, this precious blanket of air has been significantly affected by the activities of society.

11. The atmosphere constantly moves, sometimes smoothly, sometimes violently. A variety of dangerous storms can develop in association with atmospheric movements, and these cannot be predicted more than a few days in advance.

12. Climates on the Earth vary with latitude, elevation, and distance from the sea, and the position of continents with respect to climate belts has changed over the course of geologic time. On a shorter time scale, both climate and sea level can change, sometimes very significantly, so that regions which were once dry can become shallow seas, and places that were once warm can later become covered in ice. Human society, during the past couple of centuries, has become a major agent of change on our planet.

13. We have much to learn by looking upwards and examining the Universe beyond Earth. Modern instruments and spacecraft have answered many questions, but with each answer, more questions arise about our Solar System, stars, galaxies, and the array of bizarre and beautiful objects that lie at inconceivable distances from us.

14. Science comes from observation, and people make scientific discoveries based on those observations. Earth science utilizes ideas from physics, chemistry, and biology, so the study of Earth science provides an excellent means to improve science literacy.

These narrative themes serve as the *take-home message* of this book, a message that we hope students will remember long after they finish their introductory Earth Science course. In effect, the themes provide a mental framework on which students can organize and connect ideas to develop a modern, coherent image of our planet in its context.

Pedagogical Approach

Students learn best from textbooks when they can actively engage with a combination of narrative text and narrative art. Some students respond more to words, which help them to organize information, provide answers to questions, fill in the essential steps that link ideas together, and develop a personal context for understanding information. Other students respond more to images that illustrate processes. Still others respond to question-and-answer-based active learning, an approach by which students "practice" their knowledge in real time. *Earth Science* provides all three of these learning tools. The text has been crafted to be engaging and to carry students forward along a narrative arc, the art has been configured to tell a story (see "Narrative Art and *What an Earth Scientist Sees*"), the chapters are laid out to help students internalize key principles (see "Organization"), and the online activities have been designed both to engage students and to provide active feedback (see "New Guided Learning Explorations" and "Smartwork5 Online Activities"). In-text features have also been thoughtfully developed to guide students to a richer understanding of and fuller engagement with unfamiliar concepts (see "Pedagogical Features Designed to Help Students Master the Concepts").

A note about units: Scientists the world over use the metric system for describing distances, weights, temperatures, and other physical features. But many of the students using this book grew up using the English system of units. To help students visualize and appreciate units, and to see the relationship between metric and English units, we provide both. The conversions we provide are sometimes exact and sometimes approximate, to reflect the precision of a measurement provided. For example, if we describe a distance as more or less 1,000 kilometers, we'll provide the conversion as 600 miles (instead of 621.37 miles), for it would be misleading to specify one unit as an approximation and the other as a precise one.

Organization

The topics covered in this book have been arranged so that students can build their knowledge of Earth science on a foundation of overarching principles. This book contains five parts.

Part I begins by considering how the Earth formed, and how it's structured, overall, from surface to center. With this basic background, students can delve into plate tectonics, the grand unifying theory of Earth science. Plate tectonics appears early in the book, so that students can gain an appreciation for the "action" underlying geology, and they can then use plate tectonics theory as a foundation from which they can interpret and link the ideas presented in subsequent chapters. Knowledge of plate tectonics, for example, helps students understand the suite of chapters on minerals, rocks, and the rock cycle. Knowledge of plate tectonics and rocks together, in turn, provides a basis for understanding volcanoes, earthquakes, and mountains. And then, with this background, students can see how the map of the Earth and its inhabitants have changed throughout the vast expanse of geologic time. We dedicate a unique chapter to energy and mineral resources, an especially relevant topic given present-day global interest in sustainability.

Part II hones in on the landscapes of the Earth's surface, the visible part of our planet. It explains how landscapes result from a never-ending battle between uplift (ultimately driven by plate tectonics) and erosion (ultimately driven by energy from the Sun). We begin by discussing surface movements in general, most dramatically manifested by landslides. Then, we consider our planet's freshwater, the landscapes associated with water movement, and the challenges presented by floods and the loss of water supplies. We conclude this part by looking at the dramatic landscapes developed in extreme climates—the realms of glaciers and deserts.

Part III takes us offshore, into the ocean. We begin by examining the water of this realm, focusing on its characteristics, its movements, and the life within it. We conclude by considering the interface between water and the solid Earth, both beneath the oceans and along its shores.

Part IV of the book introduces students to the Earth's atmosphere. We begin by describing the character and evolution of the atmosphere and the overall composition of air. We then develop an understanding of how and why air moves, and how this movement leads to weather systems and storms. The book provides a particular focus on the devastating drama of thunderstorms, tornados, hurricanes, and mid-latitude cyclones, because of the impact these can have on society. This part culminates with an examination of the Earth's climate, and of scientific and social issues related to climate change.

Part V, the book's final part, goes beyond the confines of the Earth and heads into outer space. We'll see how new discoveries provide answers to age-old questions about where our planet came from, and what the distant future holds. Students will learn how humanity's knowledge of space has advanced over the centuries, what the other objects of our Solar System look like, and how the Sun and the stars produce energy. This part ends by taking a voyage to the farthest reaches of the Universe, which we learn about by analyzing light that began its journey to the Earth billions of years before our planet had even formed.

Although we ordered the chapters in *Earth Science* in a way that provides an overall direction to our narrative, we have ensured that this book is flexible, and that instructors can assign chapters in a different sequence. Therefore, each chapter is self-contained, and we reiterate relevant material where necessary. This flexibility allows instructors to choose their own strategies for teaching Earth science.

Special Features of This Text

Narrative Art and *What an Earth Scientist Sees*

To help students visualize topics, this book is lavishly illustrated. The figures are designed to provide a realistic context for interpreting features and phenomena without overwhelming students with extraneous detail. In this edition, many drawings and photographs have been integrated into **narrative art** that has been laid out, labeled, and annotated to tell a story—the figures are drawn to teach! Subcaptions are positioned adjacent to the relevant parts of a figure, labels point out key features, and balloons provide important detail. Subparts have been arranged to convey time progression, where relevant. The color schemes in the drawings have been tied to those of related photos, so that students can easily visualize the relationships between drawings and photos. In some examples, photographs are accompanied by annotated sketches labeled **What an Earth Scientist Sees**, which help students to be certain that they actually see the specific features that the photo was intended to show. The in-text art also serves as the foundation for the robust suite of videos, and animations and simulations that give students a fuller and more dynamic means for visualizing complicated processes that happen over long periods of time (see "*Narrative Art Videos* and Animations and Simulations").

Earth Science at a Glance

In addition to individual figures, each chapter contains at least one dramatic *Earth Science at a Glance* illustration. These illustrations either expand on a particular topic or provide a synopsis of many topics in a beautiful, artistic rendering. Renowned British artist Gary Hincks hand-painted the majority of these. Others were developed and rendered by the talented artists who produced the text's art. These illustrations provide a way for students to visualize key concepts . . . at a glance.

How can I explain? Features

Unique *How can I explain* boxes in every chapter provide simple, fun, hands-on projects that help students to better visualize and master key concepts. These exercises are particularly useful for future educators, as the projects can easily translate into effective lesson plans for pre-college classes.

Pedagogical Features Designed to Help Students Master the Concepts

Each chapter begins with a series of *learning objectives* that frame the major concepts of the chapter for the students. Every section in a chapter ends with a *Take-home message*, a brief summary that helps students identify and remember the highlight of the section before moving on to the next. *See for Yourself* panels guide students on virtual field trips, via Google Earth, to locations around the globe or in the sky where they can apply their newly acquired knowledge to interpret real-world geologic features. *How can I explain* features, described above, drive student interaction with the course material by providing simple hands-on projects that illustrate important concepts. *Did You Ever Wonder* panels prompt students to connect new information to their existing knowledge base by asking Earth science-related questions that they have probably already

Take-home message . . .

An ordinary thunderstorm has a vertical, non-rotating updraft and tends to dissipate fairly quickly. Squall-line thunderstorms form in a row along storm-generated cold pools of air or along strong cold fronts. The most violent thunderstorms, supercells, form where vertical wind shear exists. A supercell has a rotating updraft, a particularly broad, asymmetrical anvil cloud, and an overshooting top.

Quick Question

Why do supercell thunderstorms tend to survive for a relatively long time?

BOX P.3 How can I **explain** . . .

The difference between potential and kinetic energy

What are we learning?
That one kind of energy can transform into another.

What you need:
Two golf balls, two full cups of water, and paper towels

Instructions:
- Drop one golf ball into the first cup from a couple of centimeters (about an inch) above the water's surface.
- Drop the second golf ball into the second cup of water from a meter (about one yard) above the water's surface.
- Remove both balls and compare the amounts of water that remain in each cup.

What did we see?
- The ball falling from just above the water's surface displaced only a small amount of water, whereas the ball falling from a meter up hit the water much harder and therefore displaced a lot more water.
- The heights of the balls represent their potential energy, and the amount of water displaced represents an amount of kinetic energy. The golf ball that fell from the higher position had more energy to start with because by raising the ball up a meter, you provided it with more potential energy. When it dropped, that potential energy was converted into kinetic energy. The ball that fell from the lower height had less potential energy to convert to kinetic energy by the time that it arrived at the water's surface, so it hit the water more softly and displaced less water.
- The potential energy of the balls is due to gravity. An object being pulled on by gravity stores energy until it can move. The higher it is, the more energy it stores. Why? Gravity causes objects to *accelerate* (speed up) as they fall. So, the higher ball hits the water at a faster *velocity* (speed) than does the lower ball, because the higher ball falls further. Since both balls have the same mass, the difference in kinetic energy is due to their respective velocities..

1 m

2 cm

BOX 17.3 Putting **Earth Science** to Use

Interpreting a weather forecast

How do meteorologists forecast the weather? And how do they come up with statements like "Tomorrow there is a 60% chance of rain" or "Tuesday will be partly cloudy"?

Weather forecasting is a complicated process that begins with the collection of weather data worldwide from airports and cities, weather balloons, satellites, radar systems, and other instruments. The data flow continuously into national data centers, such as the National Center for Environmental Prediction in Washington, DC. These centers process the data using supercomputers, arranging the information on orderly grids that cover all or parts of the planet. Numerical models, sophisticated computer codes consisting of mathematical equations that describe the forces that govern atmospheric motion and other factors that control the behavior and evolution of the atmosphere, use these data to calculate the future state of the atmosphere as far out as 14 days. The accuracy of such predictions is strongest within a 3-day range, but models can help predict future states at longer intervals.

So how does all that relate to the chance of rain in Cincinnati at noon on Tuesday? To pinpoint that type of information, meteorologists compare, using statistics, data from the current model forecast with thousands of past model forecasts and historical data about rainfall in Cincinnati (and all other weather data relevant to people living at thousands of other locations!). These model output statistics provide the information you actually receive on your television and internet-based forecasts.

For example, when a meteorologist states that there is a 60% chance of rain tomorrow afternoon in Cincinnati, the forecaster means that, historically, rain occurred 60% of the time that the weather conditions forecasted by the model were present in the afternoon in the Cincinnati area. A meteorologist might forecast clear, partly cloudy, mostly cloudy, or cloudy weather (sometimes meteorologists use sunny and partly sunny for daytime hours—these terms have the same meaning as clear and partly cloudy). These predictions, again, result from model output statistics, which provide the forecaster with the percentage of the time clouds are expected at a specific location. Forecasters always have the option of "tweaking" the model predictions based on their local knowledge, but their forecasts always start with the computer-based predictions and model output statistics. The data are also compiled to create the weather maps and forecasts that you see on television, on the internet, or in newspapers **(Fig Bx 17.3)**.

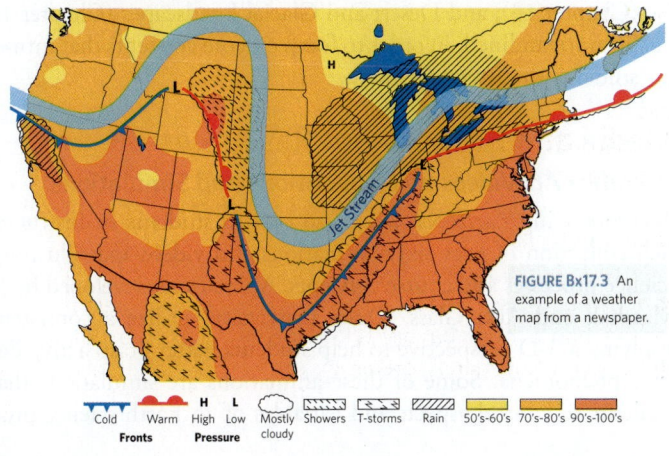

FIGURE Bx17.3 An example of a weather map from a newspaper.

Cold | Warm | H High | L Low | Mostly cloudy | Showers | T-storms | Rain | 50's-60's | 70's-80's | 90's-100's
Fronts | | **Pressure** | | | | | | | |

thought about. ***Putting Earth Science to Use*** boxes help students connect the course concepts to real-life applications. ***A Deeper Look*** boxes provide additional content to understand certain concepts more thoroughly. ***Science Toolboxes*** give students brief, accessible introductions or reviews of basic science terms and concepts that can help them better grasp the chapter material. The two-page ***Chapter Review*** spreads are useful for easy reference and as study guides for students. Each review question is tagged to one or more of the learning objectives to help students recognize what they have learned. Every review section also includes visual and applied questions designed to test basic knowledge and to stimulate critical thinking.

Up-to-Date Coverage of Current Topics

Earth Science, Second Edition, reflects the latest research in the discipline, to help students understand events and discoveries that have been featured in news headlines. Examples include the hurricanes that struck Texas, Florida, and Puerto Rico in 2017, the Kilauea eruption of 2018, the capturing of a supermassive black hole image in 2019, and the 2019 discovery of 20 new moons orbiting the planet Saturn.

More Streamlined Second Edition

Responding to reviewer feedback on the First Edition, this Second Edition is over 100 pages shorter. In streamlining content throughout the text, our aim was to make the book more accessible for students, while still providing clear, useful explanations and illustrations of the many topics covered in an Earth Science course. The following topics received notable revision:

- The rock groups material (covered in Chapters 4-6) has been reduced to be more easily covered in one or two class sessions. These chapters help students understand the relationships among the different rock groups and the processes by which they form.

- The winds and severe weather material (covered in Chapters 18-19) have been restructured to better align with how most instructors cover this material. Severe weather phenomena are now covered in one chapter (Chapter 19), where the discussion progresses from the smaller-scale phenomenon of thunderstorms and tornadoes to broader ones such as mid-latitude cyclones, before culminating in an examination of hurricanes.

- The astronomy material (Chapters 21-23) has been condensed and updated to ensure that the coverage and explanations presented reflect the most recent developments in this rapidly changing field.

- Some longer chapters in the First Edition, including Energy and Mineral Resources (Chapter 11); Streams, Lakes, and Groundwater (Chapter 13); and Desert and Glacial Landscapes (Chapter 14) have been streamlined, in order to focus on core concepts that introductory students should understand.

Media and Assessment Resources

Narrative Art Videos and Animations and Simulations

Accompanying *Earth Science,* at no cost to students or instructors, is a rich collection of over sixty animations and videos that illustrate Earth science processes and course concepts. Animations developed by Stephen Marshak and Alex Glass (Duke University) utilize a consistent style, applying a 3-D perspective to help students better grasp active Earth science phenomena. Some of these animations are simulations that allow students to control aspects and variables of an Earth science process. In

the ***Narrative Art videos***, Stephen Marshak enhances explanations of core concepts in the text by describing the processes displayed in animated versions of the book's figures. This full suite of visual resources is available to students at digital.wwnorton.com/earthscience2 and in the Learning Management System (LMS) Coursepacks. The suite is additionally available to instructors at the book's Instructor Resources page and in the searchable Interactive Instructor's Guide.

Real World Videos

In addition to the suite of original videos, and animations and simulations described above, instructors and students will have access to over a hundred ***Real World videos***, showing Earth science—and Earth scientists—in action. These videos have been carefully selected by

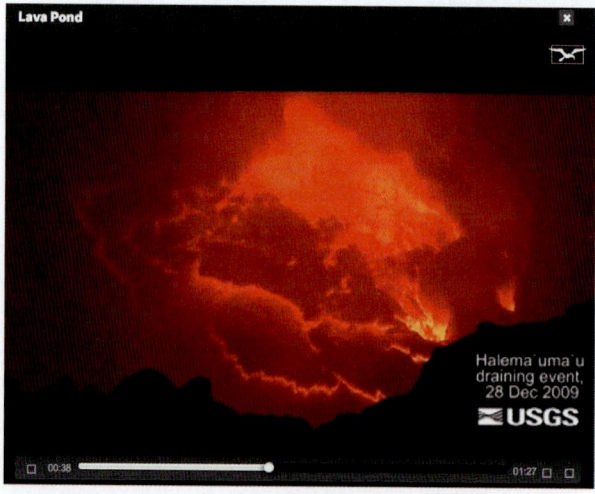

Tobin Hindle of Florida Atlantic University and Christine Clark of Eastern Michigan University. Teaching notes and classroom discussion questions to accompany the videos are available in the *Interactive Instructor's Guide*, and selected videos are available in the Smartwork5 tutorial and assessment system, with accompanying questions and feedback. Over eighty real world videos are available at digital.wwnorton.com/earthscience2, in the LMS Coursepacks, and in the online *Interactive Instructor's Guide*.

New Guided Learning Explorations

Guided Learning Explorations are formative, topical online exercises that coach students through three carefully scaffolded levels: 1) foundational concept review; 2) application questions featuring Earth science data, video and animation clips, and interactive simulations; and 3) exploratory questions that have students analyze real-world sites and make assessments using Google Earth flyover videos, maps, and other tools. Each level builds directly on the previous one to help students build a richer, more engaged understanding of the concepts. Guided Learning Explorations use game-like elements and are graded for completion, motivating students to work toward mastering each topic.

The Guided Learning Explorations cover 20 major topics taught in an introductory Earth Science course, such as plate tectonics, earthquakes, coastal processes, thunderstorms and tornadoes, and stellar evolution. These activities are available with every new book or ebook purchase, and can be set up to work within your campus LMS. They can also be purchased standalone at digital.wwnorton.com/earthscience2.

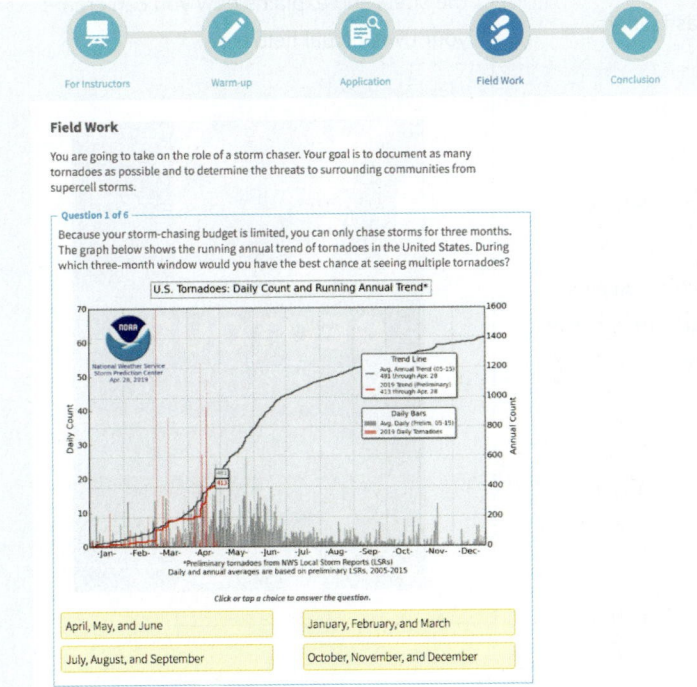

Smartwork5 Online Activities

The Smartwork5 online activities system features visual assignments that provide students with answer-specific feedback and links to the relevant section of the ebook for every question. Students get the coaching they need to work through assignments, while instructors get real-time assessment of student progress via automatic grading and detailed item analysis. Smartwork5 can also be integrated directly into instructors'

campus LMS, so that students benefit from single-sign on and instructors benefit from assignment results automatically reporting to the LMS gradebook.

Smartwork5 questions are written by and for Earth science educators, in service of the textbook's learning objectives. Among the many question types available in Smartwork5 are ranking, sorting, and labeling tasks that build directly on the text's art. Questions based on the *Narrative Art Videos*, animations and simulations, and *Real World Videos* help students engage with Earth science processes in action. New *Putting Earth Science to Use* and *How Can I Explain...* short-answer questions get students explaining Earth science phenomena and applications in their own words. And finally, Smartwork5 also provides basic reading quizzes that help students come to class better prepared, and *Geotour*-guided inquiry activities that use Google Earth. Smartwork5 comes free with all new texts in any format, or can be purchased standalone at digital.wwnorton.com/earthscience2. For this Second Edition, Christine Clark of Eastern Michigan University and Theodore Erski of McHenry County College expanded and updated the Smatwork5 course, based on the First Edition course created with Heather Lehto at Angelo State University, Tobin Hindle at Florida Atlantic University, Amy Gross at the University of North Carolina at Pembroke, Geoffrey Cook at the University of California, San Diego, and Heather Cook at California State University San Marcos.

Additional Instructor Materials
Lecture PowerPoints and Art Files

- *Lecture PowerPoints*—Designed for instant classroom use, these highly visual slides utilize photographs and line art from the book in a form that has been optimized for use in the PowerPoint environment. All art has been resized for projection. *Lecture PowerPoints* include in-class active learning prompts, class questions, *Narrative Art Videos*, and animations and simulations.

- *Labeled and Unlabeled Art PowerPoints*—These include all art from the book, formatted as JPEGs that have been pre-pasted into PowerPoints. We offer one set in which all labeling has been stripped and one set in which labeling remains.

- *Labeled and Unlabeled Art JPEGs*—We provide a complete file of individual JPEGs for art and photographs used in the book. Again, artwork is available with and without labels, so it may be used in presentations, quizzes, exams, etc.

- *Semesterly PowerPoint Update Service*—W. W. Norton & Company offers a semesterly update service that provides new, current-event-based PowerPoint slides, with instructor support, tying events in the news to core concepts from the text.

Instructor's Manual

The Instructor's Manual, prepared by Amy Gross of The University of North Carolina at Pembroke, is designed to help instructors prepare lectures, homework, and exams. Each chapter contains:

- Learning objectives

- Chapter summaries

- Complete answers to end-of-chapter Review and On Further Thought questions

- Descriptions, teaching notes, and discussion questions for all *Narrative Art Videos*, animations and simulations, and *Real World Videos*

See for yourself
Using *Google Earth*

Visiting Field Sites Identified in the Text

There's no better way to appreciate geology than to see it firsthand in the field. The challenge is that the great variety of geologic features that we discuss in this book can't be visited from any one locality. So, even if your class can take geology field trips during the semester, at most you'll see just a few geologic settings. Fortunately, Google Earth makes it possible for you to fly to spectacular geologic field sites anywhere in the world in a matter of seconds—you can take a virtual field trip electronically. Using related options, you can also look at stars in the sky.

In each chapter in this book, at least one *See for Yourself* provides sites that you can explore on your own computer (Mac or PC) using Google Earth software, or on your Apple/Android smartphone or tablet with the appropriate Google Earth app.

To get started, follow these three simple steps:

1 Check to see whether Google Earth is installed on your personal computer, smartphone, or tablet. If not, download the software from **earth.google.com** or the app from the Apple or Android app store.

2 Each *See for Yourself* site provides a thumbnail photo and brief description of the site (highlighting what you will see), as well as the latitude and longitude of the site.

3 Open Google Earth, and enter the coordinates of the site in the search window. As an example, let's find Mt. Fuji, a beautiful volcano in Japan. We specify the coordinates in the book as follows:

Latitude   35°21′41.78′N   Longitude   138°43′50.74′E

Type these coordinates into the search window as:

35 21 41.78N, 138 43 50.74E

Note that the degree (°), minute (′), and second (″) symbols are optional and can be left as blank spaces.

When you click Enter or Return, your device will bring you to the viewpoint right above Mt. Fuji illustrated by the thumbnail on the left. Note that you can use the tools built into Google Earth to vary the elevation, tilt, orientation, and position of your viewpoint. The thumbnail on the right shows the view you'll see of the same location if you tilt your viewing direction and look north.

View looking down.

View looking north.

Need More Help?

Please visit https://digital.wwnorton.com /earthscience2 to find a video showing you how to download and install Google Earth, more detailed instructions on how to find the *See for Yourself* sites, additional sites not listed in this book, links to Google Earth videos describing basic functions, and links to any hardware and software requirements. Also, notes addressing important Google Earth updates will be available at this site.

We also offer a separate book— the *Geotours Workbook*, Second Edition (ISBN 978-1-324-00096-9)—that identifies additional interesting geologic sites to visit, provides active-learning exercises linked to the sites, and explains how you can create your own virtual field trips.

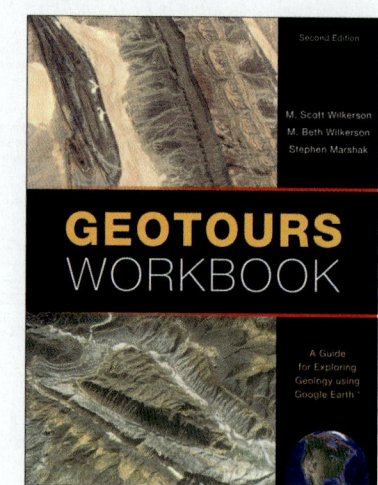

- Activity ideas to implement active learning in class
- A correlation between Marshak and Rauber's *Earth Science*, Second Edition and Tarbuck and Lutgen's *Earth Science*, Fifteenth Edition, for those looking to make the switch.

Interactive Instructor's Guide (IIG)

Searchable by chapter, learning objective, or keyword, the Interactive Instructor's Guide instantly provides multiple ideas for teaching: video clips, powerpoints, animations, and other class activities and exercises. This repository of lecture and teaching materials functions both as a course prep tool and as a means of tracking the latest ideas in teaching the Earth Science course.

Test Bank

The Test Bank, authored by Maureen Lemke of Texas State University, has been written to assess the text's learning objectives using carefully vetted and well-rounded questions. Every item in the Test Bank has been reviewed to ensure scientific accuracy. Each chapter features 50 multiple-choice questions, 10 short-answer or essay questions that test student's critical thinking and knowledge-application skills, and several art-based questions using modified images from the text. Finally, each question is tagged by text section, learning objective, difficulty level, and Bloom's taxonomy level.

Geotours Workbook, Second Edition

Created by Scott and Beth Wilkerson of DePauw University, and Stephen Marshak, *Geotours Workbook* contains active-learning exercises arranged by topic that take students on virtual field trips in Google Earth to see outstanding examples of Earth science at locations around the world. Each GeoTour is accompanied by a worksheet that includes instructions and multiple-choice questions. The *Workbook* accompanies a custom-made Google Earth kmz file created by Scott and Beth Wilkerson, available for free download by all instructors and students using *Earth Science*, Second Edition at digital.wwnorton.com/earth science2. The *Geotours Workbook* can be packaged for free with the text and includes complete user instructions and advanced instruction. Request a sample copy from your local Norton representative to preview each worksheet.

See for Yourself Google Earth Sample Sites

Earth Science users who simply want access to sample field sites for classroom presentations or distribution to students can download the sites from digital.wwnorton.com/earthscience2.

ebook

Compatible with all computers and mobile devices, the Norton ebook reader provides intuitive highlighting, note taking, and bookmarking functionalities. The Norton ebook for *Earth Science*, Second Edition includes dynamic features that engage students, such as our animations, Narrative Art Videos, and links to Google Earth See for Yourself sites. To help focus student reading, instructors can share notes with their class, including images and video. Reports on student and class-wide time-on-task allow instructors to monitor student reading and engagement. The Norton ebook reader can also be integrated into your campus learning management system. When integration is enabled, students can click on a link to the ebook from their campus LMS and be redirected.

The *Earth Science*, Second Edition ebook is available for purchase at **digital.wwnorton.com/earthscience2.**

Acknowledgments

Many people helped the authors bring this book from the concept stage to the shelf, and we greatly appreciate their roles in continuing the forward momentum of this sometimes-overwhelming project.

First and foremost, we wish to thank Kathy Marshak, who served as the book's in-home production editor. She helped coordinate material from previous books; edited and proofed text and figures; monitored the flow of manuscript and proofs on behalf of the authors; helped coordinate the art program; queried stylistic choices; monitored schedules; and served as an invaluable extra set of eyes at every stage. This book would not have happened without Kathy!

The staff of W. W. Norton & Company have, as always, been wonderful to work with, and we thank them for their insightful input and enduring patience during the development of this book. It has been a privilege to work with an employee-owned company whose editorial staff remains willing to work directly with authors. In particular, we would like to thank Thom Foley, the senior project editor for both editions of this book, who has done a Herculean job of overseeing a complicated process of managing the manuscript, chapter proofs, art, and photos, all while remaining incredibly calm. Thom invested untold hours in sorting out composition and design issues, laying out pages by hand when necessary, and making sure every word and figure was correct and ended up where it should be.

The First Edition of this book greatly benefited from the input of three supervising editors. The idea for the book originated in discussion with the late Jack Repcheck, a mentor who we will always remember. Eric Svendsen's experience and skill guided the book through the starting gate, injecting many ideas that contributed to its style and organization, connecting the project to new trends in science pedagogy and book design. Jake Schindel took over the reins during the First Edition's journey. His ability to solve problems and keep the project moving forward is impressive, and we greatly appreciate his enthusiasm, skill, and hard work throughout the revision process that led to the Second Edition.

We wish to thank Rob Bellinger, Arielle Holstein, and Jesse Singh for their innovative approach to ancillary development, and for overseeing the development of the online educational supplements. We are also very grateful to all of the instructors who have helped to develop accompanying materials. Thanks to Norma Sims Roche, who again did a spectacular job of copyediting, patiently sorting through and resolving edits upon edits upon edits; Sunny Hwang for helpful contributions to the project's development during the First Edition; Stephanie Romeo for her expert and thoughtful management of the photo program; Jane Miller for insightful photo research; Ashley Horna for coordinating the back-and-forth between the publisher and various suppliers; Debra Morton Hoyt and Rubina Yeh for overseeing the book's cover and design; Anne DeMarinis for the stunning interior design; Rachel Goodman for skillful editorial support in addressing expected and unexpected issues throughout the book's development; and Megan Schindel for expertly tracking down permissions. We are particularly grateful to the book's marketing director, Katie Sweeney, and Norton's staff of sales representatives, who have done an outstanding job of bringing this book to the attention of a broad audience.

Production of the illustrations has involved many people over many years. We thank Jan Troutt, lead artist Joanne Brummett, and the team at Troutt Visual Services who expertly provided line-art revisions and the rendering of new art for the Second Edition. Stan Maddock, as the primary artist for Norton's geoscience books over two decades, is responsible for conceptualizing the book's art style and for producing many figures.

It's also been great fun to interact with Gary Hincks, who painted many of the two-page spreads, in part using his own designs and geologic insights. Versions of several of these paintings originally appeared in *Earth Story* (produced by BBC Worldwide, 1998) and were based on illustrations conceived with Simon Lamb and Felicity Maxwell. Some of the marginal quotes used in this book were found in *Language of the Earth*, compiled by F. T. Rhodes and R. O. Stone (Pergamon, 1981).

This book and its related text, *Earth: Portrait of a Planet*, have benefited greatly from input by expert reviewers for specific chapters, by general reviewers of the entire book, and by other reviewers who have provided helpful feedback for this Edition. These reviewers include:

Mary Abercrombie, *Florida Gulf Coast University*
Kevin Barrett, *Tarrant County College, Northwest Campus*
Erica Barrow, *Ivy Tech Community College*
Karin Block, *City College of New York*
Brett Burkett, *Collin College*
Karen Busen, *Tallahassee Community College*
Marianne Caldwell, *Hillsborough Community College*
Cinzia Cervato, *Iowa State University*
Jennifer J. Charles-Tollerup, *Rowan College, Gloucester County*
Laura Chartier, *University of Alaska, Anchorage*
Robert L. Dennison, *Heartland Community College*
Melissa Driskell, *University of North Alabama*
Todd Feeley, *Montana State University*
Kerry Workman Ford, *California State University, Fresno*
Nicholas Frankovits, *University of Akron*
Kenneth Galli, *Boston College*
Bryan Gibbs, *Richland College*
Alessandro Grippo, *Santa Monica College*
Amy Gross, *University of North Carolina, Pembroke*
Mary Hall-Brown, *University of North Carolina, Greensboro*
Tobin Hindle, *Florida Atlantic University*
Mark Horrell, *Northwest Florida State College*
Ashanti Johnson, *University of Texas, Arlington*
Leslie Kanat, *Johnson State College*
Zoran Kilibarda, *Indiana University Northwest*
Michael Kozuch, *California State University, East Bay*
Michael Kruge, *Montclair State University*
Kody Kuehnl, *Franklin University*
Brett C. Latta, *Franklin University*
Kristine Larsen, *Central Connecticut State University*
Ana Larson, *University of Washington*
Rita Leafgren, *University of Northern Colorado*

Maureen Lemke, *Texas State University*
Judy McIlrath, *University of South Florida*
Jamie Mitchem, *University of North Georgia, Gainesville*
Daniel Murphy, *Eastfield College*
Jacob Napieralski, *University of Michigan, Dearborn*
Barbara O'Grady, *Montana State University*
Stacy Palen, *Weber State University*
Mark Peebles, *St. Petersburg College*
Melanie Ruberti, *Santa Fe College*
Steven H. Schimmrich, *SUNY Ulster County Community College*
Marcia Schulmeister, *Emporia State University*
Jennifer K. Sheppard, *Moraine Valley Community College*
Andrew Smith, *Vincennes University*
Steven Stemle, *Palm Beach State College*
Keith Sverdrup, *University of Wisconsin, Milwaukee*
Tawny Tibbits, *Broward College*
Anthony J. Vega, *Clarion University*
Adil M. Wadia, *The University of Akron*
Shizuko Watanabe, *Eastfield College*
Terry R. West, *Purdue University*

We also wish to thank the instructors who joined us for focus groups regarding this text and the courses it serves; their input helped shape the form and content of this book. These instructors include:

Johnsely S. Cyrus, *North Carolina Agricultural and Technical State University*
Frank DeCourten, *Sierra College*
David Douglas, *Pasadena City College*
Todd Feeley, *Montana State University*
Garry Hayes, *Modesto Junior College*
Richard W. Hurst, *California Lutheran University*
Melissa Lobegeier, *Middle Tennessee State University*
John McDavis, *Carleton College*
Rosemary Millham, *SUNY New Paltz*
Kent Murray, *University of Michigan, Dearborn*
Jeffrey Myers, *Western Oregon University*
Caroline Pew, *University of Washington*
Nancy Price, *University of Washington*
Eric Jonathan Pyle, *James Madison University*
Rene Shroat-Lewis, *University of Arkansas, Little Rock*
Carol Anne Stein, *University of Illinois, Chicago*
Rebecca Teed, *Wright State University*
Leanne Sue Teruya, *San Jose State University*
Suzanne Traub-Metlay, *Front Range Community College*

ABOUT THE AUTHORS

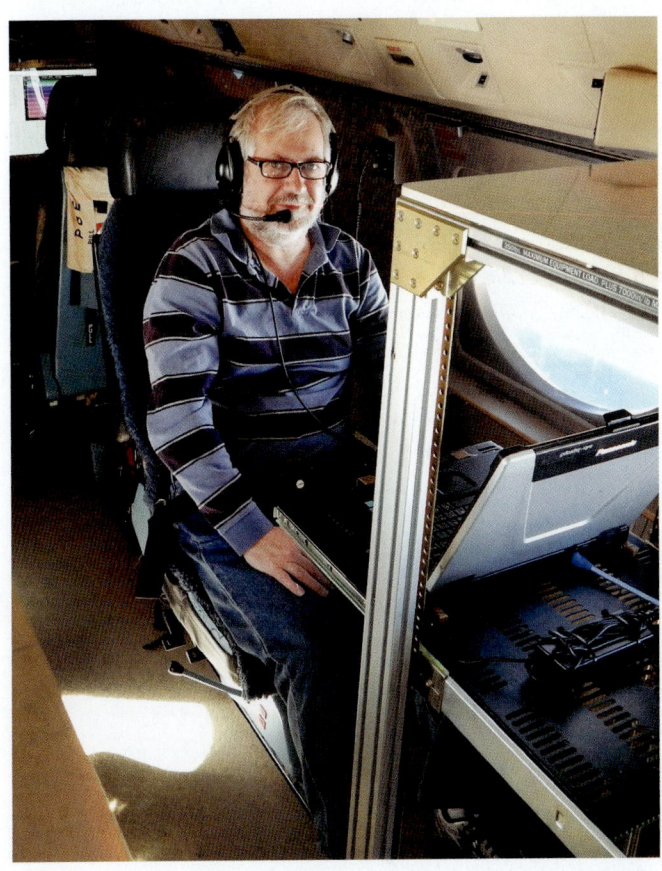

Stephen Marshak is a Professor Emeritus of Geology at the University of Illinois, Urbana-Champaign, where he also served as Department Head and as the Director of the School of Earth, Society, and Environment. He holds an A.B. from Cornell University, an M.S. from the University of Arizona, and a Ph.D. from Columbia University. Steve's research interests lie in structural geology and tectonics, and he has participated in field projects on a number of continents. Steve loves teaching and has won his college's and university's highest teaching awards. He also received the 2012 Neil Miner Award from the National Association of Geoscience Teachers (NAGT), for "exceptional contributions to the stimulation of interest in the Earth sciences." Steve is a Fellow of the Geological Society of America. In addition to research papers and *Earth Science*, he has authored *Earth: Portrait of a Planet,* and *Essentials of Geology,* and has co-authored *Laboratory Manual for Introductory Geology, Earth Structure: An Introduction to Structural Geology and Tectonics*, and *Basic Methods of Structural Geology.*

Robert Rauber is a Professor of Atmospheric Sciences at the University of Illinois, Urbana-Champaign, where he served as the Department Head and is currently the Director of the School of Earth, Society, and Environment. He holds a B.S. in Physics and a B.A. in English from the Pennsylvania State University, as well as M.S. and Ph.D. degrees in Atmospheric Science from Colorado State University. In addition to teaching and writing, Bob oversees a research program that focuses on the development and behavior of storms. To carry out this work, Bob flies into hazardous weather in specially equipped airplanes—during these flights, he's so intent on recording data that he usually doesn't notice the bouncing and lightning. Bob has won several campus teaching awards, is a Fellow of the American Meteorological Society (AMS), and served as Publication Commissioner for the AMS. In addition to research papers and *Earth Science*, his textbook writing credits include lead authorship of *Severe and Hazardous Weather: An Introduction to High Impact Meteorology* and *Radar Meteorology: A First Course.*

EARTH SCIENCE

The Earth, the Atmosphere, and Space

PRELUDE
Welcome to Earth Science!

By the end of the Prelude you should be able to . . .

A. describe the variety of disciplines covered in an Earth Science course.

B. evaluate whether and how a news story pertains to an aspect of Earth science.

C. explain the key elements of the scientific method, and distinguish between a hypothesis and a theory.

D. characterize the Earth System and its key realms.

E. state where energy that affects the Earth and other planets comes from.

F. describe a few key themes that serve as foundations for understanding the Earth and the Universe.

G. identify several practical issues that a background in Earth science can help you to understand and address.

(a) The Transantarctic Mountains separate East Antarctica from West Antarctica.

(b) The smoking summit of Mt. Erebus seen from the cockpit of a plane.

(c) A plane dropping geologists on a snowfield.

P.1 Introduction

Our C-130 transport plane rose from the frozen surface of the Ross Sea, near the coast of Antarctica, and turned south. We were heading to a field site about 250 km (155 mi) away where, with luck, we'd be able to spend the next month studying cliff exposures of some very unusual rocks (Fig. P.1). The plane climbed past the smoking summit of Mt. Erebus, the Earth's southernmost volcano, and for the next hour it flew along the Transantarctic Mountains, a long range of rugged peaks that divides the continent into two regions, East Antarctica and West Antarctica. Over millions of years, snow accumulating in Antarctica's cold climate built into *glaciers*, sheets and rivers of solid ice that slowly flow across the ground.

(d) Sledding to a field site with crates of supplies.

<< On the coast of Nova Scotia, along the Bay of Fundy, we can see interactions among many components of the Earth System—sunlight, air, water, rock, soil, and life.

The glacier covering East Antarctica attains a thickness of over 3 km (2 mi). Its top forms a high plain called the Polar Plateau.

While marveling at this stark panorama—so different from the forests, grasslands, farms, and cities of the more populated regions on our planet—we heard the engines

3

slow and felt the C-130 begin to descend. As the plane approached the glacier just below the cliff that we hoped to study, the pilot lowered the landing gear, which had been equipped with giant skis. Shouting above the engine noise, a crew member reminded us of the emergency procedure: "If you hear three short blasts of the siren, hold on for dear life!" Seconds later, the skis slammed into small frozen snowdrifts that formed wave-like ripples on the surface of the glacier. *Wham, wham, wham, wham!* It felt as though a fairy-tale giant was shaking the plane. Then, as fast as it began, the shaking stopped, for we were airborne again, looking for a softer landing surface. We finally touched down at a location where soft snow blankets the Polar Plateau. Bundled against the frigid gale generated by the plane's still-roaring propellers, we jumped out and helped unload food, fuel, stoves, tents, sledges, and snowmobiles. The instant the cargo was out, the C-130 trundled off and rose skyward.

When the glint of the plane's metal skin had passed beyond the horizon, the silence of Antarctica hit us—no dogs barked, no leaves rustled, and no traffic rumbled in this land of black rock and white snow. It would take us almost two days, even with the aid of snowmobiles, to haul sledges of food and equipment to our field site, where we would spend the next month collecting rock samples. Why go to so much effort and expense to study an exposure of rocks? The Scottish poet Walter Scott (1771–1832) asked the same question and provided a colorful answer: "Some rin uphill and down dale, knapping the chucky stanes to pieces wi' hammers, like sae mony road-makers run daft—they say it is to see how the warld was made!"

Indeed, to see how the world was made. Field expeditions like the one we've just described, along with analyses carried out with instruments in laboratories or deployed by ships and planes, calculations run on computers, and scans made with telescopes and space probes, have led to an explosion of discoveries about the land, sea, and atmosphere of the Earth, and about the countless objects that populate the rest of the **Universe** (all of space, and everything within it). These discoveries lead to an appreciation not only of our home planet, but also of its context in the broadest sense, and they underlie the subject of Earth science **(Fig. P.2)**. To help you begin your exploration of these discoveries, this Prelude briefly outlines the topics covered by a course in Earth Science, introduces themes that tie these topics together, and suggests ways that you can use what you'll learn.

FIGURE P.2 Astronauts aboard the International Space Station have this view of space beyond the Earth's horizon as they orbit the planet.

P.2 What's in an Earth Science Course?

Defining Earth Science and Its Components

For most of humanity's existence, speculation about the natural world lay in the realm of philosophy, and people attributed natural features in their surroundings to supernatural phenomena. But, beginning a few thousand years ago, and accelerating in the past few hundred years, the study of the natural world and its surroundings became the focus of **science**, the systematic analysis of natural phenomena based on observation, experiment, and calculation.

Science has evolved into several distinct disciplines, most of which may already be familiar to you. For example, if someone asks, "What is chemistry?" you might respond that it's the study of chemicals and reactions among them. Similarly, if someone asks, "What is physics?" you might respond that it concerns the study of matter and energy **(Box P.1)**, and if someone asks, "What is biology?" you might respond that it's the study of living organisms. But if someone asks, "What is Earth science?" you might be stumped.

Why can Earth science be hard to define? It's because **Earth science** combines many disciplines **(Fig. P.3)**. It includes **geology**, the study of our planet with a focus on the materials that compose it, the phenomena that change it, and its long-term history. It includes **oceanography**, the study of the water and life in the oceans, as well as the way in which ocean water moves and interacts with land and air. And it includes **atmospheric science**, the study of the air layer that surrounds the Earth. Atmospheric

FIGURE P.3 The various disciplines of Earth science.

(a) A geologist studies a rock face.

(b) An oceanographer samples the sea.

(c) An atmospheric scientist monitors and analyzes the atmosphere.

(d) An astronomer explores the cosmos.

Science Toolbox

Matter and energy

Every second of every day, you interact with matter and energy. How can we define these important terms? Let's start with matter. *Matter* is the material substance of the Universe, the stuff that occupies space. The amount of matter in an object is its *mass*, so an object with greater mass contains more matter. As we'll discuss further in Chapter 1, matter consists of tiny particles called *atoms,* which can stick or bond together to form *molecules*.

Scientists recognize four different *states of matter*, or forms in which matter exists. A *solid* can retain its shape, regardless of changes in the size of its container, because its atoms and molecules remain locked in position **(Fig. BxP.1a)**. A *liquid* can flow and conform to the shape of its container because clusters of atoms and molecules within it can move relative to one another **(Fig. BxP.1b)**. A *gas* not only can flow, but can also expand to fill its container, because its atoms and molecules are not attached to one another and can move about freely **(Fig. BxP.1c)**. We'll discuss a fourth state of matter, *plasma*, later in the book—for now, we'll simply say that it resembles a very hot gas.

From your everyday experience, you know that it's more difficult to lift a block of rock than it is to lift a block of Styrofoam of the same size. That's because the two objects have different densities. *Density* refers to the amount of matter within a given volume or, formally stated, the mass per unit volume. A volume of space that contains hardly any matter, meaning that its density approaches zero, is a *vacuum* (from the Latin *vacuus*, meaning vacant).

Energy, in the jargon of physics, represents the ability to do work or to cause a physical system to change. This means that the transfer of energy can make a car move or a light bulb glow. Physicists distinguish among several forms of energy, including *kinetic energy*, the energy of motion; *potential energy*, the energy stored in a material that can be released later; *radiant energy*, the energy carried from one location to another in the form of *electromagnetic radiation* (such as light); and *thermal energy*, or *heat*, the energy that can cause a material to warm up.

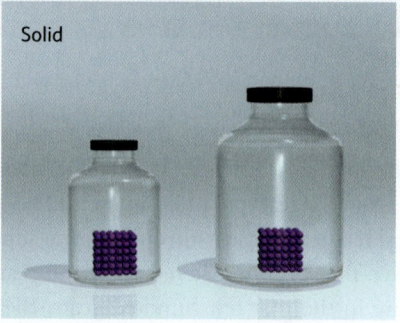

(a) A solid retains its shape and density.

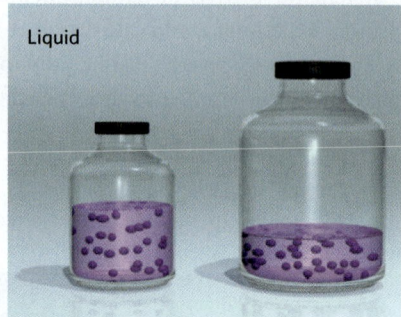

(b) A liquid conforms to its container, but retains its density.

(c) A gas fills its container, so its density can change.

FIGURE BxP.1 Three states of matter.

---

science, in turn, includes both meteorology and climate science. **Meteorology** focuses on the *weather*, meaning the condition of the atmosphere at a given location and time, as well as on the movement of air and its consequences. **Climate science** focuses on understanding factors that control *climate*—the nature of daily to seasonal variations in weather conditions for a region, averaged over many years—and how climate changes over time. High school or college courses entitled Earth Science also commonly introduce the basics of **astronomy**, or *space science*, the study of planets, stars, and other objects of the Universe, because

studying astronomy provides a framework of knowledge on which to build an understanding of the formation and history of our own planet. In sum, we see that an Earth Science course covers the nature, origin, and evolution of all of our natural surroundings. It's a broad subject, indeed!

The Scientists Who "Do" Earth Science

The information presented in this book comes from the work of scientists. Popular media often characterize scientists as awkward loners with poor taste in clothing. Who are they, really? In an Earth Science course, you'll discover that **scientists** are people who search for ideas that can explain the way our Universe operates. They do such work in many different ways. At any given time, at locations all around the world, *field scientists* scale cliffs **(Fig. P.4a)**, fly into storms **(Fig. P.4b)**, plow through stormy seas, or star-gaze from mountaintops. Meanwhile, *laboratory scientists* peer down microscopes, adjust electronic equipment,

FIGURE P.4 Research takes place in many environments.

(a) Cliff exposures in the desert of Utah provide a record of the Earth's past.

(b) To learn about severe weather, researchers fly into storms and take measurements.

(c) Laboratories provide an opportunity to carry out controlled experiments.

FIGURE P.5 Use of models in Earth science.

(a) A stream of water flowing over sand can simulate the evolution of a river.

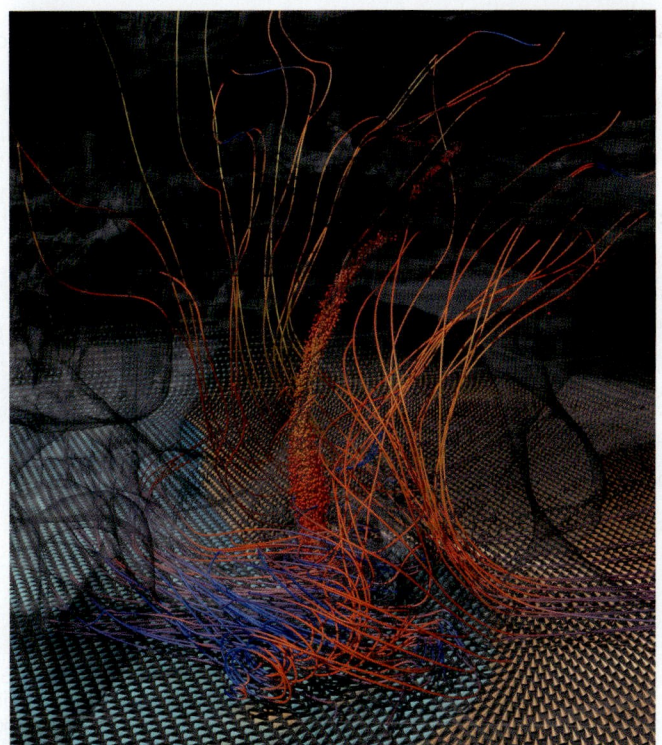

(b) The flow through a severe thunderstorm, simulated by a supercomputer.

or mix test tubes of chemicals **(Fig. P.4c)**; and *computational scientists* program computers or devise equations. In Earth science, some scientists develop computer or physical *models* (simulations) to help visualize processes that take place too slowly or too quickly to see in real time, or to characterize objects that are too small or too large to measure directly **(Fig. P.5)**.

Science Toolbox

The scientific method

Sometime during the past 100 million years, a large solid block zoomed in from space and slammed into our planet at a site in what is now the midwestern United States, a landscape that today hosts flat cornfields. The impact of this *meteorite* blasted debris skyward and carved a deep, bowl-shaped depression called a *crater*. The impact also shattered rock beneath the crater, and it caused layers of rock that had been buried deeply below the ground surface to spring upward and tilt on end. In the millions of years that followed, *erosion* (the grinding away and removal of material at the Earth's surface by flowing water or ice and blowing wind) wore down the crater until the depression disappeared entirely. But erosion did not carve deeply enough to remove the fractures and tilted rock layers that the impact had produced. Some 15,000 years ago, sand, gravel, and mud carried by a vast glacier buried the area and hid evidence of the impact from view **(Fig. BxP.2)**.

So much history beneath a cornfield! How do we know that history? It took scientific research! Scientists commonly guide their research by using the **scientific method**. Let's consider the idealized components of the scientific method and see how researchers applied them to come up with and verify the meteorite-impact story.

• *Recognizing the problem:* Any scientific project, like any detective story, generally begins by identifying a problem. The cornfield problem came to light when workers drilling a water well discovered that limestone, a rock commonly made of shell fragments, lies just below the layer of 15,000-year-old glacial deposits. In surrounding regions, the rock directly beneath the glacial sediment consists of sandstone, a rock consisting mostly of sand grains. In fact, outside of this cornfield, the limestone layer lies underneath the sandstone layer. Limestone can be used to make cement and produce agricultural lime, so workers bulldozed off the glacial sediment and dug a quarry to excavate the limestone. They were surprised to find that the rock layers exposed in the quarry were tilted steeply and contained many

FIGURE BxP.2 An ancient meteorite impact has excavated a crater and permanently changed rock beneath the surface.

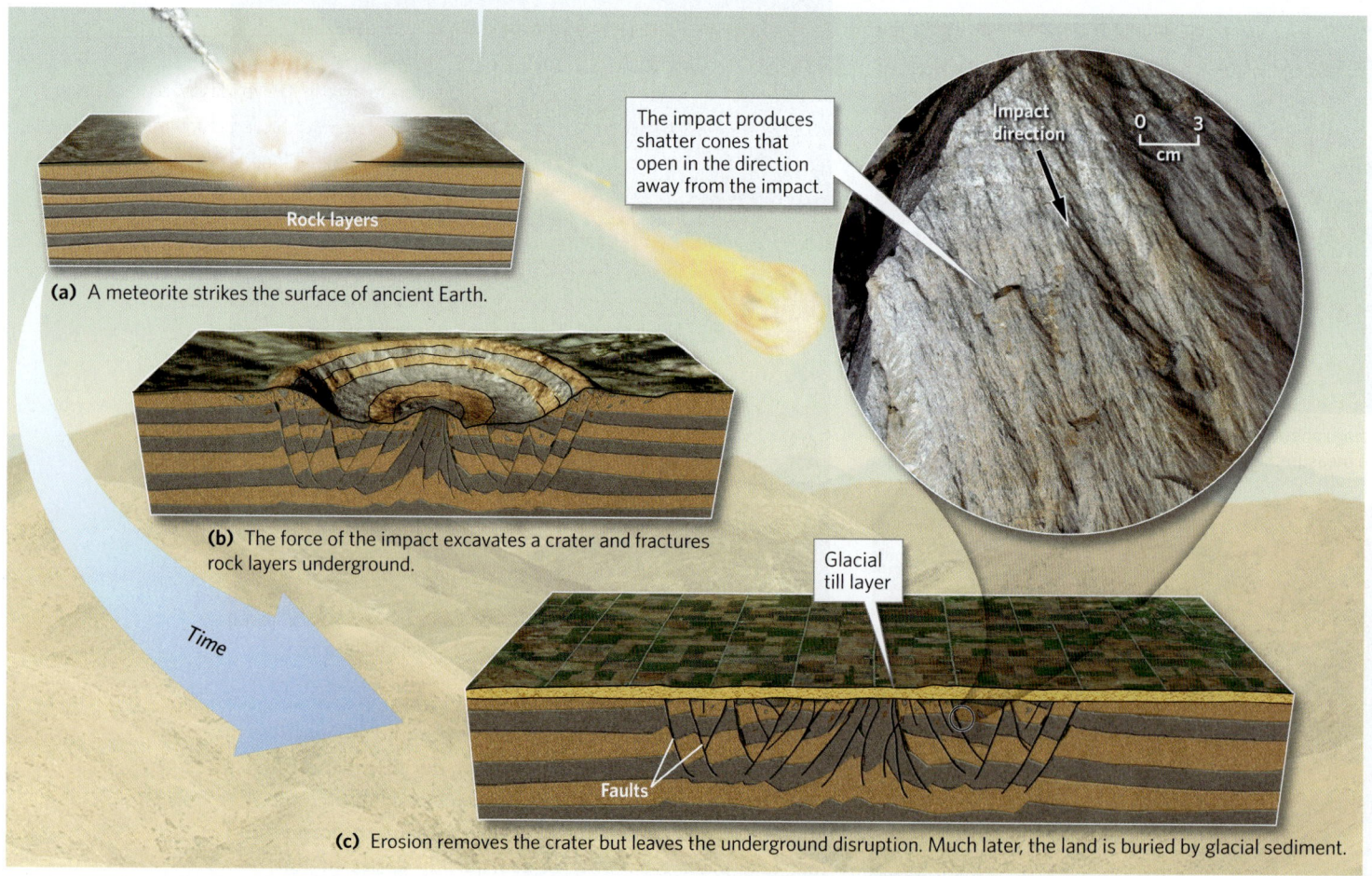

(a) A meteorite strikes the surface of ancient Earth.

The impact produces shatter cones that open in the direction away from the impact.

Impact direction

0   3 cm

(b) The force of the impact excavates a crater and fractures rock layers underground.

Time

Glacial till layer

Faults

(c) Erosion removes the crater but leaves the underground disruption. Much later, the land is buried by glacial sediment.

Rock layers

large cracks, because rock layers in surrounding regions are horizontal, like the layers in a birthday cake, and contain relatively few cracks. What phenomenon brought limestone up close to the Earth's surface, tilted the layers in the rocks, and shattered the rocks? Curious scientists journeyed to the quarry to find out.

- *Collecting data:* To better characterize and, hopefully, solve a problem, scientists collect *data*, sets of observations, measurements, or calculations. Scientists working at the quarry measured the orientation of the rock layers and *documented* (made a written or photographic record of) the fractures that broke up the rocks. Note that in studies of this quarry, data come from analyzing natural features formed long ago. In other situations, data might come from experiments or models. Observations, calculations, and experiments should be *repeatable*, in the sense that another researcher following the same procedure should obtain the same result.

- *Proposing hypotheses:* A scientific **hypothesis** is merely a possible explanation for a set of observations or calculations. Scientists may propose hypotheses before, during, or after their initial data collection. In this example, the geologists came up with two alternative hypotheses to explain the features in the quarry: (1) the features formed during an ancient volcanic explosion; and (2) the features are a consequence of an ancient meteorite impact.

- *Testing hypotheses:* Because a hypothesis is just an idea that can be either right or wrong, scientists must test a hypothesis to determine its validity. Often, such tests involve making *predictions* of what scientists will see if they make more observations or if they conduct an experiment or calculation. If a hypothesis fails these tests, scientists discard it. If a hypothesis passes these tests, it could be right.

The scientists at the quarry compared their field observations with published observations made at known sites of volcanic explosions and meteorite impacts, and with the results of models designed to simulate such events. They learned that if the features visible in the quarry were the result of volcanism, the quarry should contain specific kinds of rocks. But no such rocks were found there. If, however, the features were the result of

an impact, the rocks should contain *shatter cones*, uniquely shaped cracks. Shatter cones can be overlooked, so the scientists returned to the quarry specifically to search for them—and found them in abundance. The impact hypothesis passed this test and remained a possibility!

Scientists don't always see the entire path to solving a problem when they start working on it. In fact, at first they might not recognize the problem to be solved, so they don't always follow the components of the scientific method in a specific sequence. Furthermore, serendipity often plays a role in research, in that scientists may stumble onto a new problem or solution without planning on it. And while ideally, scientists try to confirm results by repeating observations, calculations, or experiments, in Earth science such work isn't always possible because the phenomena under study may have happened in the very distant past, or in locations that are far away. In fact, some questions just can't be answered fully using available resources and methods, so we have to accept that some interpretations remain uncertain.

In common English, the word *theory* often substitutes for the word *hypothesis*. For example, you may see sentences like, "The detective's proposal is only a theory," or "The author of the article proposes many theories," with the implication that a theory is just an idea that's as likely to be wrong as it is to be right. In scientific discussion, however, a **theory** is a scientific idea supported by abundant evidence. In other words, a theory has passed many tests and has, so far, failed none. Consequently, scientists have much more confidence in the validity of a theory than they do in the validity of a hypothesis. Continued study in the midwestern quarry eventually yielded so much evidence favoring the impact hypothesis that the proposal came to be viewed as a theory.

Scientists continue to test theories over a long time. Successful theories, those that are supported by many observations and lead to many successful predictions, eventually become part of a discipline's foundation. In some cases, scientists have been able to devise concise statements that completely describe a specific relationship or phenomenon. Such a statement, called a *scientific law*, applies without exception over a defined range of conditions. Unlike a theory, a law doesn't explain a phenomenon.

As you study Earth science, you'll have an opportunity to see how scientists conduct *scientific research*, the process of seeking to understand natural phenomena. You'll see that scientific research does not rely on subjective guesses, but rather anchors itself in the development of consistent, testable concepts by following the basic tenets of the *scientific method* **(Box P.2)**. Some scientists focus their research on defining new principles that can profoundly change our understanding of nature—think of Charles Darwin, Marie Curie, or Albert Einstein—whereas others focus on practical matters, using the results of their research to improve our lives, find our resources, and protect our environment. To drive home the point that science is a human endeavor, this book highlights where, when, and how ideas originated, so that you can answer the question, "How do we know that?"

How can I explain . . .

The difference between potential and kinetic energy

What are we learning?

That one kind of energy can transform into another.

What you need:

Two golf balls, two full cups of water, and paper towels

Instructions:

- Drop one golf ball into the first cup from a couple of centimeters (about an inch) above the water's surface.
- Drop the second golf ball into the second cup of water from a meter (about one yard) above the water's surface.
- Remove both balls and compare the amounts of water that remain in each cup.

What did we see?

- The ball falling from just above the water's surface displaced only a small amount of water, whereas the ball falling from a meter up hit the water much harder and therefore displaced a lot more water.

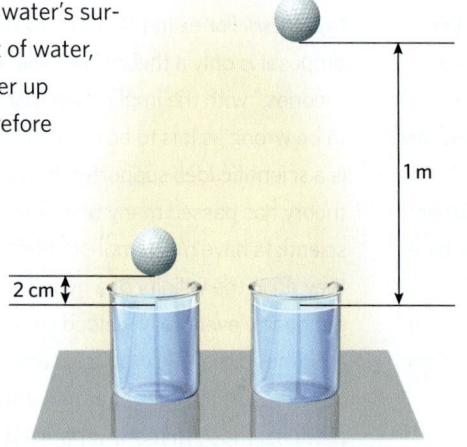

- The heights of the balls represent their potential energy, and the amount of water displaced represents an amount of kinetic energy. The golf ball that fell from the higher position had more energy to start with because by raising the ball up a meter, you provided it with more potential energy. When it dropped, that potential energy was converted into kinetic energy. The ball that fell from the lower height had less potential energy to convert to kinetic energy by the time that it arrived at the water's surface, so it hit the water more softly and displaced less water.

- The potential energy of the balls is due to gravity. An object being pulled on by gravity stores energy until it can move. The higher it is, the more energy it stores. Why? Gravity causes objects to *accelerate* (speed up) as they fall. So, the higher ball hits the water at a faster *velocity* (speed) than does the lower ball, because the higher ball falls further. Since both balls have the same mass, the difference in kinetic energy is due to their respective velocities..

Who are the scientists that make the discoveries we'll be focusing on in this book? When a headline begins with "Scientists say . . ." and then continues with "an earthquake shook Japan today," or "the supply of oil may be running out," the scientists under discussion are *geologists*. If the headline continues instead with "severe storms are approaching," or "the length of the growing season will change," the scientists referred to are *atmospheric scientists*. If the headline continues with "a shift in the pattern of ocean currents was detected," or "seawater has become more acidic in recent years," the scientists involved are *oceanographers*. And if the sentence continues, "a new planet has been found in orbit around a distant star," or "a solar storm might disrupt communications," the scientists making news are *astronomers*. In this book, for simplicity, we'll often refer to geologists, atmospheric scientists, and oceanographers together as *Earth scientists*.

Take-home message . . .

An Earth Science course encompasses geology, oceanography, atmospheric science, and astronomy. Field, laboratory, and computational scientists try to understand our surroundings using the scientific method.

Quick Question

Why is it harder to define Earth science than it is to define, say, chemistry?

P.3 Narrative Themes of This Book

To understand a novel, you need to learn names of the characters, and you need to pay attention to how the characters interact and relate to one another. Similarly, to understand a textbook, you need to learn some new words, or *terminology*, in order to be able to discuss ideas efficiently, and you need to pay attention to how ideas connect to one another. Let's now consider some of these connected ideas, or *narrative themes*, that tell the story of the Earth and its context in space to provide this book's overall take-home message.

Matter and Energy Behave in Predictable Ways, but Uncertainty Exists

The Earth, and the Universe as a whole, consist of matter and energy (see Box P.1). Throughout this book, we discuss the nature of matter and energy and the many ways that they interact. These interactions happen in predictable, natural ways **(Box P.3)**. For example, when the Sun shines on a dark-colored car, the car's surface becomes very hot; when you fill a balloon with helium gas, the balloon rises; and when you place liquid water in a freezer, it turns into solid ice. Because such phenomena happen in the same way under the same circumstances every time, scientists can predict—to some extent—how natural processes will operate.

We say "to some extent" in the previous paragraph because *unpredictability* is also a part of nature. For example, while we can predict that a boulder pushed over a cliff will tumble downslope, we can't predict exactly where it will land, for it may bang into other boulders on its way down. And when we make a measurement repeatedly, we may find that the values we obtain vary slightly, because there's a randomness to natural phenomena, or because our measurement technique isn't perfect. Scientists describe this randomness by characterizing the *degree of uncertainty* associated with a measurement or prediction. Specifically, scientists distinguish between *accuracy,* representing how close a measured value is to the true value, and *precision,* representing how close a succession of measurements of the same feature are to one another.

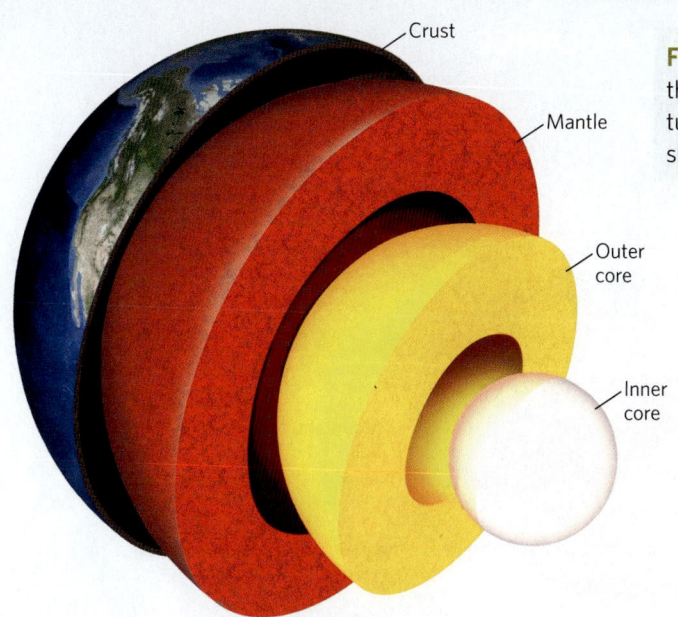

FIGURE P.6 Simplistically, the geosphere can be pictured as a set of concentric shells or layers.

Crust
Mantle
Outer core
Inner core

The Earth System Contains Many Interacting Realms

As you study the Earth, you'll discover that our home planet contains a variety of *realms*, meaning regions or volumes with a distinct character. These include:

- *Geosphere:* The part of the Earth that starts at the solid surface and goes down to the center—6,371 km (3,959 mi) below—is the **geosphere**. Geologists divide the geosphere into concentric layers, nested one within the other like the layers of an onion. We live on the surface of the *crust*, the outermost layer. The crust overlies the *mantle*, which in turn surrounds the *core* **(Fig. P.6)**.

- *Hydrosphere:* The **hydrosphere** includes all of the liquid, solid (frozen), and gaseous water of the Earth. Liquid water includes *surface water* (which occurs in oceans, lakes, streams, rivers, swamps, snow, and glaciers) and *groundwater* (which occurs in open holes and cracks underground in the upper crust).

- *Cryosphere:* Sometimes, Earth scientists refer to the frozen component (snow and ice) of the hydrosphere separately as the **cryosphere** (from *cryo–*, the Greek word for icy cold). The cryosphere includes regions where glaciers cover the land surface, where the ground remains frozen all year, and where sea ice covers the ocean.

- *Atmosphere:* The **atmosphere** is a layer of gas that extends from the Earth's surface upward. We refer to the

mixture of gases in the atmosphere as *air*. The density of air decreases with increasing elevation, so 99% of the mass of the atmosphere lies below an elevation of about 30 km (18 mi).

- *Biosphere:* All living organisms on the Earth, together with the portion of the Earth in which organisms exist, constitute the **biosphere**. The biosphere encompasses the land surface, the hydrosphere, the uppermost several kilometers of the geosphere, and the lower several kilometers of the atmosphere. The bottom of the biosphere lies at depths where groundwater approaches boiling conditions.

Earth scientists refer to all the realms that we've just described, along with the intricate ways in which those realms interact with one another over time, as the **Earth System** (**Earth Science at a Glance**, pp. 12–13). Some materials move from realm to realm in the Earth System over time. In this context, a realm can be thought of as a *reservoir*, a place that can contain a quantity of material. In some cases, a succession of movements defines a *cycle* in the Earth System, in that the material eventually ends up where it started.

The Universe and the Earth Are Very Old

Scientific research indicates that the Universe as we know it originated about 13.8 billion years ago. The Earth is much younger—its birth happened 4.54 to 4.56 billion years ago. These large numbers emphasize that our planet existed long before recorded human history began, for the oldest written symbols are only about 5,500 years old. Scientists use the term **cosmologic time** (from the word

The Earth System

Thunderhead

Lightning

External energy

Mountain uplift

Sun

Continental glacier

Rain and snow

City

Ocean

Desert

Rocky coastline

Valley

Arid mountains

Mining

Lakes

Field pattern

Deciduous forest

Forested mountains

Beach

Tropical rainforest

Coral reef

Shark

Internal energy

When you stand on the surface of the Earth, you can see the wondrous ways in which components of the Earth System interact. The geosphere consists of the solid part of our planet. You see it wherever you see exposed rock, sediment, or soil. Most of it lies underground, in the internal layers of our planet. The hydrosphere consists of all liquid water at or near the surface of the Earth. It fills oceans, lakes, and underground pores, and occurs as gas in the atmosphere. The cryosphere consists of frozen water, mostly in glaciers. The biosphere consists of living organisms, from bacteria to whales. The atmosphere is the envelope of gas that encircles the planet. Flow in the air and sea transfer heat and water around the planet.

Internal energy rising from the interior and external energy coming to the Earth from the Sun keep the Earth System dynamic, so that materials cycle from component to component over time. Human society is having a growing impact on the Earth System, by extracting resources, building and farming on the Earth's surface, and emitting waste.

Cirrus clouds

Jet stream

Moon

Aurora

Ice and snow

Wind system

Coniferous forest

Evaporation

Volcanic islands

Industrial pollution

Cold surface current

Surface waters

Delta

Swamps

Warm surface current

Twilight zone

Abyssal zone

Whale

Seafloor

Bacteria and plankton

Giant squid

Deep-sea current

Black smokers

FIGURE P.7 Time scales in Earth science.

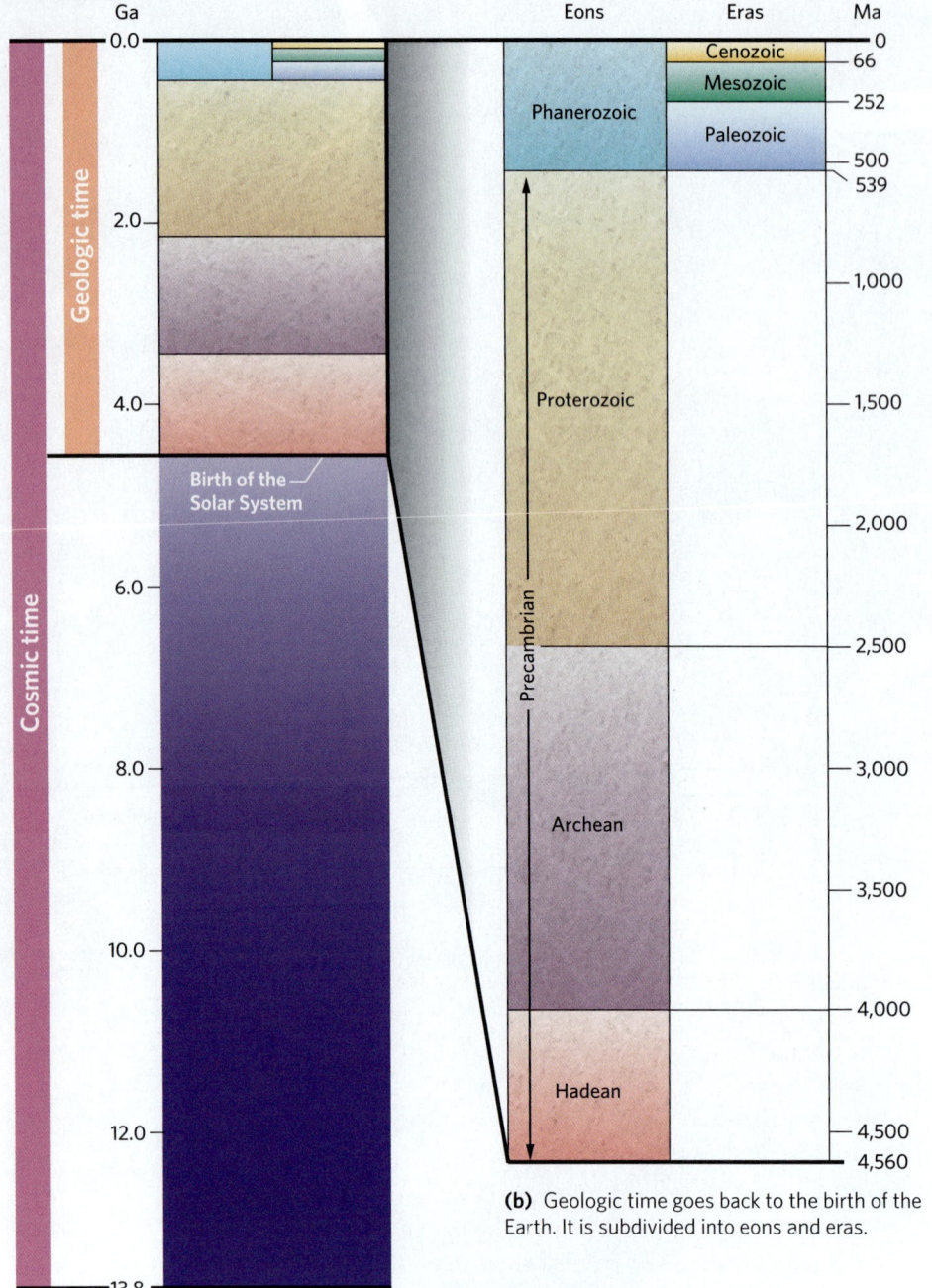

(b) Geologic time goes back to the birth of the Earth. It is subdivided into eons and eras.

(a) Cosmologic time goes back to the beginning of the Universe.

then go extinct. Because numbers used in discussing geologic or cosmologic time can be huge, researchers use the following abbreviations:

1 Ka = 1 thousand years (Ka stands for *kilo-annum*)

1 Ma = 1 million years (Ma stands for *mega-annum*)

1 Ga = 1 billion years (Ga stands for *giga-annum*)

For example, we could say: "The dinosaurs went extinct at 66 Ma;" or, "Before 2.5 Ga, the atmosphere did not contain any oxygen." To discuss geologic time, geologists use the **geologic time scale**, which divides the Earth's history into *eons* and *eras*. We'll see in Chapter 9 that eras include smaller subdivisions called *periods*.

The Theory of Plate Tectonics Explains Many Geologic Phenomena

The Earth's crust, along with the uppermost portion of the mantle, forms a rigid shell 100–150 km (60–90 mi) thick, called the *lithosphere*. This shell lies above the warmer, softer part of the mantle known as the *asthenosphere*. According to the **theory of plate tectonics** (or, simply, *plate tectonics*), the lithosphere currently consists of roughly 20 discrete pieces, called *plates*, which move relative to one another **(Fig. P.8)**. Plate interactions produce earthquakes, volcanoes, and mountain ranges.

Because of plate motions, **continents** (large land areas) move relative to one another as the **ocean basins** (places where the Earth's surface is lower and lies submerged beneath seawater) between them open and close over the course of geologic time. As a result, the map of the Earth's surface constantly changes. In fact, at certain times in the past, continents accumulated into *supercontinents,* huge continents that contained most of the Earth's land, which later broke up. For example, at 200 Ma, the Atlantic Ocean didn't exist, and a dinosaur could have walked from New York to Paris without getting its feet wet. Plate motions take place slowly—at rates of only 1–15 cm (0.5–6 in) per year—so they're almost imperceptible during the course of a human lifetime. But over geologic time, displacements can add together to become large. For example, the entire width of the Atlantic Ocean grew during the past 180 million years, the last 4% of the Earth's history.

The Earth Is a Planet, and Each Planet Is Unique

Eight planets move around our Sun. The inner planets, meaning the ones that lie closer to the Sun, are also known as *terrestrial planets* because they all have solid, rocky surfaces like the Earth, the third planet out

cosmos, a synonym for the Universe) for time intervals related to events in the history of the Universe and the term **geologic time** for time intervals related to the history of the Earth **(Fig. P.7)**. The immense span of cosmologic time means that there's more than enough time for stars to form, mature, and die. Similarly, the immense span of geologic time means that there's more than enough time for mountain ranges to rise and later be removed by erosion, and for species of living organisms to appear and

FIGURE P.8 A simplified map of the Earth's surface showing the major plate boundaries of which there are three types, indicated by different symbols [transform, divergent (ridge or rift), and convergent (trench or collision zone); see Chapter 2 for more on this]. The plates are the areas between the boundaries. The orange arrows indicate the direction and velocity of plate movement; the longer the arrow, the faster the motion.

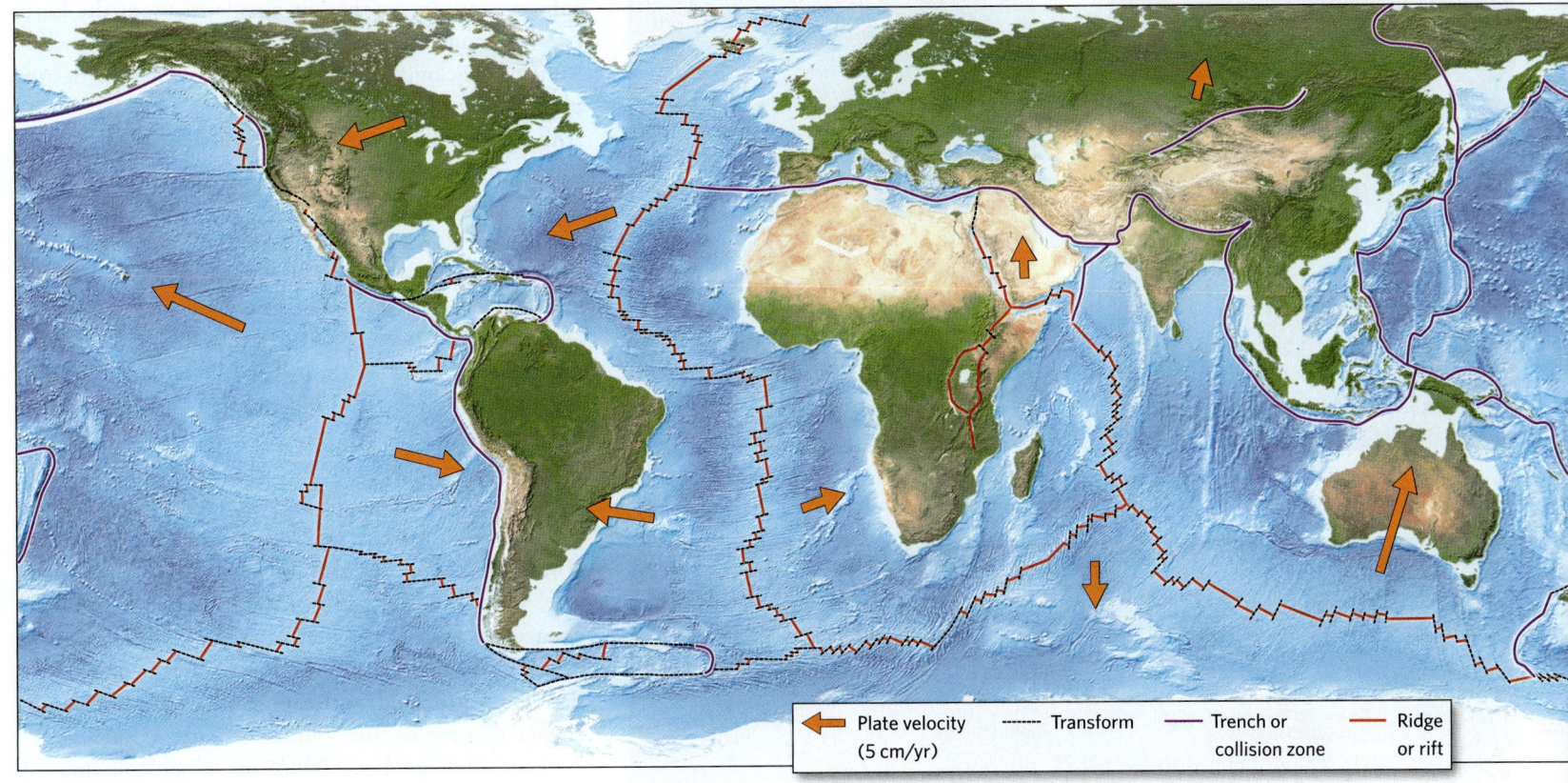

| ← Plate velocity (5 cm/yr) | ------ Transform | —— Trench or collision zone | —— Ridge or rift |

from the Sun. The outer planets are known as the *Jovian planets* because, like Jupiter, they contain vast amounts of gas and do not have rocky surfaces. No two planets look the same. Of the terrestrial planets, only the Earth has liquid water at its surface, an oxygen-rich atmosphere, and moving plates, and only the Earth, so far as we know, harbors life. In recent years, astronomers have discovered more than 4,000 *exoplanets*, planets associated with stars other than our own. Some of these planets may be Earth-like in size, but whether any host moving plates, liquid water, or life remains unknown.

Various Sources of Energy Affect the Earth and Other Planets

The Earth is a dynamic place. As we've noted, interactions among moving plates generate earthquakes, volcanoes, and mountains, while flow of water, ice, and air causes erosion of our planet's surface and redistributes materials. Three sources of energy drive this activity:

1. **Internal energy** refers to the heat contained within the Earth. Plate motion, with all its consequences,

happens because this heat keeps much of the Earth's interior soft enough to flow very slowly. Some of this heat remains from the squeezing together, or *compression*, of materials during our planet's formation, and some comes from *radioactive atoms* (atoms that spontaneously break down and release energy).

2. **Gravity**, the pull that one mass exerts on another, holds the Earth together and keeps it circling the Sun. It also causes tides to rise and fall, rivers to flow, and rocks to tumble downslope.

3. **External energy** travels to the Earth from the Sun in the form of light. It contributes to heating the ground, the atmosphere, and the oceans. Working together with gravity, external energy drives the wind and the ocean currents, which in turn transport materials from one location to another.

Internal energy, gravity, and external energy affect not just the Earth, but all objects of the Universe. But the relative amounts of energy provided by these three sources vary between different objects. We can understand many aspects of the behavior of these objects by

FIGURE P.9 The atmosphere and oceans of the Earth System are in constant motion at all time and distance scales.

(a) A satellite view shows swirling regions of clouds.

(b) In a tornado, air spirals up a funnel at immense speed.

(c) Ocean waves crash on the rocky shore of a Hawaiian island.

thinking about the interplay among the energy sources affecting them.

Change Happens

In space, stars and planets form, objects collide, stars explode, and the Universe expands. On the Earth, mountains rise, continents move, landscapes undergo erosion, the climate warms and cools, sea level rises and falls, and life evolves. Clearly, change happens everywhere—but rates of change vary immensely. For example, some stars have remained virtually unchanged since the Universe began, whereas others burned furiously for just a few million years and then blew up. On the Earth, it takes hundreds of millions of years

for continents to merge and drift apart, but a landslide may remove a mountainside in a matter of minutes, and an earthquake may last only seconds. If you look at a time-lapse satellite image of cloud patterns in the atmosphere, you can see that air circulates about the Earth over the course of days (**Fig. P.9a**), yet hidden within some of the clouds may be a tornado, which can destroy everything in its path in less than a minute (**Fig. P.9b**). And if you study the ocean, you'll see that some currents carry water across entire ocean basins over the course of months, while 600 waves may arrive at a beach every hour (**Fig. P.9c**).

Witness the aftermath of a large flood or hurricane, and it's clear that the might of natural phenomena can change human-made structures (**Fig. P.10a**). But watch a bulldozer clear a swath of forest, hear dynamite blast apart a cliff, try to sail through rafts of garbage floating in the ocean, walk in a human-made canyon between rows of skyscrapers, or measure the increasing carbon dioxide concentration in the atmosphere, and it's clear that people can also play a key role in changing our planet (**Fig. P.10b**). When discussing change on the Earth, therefore, we distinguish between **natural change**, meaning change due to geologic, meteorological, or astronomical events that would happen whether or not people inhabited the Earth, and **anthropogenic change**, meaning change that results from human activity.

For all but the last 0.000001% of geologic time, all change that took place in the Earth System was natural. In the past few centuries, however, our species has developed the ability to modify the Earth System in ways, and at rates, that do not happen naturally. During that time, we have transformed prairies and forests into fields and cities, we have moved mountains and redirected rivers, and we have synthesized a vast variety of new substances that didn't exist before. Because of human activities, the past few centuries of Earth history differ so much from previous times that some scientists refer to this time interval as the *Anthropocene*.

Take-home message . . .

Many narrative themes tie together concepts in Earth science. For example, matter and energy behave in predictable ways, the Earth is a complex system with interacting realms, the Universe and the Earth are so old that there's plenty of time for significant change to take place, plate tectonics causes geologic phenomena on the Earth, and three sources of energy affect all objects in the Universe.

Quick Question

What examples emphasize that humans now change the Earth at significant rates?

(a) The natural power of a hurricane can level a beachside community.

(b) The human-generated power of a chain saw and a bulldozer can level a forest.

P.4 Why Study Earth Science?

Hundreds of thousands of people pursue careers as professionals in various aspects of geology, oceanography, atmospheric science, and astronomy. These careers are anchored in corporations, universities, government agencies, observatories, consulting firms, and nongovernmental organizations. Nevertheless, most students reading this book don't have such careers as a goal. So why should future teachers, construction workers, executives, volunteers, lawyers, office personnel, sales managers, insurance agents, truck drivers, politicians, or doctors study Earth science? This question has many answers.

Earth Science Addresses Practical Issues

An Earth Science course may be among the most practical courses you can take. Let's consider a few examples of issues you may encounter that might be affected by what you learn in an Earth Science course.

- *Where you live:* Is the location of an apartment that you'd like to rent safe from flooding, landslides, or earthquakes **(Fig. P.11)**? Earth science can help you to assess risk due to natural hazards.

FIGURE P.11 Earth science provides insight into natural hazards.

Columbia mudslide, 2017

Indonesia earthquake, 2018

(a) Knowledge of Earth science might have prevented building homes on a dangerous slope.

(b) Knowledge of Earth science may guide the establishment of building codes that allow buildings to withstand shaking.

certain types of clouds. Has someone announced that an earthquake, meteorite, or hurricane will strike the day after tomorrow? Earth science can help you evaluate the reliability of such a prediction.

• *How to interpret societal issues:* Have you read news articles about climate change? Earth science introduces the methods used to study the issue and the basis for interpreting statements about it.

We could continue adding to this list for pages. Hopefully, by the end of your Earth Science course, you'll have insights into many issues that could affect your livelihood, your family, your country, and your planet in the future. Because the Earth is our only home, all citizens of the 21st century have a stake in its future, and the study of Earth science can help people understand how their choices may affect this future.

• *What you drink:* Does your community pump drinking water from wells? Earth science can help you evaluate whether or not a new landfill project outside of town might contaminate the water supply.

• *How to interpret the weather:* Are you worried that thunderstorms may disrupt an upcoming picnic **(Fig. P.12)**? Earth science can help you interpret a radar display to predict a storm's approach.

• *When to worry about risks:* Do you get nervous when flying in an airplane? Earth science can help you understand why planes bounce when passing through

Earth Science Helps You Understand Our Context in the Universe

This book can give you a perspective on how your home planet, you, and all your surroundings came to be. You'll learn not only about the Earth, but also about the countless other objects in the visible Universe. You will come to appreciate the immense size of the Universe, and you will realize that, although you can't feel the motion, you, and everyone you know, are taking an incredibly rapid journey on an island zipping through space. The limits of our island came into human consciousness when astronauts sent back the first photo of the Earth rising above the horizon of the Moon **(Fig. P.13)**.

FIGURE P.13 Earth science may help you understand the context of your home planet in space. This view of the Earth, taken by an astronaut orbiting the Moon, emphasizes that we live on an island in space.

(a) Coal from this mine can be burned to produce energy.

(b) Gases from this power plant can affect air quality.

Earth Science Addresses Resources and Sustainability

We take it for granted that we can construct buildings, roads, and dams out of concrete; that we can go to a store and buy metal cooking pots or paper clips; that we can run a laptop computer for hours on a built-in battery; and that we can generate the power to light a city from materials extracted from the ground (Fig. P.14a). Do you ever wonder where all this "stuff" comes from? Where do we obtain the cement in concrete, the clay in bricks, the copper of pipes and wires, or the gasoline that powers a car? All of these resources come from the Earth.

Because Earth science addresses the origin and supplies of resources, it provides a basis for understanding **sustainability**, the ability to maintain or improve our standard of living without running out of resources. In addition, you'll see why many activities of human society impact *environmental quality* (the breathability of air, the drinkability of water, and the ability of a region to support life) and may change the Earth's climate (Fig. P.14b).

Earth Science Helps You Appreciate Your Surroundings

Hopefully, this book will help you see your surroundings differently. When you drive down a highway, road cuts through rocks will no longer be merely gray cliffs, but rather open books narrating the story of the Earth's long history (Fig. P.15). When you stand on the beach, you won't just enjoy a pretty view, but may think about how the interface between land and sea constantly changes. When you gaze at clouds drifting by on a windy day, you won't just see puffs of white, but will see them as

FIGURE P.15 Geologic exploration provides not only beautiful views, but a look into the Earth's past. If you were to hike by the rock exposure shown on the left, it might look like "just a cliff of tan rock." To a geologist, however, this cliff reveals many events in the Earth's history: sand was transformed into sandstone layers (beds), and the beds were bent, due to plate motion, into curves called folds.

A cliff in California

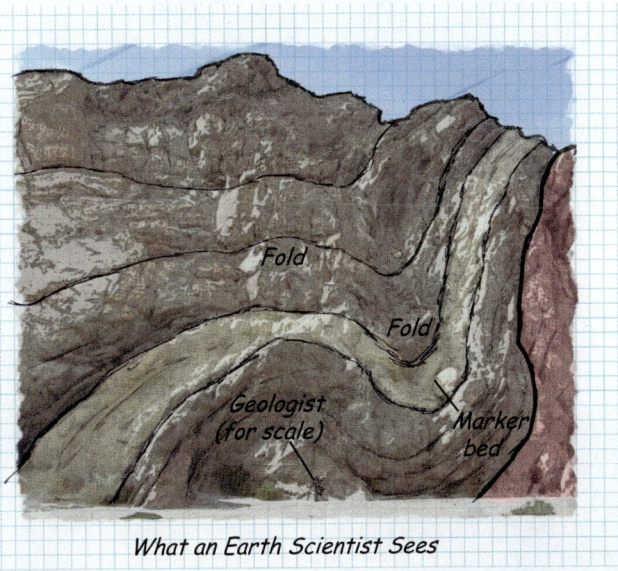

What an Earth Scientist Sees

clues to upcoming weather. And when you look at the night sky, you'll know that the stars aren't just points of light, but rather suns lighting their own solar systems, and you may wonder if distant solar systems host planets like our own.

Now, it's onward to Earth science. We'll begin the book at the beginning, with the formation of the Universe and the elements it's made of. With this background, we will then approach our own planet and describe its basic architecture before moving on to the grand unifying theory of geology, plate tectonics. Part I of the book covers topics related to the land, the materials beneath it, and the features formed on its surface. Subsequent parts address the oceans, the atmosphere, and finally, space. Read on, and you can develop a personal image of the land beneath your feet, the landscapes between you and the horizon, the sea that surrounds the land, the atmosphere that envelops the Earth, and the objects that sparkle in the night sky above.

Take-home message . . .

The study of Earth science can help you understand many practical issues, such as whether your home is subject to natural hazards. It can also help you to appreciate your surroundings and our place in the Universe, and to understand the context of sustainability and environmental issues.

Quick Question
What factors influence the safety of a home site?

ANOTHER VIEW In this snapshot of the Earth System, we can see the past and the present. These red cliffs—formed from sand dunes deposited by the wind when dinosaurs roamed the land—are interacting with the atmosphere, leading to local cloud formation. Internal processes of the Earth caused the land to rise, while wind, water, and gravity are now beveling it back down. The energy driving all this work comes from heat stored and formed inside our planet, and from radiation coming from our nearest star, the Sun.

PRELUDE REVIEW

- Earth science includes many disciplines: geology, oceanography, atmospheric science, and typically, astronomy.

- There are many kinds of scientists—some work in the field, some in a laboratory, and some at a computer. They try to interpret the natural world on the basis of observations, measurements, and calculations, generally by following the scientific method.

- The Universe—everything that exists, both on the Earth and in all of space—contains matter and energy, which behave predictably.

- Our home planet can be envisioned as a complex system, the Earth System, which includes the geosphere, hydrosphere, cryosphere, atmosphere, and biosphere.

- The interior of the Earth consists of the crust, mantle, and core.

- Scientific measurements indicate that the birth of the Universe happened 13.8 billion years ago, while the Earth formed 4.54 to 4.56 billion years ago. Geologic time, the time since the Earth's formation, has been divided into named increments.

- The outer shell of the Earth, the lithosphere, is divided into plates that move relative to one another. This theory of plate tectonics explains earthquakes, volcanoes, and mountain building.

- On the Earth, internal energy and gravity drive plate tectonics; gravity alone drives river flow and downslope movements; and external energy, together with gravity, drives the flow of oceans and the air.

- The Earth is a dynamic planet where change is the norm. Over geologic time, the planet has had time to change in major ways. Human society now causes significant change in the Earth System.

- Earth science helps provide the background needed to understand many practical issues. It will also help you appreciate and understand your surroundings, from the ground to the stars.

Key Terms

anthropogenic change (p. 16)
astronomy (p. 6)
atmosphere (p. 11)
atmospheric science (p. 5)
biosphere (p. 11)
climate science (p. 6)
continents (p. 14)
cosmologic time (p. 11)

cryosphere (p. 11)
Earth science (p. 5)
Earth System (p. 11)
external energy (p. 15)
geologic time (p. 14)
geologic time scale (p. 14)
geology (p. 5)
geosphere (p. 11)

gravity (p. 15)
hydrosphere (p. 11)
hypothesis (p. 9)
internal energy (p. 15)
meteorology (p. 6)
natural change (p. 16)
ocean basins (p. 14)
oceanography (p. 5)

science (p. 5)
scientific method (p. 8)
scientist (p. 7)
sustainability (p. 19)
theory (p. 9)
theory of plate tectonics (p. 14)
Universe (p. 4)

Review Questions

Letters in parentheses correspond to the chapter's learning objectives.

1. What subjects does an Earth science course cover? **(A)**

2. What aspect of Earth science would provide context for a news story about an earthquake? **(B)**

3. Distinguish among a hypothesis, a theory, and a scientific law. **(C)**

4. Describe the different realms of the Earth System. Name the layers of the geosphere. **(D)**

5. What kinds of energy play a role in driving the Earth System and causing aspects of the system to change over time? **(E)**

6. Describe ways you might use your understanding of Earth Science to address problems that your community faces. **(G)**

7. Explain the difference between cosmologic and geologic time. **(F)**

8. Identify practical applications of Earth science. **(G)**

9. Why does an Earth Science course discuss resources? **(G)**

On Further Thought

10. Imagine that a continent moves at 5 cm per year. How long would it take that continent to move 2,000 km? What percentage of geologic time does this length of time represent? **(F)**

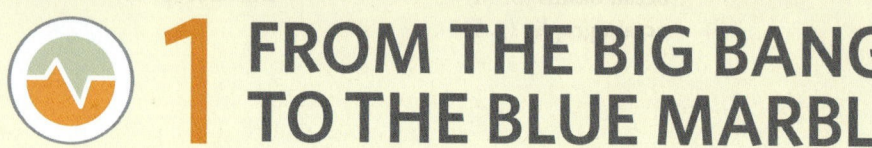

1 FROM THE BIG BANG TO THE BLUE MARBLE

By the end of the chapter you should be able to . . .

A. classify the major types of celestial objects, and indicate which ones lie within our Solar System.

B. outline the scientific explanation for the origin of the Universe and the elements within it.

C. describe how stars and planets formed and evolved.

D. interpret features that a spacecraft would detect when approaching the Earth from outer space.

E. characterize the Earth System concept, and define the realms of the Earth System.

F. distinguish major topographic features of the land and bathymetric features of the seafloor.

G. list the key materials (including the three kinds of rock) that comprise our planet.

H. draw a model of the Earth's internal layers, and describe key characteristics of these layers.

1.1 Introduction

Sometime in the distant past, more than 50,000 generations ago, our ancestors developed the capacity for complex thought. This amazing ability allowed humanity to pose broad questions about the **Universe**, everything that exists both in and on the Earth and throughout all of space beyond. Why is there land and sea? What uplifted mountains and set the courses of rivers? Why does the wind blow and the rain fall? How does the Sun produce heat and light? What are the bright objects in the sky at night? In early cultures, such musings spawned legends in which supernatural heroes sculpted the landscape or pushed stars across the heavens. In recent centuries, *science*—the systematic study of natural phenomena based on observation, experiment, and calculation—has yielded new ideas about how the Universe formed and how it has changed over time, as well as about how the Earth, our home planet, fits in.

In this chapter, we set the stage for discussing the Earth and the rest of the Universe by introducing three topics. First, we describe key concepts of modern *cosmology*, the scientific analysis of the overall structure, formation, and evolution of the Universe. Second, we provide an initial picture of the Earth's surface and its immediate surroundings by taking an imaginary journey aboard a spacecraft that zooms to this planet and then goes into orbit around it. Finally, we peek at the Earth's interior and characterize its overall layering and composition. (In this context, the term *composition* refers to the chemicals that make up a material.) You'll see that our planet contains distinct *realms* (regions or components), which interact in complex and wondrous ways. Earth scientists refer to these interrelated realms, together with the exchanges among them and the ways they influence one another, as the *Earth System*.

1.2 A Basic Image of the Universe

During the day, the Sun shines brightly, and on a cloudless night, the Moon, along with twinkling points of light, decorates the sky. City dwellers actually can see only a few of the twinkling lights because the glow from electrically lit buildings and streets makes the sky too bright. But if you head into the dark countryside on a moonless night, you'll see not only many more points of light, but you'll also

<< Research by astronomers, physicists, and geologists suggests that our Solar System formed when gravity pulled gas, dust, and ice into a disk. The central ball became our Sun. Farther out, matter accumulated into planetesimals.

FIGURE 1.1 The Milky Way, as seen in the dark night sky of Washington State.

see a hazy glowing band arcing from horizon to horizon **(Fig. 1.1)**. Our distant ancestors were mystified by this vista. What are these **celestial objects**, these objects beyond the clouds, and where do they lie in relationship to our home?

Introducing Stars, Galaxies, and Nebulae

Stare at the night sky over the course of a few hours, and you'll see that most of its twinkling points of light move at the same speed and in the same direction. In fact, if you repeat your observations over many years, you'll discover that the position of each point remains fixed relative to the others year after year. Early observers referred to such points as *stars*. Prior to the emergence of modern scientific thinking in the 16th century, people in Western cultures assumed that the stars were anchored to the surface of a distant sphere surrounding the Earth, and that this sphere, along with all other celestial objects—including the Sun—circled the Earth. In other words, people favored a *geocentric model* in which the Earth lay at the center of the Universe. Scientific observations over succeeding centuries eventually revealed that a **star** is actually an incredibly immense, amazingly hot ball that emits vast amounts of light, and that our Sun itself is a star **(Fig. 1.2a)**. We see the Sun as a sphere of light, rather than as a point, simply because it lies so close to us. Observations also demonstrated that stars cluster in vast groups, called **galaxies**, held together by the pull of gravitational force **(Box 1.1)**. Our Sun, along with over 300 billion other stars, belongs

BOX 1.1 | Science **Toolbox**

What is a force?

Simplistically, we can think of a **force** as a push or pull acting on an object. Physicists define force more formally by an equation, $F = ma$. In words, this equation states that force equals mass times acceleration, where *acceleration* means a change in the velocity of an object in a given time interval. Note that the *velocity* of an object depends on both the rate (meaning *speed,* as defined by distance ÷ time) at which the object travels, and its mass—a "change in velocity," therefore, can refer to a change in speed, a change in direction, or both. Therefore, the above equation means that the application of force can cause an object to move at a different speed and/or in a different direction than it had been moving previously. The larger the applied force, the greater the amount of change in velocity it can cause.

We can distinguish between contact forces and field forces. A *contact force* acts when one object pushes or pulls on another by touching it. To picture a contact force, imagine kicking a ball with your foot—the impact causes the ball to head off in a different direction. A *field force* (also known as an *action-at-a-distance force*) emanates invisibly from an object and can cause another object to move even if the objects aren't touching. Examples of field forces include **electromagnetism**, produced by magnets or by electric currents, and **gravity**, produced by the mass of an object. Electromagnetism can cause either *attraction* (the pulling of objects toward each other) or *repulsion* (the pushing of objects away from each other), whereas gravity can cause only attraction. In practice, the "field" of a field force refers to the region in which the force can cause a measurable consequence. The *magnetic field* surrounding a hand-held magnet can move a metal paper clip, and the *gravitational field* of the Earth keeps you from floating out of your chair.

to the **Milky Way Galaxy**, one of perhaps a trillion galaxies populating the Universe. Most of the individual stars that we see in the night sky lie within nearby parts of the Milky Way; the rest of its stars constitute the hazy glowing band visible in Figure 1.1. (Some of the points of light we see, however, are actually very distant galaxies.) Viewed from a great distance, our galaxy looks like a giant, swirling, flattened spiral similar in shape to its neighbor, Andromeda **(Fig. 1.2b)**.

The material substance, or *matter*, making up all objects in the Universe consists of atoms and molecules **(Box 1.2)**. Matter, as we saw in the Prelude, can exist in four different states: *solid*, *liquid*, *gas*, and *plasma*. Significantly, not all matter in galaxies lies within stars—if it did, we wouldn't exist. Galaxies also contain planets and moons, which we'll introduce shortly, and nebulae. A **nebula** is a huge, wispy cloud floating in space **(Fig. 1.3)**. It can contain three kinds of matter: (1) gas, mostly made up of hydrogen molecules

FIGURE 1.2 Stars and galaxies.

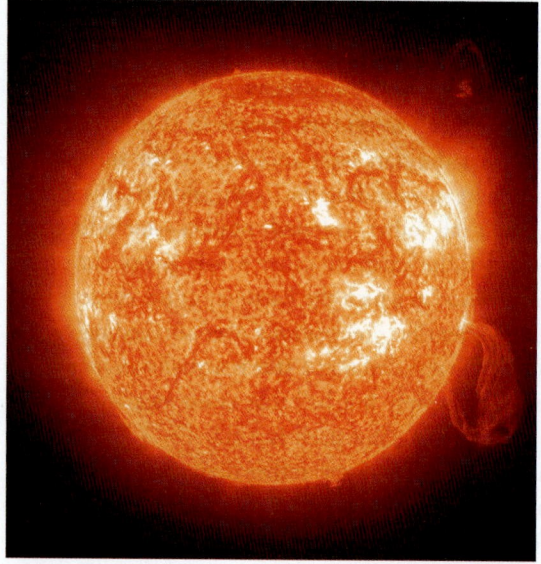

(a) The Sun, like all stars, radiates immense amounts of energy.

(b) The Andromeda Galaxy has a spiral shape much like that of the Milky Way. It contains about a trillion stars.

FIGURE 1.3 The Orion Nebula. The Hubble Space Telescope captured this image of the Orion Nebula, a cloud of gas and dust 24 light-years across, in which new stars are forming.

(H_2) and helium atoms (He); (2) **ice**, solids composed of frozen **volatile materials**—such as water (H_2O), ammonia (NH_3), carbon dioxide (CO_2), and methane (CH_4)—which can *melt* (turn into liquid) at relatively low temperatures and can even *evaporate* (turn into gas) easily under the Earth's surface conditions; and (3) **dust**, tiny particles of **refractory materials**—such as minerals and metals—which have a relatively high melting temperature and do not evaporate easily under the Earth's surface conditions. Refractory materials contain chemicals such as silica (SiO_2), iron oxide (Fe_2O_3), and magnesium oxide (MgO). Note that in the context of discussing celestial objects, the term *ice* has a broader meaning than it does in everyday English, and the term *dust* has a more restricted meaning.

Ancient observers distinguished between stars and planets when they noticed that certain celestial objects move relative to the backdrop of stars and relative to one another. Because of that movement, those objects came to be known as *planets*, from the Greek word *planan*, meaning wanderer. Modern researchers define a **planet** as a large spherical mass that follows an **orbit** (a closed path) around a star and has incorporated all the matter that lies in or near its orbit. By this definition, eight planets, including the Earth, orbit our Sun. Astronomers refer to four of these—Mercury, Venus, the Earth, and Mars—as *terrestrial planets* because they resemble the Earth by being relatively small (the Earth is the largest), and by having rocky surfaces. These planets are also known as the *inner planets* because they lie closest to the Sun. The remaining four planets—Jupiter, Saturn, Uranus, and Neptune—are the *Jovian planets,* named after

Jupiter, the largest of them. These planets, also known as the *outer planets* because they lie farther away from the Sun, are much larger than the terrestrial planets and consist mostly of gases and ices. Each planet rotates on an *axis* (an imaginary line passing through the planet's center and its two poles) like a spinning top. The Sun and the stars seem to move across the sky because of the Earth's rotation, and we see the planets move relative to the backdrop of stars because, though they are far from the Earth, the planets lie much closer to us than do the stars. The planets move relative to one another because they orbit the Sun at different distances and at different speeds.

All the planets, except Mercury and Venus, have moons. Astronomers define a **moon** as a sizable solid object that orbits a planet. The Earth has one—we call it the Moon—whereas Jupiter has 79 known moons that range in size from larger than Mercury to less than 10 km (6 mi) across. The Sun, the planets, and the moons, as well as various other smaller objects held in their orbits by the gravitational force of the Sun, together constitute the **Solar System (Fig. 1.4)**.

What lies between celestial objects? Most of this space consists of a *vacuum*, a volume that contains virtually no matter. The vacuum of *interstellar space*, the region between stars, contains only about 3 atoms or molecules per cubic meter (per 35 ft³). By comparison, the air that you breathe at sea level contains 300,000,000,000,000,000,000,000 atoms or molecules per cubic meter. A nebula may contain thousands of atoms per cubic meter, but it still has a density that is vastly less than the best vacuum produced in a laboratory on the Earth.

BOX 1.2 ▶ ## Science Toolbox

The components of matter

In order to discuss the formation of the Universe and the stars and planets within it, it's necessary to be familiar with the particles that make up matter and how they attach to one another. Here's a quick review.

Elements, Atoms, and Nuclear Bonds

An **element** is a piece of matter that cannot be subdivided into components with different properties. Copper, for example, fits the definition because we cannot subdivide copper into different materials. Salt, in contrast, is not an element because chemists can separate salt into other elements (sodium and chlorine), neither of which looks like salt. If you keep subdividing a piece of an element into smaller and smaller pieces, you end up with an *atom*, the smallest piece of an element that still has the properties of the element. Modern studies have shown that atoms are so small that a single gram (0.035 oz) of hydrogen contains about 602,000,000,000,000,000,000,000 atoms, and 5 to 10 trillion atoms fit within the area of the period at the end of this sentence. We refer to different elements using abbreviations known as *chemical symbols*: for example, H = hydrogen; He = helium; C = carbon; O = oxygen; and Fe = iron.

It wasn't until the early 20th century that researchers realized that atoms themselves can be subdivided into still smaller *subatomic particles*—namely, *protons*, which have a positive charge; *neutrons*, which have a neutral charge; and *electrons*, which have a negative charge **(Fig. Bx1.2a)**. (Simplistically, the term *charge* refers to the electrical behavior of a particle. Think of a positive charge as the "+" end of a battery and a negative charge as the "−" end of a battery. Opposite charges attract, meaning they pull toward each other, whereas like charges repel, meaning they push away from each other.) Neutrons have slightly more mass than protons, and both have much greater mass than electrons.

Protons and neutrons stick together to form a dense *nucleus* at the center of an atom (see Fig. Bx1.2a). Electrons swirl around the nucleus in an *electron cloud;* the outer edge of the electron cloud defines the surface of an atom. Considering that the nucleus of an atom is very tiny compared with the volume within the electron cloud, atoms actually consist mostly of empty space. The attractive force that holds the particles in a nucleus together is called a *nuclear bond*. Nuclear bonds act only over extremely small distances, so in order for *nuclear fusion*—the merging of two nuclei—to take place, the nuclei must be stripped of their electron shells and must collide at high velocity **(Fig. Bx1.2b)**.

FIGURE Bx1.2 Atoms and nuclear bonds.

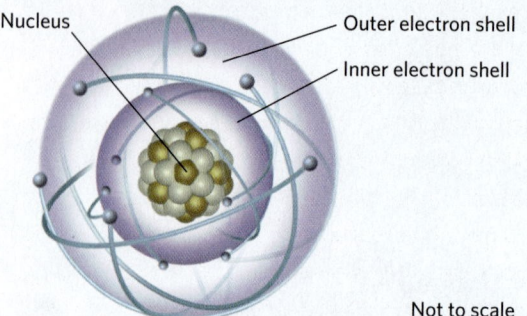

Not to scale

(a) A schematic image of an atom with a nucleus orbited by electrons.

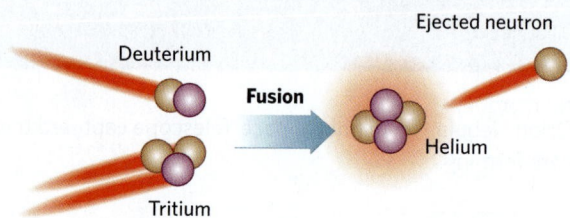

(b) Fusion happens when colliding particles combine to form a new, larger nucleus. (Deuterium and tritium are versions of hydrogen nuclei.)

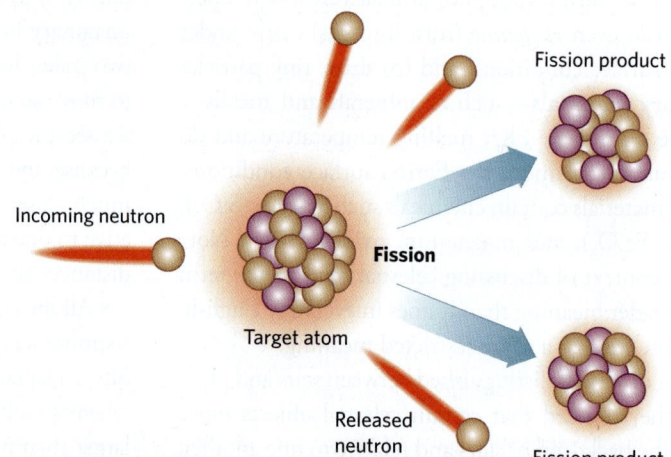

(c) Fission takes place when an atom is split by an incoming particle, producing several smaller fragments of the original atom.

Only when nuclei collide at high velocity can they get close enough for nuclear bonds to overcome electrical repulsion. We can think of *nuclear fission* as the opposite of fusion. When fission takes place, a large nucleus splits into two smaller ones **(Fig. Bx1.2c)**. Both types of *nuclear reactions* (nuclear fusion and nuclear fission) release immense amounts of energy.

Atomic Number and Atomic Mass

An atom of one element differs from an atom of another element because it contains a different number of protons. Scientists refer to the number of protons in an element's nucleus as its *atomic number*. Hydrogen has an atomic number of 1, helium has an atomic number of 2, carbon has an atomic number of 6, and iron has an atomic number of 26. Uranium, with an atomic number of 92, is the largest naturally occurring atom. When nuclear fusion takes place, two nuclei with small atomic numbers bond together to form a new nucleus with a larger atomic number; the result is an atom of a different element. When nuclear fission takes place, a large nucleus splits to form two nuclei of other elements with smaller atomic numbers.

The number of protons plus the number of neutrons approximately represents the *atomic mass* of an atom. All atoms, except for most hydrogen atoms, contain neutrons. Small atoms generally have the same number of neutrons as protons, whereas large atoms contain more neutrons than protons.

Molecules, Compounds, and Chemical Bonds

Atoms can be attached to other atoms by an attractive force called a *chemical bond*. Scientists use the word **molecule** for a particle composed of two or more atoms bonded together by chemical bonds. These bonds, unlike nuclear bonds, involve the interaction of electron clouds, not of nuclei. Some molecules contain atoms of only one element, whereas others contain atoms of two or more different elements. Chemists refer to a substance in which molecules consist of two or more elements as a *compound*.

We can specify the composition of a molecule by providing its *chemical formula*, a recipe that uses abbreviations to represent the elements in a compound and their relative proportions. For example, a hydrogen molecule (H_2) consists of two atoms of hydrogen, and a water molecule (H_2O) consists of two hydrogen atoms bonded to a single oxygen atom. Note that a molecule is the smallest piece of a compound that has the characteristics of that compound.

Measuring Distances in Space

When we see the Sun, or any star, we are seeing light that has traveled through space. (As we'll discuss later, visible light is only one of many types of *electromagnetic radiation*; other types include radio waves, X-rays, and ultraviolet light.) The distances between stars and between galaxies are so immense that, to avoid having to use colossal numbers, astronomers specify these distances in terms of light-years. One **light-year** represents the distance that light, or any electromagnetic radiation, travels in one year. Because the *speed of light* is so fast—about 300,000 km (186,000 mi) per second—one light-year equals a huge distance, roughly 9.5 trillion km (5.9 trillion mi). The nearest star to our Sun lies 4.2 light-years away, and the Andromeda Galaxy lies over 2.5 million light-years away. This means that when we look at Andromeda, we are seeing light that began its journey toward the Earth 2.5 million years ago—we are looking back in time!

FIGURE 1.4 The relative order of the planets in the Solar System. Note that the orbits all lie approximately in the same plane, which has been highlighted. Neither the distances between the planets nor their sizes are shown correctly to scale.

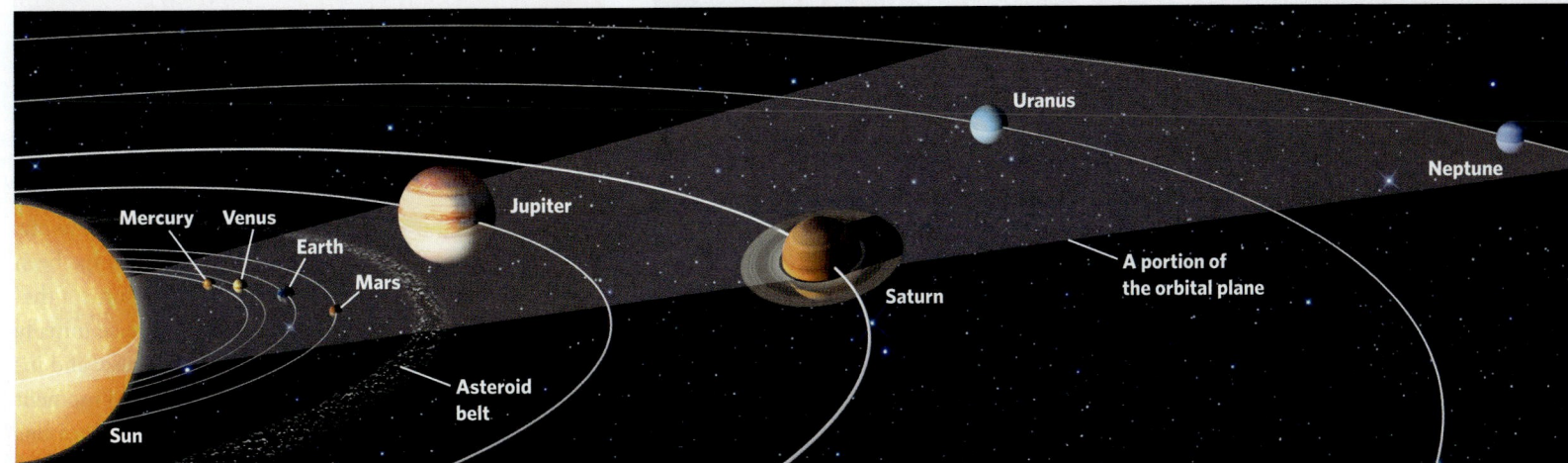

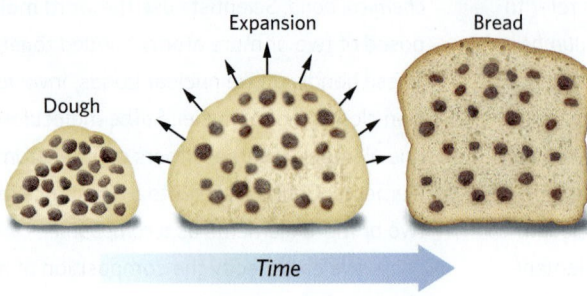

FIGURE 1.5 A raisin-bread analogy for the expanding Universe. As the dough expands, each raisin moves farther away from the others.

Dough
Expansion
Bread
Time

The Expanding Universe

In the 1920s, astronomers such as Edwin Hubble (1889–1953) braved many a frosty night gazing through telescopes at mountaintop observatories in order to chart distant galaxies too faint to be seen with the naked eye. At first, they just wanted to record the locations and shapes of the galaxies. But eventually, these astronomers learned how to determine whether the galaxies were moving toward or away from the Earth and to calculate how fast they were moving (Chapter 23 explains how).

Hubble pondered the results of this work and noted that all distant galaxies—regardless of their direction from the Earth—are moving away from us, and that the farther

FIGURE 1.6 An artist's rendering of Universe expansion from the Big Bang through the present and on into the future.

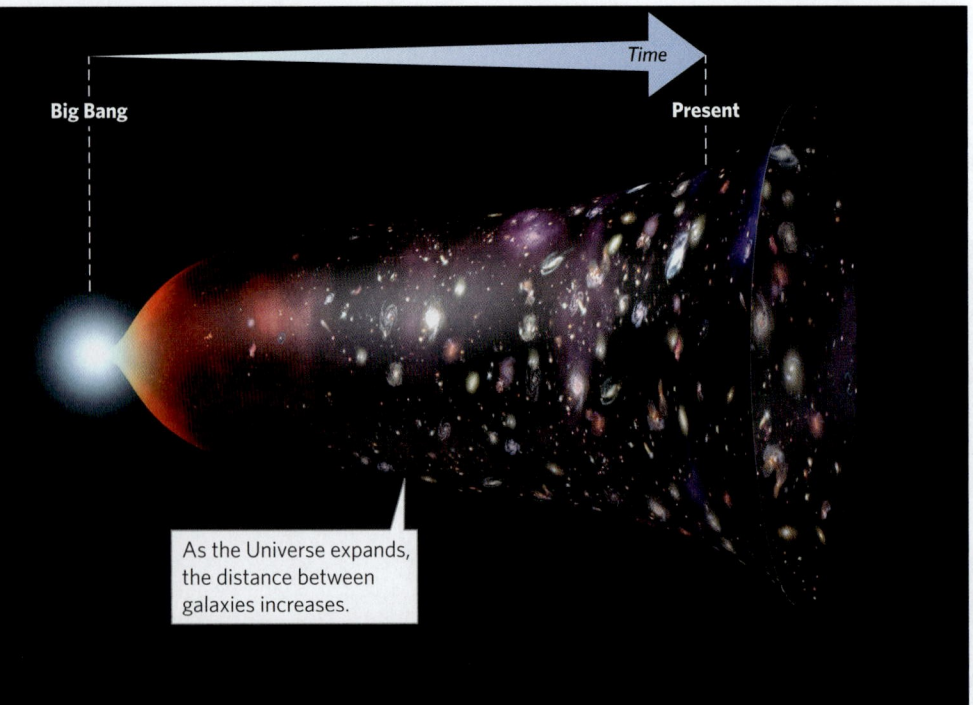

Time

Big Bang
Present

As the Universe expands, the distance between galaxies increases.

away they are, the faster they appear to be moving. This result, at first, seemed bizarre. But Hubble soon realized that it meant that the Universe must be expanding. Up until Hubble's eureka moment, astronomers had assumed that the size of the Universe remained unchanged, so his proposal was a radically new idea. Hubble's hypothesis eventually passed so many tests that it gained the status of a theory (see Box P.2), now known as the **expanding Universe theory**.

To understand Hubble's proposal and picture an expanding Universe, imagine that you're a raisin in a ball of fresh dough being baked to make raisin bread **(Fig. 1.5)**. As the dough bakes, the volume of the ball expands, and as expansion takes place, all the other raisins move away from you and from one another. Significantly, raisins that were farther from you prior to baking move away from you at a faster velocity than do raisins that were closer to you prior to baking. To see why, keep the definition of speed (speed = distance ÷ time) in mind. If the radius of the dough ball doubles during 1 hour of baking, then raisins that were initially 1 cm apart move away from each other at a speed of 1 cm/h, while raisins that were initially 4 cm apart move away from each other at 4 cm/h.

Take-home message . . .

> The Universe includes stars (such as the Sun), planets, moons, and nebulae. The Earth is one of eight planets orbiting the Sun. Our Solar System lies in the Milky Way Galaxy, one of perhaps a trillion galaxies in the Universe. As the Universe expands, galaxies move away from one another.

Quick Question

Why, when you look at the night sky, do you see the past?

1.3 Formation of the Universe and Its Elements

The Big Bang

Hubble's expanding Universe theory triggers the following question: Did the expansion of the Universe begin at a specific time in the past? If so, that time would mark the beginning of the Universe. Astronomers conclude that its expansion did indeed begin with an event known as the **Big Bang (Fig. 1.6)**. According to this idea, which has been supported by so much evidence that it has become a theory, everything that now constitutes our Universe initially existed within an extremely small, infinitely dense, and infinitely hot volume known as a *singularity*. At the moment of the Big Bang, which astronomers estimate, for reasons described later in this book, happened at about 13.8 Ga,

BOX 1.3 ## Science Toolbox

Heat and temperature

The atoms and molecules that make up a solid object do not stay rigidly fixed in place. Rather, they vibrate by moving back and forth without breaking the chemical bonds that hold them to their neighbors. Atoms in a liquid or gas not only vibrate, but can also move about with respect to one another. Their vibrations and other movements represent **thermal energy**, defined as the total kinetic energy (energy of motion) of all a substance's atoms in a given volume. When a material warms, its atoms or molecules move faster, so its thermal energy increases. Similarly, when a material undergoes *compression* (squeezing together) without losing thermal energy, it heats up. We use the term **heat** in reference to thermal energy transferred from one object or region to another.

When we say that one object is hotter or colder than another, we are describing its **temperature**, a measure of warmth relative to a known reference value. Temperature represents the average kinetic energy of atoms in the material. Heat always flows from a location of higher temperature to a location of lower temperature. To specify temperature—that is, to indicate relative hotness or coldness—scientists use the

centigrade scale (also called the *Celsius scale*), which is divided into *degrees* (°C). Scientists have calibrated the centigrade scale so that at sea level, the boiling point of water is 100°C and the freezing point is 0°C. An increment of 1°C, therefore, equals 1/100 of the temperature difference between the boiling and freezing points of water. The English system uses the *Fahrenheit scale* for temperature. On this scale, water boils at 212°F and freezes at 32°F at sea level. Note that degrees on the Fahrenheit scale are smaller than those on the centigrade scale.

Scientists sometimes use the *Kelvin scale*, named for Lord Kelvin (1824–1907, a British physicist) to specify temperature. On the Kelvin scale, the coldest possible temperature, called *absolute zero*, is designated as 0 K. (Note that numbers on the Kelvin scale do not use the degree sign.) A material simply can't get colder than absolute zero, meaning that you cannot extract any thermal energy from a substance at 0 K because all the motion of its atoms and molecules has ceased. An increment of temperature on the Kelvin scale has the same magnitude as one on the centigrade scale. 0 K equals −273.15°C, so on the Kelvin scale, the freezing point of water is 273.15 K, and the boiling point of water is 373.15 K.

---

the singularity began to inflate or expand. Of course, no human was present then, so no one actually experienced the Big Bang. But by combining clever calculations with careful observations, researchers have developed a model of how the Universe evolved once the expansion started.

During the first moments after the Big Bang, the Universe was still so dense and so hot that matter as we know it couldn't exist **(Box 1.3)**. Within a few seconds, however, the smallest atoms, hydrogen atoms, began to form. Only a few minutes later, when the Universe's temperature had fallen below 1 billion degrees centigrade (1.8 billion degrees Fahrenheit), most of the hydrogen atoms that exist today had formed. For the next several minutes, some of the hydrogen atoms collided and underwent nuclear fusion to form helium atoms as well as small quantities of other atoms with atomic numbers of less than 5. By the time the Universe was a mere 20 minutes old, this process, known as **Big Bang nucleosynthesis**, had concluded. It yielded mostly hydrogen (74% by mass) and helium (24% by mass). Later, chemical bonding of hydrogen atoms produced H_2 molecules.

Birth of the First Stars: The Nebular Theory

By the time the Universe reached its 200 millionth birthday, large volumes of H_2 and He gas had collected into

immense, slowly swirling nebulae. These first nebulae were relatively dense, because the young Universe was still fairly small, and they differed from the nebulae seen in the Universe today in that they consisted only of gas, for the elements that make up ice and dust particles hadn't formed yet. Eventually, at a great many locations, patches of nebulae became dense enough to begin collapsing due to the inward pull of gravity. When this happened, atoms and molecules of gas moved closer together, so the collapsing region became progressively denser. In a grand example of the rich getting richer, the most massive regions drew in matter from nearby, less dense regions. As explained in Box 1.3, compression of a material into a smaller volume causes its temperature to rise. So as collapsing patches within a nebula got denser and denser, they also became hotter and hotter. In addition, their initial swirling motion transformed into organized rotational motion around an axis, so that a collapsing patch of nebula took the shape of a flattened disk with a bulbous center **(Fig. 1.7a)**. Astronomers refer to such a disk as an **accretion disk** because matter accreted, or became attached, to the disk over time. Matter that was moving more slowly fell toward the central ball of the disk, while matter that was moving fast enough remained in

FIGURE 1.7 The nebular theory.

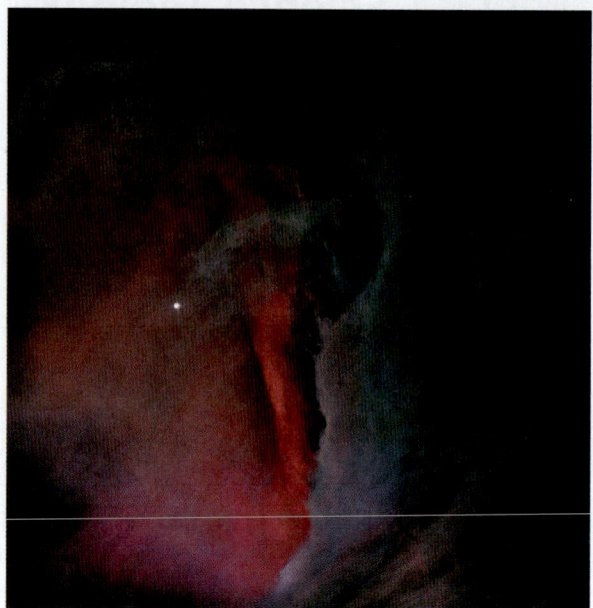

(a) Artist's rendition of a star forming as matter falls into a rotating accretion disc. As the mass collapses inward, the orbital velocity increases, as occurs when a spinning skater pulls in her arms.

(b) An early nebula might have looked like this, when the first star formed. It's mostly swirling gas.

> This truth within thy mind rehearse, That in a boundless Universe Is boundless better, boundless worse.
>
> —ALFRED TENNYSON
> (BRITISH POET, 1809–1892)

 Video
Formation of the Solar System

the flattened outer part of the disk, in orbit around the central ball.

When the central ball of an accretion disk became sufficiently massive, gravity would cause it to collapse inward even more. The resulting compression made the ball become even hotter. Hot objects radiate light, so when the ball became hot enough, it began to glow. When this happened, the central ball had become a **protostar**. What happened next depended on the amount of matter the protostar contained. That's because heat causes a gas to expand or spread out, producing an outward push called *thermal pressure*. (You can see an example of thermal pressure in action by placing an inflated balloon outside on a hot day: expansion of the warming gas in the balloon stretches the balloon until it bursts.) If a protostar contained a relatively small amount of matter, the outward push of thermal pressure would balance the inward pull of gravity before the protostar could collapse enough to become a star. But in the early Universe, which was much smaller than the Universe today, some accretion disks contained so much matter that the protostars formed from them were extremely massive. The inward pull of gravity in a very massive protostar could overcome thermal pressure, allowing these stars to continue collapsing inward until their interiors became very dense and extremely hot (about 10,000,000°C). At this stage, hydrogen nuclei in the protostar were zipping around so fast that they could come close enough to one another to slam together. Such collisions resulted in the *fusion* of hydrogen nuclei to form helium nuclei (see Box 1.2), a process

that generates prodigious amounts of energy. When fusion began in massive protostars, those bodies effectively ignited to become nuclear furnaces—true stars—and the first starlight lit the Universe **(Fig. 1.7b)**. This first generation of star formation happened at countless locations throughout the many nebulae of the young Universe, and by about 800 million years after the Big Bang, vast numbers of stars had formed and, due to the gravitational pull among them, had organized into the first galaxies. The idea that stars formed within collapsing patches of a nebula is known as **nebular theory**.

Formation of the Elements: We Are All Made of Stardust!

The Big Bang produced all of the matter that exists in the Universe, but it did not produce all of the elements that exist today. Specifically, as we noted earlier, Big Bang nucleosynthesis yielded only hydrogen, helium, and trace amounts of other elements with atomic numbers of less than 5, but the Universe today contains 92 naturally occurring elements. Where did the elements with larger atomic numbers come from? In other words, when and how did heavier elements such as carbon (atomic number 6), iron (atomic number 26), and uranium (atomic number 92) form? Heavier elements form inside stars by a process called **stellar nucleosynthesis**, during which smaller nuclei fuse together to form larger ones. In other words, stars are element factories. When a star runs out of hydrogen (its original nuclear fuel), new fusion reactions take place that produce heavier elements. In fact, in very large stars, these

FIGURE 1.8 Element factories in space.

(a) When the Moon blocks the Sun during an eclipse (see Chapter 21), we can see the gases of the solar wind streaming into space.

(b) This expanding cloud represents the remnant of a supernova explosion that happened 4,500 year ago.

reactions can also yield atoms up to and including iron (see Chapter 23).

Some of the new atoms formed within a star escape into space during the star's lifetime simply by moving fast enough to overcome the star's gravitational pull. These atoms make up *stellar wind*, a stream of charged particles that flows from a star; astronomers refer to stellar wind from our own Sun as the **solar wind (Fig. 1.8a)**. Most of the atoms, however, do not escape until the *death of a star*, the time when the star has used up its nuclear fuel and starts to undergo radical changes. As we'll see in Chapter 23, stars come in a range of sizes. A small dying star ejects matter into space after it has evolved into a *red giant*, whereas a large star becomes unstable and undergoes an immense, violent explosion, which suddenly ejects huge quantities of matter into space. Astronomers refer to a giant exploding star as a **supernova (Fig. 1.8b)**, from the Latin *nova*, meaning new, because to observers on the Earth, the explosion appears as a new, bright, but short-lived point of light in the night sky. Most of the heaviest elements form during supernova explosions, a process known as **supernova nucleosynthesis**.

First-generation stars were huge, some with a mass a hundred or more times that of our Sun. Astronomers calculate that the temperature in very massive stars becomes so extreme that fusion reactions consume nuclear fuel very quickly. As a result, a very massive star may survive only a few million years before it collapses and explodes. Supernova explosions, therefore, peppered the young Universe, and these explosions injected elements that had formed in stars, as well as elements produced during the explosion, into surrounding nebulae. So, as

first-generation stars formed and died, the atoms of heavier elements mixed with hydrogen and helium left over from the Big Bang, and the chemical composition of nebulae in the Universe changed.

Second-generation stars formed out of these new, compositionally diverse nebulae. Those second-generation stars lived and died and contributed elements to nebulae from which third-generation stars formed, and so on. Because of stellar nucleosynthesis and supernova nucleosynthesis, each generation of stars contains an increasing proportion of heavier elements. Different stars live for different lengths of time because the lifetime of a star depends on its mass, so at any given moment, the Universe contains many generations of stars. Our own Sun may be a third-, fourth-, or fifth-generation star. Think of it—since the mix of elements we find on the Earth, including within our own bodies, comes from the hearts of extinct stars, we are all made of stardust!

Did you ever wonder . . .

where the atoms that make up your body first formed?

Take-home message . . .

Research indicates that the Universe originated during the Big Bang, about 13.8 billion years ago. Big Bang nucleosynthesis produced mostly hydrogen and helium. According to the nebular theory, gas clouds composed of these elements collapsed to produce the first stars, which generated energy by nuclear fusion. These stars later exploded as supernovae. Nucleosynthesis in stars and supernovae produced larger atoms.

Quick Question

Could a terrestrial planet have formed in association with a first-generation star?

FIGURE 1.9 Formation of solid bodies in the Solar System.

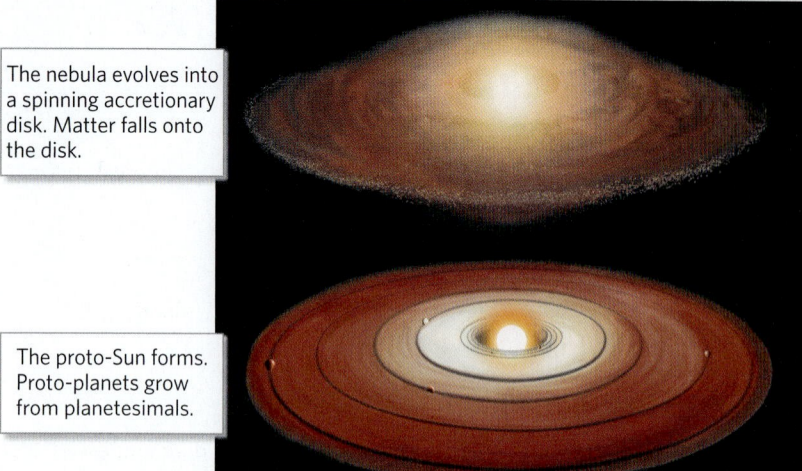

The nebula evolves into a spinning accretionary disk. Matter falls onto the disk.

The proto-Sun forms. Proto-planets grow from planetesimals.

(a) An accretion disk containing gas, ice, and dust forms. Eventually, it evolves into a protosun surrounded by a protoplanetary disk.

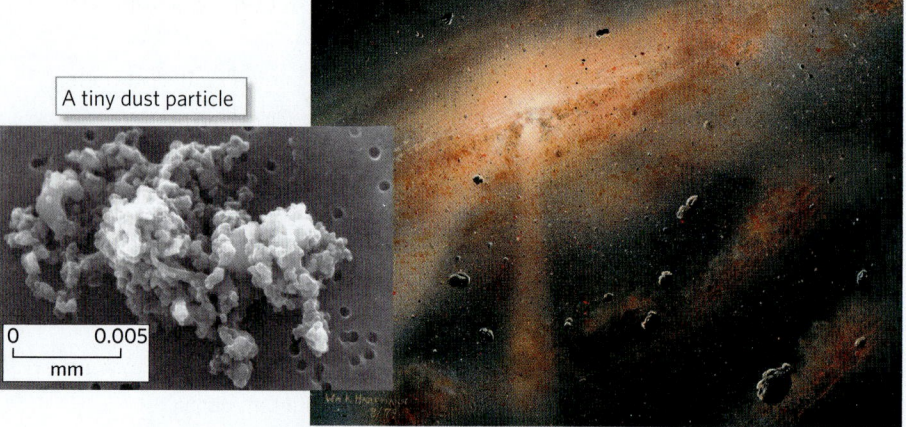

A tiny dust particle

0   0.005

mm

(b) In the protoplanetary disk, small particles of gas, ice, and dust clump together, eventually growing into planetesimals.

▶ **Video**

Formation of the Earth

(c) Some planetesimals grew into grainy masses that might have looked similar to this asteroid, Bennu, whose diameter is greater than the height of the Empire State Building.

1.4 Formation of the Earth

Condensation Theory and Planet Formation

We've seen that stars of the first generation were very massive and relatively short-lived. They had to be massive, or they couldn't have collapsed enough for fusion to begin. Smaller stars, like our Sun, burn at cooler temperatures and can survive for billions of years. How do Sun-like stars form? Their formation could take place only in nebulae that contained tiny (0.01 mm, or 0.0004 in) ice and dust particles. Such particles formed from the condensation of matter into solids, and such condensation could take place only in nebulae that contained larger atoms (such as oxygen, silicon, and iron), ones produced by stellar nucleosynthesis and supernova nucleosynthesis. Ice and dust are important because they disperse heat so efficiently that their presence decreases the thermal pressure in protostars and therefore permits smaller protostars to collapse enough to ignite and become true stars. When such stars form, gas, ice, and dust remain in a disk surrounding the star, and this disk, as we'll see below, provides the material from which planets form. The set of ideas describing how an ice- and dust-containing nebula evolves into a star surrounded by planets has come to be called **condensation theory** because of the role played by the condensation of matter into solids during the process.

With condensation theory in mind, let's examine the scientific model for the birth of our own Solar System. Its history began when an accretion disk developed within a nebula in an arm of the Milky Way Galaxy **(Fig. 1.9a)**. This disk contained not only hydrogen and helium, but also larger atoms. The central ball of this disk collapsed to become the Sun around 4.57 Ga. As this happened, the outer part of the accretion disk—the material that did not get drawn into the Sun by gravity—remained in orbit around it. This region is known as the **protoplanetary disk** because it provided the matter that would be incorporated into planets.

As soon as the Sun became a nuclear furnace, the solar wind began to blow. This wind carried volatile materials from the inner part of the protoplanetary disk into the outer part, where some froze back into ice specks, creating a boundary now called the *frost line*. Inside the frost line, closer to the Sun, terrestrial planets formed. At distances beyond the frost line, enough volatiles remained for the Jovian planets to form.

How did planetary growth take place? Specks of dust and ice acted as "seeds" to which other atoms could attach, so the specks gradually grew into soot-sized particles. Under the influence of gravity, these particles began to clump together into grains, which drifted together to form boulder-sized blocks, which in turn combined into still larger blocks **(Fig. 1.9b, c)**. Some of these blocks became big enough for their gravity to pull in other blocks and, eventually, they grew into **planetesimals**, solid bodies with diameters greater than 1 km. A few planetesimals won the competition to attract other objects in their orbits and became **protoplanets**, bodies approaching the size of today's planets. Astronomers estimate that the early Solar System may have had up to 80 such protoplanets. As these bodies grew, their gravitational pulls tugged on one another. This pull caused some to fall into the Sun, some to escape the Solar System, and some to collide. Eventually, only eight protoplanets succeeded in collecting all the material in their orbits. These bodies became the planets, including the Earth and Moon, that we observe today (Earth Science at a Glance, pp. 34–35).

The condensation theory explains key characteristics of the Solar System. Specifically, all of its planets lie in roughly the same plane because they formed within the same protoplanetary disk. Those planets in the inner orbits, where the protoplanetary disk consisted mostly of dust, became the terrestrial planets, consisting mostly of rock and metal. In contrast, those planets that grew in the outer part of the Solar System, where volatiles accumulated, became the giant Jovian planets consisting mostly of gas and ice (surrounding relatively small cores of rock and metal).

Notably, not all of the solid materials from the protoplanetary disk lie within planets. For example, large numbers of planetesimals and planetesimal fragments lie in the *asteroid belt* between the orbits of Mars and Jupiter. These objects, called **asteroids** (see Fig. 1.9c), which consist of rock and metal, may have originated either as solid fragments which never became incorporated in a planet, or as planetesimals that broke apart. Beyond the orbit of Neptune, millions of icy fragments, as well as larger spherical objects now known as *dwarf planets*, orbit the Sun. *Comets*, long-tailed glowing objects that occasionally arc across the sky, come from these distant reaches of the Solar System.

Internal Differentiation of the Earth

When a planetesimal first forms, it is fairly homogeneous. But as a planetesimal grows, it gets progressively warmer inside, for three reasons. First, as the mass of a planetesimal becomes greater, gravity squeezes its interior together more tightly, and as we have seen, compression of a material causes it to warm up. Second, during the early history of the Solar System, planetesimals frequently collided, and at each impact, kinetic energy transformed into thermal energy. (To see an example of this phenomenon, hit a nail repeatedly with a hammer, and then feel the nail.) And third, planetesimals contained lots of *radioactive atoms*, meaning atoms that can spontaneously break down and release energy. (Modern planets contain fewer radioactive atoms because many of those that existed at 4.56 Ga have since decayed.)

The interiors of large planetesimals containing refractory materials became so hot that they began to melt internally, producing droplets of molten *iron alloy* (a blend of iron and

BOX 1.4

How can I explain . . .

Differentiation of the Earth's interior

What are we learning?
Why the Earth, and other planets, separated into distinct internal layers.

What you need:
- An empty 16-fl-oz (0.5-L) glass bottle with a screw cap
- Vegetable oil (about 1½ c.; 0.350 L), to represent the mantle
- Red vinegar (about ¼ c.; 0.60 L), to represent the core

Instructions:
- Pour the oil and the vinegar into the bottle. The amounts represent the approximate proportion, by volume, of mantle material to core material in the Earth.
- Place the bottle in a refrigerator until cold. Remove the bottle and shake it hard to homogenize the contents. While the bottle remains cold, the oil and vinegar remain homogeneous (thoroughly mixed).
- Heat the bottle to a warm temperature (about 30°C, or 86°F). If it's a warm, sunny day, just put it outside; otherwise, place it in a pan of hot water.
- Watch, and you will see the oil and water separate as the denser material (vinegar) sinks to the bottom and the less dense material (oil) rises to the top. The contents of the bottle will no longer be homogeneous.

What did we see?
- A model of the Earth's differentiation.
- How warming a material increases its ability to flow more easily, so that materials of different density can separate into layers of different density.
- The difference between homogeneous and inhomogeneous materials.

other elements). These droplets were denser than the rocky material around them, so they sank downward due to the inward pull of gravity and accumulated into a ball—a *core*—at the center of the planetesimal (Box 1.4). The remaining materials formed a shell, called the *mantle*, surrounding the

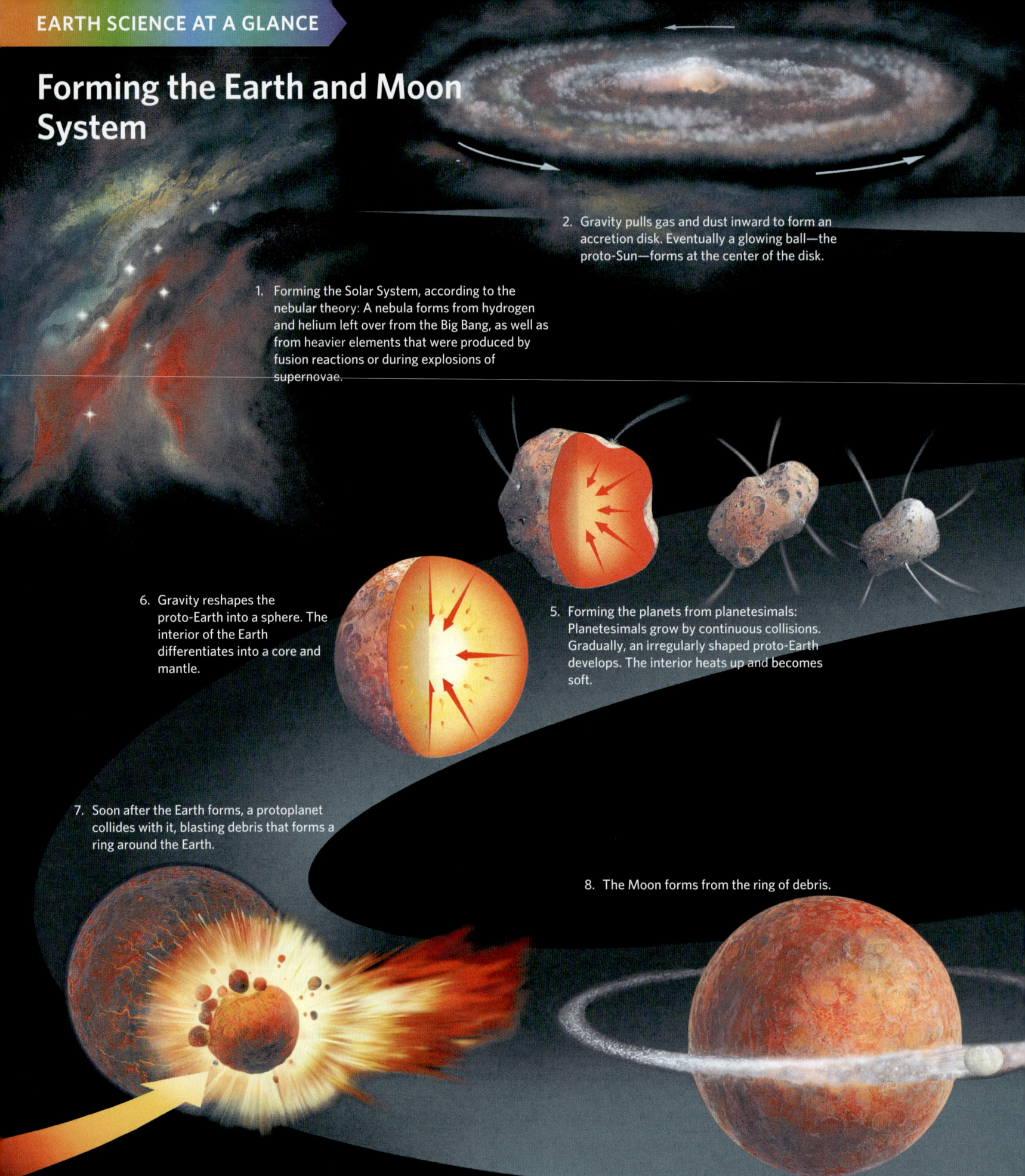

Forming the Earth and Moon System

1. Forming the Solar System, according to the nebular theory: A nebula forms from hydrogen and helium left over from the Big Bang, as well as from heavier elements that were produced by fusion reactions or during explosions of supernovae.

2. Gravity pulls gas and dust inward to form an accretion disk. Eventually a glowing ball—the proto-Sun—forms at the center of the disk.

5. Forming the planets from planetesimals: Planetesimals grow by continuous collisions. Gradually, an irregularly shaped proto-Earth develops. The interior heats up and becomes soft.

6. Gravity reshapes the proto-Earth into a sphere. The interior of the Earth differentiates into a core and mantle.

7. Soon after the Earth forms, a protoplanet collides with it, blasting debris that forms a ring around the Earth.

8. The Moon forms from the ring of debris.

3. Dust (particles of refractory materials) concentrates in the inner rings, while ice (particles of volatile materials) concentrates in the outer rings. Eventually, the dense ball of gas at the center of the disk becomes hot enough for fusion reactions to begin. When it ignites, it becomes the Sun.

4. Dust and ice particles collide and stick together, forming planetesimals.

9. Eventually, the atmosphere develops from volcanic gases. When the Earth becomes cool enough, moisture condenses and rains to produce the oceans. Some gases may be added by passing comets. The moon moves farther from the Earth as time passes.

35

FIGURE 1.10 Differentiation of the Earth's interior.

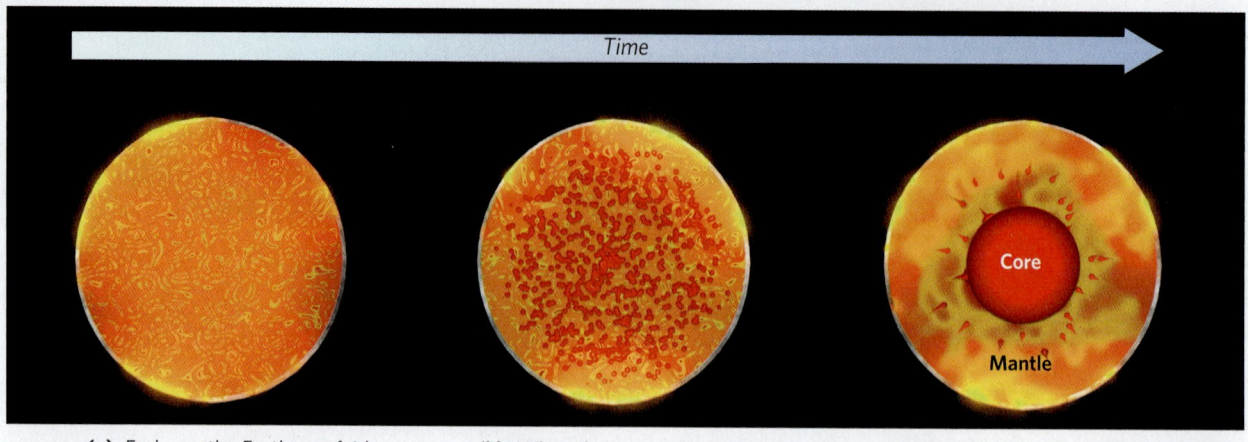

Time

(a) Early on, the Earth was fairly homogeneous inside.

(b) When the temperature got hot enough, iron began to melt.

(c) The iron accumulated at the center of the planet to form a metallic core.

core. By this process of **differentiation**, large planetesimals, including the one that eventually became the Earth, developed internal layering early in their history and became inhomogeneous **(Fig. 1.10)**. The growth of the Earth to nearly its present size, and its differentiation, occurred between 4.56 and 4.54 Ga, so this age range represents the age of the Earth. Note that when the Earth formed, the Universe was already over 9 billion years old.

Making the Earth Round

Small planetesimals had irregular shapes, as do asteroids and small moons today. Planets and large moons, on the other hand, are spherical. Why? A small planetesimal is cool and rigid, so the weak pull of gravity within it cannot cause its internal material to flow. But once a planetesimal reaches a diameter of about 400–800 km (250–500 mi), its interior has become warm and soft enough, and the force of gravity has become large enough, that material in the object's interior flows very slowly. When this happens, bulges sink and dimples rise until the planetesimal evens out into a sphere whose mass lies evenly distributed around its center, such that the force of gravity is nearly the same at all points on its surface (see Earth Science at a Glance, pp. 34–35).

Take-home message . . .

Smaller stars, like the Sun, grew from nebula that contained ice and dust, as explained by condensation theory. Planets of the Solar System formed from material in the protoplanetary disk that surrounded the newborn Sun. Gravity brought materials together into planetesimals, then protoplanets, and finally planets. In the outer part of the Solar System, large amounts of gas and ice collected to form the Jovian planets. Once they were large enough, planetesimals differentiated internally, and became spherical.

Quick Question

Why did the Jovian planets form only in the outer part of the protoplanetary disk?

1.5 The Blue Marble: Introducing the Earth

Imagine that you're on board a spacecraft flying from a distant planet toward the Earth. You've reached the final phase of your voyage, with the Earth in sight. As you approach the Earth, you'll first detect the influence of the planet on the space around it. Then, when you enter into orbit around it, you'll begin to recognize several distinct components, or realms, of the Earth System. Finally, with measurements made from space, you'll even be able to define the basic internal structure of the planet. Let's examine your discoveries.

Welcome to the Neighborhood!

From the orbit of Mars, the Earth looks like a large star with a bluish tint, but from about 30,000,000 km

FIGURE 1.11 The Earth, as seen from space. During the day, the Earth and clouds reflect sunlight. Oceans, land, and ice are all visible.

(20,000,000 mi) away, it begins to resemble a bluish glass marble. Finally, at a distance of about 1,000,000 km (600,000 mi), three times the distance to the Moon, you can see the complexities of the Earth's surface and atmosphere **(Fig. 1.11)**.

As you fly inward from the orbit of the Moon, your instruments discover that the Earth produces a magnetic field, like a signpost saying, "Welcome to the Earth!" A **magnetic field** is the region affected by the invisible force emanating from a magnet (see Box 1.1). The Earth's magnetic field, like the familiar magnetic field around a bar magnet, is a *dipole*, meaning that it has a north pole and a south pole **(Fig. 1.12a, b)**. We can portray a magnetic field by drawing curving *magnetic field*

Did you ever wonder . . .

why compasses work?

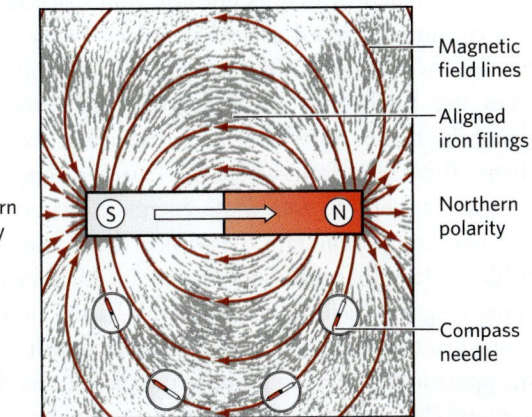

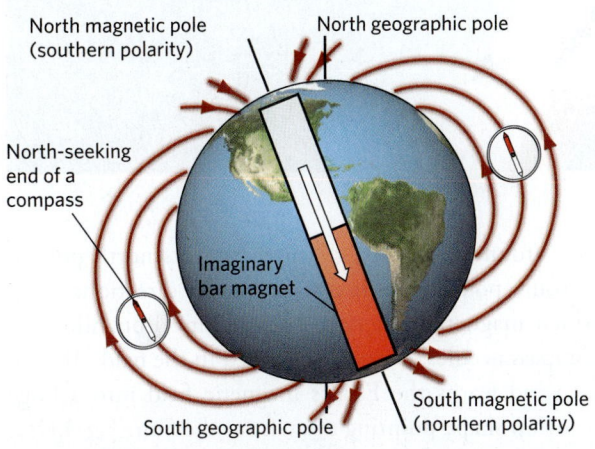

FIGURE 1.12 The Earth's magnetic field.

(a) Magnetic field lines produced by a bar magnet point into the magnet's south pole and out from the magnet's north pole. Compass needles align with the field lines.

(b) We can represent the Earth's magnetic field as an imaginary bar magnet inside the planet.

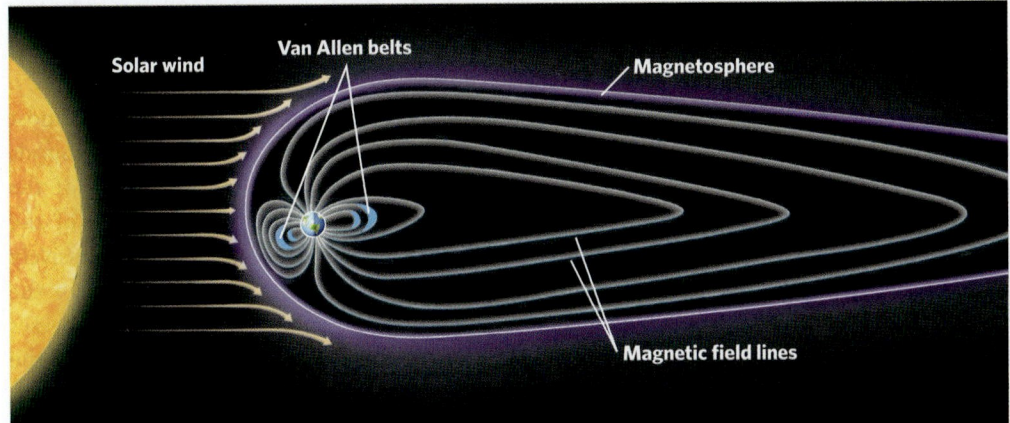

(c) The Earth behaves like a magnetic dipole, but its magnetic field lines are distorted by the solar wind. The Van Allen radiation belts trap charged particles.

(d) A view of aurorae as seen from space.

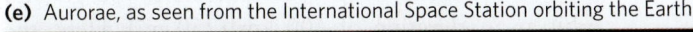

(e) Aurorae, as seen from the International Space Station orbiting the Earth.

FIGURE 1.13 An orbiting astronaut's photograph shows the haze of the atmosphere fading up into the blackness of space. The space shuttle is in the distance.

This magnetic field acts like a shield and deflects most, though not all, of the solar wind's charged particles from the Earth, for a magnetic field applies a force to charged particles. The region inside this shield is called the *magnetosphere*. High-velocity charged particles can be dangerous to organisms, so the existence of the magnetosphere contributed to making the Earth hospitable to life.

The magnetic field does not stop your spacecraft, and you continue to speed toward the planet. At a distance between 10,500 and 3,000 km (18,000 and 6,500 mi) from the Earth, you encounter the *Van Allen radiation belts,* named for the physicist who first recognized them. These belts form where the Earth's magnetic field is strong enough to trap more energetic charged particles that could penetrate the weaker, outer part of the field. Particles that make it past the Van Allen belts flow along magnetic field lines toward the Earth's north and south magnetic poles, where they interact with gases in the upper atmosphere to produce the spectacular glow of *aurorae* **(Fig. 1.12d, e)**.

lines around a magnet, running from its north pole to its south pole. These lines represent the directions along which magnetic materials, such as tiny iron filings or compass needles, align when placed in the field. The solar wind warps the Earth's magnetic field into a huge teardrop shape pointing away from the Sun **(Fig. 1.12c)**.

Introducing the Atmosphere

As you descend below an elevation of about 10,000 km (6,200 mi), the Earth more than fills your field of view, and your instruments detect that the concentration of gas outside the spacecraft has increased slightly. This change means that you've reached the uppermost limit of the **atmosphere**, the layer of gas that surrounds the Earth and represents the outermost realm of the Earth System **(Fig. 1.13)**. From the outer edge of the atmosphere on down, the density of gas progressively increases because the weight of the gas higher in the atmosphere pushes down on the gas below and squeezes its molecules closer together. As a result, 99.9% of the gas molecules in the Earth's atmosphere lie below an elevation of only 50 km (30 mi), less than 1% of the Earth's radius. To avoid entering the denser part of the atmosphere, where friction with air molecules would slow down and heat up your spacecraft, you go into orbit at an elevation of 400 km (250 mi). Here, the air barely affects your spacecraft, and you can continue your studies.

First, you analyze the composition of **air**, the mixture of gases making up the atmosphere, and find that it consists of 78% nitrogen molecules (N_2) and 21% oxygen molecules (O_2). The remaining 1%, known as trace gases, includes argon (Ar), carbon dioxide (CO_2), and methane (CH_4). Air also contains varying amounts of *water vapor* (gaseous H_2O).

Then, you measure **atmospheric pressure**, the push that air exerts on its surroundings. **Pressure** in any material (air, water, or rock) can be described

FIGURE 1.14 Characteristics of the Earth's atmosphere. Molecules pack together more tightly at the base of the atmosphere, so atmospheric pressure changes with elevation (as shown by the blue curve).

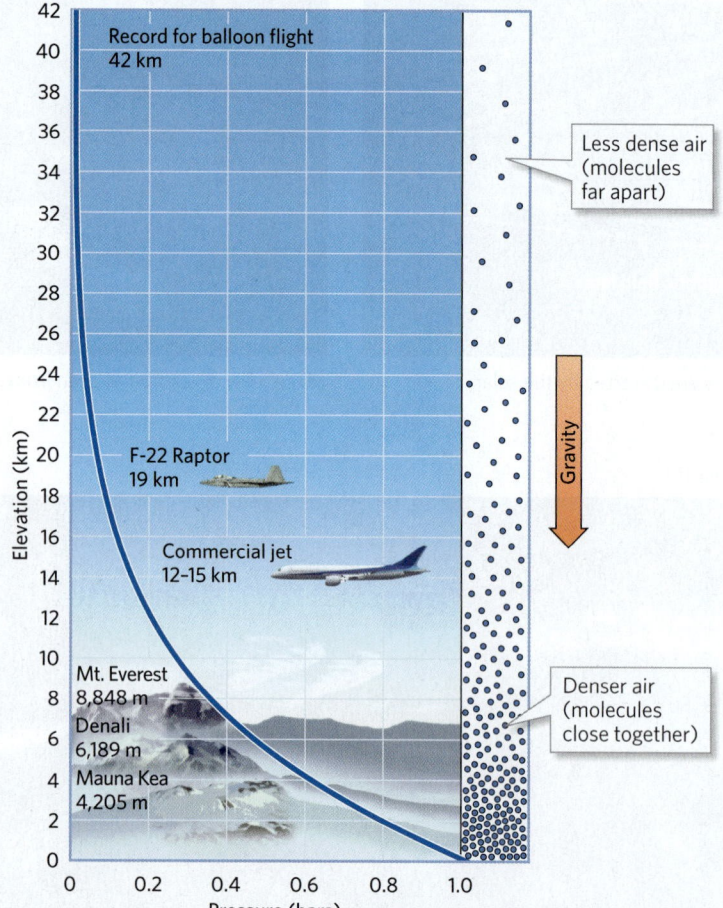

FIGURE 1.15 Clouds take many forms in the Earth's atmosphere.

in units of *force per unit area*. One such unit, a *bar*, represents the approximate average pressure of the atmosphere at sea level. Because atmospheric pressure decreases with increasing elevation, it has a value of only 0.3 bars at the peak of the Earth's highest mountain, Mt. Everest **(Fig. 1.14)**. At this elevation, 8.85 km (5.5 mi) above sea level, people can't survive long without bottled oxygen.

Next, you analyze the distribution of water vapor and measure air movement. You learn that in the lower part of the atmosphere, below an elevation of about 12 km (7.5 mi), the concentration of water vapor in the air ranges between nearly 0% and slightly over 4%. In places, water vapor condenses into *clouds*, mists of tiny liquid droplets or solid ice crystals. The shape and distribution of clouds change constantly, providing a visual clue that the air moves **(Fig. 1.15)**. In fact, you determine that the speed of *wind*, the component of air movement parallel to the Earth's surface, is generally 1–50 km/h (0.6–30 mph) near the ground—at higher altitudes (9–16 km), wind speed may become faster.

The Hydrosphere and Cryosphere

Having made a quick analysis of the atmosphere, you turn your attention to the Earth's surface and map its characteristics. Immediately, you realize that the Earth differs from all the other terrestrial planets in that water covers about 70% of its surface; the remainder is *land*, areas not covered by water. Of the water, about 97% is salty (containing about 3.5% dissolved salt) and occupies the *oceans*, huge basins filled with *saltwater*. The remainder is fresh (containing less than 0.1% dissolved salt). Of the liquid *freshwater*, only a small part occupies lakes, ponds, rivers, and streams on the surface of the land. The rest exists as **groundwater**, filling holes and cracks underground.

Your mapping also reveals that not all of the Earth's water occurs in liquid form. In polar regions and at high elevations, water freezes into ice **(Fig. 1.16)**. When ice builds up on the land surface in a thick enough layer that it lasts all year, it starts to flow very slowly due to gravity, becoming a **glacier**. In very cold regions, near-surface groundwater also freezes, producing *permafrost* (permanently frozen ground), and the sea surface freezes to form a thin layer of *sea ice*. As we noted in the Prelude, the

FIGURE 1.16 Not all of the hydrosphere is liquid water. The liquid part of the hydrosphere (in the form of ocean water) and the frozen part (glaciers of the cryosphere) come in contact along the coast of Alaska.

Ice

Land

Ocean

25 km

FIGURE 1.17 The proportions of elements making up the solid mass of the Earth. Note that iron and oxygen account for most of the mass.

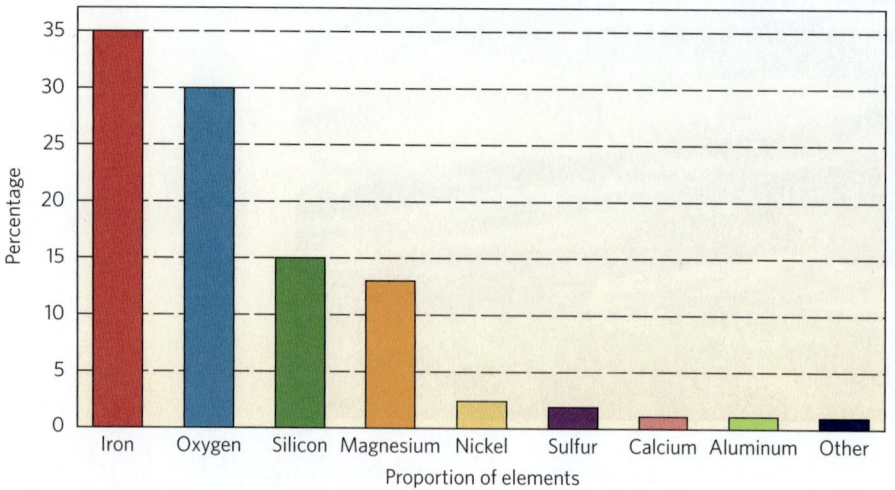

FIGURE 1.18 Materials of the geosphere.

Melt

Mineral

Glass

Sediment

Metal

Rock

frozen realm makes up the **cryosphere**, whereas liquid, frozen, and gaseous H_2O together constitute the **hydrosphere** of the Earth System.

Materials of the Geosphere

If you sent down probes to sample and analyze the composition of the solid realm of the Earth System, the **geosphere**, you would detect all 92 naturally occurring elements, but you would find that only four of these—iron, oxygen, silicon, and magnesium—make up 90% of the Earth's solid mass **(Fig. 1.17)**. You would also learn that the geosphere contains a great variety of different materials. Let's take a quick look at some basic categories of these materials **(Fig. 1.18)**—all of which will be discussed further later in the book.

- *Melts:* A **melt**, or *molten material*, forms when a solid becomes hot enough to transform into a liquid.

- *Minerals:* A **mineral** is a solid, naturally occurring substance in which atoms are arranged in an orderly pattern. We can define the proportions of elements in a mineral. A continuous piece of mineral that has naturally grown faces is a *crystal*.

- *Glasses:* A solid in which atoms are not arranged in an orderly pattern is a **glass**.

- *Grains:* Geologists refer to a relatively small fragment of a solid as a **grain**. A grain can consist of a crystal, of a part of a crystal, or of a fragment of glass or rock.

- *Sediment:* **Sediment** consists of an accumulation of loose grains that are not stuck together.

- *Soil:* Sediment that has changed due to reactions with air, water, and life at the surface of the Earth and has mixed with the remains of organisms becomes a **soil**.

- *Rocks:* A **rock** is a solid aggregate (collection) composed of mineral crystals or grains or of a mass of natural glass. Geologists recognize three principal kinds of rocks:

  1. *Igneous rock* forms by the solidification of molten material. Igneous rock can solidify underground or at the Earth's surface. Geologists refer to molten rock underground as *magma* and to molten rock that has spilled out at the Earth's surface as *lava*.

  2. *Sedimentary rock* forms by the cementing together of solid grains, or by the precipitation of minerals out of water solutions, at or near the Earth's surface.

  3. *Metamorphic rock* forms when existing rock undergoes changes, primarily due to an increase in temperature and pressure deep beneath the Earth's surface.

- *Metals:* A solid composed of metallic elements (such as iron, aluminum, copper, or tin) is a **metal**. Metals can be pounded into sheets. In a metal, some electrons flow freely, so a metal conducts electricity. An **alloy** is a blend containing more than one metallic element.

The Surface of the Geosphere

Overall, the Earth has the shape of a sphere. But it's not perfectly smooth. In detail, its surface has many ups and downs that you can detect from orbit and represent on a map by specifying vertical distances with respect to *sea level*, the surface of the ocean **(Fig. 1.19)**. An analysis of **topography**, variations in the elevation of the land surface, reveals that the land has an average elevation of about 0.8 km (0.5 mi), and an analysis of **bathymetry**, variation in the depth of water bodies, reveals that the ocean has an average depth of about 4 km (2.5 mi).

Most land on the Earth lies within large areas called **continents**—the largest of which are over 8,000 km (5,000 mi) across. (By comparison, the circumference of the Earth is 40,075 km, or about 25,000 mi.) The remainder occurs in much smaller areas, called *islands*. Mapping of the land surface reveals *plains*, broad, flat areas; *mountain ranges*, linear belts of high, rugged topography; *plateaus*, flat areas of elevated land; *volcanoes*, mountains that erupt lava; and *valleys*, elongate (long, narrow) low areas. Looking at the land surface, you see that some of it exposes rock, some has a veneer of sediment, and some has become soil. Most of the surface water on the planet, as we've already mentioned, occupies the ocean basins. As Figure 1.19 shows, the ocean basins can be divided into distinct provinces with different bathymetric characteristics.

Together, the bathymetry and topography of the Earth make it look very different from the other terrestrial planets **(Fig. 1.20)**. Most notably, the surfaces of the other planets remain pockmarked by *craters*, bowl-shaped indentations caused by the impact of meteorites (see the Prelude), whereas very few craters appear on the Earth's surface. This difference leads you to realize that some processes must be taking place that can change the Earth's surface. The changes are caused by *uplift* (movement of the land surface upward to higher elevations), **erosion** (the grinding or beveling away of the land), and **deposition** (the burying of features by sediment). In other words, unlike the other terrestrial planets, the Earth remains a dynamic place, where natural phenomena actively modify its surface.

The Biosphere

Even a cursory examination reveals that, unlike all other planets in the Solar System, the Earth System

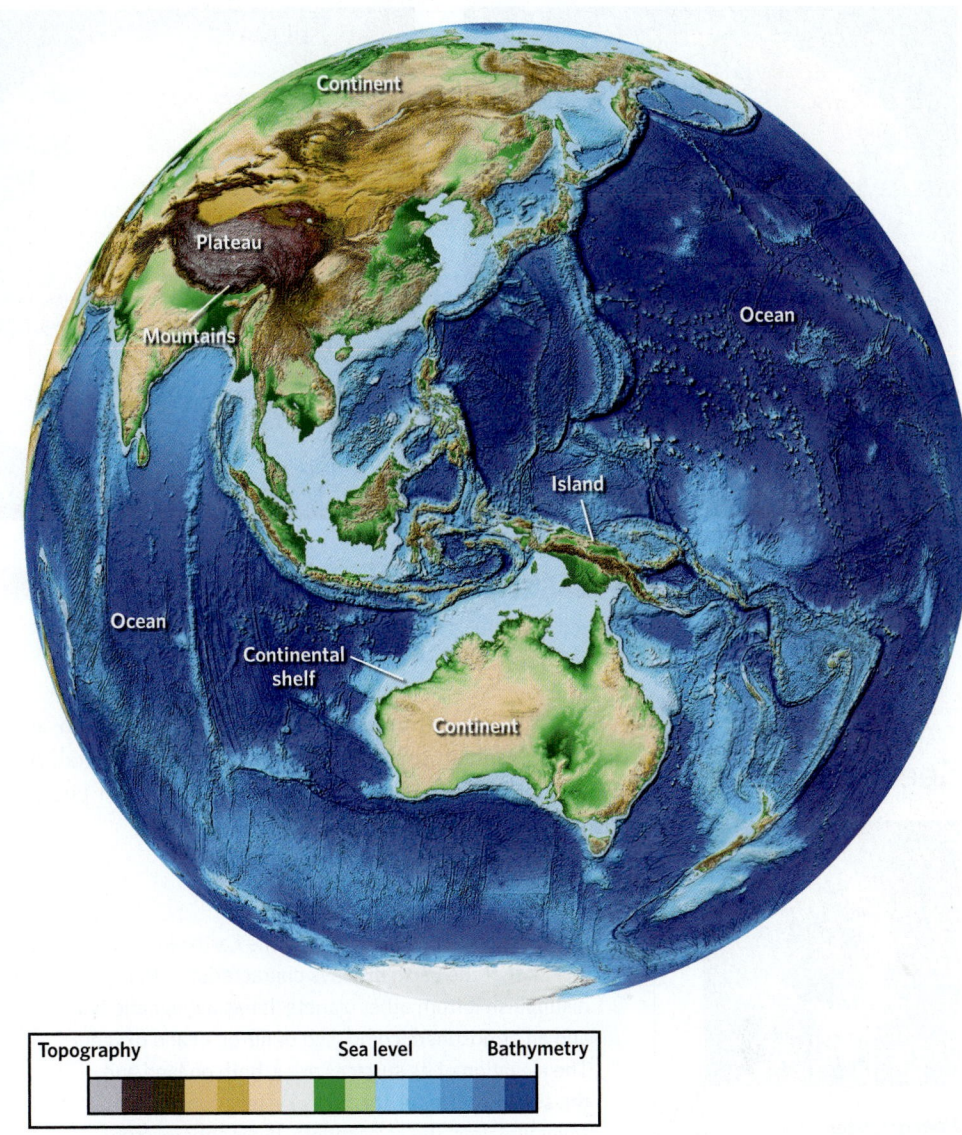

FIGURE 1.19 Bathymetry and topography. This map shows the Earth's bathymetry (variation in seafloor depth) and topography (variation in land elevation).

hosts a **biosphere**, the vast variety of organisms inhabiting our planet together with the portion of the Earth in which life survives. The biosphere extends from a few kilometers below the Earth's surface to a few kilometers above it. From your orbiting spacecraft, you can see plants and animals, and you can detect the distinct chemicals, called *organic chemicals*, produced by living organisms. Researchers estimate that the total mass of the Earth's living organisms, including *microbes* (microscopic organisms), roughly equals the weight of 10 volcanoes!

As you orbit the Earth, you can see the changes that human society has made to the planet. For example, you can see farm fields, cities and towns, highways, canals, dams, and mines, and you can detect a variety

FIGURE 1.20 Topography of three terrestrial planets compared. Comparing the topography of the Earth, Mars, and Venus shows that each planet is unique. Colors represent different elevations (blue areas are low; red is intermediate; white is high). (Not to scale.)

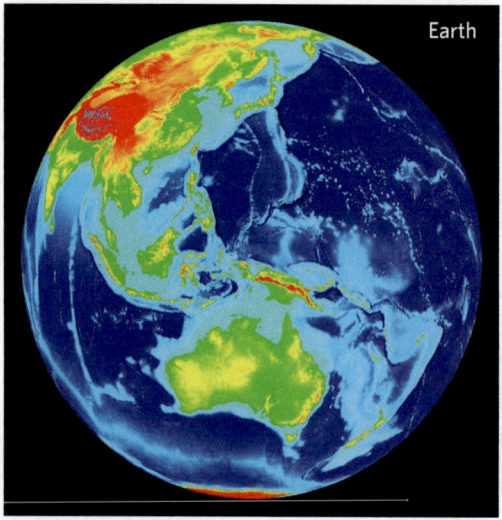
Earth

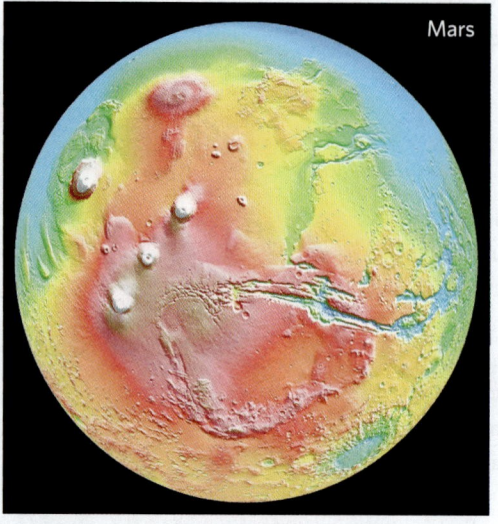
Mars

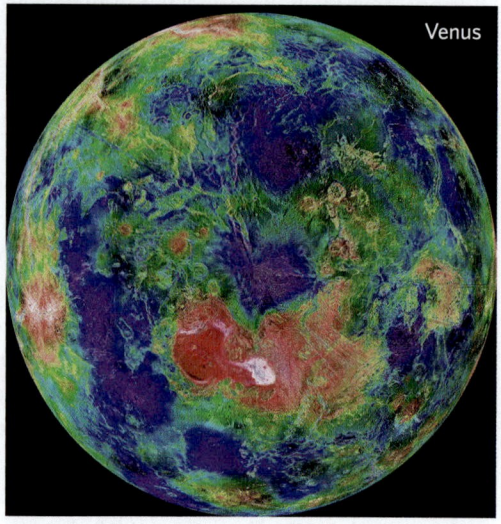
Venus

Low ── High

Altitude

See for yourself

Meteor Crater

Latitude: 35°1'37.18" N
Longitude: 111°1'20.17" W

If you look at the Moon through a telescope, you see thousands of craters, depressions that mark the locations of meteorite impacts. On the Earth, most craters have eroded away, but a few remain. Zoom to an elevation of 5.7 km (19,000 ft) and look down. You are seeing Meteor Crater, Arizona, formed by the impact of a 50-m-wide meteorite less than 50,000 years ago.

of *anthropogenic* (human-produced) materials, such as concrete, asphalt, brick, lumber, tile, and sheet metal. A view of the Earth at night, showing the lights of cities, emphasizes the reach of civilization **(Fig. 1.21)**.

Take-home message . . .

A spacecraft approaching the Earth would learn that the Earth displays characteristics that distinguish it from other planets. It has a magnetic field and an atmosphere composed of nitrogen and oxygen. The elevation of its surface varies, both on land and on the seafloor. Overall, the Earth System consists of several realms: the atmosphere, hydrosphere, cryosphere, geosphere, and biosphere.

Quick Question ──────────────

What are the various materials that make up the geosphere?

1.6 A First Glance at the Earth's Interior

Hints About What's Inside the Earth

Can you discover anything about the Earth's interior—the material between the Earth's surface and its center—without drilling a hole through it? Fortunately, the answer is yes, for the deepest hole ever drilled extends down only about 12.3 km (7.6 mi), less than 0.2% of the Earth's radius. First, by measuring the mass and volume of the whole Earth, you can calculate the planet's average density, and by sampling rocks exposed at the Earth's surface, you can measure the density of surface rocks. When you compare these two measurements, you'll find that the average density of the whole Earth is about twice that of most rocks found at its surface. Therefore, material inside the planet must be much denser than the material at its surface. Second, as we've seen, the Earth is nearly spherical, even though it spins rapidly on its axis. If the Earth were entirely molten beneath its surface, or if its mass were distributed evenly throughout, spinning would cause it to flatten into a disk. The observation that the Earth is nearly spherical means that its interior must be mostly solid and that the densest material must be concentrated near its center. Eventually, putting all these deductions together, you could come up with a model in which the Earth resembles a hard-boiled egg, with three principal layers **(Fig. 1.22)**: (1) a thin *crust* (the eggshell) composed mostly of relatively low-density rock; (2) a solid *mantle* (the white) composed of relatively high-density rock; and (3) a *core* (the yolk) consisting of very high-density metal alloy.

Refining the Picture of the Earth's Interior

Clearly, many questions would remain after your initial survey. Exactly what are the materials inside the Earth, how thick are its layers, and are the boundaries between layers

FIGURE 1.21 City lights at night, as seen from space.

sharp or gradational? These questions would be difficult to answer from your orbiting spacecraft. Geologists use other sources of data to gain insight into the composition and character of the Earth's interior and to calculate how temperature and pressure vary inside the planet **(Box 1.5)**. The study of *seismic waves*, vibrations produced by earthquakes, has been particularly helpful. The path and speed of these waves as they travel through the Earth characterize the Earth's interior somewhat as ultrasound measurements characterize the insides of a person, for seismic waves travel at different velocities through different materials, and they bend or reflect when they reach boundaries between different materials.

Here, we introduce some key characteristics of the Earth's interior layers so that we can use this information in our discussion of plate tectonics in Chapter 2. We'll provide details on studies involving seismic waves in Chapter 8. **Table 1.1** provides a simplified representation of the thickness of these layers.

THE CRUST. When you stand on the Earth's surface, you're on top of its outermost layer, the **crust (Fig. 1.23a)**.

Oceanic crust, beneath the seafloor, has a thickness of between 7 and 10 km (4 and 6 mi), whereas *continental crust*, beneath the land, has a thickness of between 25 and 70 km (15 and 45 mi). Compared with the radius of the Earth, the crust is so thin that if the Earth were the size of an inflated balloon, the crust would be about the thickness of the balloon's skin.

What is the crust made of? The top of the oceanic crust, beneath the abyssal plains, is a blanket of sediment, generally less than 1 km (0.6 mi)

(a) The hard-boiled egg analogy for the Earth's interior.

FIGURE 1.22 An early image of the Earth's internal layers.

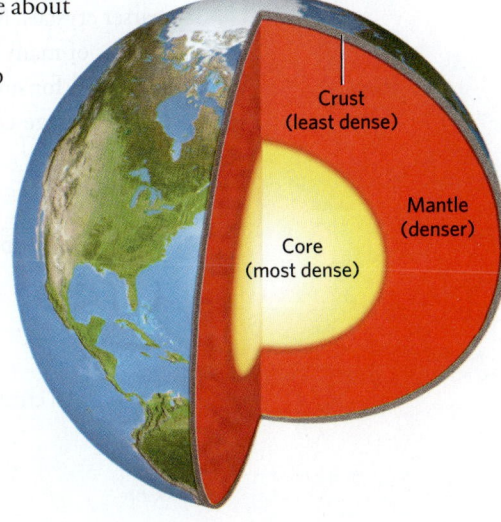

(b) In the 19th century, researchers recognized that the Earth has layers.

BOX 1.5 ▶ **A Deeper Look**

Temperatures and pressures inside the Earth

When the Earth first formed, it was probably so hot that its surface and much of its interior were molten. Over time, as heat escaped into space, the planet cooled and solidified. However, heat produced by the decay of radioactive atoms in rock keeps the interior from cooling entirely, and it still remains very hot inside. For example, while the Earth's average ground-surface temperature is about 14°C (57°F), the temperature at a depth of 40 km (25 mi) exceeds 600°C (1,100°F), and the temperature at the center may reach 6,000°C (11,000°F). Geologists refer to the rate at which temperature increases with increasing depth as the **geothermal gradient (Fig. Bx1.5)**.

Pressure also increases with depth due to the weight of overlying material. Because the materials of the geosphere are denser than the gas of the atmosphere, pressure increases more rapidly with depth in the geosphere than in the atmosphere. Geologists estimate that at a depth of 40 km, the pressure reaches 10,000 bars, and at the center of the Earth, it reaches about 3,500,000 bars.

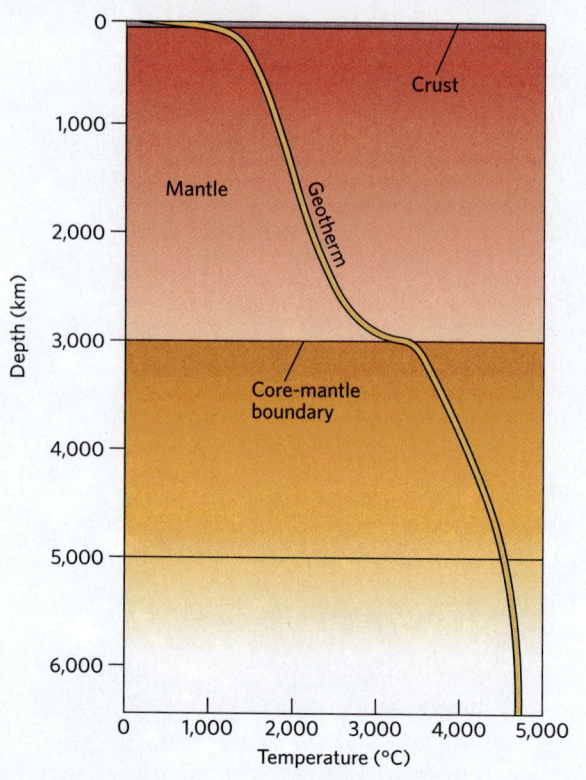

FIGURE Bx1.5 The geotherm indicates how temperature changes with depth.

thick. Beneath this blanket, the oceanic crust consists of two kinds of dark gray, relatively dense igneous rock: **basalt,** which contains tiny crystals, and **gabbro,** with coarser crystals. Most of the continental crust, in contrast, consists of many different kinds of rocks. To simplify our discussion for now, we can say that the continental crust has an average composition similar to that of **granite,** a light-colored igneous rock with a density less than that of basalt or gabbro.

THE MANTLE. The Earth's mantle, a 2,885-km (1,792-mi)-thick layer beneath the crust, represents the largest part of the Earth in terms of volume **(Fig. 1.23b)**. This layer consists entirely of a very dense igneous rock called

TABLE 1.1 Approximate Average Thicknesses of the Principal Layers Inside the Earth

Layer Name	Thickness (km)	Thickness (mi)	Composition*
Oceanic crust	7	5	Basalt and gabbro (dense rock)
Continental crust	40	25	On average, similar to granite (less dense rock)
Mantle	2,885	1,790	Peridotite (very dense rock)
Core	3,470	2,155	Iron alloy (metal)

Note: All numbers are rounded.
*Different rock types contain different proportions of chemicals.

FIGURE 1.23 A modern view of the Earth's interior layers.

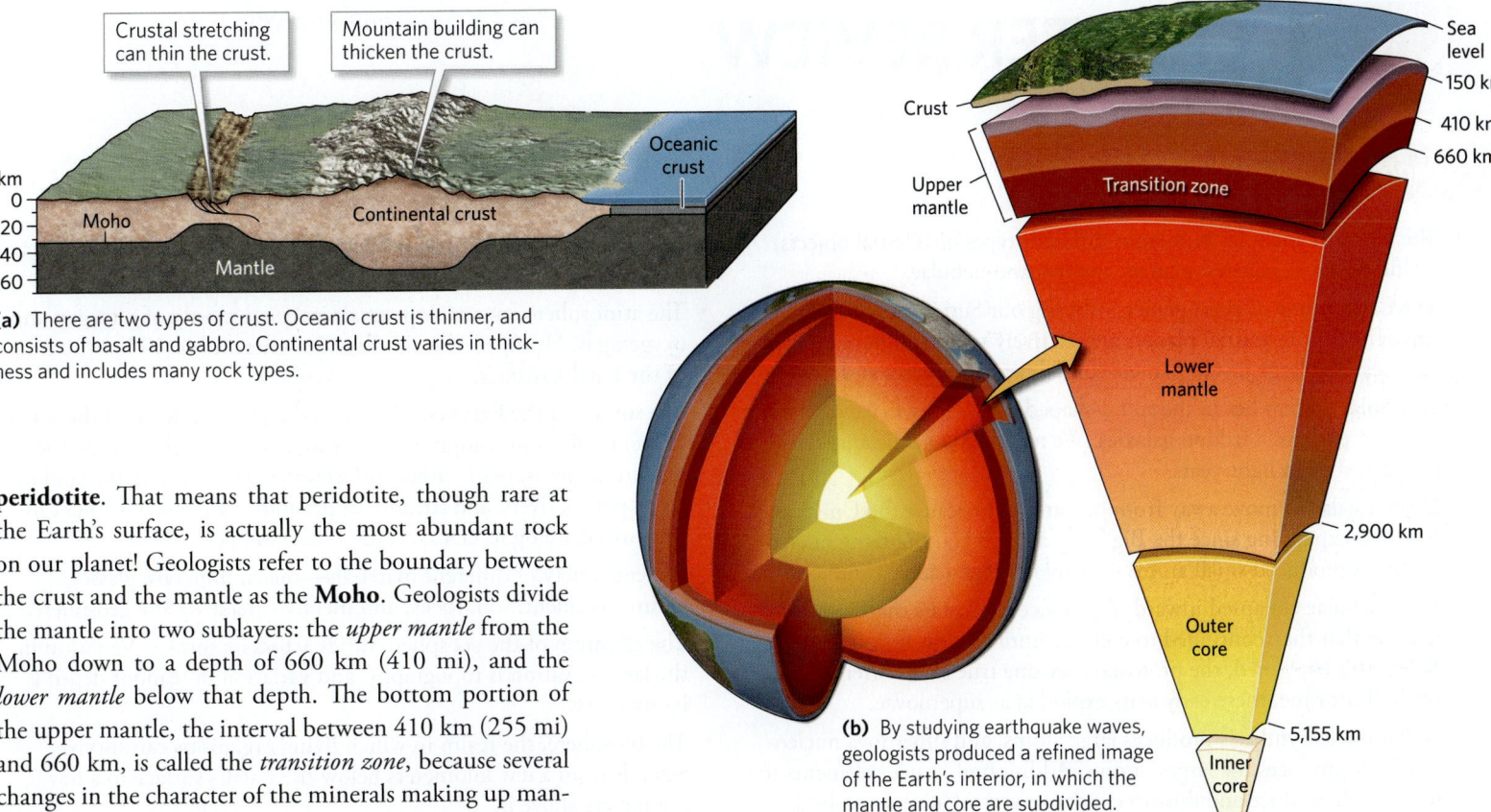

Crustal stretching can thin the crust.

Mountain building can thicken the crust.

Oceanic crust

Continental crust

Moho

Mantle

km
0
20
40
60

(a) There are two types of crust. Oceanic crust is thinner, and consists of basalt and gabbro. Continental crust varies in thickness and includes many rock types.

Crust

Upper mantle

Transition zone

Lower mantle

Outer core

Inner core

Sea level

150 km

410 km

660 km

2,900 km

5,155 km

6,371 km

(b) By studying earthquake waves, geologists produced a refined image of the Earth's interior, in which the mantle and core are subdivided.

peridotite. That means that peridotite, though rare at the Earth's surface, is actually the most abundant rock on our planet! Geologists refer to the boundary between the crust and the mantle as the **Moho**. Geologists divide the mantle into two sublayers: the *upper mantle* from the Moho down to a depth of 660 km (410 mi), and the *lower mantle* below that depth. The bottom portion of the upper mantle, the interval between 410 km (255 mi) and 660 km, is called the *transition zone*, because several changes in the character of the minerals making up mantle peridotite take place there. Nearly the entire mantle is solid, but most of it is hot enough to be able to flow somewhat like soft, but not liquid, wax, though much more slowly (at a rate of less than 15 cm, or 6 in, per year). Geologists refer to this process as *plastic flow*.

THE CORE. Early calculations suggested that the core, the central ball deep inside the Earth, has the same density as gold, so for many years people held the fanciful hope that vast riches lay at the heart of our planet. Alas, geologists eventually concluded that the core consists of a far less glamorous material: *iron alloy*, a mix of iron with 4% nickel and up to 10% oxygen, silicon, or sulfur. The *outer core* is molten, so it can flow fairly rapidly, whereas the *inner core* is solid. Flow in the outer core generates the Earth's magnetic field.

As emphasized by the above descriptions, the traditional division of the Earth's interior into layers reflects variation in the composition of the Earth with depth. In the early 20th century, geologists realized that the outer portion of the Earth could be divided in another way, by the behavior of its materials when subjected to forces. They learned that the outer layer, consisting of the crust together with the uppermost part of the mantle—a layer now called the *lithosphere*—behaves somewhat like a rigid shell, and that the layer of mantle below

the lithosphere—a layer now called the *asthenosphere*—can flow plastically. We'll refine this distinction in the next chapter and see how it became incorporated into the grand unifying theory of geology, plate tectonics.

Take-home message . . .

The Earth's interior can be divided into three layers based on their differing compositions: the crust, mantle, and core. The crust beneath the oceans differs from the crust beneath the continents. The mantle, the largest part of the Earth, consists of very dense, almost entirely solid rock, but it is so hot that it flows plastically. The core consists of iron alloy. Flow in the liquid outer core generates the magnetic field. The Earth's inner core is solid.

Quick Question
What sources of data reveal the character of the Earth's interior?

CHAPTER REVIEW

- The Universe includes numerous different types of celestial objects, including stars, galaxies, planets, moons, and nebulae.

- The Earth is one of eight planets orbiting our Sun in the Solar System. The four terrestrial planets are relatively small and have rocky surfaces. The large Jovian planets consist mostly of gases and ices.

- Our Solar System lies in the spiral-shaped Milky Way Galaxy, which is one of perhaps a trillion galaxies. We measure large distances between stars in light-years.

- Distant galaxies move away from one another because the Universe has been expanding since the Big Bang at 13.8 Ga. Big Bang nucleosynthesis produced small atoms (mostly hydrogen and helium).

- Early nebulae collapsed inward to produce protostars that were so massive that they continued to collapse until nuclear fusion began. When this happened, the protostars became true stars. After a relatively short time, these early stars exploded as supernovae.

- Stellar nucleosynthesis produces large atoms, and supernova nucleosynthesis produces the largest atoms. Addition of heavier elements to nebulae allowed ice and dust to condense as the Universe evolved.

- In our Solar System, planets formed in the portion of the accretion disk. Here, ice and dust collected into clumps, which merged to form planetesimals. Some of these planetesimals grew by collection of other material into protoplanets, then true planets.

- The terrestrial planets formed mostly from dust nearer the Sun, while the Jovian planets incorporated vast amounts of ice and gas.

- When a planetesimal composed of refractory materials became large enough for compression to make its interior very hot, it underwent differentiation into a metallic core and a rocky mantle. The heating and softening of the interior also allowed larger planetesimals to become spherical.

- A magnetic field extends into space around the Earth. The magnetic field shields the Earth from solar wind and other particles.

- The atmosphere, an envelope of air, consists mostly of nitrogen and oxygen gas. Nearly all the air—99.9%—lies within 50 km (30 mi) of the Earth's surface.

- The surface of the Earth consists of land and water. Most of the water (97%) is salty and comprises the oceans and seas of the world. The remaining freshwater is either surfacewater, which lies on the land in lakes, ponds, rivers, and streams, or groundwater, which lies beneath the surface, filling its cracks, caves, holes, and pores.

- A great variety of different materials—melts, minerals, glasses, grains, sediment, soil, rocks, and metals—make up the geosphere.

- The elevation of the geosphere varies across its surface. Variation in the land elevation is topography, and variation in seafloor depth is bathymetry.

- The biosphere, the realm in which living organisms can survive, extends from a few kilometers below the Earth's surface to a few kilometers above it.

- The interior of the Earth consists of a thin crust, a thick mantle, and a central core. A variety of data sources have led to this model.

- Pressure and temperature both increase with depth in the Earth. The rate of temperature increase is the geothermal gradient.

- The crust beneath the oceans differs in thickness and composition from the crust beneath the continents. Rocks of the crust are less dense than those of the mantle.

- The mantle can be subdivided into the upper and lower mantle, and the core, which consists of iron alloy, can be subdivided into a liquid outer core and a solid inner core. Flow in the outer core produces the Earth's magnetic field.

accretion disk (p. 29)
air (p. 38)
alloy (p. 41)
asteroid (p. 33)
atmosphere (p. 38)
atmospheric pressure (p. 38)
basalt (p. 44)
bathymetry (p. 41)
Big Bang (p. 28)
Big Bang nucleosynthesis (p. 29)
biosphere (p. 41)

celestial object (p. 23)
condensation theory (p. 32)
continent (p. 41)
crust (p. 43)
cryosphere (p. 40)
deposition (p. 41)
differentiation (p. 36)
dust (p. 25)
electromagnetism (p. 24)
element (p. 26)
erosion (p. 41)

expanding Universe theory (p. 28)
force (p. 24)
gabbro (p. 44)
galaxy (p. 23)
geosphere (p. 40)
geothermal gradient (p. 44)
glacier (p. 39)
glass (p. 40)
grain (p. 40)
granite (p. 44)
gravity (p. 24)

groundwater (p. 39)
heat (p. 29)
hydrosphere (p. 40)
ice (p. 25)
light-year (p. 27)
magnetic field (p. 37)
melt (p. 40)
metal (p. 41)
Milky Way Galaxy (p. 24)
mineral (p. 40)
Moho (p. 45)

molecule (p. 27)
moon (p. 25)
nebula (p. 24)
nebular theory (p. 30)
orbit (p. 25)
peridotite (p. 45)
planet (p. 25)

planetesimal (p. 32)
pressure (p. 38)
protoplanet (p. 32)
protoplanetary disk (p. 32)
protostar (p. 30)
refractory materials (p. 25)
rock (p. 40)

sediment (p. 40)
soil (p. 40)
Solar System (p. 25)
solar wind (p. 31)
star (p. 23)
stellar nucleosynthesis (p. 30)
supernova (p. 31)

supernova nucleosynthesis (p. 31)
temperature (p. 29)
thermal energy (p. 29)
topography (p. 41)
Universe (p. 23)
volatile materials (p. 25)

Review Questions

Letters in parentheses correspond to the chapter's learning objectives.

1. How did observers, thousands of years ago, distinguish between stars and planets in the night sky? What's the difference between a planet and a moon? **(A)**

2. How many planets does our Solar System contain, and where does the Solar System lie relative to the center of the Milky Way Galaxy? About how many galaxies are there in the Universe? **(A)**

3. What is the difference between a nebula and a vacuum? How did the first stars form according to the nebular theory? **(B)**

4. Explain the expanding Universe theory and its relationship to the Big Bang theory. According to these theories, when did the Universe form? **(B)**

5. Distinguish between Big Bang nucleosynthesis, stellar nucleosynthesis, and supernova nucleosynthesis. Which process happened in this image? Why is it fair to say that we are all made of stardust? What's the difference between gas, ice, and dust in space? **(B)**

6. Describe the formation of a star in the early Universe, according to the nebular theory, and how that process differs from the formation of the Solar System, according to the condensation theory. **(C)**

7. Distinguish among a planetesimal, a protoplanet, and a planet. **(C)**

8. Why isn't the Earth homogeneous? Why is it round? Why did the core form? **(C)**

9. What is the Earth's magnetic field? How does the magnetic field interact with the solar wind? **(D)**

10. What is the Earth's atmosphere composed of? How does the density of air change with increasing elevation? **(D)**

11. Distinguish among the various realms of the Earth System. **(E)**

12. What is the proportion of land area to ocean area on the Earth? Is the seafloor completely flat? **(F)**

13. Describe the major categories of materials constituting the geosphere. **(G)**

14. How do temperature and pressure change with increasing depth in the Earth? **(H)**

15. Label the principal layers of the Earth on this figure. **(H)**

16. What sources provide geologists with information about the character of the Earth's interior? **(H)**

17. What is the Moho? Describe basic differences between continental crust and oceanic crust. **(H)**

18. What is the mantle composed of? Is the mantle rigid and unmoving? **(H)**

19. What is the core composed of? How do the inner and outer cores differ? Which part produces the magnetic field? **(H)**

On Further Thought

20. Could the Earth have formed in association with a first-generation star? **(C)**

21. Why are the Jovian planets farther from the Sun than the terrestrial planets are? **(C)**

22. At highway speeds (100 km/h), how long would it take to drive a distance equal to the thickness of the continental crust? How about from the Earth's surface to the planet's center? **(H)**

2 THE WAY THE EARTH WORKS
Plate Tectonics

By the end of the chapter you should be able to . . .

A. discuss the evidence that Alfred Wegener used to justify his proposal that continents drift.

B. describe the process of seafloor spreading and the observations that allowed geologists to confirm that it takes place.

C. contrast the lithosphere with the asthenosphere, identify major plates of the lithosphere, and explain how the boundaries between plates can be recognized.

D. sketch the three types of plate boundaries, and describe the nature of motion that occurs across them.

E. relate types of geologic activity to types of plate boundaries, and explain how new plate boundaries can form and existing ones can cease activity.

F. outline the major ideas in the modern theory of plate tectonics, and interpret Wegener's observations in the context of this theory.

G. describe how measurements of paleomagnetism have helped to prove that plate tectonics happens.

H. explain the methods scientists use to describe and measure the velocity of plate motion.

2.1 Introduction

In September 1930, a German meteorologist, Alfred Wegener, along with several colleagues, sledged across the ice sheet of Greenland to resupply weather observers stranded at a remote camp **(Fig. 2.1a)**. After dropping off crates of food, Wegener and one companion decided to head back immediately. Sadly, they were never seen again. At the time of his death, Wegener was well known, not only to researchers studying climate, but also to geologists, because some 15 years earlier he had published a book in which he challenged geologists' long-held assumption that the locations of continents have remained fixed for all of geologic time. Wegener proposed instead that in the past, continents fit together in one vast supercontinent, like pieces of a giant jigsaw puzzle, and that this supercontinent—which he named **Pangaea** (pronounced Pan-JEE-ah; Greek for all land)—later fragmented into separate continents. Over time, according to Wegener, these continents moved apart to reach their present positions. He named this process **continental drift (Fig. 2.1b)**.

Initially, hardly any geologists accepted Wegener's hypothesis, and Wegener died without knowing that the idea would later become the foundation of a scientific revolution. Today, geologists take for granted that the map of the Earth constantly changes—continents do indeed slowly waltz around its surface, and they have combined, and then broken apart, more than once over geologic time.

The revolution that led to our modern image of a dynamic Earth began in 1960, when an American professor, Harry Hess (1906–1969), suggested that new ocean floor forms between two continents as they move apart, a process now known as *seafloor spreading*. Hess also suggested that continents can move toward each other when the old ocean floor between them sinks back down into the Earth's interior, a process now known as *subduction*. Many geologists began to explore the implications of Hess's suggestions, and by 1968, they had developed a fairly complete model describing how seafloor spreading and subduction operate. In this model, the Earth's *lithosphere*—its outer, relatively rigid shell—consists of several pieces, or *plates*, that slowly move relative to one another. Because geologists have confirmed this model by many observations, it has gained the status of a theory, called the *theory of plate tectonics*, or simply *plate tectonics*. The name comes from the Greek word *tekton*, which means builder, because plate movements "build" regional geologic features.

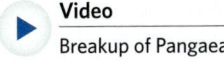

Video

Breakup of Pangaea

FIGURE 2.1 Alfred Wegener and his model of continental drift.

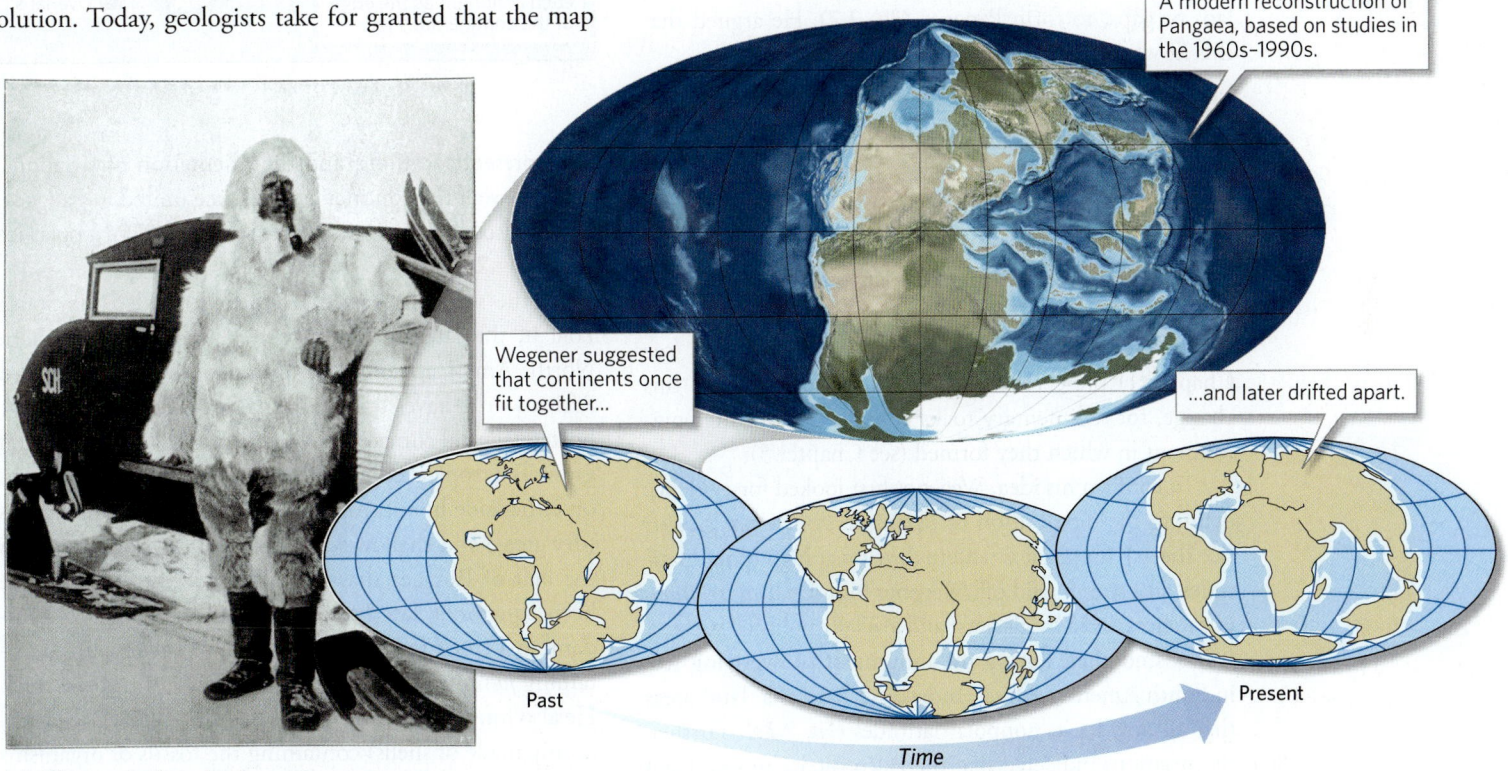

A modern reconstruction of Pangaea, based on studies in the 1960s–1990s.

Wegener suggested that continents once fit together...

...and later drifted apart.

Past

Present

Time

(a) Wegener in Greenland.

(b) Wegener's image of Pangaea and its breakup.

<< This cliff in Cyprus exposes rock that formed by seafloor spreading and was then uplifted to form part of the land along a convergent boundary. Before the theory of plate tectonics was developed, geologists could not explain the rocks on this cliff. This photo shows both pillow basalt and basalt dikes.

We begin this chapter by introducing the observations that led Wegener to propose his continental-drift hypothesis. Then we discuss subsequent discoveries that led to plate tectonics theory. Next, we describe the nature of lithosphere plates and the boundaries between them, the way that measurements of magnetism prove plate tectonics theory, and the way geologists describe plate motions. You'll see that plate tectonics provides a basis for understanding much of Earth science.

2.2 Continental Drift

Wegener's Evidence

Why did Wegener propose that Pangaea once existed and later separated into continents that then drifted apart? To find an answer, let's look at Wegener's key observations.

FIT OF THE CONTINENTS. Almost as soon as maps depicting the Atlantic Ocean became available, scholars noted that Africa and Europe looked like they would fit snugly against the Americas if the Atlantic Ocean didn't exist. Wegener took this concept further by showing that all the continents could fit together, with remarkably few overlaps or gaps, to form Pangaea **(Fig. 2.2)**. He argued that this fit was too good to be coincidence.

DISTRIBUTION OF CLIMATE BELTS IN THE PAST. Wegener knew that different climate belts occur at different latitudes. For example, polar climates lie at high latitudes, hot and dry climates at subtropical latitudes, and steamy wet climates at tropical latitudes. Wegener speculated that if a continent drifted from one latitude to another, the climate at a location on the continent would change over time, and that a succession of sedimentary-rock beds (see Chapter 1) preserved at the location would record this change, for sedimentary rocks contain clues to the environment in which they formed (see Chapter 5).

To confirm his idea, Wegener first looked for evidence defining the distribution of Paleozoic *continental glaciers* (ice sheets), for glaciers tend to develop at high (polar) latitudes. His search paid off. Wegener found *glacial striations* (scratches carved by glaciers) and glacially polished surfaces (rocks smoothed during the movement of overlying ice) in South America, Africa, India, and Australia, land areas that all now lie in nonpolar latitudes **(Fig. 2.3a)**. Further, the striations indicated that the Paleozoic ice moved from what is now ocean onto the land. (Glaciers can't move this way—instead, they must flow from the land to the sea.) Wegener also found Paleozoic sedimentary beds composed of a distinctive type of sediment left by glaciers. He realized that these Paleozoic glacial features could not have formed at the latitudes where they now occur, and suggested that

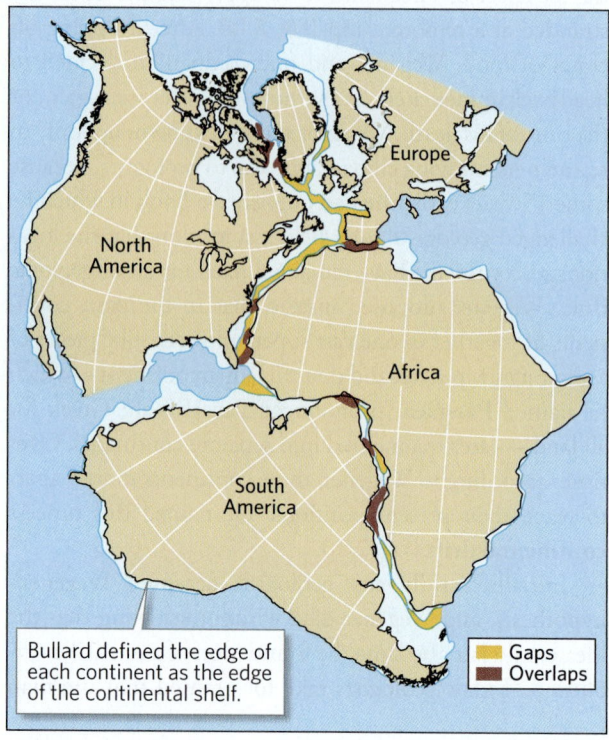

FIGURE 2.2 The "Bullard fit" of the continents. In 1965, Edward Bullard used a computer to fit the continents together and demonstrate how minor the gaps and overlaps are, although the match still isn't perfect.

Bullard defined the edge of each continent as the edge of the continental shelf.

| Gaps |
| Overlaps |

their present locations, and the orientation of striations, make sense if the continents were once united in Pangaea, with the southern part of Pangaea lying beneath a polar ice sheet **(Fig. 2.3b, c)**.

If the southern part of Pangaea straddled the South Pole in the late Paleozoic, then at that time, southern North America, southern Europe, and northwestern Africa would have straddled the equator and would have hosted tropical climates in which swamps and reefs would grow. The tropical climate belts would have been bordered on either side by subtropical climate belts hosting deserts (dry areas) and shallow seas of particularly salty water. In the belt of Pangaea that Wegener predicted would have been equatorial, late Paleozoic rock layers include abundant coal (rock formed from woody plant remains) containing *fossils* (remains preserved in rock) of tropical trees. He also found limestone (a type of sedimentary rock commonly made of shells) containing the fossils of organisms that prefer warm water. In the portions of Pangaea that Wegener predicted would be subtropical, late Paleozoic sedimentary beds include deposits formed in desert dunes or very salty water **(Fig. 2.4)**.

DISTRIBUTION OF FOSSILS. Today, different continents provide homes for different species. Kangaroos, for

FIGURE 2.3 Evidence for a supercontinent.

(a) Glacial striations of late Paleozoic age on the surface of glacially polished bedrock along the southern coast of Australia.

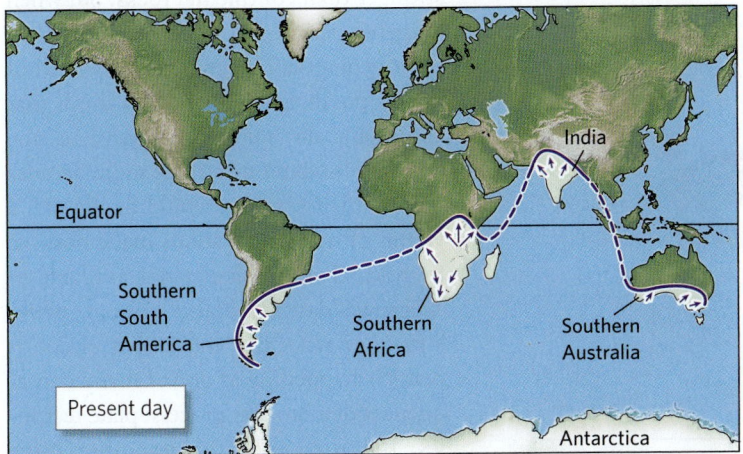

(b) A map showing the distribution of late Paleozoic glacial deposits and the orientation of associated striations.

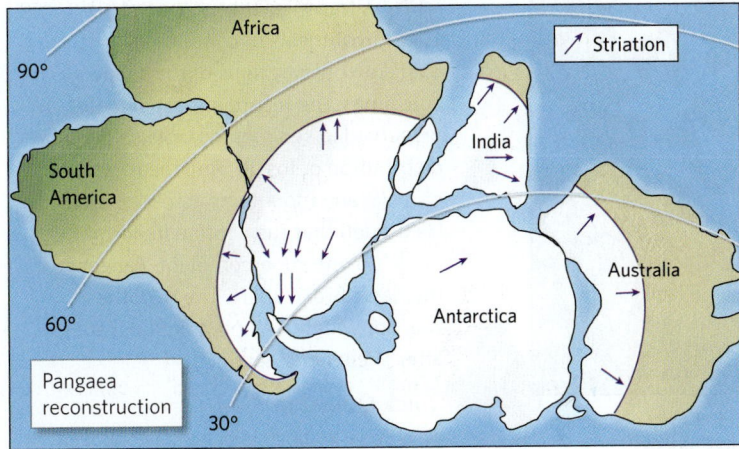

(c) In Wegener's reconstruction of Pangaea, the glaciated areas connect to outline a region of late Paleozoic southern polar ice caps.

FIGURE 2.4 Climate belts, as indicated by distinct rock types, make sense on a map of Pangaea.

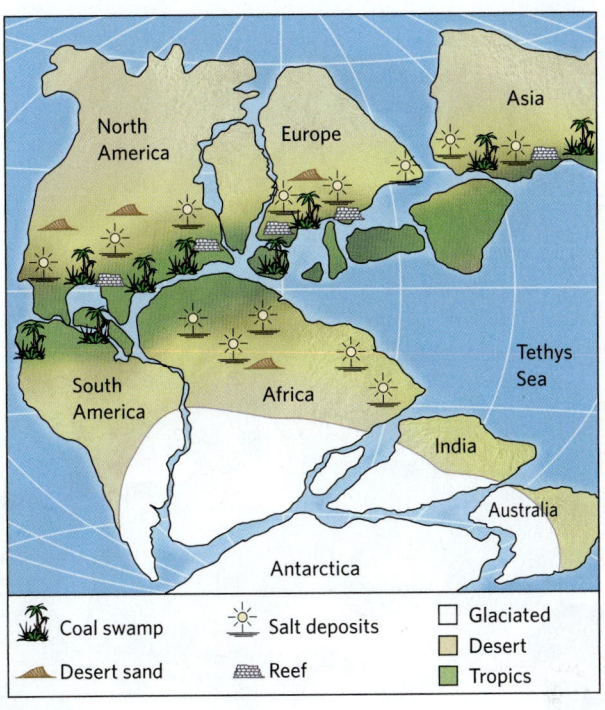

Coal swamp — Salt deposits — ☐ Glaciated

Desert sand — Reef — ☐ Desert

☐ Tropics

example, live in Australia, but nowhere else, because they cannot swim across oceans to other continents. Wegener inferred that when Pangaea existed, land animals and plants could have migrated to colonize regions that today are separated by expanses of ocean. To test this idea, he plotted locations of fossils of land-dwelling species that would have lived when Pangaea existed. As he predicted, fossils of these species occur in sedimentary beds on continents that, though they are now separate, were adjacent in his reconstruction of Pangaea **(Fig. 2.5)**.

MATCHING GEOLOGIC UNITS AND MOUNTAIN BELTS. Wegener speculated that distinct rock types should be traceable from the coast of one continent to the coast of the adjacent continent on his reconstruction of Pangaea. He was correct again. Specifically, he found belts of very old rocks in eastern South America that look just like those that occur in belts of western Africa **(Fig. 2.6a)**. In addition, he predicted that mountain belts on the coastlines of continents that were formerly connected should be similar in age. When he examined descriptions

See for yourself

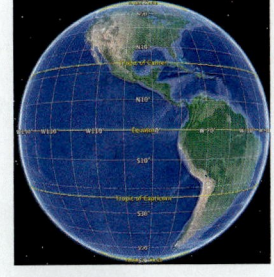

Modern Climate Belts

Latitude: 0°10′24.19″ S
Longitude: 93°7′7.22″ W

Zoom to 15,000 km (9,200 mi) and look straight down. If in Google Earth, click "grid."

Dark green areas are tropical, and brown areas are drier.

North America is now at temperate to Arctic latitudes. Wegener noted that in the late Paleozoic it was tropical, as if near the equator.

FIGURE 2.5 Fossil evidence for continental drift. The dots indicate locations of fossils for the organisms indicated.

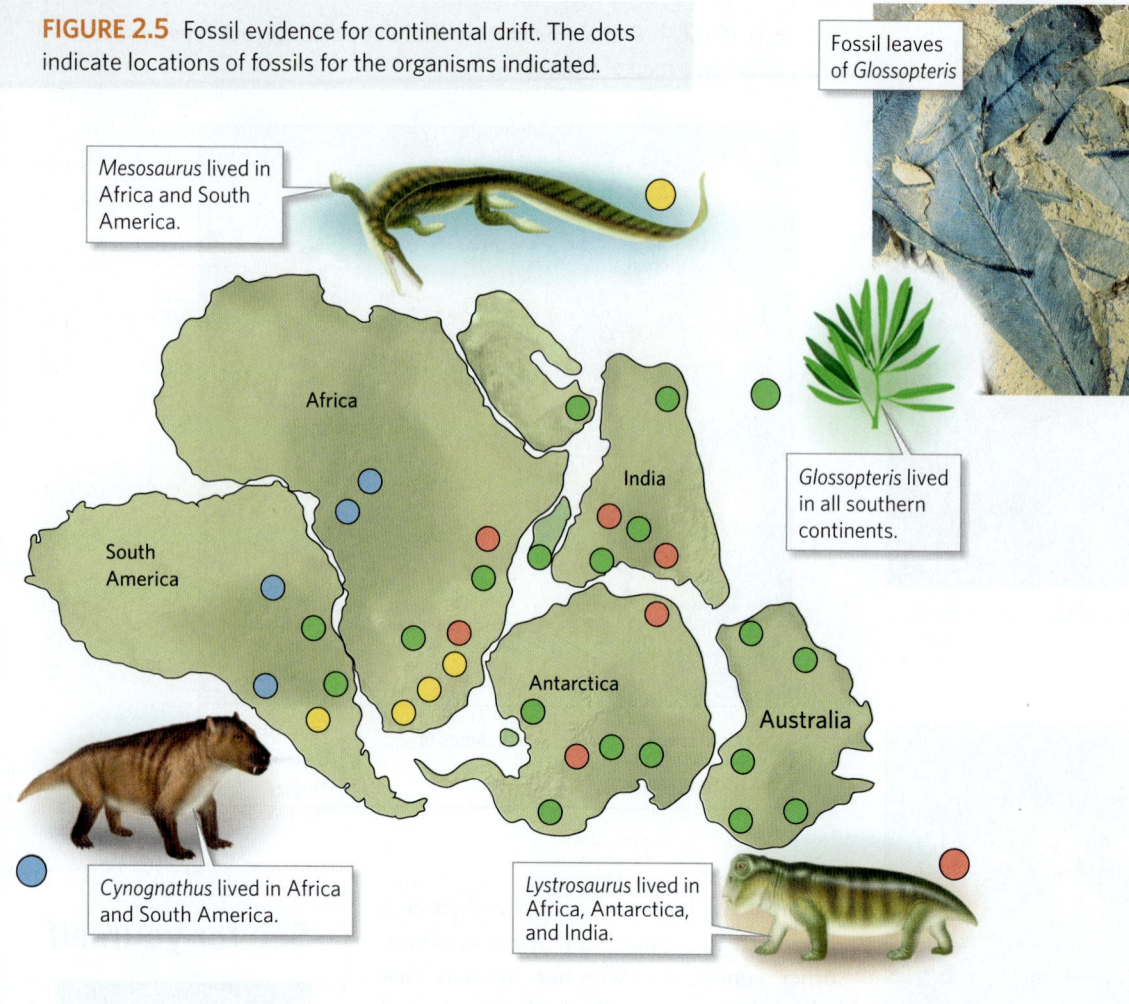

Fossil leaves of *Glossopteris*

Mesosaurus lived in Africa and South America.

Africa

India

Glossopteris lived in all southern continents.

South America

Antarctica

Australia

Cynognathus lived in Africa and South America.

Lystrosaurus lived in Africa, Antarctica, and India.

of ancient mountain belts, he found that the Appalachian Mountains of the United States and Canada align with similar-aged mountain belts of Great Britain, Scandinavia, and northwestern Africa on the map of Pangaea (Fig. 2.6b).

The Opposing View: Drift Denial

Though Wegener's evidence for continental drift seemed compelling, leading geologists of his day argued that drift, nevertheless, was impossible because no forces are strong enough to move continents. Also, Wegener had suggested that continents move by somehow plowing through rock of the ocean floor like a ship plowing through water. Other geologists emphasized that such a process can't take place, because seafloor rocks are stronger than continental rocks. So, when Wegener perished in a Greenland blizzard, most geologists remained unconvinced of his proposals and preferred to think that continents do not move relative to one another. It would take three more decades before this *fixist view* would finally die, a victim of new observations made possible by research technologies not available in Wegener's day. Geologists came to prefer a *mobilist view*, that movement of continents is happening as one of many consequences of a broader process: plate tectonics. The first step in this revolution came with the recognition of seafloor spreading.

FIGURE 2.6 Further evidence of drift: rocks on different sides of the ocean match.

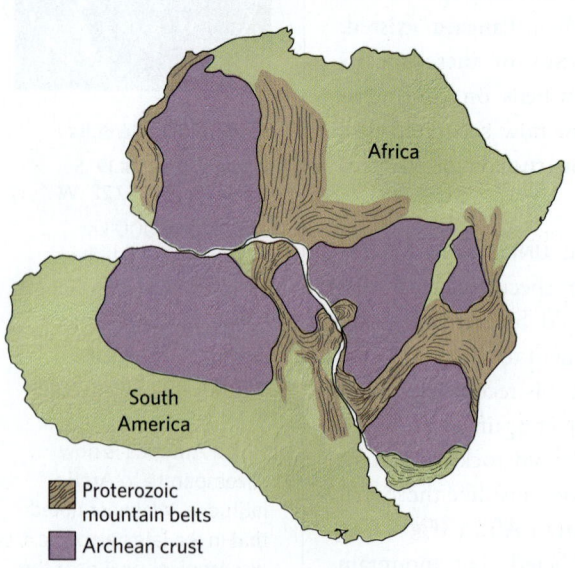

Africa

South America

■ Proterozoic mountain belts
■ Archean crust

(a) Distinctive belts of rock in South America align with similar ones in Africa without the Atlantic Ocean. Proterozoic rocks are ~2.5 to 0.5 Ga. Archean rocks are >2.5 Ga.

Europe

Greenland

North America

Africa

☐ Mountain belt

(b) If the Atlantic Ocean didn't exist, Paleozoic mountain belts on both coasts would be adjacent.

Take-home message . . .

Alfred Wegener proposed his hypothesis of continental drift based on observations of the shape of coastlines, the record of past climates preserved in sedimentary rocks, the distribution of fossils, and the matching of rocks and mountains across oceans. He argued that the continents formerly made up one supercontinent, Pangaea, that later broke apart. The hypothesis was not widely accepted until decades after Wegener's death.

Quick Question

Why were geologists of his day opposed to Wegener's hypothesis?

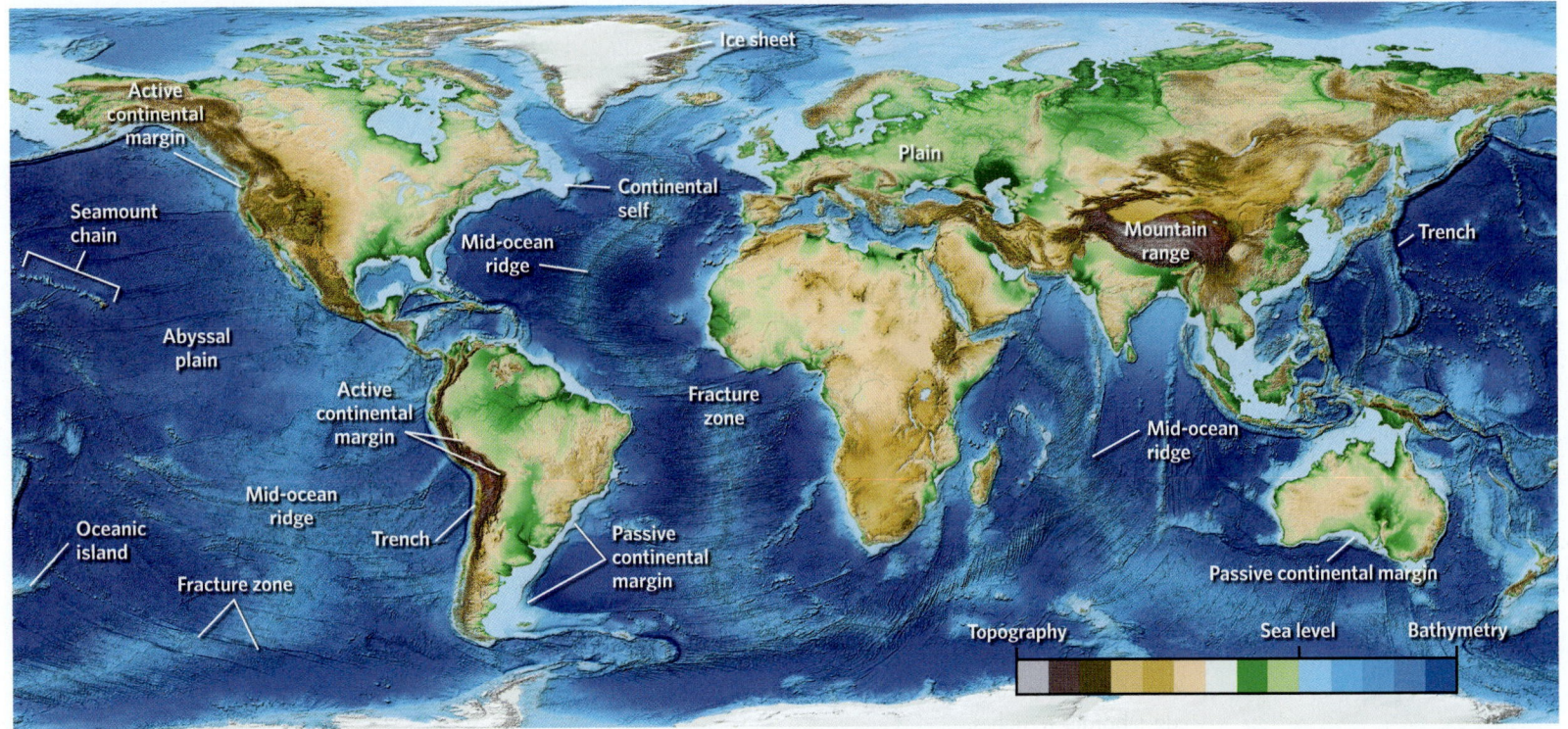

Figure labels (clockwise): Ice sheet; Active continental margin; Plain; Mountain range; Trench; Seamount chain; Continental self; Mid-ocean ridge; Abyssal plain; Active continental margin; Fracture zone; Mid-ocean ridge; Oceanic island; Mid-ocean ridge; Trench; Passive continental margin; Fracture zone; Passive continental margin; Topography; Sea level; Bathymetry

FIGURE 2.7 Bathymetry of the world's oceans.

2.3 The Discovery of Seafloor Spreading

New Geologic Observations (1930–1960)

From the time of Wegener's death in 1930 until the 1960s, geology entered an age of discovery, during which researchers developed new tools to explore the Earth System. Here, we review some of their astounding discoveries.

BATHYMETRIC FEATURES OF THE SEAFLOOR. Prior to the 20th century, the study of *bathymetry*—depth variations of the seafloor—was extremely tedious because the only way to measure depth at a location was to make a *sounding* by lowering a heavy weight attached to a cable down to the seafloor. In the deeper parts of the ocean, a single measurement could take hours. The invention of *sonar* (echo sounding) greatly sped up the process of making depth measurements. With sonar, it's possible to obtain a continuous record of the depth of the seafloor while a ship cruises, and from this record, to produce a *bathymetric profile*, a diagram that plots depth on the vertical axis against location on the horizontal axis. By obtaining many bathymetric profiles, geologists can produce a **bathymetric map** that reveals the overall shape of the seafloor. In 1977, Marie Tharp (an American researcher) and her colleagues hand-painted the first modern bathymetric map that clearly depicted important geologic features of the seafloor in a meaningful way. Modern technology allows satellites

to map bathymetry from space **(Fig. 2.7)**. Bathymetric maps have revealed the following features:

- **Ocean basins** are distinctly lower than continents. The boundary between an ocean basin and a continent is called a **continental margin**.

- Large areas of ocean basins consist of broad **abyssal plains**, flat regions that lie at depths of 4–5 km (2.5–3.1 mi) below sea level.

- **Mid-ocean ridges**, long submarine mountain ranges whose crests lie at depths of 2.0–2.5 km (1.2–1.5 mi) below sea level, occur in all ocean basins. The adjective *mid-ocean* can be a bit confusing because it gives the impression that these features all run down the center of ocean basins. This isn't always the case—some ridges lie closer to one side of an ocean basin than they do to the other. Mid-ocean ridges tend to be symmetrical, in that the bathymetry on one side of a *ridge axis* (the center line of the ridge) is a mirror image of the bathymetry on the other side. In addition, ridges are segmented, in that along their length, ridges consist of segments that do not align end to end.

- Narrow bands of broken-up oceanic crust, containing vertical fractures, cross mid-ocean ridges and trend roughly at right angles to the ridge axis. These features are known as **fracture zones**. The segments of a mid-ocean ridge terminate at fracture zones.

- Elongated troughs, known as **deep-sea trenches**, in which the ocean floor reaches depths of 7–11 km (4–7 mi), occur in or along some edges of ocean basins.

See for yourself

Trenches

Latitude: 35°37′59.86″ N
Longitude: 145°36′12.78″ E

Fly to an elevation of 5,500 km (3,500 mi) and look straight down. Zoom in. If in Google Earth, you can use the elevation tool to measure trench depth. The dark bands that you see are the trenches of the western Pacific Ocean, near Japan.

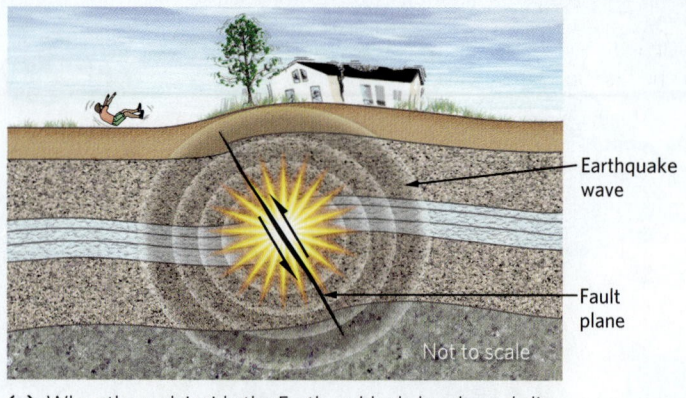

(a) When the rock inside the Earth suddenly breaks and slips, forming a fracture called a fault, it generates shock waves that pass through the Earth and shake the surface.

FIGURE 2.8 Faulting and earthquakes.

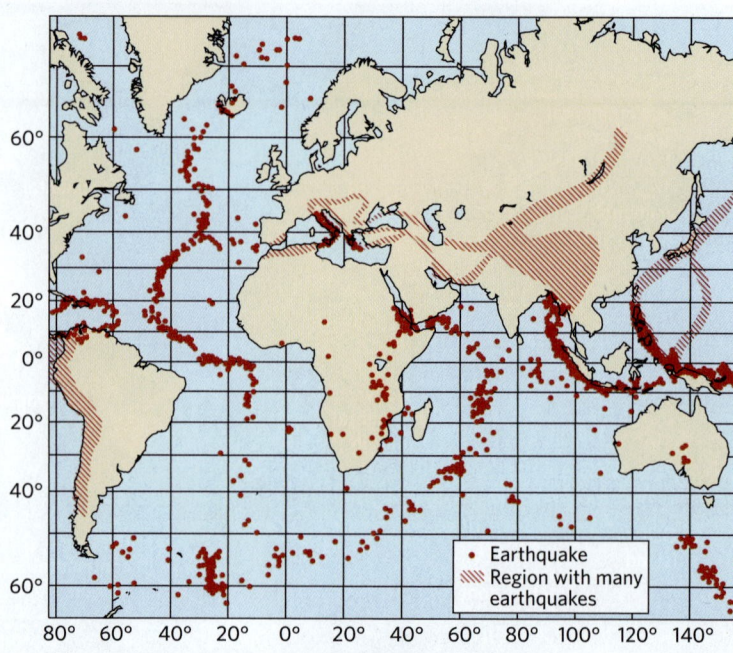

(b) A 1953 map showing the distribution of earthquake locations in the ocean basins. Note that earthquakes occur in belts.

Some trenches border continents, but some do not. Further, not all continental edges border trenches.

- At hundreds of locations in the world's oceans, volcanic eruptions have produced peaks of igneous rock that built up above the surrounding seafloor. If the top of a peak rises above sea level, it forms an *oceanic island*, whereas if its top lies below sea level, it's a *seamount*. Oceanic islands and seamounts typically occur in chains.

CHARACTER OF OCEANIC CRUST. In addition to obtaining information about the shape of the seafloor surface, geologists dredged the seafloor, drilled into the seafloor, and used various instruments to characterize subsurface layers within the oceanic crust. This work led to the realization that the upper portion of the oceanic crust beneath abyssal plains consists of a relatively thin layer of sediment that settled down from overlying seawater. Almost no sediment occurs at the ridge axis. The sediment layer thickens progressively away from a mid-ocean ridge. Beneath the sediment layer, oceanic crust consists of *basalt*, a dense gray igneous rock (see chapter-opening photo).

Researchers also measured the temperature in holes drilled into the seafloor. From these measurements, they calculated *heat flow*, the rate at which heat rises from the Earth's interior. By comparing results from many locations, researchers learned that heat flow beneath mid-ocean ridges is greater than that beneath abyssal plains.

DISTRIBUTION OF EARTHQUAKES. A **fault** is a fracture in the Earth's crust on which *slip* (sliding) takes place. Most **earthquakes**—episodes of ground shaking—are due to fault slip **(Fig. 2.8a)**. The sudden movement generates vibrations that travel through the Earth in the form of

waves, and when these waves reach the ground surface, they cause the shaking we feel as an earthquake.

When researchers learned how to find the location of an earthquake (see Chapter 8), it became possible to produce maps that show the global distribution of earthquakes. Even early versions of such maps demonstrated that earthquakes do not occur randomly, but rather cluster in distinct belts, called **seismic belts**, from the Greek word *seismos*, which means earthquake **(Fig. 2.8b)**. Seismic belts lie along trenches, along mid-ocean ridge axes, along parts of fracture zones, and on other faults.

Harry Hess and His "Essay in Geopoetry"

Initially, geologists did not see relationships among the observations described above. That changed when Harry Hess realized that the observations were clues that could ultimately help answer the question of how continents move. Hess began by thinking about the implications of seafloor-sediment thickness. He thought that if the ocean basins were as old as the Earth itself, they should be covered by thick sediment. The thinness of the sediment found on the ocean floor indicated that the ocean floor couldn't be as old as the Earth. Further, the progressive increase in thickness of the sediment away from mid-ocean ridges suggested that ridges are younger than abyssal plains. Hess concluded, therefore, that new ocean floor must be forming at the ridges, so that an ocean basin grows wider with time. But how? The association of earthquakes with mid-ocean ridges provided an answer—seafloor must be cracking and splitting apart at the ridges.

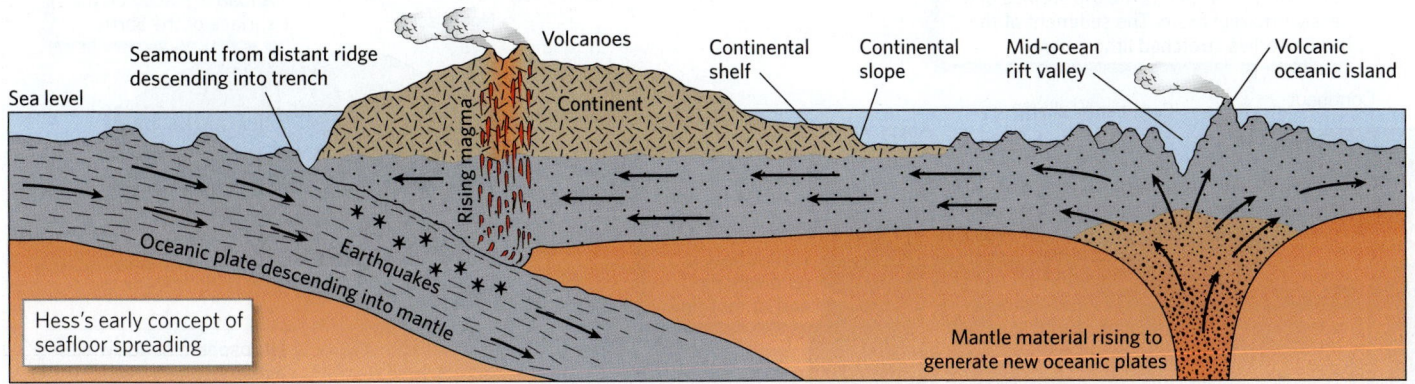

The high heat flow along mid-ocean ridge axes further suggested that hot molten rock was rising beneath the zone where this splitting takes place.

Hess put the clues together and proposed that ocean basins form by a process that came to be known as **seafloor spreading**, a name proposed by another geologist, Robert Dietz. According to Hess's proposal, seafloor stretches apart along the axis of a mid-ocean ridge, and new oceanic crust forms from magma that rises and fills the space. Once formed, new seafloor moves away from the ridge, and as a consequence, an ocean basin can grow wider over time. Hess realized that if new seafloor forms at the ridge axis, then somewhere else, old seafloor must be consumed—otherwise, the Earth would have to be expanding significantly, and no evidence indicated that it was. So he proposed that old seafloor sinks back into the mantle at deep-sea trenches, and that this movement generates the seismic belts observed along trenches (Fig. 2.9).

Hess, along with other geologists, recognized that seafloor spreading provided the long-sought explanation of how continental drift occurs, an explanation that had eluded Wegener. Simply put, continents move apart due to seafloor spreading at mid-ocean ridges, and continents move toward each other as the seafloor between them is consumed at trenches. The hypothesis seemed so elegant that Hess referred to it as "an essay in geopoetry."

Take-home message . . .

🏠 Observations of seafloor bathymetry, sediment cover, heat flow, and seismic belts led to the proposal that new seafloor forms at mid-ocean ridges, causing seafloor spreading. Old seafloor sinks back into the mantle at trenches, by the process of subduction. Continents move as a consequence of these processes.

Quick Question ──────────────
What is the relationship between seismic belts and bathymetric features of the ocean basins?

2.4 Modern Plate Tectonics Theory

The proposal of seafloor spreading set off a revolution in geologic thought. Over the course of the next decade, most geologists came to realize that their long-held interpretations of the Earth, based on the fixist premise that the positions of continents did not change and that the ocean basins that we see today date back to the birth of the Earth, were simply wrong! If seafloor spreading and subduction take place, and as a consequence, the continents move, then the outer layer of the Earth, the shell that we live on, must be quite mobile indeed.

By 1968, researchers had produced a new model of the Earth. In this model, the Earth has a rigid outer shell. (A *rigid material*, in this context, is one that bends and breaks when subjected to a force, but does not flow.) This shell is not completely intact, but rather consists of separate pieces, or *plates*, that move relative to one another. After this model passed many tests, it came to be known as the *theory of plate tectonics*, or simply **plate tectonics**. Geologists realized that the theory could explain not only all of Wegener's observations, but also many other important geologic phenomena, such as earthquakes, volcanoes, and mountain building. In effect, plate tectonics became the *grand unifying theory* of geology. In this section, we describe key aspects of the theory. We begin by clarifying what we mean by a plate.

The Concept of Lithosphere and Asthenosphere

As we noted in Chapter 1, geologists back in the 19th century concluded that the interior of the Earth consists of distinct layers (crust, mantle, and core). At first, geologists thought that the rigid shell of the Earth consisted only of the crust—in fact, that's how Hess portrayed plates in his sketch depicting seafloor spreading. But this

FIGURE 2.10 The nature and behavior of the lithosphere.

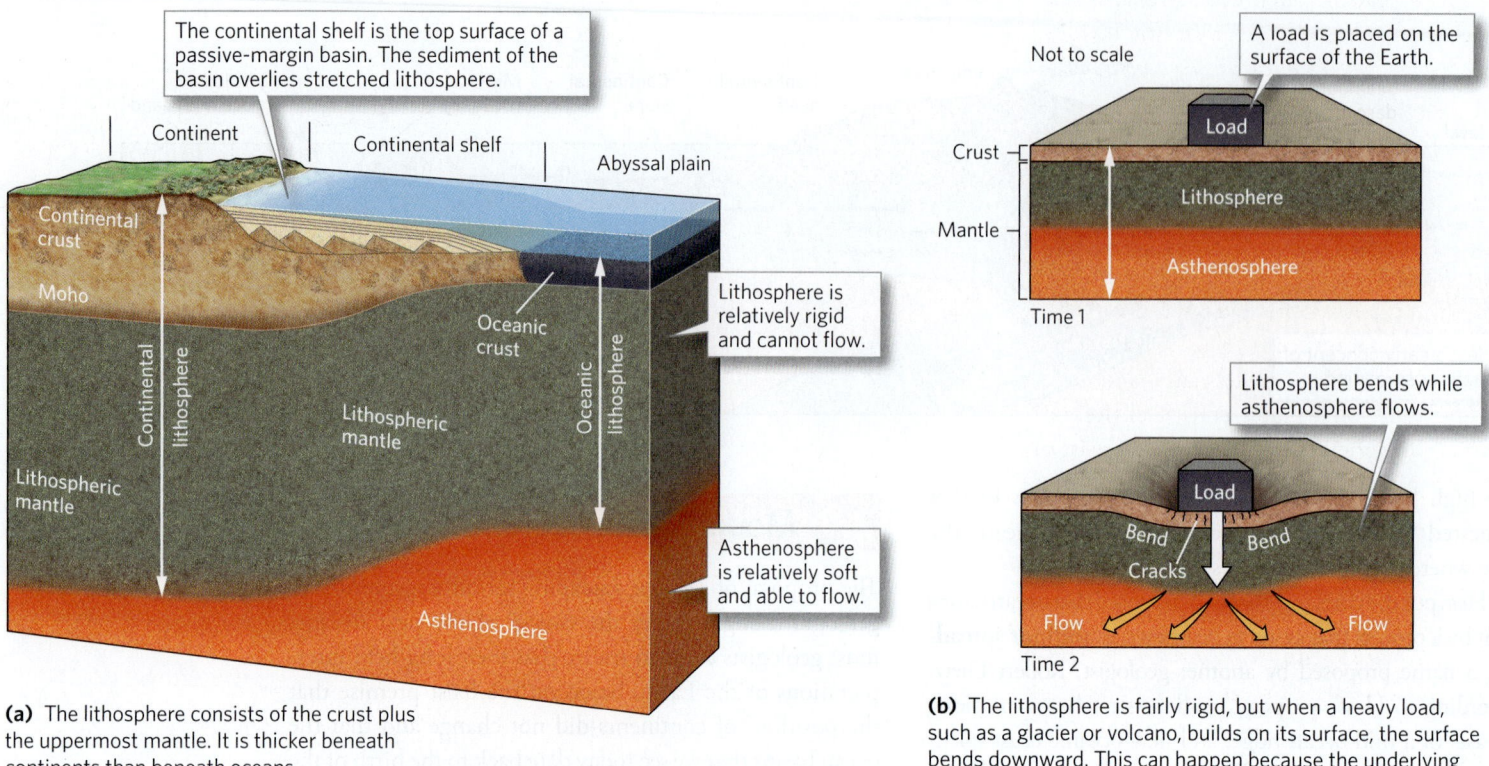

The continental shelf is the top surface of a passive-margin basin. The sediment of the basin overlies stretched lithosphere.

Continent

Continental shelf

Abyssal plain

Continental crust

Moho

Continental lithosphere

Oceanic crust

Oceanic lithosphere

Lithospheric mantle

Lithospheric mantle

Asthenosphere

Lithosphere is relatively rigid and cannot flow.

Asthenosphere is relatively soft and able to flow.

(a) The lithosphere consists of the crust plus the uppermost mantle. It is thicker beneath continents than beneath oceans.

Not to scale

A load is placed on the surface of the Earth.

Load

Crust

Mantle

Lithosphere

Asthenosphere

Time 1

Lithosphere bends while asthenosphere flows.

Load

Bend   Bend

Cracks

Flow   Flow

Time 2

(b) The lithosphere is fairly rigid, but when a heavy load, such as a glacier or volcano, builds on its surface, the surface bends downward. This can happen because the underlying plastic asthenosphere can flow out of the way.

image is not correct. The rigid outer shell of the Earth includes not only the crust, but also the top part of the upper mantle directly beneath the crust. Geologists refer to this combination—the crust plus the top part of the upper mantle—as the **lithosphere**. The crust, as we've seen, is about 7–10 km (4–6 mi) thick beneath the oceans and about 25–70 km (15–45 mi) thick beneath continents. The lithosphere, in contrast, is about 100 km (60 mi) thick beneath the oceans and 150 km (90 mi) thick beneath the continents. Geologists refer to the part of the mantle within the lithosphere as the **lithospheric mantle (Fig. 2.10a)**. The portion of the mantle that lies beneath the lithosphere is called the **asthenosphere**. In contrast to the lithosphere, the asthenosphere displays *plastic behavior*. (A *plastic material*, in this context, is a material that can flow slowly without breaking.) The asthenosphere flows plastically at rates of 1–15 cm (0.5–6 in) per year.

The distinctions between crust and mantle and between lithosphere and asthenosphere can be confusing. Crust differs from mantle by its composition, meaning that mantle rock contains a different mix of chemicals and, therefore, different rock types, than does crustal rock. Lithosphere differs from asthenosphere by its response to a force—lithosphere behaves rigidly, whereas

asthenosphere behaves plastically. If an imaginary heavy load were placed on the surface of the lithosphere, the lithosphere would bend downward and could crack, while the underlying asthenosphere would flow out of the way **(Fig. 2.10b)**.

Why is the lithospheric mantle rigid while the asthenosphere is not? This contrast in behavior exists because the Earth gets hotter with increasing depth, and as rock gets hotter, it gets softer. At the depth of the lithosphere's base, mantle rock (peridotite) reaches a temperature of about 1,280°C (2,340°F). At this temperature, and at hotter temperatures, peridotite can flow, whereas at cooler temperatures, it cannot. Put another way, the lithospheric mantle and the asthenosphere have the same composition. The difference between them is only that they behave differently (one rigidly, the other plastically) because of their respective temperatures. To picture the relationship between temperature and rigidity, imagine a simple experiment: Place one wax candle in a freezer and one in a warm oven. After a while, take them out and try to bend them. The cold candle resists bending and breaks because it's rigid, whereas the warm one changes shape without breaking because it's plastic. Heat makes rock, like wax, softer.

As we've mentioned, the Earth's lithosphere contains a number of major breaks that separate it into **plates**, which geologists also refer to as *lithosphere plates* (Fig. 2.11a). The dividing lines between plates are called *plate boundaries*, and a portion of a plate that is not near a plate boundary is called a *plate interior*. Geologists recognize about 20 plates. These include seven large plates, several smaller ones, and several very small ones known as *microplates*. Some plates have familiar-sounding names, such as the North American Plate and the African Plate, while some names, such as the Cocos Plate and the Juan de Fuca Plate, are less familiar.

Note that some plates consist entirely of oceanic lithosphere, whereas others consist of both oceanic and continental lithosphere. In addition, some continental margins are plate boundaries but some are not, so geologists distinguish between **active margins** of continents, which coincide with plate boundaries, and **passive margins**, which do not. Typically, a thick wedge of sediment accumulates along a passive margin. This wedge includes sand and mud carried to the sea by rivers as well as the shells of marine organisms. The surface of such a wedge is a **continental shelf** (see Fig. 2.10), a broad region of shallow sea between the continent and the abyssal plain (Fig. 2.11b).

Identifying Plate Boundaries

How do we recognize the location of a plate boundary? All plate boundaries are identified by the presence of *active faults*, meaning faults where slip has occurred recently and will probably occur in the future. Slip on faults at plate boundaries permits one plate to move relative to its neighbor, and this slip generates earthquakes (Fig. 2.12a). Seismic belts, therefore, define the positions of plate boundaries (Fig. 2.12b). Plate interiors, the regions away from the plate boundaries, remain relatively earthquake-free because the plates themselves are rigid.

We've stated that plates move relative to one another. Such movement can take place because the underlying plastic asthenosphere can flow. To picture this movement, imagine placing a block of wood on a layer of molasses—if you push the wood, it moves relative to the molasses below. Geologists classify plate boundaries into three types based on the motion of the plate on one side of the boundary relative to the plate on the other (Fig. 2.13). A boundary at which two plates move apart

FIGURE 2.11 Lithosphere plates and continental margins.

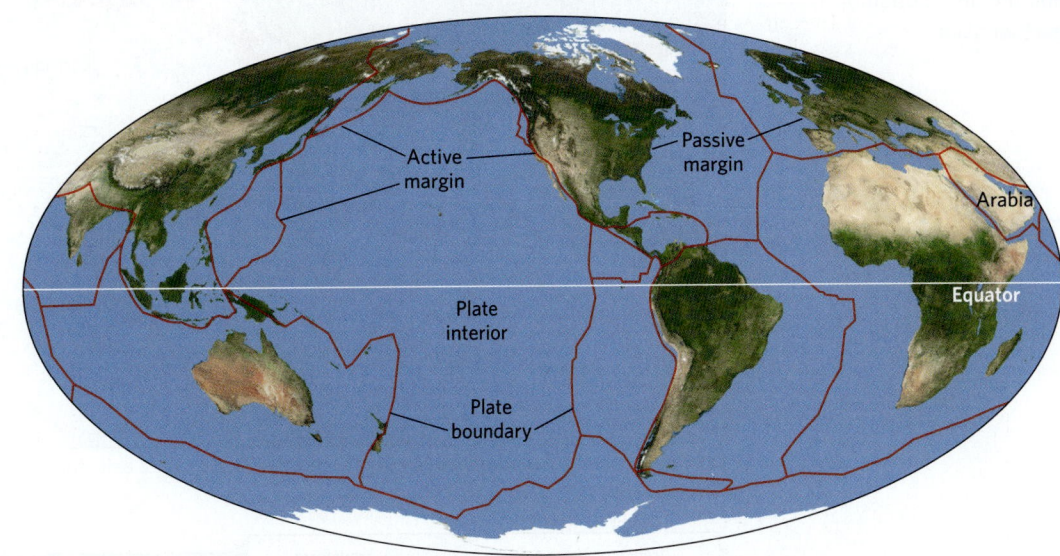

(a) An image of the globe on which some of the major plate boundaries are highlighted. Active continental margins lie along plate boundaries; passive margins do not.

(b) A bathymetric image of the continental shelf along the east coast of North America.

from each other is a *divergent boundary*, a boundary at which two plates move toward each other so that one plate sinks beneath the other is a *convergent boundary*, and a boundary at which one plate slips horizontally along the edge of another plate is a *transform boundary*. We discuss plate boundaries in more detail in the next section of this chapter.

Plate Tectonics Theory Today: A Synopsis

It took years of work by many researchers around the world to refine the theory of plate tectonics. In light of our introduction to lithosphere, asthenosphere, plates,

FIGURE 2.12 The locations of plate boundaries and the distribution of earthquakes.

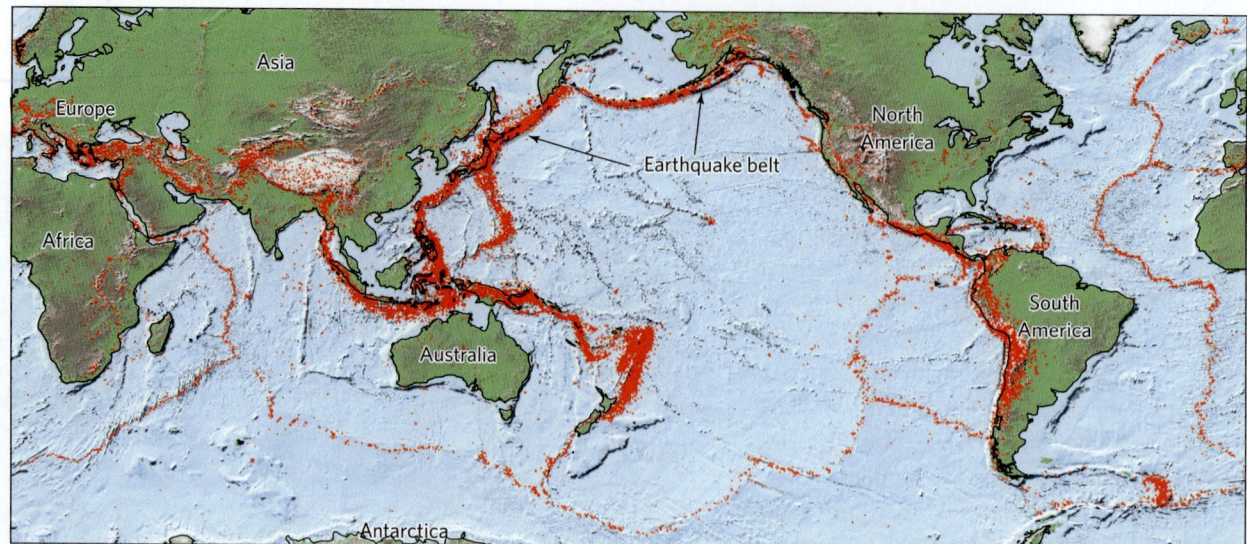

(a) The locations of earthquakes (red dots) mostly fall in distinct belts that correspond to plate boundaries. Relatively few earthquakes occur in the stabler plate interiors.

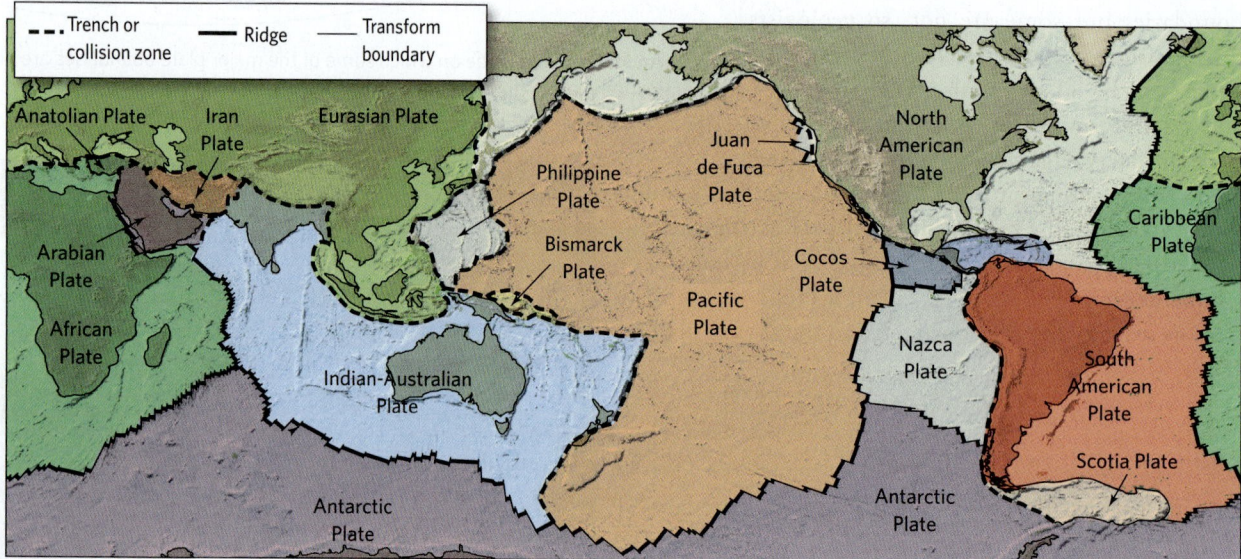

(b) A map of major plates shows that some consist entirely of oceanic lithosphere, whereas others consist of both continental and oceanic lithosphere.

FIGURE 2.13 The three types of plate boundaries.

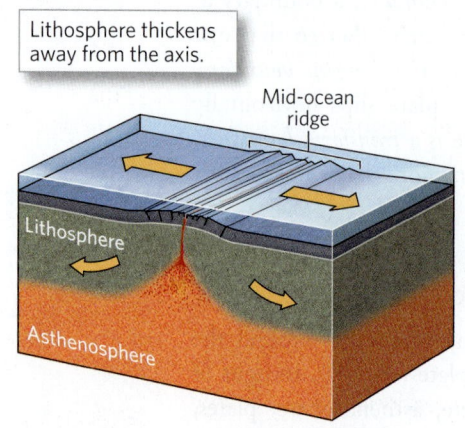

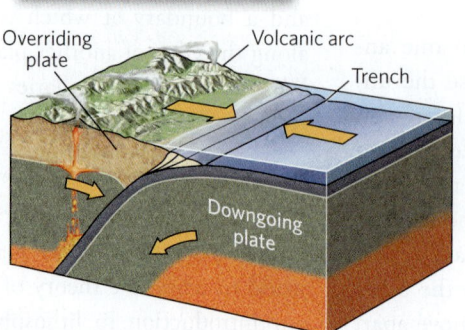

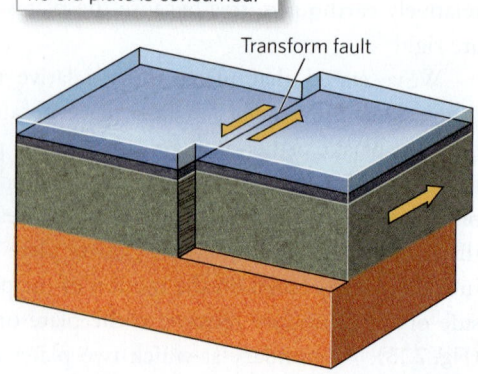

(a) At a *divergent boundary*, two plates move away from the axis of a mid-ocean ridge, where new oceanic lithosphere forms.

(b) At a *convergent boundary*, two plates move toward each other; the downgoing plate sinks beneath the overriding plate.

(c) At a *transform boundary*, two plates slide past each other on a vertical fault.

and plate boundaries, we can summarize the key components of the theory as follows:

- The Earth's lithosphere, its rigid outer shell, consists of the crust and underlying uppermost mantle. The lithosphere is divided into plates that move relative to one another.

- Plates are on the order of 100–150 km (60–90 mi) thick and range from hundreds to thousands of kilometers across. Some consist of only oceanic lithosphere, whereas others consist of both oceanic and continental lithosphere.

- Plates sit on the underlying asthenosphere, the portion of the mantle that is warm enough to flow plastically.

- As plates move, their interactions at plate boundaries cause earthquakes and other geologic phenomena, but plate interiors remain intact (**Earth Science at a Glance**, pp. 64–65). The distribution of seismic belts, therefore, defines the positions of plate boundaries.

- Geologists distinguish among three types of plate boundaries (convergent, divergent, and transform) based on relative plate motions.

- The movement of plates results in continental drift. Because of plate tectonics, the map of the Earth's surface constantly changes.

Take-home message . . .

Earth's lithosphere—its rigid outer shell—is divided into plates that move relative to one another. As a plate moves, its interior remains mostly unchanged, but slip occurs along the plate boundary, causing earthquakes. This overall concept is the theory of plate tectonics.

Quick Question
What does the lithosphere consist of? How does it differ from the asthenosphere?

2.5 Geologic Features of Plate Boundaries

Divergent Boundaries

In 1872, HMS *Challenger* set sail from England to begin the world's first research cruise. Every now and then the ship's crew took a sounding. The results hinted that a submerged mountain range ran from north to south down the center of the Atlantic Ocean basin. This feature, which rose about 2 km (1.2 mi) above the abyssal plain, came to be known as the *Mid-Atlantic Ridge* (**Fig. 2.14a, b**). This ridge extends from the latitude of far northern Greenland to the latitude of far southern South America. Significantly, the ridge is not a continuous line. Rather, along its length, it consists of distinct segments whose ends terminate at a fracture zone. Further, the segments do not line up, so the end of one segment may be offset along the fracture zone by a few kilometers to several hundred kilometers (**Fig. 2.14c**).

Bathymetric studies over the next several decades found that all the major oceans contain such ranges, which came to be known in general as *mid-ocean ridges*, although not all occupy the center line of the basin. The ridge of the east Pacific Ocean was named the *East Pacific Rise* because, in contrast to other mid-ocean ridges, it has a relatively broad smooth surface.

Until the theory of plate tectonics came along, no one knew why mid-ocean ridges existed. Plate tectonics theory, however, explains their geologic origin. A mid-ocean ridge delineates a **divergent boundary**, also known as a *spreading boundary* because at this type of plate boundary, two oceanic plates move apart by seafloor spreading. Note that during this process, no open space ever develops between diverging plates. Rather, as the plates move apart, new oceanic lithosphere forms between them (**Fig. 2.15a**). Typically, a narrow valley, bordered by faults, defines the center line of the ridge or *ridge axis* (**Fig. 2.15b**). Overall, the seafloor on each side of the ridge slopes gently away from it, reaching the abyssal-plain depth (4.5 km; 2.8 mi) at up to 800 km (500 mi) from the ridge axis. A mid-ocean ridge is roughly symmetrical, in that one side looks like a mirror image of its other side.

What happens at the ridge during seafloor spreading? As the process takes place, hot asthenosphere rises beneath the ridge. As this material rises, it begins to melt (for reasons we'll explain in Chapter 4). The resulting molten rock, or *magma*, rises and accumulates in a magma chamber below the ridge axis (see Fig. 2.15b). A *magma chamber* can be pictured as a volume filled with a mush of crystals mixed with red-hot melt. Some of the mush solidifies into gabbro along the sides of the chamber, while some rises still higher, along vertical cracks, and solidifies into wall-like sheets, called *dikes,* that are made of basalt. And some makes it all the way to the surface of the seafloor and seeps out as lava from small submarine volcanoes that follow cracks along the ridge axis. This lava cools when it comes in contact with cold seawater and forms glassy, melon-shaped blobs known as *pillows*, which accumulate into mounds of *pillow basalt* (see chapter-opening photo). The entire layer of igneous rock formed at the ridge axis, from the base of the gabbro to the top of the pillow basalt, represents new oceanic crust.

As it forms, oceanic crust moves away from the ridge axis, and as this happens, more asthenosphere rises, more

See for yourself

The Mid-Atlantic Ridge
Latitude: 2°34′26.28″ S
Longitude: 15°37′59.86″ W

Fly to an elevation of 11,300 km (6,800 mi) and look straight down to see a view of the Mid-Atlantic Ridge in the equatorial Atlantic Ocean. Zoom in to see a fracture zone.

FIGURE 2.14 Bathymetry of a mid-ocean ridge.

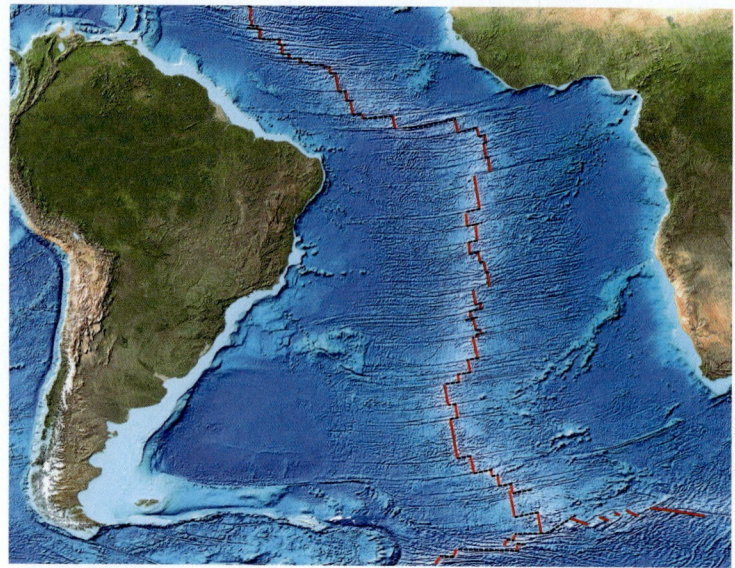

What an Earth Scientist Sees

(a) A bathymetric map of the Mid-Atlantic Ridge in the South Atlantic Ocean. The lighter shades of blue are shallower water depths. Red lines are the axis of the ridge. The sketch labels key features.

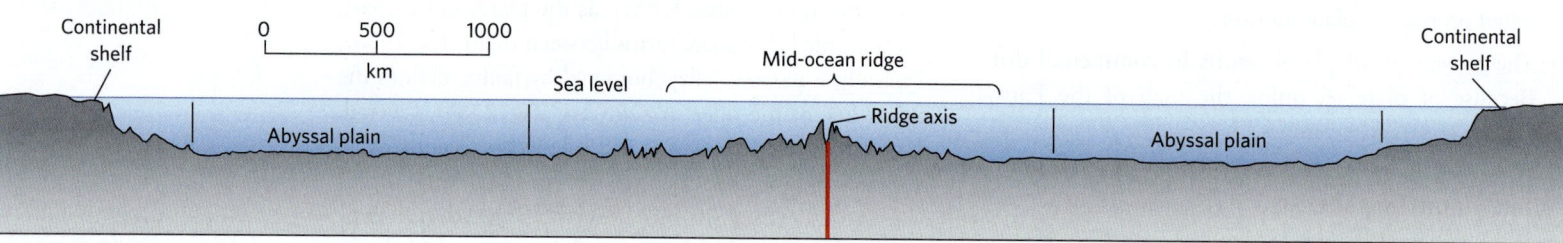

(b) A bathymetric profile (vertical slice) across the Mid-Atlantic Ridge, from continent to continent. Note that the ridge is higher than the abyssal plain. The red line indicates rising magma.

magma forms and rises, and still more crust forms. In other words, like a vast, continuously moving conveyor belt, oceanic crust forms at the ridge and then moves away from the ridge. The stretching force, or *tension*, applied to newly formed solid crust as spreading takes place breaks up this new crust, resulting in the formation of faults bordering the ridge axis. Slip on these faults causes divergent-boundary earthquakes.

At the ridge axis, a lithosphere plate consists only of new crust. As this crust moves away from the ridge axis, no additional igneous rock adds to it, so its thickness stays the same. Therefore, the crust, along with the mantle directly beneath, progressively cools as it ages. Loss of heat from the mantle causes the boundary between cooler, rigid mantle and warmer, plastic mantle to become deeper. Because this boundary defines the base of the lithosphere, the oceanic lithosphere thickens as it ages **(Fig. 2.16)**. Further, because rock becomes denser as it cools, the lithosphere, overall, gets denser as it ages. So, like a cargo ship whose keel sinks deeper into the water, and whose deck moves downward, as workers load it

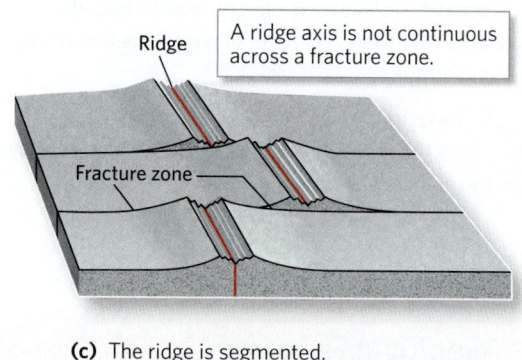

(c) The ridge is segmented.

with heavy cargo, the lithosphere moves deeper into the asthenosphere, and the seafloor becomes lower, as the lithosphere ages. As a result, abyssal plains, which are underlain by old lithosphere, are deeper than mid-ocean ridges, which are underlain by young lithosphere.

Because all seafloor forms at mid-ocean ridges, the youngest seafloor borders the ridge axis, and the oldest seafloor lies farthest from the ridge **(Fig. 2.17a)**. For this

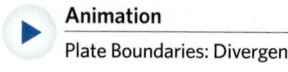

Animation

Plate Boundaries: Divergent

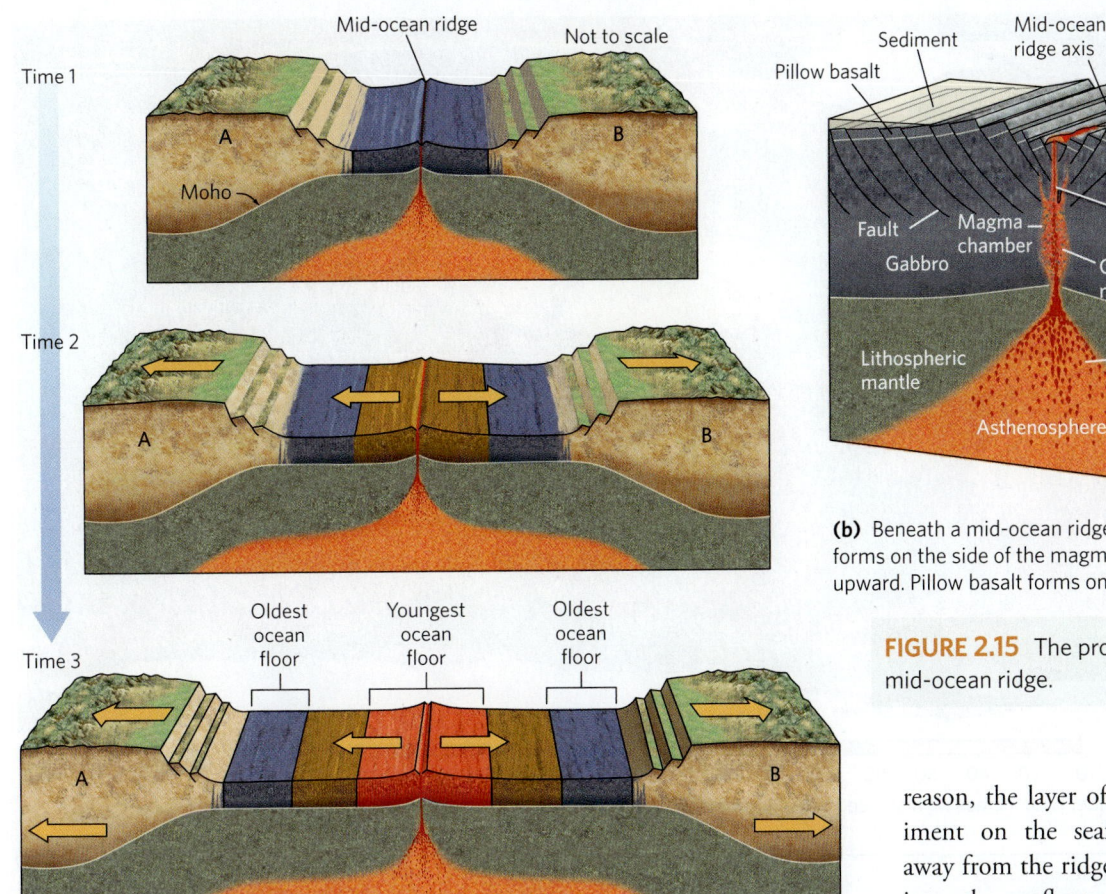

Time 1

Mid-ocean ridge

Not to scale

A

Moho

B

Time 2

A

B

Time 3

Oldest ocean floor

Youngest ocean floor

Oldest ocean floor

A

B

(a) During seafloor spreading, the ocean floor gets wider and continents on either side move apart. New oceanic crust forms at the ridge axis.

Sediment

Mid-ocean ridge axis

Fault scarp

Median valley

Pillow basalt

Fault

Gabbro

Magma chamber

Magma

Crystal mush

Dikes

Lithospheric mantle

Zone of partial melting

Asthenosphere

(b) Beneath a mid-ocean ridge is a magma chamber. Gabbro forms on the side of the magma chamber. Basalt dikes protrude upward. Pillow basalt forms on the seafloor surface.

FIGURE 2.15 The process of seafloor spreading at a mid-ocean ridge.

FIGURE 2.16 Changes accompanying the aging of lithosphere.

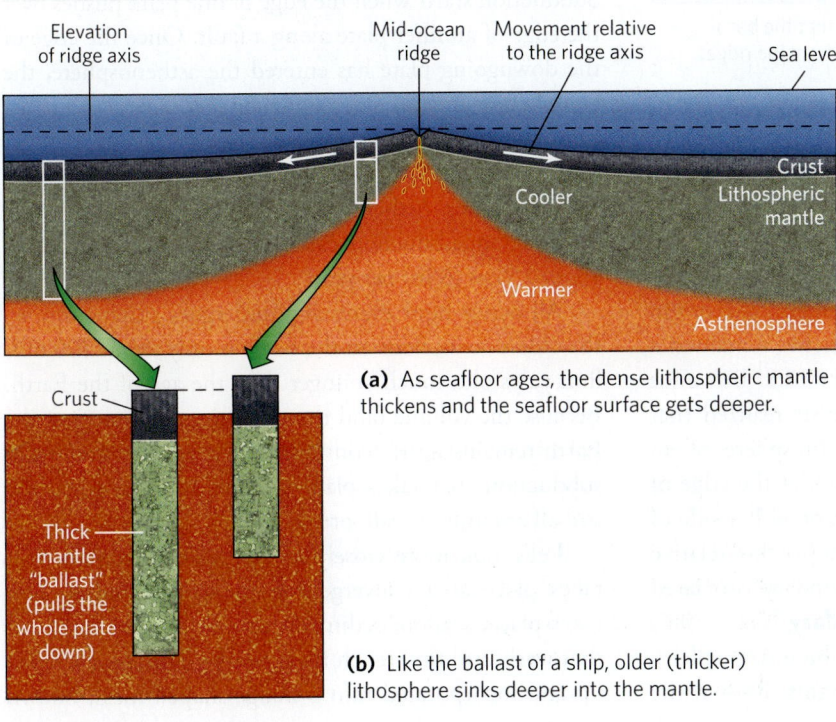

Elevation of ridge axis

Mid-ocean ridge

Movement relative to the ridge axis

Sea level

Cooler

Crust

Lithospheric mantle

Warmer

Asthenosphere

Crust

(a) As seafloor ages, the dense lithospheric mantle thickens and the seafloor surface gets deeper.

Thick mantle "ballast" (pulls the whole plate down)

(b) Like the ballast of a ship, older (thicker) lithosphere sinks deeper into the mantle.

reason, the layer of deep-sea sediment on the seafloor thickens away from the ridge axis. Drilling into the seafloor confirms that the basal (lowest) sediment layer becomes progressively older with increasing distance from the ridge. This observation serves as a proof of seafloor spreading (Fig. 2.17b).

Convergent Boundaries

During its epic voyage of discovery, HMS *Challenger* made soundings in the western Pacific, and it found a location where ocean depth exceeded 8 km (5 mi). Seventy-six years later, HMS *Challenger II* revisited the location and surveyed it with sonar, finding that the region of deep water delineated a 2,500-km (1,500-mi)-long trough, now known as the Mariana Trench, whose deepest point was almost 11 km (7 mi) below sea level—almost three times the average depth of the seafloor (Fig. 2.18a)! Similar deep-sea trenches define most of the boundary of the Pacific Plate and occur along portions of the margins of several other plates

FIGURE 2.17
Age of the
seafloor.

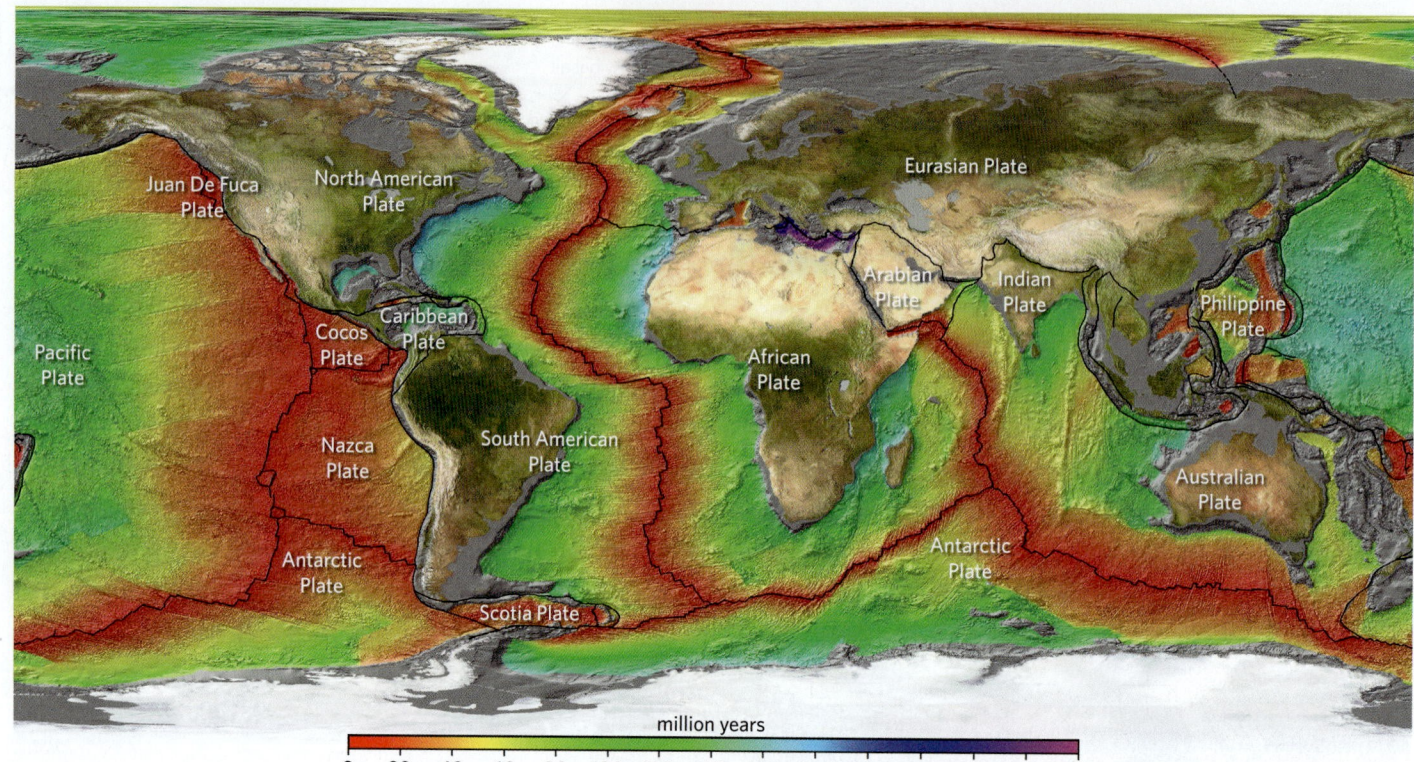

million years

0 20 40 60 80 100 120 140 160 180 200 220 240 260 280

(a) Note that the seafloor grows older with increasing distance from a mid-ocean ridge axis.

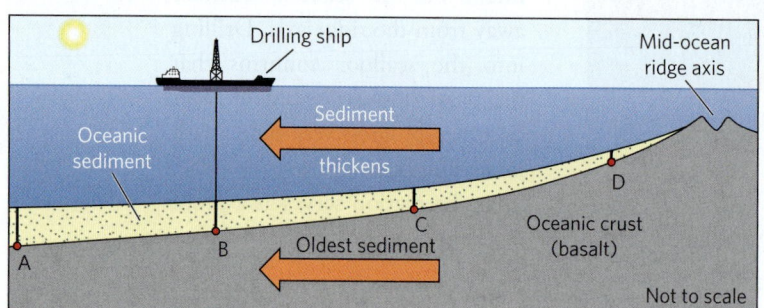

(b) Drilling into the sediment layer of the ocean floor confirms that the basal sediment in contact with basalt gets older the farther away it is from the ridge.

as well, though none are as deep as the Mariana Trench (**Fig. 2.18b, c**). **Volcanic arcs**, chains of volcanoes that typically trace out a curving arc in map view, border one side of a trench. Massive earthquakes happen with unnerving frequency beneath trenches and volcanic arcs.

As was the case with mid-ocean ridges, the origin and geologic meaning of deep-sea trenches remained a mystery before the advent of plate tectonics. But once the theory had been proposed, geologists realized that trenches mark locations where the lithosphere of an oceanic plate bends down and slides under the edge of another plate. As a consequence, plates on either side of a trench move toward each other. Due to this relative motion, geologists refer to the plate boundary associated with a trench as a **convergent boundary** (**Fig. 2.19a**). Keep in mind that, at a convergent boundary, plates don't butt into each other like angry rams. Rather, the

downgoing plate sinks into the asthenosphere beneath the *overriding plate*. Geologists refer to the sinking process as **subduction**, so a convergent boundary can also be called a **subduction zone**.

Subduction occurs for a simple reason: lithospheric mantle, once it has cooled and thickened for at least 10 million years, is denser than the underlying asthenosphere, so it can sink down through the asthenosphere. Subduction starts when the edge of one plate pushes over the edge of another plate along a fault. Once the edge of the downgoing plate has entered the asthenosphere, the plate begins to sink like an anchor (**Fig. 2.19b**). To make room for the sinking lithosphere, soft asthenosphere flows out of its way, just as water flows out of the way of a sinking anchor. Asthenosphere can flow only very slowly, however, so subduction cannot proceed any faster than about 15 cm (6 in) per year.

All ocean floor eventually undergoes subduction, so the oldest ocean floor on the Earth is only about 200 million years old, much younger than the age of the Earth. Because the volume (and therefore circumference) of the Earth remains nearly constant over time, the amount of subduction that takes place worldwide must equal the overall amount of seafloor spreading.

Let's look more closely at the geologic activity that takes place at a convergent boundary. As subduction takes place, seafloor sediment, as well as sand and mud that wash into the trench from nearby land, gets scraped up and incorporated into a wedge-shaped mass, known

FIGURE 2.18 The bathymetry of a convergent boundary.

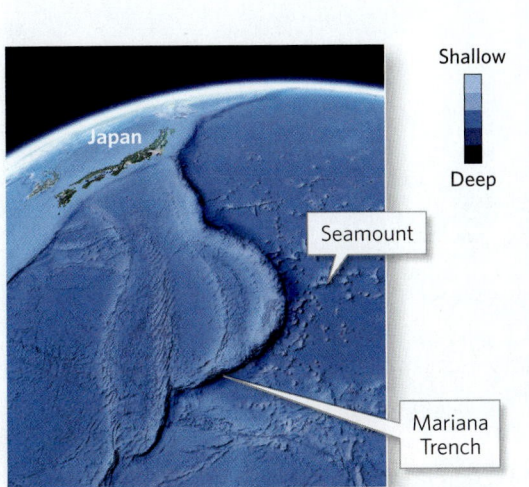

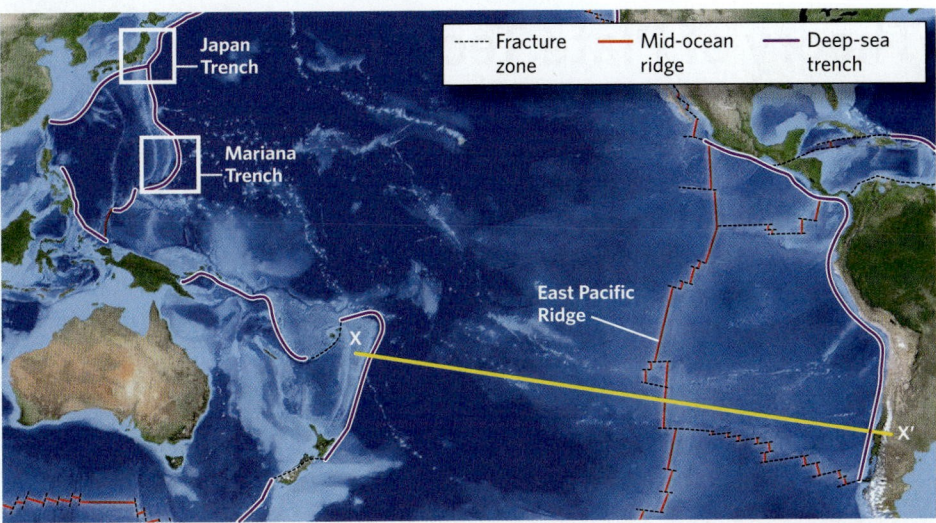

(a) A bathymetric map of the western Pacific, showing the trace of the Mariana Trench. Dark blue is very deep water.

(b) Trenches occur on both sides of the Pacific Ocean. The yellow line shows the location of the bathymetric profile in part (c).

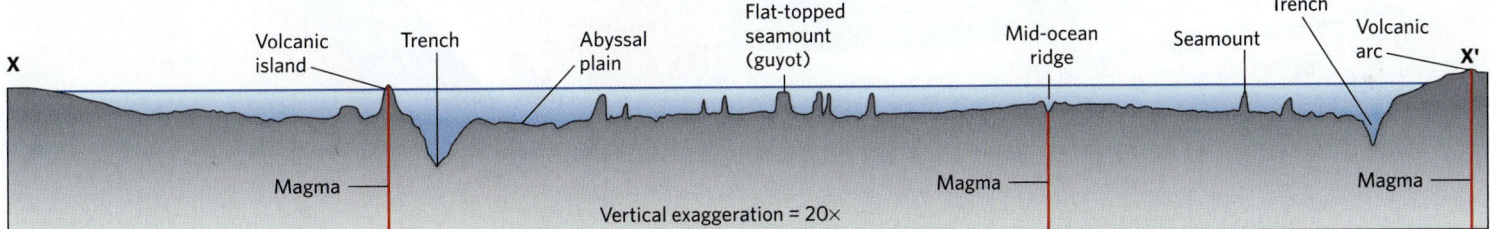

(c) A bathymetric profile showing the contrast between the depth of a trench and the depth of an abyssal plain. This profile goes from the Tonga Trench, east of Australia, to the coast of South America. Red lines indicate rising magma. The sketch labels key features.

FIGURE 2.19 During subduction, oceanic lithosphere sinks into the deeper mantle.

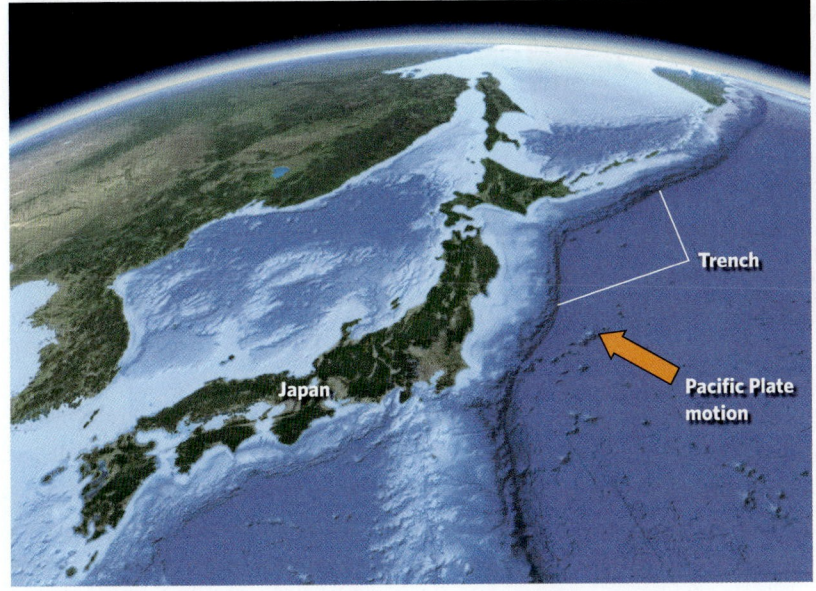

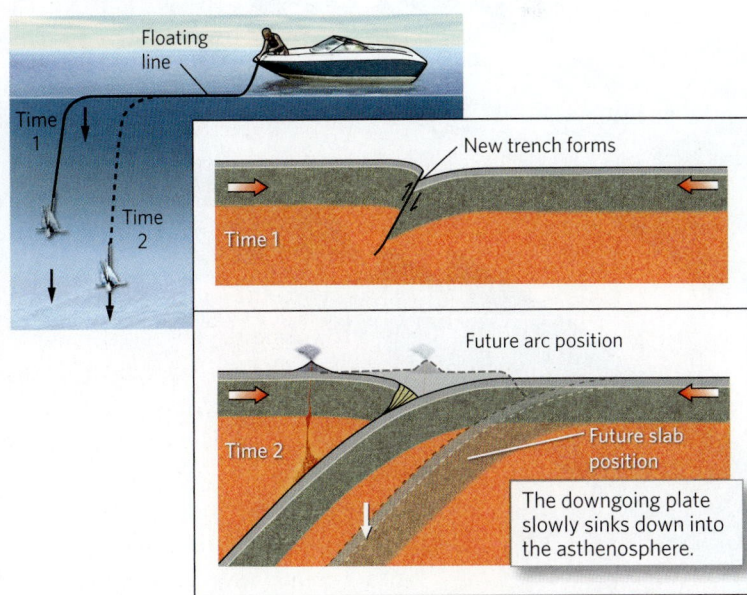

(a) A trench delineates the place where the Pacific Plate is being subducted beneath Japan.

(b) Sinking of the downgoing plate resembles the sinking of an anchor attached to a rope.

The Theory of Plate Tectonics

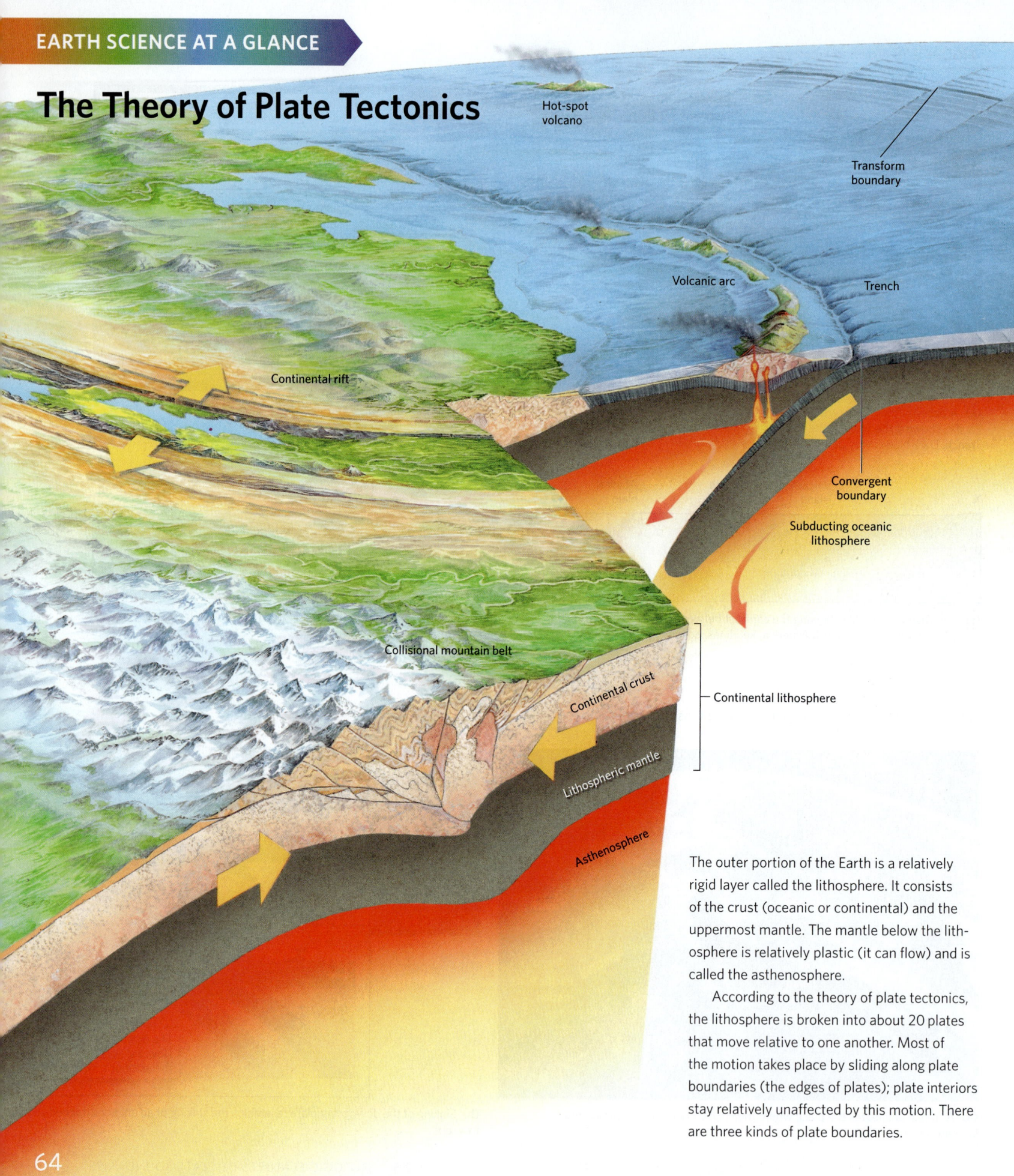

Hot-spot volcano

Transform boundary

Volcanic arc

Trench

Continental rift

Convergent boundary

Subducting oceanic lithosphere

Collisional mountain belt

Continental crust

Continental lithosphere

Lithospheric mantle

Asthenosphere

The outer portion of the Earth is a relatively rigid layer called the lithosphere. It consists of the crust (oceanic or continental) and the uppermost mantle. The mantle below the lithosphere is relatively plastic (it can flow) and is called the asthenosphere.

According to the theory of plate tectonics, the lithosphere is broken into about 20 plates that move relative to one another. Most of the motion takes place by sliding along plate boundaries (the edges of plates); plate interiors stay relatively unaffected by this motion. There are three kinds of plate boundaries.

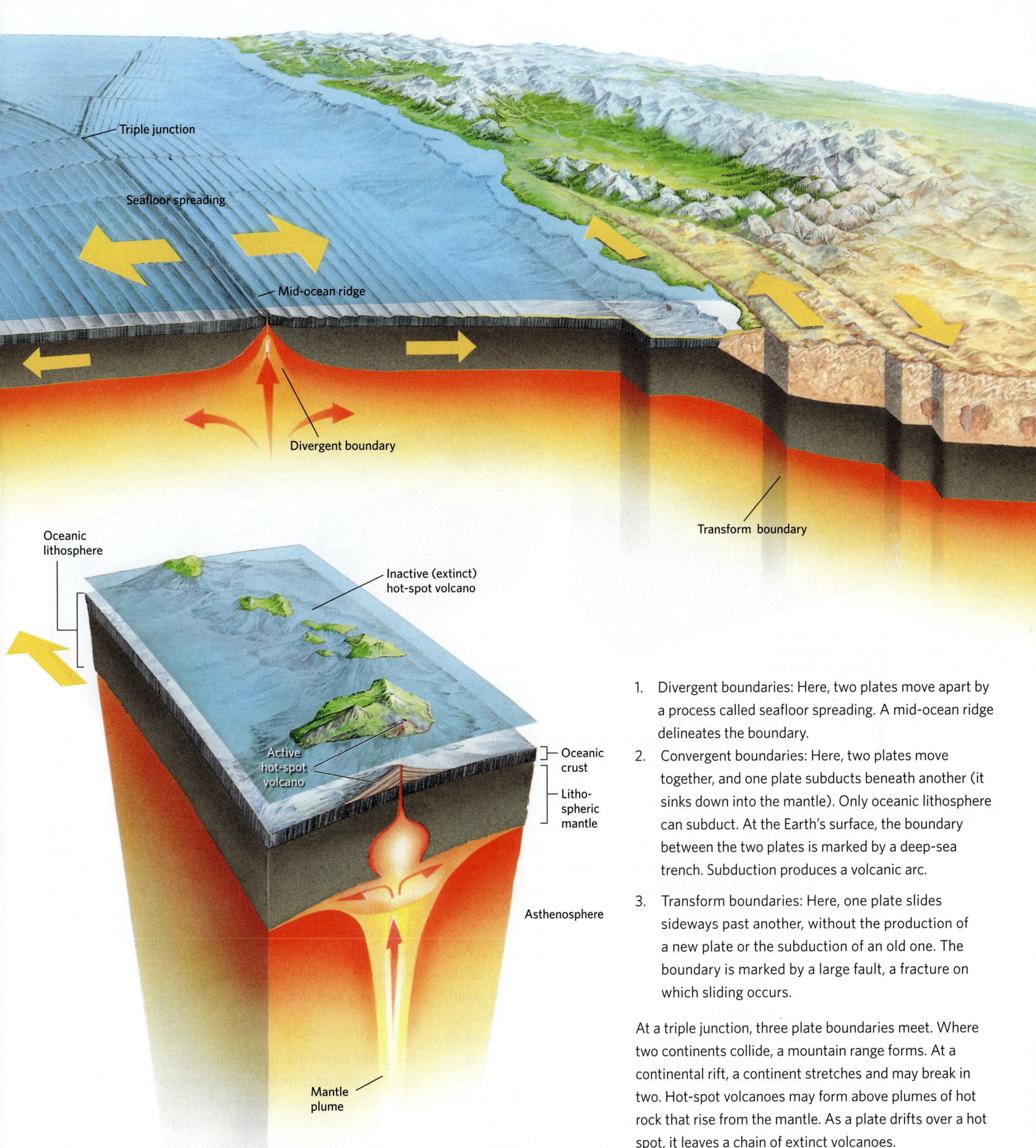

Triple junction

Seafloor spreading

Mid-ocean ridge

Divergent boundary

Transform boundary

Oceanic
lithosphere

Inactive (extinct)
hot-spot volcano

Oceanic
crust

Litho-
spheric
mantle

Active
hot-spot
volcano

Asthenosphere

Mantle
plume

1. Divergent boundaries: Here, two plates move apart by a process called seafloor spreading. A mid-ocean ridge delineates the boundary.

2. Convergent boundaries: Here, two plates move together, and one plate subducts beneath another (it sinks down into the mantle). Only oceanic lithosphere can subduct. At the Earth's surface, the boundary between the two plates is marked by a deep-sea trench. Subduction produces a volcanic arc.

3. Transform boundaries: Here, one plate slides sideways past another, without the production of a new plate or the subduction of an old one. The boundary is marked by a large fault, a fracture on which sliding occurs.

At a triple junction, three plate boundaries meet. Where two continents collide, a mountain range forms. At a continental rift, a continent stretches and may break in two. Hot-spot volcanoes may form above plumes of hot rock that rise from the mantle. As a plate drifts over a hot spot, it leaves a chain of extinct volcanoes.

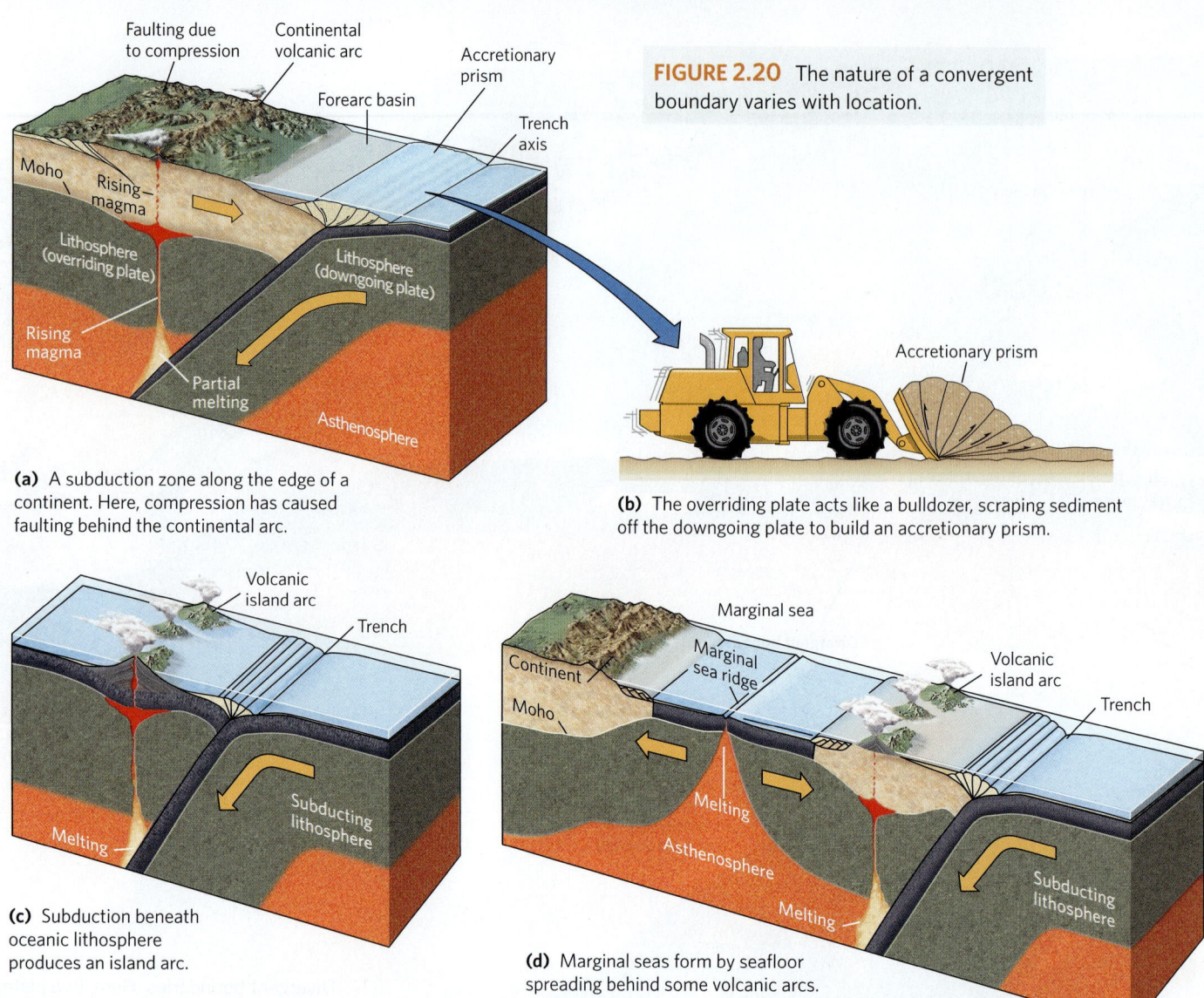

FIGURE 2.20 The nature of a convergent boundary varies with location.

(a) A subduction zone along the edge of a continent. Here, compression has caused faulting behind the continental arc.

(b) The overriding plate acts like a bulldozer, scraping sediment off the downgoing plate to build an accretionary prism.

(c) Subduction beneath oceanic lithosphere produces an island arc.

(d) Marginal seas form by seafloor spreading behind some volcanic arcs.

▶ Animation

Plate Boundaries: Convergent

as an **accretionary prism**, along the margin of the overriding plate **(Fig. 2.20a, b)**. In effect, an accretionary prism resembles a pile of sand that builds in front of a bulldozer. A volcanic arc develops on the edge of the overriding plate, on the opposite side of the accretionary prism from the trench. Note that subduction is taking place beneath the entire length of an arc at any given time, so all the volcanoes in the arc are *active volcanoes*, meaning that they all are either erupting, have recently erupted, or are likely to erupt in the future.

Where an oceanic plate subducts beneath a continent, a *continental arc* develops along the edge of the continent—the Andean arc along the coast of South America and the Cascade arc of Oregon and Washington serve as examples. At places where an oceanic plate subducts beneath another oceanic plate, volcanoes build up on the seafloor of the overriding plate and grow into a volcanic **island arc**—the Mariana arc and the Aleutian arc serve as examples **(Fig. 2.20c)**. In a few locations, a volcanic arc grows on a sliver of continental crust that has

separated from the main continent due to the growth of a small ocean basin, called a *marginal sea* **(Fig. 2.20d)**—the Japan arc is an example.

The boundary between the downgoing plate and the base of the overriding plate at a convergent boundary contains many faults, and slip on these faults generates earthquakes. Earthquakes also happen at greater depths in downgoing plates, in a band called the Wadati-Benioff zone. In fact, geologists detect these earthquakes to a depth of 660 km (410 mi) **(Fig. 2.21)**. Downgoing plates eventually sink below a depth of 660 km, but they do so without generating earthquakes. The lower mantle may be a "graveyard" for subducted plates.

Transform Boundaries

We noted earlier that mid-ocean ridges are segmented and that the segments are linked by portions of fracture zones **(Fig. 2.22a)**. Originally, researchers incorrectly assumed that the entire length of each fracture zone was an active fault. But when information about the distribution

FIGURE 2.21 Subducted lithosphere sinks into the mantle like a denser fluid sinking into a less dense fluid. The subducted plate may eventually separate into pieces as it passes through the layers of the mantle. Earthquakes occur in the subducted plate down to a depth of about 660 km (410 mi, a region known as the Wadati-Benioff zone).

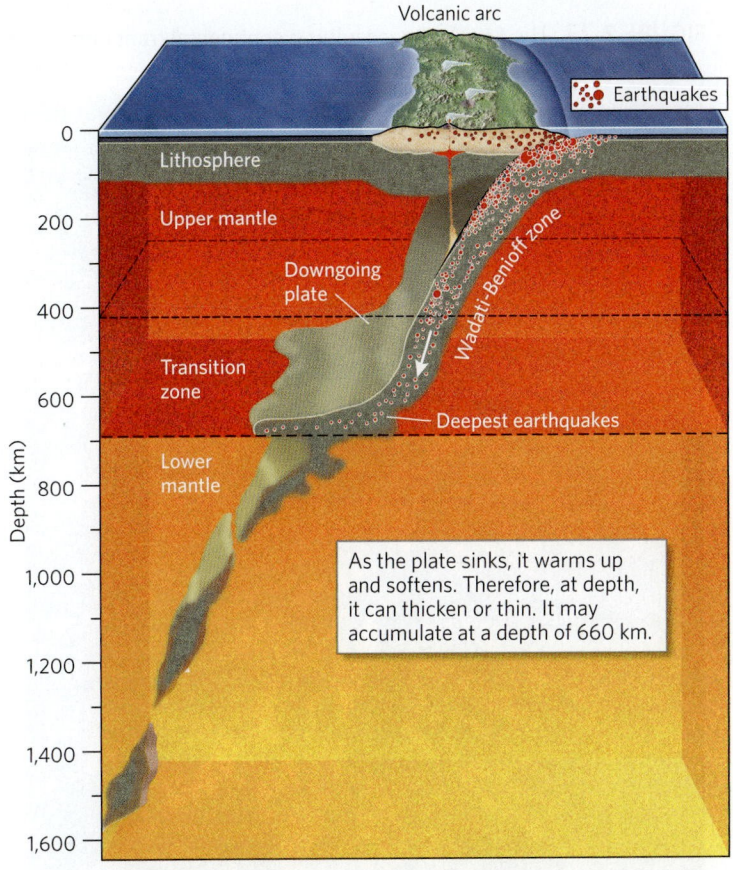

> As the plate sinks, it warms up and softens. Therefore, at depth, it can thicken or thin. It may accumulate at a depth of 660 km.

of earthquake epicenters along mid-ocean ridges became available, it became clear that slip occurs only on the segment of a fracture zone that lies between the ends of two ridge segments. The portions of fracture zones that extend beyond the ends of ridge segments, into the abyssal plain, do not generate earthquakes.

The distribution of slip along fracture zones remained a mystery until a Canadian geologist, J. Tuzo Wilson (1908–1993), began to think about fracture zones in the context of seafloor spreading. To understand Wilson's interpretation, imagine a map showing two north-south-trending mid-ocean ridge segments linked by an east-west-trending fracture zone (**Fig. 2.22b**). Plate A is moving west, and Plate B is moving east, relative to the ridge axis. The southern ridge segment intersects the fracture zone at Point X, and the northern one intersects it at Point Y. Along the portion of the fracture zone between X and Y, Plate A moves to the left relative to Plate B, so this portion must be an active fault along which earthquakes take place. In other words, this segment of the fracture zone is a plate boundary. To the west of Point X, and to the east of Point Y, however, seafloor north and south of the fracture zone moves in the same direction and at the same

FIGURE 2.22 Oceanic transform faulting

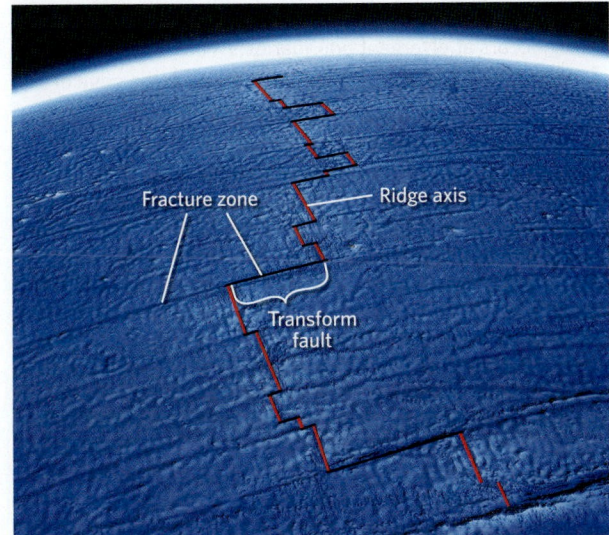

(a) Fracture zones along the Mid-Atlantic Ridge. Notice that a transform fault is only part of a fracture zone, the part between two ridge segments.

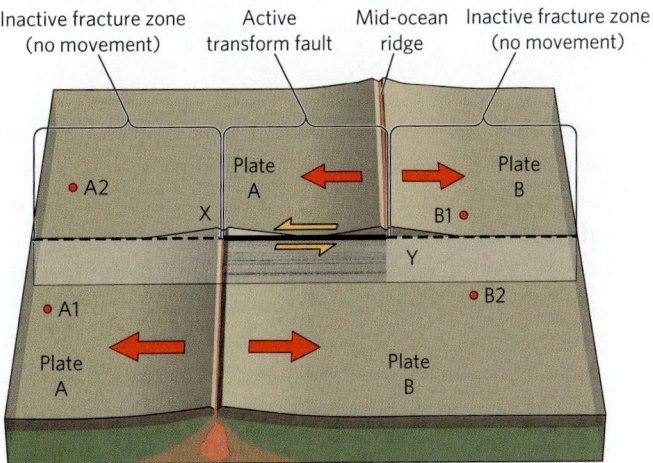

(b) Note that only the segment of the fracture zone between the two ridge segments is an active transform fault.

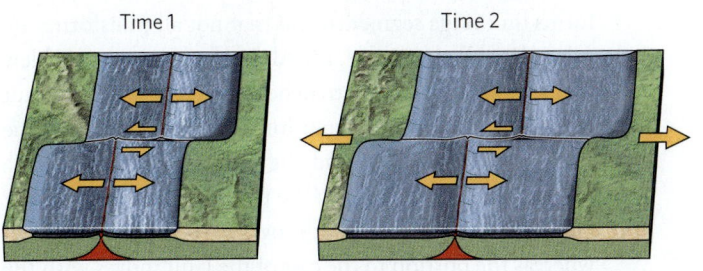

(c) Wilson's interpretation of a transform fault.

FIGURE 2.23 The San Andreas fault is a continental transform boundary.

(a) In southern California, the San Andreas fault cuts a dry landscape. The fault trace is in the narrow valley. The land has been pushed up slightly along the fault.

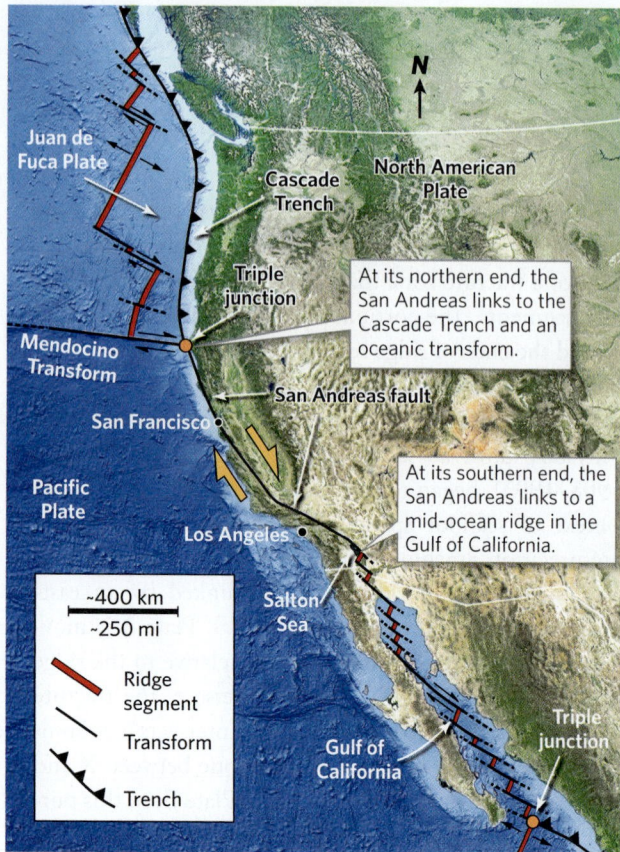

(b) The San Andreas fault is a continental transform boundary between the North American and Pacific Plates. The Pacific Plate is moving northwest relative to North America.

rate. These portions of the fracture zone, therefore, are not active faults. In fact, when you cross these segments, you stay on the same plate, although the plate on one side of the boundary is older than that on the other side.

Wilson introduced the term **transform fault** (or *transform*) for the actively slipping segment of a fracture zone between two ridge segments, and he emphasized that transform faults are a third type of plate boundary, a **transform boundary**. At such a location, one plate slips sideways relative to the other, but no new plate forms, and no old plate subducts. Transform boundaries are, therefore, vertical faults on which the slip direction parallels the Earth's surface **(Fig. 2.22c)**.

Wilson and others soon realized that not all transforms link ridge segments, and that not all transforms are submarine. For example, the Alpine fault cuts across New Zealand and links two trenches. The San Andreas fault of California serves as a transform boundary between the North American Plate and the Pacific Plate and links a ridge segment to a trench. The portion of California that lies to the west of the fault moves with the Pacific Plate, whereas the portion to the east of the fault moves with the North American Plate **(Fig. 2.23)**.

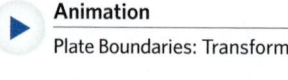
Animation
Plate Boundaries: Transform

Take-home message . . .

Geologists distinguish among three types of plate boundaries based on relative plate movement: divergent, convergent, and transform. Seafloor spreading occurs at divergent boundaries. At convergent boundaries, an oceanic plate sinks beneath the edge of another plate. At transform boundaries, one plate slips sideways along the edge of another.

Quick Question ⎯⎯⎯⎯⎯⎯⎯⎯⎯⎯⎯⎯
Why do earthquakes occur at plate boundaries?

2.6 The Birth and Death of Plate Boundaries

The configuration of plates and plate boundaries visible on our planet today has not existed for all of geologic history, and it will not stay the same in the future. Because of plate motion, new oceanic plates form, only to be consumed by subduction later on, and continents merge,

FIGURE 2.24 Continental collision.

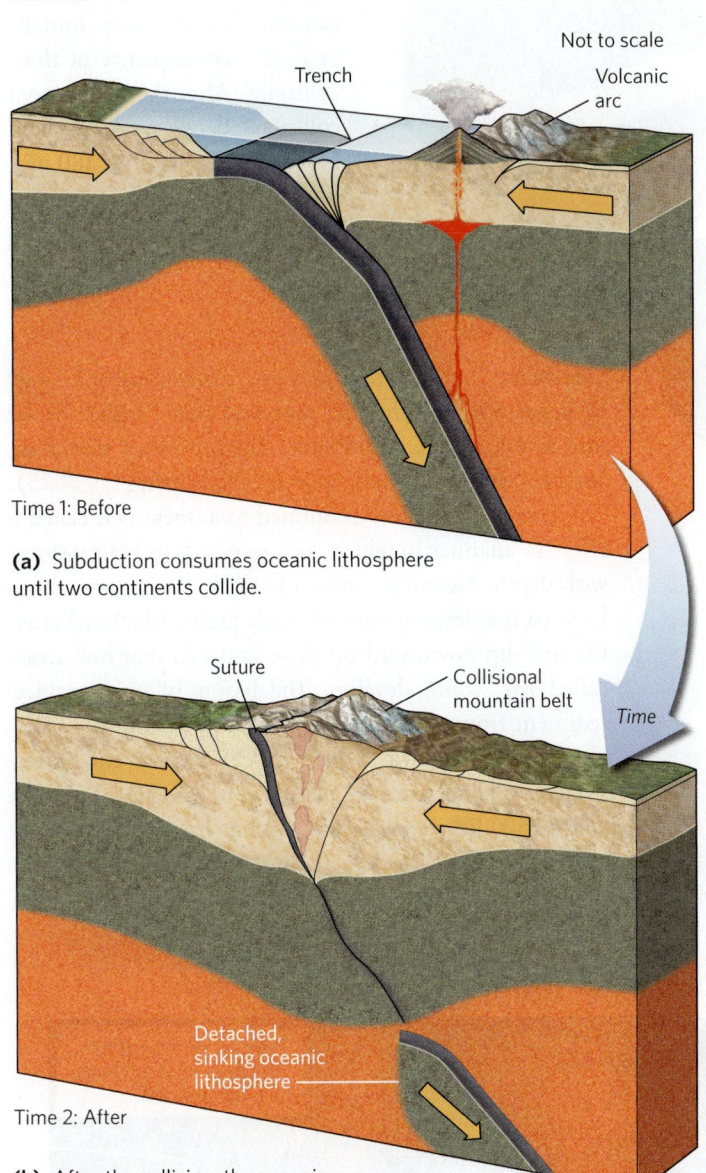

Time 1: Before

(a) Subduction consumes oceanic lithosphere until two continents collide.

Time 2: After

(b) After the collision, the oceanic lithosphere detaches and sinks into the mantle. Rock caught in the collision zone gets broken, bent, and squashed, and forms a mountain range.

(c) An oblique satellite view showing the collision of India with Asia to produce the Himalayas.

then later separate. In this section, first we examine how a convergent boundary ceases to exist when two continents collide. Then we look at how continents split apart, or *rift*, a process that may produce a new divergent boundary.

When Continents Converge: The Process of Collision

India was once a small, separate continent that lay far to the south of Asia. Over time, however, subduction consumed ocean floor between India and Asia, and India moved northward. By 40 to 50 million years ago, all the ocean floor between the two continents had been subducted, and India itself pushed into Asia. This type of event, leading to the merging of two landmasses after subduction of the intervening ocean floor, is called **collision**. The boundary between the once separate bodies is called a *suture*. If the event involves two continents, geologists refer to it as *continental collision*.

Collision occurs when two pieces of relatively buoyant crust, such as continental crust or an island arc, converge at a plate boundary, for buoyant crust cannot be subducted **(Fig. 2.24a, b)**. When collision happens, the convergent boundary that once lay between the buoyant landmasses ceases to exist. For example, when India collided with Asia, the oceanic lithosphere that was attached to the northern coast of India broke off and sank down into the mantle, so subduction along the southern coast of Asia stopped. In addition, the rock and sediment that once lay along the margins of the two landmasses, or had been scraped off the seafloor between them, underwent squeezing as if in a giant vise, resulting in the huge (approximately 8 km, or 5 mi, high) Himalayan mountain belt **(Fig. 2.24c)**. The continued push of India into Asia has caused continued growth of the Himalayas and also contributes to the uplift of the Tibetan Plateau. As continents collide, not only does the surface of the Earth go up, but the base of the crust goes down, so the continental crust becomes thicker.

Collision has yielded some of the most spectacular mountain ranges on the planet today; examples include the Himalayas and the Alps. It also produced major mountain ranges in the past, ranges that have since been mostly

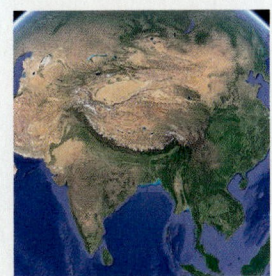

See for yourself

Himalayan Collision

Latitude: 27°59'17.95" N
Longitude: 86°55'30.71" E

Zoom to an elevation of 7,000 km (4,300 mi) and look straight down.

You're seeing the Himalayas. Zoom closer, and you'll be at the peak of Mt. Everest.

FIGURE 2.25 During the process of rifting, continental lithosphere stretches and thins. If rifting succeeds, a new mid-ocean ridge forms.

Moho

Time 1

Wide rift

Basin

Range

New passive margin

Time 2

New sediment

New mid-ocean ridge

Time 3

eroded away. For example, the Appalachian Mountains in the eastern United States initially grew as a consequence of three collisions. After the most recent collision, between Africa and North America around 280 million years ago, North America became part of Pangaea.

Stretching a Continent: The Process of Rifting

Continents, once formed, may eventually break apart—as Wegener realized when he proposed that Pangaea broke into several smaller continents. The process of stretching and breaking a continent apart is called **rifting (Fig. 2.25)**. This process tends to be confined to a linear belt called a **rift**. The manner in which rock responds to rifting varies with depth. Near the surface of the continent, stretching leads to the development of many faults. Blocks of crust tilt and slip downward on these faults so that low areas, called *rift basins*, develop. The basins fill with eroded sediment from the rift's margin as well as from the tilted blocks. Deeper down, where rock is warmer and softer, stretching takes place by plastic flow (see Fig. 2.25). As continental lithosphere thins during rifting, hot asthenosphere rises beneath the rift, so no open space develops.

FIGURE 2.26 Examples of present-day crustal rifting.

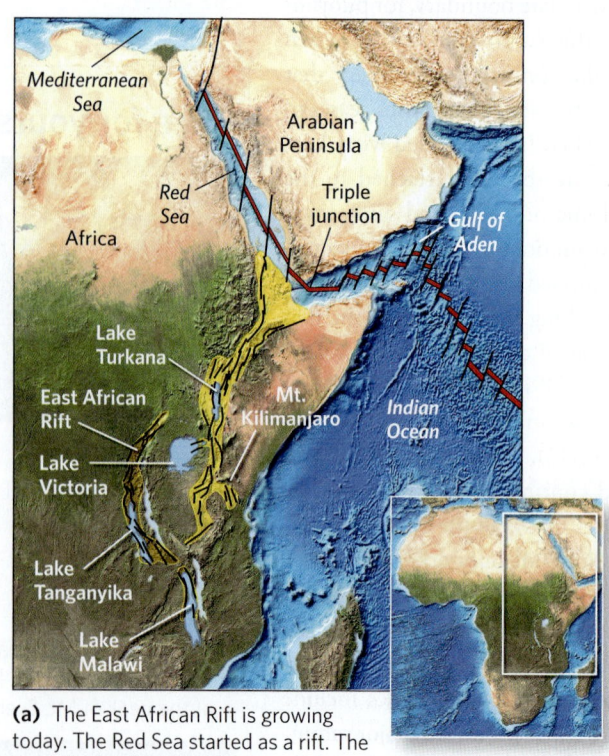

(a) The East African Rift is growing today. The Red Sea started as a rift. The inset shows the map's location.

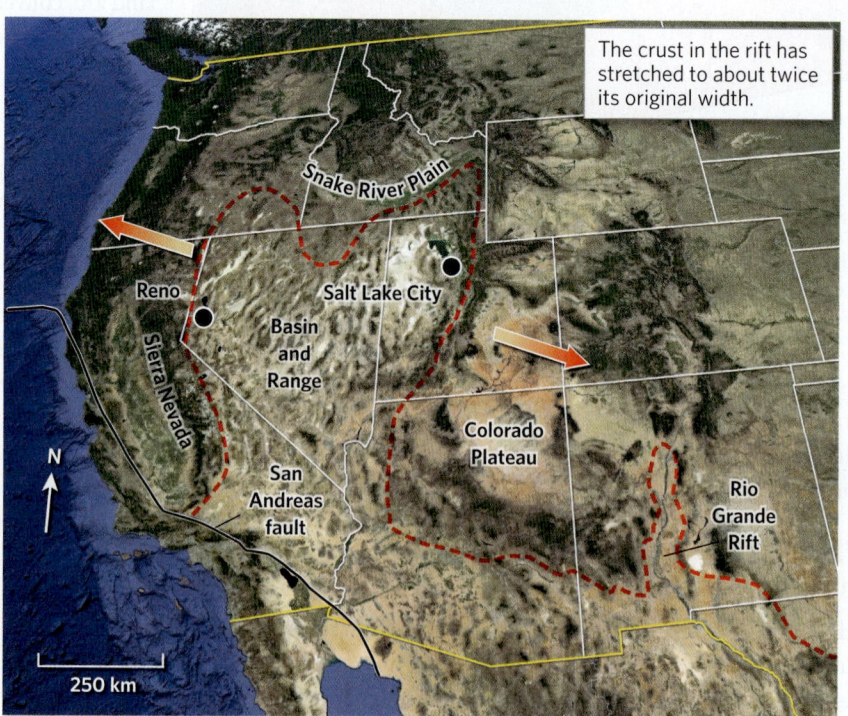

The crust in the rift has stretched to about twice its original width.

(b) The Basin and Range Province is a rift. Faulting bounds the narrow north-south-trending mountains, separated by basins. The arrows indicate the direction of stretching.

Putting Earth Science to Use

Continental shelves and exclusive economic zones

With a basic knowledge of plate tectonics, you can look at a map depicting the bathymetry of coastal areas bordering the United States and understand why the continental shelf along the Pacific coast of North America looks so different from that of the Atlantic and Gulf Coast **(Fig. Bx2.1)**. The Pacific Coast is an active margin, bordering the San Andreas fault and the Cascadia trench. In contrast, the Atlantic and Gulf coasts are passive margins, formed when North America rifted from the rest of Pangaea and the Mid-Atlantic Ridge developed. Sediment has been accumulating and building out the continental shelf along the East and Gulf coasts for almost 200 million years, so the shelf has grown very wide.

Notably, shallow seas (< 300 m, or 1000 ft deep) provide most of the fish that the commercial fishing industry catches. Therefore, continental shelves are the most important fishing grounds in the world. Because of the great width of shelves along passive margins, some governments proposed that *exclusive economic zones* (EEZs), the offshore area in which the bordering country owns all resources, be widened from 12 nautical miles (12 km) to 200 nautical miles (370 km) as measured out from the shore. The United Nations Convention of the Law of the Sea

FIGURE Bx2.1 A comparison of the continental shelves bordering the continental United States. Note that the east and Gulf coast EEZs roughly correspond to the wide continental shelves.

adopted this proposal in 1982. Sedimentary strata of passive margins also contain major oil and gas resources, so the definition of EEZs has implications for offshore drilling, as well. Because of their economic significance, boundaries of EEZs have, unfortunately, become points of conflict between nations.

Some of this asthenosphere melts, producing magma that erupts as lava from volcanoes in the rift (see Chapter 4).

Rifting takes place today at several localities on the Earth. For example, along the *East African Rift*, easternmost Africa is separating from central Africa. A fault-bounded valley, containing elongated lakes in places, delineates the axis of this rift **(Fig. 2.26a)**. The *Basin and Range Province* of the western United States is a rift containing many elongated rocky ridges (the edges of tilted crustal blocks) separated from each other by narrow, sediment-filled basins **(Fig. 2.26b)**.

A rift that continues stretching until the continent breaks apart and a new mid-ocean ridge develops is called a *successful rift*. Once seafloor spreading begins, remnants of the rift become inactive. These inactive remnants delineate the new edges of the continents. Over time, they sink, become buried by very thick accumulations of sediment, and evolve into passive margins **(Box 2.1)**. Along an *unsuccessful rift*, rifting ceases before it splits a continent. The trace of the rift remains as a scar in the crust, delineated by faults and by rift-related sedimentary and igneous rocks.

Take-home message . . .

When two buoyant pieces of crust collide, a mountain belt forms, and the convergent boundary that existed between the pieces ceases to exist. Rifting can split a continent in two and can lead to the formation of a new divergent boundary.

Quick Question

What is the difference between a successful and an unsuccessful rift?

2.7 Special Locations in the Plate Mosaic

Triple Junctions

At several localities on the Earth, three plate boundaries intersect. Each of these intersections, a *triple junction*, stands out on a global map of plate boundaries. For example,

FIGURE 2.27 Triple junctions.

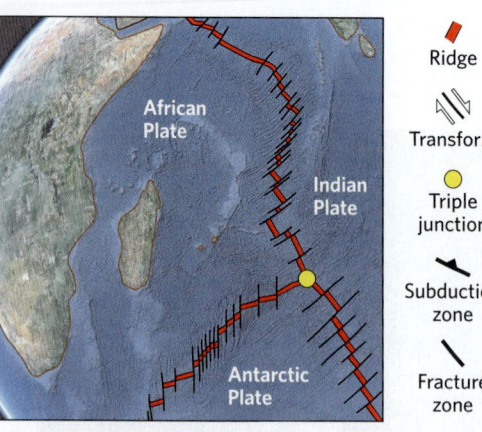

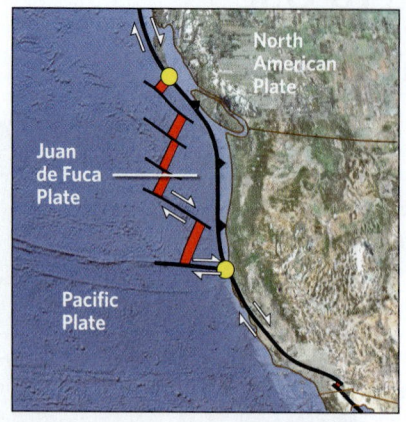

(a) A ridge-ridge-ridge triple junction occurs in the Indian Ocean.

(b) A trench-transform-transform triple junction occurs at the northern end of the San Andreas fault.

FIGURE 2.28 Hot spots.

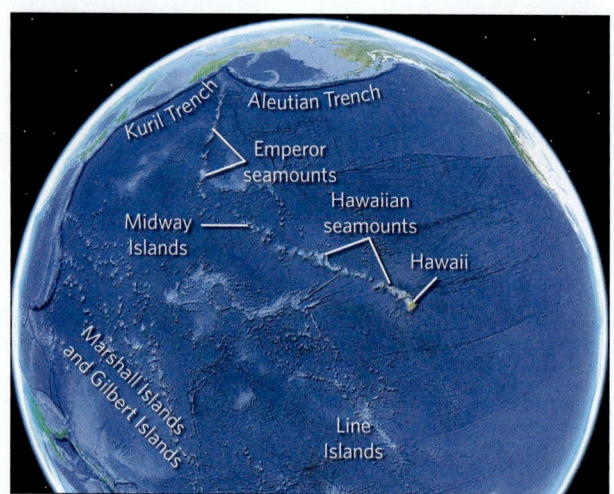

(a) A bathymetric map showing the hot-spot tracks of the Pacific Ocean.

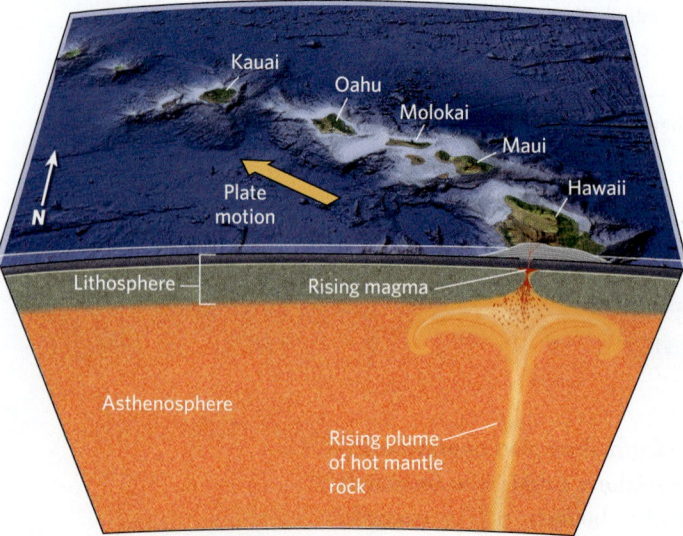

(b) According to the mantle-plume model, Hawaii lies above a rising column of hot mantle rock.

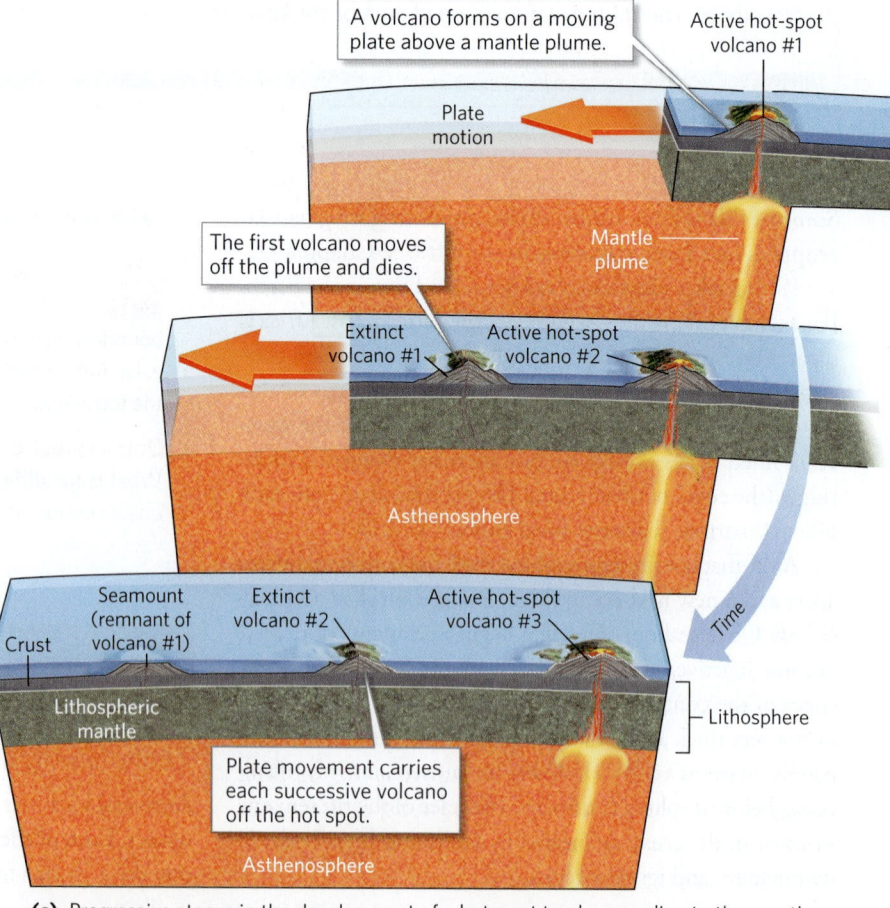

(c) Progressive stages in the development of a hot-spot track, according to the mantle-plume model.

one occurs at the point where three ridges intersect in the western Indian Ocean **(Fig. 2.27a)**. Another occurs along the coast of California, where the San Andreas fault, a transform fault, intersects the Cascade subduction zone and an oceanic transform fault **(Fig. 2.27b)**.

Hot Spots

On a global basis, most *subaerial volcanoes* (volcanoes that protrude above sea level) occur along volcanic arcs bordering subduction zones or within rifts. Similarly, most *submarine volcanoes* (volcanoes hidden by the sea) occur along mid-ocean ridges. The volcanoes of volcanic arcs and mid-ocean ridges are *plate-boundary volcanoes* in that they form as a direct consequence of interactions at a plate boundary. But not all volcanoes are plate-boundary volcanoes. Geologists have identified about a hundred locations where volcanoes do not form directly as a consequence of plate interactions. These locations are called **hot spots**.

Most hot spots occur in the interior region of a plate, away from plate boundaries. Examples include Hawaii **(Fig. 2.28a)**, a location within the Pacific Plate thousands of kilometers from the nearest plate boundary, and Yellowstone National Park, which occurs in the interior of the North American Plate. A few hot spots occur along

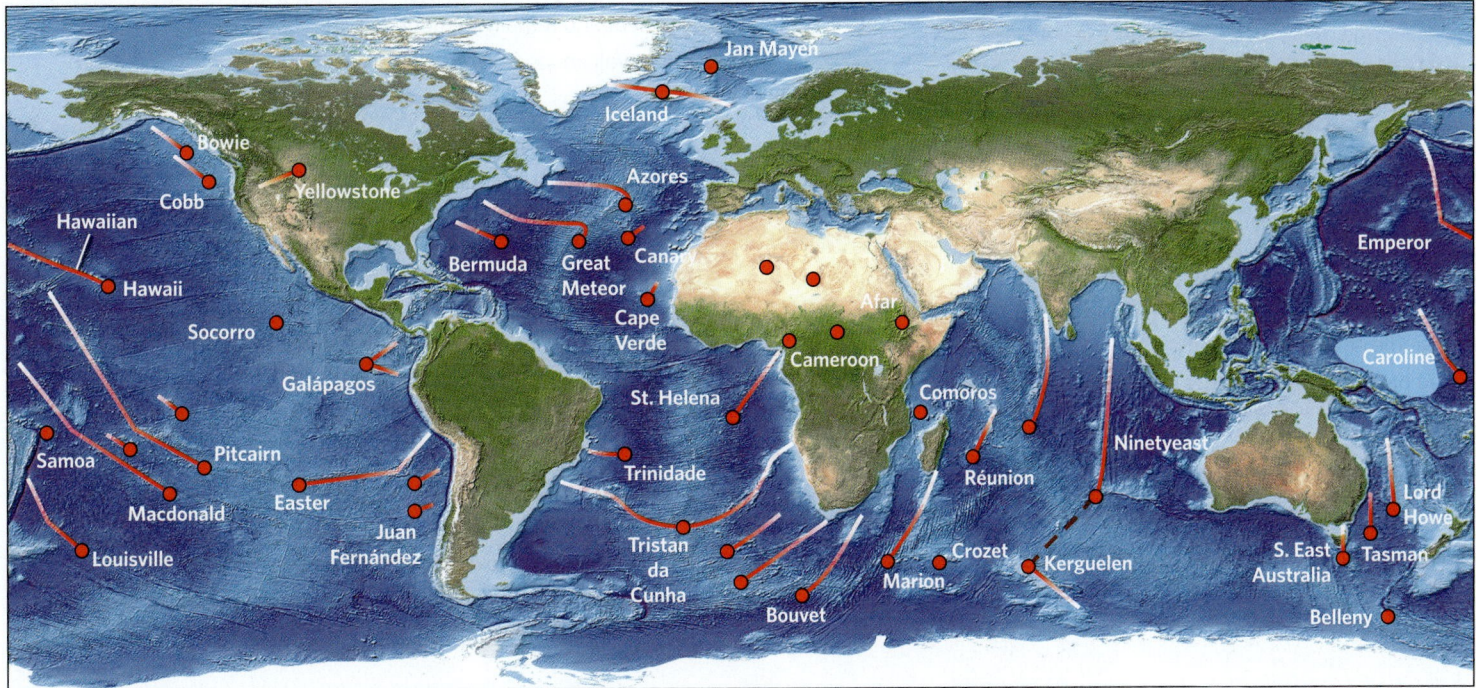

mid-ocean ridges, but the volcanoes formed at these hot spots differ from normal ridge volcanoes in that they produce much more lava than normal seafloor spreading produces. Iceland serves as an example of such a hot spot. So much lava erupts in Iceland, in fact, that it has built a plateau that rises 2 km (1.2 mi) above the rest of the Mid-Atlantic Ridge.

Why do hot spots exist? Researchers still debate this question. Most favor a model in which hot spots lie at the tops of **mantle plumes**, narrow columns of particularly hot rock that flow upward through the mantle **(Fig. 2.28b)**. When the hot peridotite in a mantle plume reaches the base of the lithosphere, it begins to melt, producing magma that seeps up through the lithosphere and erupts as a *hot-spot volcano* (see Chapter 4).

J. Tuzo Wilson noticed that most hot-spot volcanoes in plate interiors lie at the end of a chain of *extinct volcanoes*, volcanoes that will never erupt again. In the Hawaiian Islands, for example, active volcanoes erupt at the southeastern end of a chain of extinct volcanic islands and seamounts (see Fig. 2.28a). (This configuration differs from that of an island arc, in which the component volcanoes are all active.) To explain the configuration of the Hawaiian-Emperor seamount chain, Wilson proposed that the position of a hot spot stays fixed, more or less, relative to the moving plate above it. As the plate moves, the volcano forming above the hot spot eventually moves off the hot spot and becomes extinct **(Fig. 2.28c)**. A new volcano then starts to grow over the hot spot, and

this volcano remains active for a while, until it, too, gets carried off the hot spot and goes extinct **(Box 2.2)**. The process repeats for as long at the hot spot survives, which can be tens of millions of years. Extinct volcanic islands slowly sink below sea level and become seamounts. The resulting chain of inactive volcanic islands and seamounts represents a **hot-spot track** **(Fig. 2.29)**.

Take-home message . . .

Some features of the Earth occur at specific points. A triple junction is the point where three plate boundaries join. A hot spot is a point where volcanism is not the direct result of plate-boundary processes. Hot spots may form due to melting at the top of a mantle plume.

Quick Question
Distinguish a volcanic arc from a hot-spot track.

2.8 Proving Plate Tectonics Using Studies of Paleomagnetism

For a hypothesis to become a theory, it must be subjected to many tests. Two key tests of plate tectonics came from the study of paleomagnetism. Here, we introduce paleomagnetism and then show how paleomagnetic tests helped prove plate tectonics.

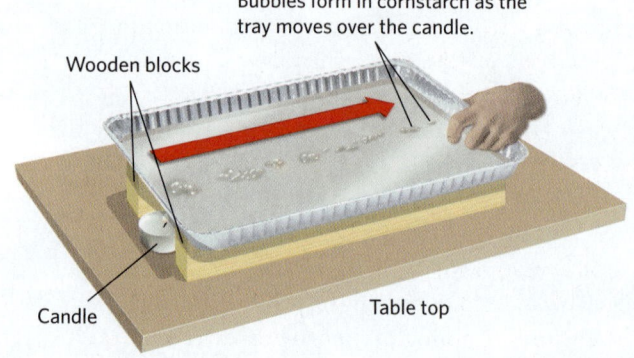

BOX 2.2 How can I explain . . .

The evolution of hot-spot tracks

What are we learning?
How a hot-spot track forms in the interior of a plate.

What you need:
- A disposable aluminum-foil baking pan
- Cornstarch and water
- A short candle and some matches
- Two wood blocks that are slightly taller than the candle

Instructions:
- Mix the water and cornstarch to make a paste.
- Cover the base of the pan with a 2-cm (1-inch)-thick layer of paste.
- Configure the blocks so that they border the candle and can support the pan.
- Light the candle.

- Slide the pan slowly over the candle. Bubbles form in the paste above the heat and remain as mounds (volcanoes) in the paste. Number the volcanoes from oldest to youngest.

Bubbles form in cornstarch as the tray moves over the candle.

Wooden blocks

Candle

Table top

What did we see?
As the paste layer (plate) moves over the fixed heat source, bubbles (representing volcanoes) form. But no bubble erupts for long, because as soon as it gets carried off the heat source, it becomes inactive (extinct).

Introducing Paleomagnetism

As we saw in Chapter 1, the flow of liquid iron alloy in the Earth's outer core generates a magnetic field that is dipolar, meaning that it has a north pole and a south pole. We can define these **magnetic poles** as the points on the Earth's surface where magnetic field lines point straight down or straight up **(Fig. 2.30a)**. Geologists, by convention, refer to the pole near the north end of the Earth as the *north magnetic pole* and to the pole at the south end as the *south magnetic pole*. We can represent the **dipole** of Earth's magnetic field with an arrow that passes through the center of the Earth and points from the Earth's north magnetic pole to its south magnetic pole. Note that the convention for labeling the Earth's magnetic poles means that the

FIGURE 2.30 Features of the Earth's magnetic field.

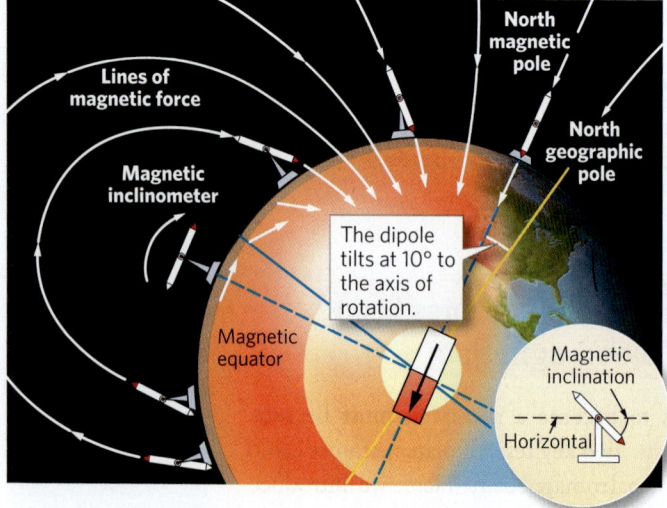

(a) The Earth's magnetic axis is not parallel to its axis of rotation.

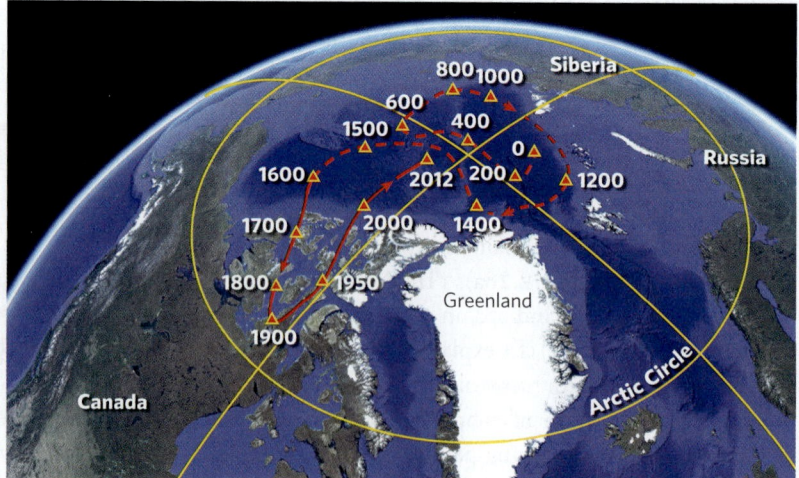

(b) A simplified map showing the changing position of the north magnetic pole over the past 2,000 years. Before about 1600, its position was not as well known, so the path is dashed.

north-seeking end of a compass needle points north when placed in the field.

At any given time, the magnetic poles of the Earth do not exactly overlie the *geographic poles*, the points where the rotational axis of the Earth intersects its surface. For example, during recorded history, the north magnetic pole has followed a circuitous route to its present location in the Arctic Ocean off the northern coast of Canada **(Fig. 2.30b)**. But magnetic poles don't seem to stray far from the geographic poles, and we can assume that, when averaged over several thousand years, the position of a magnetic pole does correspond with that of the geographic pole.

Over 1,500 years ago, Chinese sailors discovered that a piece of lodestone, when suspended from a thread, swivels until it points in a northerly direction, so it can serve as a compass. Lodestone exhibits this behavior because it consists of *magnetite*, an iron oxide mineral that is naturally magnetic. Any magnetic material, if it can move freely, will line up with magnetic field lines when placed in a magnetic field. Some rock types contain enough tiny specks of magnetite or other magnetic minerals that the rock behaves, overall, like a weak magnet. This behavior allows the rock to preserve a record of the orientation of the Earth's magnetic field, at the time the rock forms, for millions or even billions of years. This record of past magnetism is called **paleomagnetism**.

How does paleomagnetism develop? Simplistically, during the rock's formation, magnetic mineral specks in the rock align with the Earth's magnetic field. Once the rock-forming process is finished, the specks can't move, even if the orientation of the Earth's magnetic field, or of the rock, changes. For example, as basalt forms by the solidification of lava, tiny magnetite crystals grow along with other minerals. The dipoles of the magnetite crystals align with the Earth's field while the solidifying basalt remains hot. But once the new rock cools below a temperature of about 600°C, the dipole orientation of the crystals becomes locked. We can represent paleomagnetism symbolically with a *paleomagnetic dipole*, an imaginary arrow embedded in the rock.

Apparent Polar Wander

In the early 20th century, researchers developed instruments capable of detecting paleomagnetism in basalt as well as in certain types of sedimentary rock. From this work, they made a surprising discovery: in rocks that formed long ago, the paleomagnetic dipole does not point to the present-day magnetic poles. At first, they interpreted this discovery to mean that at the time the rock formed, the Earth's magnetic poles were in different locations than they are today. The location that

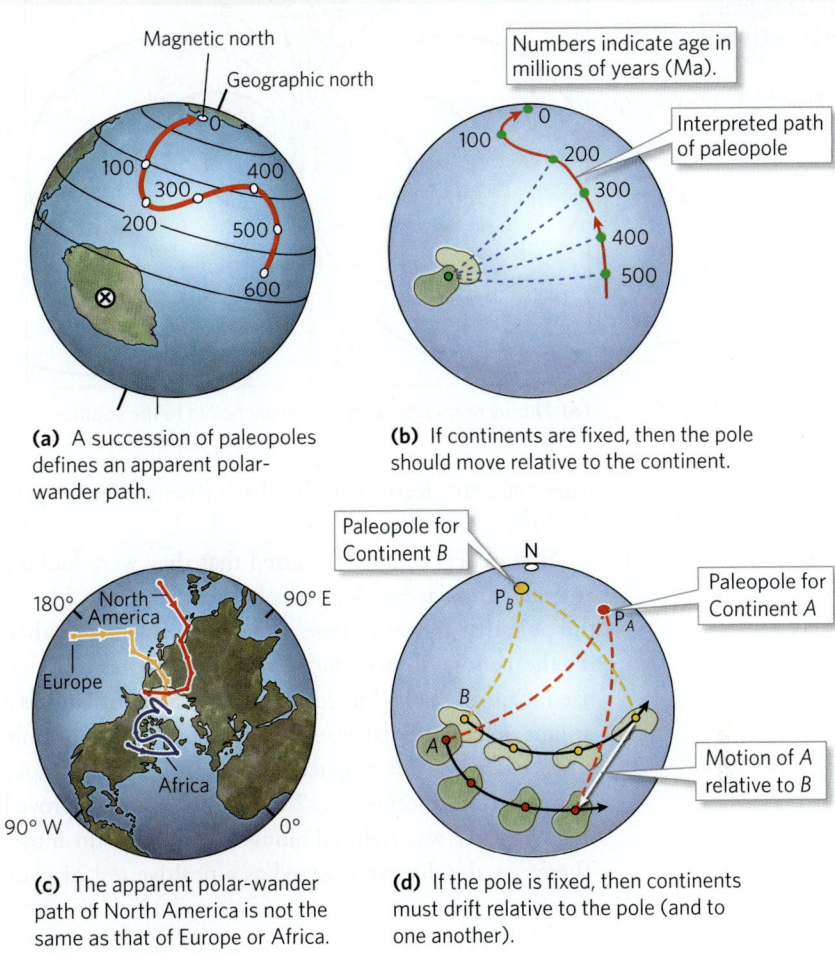

FIGURE 2.31 Apparent polar-wander paths and their interpretation.

(a) A succession of paleopoles defines an apparent polar-wander path.

(b) If continents are fixed, then the pole should move relative to the continent.

(c) The apparent polar-wander path of North America is not the same as that of Europe or Africa.

(d) If the pole is fixed, then continents must drift relative to the pole (and to one another).

a paleomagnetic dipole pointed to came to be known as a **paleopole**. Additional work showed that paleopole positions, as recorded in rocks of different ages from the same region, appeared to change over time. In fact, when plotted on a map, paleopole positions for successively younger rocks from a region trace out a curving line on the surface of the Earth. This line is called an **apparent polar-wander path (Fig. 2.31a)**.

To interpret apparent polar-wander paths, researchers first took the fixist view and assumed that the position of the continent from which a sequence of rocks had been collected had stayed the same through geologic time. If this assumption was correct, then the apparent polar-wander path would represent how the position of the Earth's magnetic pole migrated over time **(Fig. 2.31b)**. But the researchers were in for a surprise! When they obtained apparent polar-wander paths from different continents, they found that each was different from the others **(Fig. 2.31c)**. The hypothesis that continents are fixed cannot explain this observation, for if the magnetic pole moved while all the continents stayed fixed, then

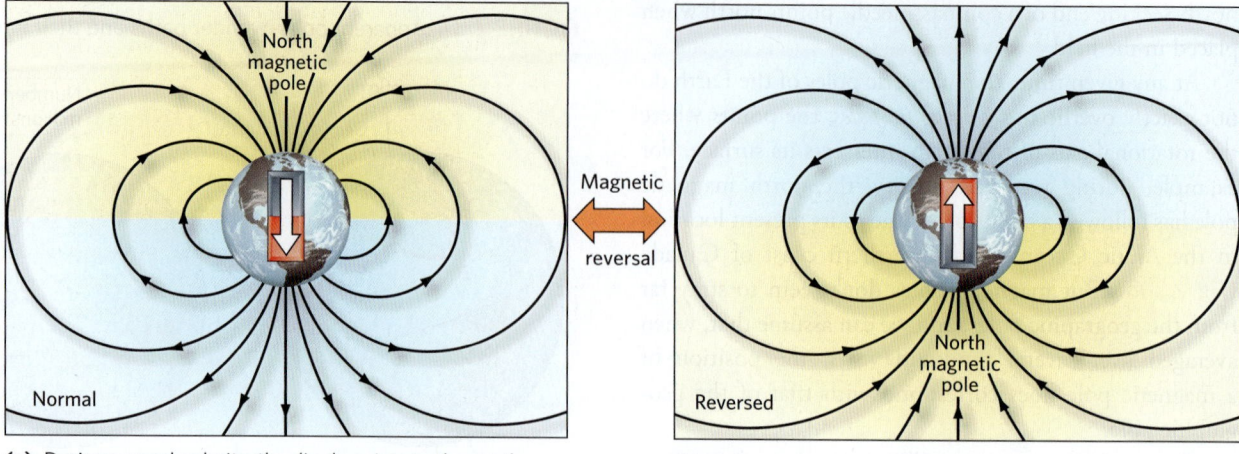

(a) During normal polarity, the dipole points to the south.

(b) During reversed polarity, the dipole points to the north.

paleomagnetic records from all continents should produce the same paths.

Researchers suddenly realized that they were looking at apparent polar-wander paths in the wrong way. It's not the pole that moves relative to fixed continents—rather, it's the continents that move relative to a more or less fixed magnetic pole! Moreover, since each continent has a unique apparent polar-wander path, continents not only move relative to the magnetic pole, but must also move relative to one another **(Fig. 2.31d)**. This discovery proved that Wegener was right all along—continents do move! Therefore, the discovery served as a positive test of plate tectonics.

Magnetic Reversals and Marine Magnetic Anomalies

Will the north-seeking end of your compass always point north? No. Study of paleomagnetism has revealed that at various times in the past, the Earth's magnetic field suddenly flipped. Sometimes the Earth has *normal polarity*, as it does today, with the north magnetic pole near the north geographic pole, and sometimes it has *reversed polarity*, with the north magnetic pole near the south geographic pole **(Fig. 2.32)**. Changes in the Earth's magnetic field from normal to reversed polarity, or from reversed back to normal polarity, are

magnetic reversals. They may be due to changes in the circulation pattern of liquid iron alloy in the outer core. Note that during a reversal, only the magnetic field polarity changes—the Earth itself does not turn upside down.

Not long before the discovery of magnetic reversals, geologists developed a new tool, called *isotopic dating* (or *radiometric dating*), that can provide the age of a rock in

FIGURE 2.33 Magnetic reversals and their chronology.

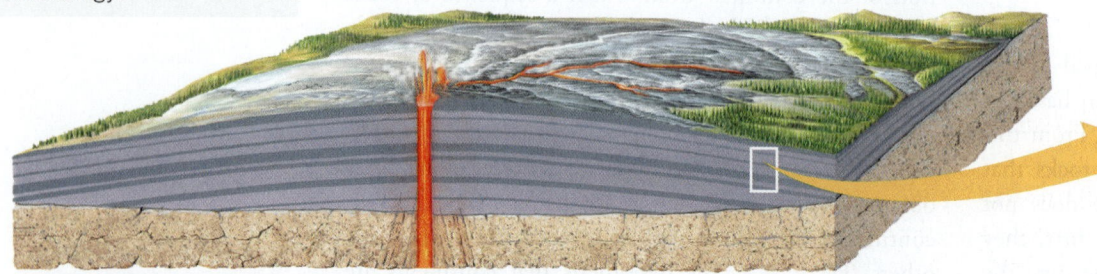

(a) Successive layers of lava build at a volcano over time. Some layers have normal polarity, whereas others have reversed polarity.

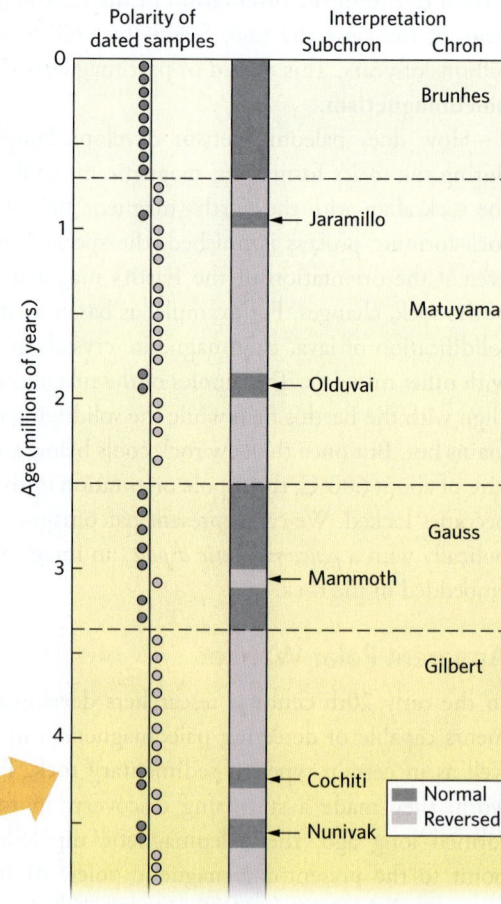

(b) Observations of such layers led to the production of a magnetic-reversal chronology with named polarity intervals, or chrons.

FIGURE 2.34 Marine magnetic anomalies.

(a) A ship towing a magnetometer detects changes in the strength of the magnetic field on the seafloor.

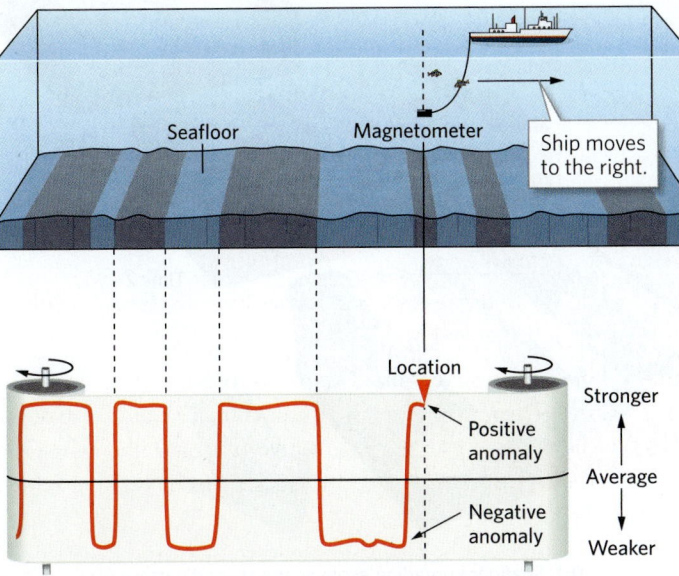

(b) On a paper record, intervals of stronger magnetism (positive anomalies) alternate with intervals of weaker magnetism (negative anomalies).

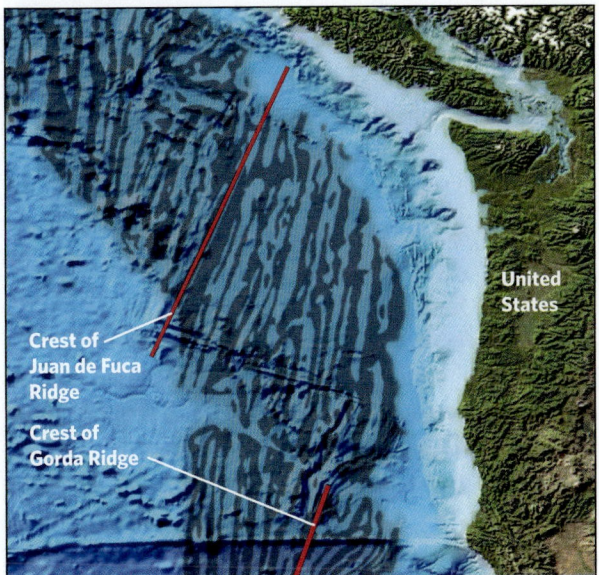

(c) A map showing areas of positive anomalies (dark) and negative anomalies (light) off the west coast of North America.

years (see Chapter 9). By recording the paleomagnetic polarity in successions of progressively older layers of basalt whose ages had been determined by isotopic dating, researchers established a *magnetic-reversal chronology*, a chart showing when reversals had happened and, therefore, the durations of polarity intervals between reversals **(Fig. 2.33a)**. A major polarity interval is called a *polarity*

chron. Some chrons include a few short-duration intervals, known as *polarity subchrons* **(Fig. 2.33b)**. Because of limitations on the precision of isotopic dating, the first magnetic-reversal chronology chart represented only the most recent 4.5 million years of Earth history.

At about the same time that some geologists were studying the timing of magnetic reversals, other geologists were measuring regional variations in the strength of the Earth's magnetic field by using an instrument called a *magnetometer*. At any given location on the surface of the Earth, the magnetic field includes two components: one produced by the main dipole of the Earth, and another produced by the magnetism of near-surface rocks. Geologists found that the strength of the magnetism that magnetometers detect varies slightly with location, and they coined the term *magnetic anomaly* for the difference between the expected strength of the Earth's main dipole field at a location and the actual observed strength of the magnetic field at that location. Field strengths stronger than expected are *positive anomalies*, whereas field strengths weaker than expected are *negative anomalies*.

When geologists towed magnetometers back and forth across the ocean and mapped variations in measured magnetic field strength above the seafloor **(Fig. 2.34a)**, they found that **marine magnetic anomalies** define a distinct pattern of alternating bands aligned parallel to mid-ocean ridge axes **(Fig. 2.34b, c)**. If we color positive anomalies dark and negative anomalies light, the pattern resembles the stripes on a zebra. Notably, variations in the width of stripes on one side of a ridge mirror those on the other side, and the relative widths of the anomalies closest to the ridge axis correspond to the relative durations of polarity chrons for the past 4.5 million years.

What does the stripe-like pattern of marine magnetic anomalies mean? In 1963, researchers realized that the striped pattern develops as magnetic reversals take place during seafloor spreading. Simply put, a positive anomaly occurs where the magnetism of basalt that makes up the seafloor has normal polarity, for in these areas, the magnetic force produced by the basalt adds to the force produced by the Earth's main dipole, so a magnetometer measures a stronger signal than expected. A negative anomaly occurs over regions of the seafloor where the basalt has reversed polarity. In these areas, the magnetic force produced by the basalt subtracts from the force produced by the Earth's main dipole, so the magnetometer measures a weaker signal **(Fig 2.35a)**.

FIGURE 2.35 The progressive development of marine magnetic anomalies.

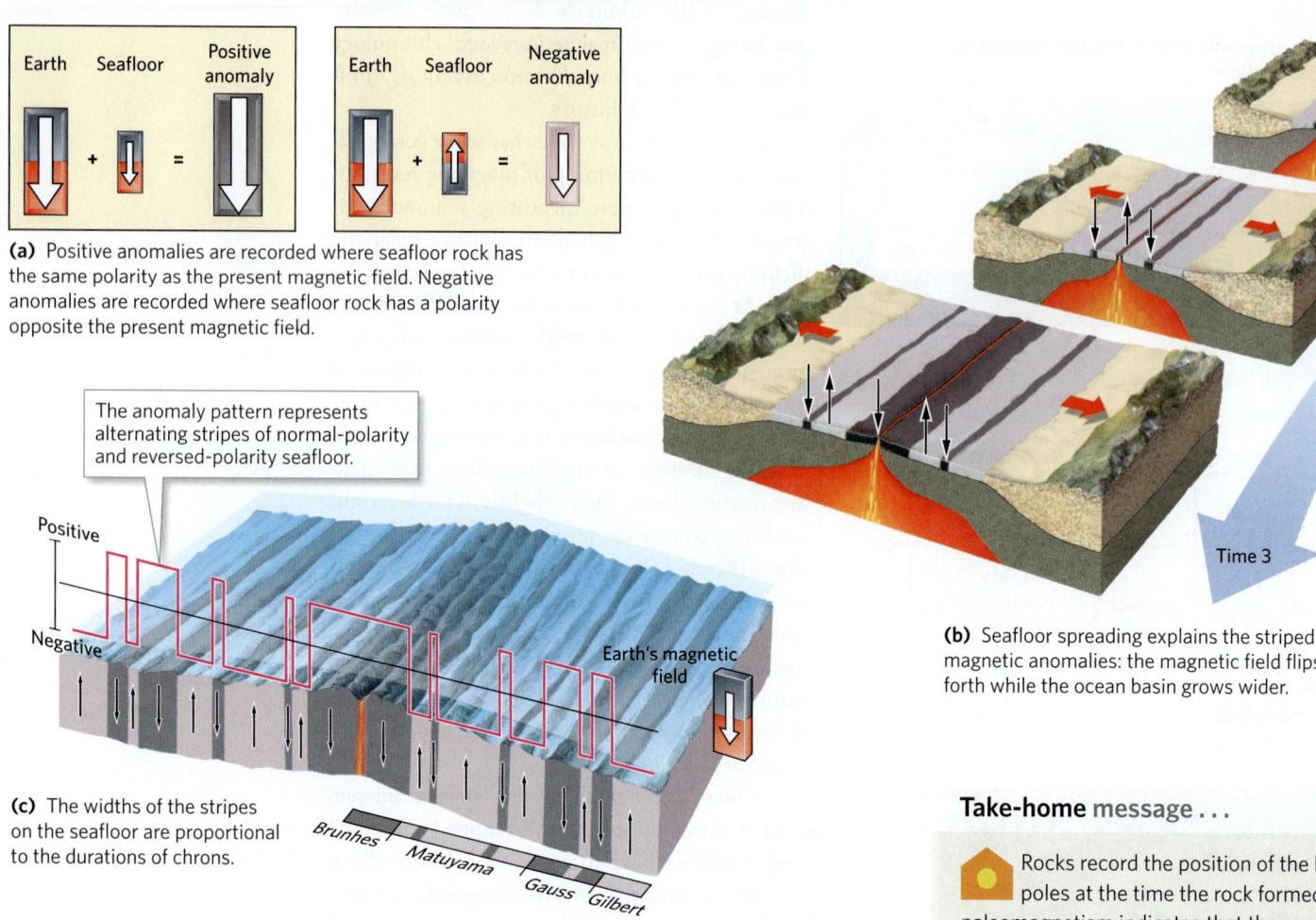

(a) Positive anomalies are recorded where seafloor rock has the same polarity as the present magnetic field. Negative anomalies are recorded where seafloor rock has a polarity opposite the present magnetic field.

The anomaly pattern represents alternating stripes of normal-polarity and reversed-polarity seafloor.

(c) The widths of the stripes on the seafloor are proportional to the durations of chrons.

Brunhes Matuyama Gauss Gilbert

(b) Seafloor spreading explains the striped pattern of magnetic anomalies: the magnetic field flips back and forth while the ocean basin grows wider.

Time 1
Time 2
Time 3

Positive
Negative
Earth's magnetic field

Take-home message . . .

Rocks record the position of the Earth's magnetic poles at the time the rock formed. The study of paleomagnetism indicates that the continents move relative to the Earth's magnetic poles. Each continent displays a different apparent polar-wander path, so the continents also move relative to one another. The polarity of the Earth's magnetic field reverses every now and then. Ocean floor formed at mid-ocean ridges records these reversals as marine magnetic anomalies, confirming the existence of seafloor spreading.

Quick Question

What determines the widths of marine magnetic anomalies?

Therefore, seafloor yielding positive anomalies formed at times when the Earth had normal polarity, whereas seafloor yielding negative anomalies formed when the Earth had reversed polarity. As seafloor spreading takes place, bands of seafloor with different polarities form and then move away from the ridge axis **(Fig. 2.35b)**. Since seafloor spreading along a given segment of ridge takes place at a fairly constant rate, the relative widths of the stripes correspond to the duration of polarity chrons **(Fig. 2.35c)**.

The discovery of the pattern of stripes and the confirmation of the relationship between the widths of the stripes and the durations of chrons serve as proof that seafloor spreading takes place, and therefore also serve as a positive test of plate tectonics. By assuming constant rates of plate motion, geologists can estimate the age of seafloor and, therefore, the times of magnetic reversals recorded by seafloor older than 4.5 Ma. These estimates indicate that the oldest seafloor on the Earth today formed only about 200 Ma, less than 5% of the age of the Earth.

> The least movement is of importance to all nature.
>
> —*BLAISE PASCAL (FRENCH MATHEMATICIAN, 1880–1930)*

2.9 The Velocity of Plate Motions

What Drives the Plates?

Because Alfred Wegener had not been able to answer the question of what makes continents drift, most geologists of his era didn't believe that continents could drift. But the minority who did agree with Wegener continued to speculate about how the continents could be mobile. In 1928, a British researcher, Arthur Holmes, suggested that continental drift may be due to convective flow in the

FIGURE 2.36 Forces driving plate motions.

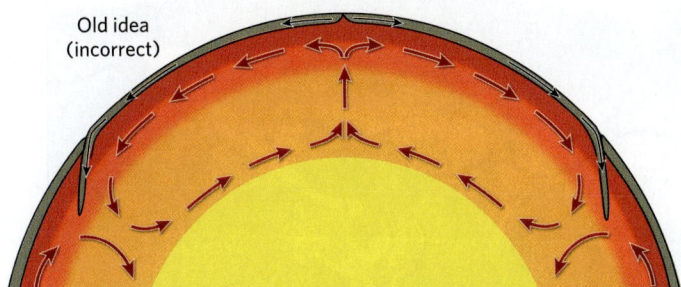

Old idea (incorrect)

(a) The old, incorrect image of simple convective cells in the asthenosphere. In this model, the plates were viewed as passive rafts carried along by convective flow in the mantle.

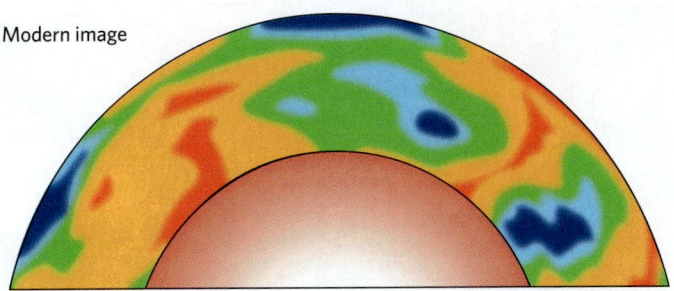

Modern image

(b) Modern studies show that convective flow in the mantle is complicated. Warm (redder) areas rise, and cooler (blue) areas sink.

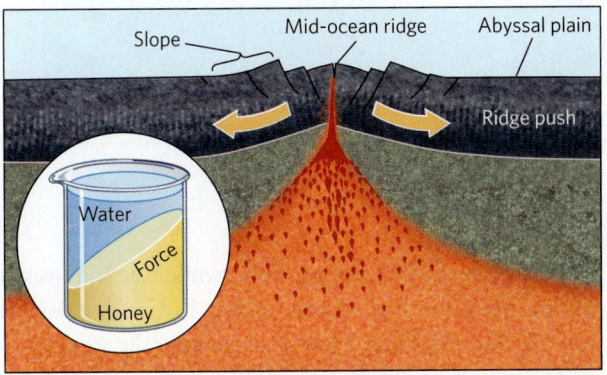

Slope Mid-ocean ridge Abyssal plain

Ridge push

Water

Force

Honey

(c) Ridge push develops because the region of a ridge is elevated. Like a wedge of honey with a sloping surface, the mass of the ridge pushes sideways.

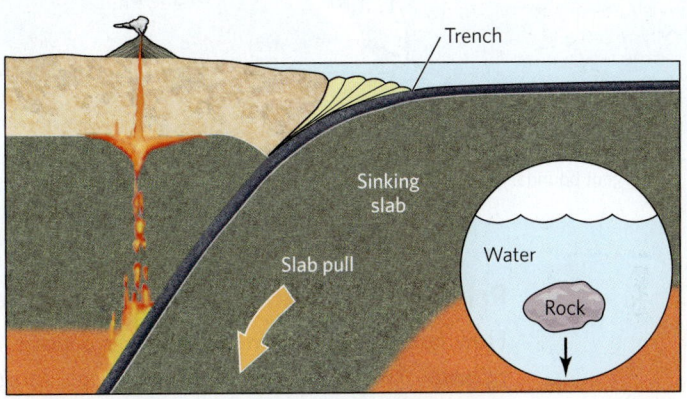

Trench

Sinking slab

Slab pull

Water

Rock

(d) Slab pull develops because lithosphere is denser than the underlying asthenosphere and sinks like a rock in water (though much more slowly).

mantle. **Convective flow**, in general, refers to a process of circulation and heat transfer that happens when a deeper layer of material warms up and, therefore, expands and becomes less dense than the overlying layer of cooler material. As a result, the deeper material becomes buoyant, and if the material is weak enough to flow, it rises. When this happens, cooler material sinks to take the place of the rising warm material. The overall pattern of convective flow is called a *convective cell*. Holmes suggested that convective flow could happen in the mantle because mantle rock is hot enough to behave plastically.

When the continental-drift hypothesis evolved into the theory of plate tectonics, geologists initially attributed the motion of plates entirely to convection, and they pictured plates as rafts being carried by simple convective cells in the asthenosphere. As a result, early drawings depicting plate motion implied that warm rock rose at the mid-ocean ridge and cool rock sank at subduction zones **(Fig. 2.36a)**. But it's not that simple **(Fig. 2.36b)**! Eventually, geologists realized that it's impossible to draw a global system of convective cells that can explain the complexity of real plate-boundary configurations. So, while

convective flow in the mantle does happen, and probably does influence the long-term movement of plates, the details of plate motion are controlled by forces applied along plate boundaries. Here, we describe two of these forces: ridge-push force and slab-pull force.

Ridge-push force develops because the lithosphere beneath mid-ocean ridges sits higher than the lithosphere beneath adjacent abyssal plains. The elevated lithosphere tries to spread sideways due to gravity, much as a mound of honey spreads sideways and moves horizontally across a table. This motion drives plates in a direction pointing away from the ridge axis **(Fig. 2.36c)**. **Slab-pull force** develops because oceanic lithospheric mantle is cooler, and therefore denser, than the warmer asthenosphere below. So, once a plate starts to subduct, it sinks, like a rock or anchor into water, though much more slowly. The subducted section slowly pulls the rest of the plate behind it **(Fig. 2.36d)**. Geologists still debate the relative roles of the different forces acting on plates. Most likely, the motions that we see reflect a complex combination of forces. A growing consensus, however, suggests that slab pull provides the dominant force applied to plates.

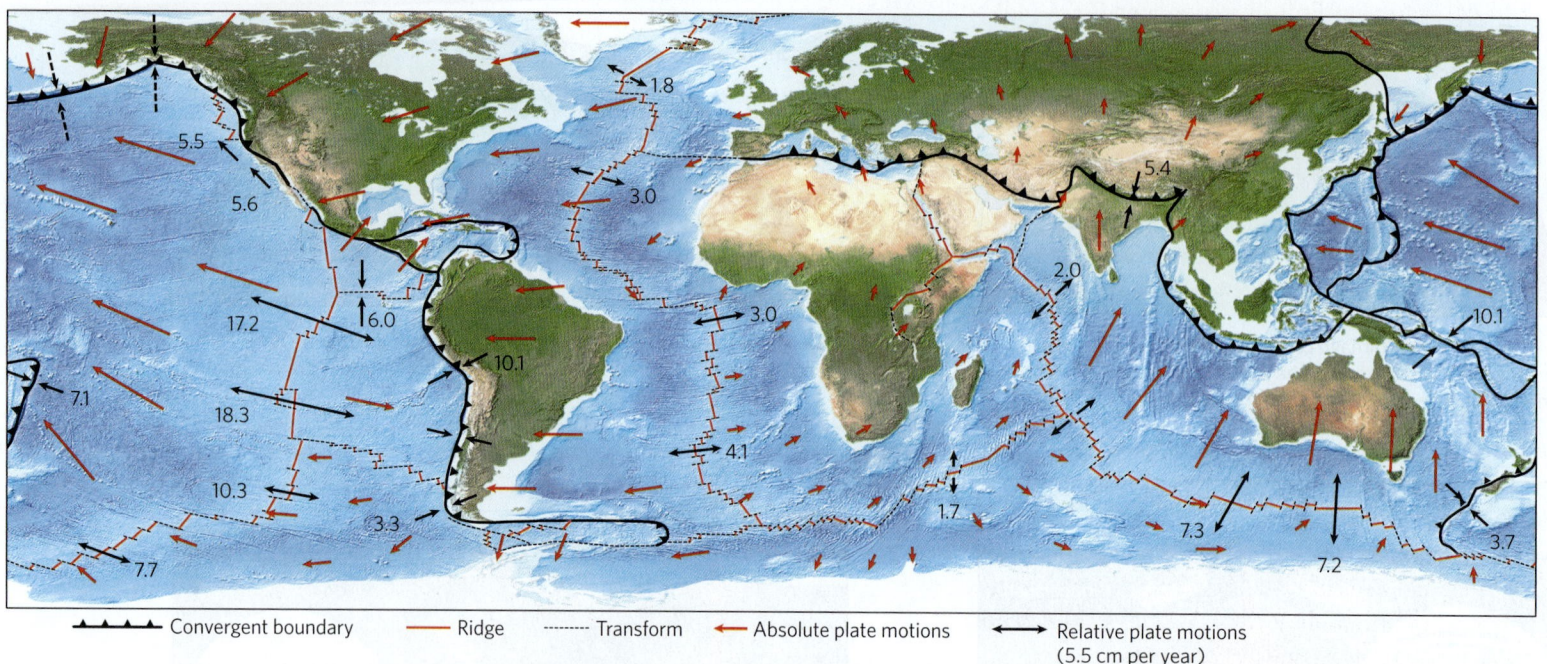

Convergent boundary — Ridge ------- Transform → Absolute plate motions ←→ Relative plate motions (5.5 cm per year)

Relative versus Absolute Plate Velocities

How fast do plates move? The answer depends on your *reference frame*. To illustrate this concept, imagine two cars speeding in the same direction down the highway. From the viewpoint of a tree along the side of the road,

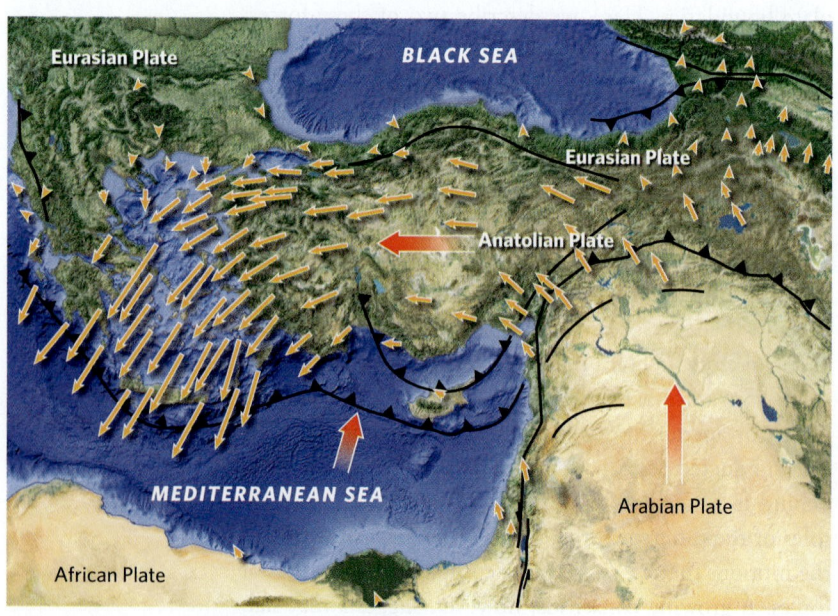

FIGURE 2.38 GPS measurements can now track modern-day plate motions accurately. The length of each arrow on this map of the eastern Mediterranean region indicates the velocity at which a point on the crust at the end of the arrow is moving. Turkey is moving westward, and the Greek islands are moving southwestward.

Car A zips by at 100 km per hour, while Car B moves at 80 km per hour. But relative to Car B, Car A moves at only 20 km per hour. How you describe the velocity of a car depends on whether the reference frame is another moving car or a fixed tree. Similarly, geologists use two different reference frames for describing plate velocity. If we describe the movement of Plate A relative to Plate B, then we are talking about **relative plate velocity**. But if we describe the movement of both plates relative to a fixed reference point that is not on one of the plates, then we are speaking of **absolute plate velocity (Fig. 2.37)**.

Note that to completely specify velocity, you need to indicate both the rate of movement and the direction of movement. Physicists define rate of movement by the equation: rate = distance ÷ time. We can determine the rate of relative motions across a mid-ocean ridge by measuring the distance of marine magnetic anomalies of a known age from the ridge axis. If the spreading rate at the ridge is constant, then the rate of a point on the plate relative to the ridge axis can be calculated from the equation above. The rate of the plate on one side of the ridge relative to the plate on the other is twice this value. The direction of motion is perpendicular to the ridge axis.

To measure absolute plate velocities, at least approximately, we can measure the movement of a plate relative to a hot spot. Specifically, if we assume that the position of a hot spot does not change much for a long time, then a hot-spot track on a plate moving over a mantle plume provides a record of the plate's rate and direction of movement. For example, if we assume that the Hawaiian hot spot's position is fixed, then the orientation of the Hawaiian-Emperor seamount chain indicates the absolute direction of Pacific Plate motion, and by dating the rocks of each volcano in the chain, we can calculate the rate of plate movement. Note that the Hawaiian chain runs northwest, whereas the Emperor chain curves north-northwest (see Fig. 2.28a). Isotopic dates of volcanic rocks from the bend indicate that they formed about 43 million years ago, suggesting that the direction in which the Pacific Plate was moving changed at that time.

GPS: Observing Plate Motions in Real Time

Using the methods described above, geologists determined that relative plate motions on Earth today occur at rates of about 1–15 cm (0.4–6 in) per year, about the rate at which your fingernails grow. These rates, though small, can yield large displacements during the immensity of geologic time. For example, in 1 million years, a plate moving at 1 cm per year travels 10 km (6 mi).

Can we detect very slow plate motions by direct observation? Until the 1990s, the answer was no. But now, by using the **global positioning system** (**GPS**)—the same technology that drivers use to find their destinations—geologists can detect displacements as small as a few millimeters per year **(Fig. 2.38)**. With such resolution, it's actually possible to see movement of plates over the course of a year! Our ability to observe plate motions serves as the ultimate proof of plate tectonics.

Paleogeography

Knowledge of the positions of marine magnetic anomalies allowed geologists to reconstruct the opening of modern oceans back to about 200 Ma. Defining the relative positions and latitudes of continents earlier in Earth history, however, requires the use of other data sources. For example, paleomagnetic studies can indicate the latitudes of continents at times in the past, and they can indicate whether two continents were moving as one or were moving separately (demonstrating that they were apart). Study of rock units and fossils also helps determine when continents were merged and when they were separate.

By taking such data into account, geologists have refined the image of continental drift as Wegener had imagined it, and they can describe the complex ways in which the map of our planet's surface has changed over time **(Fig. 2.39)**. This work shows that Pangaea was not the only supercontinent during Earth history. In fact, continents merged to form supercontinents, which later broke up, at least four times during the past few billion years.

FIGURE 2.39 Due to plate tectonics, the map of the Earth's surface slowly changes. Here we see the breakup of Pangaea over the past 250 million years.

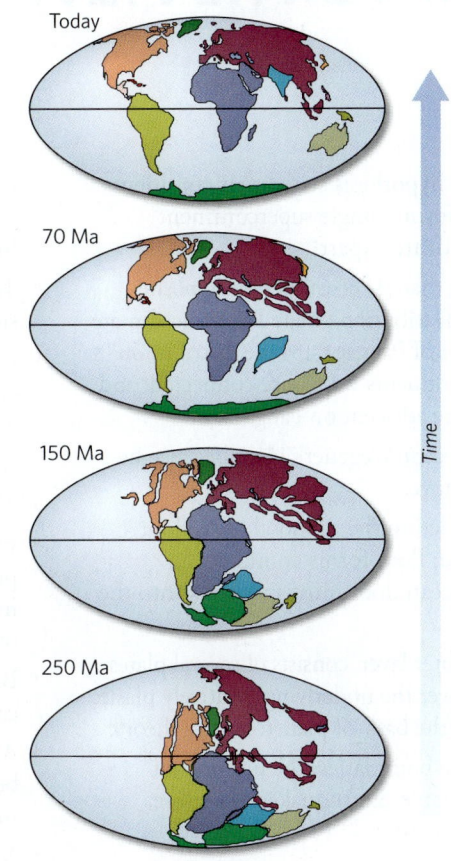

Plate interactions have produced a multitude of different rock types, as we will see in the next few chapters, as well as many mountain belts and other geologic features, which we will discuss later in the book.

Take-home message . . .

Plate motion takes place because plates are acted on by ridge push, slab pull, and convective flow. This motion happens at rates of 1 to 15 cm per year. Relative plate velocity refers to the rate at which a plate moves relative to its neighbor, whereas absolute plate velocity refers to the rate at which a plate moves relative to a fixed reference point. GPS measurements can now detect plate motions directly. Geologists have tracked the movement of continents over the Earth's history.

Quick Question

Was Pangaea the only supercontinent in history?

2 CHAPTER REVIEW

- Alfred Wegener's continental-drift hypothesis stated that continents had once been joined together to form a single supercontinent (Pangaea), and then subsequently drifted apart.

- Wegener argued for drift by noting that (1) coastlines on opposite sides of the oceans match; (2) the distribution of late Paleozoic climate belts is compatible with the concept of Pangaea; (3) the distribution of fossil species suggests that the continents were once connected; and (4) distinctive rock assemblages were adjacent on Pangaea.

- Most geologists did not initially accept Wegener's ideas because he couldn't explain how continents move.

- Research in the mid-20th century led to the proposal of seafloor spreading—the idea that new ocean floor forms at mid-ocean ridges—and to the idea that old ocean floor must sink back into the mantle at trenches.

- The lithosphere, the Earth's rigid outer layer, consists of several plates that move relative to one another over the underlying, relatively plastic asthenosphere. This concept forms the basis of plate tectonics theory.

- Plate interactions occur along plate boundaries; the interiors of plates remain relatively rigid and stable. Earthquakes, therefore, delineate the position of plate boundaries.

- There are three types of plate boundaries—divergent, convergent, and transform—distinguished by the movement of the plate on one side of the boundary relative to the plate on the other side.

- Divergent boundaries, where seafloor spreading takes place, are marked by mid-ocean ridges. As a consequence of seafloor spreading, continents on either side of an ocean drift apart.

- Convergent boundaries are delineated by deep-sea trenches and adjacent volcanic arcs. At a convergent boundary, oceanic lithosphere subducts and sinks into the mantle.

- Transform boundaries are large faults along which one plate slides sideways past another.

- A convergent boundary ceases to exist when two buoyant pieces of crust converge at the subduction zone and collision occurs.

- A large continent can split into two smaller ones by the process of rifting. During rifting, continental lithosphere stretches and thins.

- Three plate boundaries intersect at a triple junction.

- Hot spots are places where volcanism not related to plate-boundary processes occurs. As a plate moves over the hot spot, the volcano moves off the hot spot and becomes extinct. Hot spots may form over mantle plumes, columns of rising asthenosphere.

- By measuring paleomagnetism in successively older rocks, geologists can define an apparent polar-wander path.

- Apparent polar-wander paths are different for different continents because continents move relative to one another, while the Earth's magnetic poles remain roughly fixed.

- Marine magnetic anomalies on the seafloor develop because magnetic reversals happen while seafloor spreading is taking place.

- Plate motion is probably the result of several forces, including ridge push, slab pull, and convective flow in the asthenosphere. Plates move at rates of about 1–15 cm (0.4–6 in) per year.

- Modern GPS methods can measure plate motions in real time.

Key Terms

absolute plate velocity (p. 80)
abyssal plain (p. 53)
accretionary prism (p. 66)
active margin (p. 57)
apparent polar-wander path (p. 75)
asthenosphere (p. 56)
bathymetric map (p. 53)
collision (p. 69)
continental drift (p. 49)
continental margin (p. 53)
continental shelf (p. 57)
convective flow (p. 79)
convergent boundary (p. 62)

deep-sea trench (p. 53)
dipole (p. 74)
divergent boundary (p. 59)
earthquake (p. 54)
fault (p. 54)
fracture zone (p. 53)
global positioning system (p. 81)
hot spot (p. 72)
hot-spot track (p. 73)
island arc (p. 66)
lithosphere (p. 56)
lithospheric mantle (p. 56)
magnetic pole (p. 74)

magnetic reversal (p. 76)
mantle plume (p. 73)
marine magnetic anomaly (p. 77)
mid-ocean ridge (p. 53)
ocean basin (p. 53)
paleomagnetism (p. 75)
paleopole (p. 75)
Pangaea (p. 49)
passive margin (p. 57)
plate (p. 57)
plate tectonics (p. 55)
relative plate velocity (p. 80)
ridge-push force (p. 79)

rift (p. 70)
rifting (p. 70)
seafloor spreading (p. 55)
seismic belt (p. 54)
slab-pull force (p. 79)
subduction (p. 62)
subduction zone (p. 62)
transform boundary (p. 68)
transform fault (p. 68)
volcanic arc (p. 62)

Letters in parentheses correspond to the chapter's learning objectives.

1. What was Wegener's continental-drift hypothesis? What was his evidence? How did he construct this map? **(A)**

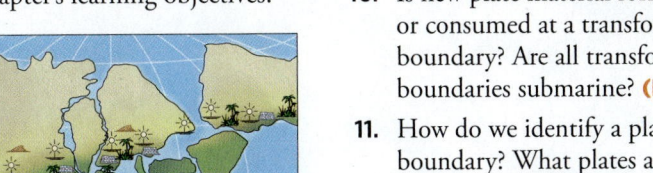

2. Describe discoveries made from the 1930s to the 1960s about the bathymetry of the seafloor. **(B)**

3. Do earthquakes occur randomly, or are they associated with bathymetric and topographic features? Does heat flow from the Earth's interior vary randomly, or are variations associated with bathymetric features? **(B)**

4. What is the hypothesis of seafloor spreading as defined by Harry Hess? How did Hess explain how seafloor spreading could take place without an increase in the Earth's circumference? **(B)**

5. How did drilling into the seafloor help prove seafloor spreading? **(B)**

6. What are the characteristics of a lithosphere plate? Is it composed of crust alone? What characteristic defines the boundary between lithosphere and asthenosphere? **(C)**

7. How do active and passive continental margins differ? **(C)**

8. How does oceanic lithosphere differ from continental lithosphere, and how does this difference explain the existence of ocean basins? **(C)**

9. Describe the three types of plate boundaries. For each, be sure to indicate the nature of relative plate motion, and name a bathymetric or topographic feature associated with that type of plate boundary. **(D)**

10. Is new plate material formed or consumed at a transform boundary? Are all transform boundaries submarine? **(D)**

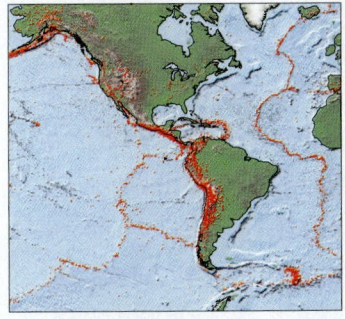

11. How do we identify a plate boundary? What plates appear on this map? **(D)**

12. How does oceanic crust form along a mid-ocean ridge? How does oceanic lithospheric mantle form? **(E)**

13. Identify the major geologic features of a convergent boundary on this drawing. **(E)**

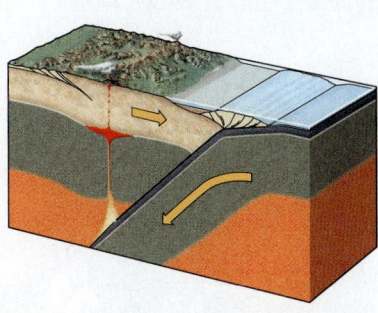

14. Describe the process of continental collision and give examples of where this process has occurred. **(E)**

15. Describe the characteristics of a rift and give examples of where rifting takes place today. **(E)**

16. What is a marine magnetic anomaly? How is it detected? **(G)**

17. How is a hot-spot track produced, and how can hot-spot tracks be used to determine the past absolute motion of a plate? **(H)**

18. What is paleomagnetism? How did the discovery of apparent polar-wander paths serve as a proof that continents move? **(G)**

19. How did the observed pattern of marine magnetic anomalies form, and how did its existence help prove plate tectonics? **(G)**

20. Discuss the major forces that move lithosphere plates. **(F)**

21. Explain the difference between relative plate velocity and absolute plate velocity. Can we measure plate velocity directly? **(H)**

22. Explain the bend in the Hawaiian-Emperor seamount chain in terms of the absolute motion of the Pacific Plate. **(F)**

23. How does the existence of the Appalachian Mountains indicate that there was an ocean separating North America from Africa and Europe prior to the formation of Pangaea? **(E)**

3 WHAT IS THE EARTH MADE OF?
Introducing Minerals and the Nature of Rock

By the end of the chapter you should be able to . . .

A. define the word *mineral* in the context of geology, and describe how minerals form.

B. describe key physical characteristics of minerals, and identify common minerals.

C. sort common minerals into classes based on their composition.

D. describe characteristics that distinguish rocks from other materials.

E. explain why a chunk of rock doesn't easily break apart into separate grains.

F. name the three rock groups, and provide the basis for this classification.

3.1 Introduction

The track got narrower and bumpier as the caravan of four-wheel-drive vehicles moved slowly uphill, carrying students hoping to visit an abandoned mine thought to contain beautiful minerals. Finally, the group reached the mine—but it wasn't abandoned! A one-armed watchman glared out from behind an old refrigerator door on which he'd painted the words, "Is there life after death? Pass this sign and find out!" After a few moments of awkward silence, one of the students spoke up. "We're just poking around looking for minerals," she said. The watchman replied, "I know . . . I've seen you with my binoculars." Undeterred, the student quietly asked, "Any chance we can take a look in the mine?" There was another long silence. Then, the man whispered, "You forgot the magic word!" In unison, all the students looked up and said, "Please?" With a grin, the watchman said, "What the heck, I could use the conversation. You can look but don't collect." He led them into the mine, a tunnel that had been carved into a rock cliff. With headlamps on and eyes wide open, the students explored. On some walls the rocks were streaked with bands of bright blue and green (Fig. 3.1), and on others the rocks contained shiny, bronze-like crystals. The students spent a wonderful hour marveling at the great variety of shapes and colors their lamps illuminated. The trip had not been wasted!

We briefly introduced minerals and rocks in Chapter 1, in the context of describing the materials of the geosphere, the solid substrate of the Earth System. In this chapter, we consider minerals in greater depth, with the goal of providing a more complete picture of what they are and how they form. We begin by clarifying the definition of a mineral. Next, we discuss how to distinguish among different types of minerals and how to classify them. We finish our discussion by examining why people consider some minerals to be gems. After introducing minerals, we turn our attention to rocks, first by clarifying the distinction between a mineral and a rock, then by introducing the basis for classifying rocks into three groups, and finally by describing the various methods that geologists use to study rocks. Chapter 3 sets the stage for the next three chapters, in which we explore the great variety of rocks that occur on the Earth and learn how to interpret them.

<< A museum specimen of red quartz crystals, colored by trace amounts of iron oxide.

FIGURE 3.1 Colorful minerals found on the wall of a mine. These are copper-bearing minerals.

5 cm

3.2 What Is a Mineral?

Defining Minerals

In everyday English, the word *mineral* has a rather vague meaning because people commonly refer to a great variety of different materials as minerals. For example, when playing the game 20 Questions, you may think of a mineral simply as anything that can't be considered an animal or plant. To a geologist, however, the term has a more restricted definition. In geologic discussion, a **mineral** is a naturally occurring, homogeneous, crystalline solid that has a definable chemical composition and, in most cases, is inorganic. Let's pull apart this mouthful of a definition and examine each component in detail:

- *Naturally occurring:* Real minerals form in nature, not in factories. We need to emphasize this point because in recent decades, chemists have synthesized materials with characteristics identical to those of real minerals. These new materials can be called *synthetic minerals*.

- *Homogeneous:* Ideally, a piece of a mineral has the same composition and structure throughout.

- *Solid:* Liquids (such as oil or water) and gases (such as air) are not minerals.

- *Crystalline solid:* If the atoms in a material are not distributed randomly, but rather remain fixed in a specific, orderly pattern, we can refer to the material as a **crystalline solid** (Fig. 3.2a). All minerals have this characteristic. The internal architecture of a mineral, simplistically, resembles a grid of scaffolding.

FIGURE 3.2 What is a crystalline solid?

A crystal face

A crystal structure resembles scaffolding.

Quartz crystal

Washington Monument

(a) Crystalline solids, such as quartz, contain an orderly arrangement of atoms. The geometry of this arrangement defines the crystal structure.

(b) A "cut crystal" bowl is not actually crystal, but rather is made of glass. The atoms within do not have an orderly arrangement.

Animation

Minerals

- *Definable chemical composition:* This phrase emphasizes that a given mineral contains specific elements in specific proportions. As a result, we can write a chemical formula for a mineral **(Box 3.1)**. For example, a sample of calcite has the formula $CaCO_3$, and a sample of quartz has the formula SiO_2.

- *Inorganic:* To understand this aspect of the definition, we must first distinguish between organic and inorganic chemicals. **Organic chemicals** are carbon-containing compounds that either occur in living organisms or resemble compounds produced in living organisms. In many organic chemicals, the carbon atoms are arranged in chains or rings and are bonded to hydrogen, nitrogen, or oxygen atoms. Almost all minerals are inorganic. Therefore, sugar, which is an organic chemical, isn't a mineral even though it's a crystalline solid. But we have to add the qualifier "almost all" because *mineralogists* (researchers who study minerals) consider about 50 organic substances that form when crystals grow naturally in organic debris to be minerals.

With the formal definition of a mineral in mind, we can make an important distinction between a mineral and a glass. A **glass** is an inorganic solid that does not have a crystal structure **(Fig. 3.2b)**. In other words, while the atoms, ions, or molecules in a mineral lock into a regular arrangement like soldiers standing in formation, those in a glass are arranged in a semi-chaotic way, like people standing around at a party. Of course most of the glass that you see in daily life has been manufactured by people. But, as we'll see in the next chapter, geologic processes at some volcanoes can produce glass. Even though volcanic glass is a natural inorganic solid, mineralogists don't consider it to be a mineral because its molecules aren't arranged in an orderly way.

If you ever need to figure out whether a substance is a mineral or not, just check it against the criteria of the definition. Is gasoline a mineral? No—it consists of organic chemicals and is a liquid. Is table salt a mineral? Yes—it's a solid crystalline compound with the formula NaCl, and it's formed by a geologic process.

Beauty in Patterns: Crystals and Their Structure

We've mentioned already that minerals, by definition, are crystalline. As a result, they can occur in the form of a crystal. As is the case for the word *mineral*, geologists use a more limited definition of *crystal* than we use in daily life. Specifically, in a geologic context, a **crystal** is a single, continuous (meaning uninterrupted) piece of solid material inside which atoms, molecules, or ions are fixed in an orderly arrangement. Geologists use another word, *grain*, in a more general sense for any small, natural solid particle. Some grains are crystals that grew into their present shape, but others are pieces of larger crystals, fragments containing many tiny crystals, or even shards of glass.

Some crystals, known as *euhedral crystals*, have distinctive geometric shapes defined by smooth, flat surfaces, known as **crystal faces**, that intersect at sharp

BOX 3.1

Science Toolbox

Basic chemistry terminology

To describe minerals, we need to use many words defined by chemists. If you've taken a chemistry course, you may know these terms already. But to avoid any misunderstandings, we define the most common terms here for ready reference. We list the terms in an order that allows us to use previous terms to define later ones.

- *Element:* A pure substance that cannot be separated into other elements. Chemists represent each element by a symbol. For example, H = hydrogen; C = carbon; O = oxygen; and Fe = iron.
- *Atom:* The smallest piece of an element that retains the characteristics of that element. An atom consists of a nucleus surrounded by an electron cloud. The nucleus of an atom contains both protons and neutrons. (There's one exception to the previous statement: hydrogen's nucleus contains only a proton.) Electrons have a negative charge, protons have a positive charge, and neutrons have a neutral charge. Atoms have the same number of electrons as protons, making them electrically neutral. If atoms gain or lose electrons, they become positive or negative ions.
- *Atomic number:* The number of protons in an atom.
- *Atomic mass:* Approximately the number of protons plus neutrons in an atom.
- *Ion:* An atom that is not neutral. An *anion* has a negative charge because it has more electrons than protons, while a *cation* has a positive charge because it has more protons than electrons. We indicate charge with a superscript sign. For example, Cl^- (chloride anion) has gained one extra electron, whereas Fe^{2+} (ferrous cation) has lost two electrons.
- *Chemical bond:* An attractive force that holds two or more atoms together. There are several types. *Covalent bonds* form when atoms share electrons. *Ionic bonds* form when a cation and an anion get close together and attract each other because opposite charges attract. In metals, materials with *metallic bonds*, some electrons can move freely.
- *Molecule:* Two or more atoms bonded together. The atoms may be of the same element or of different elements. For example, H_2 is a molecule of hydrogen, and H_2O is a molecule of water.
- *Compound:* A substance in which molecules contain two or more elements.
- *Chemical:* A general name used for a pure substance (either an element or a compound).
- *Chemical formula:* A shorthand recipe that itemizes the various elements in a chemical and specifies their relative proportions. For example, an SiO_2 molecule contains two atoms of oxygen (O) for every atom of silicon (Si).
- *Chemical reaction:* A process that involves the breaking or forming of chemical bonds. Chemical reactions can break molecules apart or produce new molecules.
- *Mixture:* A combination of two or more elements or compounds that can be separated without a chemical reaction. For example, a cereal composed of bran flakes and raisins is a mixture—you can separate raisins from flakes without destroying either.
- *Solution:* A type of material in which one chemical (the *solute*) dissolves in another (the *solvent*). A solute may separate into ions during the process of dissolution. For example, when salt (NaCl) dissolves in water, it separates into sodium (Na^+) and chloride (Cl^-) ions. In a solution, atoms or molecules of the solvent surround atoms or molecules of the solute.
- *Precipitate* (noun): A solid formed when ions in a liquid solution join together to form solid grains that separate from and settle out of a solution. Salt remaining when seawater evaporates is a precipitate.
- *Precipitate* (verb): The process of forming a solid that separates out of a solution. When saltwater evaporates, salt crystals precipitate from the water.

edges. Crystal faces form naturally during the crystal's formation. Each face of a euhedral crystal of a given mineral forms a specific angle relative to its neighbor. Euhedral crystals come in many fascinating shapes—cubes, trapezoids, pyramids, octahedrons, hexagonal columns, prisms, blades, needles, columns, and obelisks—that look like they belong in the pages of a geometry book **(Fig. 3.3)**. The faces of a euhedral crystal grew to their present shape—they are not formed by breaking, cutting, or grinding. Therefore, if you purchase "crystal glassware" for your dining room, you're not buying natural mineral crystals, for two reasons: first, glassware consists of glass, so it does not have a crystal structure, and second, the flat faces of the glassware were produced by artisans using a grinding wheel, not by growth of the glass. Significantly, not all crystals are euhedral. Natural

Did you ever wonder . . .

whether there's a difference between a crystal of quartz and "crystal glassware"?

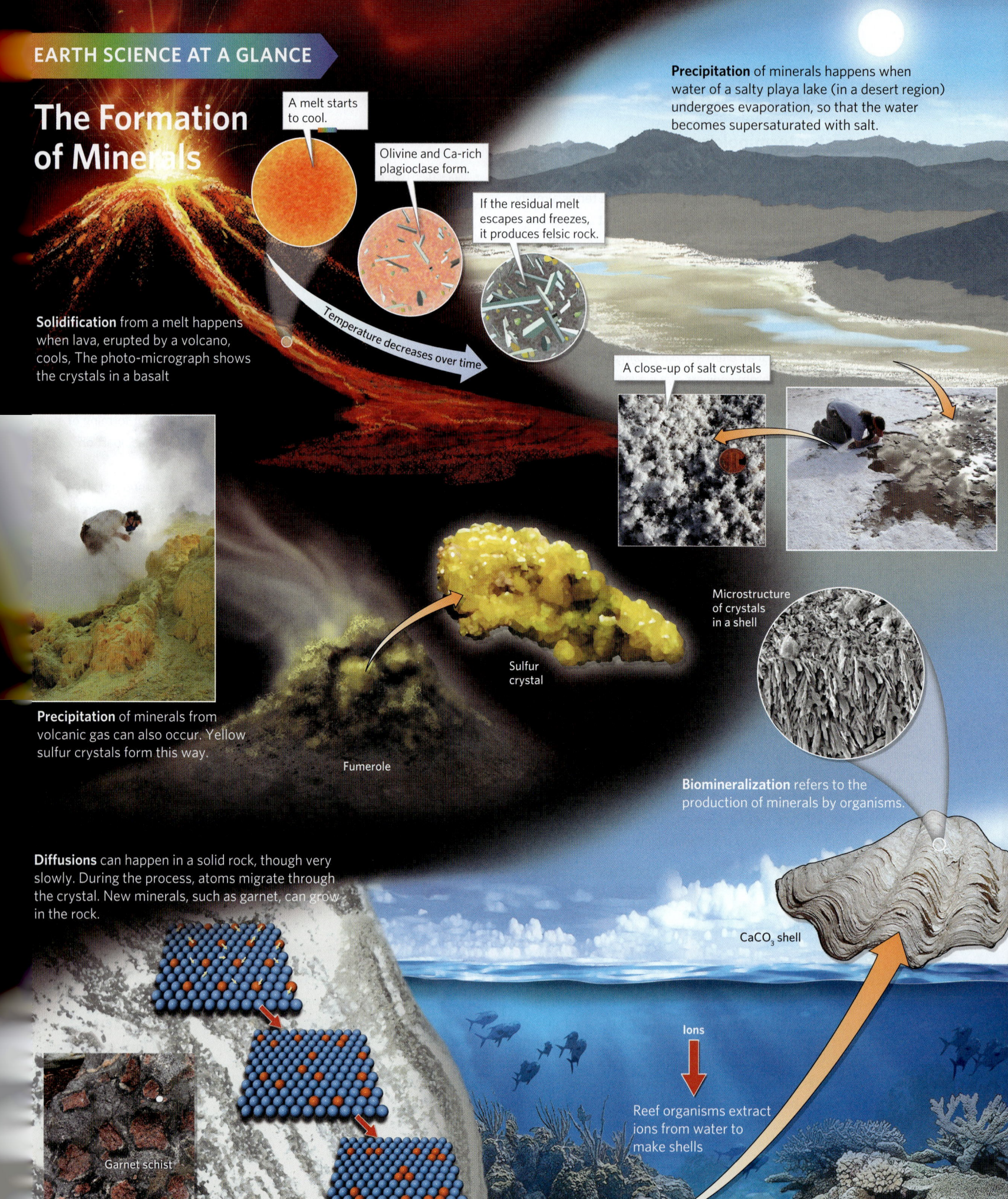

The Formation of Minerals

A melt starts to cool.

Olivine and Ca-rich plagioclase form.

If the residual melt escapes and freezes, it produces felsic rock.

Temperature decreases over time

Solidification from a melt happens when lava, erupted by a volcano, cools, The photo-micrograph shows the crystals in a basalt

Precipitation of minerals happens when water of a salty playa lake (in a desert region) undergoes evaporation, so that the water becomes supersaturated with salt.

A close-up of salt crystals

Microstructure of crystals in a shell

Sulfur crystal

Fumerole

Precipitation of minerals from volcanic gas can also occur. Yellow sulfur crystals form this way.

Biomineralization refers to the production of minerals by organisms.

Diffusions can happen in a solid rock, though very slowly. During the process, atoms migrate through the crystal. New minerals, such as garnet, can grow in the rock.

$CaCO_3$ shell

Ions

Reef organisms extract ions from water to make shells

Garnet schist

processes, as we'll see later, can also produce *anhedral crystals*, which have irregularly shaped surfaces.

The orderly arrangement of atoms inside a crystal—its **crystal structure**—is one of nature's most spectacular examples of *pattern*, the repetition of a feature in a geometric arrangement. Just as the regular spacing and angular relationships of a design motif define a pattern in wallpaper, the regular spacing and angular relationships of atoms or molecules define the pattern in a crystal **(Fig. 3.4a, b)**. If you picture atoms or ions in minerals as tiny balls held in place by chemical bonds, the geometric grid defining the relative positions of atoms in the crystal can be called a *crystal lattice*.

As an example of crystal structure, let's consider models depicting the arrangement of ions in *halite* (NaCl), the salt you use to season food. Within a crystal of halite, Cl⁻ alternate with Na⁺ in a lattice that looks like an array of lines intersecting one another at a 90° angle, so halite crystals resemble a cube **(Fig. 3.4c)**. Note that the pattern of atoms or ions in a mineral displays **symmetry**, meaning that the shape of one part of a mineral is a mirror image of the shape of another part **(Fig. 3.4d)**.

How Do Minerals Form?

While some mineral grains have survived for millions to billions of years, at any given time in our dynamic Earth System, new mineral grains are forming. Mineralogists recognize five different mineral-forming processes. Which process happens at a given place and time depends on the geologic setting of the mineral-forming environment (**Earth Science at a Glance**, p. 88):

- *Solidification of a melt:* The process by which a melt transforms into a solid is called *solidification* or *freezing*. During solidification, atoms or ions moving about in the melt come in contact with a crystal and bond to its surface. As the solid mineral grows, the amount of melt decreases, until transformation has become complete. The growth of water ice in a freezer represents this process.

- *Precipitation from a liquid solution:* During precipitation, ions of a solute dissolved in a solvent bond together to form a solid crystal that separates from the solvent. The whitish salt crust, left when a pan of seawater evaporates, forms by precipitation.

- *Biomineralization:* Some minerals grow at the interface between the physical and biological components of the Earth System. This process, called *biomineralization*, takes place when metabolism in a living organism causes atoms to precipitate either within or on the organism's cells or immediately adjacent to its cells. Biomineralization by clams produces calcite

Halite   Diamond   Staurolite   Quartz

Garnet   Stibnite   Calcite   Kyanite

FIGURE 3.3 Crystals come in a variety of shapes. Each example shown here illustrates an ideal crystal of the specified mineral.

FIGURE 3.4 The internal structure of minerals.

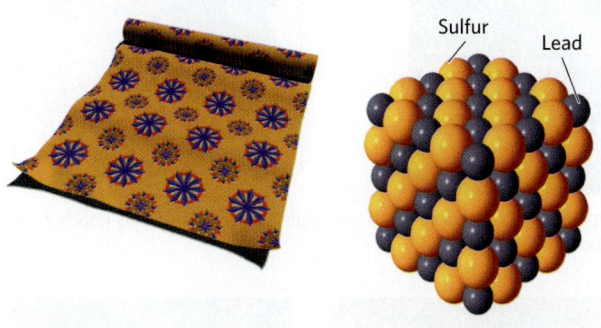

Sulfur   Lead

(a) The repetition of a flower motif on wallpaper.

(b) The repetition of alternating sulfur and lead atoms in the mineral galena (PbS).

Animation
Minerals: Crystalline Structure

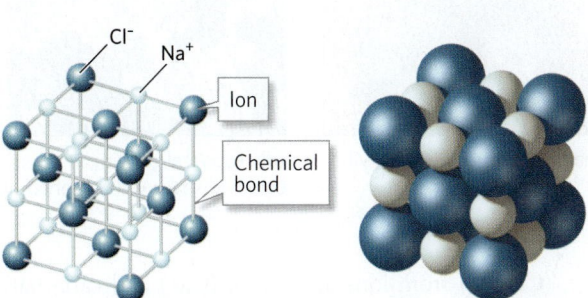

Cl⁻   Na⁺   Ion   Chemical bond

(c) Models of halite (NaCl): The model on the left portrays ions as balls and chemical bonds as sticks. The model on the right gives a sense of how ions fit together.

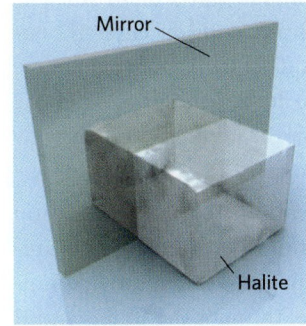

Mirror   Mirror   Halite   Snowflake

(d) Minerals display symmetry. One half of a halite crystal or a snowflake (an ice crystal) is a mirror image of the other.

Animation
Minerals: Geologic Processes

FIGURE 3.5 The growth of crystals.

Atoms attach to the crystal face.

Time

(a) A crystal grows as atoms in the surrounding material attach themselves to its faces. As the crystal grows, its faces move outward into the available space. This example shows precipitation from a water solution.

(b) Crystals that grow into an open space, like the purple quartz crystals (amethyst) in this geode from Brazil, develop a euhedral form (see the inset).

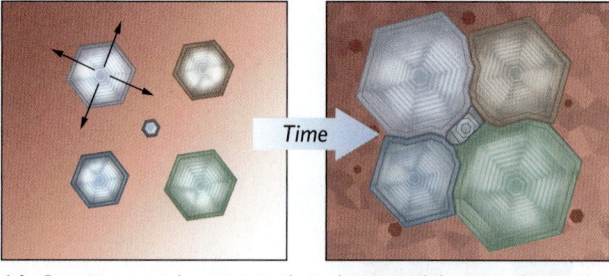

Time

(c) Growing crystals maintain their shape until they interfere with one another.

(d) A crystal growing in a confined space will be anhedral.

minerals involves diffusion of atoms through a solid, a process that happens much more slowly than does diffusion through air or water. Diffusing atoms can rearrange locally within a solid or move to a new location within a solid. Both processes result in new crystal growth. For example, garnet—a purple mineral whose crystals have 12 sides—can grow by diffusion in solid rock when the rock has been warmed to a high temperature.

How do crystals grow (Box 3.2)? The process starts with the chance appearance of a *seed,* meaning an extremely tiny crystal. Once a seed exists, other atoms in the surrounding material attach themselves to the faces of the seed, and the seed gets large enough to be called a crystal. As the crystal grows, its faces move outward, so that at any given time, the youngest part of the crystal lies at the crystal face (Fig. 3.5a). Crystal faces maintain the same angle relative to each other as they grow—for example, the angle between two adjacent faces of a quartz crystal is always 120°. A crystal's shape (needle-like, sheet-like, blade-like) depends both on the geometry of the crystal lattice and on whether or not the crystal can grow faster in one direction than in others.

If a mineral grows without being inhibited by its surroundings, it becomes a euhedral crystal. For example, crystals protruding from the walls of a **geode**, a mineral-lined open cavity in rock, commonly have a euhedral form (Fig. 3.5b). Alternatively, crystals that grow within an enclosed space commonly become anhedral (Fig. 3.5c, d) because the walls of the space inhibit the formation of crystal faces.

Take-home message . . .

Minerals are homogeneous, naturally occurring solids with a crystal structure and a definable chemical formula; most are inorganic. They grow in many ways, including solidification from a melt and precipitation from a solution. Crystal faces of a given mineral have a specific orientation relative to one another.

Quick Question

Do all crystals have the same shape? Why or why not?

$(CaCO_3)$ from ions of calcium (Ca^+) and carbonate (CO_3^{2-}) that the clams extract from water; the clams use the calcite to form their hard shells.

- *Precipitation from a gas:* We usually think of precipitation as taking place in liquid. But around volcanic vents, where volcanic gases and steam enter the atmosphere and cool very quickly, minerals precipitate directly from the gases to form deposits around the vent. Bright yellow deposits of sulfur crystals grow by this mechanism.

- *Diffusion in solids:* Diffusion involves the migration of atoms or molecules through a material. For example, if you drip a drop of ink into water, the ink molecules diffuse, or disperse, through the water, and if you spray perfume into the air, the perfume molecules diffuse through the air. Formation of certain

3.3 How Can You Tell One Mineral from Another?

Identifying Minerals by their Physical Properties

Researchers have identified and named almost 4,000 different minerals so far, and new ones are discovered every year. Minerals can be practical, beautiful, or even hazardous (Box 3.3). They differ from one another because they have

Crystal growth

What are we learning?

Crystals grow over time at rates affected by environmental conditions.

What you need:

• Water and salt
• A 0.5-L (16-fl-oz) glass jar
• A spoon and a string

Instructions:

• Fill the jar with hot water and dissolve as much salt in the water as it can hold.
• Tie the string to the spoon and place the spoon across the top of the jar, so the string hangs down into the water.
• Place the jar, uncovered, in a refrigerator, and let the water evaporate over a period of days. Salt crystals will form on the base of the jar and on the string.
• Remove the string to examine the crystals with a magnifying glass.

• Repeat the experiment, only this time place the jar in a warm-water bath on a hot plate. Let the water in the jar evaporate quickly.

What did we see?

As the water evaporates, the solution becomes saturated, so that it has no more capacity to hold dissolved salt. The Na^+ and Cl^- ions combine to form salt molecules, and crystals grow as those molecules precipitate. Under magnification, the cubic shape of the crystals may be visible. The rate of evaporation affects crystal size.

different chemical compositions, different crystal structures, or both. Amateurs and professionals alike get a kick out of recognizing distinctive minerals. They might hover around a display case in a museum and identify *mineral specimens* (intact pieces of one or more crystals or grains of an individual mineral) without bothering to look at the labels. How do they do it? The skill lies in learning to recognize the basic physical properties—the visual and material characteristics—that distinguish one mineral from another. You can characterize some physical properties of minerals, such as shape and color, from a distance. Others, such as hardness and magnetization, can be determined only by handling the specimen or by performing an identification test on it. *Identification tests* include scratching the mineral against another object, placing it near a magnet, lifting it, tasting it, or placing a drop of acid on it. Let's examine some of the physical properties used in mineral identification.

COLOR. Color results from the way a mineral interacts with light. The color you see when looking at a mineral represents the colors of the visible light spectrum that the mineral doesn't absorb. Some minerals always have the same color, but others come in a range of colors **(Fig. 3.6a)**. In some cases, color variations in a mineral are due to impurities in the crystal. The presence of iron atoms in quartz, for example, makes the quartz red or purple.

STREAK. The **streak** of a mineral refers to the color of a powder produced by scraping the mineral against an unglazed ceramic plate **(Fig. 3.6b)**. Streaks for a given mineral tend to be less variable than the color of whole specimens, so they provide a more reliable clue to a mineral's identity.

LUSTER. **Luster** refers to the way a mineral's surface reflects light. You can describe luster by comparing the appearance of the mineral with the appearance of a familiar substance. For example, minerals that look like metal have a *metallic luster*, and those that do not have a *nonmetallic luster* **(Fig. 3.6c, d)**. Various adjectives used for different types of nonmetallic luster include silky, glassy, satiny, resinous, pearly, and earthy.

HARDNESS. **Hardness** indicates the relative ability of a mineral to resist scratching. Hard minerals can scratch soft minerals, but soft minerals cannot scratch hard ones. In the early 1800s, a researcher named Friedrich Mohs listed some minerals in sequence of relative hardness and assigned numbers (1 to 10) to those minerals. In the resulting **Mohs hardness scale**, a mineral with a hardness of 5, for example, can scratch all minerals with a hardness of less than 5. We can add familiar objects (fingernails, copper coins, window glass, and steel knives) to the scale to provide context for judging

BOX 3.3 ▶

Putting Earth Science to Use

Hazardous minerals

Most minerals are harmless, and some we ingest routinely. For example, we use halite to salt food, calcite as an antacid, and certain types of clay to thicken milkshakes. But some minerals pose hazards to human health due to their composition, crystal habit, or grain size. Let's consider a couple of examples.

Some minerals contain chemicals that can poison you. For example, a version of pyrite (FeS_2), called arsenopyrite (FeAsS), contains arsenic, a deadly poison. Arsenopyrite can reside unchanged in rock buried deep in the Earth for millions of years. But when exposed to oxygen-bearing groundwater or to the air, the mineral undergoes a chemical reaction with oxygen to become a chemical that can dissolve in water. Drinking arsenic-bearing groundwater from wells can cause arsenic poisoning.

Concerns about asbestos have been featured in the news for decades. That's because asbestos was once used widely to manufacture insulation, brakes, roof shingles, and floor tiles, so people commonly come in contact with it. *Asbestos* refers to a group of several silicate minerals that grow in clumps of very fine, flexible needles or fibers **(Fig. Bx3.3a, b)**. Minerals in this group have a very high melting temperature and are very strong—that's why asbestos floor tiles and shingles were popular through the mid-20th century. Unfortunately, some (not all) kinds of asbestos, if inhaled, become embedded in human lungs and can cause cancer. For this reason, asbestos has been banned for most applications, and most remodeling or demolition projects for pre-1980s buildings require contractors to follow strict rules for *asbestos abatement*. They must enclose the areas where asbestos is being removed and must train workers to wear protective clothing and minimize production of asbestos dust **(Fig. Bx3.3c)**.

Quartz and feldspar, and many other silicate minerals, are safe under most circumstances, but can pose a danger if pulverized. Specifically, the dust of pulverized silicate minerals can, if inhaled, become embedded in lungs and cause *silicosis*. Because these minerals are very stable, neither metabolic processes in the lungs nor coughing can remove the dust. The irritation caused by the dust triggers the growth of fibrous masses that eventually block access to air and make breathing difficult. Everyone breathes a little dust now and then with no ill effects, but prolonged exposure to dust can be dangerous.

FIGURE Bx3.3 Asbestos has characteristics that can make it both useful and hazardous.

(a) Asbestos is a fibrous mineral.

(b) At high magnification, asbestos fibers look like twigs and vines.

(c) Workers removing asbestos must wear protective gear, and the work area must be sealed off .

FIGURE 3.6 Physical properties of minerals.

(a) Color is diagnostic of some minerals, but not all. For example, quartz can come in many colors.

(b) To obtain the streak of a mineral, rub it against an unglazed porcelain plate. The streak consists of mineral powder.

(c) Pyrite has a metallic luster because it gleams like metal.

(d) Feldspar has a nonmetallic luster.

(e) Calcite reacts with hydrochloric acid to produce carbon dioxide gas.

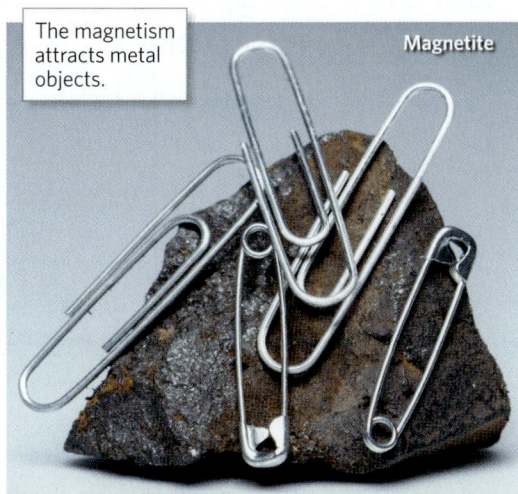

(f) Magnetite is magnetic.

(g) Crystal habit refers to the shape or character of a crystal. These giant gypsum crystals, from a cave in Mexico, are columnar.

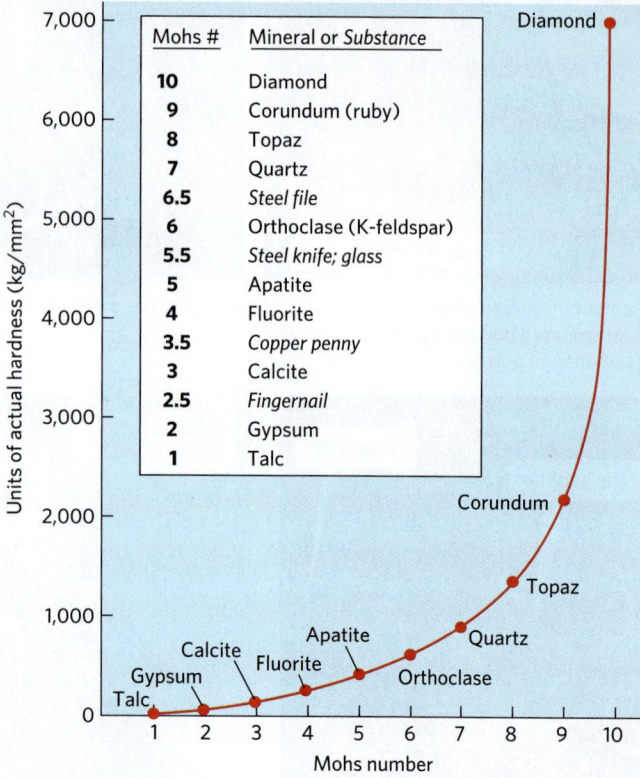

TABLE 3.1 Mohs hardness scale. Note that Mohs's numbers are relative—in reality, diamond is 3.5 times harder than corundum, as the graph shows.

Mohs #	Mineral or *Substance*
10	Diamond
9	Corundum (ruby)
8	Topaz
7	Quartz
6.5	*Steel file*
6	Orthoclase (K-feldspar)
5.5	*Steel knife; glass*
5	Apatite
4	Fluorite
3.5	*Copper penny*
3	Calcite
2.5	*Fingernail*
2	Gypsum
1	Talc

the hardness of unfamiliar minerals **(Table 3.1)**. For example, if glass has a hardness of about 5.5, then any mineral that can scratch glass has a Mohs hardness higher than 5.5, and any mineral the glass can scratch has a hardness of less than 5.5.

SPECIFIC GRAVITY. Specific gravity represents the density of a mineral, as specified by the ratio between the weight of a volume of the mineral and the weight of an equal volume of water. For example, 1 cm³ of quartz has a weight of about 2.7 g, whereas 1 cm³ of water has a weight of 1.0 g, so the specific gravity of quartz is 2.7. In practice, you can develop a feel for specific gravity by hefting mineral specimens in your hand.

SPECIAL PROPERTIES. Some minerals have distinctive properties that readily distinguish them from other minerals. For example, calcite ($CaCO_3$) and dolomite [$CaMg(CO_3)_2$] fizz in dilute hydrochloric acid (HCl) to produce carbon dioxide (CO_2) gas **(Fig. 3.6e)**; graphite makes a gray mark on paper; magnetite attracts iron objects **(Fig. 3.6f)**; halite tastes salty; and plagioclase has striations (thin parallel corrugations or stripes) on its surface.

CRYSTAL HABIT. The **crystal habit** of a mineral refers to the shape of a single euhedral crystal, or to the character of an aggregate of many well-formed crystals that grew together as

a group **(Fig. 3.6g)**. Crystal habit depends on the internal arrangement of atoms in the crystal and on whether or not the crystal grows in all directions at the same rate. A description of crystal habit generally includes adjectives that define the relative dimensions and the geometric shape of the crystal. For example, crystals that have roughly the same length in all directions are called blocky; those that are much longer in one dimension than in others are columnar or, in the extreme, needle-like. Similarly, specimens shaped like sheets of paper are platy, and those shaped like knives are bladed.

FRACTURE AND CLEAVAGE. Different minerals fracture (break) in different ways, depending on the internal arrangement of their atoms. If a mineral breaks to form distinct planar surfaces that have a specific orientation relative to the crystal structure, then we say that the mineral has **cleavage** and refer to each surface as a *cleavage plane*. Cleavage planes form in the directions where the bonds holding atoms together in the crystal are the weakest **(Fig. 3.7a–f)**. Some minerals have only one direction of cleavage. For example, mica has very weak bonds in one direction but strong bonds in the other two directions, so it splits easily into parallel sheets, each bounded by a cleavage plane. Other minerals have two or three directions of cleavage that intersect at a specific angle. In halite, for example, three directions of cleavage intersect at right angles, so halite crystals break into little cubes. Cleavage planes are not crystal faces, though sometimes the two types of surfaces can be mistaken for one another. By examining carefully, you can generally distinguish between cleavage planes and crystal faces—cleavage planes of the same orientation are repeated throughout the specimen, while crystal faces are not **(Fig. 3.7g)**. Minerals that have no cleavage at all break either by forming irregular fractures or by forming smoothly curving, clamshell-shaped **conchoidal fractures (Fig. 3.7h)**.

Classifying Minerals by Their Composition

THE PRINCIPAL CHEMICAL CLASSES. Figuring out a way to classify minerals—to group them in a meaningful way that emphasizes common features—wasn't easy because, at first glance, so many characteristics could conceivably provide a basis for classification. If you were given a tray of 20 random mineral specimens, you might try to classify them based on obvious physical characteristics such as color—but you would soon find that such a basis isn't useful, for many minerals with the same color differ from one another in important ways.

A Swedish chemist, Jöns Berzelius (1779–1848), proposed that minerals be grouped by similarities in their chemical composition. Mineralogists have found this approach helpful, and they now distinguish several principal *chemical classes* of minerals that differ from one another

FIGURE 3.7 The nature of mineral cleavage and fracture.

(a) Mica has one strong plane of cleavage and splits into sheets.

(b) Pyroxene has two planes of cleavage that intersect at 90°.

(c) Amphibole has two planes of cleavage that intersect at 60°.

Halite breaks into cubes.

(d) Halite has three mutually perpendicular planes of cleavage.

Calcite breaks into rhombs.

(e) Calcite has three planes of cleavage, one of which is inclined.

(f) Diamond has four planes of cleavage, each inclined to the others.

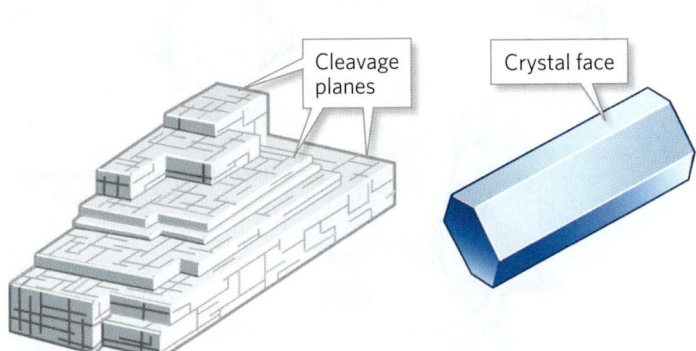

Cleavage planes

Crystal face

(g) How do you distinguish between cleavage planes and crystal faces? Cleavage planes can be repeated, whereas a crystal face is a single surface.

Irregular fracture

Crystal face

Conchoidal fracture

Quartz

Garnet

(h) Minerals without cleavage can develop irregular or conchoidal fractures.

in terms of the anion (negative ion) or anionic group (negatively charged molecule) that the minerals contain. Different minerals within a class differ from one another in terms of the cations (positively charged ions) that they contain or in terms of their crystal structures, or both. Some of the major mineral classes are outlined below.

- *Silicates:* All *silicate* minerals contain the SiO_4^{4-} anionic group. Examples of minerals in this class include quartz (SiO_2) and feldspars (such as $KAlSi_3O_8$). We will learn more about silicates in the next section.

- *Sulfides:* Sulfides consist of a metal cation bonded to a sulfide anion (S^{2-}). Examples include galena (PbS) and pyrite (FeS_2).

- *Oxides:* Oxides consist of metal cations bonded to oxygen anions. Typical oxide minerals include hematite (Fe_2O_3) and magnetite (Fe_3O_4).

- *Halides:* The anion in a halide is a halogen, such as chloride (Cl^-) or fluoride (F^-). Halite, or rock salt (NaCl), and fluorite (CaF_2), a source of fluoride, are common examples of halides.

- *Carbonates:* In carbonate minerals, the molecule CO_3^{2-} serves as the anionic group. Examples of carbonates include calcite ($CaCO_3$) and dolomite [$CaMg(CO_3)_2$].

- *Native metals:* Native metals consist of pure masses of a single metal. The metal atoms are bonded by metallic bonds. Copper and gold, for example, may occur as native metals.

FIGURE 3.8 The structure of silicate minerals.

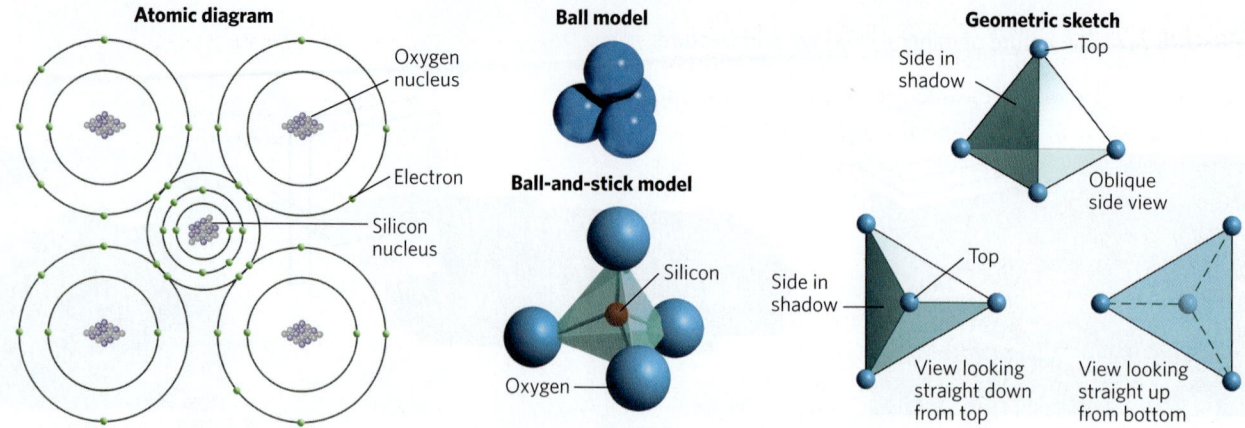

Atomic diagram

Oxygen nucleus

Electron

Silicon nucleus

Ball model

Ball-and-stick model

Silicon

Oxygen

Geometric sketch

Side in shadow

Top

Oblique side view

Side in shadow

Top

View looking straight down from top

View looking straight up from bottom

(a) The fundamental building block of a silicate mineral is the silicon-oxygen tetrahedron. Oxygen atoms occupy the corners of the tetrahedron, and a silicon atom lies at the center. Geologists portray the tetrahedron in a number of different ways.

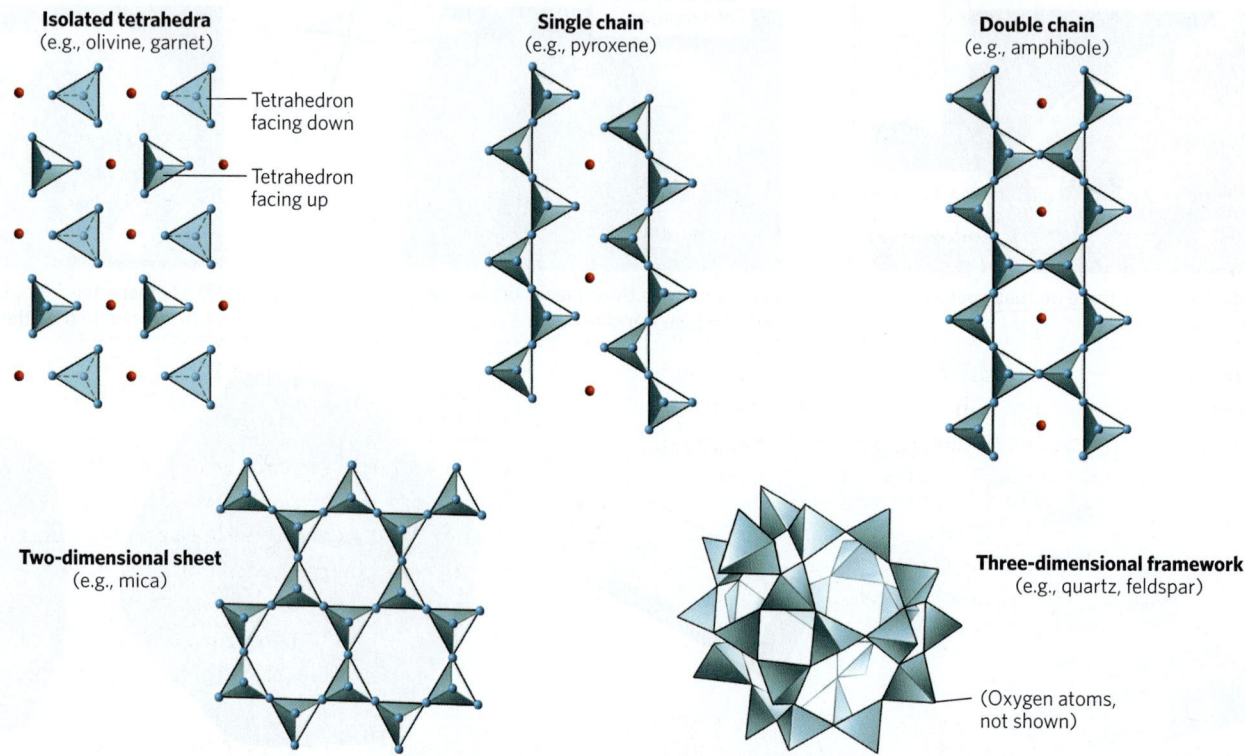

Isolated tetrahedra
(e.g., olivine, garnet)

Tetrahedron facing down

Tetrahedron facing up

Single chain
(e.g., pyroxene)

Double chain
(e.g., amphibole)

Two-dimensional sheet
(e.g., mica)

Three-dimensional framework
(e.g., quartz, feldspar)

(Oxygen atoms, not shown)

(b) The classes of silicate minerals differ from one another by the way in which the silicon-oxygen tetrahedra are linked. Where the tetrahedra link, they share an oxygen atom. Oxygen atoms are shown in blue. Positive ions (not shown) occupy spaces between tetrahedra.

- *Sulfates:* Sulfates consist of metal cations bonded to SO_4^{2-} anionic groups. Many sulfates form by precipitation out of water at or near the Earth's surface. Gypsum ($CaSO_4 \cdot 2H_2O$) forms by evaporation of saltwater.

THE SILICATE MINERALS. Of the various mineral classes, **silicate minerals** account for the largest proportion of the Earth. In fact, over 95% of rocks in the continental crust consist of silicate minerals, and the rocks that make up oceanic crust and the Earth's mantle consist almost entirely of silicate minerals. Internally, silicate minerals contain various arrangements of an anionic group called the **silicon-oxygen tetrahedron** (SiO_4^{4-}),

sometimes informally called the *silica tetrahedron*. Each silica tetrahedron consists of a silicon atom surrounded by four oxygen atoms, so it has a pyramid-like shape with four triangular faces **(Fig. 3.8a)**. In a silicate mineral, tetrahedra can bond together by sharing oxygen atoms.

Mineralogists distinguish among different groups of silicate minerals based on how the tetrahedra link together, which in turn determines the ratio of silicon to oxygen in the mineral **(Fig. 3.8b)**. In olivine, the tetrahedra do not share any oxygen atoms and are held together only by the attraction of cations, whereas in pyroxene, tetrahedra link to form a single chain; in amphibole, they link to form double chains; in mica, they link to form two-dimensional sheets; and in quartz or feldspar, they link to form a three-dimensional jungle gym–like network. (Notably, geologists distinguish between two types of feldspar based on composition: plagioclase and orthoclase.)

Take-home message . . .

We can distinguish among different minerals by physical properties such as luster, color, cleavage, hardness, and specific gravity. Minerals can be grouped into classes on the basis of their chemical composition. Silicate minerals are the most abundant minerals in the Earth.

Quick Question ————————————
What is the difference between a cleavage plane and a crystal face in a mineral?

3.4 Something Special: Gems

Mystery and romance follow famous **gems**, mineral specimens that are particularly beautiful or valuable and have been prepared in a way that enhances their beauty (**Fig. 3.9**). Some gems are unique minerals, but others are merely pretty and rare versions of more common minerals. For example, ruby is a special version of the mineral corundum, and emerald is a special version of

FIGURE 3.10 The process of creating facets on gems.

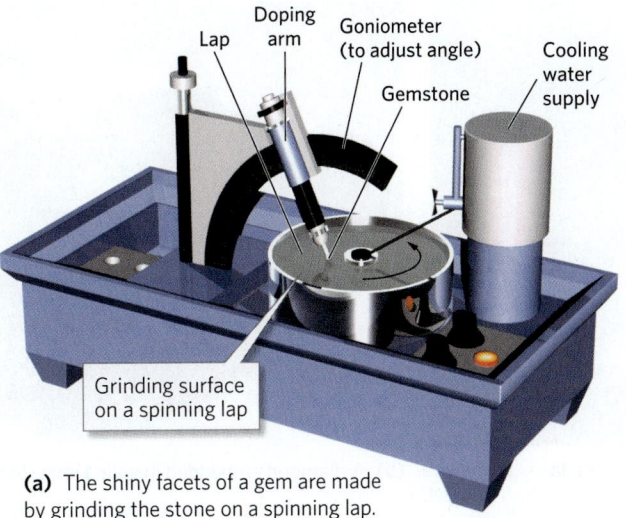

(a) The shiny facets of a gem are made by grinding the stone on a spinning lap.

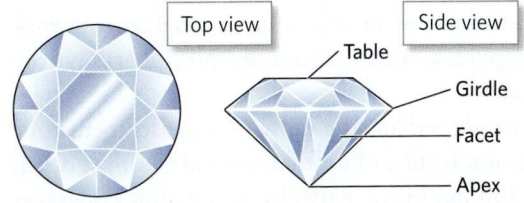

(b) There are many different "cuts" for a gem. Here we see the top and side views of a brilliant-cut diamond with 57 facets.

Did you ever wonder . . .

how gems in jewelry get all those shiny faces?

FIGURE 3.9 Gem-quality versions of corundum (Al_2O_3) can be fashioned into gems of many shapes and colors. Ruby is red corundum. Typical sapphires are blue corundum.

the mineral beryl. The beauty of a gem lies in its color and, in the case of transparent gems, its "fire"—the way that light bends and reflects within the gem. The quality of a transparent gem depends on its color as well as on the concentration of *inclusions* (tiny flecks of other minerals, or of bubbles filled with air or water). A gem containing many inclusions won't be clear and will probably have less value.

The gems used in most jewelry are cut stones, meaning that they have many smooth, shiny faces that form sharp angles with their neighbors. The smooth faces, technically known as **facets**, are not natural cleavage planes, nor are they crystal faces. Rather, they are produced when gem cutters carefully grind and polish specimens by using a faceting machine (**Fig. 3.10**). To produce a facet, the gem cutter sticks a gemstone to the end of a *doping stick* and then holds it against a spinning grinding plate called a *lap* until the facet is complete. Then, the gem cutter rotates the gemstone by a specific angle and grinds away another face.

A *gemstone*, a mineral specimen that will be a gem when cut and polished, can form in many ways, like any other mineral. Some solidify from a melt, some form by diffusion, some precipitate from a water solution, and some are a consequence of the chemical interaction of rock with water near the Earth's surface.

See for yourself

Kimberley Diamond Mine

Latitude: 28°44'17.06" S
Longitude: 24°46'30.77" E

Zoom to an elevation of 13 km (~8 mi) and look straight down.

The field of view shows the town of Kimberley, South Africa, and its inactive diamond mine. The mine looks like a circular pit. You can also see the tailings pile of excavated rock debris.

FIGURE 3.11 Where diamonds come from.

(a) Surface diamond mines in northern Canada.

(b) A diamond embedded in solid kimberlite.

(c) Sifting for diamonds in a streambed.

See for yourself

Ekati Diamond Mine, Canada

Latitude: 64°43′15.44″ N
Longitude: 110°36′56.27″ W

Zoom to an elevation of 100 km (~62 mi) and look straight down.

The Ekati Diamond Mine is in a remote, largely uninhabited region of the Northwest Territories of Canada. Prospectors found diamonds here in the early 1990s after a 20-year search. The mine opened in 1998 and within 10 years had produced more than 40 million carats (8,000 kg) of diamonds. Zoom down to 10 km (~6 mi) to see details of the mining operation.

Some gemstones occur only in an unusual igneous rock called *pegmatite*, which forms by precipitation from steamy melts.

Diamond, perhaps the most widely used gemstone today, grows from carbon that was carried to a depth of over 100 km in the Earth by subduction. The great pressure at such depth causes carbon atoms to arrange themselves into a compact and very strong mineral, diamond. (At shallow depths, carbon crystalizes into graphite.) In fact, diamond is the hardest mineral known. Geologists speculate that the diamonds return to the surface when rifting thins the crust and causes some of the underlying mantle to melt, and the rising magma carries the diamonds up with it. Solidification of diamond-containing magma produces a special kind of igneous rock called *kimberlite* (named for Kimberley, South Africa, where it was first found). Diamonds in solid kimberlite can be obtained by digging up the kimberlite and crushing it **(Fig. 3.11a, b)**. But nature can also break diamonds free from the Earth, and these diamonds remain as solid grains in river gravel. Such diamonds are obtained by separating them from deposits of the gravel **(Fig. 3.11c)**.

Take-home message . . .

Gemstones are particularly rare and beautiful minerals. The gems found in jewelry have been faceted using a lap. The facets are not natural crystal faces or cleavage surfaces. The fire of a gem comes from the way it reflects light internally.

Quick Question

What's the difference between a facet on a gem and a crystal face?

3.5 Introducing Rocks

What Is a Rock?

In the mid-19th century, as populations along the west coast of the United States began to grow, railroad companies decided that the time had come to complete a transcontinental railroad. They brought in thousands of workers who faced death all too frequently as they blasted and chiseled their way through the Sierra Nevada range of California to make a path for the tracks. If you had asked those workers to define a rock, they might have responded with the obvious: a rock is a hard, heavy, solid mass that's difficult to dig into. Geologists use a more precise definition: a **rock** is a coherent, naturally occurring solid consisting of an aggregate of mineral grains or, less commonly, a mass of glass. To clarify this definition, let's look at its components:

- *Coherent:* A rock holds together as a solid mass. It's not a pile of loose grains.

- *Naturally occurring:* Rocks form only during geologic processes. Manufactured materials, such as concrete and brick, are not rocks, although they contain materials similar to those in rocks.

- *Aggregate of mineral grains or a mass of glass:* The vast majority of rocks consist of an aggregate, or collection, of many mineral grains that are attached to one another. Some rocks contain only one kind of mineral, whereas others contain several different kinds. A few rock types, formed only at volcanoes, consist of glass (see Chapter 4).

Why does a rock hold together? Most rocks are a coherent mass either because **cement**, a binder composed of mineral crystals that precipitate from water in the space

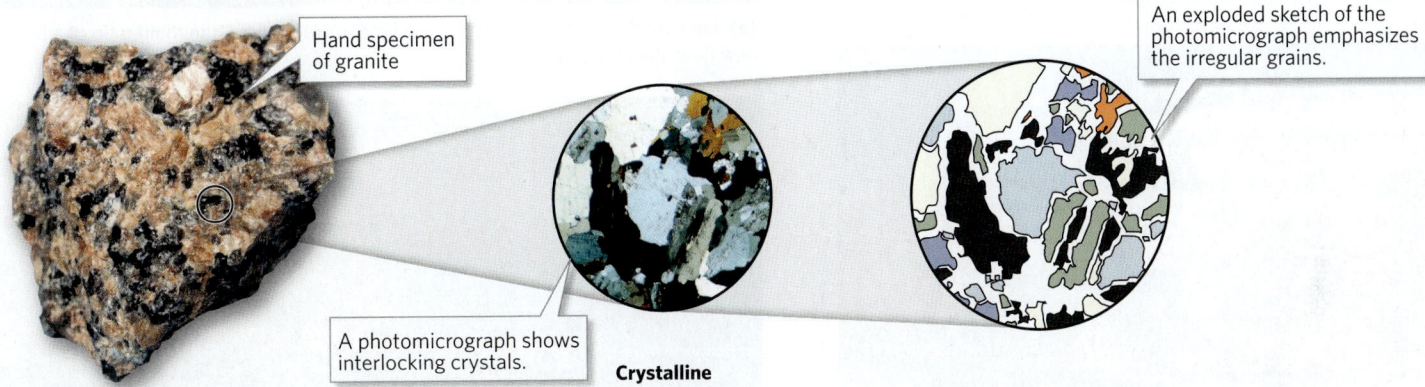

Hand specimen of sandstone

A photomicrograph shows grains held together by cement.

An exploded sketch of the photomicrograph distinguishes the grains from the cement.

Cement

Sand grain

Clastic

FIGURE 3.12 The two different textures of rock.

(a) Clastic texture is illustrated by the grains and cement in limestone.

Hand specimen of granite

An exploded sketch of the photomicrograph emphasizes the irregular grains.

A photomicrograph shows interlocking crystals.

Crystalline

(b) Crystalline texture is illustrated by the interlocking crystals in a granite.

FIGURE 3.13 What is bedrock, and what isn't? The loose boulders on the ground at the base of this cliff in Utah are not bedrock, but the top of the cliff behind them is.

between grains, holds the grains together **(Fig. 3.12a)**, or because grains in the rocks grew to interlock with one another, like pieces in a jigsaw puzzle **(Fig. 3.12b)**. Rocks made of grains held together by cement have a **clastic texture**, while rocks whose grains grew to interlock with one another have a **crystalline texture**. Rocks made of glass hold together because they originated as a continuous mass (meaning that they have no separate grains) or because they consist of glass grains that welded together while hot.

At the surface of the Earth, rock occurs either as **bedrock**, meaning rock that's still attached to the Earth's crust below, or as loose pebbles, cobbles, blocks, or boulders that have broken free from bedrock and have moved from their point of origin **(Fig. 3.13)**. Geologists refer to an exposure of bedrock as an **outcrop**. An outcrop may be a knob out in a field or in the woods, a cliff or ridge along a mountain, or the ledges that form the wall or bed of a stream **(Fig. 3.14a)**. In the past few centuries, people have created outcrops by carving road cuts, rail cuts, and other excavations **(Fig. 3.14b)**. To people who live in cities or forests or on farmland, bedrock outcrops may be unfamiliar because vegetation, sand, mud, gravel, soil, water, asphalt, concrete, or buildings cover them. But in

Did you ever wonder . . .

if the brick used in building your house is a kind of rock?

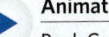 **Animation**

Rock Groups: Definition

FIGURE 3.14 Common types of outcrops.

(a) A stream cut near Catskill, New York, exposes rock that would otherwise be hidden by trees.

(b) To keep the slope (grade) of this highway gentle, engineers cut an artificial canyon through rock west of Denver.

(c) Natural cliff exposures on a mountain face near Salt Lake City, Utah. The cliff slopes are too steep for trees to take root.

FIGURE 3.15 The three basic rock groups.

Igneous

(a) Lava (molten rock) freezes to form igneous rock. Here, the molten tip of a brand-new flow still glows red.

Sedimentary

(b) Sand, formed from grains eroded from the cliffs above, collects on the beach. If buried, and turned to rock, it becomes layers of sandstone, like those making up the cliff.

Metamorphic

(c) Metamorphic rock forms when pre-existing rocks endure change in environmental conditions that cause new minerals and textures to form.

mountainous areas, outcrops may be nearly everywhere (Fig. 3.14c).

The Basics of Rock Classification

Beginning in the 18th century, geologists struggled to develop a sensible way to classify rocks. Eventually, it became clear that a *genetic scheme*—a scheme that focuses on the origin (genesis) of rocks—provides the best approach for classifying rocks, and this is the approach that we continue to use today.

Using the modern genetic classification of rocks, geologists recognize three basic rock groups:

1. **Igneous rocks** form by the freezing (solidification) of molten rock, or *melt* (Fig. 3.15a). The process of solidification can take place underground or at the Earth's surface.

2. **Sedimentary rocks** form either by the cementing together of grains broken off pre-existing rocks or by the precipitation of mineral crystals out of water solutions at or near the Earth's surface (Fig. 3.15b). The material making up sedimentary rocks is called *sediment*.

3. **Metamorphic rocks** form when pre-existing rocks undergo changes in response to a modification of their environment, without first melting. For example, metamorphism takes place in response to the rise in temperature and pressure that happens when rock that was near the Earth's surface ends up deep beneath a mountain belt. Metamorphism produces new minerals and new arrangements of crystals (Fig. 3.15c).

Each of the three basic rock groups contains many different individual rock types, each of which has a name. Some of these names may be familiar to you (such as granite and sandstone), but some may not (such as peridotite and arkose). Rock names come from various sources—some reflect the dominant component making up the rock, some indicate a region where the rock was first discovered, some derive from a root word of Latin origin, and some come from local traditional names. All told, there are hundreds of different rock names. In this book, we introduce only about 30 more common names.

Fundamental Characteristics of Rocks

In order to describe and identify rocks, geologists not only need to identify the minerals within a rock, but also to characterize the overall physical properties of the rock. Key characteristics of rocks include the following:

- *Grain size and shape:* In some rocks, the grains are so small that they can't be seen without a microscope,

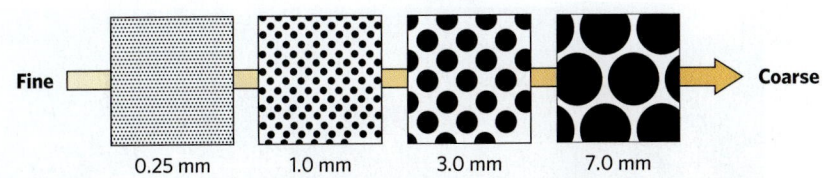

FIGURE 3.16 Describing grains in rock.

(a) Geologists define grain size by using a comparison chart like this one.

Fine | 0.25 mm | 1.0 mm | 3.0 mm | 7.0 mm | Coarse

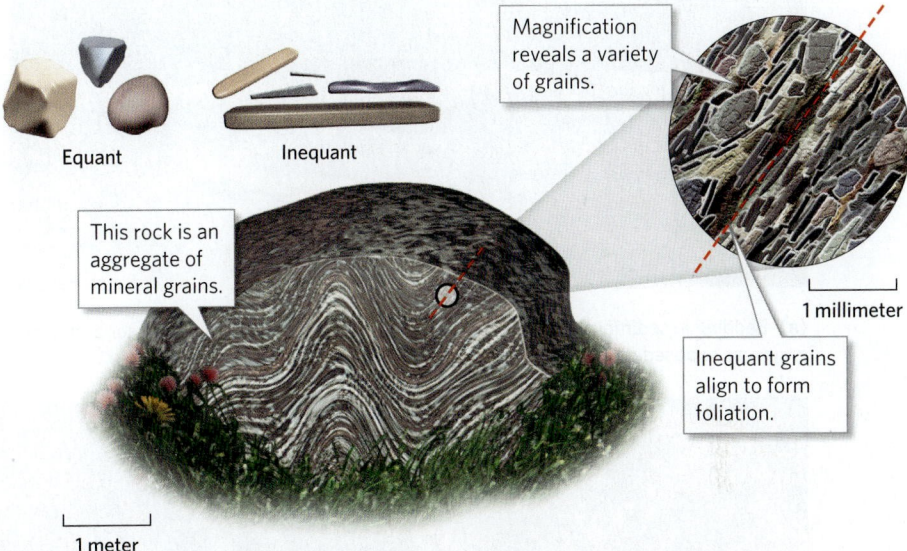

Magnification reveals a variety of grains.

Equant   Inequant

This rock is an aggregate of mineral grains.

1 millimeter

Inequant grains align to form foliation.

1 meter

(b) Grains in rock come in a variety of shapes. Some are equant, whereas others are inequant. In this example of metamorphic rock, inequant grains align to define metamorphic foliation.

whereas in others, they may be bigger than a fist (Fig. 3.16a). In some rocks the grains are **equant**, meaning that they have similar dimensions in all directions, whereas in others they are **inequant**, meaning that their dimensions are not the same in all directions (Fig. 3.16b). And in some rocks, all the grains have the same size, whereas other rocks contain a variety of different-sized grains.

- *Composition:* The term *composition* refers to the proportions of different chemicals making up a rock. A rock's composition depends on the minerals it contains because minerals differ from one another in terms of the chemicals they contain.

- *Texture:* The term **texture** refers to both the arrangement of grains in a rock (the way grains connect to one another) and the degree to which grains are equant or inequant. In rocks containing inequant grains, texture also refers to the degree to which the grains align parallel to one another. As we've seen, in rocks with a clastic texture, grains are cemented to one another, whereas in rocks with a crystalline texture, grains interlock with one another.

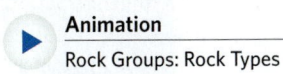

Animation
Rock Groups: Rock Types

FIGURE 3.17 Layering in rock.

(a) Bedding in sedimentary rock, here defined by alternating layers of coarser and finer grains, as exposed on a cliff in northwestern China.

Foliation plane

(b) Foliation in this outcrop of metamorphic rock near Mecca, California, is defined by alternating light and dark layers. The color of each layer depends on the minerals it contains.

- *Layering:* Some rock bodies appear to contain distinct layering, defined either by bands of different compositions or textures, or by the alignment of inequant grains. Layering in sedimentary rocks is called *bedding* **(Fig. 3.17a)**, whereas layering in metamorphic rocks is called *metamorphic foliation* **(Fig. 3.17b)** (see also Fig. 3.16b).

Studying Rocks

Ideally, the study of a rock begins at the outcrop. If the outcrop is big enough, you'll be able to see layering, and you can study relationships between the rock you're interested in and other rocks around it **(Fig. 3.18a)**. To gain insight into the composition and texture of the rock, you may need to break off a **hand specimen**, a fist-sized

FIGURE 3.18 Studying rocks in the field.

(a) A field geologist examining a hand specimen (on a cold day).

(b) A hand specimen.

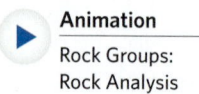

Animation

Rock Groups: Rock Analysis

piece that you can look at more closely with a magnifying glass **(Fig. 3.18b)**. Such observations enable you to identify individual mineral grains and see connections between grains.

In rocks that have very small grains, you may need to prepare a **thin section**, a very thin slice (about 0.03 mm thick, the thickness of a human hair) glued to a glass slide, from a hand specimen **(Fig. 3.19a)**. By placing the thin section beneath the lens of a *petrographic microscope* **(Fig. 3.19b)**, a special microscope used to study rocks, you can see grains at high magnification. A petrographic microscope differs from an ordinary microscope in that it illuminates the thin section with transmitted *polarized light*. This means that the illuminating light consists of light waves aligned in the same direction, and that the light passes up through the thin section from underneath. When viewed with this type of lighting, each type of mineral displays a unique suite of colors **(Fig. 3.19c)**. The specific color the observer sees depends on the specific mineral making up the grain, the orientation of the crystal lattice relative to the light beam, and the thickness of the grain. The brilliant colors and strange shapes in a thin section rival the beauty of stained glass. To record your observations, you can obtain a **photomicrograph**, a photograph taken through the lens of the microscope.

Since the 1950s, new high-tech electronic instruments have enabled geologists to examine rocks on an even finer scale than is possible with a petrographic microscope. Modern research laboratories may use *electron microprobes*, which focus a beam of electrons on a small part of a grain and reveal the chemical composition of the mineral **(Fig. 3.20)**; *mass spectrometers*, which analyze the proportions of different types of atoms in a sample; and *X-ray diffractometers*, which help identify minerals by detecting how an X-ray beam interacts with crystal structures.

Take-home message . . .

Geologists classify rocks into three groups—igneous, sedimentary, and metamorphic—based on the way the rock forms. Each group includes many different kinds of rocks, distinguished from one another by physical characteristics. You can study rocks both at outcrops in the field and by taking samples back to the lab for analysis with petrographic microscopes and high-tech instruments.

Quick Question

Considering the definition of a rock, why isn't sand a rock? Why isn't concrete?

FIGURE 3.19 Studying rocks in thin section.

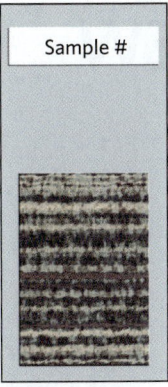
Sample #

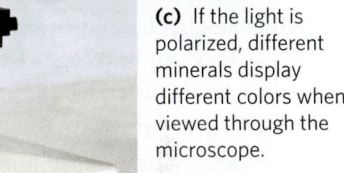
500 μm

(a) The geologist glues a chip of rock to a glass slide and grinds it down until it is so thin that light can pass through it.

(b) With a petrographic microscope, it's possible to view thin sections with light that shines through the sample from below.

(c) If the light is polarized, different minerals display different colors when viewed through the microscope.

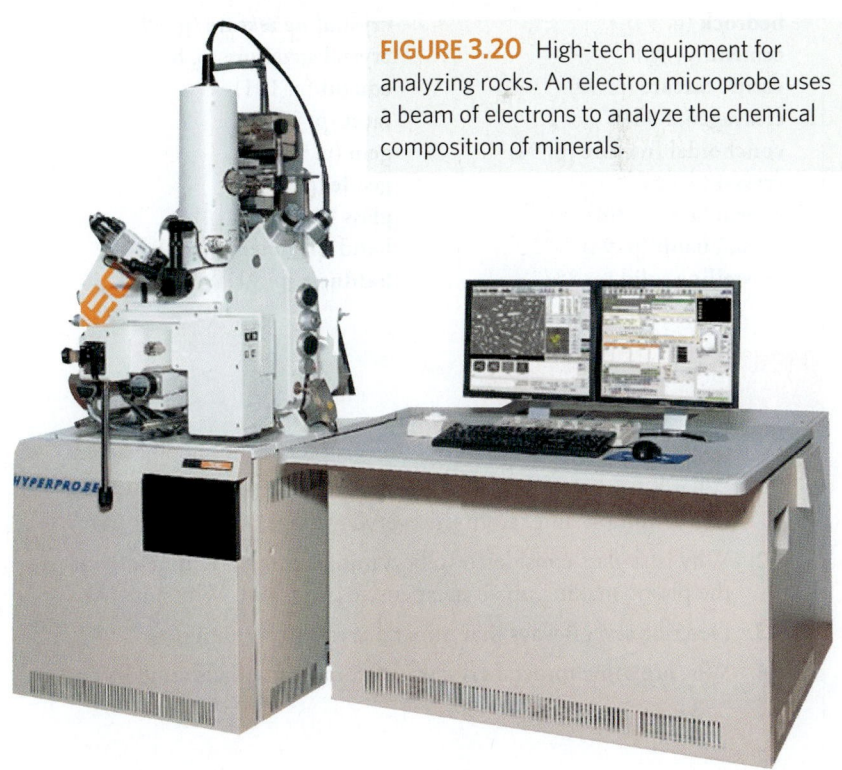

FIGURE 3.20 High-tech equipment for analyzing rocks. An electron microprobe uses a beam of electrons to analyze the chemical composition of minerals.

3 CHAPTER REVIEW

Chapter Summary

- Minerals are naturally occurring, homogeneous crystalline solids formed by geologic processes. They have a definable chemical composition and a crystal structure characterized by an orderly arrangement of atoms, ions, or molecules. Most minerals are inorganic.

- The crystal structure of a mineral is defined by the orderly arrangement of atoms in a specific pattern.

- Minerals can form by solidification of a melt, precipitation from a water solution, diffusion through a solid, metabolism of organisms, or precipitation from a gas.

- Mineralogists have identified about 4,000 different types of minerals. They can be recognized by their physical properties, such as color, streak, luster, hardness, specific gravity, crystal habit, cleavage, magnetism, and reactivity with acid.

- Minerals can be grouped into classes (such as silicates, oxides, sulfides, and carbonates) based on their chemical composition.

- Silicate minerals, the most common minerals on the Earth, consist of networks of silicon-oxygen tetrahedra. Groups of silicate minerals can be distinguished from one another by the way in which the tetrahedra link together.

- Gemstones are minerals known for their beauty and rarity. The facets on cut gems used in jewelry are made by grinding and polishing the stones with a faceting machine.

- Rocks are coherent, naturally occurring aggregates of minerals or masses of glass. They can be classified into three groups—igneous, sedimentary, and metamorphic—based on how they formed.

- Each rock group contains many different kinds of rocks, each with a name. Rocks can be identified based on physical characteristics such as grain size, texture, and composition.

- Rocks are exposed in outcrops, from which they can be collected for detailed studies that use petrographic microscopes and a variety of high-tech analytic equipment.

Key Terms

bedrock (p. 99)
cement (p. 98)
clastic texture (p. 99)
cleavage (p. 94)
conchoidal fracture (p. 94)
crystal (p. 86)
crystal face (p. 86)
crystal habit (p. 94)
crystalline solid (p. 85)

crystalline texture (p. 99)
crystal structure (p. 89)
equant (p. 101)
facet (p. 97)
gem (p. 97)
geode (p. 90)
glass (p. 86)
hand specimen (p. 102)
hardness (p. 91)

igneous rock (p. 101)
inequant (p. 101)
luster (p. 91)
metamorphic rock (p. 101)
mineral (p. 85)
Mohs hardness scale (p. 91)
organic chemicals (p. 86)
outcrop (p. 99)
photomicrograph (p. 103)

rock (p. 98)
sedimentary rock (p. 101)
silicate mineral (p. 96)
silicon-oxygen tetrahedron (p. 96)
specific gravity (p. 94)
streak (p. 91)
symmetry (p. 89)
texture (p. 101)
thin section (p. 103)

Review Questions

Letters in parentheses correspond to the chapter's learning objectives.

1. What is a mineral, as geologists understand the term? How does this definition differ from the everyday usage of the word? **(A)**

2. Why isn't glass considered to be a mineral? Salt is a mineral, but the plastic making up an inexpensive pen is not. Why not? **(A)**

3. Describe several ways that mineral crystals can form. **(A)**

4. Why are some mineral crystals euhedral, and others anhedral? **(A)**

5. How can you determine the hardness of a mineral? What's the Mohs hardness scale? **(B)**

6. List and define the principal physical properties used to identify a mineral. Which mineral reacts with acid to produce CO_2, as shown in the figure? **(B)**

7. What characteristic provides the basis for separating minerals into classes? **(C)**

8. How do you distinguish cleavage planes from crystal faces on a mineral? Which do each of the accompanying pieces of art show? Explain how each type of surface forms. **(B)**

9. What is a silicon-oxygen tetrahedron? On what basis do mineralogists organize silicate minerals into distinct groups? **(C)**

10. Why are some minerals considered gemstones? How do jewelers make the facets on a gem? **(C)**

11. What holds the grains together in a rock? Identify the textures shown in the accompanying figures. **(E)**

12. On what basis can you distinguish rock from loose sediment? **(D)**

13. On what basis do geologists classify rocks into three groups? What are those groups? **(F)**

14. What is an outcrop? Provide examples of several different types of outcrops. **(E)**

15. What tools can geologists use to study rocks in the field or lab? **(D)**

On Further Thought

16. Imagine that you find two milky-white crystals, each about 2 cm across. One consists of plagioclase and the other of quartz. How can you determine which is which? **(B)**

17. Could you use crushed calcite to grind facets on a diamond? Why or why not? **(B)**

18. You are an architect who has been hired to build a decorative plaza of rock. Heavy traffic will be driving over the plaza frequently, so the rock needs to be very durable. Which would make the strongest and most durable surface for the plaza, limestone or granite? (Hint: Think about the minerals that make up these rocks. Limestone contains calcite and granite contains quartz and feldspar.) **(B)**

ANOTHER VIEW Calcite crystals peak through a window in a geode.

4 VOLCANISM AND IGNEOUS ROCKS

By the end of the chapter you should be able to . . .

A. explain why melting happens inside the Earth and how molten rock becomes igneous rock.

B. differentiate between extrusive and intrusive igneous rock, and describe types of igneous intrusions.

C. recognize different kinds of igneous rocks, explain why they exist, and discuss how to identify and interpret them.

D. distinguish among the various kinds of volcanic eruptions, explain why they differ, and describe the materials produced by a volcano.

E. interpret the hazards that develop during or after a volcanic eruption.

F. evaluate how we can predict and mitigate volcanic hazards.

G. relate occurrences of igneous activity and volcanism to plate tectonics.

4.1 Introduction

For many years, the Roman town of Pompeii prospered at the foot of Mt. Vesuvius, in what is now southern Italy. Pompeiians thought that Vesuvius, which at the time towered 3 km (10,000 ft) above the sea, was just another scenic mountain. They were wrong, very wrong, for Mt. Vesuvius is a **volcano**. Geologists use the term volcano both for a *vent* (opening) from which **melt** (molten rock), fragments of solidified melt, and gas emerge from below a planet's surface, and for a hill or mountain built from the materials that come out of a vent. For several weeks, in the spring of 79 C.E., earthquakes had jolted Pompeii with unnerving frequency, and Vesuvius had grumbled like distant thunder. But people browsing in the town's markets paid little heed until 1:00 P.M. on August 24, when Vesuvius suddenly roared, and a dark, mottled cloud churned out of its summit. This was no normal cloud. Instead of just steam, it contained sulfurous gas, *ash* (composed of tiny flakes of glass and pulverized rock), and marble-sized rock fragments. As lightning sparked in its crown, the cloud spread over Pompeii, turning day into night. Choking fumes filled the air, ash sifted down like heavy snow, and rock fragments fell like hail. Panic ensued as inhabitants rushed to escape. Sadly, for most it was too late. Vesuvius suddenly exploded, blasting out immensely more hot gas and debris. Much of this material swirled high into the atmosphere, but some rushed downslope in a scalding avalanche of hot ash that swept over and buried Pompeii minutes later (Fig. 4.1a).

The blanket of volcanic debris that spread over Pompeii protected the ruins of the town so well that 18 centuries later, archaeologists excavating the debris found an amazingly complete record of daily life in the Roman Empire (Fig. 4.1b). Occasionally, they also found open spaces in the debris that, when filled with plaster, provided casts of Pompeii's inhabitants, twisted in agony or huddled in despair, at their moment of death (Fig. 4.1c).

The danger posed by volcanoes like Mt. Vesuvius is hardly a thing of the past. In 2018, a major eruption took place on the *flank* (side) of Kilauea, a volcano on the Big Island of Hawaii. During this event, known as the lower Puna eruption, glowing red molten rock rose from depth and spurted or seeped out of vents (Fig. 4.2a). These vents fed a stream of lava, called a *lava flow*. In places where lava reached the shore, it spilled into the sea, instantly

<< The drama of an eruption—in this case, at the peak of the Santiaguito Volcano, in Guatemala. Behind a cloud of ash, molten rock that has risen from the mantle spurts out explosively from the summit.

FIGURE 4.1 The eruption of Vesuvius buried Pompeii in 79 C.E.

(a) A painting depicting the massive eruption.

You can see ruts carved into streets by chariots.

(b) Streets and buildings of Pompeii are well preserved.

(c) Plaster casts of unfortunate victims who were buried beneath volcanic ash and debris.

107

FIGURE 4.2 Lava flows from the 2018 eruption of Kilauea, Hawaii.

(a) During the lower Puna eruption, lava spurted out of a vent and flowed through a neighborhood.

(b) Lava spilling into the sea causes water to instantly turn into steam.

Glowing waves rise and
flow, burning all life on
their way, and freeze
into black, crusty rock
which . . . builds the land,
thereby adding another
day to the geologic past . . .
I became a geologist
forever, by seeing with
my own eyes—the Earth
is alive!

—*HANS CLOOS (GEOLOGIST,
1886–1951), ON SEEING AN
ERUPTION OF MT. VESUVIUS*

turning the water to billows of steam **(Fig. 4.2b)**. Sadly, the 2018 flow not only burned through forests and fields, but also covered houses and roads in its path. As the lava flow cooled, its surface darkened and formed a hardened rind. Eventually, the melt cooled completely and became new igneous rock.

What happened on that fateful day in Pompeii in 79 C.E. and more recently in Hawaii? Both regions experienced a violent **volcanic eruption**, defined as an event during which molten rock flows or sprays, or solid debris blasts, out of a vent. The ancient Romans thought that eruptions happened when Vulcan, the god of fire, fueled his subterranean forges to manufacture weapons. No one believes the Roman myth anymore, but the god's name has been immortalized as the root of the English word *volcano*. Research demonstrates that volcanic eruptions are a manifestation of **igneous activity**, the name geologists use for the overall process during which rock deep underground melts, producing molten rock that rises up into the crust and, in some cases, makes it all the way to the Earth's surface. Geologists refer to molten rock underground as **magma** and to molten rock erupted at the Earth's surface at a volcano as **lava**.

Any rock whose formation involves solidification of a melt is an **igneous rock**. Considering the fiery heat of the melt from which igneous rocks solidify, the name *igneous*, from the Latin *ignis*, meaning fire, makes sense. Igneous rocks can be found in many places, for they make up the entire oceanic crust and much of the continental crust. It may seem strange to speak of "freezing" in regard to rock formation. Most people think of freezing as a change from liquid water to solid ice at a temperature below 0°C (32°F). Nevertheless, the transformation of a magma or lava into an igneous rock represents the same

phenomenon: cooling a liquid until its molecules lock into place so it transforms, overall, into a solid. In contrast to water, however, molten rock freezes at high temperatures. For example, the Hawaiian lava started to solidify at about 1,100°C (2,000°F)—to put this temperature in perspective, keep in mind that your home oven reaches a top temperature of only 260°C (500°F).

In this chapter, we examine the process, products, and consequences of igneous activity. We begin with an explanation of why rocks melt inside the Earth to produce magma in the first place, followed by a description of various types of igneous rocks and how characteristics of these rocks provide clues to the environment in which the rocks solidified. Next, we focus on volcanic eruptions themselves, and we discuss how human society can be affected by them. We then conclude this chapter by relating igneous activity to plate tectonics.

4.2 Melting and the Production of Magma

Causes of Melting Inside the Earth

Popular media give the impression that the mantle beneath the Earth's crust consists entirely of molten rock, and that igneous activity happens whenever a conduit connects this "underground magma sea" to the Earth's surface. This image is wrong! The crust and mantle both consist mostly of solid rock. You can see why by examining a graph that shows how the **geotherm**, the line representing how temperature changes with depth in the Earth, compares with the *melting curve*, the line representing the pressure and temperature conditions at which melting starts **(Fig. 4.3a)**. Rock can remain

FIGURE 4.3 Decompression melting.

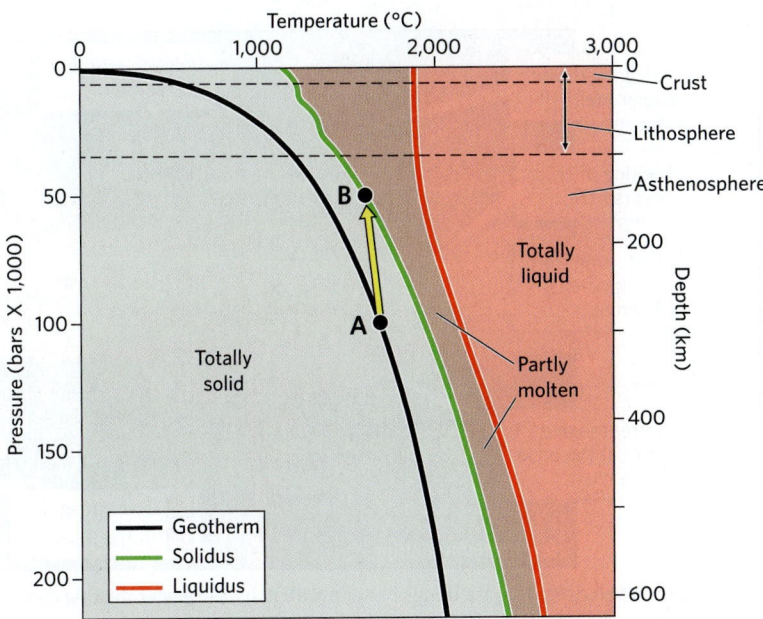

(a) Decompression takes place when the pressure acting on hot rock decreases. As this graph of pressure and temperature conditions in the Earth shows, when rock rises from point A to point B, the pressure decreases a lot, but the rock cools only a little, so the rock begins to melt. The *solidus* line indicates conditions at which melt begins to form, and the *liquidus* line indicates conditions at which the rock melts completely.

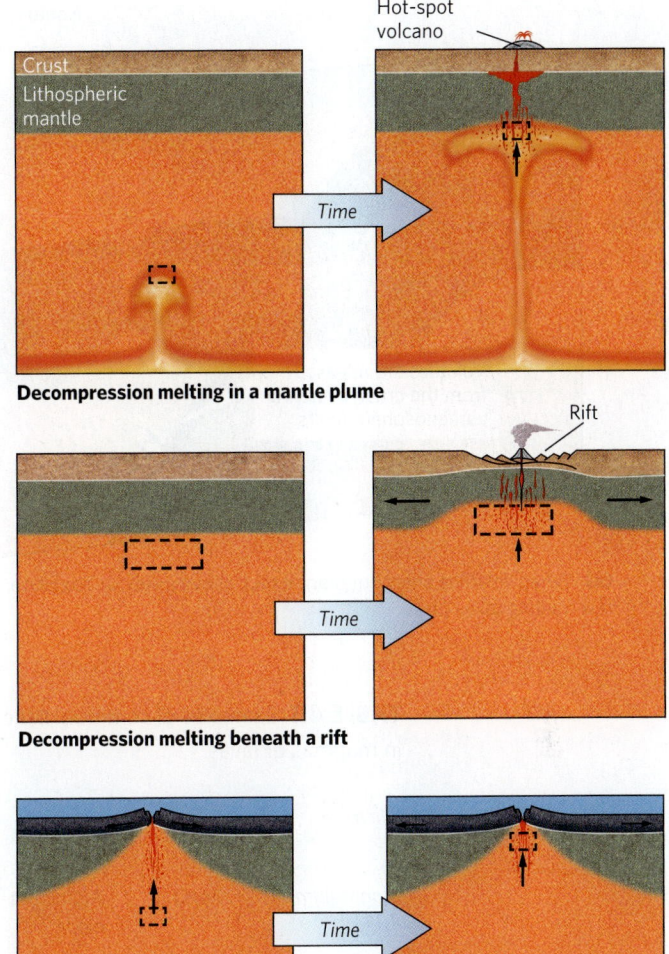

Decompression melting in a mantle plume

Decompression melting beneath a rift

Decompression melting beneath a mid-ocean ridge

(b) The conditions leading to decompression melting occur in several different geologic environments. In each case, a volume of hot asthenosphere (outlined by dashed lines) rises to a shallower depth, and magma (red dots) forms.

solid at great depth in the Earth, despite the high temperatures, because pressure also increases with depth in the Earth. Why? While high temperature causes atoms to vibrate faster, working to break free of crystals in rock to form a liquid, high pressures have the affect of holding atoms together and preventing them from breaking free, keeping crystals solid. Yet magma does form in some locations. Why? Melting of pre-existing rock to produce magma takes place in response to local changes in temperature, pressure, or chemical composition. Let's look at each of these causes individually.

- *Melting due to a decrease in pressure:* At lower pressures, atoms are held together less tightly, so bonds between them break more easily. Therefore, if the pressure squeezing very hot rock decreases, melting may begin. Such **decompression melting** happens when hot rock from deeper in the mantle rises to shallower depths without cooling significantly **(Fig. 4.3b)**.

- *Melting due to the addition of volatiles:* Volatile materials, such as H_2O and CO_2, are compounds that can easily exist as gases near the Earth's surface. When volatiles seep into very hot rock, the rock begins to melt because these compounds promote the separation of molecules from crystals. Geologists refer to this process as **flux melting (Fig. 4.4a)**; in this context, a *flux*

is a chemical that, when added, lowers another material's melting temperature.

- *Melting due to heat transfer:* Not all rock types start to melt at the same temperature. For example, mantle rock begins to melt at about 1,200°C (2,200°F), whereas typical continental crustal rock begins to melt at about 700°C (1,300°F). Therefore, magma rising from the mantle can be much hotter than the melting temperature of continental crust it passes through. As a result, heat from the magma can increase the temperature of surrounding crustal rocks enough to cause them to melt **(Fig. 4.4b)**. To picture the process, imagine injecting hot fudge into a ball of ice cream—the fudge raises the temperature in the ice cream and causes it to melt. We call this process **heat-transfer melting**.

Did you ever wonder . . .

where the molten rock erupting from a volcano comes from?

FIGURE 4.4 Flux melting and heat-transfer melting.

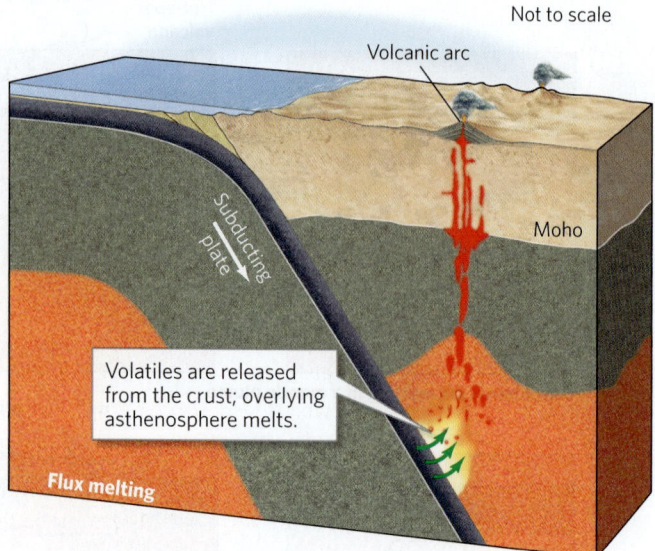

Not to scale

(a) When volatiles enter hot mantle rock above a subducting plate, flux melting takes place.

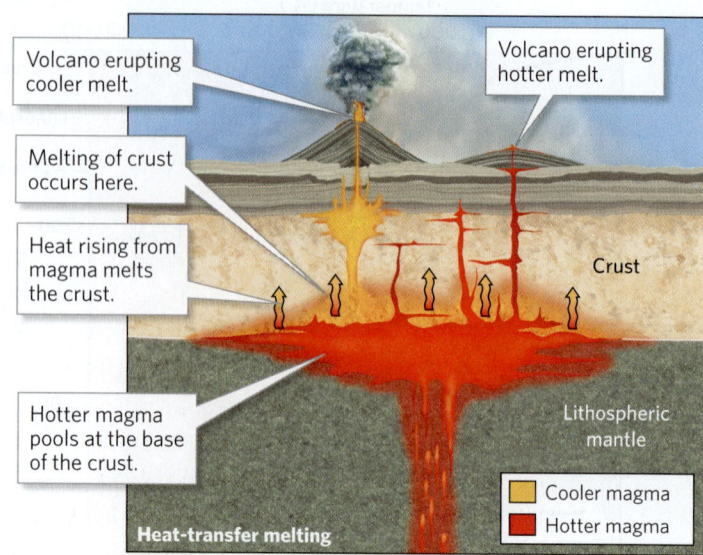

(b) When rising magma brings heat up with it, heat-transfer melting can take place in the overlying or surrounding rock.

TABLE 4.1 Examples of common oxide compounds in magmas or lavas

Silica	SiO_2	Iron oxide	FeO or Fe_2O_3
Magnesium oxide	MgO	Aluminum oxide	Al_2O_3
Sodium oxide	Na_2O	Calcium oxide	CaO

Chemical Types of Magma and Lava

Video
Partial Melting

Animation
Lava Composition
(Interactive)

Water, a liquid that you're very familiar with, consists entirely of one chemical, H_2O. In contrast, magma or lava typically contains several different chemicals in varying proportions. We can represent these chemicals, simplistically, as *oxide compounds*, whose molecules consist of metal atoms bonded to oxygen **(Table 4.1)**. Not all melts have the same composition, meaning that

TABLE 4.2 The four major types of magma

Type	Ratio of silica to iron and magnesium oxide	Silica (%)	Temperature (°C)	Viscosity
Felsic	High	66–76	700	High
Intermediate		52–66	900	
Mafic		45–52	1,100	
Ultramafic	Low	38–45	1,300	Low

different melts contain different proportions of different oxides. Geologists distinguish four major types of melts—*felsic, intermediate, mafic, and ultramafic*—by the proportion of silica relative to the sum of iron and magnesium oxide that the molten rock contains **(Table 4.2)**. Felsic melts contain the most silica and ultramafic ones contain the least. The specific composition of a molten rock found at a given location depends on many factors **(Box 4.1)**.

Significantly, the proportion of silica in molten rock affects the temperature at which the melt can remain liquid, as well as the melt's **viscosity**, or ease of flow (see Table 4.2). For example, felsic melts can still be liquid at temperatures hundreds of degrees lower than those at which ultramafic melts will already have become solid. And felsic melts have high viscosity, so they can't flow easily, whereas ultramafic melts have relatively low viscosity, so they can flow relatively easily. The proportion of silica affects viscosity because silica tends to form chain-like molecules that tangle with one another and slow the flow.

Gases in Magma

So far, we've focused our attention on the chemicals that make up the molten rock of a magma. Magma generally also contains dissolved volatiles, such as H_2O, CO_2, sulfur dioxide (SO_2), and hydrogen sulfide (H_2S). Such volatiles exist as ionic groups (such as OH) bonded to minerals in solid rock. When rock melts, however, the volatiles separate from solid minerals and become separate molecules. Under the high pressure at the depth of

A Deeper Look

Why does magma composition vary?

Magma varies significantly in composition from place to place, as reflected by the rock that eventually solidifies from it. Many factors control magma composition:

- *Composition of the magma source:* The chemicals available to go into a melt depend on the chemicals within the *magma source,* the rock from which the magma was extracted. Not all magmas are derived from the same magma source, so not all have the same composition.
- *Partial melting:* Under the temperature and pressure conditions that occur in our planet, only a small percentage of the rock at a given location actually melts before the resulting magma migrates away. In other words, magma tends to be the product of **partial melting**. The percentage of a magma source that melts to form magma affects the proportions of chemicals in the magma. Significantly, partial melting yields magma that is more felsic than the magma source because a higher proportion of the chemicals needed to form felsic minerals migrate into the magma at lower temperatures. For this reason, partial melting of ultramafic rock in the mantle produces mafic magma.
- *Chemical interaction with surroundings:* Magma may incorporate chemicals dissolved from the solid rock through which it migrates or from blocks of rock that fall into the magma. This process is called **assimilation**.

- *Fractional crystallization:* Since magma contains many different chemical compounds, many different minerals can form when it freezes. In the 1920s, an American geologist named Norman Bowen discovered that not all of these minerals form at the same time. Specifically, when a mafic magma starts to freeze, crystals of olivine and pyroxene, minerals containing a high proportion of iron and magnesium oxide, grow first **(Fig. Bx4.1a)**. Their formation removes iron and magnesium from the melt. Therefore, as cooling continues, the next minerals to form contain relatively less iron and magnesium oxide. Bowen observed that amphiboles form next, and micas, quartz, and orthoclase (potassium feldspar) form last. While all these minerals are forming, plagioclase crystals also grow. As the melt progressively cools, the composition of the plagioclase crystals changes—early-formed plagioclase contains more CaO, while later-formed plagioclase contains more NaO. The overall sequence of crystallization as molten rock cools is known as **Bowen's reaction series (Fig. Bx4.1b)**, and the process of crystal formation and associated removal of chemicals from the melt is called **fractional crystallization**. Because fractional crystallization progressively extracts iron and magnesium from the magma, the composition of a magma reflects the degree of fractional crystallization.

FIGURE Bx4.1 Bowen's reaction series is the sequence of mineral crystallization in cooling magma.

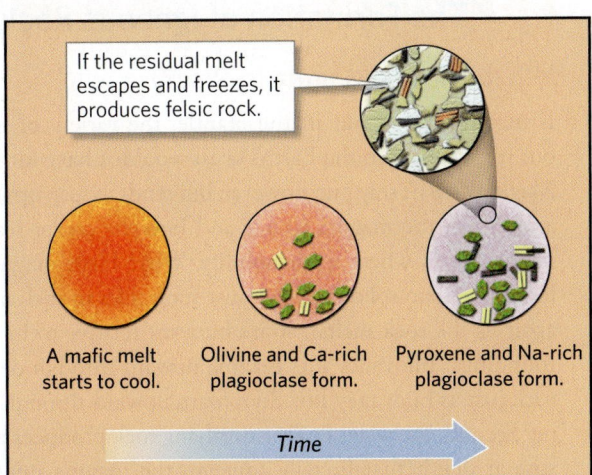

If the residual melt escapes and freezes, it produces felsic rock.

A mafic melt starts to cool.

Olivine and Ca-rich plagioclase form.

Pyroxene and Na-rich plagioclase form.

Time

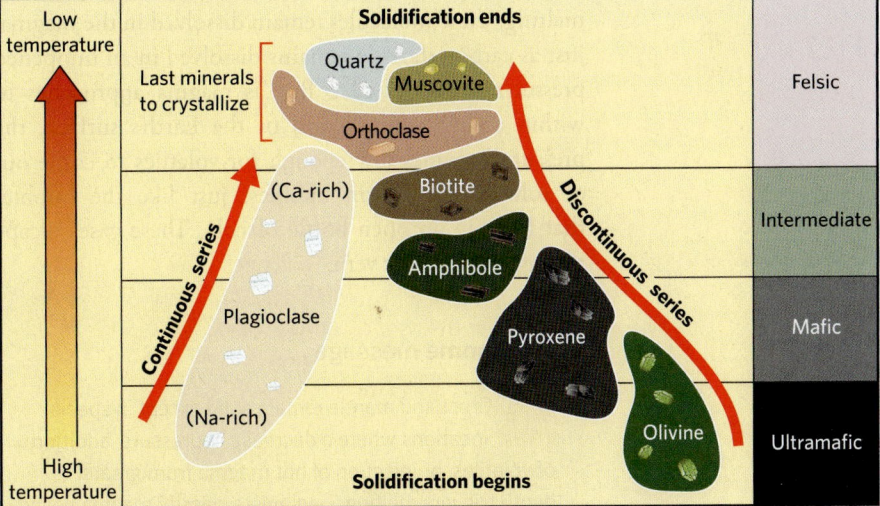

(a) Bowen discovered that with decreasing temperature, fractional crystallization begins, and the composition of the remaining magma becomes increasingly felsic.

(b) This chart displays the discontinuous and continuous sides of Bowen's reaction series. Rocks formed from minerals at the top of the series are mafic, whereas rocks formed from the bottom of the series are felsic. Here, the term *discontinuous* indicates that a succession of different minerals forms, whereas the term *continuous* means one mineral (plagioclase), with varying compositions (Ca-rich vs. Na-rich), forms. Orthoclase is a potassium feldspar.

FIGURE 4.5 Formation of extrusive and intrusive igneous rocks.

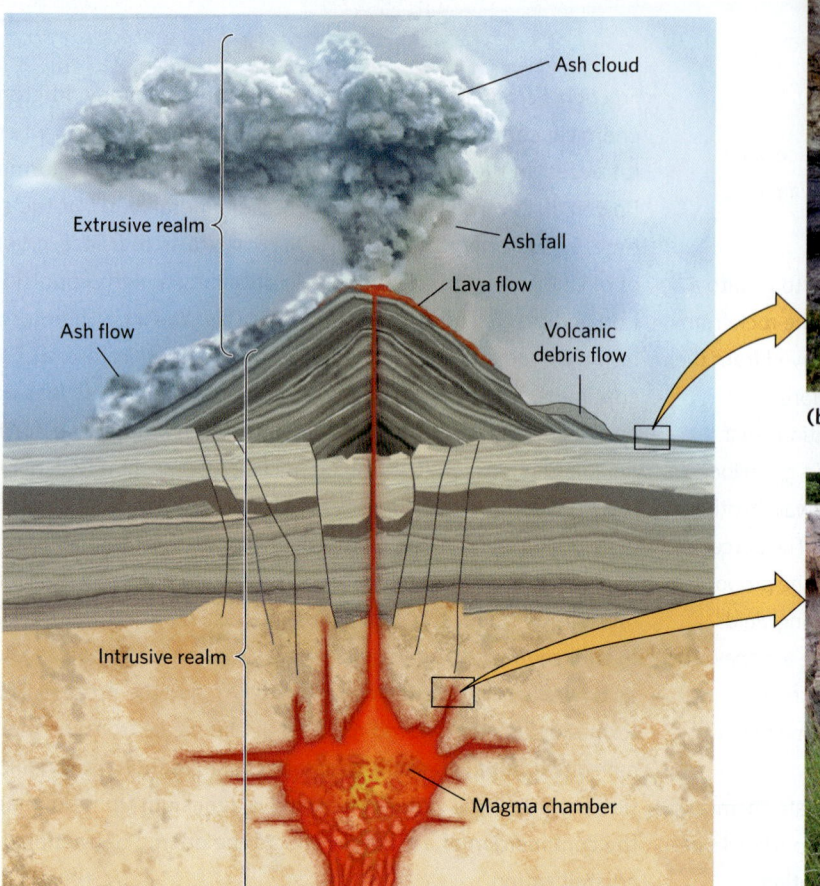

(b) Extrusive rocks include lava flows and pyroclastic layers.

(a) The intrusive realm lies underground and the extrusive realm lies aboveground. Lava flows, as well as various types of ash eruptions, all produce extrusive rocks.

(c) An intrusion of basalt (dark rock) cuts across a mass of granite (light rock).

melting, these molecules remain dissolved in the magma, just as carbon dioxide remains dissolved in an unopened pressurized can of cola. But as magma approaches to within about 5 km (3 mi) of the Earth's surface, the pressure becomes low enough for volatiles to come out of solution and form bubbles, just like the bubbles that form in an open bottle of cola. These gases escape through a volcano's vent.

Take-home message . . .

Crust and mantle remain solid except in special locations where a decrease in pressure, addition of volatiles, or injection of hot magma from greater depth triggers melting. Geologists classify magma based on the proportion of silica that it contains. Magma can contain dissolved gases.

Quick Question

What are some factors that can influence magma composition?

4.3 The Formation of Igneous Rock

What Causes Magma to Rise?

If magma stayed put in the mantle, the variety of igneous rocks found in the Earth's crust wouldn't have formed. Magma doesn't stay put, however, but tends to seep upward, away from the magma source and into the cooler realms of the Earth, where it can solidify. Why? First, magma is less dense than solid rock (because molecules in rock move apart as the rock melts), so it's buoyant relative to its surroundings. Buoyancy lifts magma upward through denser rock just as buoyancy lifts Styrofoam upward through water. Second, the weight of the overlying rock produces pressure at depth that literally squeezes the magma upward, somewhat as the weight of your foot squeezes mud up between your toes when you step in a puddle.

Extrusive versus Intrusive Igneous Rock

Most people have seen photos or movies of volcanic eruptions. Such dramatic images highlight floods of lava

FIGURE 4.6 Columnar jointing.

The vertical lines are columnar joints.

(a) Huge columnar joints in Devils Tower, Wyoming.

Sky

Rubbly top of flow
Columnar-jointed interior
Rubbly base of flow
Older flow

(b) Columnar jointing in a lava flow in Yellowstone National Park.

or clouds and avalanches of fragmented debris erupting from a vent. New igneous rock that forms when lava flowing from a volcano solidifies after coming in contact with air or water, or when the debris ejected by a volcano accumulates and either welds together or becomes cemented together, is called **extrusive igneous rock (Fig. 4.5a, b)**.

Not all igneous rock, however, originates after extrusion from a volcano. In fact, a vastly greater proportion of the Earth's igneous rock forms by solidification of magma underground, out of view **(Fig. 4.5c)**. We refer to rock produced by the freezing of magma underground, after it has *intruded* (pushed its way into) cooler overlying rock, as **intrusive igneous rock**; the pre-existing rock surrounding an intrusion is called **wall rock**.

Transforming Melt into Rock

How long it takes for a melt to solidify depends on how fast heat transfers from the melt into its surroundings. For example, an intrusion that cools deep in the crust, surrounded by warm wall rock, cools slowly, like coffee in a thermos bottle, for rock, like a thermos bottle, serves as a good insulator. In contrast, lava flowing onto the cold surface of the Earth cools quickly because it is not well insulated. In fact, if lava comes in contact with water, it cools especially fast, because water removes heat efficiently. The cooling rate also depends on the shape and size of a body of melt. Because the body loses heat to its surroundings only at its surface, bodies with a large surface area per unit volume cool faster. Therefore, a pancake-shaped body cools faster than a spherical body of the same volume, a shoebox-sized intrusion cools more quickly than a building-sized intrusion, and small drops of lava sprayed into the air cool faster than a thick lava flow.

Igneous rock remains hot immediately after it solidifies. It takes additional time to cool down to the temperature of its surroundings. During the final stages of cooling, igneous rock shrinks, because as its temperature decreases, its volume decreases. When it shrinks, the rock tends to crack. In some cases, the cracks outline roughly hexagonal columns, producing a visually striking pattern called **columnar jointing (Fig. 4.6)**.

Take-home message . . .

Magma rises because it is buoyant and because pressure from the weight of overlying rocks squeezes it upward. When molten rock enters a cooler environment, it freezes. Intrusive igneous rock solidifies underground, whereas extrusive igneous rock forms on the Earth's surface. The rate of cooling depends on the environment and on the size and shape of the melt body.

Quick Question
Which cools faster, a large blob of magma intruded at depth or a thin flow of lava extruded at the Earth's surface?

4.4 The Products of Igneous Activity

Igneous Intrusions

We can't see an **igneous intrusion**, a body of rock that solidifies underground, while it's forming. But over time, erosion of overlying rock can expose intrusions, allowing us to study their characteristics and distinguish among different types of intrusions. (As discussed in Chapter 11, understanding intrusions has important practical applications, for much of

FIGURE 4.7 Igneous sills and dikes; examples of tabular intrusions.

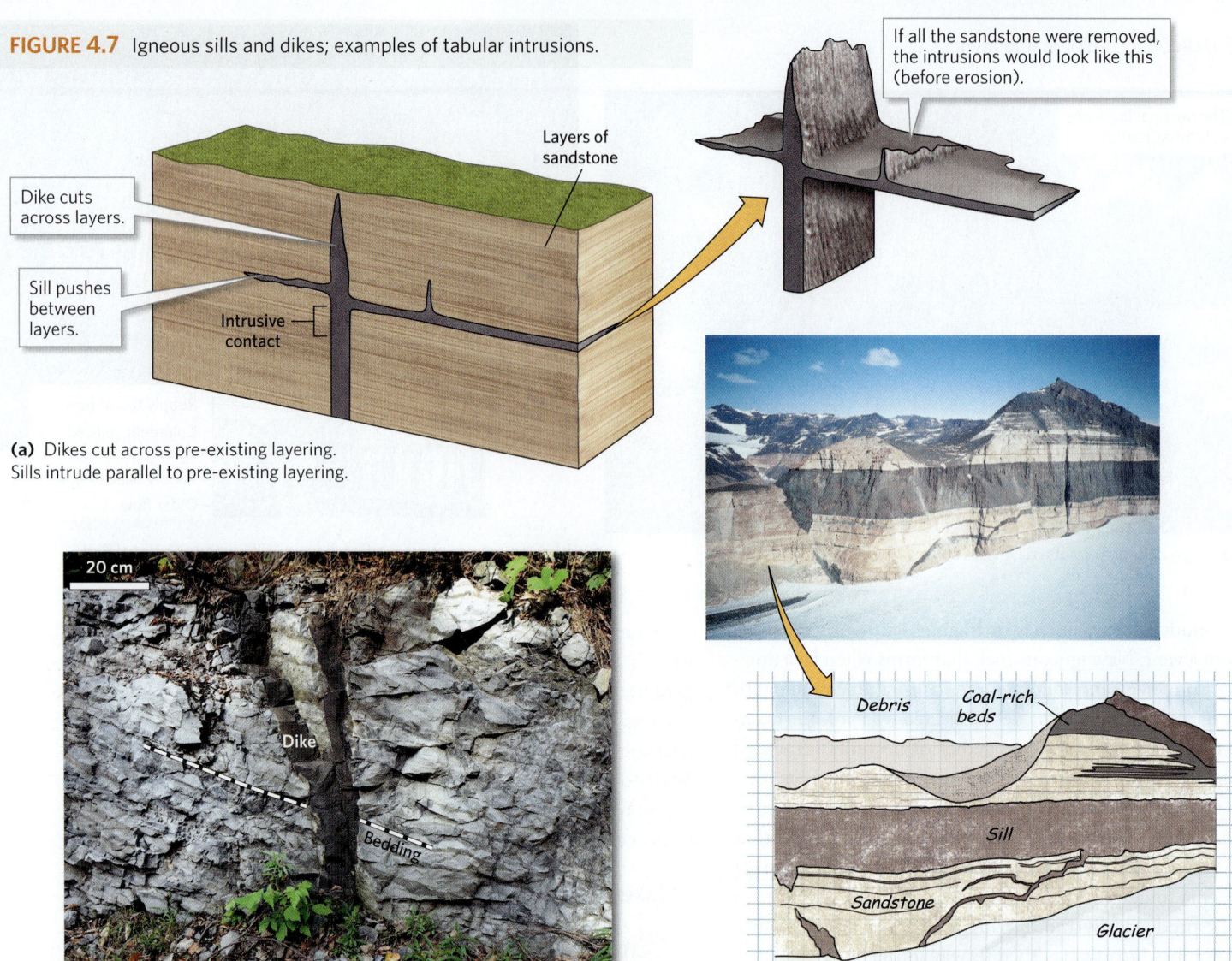

Dike cuts across layers.

Sill pushes between layers.

Intrusive contact

Layers of sandstone

If all the sandstone were removed, the intrusions would look like this (before erosion).

(a) Dikes cut across pre-existing layering. Sills intrude parallel to pre-existing layering.

20 cm

Dike

Bedding

Debris   Coal-rich beds

Sill

Sandstone

Glacier

What an Earth Scientist Sees

(b) An example of an igneous dike cutting across sedimentary rocks in Montreal, Canada.

(c) Large sills of basalt intruded sandstone beds in Antarctica, here exposed at Finger Mountain.

the world's supply of valuable metals formed in association with intrusions.)

Tabular intrusions, or *sheet intrusions,* have relatively planar surfaces and somewhat uniform thickness. They typically range in length from meters to tens of kilometers, and in thickness from centimeters to tens or even hundreds of meters. Geologists distinguish between two types, based on the orientation of the intrusion relative to its wall rock. A **dike** cuts across pre-existing layering (bedding or foliation) of wall rock, whereas a **sill** intrudes parallel to pre-existing layering **(Fig. 4.7)**. By convention, in places where wall rock does not have layering, geologists still refer to steep or vertical tabular intrusions as dikes and to near-horizontal bodies as sills. In some locations, a blister-shaped intrusion, called a *laccolith,* forms.

Plutons are blob-shaped intrusions that range from tens of meters across to several kilometers across. Some plutons may form from the solidification of magma that filled a **magma chamber**, an underground space containing a high proportion of magma **(Fig. 4.8a)**. Alternatively, plutons may be built by successive intrusions of adjacent sills, each intruding beneath the previous one **(Fig. 4.8b)**. We can see plutons where they have been exposed by erosion **(Fig. 4.8c)**. When numerous plutons form within a region, they merge into a vast igneous body, known as a **batholith**, that can be hundreds of kilometers long and up to a hundred kilometers wide **(Fig. 4.9)**. For example, plutons that intruded between 145 and 80 million years ago form the *Sierra Nevada batholith* of California; the spectacular cliffs of Yosemite National Park expose rock of this batholith.

FIGURE 4.8 Plutons and batholiths.

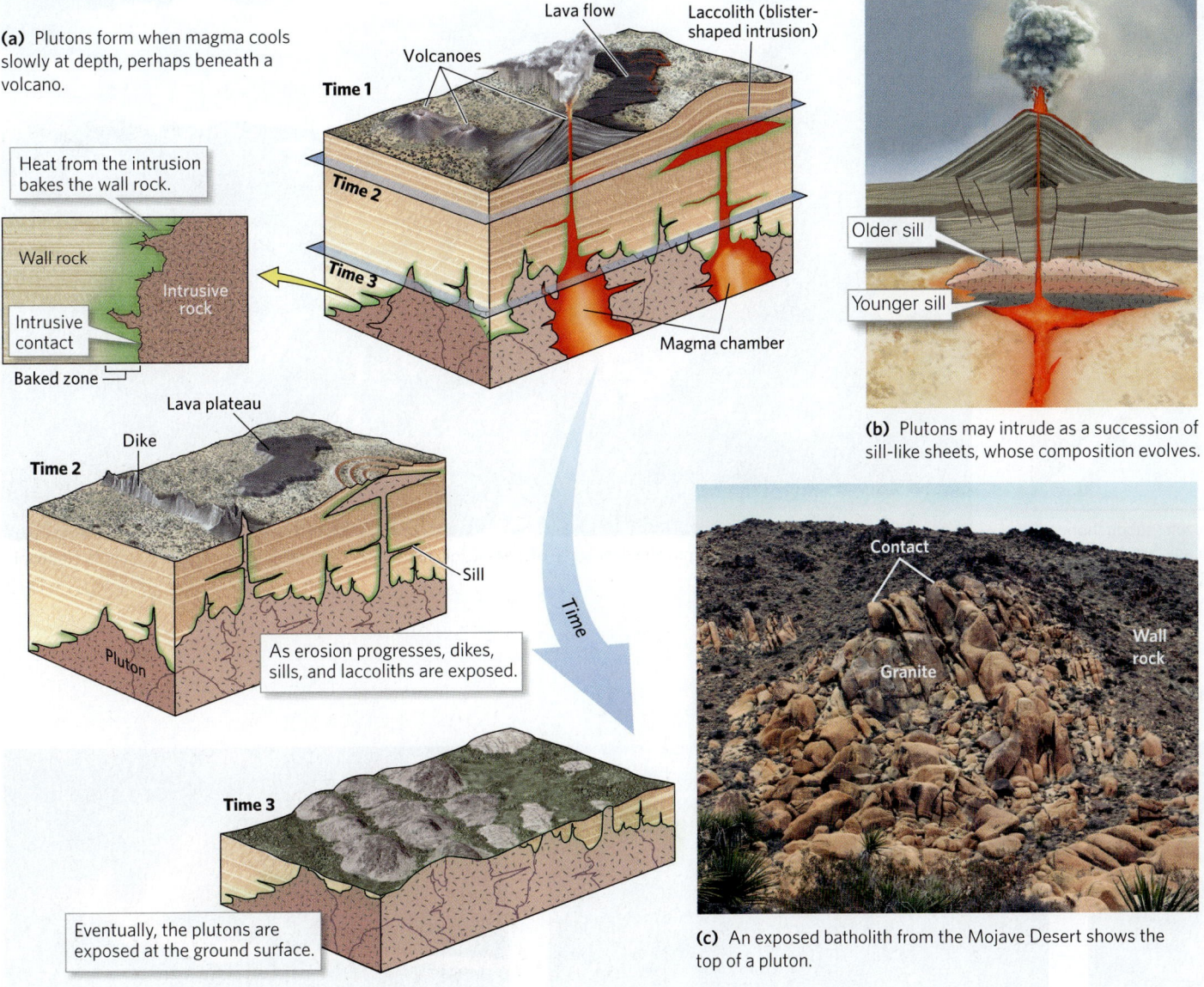

(a) Plutons form when magma cools slowly at depth, perhaps beneath a volcano.

Heat from the intrusion bakes the wall rock.

Wall rock

Intrusive rock

Intrusive contact

Baked zone

Time 1

Volcanoes

Lava flow

Laccolith (blister-shaped intrusion)

Time 2

Time 3

Magma chamber

Lava plateau

Dike

Time 2

Sill

Pluton

As erosion progresses, dikes, sills, and laccoliths are exposed.

Time

Time 3

Eventually, the plutons are exposed at the ground surface.

Older sill

Younger sill

(b) Plutons may intrude as a succession of sill-like sheets, whose composition evolves.

Contact

Granite

Wall rock

(c) An exposed batholith from the Mojave Desert shows the top of a pluton.

Lava Flows

Geologists use the term **lava flow** both for molten lava moving over the Earth's surface and for the layer of solid igneous rock formed when the lava freezes. The characteristics of a lava reflect both the viscosity and volume of a flow **(Box 4.2)**. For example, a relatively cool lava with a felsic composition has such high viscosity that it builds into a bulbous mound, called a **lava dome**, that does not extend far from the vent. Intermediate-composition lava extrudes in short, blocky flows. Hot, mafic lava has such low viscosity that it can form long, relatively thin flows **(Fig. 4.10a)**. Cooling eventually causes the surface of a mafic flow to crust over while the internal part of the flow remains molten and can move

(Fig. 4.10b). In some cases, the lava eventually flows only through **lava tubes**, tunnel-like conduits in the interior of a larger lava flow **(Fig. 4.10c)**. In some cases, lava tubes eventually drain and become empty tunnels **(Fig. 4.10d)**.

The surface texture of a mafic lava flow depends on the volume and rate of flow of the lava. Slower-moving low-volume flows have soft, pasty surfaces and wrinkle into smooth, glassy, rope-like ridges; geologists refer to such flows by their Polynesian name, **pahoehoe** (pronounced pa-hoy-hoy) **(Fig. 4.11a, b)**. Flows containing a larger volume of faster-moving lava crust over while the flow is still moving; the solidified layer breaks into a jumble of jagged fragments, forming a rubbly flow also known by its Polynesian name, **a'a'** (pronounced ah-ah) **(Fig. 4.11c)**.

FIGURE 4.9 Batholiths in the western United States.

(a) During the Mesozoic, subduction produced a huge volcanic arc. In the crust beneath the arc, large granite batholiths formed. Erosion exposed them.

(b) The Sierra Nevada of California provide exposures of a Mesozoic batholith. The huge granite cliffs of Yosemite National Park are parts of this batholith.

FIGURE 4.10 Mafic lava flows.

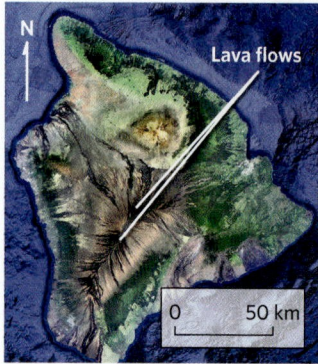

(a) Numerous lava flows (dark streaks) are visible on Hawaii.

(b) A highway cuts through a mafic lava flow on the Big Island of Hawaii. (Cinder cones appear in the background.)

(c) Flowing lava, sourced from a lava fountain on Réunion, an island in the Indian Ocean, lights the night sky.

(d) In a lava tube, molten lava flows under a crust of solid rock.

(e) A drained lava tube exposed by a road cut on Hawaii.

FIGURE 4.11 Surface textures of mafic lava flows.

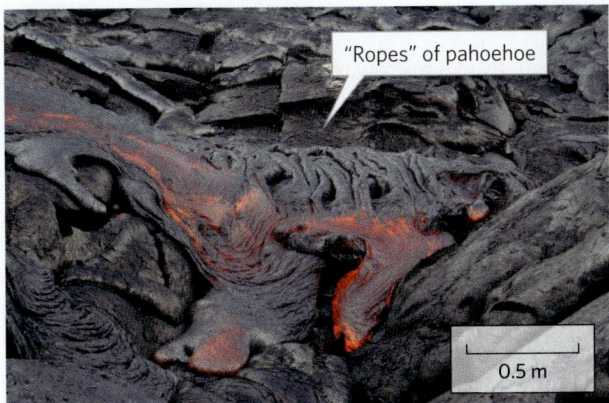

"Ropes" of pahoehoe

0.5 m

(a) Pahoehoe developing on a new lava flow in Hawaii.

"Ropes" of pahoehoe

0.5 m

(b) The surface of a pahoehoe flow after it has cooled and solidified. Note the coin for scale.

(c) The rubbly surface of an a'a' flow, Sunset Crater, Arizona.

Older pillows

(d) Pillow basalt develops when lava erupts underwater. Later uplift may expose pillows above sea level, as in this Oregon outcrop.

▶ **Animation**
Volcanoes and Lava Flows: Flows

Mafic flows that erupt underwater look different from those that erupt on land because the lava cools particularly quickly when in contact with water. Moments after extrusion, submarine mafic lava forms a glass-encrusted blob, or *pillow*. The rind of the pillow stops the flow's advance until the pressure of the lava squeezing into the pillow breaks the rind, and then a new blob of lava squirts out, which itself freezes into a pillow. Geologists refer to layers composed of these blobs as **pillow basalt (Fig. 4.11d)**.

Pyroclastic Deposits

In 1943, as Dionisio Pulido prepared to sow his field west of Mexico City, an earthquake jolted the ground, and the surface of the field bulged upward and cracked. Fragments of rock, ranging from dust-sized to melon-sized, filled the air, and Dionisio fled. By the following morning, the field lay buried beneath a 40-m (130-ft)-high mound of gray-black debris. Dionisio had witnessed the birth of a new volcano, soon named Paricutín.

Geologists refer to all kinds of fragments produced by an eruption, such as that ejected by Paricutín, as **pyroclastic debris**, from the Greek word for fire, or as *tephra*. Some pyroclastic debris forms when already solidified or partially solidified lava in a volcano's vent breaks apart and sprays upward, or when pre-existing igneous rock surrounding the volcanic vent gets blasted apart. Some of this debris rises skyward to high elevations **(Fig. 4.12a)**, but some avalanches down the side of the volcano in a **pyroclastic flow**; such a flow contributed to the burial of Pompeii. Pyroclastic debris also forms when lava spews up into the air in a **lava fountain**, producing drops or blobs of lava that freeze in midair or shortly after falling on the ground **(Fig. 4.12b)**. Different names apply to different sizes of pyroclastic debris: **Ash** consists of tiny flakes or slivers, commonly composed of glass **(Fig. 4.12c)**. **Lapilli** are marble- to golf-ball-sized pieces that can be formed from fragmented rock, from ash that clots together when mixed with water **(Fig. 4.12d)**, or from clots of lava, in which case they're called *cinders* **(Fig. 4.12e)**. **Blocks** are

The meaning of viscosity

What are we learning?

The viscosity of a magma or lava affects how it flows and how easily gas can escape from it.

What you need:

- A squeeze bottle of ketchup and a jar of smooth peanut butter
- Large plastic cutting board; stopwatch; scoop or spoon; cup; plastic straws; microwave oven

Instructions:

- Place the cutting board on a counter with one end propped up so that the surface of the board tilts at an angle of 45° into a sink or tray.
- Place a scoop of one of the food substances near the upslope end of the cutting board. See how far it flows down the cutting board in a given amount of time, and estimate its ending thickness. Repeat with the second substance. Which is more viscous?
- Based on your reading, which substance represents felsic magma?
- Insert one straw into a cup of ketchup and another into a jar of peanut butter, then blow hard into each one. In which substance do bubbles form and rise more easily? What does this mean in terms of how viscosity affects the escape of gas?

- Heat the substances briefly in a microwave oven. Repeat the experiment on the tilted cutting board, using the warmed products. How does temperature affect viscosity?

What did we see?

This exercise illustrates the concept of using physical models to represent the behavior of real systems. In detail, it illustrates the contrast between high viscosity and low viscosity and the effect of temperature on viscosity.

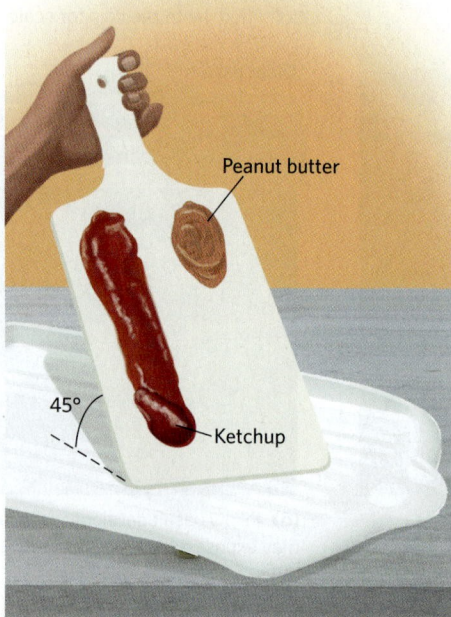

Peanut butter

45°

Ketchup

Bubbles in ketchup

Bubbles don't form easily in peanut butter

---

angular larger chunks **(Fig. 4.12f)**, and **volcanic bombs** are blocks that were still soft during eruption so that they became streamlined as they fell, or flattened into a pie shape when they landed **(Fig. 4.12g)**.

Take-home message . . .

Igneous rocks form both underground and at the Earth's surface. Intrusive igneous rocks form when magma solidifies underground in dikes, sills, or plutons. Extrusive igneous rocks include lava flows and pyroclastic debris (also known as tephra).

Quick Question

Distinguish among a dike, a sill, and a pluton.

4.5 Classifying Igneous Rocks

Because melts can have a variety of compositions, and because they can freeze to form igneous rocks in many different environments above and below the surface of the Earth, a wide range of different igneous rock types occur on the Earth. Geologists organize these rocks into classes on the basis of two characteristics: texture and composition. A description of texture indicates whether an igneous rock is *crystalline* (consists of interlocking crystals), *fragmental* (consists of pieces that are either welded or cemented to one another), or *glassy* (consists mostly of glass) **(Fig. 4.13)**. A description of composition focuses on the relative proportions of silica to magnesium and iron oxides and, therefore,

FIGURE 4.12 Pyroclastic debris.

(a) Pyroclastic debris billowing from an eruption of Chaitén, in Chile.

(b) Clots of lava being ejected in a lava fountain.

(c) Electron photomicrograph of ash.

(d) Accretionary lapilli.

(e) Cinders (tephra).

(f) Fragments of rock blasted off the volcano form angular blocks.

(g) Clots of lava that freeze in the air form bombs.

Crystals (a) Granite is crystalline.

Pyroclastic grains (b) Tuff is fragmental.

Glass (c) Obsidian is glassy.

distinguishes among ultramafic, mafic, intermediate, and felsic rocks. An igneous rock's texture tells us about the environment in which it formed, and its composition tells us about the magma source and the way in which the magma evolved.

Crystalline Igneous Rocks

A **crystalline igneous rock** consists of mineral crystals that grew in a melt and fit together like pieces in a jigsaw puzzle. Geologists distinguish between fine-grained (*aphanitic*) igneous rocks, in which the crystals are too small to be identified without a microscope, and coarse-grained (*phaneritic*) igneous rocks, in which the minerals making up individual grains can be distinguished with the naked eye. Some igneous rocks have a uniform texture, in that they contain mostly grains of about the same size, but some contain large grains surrounded by tiny grains. In such *porphyritic* igneous rocks, the larger grains are called *phenocrysts* and the surrounding finer grains constitute *groundmass*. Simplistically, the grain size of a crystalline igneous rock reflects the cooling rate: finer-grained rocks cooled quickly, whereas coarser-grained rocks cooled slowly. A porphyritic rock develops when a magma starts cooling slowly, so that early-forming crystals become large, and then cools quickly, so that the remaining melt forms a fine-grained groundmass.

There are many different names for crystalline igneous rocks. The name applied to a given rock sample depends on both the composition and the grain size of the sample **(Fig. 4.14)**. For example, a fine-grained felsic rock is a *rhyolite* while a coarse-grained one is a *granite*; a fine-grained intermediate rock is an *andesite* while a coarse-grained one is a *diorite*; a fine-grained mafic rock is a *basalt* while a coarse-grained one is a *gabbro*; and a fine-grained ultramafic rock is a *komatiite* (an uncommon rock) while

a coarse-grained one is a *peridotite*. Note that the two members of each of these pairs have the same chemical composition but cooled at different rates. The color and density of an igneous rock provide clues to its composition. For example, samples of felsic rock tend to be tan or pink/maroon and have a relatively low density, whereas samples of mafic rock tend to be black or dark gray and have a relatively high density.

Glassy and Fragmental Igneous Rocks

In some cases, lava cools so quickly that atoms or molecules within it do not attain the orderly arrangement that they have in minerals, but rather freeze in a disordered configuration. This process yields a **glassy igneous rock**. Some glassy rocks consist of a solid mass of glass through and through, and others contain tiny crystals surrounded by a groundmass of glass. A mass of solid, felsic glass is called **obsidian** (see Fig. 4.13c). This rock tends to be black or brown and breaks conchoidally, so it can be fashioned into arrowheads. Commonly, rapidly cooling lava contains gas bubbles, and if the lava freezes before the bubbles have a chance to escape, the bubbles remain as open spaces, known as **vesicles**, within the solid rock. **Pumice** is a light-colored felsic glassy rock that contains an abundance of very tiny vesicles; it forms from the cooling of frothy lava **(Fig. 4.15a)**. **Scoria** is a mafic volcanic rock that also contains abundant vesicles, but the vesicles in scoria are bigger than those in pumice, and the walls between vesicles are thicker **(Fig. 4.15b)**.

Fragmental igneous rock forms either when hot fragments of pyroclastic debris weld together while still extremely hot or when cooled fragments accumulate and later undergo compaction and cementation. Geologists use the term **pyroclastic rock** for fragmental igneous rocks formed directly and exclusively from the products of an eruption. Examples include **tuff (Fig. 4.16a)**, a fine-grained rock composed mostly of volcanic ash (**Fig. 4.16b**), and **volcanic breccia**, a coarser rock composed of volcanic blocks.

Take-home message . . .

Igneous rocks form from material that solidified from a melt. They can be crystalline, glassy, or fragmental. Grain size in a crystalline rock reflects the rock's cooling rate. The name used for a given sample of crystalline rock depends on its grain size and composition. Fragmental igneous rocks form from pyroclastic debris.

Quick Question

What rock type forms from felsic magma cooled in a large pluton at depth?

4.6 The Nature of Volcanoes

Volcanic eruptions are the most visible manifestations of igneous activity. Some volcanoes can explode catastrophically, like Mt. Vesuvius, but others, such as Kilauea, don't. Now that we have introduced igneous materials and their origins, let's focus our attention on volcanoes and their behavior. We begin by examining the different ways in which volcanoes erupt.

Will It Flow, or Will It Blow?

The character of a volcanic eruption—whether it emits flows of lava or blasts out billows of pyroclastic debris—defines a volcano's **eruptive style**. Geologists have traditionally assigned names to different eruptive styles based on well-known examples (**Earth Science at a Glance**, pp. 128–129). Here, we distinguish between two main eruptive styles.

During an **effusive eruption**, or a *lava-dominated eruption*, low-viscosity mafic lava spurts or spills out of the vent and flows downslope as a lava flow (**Fig. 4.17a, b**). If the lava coming from an effusive eruption is under sufficient pressure, it may spurt skyward out of the vent as a *lava fountain* (**Fig. 4.17c**), rising as high as a few hundred meters above the vent opening. The pressure driving a fountain develops because gases in the lava expand as it rises—bursting of the gas bubbles may contribute to breaking the lava into clots. Sometimes lava from an effusive eruption pools around a vent, producing a *lava lake* (**Fig. 4.17d**). Cooling causes the surface of a lava lake to crust over with plates of dark rock, which move and crack due to convection in the still-molten lava beneath. Only mafic lava has low enough viscosity to erupt effusively, so effusive eruptions produce basalt.

As the name emphasizes, an **explosive eruption** involves an energetic blast that forcefully ejects material

FIGURE 4.14 Classification of crystalline igneous rocks.

(a) Types of crystalline igneous rocks, arranged by grain size and composition.

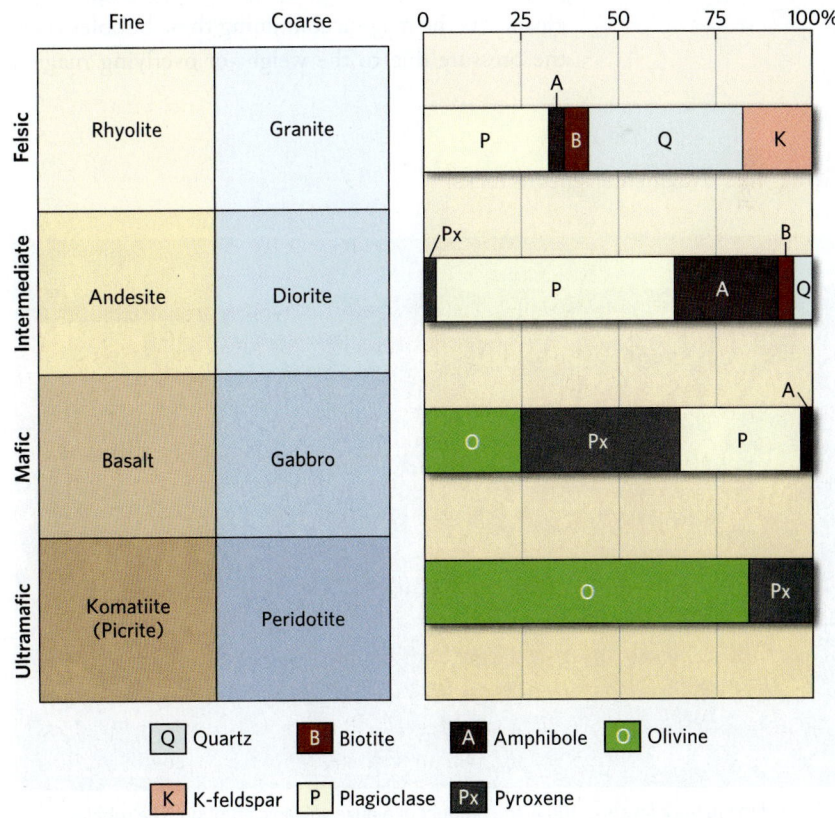

(b) The mineral composition of different types of crystalline igneous rocks. The chart at the right shows typical percentages of the minerals in the rock types on the left.

FIGURE 4.15 Igneous rocks containing vesicles.

(a) Pumice is so light that paper can hold it up.

(b) Scoria looks like a dark sponge, though it is very hard.

Did you
ever wonder . . .

whether all volcanoes look
the same?

from a volcano. Such eruptions, which occur in a range of sizes **(Box 4.3)**, produce clouds and avalanches of pyroclastic debris. Why do explosive eruptions take place? Let's consider two of the most common reasons:

1. *Expanding volcanic gas bubbles:* Bubbles of volcanic gas can become trapped in intermediate or felsic magma that is rising in a volcano. Such magma is so viscous that the bubbles can't rise through it and escape. As the magma containing these bubbles rises, the pressure due to the weight of overlying magma

decreases, so the gas bubbles try to expand. But due to the strength of the magma, or because the magma has already solidified into glass, the bubbles can't grow. As a result, the outward pressure within the bubbles increases. If a lava dome at the crest of the volcano, or the rock forming the flank of the volcano, suddenly breaks or slumps away, the inward pressure acting on the magma suddenly decreases, so the bubbles suddenly expand even more, and the pressure within them becomes so great that it breaks the glassy walls surrounding them. Instantly, the gas expands explosively, blasting the shattered bubble walls and adjacent rock out of the vent. When this happens, an *eruptive jet* driven by the pressure of the eruption sends debris up several hundred meters **(Fig. 4.18a, b)**. The hot ash in the jet mixes with and warms the surrounding air. The resulting cloud of

FIGURE 4.16 Fragmental igneous rocks.

(a) Each of these thick layers of tuff is the product of a huge volcanic eruption. The rubble-strewn slopes that cover the lower two-thirds of the cliff consist of debris that fell from the cliff.

A golf-ball-sized fragment of pumice

(b) A close-up of the tuff, which consists mainly of ash, but also includes other fragments such as pumice lapilli.

FIGURE 4.17 Examples of effusive eruptions.

(a) An effusive eruption on Hawaii. Lava spills out of a small crater and collects in a fast-moving flow.

(b) A lava flow from the 2018 lower Puna eruption on the Big Island of Hawaii.

(c) A lava fountain from the 2018 lower Puna eruption on the Big Island of Hawaii.

(d) An example of a lava lake. Note that a thin layer of dark new rock has formed over the molten lava.

hot ash and hot air becomes buoyant relative to the surrounding cold air, so it continues to rise as a convective column to stratospheric heights **(Fig. 4.18c)**. Part of the column of erupting material, however, collapses due to gravity and avalanches down the volcano to form a pyroclastic flow. In some cases, explosive eruptions may have enough force to blow the top off the volcano and rapidly drain a magma chamber.

2. *Interaction of water with hot rock or magma:* Some explosive eruptions occur when water comes in contact with hot magma in or below the vent, which converts it into high-pressure steam that blasts pyroclastic debris out of the vent. Such *phreatic explosions* can send ash and cinders skyward without erupting any lava. This happened during the 2018 lower Puna eruption in Hawaii, when groundwater entered a debris-filled vent. It can also occur during the eruption of a volcano whose vent lies just below the sea surface **(Fig. 4.18d)**.

The Architecture of Volcanoes

All volcanoes share the same basic components. Beneath the volcano, a mush of crystals and melt accumulates in a magma chamber. (In general, the amount of liquid the chamber contains varies from 3% to 30%; the remainder consists of crystals.) Not all magma chambers lie at the same depth—shallower ones occur 3–20 km (2–12 mi) below the surface, whereas deeper ones lie at 20–50 km (12–30 mi). Magma and gas rise from the chamber through a conduit to the Earth's surface and erupt from a vent. Over time, solidified lava and pyroclastic debris build up around the vent to produce a hill or mountain, known technically as the *volcanic edifice,* with a peak or *summit,* and

FIGURE 4.18 Examples of explosive eruptions.

(a) A large explosive eruptive cloud (Plinean type) contains several components.

(b) The eruptive cloud of an eruption on Mt. Etna, Italy.

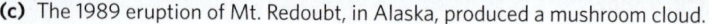

(c) The 1989 eruption of Mt. Redoubt, in Alaska, produced a mushroom cloud.

(d) A phreatic eruption near Tonga in the western Pacific.

sides or *flanks*. Dikes may form when magma intrudes into cracks that cut through rock of the edifice.

In some volcanoes, the conduit through which magma and gas reach the Earth's surface has a chimney-like shape and is topped by a circular depression, or **crater**, that resembles a bowl (Fig. 4.19a). Craters, which may be up to 500 m (1,600 ft) across and 200 m (650 ft) deep, develop in two ways: some form during eruptions when pyroclastic debris builds up around the vent, whereas others form after an eruption when the volcano's summit collapses into the drained conduit. Not all eruptions come from a single crater, however. During **fissure eruptions**, curtains of lava spew from a

vent that has the shape of an elongate crack, or *fissure*, that cuts across the flank of the volcano (Fig. 4.19b). Though eruptions commonly take place from a *summit vent* at the peak of a volcano, not all do—some eruptions come from *flank vents* on the side slope of a volcano.

The overall shapes of volcanoes vary as well. Geologists distinguish among three shapes. **Shield volcanoes**, so named because they resemble a soldier's shield lying on the ground, are broad, gentle domes. They form from layer upon layer of low-viscosity mafic lava produced by effusive eruptions (Fig. 4.20a). **Cinder cones** consist of symmetrical, cone-shaped piles of volcanic cinders ejected by a lava fountain during an effusive eruption (Fig. 4.20b).

Animation

Volcanoes and Lava
Flows: Types

FIGURE 4.19 Crater eruptions and fissure eruptions come from conduits of different shapes.

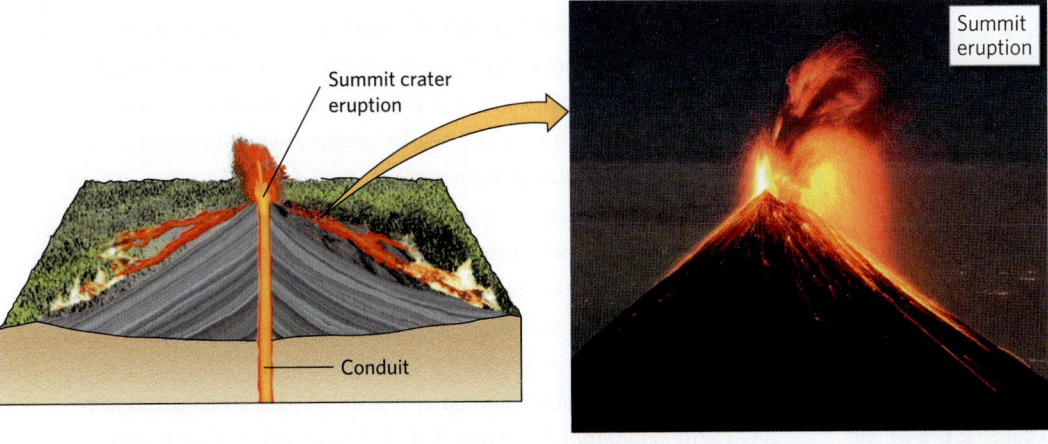

Summit crater eruption

Conduit

Summit eruption

Crater between eruptions

(a) In a crater eruption, lava spouts from a chimney-shaped conduit.

Lava curtain

Fissure

Fissure eruption

(b) In a fissure eruption, lava comes out in a curtain along the length of a crack.

FIGURE 4.20 Volcano shapes.

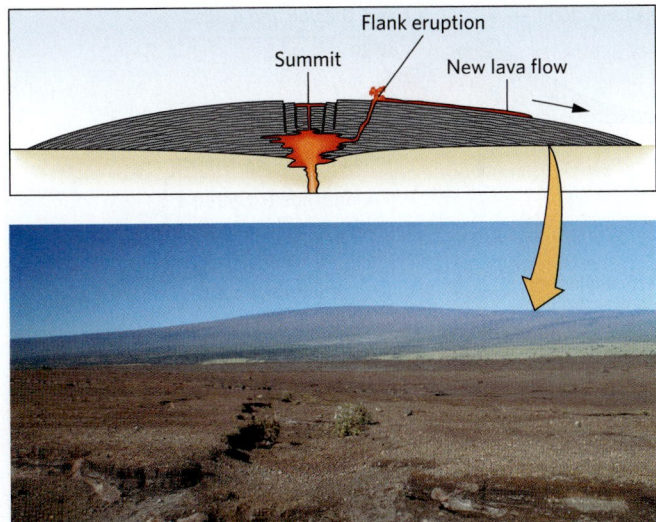

Flank eruption

Summit

New lava flow

(a) A 1.2 km-wide shield volcano, made from successive layers of low-viscosity basaltic lava, has very gentle slopes.

(b) A 1.2 km-wide cinder cone in Arizona. A lava flow covers the land surface in the distance. Note the road for scale.

BOX 4.3

A Deeper Look

Volcanic explosions to remember

Mt. St. Helens, a snow-crested volcano in the Cascade Range of the northwestern United States, had not erupted since 1857. However, geologic evidence suggested that the mountain had a violent past, punctuated by explosive eruptions. On March 20, 1980, an earthquake announced that the volcano was awakening once again. A week later, the summit began emitting gas and pyroclastic debris. Geologists who set up monitoring stations to observe the volcano noted that its north side was beginning to bulge markedly, suggesting that the magma chamber was filling with magma and causing the whole volcano to expand. Concern that an eruption was imminent led local authorities to evacuate people from the area.

The main eruption came suddenly. At 8:32 A.M. on May 18, David Johnston, a US Geological Survey geologist monitoring the volcano from a station 10 km away, shouted over his two-way radio, "Vancouver, Vancouver, this is it!" An earthquake had triggered a huge landslide that caused 3 km³ (0.7 mi³) of the volcano's weakened north side to slide away. The landslide released pressure on the magma in the volcano, causing a sudden and violent expansion of gases that blasted through the side of the volcano (Fig. Bx4.3a). Rock, steam, and ash screamed north at the speed of sound and flattened a forest and everything in it over an area of 600 km² (140 mi²) (Fig. Bx4.3b). Tragically, Johnston, along with 60 other people, vanished forever beneath the debris.

Seconds after the sideways blast, a vertical column carried over 500 million tons of ash (about 1 km³) up to the stratosphere, where strong high-altitude winds eventually transported it around the globe. In towns near the volcano, a blizzard of ash buried fields. Water-saturated ash flooded river valleys, carrying away everything in their path. When the eruption was finally over, the peak of Mt. St. Helens had disappeared—the summit now lay 440 m (1,450 ft) lower, and the once snow-covered mountain was a gray mound with a large gouge in one side.

An even greater explosion happened in 1883. Krakatau (also known as Krakatoa), a volcano in the sea between Java and Sumatra, where the Indian Ocean floor subducts beneath Southeast Asia, had grown to become a 9-km (6-mi)-long island rising 800 m (2,600 ft) above the sea. On May 20, the island began to erupt with a series of large explosions, yielding ash that settled as far as 500 km (300 mi) away. Smaller explosions continued through June and July, and steam and ash rose from the island, forming a huge black cloud that rained ash into the surrounding straits. Krakatau's demise came at 10:00 A.M. on August 27, perhaps when the volcano cracked and the magma

chamber flooded with seawater that flashed to steam. The resulting blast, five thousand times greater than the Hiroshima atomic-bomb explosion, could be heard as far as 4,800 km (3,000 mi) away. Tsunamis pushed out by the explosion raced across the sea, killing over 36,000 people when they slammed into nearby coastal towns. When the air finally cleared, Krakatau was gone, replaced by a depression some 300 m (1,000 ft) deep. All told, the eruption sent 20 km³ (about 5 mi³) of rock skyward. Ash that reached stratospheric elevations caused spectacular sunsets during the next several years.

Both of the volcanic explosions we've just described pale in comparison to the largest ones in the Earth's history. In fact, even the largest observed eruption (Tambora, in 1815) was small compared with an explosion that took place over 600,000 years ago in what is now Yellowstone National Park, Wyoming (Fig. Bx4.3c). Geologists use the term **supervolcano** for a volcano that yields an explosive eruption that ejects more than 1,000 km³ (240 mi³) of debris. At least 10 such events have been identified in Yellowstone from the occurrence of thick and widespread tuffs. The most recent supervolcanic eruption occurred at New Zealand's Taupo Volcano about 26,500 years ago. Geologists specify the size of explosive eruptions by specifying a number (from 0 to 8) on a scale called the volcanic explosivity index. The 79 C.E. eruption of Vesuvius and the 1980 eruption of Mt. St. Helens both have a rating of 5, whereas supervolcanoes such as Yellowstone and Taupo have a rating of 8.

FIGURE Bx4.3 Memorable explosive eruptions.

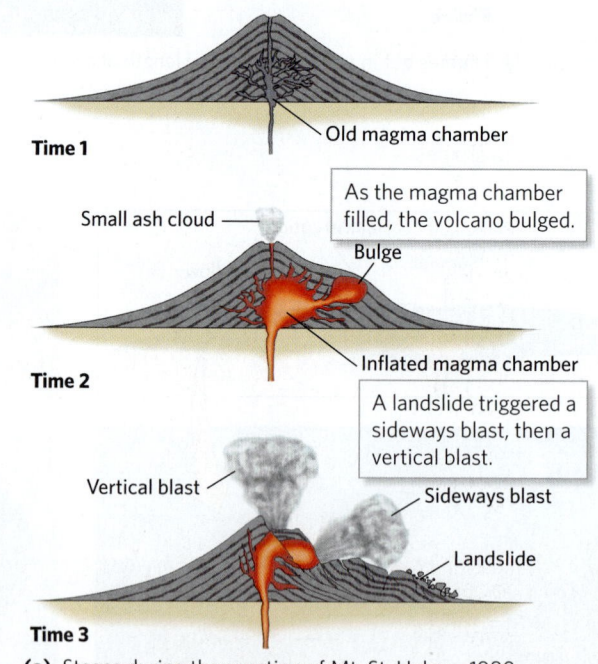

(a) Stages during the eruption of Mt. St. Helens, 1980.

- Time 1 — Old magma chamber
- Small ash cloud — As the magma chamber filled, the volcano bulged. — Bulge
- Time 2 — Inflated magma chamber
- A landslide triggered a sideways blast, then a vertical blast.
- Vertical blast — Sideways blast
- Time 3 — Landslide

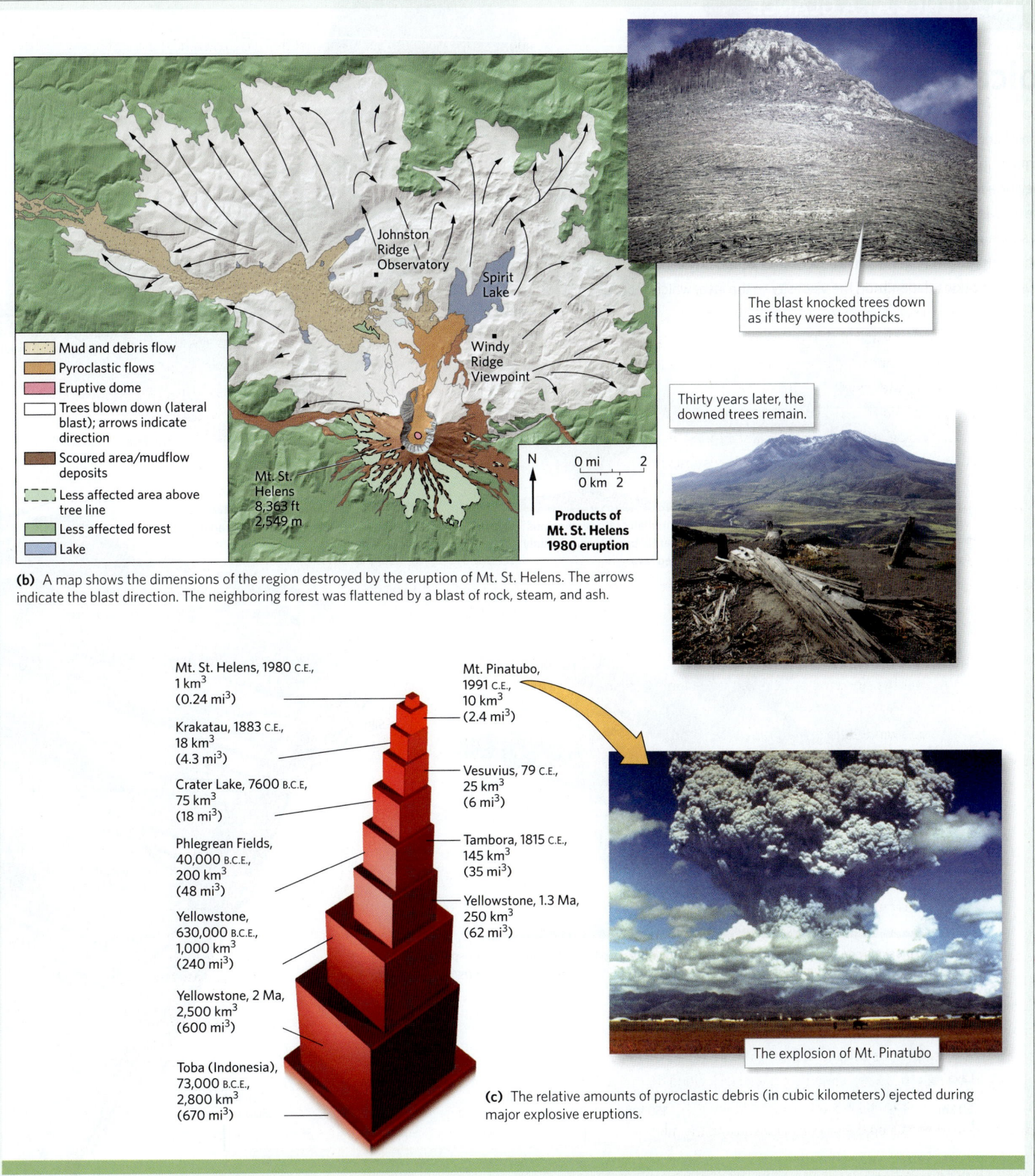

Mud and debris flow

Pyroclastic flows

Eruptive dome

Trees blown down (lateral blast); arrows indicate direction

Scoured area/mudflow deposits

Less affected area above tree line

Less affected forest

Lake

Johnston Ridge Observatory

Spirit Lake

Windy Ridge Viewpoint

Mt. St. Helens 8,363 ft 2,549 m

N

0 mi 2
0 km 2

Products of Mt. St. Helens 1980 eruption

The blast knocked trees down as if they were toothpicks.

Thirty years later, the downed trees remain.

(b) A map shows the dimensions of the region destroyed by the eruption of Mt. St. Helens. The arrows indicate the blast direction. The neighboring forest was flattened by a blast of rock, steam, and ash.

Mt. St. Helens, 1980 C.E., 1 km^3 (0.24 mi^3)

Krakatau, 1883 C.E., 18 km^3 (4.3 mi^3)

Crater Lake, 7600 B.C.E, 75 km^3 (18 mi^3)

Phlegrean Fields, 40,000 B.C.E., 200 km^3 (48 mi^3)

Yellowstone, 630,000 B.C.E., 1,000 km^3 (240 mi^3)

Yellowstone, 2 Ma, 2,500 km^3 (600 mi^3)

Toba (Indonesia), 73,000 B.C.E., 2,800 km^3 (670 mi^3)

Mt. Pinatubo, 1991 C.E., 10 km^3 (2.4 mi^3)

Vesuvius, 79 C.E., 25 km^3 (6 mi^3)

Tambora, 1815 C.E., 145 km^3 (35 mi^3)

Yellowstone, 1.3 Ma, 250 km^3 (62 mi^3)

The explosion of Mt. Pinatubo

(c) The relative amounts of pyroclastic debris (in cubic kilometers) ejected during major explosive eruptions.

Volcanoes

Beneath a volcano, magma rises to fill a pervasively cracked region of crust and forms a magma chamber. Some of the magma erupts at a surface vent. Once molten rock has erupted at the surface, it becomes lava. Some lava spills down the sides of the volcano in lava flows. Some fountains out of a vent to form scoria fragments that pile up in a cone around the vent. Eruptions may eject larger chunks as blocks or bombs. The nature of eruptions depends on the viscosity of the lava, which

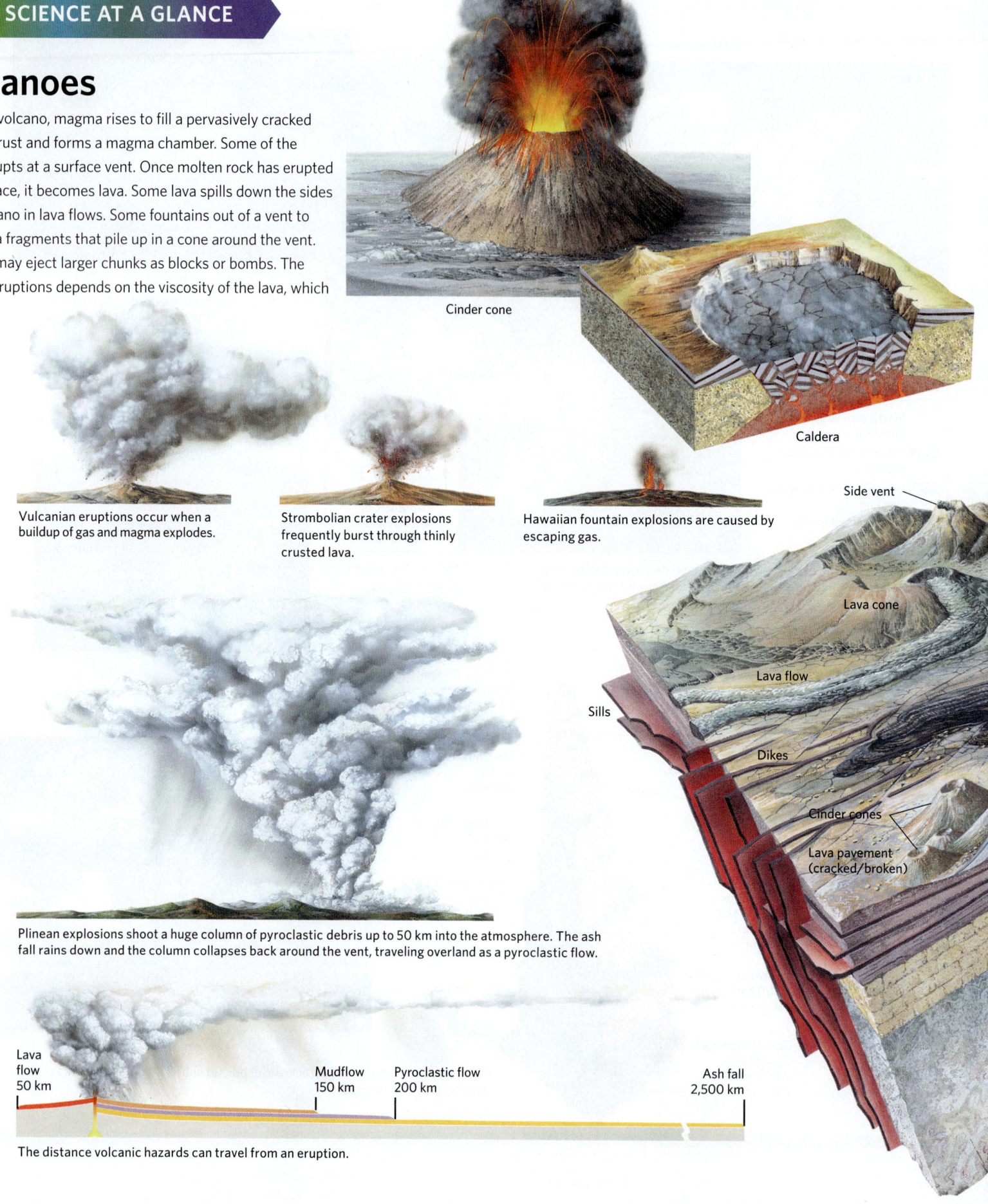

Cinder cone

Caldera

Vulcanian eruptions occur when a buildup of gas and magma explodes.

Strombolian crater explosions frequently burst through thinly crusted lava.

Hawaiian fountain explosions are caused by escaping gas.

Side vent

Lava cone

Lava flow

Sills

Dikes

Cinder cones

Lava pavement (cracked/broken)

Plinean explosions shoot a huge column of pyroclastic debris up to 50 km into the atmosphere. The ash fall rains down and the column collapses back around the vent, traveling overland as a pyroclastic flow.

Lava flow 50 km

Mudflow 150 km

Pyroclastic flow 200 km

Ash fall 2,500 km

The distance volcanic hazards can travel from an eruption.

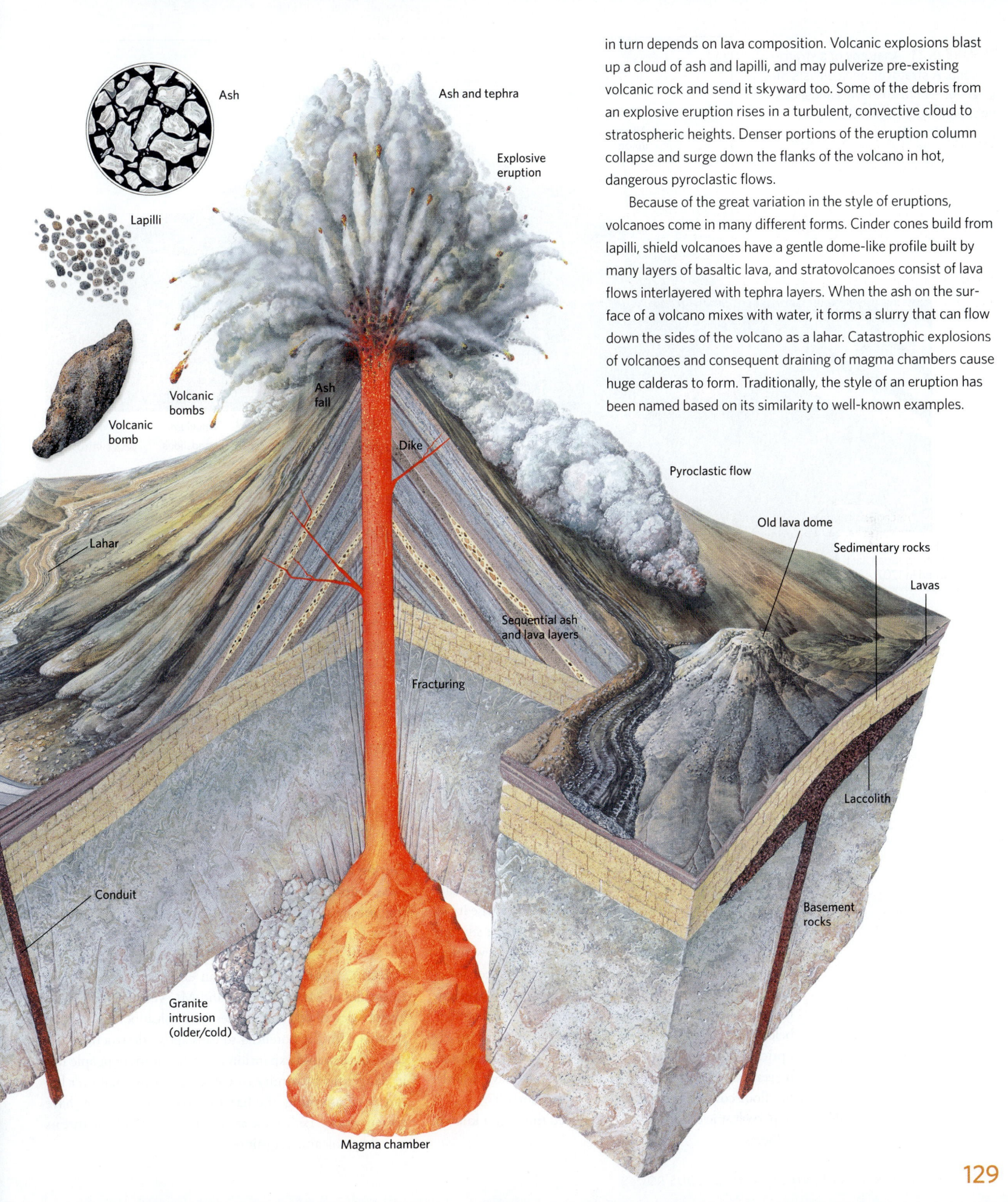

in turn depends on lava composition. Volcanic explosions blast up a cloud of ash and lapilli, and may pulverize pre-existing volcanic rock and send it skyward too. Some of the debris from an explosive eruption rises in a turbulent, convective cloud to stratospheric heights. Denser portions of the eruption column collapse and surge down the flanks of the volcano in hot, dangerous pyroclastic flows.

Because of the great variation in the style of eruptions, volcanoes come in many different forms. Cinder cones build from lapilli, shield volcanoes have a gentle dome-like profile built by many layers of basaltic lava, and stratovolcanoes consist of lava flows interlayered with tephra layers. When the ash on the surface of a volcano mixes with water, it forms a slurry that can flow down the sides of the volcano as a lahar. Catastrophic explosions of volcanoes and consequent draining of magma chambers cause huge calderas to form. Traditionally, the style of an eruption has been named based on its similarity to well-known examples.

Ash

Lapilli

Volcanic bomb

Ash and tephra

Explosive eruption

Volcanic bombs

Ash fall

Lahar

Dike

Pyroclastic flow

Old lava dome

Sedimentary rocks

Lavas

Sequential ash and lava layers

Fracturing

Laccolith

Conduit

Basement rocks

Granite intrusion (older/cold)

Magma chamber

FIGURE 4.21 Anatomy of a strato-volcano. A stratovolcano consists of layers of tephra and lava. Landslides occasionally occur and transport material downslope.

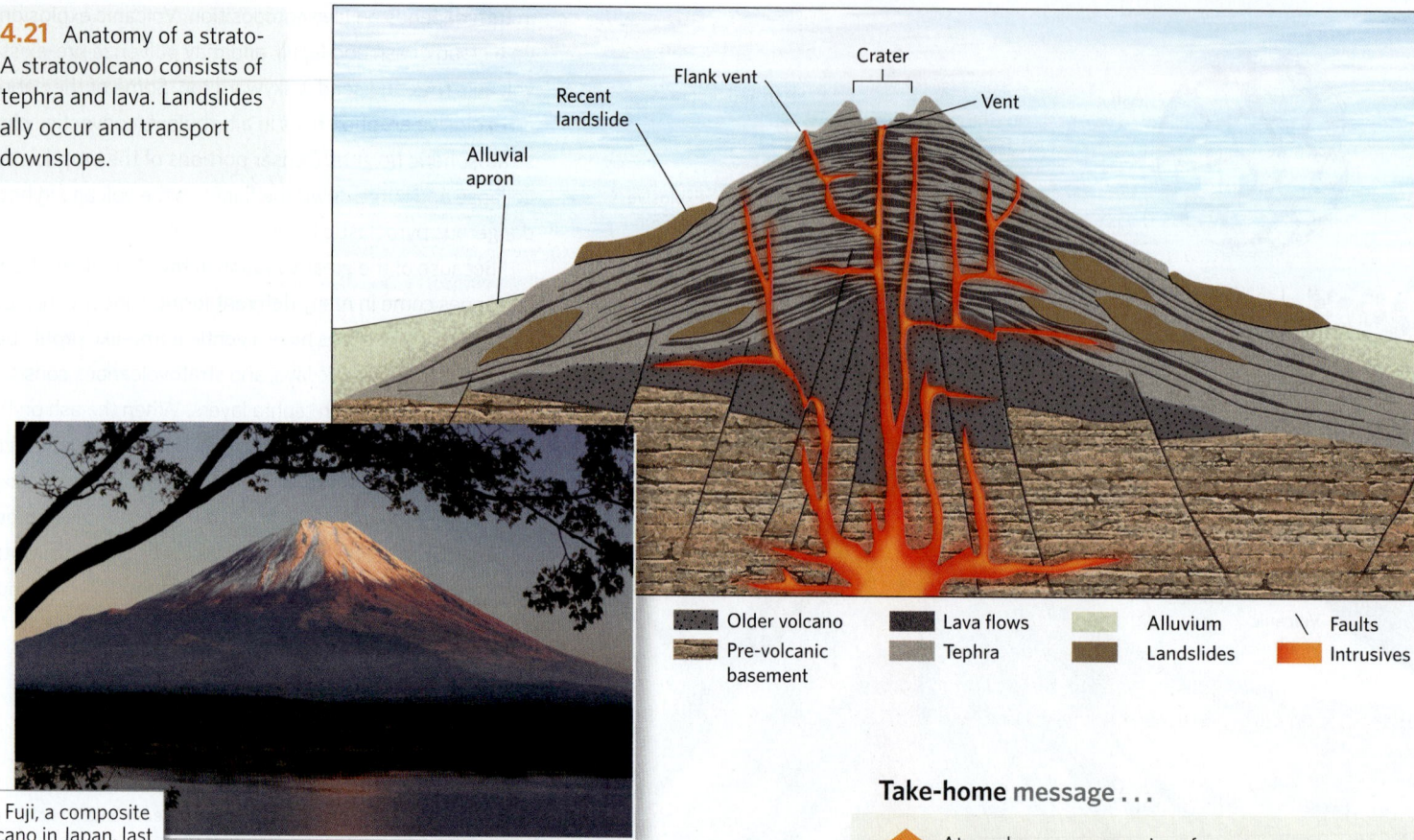

Mt. Fuji, a composite volcano in Japan, last erupted in 1707.

Stratovolcanoes, also known as *composite volcanoes*, tend to be large (up to a few kilometers high and 25 km, or 15 mi, across), cone-shaped mountains made from alternating layers of lava and pyroclastic debris erupted during alternating effusive and explosive eruptions **(Fig. 4.21)**. The prefix *strato–* emphasizes the layered character of the igneous rock that makes up the volcano. Their shape, exemplified by Japan's Mt. Fuji, serves as the classic image most people have of a volcano. In most examples, the upper, steeper part of the cone consists of erupted layers of lava, tephra, and tuff that collected in place, whereas the lower, less steep part consists of deposits formed when pyroclastic debris or broken-up lava either slid or got transported by water downslope. Stratovolcanoes commonly erupt intermediate lava, which doesn't flow far from the summit because it's so viscous. Large explosions occasionally blast off the summit or side of a stratovolcano. When this happens, a new volcanic edifice may grow over the remnants of its predecessor.

After major eruptions, the summit area of a volcano may collapse into the large, drained magma chamber below, producing a **caldera**. Calderas are circular to elliptical depressions, up to tens of kilometers across and several hundred meters deep, that generally have a fairly flat floor covered by lava or pyroclastic debris **(Fig. 4.22)**. Supervolcanic eruptions have produced the largest known calderas.

Take-home message . . .

At a volcano, magma rises from a magma chamber and erupts from a chimney-like conduit or a crack-like fissure. Eruptive style varies—during effusive eruptions, mostly lava erupts, whereas during explosive eruptions, large volumes of pyroclastic debris blast skyward. The shapes of volcanoes include shields, cinder cones, and stratovolcanoes. Collapse following a major explosion can produce a caldera.

Quick Question
What phenomena can trigger a volcanic explosion?

4.7 Beware! Volcanoes Are Hazards

During the eruption of 2018, on Hawaii, lava fountains erupted from fissures that cut right through housing subdivisions, fields, and forests. Lava flows, some up to 6 m (20 ft) thick and traveling at speeds of up to 25 km/h (15 mph), eventually buried almost 35 km² (15 mi²). Clearly, the Hawaiian eruption, as well as the calamities at Mt. St. Helens and Mt. Vesuvius that we've already described, emphasizes that volcanoes are natural hazards with the potential to cause great destruction. Because of the rapid expansion of cities, far more people live in dangerous proximity to volcanoes today than ever before, so if anything, the hazards posed by volcanoes have gotten worse. Let's look at the different kinds of threats posed by volcanic eruptions.

Hazards Due to Lava and Ash

When you think of an eruption, perhaps the first threat that comes to mind is the lava that flows from a volcano. Indeed, on many occasions, lava has overwhelmed towns (Fig. 4.23a–d). Basaltic lava from effusive eruptions poses the greatest threat because it can flow relatively quickly and spread over a broad area. Lava flows have overrun roads, housing developments, and vehicles. Usually people have time to get out of the way of such flows, but not necessarily with their possessions. Sometimes, buildings burst into flames from the intense heat of a nearby flow.

Catastrophic explosions blast large quantities of pyroclastic debris into the air (Fig. 4.23e–g). Close to the volcano, lapilli and blocks tumble from the sky, smashing through buildings or crushing them beneath a blanket up to several meters thick. Winds can carry fine ash over a broad region. In the Philippines, for example, a typhoon spread ash from the 1991 eruption of Mt. Pinatubo over a 4,000-km² (1,500 mi²) area. When this ash sifts down from the sky, it buries crops, poisons soil, and insidiously infiltrates machinery, causing moving parts to wear out. A

pyroclastic flow, racing down the flank of a volcano, can be so hot and poisonous that it means instant death to anyone caught in its path.

The explosive force of an eruption can flatten buildings and forests and, if it occurs in an island, can generate *tsunamis* (sea waves that are not a consequence of the wind, but rather due to sudden displacement of water). Pyroclastic flows produced by such eruptions can be particularly dangerous. For example, in 1902, when a lava dome at the summit of Mt. Pelée, on the island of Martinique, suddenly broke and released the pressure inside the volcano, the ensuing explosion produced a pyroclastic flow that rushed down the side of the mountain and swept through the city of St. Pierre, 10 km (6 mi) away. All but two of the city's 28,000 inhabitants perished.

If ash from an explosive eruption mixes with water, it can start to flow down river valleys in a viscous slurry called a **lahar** (Fig. 4.23h). Because lahars are denser than water, they pack more force than clear water flowing at the same velocity and can carry away everything in their path, including bridges, boulders, and huge trees. Perhaps the most

FIGURE 4.22 Formation of a volcanic caldera.

Time

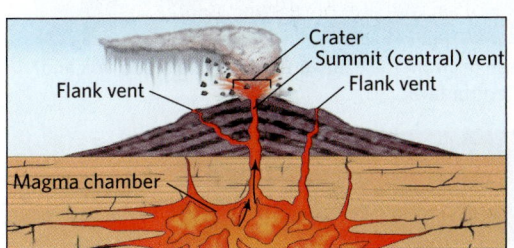

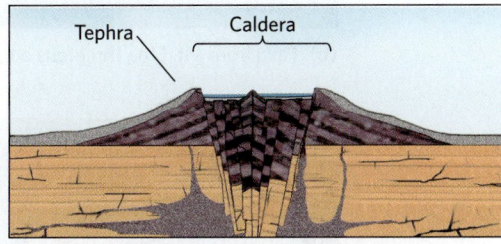

(a) As an eruption begins, the magma chamber inflates with magma. There may be a central vent and one or more flank vents.

(b) During the eruption, the magma chamber drains, and the central portion of the volcano collapses downward.

(c) The collapsed area becomes a caldera. Later, a new volcano may begin to grow within the caldera.

(d) This caldera in Oregon formed about 7,700 years ago. Afterward, it filled with water to become Crater Lake. Wizard Island, protruding from the lake, is a small volcano that grew on top of the caldera floor.

FIGURE 4.23 Volcanic hazards: lava and ash.

Lava Flows

(a) A lava flow reaches a house in Hawaii and sets it on fire.

(b) Lava from Mt. Etna threatens a town and olive grove in Sicily.

(c) Residents rescue household goods after a lava flow filled the streets of Goma, along the East African Rift.

(d) This empty school bus was engulfed by lava in Hawaii.

Pyroclastic Debris

(e) A pyroclastic flow rushes down the slope of the Soufriere Hills Volcano, on the island of Montserrat.

(f) A blizzard of ash fell from the cloud erupted by the Kelud Volcano eruption in Solo, Central Java.

(g) Lapilli falls from an eruption in Iceland.

(h) A lahar submerges farmland in Columbia.

destructive lahar of recent times accompanied the eruption of a snow-crested volcano in Colombia in 1985. The lahar surged down the mountainside into a valley, burying the sleeping town of Armero, 60 km (40 mi) away, with a 5-m (16-ft)-thick layer of mud and entombing the town's 25,000 citizens. Lahars may take place during or soon after an explosive eruption if heavy rain falls on recently deposited ash. They may also occur when snow and ice on a volcano melts and mixes with previously deposited ash.

Volcanic ash and gas that rise to stratospheric heights can be a hazard to airplanes. Ash and acidic aerosols (produced when sulfurous gas dissolves in water) scratch windows and damage the fuselage. Ash sucked into a jet engine melts to produce a glassy coating that restricts air flow. As a consequence, temperature sensors indicate that the engine is overheating, making the engine shut down. This happened to a British Airways 747 that flew through the ash cloud over a volcano on Java in 1982. All four engines failed, and for 13 minutes, the plane glided downward from its cruising altitude of 11.5 km (37,000 ft) as the pilots frantically tried to restart the engines. Finally, at 3.7 km (12,000 ft), the engines cooled sufficiently to roar back to life. The plane headed to Jakarta, where, without functioning instruments, the pilot brought the 263 passengers and crew in for a safe landing. Because of the risk to air travel that ash represents, all airspace in Europe was closed for days following the 2010 ash-rich eruption of Iceland's Eyjafjallajökull volcano. The closure stranded millions of passengers and disrupted the global economy.

Hazards Due to Volcanic Gas

Volcanoes emit significant amounts of gas. This gas plays an important role in the Earth System because it includes H_2O, essential to life, and CO_2, which has played an important role in regulating the atmosphere's temperature over geologic time. On a few occasions, however, accumulations of CO_2 have proved deadly. In 1986, this happened in Cameroon, when CO_2 seeping from a volcanic vent dissolved in the cold water of a lake filling the volcano's crater. A landslide disturbed the cold water, which released the gas suddenly. An invisible cloud of CO_2 silently drifted down the flank of the volcano and suffocated the inhabitants of villages downslope. Sulfur-bearing gases are also problematic. They can form choking clouds that make breathing difficult near the vent, and they can dissolve in atmospheric water to produce acidic aerosols. If the weather conditions are appropriate, sulfur aerosols can accumulate to produce *vog*, polluted air that is very hazy.

Volcanoes and Climate

In 1783, Benjamin Franklin, who was serving as the American ambassador to France, noticed that the summer weather seemed to be unusually cool and hazy. Franklin,

an accomplished scientist as well as a statesman, couldn't resist seeking an explanation for this phenomenon. He learned that in June of 1783, a huge volcanic eruption had taken place in Iceland. He wondered if the "smoke" from the eruption had prevented sunlight from reaching the Earth. Franklin reported this idea at a scientific meeting, and by doing so, was the first scientist to suggest a link between volcanic eruptions and climate.

Franklin's idea was confirmed in 1815, when Mt. Tambora in Indonesia exploded, making the sky so hazy that temperatures dipped by several degrees. It remained so cold in the northern hemisphere that 1816 became known as "the year without a summer." The dreary weather confined Mary Shelley and her literary companions indoors, where they read ghost stories. Lord Byron challenged the companions to write their own ghost stories. Mary Shelley took the task to heart and produced *Frankenstein*. The weather had far-reaching effects, as it caused crops to fail, leading to starvation and emigration. Studies of atmospheric conditions following the 1991 eruption of Mt. Pinatubo in the Philippines provide clear documentation of the short-term effects that an eruption can have on global atmospheric temperature.

How does volcanic activity cool the climate? When a large explosive eruption takes place, ash and aerosols enter the stratosphere and, within weeks, encircle the planet. This material remains suspended in the atmosphere for as long as a few years, and the haze it produces reflects sunlight back to space during the day, so the Earth's surface doesn't warm as much as it would if the eruption hadn't occurred.

Did you ever wonder . . .

whether volcanic eruptions affect air travel?

Take-home message . . .

Volcanoes can be dangerous! Lava flows, pyroclastic flows, ash clouds, explosions, lahars, landslides, and tsunamis produced during eruptions can destroy cities and farmland. Ash in the air can be a hazard for air travel. Ash and aerosols thrown high into the atmosphere reflect sunlight and can cause a temporary global drop in temperature.

Quick Question
Why can a lahar do so much more damage than a similarly sized flood of clear water?

4.8 Protecting Ourselves from Vulcan's Wrath

The Bronze Age Minoan culture thrived on the islands of the eastern Mediterranean beginning around 3650 B.C.E. Around 1645 B.C.E., the Minoan city of Akrotiri vanished, buried beneath thick layers of ash—some of the material that had comprised the Santorini volcano until it exploded. All

FIGURE 4.24 The shape of a volcano changes as it erodes over time.

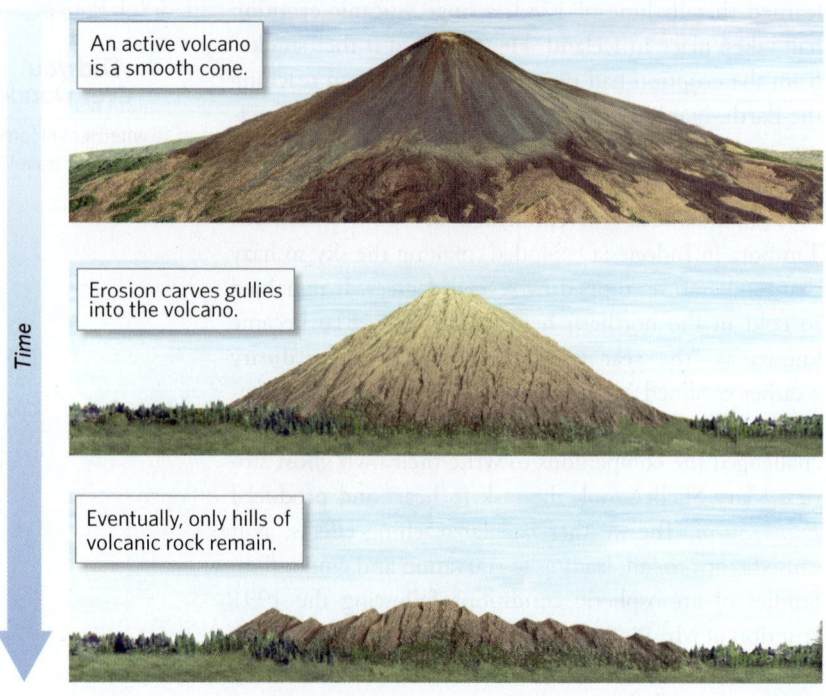

An active volcano is a smooth cone.

Erosion carves gullies into the volcano.

Eventually, only hills of volcanic rock remain.

Time

eruptions (see Chapter 9). The landscape on a volcano's surface also provides clues to activity: a volcano whose surface exposes lava or pyroclastic debris that has not been vegetated or eroded by streams has probably erupted recently; a volcano whose surface has become vegetated and incised by streams has not erupted for a while; and a volcano that has completely eroded away, so that underlying intrusions crop out, has probably long been extinct **(Fig. 4.24)**.

Predicting Eruptions

With data on the timing of the past several eruptions at an active volcano, geologists can calculate the **recurrence interval** of activity at the volcano, meaning the average time between eruptions. Knowledge of the recurrence interval provides a rough estimate of eruption frequency. If, for example, the recurrence interval is 50 years, then the volcano will probably erupt a few times over the course of a couple of centuries, but if the recurrence interval is 5,000 years, the likelihood of an eruption taking place in your lifetime is slim, though it's not impossible.

Because the recurrence interval merely gives the average time between eruptions, it's not possible to predict the exact timing of an eruption years in the future. However, short-term (weeks to months) predictions of impending volcanic activity may be feasible. Some volcanoes send out distinct warning signals of impending eruptions that geologists can monitor:

- *Earthquake activity:* When magma flows into a magma chamber, rocks surrounding the magma chamber break and slip, and small explosions may take place. All this movement causes earthquakes, so in the days or weeks preceding an eruption, the volcano and its surroundings become seismically active.

- *Changes in heat flow:* The intrusion of hot magma into a volcano increases local *heat flow*, the amount of heat passing through rock. In some cases, the increase in heat flow melts snow or ice on the volcano.

- *Increases in gas and steam emission:* Even though magma remains below the surface, gases bubbling out of the magma, or steam formed by the heating of groundwater by the magma, may percolate upward through cracks in the Earth and rise from the volcanic vent **(Fig. 4.25a)**. So an increase in the volume of gas emission, or the formation of new hot springs, indicates that magma has entered the ground below.

- *Changes in shape:* As magma fills the magma chamber inside the volcano, it pushes outward and can cause the surface of the volcano to bulge **(Fig. 4.25b)**.

Mitigating Volcanic Hazards

There's no way to stop an eruption, so what can we do to prevent the loss of life and property near an

that remains of Santorini today is a huge caldera whose rim projects above sea level as the Greek island of Thera. Damage caused by ash clouds and tsunamis generated by the eruption may have contributed to the eventual demise of Minoan civilization and may have been the basis of disaster legends of the ancient Greeks and other cultures. Clearly, volcanic eruptions can be natural hazards of extreme danger. In this section, we examine the evidence that geologists use to determine whether a volcano has the potential to erupt and the steps people can take to mitigate its destructive effects.

Did you ever wonder . . .

whether geologists can predict volcanic eruptions?

Active or Extinct?

The first step in determining whether a volcano may be a hazard is to determine whether it has the potential to erupt. Geologists refer to volcanoes that have erupted during the last several thousand years and might erupt in the future as **active volcanoes**. In contrast, **extinct volcanoes** are those that have shut off entirely and can never erupt again because the geologic conditions leading to eruption no longer exist. Geologists use the term **dormant volcanoes** for active volcanoes that are not currently erupting. Extinct volcanoes are not a worry—active volcanoes are.

To decide whether a volcano is active or extinct, it's necessary to determine when past eruptions took place. If a volcano erupted historically, it's still active. In regions where the historical record doesn't extend back more than a few centuries, geologists use isotopic dating (radiometric dating) to determine the ages of volcanic rocks or of wood buried by pyroclastic debris, and they assign dates to past

Reading a volcanic hazard map

Geologists develop volcanic hazard maps to show the potential impact of a volcanic eruption on the surrounding region. **Figure Bx4.4** shows an example of such a map for the area surrounding Mt. Rainier in Washington State. While lava flows and pyroclastic flows present a significant threat close to the volcano, lahars from a Rainier eruption could travel as far as Tacoma, some 95 km (59 mi) away. The United States Geological Survey (USGS) publishes long-term hazard assessment reports that assess the types and likelihood of hazards at specific volcanic sites and where the affected areas would be in those regions. You can search these reports by state or by specific volcano at https://volcanoes.usgs.gov/vhp/hazard_assessments .html. Do you reside (or attend school) in a region that could be affected by a volcanic eruption? If so, what hazards is your region most vulnerable to? Municipalities that lie within a volcano's impact area, such as Pierce County in Washington State, often have hazard maps, emergency plans, and evacuation routes posted on their websites. It's a good idea to familiarize yourself with these resources if you live in a vulnerable area.

FIGURE Bx4.4 A volcanic hazard map for Mt. Rainier, Washington.

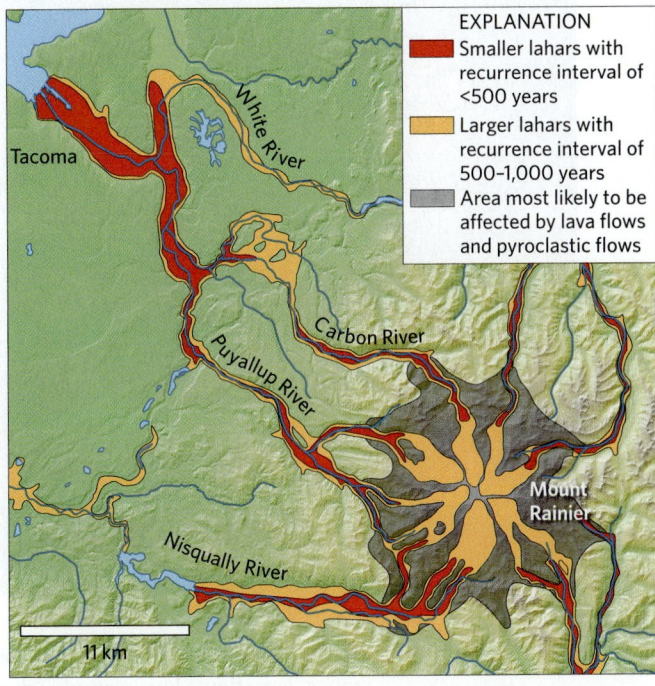

EXPLANATION

- Smaller lahars with recurrence interval of <500 years
- Larger lahars with recurrence interval of 500–1,000 years
- Area most likely to be affected by lava flows and pyroclastic flows

11 km

active volcano? As an important first step, geologists compile a **volcanic hazard map** (Box 4.4), which delineates areas that lie in the path of potential lava flows, lahars, or pyroclastic flows. If an eruption is imminent, people within these danger zones should evacuate. Unfortunately, because of the uncertainty of predicting eruptions, the decision about whether or not to evacuate can be difficult.

FIGURE 4.25 Monitoring the activity of volcanoes.

(a) Monitoring volcanic activity is a risky business, for the gases emitted by volcanoes can be deadly.

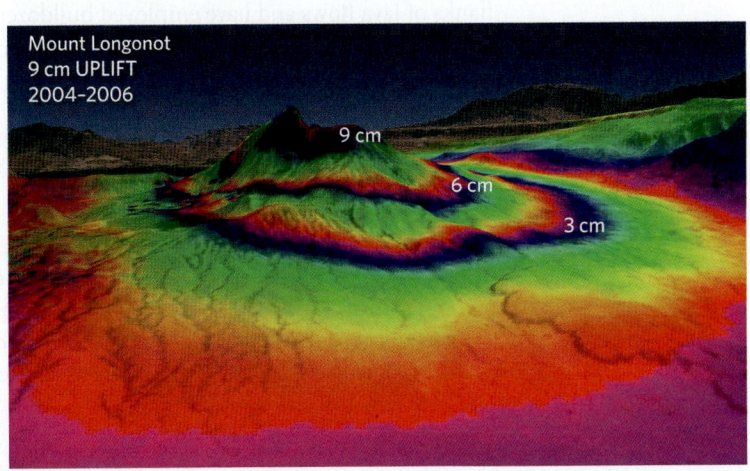

Mount Longonot
9 cm UPLIFT
2004–2006

9 cm
6 cm
3 cm

(b) Ground movements may reflect the intrusion of magma below a volcano. Each color-spectrum band represents 3 cm (1 in) of uplift. The measurements were made by satellite radar.

In a very few cases, people have tried to divert or stop a lava flow. For example, during a 1669 eruption of Mt. Etna, an active volcano on the island of Sicily, a basaltic lava flow approached the town of Catania. Fifty townspeople boldly hacked through the solidified side of the flow to create an opening through which lava could exit. They hoped to cut off the supply of lava feeding the end of the flow that was approaching their homes. Their strategy worked, but unfortunately, the diverted flow began to move toward the neighboring town of Paterno. Five hundred Paternoans chased away the Catanians. The hole was closed, and the flow headed back toward Catania, eventually overrunning part of the town. More recently, people have used high explosives to blast breaches in the flanks of lava flows and have employed bulldozers to build dams and channels to divert lava. Efforts to divert flows from eruptions of Mt. Etna in 1983 and in 1992 were successful (Fig. 4.26).

Take-home message . . .

Geologists can distinguish between active and extinct volcanoes, and they can provide near-term predictions of eruptions, allowing people to take precautions. Volcanic hazard maps help determine what areas should be evacuated. Rarely, it's possible to divert lava flows.

Quick Question

What evidence suggests that a volcano may be about to erupt?

4.9 Where Does Igneous Activity Occur?

Now that we've learned about the importance of igneous activity—both extrusions at volcanoes and intrusions underground—to many aspects of the Earth System as well as to society, we return to the fundamental question of why igneous activity happens where it does. As noted earlier, igneous activity happens only at particular locations on the Earth. Why? Before the theory of plate tectonics was proposed, geologists didn't have a good answer to this question. We now realize that most igneous activity occurs along divergent and convergent boundaries, but some occurs at hot spots and in rifts (Fig. 4.27). In this section, we relate the character of igneous activity to different geologic settings.

Igneous Activity at Mid-Ocean Ridges

During seafloor spreading at a mid-ocean ridge, underlying hot asthenosphere rises from deeper down in the mantle up to shallower depths (see Fig. 4.3b). When the peridotite of the asthenosphere rises above a depth of about 100 km (about 60 mi), it undergoes decompression melting and produces mafic magma, which rises still farther. Some of this magma accumulates in a magma chamber at a depth of 7–4 km (4–2.5 mi) beneath the ridge axis. Slow solidification of some of the crystal mush within this chamber produces gabbro. But some magma continues its upward movement, filling vertical cracks and solidifying to form basalt dikes, and some reaches the seafloor and erupts to form pillow basalt, which forms a layer on the seafloor.

We don't generally see the volcanic activity of mid-ocean ridges because most of it occurs beneath 2 km (1.2 mi) of seawater. Geologists in submersibles, however, have been able to examine these submarine eruptions in a few locations and have found that they take place along fissures parallel to the ridge axis (Fig. 4.28a, b). An individual fissure, which remains active for tens to hundreds of years before becoming inactive, yields elongate mounds of pillow basalt. New oceanic crust along ridges contains many cracks into which seawater can percolate. This water, warmed by magma, dissolves minerals in the crust before reentering the sea at *hydrothermal vents*, cracks that emit hot-water solutions from below the seafloor. When these solutions come in contact with cold seawater, they cool, and the dissolved minerals they contain precipitate to form a dark cloud of finely suspended grains. As a consequence, hydrothermal vents are known as **black smokers** (Fig. 4.28c).

Igneous Activity at Convergent Boundaries

Most *subaerial volcanoes* (those that erupt above sea level) occur along convergent boundaries. The magma that feeds these volcanoes forms because volatile compounds seep from the subducting plate into the overlying asthenosphere and trigger flux melting in the asthenosphere (see Fig. 4.4a). These volatiles come from subducted oceanic crust, which contains volatile compounds due to hydrothermal activity at mid-ocean ridges. When an oceanic plate sinks into the mantle at a convergent boundary, these volatiles are carried with it. At a depth of about 150 km (90 mi), the subducted plate becomes hot enough for bonds holding the volatiles to the crustal minerals to break, and the volatiles escape into the overlying asthenosphere.

Once magma forms in the asthenosphere above the subducted plate, it rises and eventually seeps through the lithosphere of the overriding plate. Some magma freezes underground in dikes, sills, or plutons, but some rises to the Earth's surface and erupts from a chain of volcanoes called a **volcanic arc**. If the overriding plate consists of oceanic lithosphere, the chain is an *island arc* **(Fig. 4.29a)**, whereas if the overriding plate consists of continental lithosphere, the chain is a *continental arc* **(Fig. 4.29b)**.

Many different kinds of rock form at volcanic arcs. If magma rises directly from the mantle, it has a mafic composition and produces basalt. Rising magma, however, may evolve and react with the crust as it moves upward, and it may trigger heat-transfer melting of the crust itself. As a result, volcanic arcs commonly erupt andesite. Typically, both effusive and explosive eruptions take place in volcanic arcs, producing stratovolcanoes. Large volumes

Did you ever wonder . . .

whether we can see all the volcanic eruptions taking place on the Earth today?

FIGURE 4.27 The geologic settings of volcanoes. The map shows the distribution of volcanoes around the world. The diagrams above the map show the five basic types of geologic settings in which volcanoes form, in the context of plate tectonics theory.

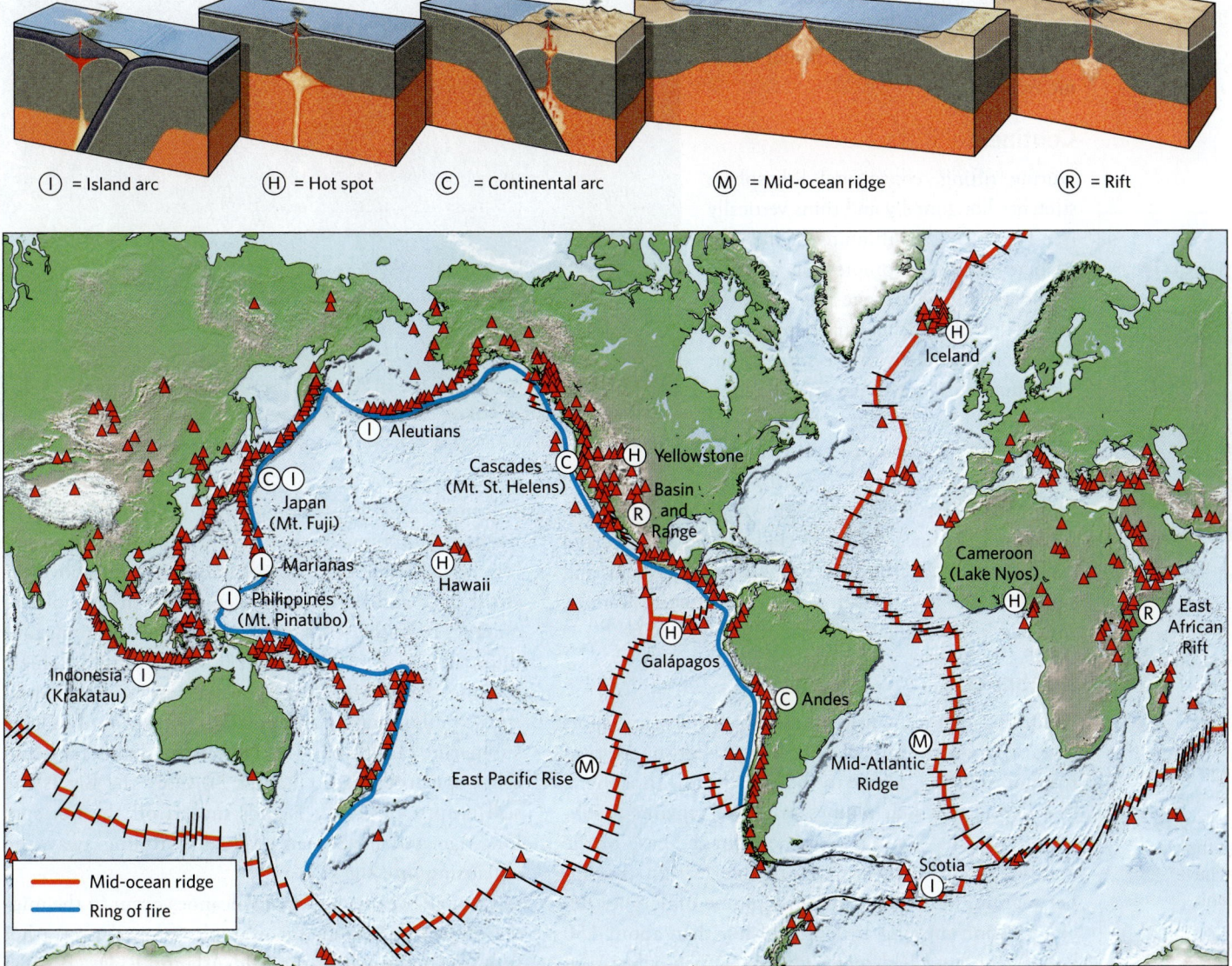

Ⓘ = Island arc   Ⓗ = Hot spot   Ⓒ = Continental arc   Ⓜ = Mid-ocean ridge   Ⓡ = Rift

— Mid-ocean ridge
— Ring of fire

FIGURE 4.28 Igneous activity along a mid-ocean ridge.

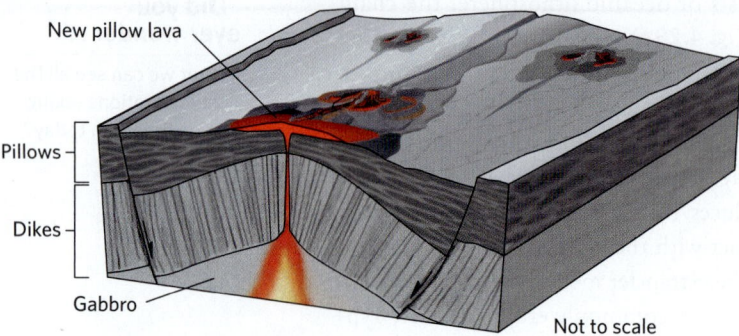

New pillow lava

Pillows

Dikes

Gabbro

Not to scale

(a) Mounds of pillow basalt erupt along fissures.

(b) Pillow basalt on the seafloor along the Juan de Fuca Ridge.

(c) An example of an active black smoker along the Mid-Atlantic Ridge.

of granite and diorite form plutons in the crust beneath continental arcs; in some cases, so much intrusion takes place that large batholiths form beneath the arcs. The Sierra Nevada batholith, for example, formed beneath a continental volcanic arc.

Continental Rifts

During rifting, continental lithosphere stretches horizontally and thins vertically (see Fig 4.3b). The thinning of the lithosphere causes decompression melting of the underlying asthenosphere and the production of mafic magma. Some of this magma rises straight to the surface, but some undergoes fractional crystallization during its rise, or causes heat-transfer melting within the crust, to yield felsic magma. As a consequence, rifts host both basaltic fissure eruptions, in which curtains of lava fountain up or linear chains of cinder cones develop, and explosive rhyolitic eruptions, which produce calderas and vast sheets of tuff. In a few locations, stratovolcanoes, such as Mt. Kilimanjaro, grow in rifts **(Fig. 4.30)**.

Hot Spots

According to the mantle-plume hypothesis, hot spots form where a column of very hot asthenosphere rises from deep in the mantle to the base of the lithosphere. The depth at which plumes originate remains a subject of debate—some may come from the base of the mantle, whereas others may originate within the upper mantle. Hot rock rising in a plume undergoes decompression melting at depths of less than about 150 km (90 mi) to yield mafic magma. When a hot-spot

volcano begins to form on oceanic lithosphere, basaltic magma erupts underwater and yields a mound of pillow lava. With time, the volcano grows up above the sea surface and becomes an island. When the volcano emerges from the sea, successive effusive eruptions produce thousands of thin basalt flows that build a shield volcano **(Fig. 4.31)**. As the volcano grows, it succumbs to the pull of gravity and portions slip seaward in large submarine landslides. The Hawaiian Islands represent a hot-spot track (see Chapter 2). Only the Big Island still erupts; the other islands moved off the hot spot long ago and have been gradually eroding away and collapsing into the sea.

Not all oceanic hot-spot volcanoes occur in the middle of plates. Iceland, for example, formed over a hot spot under the axis of the Mid-Atlantic Ridge. Because

---

See for yourself

Mauna Loa, Hawaii

Latitude: 19°27′0.50″ N
Longitude: 155°36′2.94″ W

Look straight down from 25 km (~15.5 mi).

You can see the NNE-trending crest of Mauna Loa, a shield volcano associated with the Hawaiian hot spot. A large, elliptical caldera dominates the view. Basaltic lava flows have spilled down its flank.

FIGURE 4.29 Volcanic arcs at convergent boundaries.

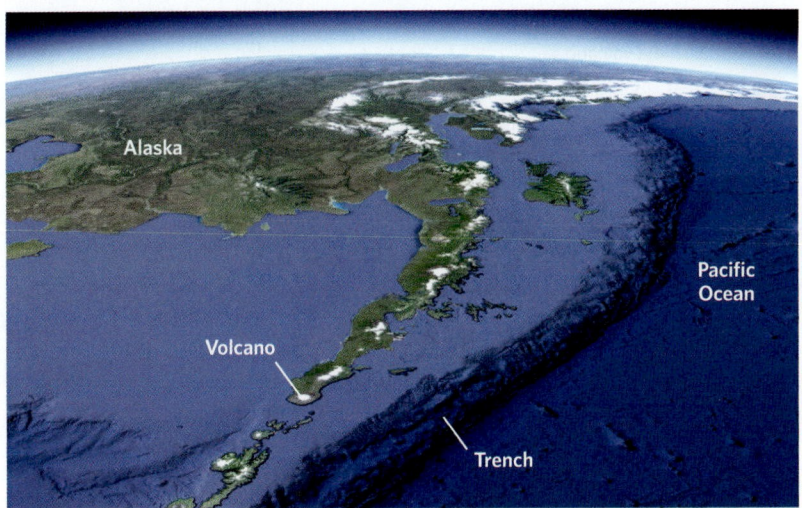

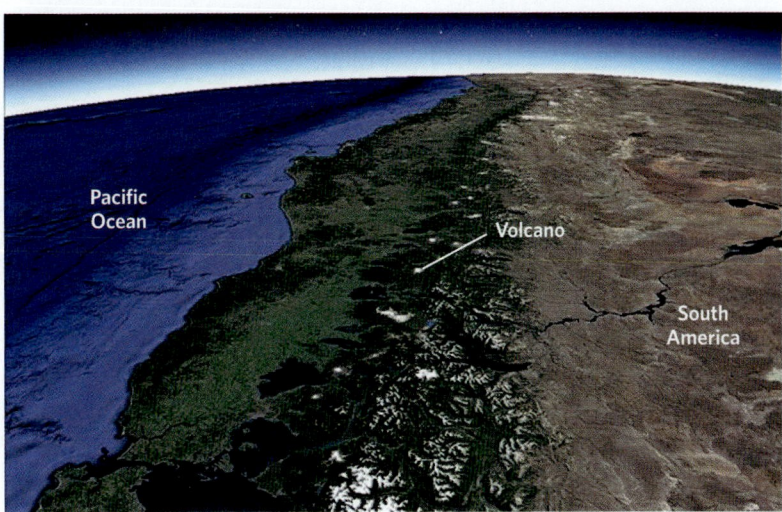

(a) The Aleutian Islands of Alaska, viewed looking northeast. The white peaks are volcanoes. This volcanic island arc formed at a convergent boundary.

(b) The Andes lie near the western coast of South America. This mountain chain is a continental volcanic arc.

of this hot spot, far more magma erupts in Iceland than at other places along the ridge, and these eruptions have built a broad plateau of basalt **(Fig. 4.32a)**. Iceland straddles a divergent boundary, so plate motion actively stretches the island apart. Indeed, the central part of the island is a narrow rift—the trace of the Mid-Atlantic Ridge **(Fig. 4.32b)**.

So far, we've focused on hot spots in oceanic lithosphere. Hot spots also occur in continental lithosphere.

The best-known example, Yellowstone National Park, lies at the northeastern end of a string of calderas **(Fig. 4.33)**. Unlike the Hawaiian hot spot, the Yellowstone hot spot has erupted both basaltic and rhyolitic lava and pyroclastic debris, because basaltic magma rising from the mantle causes heat-transfer melting in the continental crust, a process that yields felsic magma. As we've noted, huge supervolcanic explosive eruptions have occurred in association with the Yellowstone hot spot (see Box 4.3). These

Animation

Hot-Spot Formation

FIGURE 4.30 Igneous activity in continental rifts.

(a) These recent cinder cones have been cut by faults associated with rifting.

(b) Mt. Kilimanjaro, the highest mountain in Africa (5.9 km, or 3.7 mi high), is a stratovolcano that has formed in a rift zone.

FIGURE 4.31 The structure of an oceanic hot-spot volcano is complicated.

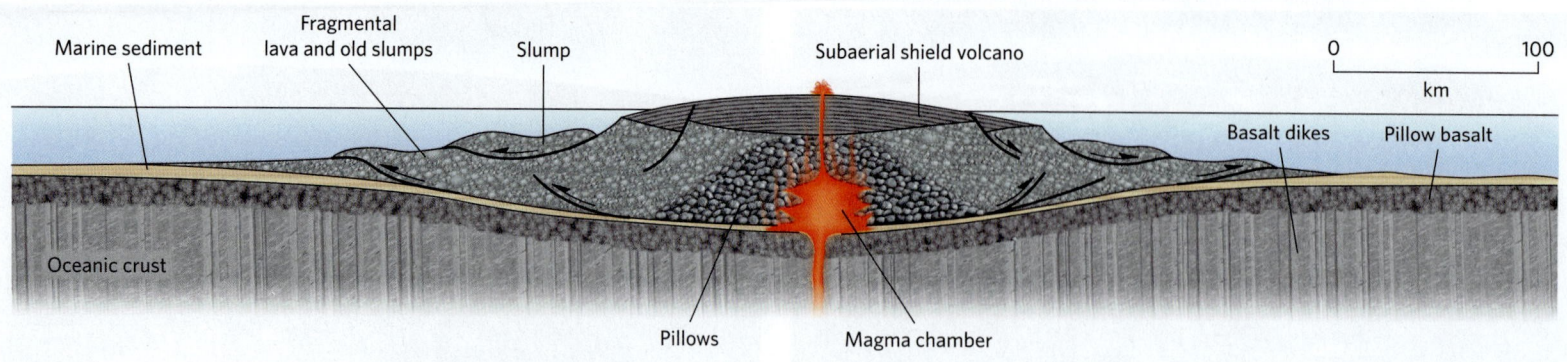

eruptions formed the thick yellow, tan, and orange layers of tuff that led to the park's name and produced huge calderas.

Large Igneous Provinces

In several locations around the world, particularly large volumes of low-viscosity mafic lava have erupted over a relatively short time and spread out in vast flows, some of which extend 500 km (300 mi) from the vent. Geologists refer to the rock formed from these flows as **flood basalt (Fig. 4.34a)**. Successive flood-basalt eruptions build a broad *basalt plateau* whose total volume may be immense (up to 175,000 km³, or 42,000 mi³). Geologists refer to such a region as a **large igneous province** or **LIP**. An example of a LIP forms the Columbia River Plateau of the northwestern United States—here,

a 3.5-km-thick succession of flood-basalt layers accumulated at 15 Ma. Several immense submarine basalt plateaus have also been identified on the ocean floor. The largest of these *oceanic plateaus*, the Ontong-Java Plateau in the western Pacific, has an area of 2,000,000 km² (770,000 mi²).

What causes LIP flood-basalt eruptions? According to one hypothesis, flood basalts form when a large mantle plume rises beneath a region that is undergoing rifting **(Fig. 4.34b)**. As the bulbous head of the plume reaches the base of the lithosphere, a particularly large volume of mantle undergoes decompression, and because rock in the plume is particularly hot, more partial melting takes place than is usual. When conduits form in the rift and provide access for the melt to reach the surface, immense fissure eruptions take place.

FIGURE 4.32 Iceland lies over a hot spot on the Mid-Atlantic Ridge.

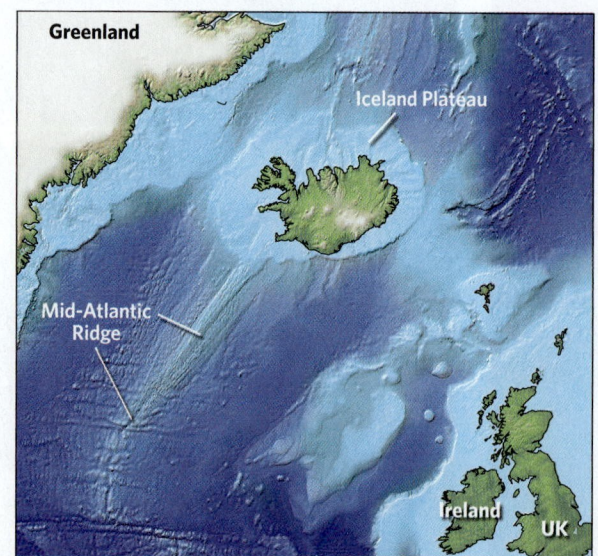

(a) A bathymetric map shows that Iceland sits atop a huge oceanic plateau straddling the Mid-Atlantic Ridge. Light blue is shallower water; dark blue is deeper.

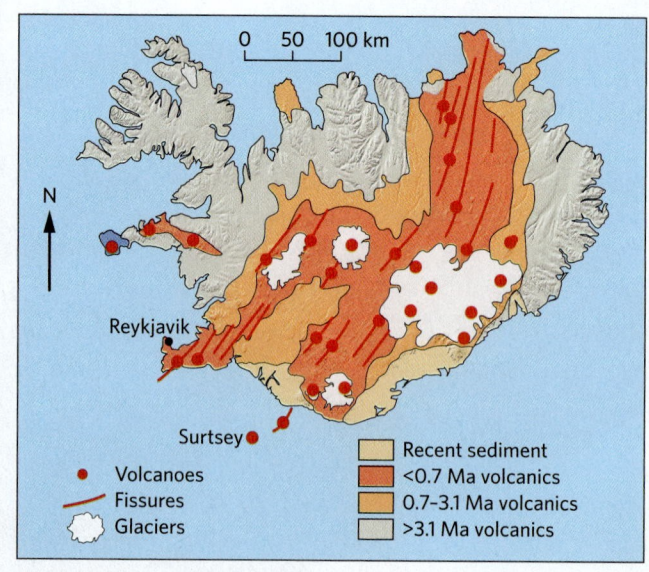

(b) A geologic map of Iceland shows that the youngest volcanoes occur in the central rift, which is effectively the on-land portion of the Mid-Atlantic Ridge.

A resurgent dome is a bulge formed when a magma chamber inflates.

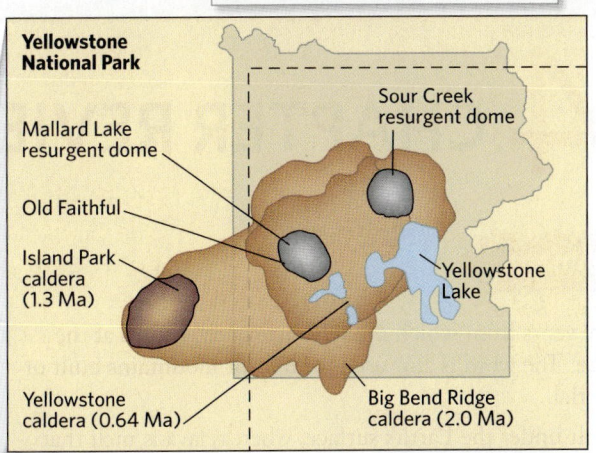

(b) Yellowstone overlies a huge caldera. Immense eruptions occurred here at least three times during the past 2 million years.

FIGURE 4.33 Volcanic activity at the Yellowstone hot spot.

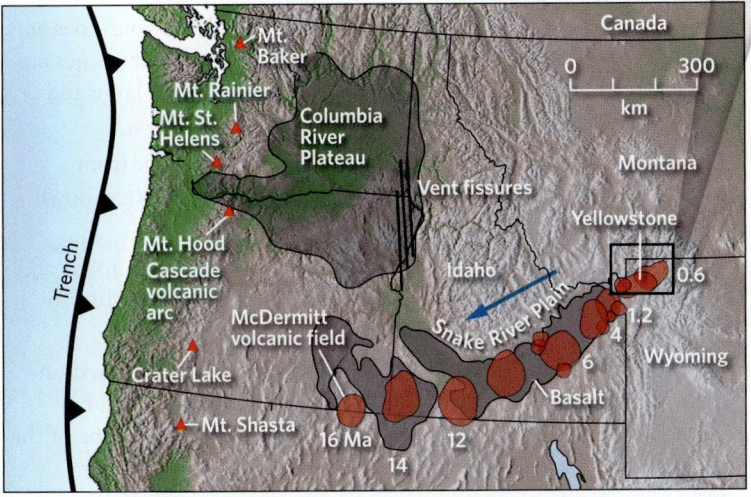

(a) Yellowstone National Park lies at the end of a continental hot-spot track. Progressively older calderas follow the Snake River Plain to the west. The blue arrow indicates plate motion.

(c) Felsic tuffs form the colorful walls of Yellowstone Canyon.

FIGURE 4.34 Large igneous provinces (LIPs).

(a) Flood basalts form the layers exposed in Palouse Canyon, Washington.

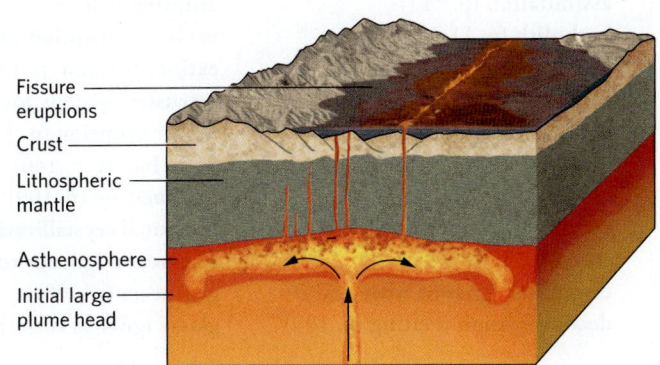

(b) The plume model for formation of flood basalts.

4 CHAPTER REVIEW

Chapter Summary

- Volcanoes are vents from which molten rock (melt) erupts at the Earth's surface. The word is also used for hills or mountains built of erupted material.

- Magma is melt under the Earth's surface, whereas lava is melt that has erupted from a volcano at the Earth's surface.

- Magma forms under special conditions that cause hot rock in the Earth to melt. Melting occurs due to decompression, addition of volatiles, or heat transfer. Generally, only a small proportion of a magma source rock melts to become magma.

- Magma composition reflects the composition of the magma source and the way the magma evolves.

- Once formed, magma rises because of its buoyancy and because of pressure caused by the weight of overlying rock.

- Intrusive igneous rocks form when magma intrudes into pre-existing rock underground. Extrusive igneous rocks form when lava or debris extrudes from a volcano at the Earth's surface.

- The rate at which magma cools depends on the depth, size, and shape of an intrusion.

- Tabular intrusions include dikes and sills. Blob-shaped intrusions are called plutons. Huge composites of many plutons are batholiths.

- The viscosity of lava depends largely on its composition. Viscous felsic lava flows tend to pile into domes at a volcano's vent, whereas less viscous mafic lava can flow great distances.

- Molten rock may be extruded and solidify in lava flows, or it may explode into the air to form ash, lapilli, and other pyroclastic debris.

- Igneous rocks are classified according to texture and composition. Some igneous rocks consist of glass or of pyroclastic debris. Crystalline igneous rocks contain interlocking crystals, whereas fragmental igneous rocks consist of fragments welded or cemented together.

- Effusive volcanic eruptions are dominated by lava flows or fountains, whereas explosive volcanic eruptions blast large quantities of pyroclastic debris skyward.

- A volcano's shape depends on its eruptive style. Shield volcanoes are broad, gentle domes formed by effusive eruptions. Cinder cones are symmetrical hills of lapilli. Stratovolcanoes can become large and consist of alternating layers of pyroclastic debris and lava.

- Eruptions may occur in a crater at a volcano's summit or from fissures on its flanks. Collapse of a volcano produces a huge, bowl-shaped depression called a caldera.

- Volcanic eruptions pose many hazards: lava flows overrun roads and towns, ash falls blanket the landscape, pyroclastic flows incinerate everything in their path, and lahars bury the land surface.

- We can distinguish between active and extinct volcanoes based on their eruption history. Imminent eruptions can be predicted by earthquake activity, changes in heat flow, changes in the shape of the volcano, and emissions of gas and steam.

- Plate tectonics theory can explain where igneous activity happens. Magma forms at convergent boundaries due to flux melting, and at divergent boundaries, rifts, and hot spots due to decompression melting. In continental crust, heat-transfer melting can also take place.

Key Terms

a'a' (p. 115)
active volcano (p. 134)
ash (p. 117)
assimilation (p. 111)
batholith (p. 114)
black smoker (p. 136)
block (p. 117)
Bowen's reaction series (p. 111)
caldera (p. 130)
cinder cone (p. 124)
columnar jointing (p. 113)
crater (p. 124)
crystalline igneous rock (p. 120)
decompression melting (p. 109)

dike (p. 114)
dormant volcano (p. 134)
effusive eruption (p. 121)
eruptive style (p. 121)
explosive eruption (p. 121)
extinct volcano (p. 134)
extrusive igneous rock (p. 113)
fissure eruption (p. 124)
flood basalt (p. 140)
flux melting (p. 109)
fractional crystallization (p. 111)
fragmental igneous rock (p. 120)
geotherm (p. 108)
glassy igneous rock (p. 120)

heat-transfer melting (p. 109)
igneous activity (p. 108)
igneous intrusion (p. 113)
igneous rock (p. 108)
intrusive igneous rock (p. 113)
lahar (p. 131)
lapilli (p. 117)
large igneous province (LIP) (p. 140)
lava (p. 108)
lava dome (p. 115)
lava flow (p. 115)
lava fountain (p. 117)
lava tube (p. 115)

magma (p. 108)
magma chamber (p. 114)
melt (p. 107)
obsidian (p. 120)
pahoehoe (p. 115)
partial melting (p. 111)
pillow basalt (p. 117)
pluton (p. 114)
pumice (p. 120)
pyroclastic debris (p. 117)
pyroclastic flow (p. 117)
pyroclastic rock (p. 120)
recurrence interval (p. 134)
scoria (p. 120)

shield volcano (p. 124)
sill (p. 114)
stratovolcano (p. 130)
supervolcano (p. 126)

tuff (p. 120)
vesicle (p. 120)
viscosity (p. 110)
volcanic arc (p. 137)

volcanic bomb (p. 118)
volcanic breccia (p. 121)
volcanic eruption (p. 108)
volcanic hazard map (p. 135)

volcano (p. 107)
wall rock (p. 113)

Review Questions

Letters in parentheses correspond to the chapter's learning objectives.

1. Describe the three processes that are responsible for causing rock inside the Earth to melt. **(A)**

2. Why does molten rock rise from depth to the surface of the Earth, and why does it solidify? **(A)**

3. Explain the difference between an intrusive and an extrusive igneous rock. Why do they have different characteristics? **(B)**

4. What is the difference between a sill and a dike, and how do both differ from a pluton or batholith? Identify the intrusions shown in the figure. **(B)**

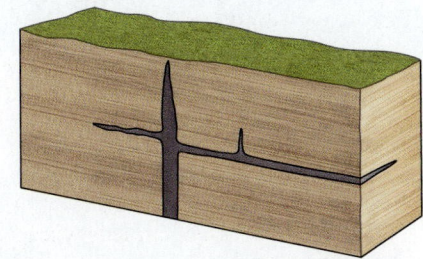

5. Why are there so many different compositions of magma? Does partial melting produce magma with the same composition as the magma source from which it was derived? **(C)**

6. Explain the process of fractional crystallization. **(C)**

7. What factors control the cooling rate of molten rock, and how does cooling rate affect texture? **(C)**

8. What factors control the viscosity of a melt, and how does viscosity affect the behavior of magma or lava? **(D)**

9. Describe the different kinds of material that can erupt from a volcano. Identify them on the figure. **(D)**

10. Describe the differences among shield volcanoes, stratovolcanoes, and cinder cones. What factors determine which type of volcanic edifice develops? **(D)**

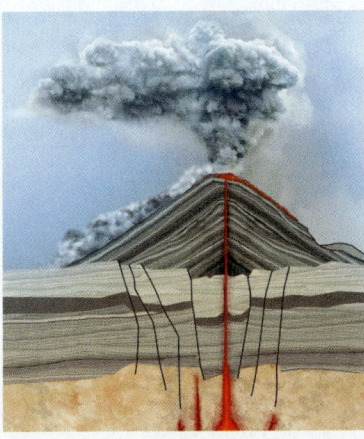

11. Identify some of the major volcanic hazards and explain how they develop. **(E)**

12. To what extent can geologists predict volcanic eruptions, and what observations provide the basis for a prediction? **(F)**

13. Explain how steps can be taken to protect people from the effects of eruptions. **(F)**

14. Why does melting take place beneath the axis of a mid-ocean ridge? **(G)**

15. Why do magmas form in association with subduction? **(G)**

16. What process in the mantle may be responsible for causing hot-spot volcanoes to form? **(G)**

17. Describe how magmas are produced at continental rifts. Why can you find both basalt and rhyolite in such settings? **(G)**

18. What is a large igneous province (LIP), and how might it form? **(G)**

On Further Thought

19. Do people living near the volcanoes of Hawaii face the same kinds of volcanic hazards as do people living near Mt. Rainier in the northwestern United States? **(E)**

20. The Cascade Range of the northwestern United States is only about 800 km (500 mi) long. The volcanic chain of the Andes is several thousand kilometers long. Look at a map showing the Earth's plate boundaries and explain why the Andes are so much longer than the Cascade Range. **(G)**

5 A SURFACE VENEER
Sedimentary Rocks

By the end of the chapter you should be able to …

A. describe the process of weathering and sediment production, and distinguish between physical and chemical weathering.

B. explain the difference between sediment and soil, and describe how soil forms and why conserving it has become a challenge.

C. outline the stages involved in the transformation of sediment into clastic sedimentary rock, and recognize the major types of clastic sedimentary rocks.

D. provide examples of biochemical, chemical, and organic sedimentary rocks, and discuss how each forms.

E. relate examples of sedimentary structures to the environments in which they develop.

F. associate kinds of sedimentary strata and structures with specific depositional settings.

G. explain why layers of marine strata lie beneath what is now dry land in continental interiors, and why the thickness of sedimentary strata varies from place to place.

5.1 Introduction

What lies beneath the floor of the Mediterranean Sea? No one really knew until the 1970s, when researchers drilled a 1-km (0.6-mi)-deep hole into the sea's floor. They expected to find layers of clay, sand, and gravel (materials that had been carried to the sea by rivers), as well as layers of shells (materials left behind when marine organisms died). The researchers did find these layers, but they also found thick layers of halite and gypsum, salts that crystallize when seawater dries up **(Fig. 5.1a)**. The presence of these salts puzzled the researchers, because to form them, the Mediterranean would have had to evaporate almost completely multiple times. Could that really have happened?

The answer proved to be yes. When the African Plate started to collide with the Eurasian Plate several million years ago, land blocked off the Straits of Gibraltar, the only opening between the Atlantic Ocean and the Mediterranean. Since rivers supply only 10% of the water needed to keep the Mediterranean full, cutting off inflow from the Atlantic indeed caused the Mediterranean to dry up until it became a 2-km (1.2-mi)-deep depression, or *basin*, whose surface was a salt-encrusted desert **(Fig. 5.1b)**. This basin remained dry until sea level in the Atlantic rose sufficiently for water to spill over the land bridge between Africa and Europe, causing the Straits of Gibraltar to become a gigantic waterfall that refilled the Mediterranean. This process of filling and drying up was repeated several times as sea level rose and fell over time.

Geologists refer to materials such as those found beneath the floor of the Mediterranean—clay, gravel, sand, shell accumulations, and salt—as sediment. More precisely, **sediment** consists of one or more of the following: loose fragments of rocks or minerals; shells and shell fragments; or mineral crystals that precipitate directly from water. A loose fragment of pre-existing rock, mineral, or shell is called a **clast** (from the Greek *klastos*, broken). Clasts come in a variety of sizes, ranging from too small to see, even with a microscope, to car-sized or even larger. Each clast size category has a name **(Table 5.1)**. Geologists commonly use the term **grain** as a synonym for a small *clast*, as well as for a single crystal of a mineral.

Researchers who drill deeply into the seafloor find that the character of seafloor sediments changes progressively

<< Ledges of sandstone and siltstone crop out on a cliff in northwestern China. New sediment (sand and gravel) has accumulated at the base of the cliff. The wind has sculpted the sand surface into ripples.

FIGURE 5.1 Evidence that the Mediterranean Sea was once dry.

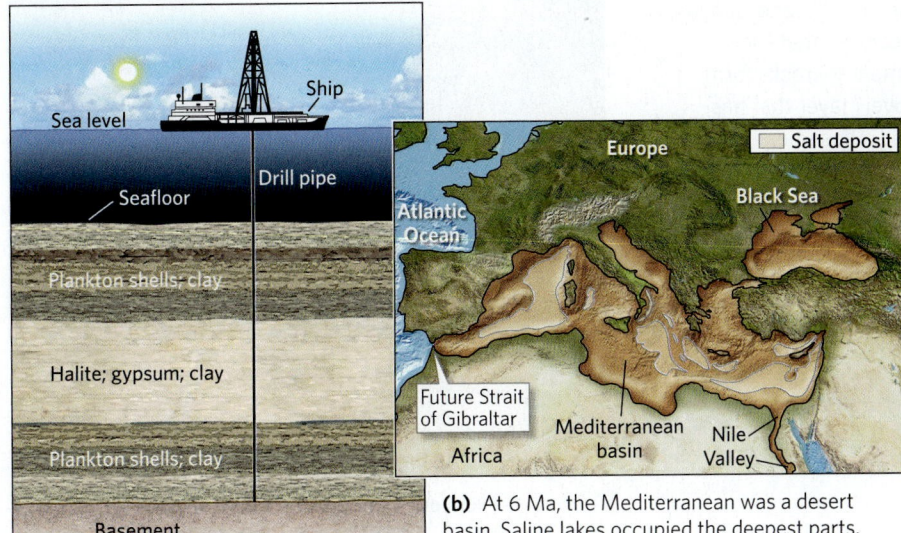

(a) A cross section displaying the sediment layers beneath the floor of the Mediterranean Sea. (Not to scale)

(b) At 6 Ma, the Mediterranean was a desert basin. Saline lakes occupied the deepest parts.

with depth. In the upper few hundred meters, sediment is *unconsolidated*, meaning that it easily separates into grains. Deeper down, sediment becomes *consolidated*, meaning that its grains pack together tightly. Deeper still, minerals that have precipitated from water fill gaps between the grains and bind them together. This change represents the transformation of a sediment into solid *sedimentary rock*, a process called *lithification*. More broadly defined, **sedimentary rock** is rock that forms at or near the surface of the Earth in one of the following ways: (1) by the compacting and cementing together of clasts produced by fragmentation of pre-existing rock; (2) by the growth of shell masses or by the cementing together of shells and shell fragments; (3) by the accumulation and alteration of organic matter left after the death of plants or plankton; or (4) by the precipitation of minerals directly from

TABLE 5.1 Names for Clasts of Different Sizes

Clast Name	Diameter
Boulder	More than 256 mm (10 in)
Cobble	Between 64 mm (2.5 in) and 256 mm
Pebble	Between 2 mm (0.0787 in) and 64 mm
Sand	Between $\frac{1}{16}$ mm (0.0025 in) and 2 mm
Silt	Between $\frac{1}{256}$ mm (0.00015 in) and $\frac{1}{16}$ mm
Clay	Less than $\frac{1}{256}$ mm

FIGURE 5.2 Standing at the edge of the inner gorge in the Grand Canyon, you can see that the sedimentary rocks form a "cover" layer that has buried a basement of old metamorphic rocks.

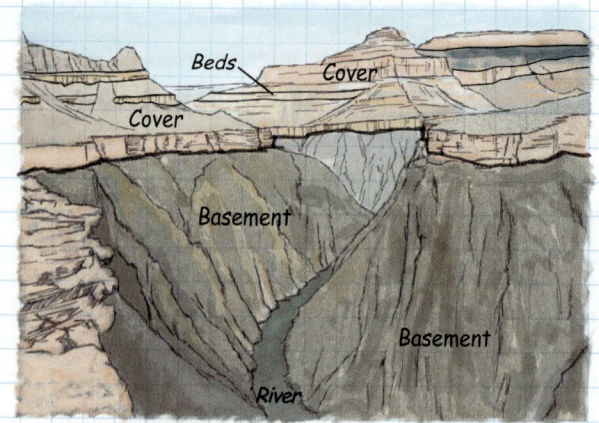

What an Earth Scientist Sees

water solutions. Sediments, and sedimentary rocks, occur in distinct horizontal layers called *beds*, which form a "cover," or veneer, that overlies a *basement* of igneous or metamorphic rock **(Fig. 5.2)**.

Like pages from a history book, beds of sediments or sedimentary rocks record ancient events and ancient environments—the story of the Earth's past. To help you understand how to read this record, we begin this chapter by discussing the production of sediment by *weathering*, a process that breaks down pre-existing rock to form clasts and dissolved ions. Then we turn our attention to the formation of sedimentary rock and show how the study of sedimentary rock can help us to characterize ancient environmental conditions. Finally, in this context, we introduce different classes of sedimentary rock and explain how and where they form.

FIGURE 5.3 Evidence of weathering.

(a) A statue that shows the effects of weathering.

(b) This outcrop shows the contrast between weathered and fresh granite. Note the hammer for scale. The white dashed line is the boundary between weathered and fresh.

5.2 Weathering and the Formation of Sediment

Have you ever seen an old marble statue that has been standing outside for a long time? Look closely and you'll notice that the details the sculptor worked so hard to create have been blunted and that cracks have formed in the stone **(Fig. 5.3a)**. The statue is undergoing **weathering**, the combination of processes that gradually modify and weaken rock when exposed to air and water at or just below the Earth's surface. We refer to the cracking and breaking up of rock as **physical weathering** (or *mechanical weathering*) and to the chemical reaction of rock with air and water as **chemical weathering**. Weathering ultimately causes *fresh rock*, meaning unweathered rock, to disintegrate and turn into new sediment composed of loose grains **(Fig. 5.3b)** as well as ions dissolved in water. Significantly, weathering makes rock more susceptible to **erosion**, the process by which glaciers, rivers, wind, or waves remove sediment and carry it away.

Prying Rock Apart: Physical Weathering

The process of physical weathering begins with the formation of natural cracks, or **joints**, in rock. Joints develop for many reasons. For example, when erosion removes *overburden* (overlying rock), rock that had been many kilometers below the surface of the Earth rises to a shallower depth. In the process, the rock undergoes slight expansion and cooling because pressures and temperatures progressively decrease toward the surface. The resulting change in shape can cause the rock to crack.

Almost all rock outcrops contain joints, which are generally spaced from centimeters to meters apart. Sedimentary rocks commonly break along planar joints oriented perpendicular to *bedding planes* (the boundaries between beds), so that beds break into rectangular blocks **(Fig. 5.4a)**. In bodies of homogeneous crystalline rock, such as granite, *exfoliation joints* can split the rock into onion-like sheets oriented roughly parallel to the ground surface **(Fig. 5.4b, c)**. Joints may also be irregular

FIGURE 5.4 Jointing breaks rocks into blocks or sheets that can separate from an outcrop.

100-m-high sandstone cliff, Ireland

Joint

Bedding

Blocks

(a) Vertical joints break beds of sedimentary rock into blocks.

Joints in granite, California

(b) Exfoliation joints on a sloping granite mountain face.

Different rock types display different types of joints.

Sandstone

Granite

Bedding plane

Vertical joints

Exfoliation joints

Acadia National Park, Maine

Exfoliation joint surface

Vertical joint trace

(c) Nearly horizontal exfoliation joints exposed in a road cut.

FIGURE 5.5 Wedging is one type of physical weathering.

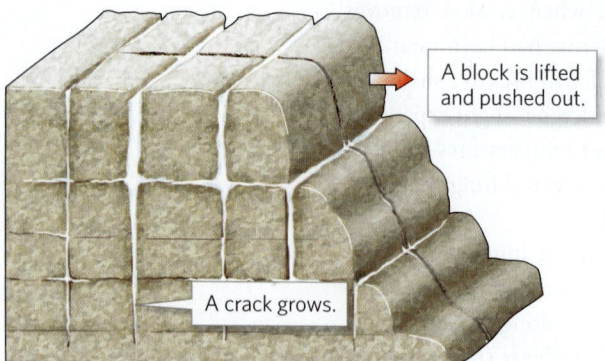

(a) Frost wedging occurs when water fills joints, freezes, expands, and wedges the joints open.

(b) Root wedging pushes open a joint, slowly separating a block from the cliff.

surfaces that break rocks into jagged pieces. A number of phenomena push joints open, causing them to widen and lengthen. For example, during *frost wedging*, water trapped in a joint freezes and expands **(Fig. 5.5a)**, and during *root wedging*, a tree root grows and expands **(Fig. 5.5b)**. Development of joints transforms formerly intact bedrock into separate blocks, which eventually break free from the bedrock, tumble downslope, and break into smaller blocks.

Physical weathering can also act at the grain scale. For example, in dry climates or along the seashore, salty water seeps into open spaces, known as pores, within the rock. As the water evaporates, salt crystals precipitate in the pores, causing *salt wedging* by pushing apart the adjacent grains.

Altering Rock: Chemical Weathering

During chemical weathering, molecules in air and water react with minerals in rock by breaking chemical bonds. A variety of different chemical reactions can affect rock. For example, during **dissolution**, chemical components of minerals become ions in a water solution. This reaction primarily affects relatively soluble salts and carbonate minerals **(Fig. 5.6)**, but even quartz grains eventually dissolve slightly. During *hydrolysis*, reactions with water transform existing minerals into new ones. For example, hydrolysis transforms feldspar and many other silicate minerals into clay. Iron-bearing minerals can undergo *oxidation*, during which the minerals react with atmospheric oxygen or with oxygen dissolved in water. In effect, oxidation causes rocks to "rust," yielding

FIGURE 5.6 Dissolution is a form of chemical weathering.

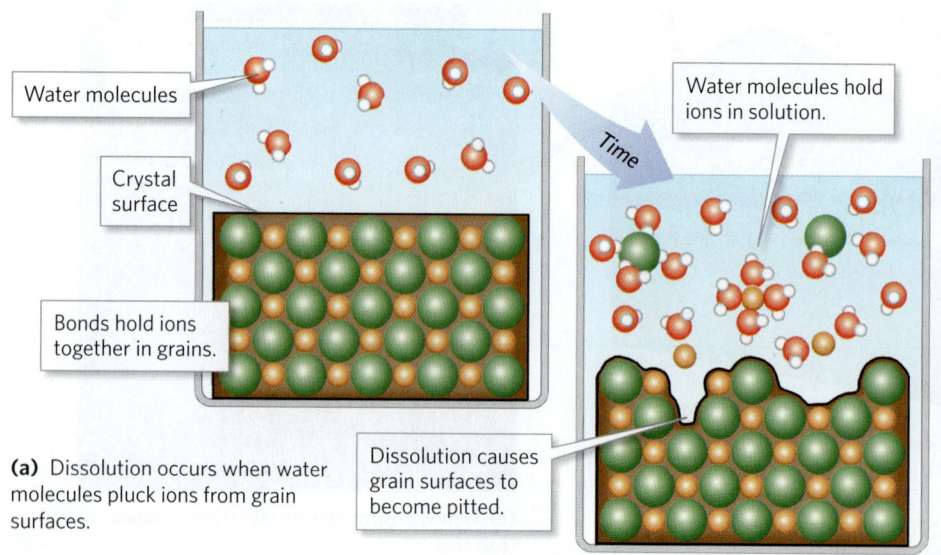

(a) Dissolution occurs when water molecules pluck ions from grain surfaces.

(b) Dissolution along joints intersecting the surface of limestone bedrock in Ireland produced these troughs.

a reddish-brown mixture of weak iron oxide and iron hydroxide minerals. Note that different minerals undergo different chemical weathering reactions. For example, feldspar weathers to clay, but quartz doesn't, so weathering of granite eventually produces both quartz grains and clay.

Recent research emphasizes that organisms can play a major role during chemical weathering. For example, roots of plants, fungi, and lichens secrete organic acids that help dissolve minerals in rocks. And some bacteria and archaea, tiny single-celled organisms, literally eat minerals for lunch, in that they pluck molecules from minerals and use the energy from the molecules' chemical bonds to support their metabolism.

Physical and Chemical Weathering Working Together

So far, we've considered the processes of chemical and physical weathering separately. In the natural world, however, they work together. Physical weathering speeds up chemical weathering because, by breaking up rock, it provides more surface area that can come in contact with air or water (Fig. 5.7a). In turn, chemical weathering speeds up physical weathering, either by dissolving cements that hold rock grains together or by transforming hard minerals such as feldspar into soft minerals such as clay. Both processes weaken rock, allowing it to disintegrate more easily (Fig. 5.7b). In some environments, the products of weathering become the raw materials from which soil can form (Box 5.1).

Not all rock types weather at the same rates. We say that rocks that weather more slowly are *resistant* to weathering, whereas rocks that weather more rapidly are *nonresistant*. As a result of such *differential weathering*, cliffs composed of alternating resistant and nonresistant rock layers take on a stair-step or sawtooth shape (Fig. 5.8a). You can easily see the consequences of differential weathering in a graveyard—inscriptions on granite

headstones remain sharp and clear for centuries, whereas those on marble headstones become blunted within decades (Fig. 5.8b), for granite consists of quartz and other durable silicate minerals, whereas marble consists of calcite, a carbonate mineral that dissolves relatively easily in acidic rainwater. Notably, weathering tends to blunt edges and round corners over time because weathering attacks a flat rock face from only one direction, an edge from two directions, and a corner from three directions (Fig. 5.9).

Video
Forming Soil

In every grain of sand, there is the story of the Earth.

—RACHEL CARSON (AMERICAN NATURALIST, 1907–1964)

Take-home message . . .

During physical weathering, rock breaks into pieces, and during chemical weathering, air and water chemically react with minerals in rock. Working together, chemical and physical weathering cause rock to disintegrate.

Quick Question
Why do the corners of blocks of rock tend to become blunted over time?

FIGURE 5.7 Physical and chemical weathering processes work together.

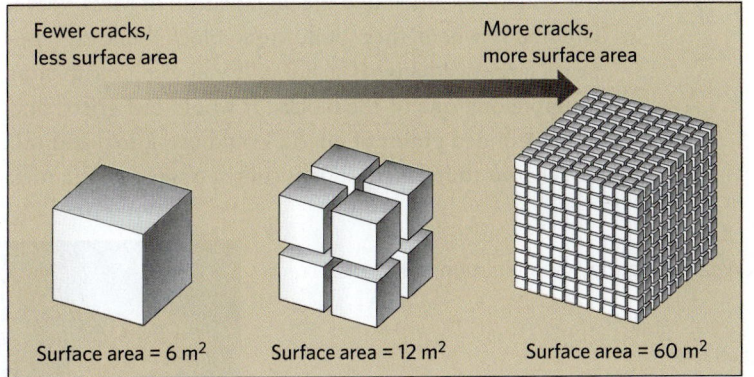

Fewer cracks, less surface area → More cracks, more surface area

Surface area = 6 m²   Surface area = 12 m²   Surface area = 60 m²

(a) As rock breaks apart due to physical weathering, the surface area increases relative to the volume; therefore, more rock surface is exposed to chemical weathering.

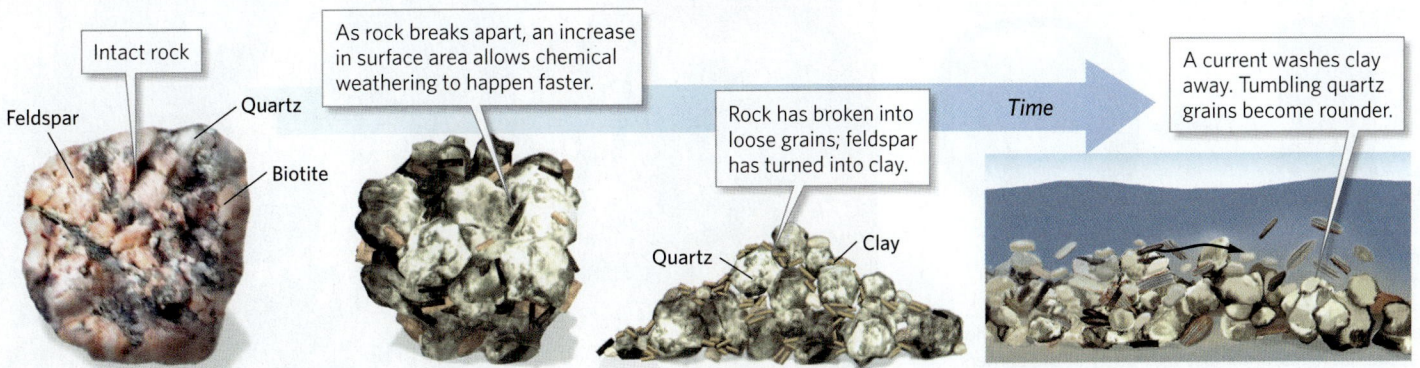

Intact rock

As rock breaks apart, an increase in surface area allows chemical weathering to happen faster.

Rock has broken into loose grains; feldspar has turned into clay.

A current washes clay away. Tumbling quartz grains become rounder.

Time

Feldspar   Quartz   Biotite   Quartz   Clay

(b) Chemical weathering weakens rock, so it breaks apart. As this happens, the surface area increases, so chemical weathering happens still faster. Eventually, the rock completely disintegrates to form sediment. Weathering of granite produces quartz sand and clay.

FIGURE 5.8 Differential weathering.

(a) Sawtooth weathering profiles developed alternating strong and weak layers on this exposure in New Mexico. Weak layers are indented.

(b) Inscriptions on a granite headstone (left) last for centuries, but those on a marble headstone (right) may weather away in decades. These gravestones are in the same cemetery and are about the same age.

5.3 Clastic Sedimentary Rocks

Processes That Form Clastic Sedimentary Rocks

Nine hundred years ago, a thriving community of Native Americans inhabited the high plateau of Mesa Verde, Colorado, where they built stone-block buildings beneath large overhangs (Fig. 5.10). If you were to rub your thumb along one of the blocks, it would feel gritty, and small rounded grains of quartz would break free and roll under your thumb. The blocks consist of *sandstone*, a rock made up of sand grains (small clasts) cemented together. Geologists refer to all rocks formed from once-separate clasts that have been packed together and then cemented to one another as **clastic sedimentary rocks**. Formation of clastic sedimentary rock involves the following steps (Fig. 5.11):

- *Weathering:* Clasts form by disintegration of bedrock due to weathering.

- *Downslope movement:* Gravity pulls clasts from higher to lower elevations. They may tumble individually or as part of a landslide.

FIGURE 5.9 Weathering may result in rounded forms.

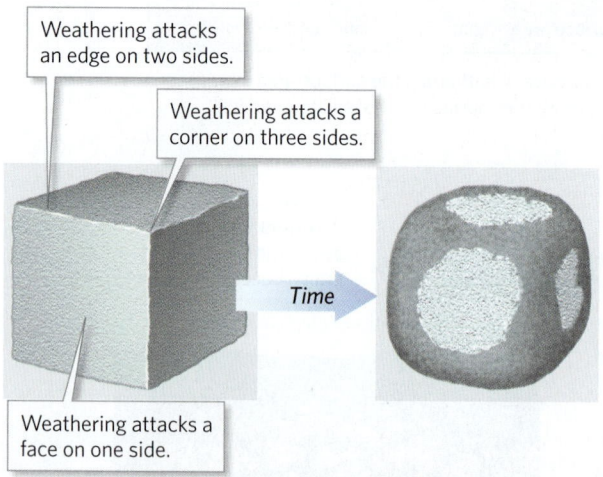

Weathering attacks an edge on two sides.

Weathering attacks a corner on three sides.

Time

Weathering attacks a face on one side.

(a) Weathering attacks a block of rock more vigorously at the edges than at the flat faces, and most vigorously at the corners. Thus, homogeneous rocks tend to weather into rounded blocks.

(b) Weathering along cracks in this granite in the Mojave Desert led to the formation of rounded blocks.

FIGURE 5.10 Between 600 and 1300 c.e., Pueblo people inhabited these dwellings made of sandstone blocks beneath a protective ledge of sandstone bedrock. The inset shows the grainy character of the rock.

Sandstone layer

- *Erosion:* Moving air, water, or ice (a *transporting medium*) can carve into bedrock or debris accumulations, pick up clasts, and carry them away, a process known as erosion.

- *Transportation:* Clasts can be carried far from their origin. The ability of the transporting medium to carry sediment depends on the medium's viscosity and velocity. For example, fast-moving, muddy water can transport large clasts, whereas slow-moving, clear water can transport only small ones.

- *Deposition:* The accumulation of clasts to form layers of sediment at one location is the process of **deposition**. It happens when a transporting medium loses its ability to carry clasts. For example, when flowing water slows, the clasts it carries settle out.

- *Lithification:* **Lithification**, the transformation of unconsolidated sediment into solid rock, involves a few stages. First, sediment undergoes *burial*, meaning that new sediment layers accumulate above it.

FIGURE 5.11 Steps in clastic sedimentary rock formation.

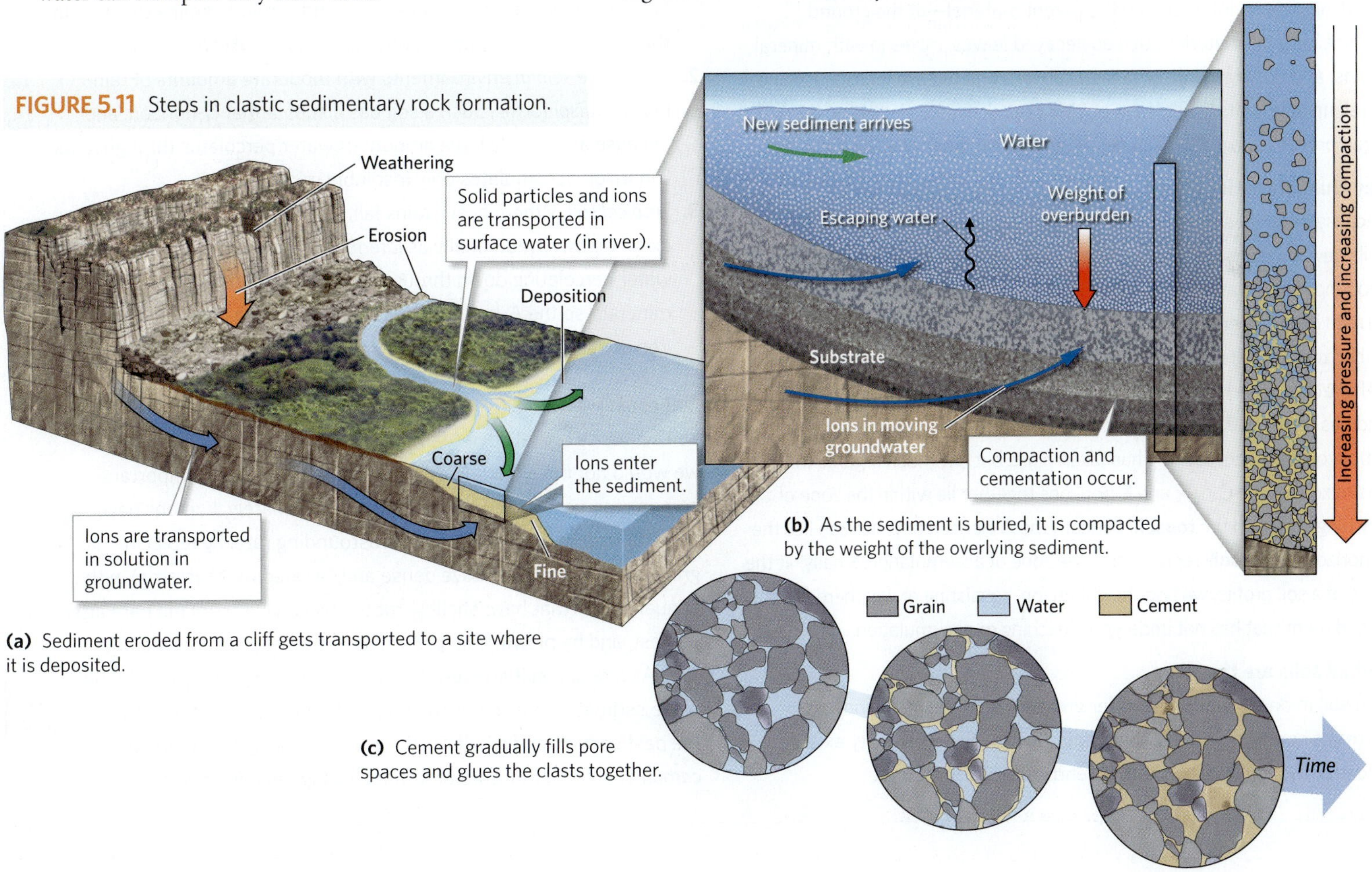

Weathering

Erosion

Solid particles and ions are transported in surface water (in river).

Deposition

Coarse

Ions enter the sediment.

Fine

Ions are transported in solution in groundwater.

(a) Sediment eroded from a cliff gets transported to a site where it is deposited.

New sediment arrives

Water

Escaping water

Weight of overburden

Substrate

Ions in moving groundwater

Compaction and cementation occur.

(b) As the sediment is buried, it is compacted by the weight of the overlying sediment.

Increasing pressure and increasing compaction

Grain   Water   Cement

(c) Cement gradually fills pore spaces and glues the clasts together.

Time

BOX 5.1 ▶ **Putting Earth Science to Use**

Soil and its formation

If you've ever had the chance to dig in a garden, field, or forest floor, you've seen firsthand that the soil in which flowers, crops, and trees grow looks and feels different from rock, beach sand, or potter's clay. **Soil** consists of sediment that has been modified over time by interaction with rainwater, air, organisms, and decaying organic matter. Because of soil's significance to society, it's important to understand how it forms, why its character varies with location, and why people should be concerned about conserving it.

Formation of soil

Three key processes contribute to soil formation. First, chemical and physical weathering break up and alter pre-existing rock or sediment at the Earth's surface. This process yields loose debris known as *parent material*. Second, rainwater falls on the parent material, sinks in, and percolates downward. As this water moves, it incorporates dissolved ions and picks up clay flakes. The level from which downward-percolating water removes ions and clay is called the *zone of leaching*. Farther underground, new mineral crystals precipitate from the water, and clay stops moving. The level in which new mineral growth takes place, and transported clay collects, is called the *zone of accumulation* **(Fig. Bx5.1a)**. Third, organisms interact with the parent material—at the ground surface, organic material, such as decayed leaves, mixes in with mineral grains, and underground, organisms absorb and release ions, churn and break up aggregates, and leave behind organic waste. Following these three processes, what was once just rock or sediment has become soil.

Key characteristics of soil

Because different soil-forming processes operate at different depths, soils develop distinct zones, known as **soil horizons**, arranged in a vertical sequence known as a **soil profile**. Let's look at an example of a soil profile formed in a temperate forest **(Fig. Bx5.1b)**. The highest horizon, the O-horizon, consists mostly of *humus* (decayed plant debris). Below the O-horizon, we find the A-horizon, in which humus has decayed further and has mixed with mineral grains. In some locations, you may also find an E-horizon, in which leaching has taken place but not much organic matter has mixed in. The O-, A-, and E-horizons together lie within the zone of leaching and form the **topsoil**. Beneath the A-horizon or E-horizon lies the B-horizon, or **subsoil**, representing the zone of accumulation. Finally, at the base of a soil profile, we find the C-horizon, consisting of weathered rock or sediment that has not undergone leaching or accumulation.

Not all soils are the same

The soil in one locality may differ greatly from the soil in another in terms of composition, thickness, and texture. Such diversity exists because the makeup of a soil depends on several factors:

- *Climate:* Rainfall and temperature affect the rate of chemical weathering, the amount of leaching, and the abundance of organisms in soil.
- *Composition:* Soils formed from different parent materials have different chemical compositions.
- *Slope steepness:* Thick soil can accumulate on flat land, but on a steep slope, soil washes or slumps away before it can become thick. So, soil thickness increases as slope angle decreases.
- *Time:* Soil takes time to form, so a young soil tends to be thinner and less developed than an old soil. A thick soil may take millennia to form.
- *Vegetation type:* Different kinds of plants extract or add different nutrients and quantities of organic matter and have different kinds of root systems, all of which affect soil character.

With the effects of these factors in mind, researchers have developed classification schemes for soils, using specific names for different types. **Figure Bx5.1c** shows some examples:

1. *Desert soil:* in deserts, where hardly any rain falls and little or no vegetation grows, an *aridisol* forms. Such a soil has no O-horizon because no organic matter accumulates, and it has almost no A-horizon because hardly any leaching takes place. Not enough rain falls to remove calcite, so it remains and cements grains together in the B-horizon to produce a hard mass called *caliche*.
2. *Temperate soil:* In environments with moderate amounts of rainfall, an *alfisol* forms. Such a soil has a substantial O-horizon, and because a relatively large amount of water percolates through it, its B-horizon accumulates only insoluble minerals.
3. *Tropical soil:* Where heavy rains fall, a distinct brick-red soil called an *oxisol* (known traditionally as *laterite*) develops. The large volume of water percolating down through an oxisol leaches out most soluble minerals, so the A-horizon consists mostly of an insoluble residue of iron and aluminum oxide, which gives the oxisol its brick-red color.

Soil conservation

Soil serves as one of society's most valuable resources—without it, we would not have food or wood. Soil also serves as an important carbon reservoir in the Earth System. Unfortunately, humans have been damaging this resource at an astounding rate. By replacing native grasses and trees (that have dense and/or deep root systems) with annual crops (that have shallow roots), by leaving farmland bare after harvest, and by plowing deeply, we have exposed soil to wind and rain. As a result, **soil erosion** (the process of carrying away soil), by some estimates, has removed about half of the Earth's topsoil over the past century and a half. Furthermore, in the soil that remains, *soil carbon* content, a measure of fertility, has greatly diminished. In some

Rain enters ground.

Plant debris accumulates.

Worms churn.

Microbes and fungi metabolize.

Roots weather minerals.

Downward-percolating water transports ions and clay.

Ions and clay accumulate.

Zone of leaching

Zone of accumulation

~10 cm

O
A
E
B
C

Horizon designation

Topsoil

Transition

Subsoil

Weathered bedrock

Solid bedrock

This soil in a temperate realm hosts a forest. It displays a good O-horizon.

(a) Soil formation involves many processes.

(b) Soils develop distinct horizons. This example is from a temperate forest.

Aridisol forms in deserts. Rainfall is so low that no O-horizon forms, and soluble minerals accumulate in the B-horizon.

Alfisol forms in temperate climates. An O-horizon forms, and relatively insoluble materials accumulate in the B-horizon.

Oxisol forms in tropical climates, where percolating rainwater leaches all soluble minerals, leaving only iron- and aluminum-rich residues.

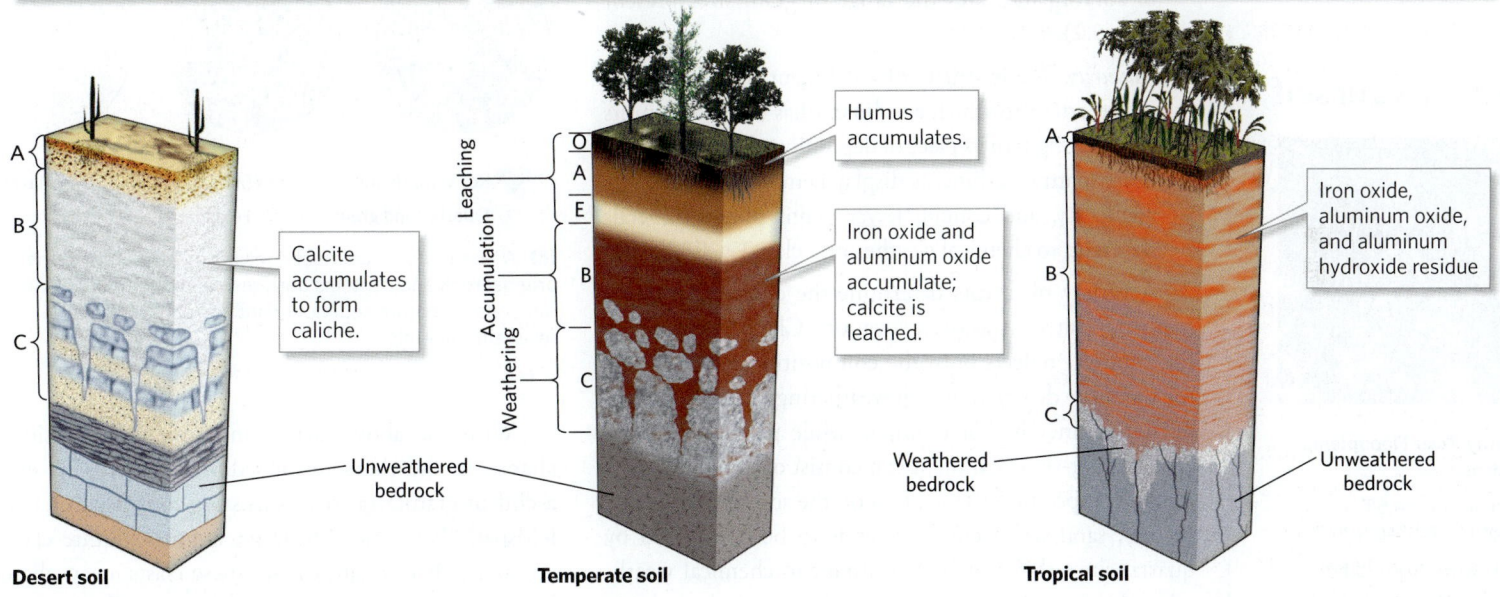

Calcite accumulates to form caliche.

Unweathered bedrock

Desert soil

Humus accumulates.

Iron oxide and aluminum oxide accumulate; calcite is leached.

Leaching
Accumulation
Weathering

Temperate soil

Iron oxide, aluminum oxide, and aluminum hydroxide residue

Weathered bedrock

Unweathered bedrock

Tropical soil

(c) Examples of soil classification.

places, soil carbon has dropped by 50% because harvesting takes away plant debris that would otherwise mix back into soil. As a result, farmers must add ever-increasing amounts of fertilizer to keep soil usable. Clearly, *soil conservation*—saving and restoring soil—must take place, or society may run out of sufficient food. Key methods include:

protecting soil with a cover crop or with mulch, after harvest, in order to prevent soil from drying out and from being disturbed by wind or rain; adopting no-till farming practices; decreasing runoff by terracing farmland, contour plowing, or installing erosion barriers; planting wind-breaks of hedges or trees; and mixing organic carbon back into the soil.

The weight of the overlying sediment squeezes out water or air that had been trapped in pores between clasts. This process of **compaction** causes clasts to fit together more tightly. Finally, minerals precipitate from groundwater in the pores and, like glue, bind clasts together to make an overall solid rock. This binding process is called **cementation**.

Characteristics of Clastic Sediment

Sediment, the loose clasts produced by weathering and erosion, varies greatly in character. To describe sediment, geologists focus on the following characteristics:

- *Clast size:* Clasts range from very large to microscopic. Table 5.1 provides the names assigned to different clast sizes.

- *Clast composition:* Larger clasts tend to be rock fragments containing a variety of different minerals. Smaller clasts (also called *grains*) tend to be individual mineral crystals, or parts of crystals.

- *Angularity:* Clasts that have sharp edges and corners are called angular, whereas those with blunted and smooth surfaces are called rounded **(Fig. 5.12a)**.

- *Sorting:* In a well-sorted sediment, clasts are nearly the same size, whereas a poorly sorted sediment contains clasts of many different sizes **(Fig. 5.12b)**. The degree of **sorting** indicates the range of grain sizes present **(Box 5.2)**.

- *Maturity:* The **maturity** of a sediment is an indication of the extent to which a sediment has evolved from its place of origin to its place of deposition. As they become more mature, sediments display better sorting and less angularity, and contain fewer grains of the minerals that tend to chemical weather into clay **(Fig. 5.12c)**.

A variety of factors determine the character of sediment deposited at a given location. Clast composition, for example, reflects both the composition of the source rock and the degree to which weathering has altered nonresistant minerals. For example, while pebble-, cobble-, and boulder-sized sediment can consist of rock fragments of any composition (depending on the source of the fragments), sand-sized sediment tends to be dominated by quartz, since this mineral is resistant to chemical weathering. Feldspar, for example, turns into clay when chemically weathered, and clay washes away. Consequently, feldspar remains as a component of sediment only in settings where such weathering hasn't taken place, and very fine sediment tends to be composed of clay, a product of chemical weathering. Notably, clast size depends on the nature of the transporting medium—ice and fast-moving water can transport large clasts, moderate moving water can transport sand, and wind or slow-moving water can transport only silt and clay.

See for **yourself**

Murray River Floodplain, Australia

Latitude: 34°7′4.68″ S
Longitude: 140°46′9.75″ E

Look down from 15 km (~9.3 mi).

The Murray River meanders across the flat-lying landscape of South Australia. In this view, the water is restricted to the channel, exposing the sand and silt deposits that have covered its floodplain.

FIGURE 5.12 Grain characteristics change with transportation.

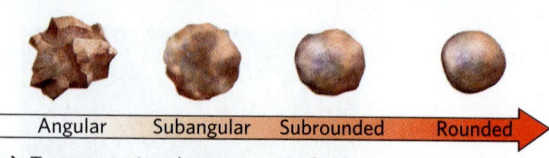

Angular · Subangular · Subrounded · Rounded

(a) Transportation decreases angularity.

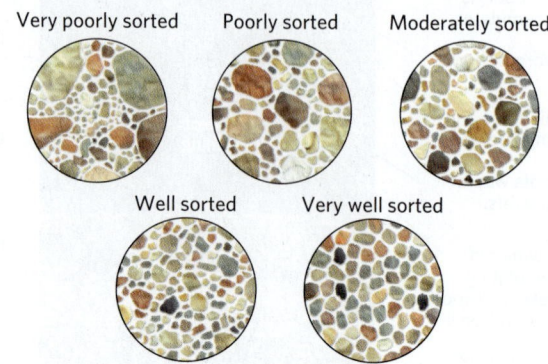

Very poorly sorted · Poorly sorted · Moderately sorted

Well sorted · Very well sorted

(b) Transportation sorts grains by size.

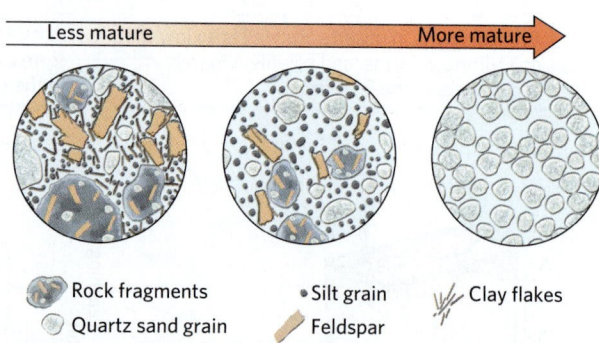

Less mature → More mature

Rock fragments · Silt grain · Clay flakes
Quartz sand grain · Feldspar

(c) A less mature sediment consists of fragments of the original rock and contains both resistant and nonresistant minerals. A mature sediment contains only well-sorted resistant minerals.

With the above factors in mind, let's consider the change in the character of sediment derived by eroding a cliff of granite (a rock consisting mostly of quartz and feldspar). Near the cliff, large, angular granite clasts accumulate. If a stream carries these clasts away, the clasts bang against each other, break into smaller pieces, and become rounded. If the granite breaks down into sand-sized grains, the sand can contain feldspar and other silicate minerals, in addition to quartz, only if chemical weathering was minimal. If chemical weathering transforms feldspars and other minerals into clay, transport by wind or moving water will carry the clay away to be deposited elsewhere, leaving quartz sand behind. Note that sediment maturity increases with increasing amounts of weathering and transport.

BOX 5.2 ▸ How can I explain . . .

Sediment sorting

What are we learning?

The size of sediment that a medium such as water or wind can transport depends on the velocity of the fluid. Therefore, fluid flow sorts sediment—without sorting, we wouldn't have distinctive shale, sandstone, and conglomerate beds. It's difficult to demonstrate water sorting without access to a flume (a long tank in which water can flow). But you can demonstrate wind sorting using a common blow drier.

What you need:

- A large cookie sheet with raised edges
- A cup of mixed sand and small pebbles
- A blow drier, set on cool; eye protection

Instructions:

- Spread the sand/pebble mixture across the sheet, near one end. Give the sheet a tap so the layer becomes smooth.
- Turn on the drier at low speed, and bring it just close enough to the end of the sheet with the sediment until sediment starts moving.
- Gently blow back and forth and watch how the sand moves to the other end of the pan, while the pebbles remain close to where they started.
- Increase the fan speed to high; the pebbles start to move.

Mixed sand and pebbles

Before

Sand   Pebbles

After

What did we see?

- At low speed, the wind sorts the sand from the pebbles. This process resembles the way in which wind or a stream carries away sand and leaves behind gravel.
- When speed increases, even the pebbles move. This observation emphasizes that faster wind (or faster currents) move larger clasts.

Types of Clastic Sedimentary Rocks

When assigning a name to a clastic sedimentary rock, geologists focus first on the rock's clast size. Other factors, such as angularity and composition, then come into play, in order to distinguish among clastic rocks of a given grain size. Table 5.2 lists the principal types of clastic sedimentary rocks, from the coarsest down to the finest. Note that a rock formed by lithification of gravel (pebbles, cobbles, or boulders) yields a *breccia* (Fig. 5.13a) if the clasts are angular, or a *conglomerate* if the clasts are rounded (Fig. 5.13b). A rock formed by lithification of sand yields *sandstone* (Fig. 5.13c), if composed primarily of quartz, or *arkose* if it contains quartz and feldspar (Fig. 5.13d). *Siltstone* forms by lithification of silt (generally tiny quartz grains). Lithification of *mud* (a mixture of clay minerals, extremely tiny quartz grains, and water) produces *shale* if the rock splits into thin sheets (Fig. 5.13e), and *mudstone* if it does not.

TABLE 5.2 Clastic Sedimentary Rocks

Rock Name	Clast size	Sediment	Clast character
Breccia	Very coarse	Gravel*	Angular
Conglomerate	Very coarse	Gravel*	Rounded
Sandstone	Medium	Sand	Mostly quartz
Arkose	Medium	Sand	Quartz and feldspar
Siltstone	Fine	Silt	Mostly quartz
Shale**	Very fine	Mud	Mostly clay
Mudstone**	Very fine	Mud	Mostly clay

* Gravel can include pebbles, cobbles, and/or boulders.
** Shale splits into thin sheets; mudstone does not.

FIGURE 5.13 Different kinds of sediments lithify into different kinds of sedimentary rocks.

Sediment $\xrightarrow{\text{Lithification}}$ Sedimentary rock

(a) Lithification of an accumulation of angular clasts yields breccia.

(b) Rounded gravel lithifies into conglomerate.

(c) Sediment deposited close to its source can be rich in feldspar. Lithification of this sediment yields arkose.

(d) Layers of sand lithify into sandstone.

(e) Layers of mud lithify to form shale.

Take-home message . . .

5.4 Biochemical, Chemical, and Organic Rocks

So far, we've discussed the process of forming new sedimentary rocks from solid grains (clasts) derived from pre-existing rocks. But, as we've noted, weathering not only yields such grains, but also produces ions in solution. These ions can be extracted by organisms to form *biochemical sedimentary rocks*, or can precipitate directly to form *chemical sedimentary rocks*. Sedimentary rocks can also incorporate organic material from the biosphere, to produce *organic sedimentary rocks*. Let's look at how these diverse types of rocks form **(Table 5.3)**.

Biochemical Sedimentary Rocks

Some organisms have the ability to extract dissolved ions from solution to make shells, and when they die, the mineral material in their shells remains. Accumulations of this material, when lithified, become **biochemical sedimentary rock**, of which geologists recognize several different types.

BIOCHEMICAL LIMESTONE. A snorkeler gliding above a reef sees an incredibly diverse community of corals and algae, around which creatures such as clams, oysters, and snails live, and above which plankton float **(Fig. 5.14a)**. These organisms share an important characteristic: they make shells of calcite ($CaCO_3$). When the organisms die, the shells remain. Any rock formed predominantly from $CaCO_3$ is called a **limestone**, and because the anionic group in limestone is CO_3^{2-} (carbonate), geologists consider limestone to be a type of **carbonate rock**. Not all limestones contain shells or shell fragments; those that do are called *biochemical limestone*. This category includes *coquina*, poorly cemented shell fragments **(Fig. 5.14b)**; *fossiliferous limestone*, which is full of visible fossil shells or shell fragments **(Fig. 5.14c)**; *micrite*, consisting of very fine carbonate grains; and *chalk*, consisting of calcite plankton shells. In outcrops, limestone typically looks like a massive light gray to dark bluish-gray rock, with a somewhat crystalline texture, that breaks

FIGURE 5.14 The formation of limestone, a type of carbonate rock.

(a) Corals and other organisms living on this reef make their shells out of calcium carbonate.

(b) Shells on a beach. If cemented, this material becomes coquina.

(c) Broken shells were cemented together to form this 415-Ma fossiliferous limestone.

into chunky blocks, rather than a pile of shell fragments **(Fig. 5.14d)**. That's because underground processes change the texture of the rock over time. For example, water passing through the rock dissolves some grains, causes new grains to grow, and precipitates cement.

BIOCHEMICAL CHERT. Along the California coast, just northwest of San Francisco, road cuts expose a reddish, almost porcelain-like rock occurring in layers between 3–15 cm (1–6 in) thick **(Fig. 5.15a)**. Hit it with a hammer, and the rock breaks to form smooth *conchoidal* (clamshell-shaped) fractures. Geologists refer to any rock made from quartz grains that are too small to be seen without extreme magnification as **chert**. The chert of coastal California formed from the shells of silica-secreting plankton that accumulated on the seafloor, so it can be called *biochemical chert*.

(d) Ancient limestone tends to occur in gray, blocky layers, like those exposed in this road cut in New York State.

(a) This chert developed from deep-sea sediment made up of the shells of silica-secreting plankton.

Sandstone and shale

Coal layer

(b) Coal is deposited in layers (beds), just like other kinds of sedimentary rocks.

FIGURE 5.15 Examples of biochemical and organic sedimentary rocks.

TABLE 5.3 Biochemical, Chemical, and Organic Sedimentary Rocks

	Rock Name	Minerals contained	Category
Carbonates	Limestone	Calcite	Biochemical or chemical
	- Coquina	Calcite	Biochemical
	- Fossiliferous limestone	Calcite	Biochemical
	- Micrite	Calcite	Biochemical
	- Travertine	Calcite	Biochemical
	Dolostone	Dolomite	Biochemical or chemical
Evaporites	Rock salt	Halite	Chemical
	Gypsum	Gypsum	Chemical
	Chert	Cryptocrystalline quartz	Biochemical or chemical
	Coal	Carbon molecules	Organic
	Organic shale	Clay and hydrocarbons	Clastic and organic

Did you ever wonder . . .

how the chert used for arrowheads first formed?

Organic Sedimentary Rocks

We've seen how shells become biochemical sedimentary rock. What becomes of the cellulose, fats, carbohydrates, or proteins of organisms? Commonly, this material either gets eaten or decomposes at the Earth's surface. But in some environments, organic debris accumulates with other sediment and undergoes burial before completely decomposing. When lithified, such sediment becomes **organic sedimentary rock**. The most common example, **coal**, formed from woody plant remains (**Fig. 5.15b**). *Organic shale*, a source of hydrocarbons (oil and gas), forms when the remains of plankton mix with clay to make mud, which then becomes shale.

Chemical Sedimentary Rocks

The colorful terraces that grow around the vents of hot springs and the layers of salt that underlie the floor of the Mediterranean Sea have something in common—they consist of rock formed by the direct precipitation of minerals from water solutions. We call such rocks **chemical sedimentary rocks**. Geologists divide these rocks into several different compositional types.

EVAPORITES. In 1965, Craig Breedlove became the first person to drive a vehicle on land at a speed of 600 mph (966 km/h). Such high-speed trials must take place on extremely long, flat surfaces. Not many places provide such conditions—but the Bonneville Salt Flats of Utah do. They formed when an ancient salt lake evaporated. During the process of *evaporation*, water molecules escape from a water body and drift up into the atmosphere, but the mineral salts that had been dissolved in the water remain. As evaporation progresses, the saltwater becomes *supersaturated*, meaning that it has exceeded its capacity to contain dissolved ions. In a supersaturated solution, ions bond together to form solid grains that either settle out of the water or grow on the floor of the water body. This process takes place not only in desert salt lakes, but also along the edges of restricted seas in hot climates, and it produces layers of rock salt (halite) and gypsum with a crystalline texture (**Fig. 5.16**). Because of the way these rocks form, geologists refer to them as **evaporites**.

TRAVERTINE. *Travertine* consists of crystalline calcite formed by precipitation from groundwater that has seeped out of the ground. Emergence of groundwater from hot springs leads to travertine precipitation because when the water degasses (loses some dissolved CO_2), evaporates, and cools, it becomes supersaturated. Travertine produced at hot springs may build terraces and mounds (**Fig. 5.17a**). Travertine also forms where water seeps out of the walls or ceilings of caves, producing *speleothems* (**Fig. 5.17b**). In

FIGURE 5.16 The formation of evaporite deposits.

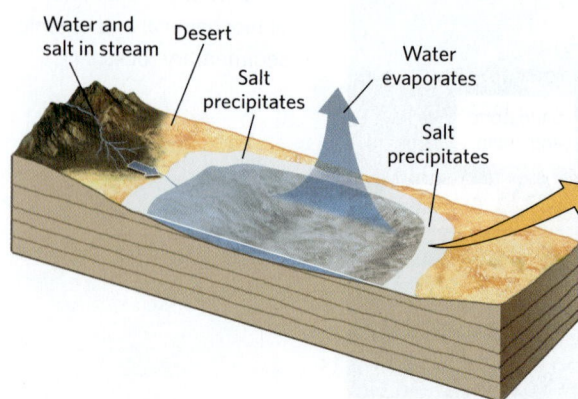

(a) When saltwater evaporates, various mineral salts, including halite and gypsum, precipitate.

(b) Under the desert sun, a salt lake on the floor of Death Valley has dried up, leaving a white crust of new salt crystals.

FIGURE 5.17 Examples of travertine (chemical limestone) deposits.

Terrace of new travertine

(a) Travertine accumulates in terraces at Mammoth Hot Springs in Yellowstone National Park, Wyoming.

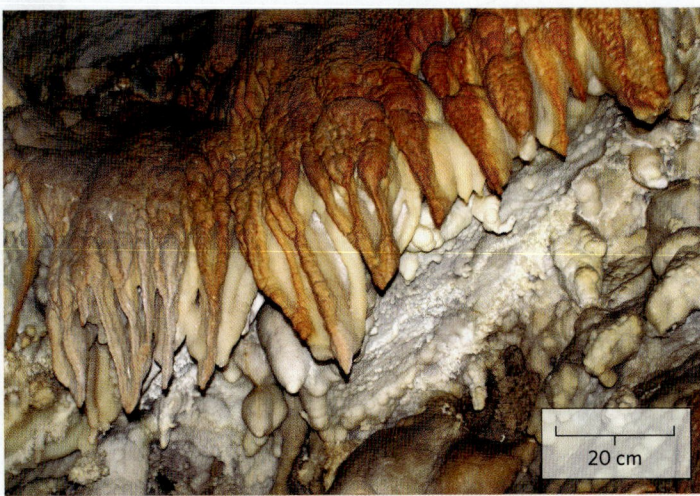

(b) Travertine speleothems form as calcite-rich water drips from the ceiling of Timpanogos Cave in Utah.

some environments, the metabolism of microbes accelerates calcite precipitation.

DOLOSTONE. Significantly, some carbonate rocks contain a large proportion of *dolomite*, a mineral with the chemical formula $CaMg(CO_3)_2$. These rocks are known as **dolostone** or dolomite. They form when limestone reacts with magnesium-bearing groundwater. This change can take place soon after the limestone has formed or long after the limestone has been buried deeply.

REPLACEMENT CHERT. As we've seen, some chert forms from the shells of silica-secreting plankton. Chert can also form when silica chemically precipitates from groundwater and replaces other minerals within a limestone. Geologists refer to this rock as *replacement chert*. Replacement chert occurs in lumpy blobs called *nodules* **(Fig. 5.18)**, which may form a layer parallel to a bedding plane.

Take-home message . . .

Biochemical sedimentary rocks incorporate minerals produced by organisms in their shells. Sediment derived from calcite shells produces various types of limestone, and sediment from silica-shelled plankton produces bedded chert. Precipitation from solutions produces a variety of chemical sedimentary rocks. Organic debris produces coal and organic shale.

Quick Question

Why are limestone and dolostone referred to as carbonate rocks, while rock salt and gypsum are referred to as evaporites?

FIGURE 5.18 Replacement chert occurs as a layer of black nodules in this chalk cliff in southern England.

5.5 Sedimentary Structures

Sedimentary rock isn't just a homogeneous, featureless mass. Rather, it contains layering as well as fascinating textures and shapes that may form during or immediately after sediment deposition. Geologists refer to such features as **sedimentary structures**. These structures can provide clues to the environment in which the sediments were deposited and, therefore, help geologists interpret Earth history. Let's look at a few important examples of sedimentary structures.

Bedding and Stratification

If you look at an outcrop of sedimentary rock, you'll see parallel planes that separate the rock into distinct layers.

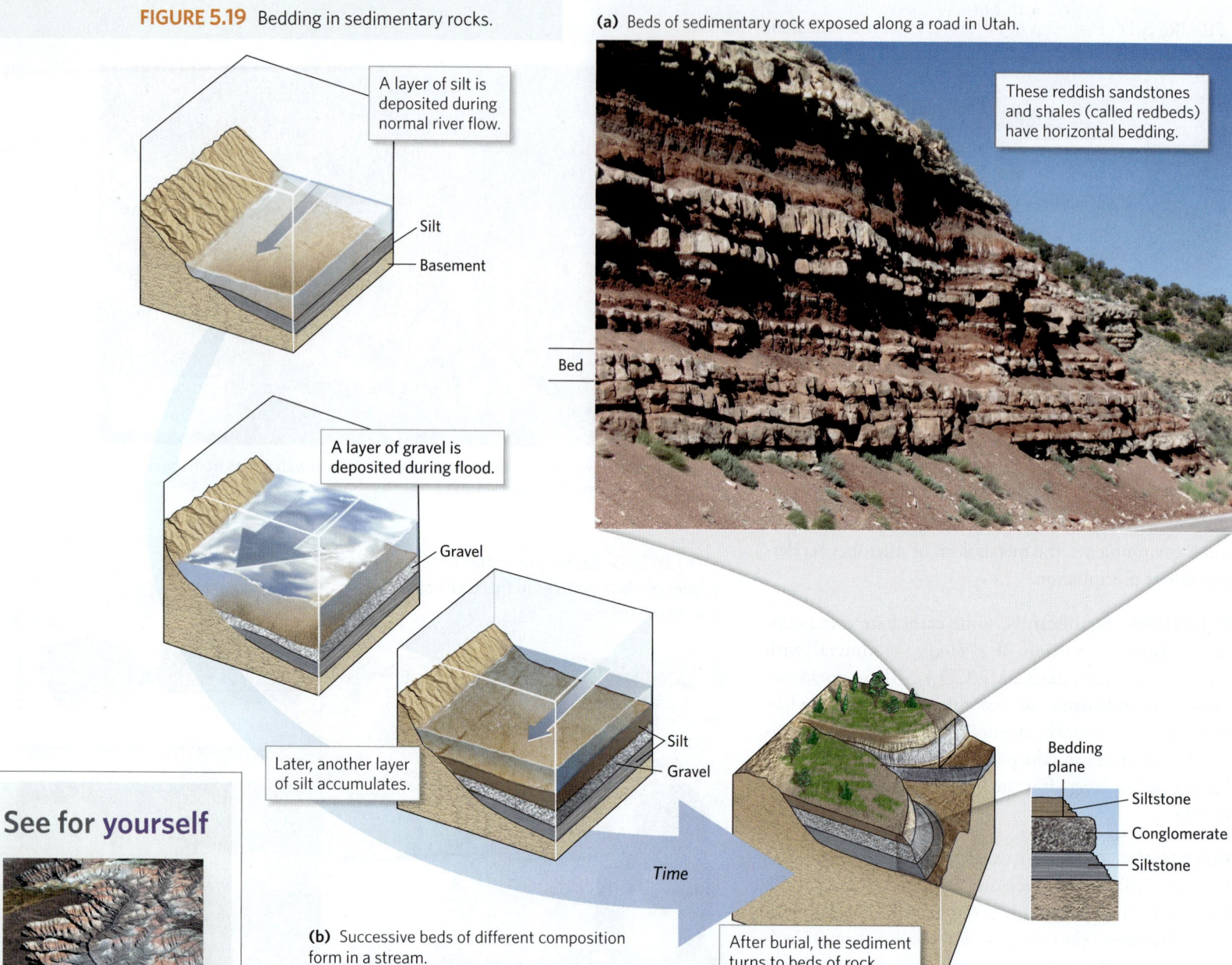

FIGURE 5.19 Bedding in sedimentary rocks.

(a) Beds of sedimentary rock exposed along a road in Utah.

A layer of silt is deposited during normal river flow.

Silt
Basement

These reddish sandstones and shales (called redbeds) have horizontal bedding.

Bed

A layer of gravel is deposited during flood.

Gravel

Later, another layer of silt accumulates.

Silt
Gravel

Bedding plane

Siltstone
Conglomerate
Siltstone

Time

(b) Successive beds of different composition form in a stream.

After burial, the sediment turns to beds of rock.

See for yourself

Grand Canyon, Arizona

Latitude: 36°8'8.94" N
Longitude: 112°15'48.56" W

Look straight down from an elevation of 15 km (~9.3 mi).

You can see the spectacular color banding caused by the succession of stratigraphic formations exposed on the walls of the Grand Canyon.

Recall that a single layer of sedimentary rock is a *bed*, and the boundary between two beds is a *bedding plane*. Geologists commonly use the term **strata** for a succession of several beds, **bedding** for the occurrence of beds, and *stratigraphy* for the study of strata **(Fig. 5.19a)**. A **stratigraphic formation** (or, simply, a *formation*) is a set of beds that can be recognized and traced for a significant distance and can be distinguished from the set of beds above or below (see Chapter 9).

Why does bedding form? To answer this question, remember how sediment accumulates. The sediment deposited at any given time reflects the composition of the sediment source and the conditions of deposition. Therefore, changes in the sediment source or the conditions of deposition can change the type of sediment deposited, and this change can delineate bedding. As an example, picture the sediment layers deposited by a river **(Fig. 5.19b)**. On a normal day, the river flows slowly and transports only silt, so layers of silt collect along the river bottom. After a heavy rainstorm, the river flows faster and transports pebbles, so a layer of gravel accumulates over the silt layer. Later, when the river slows again, another layer of silt buries the gravel. This succession of sediments, when lithified, becomes alternating beds of siltstone and conglomerate.

Current-Related Structures

In quiet water or calm air, sediment settles like snow, and when it does, the sediment layer that forms has a smooth surface and a homogeneous texture. In contrast, sediment layers deposited by flowing water or air tend to develop distinctive sedimentary structures both on the bed surface and within the bed, as we will now see.

RIPPLE MARKS, DUNES, AND CROSS BEDS. The top surface of a sediment layer deposited beneath flowing air or water typically displays wave-like ridges and troughs, which can be preserved **(Fig. 5.20)**. These **ripple marks** tend to be relatively small (less than 10 cm, or 4 in, high) and aligned perpendicular to the current flow. Burial of ripple marks by another layer of sediment preserves them, so you can find them on bedding planes of ancient rocks. **Dunes** also form beneath wind or flowing water, but are larger than ripple marks **(Fig. 5.21a, b)**.

FIGURE 5.20 Ripples and ripple marks.

(a) Modern ripples, exposed at low tide, on a sandy beach on Cape Cod, Massachusetts.

(b) Ripples on a tilted, 145-Ma sandstone bed.

FIGURE 5.21 The formation of dunes and cross beds.

(a) A small dune developing in Death Valley. Small ripples have formed on the top of the dune.

(b) Large sand dunes in a windstorm.

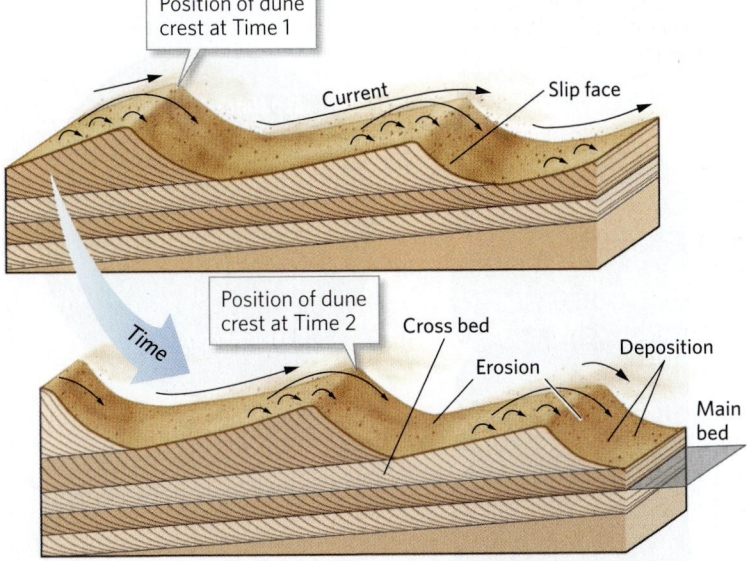

(c) Cross beds form when a wind current carries sand up the windward side of a ripple or dune, where it builds up until it falls down the slip face.

What an Earth Scientist Sees

(d) This sandstone exposed in a cliff face in Zion Canyon, Utah, contains large cross beds formed 200 Ma.

If you slice into a ripple mark or dune, you'll find distinct *laminations* (thin layers within a thicker layer) known as **cross beds**, inclined at an angle to the top of the sediment layer. To see how cross beds develop, imagine a current of air or water moving uniformly in one direction. The current picks up sand from the upstream or windward face of the ripple or dune and deposits it on the downstream or leeward face **(Fig. 5.21c)**. Sand builds up until the downstream slope becomes too steep and gravity causes the sand to tumble down the leeward face, or *slip face*. As a dune or ripple builds downstream, more sand buries each successive slip face, preserving the previous one as a cross bed. When new layers of dunes or ripples build over existing ones, a succession of cross-bedded layers, separated from each other by main bedding planes, accumulates **(Fig. 5.21d)**.

GRADED BEDDING. When unconsolidated sediment slides down an underwater slope, it mixes with water to produce a murky, turbulent cloud. This cloud of suspended sediment is denser than clear water, so it flows downslope as an underwater avalanche known as a **turbidity current (Fig. 5.22)**. At the base of the slope, the turbidity current slows, so the sediment that it carries settles out to produce a submarine fan. Larger grains sink faster than smaller grains, so the coarsest sediment accumulates first, and progressively finer grains accumulate on top, with the finest sediment (clay) settling out last. This process forms a **graded bed**, a layer of sediment in which grain size varies from coarse at the bottom to fine at the top.

Sedimentary Structures in Drying Mud

In some environments, mud becomes exposed to air after deposition. If it rains before the mud dries completely, small impact craters called *raindrop impressions* form where raindrops strike the mud surface. If the mud dries out, it shrinks and breaks into hexagonal plates that warp up at their edges. Gaps, known as **mudcracks**, develop between the plates **(Fig. 5.23)**.

FIGURE 5.22 The development of graded beds.

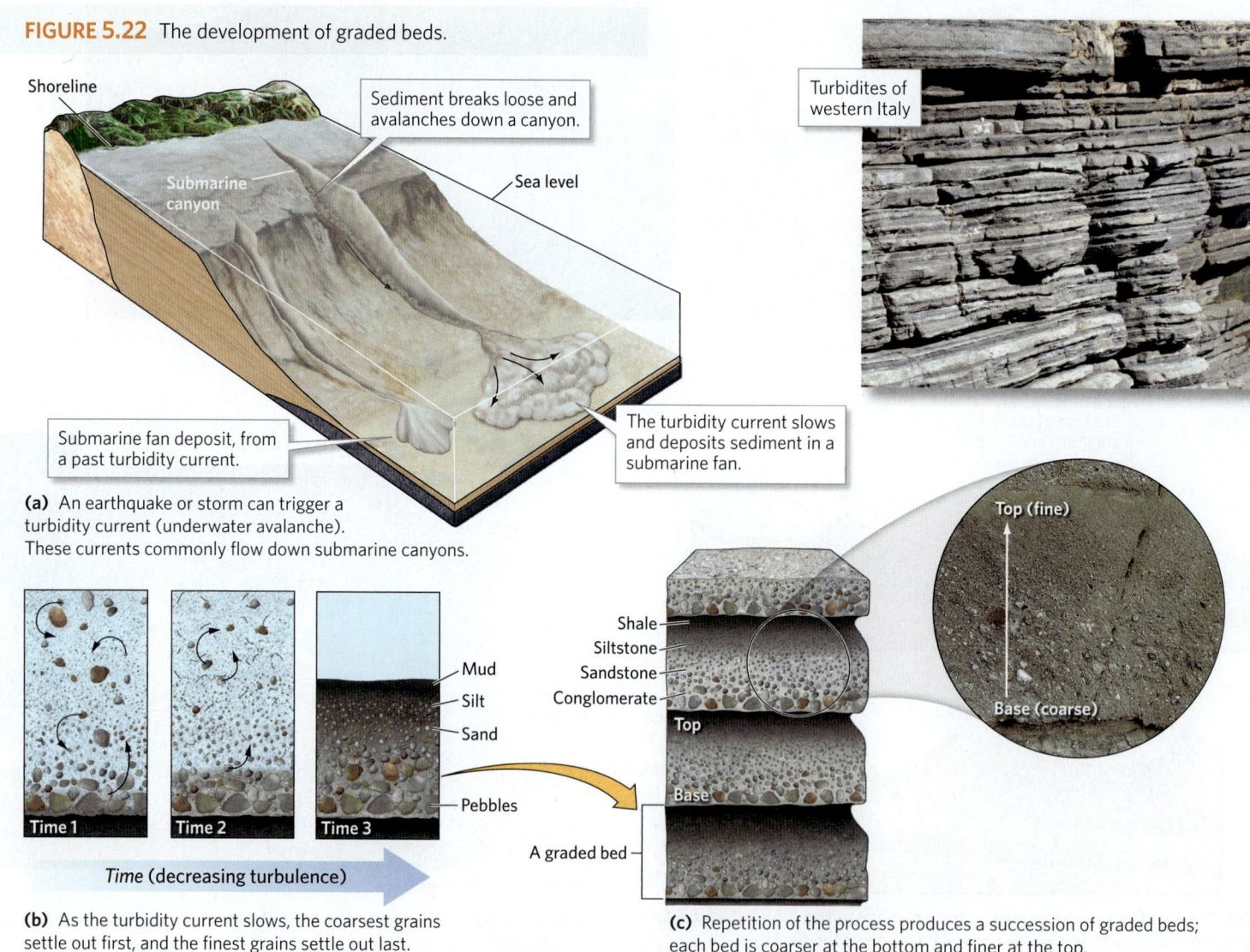

Turbidites of western Italy

Shoreline

Sediment breaks loose and avalanches down a canyon.

Submarine canyon

Sea level

Submarine fan deposit, from a past turbidity current.

The turbidity current slows and deposits sediment in a submarine fan.

(a) An earthquake or storm can trigger a turbidity current (underwater avalanche). These currents commonly flow down submarine canyons.

Mud
Silt
Sand
Pebbles

Time 1   Time 2   Time 3

Time (decreasing turbulence)

(b) As the turbidity current slows, the coarsest grains settle out first, and the finest grains settle out last.

Shale
Siltstone
Sandstone
Conglomerate

Top
Base

A graded bed

Top (fine)

Base (coarse)

(c) Repetition of the process produces a succession of graded beds; each bed is coarser at the bottom and finer at the top.

FIGURE 5.23 Formation of mudcracks.

(a) Mudcracks in red mud at Bryce Canyon, Utah. As the mud dries, it contracts, and the cracks form.

(b) Mudcracks preserved in a 410-Ma bed exposed on the base of a cliff in New York.

See for yourself

Alluvial Fans and Evaporites, Death Valley, California

Latitude: 36°7'35.28" N
Longitude: 116°45'24.44" W

Look obliquely east from 1.5 km (1 mi) up.

Death Valley is a narrow rift whose floor lies below sea level. You can see alluvial fans and white evaporites at many localities along the length of the valley.

Take-home message . . .

Sedimentary structures include bedding, ripple marks and dunes, cross bedding, mudcracks, and graded beds, all formed during or soon after deposition of sediment. Some of these features form on the surfaces of beds, some within beds.

Quick Question

What role does a current play in developing sedimentary structures?

5.6 Depositional Settings and Basins

By looking at characteristics of strata, can we determine the **depositional setting** (environmental conditions at the time of deposition) in which sediments accumulated? Fortunately, the answer is yes. To do this, geologists, like crime-scene investigators, look for clues. Detectives check fingerprints and DNA to identify a culprit, while geologists examine grain size, composition, sorting, sedimentary structures, and *fossils* (relics of organisms buried with the sediment) to identify a depositional setting. We group these settings into two main categories: *non-marine settings* occur on land or in lakes and streams that flow on land, whereas *marine settings* occur in the oceans or along the coasts of oceans. Let's consider a few examples of each (**Earth Science at a Glance**, pp. 166–167).

Non-Marine Deposition

Since ice is solid, glaciers can carry clasts of any size, so when the ice melts away, the sediment left behind is unsorted. Therefore, if you see sediment composed of large clasts surrounded by finer-grained sediment, you may be seeing glacial deposits **(Fig. 5.24a)**. Flowing water, in contrast, can sort sediment. In streams, coarser deposits (gravel) typically collect in the faster water of the stream channel, while finer sediment gets carried further

FIGURE 5.24 Examples of terrestrial depositional environments.

(a) Glacial till deposited by a glacier in France.

(b) Sediment in a Maryland stream.

(c) Alluvial fan in Death Valley, California.

FIGURE 5.25 Examples of shallow-marine depositional environments.

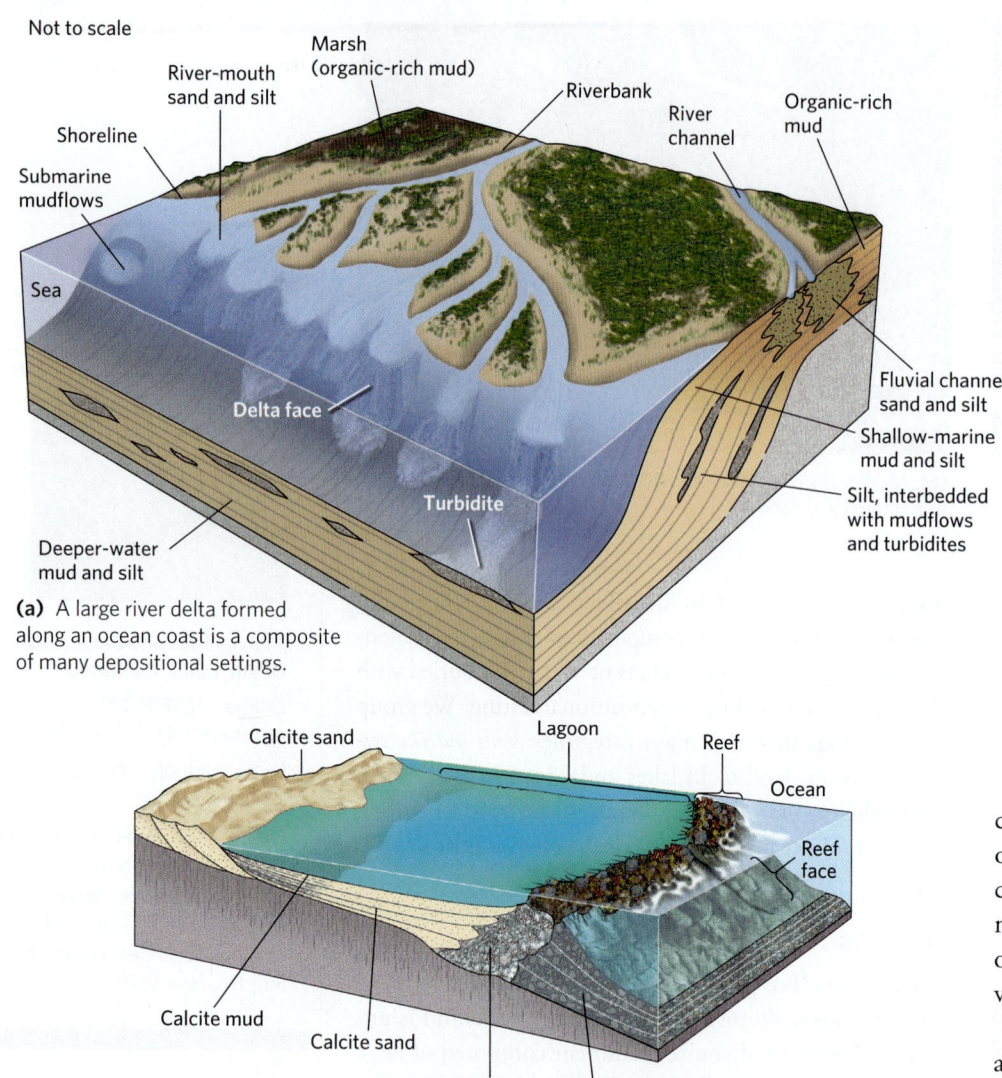

Not to scale

Marsh (organic-rich mud)

River-mouth sand and silt

Shoreline

Riverbank

River channel

Organic-rich mud

Submarine mudflows

Sea

Delta face

Fluvial channel sand and silt

Shallow-marine mud and silt

Turbidite

Silt, interbedded with mudflows and turbidites

Deeper-water mud and silt

(a) A large river delta formed along an ocean coast is a composite of many depositional settings.

Calcite sand

Lagoon

Reef

Ocean

Reef face

Calcite mud

Calcite sand

Reef buildup

Broken fragments of reef

(b) Carbonate reefs form along shorelines where the water is clear and warm.

FIGURE 5.26 Photo through a microscope of calcite plankton shells.

5 μm

downstream. If there's a flood, however, the water level in the stream may rise to flow over the stream's *floodplain*, the flat land on either side of the channel, where it slows down and deposits silt and mud. Therefore, if you see strata in which thin layers of conglomerate are surrounded by layers of siltstone and shale **(Fig. 5.24b)**, you may be looking at stream deposits. Finally, out in a desert, wind sorts and rounds sand, and may build dunes, and at locations where, after a heavy downpour, temporary streams flow out of mountain canyons and onto the adjacent desert plains, sand and gravel accumulate in a wedge called an *alluvial fan* **(Fig. 5.24c)**. Since sediment in an alluvial fan has not travelled very far from its source, it may still contain feldspar and other silicate minerals that did not have the chance to weather. Therefore, if you see sandstone deposits with huge cross-bedded layers, you're

likely seeing sediments that were deposited in dunes (see Fig. 5.21d), and if you see layers of arkose and conglomerate, you could be seeing deposits of an alluvial fan. Notably, non-marine deposits may develop a red color during lithification, because in the presence of oxygen, an iron-oxide mineral called hematite precipitates in cement. Such deposits, regardless of grain size, are known informally as *redbeds* (see Fig. 5.19).

Marine Deposition

Where large rivers empty into the sea, a *marine delta* develops. These deltas host various depositional environments, including swamps, channels, floodplains, and submarine slopes **(Fig. 5.25a)**. Ocean currents and waves pick up and transport some of the clastic sediment brought to the sea. Along the shore, the intensity of wave action controls the type of sediment deposited. Where sand washes back and forth in the surf, for example, it becomes well sorted and well rounded, and sand beds may host ripple marks. In broad tidal flats, lagoons, and in deeper water offshore—regions unaffected by waves—silt and mud can also accumulate.

In warm shallow-marine settings without a supply of clastic sediment, the clear, nutrient-rich water can nurture organisms that produce carbonate shells. Nearby beaches collect sand made of small shell fragments, lagoons accumulate carbonate mud, and on reefs, corals build mounds of carbonate **(Fig. 5.25b)**. These deposits transform into various types of limestone.

On the slope that occurs between continental shelves and abyssal plains or trenches, turbidity currents carve deep submarine canyons. Where these canyons empty out, submarine fans of clastic sediment accumulate. On abyssal plains, far from land, only fine clay and plankton are available to settle out of the ocean, so deposits of deep-marine sediment, when buried, yield finely laminated mudstones, chert, or chalk **(Fig. 5.26)**.

Transgression and Regression, and Sedimentary Basins

The depositional environment at a specific location on or bordering continents and islands doesn't remain the same over geologic time. When relative sea level rises, the shoreline migrates inland, a process called **transgression**, and when relative sea level falls, the shoreline migrates seaward, a process called **regression** **(Fig. 5.27)**. During a transgression, terrestrial deposits may be buried by shallow marine deposits, which in turn are covered by deeper marine deposits. These processes can lead to the eventual formation of sediment blankets over a broad region. At times in the Earth's

FIGURE 5.27 The effects of transgression and regression during deposition of a sedimentary sequence.

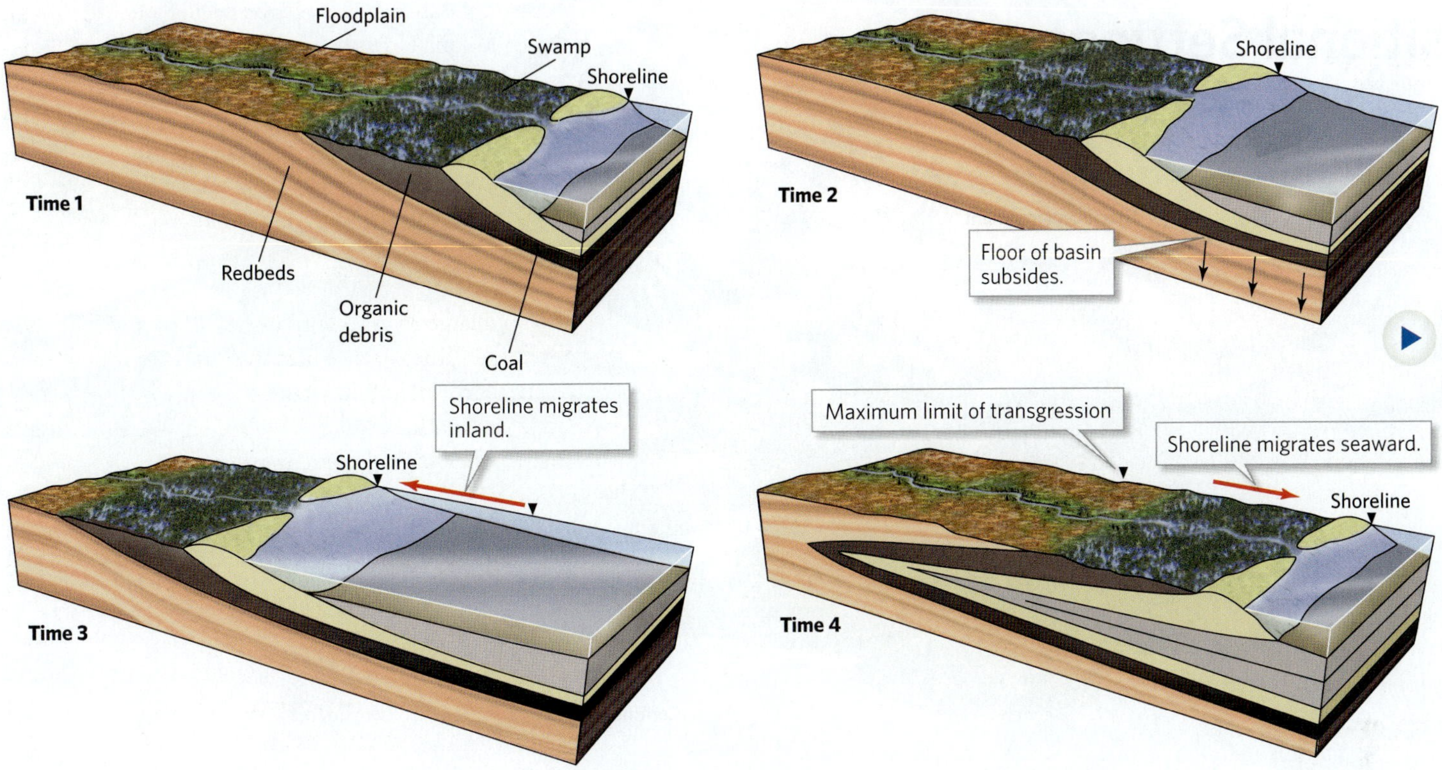

Floodplain

Swamp

Shoreline

Time 1

Redbeds

Organic debris

Coal

Shoreline

Time 2

Floor of basin subsides.

▶ **Video**
Transgression and Regression

Shoreline migrates inland.

Shoreline

Time 3

Maximum limit of transgression

Shoreline migrates seaward.

Shoreline

Time 4

history, sea level rose so much that large areas of continental interiors became submerged. Later, when sea level dropped, regression took place and the continents became dry land again.

The sedimentary veneer on the Earth's surface varies from 0 to 20 km (up to 12 mi) in thickness. Particularly thick sediment accumulations develop only in certain locations where the Earth's crust sinks, or undergoes **subsidence**, forming a depression in which sediment collects. Geologists distinguish among several types of **sedimentary basins**, as identified in **Figure 5.28**.

Take-home message . . .

🏠 Different types of sedimentary rocks form in different depositional environments. For example, strata deposited along a river differ from strata deposited by ocean waves or in the deep sea. Geologists use clues such as the character of sedimentary structures to figure out the depositional environment in which the sediments under study were deposited. Thick accumulations of sediment develop in sedimentary basins.

Quick Question ⎯⎯⎯⎯⎯⎯⎯⎯⎯⎯
Distinguish between transgression and regression.

See for yourself

Shallow Marine Environments, Red Sea, Egypt

Latitude: 22°38′17.70″ N
Longitude: 36°13′21.27″ E

Look straight down from an elevation of 4 km (~2.5 mi).

The desert sands of the Sahara abut the blue waters of the Red Sea. What types of sedimentary rocks would form if the sediments visible in this view were to be buried and preserved?

FIGURE 5.28 The geologic settings of sedimentary basins.

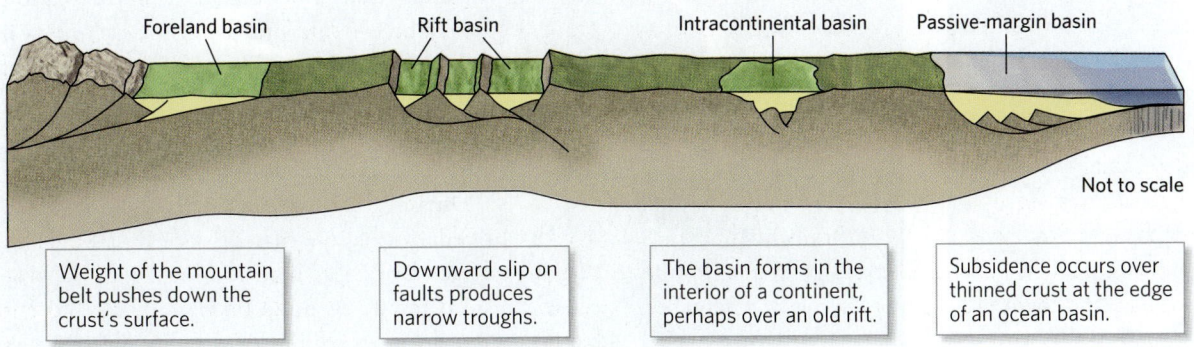

Foreland basin

Rift basin

Intracontinental basin

Passive-margin basin

Not to scale

Weight of the mountain belt pushes down the crust's surface.

Downward slip on faults produces narrow troughs.

The basin forms in the interior of a continent, perhaps over an old rift.

Subsidence occurs over thinned crust at the edge of an ocean basin.

Depositional Settings

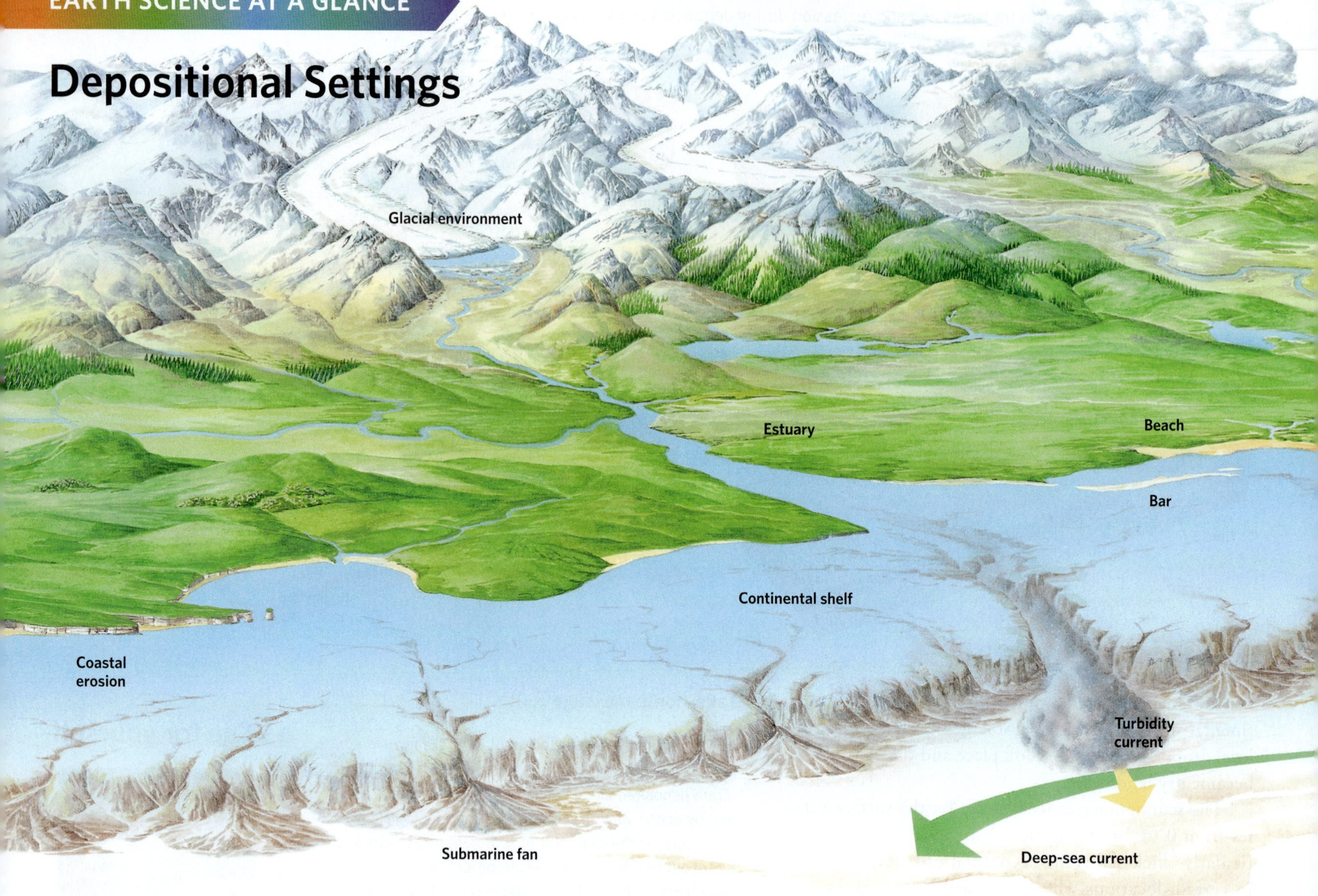

Glacial environment

Estuary

Beach

Bar

Continental shelf

Coastal erosion

Turbidity current

Submarine fan

Deep-sea current

Beach sand

Reef

Coral reef

Desert environment

Saline lake

Fluvial environment

Sand dunes

Coastal environment

Coastal swamp

Reef

Delta

Shale

Siltstone

Sandstone

Conglomerate

Fossiliferous limestone

The character of a sediment depends on the composition of the sediment and on the depositional environment in which it accumulated. Glaciers carry sediment of all sizes, so they leave deposits of poorly sorted till. Streams deposit coarser sands and gravels in their channels and finer ones on floodplains. In the quiet water of lakes, fine-grained mud accumulates. In desert environments, sand builds into dunes, evaporates precipitate in saline lakes, and gravel and sand gather in alluvial fans. Wave action along coastal beaches leaves behind well-sorted sand. In swampy areas, large volumes of plant matter accumulate. Where a river flows into the sea, its water slows and deposits a large delta. In warmer coastal marine environments, carbonate sediments, from the shells of organisms, are deposited. Locally, corals and other organisms build carbonate reefs.

Offshore, submarine canyons channel avalanches of sediment, or turbidity currents, out to the deep seafloor. Far from shore, pelagic marine sediment, commonly from the shells of plankton, slowly settles out. Burial and lithification of these various types of sediments turn them into examples of the different classes of sedimentary rocks— clastic, chemical, biochemical, and organic.

5 CHAPTER REVIEW

- Rocks at or near the Earth's surface undergo physical and chemical weathering and, as a result, eventually disintegrate into loose clasts. Weathering also produces ions in solution.

- Sedimentary rock forms at or near the Earth's surface by cementation of clasts from pre-existing rock, by growth of shell masses or cementation of shell fragments, by accumulation of organic matter, or by precipitation from water solutions.

- Soil forms when downward-percolating rainwater and living organisms interact with rock and debris. Leaching takes place nearer the surface and accumulation farther down.

- The character of soil depends on climate, time, sediment source composition, and other factors. Soil develops distinct horizons. Climate affects the nature of these horizons.

- Clastic sedimentary rocks form when sediment, produced by weathering and erosion, undergoes transportation deposition, burial, and finally, lithification. Such rocks are classified by grain size. Common clastic sedimentary rocks include shale, siltstone, sandstone, arkose, conglomerate, and breccia.

- Biochemical sedimentary rocks develop from the shells of organisms. Rocks made mostly of calcite shells are called limestone. Biochemical chert forms from silica shells of plankton Organic sedimentary rocks consist of organic remains of organisms, such as plant debris.

- Chemical sedimentary rocks precipitate directly from water solutions.

- Sedimentary structures formed during deposition include bedding, cross bedding, graded bedding, ripple marks, dunes, and mudcracks. These structures serve as clues to depositional settings.

- Glaciers, streams, alluvial fans, deserts, rivers, lakes, deltas, beaches, shallow seas, and deep seas each accumulate a different, distinctive assemblage of sedimentary strata.

- Transgressions occur when sea level rises and the shoreline migrates inland. Regressions occur when sea level falls and the shoreline migrates seaward.

- Thick layers of sedimentary rocks accumulate in sedimentary basins.

Key Terms

bedding (p. 160)
biochemical sedimentary rock (p. 156)
carbonate rock (p. 156)
cementation (p. 154)
chemical sedimentary rock (p. 158)
chemical weathering (p. 147)
chert (p. 157)
clast (p. 145)
clastic sedimentary rock (p. 150)
coal (p. 158)
compaction (p. 154)

cross bed (p. 162)
deposition (p. 151)
depositional setting (p. 163)
dissolution (p. 148)
dolostone (p. 159)
dune (p. 161)
erosion (p. 147)
evaporite (p. 158)
graded bed (p. 162)
grain (p. 145)
joint (p. 147)
limestone (p. 156)

lithification (p. 151)
maturity (p. 154)
mudcrack (p. 162)
organic sedimentary rock (p. 158)
physical weathering (p. 147)
regression (p. 164)
ripple mark (p. 161)
sediment (p. 145)
sedimentary basin (p. 165)
sedimentary rock (p. 145)
sedimentary structure (p. 159)
soil (p. 152)

soil erosion (p. 152)
soil horizons (p. 152)
soil profile (p. 152)
sorting (p. 154)
strata (p. 160)
stratigraphic formation (p. 160)
subsidence (p. 165)
subsoil (p. 152)
topsoil (p. 152)
transgression (p. 164)
turbidity current (p. 162)
weathering (p. 147)

Letters in parentheses correspond to the chapter's learning objectives.

1. Explain why weathering is an important step in the production of sediment, and indicate how physical and chemical weathering differ from each other. What are the features shown in the illustration, and what role do they serve during weathering? **(A)**

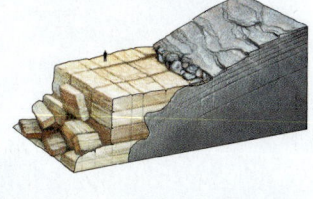

2. Provide examples of chemical reactions that can be involved in weathering. What happens to feldspar during chemical weathering? What happens to quartz? **(A)**

3. What characteristics make a soil different from a sediment? What are soil horizons, and how do they form? How does climate influence soil development? **(B)**

4. How do human activities affect soil erosion rates? What can be done to conserve soil? **(B)**

5. Describe the steps leading to the production of a clastic sedimentary rock from fresh bedrock. What processes contribute to lithification? **(C)**

6. How do grain size, sorting, and angularity change during the transportation and weathering of sediment? **(C)**

7. Name the rock shown in the illustration. How did the clasts become rounded? **(C)**

8. Shale and mudstone form from the lithification of what type of sediment? How does a shale differ from a mudstone? **(C)**

9. Explain the processes involved in forming biochemical sedimentary rocks. How do these processes differ from those forming chemical sedimentary rocks? **(D)**

10. Provide examples of different types of limestone, and indicate which are biochemical and which are chemical. Why are limestones considered a type of carbonate rock? Name another type of carbonate rock and explain how it forms. **(D)**

11. What is chert, and how does it form? **(D)**

12. What kind of rock forms from the sediment deposited in a coral reef? **(D)**

13. Explain how bedding develops in sedimentary strata. What visual features can you use to recognize bedding? What is a stratigraphic formation? **(E)**

14. If the surface of a bed has ripple marks, was the sediment in the bed deposited in a current or in still water? What does the presence of mudcracks in a layer of shale imply? **(E)**

15. How do cross beds form? Draw a sketch to indicate how the orientation of cross beds serves as a clue to the direction of current flow at the time of deposition. **(E)**

16. Explain how sediment containing graded bedding develops, and give an example of a setting where such sediment might accumulate. **(E)**

17. How does the sediment deposited in a glacial depositional setting differ from sediment deposited along a river? What is the difference between an alluvial fan and a marine delta? What kind of sedimentary rock might form from alluvial-fan deposits? **(F)**

18. Explain the difference between transgression and regression. **(G)**

19. What process must happen in order for a very thick succession of sedimentary strata to accumulate in a sedimentary basin? **(G)**

On Further Thought

20. The top three formations exposed on the walls of the Grand Canyon are, from oldest to youngest: the Coconino Sandstone (composed of quartz sand in beds that contain large cross beds, similar to those shown in Figure 5.21); the Toroweap Formation (composed of shale and gypsum beds); and at the top, the Kaibab Limestone (composed of grey limestone containing marine fossils). All three formations are late Paleozoic, deposited between about 250 and 260 Ma. Indicate which depositional setting existed at the time when each unit was deposited. What might have caused the change in depositional setting between the time when the Coconino Sandstone was deposited, and the time when the Kaibab Limestone was deposited? **(G)**

21. Fly to latitude 23°58'40.98"N and longitude 77°30'20.37"W using Google Earth. You will see the shore of an island in the Bahamas. This location is far from any source of clastic grains composed of silicate minerals, because it is far from any rivers carrying such sediment into the sea. The climate of the Bahamas is tropical. What does the sand on the nearby beaches consist of? If buried and lithified, what type of sedimentary rock will this become? Zoom down to low elevation (< 5 km), and you can see large ripples in the sand. What caused these ripples to form? **(D, E)**

6 A PROCESS OF CHANGE
Metamorphism and the Rock Cycle

By the end of the chapter you should be able to . . .

A. define metamorphism, and characterize the key features of metamorphic rocks.

B. describe processes that cause mineral assemblages and textures to change during metamorphism.

C. explain how geologists classify metamorphic rocks, and describe the major metamorphic rock types.

D. distinguish among different metamorphic grades, and explain why they exist.

E. characterize geologic phenomena in the Earth that cause metamorphism, and provide examples of places where you might find metamorphic rocks.

F. sketch a model illustrating pathways through the rock cycle, and discuss the energy driving the cycle.

6.1 Introduction

After a caterpillar grows to full size, it hangs from a branch and becomes encased in a chrysalis. Inside, the organic material of the caterpillar reorganizes, and a butterfly takes shape (Fig. 6.1a). Biologists refer to this amazing change as *metamorphosis*, from the Greek words *meta*, meaning change, and *morphe*, meaning form. In geologic environments where temperatures and pressures become relatively high, rocks undergo changes every bit as dramatic as those that happen when a caterpillar turns into a butterfly. Geologists use a similar word, **metamorphism**, for changes that take place in rocks at high temperatures and pressures, and they use the term *metamorphic rock* for the product of this change (Fig. 6.1b). Specifically, a **metamorphic rock** is one that forms when a **protolith** (pre-existing rock) changes due to the growth of new minerals or the development of a new *texture* (as defined by the shape, size, and arrangement of mineral grains) without the involvement of melting or chemical weathering.

To help you understand metamorphism and its products, this chapter begins by describing the causes of metamorphism and the basis for classifying metamorphic rocks. We then characterize the geologic settings in which these rocks form in the context of plate tectonics theory. We conclude this chapter—and this book's introduction to Earth materials overall—by examining the *rock cycle*, the progressive transition of materials from one rock type to another rock type that can take place on our dynamic planet over geologic time. You'll see that metamorphism represents but one avenue of transition in the rock cycle.

6.2 Causes and Consequences of Metamorphism

What Is a Metamorphic Rock?

If someone were to put a hand specimen of rock on a table in front of you, how would you know that it's metamorphic? First, metamorphic rocks may possess **metamorphic minerals**, new minerals that grow within solid rock under metamorphic conditions. In fact,

>> This outcrop, on a hill in Scotland, reveals metamorphic rock that formed when sedimentary rock was subjected to high temperatures and pressures after being buried very deeply during mountain building. In response to a change in the rock's environment, a new group of metamorphic minerals grew.

FIGURE 6.1 Dramatic transformations.

(a) A caterpillar undergoes metamorphosis to become a butterfly.

(b) A shale undergoes metamorphism to become a garnet schist.

metamorphism typically produces a *metamorphic mineral assemblage*, meaning a group of minerals that were not all present in the protolith. Second, metamorphic rocks can display **metamorphic textures**, distinctive arrangements and orientations of mineral grains not found in the protolith. All metamorphic rocks have *crystalline textures* (composed of interlocking grains), regardless of whether the protolith did. And some metamorphic rocks contain *metamorphic foliation*, a type of layering not found in sedimentary or igneous rocks. Development of metamorphic minerals and textures can make a metamorphic rock look as different from its protolith as a butterfly does from a caterpillar.

Nothing in the world lasts, save eternal change.

—*HONORAT DE BUEIL*
(FRENCH POET, 1589–1650)

Changes Involved in Metamorphism

If metamorphism does not involve melting (the formation of magma) or chemical weathering (the reaction of minerals with air and water at or near the Earth's surface), then how does it take place? To understand metamorphism, even at a basic level, we need to zoom in on the interior of a mineral grain to picture atoms and the chemical bonds that hold them together (see Chapter 3). During metamorphism, **diffusion**, the process by which atoms migrate slowly through a material, takes place. For diffusion to happen, the bonds holding an atom to its neighbor must break, and the atom must move slightly before it reattaches to new neighbors. When this process repeats countless times for countless atoms, atoms effectively rearrange themselves so that new minerals grow while pre-existing ones disappear. (Other processes, discussed in more advanced books, can also be involved in the changes taking place during metamorphism.) When diffusion, and other atomic-scale processes, take place, a variety of changes can occur.

Animation

Metamorphic Change: Recrystallization

- *Recrystallization:* The process of **recrystallization** changes the shape and size of mineral grains in a rock, and in some cases the way in which the grains are held together, without changing the identity of the minerals involved **(Fig. 6.2a)**. For example, recrystallization transforms sandstone, composed of rounded quartz grains held together by quartz cement, into quartzite, a metamorphic rock composed of interlocking quartz crystals.

- *Phase change:* If a mineral in a rock transforms into another mineral that has the same chemical composition but a different crystal structure, we say that the mineral has undergone a **phase change**. The production of diamond from graphite represents a phase change: both minerals consist entirely of carbon, but the arrangement of carbon atoms in graphite differs from that in diamond (see Chapter 3).

Animation

Metamorphic Change: Phase Change

- *Metamorphic reactions:* During metamorphism, a shale (a rock composed of clay) may change into a rock composed of mica and garnet. Such a change, involving the growth of new minerals that were not present in the protolith, exemplifies a **metamorphic reaction (Fig. 6.2b)**. During a metamorphic reaction, atoms separate from pre-existing minerals, diffuse through the solid mineral, then bond in a different configuration to form new minerals.

- *Pressure solution:* If a water film coats the surfaces of mineral grains, portions of the grains can dissolve at contact points where they are being squeezed against neighboring grains. The resulting ions migrate through the water film to another location in the rock, where they precipitate. This process, known as *pressure solution*, modifies the texture of a rock **(Fig. 6.2c)**.

Animation

Metamorphic Change: Pressure Solution

- *Plastic deformation:* At elevated temperatures and pressures, mineral grains can change shape very slowly without breaking or dissolving **(Fig. 6.2d)**. Such *plastic deformation* allows grains to flatten or become elongate.

FIGURE 6.2 Metamorphic processes as seen through a microscope.

Protolith ⟶ Metamorphic rock

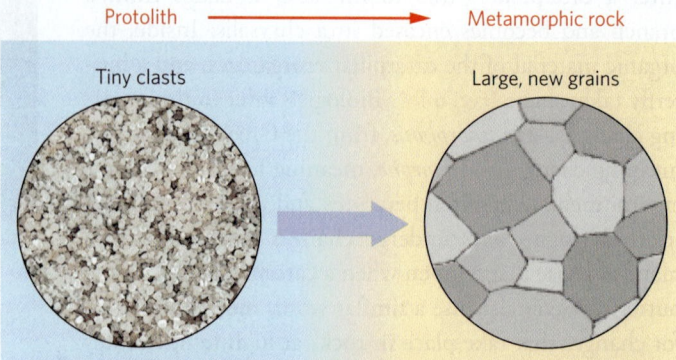

(a) Mineral grains can recrystallize to form new, interlocking grains of the same mineral. Typically, the new grains are larger.

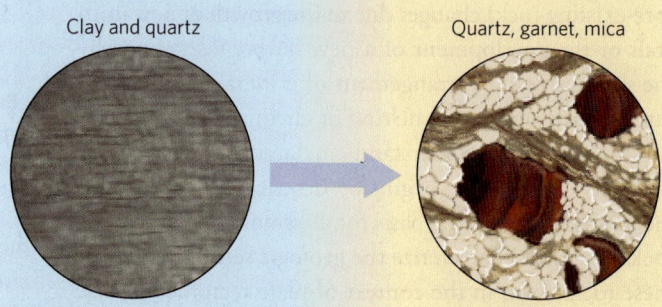

(b) Metamorphic reactions can change the original mineral assemblage into a new, metamorphic mineral assemblage.

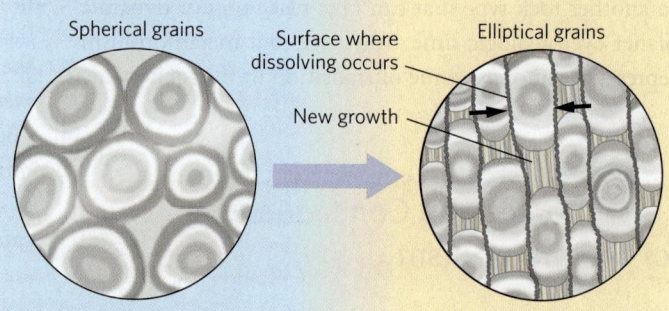

(c) Pressure solution can dissolve grains on the sides being pressed together more tightly. Black arrows indicate the squeezing direction.

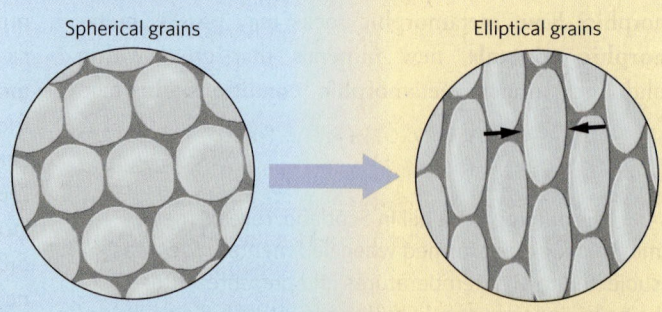

(d) Plastic deformation can change the shape of grains without breaking them. Black arrows indicate the compression direction.

Agents of Metamorphism

You've seen diffusion when you've poured milk into coffee and watched it disperse. But in solid rock, the process can take place so slowly that a rock can survive for billions of years, largely unchanged, unless subjected to one or more *agents of metamorphism*, phenomena that drive metamorphism:

INCREASE IN TEMPERATURE. Think about what happens to atoms in a mineral grain as the grain becomes warmer. Heat energy makes atoms vibrate more rapidly, so the chemical bonds locking atoms to their neighbors stretch and bend. If a bond stretches or bends too much, it breaks, and the atom detaches from its neighbor. The atom can then move slightly and bond to a different atom. As we've seen, repetition of this process allows many atoms to migrate slowly through a solid, enabling recrystallization and metamorphic reactions. Increases in temperature play the most important role in driving metamorphism, because unless atoms can move and vibrate fast enough for chemical bonds to break, metamorphism won't happen.

INCREASE IN PRESSURE. When you swim underwater, water squeezes against you equally from all sides—in other words, your body feels *pressure*. The deeper you go, the greater the pressure you experience, because the weight of overlying water increases. Pressure also increases with depth in the solid rock of the Earth's crust. The squeeze caused by increasing pressure drives atoms to fit together more tightly than they did in protolith minerals. In Chapter 1, we introduced the *bar* as a unit of pressure. Pressures at which metamorphism takes place are so large that geologists specify them in *kilobars* (1 kbar = 1,000 bar). For example, at a depth of 20 km (about 12 mi) in the crust, pressures reach about 5.5 kbar.

INCREASE IN BOTH TEMPERATURE AND PRESSURE. So far, we've looked separately at changes caused by pressure and by temperature. In the Earth, however, pressure and temperature both increase with increasing depth (**Fig. 6.3**). Experiments show that the specific metamorphic reactions taking place in a rock depend on both temperature and pressure.

INTERACTION WITH HYDROTHERMAL FLUIDS. Sometimes metamorphism takes place in the presence of very hot water solutions, known as **hydrothermal fluids**. The presence of hydrothermal fluids speeds up metamorphic reactions because the atoms involved in such reactions can diffuse faster through the fluid than they can through a solid mineral grain. In addition, hydrothermal fluids passing through a rock may pick up some dissolved ions and drop others off, just as a bus picks up and drops off passengers.

So, interaction with hydrothermal fluids can change the overall chemical composition of a rock during metamorphism. Geologists refer to compositional changes that take place during metamorphism as **metasomatism**.

APPLICATION OF DIFFERENTIAL STRESS. We noted above that pressure refers to an equal push in all directions. Pressure can cause the volume of an object to change, as would happen if you carry a balloon with you to the bottom of a pool, but it won't affect the object's shape. In solids, however, the push in one direction may be greater than the push in another direction. You can picture why if you put a ball of dough on the floor, lay a book on it, and step on the book—the ball flattens into a pancake, oriented parallel to the floor, because the downward push you apply with your foot exceeds the push provided by air in other directions (**Fig. 6.4a**). If you press the ball of dough between a wall and a vertical book, the dough will flatten into a pancake shape parallel to the wall (**Fig. 6.4b**). Note that the orientation of flattening depends on the direction of the greatest push (**Box 6.1**).

Pushing or squeezing of an object is called *compression*. Not all stress is compressive. For example, if you grab each side of the dough ball with your hands and pull, you're applying *tension* to the ball, so it stretches and becomes longer. And if you place the dough on a table, set your hand on top of it, and move your hand parallel to the table, you're applying *shear* to the ball. Shear is a push applied at an angle that causes one part of a material to move sideways relative to another part (**Fig. 6.4c**). Compression, tension, and shear are all examples of **stress**, the force applied per unit area at a location. Geologists use the term **differential stress** to distinguish conditions in which the push or pull in one direction differs from that in another, or in which shearing is taking place. Note that *pressure* also represents the application of stress, but as noted above, it represents a special case in which the amount of stress applied to a body has the same value in all directions. (We'll discuss the general concept of stress further in Chapter 7.)

At the pressures and temperatures that exist below a depth of 10–15 km (6–9 in) in the crust, slowly subjecting a rock to differential stress can cause the rock's texture to change. Not only can it cause some grains to become *platy* (pancake-shaped) or *elongate* (cigar-shaped), but it can cause such grains to align in the same direction; these changes can be due to pressure solution or plastic deformation. When platy or elongate grains in a rock become aligned, we say that they display a **preferred orientation** (**Fig. 6.4d**). Note that platy and elongate grains are *inequant*, meaning they have different dimensions in different directions—we distinguish inequant grains from *equant* grains, which have the same dimensions in all directions (**Fig. 6.4e**).

FIGURE 6.3 Approximate pressure and temperature gradients in the Earth's crust.

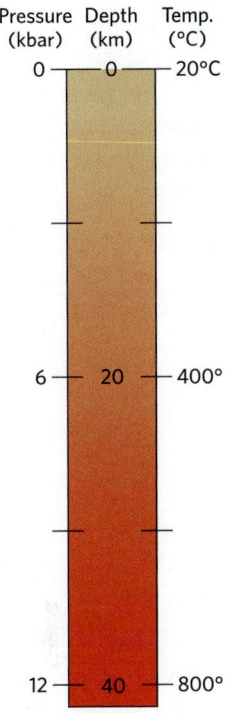

Pressure (kbar) / Depth (km) / Temp. (°C)

0 — 0 — 20°C

6 — 20 — 400°

12 — 40 — 800°

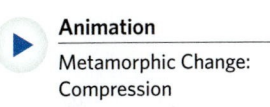
Animation
Metamorphic Change: Compression

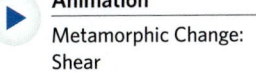
Animation
Metamorphic Change: Shear

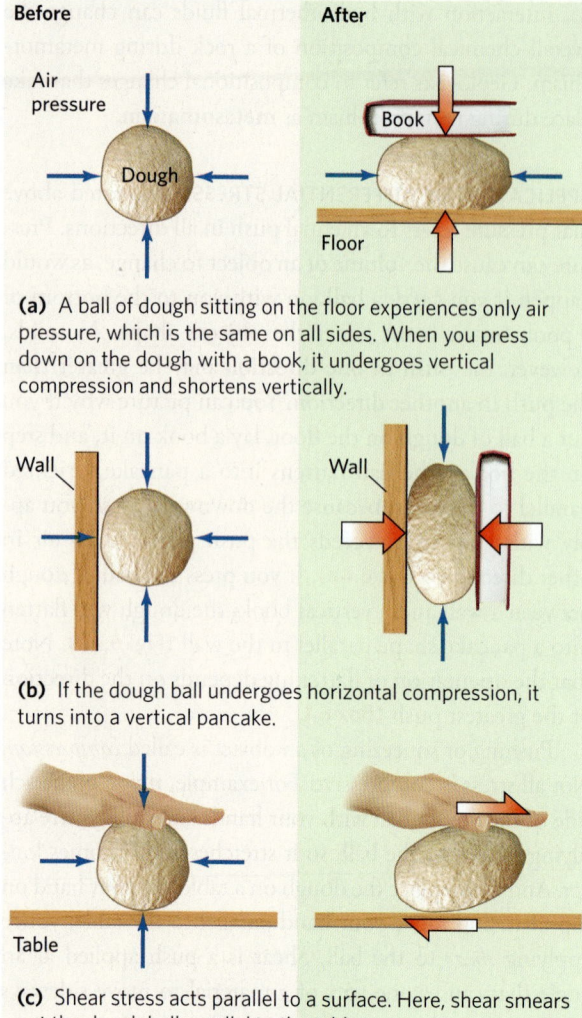

(a) A ball of dough sitting on the floor experiences only air pressure, which is the same on all sides. When you press down on the dough with a book, it undergoes vertical compression and shortens vertically.

(b) If the dough ball undergoes horizontal compression, it turns into a vertical pancake.

(c) Shear stress acts parallel to a surface. Here, shear smears out the dough ball parallel to the table.

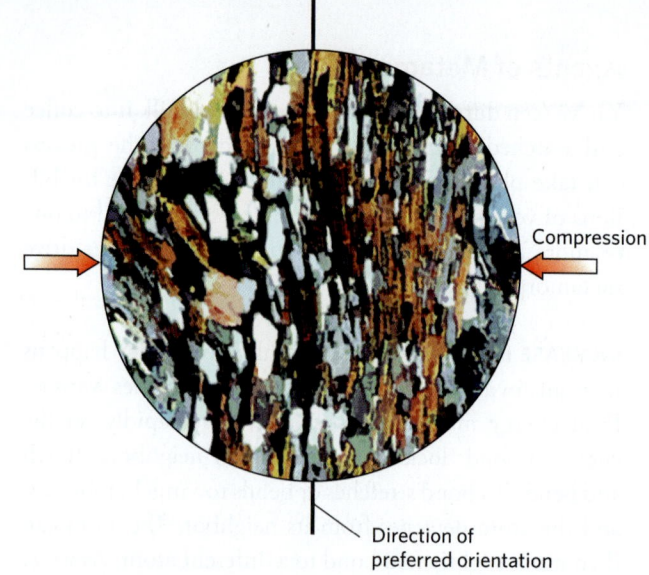

(d) In metamorphic rock, inequant grains may be aligned to form a preferred orientation. As seen through a microscope, the flat planes of grains are perpendicular to the compression direction.

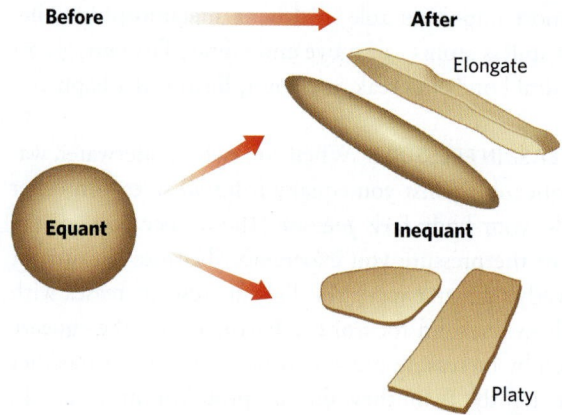

(e) Compression and shear can transform equant grains into inequant grains. Inequant grains can be elongate (cigar-shaped) or platy (pancake-shaped).

Take-home message . . .

A metamorphic rock contains mineral assemblages and textures that differ from those found in its protolith. Many processes, such as recrystallization, metamorphic reactions, phase changes, pressure solution, and plastic deformation can take place during metamorphism. These processes happen in response to changes in temperature and pressure, interaction with hydrothermal fluids, or by the application of differential stress.

Quick Question

Can the overall composition of rock change during metamorphism?

6.3 Types of Metamorphic Rocks

Now that we've introduced the processes involved in metamorphism and the conditions that cause metamorphism, let's examine the various types of rocks formed by metamorphism. Coming up with a way to classify and name the great variety of metamorphic rocks on the Earth has not been easy. After decades of debate, geologists ended up dividing metamorphic rocks into two fundamental classes—foliated and nonfoliated—based on whether or not the rock contains **metamorphic foliation**, layering formed in response to metamorphism. We can distinguish among different types of foliated rocks based on the character of the foliation, and we can distinguish among different types of nonfoliated rocks based on their composition.

Foliated Metamorphic Rocks

The word *foliation* comes from the Latin *folium*, meaning leaf. Metamorphic foliation can give metamorphic rock a striped or streaked appearance in an outcrop, and in some cases it makes a rock susceptible to splitting into thin sheets. Not all metamorphic foliation looks the same—its character depends on the rock's grain size and on whether the planes of foliation represent the alignment of inequant grains or the existence of alternating

How can I explain . . .

The meaning of differential stress

What are we learning?

- That stress is the application of force to a specified area.
- That different orientations of stress have different consequences.
- That application of stress can produce a preferred orientation.

What you need:

- A baseball-sized mass of Play-Doh or Plasticine modeling material
- A heavy book, six coins, and a large nail

Instructions:

- Mold the modeling material into a spherical ball.
- Insert the coins, each perpendicular to the ball's surface, at different locations along a circumference on the surface of the ball, as shown in the figure. If you slice the ball carefully, you can see the orientation of the coins.

- Place the ball on a table, place the book horizontally on the surface of the ball, and press down, if the book isn't heavy enough to change the shape of the ball.
- Examine the shape of the ball and the orientation of the coins.
- Repeat the experiment, but this time insert a large nail partway into the ball between the horizontal book and the surface of the ball. The nail should be vertical. The weight of the book should push the nail deep into the ball
- Repeat the experiment (not shown), but this time squeeze the ball between two vertical books, or between the book and a wall.

What did we see?

- A preferred orientation of platy minerals can develop due to application of differential stress.
- A force applied over a broad area does not have the same consequences as the same force applied to a small area.
- The changes caused by applying a differential stress depend on the orientation of the stress.

Dough ball   Sliced ball   Coin

---

layers of different metamorphic minerals. Here, we describe the most common foliated metamorphic rocks.

- *Slate:* The finest-grained foliated metamorphic rock, **slate**, forms by metamorphism of clay-rich rock (shale or mudstone) when it is subjected to horizontal compression, a differential stress, at relatively low pressures and temperatures. Under these conditions, clay flakes (very tiny, platy grains) develop a preferred orientation by becoming aligned parallel to one another and perpendicular to the direction of maximum compression **(Fig. 6.5a)**. The result is foliation called **slaty cleavage**; it forms partly due to rotation of pre-existing clay flakes and partly due to the growth of new grains. Slate splits into thin sheets parallel to slaty cleavage, so the rock can be easily used to make roofing shingles **(Fig. 6.5b)**.

- *Phyllite:* In Greek, the word *phyllon* means leaf—that's why Greek pastry dough that comes in very thin sheets is called phyllo. Geologists use a similar word, **phyllite**, for a fine-grained metamorphic rock

containing foliation defined by the preferred orientation of fine-grained, light-colored translucent mica. The mica gives phyllite a silky luster **(Fig. 6.6a)**. Phyllite is formed by the metamorphism of slate at a temperature high enough to cause clay to recrystallize into mica when subjected to differential stress.

- *Schist:* Rock that has medium to coarse grains and possesses *schistosity*, foliation defined by the preferred orientation of large crystals of mica (generally muscovite or biotite), is called **schist (Fig. 6.6b)**. Schist forms at temperatures higher than those that produce phyllite, for at higher temperatures, mica can grow into larger crystals. Other minerals, such as garnets, can also grow in schist. The existence of schistosity means that the rock metamorphosed while being subjected to differential stress.

- *Gneiss:* A **gneiss** is a metamorphic rock consisting of alternating layers of dark-colored and light-colored minerals **(Fig. 6.7a, b)**. Such *compositional layering*, or **gneissic banding**, can form in several ways. In some

Did you ever wonder . . .

why slate makes such nice roofing shingles?

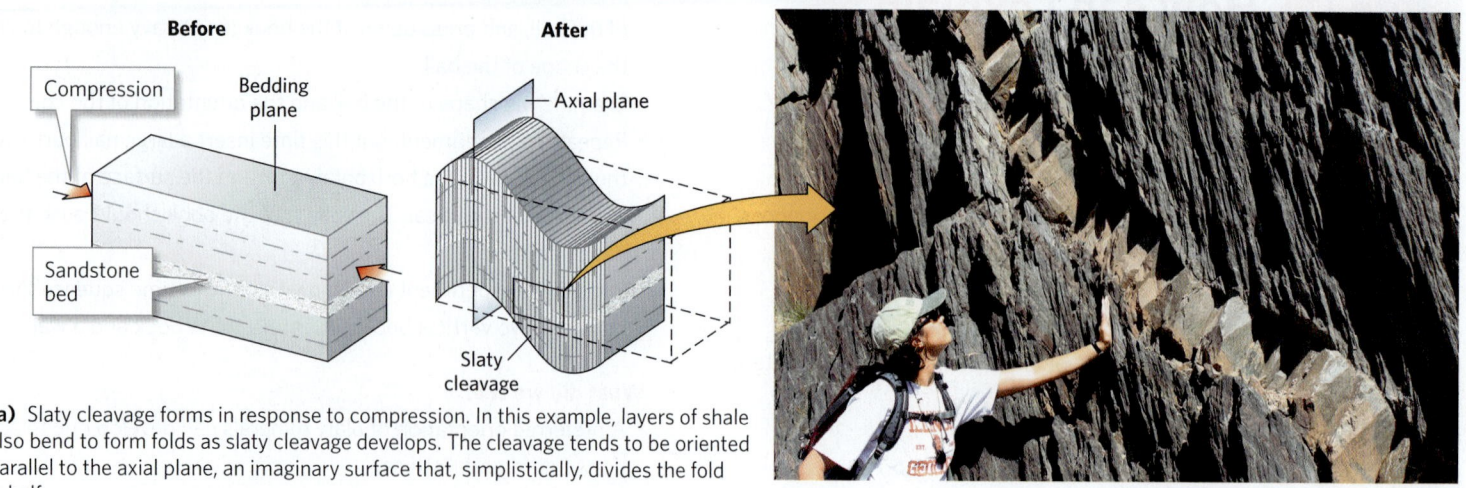

Before   After

(a) Slaty cleavage forms in response to compression. In this example, layers of shale also bend to form folds as slaty cleavage develops. The cleavage tends to be oriented parallel to the axial plane, an imaginary surface that, simplistically, divides the fold in half.

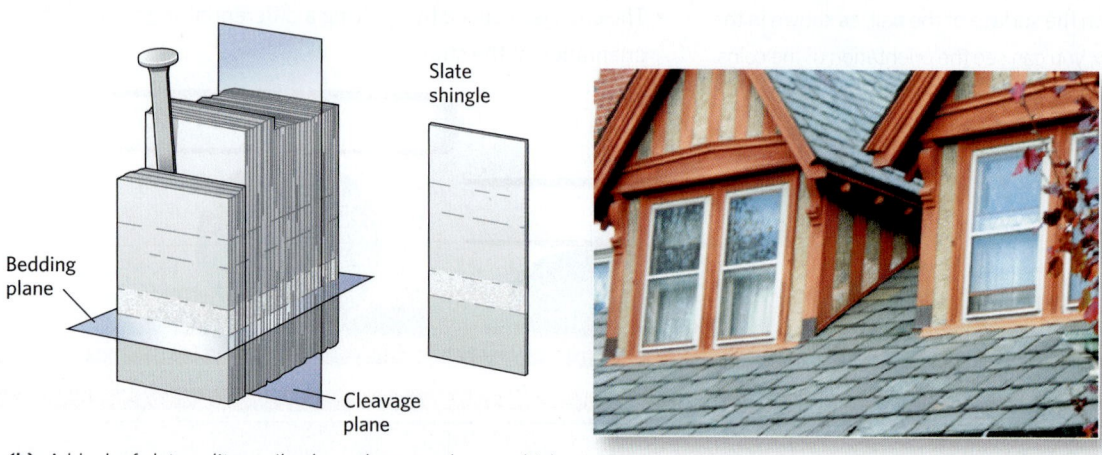

(b) A block of slate splits easily along cleavage planes, which may be at a high angle (shown here at 90°) to the bedding planes. Workers split slate to produce roof shingles that, when overlapped, make a watertight surface (inset photos).

FIGURE 6.6 Examples of foliated metamorphic rocks.

(a) During the formation of this phyllite, clay recrystallized to form tiny mica flakes that reflect light, giving the rock a silky sheen. At this locality, in Brazil, compression wrinkled the foliation.

(b) This schist, in New York City, contains coarse mica flakes along with other metamorphic minerals.

FIGURE 6.7 The nature of gneiss and the formation of gneissic banding.

(a) In this block of gneiss, the layers are centimeters across.

(b) An outcrop of 2.7-Ga gneiss in Ontario, Canada, displays curving foliation.

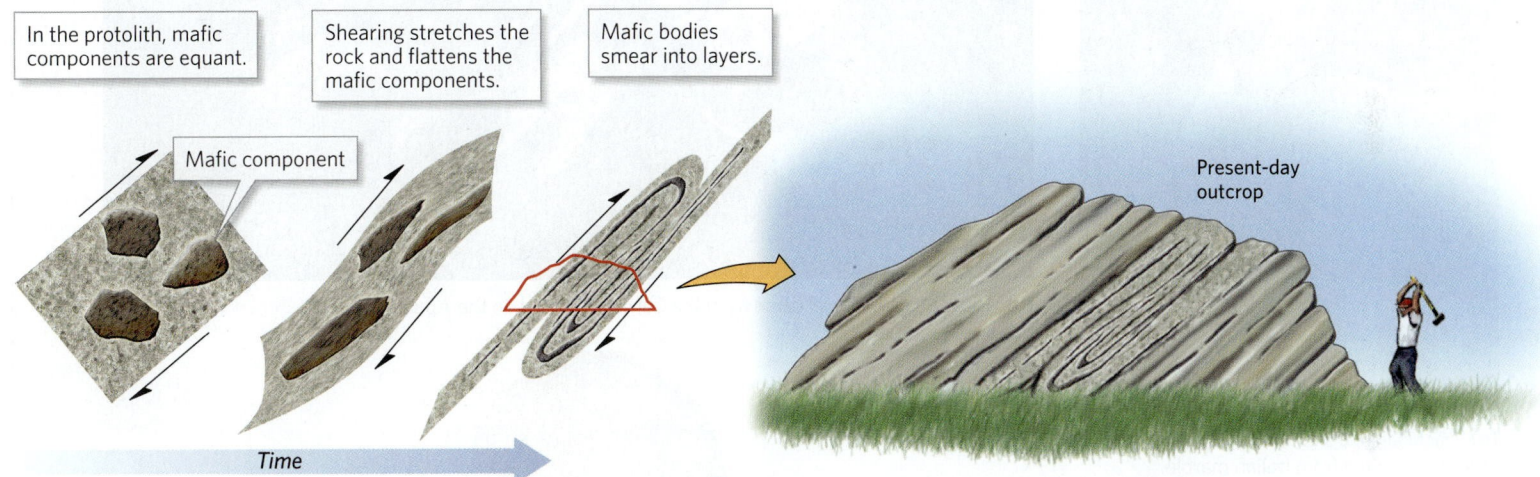

In the protolith, mafic components are equant.

Shearing stretches the rock and flattens the mafic components.

Mafic bodies smear into layers.

Mafic component

Present-day outcrop

Time

(c) Formation of gneiss, in some cases, involves extreme shear. In this example, bodies of mafic rock are embedded in a felsic rock. Due to shear, original contrasting rock types are smeared into parallel layers.

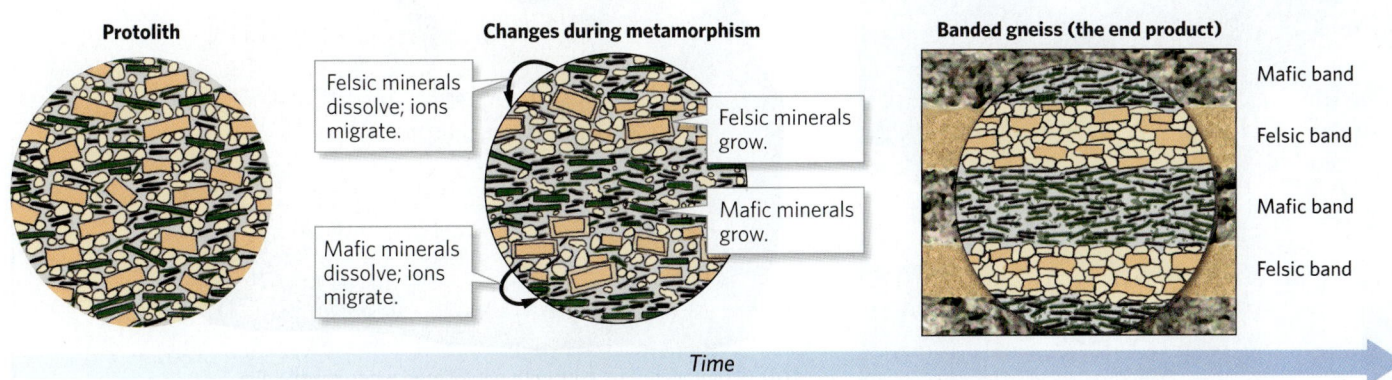

Protolith

Changes during metamorphism

Banded gneiss (the end product)

Felsic minerals dissolve; ions migrate.

Felsic minerals grow.

Mafic minerals grow.

Mafic minerals dissolve; ions migrate.

Mafic band

Felsic band

Mafic band

Felsic band

Time

(d) Gneiss may also form by metamorphic differentiation, during which metamorphic reactions cause felsic and mafic mineral crystals to grow in distinct, separate layers.

Putting Earth Science to Use

The sculptures of Michelangelo

Believe it or not, deepening your understanding of different rock types and how they form can enrich your experience when viewing and assessing great works of art. Michelangelo (1475–1564), one of the most famous Italian artists of the Renaissance, once said,

"Every block of stone has a statue inside it, and it is the task of the sculptor to discover it." Michelangelo's fame as a sculptor came from the drama and realism of his creations in marble **(Fig. Bx6.2a)**. *David*, a statue that he started when he was only 26, rises to a height of over 5 m (17 ft) and weighs an estimated 5,600 kg (12,500 lb).

FIGURE Bx6.2 Michelangelo's marble.

(a) Michelangelo's famous sculpture *David* was carved from Italian marble.

(b) The marble for *David* came from the Carrara quarries, in the Apuan Alps of northwestern Italy.

(c) An example of an Italian marble quarry sliced into a mountainside.

(d) Sculptors like to work with nonfoliated white marble due to its softness and uniform texture. Its homogeneous texture of interlocking crystals makes Carrara marble, specifically, resistant to shattering during carving.

Michelangelo did not use just any blocks of marble for his works. The marble in *David* came from the Carrara quarries, perched at the peak of a 1.6-km (1-mi)-high mountain range called the Apuan Alps in northwestern Italy. There, quarries originally established during the Roman Empire expose immense cliffs of nearly pure white stone **(Fig. Bx6.2b)**. How did the Carrara marble form? Its story began over 200 million years ago, when the region hosted shallow seas in which marine organisms with calcite shells lived and died. The shells were buried and lithified to form a very pure limestone. During Cenozoic mountain building, the limestone ended up at a depth of 15–18 km (9–11 mi) below the Earth's surface, where it underwent metamorphism at a temperature of around 400°C (750°F) and recrystallized. The lack of minerals other than calcite in the protolith meant that the resulting marble was very pure. Continued mountain building, together with erosion, eventually brought the marble back to the Earth's surface.

For over two millennia, people have sweated and struggled to break chunks of this marble free from bedrock **(Fig. Bx6.2c)**. In Michelangelo's day, workers wedged blocks of rock free from cliffs by pounding a row of chisels into the rock to produce a crack. Beginning in the 19th century, wire saws became popular tools. These saws consist of an abrasive wire that revolves around two pulleys. Movement of the wire slowly grinds a slot into the rocks. Because the quarries are high on the sides of the mountains, blocks could be towed by oxen down the mountainside to waiting ships. Today, trucks and trains carry the stone to a port, and ships transport it worldwide.

Why did Michelangelo choose Carrara marble to sculpt? The rock has a very homogeneous texture of interlocking crystals, making the marble resistant to shattering during carving **(Fig. Bx6.2d)**, and because of its purity, it has a uniform color. The softness of the calcite (3 on the Mohs hardness scale) making up the rock allows it, like any marble, to be highly polished. Light can penetrate slightly into Carrara marble, so the surface of sculptures made from it are slightly translucent. This quality gives the sculptures a waxy glow similar to that of human skin.

cases, it develops when a protolith of one rock type that contains bodies of a different rock type undergoes extreme shear stress. As a consequence, plastic deformation smears out the different rock types into aligned sheets **(Fig. 6.7c)**. Banding can also develop by an incompletely understood process known as *metamorphic differentiation*, during which metamorphic reactions segregate different mineral types into different layers **(Fig. 6.7d)**.

Nonfoliated Metamorphic Rocks

In a nonfoliated metamorphic rock, metamorphic minerals are either equant or, if inequant, have a random orientation. Common nonfoliated rocks include the following:

- *Hornfels:* When a protolith undergoes heating without being subjected to differential stress, it undergoes metamorphism to become a **hornfels**, a fine- to medium-grained nonfoliated rock that contains a variety of metamorphic minerals. In a hornfels, inequant grains do not have a preferred orientation.

- *Quartzite:* During metamorphism of a pure quartz sandstone, sand grains and cements recrystallize. This process destroys the distinction between cement and grains, and it eliminates most pore space, resulting in a metamorphic rock called **quartzite**, composed of interlocking quartz crystals **(Fig. 6.8a)**. Notably, though we introduce quartzite here as a nonfoliated rock, differential stress applied to a quartzite can produce foliation in the rock.

- *Marble:* Subjecting limestone to metamorphism yields **marble**, a metamorphic rock composed almost entirely of interlocking calcite crystals **(Box 6.2)**. During the formation of marble, grains in the protolith recrystallize, so fossil shells, pores, and the distinction between grains and cement all disappear **(Fig. 6.8b)**. As is the case for quartzite, not all marble lacks foliation—differential stress during metamorphism can cause the rock to flow plastically, producing marble that has beautiful color banding **(Fig. 6.8c)**.

Metamorphic Grade

Not all metamorphism takes place under the same **metamorphic conditions**, meaning the temperature and pressure at which metamorphic minerals and textures form. For example, rock at a depth of 40 km (24 mi) beneath a mountain range undergoes metamorphism at a higher temperature and pressure than does rock at a depth of 20 km (12 mi). Geologists use the concept of **metamorphic grade** in an informal way to characterize the metamorphic conditions in which a given rock underwent metamorphism. Temperature plays the dominant role in determining the nature of the metamorphic reactions that take place, so grades are distinguished from one another primarily by the temperature of metamorphism. Specifically, *low-grade* metamorphic rocks form at temperatures of 200°C–400°C (480°F–750°F), *intermediate-grade*

FIGURE 6.8 Examples of nonfoliated metamorphic rocks.

Quartz sandstone—the protolith of quartzite

(a) In this sandstone from Kuwait (left), the sand grains stand out. In contrast, this nonfoliated maroon quartzite (right) looks glassy. Note the contrast in texture revealed by the photomicrographs.

(b) In this unmetamorphosed limestone from New York (left), you can see bedding as well as fossils and shell fragments. Such distinctions are not visible in the white marble (right) exposed in an Italian quarry. Note the texture contrasts in the photomicrographs.

(c) The marble tiles in this floor display color banding due to imperfections inherited from original bedding.

FIGURE 6.9 Intensity of metamorphism is indicated by metamorphic grade.

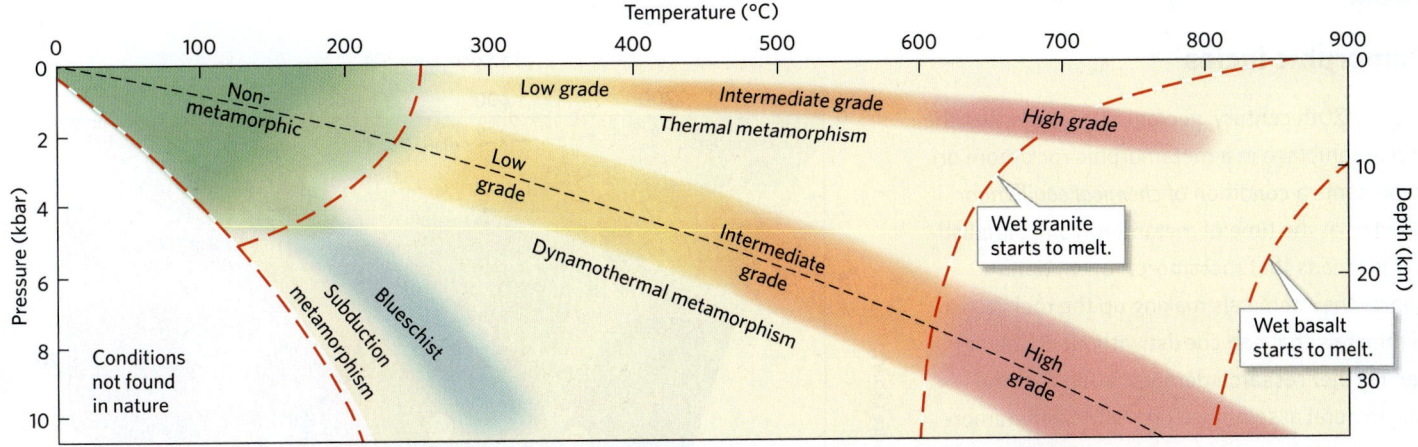

(a) This graph depicts the approximate temperatures and pressures of metamorphic grades. Different metamorphic conditions occur in different geologic settings. *Wet* means that the rock contains water and therefore melts at a lower temperature than does dry (water-free) rock.

Grade	PROTOLITH	LOW GRADE	INTERMEDIATE GRADE	HIGH GRADE

(b) The metamorphic minerals that form in a given rock depend on metamorphic grade and protolith composition. This chart lists the metamorphic minerals that form from a shale protolith at different metamorphic grades.

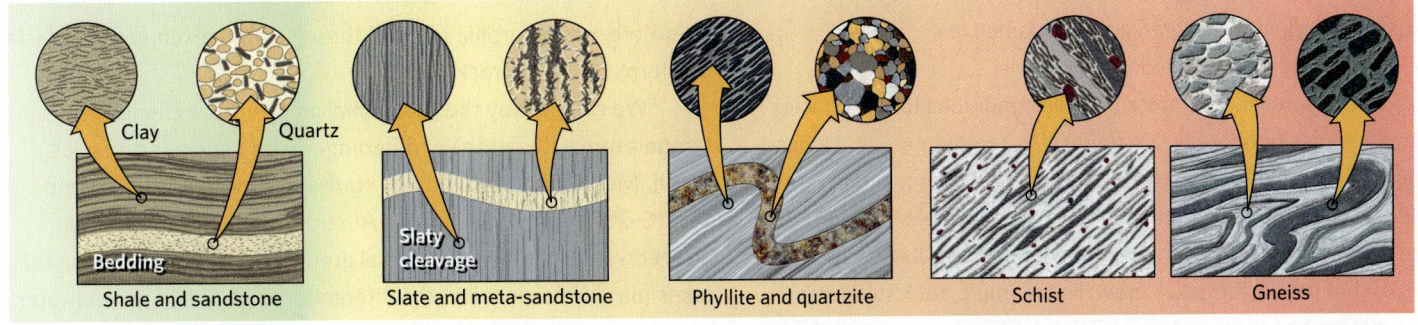

(c) Here we see the consequences of the progressive metamorphism of shale and sandstone during mountain building. The background colors indicate the range of grades in which key metamorphic minerals form. Note how textures change with increasing grade.

metamorphic rocks at 400°C–600°C (750°F–1,100°F), and *high-grade* metamorphic rocks at 600°C–850°C (1,100°F–1,560°F) **(Fig. 6.9a)**. Geologists consider slate and phyllite to be examples of low-grade metamorphic rocks, most schist and some gneiss to be intermediate-grade metamorphic rocks, and some schist and most gneiss to be high-grade metamorphic rocks. Notably, metamorphic rocks formed at different grades contain different minerals (with the exception of quartzite and marble, which consist only of quartz and calcite, respectively, regardless of grade), and the texture of metamorphic rocks changes with increasing grade **(Fig. 6.9b, c)**. A more formal characterization of metamorphic conditions uses the concept of *metamorphic facies* **(Box 6.3)**.

BOX 6.3

A Deeper Look

Metamorphic facies

In the early 20th century, geologists realized that the mineral assemblage in a metamorphic rock more or less represents a condition of *chemical equilibrium* that existed at the time of metamorphism. Simplistically, this means that metamorphic reactions reorganize the chemicals making up the rock into a set of minerals that can coexist without changing further. Further research demonstrated that the specific mineral assemblage present in a metamorphic rock depends on temperature and pressure conditions and also on the composition of the protolith. This discovery led the geologists to propose the concept of *metamorphic facies.*

A **metamorphic facies** is a set of metamorphic mineral assemblages that form under a certain range of pressure and temperature. Each specific mineral assemblage in a facies reflects the original protolith composition. According to this definition, a given metamorphic facies includes several different kinds of rocks that differ from one another in terms of chemical composition and, therefore, mineral content. But all the rocks of a given facies formed under roughly the same temperature and pressure conditions. Geologists recognize several facies, including zeolite, prehnite-pumpellyite, hornfels, greenschist, amphibolite, blueschist, eclogite, and granulite—all named for a rock type or mineral associated with the facies.

We can represent the metamorphic conditions for each facies on a graph whose horizontal axis represents temperature and whose vertical axis represents pressure **(Fig. Bx6.3)**. Each area on the graph labeled with a facies name represents the approximate range of temperatures and pressures in which mineral assemblages characteristic of that particular facies grow. For example, rock subjected to the pressure and temperature at Point A (4.5 kbar and 400°C, or 750°F) develops a mineral assemblage characteristic of the greenschist facies. As Figure Bx6.3 implies, the boundaries between facies are gradational and approximate. Note that some amphibolite-facies rocks and all granulite-facies rocks form at pressure-temperature conditions under which a granite containing water will melt.

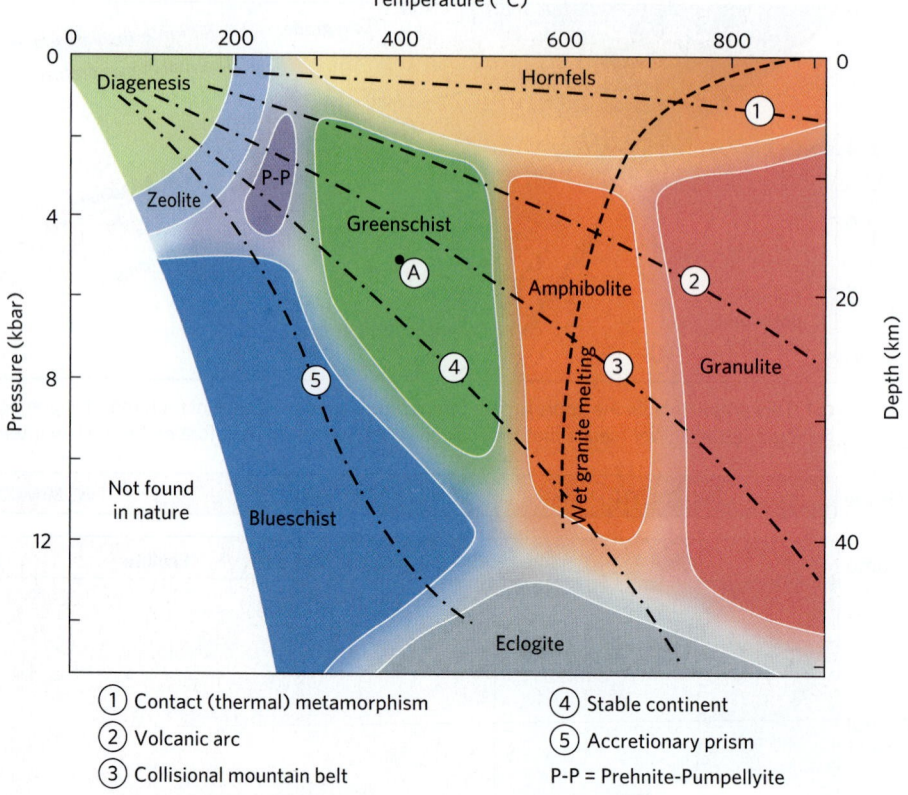

① Contact (thermal) metamorphism
② Volcanic arc
③ Collisional mountain belt
④ Stable continent
⑤ Accretionary prism
P-P = Prehnite-Pumpellyite

FIGURE Bx6.3 The common metamorphic facies. The numbers refer to different geothermal gradients.

Therefore, metamorphic rocks of these very high-temperature facies can form only if the rock is dry.

We can portray the *geothermal gradient* (the change in temperature with depth) of different crustal regions on a facies graph. Metamorphic conditions start at temperatures of around 200°C–250°C (400°F–482°F). At temperatures less than that, any recrystallization, new mineral growth, or pressure solution that takes place in response to the interaction of rock and groundwater is called *diagenesis*. Note that beneath mountain belts, the geothermal gradient passes through the zeolite, prehnite-pumpellyite, greenschist, amphibolite, and granulite facies. In contrast, temperature increases more slowly with depth in the accretionary prisms that form during subduction, where blueschist-facies rocks form. Hornfels forms at high temperatures and relatively low pressures.

Considering the temperature range at which metamorphic minerals grow, geologists estimate that metamorphic conditions exist at depths between 10 km and 40 km (6–25 mi) in the crust, depending on geothermal gradient in a particular geologic setting. More precisely, however, the melting temperature of a rock depends on its mineral composition and water content. In some cases, a felsic melt starts to form in a rock subjected to high-grade metamorphism while the mafic components of the rock remain solid. If the melt solidifies before migrating away, a rock known as a *migmatite*, effectively a mixture of igneous and metamorphic rock, forms **(Fig. 6.10)**.

Take-home message . . .

> Geologists divide metamorphic rocks into classes based on whether the rock contains foliation. Foliated rocks include slate, schist, and gneiss. Nonfoliated rocks include marble and quartzite. The type of metamorphic rock that forms depends on the conditions of metamorphism.
>
> **Quick Question** ⎯⎯⎯⎯⎯⎯⎯⎯
> How do geologists describe the intensity of metamorphism at a location?

FIGURE 6.10 This road cut in Ontario contains a nearly vertical layer of migmatite. The light layers consist of felsic igneous rock, and the dark layers are mafic metamorphic rock.

Light (felsic) bands formed from melt.

Dark (mafic) bands remained solid.

6.4 Where Does Metamorphism Occur?

We've seen that metamorphism occurs at temperatures that are hotter than those that cause lithification of sedimentary rocks and cooler than those that form igneous rocks. Let's look at some of the geologic environments in which these metamorphic conditions develop.

Thermal or Contact Metamorphism

Imagine a place where hot magma has intruded cooler wall rock. Heat flows from the magma into the wall rock because heat always flows from hotter to colder materials. Consequently, the magma cools and solidifies while the wall rock heats up. As this happens, hydrothermal fluids may be circulating through both the intrusion and the wall rock. Heat and hydrothermal circulation cause the wall rock to undergo metamorphism. The highest-grade metamorphic rock forms immediately adjacent to the intrusion, where the temperatures are highest, and progressively lower-grade rocks form farther away. The band of metamorphic rock that forms around an igneous intrusion is called a **metamorphic aureole**—the word *aureole* comes from the Latin *aureola*, meaning crown or halo **(Fig. 6.11a, b)**. The width of a metamorphic aureole depends on the amount of heat the intrusion releases, which

in turn depends on the temperature, size, and shape of the intrusion, as well as on the amount of hydrothermal circulation that takes place.

We can refer to the local metamorphism caused by heat from an igneous intrusion as either **thermal metamorphism**, to emphasize that it develops in response to heat without a change in pressure and without differential stress, or as **contact metamorphism**, to emphasize that it develops adjacent to the contact between an intrusion and its wall rock. Because this type of metamorphism takes place without differential stress, no preferred orientation develops, so metamorphic aureoles typically contain hornfels. As explained by the theory of plate tectonics, magma intrudes the crust at convergent boundaries, in rifts, and during certain stages of mountain building, so these are the geologic environments where contact metamorphism occurs.

Dynamothermal or Regional Metamorphism

Along convergent boundaries or during continental collision, large slices of continental crust slip along faults and move up and over other portions of the crust. As a consequence, a protolith that was once near the Earth's surface along the margin of a continent can end up at great depth beneath a mountain range **(Fig. 6.11c, d)**. In this environment, three processes affect the protolith: (1) it heats up, because temperature increases with depth and because igneous activity may happen nearby; (2) it endures greater pressure,

FIGURE 6.11 Mechanisms of metamorphism.

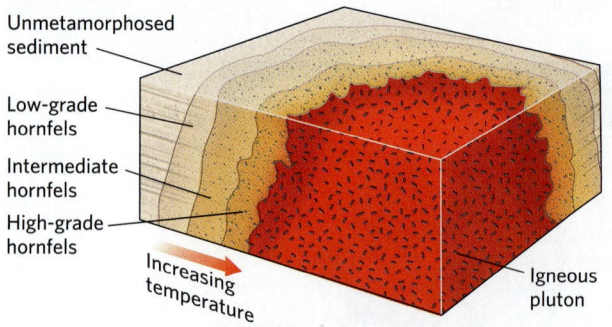

Unmetamorphosed
sediment

Low-grade
hornfels

Intermediate
hornfels

High-grade
hornfels

Increasing
temperature

Igneous
pluton

(a) Thermal metamorphism occurs when heat from a large pluton produces a metamorphic aureole in which hornfels develops. Metamorphic grade decreases progressively away from the pluton.

Hornfels

Blueschist

(b) Blueschist forms at the base of an accretionary prism at a convergent boundary.

At Point A, temperature = 20°C, pressure = 1 bar

Before

Point A starts out as sediment near the Earth's surface.

Time

At Point A, temperature = 450°C, pressure = 6 kbar

After

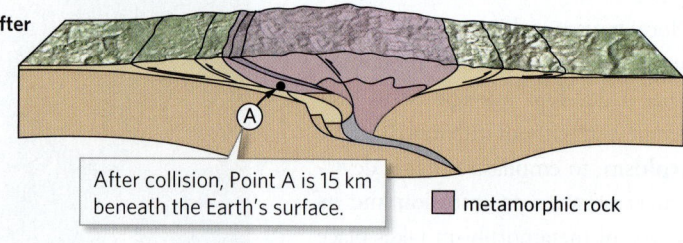

After collision, Point A is 15 km beneath the Earth's surface.

■ metamorphic rock

(c) Dynamothermal metamorphism occurs during the development of mountain belts. The process is also called regional metamorphism.

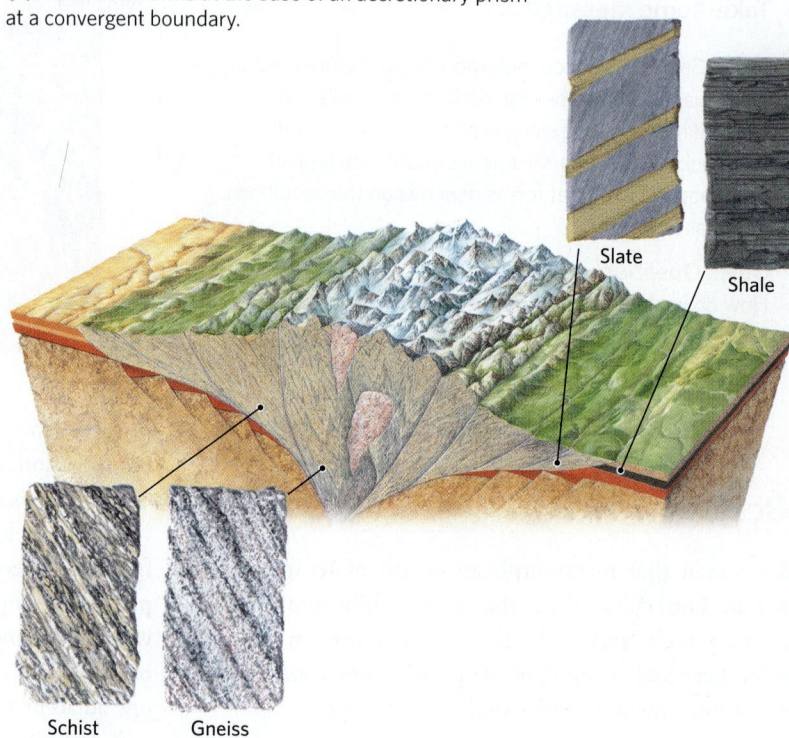

Slate

Shale

Schist   Gneiss

(d) A variety of dynamothermal metamorphic rocks form in a collisional mountain belt.

because the weight of the overburden increases with depth; and (3) it is subjected to differential stress (compression and shear). As a result, a protolith that ends up at depth beneath a mountain belt transforms into foliated metamorphic rock. The type of foliated rock that forms depends on the grade of metamorphism: slate forms at shallower depths, whereas schist and gneiss form at greater depths. Since the metamorphism we've just described involves not only changes due to elevated temperature, but also changes due to differential stress, geologists refer to it as **dynamothermal metamorphism** (from the Greek word *dynamis*, meaning power). Typically, such metamorphism affects a large region—often the length and breadth of a mountain belt—so it's also known as **regional metamorphism**.

A special type of dynamothermal metamorphism occurs in the accretionary prisms that form along convergent plate boundaries. Here, some sediment and basalt move with the subducting plate to the base of the prism before it is scraped off and incorporated into the prism. Since the subducting plate is relatively cold, rocks at the base of the prism are subjected to high pressures at relatively low temperatures, and they also endure shear. These conditions produce *blueschist*, a rock whose color comes from the presence of a bluish amphibole (see Fig. 6.11b).

Where Do We Find Metamorphic Rocks Today?

When you stand on an outcrop of metamorphic rock, you are standing on rock that once lay many kilometers beneath the surface of the Earth. How did such rock return to the Earth's surface? Geologists refer to the overall process by which deeply buried rocks end up back at the surface as **exhumation**. For exhumation to take place, rock from depth must undergo uplift, and rock at the Earth's surface must be eroded away.

Keeping in mind the geologic processes that form metamorphic rock, and those that later cause its exhumation, let's ask the question, Where can you find exposed metamorphic rocks today? You can start your quest by hiking into a mountain range. Because the processes involved in mountain building produce metamorphic rocks, the towering cliffs in the interior of a mountain range typically consist of schist, gneiss, quartzite, and marble. You can also find metamorphic rocks where deep canyons have been cut into basement **(Fig. 6.12a)**. You can find even more metamorphic rocks by walking across a **shield**, a broad expanse of Precambrian continental crust that includes rocks that were metamorphosed during the many mountain-building events that took place over a billion years ago and remained long after the peaks of the resulting mountain ranges eroded away **(Fig. 6.12b, c)**.

See for yourself

Wind River Mountains, Wyoming

Latitude: 43°6'7.22" N
Longitude: 109°21'36.45" W

Look down, obliquely from 20 km (~12.5 mi).

Faulting uplifted the Precambrian basement of western Wyoming 40 to 80 million years ago. This rock underwent high-grade metamorphism over 2 billion years ago. You can see that overlying sedimentary strata have warped into a huge fold.

FIGURE 6.12 Exposures of metamorphic rock.

(a) The Gunnison River has carved a deep canyon into the Precambrian rock in Colorado.

(b) A photograph from an airplane window of the flat landscape of the eastern Canadian Shield. The light areas are water.

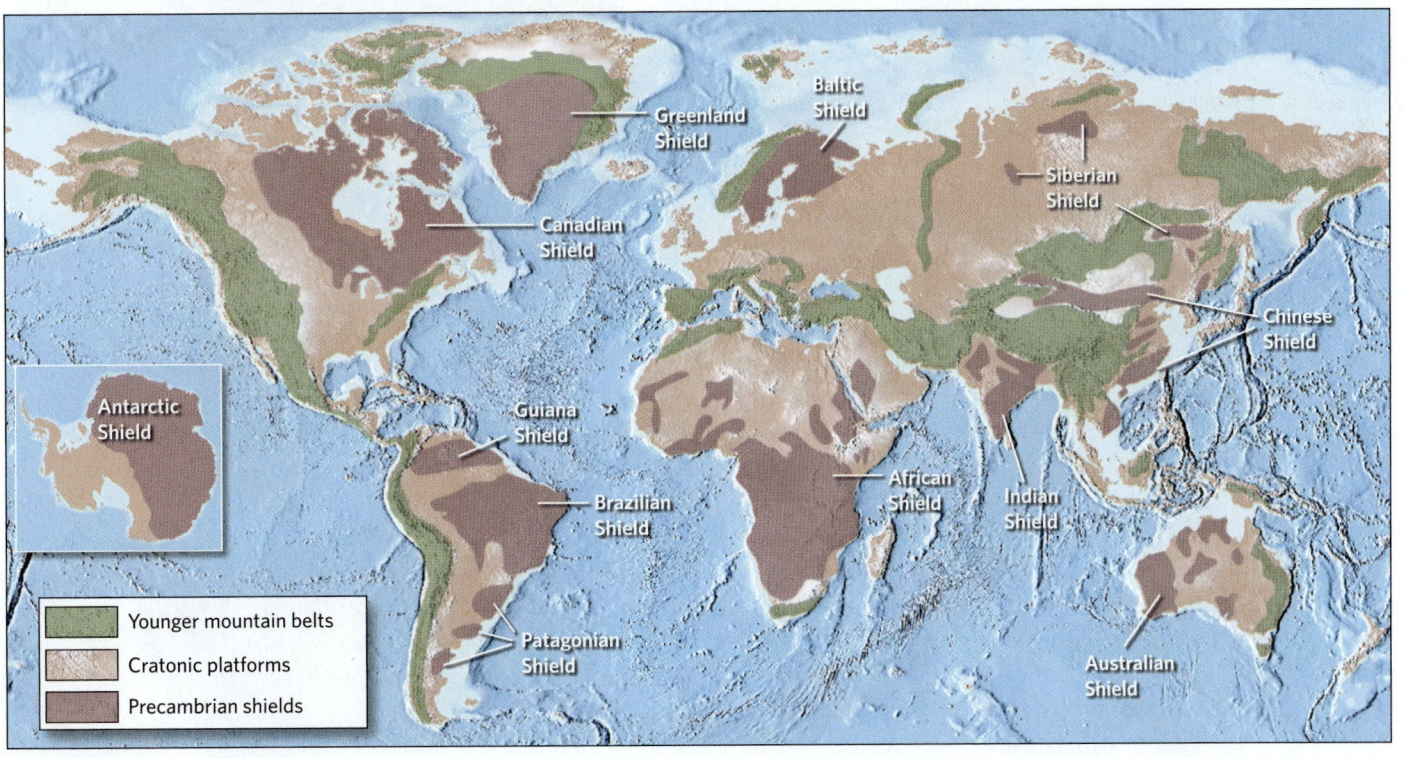

Younger mountain belts
Cratonic platforms
Precambrian shields

Antarctic Shield
Greenland Shield
Baltic Shield
Siberian Shield
Canadian Shield
Chinese Shield
Guiana Shield
African Shield
Indian Shield
Brazilian Shield
Patagonian Shield
Australian Shield

(c) A map showing the distribution of shields, areas where broad expanses of Precambrian crust, including Precambrian metamorphic rocks, crop out.

See for yourself

Canadian Shield, East of Hudson Bay

Latitude: 61°09′50.15″ N
Longitude: 76°42′43.88″ W

Look down from 225 km (140 mi).

The Canadian Shield at this locality is covered by sparse vegetation. Differential erosion of the gneiss makes the foliation stand out. The east-west band of foliation delineates a zone of dynamic metamorphism associated with a fault zone.

Take-home message . . .

Contact metamorphism develops around igneous intrusions due to heat from the intrusion. Regional metamorphism develops beneath mountain ranges where rock undergoes compression and shear at high temperatures and pressures. Most exposures of metamorphic rocks occur in mountain ranges and shields.

Quick Question

Why can we see metamorphic rocks at the Earth's surface today?

6.5 The Rock Cycle

Even though rock seems to be the embodiment of strength and durability, in the time frame of Earth history, it doesn't last forever. Due to a great variety of

geologic processes, the atoms making up minerals of one rock type may eventually be released, rearranged, relocated, and rebonded to become incorporated in other minerals nearby or far away (**Earth Science at a Glance**, pp. 188–189). As a result of these processes, material that starts out in one rock type may end up in another rock type. Geologists refer to the progressive transformations that result in the passage of atoms through different rock types over geologic time as the **rock cycle** (**Fig. 6.13**). A discussion of the rock cycle illustrates the relationships among the rock types described in this chapter and in the previous two chapters and provides an example of how the Earth System evolves over time.

Pathways through the Rock Cycle

By following the arrows in Figure 6.13, you can see many pathways through the rock cycle. To picture a complete path around the cycle, imagine that magma intrudes the crust, cools, and becomes a new igneous

> **Did you ever wonder . . .**
>
> whether rocks, once formed, last forever?

FIGURE 6.13 The rock cycle.

Erosion, transportation, and deposition

Erosion, transportation, and redeposition

Heating and remelting

Sedimentary rock

Heating and melting

Erosion, transportation, and deposition

Burial and heating

Burial and/or heating

Igneous rock

Input of new melt into the crust from the mantle

Return of material to the mantle by subduction

Melting

Metamorphic rock

Burial, heating, and remetamorphism

FIGURE 6.14 Examples of the three classes of rock.

(a) Igneous rock forming, Hawaii.

(b) Sedimentary strata, Utah.

(c) Metamorphic rock, Brazil.

rock **(Fig. 6.14a)**. If that igneous rock later undergoes weathering and erosion, it becomes sediment that eventually undergoes transport, deposition, burial, and lithification to produce new sedimentary rock **(Fig. 6.14b)**. If that sedimentary rock becomes buried very deeply, it turns into metamorphic rock **(Fig. 6.14c)**. At even higher temperatures, the rock may partially melt and produce magma, which may rise up in the crust and solidify to form new igneous rock. The path we've just described isn't the only path through the rock cycle,

however, for there can be shortcuts. For example, metamorphic rock may be uplifted and eroded to form new sediment that later undergoes burial and lithification to form new sedimentary rock. Likewise, igneous rock may be metamorphosed directly, without first weathering into sediment.

Not all materials pass through the rock cycle at the same rate, and for that reason, we find rocks of many different ages at the surface of the Earth. Some rocks remain in one form for less than a few million years, while others have stayed unchanged for most of Earth history. For example, a rock in the Appalachian Mountains has passed through stages of the rock cycle many times during the past billion years because the eastern margin of North America has been subjected to multiple episodes of rifting, basin formation, and mountain building. In contrast, a 3-Ga igneous rock found to the north, in the Canadian Shield, has not yet passed the first stage of the rock cycle.

What Energy Drives the Rock Cycle?

The rock cycle happens because the Earth is a dynamic, ever-changing planet where many rock-forming environments exist. External energy (solar radiation), internal energy (the Earth's internal heat), and gravitational energy all play a role in driving the rock cycle by keeping the mantle, crust, atmosphere, and oceans in motion. Specifically, our planet's internal heat and gravitational field work together to drive plate movements and mantle plumes. Plate interactions cause the uplift of mountain ranges, a process that leads to weathering, erosion, and sediment production, as well as to metamorphism and igneous activity. Circulation of the atmosphere—driven by a combination of solar radiation and gravity—produces rain, snow, and wind. The rain feeds streams, and the snow accumulates to form glaciers, which, along with wind and wind-driven waves, act as agents of erosion that produce sediment. In the Earth System, life also plays a key role by contributing to weathering and mineral precipitation.

 Video
Rock Cycle

See for **yourself**

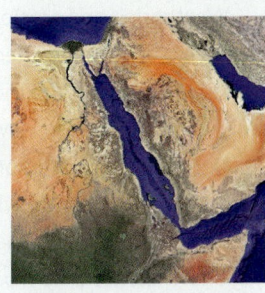

Coast of the Red Sea, Egypt

Latitude 20°29′14.04″ N
Longitude 38°16′32.06″ E

Look down from an altitude of 4,500 km (~2,796 mi).

Below, you can see the color contrast between the dark Precambrian metamorphic rocks and the lighter-colored younger rocks. The Precambrian rocks were uplifted during rifting.

Take-home message . . .

🏠 Rocks don't last forever. Melting, weathering, burial, and metamorphism can change rock so that the atoms in one rock type end up in another and, later, in another still. This progression, called the rock cycle, is driven by energy coming from both inside the Earth and from the Sun, as well as from gravity.

Quick Question

How do steps in the rock cycle relate to plate interactions?

Rock-Forming Environments and the Rock Cycle

Rocks form in many different environments. Igneous rocks develop where melt rises from depth and cools. Intrusive igneous rocks form where magma cools underground; extrusive igneous rocks form where lava and ash erupt at the surface.

Weathering and erosion break up existing rock and produce sediment. Different kinds of sediments develop in different places, reflecting both the composition of the source and the setting in which the sediment accumulates. When this sediment eventually gets buried and undergoes lithification, new sedimentary rocks form.

Under certain conditions, pre-existing rocks can undergo change in the solid state—metamorphism—which produces metamorphic rocks. Contact metamorphism is due to heat released by an intrusion of magma. Regional metamorphism occurs where tectonic processes cause rocks from the surface to be buried very deeply.

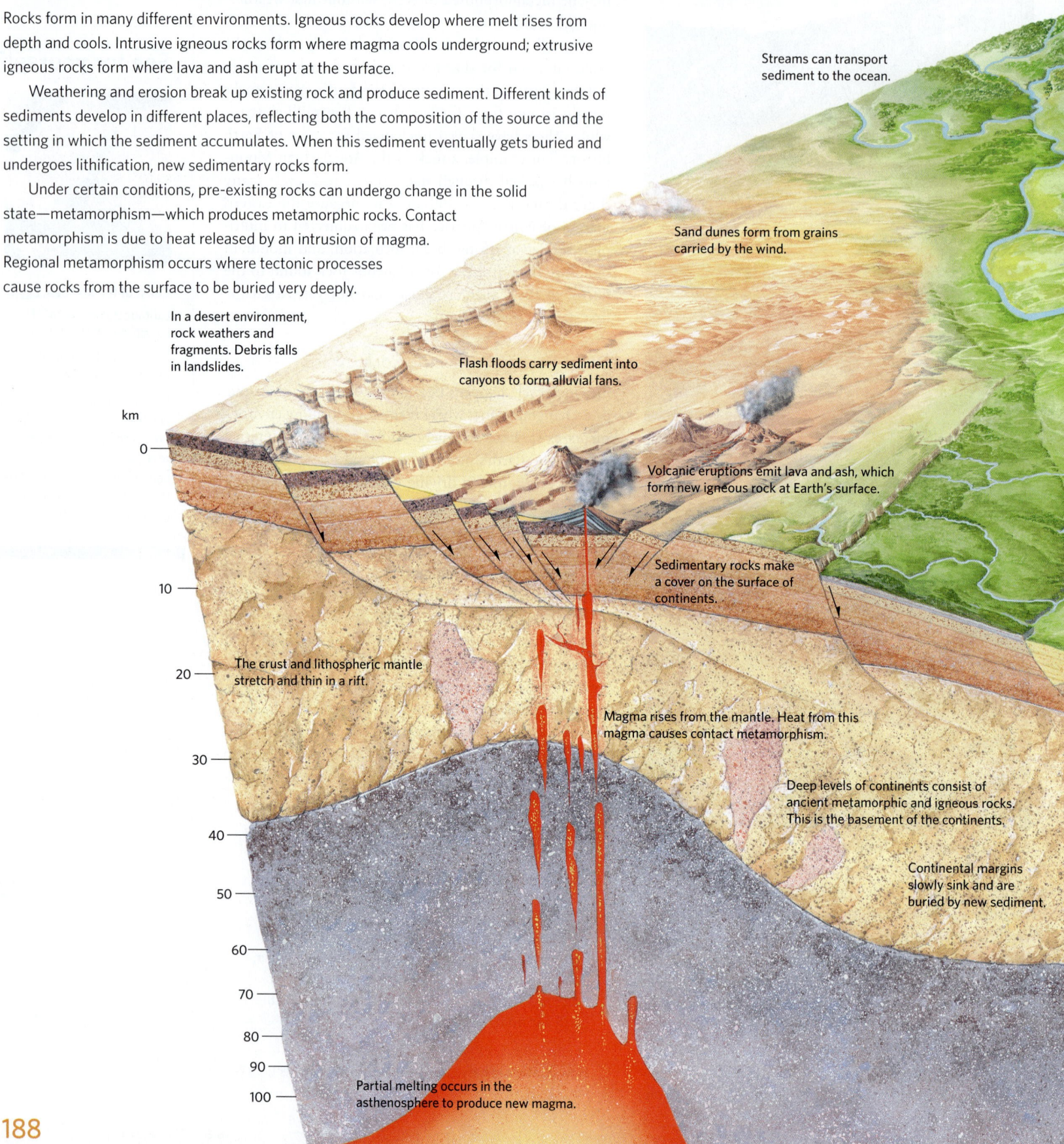

Streams can transport sediment to the ocean.

Sand dunes form from grains carried by the wind.

In a desert environment, rock weathers and fragments. Debris falls in landslides.

Flash floods carry sediment into canyons to form alluvial fans.

Volcanic eruptions emit lava and ash, which form new igneous rock at Earth's surface.

Sedimentary rocks make a cover on the surface of continents.

The crust and lithospheric mantle stretch and thin in a rift.

Magma rises from the mantle. Heat from this magma causes contact metamorphism.

Deep levels of continents consist of ancient metamorphic and igneous rocks. This is the basement of the continents.

Continental margins slowly sink and are buried by new sediment.

Partial melting occurs in the asthenosphere to produce new magma.

km
0
10
20
30
40
50
60
70
80
90
100

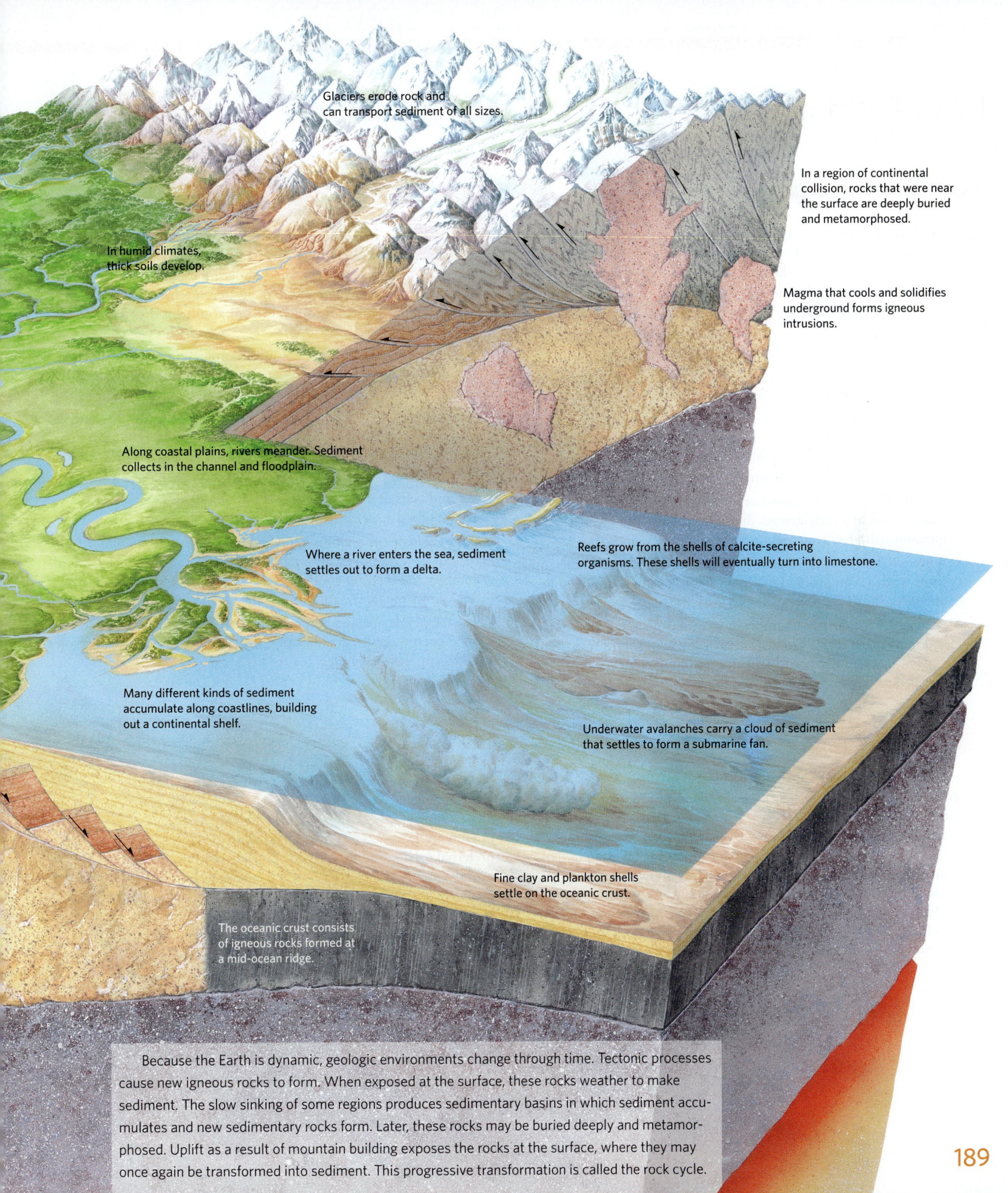

Glaciers erode rock and can transport sediment of all sizes.

In a region of continental collision, rocks that were near the surface are deeply buried and metamorphosed.

In humid climates, thick soils develop.

Magma that cools and solidifies underground forms igneous intrusions.

Along coastal plains, rivers meander. Sediment collects in the channel and floodplain.

Reefs grow from the shells of calcite-secreting organisms. These shells will eventually turn into limestone.

Where a river enters the sea, sediment settles out to form a delta.

Many different kinds of sediment accumulate along coastlines, building out a continental shelf.

Underwater avalanches carry a cloud of sediment that settles to form a submarine fan.

Fine clay and plankton shells settle on the oceanic crust.

The oceanic crust consists of igneous rocks formed at a mid-ocean ridge.

Because the Earth is dynamic, geologic environments change through time. Tectonic processes cause new igneous rocks to form. When exposed at the surface, these rocks weather to make sediment. The slow sinking of some regions produces sedimentary basins in which sediment accumulates and new sedimentary rocks form. Later, these rocks may be buried deeply and metamorphosed. Uplift as a result of mountain building exposes the rocks at the surface, where they may once again be transformed into sediment. This progressive transformation is called the rock cycle.

189

6 CHAPTER REVIEW

Chapter Summary

- Metamorphism refers to change that takes place in rock at high temperatures and pressures to produce new minerals or textures in the rock without melting or chemical weathering. The product of this change is a metamorphic rock.

- The protolith from which a metamorphic rock forms can be igneous, sedimentary, or pre-existing metamorphic rock.

- Metamorphism involves recrystallization, phase changes, metamorphic reactions, pressure solution, and/or plastic deformation. These processes occur by diffusion (migration of atoms) and other processes at the atomic scale.

- Metamorphism is driven by various agents, including increases in temperature and pressure, application of differential stress, and interaction of rock with hydrothermal fluids.

- If hydrothermal fluids deposit or remove ions, so that the chemical composition of a rock changes during metamorphism, we say that metasomatism has occurred.

- A preferred orientation of grains in a rock develops where differential stress (such as compression that is greater in one direction than in others, or shear) aligns inequant grains.

- Geologists separate metamorphic rocks into two classes: foliated rocks and nonfoliated rocks. Foliated rocks include slate, phyllite, schist, and gneiss. Nonfoliated rocks include hornfels, quartzite, and marble.

- Metamorphic conditions vary with location in the crust. Metamorphic rocks formed under relatively low temperatures are low-grade, whereas those formed under high temperatures are high-grade. Intermediate-grade rocks develop between these two extremes. Different minerals form at different metamorphic grades.

- A metamorphic facies is a set of metamorphic mineral assemblages that develop under a specified range of temperature and pressure conditions.

- Thermal (contact) metamorphism occurs in a metamorphic aureole surrounding an igneous intrusion. Dynamothermal (regional) metamorphism results when rocks undergo heating and shearing during mountain building.

- We find belts of metamorphic rocks in mountain ranges. Shields expose broad areas of Precambrian metamorphic rocks.

- Over geologic time, atoms in one rock type may end up in different rock types. This progressive transfer of atoms from one rock type to another, and then yet another, is called the rock cycle. Not all material follows the same path through the rock cycle.

- The rock cycle operates because the Earth System remains dynamic. Solar energy, the Earth's internal energy, gravity, and life all play roles in driving the rock cycle.

Key Terms

contact metamorphism (p. 183)
differential stress (p. 173)
diffusion (p. 171)
dynamothermal metamorphism (p. 184)
exhumation (p. 185)
gneiss (p. 175)
gneissic banding (p. 175)
hornfels (p. 179)
hydrothermal fluid (p. 173)

marble (p. 179)
metamorphic aureole (p. 183)
metamorphic conditions (p. 179)
metamorphic facies (p. 182)
metamorphic foliation (p. 174)
metamorphic grade (p. 179)
metamorphic mineral (p. 171)
metamorphic reaction (p. 172)
metamorphic rock (p. 171)
metamorphic texture (p. 171)

metamorphism (p. 171)
metasomatism (p. 173)
phase change (p. 172)
phyllite (p. 175)
preferred orientation (p. 173)
protolith (p. 171)
quartzite (p. 179)
recrystallization (p. 172)
regional metamorphism (p. 184)

rock cycle (p. 186)
schist (p. 175)
shield (p. 185)
slate (p. 175)
slaty cleavage (p. 175)
stress (p. 173)
thermal metamorphism (p. 183)

Letters in parentheses correspond to the chapter's learning objectives.

1. How are metamorphic rocks different from igneous and sedimentary rocks? **(A)**

2. What is a protolith? Do metamorphic rocks necessarily contain the same minerals that are present in a protolith? **(A)**

3. What processes can cause metamorphism? How does recrystallization differ from a metamorphic reaction? **(B)**

4. Describe the various agents of metamorphism. Why is diffusion important during metamorphism? **(B)**

5. What is metamorphic foliation, and how does it form? How does differential stress differ from pressure? What is the relationship between foliation and differential stress? Draw the orientation of foliation on this thin section. **(C)**

6. How does slate differ from phyllite? How does phyllite differ from schist? How does schist differ from gneiss? What is a migmatite? **(C)**

7. Why is hornfels nonfoliated? **(C)**

8. How does marble differ from quartzite? **(C)**

9. What is meant by metamorphic grade, and how can it be determined? **(D)**

10. What is a metamorphic facies? Identify the axes of the graph, and identify areas representing conditions in which low-, intermediate-, and high-grade metamorphic rocks form. How does grade vary with depth beneath a mountain range? **(D)**

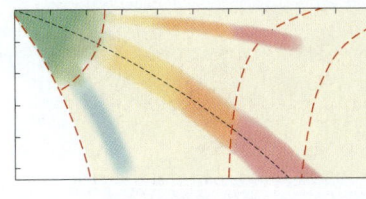

11. Describe the geologic settings where thermal and dynamothermal metamorphism take place. **(E)**

12. In the diagram, label the metamorphic aureole, and label the highest-grade metamorphic rock in the aureole. **(E)**

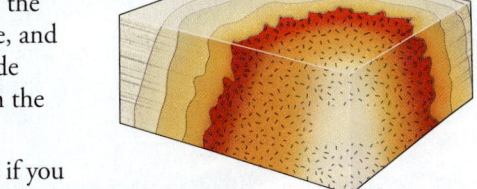

13. Where would you go if you wanted to find exposed metamorphic rocks? How did such rocks return to the surface of the Earth after undergoing metamorphism at depth in the crust? **(E)**

14. Once formed, does a rock necessarily last for all of Earth history? Explain your answer. **(F)**

15. Describe a pathway through the rock cycle. Why does the rock cycle happen on the Earth? **(F)**

16. Have all continental rocks passed through the rock cycle the same number of times? Explain your answer. **(F)**

17. Do you think that you would be likely to find a broad region (hundreds of kilometers across) in which outcrops consist of high-grade hornfels? Why or why not? **(D)**

18. Could you find a layer of metamorphic rock between the layers of sedimentary rock in a sedimentary basin? Why or why not? **(B, E)**

19. The geothermal gradient of Mars is 8°C/km (i.e., temperature increases by 8°C per kilometer of depth), and the crust of Mars is 30 km thick. Do high-grade metamorphic rocks form in this crust? **(D)**

20. Does the basalt of the oceanic crust pass through stages of the rock cycle as it moves from a mid-ocean ridge to a trench? **(F)**

21. Does the rock cycle happen on the Moon? **(F)**

ANOTHER VIEW The owners of a mansion in Newport, Rhode Island, chose a polished slice of this marble, featuring complex swirls, to decorate a fireplace. The swirls formed deep in the crust, when the rock was warm enough and soft enough to flow.

7 CRAGS, CRACKS, AND CRUMPLES
Mountain Building and Geologic Structures

By the end of the chapter you should be able to . . .

A. describe how rocks can deform (crack, bend, or flow) in response to stress.

B. identify basic geologic structures (joints, faults, folds, foliation), and sketch them.

C. explain why mountain belts grow and how their formation relates to plate tectonics.

D. draw connections between the formation of certain rock types and processes that build mountains.

E. explain the concept of isostasy and how it relates to mountain uplift.

F. relate mountain landscapes to erosion.

G. identify a craton and distinguish it from a mountain belt.

7.1 Introduction

Geographers call the peak of Mt. Everest "the top of the world," for this mountain, which lies in the Himalayas of south-central Asia, rises higher than any other on Earth. The ice on Mt. Everest's summit glistens at an elevation 8.85 km (29,029 ft) above sea level, almost the cruising height of modern jets. In 1953, Sir Edmund Hillary, from New Zealand, and Tenzing Norgay, a Nepalese guide, became the first to reach the summit. Since then, thousands of other people have also succeeded, but over 250 have died trying. Mountains appeal to nonclimbers as well—almost everyone loves a vista of snow-crested peaks. The stark cliffs, clear air, meadows, forests, streams, and glaciers of mountainous landscapes provide a refuge from the mundane.

With the exception of large volcanoes formed over hot spots, mountains do not occur in isolation, but rather in ranges of elevated land called **mountain belts**, or **orogens** (from the Greek words *oros*, meaning mountain, and *genesis*, meaning birth). To build such belts, thousands of cubic kilometers of rock must be pushed skyward against the pull of gravity. Clearly, the existence of mountains serves as one of the most obvious indications of the Earth's dynamic activity! A map of the present-day Earth reveals about a dozen major mountain belts and numerous smaller ones **(Fig. 7.1)**. In addition, many places that are now lowlands were mountains in the distant geologic past. So **orogeny**, the process of mountain-belt formation, has happened many times and in many places over the Earth's history.

Orogeny not only leads to **uplift**, the vertical rise of the land surface and the rock beneath, but also causes rocks to undergo **deformation**, a process during which rocks bend, break, or flow. **Geologic structures**, the features produced by deformation, include *joints* (natural cracks), *faults* (fractures on which one body of rock slides past another), *folds* (bends, curves, or wrinkles of rock layers), and *foliation* (the layering in rock that develops when metamorphism happens during deformation).

A given mountain-building event may last for tens of millions of years. As soon as land starts to rise into a mountain belt, however, *erosion* begins to grind it away, producing sediment and sculpting spectacular, jagged landscapes. Erosion is the force that carves the cliffs and valleys and sharp peaks of mountains—without it, a mountain belt might be just a region of elevated land. Erosion and uplift battle as a mountain belt forms. When uplift ceases, erosion can eventually shave a mountain range down to near sea

<< The Wasatch Mountains rise above the clouds in Utah. Tectonic processes uplifted the land. Erosion is sculpting the peaks and valleys.

level. But even after the high peaks are gone, a low-lying belt of fractured, contorted, and metamorphosed rock remains. These scars serve as a permanent monument to what was once a region of high peaks.

In this chapter, we first learn about the processes that deform rocks and how to describe and interpret the geologic structures that result from their deformation. Next, we turn our attention to the question of why mountain belts form and find that answers come from plate tectonics theory. Finally, we consider the phenomena that sculpt regions of uplift to form the landscapes that we can see when visiting mountainous terrains.

7.2 Rock Deformation

What Is Deformation?

To get a visual sense of what geologists mean by *deformation*, let's contrast rock that has not been affected by mountain building with rock that has been affected. Our undeformed example comes from a road cut in the interior plains of North America, and our deformed example comes from a cliff in the Alps, a mountain belt of Europe.

The road cut exposes nearly horizontal beds of sandstone, shale, and limestone cut by a few joints **(Fig. 7.2a)**. These beds have the same orientation that they had when first deposited. The sand grains in the sandstone of this outcrop are nearly spherical (the same shape as when they were deposited), and the clay flakes in the shale lie roughly parallel to the bedding plane. When we say that this outcrop exposes *undeformed rock*, we mean that it contains no geologic structures, other than a few joints (visible as thin vertical cracks cutting across the bedding).

The rocks of the Alpine cliff look very different **(Fig. 7.2b)**. Here, we find layers of quartzite, slate, and marble (the metamorphic equivalents of sandstone, shale, and limestone) that have been contorted into folds, so that the rock layers, instead of being horizontal, trace out wave-like shapes. On microscopic examination, we may find that the grains in the quartzite are not spherical, but have been flattened into elliptical shapes. And in the slate, the clay flakes no longer align with the bedding plane, but rather align in a new orientation that defines slaty cleavage, a type of foliation (see Chapter 6). Finally, we see that the quartzite and slate layers end abruptly at a sloping fault, below which the outcrop consists of folded marble. Slip on this fault brought the slate and quartzite up from a deeper level in the range and juxtaposed it against the marble. The folds, faults, foliation, and flattened grains in the Alpine cliff are all manifestations of deformation, so we say that this outcrop exposes *deformed rock*. The character of the geologic structures visible in the cliff emphasizes that during deformation, rocks can undergo

FIGURE 7.1 This digital map of world topography shows the locations of major mountain ranges.

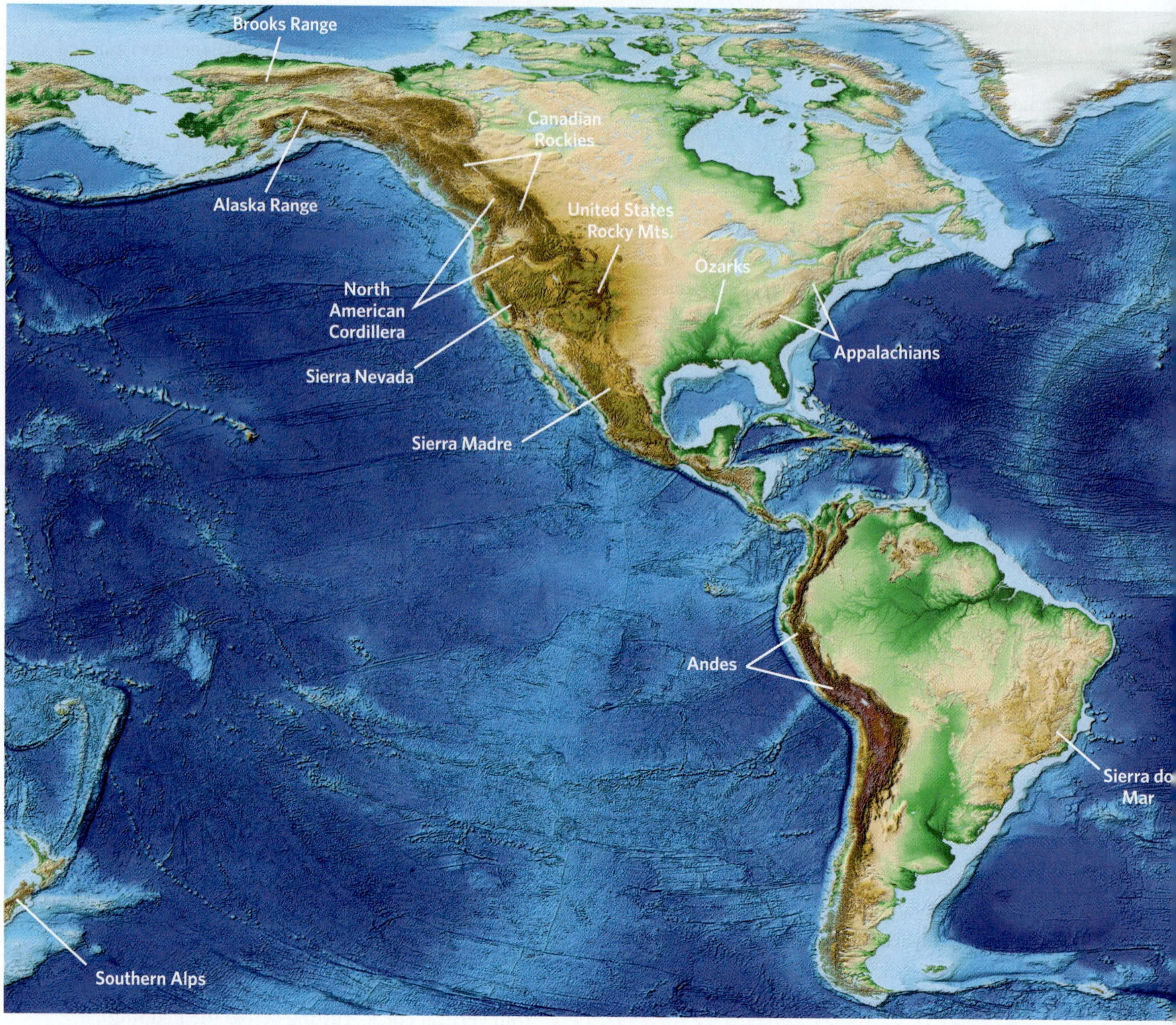

Brooks Range

Canadian Rockies

Alaska Range

United States Rocky Mts.

Ozarks

North American Cordillera

Sierra Nevada

Appalachians

Sierra Madre

Andes

Sierra do Mar

Southern Alps

FIGURE 7.2 Deformation changes the character and configuration of rocks.

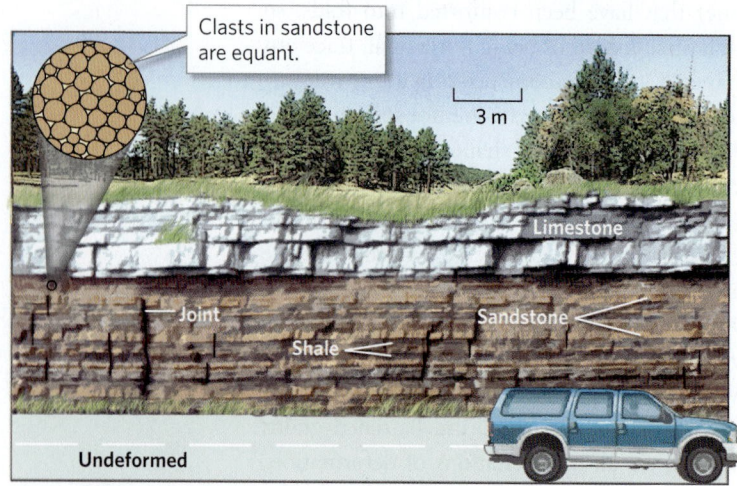

Clasts in sandstone are equant.

3 m

Limestone

Joint

Sandstone

Shale

Undeformed

(a) These flat-lying beds along a highway in the interior plains of North America are essentially undeformed. A few joints, formed when overlying rock eroded away, are visible.

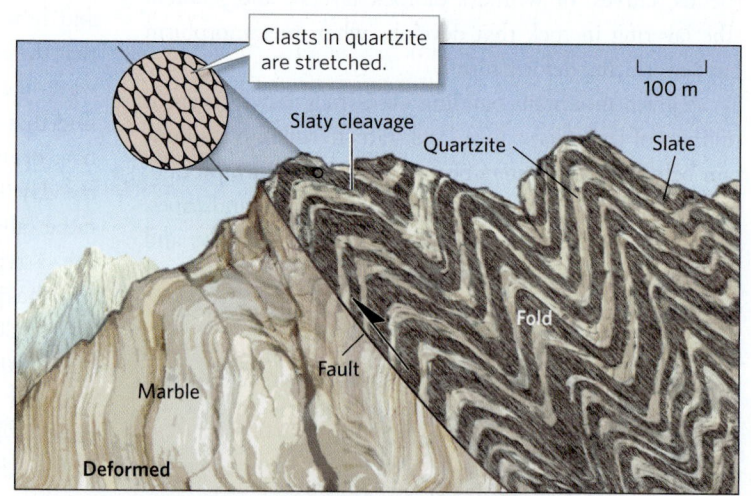

Clasts in quartzite are stretched.

100 m

Slaty cleavage

Quartzite

Slate

Fold

Fault

Marble

Deformed

(b) In this Alpine cliff, deformation has caused rock layers to undergo folding and faulting. In addition, foliation (in this case, slaty cleavage) has developed, and clasts have been stretched.

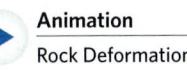

one or more of the following changes (Fig. 7.3): (1) a change in location, or *displacement*; (2) a change in orientation, or *rotation*; and (3) a change in shape, or *distortion*.

Geologists refer to the amount of distortion that a body of rock or a region of crust develops during deformation as its **strain**. We can distinguish among different kinds of strain according to the nature of the overall shape change. Specifically, if a layer becomes longer, it has undergone *stretching*, or extension, whereas if the layer becomes shorter, it has undergone *shortening* (Fig. 7.4a–c). And if a change in shape involves the movement of one part of a rock body sideways past another, it has undergone *shear strain* (Fig. 7.4d, e).

Brittle versus Plastic Deformation

Drop a porcelain plate on a hard floor, and it shatters into pieces. You've just observed **brittle deformation**, the process by which a material cracks or fractures to form pieces that no longer hold together (Fig. 7.5a, b). The formation of joints and faults serves as a geologic example of brittle deformation in rocks. Now, squeeze a ball of soft dough between a book and a tabletop, or bend a stick of chewing gum. You've just observed **plastic deformation**, the process during which objects change shape without visibly breaking (Fig. 7.5c, d). The formation of folds serves as a geologic example of plastic deformation. Everyday examples of deformation, such as breaking plates or squeezing dough, take place fast enough for you to see. In contrast, geologic deformation in mountain ranges generally takes place so slowly (at rates of less than a couple of millimeters to a few centimeters a year) that you can't see it happen in the course of a human lifetime.

What actually happens within rock during deformation? Recall that the atoms that make up mineral grains are connected to one another by chemical bonds. During brittle deformation, many bonds break and stay broken, leading to the formation of a permanent crack across which material no longer connects. During plastic deformation, in contrast, some bonds break, but new ones

Animation
Rock Deformation

FIGURE 7.3 The components of deformation include displacement, rotation, and distortion.

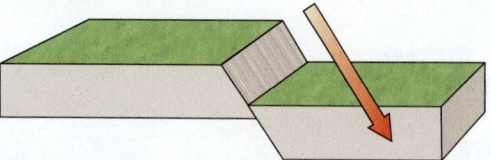

Slip on a fault transported this rock down from a higher position.

>2.7-Ga greenschist

~2.2-Ga quartzite

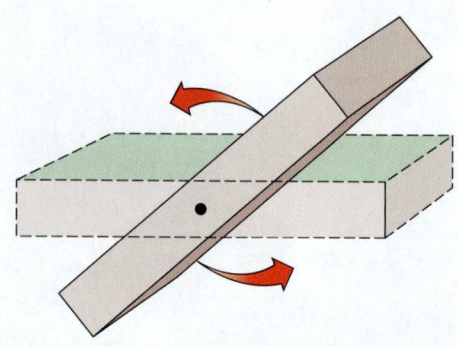

(a) Displacement occurs when a block of rock moves from one location to another.

(b) Rotation occurs when a body of rock undergoes tilting.

These beds were horizontal when deposited. They are now tilted.

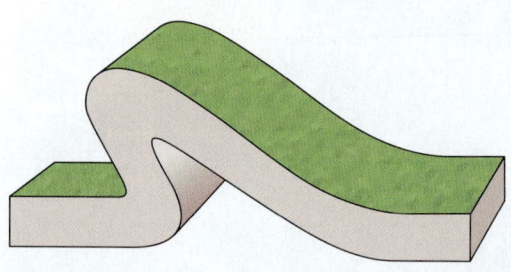

These beds were once horizontal and of constant thickness. They are now distorted.

(c) Distortion occurs when a body of rock changes shape. The development of a fold represents one type of distortion.

quickly form. In this way, the atoms within a rock are rearranged, and the rock gradually changes shape without breaking into pieces.

What determines whether rock deforms brittlely or plastically? The deformation behavior of a rock depends on several factors:

- *Temperature:* Heat makes materials softer, so warmer rocks tend to deform plastically, whereas colder rocks tend to deform brittlely. To see the effect of temperature for yourself, compare the behavior of a candle warmed in an oven with that of one cooled in a freezer when you bend it with your hands—the former easily changes shape, but the latter snaps in two.

- *Pressure:* Under great pressures deep in the Earth, rock behaves more plastically than it does under the lower pressures near the surface. Pressure effectively prevents rock from separating into fragments as it deforms.

- *Deformation rate:* A sudden change in shape can cause brittle deformation, whereas a slow change in shape can cause plastic deformation. For example, if you hit a marble bench with a hammer, it shatters, but if you leave the bench alone for a century, it gradually sags without breaking.

Pressure and temperature both increase with depth in the Earth. As a consequence, in typical continental crust, rocks generally behave brittlely above a depth of 10–15 km (6–9 mi) and plastically below that depth.

Differential Stress: The Cause of Deformation

Up to this point, we've focused on picturing the consequences of deformation. Describing the causes of deformation is a bit more challenging. Museum displays about mountain building typically bypass the issue by saying something like, "The mountains were produced by forces

FIGURE 7.4 Strain is a measure of the distortion, or change in shape, that takes place in rock during deformation.

(a) An unstrained cube and an unstrained fossil brachiopod shell.

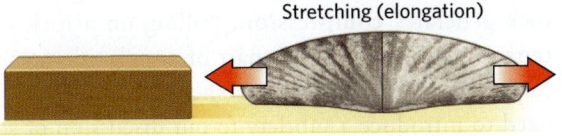

(b) Horizontal stretching changes the cube into a horizontal brick and elongates the shell.

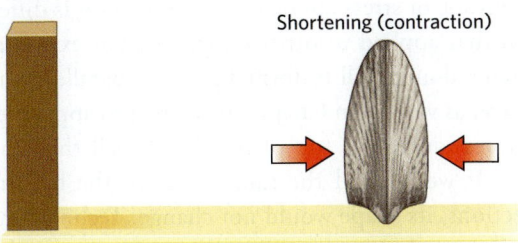

(c) Horizontal shortening changes the cube into a vertical brick and makes the shell narrower.

(d) Shear strain transforms the cube into a parallelogram and changes angular relationships in the shell.

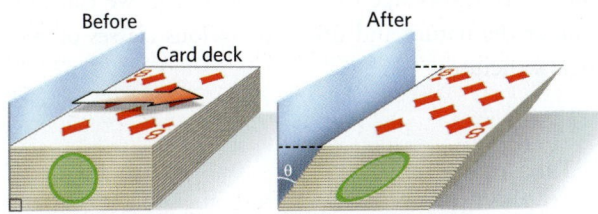

(e) You can simulate shear strain by moving a deck of cards so that each card slides a little with respect to the one below.

deep within the Earth." What does this mean? Isaac Newton stated that a **force** can cause an object to speed up, slow down, or change direction. In the context of geology, plate interactions, such as continental collisions, apply forces to rocks and cause them to change location, orientation, or shape. In other words, the application of forces in the Earth ultimately drives deformation.

FIGURE 7.5 Brittle versus plastic deformation.

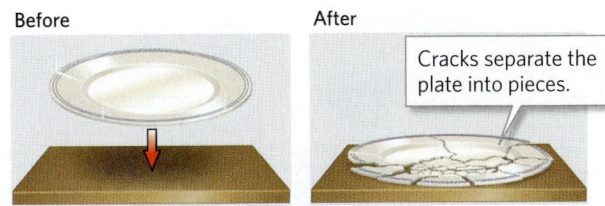

(a) Brittle deformation occurs when you drop a plate and it shatters.

(b) The cracks in these sandstone beds formed by brittle deformation.

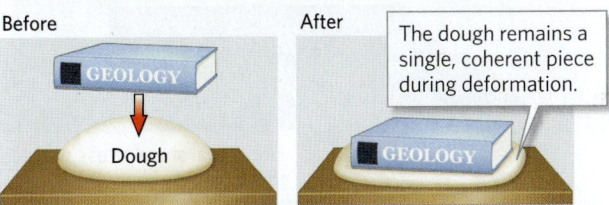

(c) Plastic deformation occurs when you squash a ball of dough.

(d) The quartzite cobbles in this conglomerate were flattened plastically.

FIGURE 7.6 Types of stress.

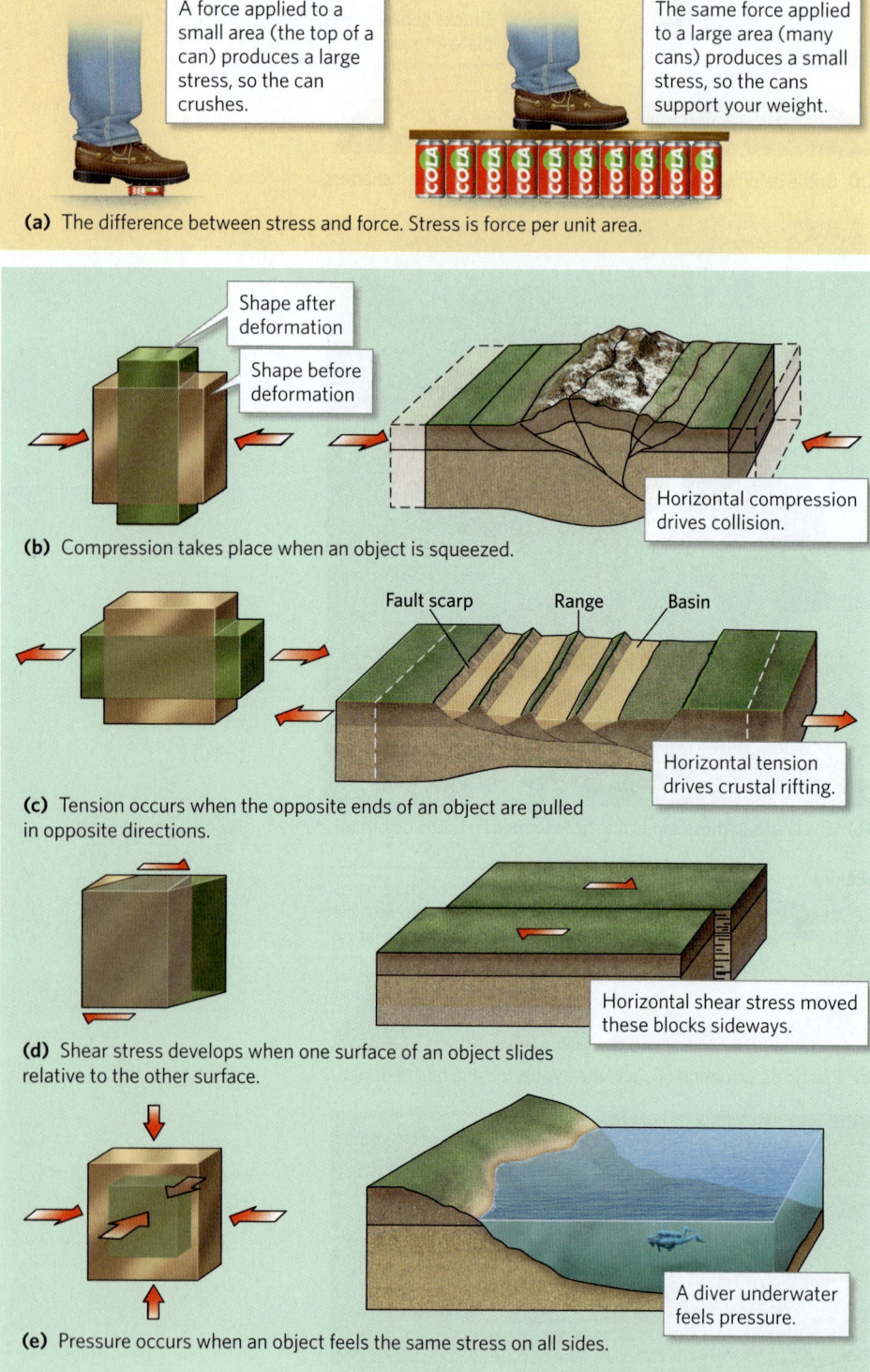

A force applied to a small area (the top of a can) produces a large stress, so the can crushes.

The same force applied to a large area (many cans) produces a small stress, so the cans support your weight.

(a) The difference between stress and force. Stress is force per unit area.

Shape after deformation

Shape before deformation

Horizontal compression drives collision.

(b) Compression takes place when an object is squeezed.

Fault scarp   Range   Basin

Horizontal tension drives crustal rifting.

(c) Tension occurs when the opposite ends of an object are pulled in opposite directions.

Horizontal shear stress moved these blocks sideways.

(d) Shear stress develops when one surface of an object slides relative to the other surface.

A diver underwater feels pressure.

(e) Pressure occurs when an object feels the same stress on all sides.

the force acts. For example, the force caused by your body weight applied to the top of a single soda can crushes the can, but the same force spread over the tops of 100 cans does not **(Fig. 7.6a)**. How does the concept of stress apply to geology? During mountain building, the force of one plate interacting with another is spread across the contact between the two plates, so the deformation resulting at any specific location actually depends on the stress developed at that location, not on the overall force.

Different types of stress develop in the Earth's crust in different geologic settings **(Fig. 7.6b–d)**. Squeezing a rock generates **compression**, pulling on a rock yields **tension**, and shoving one part of a rock sideways with respect to another produces **shear**. At this point, we need to introduce a subtlety to our discussion of stress. Specifically, in order to cause deformation, rocks must be subjected to **differential stress**, meaning that the magnitude of stress applied in one direction is different from that applied in other directions. For example, to flatten a dough ball to form a pancake parallel to a flat surface, as we saw in Chapter 6, we had to apply greater compression to the top of the dough ball than to the sides. If we applied the same stress to the ball in all directions, its shape would not change. Technically, the familiar word *pressure* refers to a special stress condition that happens when the stress acting on all sides of an object is the same **(Fig. 7.6e)**.

Note that *stress* and *strain*, in the context of geology, have very different meanings, even though, in everyday English, the terms may seem interchangeable. In geology, *stress* refers to compression, tension, or shear, whereas *strain* is a measure of the shape changes resulting from deformation. Put simply, stress causes strain. For example, compression causes shortening, tension generates stretching, and shear produces shear strain. With this knowledge of stress and strain, we can now look at the nature and origin of various classes of geologic structures.

Take-home message . . .

Because of differential stress that develops in the Earth, rocks may change their shape, position, or orientation. All of these changes are manifestations of deformation. Strain is a measure of the change in shape. During brittle deformation, rocks break, whereas during plastic deformation, rocks bend and distort without breaking.

Quick Question
When you sit on a chair, does the weight of your body apply vertical compression or vertical tension?

Geologists, however, use the word *stress* instead of force when talking about generation of deformation, where we define **stress** as force per unit area. We must distinguish between stress and force because the actual consequences of applying a force depend not only on the amount of force applied, but also on the area over which

FIGURE 7.7 Examples of joints.

(a) Prominent vertical joints cut these red sandstone beds in Arches National Park, Utah.

(b) Vertical joints in shale on a cliff face near Ithaca, New York.

7.3 Geologic Structures Formed by Brittle Deformation

Joints: Natural Cracks in Rocks

If you examine just about any outcrop of rock, you'll find cracks, meaning surfaces across which rock broke so that the pieces on either side of the break no longer connect. Geologists refer to a natural crack in rock as a **joint** if there has been no movement of the rock on either side of the crack (**Fig. 7.7**). Because rock is not connected across a joint, these cracks are planes of weakness, so their presence can affect the stability of cliffs.

Joints develop in response to tension in brittle rock. In other words, a rock cracks and a joint forms because the rock has been pulled slightly apart. The tension that causes joints may exist for a variety of geologic reasons. For example, some joints develop when the rock of a region cools and shrinks—the same process happens when you heat a glass plate and then plunge it into cold water. Other joints develop when rock formerly at depth undergoes a decrease in compression as overlying rock erodes away, and therefore changes shape slightly. (Note that joints can form in response to cooling or a decrease in compression even in regions where no mountain building has occurred—that's why we see the traces of joints in the undeformed outcrop depicted in Figure 7.2a.) Finally, some joints form when rock layers bend and stretch during mountain building.

Faults: Surfaces of Slip

After the San Francisco earthquake of 1906, geologists found a rupture that had torn through the land surface near the city. Where this rupture crossed orchards, it offset rows of trees, and where it crossed a fence, it broke the fence in two—the western side of the fence moved northward by about 2 m (6 ft) relative to the eastern side. The rupture represents the trace of the San Andreas fault. A **fault**, as we have seen, is a fracture surface on which sliding occurs. **Slip**, the sliding movement along a fault, can generate earthquakes. Geologists refer to the amount of movement that takes place across a fault as the fault's **displacement**.

Faults have formed throughout Earth history. *Active faults* are those on which sliding has been occurring in recent geologic time, whereas *inactive faults* ceased to slip long ago. Some faults, such as the San Andreas, displace the ground surface when they move. Others slip only at depth and remain invisible at the surface unless they are later exposed by erosion.

Fault Classification

Not all faults result in the same kind of crustal deformation. Specifically, some accommodate shortening, some accommodate stretching, and some accommodate horizontal shear. It's important for geologists to distinguish among different kinds of faults in order to interpret their geologic significance. Fault classification focuses on two

FIGURE 7.8 The principal categories of faults.

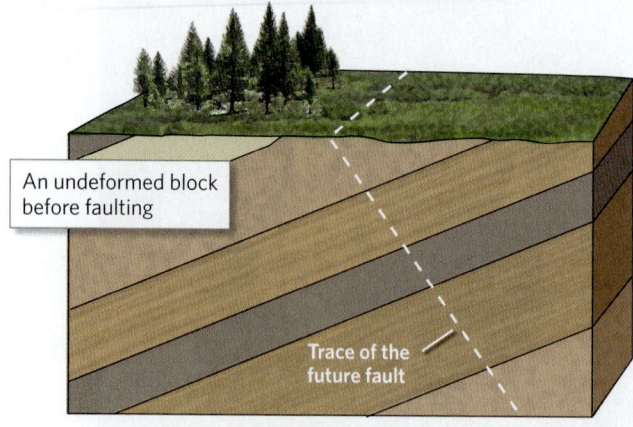

An undeformed block before faulting

Trace of the future fault

Weathered fault scarp

Dip line

Half arrows indicate slip.

Hanging-wall block

Footwall block

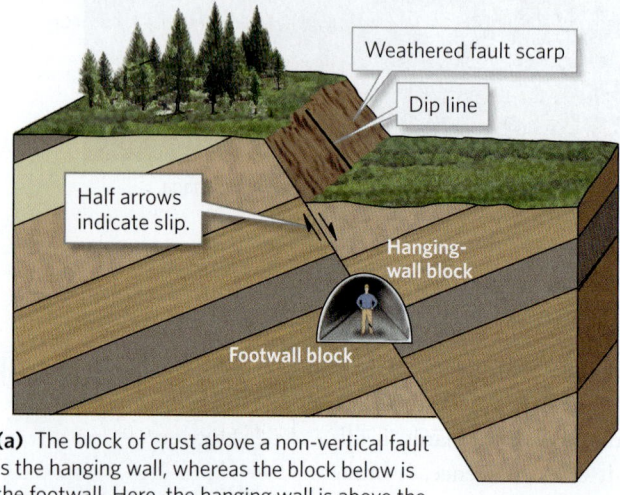

(a) The block of crust above a non-vertical fault is the hanging wall, whereas the block below is the footwall. Here, the hanging wall is above the man's head, and the footwall is below his feet.

Dip-slip faults

Hanging wall up

Reverse (steep slope)   Thrust (gentle slope)

Hanging wall down

Normal

Hanging wall

Footwall

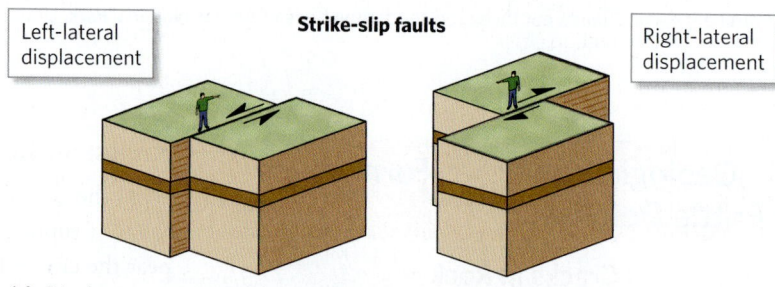

(b) Displacement on a dip-slip fault is parallel to the fault's slope. Reverse faults and thrust faults cause crustal shortening, whereas normal faults cause crustal stretching (red arrows).

Left-lateral displacement

Strike-slip faults

Right-lateral displacement

(c) Displacement on a strike-slip fault moves one block horizontally with respect to the other. There is no up-and-down motion.

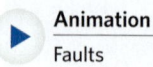

Animation

Faults

characteristics of faults: (1) the *dip,* or slope, of the fault plane **(Box 7.1)** and (2) the shear sense across the fault. By *shear sense,* we mean the direction in which material on one side of the fault moves relative to the material on the other side. With this concept in mind, let's consider the principal kinds of faults.

- *Dip-slip faults:* On a **dip-slip fault**, slip takes place parallel to a line pointing down the slope of the fault surface (the dip line; see Box 7.1), so the *hanging-wall block,* meaning the rock above the fault surface, slides up or down relative to the *footwall block,* the rock below the fault surface **(Fig. 7.8a)**. If the hanging-wall block slides up, the fault is a **reverse fault**. A reverse fault with a gentle dip (meaning a slope of less than 30°) is also known as a **thrust fault**. Reverse and thrust faults accommodate shortening of the crust, as happens during continental collision. If the hanging-wall block slides down, the fault is a **normal fault**. Normal faults accommodate stretching of the Earth's crust, as happens during rifting **(Fig. 7.8b)**.

- *Strike-slip faults:* A **strike-slip fault** is a fault on which the slip direction is parallel to a horizontal line (the *strike line;* see Box 7.1) on the fault surface. This means that the block on one side of the fault slips sideways relative to the block on the other side, and there is no upward or downward motion. Most strike-slip faults have a steep to vertical dip. Geologists distinguish between two types of strike-slip faults based on the shear sense as viewed when you are facing the fault and looking across it. If the block on the far side has slipped to your left, the fault is a *left-lateral* strike-slip fault, and if the block has slipped to your right, the fault is a *right-lateral* strike-slip fault **(Fig. 7.8c)**.

Recognizing Faults

How do you recognize a fault when you see one? In many cases, faulting visibly shifts distinct layers in rocks so that layers on one side of a fault are not continuous with layers

BOX 7.1 ▶ **Science Toolbox**

Describing the orientation of geologic structures

When discussing geologic structures, it's important to be able to communicate information about their orientation. For example, does a fault exposed in an outcrop at the edge of town continue beneath the nuclear power plant 3 km to the north, or does it go beneath the hospital 2 km to the east? If we know the fault's orientation, we may be able to answer such questions. To describe the orientation of a geologic structure, geologists picture the structure as a simple geometric shape, then specify the angles that the shape makes with respect to a horizontal plane (a flat surface parallel to sea level), a vertical plane (a flat surface perpendicular to sea level), and the north direction (a line of longitude).

Let's start by considering planar structures such as faults, beds, joints, and foliation. We call these *planar structures* because they are two-dimensional surfaces, so a portion of them can be represented as a geometric plane. A planar structure's orientation can be specified by its *strike and dip*. The **strike** is the angle between an imaginary horizontal line, the *strike line*, on the planar structure and an imaginary horizontal line pointing to true north **(Fig. Bx7.1a, b)**. The **dip** is the angle of the planar structure's slope, or more precisely, the angle between a horizontal plane and the *dip line* (an imaginary line parallel to the steepest slope on the structure), as measured in a vertical plane perpendicular to the strike. A horizontal

planar structure has a dip of 0°, whereas a vertical one has a dip of 90°. We can measure the strike with a special type of compass **(Fig. Bx7.1c)** and the dip with a clinometer, a type of protractor that uses a bubble to indicate horizontal, and we can represent strike and dip on a map using the symbol shown in Figure Bx7.1b.

We can envision a *linear structure* as a geometric line rather than a plane. Examples of linear structures include scratches or grooves on a rock surface. Geologists specify the orientation of a linear structure by giving its *plunge and bearing* **(Fig. Bx7.1d)**. The **plunge** is the angle between a linear structure and an imaginary horizontal line in the vertical plane that contains the structure. A horizontal linear structure has a plunge of 0°, and a vertical one has a plunge of 90°. The **bearing** is the angle between due north and the projection of the linear structure on a horizontal plane. (A projection is the shadow of a line on a horizontal plane if lit from above.)

FIGURE Bx7.1 Specifying the orientation of planar and linear structures.

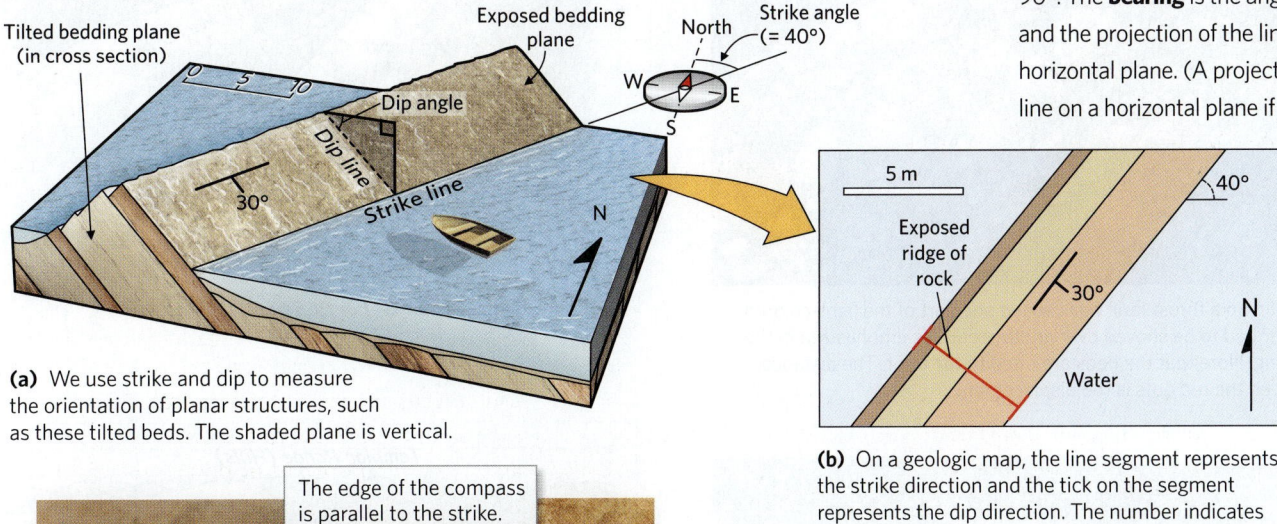

(a) We use strike and dip to measure the orientation of planar structures, such as these tilted beds. The shaded plane is vertical.

(b) On a geologic map, the line segment represents the strike direction and the tick on the segment represents the dip direction. The number indicates the dip angle as measured in degrees.

(c) Geologists use a special compass to measure strike. The compass includes an inclinometer.

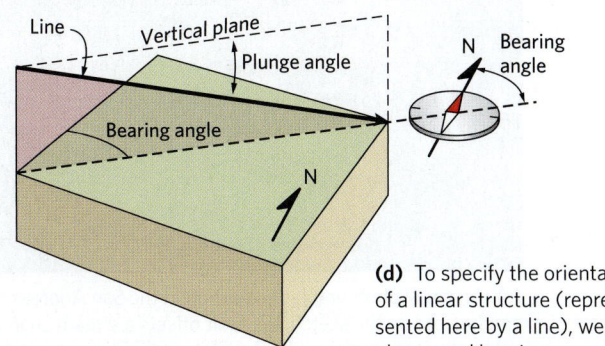

(d) To specify the orientation of a linear structure (represented here by a line), we use plunge and bearing.

FIGURE 7.9 Recognizing fault displacement in the field.

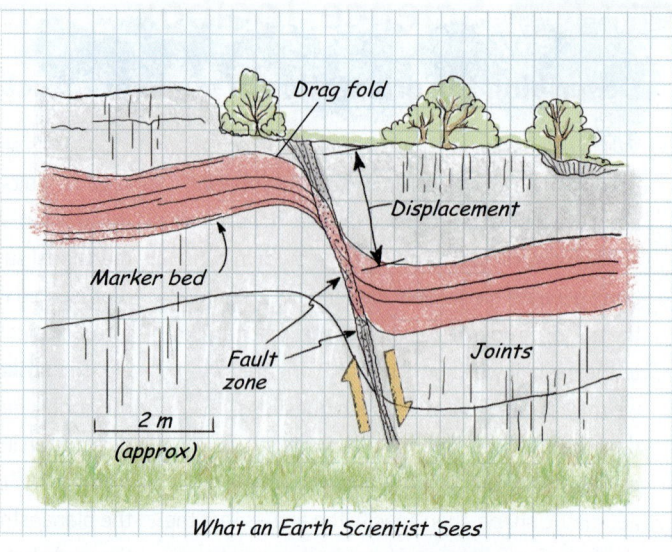

What an Earth Scientist Sees

(a) A steep normal fault has displaced a distinctive redbed (a marker layer). Note that the faulting has formed a 0.5-m-wide fault zone of broken rock. Folds have developed adjacent to the fault.

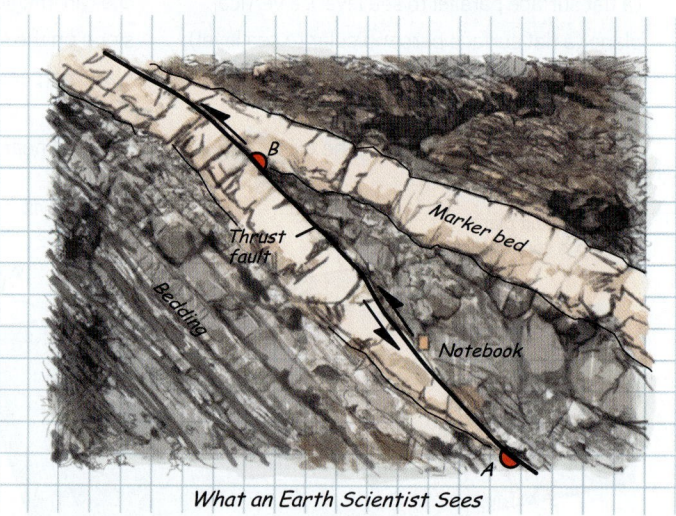

What an Earth Scientist Sees

(b) Slip on a thrust fault has caused one part of the light-colored marker bed to be shoved over another part, as emphasized in the drawing. Note that the beds are tilted to the right. The distance between the red dots is the displacement.

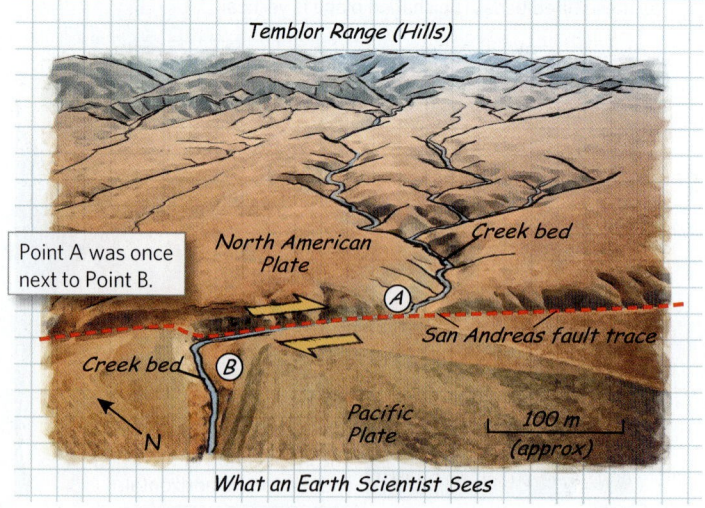

What an Earth Scientist Sees

(c) An aerial photograph of a portion of the San Andreas fault. As emphasized by the sketch, the fault offsets a stream channel in a right-lateral shear sense.

FIGURE 7.10 Features of exposed fault surfaces.

A fault scarp formed by slip on a normal fault in Nevada.

(a) This normal fault has displaced the ground surface. It formed during an earthquake in 1954.

(b) This fault breccia consists of irregular fragments of light-colored rock embedded in a dark matrix of fault gouge.

Striation orientation

(c) Slip lineations or fault striations on the surface of a strike-slip fault may look like grooves or scratches.

on the other side **(Fig. 7.9a, b)**. If a fault intersects the ground surface, slip on the fault can displace natural landscape features such as stream valleys or glacial moraines **(Fig. 7.9c)**, or human-made features such as highways, fences, or rows of trees in orchards. Displacement on a fault that offsets the ground surface produces a step, known as a **fault scarp**, on the ground surface **(Fig. 7.10a)**.

Faulting under brittle conditions may crush or break rock in a band or zone bordering the fault. If this shattered rock consists of visible angular fragments, it's called *fault breccia* **(Fig. 7.10b)**, but if it consists of a fine powder, it's called *fault gouge*. Some fault surfaces are polished and grooved by movement on the fault. Polished fault surfaces are called **slickensides**, and linear grooves on fault surfaces are *slip lineations* **(Fig. 7.10c)**. Because faults tend to break up and weaken rock, the *fault trace*, the line of intersection between the fault and the ground surface, may preferentially erode to become a linear valley.

Take-home message . . .

Geologic structures formed by brittle deformation include joints (natural cracks) and faults (fractures on which sliding occurs). Geologists distinguish among different kinds of faults based on the relative displacement across the fault. Faulting can break up rock, and fault surfaces themselves may be polished and grooved due to slip.

Quick Question

How can you distinguish between a joint and a fault in the field?

7.4 Folds and Foliation

Basic Fold Shapes

Imagine a carpet lying flat on the floor. Push on one end of the carpet, and it will wrinkle or contort into a series of wave-like curves. Stresses developed during mountain building can cause bedding, foliation, or other features in rock to warp or bend in a similar manner. The result—a curve in the shape of a rock layer—is called a **fold** (see Fig. 7.2b).

Not all folds look the same: some look like arches, some look like troughs, and some have other shapes. To describe these shapes, we must first label the parts of a fold **(Fig. 7.11a)**. The **limbs** are the sides of the fold that have the least curvature, and the **hinge** refers to a line along which the fold has the greatest curvature. The *axial plane*, an imaginary plane that contains the hinges of successive layers, effectively divides the fold into two halves. With these terms in hand, we can distinguish among the following types of folds:

- *Anticlines, synclines, and monoclines:* Folds that have an arch-like shape in which the limbs dip away from the hinge are called **anticlines** (see Fig. 7.11a), whereas folds with a trough-like shape in which the limbs dip toward the hinge are called **synclines** **(Fig. 7.11b)**. A **monocline** has the shape of a carpet draped over a stair step **(Fig. 7.11c)**.

- *Nonplunging and plunging folds:* If the hinge is horizontal, the fold is called a *nonplunging fold*, but if

Animation
Folds

FIGURE 7.11 Geometric characteristics of folds.

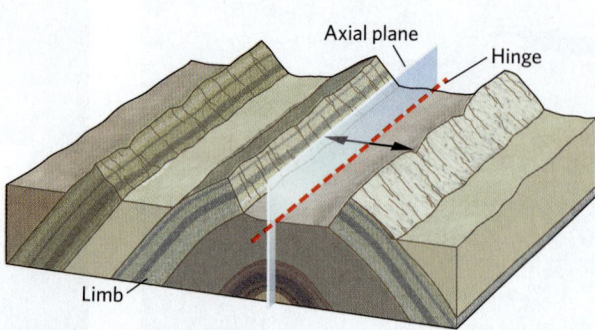

(a) An anticline looks like an arch. The limbs dip away from the hinge, as indicated by the outward-pointing arrows. Note that because of erosion, the oldest strata are exposed along the hinge.

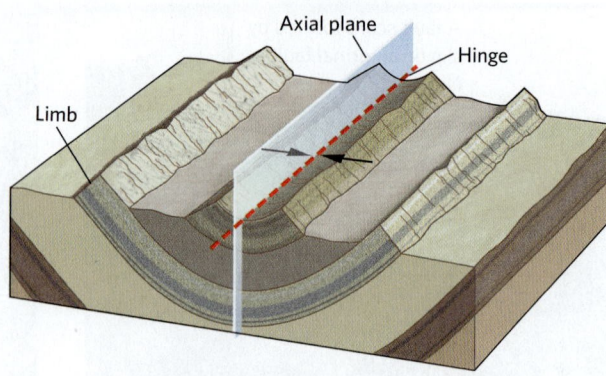

(b) A syncline looks like a trough. The limbs dip toward the hinge, as indicated by the inward-pointing arrows. Note that because of erosion, the youngest strata are exposed along the hinge.

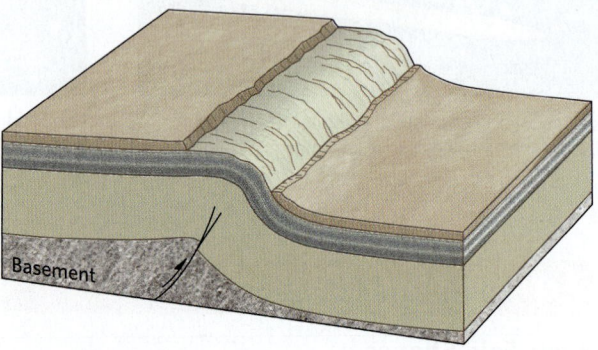

(c) A monocline looks like a stair step and is commonly draped over a fault block.

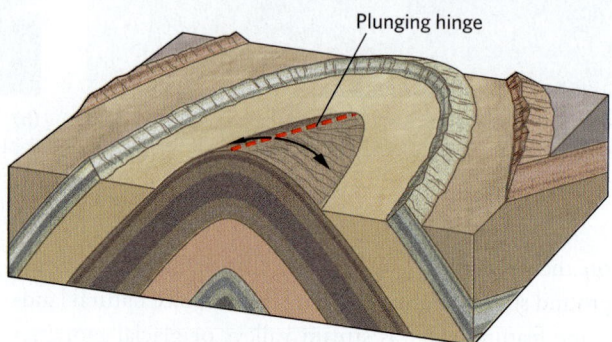

(d) A plunging anticline has a tilted hinge.

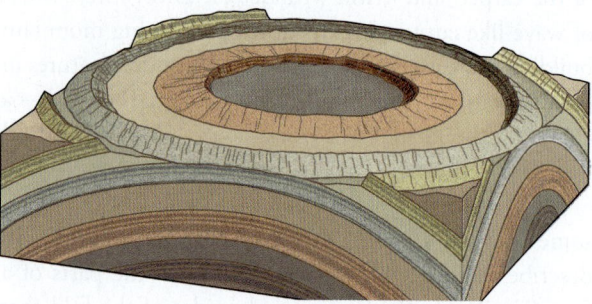

(e) A dome has the shape of an overturned bowl.

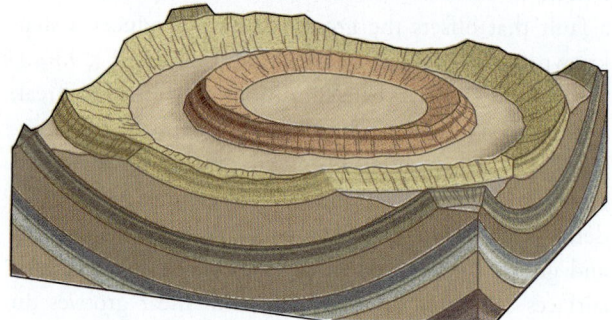

(f) A basin has the shape of an upright bowl.

See for yourself

Appalachian Fold Belt

Latitude: 40°53′24.28″ N
Longitude: 77°4′28.11″ W

Zoom to 60 km (37 mi). Tilt the view and look west.

These ridges are relicts of large folds formed when Africa collided with North America, producing the Appalachian orogen about 280 Ma. Erosion has removed the high mountains. Now, the region is the Valley and Ridge Province; hard sandstone beds resistant to erosion underlie the ridges.

the hinge is tilted, the fold is called a *plunging fold* (**Fig. 7.11d**).

- *Domes and basins:* A fold with the shape of an over-turned bowl is called a **dome**, whereas a fold shaped like an upright bowl is called a **basin** (**Fig. 7.11e, f**). Domes and basins both form circular patterns that look like bull's-eyes on a map if the land surface cuts across the structure. If domes and basins involve sedimentary beds, the oldest beds will crop out in the center of a dome, whereas the youngest beds will crop out in the center of a basin.

Using these terms, you can identify the various folds shown in **Figure 7.12**.

Formation of Folds

Why do folds form? Some rock layers wrinkle up, or *buckle*, in response to end-on compression. In such cases, a "train" of anticlines and synclines may form, with their hinges perpendicular to the direction of maximum compression (**Fig. 7.13**). Other folds form where shear gradually shifts one part of a rock body relative to another—layers caught up in the zone of shear undergo folding as

FIGURE 7.12 Examples of folds on outcrops and in the landscape.

(a) This anticline, exposed in a road cut near Kingston, New York, involves beds of Paleozoic limestone.

(b) This syncline, exposed in a road cut in Maryland, involves beds of Paleozoic sandstone and shale.

(d) This train of folds, exposed in sea cliffs in eastern Ireland, includes anticlines and synclines. The folds involve beds of Paleozoic sandstone and shale.

(c) This fold, exposed along the coast of Brazil, occurs in Precambrian gneiss.

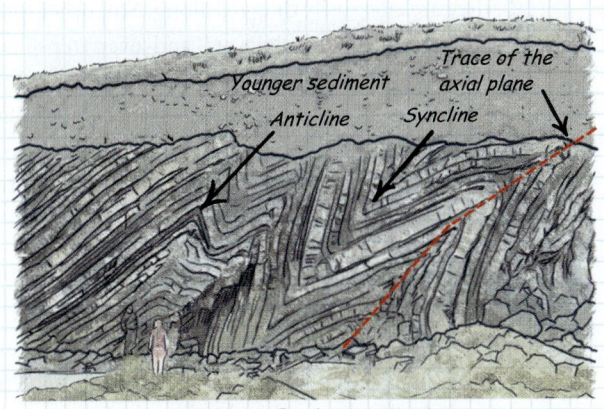

What an Earth Scientist Sees

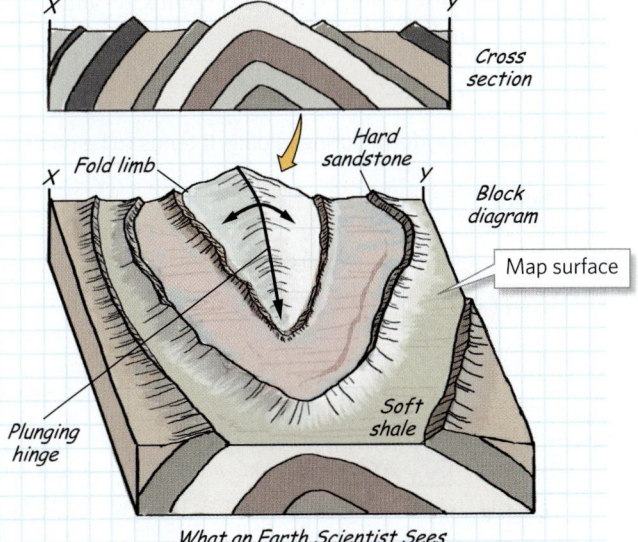

What an Earth Scientist Sees

(e) The plunging anticline of Sheep Mountain, Wyoming, is easy to see because of the lack of vegetation. Resistant rock layers (sandstone) stand out as ridges, whereas weaker rock layers (shale) erode away. A block diagram shows how the surface exposures relate to underground structures.

FIGURE 7.13 Folding can be caused by many different processes, illustrated here in cross section.

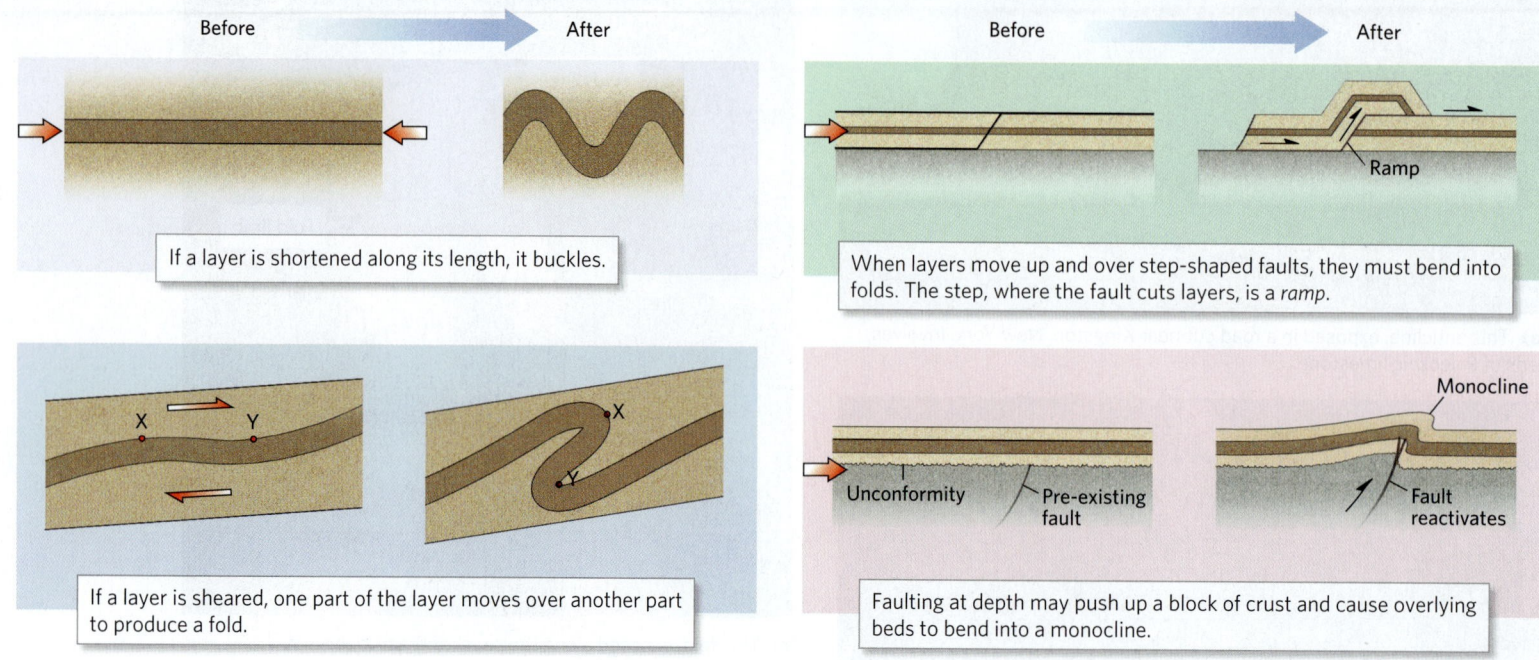

Before → After

If a layer is shortened along its length, it buckles.

Before → After

When layers move up and over step-shaped faults, they must bend into folds. The step, where the fault cuts layers, is a *ramp*.

Ramp

If a layer is sheared, one part of the layer moves over another part to produce a fold.

Monocline

Unconformity — Pre-existing fault — Fault reactivates

Faulting at depth may push up a block of crust and cause overlying beds to bend into a monocline.

they are effectively dragged along. Still other folds develop where rock layers move up and over step-like bends in a thrust fault, and must bend to conform to the fault's shape. Finally, some folds form when slip on a fault causes a block of basement, overlain by beds of sedimentary rock, to move upward relative to a neighboring block; when this happens, the beds bend.

Foliation Produced by Deformation

As we discussed in Chapter 6, *foliation*—parallel surfaces or layering such as slaty cleavage, schistosity, and gneissic banding—can develop during metamorphism. During the formation of foliation, differential stress can cause elongate or platy grains to align parallel to one another (**Fig. 7.14**).

Take-home message...

Folds, such as anticlines and synclines, are bends or curves in rock layers that form in response to differential stress. Stress can also cause the shape and orientation of grains to change, yielding foliation.

Quick Question

What processes cause rock layers to undergo folding?

7.5 Causes of Mountain Building

Before plate tectonics theory became established, geologists were just plain confused about why mountains formed. In the context of the new theory, however, the many processes driving mountain building became

FIGURE 7.14 Slaty cleavage may form during folding. It tends to be parallel to the axial plane.

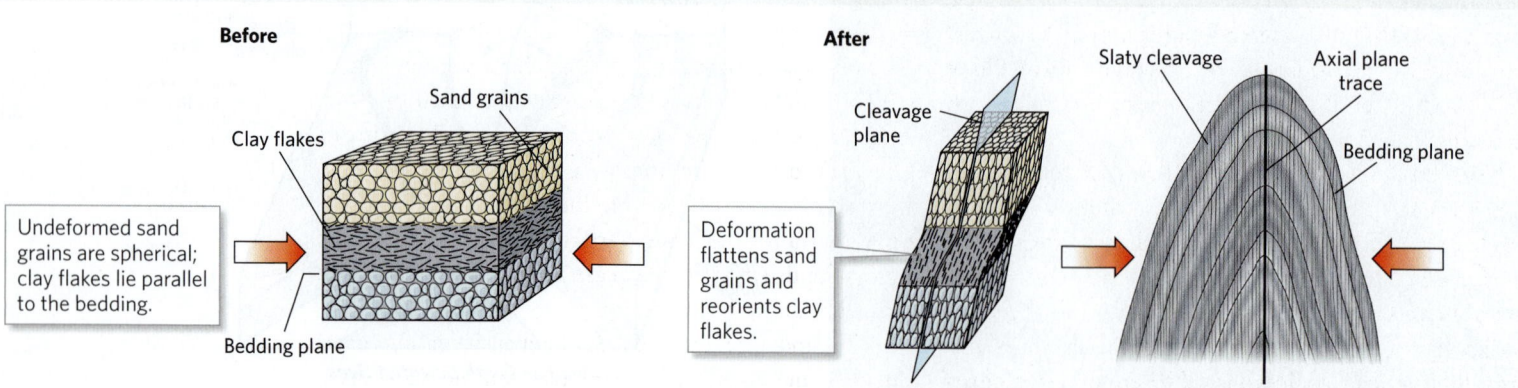

Before

Sand grains

Clay flakes

Undeformed sand grains are spherical; clay flakes lie parallel to the bedding.

Bedding plane

After

Cleavage plane

Deformation flattens sand grains and reorients clay flakes.

Slaty cleavage

Axial plane trace

Bedding plane

FIGURE 7.15 Convergent-boundary orogeny.

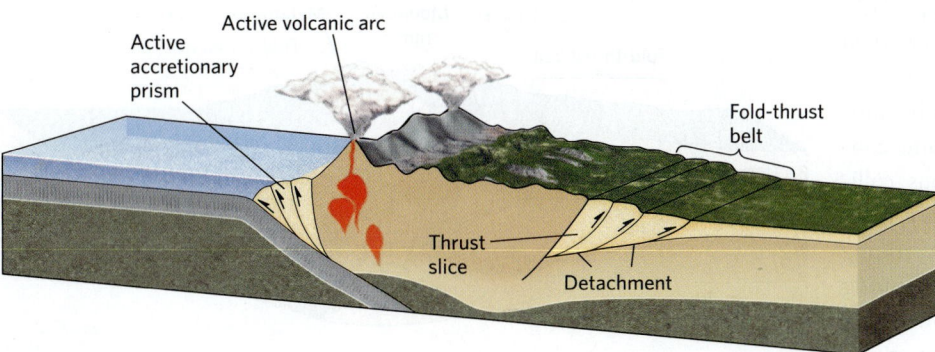

(a) At some convergent boundaries, compression uplifts a mountain belt.

(b) These mountains in Chile are part of the Andes, which formed along a convergent boundary.

clear: mountains form primarily in response to convergent-boundary deformation, continental collision, and rifting. In this section, we look at three different geologic settings and the types of mountains and geologic structures that develop in each one.

Mountain Belts Related to Subduction

At convergent boundaries, the oceanic lithosphere of the downgoing plate subducts and sinks into the mantle beneath the overriding plate. This process, as we saw in Chapter 2, triggers melting in the mantle and growth of a volcanic arc along the edge of the overriding plate. In many locations where the overriding plate consists of continental crust, the interaction between the downgoing plate and the overriding plate produces compression in the continental crust. This compression, in turn, causes the crust to shorten and produces a **convergent-boundary orogen** (Fig. 7.15a).

Many types of geologic structures form within convergent-boundary orogens. For example, in the very warm crust at depth in the orogen, plastic deformation and associated metamorphism cause folds, foliation, and reverse faults to form. This deformation also moves rocks from deeper levels up toward the Earth's surface, producing a belt of uplifted land that undergoes erosion by rivers and glaciers to produce sharp peaks. The Andes serve as an example of convergent-boundary deformation, for they have formed due to the subduction of Pacific Ocean floor beneath South America (Fig. 7.15b).

Notably, along the continental side of such a belt, at shallower depths, numerous thrust faults form, each carrying a wide, and relatively thin, sheet of rock called a *thrust slice*. As Figure 7.15a shows, each thrust slice moves up and over the one in front of it, so the slices overlap like shingles. Rocks within thrust slices undergo folding as the slices move. Therefore, geologists

refer to the overall region containing thrust faults and associated folds as a **fold-thrust belt**. Note that the thrust faults intersect a regional, nearly horizontal fault, called a *detachment*, at depth.

Mountain Belts Related to Collision

Once the oceanic lithosphere between two blocks of relatively buoyant continental crust subducts completely, the blocks, which were once separated by an ocean, collide with each other and produce a **collisional orogen** (Fig. 7.16a, b). The boundary between what had been separate blocks is called a **suture**. Large thrust faults form during a collisional orogeny as the edge of one block slips up and over the margin of the other. As a result, rocks in the footwall block may end up tens of kilometers below the land surface, where they undergo folding and metamorphism. Typically, these rocks develop foliation. The most intense deformation and metamorphism in a collisional mountain belt occurs in the internal or central zone of the orogen. On both sides of the internal metamorphic zone, fold-thrust belts tend to develop.

Because of the deformation that takes place within a collisional orogen, the crust thickens substantially. In fact, in some examples, the crust may thicken to 70 km (43 mi), almost twice the thickness of normal continental crust. Also during collision, the land may be uplifted by several kilometers. Due to folding and thrust faulting in the interior of the orogen, rocks that metamorphosed at great depth eventually squeeze upward and may be exposed at the land surface.

The most intense collisions happen when two continents come together. For example, continental collision

See for **yourself**

Folds, Central Australia

Latitude: 24°18′44.08″ S
Longitude: 132°10′34.02″ E

Zoom to 30 km (~18.5 mi) and look down.

Alternating beds of resistant and nonresistant sedimentary strata have been warped into plunging folds.

 Video
Continental Collision

yielded the Himalayas and the Alps during the Cenozoic (**Fig. 7.16c**; **Earth Science at a Glance**, pp. 212–213). But collisions also occur when small continental blocks, island arcs, or oceanic plateaus collide with continents or each other. Along some convergent boundaries, numerous collisions over geologic time suture a series of blocks to the edge of the overriding plate. Geologists refer to the suturing of smaller blocks to a larger one as **accretion**. An incoming buoyant crustal block is called an *exotic terrane* when it lies offshore and an *accreted terrane* once it has been sutured to the overriding plate (**Fig. 7.17a**). The process of accretion can add a substantial amount of new crust to the edge of a continent (**Fig. 7.17b**).

Mountain Belts Related to Continental Rifting

A *continental rift*, as we saw in Chapter 2, is a place where a continent stretches. This movement causes normal faults to develop in the upper crust (**Fig. 7.18a**). Movement on these faults causes blocks of crust to drop down, and in the process, strata in the blocks undergo tilting. Commonly, a nearly horizontal detachment underlies all of the blocks. The low areas between the tilted fault blocks fill with sediment eroded from the blocks. As a result, rifts typically contain sets of narrow, elongate mountain ranges composed of tilted fault blocks, separated by deep, sediment-filled rift basins. Because these ranges form due to slip on faults, they are sometimes called *fault-block mountain ranges*. Rifting produced the Basin and Range Province of Utah, Nevada, and Arizona (**Fig. 7.18b**). The Basin and Range Province lies within the North American Cordillera, one of North America's mountain belts (**Box 7.2**).

Measuring Mountain Building in Progress

The rumblings of earthquakes and the eruptions of volcanoes in some ranges attest to present-day, continuing

FIGURE 7.16 Collisional orogeny.

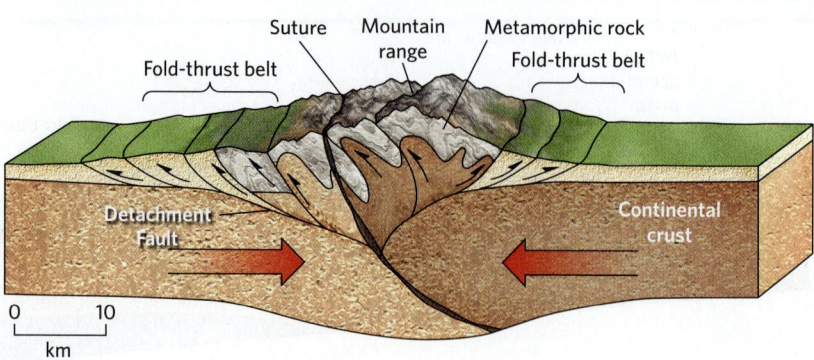

(a) During continental collision, continents squeeze together and deform. Thrust faulting brings metamorphic rock up from beneath the mountain belt to shallower levels.

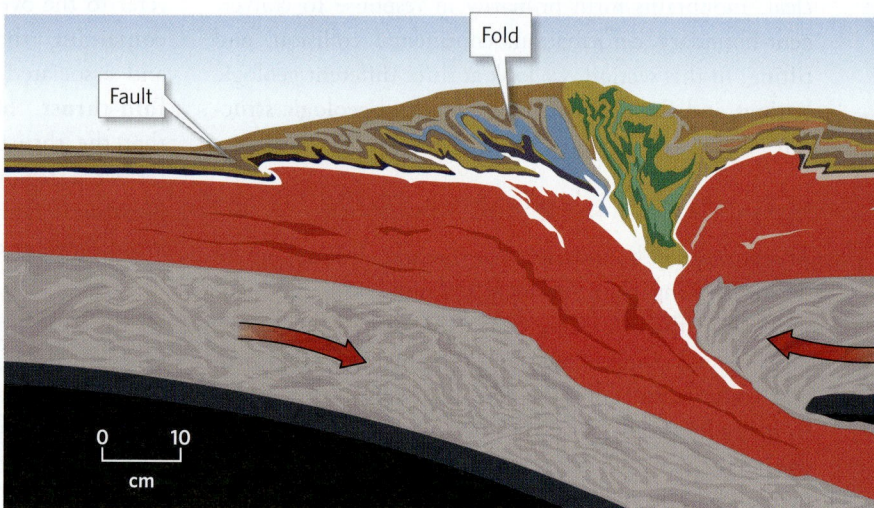

(b) Geologists can simulate collisional orogeny in the laboratory using layers of colored sand. Dragging the left side of such a model under the right side produces geologic structures and uplift, as shown in this sketch.

(c) This view of the Himalayas from space emphasizes that these mountains were uplifted when India collided with and pushed into Asia.

FIGURE 7.17 The accretion of exotic terranes can widen a mountain belt.

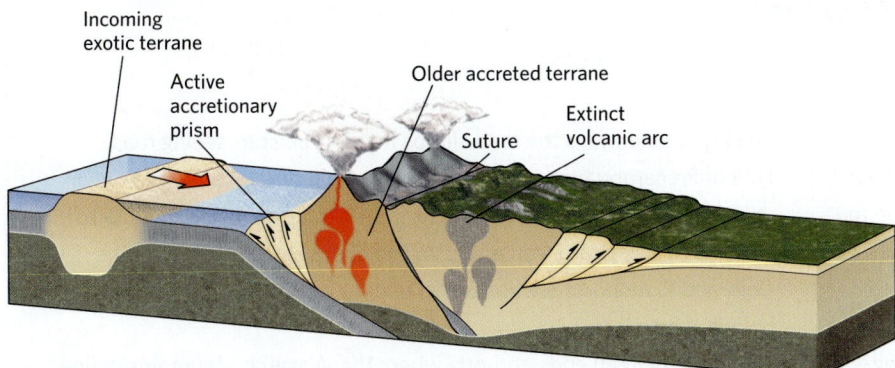

Incoming exotic terrane
Active accretionary prism
Older accreted terrane
Suture
Extinct volcanic arc

(a) An example of exotic terranes attaching to a continent as a consequence of subduction.

0   500
km

Arctic Ocean

USA
Canada

Eastern edge of North American Cordillera

Pacific Ocean

After Wrangellia attached to North America, strike-slip faults broke it into pieces.

N

Canada
USA

Accreted terranes
Wrangellia
North American crust
Terrane boundary

USA
Mexico

(b) The western portion of North America consists of accreted terranes that attached to the continent during the Mesozoic. A distinct terrane called Wrangellia (highlighted in red) was sliced into pieces that were displaced by strike-slip faults.

movements that geologists can measure through field studies and satellite technology. For example, researchers can determine where coastal areas have been rising relative to sea level by locating ancient beaches that now lie high above the water (Fig. 7.19a). And they can locate where the land surface has risen relative to a river by identifying places where a river has recently carved a new valley. In addition, GPS (the global positioning system) allows direct measurement of uplift and horizontal shortening rates. While standard

FIGURE 7.18 Rift-related orogeny.

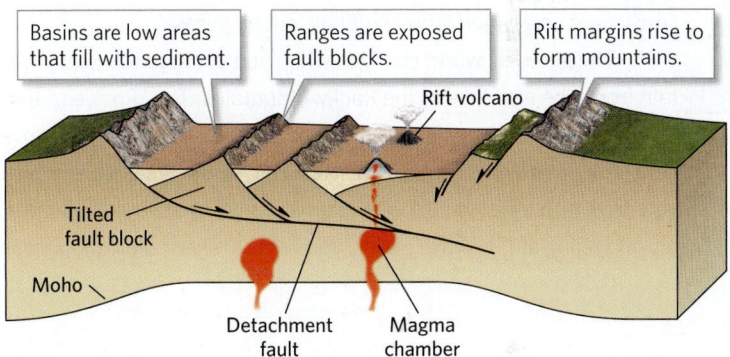

Basins are low areas that fill with sediment.
Ranges are exposed fault blocks.
Rift margins rise to form mountains.
Rift volcano
Tilted fault block
Moho
Detachment fault
Magma chamber

(a) Rifting leads to the development of numerous narrow mountain ranges separated by rift basins.

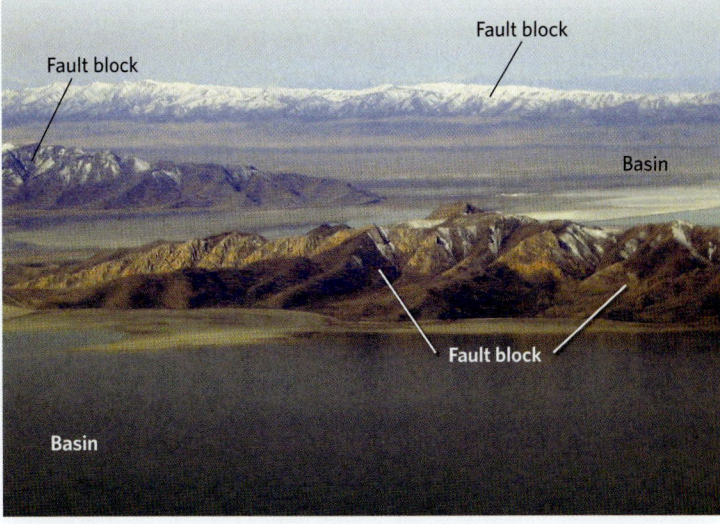

Fault block
Fault block
Basin
Fault block
Basin

(b) The Basin and Range Province of the western United States formed during Cenozoic rifting.

BOX 7.2

Putting Earth Science to Use

Seeing the mountain belts of North America

If you ever have the opportunity to drive across the middle of the USA, at the latitude of Washington DC or San Francisco, take it . . . and don't ignore the road cuts and landscapes that you pass! If you begin on the east coast and head west, you'll cross the remnants of the Appalachian orogen (Fig. Bx7.2). During the final phase of deformation that produced this orogen, North America collided with Africa and South America. What's the evidence for this collision? West of Washington, you'll cross the Valley and Ridge Province, the eroded remnant of a fold-thrust belt that was a product of compressional stress (see Fig. 7.15a). Watch how the bedding dip changes from one road cut to the next, as strata curve around folds. The locations of resistant and non-resistant rocks, controlled by folding, determine where erosion carved valleys or left ridges. You know you've reached the western edge of the Valley and Ridge, and have entered the Appalachian Plateau, when road cuts expose horizontal strata. As you cross the interior plains, part of North America's craton, the bedrock in most outcrops consists of nearly horizontal bedding, cut by joints.

The geologic world changes when you come to the Rocky Mountain front, the east face of the Rocky Mountains. The "Rockies," the eastern portion of the North American Cordillera, rose due to convergent-margin tectonics in the late Mesozoic through early Cenozoic. As you drive into the foothills, you observe that strata tilt to a near-vertical dip. Traveling at highway speeds, you cross these steeply dipping strata

quickly. Suddenly, as the road climbs higher, you start seeing road cuts of Precambrian metamorphic rocks. Near here, you cross a major reverse fault—slip on this fault contributed to uplifting the mountains. Continuing westward, you arrive on the relatively flat surface of the Colorado Plateau, where horizontal beds eroded into flat-topped mesas bounded by joints.

The plateau ends abruptly where the Wasatch Mountains define the boundary between the Colorado Plateau and the Basin and Range rift. As the crust stretched, the rift floor dropped down, so as you drive down the west flank of the mountains, climate changes from alpine to desert. For the next day of driving, you'll cross desert plains, underlain by grabens or half-grabens, alternating with narrow, rugged mountains (the unburied tips of tilted fault blocks; see Fig. 7.18). The landscape and rocks change abruptly once again when you reach the front of the Sierra Nevada, at the edge of the Basin and Range. You now enter a province where road cuts expose Mesozoic granite that once lay beneath a continental volcanic arc. From here to the west coast, outcrops contain crust that was accreted to the continent during Mesozoic and early Cenozoic convergent-margin tectonics. Just before reaching the Pacific, you'll cross a present-day plate boundary—the San Andreas fault—and pass from the North American Plate to the Pacific Plate. Your road trip traversed over 4,000 km (2,500 mi) and introduced you to a remarkable record of the Earth's long history!

FIGURE Bx7.2 A traverse across the United States, from Washington, DC to San Francisco, crosses many geologic provinces. Geologic features visible at various locations speak to the long geologic history of the North American continent.

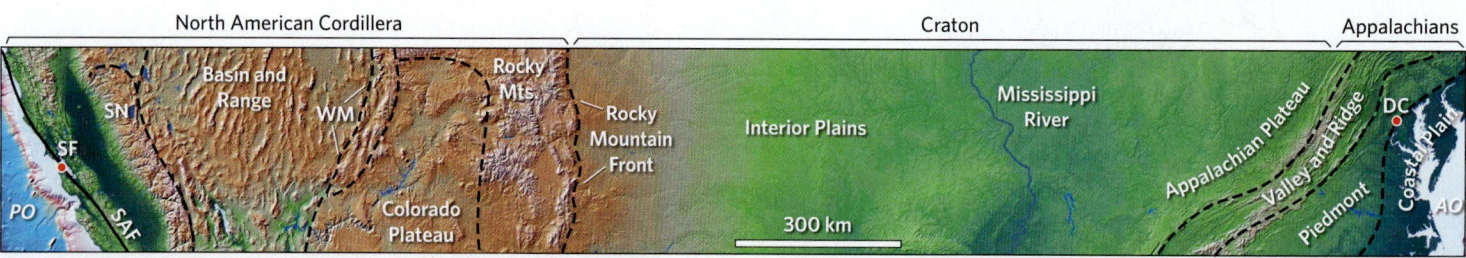

Note: SF = San Francisco; SAF = San Andreas fault; PO = Pacific Ocean; SN = Sierra Nevada; WM = Wasatch Mts.; DC = Washington, DC; AO = Atlantic Ocean

Colorado Plateau strata

Rocky Mountain Front

Valley and Ridge dipping strata

FIGURE 7.19 Evidence of present-day mountain building.

(a) Wave erosion cut these terraces along a beach. During earthquakes, the terraces were uplifted above sea level.

Uplift due to 1703 earthquake
Uplift due to 1923 earthquake
Present sea level

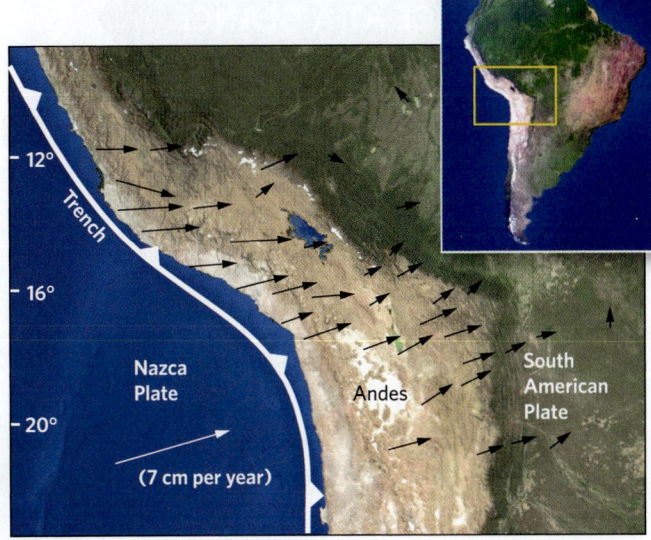

(b) GPS measurements in the Andes. The lengths of the arrows indicate the velocities of locations in the Andes relative to the interior of South America. The white arrow indicates relative plate motion.

hand-held GPS devices provide locations with accuracies of only about ±2 m (6 ft), research-quality GPS devices can specify locations to within ±2 mm (0.008 in). By comparing the precise positions of two different locations over a period of a few years, geologists can detect crustal motion. In effect, we can "see" the Andes shorten horizontally at a rate of a couple of centimeters per year as the subducting oceanic lithosphere compresses the continental crust of South America (Fig. 7.19b).

Take-home message . . .

Mountain belts form in association with convergent boundaries, continental collisions, and rifting. During convergent-boundary and collisional orogeny, crust thickens, thrust faults and folds form, and metamorphism takes place. Rifting yields fault-block mountains separated by basins.

Quick Question

How does the stress in a rift differ from stress in a collisional orogen?

7.6 Other Consequences of Mountain Building

Formation of Rocks In and Near Mountains

Mountain building produces conditions that can form a great variety of rock types. In fact, rocks of all three basic groups can form in mountain belts (Fig. 7.20a):

- *Igneous:* Melting takes place in the mantle and lower crust in regions where mountains are forming. The resulting magma rises to form plutons in continental

FIGURE 7.20 Rocks of all three basic groups may be formed by orogeny.

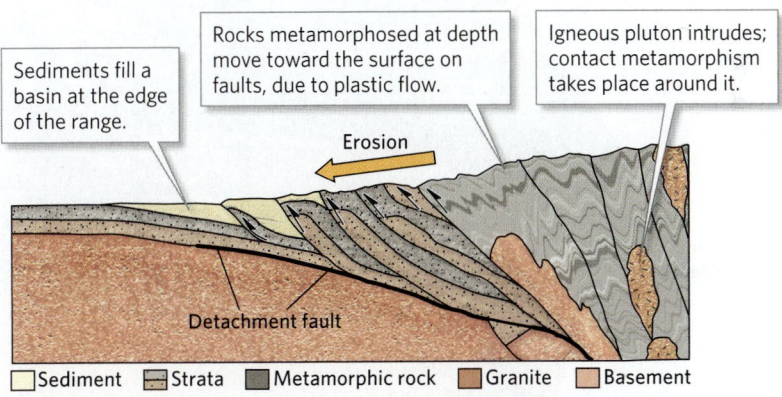

Sediments fill a basin at the edge of the range.

Rocks metamorphosed at depth move toward the surface on faults, due to plastic flow.

Igneous pluton intrudes; contact metamorphism takes place around it.

Erosion

Detachment fault

☐ Sediment ☐ Strata ☐ Metamorphic rock ☐ Granite ☐ Basement

(a) In the internal zone of a mountain range, metamorphic rocks and igneous rocks form. At the edge of the range, sedimentary rocks form.

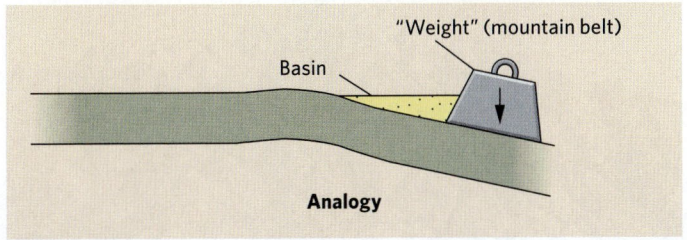

"Weight" (mountain belt)

Basin

Analogy

(b) A sedimentary basin develops because the mountain range acts as a weight that pushes down the surface of the lithosphere.

crust. When the magma in orogens freezes, it becomes igneous rock.

- *Sedimentary:* Weathering and erosion in mountain belts generate vast quantities of sediment. This sediment accumulates in low areas. In some locations, the weight of the mountain belt itself pushes down the surface of the nearby lithosphere, thereby producing

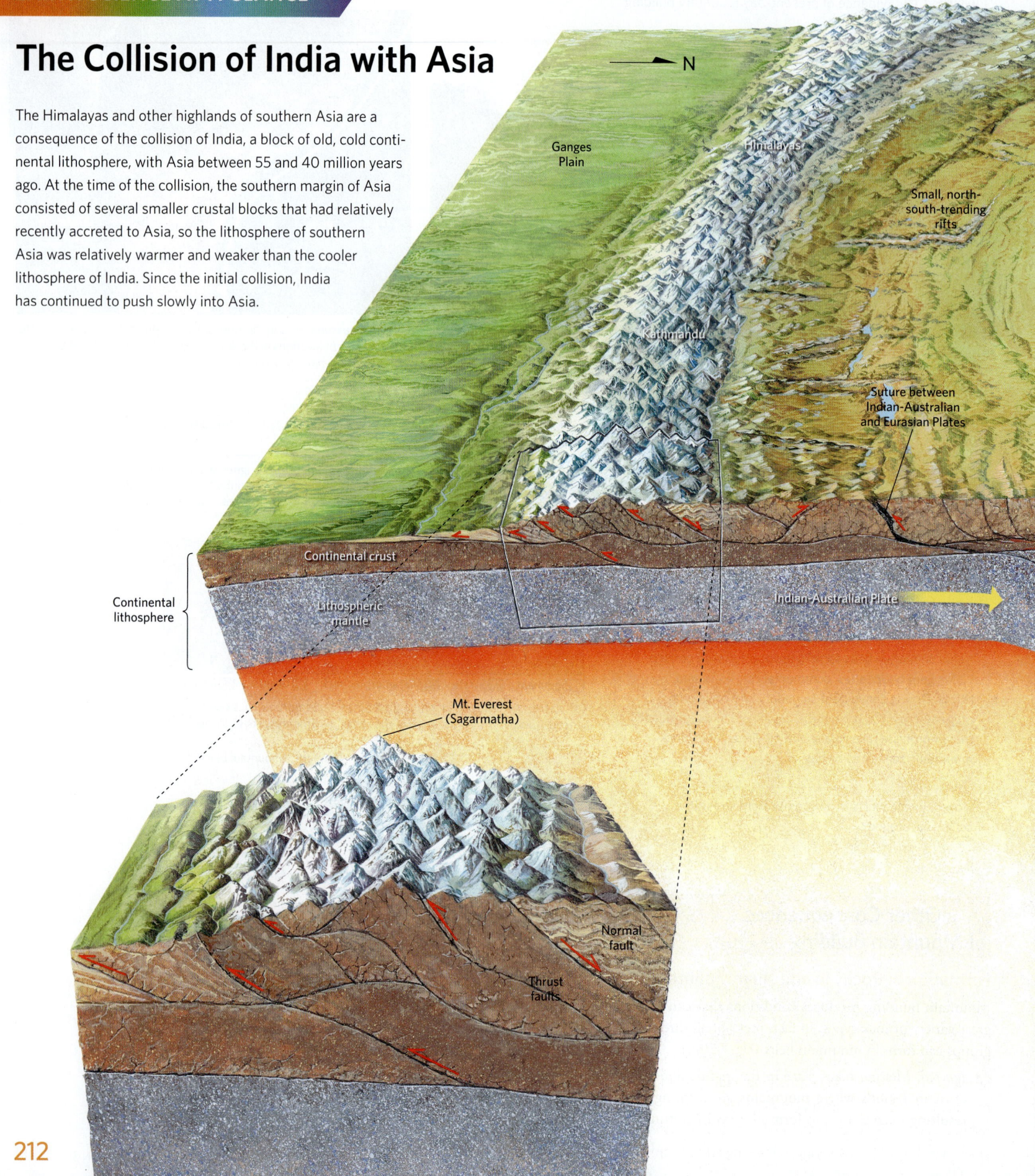

The Collision of India with Asia

The Himalayas and other highlands of southern Asia are a consequence of the collision of India, a block of old, cold continental lithosphere, with Asia between 55 and 40 million years ago. At the time of the collision, the southern margin of Asia consisted of several smaller crustal blocks that had relatively recently accreted to Asia, so the lithosphere of southern Asia was relatively warmer and weaker than the cooler lithosphere of India. Since the initial collision, India has continued to push slowly into Asia.

N

Ganges Plain

Himalayas

Small, north-south-trending rifts

Kathmandu

Suture between Indian-Australian and Eurasian Plates

Continental crust

Continental lithosphere

Lithospheric mantle

Indian-Australian Plate

Mt. Everest (Sagarmatha)

Normal fault

Thrust faults

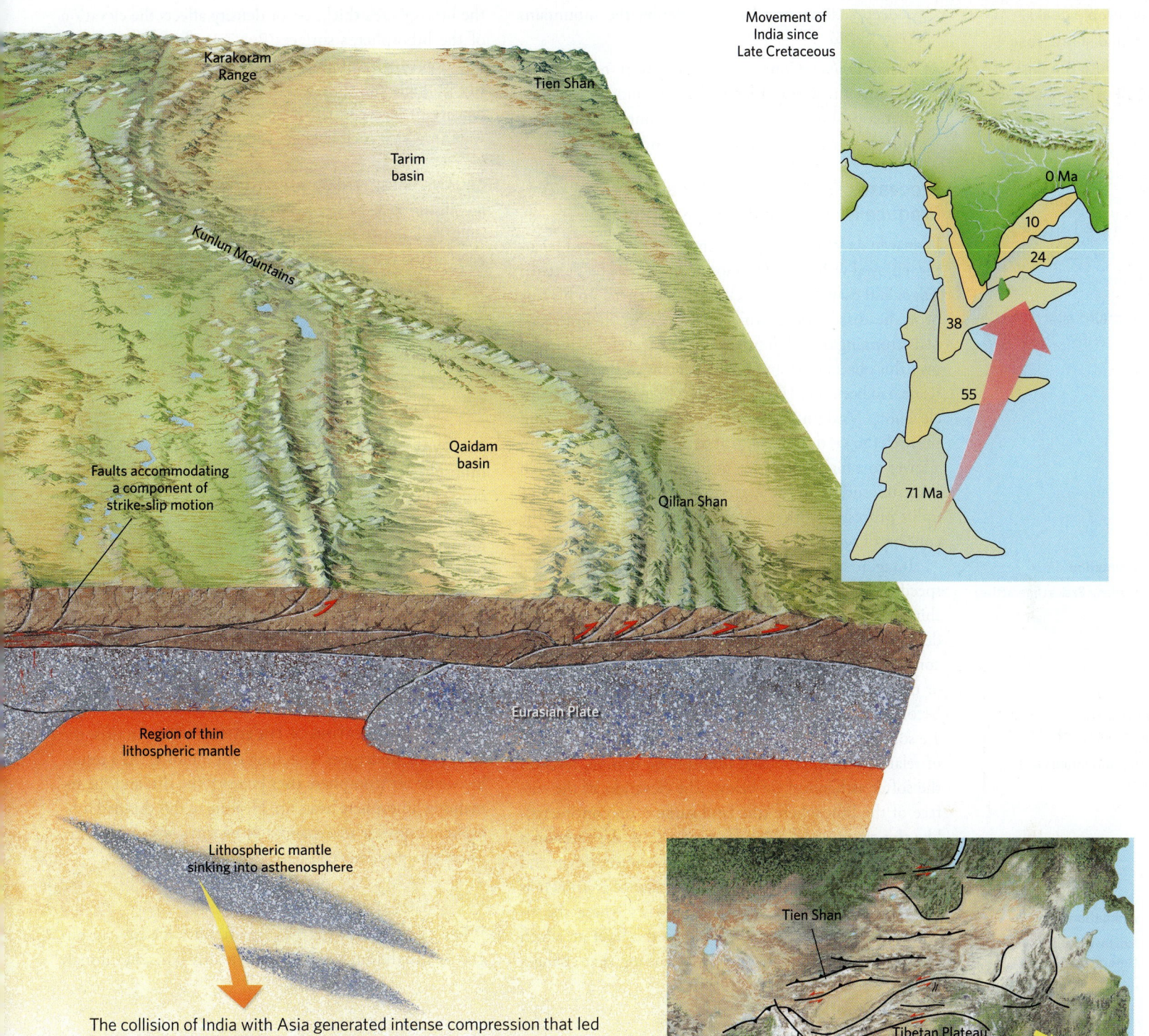

Movement of India since Late Cretaceous

0 Ma
10
24
38
55
71 Ma

Karakoram Range

Tien Shan

Tarim basin

Kunlun Mountains

Faults accommodating a component of strike-slip motion

Qaidam basin

Qilian Shan

Eurasian Plate

Region of thin lithospheric mantle

Lithospheric mantle sinking into asthenosphere

Tien Shan

Tibetan Plateau

Himalayas

The collision of India with Asia generated intense compression that led to the uplift of the Himalayas and the Tibetan Plateau. A variety of geologic structures have formed in the region. For example, the southern edge of the Himalayas is a fold-thrust belt in which thrust sheets are moving southward over India. Portions of China and Southeast Asia have slipped eastward on strike-slip faults. And reverse faults in central Asia have become active, leading to growth of ranges such as the Tien Shan. Why the broad plateau of Tibet rose remains something of a mystery. In part, the plateau's uplift may be a consequence of the crust thickening as it is being squashed horizontally and, as indicated by GPS measurements, spreading sideways to the east. Uplift may also be due to the heating of the region. If slabs of the underlying lithospheric mantle dropped off and sank, hot asthenosphere would rise to fill the space. Heating of the overlying, remaining lithosphere would cause it to rise.

213

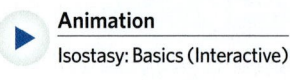

Animation
Isostasy: Basics (Interactive)

Did you ever wonder . . .

whether mountains last forever?

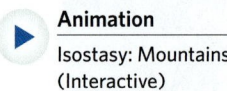

Animation
Isostasy: Mountains (Interactive)

a deep sedimentary basin bordering the mountains (Fig. 7.20b).

- *Metamorphic:* Contact metamorphism occurs near plutons intruded during mountain building. Regional metamorphism occurs where mountain building thrusts one part of the crust over another.

Processes That Cause Uplift and Produce Mountainous Topography

Leonardo da Vinci, the great Renaissance artist and scientist, enjoyed walking in the mountains, where he sketched ledges and examined the rocks he found there. To his surprise, he discovered marine shells (fossils) in limestone beds cropping out a kilometer above sea level. He puzzled over this observation and finally concluded that the limestone had been lifted from below sea level up to its present elevation. Modern geologists agree with Leonardo, and they refer to vertical movement of the Earth's surface from a lower to a higher elevation as *uplift*. What processes can cause the surface of the Earth to rise? To understand how uplift processes work, we must begin by introducing the concept of *isostasy*.

Imagine a ship anchored in a port. Its deck lies at a specific elevation above the sea surface. If we make the ship heavier by adding cargo, or shorter by removing a deck, the position of its top surface becomes lower. In contrast, if we make the ship lighter by removing cargo, or taller by adding a deck, the position of its top surface becomes higher. Vertical movements of the lithosphere are somewhat similar, for the lithosphere, which consists of relatively rigid crust and lithospheric mantle, rests on the softer asthenosphere below. The elevation of the surface of the lithosphere (the land surface, on continents), like the elevation of the surface of a ship, depends on a balance between forces pulling the lithosphere down and forces pushing it up (Fig. 7.21). This balance is known as **isostasy**. Put another way, isostasy exists where the elevation of the Earth's surface reflects the level at which the lithosphere naturally "floats." Anything that changes

the lithosphere's thickness or density affects the elevation of the lithosphere's surface (Box 7.3). So, to answer the question of why mountain belts can rise, we must identify geologic processes that can change the thickness or density of layers in the lithosphere. Let's consider some ways in which such changes can take place.

CRUSTAL SHORTENING AND THICKENING. The crust in a collision zone shortens horizontally and thickens vertically. As a result, the surface of the crust goes up, and the base of the crust goes down; the base of the lithosphere goes down as well (Fig. 7.22). For example, in the Himalayas, the surface of the crust has risen to an elevation of over 8 km (5 mi), and the base of the crust now lies at a depth of over 60 km (37 mi). This downward protrusion of crust is called a *crustal root*, and the downward protrusion of the lithosphere is called the *lithospheric mantle root*.

REMOVAL OF LITHOSPHERIC MANTLE. The weight of the lithospheric mantle (composed of denser rock than the crust) effectively pulls the lithosphere down, just as heavy cargo makes a ship settle deeper into the water. Therefore, removing some or all of the lithospheric mantle from the bottom of a plate causes the surface of the remaining lithosphere to rise in order to maintain isostasy, even if the crust's thickness remains unchanged. Such removal—a process that geologists call *delamination*—resembles removal of cargo or ballast from the hold of a ship: as the weight of the cargo decreases, the deck of the ship rises.

THINNING AND HEATING OF THE LITHOSPHERE. In rifts, the lithosphere undergoes stretching and thinning. As a result, less dense asthenosphere rises beneath the rift, and the remaining lithosphere heats up. Heating of this thinned lithosphere causes it to expand and become even less dense. These two processes cause the overall region to rise.

What Goes Up Must Come Down

When the land surface rises significantly for any reason, gravity and the Sun's energy begin to drive erosion through

FIGURE 7.21 A ship analogy for the concept of isostasy. The elevation of a ship's deck above sea level changes when a load is added to the ship. Water flows out of the way as the ship moves downward.

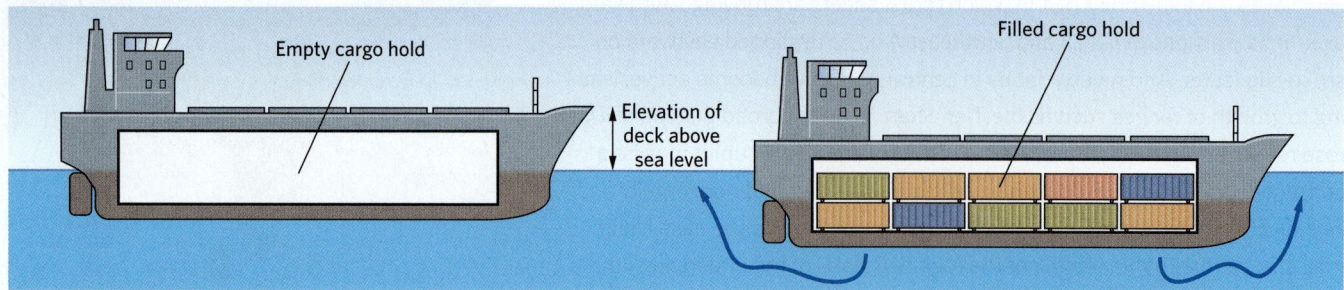

Empty cargo hold

Filled cargo hold

Elevation of deck above sea level

How can I **explain** . . .

Isostasy

What are we learning?
How a change in density and a change in thickness can be accommodated to maintain isostasy.

What you need:
- A tub of water
- Wooden blocks, some of dense hardwood, like oak, and some of less dense wood, like pine (begin with two pine blocks and one oak block, all of the same thickness)

Instructions:
- Place a pine block in the water, and let it come to equilibrium, meaning that isostasy has been achieved. Use a ruler to measure the distance between the water surface and the top of the block. Record your result.
- Add a second pine block on top of the first. (You may have to prop the blocks against the corner of the tub to keep them from tipping.) With the second block in place, measure the distance between the water surface and the top of the top block. You will see that this distance has increased.
- Now, place a block of oak in the water. Using the ruler, determine whether the distance from the top of this block to the water surface is greater or less than the distance you measured for the single pine block.
- By lining up a row of floating blocks of different thickness but the same density, you can see how the high elevations of a collisional mountain belt come to overlie the thickest crust.

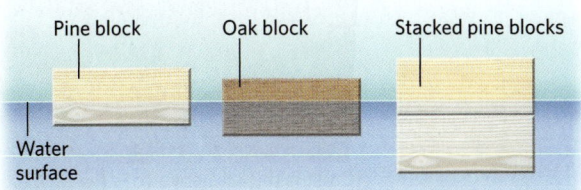

Pine block   Oak block   Stacked pine blocks

Water surface

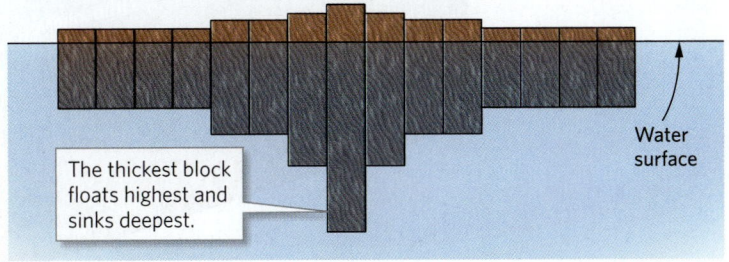

The thickest block floats highest and sinks deepest.

Water surface

What did we see?
The distance of a block's top surface from the water surface depends on both the thickness and the density of the block. Increase the block's thickness (by adding a second block), and the surface is higher. Increase the block's density (by replacing a block with one of denser wood), and the surface is lower. If you were able to measure thicknesses, volumes, and densities, you would see that a wood block sinks until the mass of the block below the water is equal to the mass of the water displaced. This relationship, explaining buoyancy, bears the name *Archimedes' principle*, recognizing its discoverer, an ancient Greek scientist.

---

the action of wind and precipitation. As a slope steepens, for example, landslides cause rock and debris to tumble down the slope. When winds blow clouds over the mountains, rain provides water that collects in streams whose flow carries away debris and sculpts valleys and canyons. If temperatures remain cold enough during the year, glaciers grow and slowly flow, carving peaks and deepening valleys. The net effect of all these processes is to grind away elevated areas and produce the jagged landscapes that we associate with mountainous terrain **(Fig. 7.23)**. It's important to keep in mind that uplift and erosion happen simultaneously in active mountain belts, so for the elevation of a range to increase over time, the range must uplift faster than it erodes. If the uplift rate becomes less than the erosion rate, the elevation of the range decreases.

Take-home message . . .

Mountain building is typically accompanied by igneous activity and metamorphism, as well as by deposition of sediment. Beneath some mountain belts, the continental crust has thickened, so the surface of the crust "floats" higher. Once uplift has occurred, erosion sculpts rugged topography.

Quick Question
How can thinning of the lithosphere cause uplift?

FIGURE 7.22 The concept of isostasy as applied to collisional mountain building.

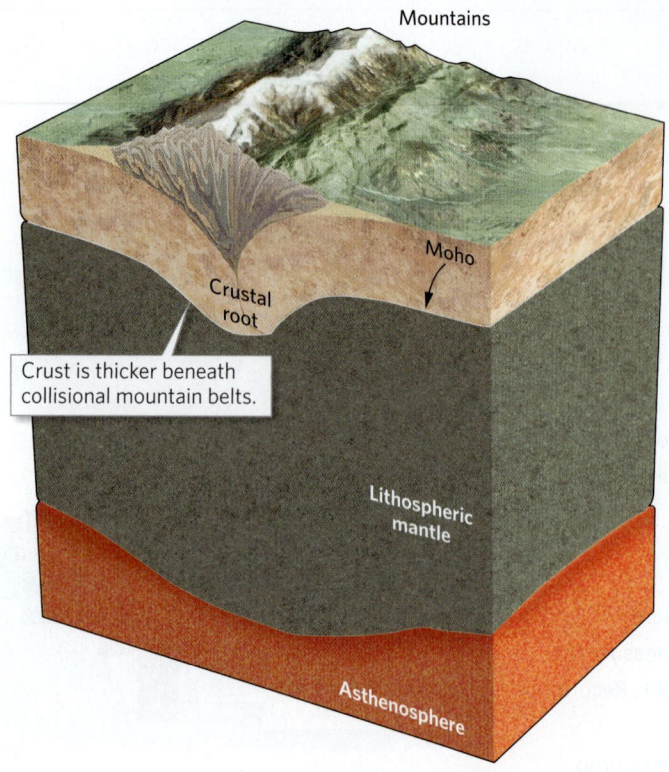

Mountains

Moho

Crustal root

Crust is thicker beneath collisional mountain belts.

Lithospheric mantle

Asthenosphere

7.7 Basins and Domes in Cratons

So far in this chapter, we've discussed deformation caused by dramatic mountain-building events. A **craton** consists of continental lithosphere that has not been affected by mountain-building events for at least the last 1 billion years. Cratons, therefore, tend to have relatively subdued topography. The interior of North America, the region between the North American Cordillera to the west and the Appalachians to the east, is a craton (**Fig. 7.24**). Cratons form when crust has been able to cool substantially and has, therefore, become relatively strong and stable. Geologists divide cratons into **shields**, in which Precambrian metamorphic and igneous rocks are exposed at the ground surface, and **cratonic platforms**, where a relatively thin layer of Paleozoic and Mesozoic strata cover the Precambrian rocks (see Fig. 7.24).

In the shield areas of cratons, we find widespread exposures of intensively deformed metamorphic rocks with abundant examples of folds and foliation. That's because the crust making up the cratons was deformed during multiple mountain-building events in the distant past, more than a billion years ago. These mountain-building events are so old that erosion has worn away the original mountains, in the process exhuming deep crustal rocks.

In cratonic platforms, the sedimentary beds typically define regional domes and basins—broad areas (over a few hundred kilometers across) that gradually sank or uplifted, respectively, over geologic time (**Fig. 7.25**). For example, in Missouri, strata arch across a broad uplift, the Ozark dome, whose diameter is 320 km (200 mi). The thickness of strata decreases toward the top of the dome because less sediment accumulated within the dome than in adjacent basins. When erosion removes the top of the dome, contacts between stratigraphic formations display a bull's-eye shape, and the oldest

FIGURE 7.23 As soon as land rises, water and ice begin eroding it.

(a) Glaciers carved these rugged peaks in Switzerland.

(b) Streams cut these valleys into weathered bedrock in Brazil.

FIGURE 7.24 North America's craton consists of a shield, where Precambrian rock is exposed, and a cratonic platform, where Paleozoic sedimentary rock covers the Precambrian rock.

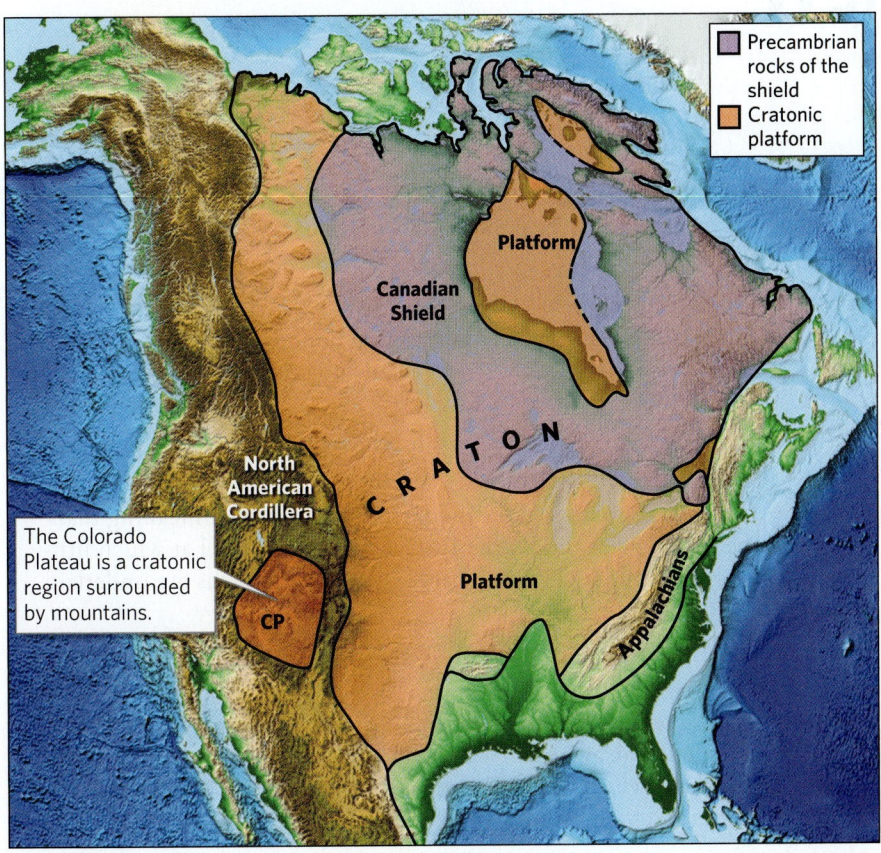

Legend:
- Precambrian rocks of the shield
- Cratonic platform

FIGURE 7.25 Domes and basins of the North American cratonic platform.

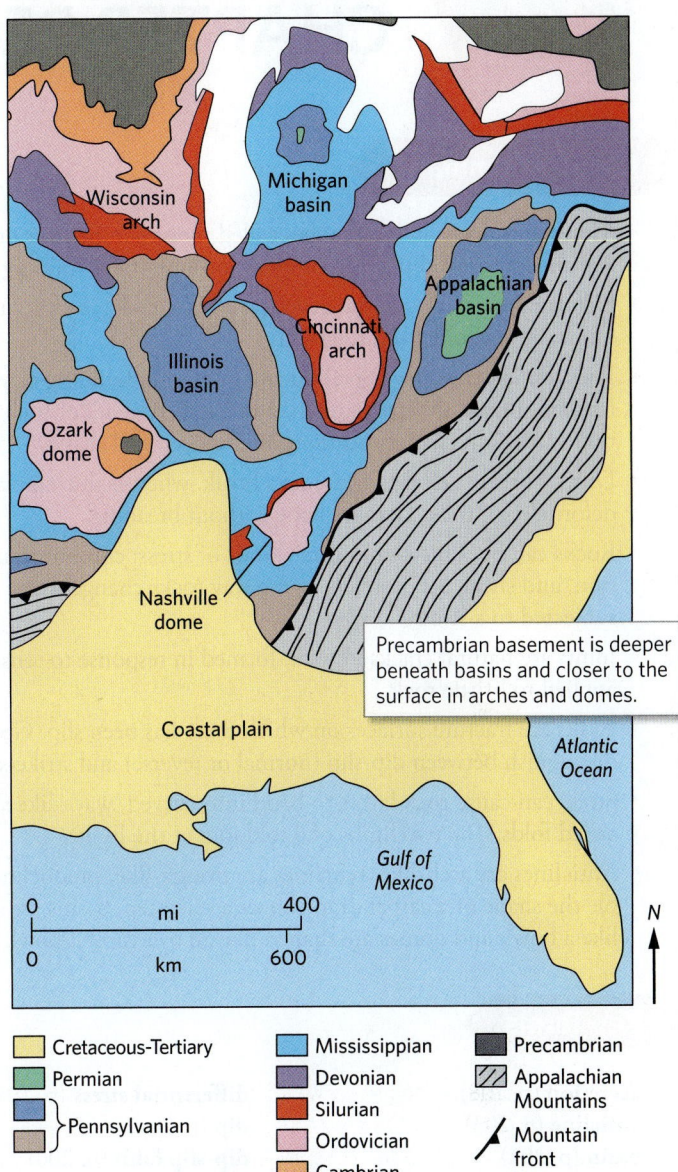

Precambrian basement is deeper beneath basins and closer to the surface in arches and domes.

Legend:
- Cretaceous-Tertiary
- Permian
- Pennsylvanian
- Mississippian
- Devonian
- Silurian
- Ordovician
- Cambrian
- Precambrian
- Appalachian Mountains
- Mountain front

(a) A geologic map of the mid-continent platform region, showing the locations of basins and domes. The terms in the legend refer to geologic time periods (see Chapter 9 for more information).

exposed rocks lie at the center of the dome at the ground surface. In the Illinois basin, sedimentary strata warp downward into a huge bowl that is also about 300 km (180 mi) across. The strata get thicker toward the basin's center, indicating that the floor of the basin was sinking, providing space, at the same time sediment was accumulating. The basin, like the dome discussed above, displays a bull's-eye shape, but here the youngest strata are exposed in the center.

Take-home message . . .

Cratons are portions of continents that consist of old (Precambrian) and relatively stable crust. Parts of cratons may be covered by Phanerozoic sedimentary strata. A basin occurs where the strata are warped down into a bowl shape, and a dome occurs where the strata are warped up.

Quick Question

Why are strata thicker beneath the center of a basin than beneath the center of a dome?

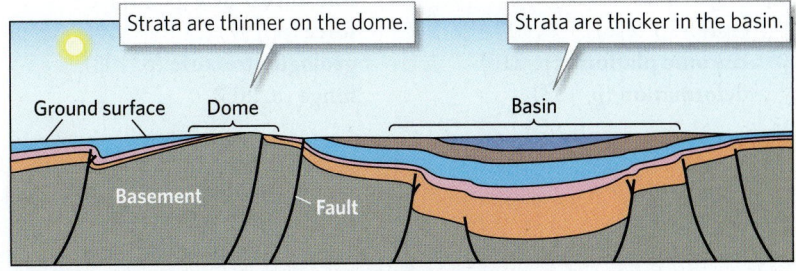

Strata are thinner on the dome.

Strata are thicker in the basin.

(b) A cross section showing how strata thin toward the crest of a dome and thicken toward the center of a basin. The cross section is vertically exaggerated.

7 CHAPTER REVIEW

- Mountains occur in long ranges called mountain belts or orogens. An orogen forms during an orogeny, or mountain-building event.

- Deformation results in the development of geologic structures: joints, faults, folds, and foliation.

- Differential stress developed during mountain building causes rocks to bend, break, shorten, stretch, and shear. Because of such deformation, rocks can change location, orientation, and shape.

- During brittle deformation, rocks break, whereas during plastic deformation, rocks change shape without breaking.

- Rocks can be subjected to three kinds of stress: compression, tension, and shear. Strain refers to the way rocks change shape when subjected to a stress.

- Joints are natural cracks in rock, formed in response to tension applied to brittle rock.

- Faults are fracture surfaces on which there has been slip. Geologists distinguish between dip-slip (normal or reverse) and strike-slip faults.

- Stress can cause rock layers to bend into curved, wave-like shapes called folds. The two limbs of a fold join at the hinge.

- Anticlines are arch-like, synclines are trough-like, monoclines resemble the shape of a carpet draped over a stair step, basins are shaped like a bowl, and domes are shaped like an overturned bowl.

- Subduction at convergent boundaries and collision between blocks of crust can produce mountains in which compression produces folds and thrusts.

- Rifting produces fault-block mountains.

- Mountain belts formed at convergent boundaries may incorporate accreted terranes. Accretion can add crust to the edge of a continent.

- With modern GPS technology, it is now possible to measure shortening and uplift in mountain belts directly.

- Uplift over broad regions is controlled by isostasy, meaning that the elevation of the Earth's surface reflects the level at which lithosphere naturally "floats." Large collisional orogens are underlain by roots.

- Once mountains are uplifted, erosion sculpts them into jagged peaks.

- Cratons are the old, relatively stable parts of continental crust. They include shields, where Precambrian rocks are exposed, and cratonic platforms, which have a cover of sedimentary strata. Broad regional domes and basins form in platform areas.

accretion (p. 208)
anticline (p. 203)
basin (p. 204)
bearing (p. 201)
brittle deformation (p. 195)
collisional orogen (p. 207)
compression (p. 198)
convergent-boundary orogen (p. 207)
craton (p. 216)
cratonic platform (p. 216)
deformation (p. 193)

differential stress (p. 198)
dip (p. 201)
dip-slip fault (p. 200)
displacement (p. 199)
dome (p. 204)
fault (p. 199)
fault scarp (p. 203)
fold (p. 203)
fold-thrust belt (p. 207)
force (p. 197)
geologic structure (p. 193)
hinge (p. 203)

isostasy (p. 214)
joint (p. 199)
limb (of fold) (p. 203)
monocline (p. 203)
mountain belt (p. 193)
normal fault (p. 200)
orogen (p. 193)
orogeny (p. 193)
plastic deformation (p. 195)
plunge (p. 201)
reverse fault (p. 200)
shear (p. 198)

shield (p. 216)
slickenside (p. 203)
slip (p. 199)
strain (p. 195)
stress (p. 198)
strike (p. 201)
strike-slip fault (p. 200)
suture (p. 207)
syncline (p. 203)
tension (p. 198)
thrust fault (p. 200)
uplift (p. 193)

Review Questions

Letters in parentheses correspond to the chapter's learning objectives.

1. What changes do rocks undergo during mountain building? **(A)**

2. Contrast brittle and plastic deformation. **(A)**

3. What factors determine whether a rock will behave in brittle or plastic fashion? **(A)**

4. How do force, stress, and strain differ from each other? **(A)**

5. How is a fault different from a joint? **(B)**

6. Compare normal, reverse, and strike-slip faults. Which type of fault does this diagram show? **(B)**

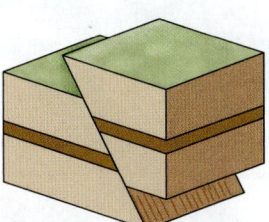

7. How can you recognize faults in the field? **(B)**

8. Describe the differences among an anticline, a syncline, and a mono-cline. Which type of fold does this figure show? **(B)**

9. How do domes and basins differ from each other, and from anticlines and synclines? **(B)**

10. Discuss the relationship between foliation and deformation. **(B)**

11. Discuss the processes by which mountain belts form at convergent boundaries, during continental collisions, and in continental rifts. **(C)**

12. Can we measure mountain building in progress? If so, how? **(C)**

13. Describe the principle of isostasy. **(E)**

14. Why do mountain belts develop rugged topography over time? **(F)**

15. Why do sedimentary basins form along the margins of orogens? Where might you find metamorphic and igneous rocks in an orogeny? **(D)**

16. How are the structures affecting the sedimentary strata of a craton different from those of a mountain belt? Which type of structure does this diagram show? **(G)**

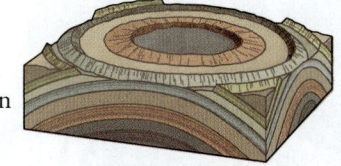

On Further Thought

17. Imagine that a geologist sees two outcrops containing the same resistant sandstone, as depicted in the cross-section sketch below. The region between the outcrops is covered by soil. The curving lines in the bed indicate the shape of the cross beds in the sandstone. Keeping in mind how cross beds form, sketch the connection of the cross-bedded layer from one outcrop to the other, before erosion. What geologic structure have you drawn? **(B)**

18. The Pyrenees, an east-west-trending mountain range along the border between Spain and France, uplifted during the Cenozoic. On both sides of the range, fold-thrust belts have developed. In the interior of the range, metamorphic rocks are exposed. What kind of stress caused the range to form? In what direction did Spain move, relative to France, to cause this stress? **(C)**

8 A VIOLENT PULSE
Earthquakes

By the end of the chapter you should be able to . . .

A. explain what an earthquake is, what processes cause earthquakes, and how earthquake energy moves through the Earth.

B. describe the basic concepts behind the methods used to measure and locate earthquakes.

C. interpret the meaning of an earthquake's magnitude and intensity as discussed in news media.

D. relate earthquakes to specific geologic settings in the context of plate tectonics theory.

E. distinguish among the different ways in which earthquakes trigger damage, and differentiate between a tsunami and a storm wave.

F. critique a prediction of an earthquake, given the real limitations on the predictability of earthquakes.

G. list the steps that people can take to prevent earthquake damage and casualties.

H. explain how seismic waves behave as they pass through the Earth's interior.

I. discuss what the study of seismic waves can tell us about layering inside the Earth.

8.1 Introduction

It was mid-afternoon on March 11, 2011, and in many seaside towns along the Pacific coast of Honshu, the largest island of Japan, fishing fleets unloaded their catch, factories churned out goods, shoppers browsed the stores, and office workers tapped at computers. No one realized that their surroundings would soon change forever. Honshu lies near a convergent boundary where the Pacific Plate slips beneath the edge of Japan and sinks back into the mantle. Averaged over time, this movement takes place at about 8 cm (3 in) per year. But the movement doesn't happen smoothly. Rather, for a while, rocks adjacent to the boundary quietly and subtly bend and warp. Then, suddenly, just as a wooden stick snaps after you've bent it too far, a measurable amount of movement takes place in a matter of seconds to minutes **(Fig. 8.1)**. On March 11, at 2:46 P.M., the "snap" started at a point located about 130 km (80 mi) east of Japan's coast and about 24 km (15 mi) below the Earth's surface. When it happened, Japan lurched eastward by a few meters and a portion of seafloor rose vertically by several centimeters. The stage had been set for a disaster.

The instant that movement took place, vibrations began to pass through the surrounding rock, just like the vibrations that pass along the two pieces of the bent stick when you snap it between your hands. The vibrations traveled at an average speed of 11,000 km (7,000 mi) per hour—10 times the speed of sound in air—and transported energy from the site of the snap outward. When they reached the surface of the Earth, they caused an episode of ground shaking. Geologists refer to the event generating these vibrations, and to the ground motions that result, as an **earthquake**.

So much energy was released by the March 11 event that the resulting earthquake—named the Tōhoku earthquake, after a province in Honshu—had historic consequences. In the area affected by the earthquake, shaking caused people to lose their balance and fall. Bottles and plates flew off shelves and crashed to the floor, bookshelves tipped over, and furniture danced around rooms. Buildings twisted and swayed, and in some cases their ceilings and facades collapsed in a shower of debris **(Fig. 8.2a)**. Outside, dust rose from the ground and landslides tumbled down hillslopes. In addition, gas pipelines broke, sending flammable vapors into the air. Some of the gas ignited in billows of flame.

<< Rescuers are searching a building that tipped over during a 2016 earthquake in Taiwan.

FIGURE 8.1 What happens during an earthquake?

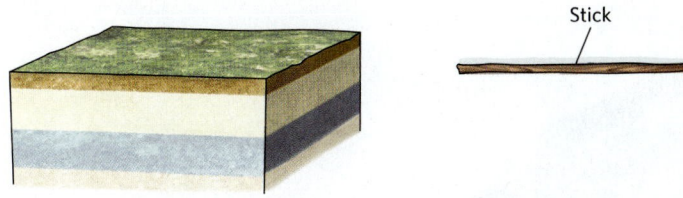

(a) Before deformation, the rock layers in this example are not bent.

The layer of rock bends elastically (exaggerated here).

(b) As deformation occurs, rock bends elastically, like a stick that you arch between your hands. The drawing exaggerates the amount of bending.

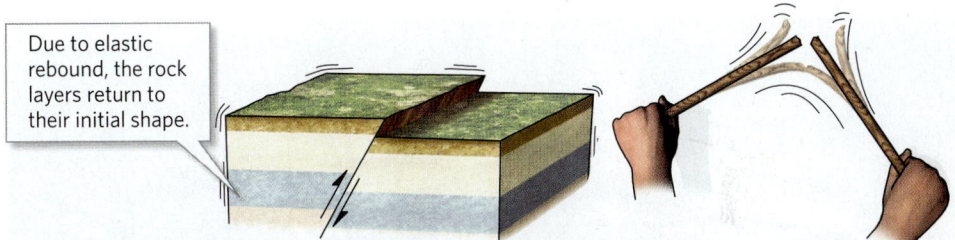

Due to elastic rebound, the rock layers return to their initial shape.

(c) Eventually, the rock breaks, and sliding suddenly occurs on a fault. This break generates vibrations like those you feel when you break a stick.

Meanwhile, the sudden displacement of the seafloor pushed up the surface of the ocean itself. This movement produced a *tsunami*, a very broad, fast-moving wave. When this wave approached the shore, it grew to a height of over 10 m (33 ft). It overtopped seawalls and battered the landscape as far as several kilometers inland **(Fig. 8.2b, c)**.

Earthquakes are a fact of life on our dynamic planet. Fortunately, most cause no damage or casualties, either because they are small or because they take place in unpopulated areas. But a few hundred earthquakes per year rattle the ground sufficiently to crack or topple buildings and injure their occupants, and every 5 to 20 years, on average, a great earthquake, such as the Tōhoku event, becomes a horrific calamity.

What geologic phenomena trigger earthquakes? Why do earthquakes take place where they do? How do they cause damage? Can we predict when earthquakes will happen or even prevent them from happening? What can they tell us about the interior of the Earth? *Seismologists* (from the Greek word *seismos*, meaning shock or earthquake), researchers who study earthquakes, have addressed many of these questions. In this chapter, we present some of their answers.

FIGURE 8.2 The Tōhoku earthquake and tsunami, Japan, 2011.

(a) Some buildings collapsed due to ground shaking.

(b) The rising water of the tsunami filled harbors and spilled over seawalls.

Coast

(c) An aerial view shows the tsunami advancing across the shore.

8.2 What Causes Earthquakes?

Ancient cultures offered a variety of explanations for **seismicity** (earthquake activity), most of which involved the restless gyrations of mythical subterranean animals. Modern research has replaced these supernatural explanations with the view that, while volcanic eruptions and nuclear explosions can trigger some earthquakes, most take place when one body of rock suddenly moves past another on a **fault**, a fracture surface on which sliding, or *slip*, takes place **(Fig. 8.3)**. The amount of slip on a fault is the fault's *displacement* **(Fig. 8.4)**. The energy released by an earthquake travels through the Earth as **seismic waves** or *earthquake waves*.

FIGURE 8.3 An example of a fault surface in Arizona that became exposed because the rock above the fault surface has eroded away. Slip lineations formed on this fault surface during movement.

Slip lineations

Fault surface

Fault trace

What an Earth Scientist Sees

How Does Faulting Generate Seismic Waves?

As we discussed in Chapter 7, slip on a fault displaces the rock on one side of the fault relative to the rock on the other side. You can find faults in many locations—but don't panic! Very few faults act as a source of earthquakes at any particular time, so geologists distinguish between *active faults*, which have moved relatively recently or might move in the future, and *inactive faults*, which moved in the distant geologic past, but probably won't move again in the future. Faulting produces seismic waves in two ways: (1) when previously intact rock suddenly breaks, forming a new fault on which slip takes place; and (2) when a pre-existing fault suddenly slips again.

To picture the rupture of intact rock, imagine that you grip each side of a brick-shaped block of rock with a clamp. Now, apply a slight upward push on one of the clamps and a slight downward push on the other. This action generates differential stress in the rock (see Chapter 7), and the rock bends slightly, but doesn't break. If you stop pushing, and therefore decrease the stress, the rock "relaxes" and returns to its original shape. Geologists refer to a change in shape that can be reversed by the removal of stress as **elastic deformation**—the same phenomenon happens when you bend a stick and then let it go. Now repeat the experiment, pushing one side of the block up and the other side down even more, so that the stress becomes greater. This time, small cracks start to develop in the rock **(Fig. 8.5a)**. With continued pushing, the cracks begin to connect to one another until, suddenly, a fracture cuts across the entire block of rock **(Fig. 8.5b)**. The instant that happens, the block breaks in two, and rock on one side of the fracture slides past rock on the other side. Because sliding takes place, the fracture becomes a fault.

FIGURE 8.4 A wooden fence built across the San Andreas fault was offset during the 1906 San Francisco earthquake.

Displacement   *Fault trace*

What an Earth Scientist Sees

If you look closely at Figure 8.5b, you'll notice that the portion of the rock in which bending takes place, and elastic deformation accumulates, is wider than the fault itself. When the new fault suddenly forms and slips, the bent rock on either side of the fault straightens out, or *rebounds*, so the accumulated elastic deformation relaxes and the stress decreases **(Fig. 8.5c)**. When relaxation takes place, however, the rock doesn't move smoothly back to its original shape; rather, it "twangs" back and forth, like a bouncing spring. It's this back-and-forth movement of rock on

Time

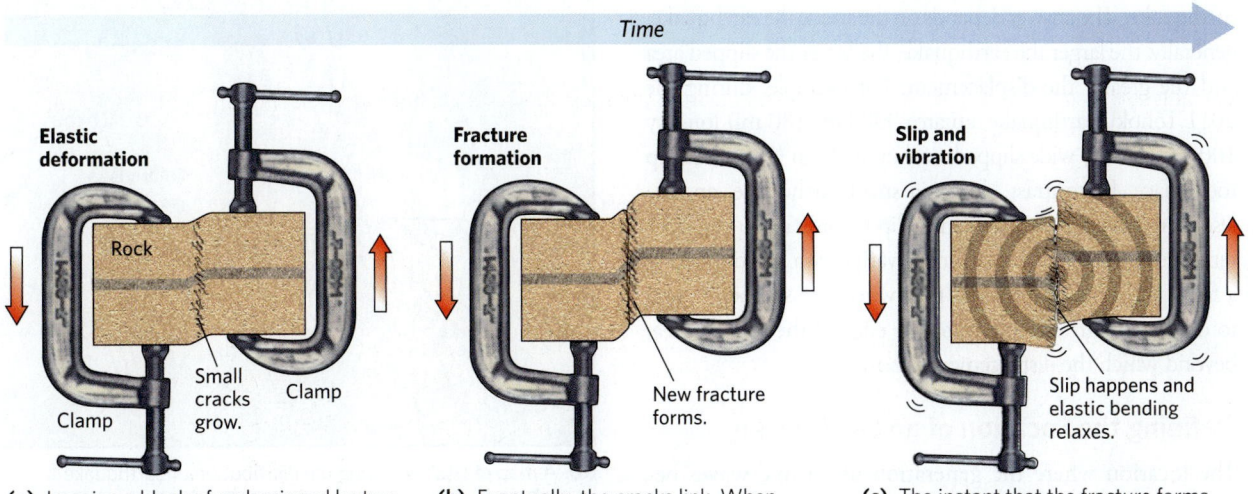

FIGURE 8.5 A model representing the development of a new fault.

Elastic deformation

Rock

Small cracks grow.   Clamp

Clamp

(a) Imagine a block of rock gripped by two clamps. Move one clamp up, and the rock starts to bend. Small cracks develop along the bend.

Fracture formation

New fracture forms.

(b) Eventually, the cracks link. When this happens, a fracture cuts completely through the rock.

Slip and vibration

Slip happens and elastic bending relaxes.

(c) The instant that the fracture forms, the rock breaks into two pieces that slide past each other. The energy that is released generates vibrations (seismic waves).

either side of the fault that generates vibrations, known as *seismic waves*. You've seen the same action when you bend a stick until it snaps. Seismologists refer to this concept as the **elastic-rebound theory**. Note that rebound does not return the rock to its original position, for as we've seen, faulting causes displacement. Furthermore, once a fault forms, it doesn't continue to slip forever because *friction*—the resistance to sliding caused by bumps and irregularities on a surface—eventually slows and stops the movement.

To understand why pre-existing faults slip again, keep in mind that once a fault has formed, it remains weaker than the surrounding, intact crust. So when stress builds up in the crust, slip can take place on a pre-existing fault before the stress becomes large enough to generate a new fault. Notably, a pre-existing fault won't slip again until the stress overcomes the resistance of friction. Therefore, as the magnitude of stress increases, rock on either side of the fault undergoes elastic deformation, so when the pre-existing fault finally slips, elastic rebound takes place. Seismologists refer to the cycle of stress buildup, then sliding and stress release on a pre-existing fault, as **stick-slip behavior**.

The main rupturing event along a fault produces the **mainshock**. Commonly, a number of smaller earthquakes, called **foreshocks**, precede the mainshock. They may be due to the development and growth of cracks in the zone that will become the principal surface of rupturing. In the days to months following a large earthquake, the region affected by that earthquake endures a series of **aftershocks**. The largest aftershock tends to be 10 times smaller than the mainshock, and most are much smaller than that. Aftershocks happen because slip during the mainshock doesn't relax all the elastic deformation in rock adjacent to the fault.

The Area and Amount of Slip during an Earthquake

How large is the area of a fault surface that slips during an earthquake? The answer depends on the size of the earthquake: generally, the larger the earthquake, the larger the slipped area and the greater the displacement. For example, during the 2011 Tōhoku earthquake, an area 300 km (180 mi) long by 100 km (60 mi) wide slipped, and up to 30 m (100 ft) of slip took place. In contrast, during a small earthquake, an area less than a square kilometer may slip by only a few centimeters. The amount of displacement varies with location along a fault; it tends to be greatest near where the slip begins and to die out progressively toward the edge of the slipped area, beyond which the displacement is zero.

Defining the Location of an Earthquake

The location where the generation of seismic waves begins is called the **focus**, or *hypocenter*, of an earthquake **(Fig. 8.6a)**. An earthquake whose focus lies at a depth of less than 70 km (43 mi) is a *shallow-focus earthquake*,

one that occurs at a depth of between 70 km and 300 km (185 mi) is an *intermediate-focus earthquake*, and one whose focus lies between 300 and 660 km (410 mi) is a *deep-focus earthquake*. Because earthquake foci do not lie on the Earth's surface, we can't plot their positions directly

FIGURE 8.6 Earthquake foci and epicenters.

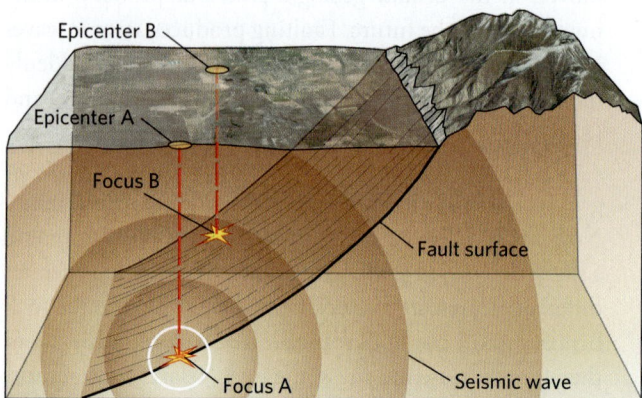

(a) The focus is the point on a fault where slip begins and from which seismic waves begin radiating. The epicenter is the point on the Earth's surface directly above the focus.

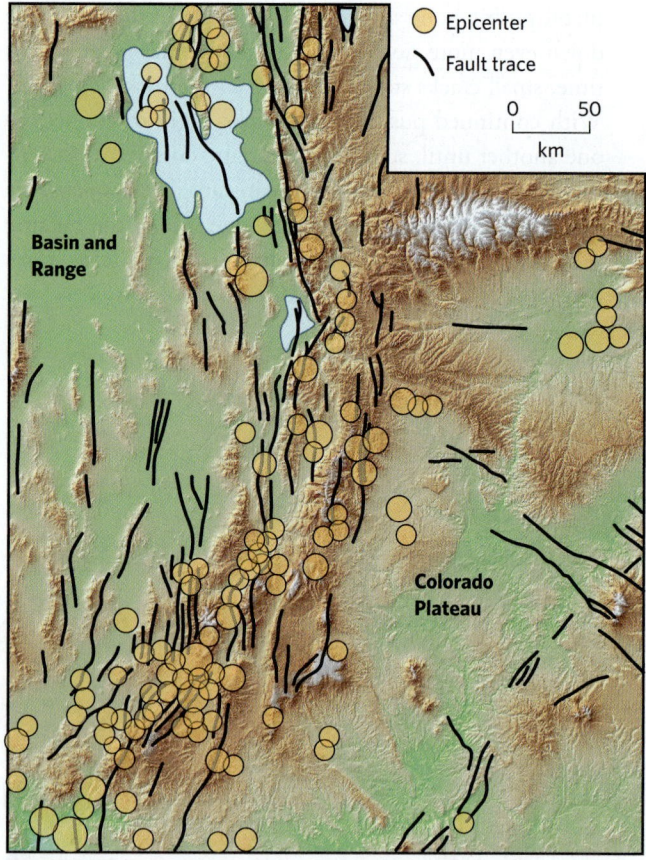

(b) A map of Utah, showing the distribution of earthquake epicenters recorded over several years. The distribution indicates that seismic activity happens mostly in a distinct belt following the boundary between the Basin and Range Province and the Colorado Plateau.

See for yourself

San Andreas fault, near San Francisco, California

Latitude: 37°31′12.66″ N
Longitude: 122°21′41.04″ W

Zoom to 2 km (~1.2 mi) and look NE.

Very narrow lakes fill a narrow valley south of San Francisco. The valley is the trace of the San Andreas fault.

on a map. What you are actually looking at on a map when you see a dot representing the position of an earthquake is the **epicenter** of the earthquake, the point on the surface of the Earth that lies vertically above the focus **(Fig. 8.6b)**.

Take-home message . . .

Most earthquakes happen when stress causes either sudden formation of a new fault or slip on a pre-existing fault. When the slip takes place, elastically deformed rock on either side of the fault rebounds and generates seismic waves. The focus of an earthquake is the point within the Earth where slip begins, and the epicenter is the point on the Earth's surface that is vertically above the focus.

Quick Question ────────────────

How does stick-slip behavior operate?

8.3 Seismic Waves and Their Measurement

The Different Types of Seismic Waves

We've mentioned that the energy produced by slip on a fault moves through the Earth in the form of seismic waves. Seismologists distinguish between two general categories of seismic waves based on where the waves move: **body waves** pass through the interior of the Earth, whereas **surface waves** travel along the Earth's surface.

Body waves cause rock to vibrate in two different ways. Seismologists refer to waves that cause back-and-forth vibrations *parallel* to the direction in which the wave itself moves as **compressional waves (Fig. 8.7a)**. As a compressional wave passes through a material, it first contracts (squeezes together), then dilates (expands). To see this kind of motion in action, push on the end of a spring to compress the coils together, then watch as the pulse of contraction moves along the length of the spring. Body waves that cause up-and-down vibration perpendicular to the direction of wave motion are known as **shear waves (Fig. 8.7b)**. To see shear-wave motion, jerk the end of a rope up and down repeatedly and watch how the up-and-down motion travels along the rope. Surface waves also have two different forms: some cause the ground surface to move up and down in rolling undulations, whereas others cause the ground surface to shimmy back and forth sideways, like a snake **(Fig. 8.7c)**. Seismologists assign names to the different kinds of waves we've just described:

- *P-waves* (for primary) are compressional body waves.
- *S-waves* (for secondary) are shear body waves.

- *R-waves* (for Rayleigh) are surface waves that cause the ground to undulate up and down.
- *L-waves* (for Love) are surface waves that cause the ground to shimmy back and forth.

The different types of seismic waves travel at different velocities. P-waves move fastest, which is why they are called primary waves. S-waves travel at about 60% of the speed of a P-wave, which is why they are called secondary waves. Surface waves are slower than body waves, so both R-waves and L-waves arrive at a location on the Earth's surface substantially after the body waves have arrived there.

Using Seismometers to Record Earthquakes

If you're standing at the epicenter of a significant earthquake, you'll certainly feel it. The same earthquake may be imperceptible to a person farther away. That's because seismic waves weaken as they travel, for Earth materials act like shock absorbers, dampening vibrations as they pass. (You've seen a similar loss in energy if you've thrown a pebble in a pond—the waves get smaller as they move farther away from the point of impact.) In order to detect and study earthquakes of all sizes, even far from the epicenter, seismologists use a **seismometer**, an instrument that can measure the ground motion produced by an earthquake, even if that motion is very tiny.

Seismometers can be configured in two ways: a vertical-motion seismometer detects up-and-down ground motion **(Fig. 8.8a)**, whereas a horizontal-motion seismometer detects back-and-forth ground motion **(Fig. 8.8b)**. The heart of a traditional mechanical seismometer consists of a heavy weight suspended from a spring. The spring, in turn, hangs from a sturdy frame that has been bolted to the ground. A pen extends from the weight and touches a revolving paper-covered cylinder that is also attached to the frame. Before an earthquake, when the ground is steady, the pen traces out a straight reference line on the paper as the cylinder turns, but when a seismic wave arrives and causes the ground surface to move, the seismometer frame—along with the paper-covered cylinder—moves with it **(Fig. 8.8c)**. Because of its *inertia* (the tendency of an object at rest to remain at rest), however, the weight, with its attached pen, remains fixed. As the revolving cylinder moves with respect to the fixed pen, the pen traces a line on the paper that represents the ground motion. If the cylinder were not revolving, the pen would go back and forth, or up and down, in place. But because the paper moves under the pen, the pen traces out a line that looks somewhat like a wave **(Box 8.1)**. Modern electronic seismometers work on the same principle, but the weight is a magnet that moves relative to a wire coil, thereby producing an electrical signal that can be recorded digitally.

FIGURE 8.7 Different types of seismic waves.

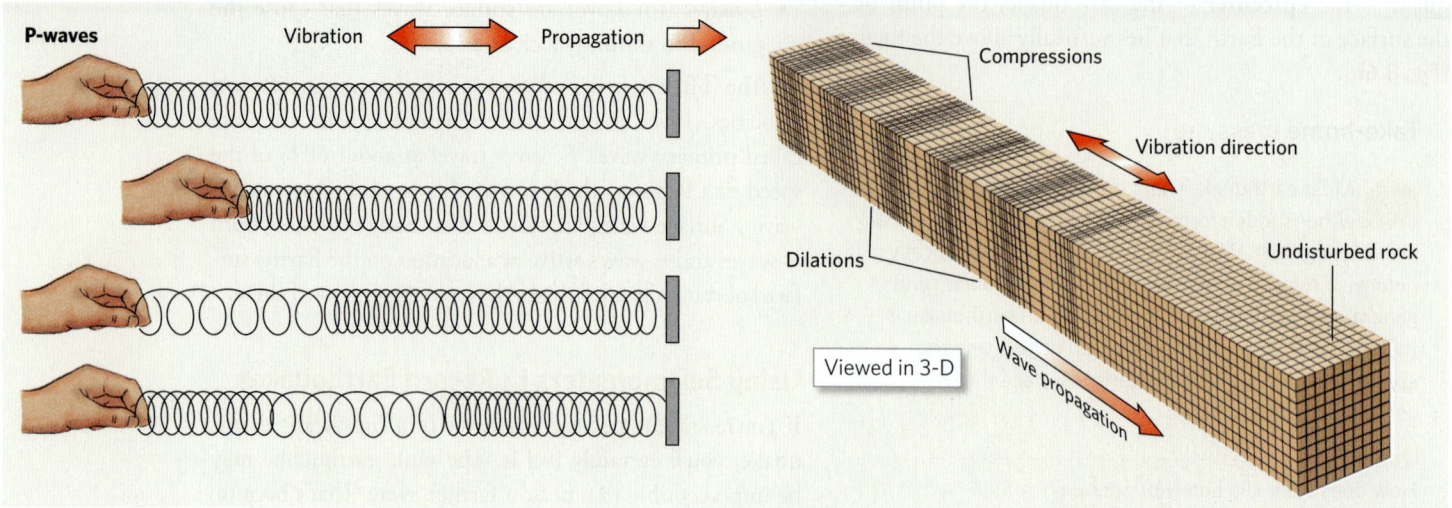

(a) Compressional waves can be generated by pushing and pulling on the end of a spring. The vibration direction is parallel to the direction of wave propagation.

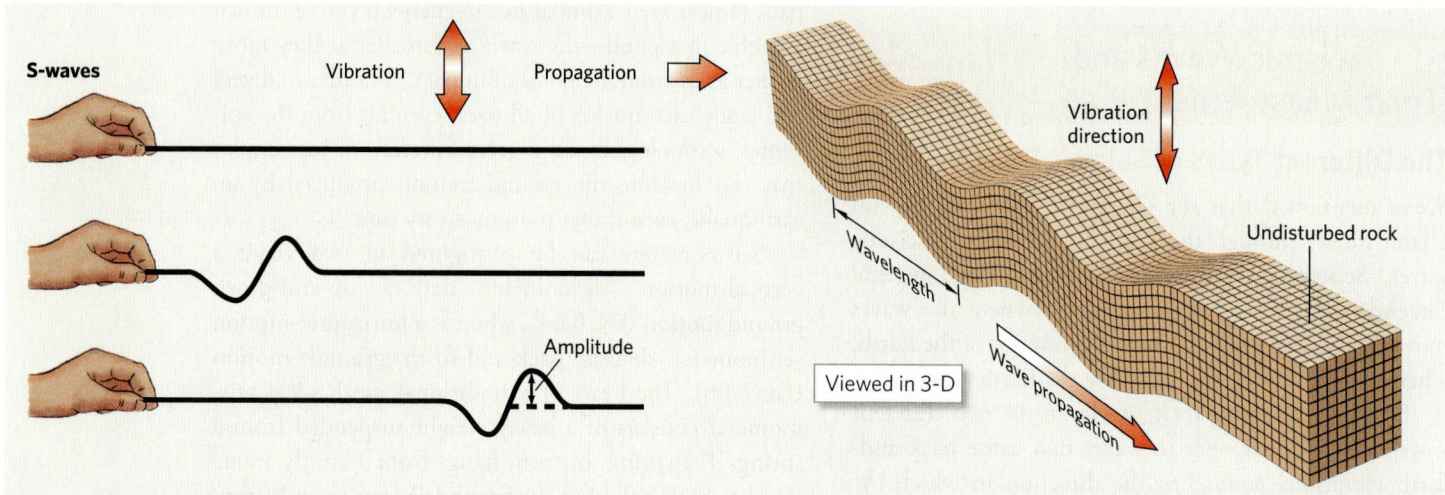

(b) Shear waves can be produced by moving the end of a rope up and down. The vibration direction is perpendicular to the direction of wave propagation.

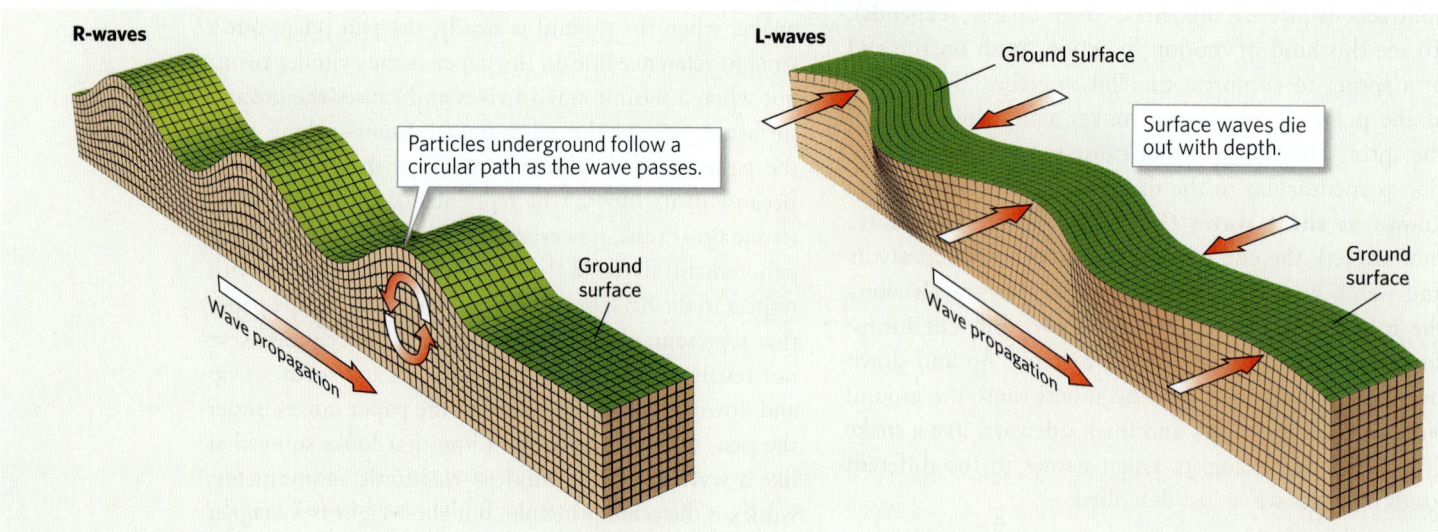

(c) There are two types of surface waves. R-waves make the ground surface go up and down. When an L-wave passes, the ground surface moves back and forth like a slithering snake.

FIGURE 8.8 The basic operation of a seismometer.

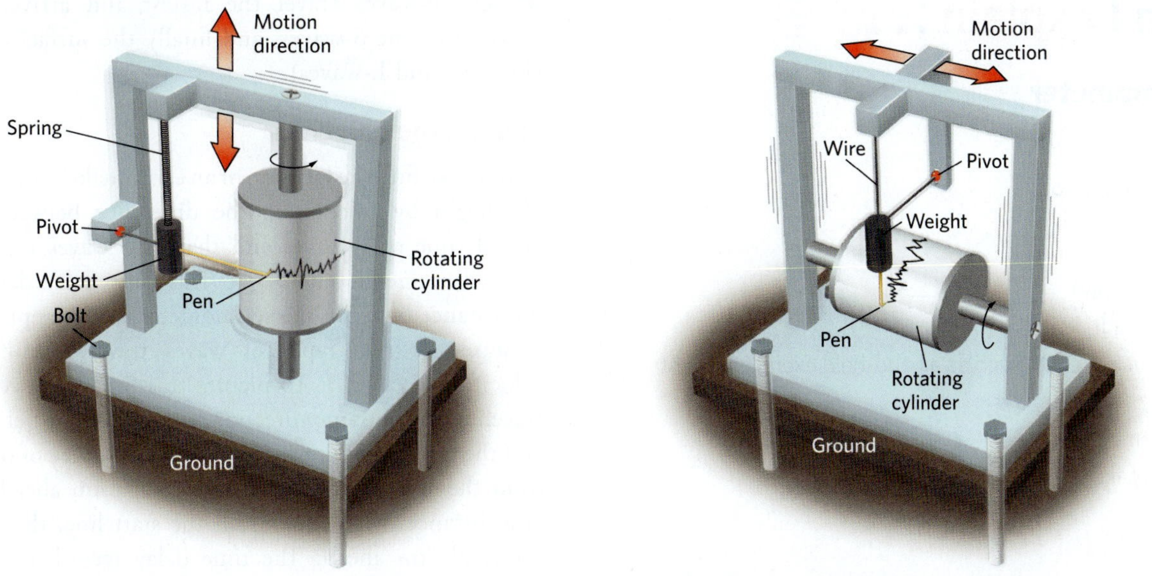

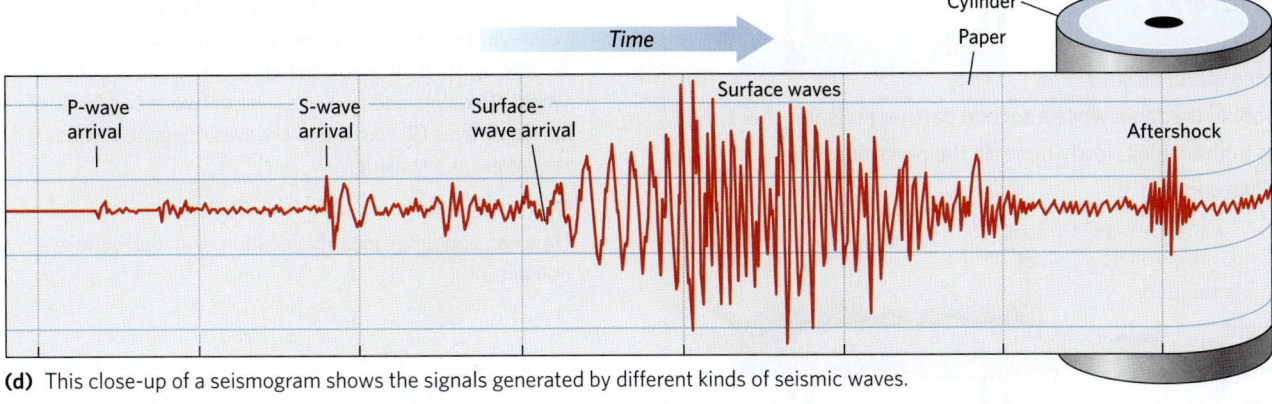

Motion direction

Spring
Pivot
Weight
Bolt
Pen
Rotating cylinder
Ground

(a) A vertical-motion seismometer records up-and-down ground motion.

Motion direction

Wire
Pivot
Weight
Pen
Rotating cylinder
Ground

(b) A horizontal-motion seismometer records back-and-forth ground motion.

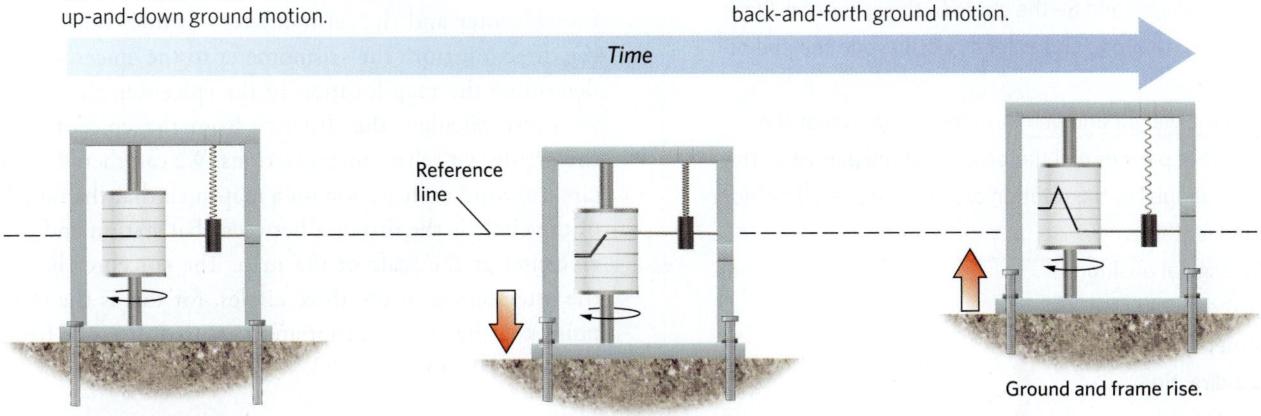

Time

Reference line

Before earthquake

Ground and frame sink.

Ground and frame rise.

(c) Before an earthquake, the pen traces a straight line. During an earthquake, the paper-covered cylinder moves up and down while the pen stays in place.

Time

Cylinder
Paper

P-wave arrival
S-wave arrival
Surface-wave arrival
Surface waves
Aftershock

(d) This close-up of a seismogram shows the signals generated by different kinds of seismic waves.

An earthquake record produced by a seismometer is called a **seismogram** (Fig. 8.8d). At first glance, a typical seismogram looks like a messy squiggle of lines, but to a seismologist it contains a wealth of information.

The horizontal axis on a seismogram represents time, and the vertical axis represents the **amplitude** (technically, one-half the wave height) of seismic waves. We refer to the instant at which a seismic wave appears at

BOX 8.1

How can I explain . . .

The workings of a seismometer

What are we learning?

How a seismometer detects ground motion.

What you need:

- A small table (preferably a bit wobbly)
- A strong string (about 60 cm, or 2 ft, long)
- A 500- to 1,000-g weight (about 1 to 2 lb), such as a hand-held exercise weight
- A medium-point felt-tip pen
- Adhesive tape, or something comparable
- A long sheet of paper; you can make this by taping together five or six 8.5" × 11" sheets

Instructions:

- Tape the pen to one end of the weight, and tie the string to the other end. Place the sheet of paper on the table so that one end extends just beyond the end of the table.
- Hold the string, and suspend the weight and pen from the string so that the pen touches the paper. Have your partner pull the paper past the pen while the table remains stationary. This simulates the rotation of a paper-covered cylinder under a pen when the ground is stable.
- Now return the paper to its original position. This time, keep the paper fixed by taping it to the table. Have your partner wobble the table back and forth in a direction perpendicular to the strip of paper. The wobbling simulates an earthquake. The paper's lack of motion simulates a situation in which the cylinder of a seismometer does not move.

- Repeat the experiment once more, but this time, have one partner wobble the table while a second partner pulls the sheet of paper (no longer taped to the table) slowly beneath the pen's position. The pen traces out wave-like curves.

What did we see?

A seismometer works because the pen, which is connected to a weight, stays fixed due to inertia, while the cylinder and frame of the seismometer move with the shaking Earth. The cylinder must rotate so that the pen's trace doesn't keep overprinting itself.

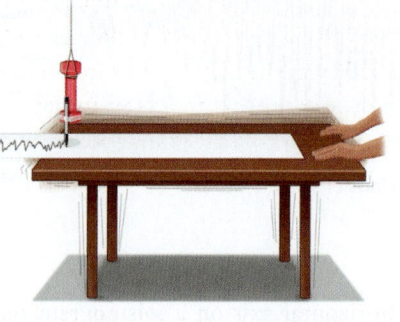

a seismometer station as the **arrival time** of the wave. The first squiggles on the record represent P-waves because P-waves travel the fastest and arrive first. Next come the S-waves, and finally the surface waves (R-waves and L-waves).

Finding the Epicenter

How do we find the location of an earthquake's epicenter? We begin by calculating the difference between the arrival time of P-waves and that of S-waves recorded at a seismometer **(Fig. 8.9a)**. This time delay depends on distance between the seismometer and the epicenter because the two types of waves travel at different velocities **(Fig. 8.9b)**. To picture this, imagine two cars traveling in the same direction, one at 60 km per hour and the other at 50 km per hour. At a distance of 60 km from the start line, the faster car is 10 km ahead, but at a distance of 120 km from the start line, the faster car is 20 km ahead. The time delay recorded at one seismometer, however, tells us only the distance between the epicenter and the seismometer—it does not tell us the direction from the seismometer to the epicenter. To determine the map location of the epicenter, therefore, we must calculate the distance from the epicenter to three different seismometer stations. We can then draw a circle around each station on a map such that the radius of the circle is the distance between that station and the epicenter at the scale of the map. The epicenter lies at the intersection of the three circles, for this is the only point that has the appropriate measured distance from all three stations **(Fig. 8.9c)**.

Take-home message . . .

Earthquake energy travels as seismic waves. Body waves travel through the interior of the Earth, whereas surface waves travel along the ground surface. Seismologists distinguish between two kinds of body waves (P-waves and S-waves) and between two kinds of surface waves (R-waves and L-waves). Seismometers record ground shaking.

Quick Question

How can you determine the location of an earthquake's epicenter?

8.4 Defining the Size of Earthquakes

During a typical earthquake, the ground shakes when body waves coming up through the solid Earth reach the surface, as well as when surface waves moving along the ground surface pass by. At a given location, some

FIGURE 8.9 Locating an earthquake's epicenter.

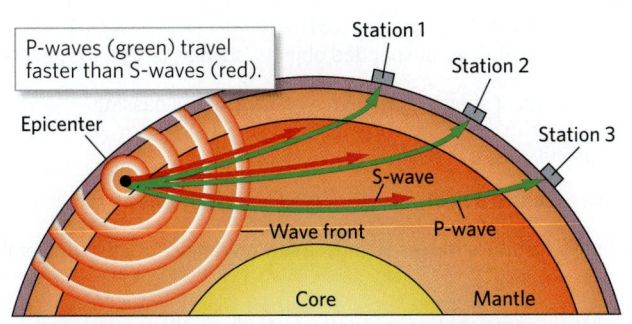

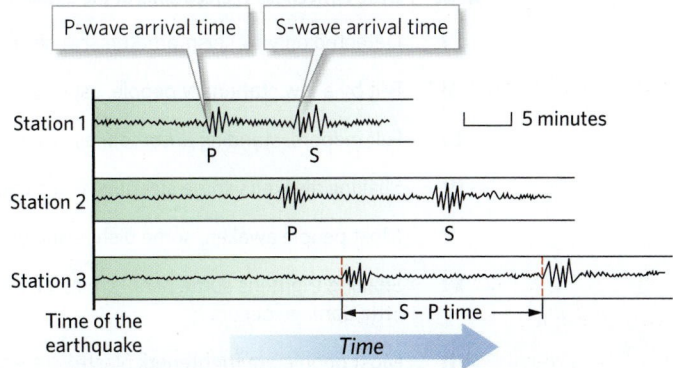

(a) Seismic waves of different types travel at different velocities. The greater the distance between the epicenter and the seismometer station, the longer it takes for seismic waves to arrive there, and the greater the delay between the P-wave and S-wave arrival times. This difference is called the S − P time.

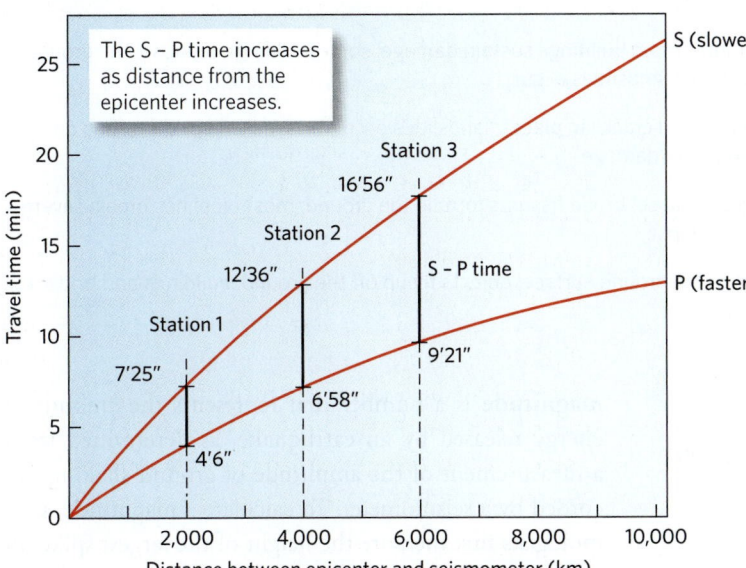

(b) We can represent the different arrival times of P-waves and S-waves on a graph. The red lines are called travel-time curves. The S − P time at a given distance from the epicenter is represented by the vertical distance between the S-wave travel-time curve and the P-wave travel-time curve.

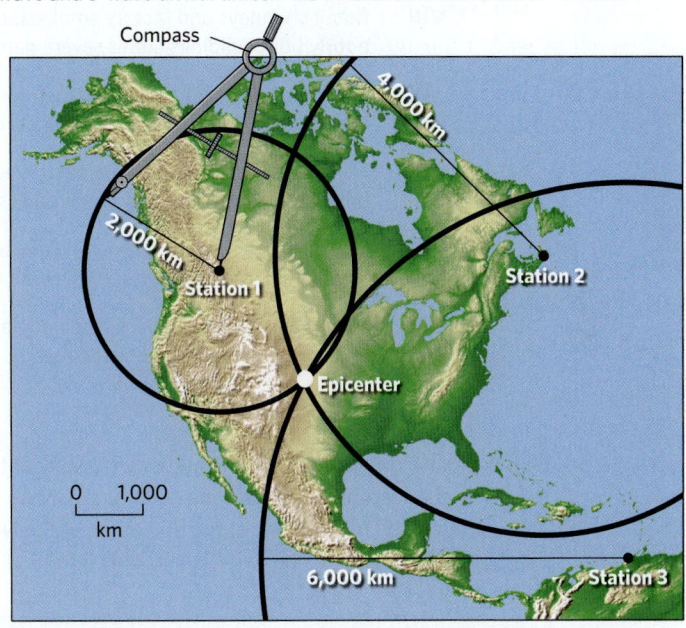

(c) If an earthquake epicenter lies 2,000 km from Station 1, we draw a circle with a radius of 2,000 km around that station at the scale of the map. We repeat for the other two stations. The intersection of the three circles is the epicenter.

earthquakes are "large," in that they shake the ground violently, whereas others are "small," in that they can barely be felt. Seismologists have developed two different scales—the intensity scale and the magnitude scale—to define earthquake size in a uniform way so that we can compare different earthquakes.

Modified Mercalli Intensity Scale

Earthquake **intensity** refers to the degree of ground shaking at a locality. In 1902, an Italian scientist named Giuseppe Mercalli devised a scale for defining earthquake intensity by assessing both the damage that the earthquake caused and people's perception of the shaking. A version of this scale, called the **Modified Mercalli**

Intensity (MMI) scale, continues to be used today **(Table 8.1)**. We represent earthquake intensities on this scale by Roman numerals. Because earthquake waves lose energy as they travel, the intensity value tends to be greatest near the epicenter and to decrease progressively away from the epicenter. Seismologists can draw a map to show how the intensity of an earthquake varies over a region **(Fig. 8.10)**.

Earthquake Magnitude Scales

When you hear a report of an earthquake disaster in the news, you will probably encounter a phrase like, "An earthquake with a magnitude of 7.2 struck the city yesterday." What does this phrase mean? Earthquake

 Animation

Earthquake Waves and Epicenter Location (Interactive)

TABLE 8.1 Modified Mercalli Intensity Scale

MMI	Destructiveness (Perceptions of the Extent of Shaking and Damage)
I	Detected only by seismometers; causes no damage.
II	Felt by a few stationary people, especially in upper floors of buildings; suspended objects, such as lamps, may swing.
III	Felt indoors; standing automobiles sway on their suspensions; it seems as though a heavy truck is passing.
IV	Shaking awakens some sleepers; dishes and windows rattle.
V	Most people awaken; some dishes and windows break; unstable objects tip over; trees and poles sway.
VI	Shaking frightens some people; plaster walls crack; heavy furniture moves slightly; a few chimneys crack, but overall, little damage occurs.
VII	Most people are frightened; plaster cracks; windows break; some chimneys topple; unstable furniture overturns; poorly built buildings sustain considerable damage.
VIII	Many chimneys and factory smokestacks topple; heavy furniture overturns; substantial buildings sustain damage; poorly built buildings suffer severe damage.
IX	Frame buildings separate from their foundations; most buildings sustain damage; some buildings collapse; the ground cracks; underground pipes break; rails bend; some landslides occur.
X	Most masonry structures are destroyed; the ground cracks in places; landslides occur; bridges collapse; facades on buildings collapse; railways and roads suffer severe damage.
XI	Few masonry buildings remain; many bridges collapse; broad fissures form in the ground; most pipelines break; severe liquefaction of sediment occurs; some dams collapse.
XII	Earthquake waves cause visible undulations of the ground surface; objects fly up off the ground; buildings and bridges of all types are completely destroyed.

FIGURE 8.10 Modified Mercalli Intensity (MMI) contours for the 1886 Charleston, South Carolina earthquake. Note that ground shaking reached an MMI of X at the epicenter, but in New York City, ground shaking reached an MMI of only II to III.

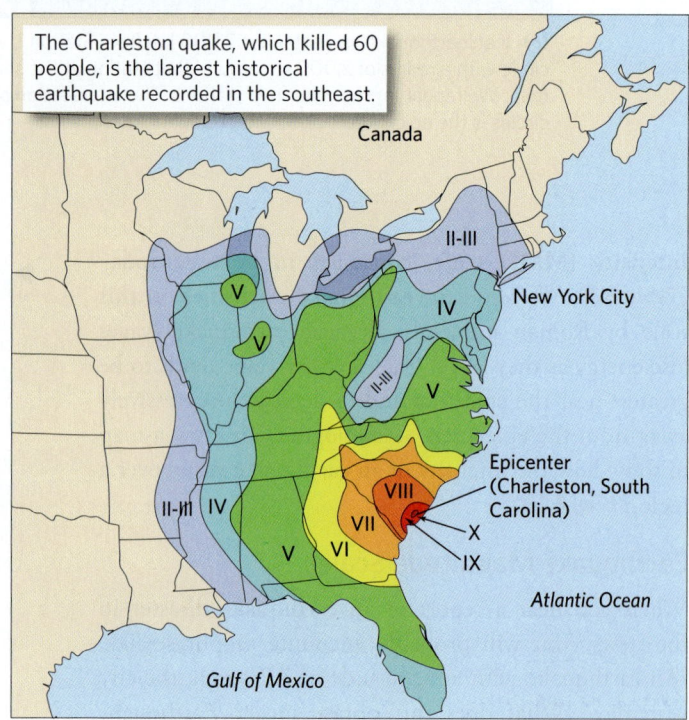

The Charleston quake, which killed 60 people, is the largest historical earthquake recorded in the southeast.

magnitude is a number that represents the amount of energy released by an earthquake, as determined from a measurement of the amplitude of ground shaking, recorded by a seismometer. To calculate a magnitude, seismologists first measure the height of the largest spike on a seismogram, for it represents the maximum amplitude of the ground motion. Then, after they have determined the distance between the epicenter and the seismometer, they adjust the measurement to be equivalent to the maximum amplitude that would be recorded by a seismometer positioned a certain distance—the *reference distance*—from the epicenter. Because of this adjustment, seismologists obtain the same magnitude for a particular earthquake from a measurement at any seismometer. In other words, a given earthquake has only one magnitude number, so, unlike intensity, magnitude does not depend on distance from the epicenter. We specify magnitude (*M*) using Arabic numerals.

In 1935, an American seismologist, Charles Richter, developed a scale for defining earthquake magnitude. This scale, now known as the **Richter scale**, is logarithmic, meaning that an increase of one unit of magnitude represents a tenfold increase in the maximum amplitude of ground motion. So, a magnitude 8 earthquake results in ground motion that is 10 times greater than that of a magnitude 7 earthquake and 1,000 times greater than

FIGURE 8.11 Using the Richter magnitude scale.

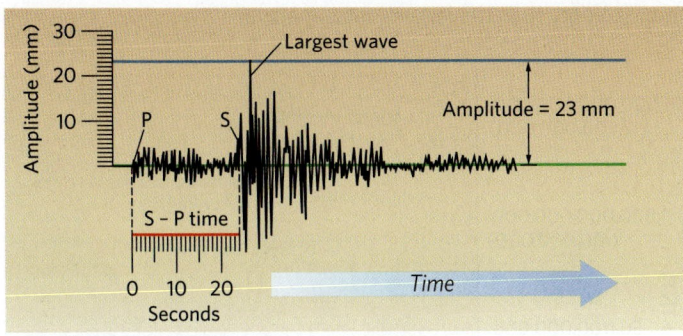

(a) To calculate the Richter magnitude from a seismogram, we first measure the amplitude of the largest wave. We then calculate the difference between the arrival times of S-waves and P-waves (S − P time) to determine the distance to the epicenter (see Fig. 8.9).

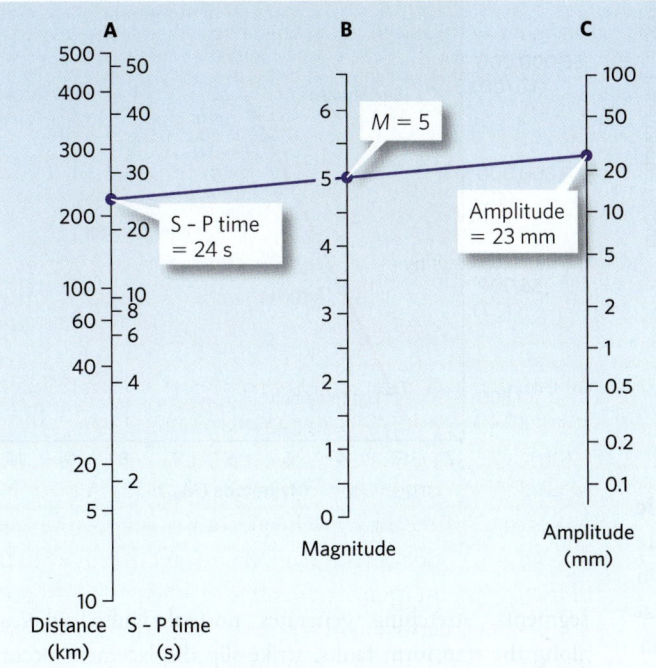

(b) Next we draw a line from the point on Column A representing the S − P time (or distance to the epicenter) to the point on Column C representing the wave amplitude. We can then read the Richter magnitude off Column B.

TABLE 8.2 Adjectives for Describing Earthquakes

Adjective	Magnitude	Intensity at Epicenter	Effects
Great	>8.0	X to XII	Total destruction
Major	7.0 to 7.9	IX to X	Extreme damage
Strong	6.0 to 6.9	VII to VIII	Moderate to serious damage
Moderate	5.0 to 5.9	VI to VII	Slight to moderate damage
Light	4.0 to 4.9	IV to V	Felt by most; slight damage
Minor	<3.9	III or smaller	Felt by some; hardly any damage

that of a magnitude 5 earthquake. Richter used 100 km (62 mi) as the reference distance. Since there's not necessarily a seismometer at exactly 100 km from an epicenter, he developed a simple chart to adjust for the distance of a seismometer station from the epicenter (Fig. 8.11).

These days, seismologists use a somewhat different, more accurate scale, called the **moment magnitude scale**, to represent an earthquake's size. To calculate an earthquake's moment magnitude (abbreviated M_w), seismologists take into account the amplitudes of several different seismic waves, the dimensions of the slipped area on the fault, and the displacement that occurred. The moment magnitude scale, like the Richter scale, is logarithmic.

To make discussion of earthquakes easier, seismologists use familiar adjectives to describe an earthquake's magnitude (Table 8.2). Earthquakes that most people can feel have a magnitude greater than M_w 4, and earthquakes that can cause moderate damage have a magnitude greater than M_w 5. The catastrophic 2011 Tōhoku earthquake registered as M_w 9.0, and the largest recorded earthquake in history, the 1960 Chilean quake, registered as M_w 9.5. News reporters sometimes incorrectly state that the magnitude scale "goes from 1 to 10." In fact, microearthquakes (M_w –1 or M_w –2) can be detected by seismometers positioned close to the epicenter, and there is no defined upper limit to the magnitude scale. That said, seismologists estimate that M_w 9.5 is about as big as an earthquake can get, given the known dimensions of faults on the Earth.

Energy Release by Earthquakes

To give a sense of the amount of energy released by an earthquake, seismologists compare earthquakes to other energy-releasing events. For example, according to some estimates, an M_w 5.3 earthquake releases about as much energy as the Hiroshima atomic bomb, and an M_w 9.0 earthquake releases significantly more energy than the largest hydrogen bomb ever detonated. Notably, although an increase of one unit of magnitude represents a tenfold increase in the maximum amplitude of ground motion, it represents a 32-fold increase in energy release. Therefore, an M_w 8 earthquake releases about 1 million times more energy than an M_w 4 earthquake (Fig. 8.12). Fortunately, very large earthquakes occur much less frequently than small earthquakes.

Take-home message . . .

 We can specify earthquake size by intensity (a measure based on perception of damage caused and shaking felt) or by magnitude (a representation of energy released by the earthquake, based on a measurement of ground motion). An increase of one magnitude unit represents a tenfold increase in the amplitude of shaking and a 32-fold increase in energy. A great earthquake releases vastly more energy than a giant hydrogen bomb.

Quick Question

Can you specify the size of an earthquake by giving just one intensity number? How about a single magnitude number?

Did you ever wonder . . .

whether an earthquake will happen near where you live?

▶ **Animation**
Earthquake Basics

8.5 Where and Why Do Earthquakes Occur?

Earthquakes do not take place everywhere on the globe. By plotting the distribution of earthquake epicenters on a map, seismologists have found that most, but not all, earthquakes occur in **seismic belts** (Fig. 8.13). Most seismic belts correspond to plate boundaries, and earthquakes within these belts are known as *plate-boundary earthquakes*. Those earthquakes that occur away from plate boundaries are called *intraplate earthquakes*.

Plate-Boundary Earthquakes

Plates move relative to their neighbors at rates of 1 to 15 cm (0.4 to 6 in) per year. So, over a period of decades to centuries, large stresses build along plate boundaries, and these stresses cause sudden slip on faults. As we saw in Chapter 7, geologists distinguish among different types of faults. The fault type that occurs at a given plate boundary depends on the relative motion across that boundary (**Earth Science at a Glance**, pp. 236–237).

DIVERGENT-BOUNDARY SEISMICITY. At a divergent boundary (a mid-ocean ridge), two plates form and move apart. Divergent boundaries are broken into ridge segments linked by transform boundaries. Therefore, two kinds of faults develop at divergent boundaries: along the ridge

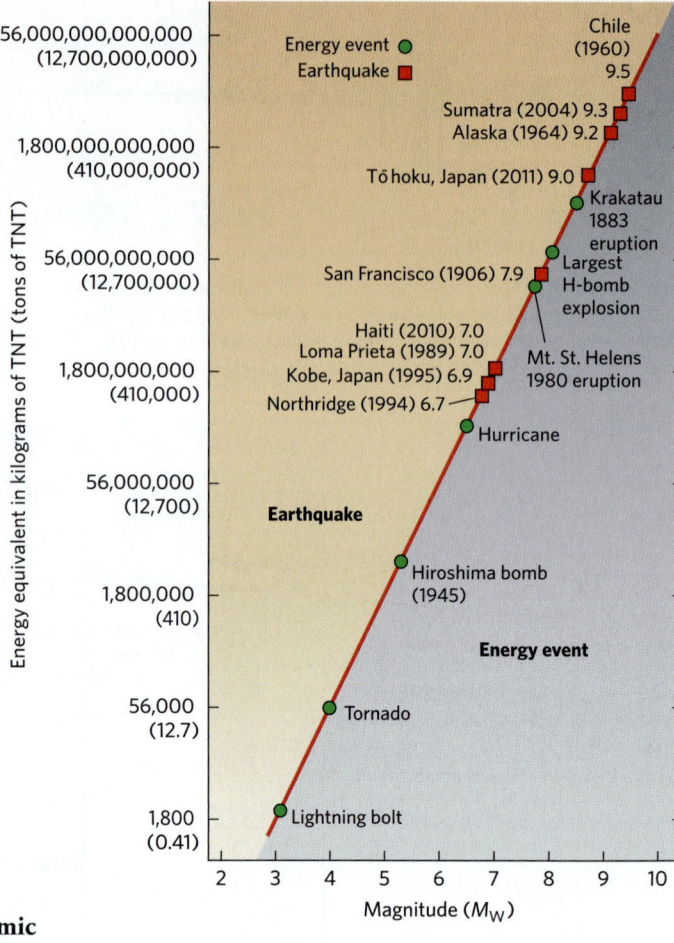

FIGURE 8.12 The energy released by earthquakes increases dramatically with magnitude. Great earthquakes release vastly more energy than the largest nuclear bombs.

segments, stretching generates normal faults, whereas along the transform faults, strike-slip displacement occurs (Fig. 8.14). Earthquakes on these faults have shallow foci. Most of these earthquakes are not responsible for major damage because they occur far from population centers.

TRANSFORM-BOUNDARY SEISMICITY ON CONTINENTS. At transform boundaries, strike-slip displacement takes place. The majority of transform faults in the world link segments of mid-ocean ridges, as we've seen. But a few, such as the San Andreas fault of California, the Alpine fault of New Zealand, and the Anatolian faults in Turkey, cut through continental crust. Large earthquakes on continental transform faults can be very destructive, for they have shallow foci. This means that the seismic waves they produce can still carry a large amount of energy when they reach the ground surface. Further, because they're on land, they may lie near population centers.

FIGURE 8.13 A map of epicenters over a period of several decades emphasizes that most earthquakes occur in distinct seismic belts along plate boundaries. The color of an epicenter dot represents the depth of the focus of the earthquake.

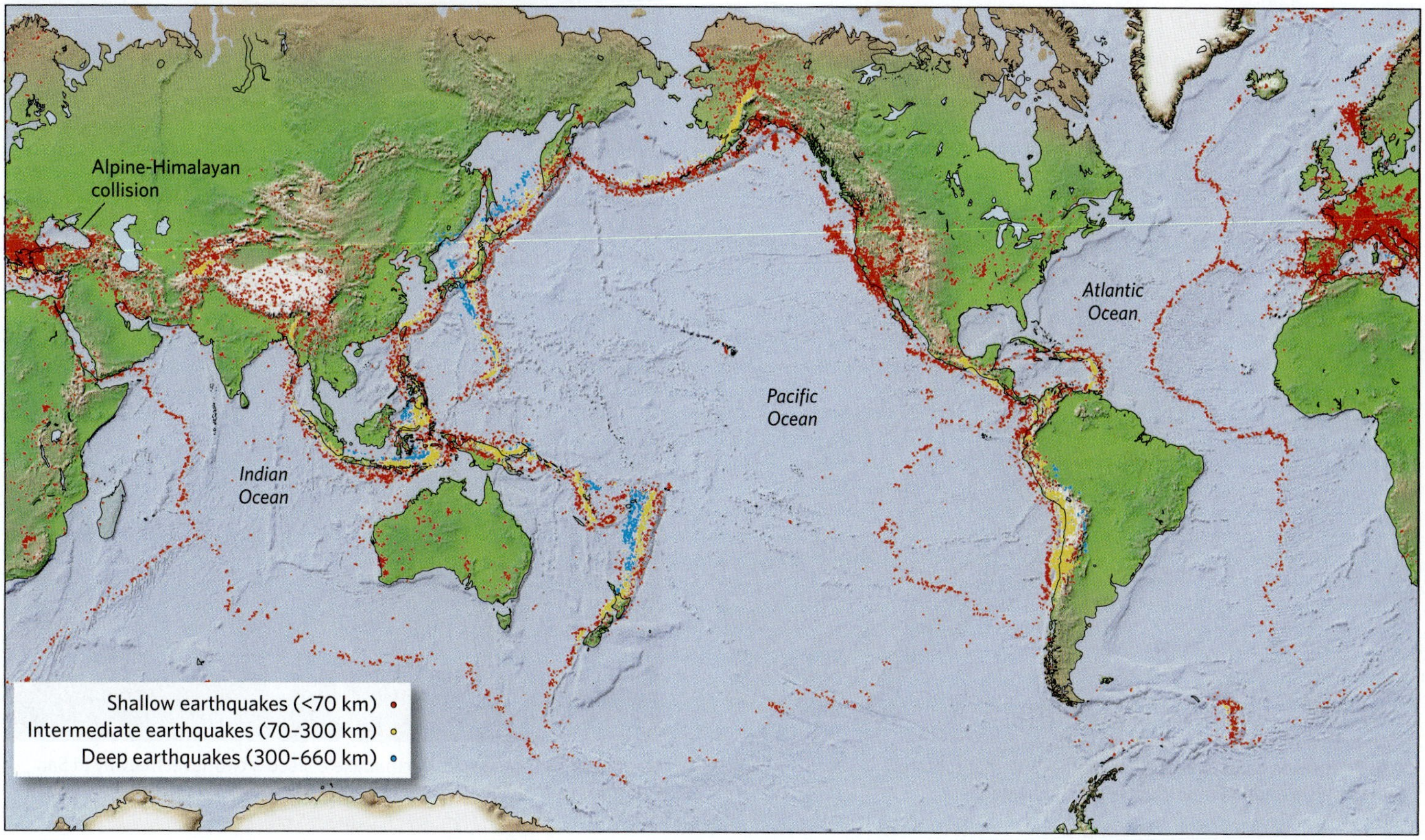

Alpine-Himalayan collision

Atlantic Ocean

Pacific Ocean

Indian Ocean

- Shallow earthquakes (<70 km) •
- Intermediate earthquakes (70–300 km) ○
- Deep earthquakes (300–660 km) •

Earthquakes along the San Andreas fault, on which the Pacific Plate slides past the North American Plate at an average rate of 6 cm (2.4 in) per year, typify the nature of seismic activity on a transform boundary. Stick-slip behavior causes numerous earthquakes to happen along this fault and on subsidiary faults near it (**Fig. 8.15a**). Some of these have been large and destructive. In 1906, for example, an M_w 7.9 earthquake struck San Francisco when a 477-km (300-mi)-long segment of the fault suddenly slipped by up to 7 m (23 ft). As streets in the city undulated, buildings swayed and collapsed. Toppled stoves and lamps sparked a fire that consumed the ruins (**Fig. 8.15b**). The largest earthquake near San Francisco since then, an M_w 6.9 event, struck Loma Prieta, 100 km to the south, in 1989. This event caused the tragic collapse of a double-decked freeway and disrupted a World Series baseball game (**Fig. 8.15c**).

CONVERGENT-BOUNDARY SEISMICITY. Convergent boundaries are complicated regions where many earthquakes at a variety of depths take place (**Fig. 8.16a**). The most dangerous

FIGURE 8.14 The distribution of earthquakes at a divergent boundary. Note that normal faults occur along the ridge axis and strike-slip faults occur along active transform faults. Earthquakes do not occur along inactive fracture zones.

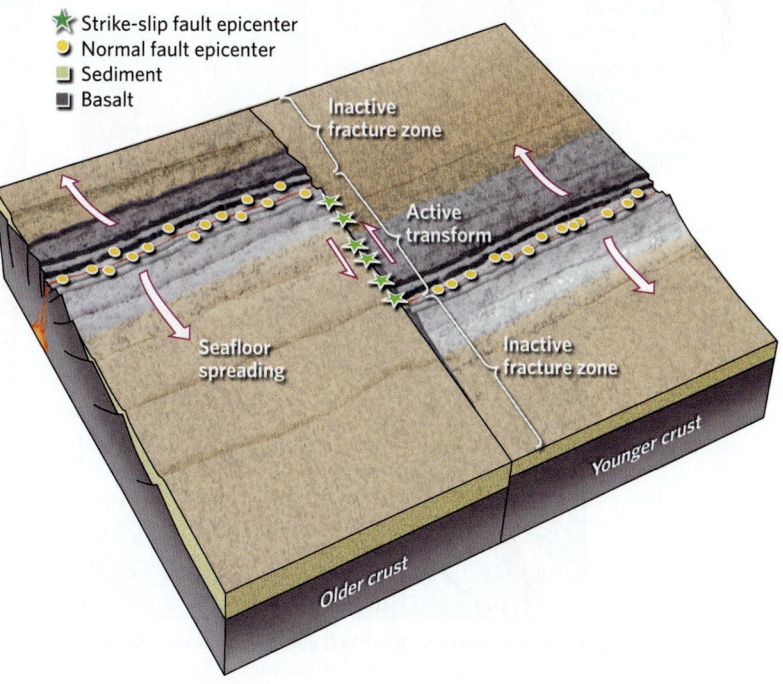

☆ Strike-slip fault epicenter
● Normal fault epicenter
▢ Sediment
▩ Basalt

Inactive fracture zone

Active transform

Seafloor spreading

Inactive fracture zone

Younger crust

Older crust

FIGURE 8.15 Seismicity on a continental transform fault.

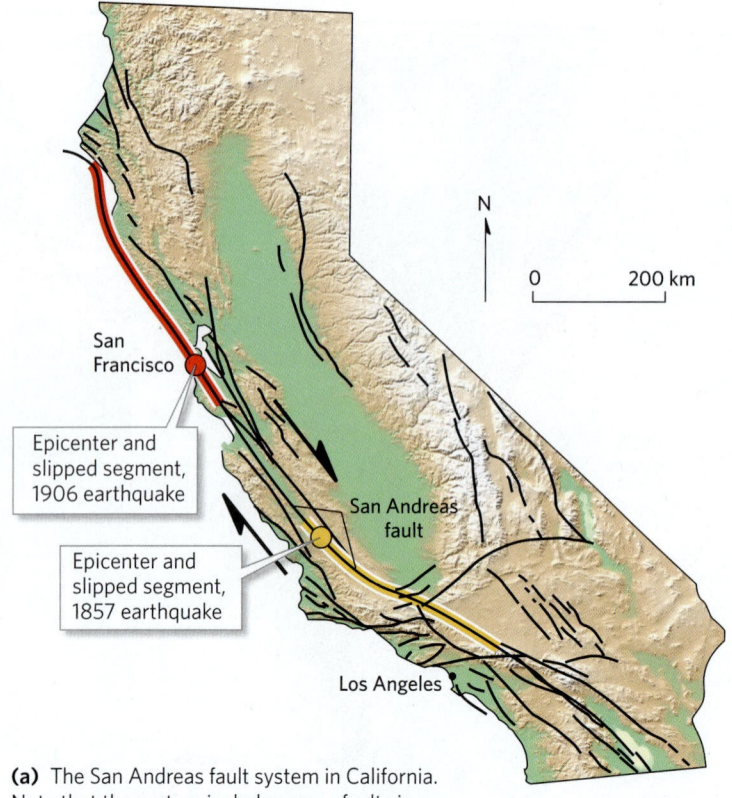

(a) The San Andreas fault system in California. Note that the system includes many faults in a 100-km-wide band. The portions of the fault that ruptured in the 1906 and 1857 earthquakes are indicated.

San Francisco

Epicenter and slipped segment, 1906 earthquake

Epicenter and slipped segment, 1857 earthquake

San Andreas fault

Los Angeles

N

0   200 km

(b) A street in San Francisco after the 1906 earthquake.

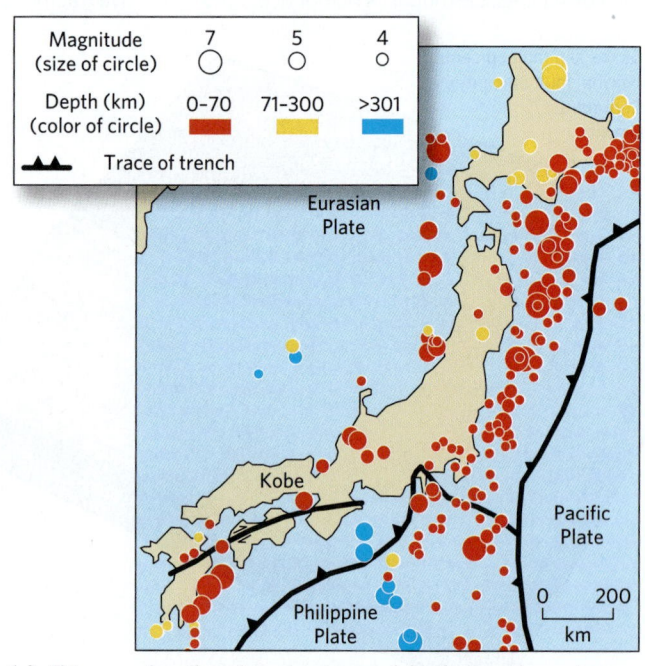

(c) The 1989 Loma Prieta earthquake caused a two-level freeway in San Francisco to collapse.

FIGURE 8.16 Convergent-boundary seismicity.

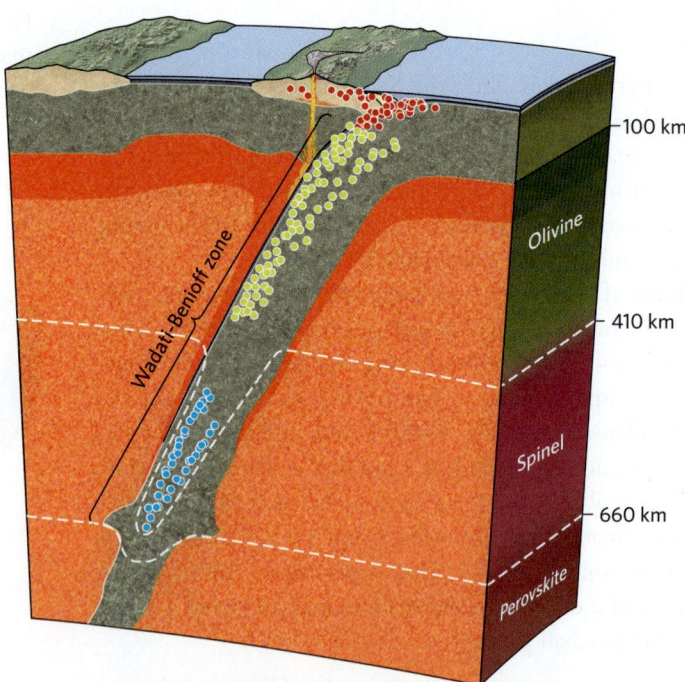

Magnitude (size of circle): 7   5   4

Depth (km) (color of circle): 0–70   71–300   >301

Trace of trench

Eurasian Plate

Pacific Plate

Philippine Plate

Kobe

0   200
km

(a) This map of earthquake epicenters and depths in and near Japan shows how complex convergent boundaries can be.

Wadati-Benioff zone

Olivine

Spinel

Perovskite

100 km
410 km
660 km

(b) At a convergent boundary, shallow earthquakes (red dots) occur along the contact between the downgoing plate and the overriding plate, as well as in the two plates themselves. Intermediate-focus (yellow dots) and deep-focus (blue dots) earthquakes define the Wadati-Benioff zone.

FIGURE 8.17 A composite diagram portraying the geologic settings in which earthquakes occur in continental lithosphere. (Subduction-related earthquakes in continental crust are not shown.)

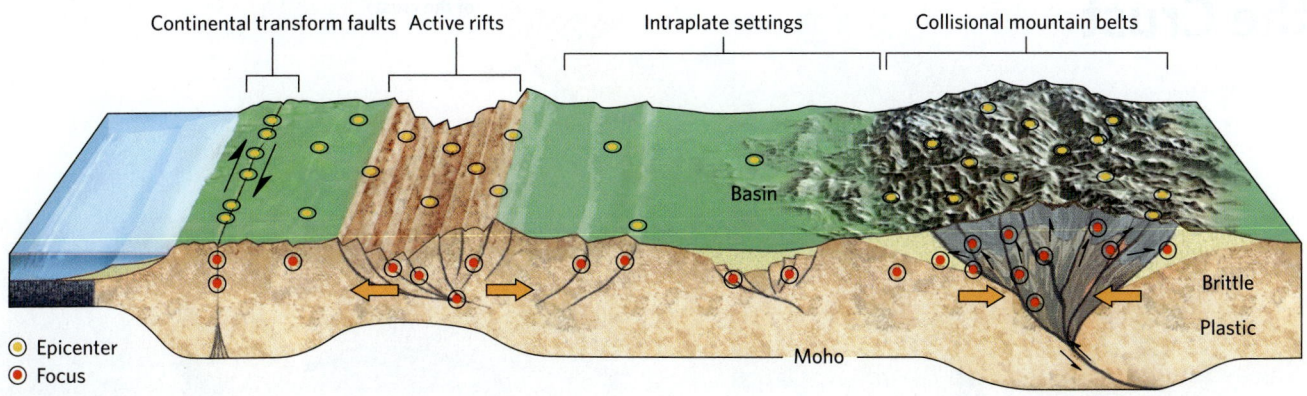

ones occur along relatively shallow thrust faults that delineate the boundary between the base of the overriding plate and the top of the subducting plate. Because the foci of these earthquakes are so shallow, most of the seismic energy produced by elastic rebound reaches the ground surface. Notable examples of devastating convergent-boundary earthquakes include the largest recorded earthquakes: the 1960 M_w 9.5 earthquake in Chile, the 1964 M_w 9.2 Good Friday earthquake in Alaska, the 2004 M_w 9.3 earthquake in Sumatra, and the 2011 M_w 9.0 Tōhoku earthquake in Japan.

Unlike divergent or transform boundaries, convergent boundaries also host intermediate- and deep-focus earthquakes. A plot of the foci of such earthquakes on a cross section through the Earth defines a sloping band of seismicity called a *Wadati-Benioff zone* (named for the seismologists who first recognized it), which extends to a depth of 660 km (410 mi). Earthquakes of this zone occur within subducted lithosphere as it sinks down through the asthenosphere **(Fig. 8.16b)**. The cause of these earthquakes remains a topic of research.

Earthquakes within Continents

Not all earthquakes on continents are associated with plate boundaries. Here, we consider a few other geologic settings within continents in which earthquakes occur **(Fig. 8.17)**.

CONTINENTAL RIFTS. The stretching of crust at a continental rift generates normal faults, and slip on these faults produces earthquakes. Active rifts today include the East African Rift, the Basin and Range Province, and the Rio Grande Rift of New Mexico. In all these places, shallow-focus earthquakes occur, some of which cause major damage.

CONTINENTAL COLLISION ZONES. In 2015, a large earthquake shook the ground in the mountainous landscape of Nepal **(Fig. 8.18)**. Monuments that had stood for

FIGURE 8.18 The 2015 Nepal earthquake destroyed many buildings. For example, this elaborate brick tower, built in 1832, crumbled to rubble.

Faulting in the Crust

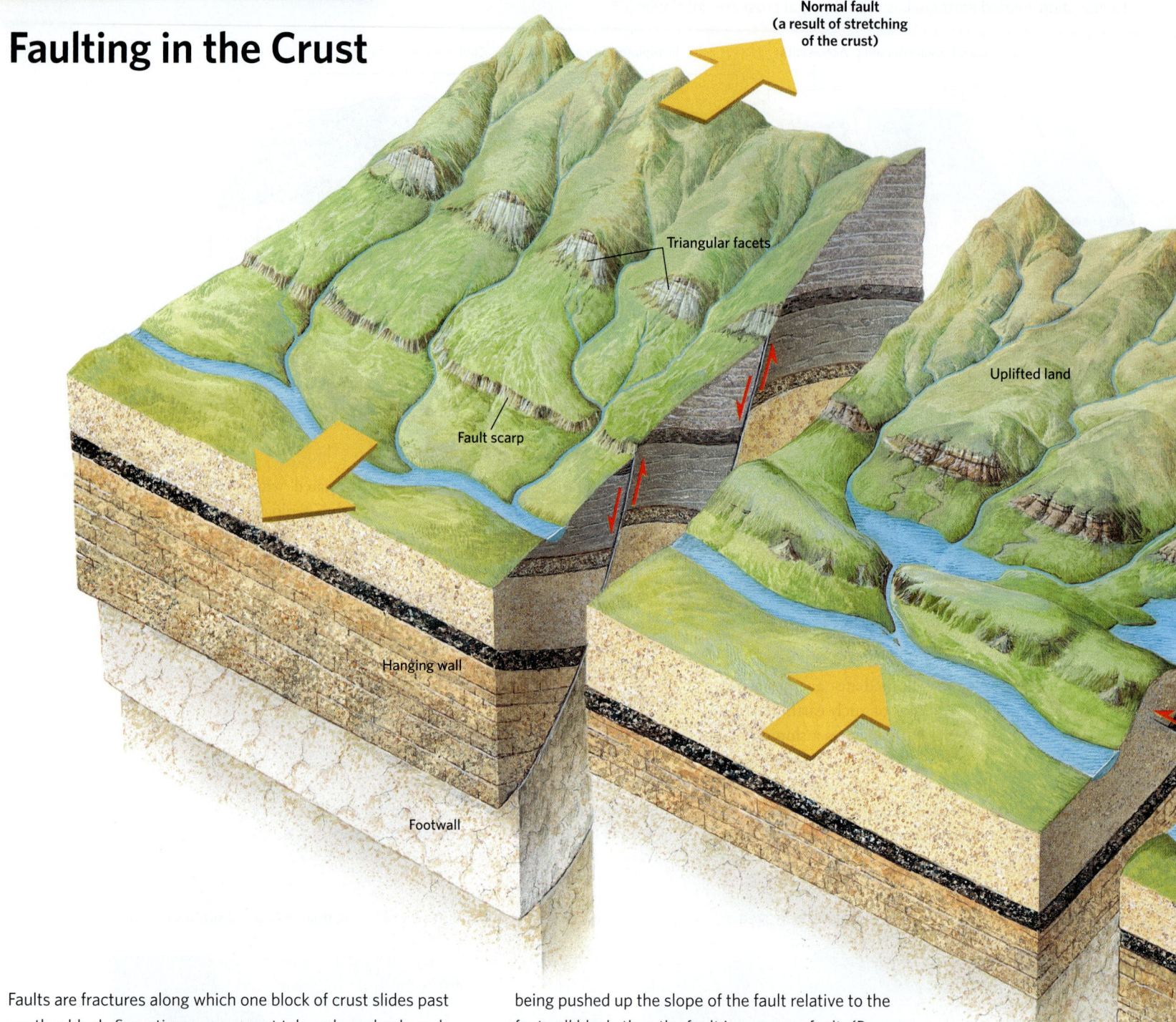

Normal fault
(a result of stretching
of the crust)

Triangular facets

Fault scarp

Uplifted land

Hanging wall

Footwall

Faults are fractures along which one block of crust slides past another block. Sometimes movement takes place slowly and smoothly, without earthquakes, but other times the movement happens suddenly, and rocks break as a consequence. The sudden breaking of rock sends seismic waves through the crust, creating vibrations at the Earth's surface—an earthquake.

Geologists recognize three types of faults. If the hanging-wall block (the rock above a fault plane) slides down the fault's slope relative to the footwall block (the rock below the fault plane), the fault is a normal fault. (Normal faults form where the crust is being stretched apart, as in a continental rift.) If the hanging-wall block is

being pushed up the slope of the fault relative to the footwall block, then the fault is a reverse fault. (Reverse faults develop where the crust is being compressed or squashed, as in a collisional mountain belt.) If one block of rock slides past another and there is no up or down motion, the fault is a strike-slip fault. Strike-slip fault planes tend to be nearly vertical. If a fault displaces the ground surface, it creates a ledge called a fault scarp. Where a fault scarp cuts a system of rivers and valleys, it truncates ridges and produces triangular facets. Strike-slip faults may offset ridges and streams sideways.

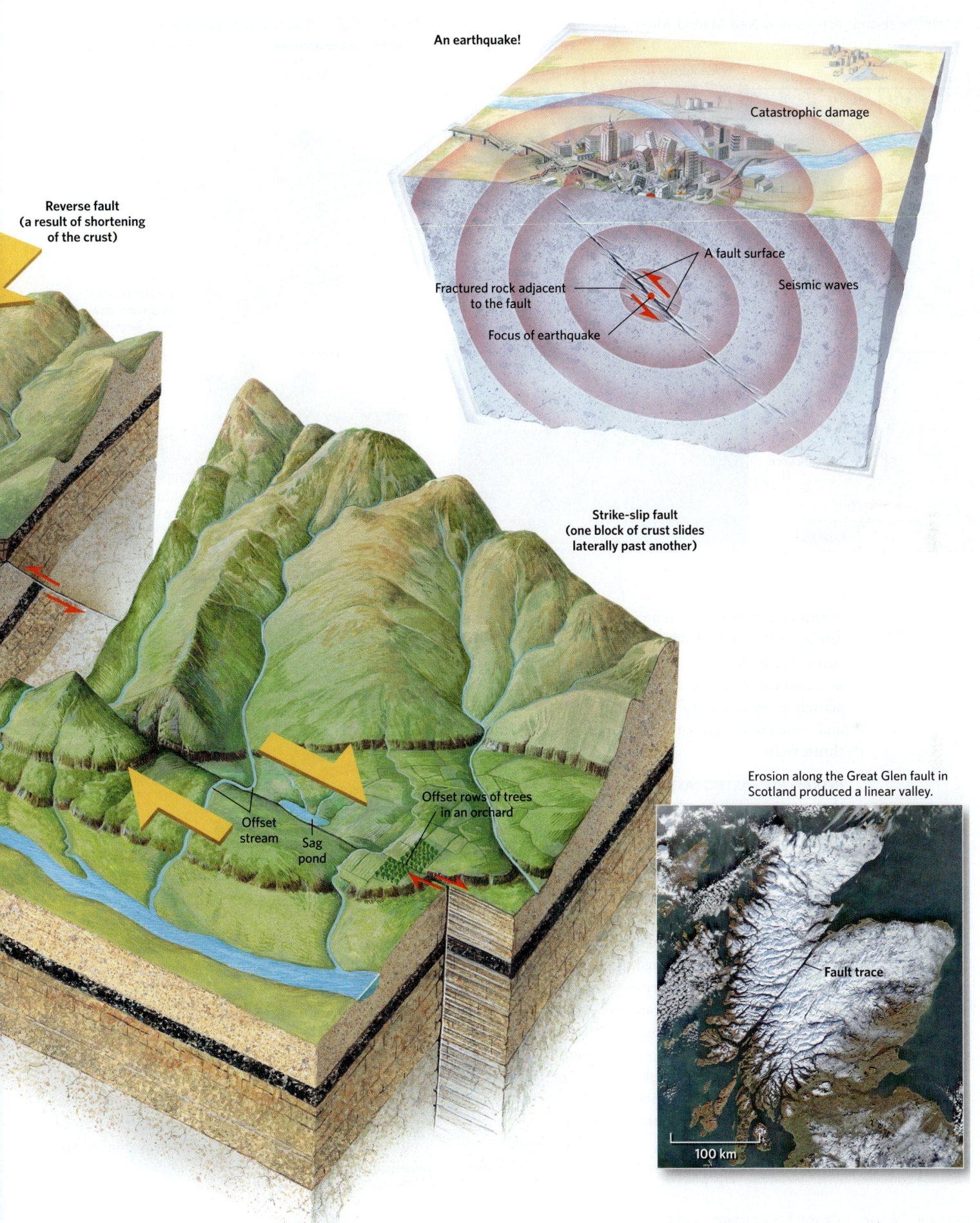

An earthquake!

Catastrophic damage

Reverse fault
(a result of shortening
of the crust)

A fault surface

Fractured rock adjacent
to the fault

Seismic waves

Focus of earthquake

Strike-slip fault
(one block of crust slides
laterally past another)

Offset rows of trees
in an orchard

Offset
stream

Sag
pond

Erosion along the Great Glen fault in
Scotland produced a linear valley.

Fault trace

100 km

FIGURE 8.19 Intraplate seismic activity near New Madrid, Missouri.

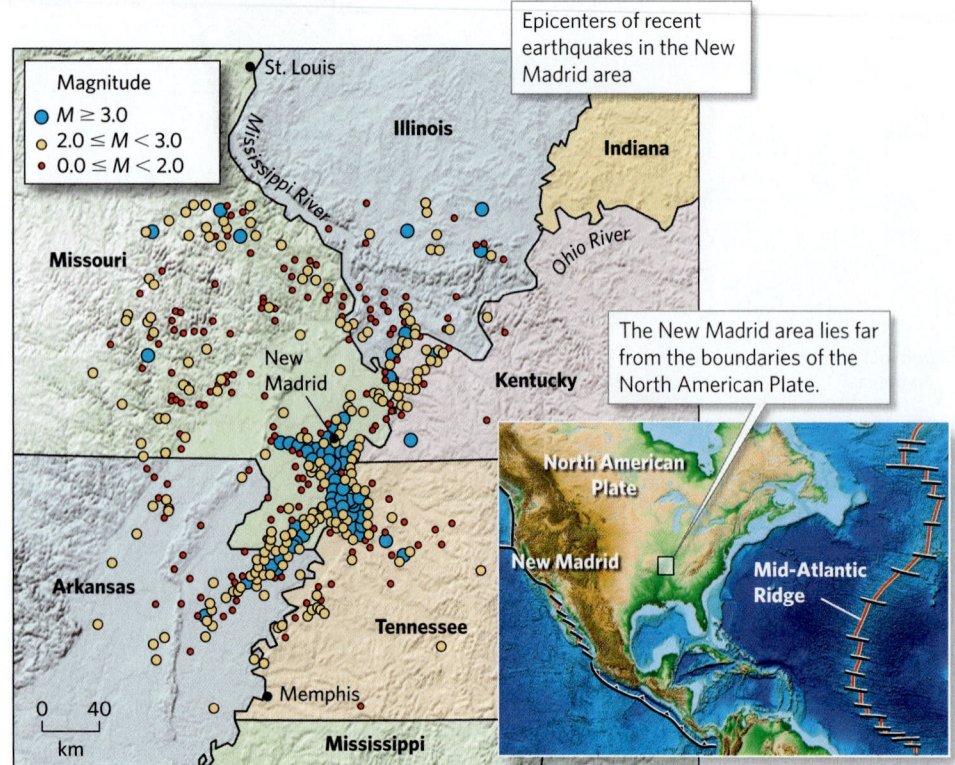

Epicenters of recent earthquakes in the New Madrid area

The New Madrid area lies far from the boundaries of the North American Plate.

FIGURE 8.20 During an earthquake, the ground can shake in many ways at once, causing surface structures to move.

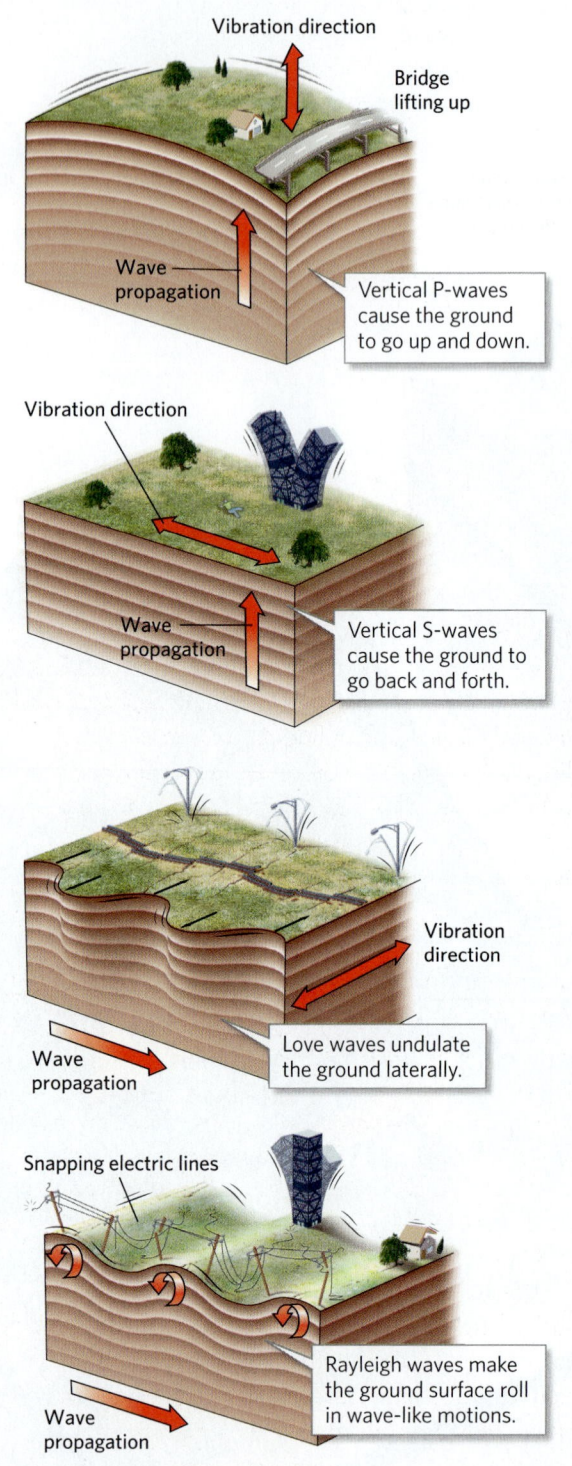

Vibration direction

Bridge lifting up

Wave propagation

Vertical P-waves cause the ground to go up and down.

Vibration direction

Wave propagation

Vertical S-waves cause the ground to go back and forth.

Vibration direction

Wave propagation

Love waves undulate the ground laterally.

Snapping electric lines

Rayleigh waves make the ground surface roll in wave-like motions.

Wave propagation

centuries collapsed in Kathmandu, and throughout the remote countryside, landslides buried communities and roads. Earthquakes are fairly frequent in Nepal, as well as elsewhere along the Himalayas, because the range is actively growing due to the collision of India with Asia, and compression caused by this collision drives slip on thrust faults.

INTRAPLATE EARTHQUAKES. In any given year, 95% of the seismic energy produced on the Earth comes from plate-boundary, collision, or rift-related earthquakes. The remaining types, which occur in the interiors of plates, are known as **intraplate earthquakes**. The foci of almost all intraplate earthquakes lie at depths of less than 25 km (15 mi). These earthquakes may happen when weak pre-existing fault zones, some of which initially formed over a billion years ago, undergo slip in response to stresses applied to continents.

Intraplate earthquakes are not uniformly distributed. In North America, most take place in southeastern Missouri, eastern Tennessee, eastern South Carolina, and southern Quebec. The largest historical intraplate earthquakes to affect the continental United States struck near New Madrid, in southernmost Missouri. During the winter of 1811–1812, three M_w 7.0 to 7.4 earthquakes, resulting from slip on faults beneath the Mississippi Valley,

shook the region. Displacement of the ground surface temporarily reversed the flow of the Mississippi River, and ground movement toppled cabins. The region remains seismically active (**Fig. 8.19**).

8.6 How Do Earthquakes Cause Damage?

Ground Shaking

An earthquake starts suddenly and may last from a few seconds to a few minutes. The duration of ground shaking at a given locality depends on how long it took for sliding to occur and on the distance between the locality and the earthquake's focus. The second factor reflects the fact that not all earthquake waves travel at the same velocity, so they don't all arrive at the same time.

Different kinds of seismic waves cause different kinds of ground motion **(Fig. 8.20)**. The severity of the shaking at a given location depends on (1) the magnitude of the earthquake, because larger-magnitude events release more energy; (2) the distance from the focus, because seismic energy decreases as seismic waves pass through the Earth; and (3) the strength of the materials just beneath the ground surface, because seismic waves tend to cause more motion in weaker materials.

If you're out in an open field during an earthquake, ground motion alone won't kill you. You may be knocked off your feet and bounced around a bit, but your body is too flexible to break. Human-made structures aren't so lucky, however **(Fig. 8.21)**. When seismic waves pass, roads, rail lines, and pipelines buckle or rupture. Buildings sway, twist back and forth, or lurch up and down, so windows may shatter, roofs fail, and building facades crash to the ground. If the floors of a multistory building or the deck of a bridge bounce up, they slam back down on their support

FIGURE 8.21 Examples of earthquake damage due to ground shaking.

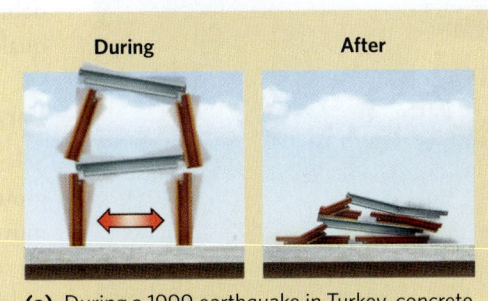

(a) During a 1999 earthquake in Turkey, concrete buildings collapsed when supports gave way and floors piled on one another like pancakes.

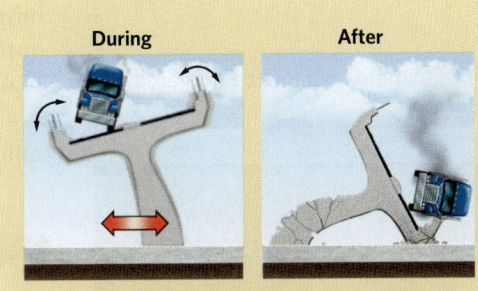

(b) An elevated bridge tipped over during the 1995 earthquake in Kobe, Japan.

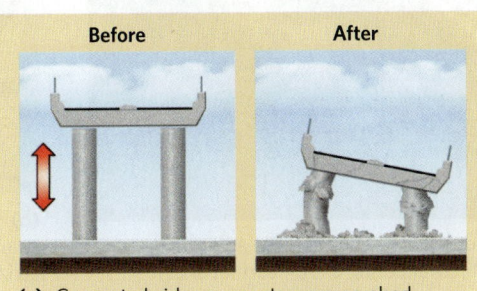

(c) Concrete bridge supports were crushed during the 1994 earthquake in Northridge, California, when the overlying bridge bounced up and slammed back down.

(d) A neighborhood of masonry buildings in Armenia collapsed during a 1999 earthquake because the walls broke apart.

FIGURE 8.22 During the 1994 Northridge earthquake, a steep slope along the coast of California collapsed, carrying part of a home with it.

Did you ever wonder . . .

whether California will fall into the sea during an earthquake?

FIGURE 8.23 Examples of sediment liquefaction triggered by earthquakes.

(a) Liquefaction under their foundations caused these apartment buildings in Niigata, Japan to tip over during a 1964 earthquake.

(c) An entire neighborhood on the Indonesian island of Sulawesi collapsed when the sediment beneath liquified.

columns, generating enough downward force to crush the columns and cause the structures to collapse. Falling debris from building failure causes the majority of earthquake-related deaths and injuries.

Landslides

Seismic shaking can cause ground on steep slopes or ground underlain by weak sediment to give way. This movement results in a *landslide*, the tumbling or flow of soil and rock downslope (see Chapter 12). Landslides are often a major cause of earthquake damage. Earthquake-triggered landslides along the coast of California may make headlines, for when steep cliffs facing the Pacific collapse, expensive homes tumble to the beach below **(Fig. 8.22)**. Such events lead to the misperception that "California will fall into the sea" during an earthquake. Although small portions of the coastline do indeed tumble down to the beach, the state as a whole remains firmly attached to the continent, despite what Hollywood scriptwriters say.

(b) During the 2011 Christchurch earthquake, liquefied sand spurted out and spread over the pavement, so the pavement collapsed to form a sinkhole.

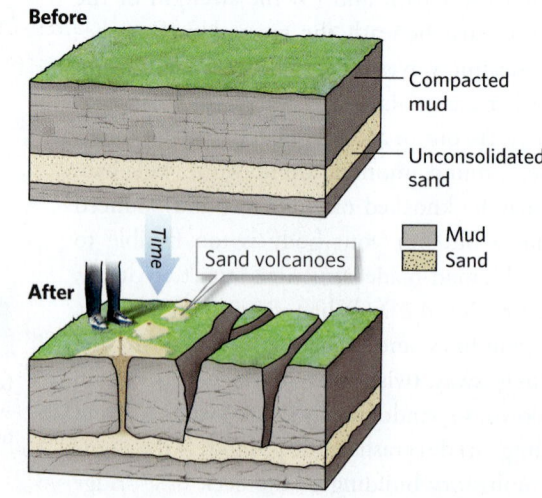

(d) Liquefaction of a sand layer causes the ground to crack and sand volcanoes to erupt.

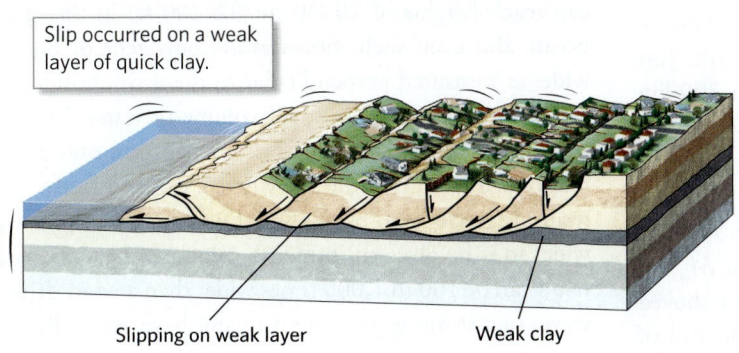

FIGURE 8.24 The 1964 Turnagain Heights disaster.

Slip occurred on a weak layer of quick clay.

Slipping on weak layer   Weak clay

Sediment Liquefaction

In 1964, an M_w 7.5 earthquake struck Niigata, Japan, a city that had been partly built on land underlain by wet sand. During the ground shaking, the foundations of over 15,000 buildings sank into the ground, causing walls and roofs to crack. In fact, several buildings tipped over (Fig. 8.23a)! In 2011, an earthquake in Christchurch, New Zealand, caused sand to erupt and produce small, cone-shaped mounds on the ground surface. When the sand moved from underground onto the surface, pit-like depressions developed nearby, some even large enough to swallow cars (Fig. 8.23b). During a 2018 earthquake in Indonesia, in places the ground turned to liquid and literally flowed, carrying away all the buildings that had been built on it (Fig. 8.23c).

The above examples illustrate the process of **sediment liquefaction**. Liquefaction in wet sand happens when shaking causes the sand grains to try to settle together more tightly. But this movement creates an increase in the pressure of the groundwater, and that force, in turn, pushes the grains further apart. As a result, what had been stable, load-bearing sand turns into a sand-water slurry, incapable of supporting weight. If cracks open up between the liquefied sand layer and the ground surface, pressure caused by the weight of overlying sediment squeezes the wet sand upward and out onto the ground surface, building cone-shaped mounds called *sand volcanoes* (Fig. 8.23d). The settling of sediment layers into a liquefied layer can also disrupt bedding and can lead to formation of open fissures and pits on the land surface.

A similar phenomenon happens in a special type of clay called *quick clay*: ground shaking transforms what had been a gel-like solid mass into a liquid, slippery mud by destroying cohesion among the grains. This phenomenon happened during the 1964 Good Friday earthquake in southern Alaska, beneath the Turnagain Heights neighborhood of Anchorage. When ground shaking began, a layer of quick clay beneath the development liquefied, and the overlying layers of sediment, along with the houses built on top of them, slumped downslope, turning the landscape into a chaotic jumble (Fig. 8.24).

Fire

The shaking during an earthquake can make lamps, stoves, or candles with open flames tip over, and it may break wires or topple power lines, generating sparks. As a consequence, areas already turned to rubble, and even areas that are not so badly damaged, may be consumed by fire. Ruptured gas pipelines and oil tanks feed the flames, sending columns of fire erupting skyward (Fig. 8.25a). Once a fire starts to spread, it may become an unstoppable inferno, especially since debris-filled streets block fire-fighting equipment and broken pipes make fire hydrants inoperable. When a large earthquake hit Tokyo in 1923, coals from cooking stoves set buildings made of wood and paper alight. The fire spread rapidly and warmed the air above the city. As hot air rose, cool air rushed in from outside the burning area, generating wind gusts of over 160 km/h (100 mph). The winds stoked the blaze, producing a firestorm that consumed the city (Fig. 8.25b).

Did you ever wonder . . .

how long an earthquake lasts?

 Animation
Tsunami Initiation and Arrival

FIGURE 8.25 Fire sometimes follows an earthquake.

(a) Broken gas tanks erupt in fountains of flame after the 2011 Tōhoku, Japan, earthquake.

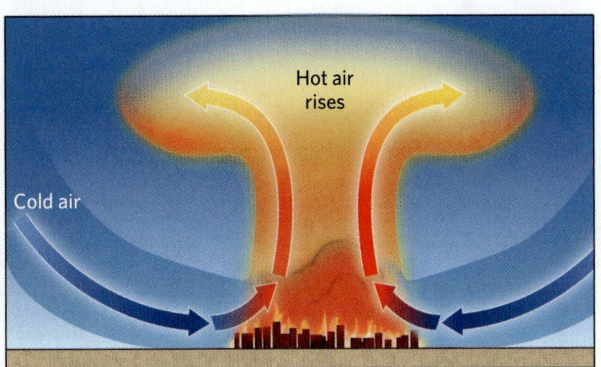

Hot air rises

Cold air

(b) A firestorm develops when cool air rushes in to replace rising hot air above a huge fire. The cool air stokes the blaze.

Tsunamis

The azure waters and palm-fringed islands of the Indian Ocean's eastern coast hide the Sunda Trench, one of the most seismically active plate boundaries on Earth. Just before 8:00 A.M. on December 26, 2004, a 1,300-km (800-mi)-long by 100-km (62-mi)-wide area of a thrust fault along this convergent boundary broke, and the overriding plate lurched westward by as much as 15 m (50 ft), pushing the seafloor up by tens of centimeters. Elastic rebound from this slip triggered the great M_w 9.3 Sumatra earthquake, and the rise of the seafloor shoved the overlying ocean surface upward. Because the area of seafloor that rose was so broad, the volume of water it displaced was immense. Gravity caused the uplifted water to collapse downward and spread out (Fig. 8.26a). The uplift and collapse of the broad mound of water generated several broad waves that traveled outward at a velocity of about 800 km per hour (500 mph)—almost the speed of a jet plane (Fig. 8.26b).

Geologists use the term **tsunami** for a wave produced by displacement of the seafloor. The displacement can be due to an earthquake, an explosive volcanic eruption, or, as we'll see in Chapter 12, a submarine landslide. *Tsunami* is a Japanese word that translates literally as harbor wave, an apt name because tsunamis can be particularly damaging to harbor towns.

Tsunamis differ significantly from familiar, wind-driven storm waves (Fig. 8.27). Large wind-driven waves can reach heights of 10–30 m (32–100 ft) in the open ocean. But even such monsters are only tens of meters wide, as measured perpendicular to the wave motion, so they contain a relatively small volume of water. In contrast, although the sea surface rises by at most only a few tens of centimeters at the site where a tsunami forms, the resulting wave may be tens to hundreds of kilometers wide, so it involves an immense volume of water. A tsunami can be 100 to 1,000 times wider than a wind-driven wave, so a storm wave and a tsunami have very different consequences.

When any wave approaches the shore, friction between the base of the wave and the seafloor slows the bottom of the wave, so the back of the wave catches up to the front, and the added water builds the wave higher. Near the beach, the top of the wave may fall over the front of the wave, forming a breaker. In the case of a wind-driven wave, the breaker may be tall when it washes onto the beach, but because the wave doesn't contain much water, the flow of water stops at the landward edge of the beach. In the case of a tsunami, however, when friction slows the wave and increases its height near the shore, the wave doesn't run out of water after it has crossed the beach. So if the coastal land is low-lying, the tsunami keeps moving inland, eventually submerging a huge area. The largest tsunamis can grow to heights of 30 m (100 ft) and can submerge land several kilometers inland from the shore. Keep in mind that we define a tsunami by its cause and its width, not by its height. Some tsunamis are only tens of centimeters to a meter high and may not even be noticed.

The 2004 Indian Ocean tsunami, the first to be recorded on video, alerted the world to how catastrophic tsunami damage can be (Fig. 8.28). When the tsunami

FIGURE 8.26 Formation of a tsunami.

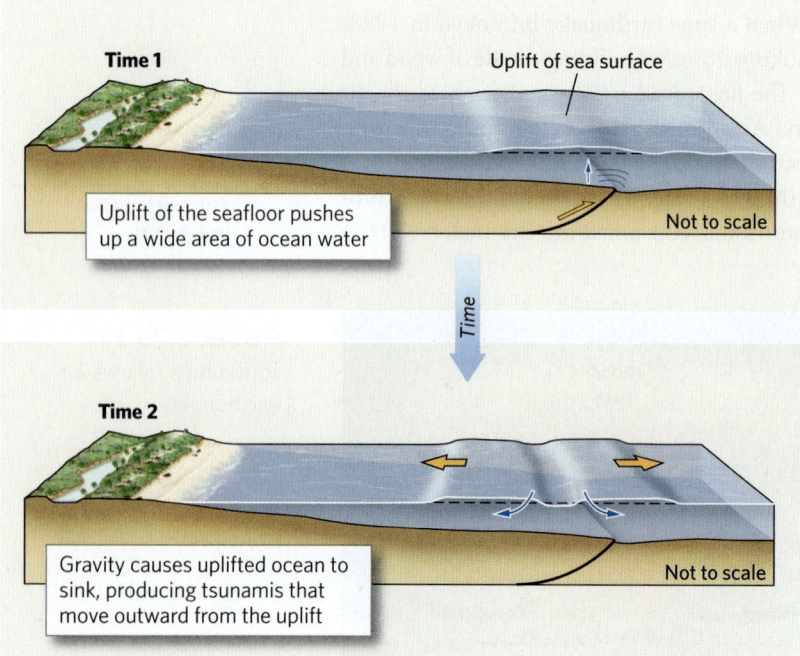

(a) After initial uplift of a mound of water, gravity causes the wave to spread out. A tsunami travels almost as fast as a jet plane.

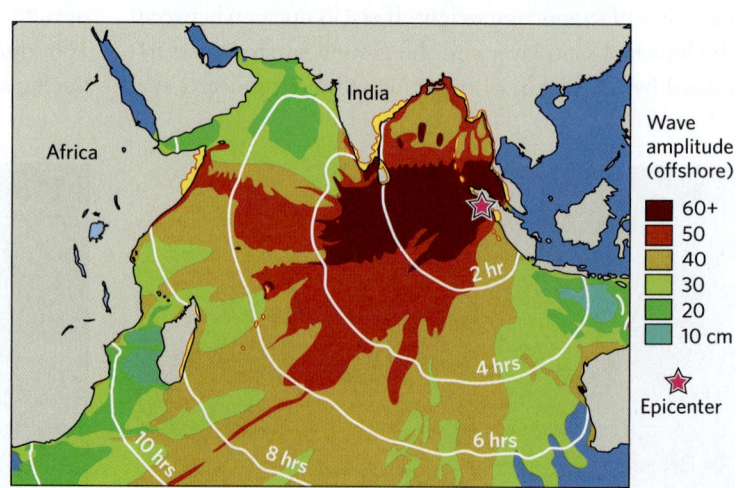

(b) A map showing the height of the 2004 tsunami as it crossed the Indian Ocean. The white lines indicate the position of the wave front at the time specified.

FIGURE 8.27 The difference between a storm wave and a tsunami. Storm waves can be high, but because they are not very wide, they contain relatively little water. A tsunami is not very high out in the open ocean, but as it approaches the land, the rear catches up and the wave grows higher. Each wave is very wide.

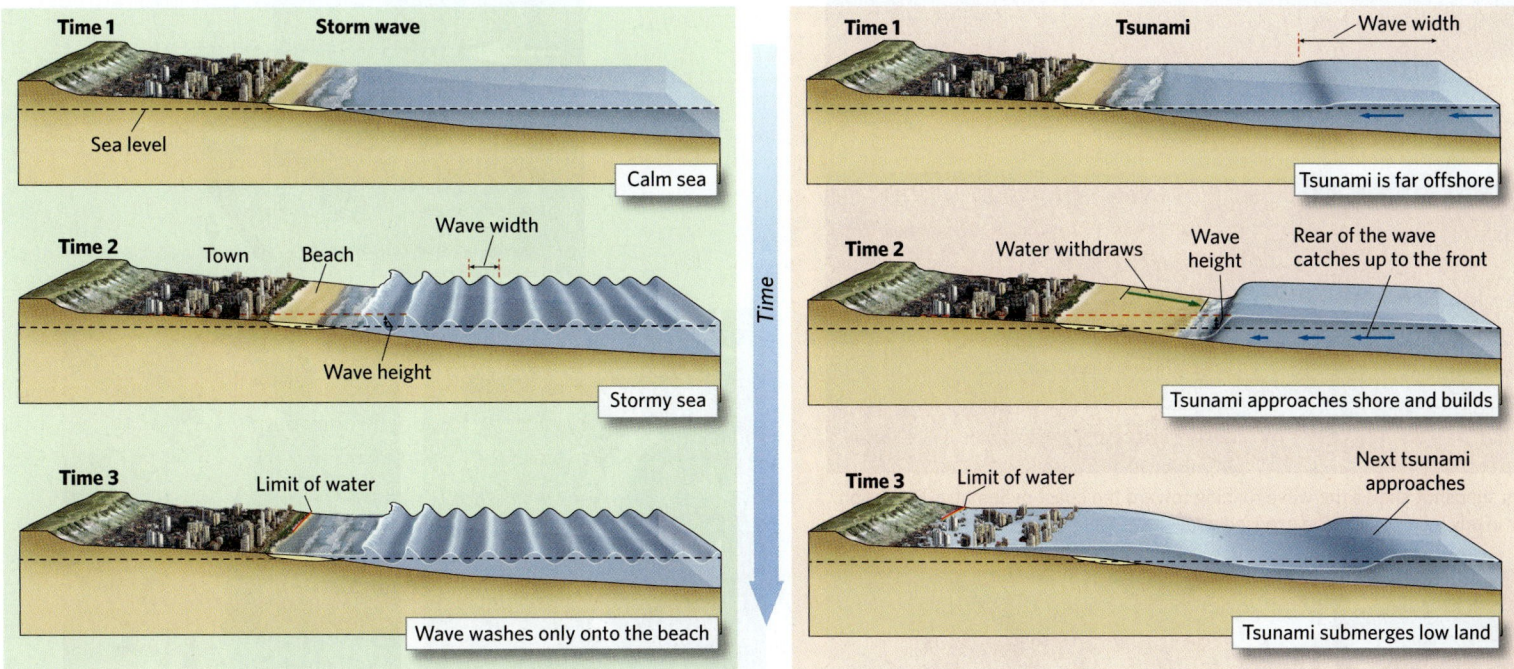

approached Banda Aceh, a city at the northern end of the island of Sumatra, the sea receded much farther than anyone had ever seen, exposing large areas of reefs that normally remained submerged even at low tide. People walked out onto the exposed reefs in wonder. Then, a wall of frothing water appeared offshore. With a rumble that grew to a roar, the tsunami approached. Friction with the seafloor slowed it to less than 30 km/h (20 mph), but it still moved faster than people could run. In places, the wave front reached heights of 15–30 m (45–100 ft) as it overwhelmed the beach. The impact of the water ripped boats from their moorings, snapped trees, battered buildings into rubble, and tossed cars and trucks like toys. Water just kept coming, eventually flooding land as far as 7 km (4 mi) inland. When gravity finally pulled the water back seaward, it carried with it a massive jumble of flotsam as well as the bodies of its unfortunate victims. Sadly, the horror of Banda Aceh was just a preamble to the devastation that would soon visit other stretches of Indian Ocean coast. Tsunamis crossed the Indian Ocean, striking southern India and Sri Lanka within two hours. Then, continuing on, the tsunamis crashed into the east coast of Africa about seven hours after the initial quake. In the end, more than 230,000 people died that day.

The tsunami that struck Japan soon after the 2011 Tōhoku earthquake also generated a new level of international awareness. The 10-m (30-ft)-high seawalls built to protect Japan's coast were no match for the advance of the wave, which rose to heights of 10–33 m (30–100 ft) when it reached shore. Racing inland, the wave picked up dirt and debris and evolved into a viscous slurry with the consistency of wet mud, so nothing could withstand its impact. It devastated coastal towns so completely that they looked as though they had been flattened by nuclear bombs (Fig. 8.29).

But even when the Tōhoku tsunami receded, the catastrophe was not over. Rising water inundated the Fukushima Daiichi nuclear power plant, where it not only destroyed power lines, cutting the plant off from the electrical grid, but also drowned the backup diesel generators. As a result, the water pumps driving the plant's cooling system stopped functioning, and the water surrounding the plant's hot reactor cores boiled away. Some of the water became so hot that H_2O molecules separated into H_2 and O_2 gas, which, when ignited by a spark, exploded. Explosions blew the tops off three of the plant's four containment buildings, contaminating the surroundings with radioactivity.

Disease

Once the ground shaking and fires have ceased, disease may still threaten lives in an earthquake-damaged region. Ground movement breaks water and sewer lines, thereby destroying clean-water supplies and exposing the public to pathogens (Box 8.2). Ground rupture and landslides

FIGURE 8.28 The great Indian Ocean tsunami of 2004.

(a) This snapshot shows the wave rushing toward the coast of Sumatra. Recession of water in advance of the wave exposed a reef.

(b) Here, the wave blasts through a grove of palm trees as it strikes Thailand.

At its highest, the tsunami's front was 15–30 m high.

(c) Satellite photos of the Indonesian province of Aceh before and after the tsunami struck. Note that the city has been washed away and the beach has vanished. The inset shows the relative size of the wave as it struck Banda Aceh.

FIGURE 8.29 Damage along the coast of Japan caused by the 2011 tsunami.

may cut transportation lines, preventing food and medicine from reaching the stricken area. The severity of such problems may exceed the ability of emergency services to cope, so it may take months to years for daily life in a damaged area to return to normal.

Take-home message . . .

Earthquakes cause devastation in many ways. Ground shaking, landslides, and sediment liquefaction disrupt landscapes and cause buildings to crumble. Though ground shaking itself cannot kill people, falling debris can. In coastal areas, tsunamis may wash over broad areas of low land. Fire and disease may follow earthquakes, causing even more loss of life.

Quick Question

How does a tsunami differ from a storm wave?

BOX 8.2 ▶ A Deeper Look

The 2010 Haiti catastrophe

Haiti sits astride the transform boundary along which the North American Plate moves westward at about 2 cm (0.8 in) per year relative to the Caribbean Plate (Fig. Bx8.2a). Therefore, earthquakes in Haiti are inevitable. But the last major earthquakes on this plate boundary had happened over 200 years ago, so by 2010, elastic deformation and associated stress had been building for quite some time. On the sunny afternoon of January 12, at 4:53 P.M., a 70-km (45-mi)-long segment of a large strike-slip fault suddenly slipped as much as 4 m (12 ft). The motion began at 25 km (15 mi) west-southwest of Port-au-Prince, the capital of Haiti, and 13 km (8 mi) beneath the ground surface (Fig. Bx8.2b). The seismic waves of the resulting M_w 7 earthquake caused the ground to lurch violently for about 35 seconds.

The impact of an earthquake on society depends not only on the earthquake's size, but also on the nature of the ground materials, on the steepness of the slopes, on construction practices, and on the quality of emergency services in the affected area. Much of Port-au-Prince sits on a basin of weak sediment, which amplified ground movements—in fact, beneath the harbor, sediment liquefied, causing wharfs to sink into the sea. Sadly, the city's buildings were not designed to withstand ground shaking, so many collapsed into rubble (Fig. Bx8.2c). In addition, some of the city's neighborhoods were perched on steep slopes, which slid downhill during the quake, carrying the neighborhoods with them (Fig. Bx8.2d). When the shaking finally stopped, most of Port-au-Prince had collapsed. As a dense cloud of white dust slowly rose over the rubble, survivors began the frantic scramble to dig out victims, a task made more hazardous by aftershocks, of which over 50 had magnitudes between 4.5 and 6.1. The renewed shaking caused still-standing but weakened structures to collapse on rescuers. Some estimates place the death toll at 230,000. In the days that followed, local emergency services were overwhelmed, and access to the victims was nearly impossible. An air caravan of aid arrived to help out, but even so, in the months that followed, cholera spread. Years after the event, recovery from the earthquake remains incomplete.

FIGURE Bx8.2 The January 2010 earthquake in Haiti and its geologic setting.

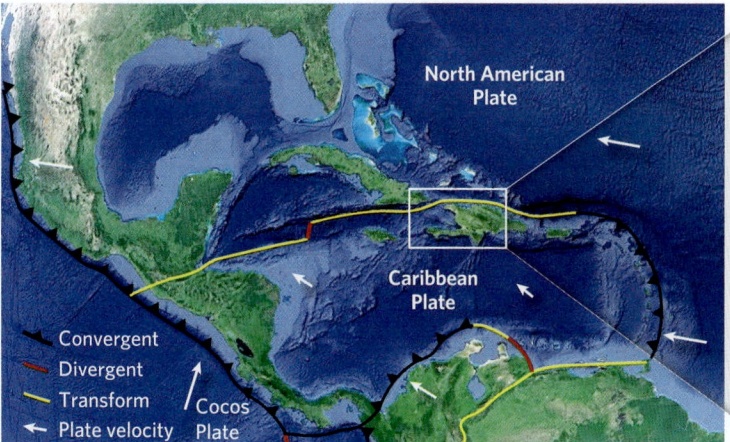

(a) The Caribbean Plate has complex boundaries delineated by bathymetric features. The white rectangle shows the location of Haiti.

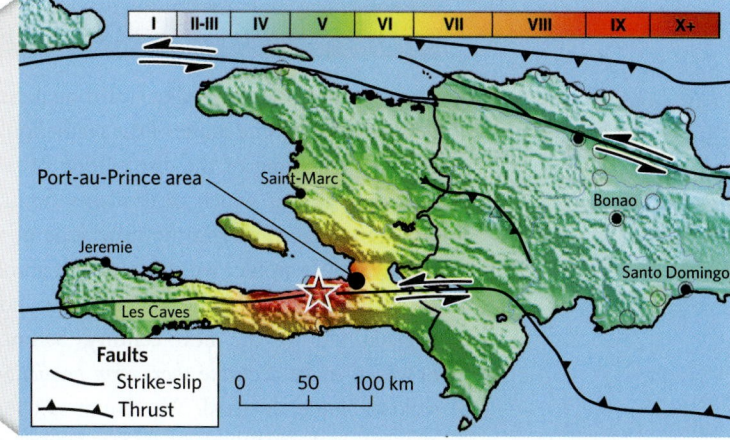

(b) A map showing the intensity of shaking in Haiti on the MMI scale. The star marks the epicenter of the quake.

(c) Survivors salvage what they can in Port-au-Prince, the capital of Haiti, after the devastating earthquake.

(d) Ground shaking during the earthquake caused most of the houses in this residential neighborhood to collapse.

FIGURE 8.30 Evidence of past earthquakes in the geologic record is used to determine recurrence intervals. The block shows subsurface sediment layers near an active fault. Buried sand volcanoes and disrupted beds (indicated by numbers) reveal when earthquakes happened in the past.

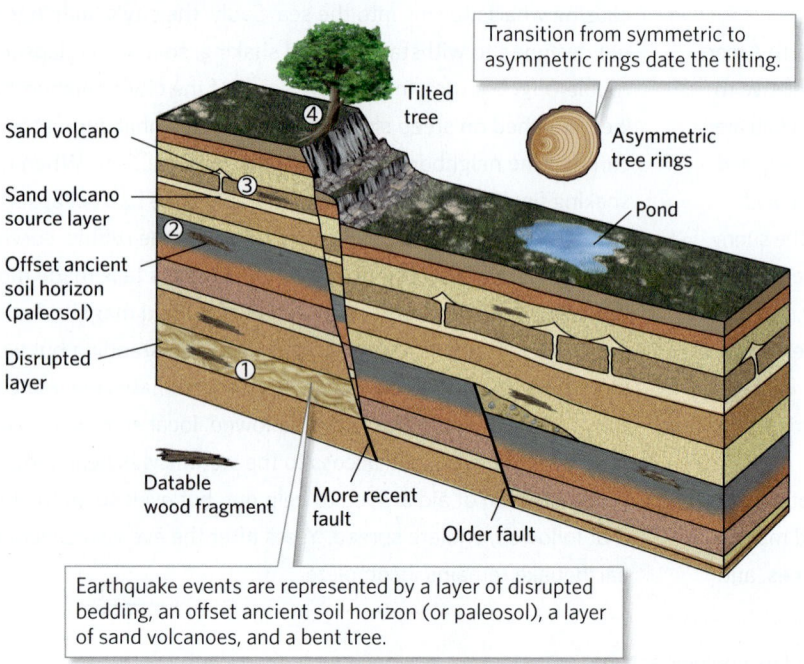

Transition from symmetric to asymmetric rings date the tilting.

Tilted tree

Asymmetric tree rings

Sand volcano

Sand volcano source layer

Pond

Offset ancient soil horizon (paleosol)

Disrupted layer

Datable wood fragment

More recent fault

Older fault

Earthquake events are represented by a layer of disrupted bedding, an offset ancient soil horizon (or paleosol), a layer of sand volcanoes, and a bent tree.

8.7 Can We Predict the "Big One?"

> We learn geology the morning after the earthquake.
>
> —RALPH WALDO EMERSON
> (AMERICAN POET, 1803–1882)

Can seismologists predict earthquakes? The answer depends on the time frame of the prediction. With our present understanding of the distribution of seismic belts and the frequency at which earthquakes occur, we can make *long-term predictions* (on the time scale of decades to centuries). For example, we can say with some certainty that an earthquake will rattle Istanbul, but not north-central Canada, during the next century. Seismologists cannot, however, make accurate *short-term predictions* (on the time scale of hours to years). We cannot say, for example, that an earthquake will happen in San Francisco 40 days from now. New technologies, however, allow seismologists to give people a warning seconds to minutes before seismic waves strike. In this section, we look at the scientific basis of long-term predictions and of earthquake early-warning systems.

Long-Term Predictions

A long-term prediction estimates the probability, or likelihood, that an earthquake will happen during a specified time period. For example, a seismologist may say, "The probability of a major earthquake occurring in the next 50 years in this state is 20%." This sentence implies that there's a 1-in-5 chance that an earthquake will happen before 50 years have passed. Seismologists refer to studies leading to long-term predictions as *seismic risk assessments*.

Urban planners use such assessments to design building codes for a region: codes requiring stronger buildings make sense for regions with greater seismic risk, for the chance that a building will be shaken during its lifetime is greater.

The basic premise of seismic risk assessment can be stated as follows: A region where many earthquakes have occurred in the past is likely to experience earthquakes in the future. Seismic belts are, therefore, regions of higher seismic risk (see Figure 8.13). This doesn't mean that disastrous earthquakes can't happen far from a seismic belt—they can and do—but the probability that an earthquake will happen in such places in any given time window is less.

To provide a more specific sense of earthquake likelihood, seismologists try to specify the **recurrence interval**, the average time between successive events, for earthquakes of a given size in a region. Since earthquakes do not happen periodically, meaning at predictably spaced time intervals, an earthquake recurrence interval does not give the exact time between successive events. Because the idea of recurrence intervals can be confusing, seismologists prefer to specify the *annual probability*, a calculation of the likelihood that an earthquake will happen in a given year, where

$$\text{Annual probability} = 1 \div \text{recurrence interval}$$

For example, if the recurrence interval for an M_w 7 earthquake in a region is 100 years, then the annual probability of such an earthquake is 1/100, or 1%. This means that there is a 1 in 100 chance of an M_w 7 earthquake happening in a given year. Note, however, that because stress builds up over time on a fault, the elastic-rebound theory hints that the annual probability of an earthquake may progressively increase as time passes.

To determine the recurrence interval for large earthquakes within a given seismic belt, seismologists must determine when previous earthquakes happened within that belt. For places where the historical record does not extend far enough back in time to reveal multiple large events, researchers look for evidence of large earthquakes preserved in the geologic record. For example, in places where sedimentary strata have accumulated in a basin over a fault, researchers may dig a trench and look for buried layers of sand volcanoes or disrupted bedding in the stratigraphic record. Each such layer, whose age can be determined by using radiocarbon dating of plant fragments (see Chapter 9), records the time of an earthquake **(Fig. 8.30)**. Seismologists calculate the number of years between successive events and then calculate the average to obtain the recurrence interval. Information on recurrence intervals allows seismologists to refine regional maps illustrating seismic hazards **(Fig. 8.31)**.

Paleoseismology studies demonstrated that the *Cascadia seismic zone*, along the coast of Washington and Oregon (the

FIGURE 8.31 Examples of seismic hazard maps.

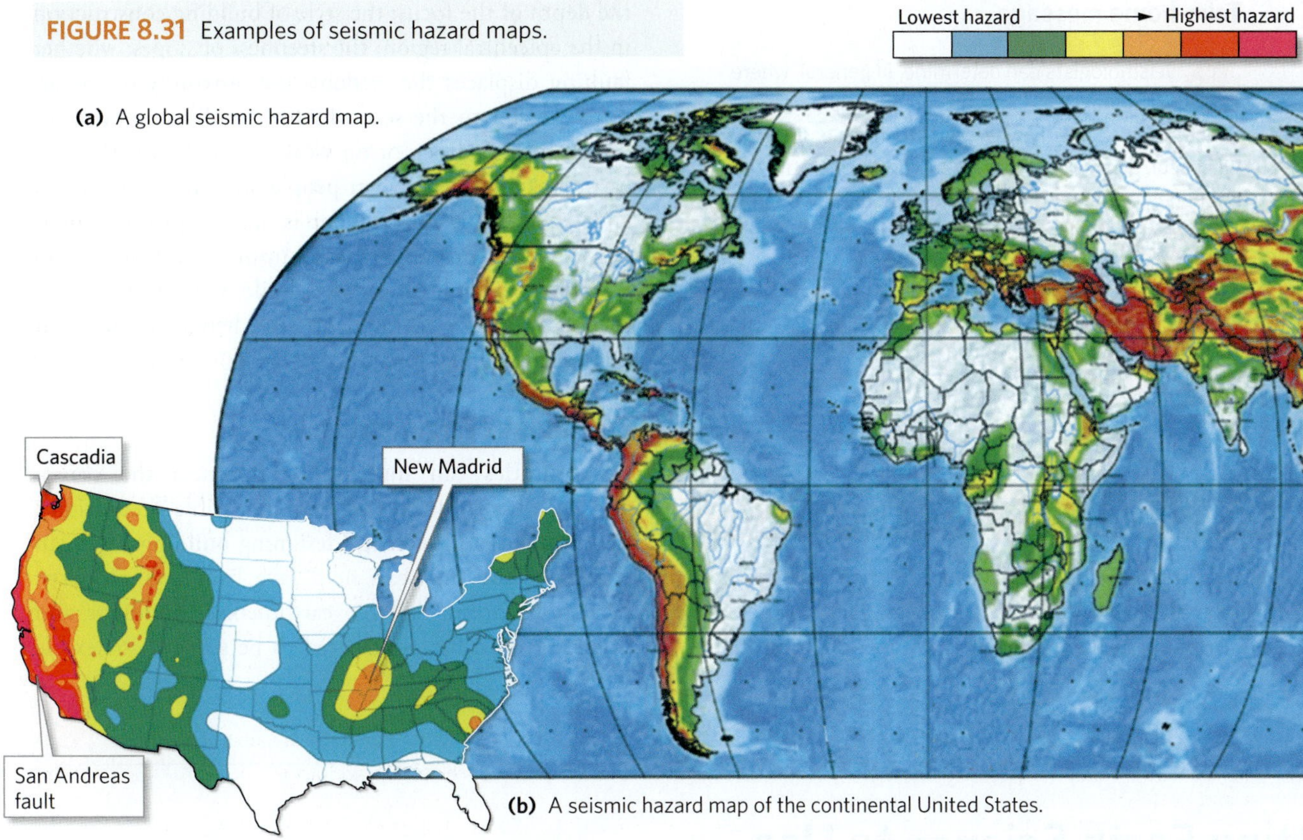

Lowest hazard ──────▶ Highest hazard

(a) A global seismic hazard map.

Cascadia

New Madrid

San Andreas fault

(b) A seismic hazard map of the continental United States.

Pacific Northwest region, west of the Cascade Mountains), hosted an M_w 9 earthquake a few hundred years ago. The evidence came from the dating of a "ghost forest," an area of trees that died when the land subsided and was submerged by a tsunami. Researchers found that a distinct sand layer was deposited by the tsunami, directly on what had been the forest floor. There is no written historical record of this event in English, but the event was recorded in the oral history of Native Americans and was documented by Japanese historians who recorded the arrival time of the far-field tsunami that had crossed the Pacific. The tsunami is now attributed to a *megathrust earthquake*, a giant earthquake produced when a large area of the fault separating a subducting plate from an overriding plate suddenly slips, that struck the Pacific Northwest on January 26, 1700. (The Tōhoku, Sumatra,1964 Alaska, and 1960 Chile quakes are also megathrust earthquakes.) Further paleoseismology studies indicate that megathrust earthquakes have a recurrence interval of 200 to 500 years in the Cascadia region.

Earthquake Early-Warning Systems

Short-term predictions, specifying that an earthquake will happen on a given date or within a time window of days to years, are bogus. In fact, such predictions will probably never be reliable. The concept of short-term prediction should not be confused, however, with the concept of an *earthquake early-warning system*, which is based on a real signal and can potentially save lives. An early-warning system works as follows: When an earthquake happens, the seismic waves it produces start traveling through the Earth. Seismometers positioned between the epicenter and a city will detect seismic waves before they have had time to reach the city. The instant that these seismometers detect the earthquake, a transmitter sends a signal to a control center, which automatically broadcasts emergency signals to the city. These signals, which travel at the speed of light, reach the city several seconds to a minute before the seismic waves. The arrival of the signals can trigger automatic shutdowns of gas pipelines, trains, nuclear reactors, and power lines. The signals can also set off sirens and trigger broadcasts on radio, TV, and cellular networks, alerting people to take precautions.

Because tsunamis are so dangerous, predicting their arrival can save thousands of lives. At the tsunami early-warning center in Hawaii, observers keep track of earthquakes around the Pacific and use data relayed from tide gauges, buoys, and seafloor pressure gauges to determine whether a particular earthquake has generated a tsunami **(Fig. 8.32)**. If the observers detect a tsunami, they flash warnings to authorities around the Pacific. Sirens along the shore alert people to evacuate to higher ground.

FIGURE 8.32 Buoys can detect a tsunami in the open ocean so that people on land can be warned to evacuate low-lying coastal areas.

8.8 Preparing for Earthquakes

The destruction and death caused by an earthquake of a given size depend on a number of factors, including the proximity of the epicenter to a population center; the depth of the focus; the style of building construction in the epicentral region; the steepness of slopes; whether faulting displaces the seafloor; the proximity of the affected region to the sea; whether building foundations are on solid bedrock or on weak materials; whether the earthquake happens when people are outside or inside; and whether the government is able to provide emergency services promptly. To minimize the calamity of an earthquake, people can strive to build stronger structures, choose safer sites to build on, and better prepare their homes and families for such an event (Box 8.3).

Earthquake Engineering

Communities can mitigate or diminish the consequences of earthquakes by taking sensible precautions. **Earthquake engineering** (designing buildings that can withstand shaking) can help save lives and property. In regions prone to large earthquakes, buildings and bridges should be constructed to be somewhat flexible

---

BOX 8.3 ▸ **Putting Earth Science to Use**

Staying safe in earthquake country

Because so many communities are built on or near seismic belts, many people will endure an earthquake at some point in their lives. Therefore, everyone needs to take personal responsibility for earthquake preparedness. As a starting point, pay attention to maps that convey information about land stability. For example, you may wish to avoid living in a house or apartment that straddles a fault zone or sits on land that may be susceptible to landslides, liquefaction, or inundation due to dam failure or tsunami. Check with local officials and/or visit your town's website to ensure that satisfactory emergency plans exist, that evacuation routes are clear, and that your region has access to supplies for rescue and recovery.

In your own home, you can proactively determine whether simple earthquake retrofitting efforts have been completed by asking, Is the building properly anchored to its foundation? Are floors properly anchored to support columns? Are there diagonal support beams in walls? You should also take basic precautions, including bolting or strapping hot-water heaters, furnaces, and bookshelves to the walls, so that they don't topple over; installing latches on cabinets so they don't pop open during shaking; and knowing how to shut off gas,

FIGURE Bx8.3 If an earthquake strikes, consider taking cover under a sturdy table near a wall.

water, and electricity. Hold an earthquake preparedness drill for your family, and ask your local schools, offices, and factories to do the same. As part of your drill, know where to go if shaking begins. In general, it's best to drop to the floor on your hands and knees and seek cover beneath or next to a very sturdy piece of furniture that won't get crushed if the ceiling falls, and that isn't adjacent to glass windows (Fig. Bx8.3).

As long as lithosphere plates continue to move, earthquakes will continue to shake us. But you can reduce the chances of damage or injury by being prepared.

FIGURE 8.33 Preventing earthquake damage and casualties.

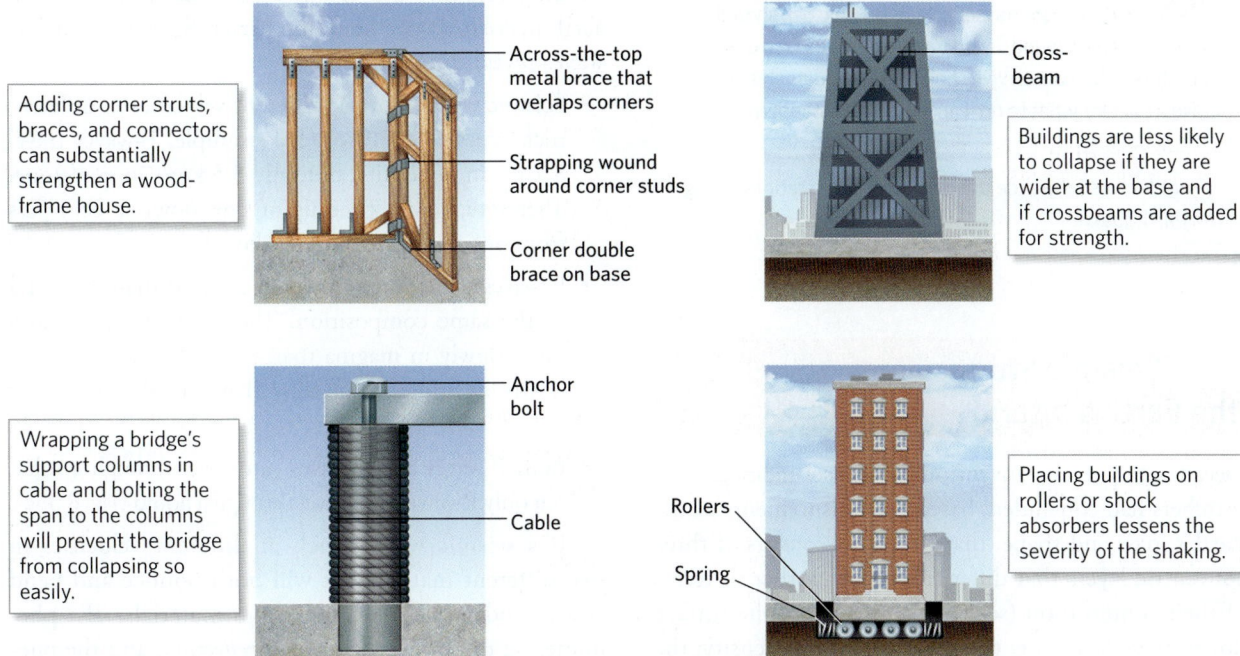

Adding corner struts, braces, and connectors can substantially strengthen a wood-frame house.

Across-the-top metal brace that overlaps corners

Strapping wound around corner studs

Corner double brace on base

Cross-beam

Buildings are less likely to collapse if they are wider at the base and if crossbeams are added for strength.

Wrapping a bridge's support columns in cable and bolting the span to the columns will prevent the bridge from collapsing so easily.

Anchor bolt

Cable

Rollers

Spring

Placing buildings on rollers or shock absorbers lessens the severity of the shaking.

(a) Damage to structures can be prevented if those structures are designed to withstand shaking.

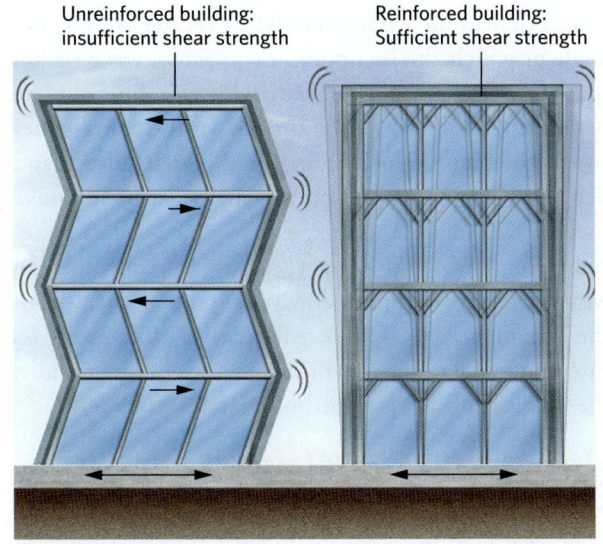

Unreinforced building: insufficient shear strength

Reinforced building: Sufficient shear strength

(b) An unreinforced building will shear side to side in a way that causes floors to shift out of alignment.

so that ground shaking cannot crack them, but they should have sufficient bracing to dampen their movements **(Fig. 8.33)**. In addition, supports should be strong enough to stand up to the weight of floors that may drop down after having bounced upward. In some cases, simple changes in construction practices can make a building stronger. For example, wrapping steel cables around bridge support columns makes them many times stronger; bolting bridge spans to the tops of support columns prevents the spans from bouncing off; and adding diagonal braces to frames keeps them from twisting and shearing too much.

In regions with significant seismic risk, certain kinds of construction should be avoided. For example, concrete-block, unreinforced concrete, and unreinforced brick buildings crack and tumble under conditions in which wood-frame, steel-girder, or reinforced concrete buildings remain standing. Traditional heavy, brittle tile roofs can shatter and bury the inhabitants inside, whereas lightweight sheet-metal roofs do not. Inadequate structures can be made safer by *seismic retrofitting*, the process of strengthening existing structures.

Earthquake Zoning

Urban planners in seismically active areas can decrease hazard by **earthquake zoning**. Zoning relies on an assessment of where land is stable and where it is not. Zoning should lead to regulations that discourage building on land underlain by weak mud or wet sand that could liquefy. Similarly, zoning should discourage building on top of, on, or at the base of steep escarpments where landslides might take place, or where the rupture of a dam could cause a flood. And zoning should prevent construction in areas prone to flooding by tsunamis. Finally, zoning should prevent construction of critical buildings (schools, hospitals, fire stations, communications centers, power plants) over active faults, for fault slip could crack and destroy the buildings.

Did you ever wonder . . .

if buildings can be designed to survive earthquakes?

Take-home message . . .

Earthquakes are a fact of life on this dynamic planet. People in regions facing high seismic risk should build on stable ground, avoid unstable slopes, and design structures that can survive shaking. Individuals should have a plan for what to do if an earthquake happens.

Quick Question

What factors influence the degree of devastation during an earthquake?

8.9 Seismic Study of the Earth's Interior

FIGURE 8.34 Simplified image of the Earth's interior, first proposed in the 19th century. The numbers in the figure are the measurements of density.

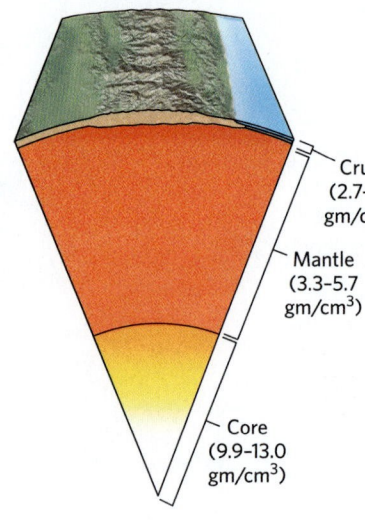

Crust (2.7–3.3 gm/cm³)

Mantle (3.3–5.7 gm/cm³)

Core (9.9–13.0 gm/cm³)

Decades before the invention of the seismometer, researchers had concluded, based on measurements of the Earth's mass and shape, that the Earth consists of three concentric layers that differ from one another in terms of their composition (see Chapter 1). From the surface down, these layers are: the crust, with a low density; the mantle, with an intermediate density; and the core, with a high density **(Fig. 8.34)**. To go beyond this basic understanding to define the specific depths of the boundaries between layers and to characterize the properties of the layers, researchers needed a tool that would allow them to "see" inside the Earth. The study of seismic waves provided such a tool. By measuring how fast seismic waves travel through the Earth, and how they reflect or bend as they travel, seismologists have been able to provide a much more refined image of the Earth's interior.

Controls on the Velocity and Bending of Seismic Waves

The ability of a seismic wave to travel through a material, as well as the velocity at which the wave travels, depends on several characteristics of the material. Factors such as *density* (mass per unit volume), *rigidity* (how stiff, or resistant to bending, a material is), and *compressibility* (how easily a material's volume changes in response to squashing) all affect seismic-wave velocity and therefore the **travel time** of a set of waves, meaning the time it takes them to move from the epicenter of an earthquake to a specific seismometer. Studies of seismic waves reveal the following:

- Seismic waves move at different velocities in different rock types **(Fig. 8.35a)**. For example, P-waves travel 8 km/s in peridotite, but only 3.5 km/s in sandstone. Therefore, waves speed up or slow down as they pass from one rock type into another.

- P-waves travel more slowly in a liquid than in a solid of the same composition. Therefore, P-waves travel more slowly in magma than in solid rock, and more slowly in molten iron alloy than in solid iron alloy **(Fig. 8.35b)**.

- Both P-waves and S-waves can travel through a solid, but only P-waves can travel through a liquid **(Fig. 8.35c)**.

If a seismic wave travels at different velocities in two different materials, it will both bounce and bend at a boundary between those two materials. The phenomenon of bouncing is called *reflection*, and the phenomenon of bending is called *refraction*. (You can see reflection and refraction by shining a light at the surface of a body of water.) The angle at which a reflected wave bounces off a boundary is always the same as the angle at which the incoming, or *incident*, wave strikes that boundary. But the angle by which a refracted wave bends at a boundary depends both on the contrast between the wave velocities in the materials above and below the boundary and on the angle at which a wave hits the boundary. As a rule, as waves pass from a material through which they travel rapidly into one through which they travel more slowly, they bend away from the boundary **(Fig. 8.36a)**. Alternatively, if waves pass into a material through which they travel more rapidly, the waves bend toward the boundary **(Fig. 8.36b)**.

Identifying the Earth's Layers, Seismically

Studies of seismic-wave velocity, refraction, and reflection have been used to locate the major boundaries between layers inside the Earth. Let's see how.

FIGURE 8.35 The propagation of earthquake waves.

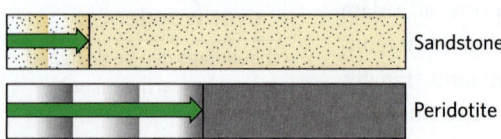

Sandstone

Peridotite

(a) Seismic waves travel at different velocities in different rock types. After a given time, a wave will have traveled farther in peridotite than in sandstone.

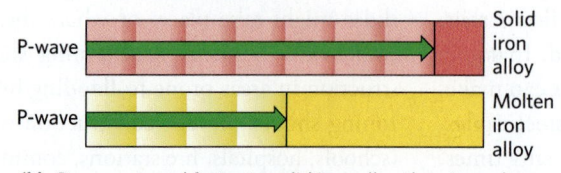

P-wave — Solid iron alloy

P-wave — Molten iron alloy

(b) P-waves travel faster in solid iron alloy than in molten iron alloy.

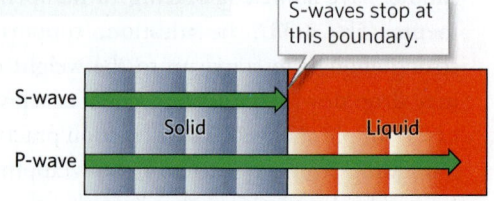

S-waves stop at this boundary.

S-wave — Solid — Liquid

P-wave

(c) Both P-waves and S-waves can travel through a solid, but only P-waves can travel through a liquid.

FIGURE 8.36 Refraction and reflection of seismic waves. The angle at which a wave is reflected by a boundary between two materials is always the same as the angle at which the incident wave strikes that boundary.

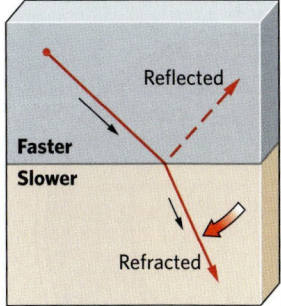

(a) A wave that enters a material through which it travels more slowly bends away from the boundary.

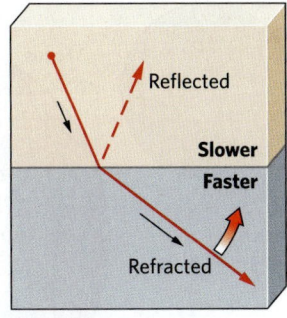

(b) A wave that enters a material through which it travels more rapidly bends toward the boundary.

FIGURE 8.37 Discovery of the Moho.

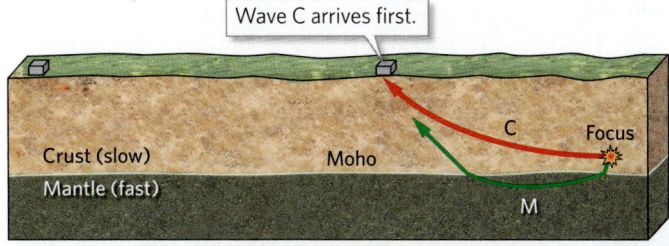

(a) P-waves traveling mostly through the crust reach a nearby seismometer first.

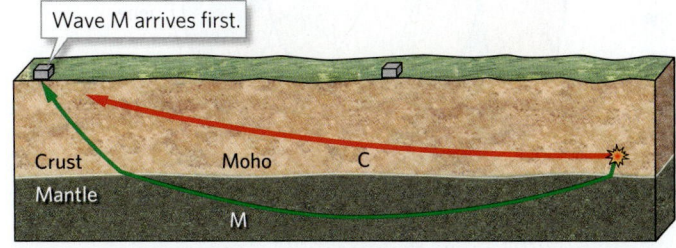

(b) P-waves traveling for most of their path through the mantle reach a distant seismometer first.

DISCOVERING THE CRUST-MANTLE BOUNDARY. In 1909, Andrija Mohorovičić, a Croatian seismologist, noted that P-waves arriving at seismometer stations less than 200 km (125 mi) from the epicenter of an earthquake traveled at an average speed of 6 km (4 mi) per second, whereas P-waves arriving at seismometers more than 200 km from the epicenter traveled at an average speed of 8 km/h (5 mph). To explain this observation, he suggested that P-waves reaching nearby seismometers followed a shallow path that kept them entirely within the crust, in which they traveled relatively slowly, whereas P-waves reaching distant seismometers traveled through the mantle for part of their route, and that their velocity was higher in the mantle. He realized that the waves refracted as they crossed the crust-mantle boundary **(Fig. 8.37)**. Mohorovičić was able to calculate the depth of the crust-mantle boundary from this observation, and he proposed that beneath continents, it occurred at a depth of 35–40 km (22–25 mi). Later studies showed that the depth of the crust-mantle boundary beneath continents varies from 25–70 km (15–45 mi), and beneath oceans from 7–10 km (4–6 mi). The crust-mantle boundary is now called the **Moho**, in honor of Mohorovičić.

DISCOVERING THE STRUCTURE OF THE MANTLE. Seismologists have determined that seismic waves travel at different speeds at different depths in the mantle **(Fig. 8.38)**. Specifically, between depths of about 100–200 km (60–125 mi) beneath the seafloor, seismic velocities are lower than in the overlying portion of the mantle This 100- to 200-km-deep layer is now known as the **low-velocity zone (LVZ)**. Researchers suggest that the LVZ corresponds to a layer in which mantle rock has undergone partial melting. Because seismic waves travel more slowly through liquids

than through solids, even a tiny bit of melt slows them down. Simplistically, the top of the LVZ delineates the base of the lithosphere and the top of the asthenosphere beneath oceanic plates. Seismologists do not find a well-developed LVZ beneath continents.

FIGURE 8.38 The velocity of P-waves changes with depth in the mantle because physical properties of the mantle change with depth.

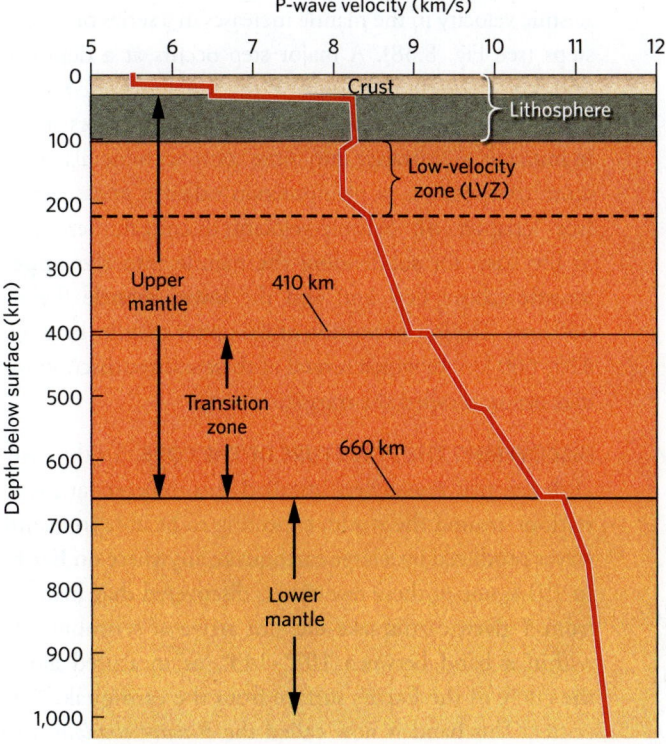

FIGURE 8.39 Shadow zones and the discovery of the Earth's core. (Note that the circumference of a circle is 360°, so it is 180° from a given point to a point on the other side of the planet.) The black arrows indicate the various paths that seismic waves propagating from the focus can take.

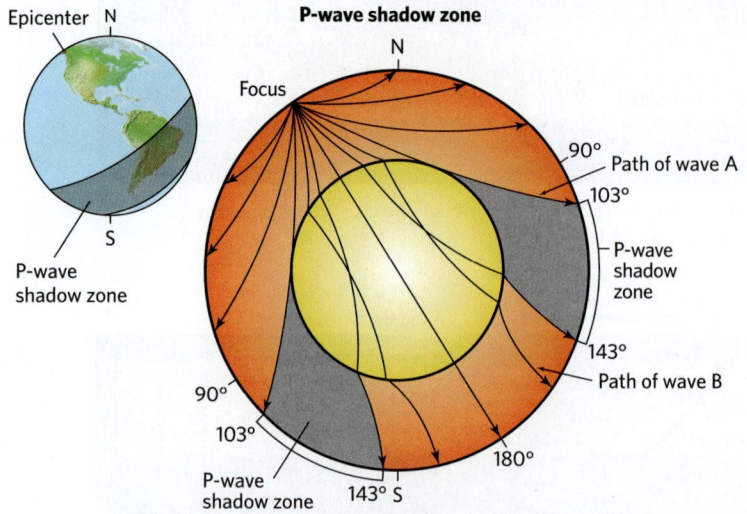

(a) P-waves do not arrive in the P-wave shadow zone because they are refracted at the core-mantle boundary.

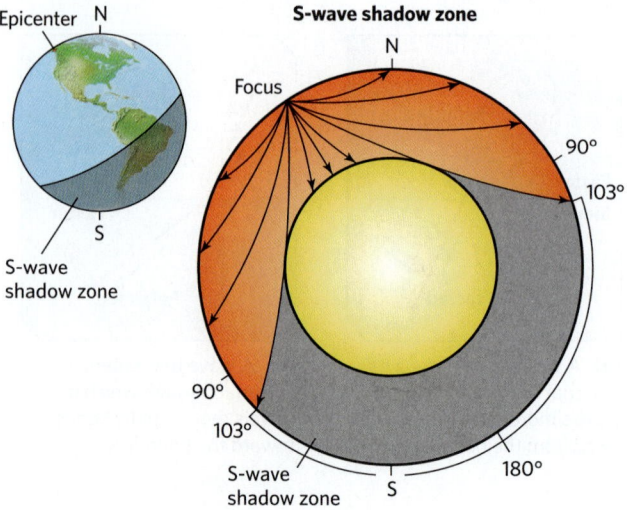

(b) S-waves do not arrive in the S-wave shadow zone because they cannot pass through the liquid outer core.

Below about 200 km, seismic-wave velocities in the mantle, beneath both continents and oceans, increase with depth (see Fig. 8.38). Seismologists interpret this increase in velocity with depth to mean that mantle peridotite becomes progressively less compressible, more rigid, and denser with depth. This interpretation makes sense, considering that the weight of overlying rock increases with depth, and that as pressure increases, the atoms making up minerals squeeze together more tightly and are less free to move.

At depths between 410 km–660 km (255–410 mi), seismic velocity in the mantle increases in a series of abrupt steps (see Fig. 8.38). A major step occurs at a depth of 660 km. Experiments suggest that such *seismic-velocity discontinuities* occur at depths where pressure causes atoms in minerals to rearrange into more compact minerals of the same composition, a phenomenon called a *phase change* (see Chapter 6). Seismic-velocity discontinuities serve as the basis for subdividing the mantle into the **upper mantle** (above 660 km) and the **lower mantle** (below 660 km). The lowest portion of the upper mantle, between 410–660 km, in which several of these seismic discontinuities occur, is called the **transition zone**.

DISCOVERING THE STRUCTURE OF THE CORE. In the early 20th century, researchers installed seismometers at many stations around the world, expecting to be able to record waves produced by a large earthquake anywhere on Earth. In 1914, one of these researchers discovered that P-waves from a given earthquake did not arrive at seismometers within a band between 103°–143°, as measured along the curve of the Earth's surface from the earthquake epicenter. This band is now called the *P-wave shadow zone*

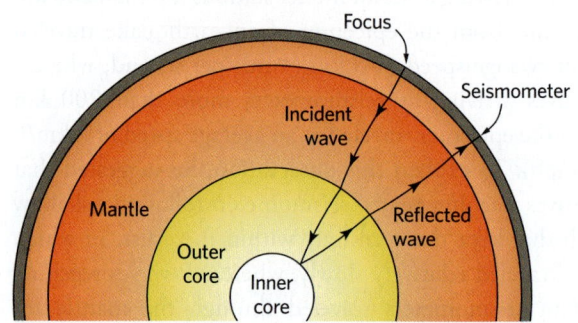

(c) A P-wave reflects off the inner core-outer core boundary.

(Fig. 8.39a). The presence of the P-wave shadow zone means that deep in the Earth, a major boundary exists where seismic waves abruptly refract downward because their velocity suddenly decreases. The dimensions of the shadow zone allowed seismologists to calculate that this boundary lies at a depth of 2,900 km (1,800 mi), and they consider it to be the **core-mantle boundary**.

Seismologists also found that S-waves did not arrive at seismometer stations located between 103°–180° from the epicenter (a band called the *S-wave shadow zone*)—which means that S-waves cannot pass through the core at all. If they could, an S-wave headed straight down through the Earth should reach the ground surface on the other side of the planet. Recall that S-waves cannot pass through liquid, so the fact that S-waves do not pass through the core means that the core, or at least part of it, consists of liquid (Fig. 8.39b).

At first, seismologists thought that the entire core might be liquid iron alloy. But in 1936, a Danish

FIGURE 8.40 The velocity-versus-depth curve is a graph showing how velocities of P-waves and S-waves vary with increasing depth in the Earth. Note that the graph does not show a velocity for S-waves in the outer core because S-waves cannot travel through molten iron (a liquid).

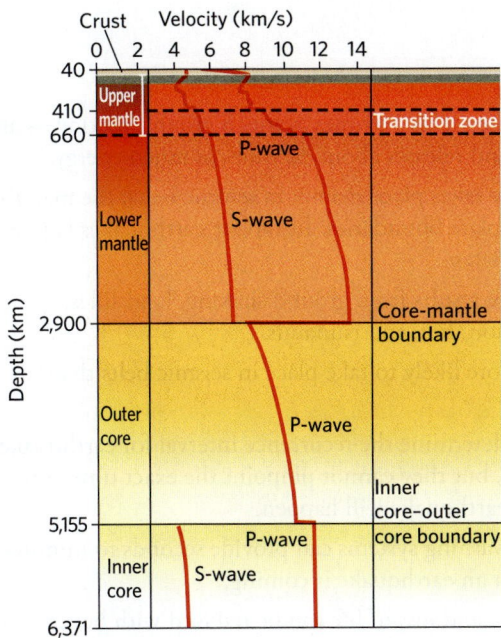

with location at a given depth. The regions of lower velocity probably represent warmer mantle material, and the regions of higher velocity probably represent cooler mantle material, for as rock gets hotter and softer, it transmits seismic energy less rapidly. The occurrence of warmer and cooler regions is a consequence of convection in the mantle **(Fig. 8.41b)**. The Earth's interior is indeed a dynamic place!

Take-home message . . .

By studying the velocity at which seismic waves travel through the Earth's interior, and by determining the depths at which seismic waves reflect or refract, seismologists have been able to locate the boundaries between Earth's layers, such as the Moho between the crust and mantle. Such work has also demonstrated that the outer core is molten. Using modern techniques, researchers can even identify patterns of convection in the mantle.

Quick Question
What is the S-wave shadow zone, and why does it exist?

FIGURE 8.41 Modern images of the Earth's interior.

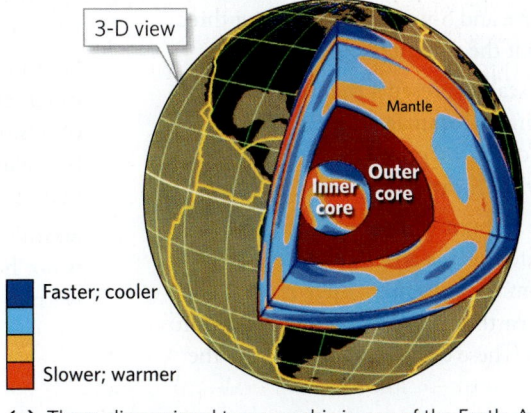

(a) Three-dimensional tomographic image of the Earth. Areas of lower seismic velocity may be relatively warmer than their surroundings, and areas of higher velocity may be relatively cooler.

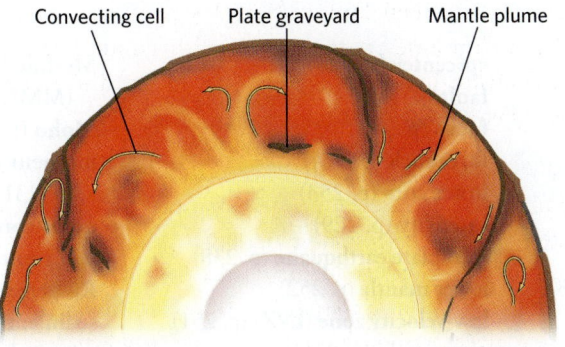

(b) A conceptual image of what the Earth's interior looks like. Hotter mantle material undergoes upwelling, and cooler mantle material sinks. Plates form at the surface and sink back down into the mantle.

seismologist, Inge Lehmann, discovered that P-waves passing through the core reflected off a boundary within the core **(Fig. 8.39c)**. She proposed that the core includes two parts: an **outer core** consisting of liquid iron alloy, and an **inner core** consisting of solid iron alloy. The depth of the boundary between the inner and outer core was eventually located at a depth of about 5,155 km (3,203 mi).

A MODERN IMAGE OF EARTH'S LAYERS. Through painstaking effort, seismologists have compiled data on seismic-wave travel times to develop a graph, known as a *velocity-versus-depth curve*, that shows the average depths at which seismic-wave velocity suddenly changes and the average amounts of change. Depths at which major changes take place define the principal layers and sublayers of the Earth down to its center **(Fig. 8.40)**.

More detailed studies in recent years have shown that the onion-like layered model of the Earth we've described so far is an oversimplification. Using a technique called *seismic tomography*, seismologists can produce three-dimensional images of seismic-velocity variation in the Earth's interior, just as doctors produce three-dimensional CT (computerized tomography) scans of the human body **(Fig. 8.41a)**. Tomographic studies allow seismologists to identify regions in the mantle where seismic waves travel faster or slower than expected, and these studies have led to the realization that the velocities of seismic waves vary significantly

8 CHAPTER REVIEW

- An earthquake is an event generating vibrations in the Earth that can lead to ground shaking at the Earth's surface.

- Earthquakes are generally a consequence of slip on a fault. Prior to slip, rock deforms elastically. During slip, the rock rebounds to its original shape, sending out vibrations.

- Some earthquakes are due to formation of new faults, but most happen when stress overcomes friction on a pre-existing fault and the fault slips. Faults typically exhibit stick-slip behavior.

- The place where an earthquake begins is called the focus of the earthquake, and the point on the ground surface directly above the focus is its epicenter.

- Earthquake energy travels in the form of seismic waves. Body waves, which pass through the interior of the Earth, include P-waves and S-waves. Surface waves pass along the surface of the Earth.

- A seismometer can detect seismic waves. Seismograms demonstrate that different types of seismic waves travel at different velocities. Using the difference between P-wave and S-wave arrival times at three locations, seismologists can pinpoint the epicenter.

- The Modified Mercalli Intensity scale represents the size of an earthquake at a locality by assessing the damage caused by an earthquake and people's perception of the ground shaking. Earthquake intensity decreases with distance from the epicenter.

- Magnitude scales characterize the amount of energy released at the source of an earthquake, as indicated by the amplitude of ground motion at a reference distance from the epicenter. There is one magnitude number for any given earthquake. The Richter scale is an early version of a magnitude scale. These days, seismologists use the moment magnitude (M_w) scale.

- An M_w 8 earthquake yields 10 times as much ground motion as an M_w 7 earthquake and releases about 32 times as much energy.

- Earthquake activity takes place mainly in seismic belts, the majority of which lie along plate boundaries. Intraplate earthquakes happen in the interior of plates.

- Earthquake damage results from ground shaking, landslides, sediment liquefaction, fire, and tsunamis.

- Earthquakes are more likely to take place in seismic belts than elsewhere.

- Seismologists can determine the recurrence interval for earthquakes in a particular belt, but they cannot pinpoint the exact time and place at which an earthquake will happen.

- Earthquake early-warning systems can provide seconds to minutes of advance notice that an earthquake is coming.

- Earthquake damage and loss of life can be reduced with better construction practices and zoning and by educating people about what to do during an earthquake.

- Seismic waves reflect or refract at boundaries between layers of different materials inside the Earth. By studying the movements of seismic waves, seismologists can identify the depths at which boundaries between layers occur, and they can identify where material is solid or molten.

- Seismic tomography studies show that seismic velocity in the mantle is not homogeneous. This variation indicates that some regions are warmer and softer than others.

Key Terms

aftershock (p. 224)
amplitude (p. 227)
arrival time (p. 228)
body wave (p. 225)
compressional wave (p. 225)
core-mantle boundary (p. 252)
earthquake (p. 221)
earthquake engineering (p. 248)
earthquake zoning (p. 249)
elastic deformation (p. 223)
elastic-rebound theory (p. 224)

epicenter (p. 225)
fault (p. 222)
focus (p. 224)
foreshock (p. 224)
inner core (p. 253)
intensity (p. 229)
intraplate earthquake (p. 238)
lower mantle (p. 252)
low-velocity zone (LVZ) (p. 251)
magnitude (p. 230)
mainshock (p. 224)

Modified Mercalli Intensity (MMI) scale (p. 229)
Moho (p. 251)
moment magnitude scale (p. 231)
outer core (p. 253)
recurrence interval (p. 246)
Richter scale (p. 230)
sediment liquefaction (p. 241)
seismic belt (p. 232)
seismicity (p. 222)

seismic wave (p. 222)
seismogram (p. 227)
seismometer (p. 225)
shear wave (p. 225)
stick-slip behavior (p. 224)
surface wave (p. 225)
transition zone (p. 252)
travel time (p. 250)
tsunami (p. 242)
upper mantle (p. 252)

Letters in parentheses correspond to the chapter's learning objectives.

1. What is an earthquake? How does earthquake energy travel through the Earth? **(A)**

2. What is a fault? Are all faults likely to generate earthquakes? Distinguish among foreshocks, the mainshock, and aftershocks. How does the size of an earthquake relate to the area of a fault that slips, and the amount of slip that occurs? **(A)**

3. Do all earthquakes require that a new fault be formed? Describe elastic-rebound theory and the concept of stick-slip behavior. **(A)**

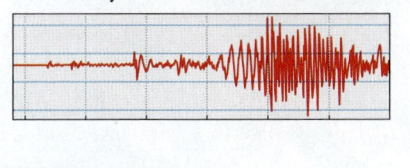

4. Label the focus and epicenter of the earthquakes shown in the diagram. **(A)**

5. What is the difference between a body wave and a surface wave? How do P-waves and S-waves differ from one another? How do R-waves and L-waves differ from one another? Label P, S, and surface waves on the seismogram. **(A)**

6. Explain how the ground movements produced by an earthquake are detected by a seismometer. **(B)**

7. How can seismologists determine the location of an earthquake's epicenter? **(B)**

8. Explain the differences among the scales used to describe the size of an earthquake. **(C)**

9. How does seismicity on mid-ocean ridges compare with seismicity at convergent or transform boundaries? Do all earthquakes occur at plate boundaries? **(D)**

10. What is a Wadati-Benioff zone? Why do intermediate-focus and deep-focus earthquakes occur there? **(D)**

11. Describe some non-plate-boundary geologic settings where earthquakes can occur. **(D)**

12. Describe the types of damage caused by earthquakes. Is all damage due to ground shaking? **(E)**

13. What is sediment liquefaction, and how can it cause damage during an earthquake? **(E)**

14. What is a tsunami, and why do tsunamis form? How do they differ from storm waves? **(E)**

15. How are long-term earthquake predictions made? What is the basis for determining a recurrence interval? **(F)**

16. What is an earthquake early-warning system? **(F)**

17. What types of structures are most prone to collapse in an earthquake? What types are most resistant to collapse? What causes most loss of life during an earthquake? **(G)**

18. Why do seismic waves refract at specific depths within the Earth?

19. What clue led to the definition of the Moho? **(I)**

20. What causes P-wave and S-wave shadow zones, and what does their presence imply? Which shadow zone does the figure show? **(H)**

21. What does a tomographic image of the Earth's interior show? **(I)**

22. Is seismic risk greater in a town on the western coast of South America or in one on the eastern coast? Explain your answer. **(D)**

23. On the seismogram of an earthquake recorded at a seismometer station in Paris, France, the S-wave arrives 6 minutes after the P-wave. On the seismogram obtained by a station in Mumbai, India, for the same earthquake, the difference between the P-wave and S-wave arrival times is 4 minutes. Which station is closer to the epicenter? From the information provided, can you pinpoint the location of the epicenter? Explain. **(C)**

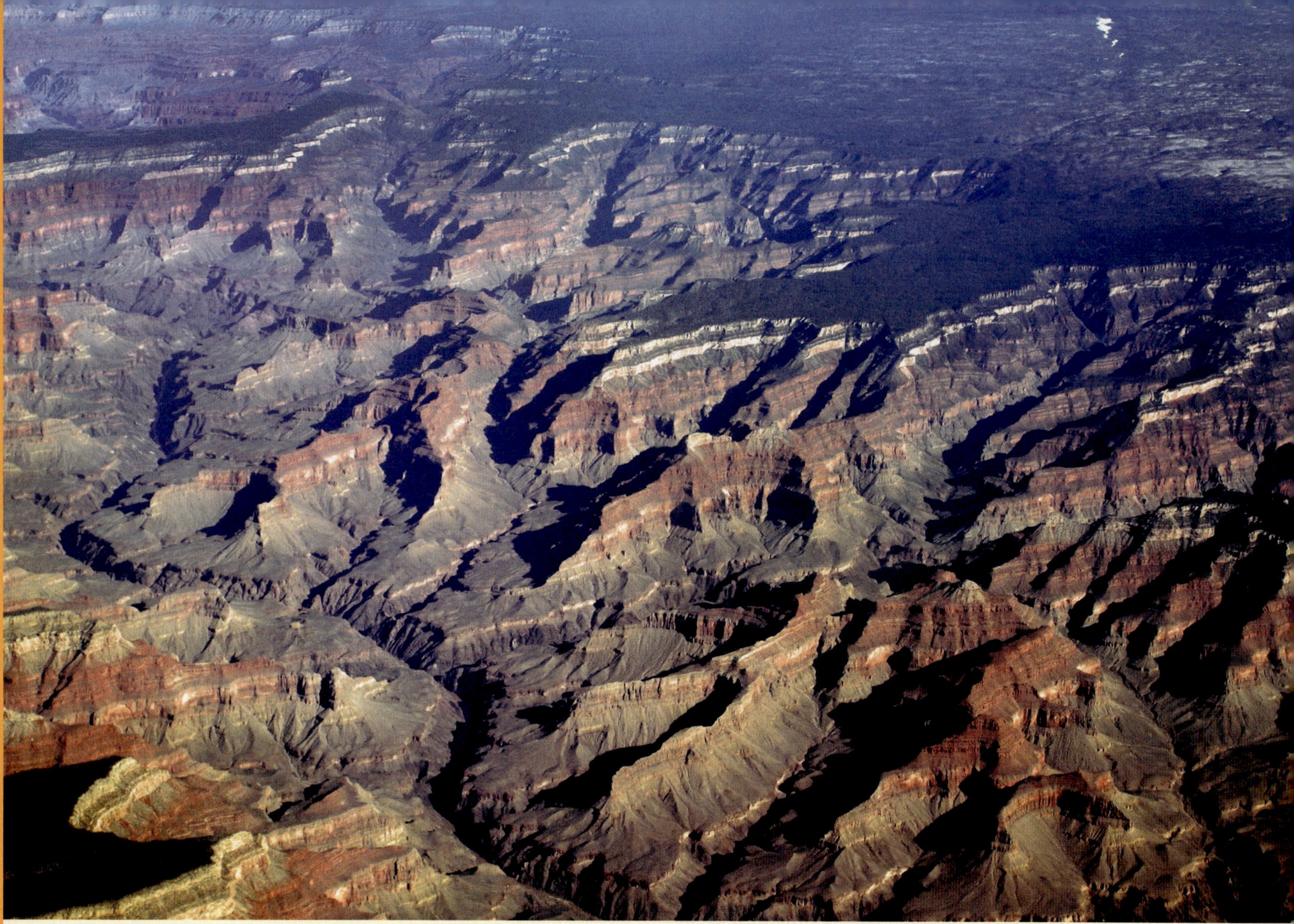

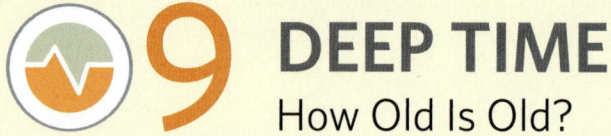

9 DEEP TIME
How Old Is Old?

In 1869, a one-armed Civil War veteran named John Wesley Powell, along with nine companions, set out to explore the Grand Canyon, the greatest gorge on the Earth. For three months, they drifted down the Colorado River, which flows down the floor of the canyon. During their voyage, seemingly insurmountable walls of rock both imprisoned and amazed the explorers and led them to pose important questions about the Earth and its history, the same questions that tourists visiting the canyon may ponder today: How long did it take to carve the canyon? When did the rocks making up the walls of the canyon form? Was there a time before these rocks accumulated? Thinking about such questions opens the door to thinking about **geologic time**, the span of time since the Earth's formation.

Our modern understanding of geologic time stems from 19th-century work that established principles for describing the relative age of geologic features. By *relative age*, we mean whether one feature is older or younger than another. With this concept in mind, we begin this chapter by discussing relative-age determination. We next introduce fossils, which can help to sort out age relationships, and we describe what they tell us about the evolution of life on the Earth. Understanding relative ages sets the stage for developing the *geologic column*, a chart that divides geologic time into intervals. When geologists developed methods for *isotopic dating* (radiometric dating) in the mid-20th century, it became possible to define the *numerical age*—the age in years—of rocks. Using isotopic dating, geologists determined numerical ages of intervals on the geologic column. This tool led to the production of the *geologic time scale* and, ultimately, to an estimate of when the Earth itself formed.

With the concept of geologic time in hand, a hike down a trail into the Grand Canyon becomes a trip through what popular authors call *deep time*. The geologic discovery that our planet's history extends billions of years before the start of human history changed humanity's perception of time and the Universe as profoundly as the astronomical discovery that the limit of deep space extends billions of light-years beyond the edge of our Solar System.

<< A hike down through the Grand Canyon of Arizona, seen here from an airplane, takes you on a journey back through deep time. Rocks at the base of the Canyon are about 1.8 billion years old.

9.2 Geologic Principles and Relative Age

The discovery of physical laws by Sir Isaac Newton helped spark the Age of Enlightenment in Europe, a time when scientists began to seek natural, rather than supernatural, explanations for the features and phenomena in the world around them. By the 1850s, geologists had established several principles that ultimately provided a foundation for developing the concept of geologic time.

Uniformitarianism

While wandering in the highlands of Scotland, James Hutton (1726–1797) noticed that many of the features he found in sedimentary rocks resembled features that he could see forming in modern depositional environments. For example, the surfaces of some sandstone beds displayed ripple marks identical to those he saw on a modern beach. These observations led Hutton to speculate that ancient rocks and landscapes were a long-ago product of the same natural processes operating in modern times.

Hutton's idea has come to be known as the principle of **uniformitarianism**. This principle means that physical processes operating in the modern world also operated in the past, at roughly the same rates (Fig. 9.1). More concisely, uniformitarianism means that "the present is the key to the past." Hutton further deduced that not all the rocks he observed, or the structures that affected them, could have formed at the same time. Because no one can see the entire process of sediment first turning into rock and then later rising into mountains, Hutton concluded that the production of rocks is very slow, and that the Earth's history must go back a long time before human history began. This proposal, along with several others, became the basis for so much geologic thinking that modern geologists consider Hutton "the father of geology."

Determining Relative Ages and Geologic History

The time at which one event happens with respect to others, in a sequence, defines the **relative age** of the events. For example, when a historian says that World War II happened after World War I, without stating the years during which the wars happened, they are specifying the relative ages of the two wars. Historians can figure out relative ages of these wars by examining evidence indicating that one war led to the other. When geologists say that one rock is older than another, they are specifying the relative ages of the two rocks. The following principles provide the foundation for determining the relative ages of rocks as well as of other geologic features and events:

FIGURE 9.1 The principle of uniformitarianism.

(a) Mudcracks form when a modern-day mud puddle dries up. Note the pen for scale.

(b) A 300-million-year-old rock containing preserved mudcracks. According to the principle of uniformitarianism, they formed the same way as modern mudcracks.

- The principle of **original horizontality** states that layers of sediment, when first deposited, are roughly horizontal **(Fig. 9.2a)**. Why? Sediments accumulate on relatively flat surfaces, such as floodplains or the seafloor. If they collect on a steep slope, they slide downslope before they can be buried and lithified. With this principle in mind, geologists conclude that folds and tilted beds represent deformation events that happened after deposition.

- The principle of **superposition** states that each layer of sedimentary rock must be younger than the one below it, for a layer of sediment cannot accumulate unless there is already a surface on which it can collect. Therefore, in a sedimentary sequence, the oldest layer lies at the bottom and the youngest at the top **(Fig. 9.2b)**.

- The principle of **cross-cutting relations** states that if one geologic feature cuts across another, the feature that has been cut is older. For example, if an igneous dike cuts across a sequence of sedimentary beds, the beds must be older than the dike **(Fig. 9.2c)**. If a layer of sediment buries the dike, the sediment must be younger than the dike; and if a fault cuts across and displaces layers of sedimentary rock, then the fault must be younger than the layers.

- The principle of **baked contacts** states that an igneous intrusion "bakes" (or thermally metamorphoses) wall rock, so the rock that contains a metamorphic aureole must be older than the intrusion **(Fig. 9.2d)**.

- The principle of **inclusions** states that a rock containing an *inclusion* (fragment of another rock) must be younger than the inclusion. So a conglomerate containing basalt pebbles is younger than the basalt, whereas a basalt containing sandstone fragments must be younger than the sandstone **(Fig. 9.2e)**.

Animation

Relative Age Dating

Geologists apply the above principles to determine the relative ages of geologic features (rocks, structures, erosional features), each of which is the consequence of a specific *geologic event*. Examples of geologic events include deposition, erosion, intrusion or extrusion of igneous rocks, and deformation (folding or faulting). The sequence of events, in terms of relative age, defines the **geologic history** of the region.

To visualize how to unravel the geologic history of a region, let's decipher the relative ages of the geologic events depicted in **Figure 9.3**. Based on the principle of superposition, deposition of Bed 1 happened first, followed by deposition of Beds 2 through 8. We know that the sill intruded after the deposition of Bed 5 because it contains sandstone inclusions. Then all the beds, together with the sill, underwent folding. The granite intruded after the folding, because we see that the pluton cuts the fold and that the folded rocks have been metamorphosed along the contact. The fault then slipped, because it cuts and offsets the pluton. The dike intruded after the fault, because it cuts the fault. Finally, erosion formed the present ground surface, which cuts across all other features. Note that Bed 8 has been completely eroded away.

Take-home message . . .

The principle of uniformitarianism—the present is the key to the past—provides a basis for interpreting geologic features and implies that the Earth must be old. This principle, together with others derived from it, allow geologists to determine the relative ages of features and construct the geologic history of a region.

Quick Question

What observations led Hutton to propose uniformitarianism?

FIGURE 9.2 Geologic principles used for determining relative ages.

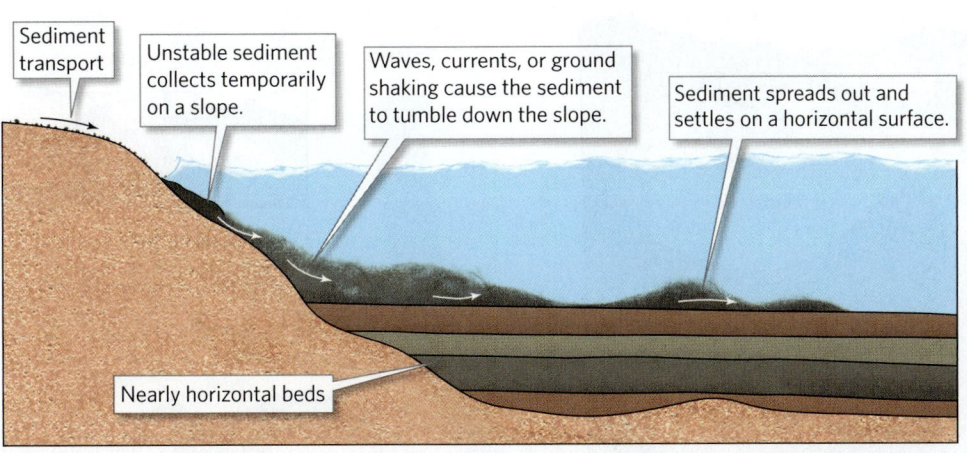

Horizontal sandstone beds in Wisconsin

Bedding plane

Cross beds

What an Earth Scientist Sees

(a) Original horizontality: Gravity causes sediment to accumulate in horizontal sheets. So, when we see horizontal beds of rock in Wisconsin (right), we assume that the beds are in their original orientation.

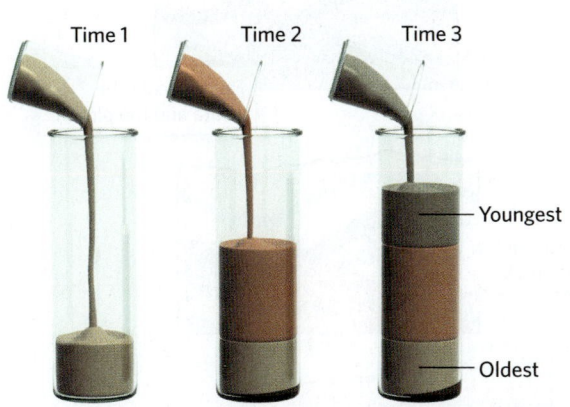

Time 1 Time 2 Time 3

Youngest

Oldest

(b) Superposition: In a sedimentary sequence, the oldest bed is on the bottom and the youngest on the top. Pouring sand into a glass cylinder illustrates this point. In the photo on the right, the beds get younger going up.

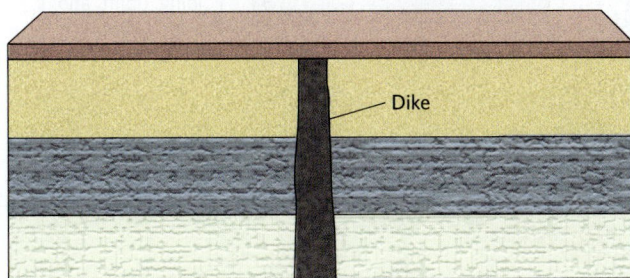

Dike

(c) Cross-cutting relations: The dike is younger than the beds it cuts across. The sediment layer that buries the dike is younger than it.

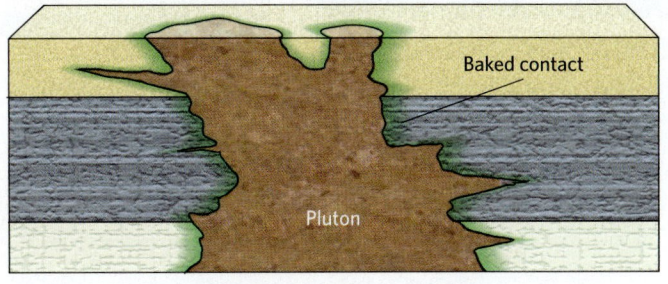

Baked contact

Pluton

(d) Baked contacts: A metamorphic aureole (green area) surrounds a pluton. The pluton is younger than the rock it has baked.

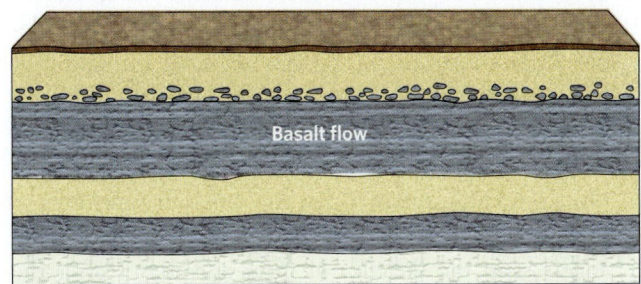

Basalt flow

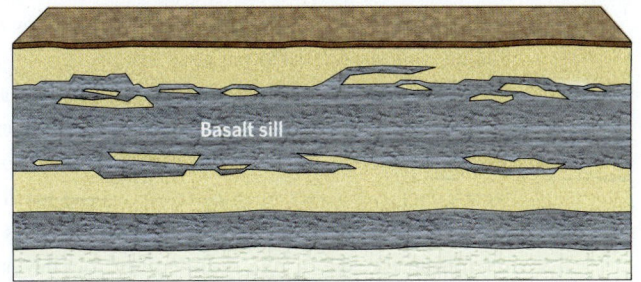

Basalt sill

(e) The pebbles derived from the underlying basalt flow must be older than the conglomerate that contains them (left). The xenoliths of sandstone incorporated in a basalt sill must be older than the sill that contains them (right).

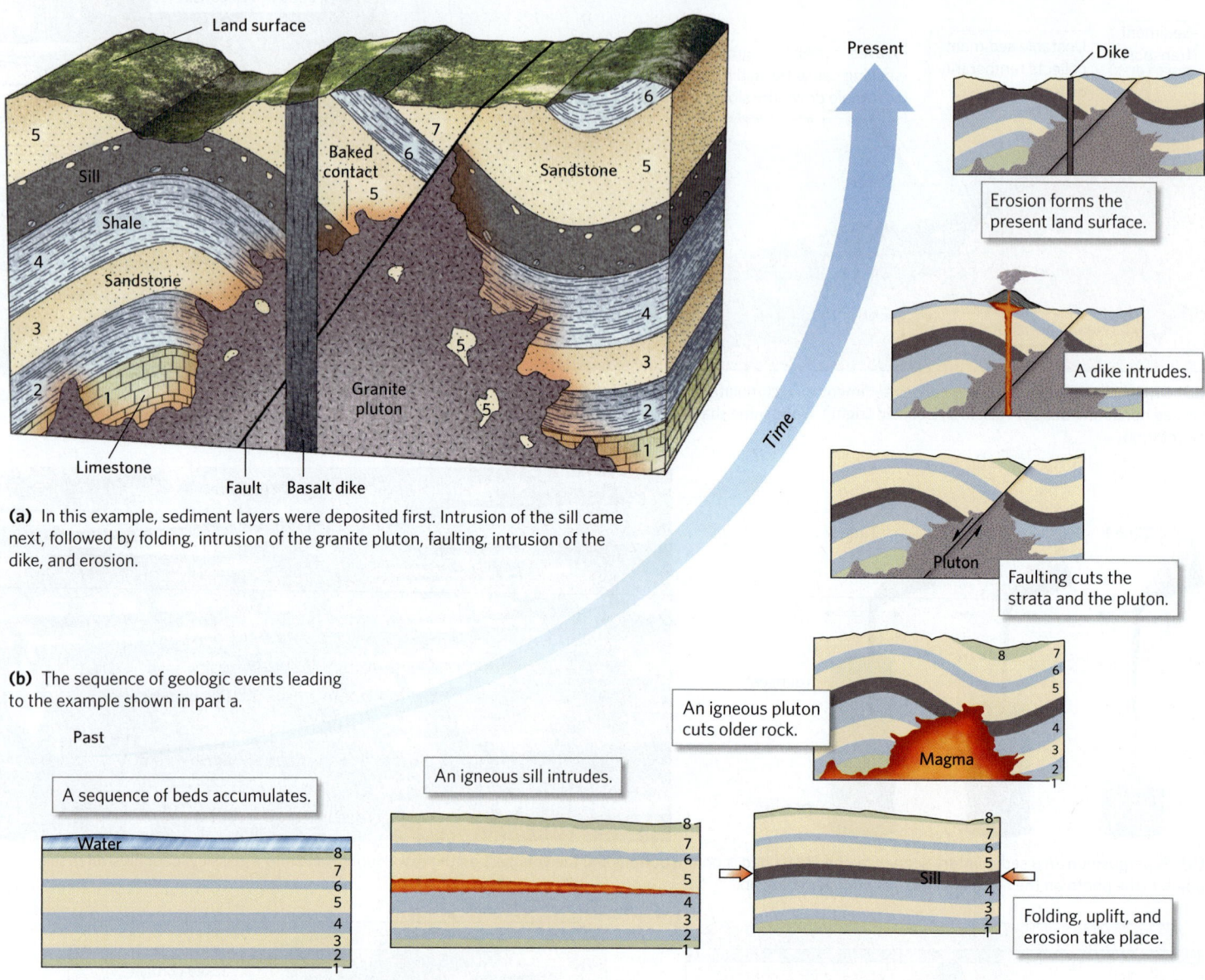

(a) In this example, sediment layers were deposited first. Intrusion of the sill came next, followed by folding, intrusion of the granite pluton, faulting, intrusion of the dike, and erosion.

(b) The sequence of geologic events leading to the example shown in part a.

Erosion forms the present land surface.

A dike intrudes.

Faulting cuts the strata and the pluton.

An igneous pluton cuts older rock.

An igneous sill intrudes.

A sequence of beds accumulates.

Folding, uplift, and erosion take place.

9.3 Memories of Past Life: Fossils and Evolution

If you look at sedimentary strata, you may find shapes that resemble shells, bones, leaves, or footprints (Fig. 9.4a). Researchers consider such **fossils** (from the Latin word *fossilis*, which means dug up) to be remnants or traces of ancient living organisms preserved in rock.

The 19th century saw **paleontology**, the study of fossils, ripen into a science as museum drawers filled with specimens (Fig. 9.4b). As we'll see, *paleontologists* (researchers who specialize in studying the fossil record) eventually learned how to use fossils as a basis for determining the age of one sedimentary rock layer relative to another, so fossils have become an indispensable tool for studying geologic history and the evolution of life (Fig. 9.4c).

Formation and Preservation of Fossils

Fossils form when organisms die and become buried by sediment or ash, or when organisms travel over or through sediment and leave imprints or debris. Paleontologists refer to the process of fossil formation as **fossilization**. To see how a typical fossil develops in sedimentary rock, let's follow the fate of an old dinosaur as it searches for food along a muddy riverbank (Fig. 9.5). On a scorching summer day, the dinosaur succumbs to the heat and collapses dead in the mud. Soon after, scavengers strip the skeleton of meat and may scatter the

FIGURE 9.4 Examples of fossils and fossil collections.

(a) Fossil skeleton in 200-million-year-old sandstone.

(b) A drawer of fossil specimens in a museum.

(c) A paleontologist collecting specimens.

bones. But before the bones have had time to weather away, the river floods and buries the bones, along with the dinosaur's footprints, under a layer of silt. More sediment buries the bones and prints still deeper, until eventually, the sediment containing the bones and footprints turns to rock (siltstone and shale). The footprints remain outlined by the boundary between the siltstone and the shale, while the bones reside within the siltstone. Minerals precipitated from groundwater passing through the siltstone gradually replace some of the chemicals constituting the bones, until the bones themselves become rock-like. The buried bones and footprints

are now fossils. A hundred million years later, uplift and erosion expose the dinosaur's grave, and a lucky paleontologist finds them and excavates them. The dinosaur rises again, but this time in a museum.

Not all living organisms become fossils when they die. In fact, only an exceedingly small number do, for it takes special circumstances to produce a fossil and allow it to survive:

- *Death in an anoxic environment:* A dead squirrel on a road won't become a fossil. As time passes, scavengers come along and eat the carcass. If that doesn't happen, microbes infest the carcass and gradually digest it, or oxidation (chemical reaction with oxygen) breaks it down into gases. A carcass has a better chance of being preserved if it settles in an anoxic (oxygen-poor) environment, where oxidation happens slowly, scavenging organisms aren't abundant, and microbial metabolism takes place very slowly.

- *Rapid burial:* If an organism dies in a depositional environment where sediment accumulates rapidly, it has a better chance of being buried before disintegrating.

- *The presence of hard parts:* Organisms without durable shells, skeletons, or other hard parts usually won't be fossilized, for soft flesh decays long before hard parts under most depositional conditions. For this reason, paleontologists have identified many more fossils of oysters, for example, than of jellyfish.

By carefully studying modern organisms, paleontologists have been able to estimate the **preservation potential** of organisms, meaning the likelihood that an organism will be buried and transformed into a fossil. In a typical modern-day shallow-marine environment, only a small fraction of the organisms have a high preservation potential. Of these organisms, only a few die in a depositional setting where they actually become fossilized. So fossilization is the exception rather than the rule.

The Many Different Kinds of Fossils

Perhaps when you think of a fossil, you picture either a dinosaur bone or the imprint of a seashell in rock. In fact, paleontologists distinguish many different kinds of fossils according to the specific way in which the organisms were fossilized. Let's look at examples of these categories.

- *Frozen or dried body fossils:* In a few environments, whole bodies of organisms may be preserved. Most of these fossils are fairly young by geologic standards—their ages can be measured in thousands of years. Examples include woolly mammoths that became incorporated in the permafrost (permanently frozen ground) of Siberia (Fig. 9.6a) or "mummified" fossils preserved in desert caves.

Animation

Fossils and Fossilization: Making of a Fossil

FIGURE 9.5 The stages in fossilization of a dinosaur.

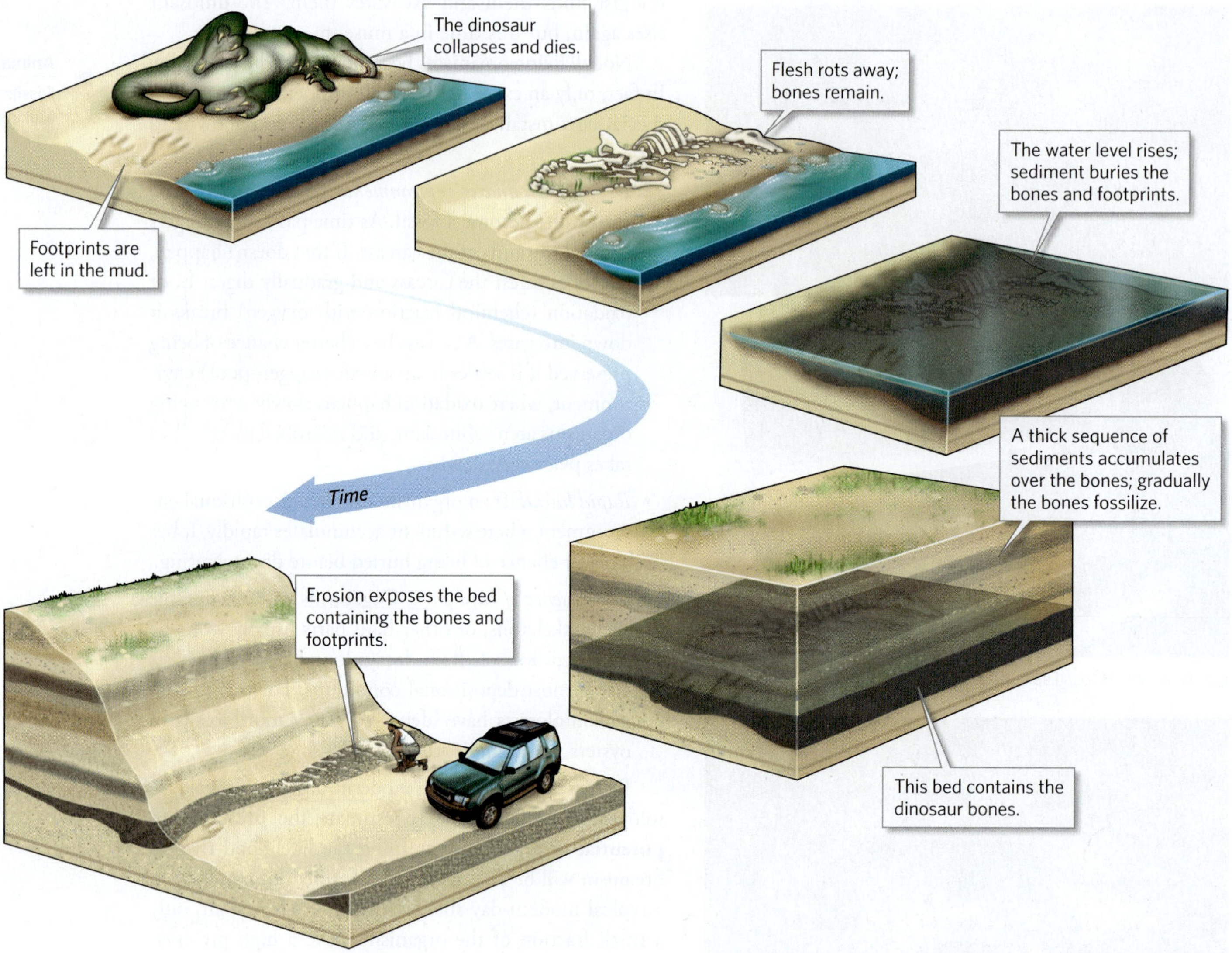

The dinosaur collapses and dies.

Footprints are left in the mud.

Flesh rots away; bones remain.

The water level rises; sediment buries the bones and footprints.

Time

A thick sequence of sediments accumulates over the bones; gradually the bones fossilize.

Erosion exposes the bed containing the bones and footprints.

This bed contains the dinosaur bones.

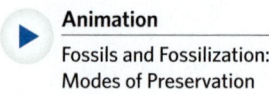

Animation

Fossils and Fossilization: Modes of Preservation

- *Body fossils preserved in amber or tar:* Insects landing on trees may become trapped in the sticky sap that trees produce. This sap envelops the insects and over time hardens into amber. Amber can preserve insects for 40 million years or more **(Fig. 9.6b)**.

- *Preserved or replaced bones, teeth, and shells:* Bones, teeth, and shells consist of durable minerals that may survive in tar or rock **(Fig. 9.6c)**. Some bone, tooth, or shell minerals are not stable and recrystallize over time. But even when this happens, the shape of the original item may be preserved in the rock.

- *Molds and casts:* As sediment compacts around a shell or body, it conforms to the shape of that item. If the shell or body later disappears because of weathering and dissolution, a cavity called a *mold* remains. If sediment later fills the mold, it too preserves the organism's

shape **(Fig. 9.6d)**. The resulting *cast* protrudes from the surface of the adjacent bed. Usually only hard parts turn into molds or casts. Rarely, the shapes of soft parts may be preserved, forming *extraordinary fossils* **(Fig. 9.6e)**.

- *Carbonized impressions of bodies:* Impressions are flattened molds created when soft or semisoft organisms or their parts (leaves, insects, invertebrates, sponges, feathers, jellyfish) get pressed between layers of sediment. Chemical reactions eventually remove most organic material, leaving only a thin film of carbon on the surface of the impression **(Fig. 9.6f)**.

- *Permineralized organisms: Permineralization* refers to the process by which minerals precipitate from groundwater that has seeped into the pores of porous

FIGURE 9.6 Examples of different kinds of fossils.

(a) This 1-m-long baby mammoth, found in permafrost in Siberia, died 37,000 years ago.

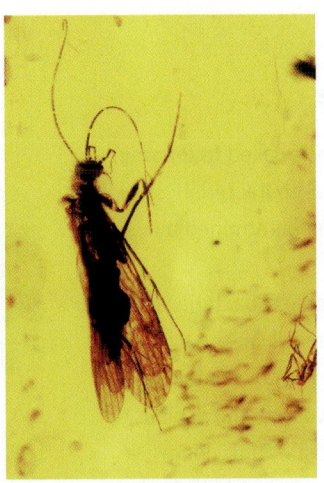

(b) This insect became embedded in amber about 200 million years ago.

(c) A fossil skeleton of a 2-m-high giant ground sloth from the La Brea Tar Pits in California.

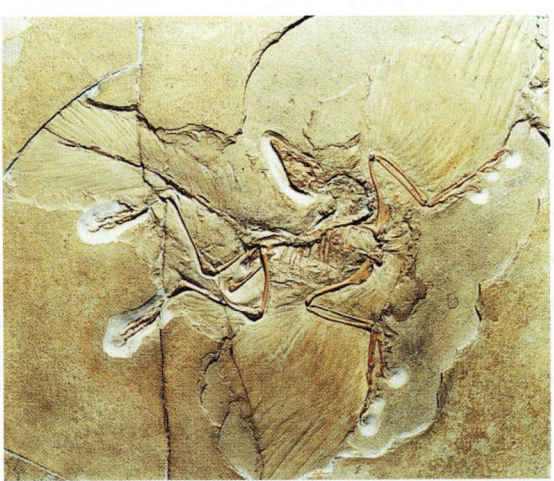

(d) A mold of *Archaeopteryx*, a very early bird, from the 150-million-year-old Solnhofen Limestone of Germany. The imprints of feathers are clearly visible.

(e) The hard parts of invertebrates are the most likely to be preserved. Here we see casts of brachiopod shells.

(f) Carbonized impressions of fern fronds in shale.

(g) Petrified wood from Arizona. Chert replaced the original wood.

(h) These dinosaur footprints from Connecticut are one form of trace fossil.

(i) Very tiny fossils, formed from plankton shells, are called microfossils.

A Deeper Look

Taxonomy: The organization of life

The study of how to identify and name organisms is called **taxonomy**. Traditionally, this was done by comparing physical characteristics. Now, relationships among some organisms can be determined by comparing DNA.

Biologists divide life into three domains (Archaea, Bacteria, and Eukarya). Archaea and Bacteria are microscopic unicellular organisms whose cells are prokaryotic (do not have a central nucleus containing DNA). Eukarya, whose cells are eukaryotic (do have a central nucleus), have traditionally been sorted into kingdoms (Protista, Fungi, Plantae, and Animalia), each of which consists of one or more phyla. Each phylum is divided into one or more classes, each class consists of one or more orders, each order of one or more genera, and each genus of one or more species. Taxonomic relationships among organisms can be portrayed on a *phylogenetic tree*, informally known as the *tree of life*, which shows which organisms radiate from a common ancestor **(Fig. Bx9.1)**. The configuration of the "tree" remains very controversial.

How are modern humans classified according to the phylogenetic tree? All humans are assigned to the genus *Homo* and the species *sapiens*. Humans, together with apes and monkeys, make up the order

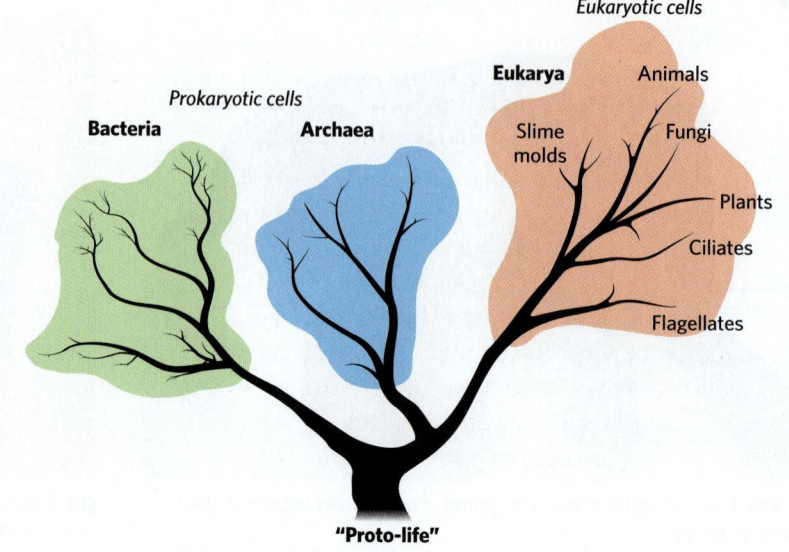

FIGURE Bx9.1 A simplified phylogenetic tree depicting taxonomic relationships among organisms.

Primates, which, together with other warm-blooded organisms with hair, constitute the class Mammalia. Mammals, along with all animals with backbones, are in the phylum Chordata, and all animals of any type are in the kingdom Animalia.

---

material, such as wood or bone. Petrified wood, for example, forms by permineralization of wood, which transforms the wood into chert **(Fig. 9.6g)**.

- *Trace fossils:* Trace fossils are marks or debris that organisms leave behind in sediment. Examples include footprints, feeding traces, burrows, and dung (coprolites) **(Fig. 9.6h)**.

- *Chemical fossils:* Living things consist of complex organic chemicals. Over geologic time, most of these chemicals break down. Some become different, but distinctive, chemicals. A distinctive chemical derived from an organism and preserved in rock is called a *chemical fossil* or *biomarker.*

Paleontologists also find it useful to distinguish among different fossils on the basis of their size. *Macrofossils* (like those in Fig. 9.6a–h) are fossils large enough to be seen with the naked eye. But some rocks and sediments also contain abundant *microfossils*, which can be seen only with a microscope **(Fig. 9.6i)**. Microfossils include remnants of plankton, bacteria, and pollen.

Classifying Fossils

Paleontologists classify fossils using the same principles that biologists use to classify modern organisms **(Box 9.1)**. There's nothing magical about classifying fossils. You can recognize common fossils in the field by examining their *morphology*, meaning their form or shape **(Fig. 9.7)**. If the fossil is complete and has distinctive features, the process can be straightforward, but if it's broken into fragments or if parts are missing, identification can be a challenge. Many fossil organisms resemble modern ones, so it is relatively easy to figure out how to classify them. For example, a fossil clam looks like a clam and not like, say, a snail. To classify a fossil more specifically, down to genus or species level, identification may involve recognizing such details as the number of ridges on the surface of its shell. Some fossils do not resemble known living organisms, however, so determining their taxonomic relationships can be difficult. Therefore, building a museum display that depicts fossil organisms as they appeared when alive requires imagination **(Fig. 9.8)**.

The Concept of Extinction

In the 18th century, paleontologists recognized that not all fossils represented the remains of observed living species. But they tacitly assumed that since the world had not been explored completely, all fossils represented species living somewhere on the planet. By the 19th century, it became clear that this interpretation could not be true, for explorers had not found giant animals, such as mastodons or dinosaurs, anywhere. Based on this realization, the French paleontologist Georges Cuvier (1769–1832) argued that some fossil species had gone **extinct**, meaning that all individuals of those species had died. We take the phenomenon of extinction for granted today, since we have seen numerous animals become extinct in historic time, but Cuvier's proposal was revolutionary in his day.

Using Fossils to Determine Relative Ages: Fossil Succession

As Britain entered the industrial revolution in the late 18th century, factories demanded coal to fire their steam engines and needed an inexpensive means to transport raw materials and manufactured goods. Investors decided to construct a network of canals, and they hired an engineer named William Smith (1769–1839) to survey some of the excavations. Canal digging provided fresh exposures of bedrock that had previously been covered by vegetation. Smith learned to recognize distinct layers of sedimentary rock and to identify the **fossil assemblage** (the group of fossil species) that each layer contained. He also realized that a particular assemblage could be found in only a limited sequence of strata (or beds) and not above or below. In other words, once a fossil species disappears at a level in a sequence of beds, it never reappears higher in the sequence. Extinction is forever.

Smith's observation, which has been repeated at millions of locations around the world, has been codified as the principle of **fossil succession**. To see how this principle works, examine Figure 9.9, which depicts a sequence of strata. Bed 1 at the base contains Species A, Bed 2 contains Species A and B, Bed 3 contains B and C, Bed 4 contains C, and so on. From these data, we can define the *range* for each species, meaning the interval in the sequence in which fossils of that species occur. In Figure 9.9, the succession of fossils, from oldest to youngest, is A, B, C, D, E, F. Note that the range of one species may overlap with that of another, and that when a species goes extinct, it does not reappear in higher beds.

Once the relative ages of fossil species have been determined, those species can be used to determine the relative ages of the beds containing them. For example, if a bed contains Fossil A (from Fig. 9.9), geologists can say that the bed is older than a bed containing Fossil F, even if the two beds do not crop out in the same area.

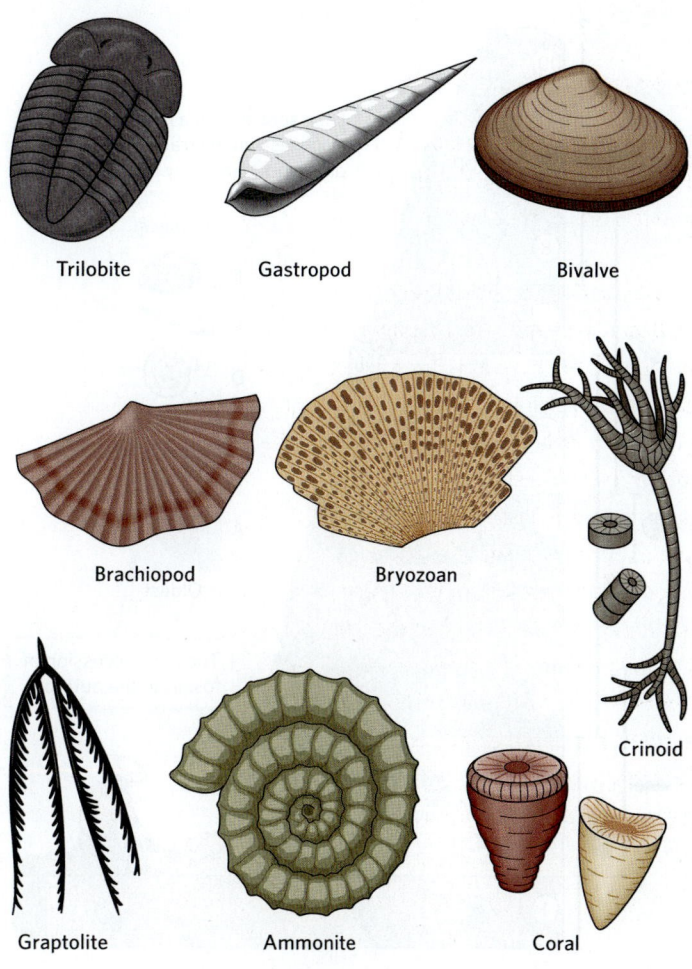

FIGURE 9.7 Common types of invertebrate fossils.

Trilobite   Gastropod   Bivalve

Brachiopod   Bryozoan   Crinoid

Graptolite   Ammonite   Coral

FIGURE 9.8 An artist's portrayal of what some fossil organisms looked like when alive. The painting is based on fossils found in the Burgess Shale of Canada.

FIGURE 9.9 The principle of fossil succession.

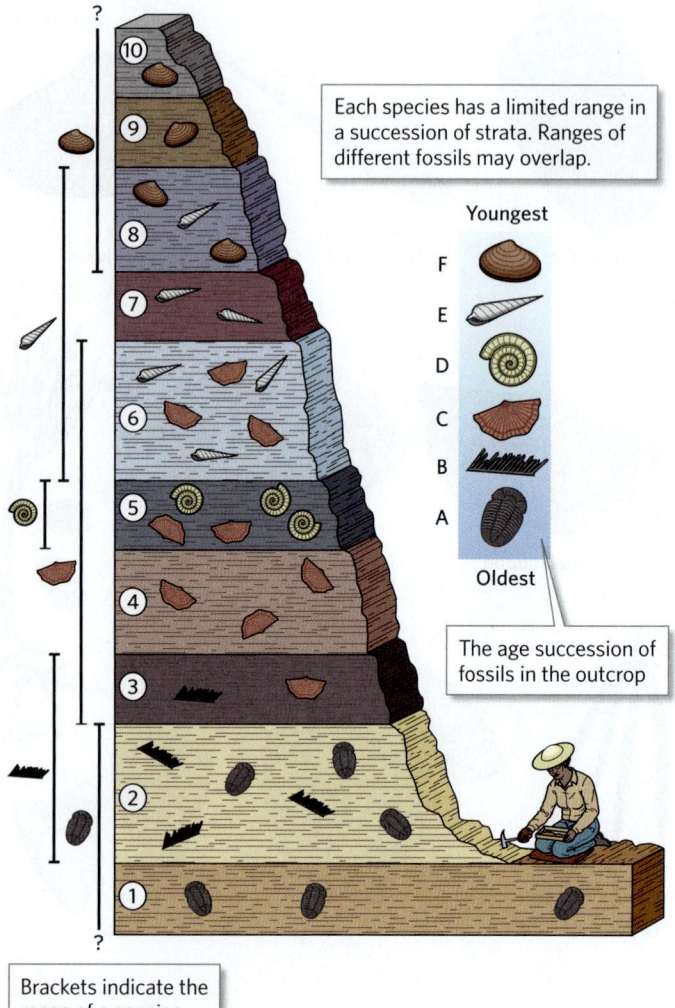

Each species has a limited range in a succession of strata. Ranges of different fossils may overlap.

Youngest

F
E
D
C
B
A

Oldest

The age succession of fossils in the outcrop

Brackets indicate the range of a species.

▶ Animation

Biostratigraphy (Interactive)

Tracing the History of Life

By some estimates, paleontologists have collected more than 250,000 different species of fossils during the past two centuries. This *fossil record* has allowed paleontologists to reconstruct the history of life over time. Paleontologists realized that, to explain fossil succession, something must have led to the extinction of some species and the appearance of new species.

The existence of evolution is an observation based on studies of the fossil record. But the observation doesn't explain how evolution happens. Charles Darwin (1809–1882) and Alfred Wallace (1823–1913) addressed this problem in an 1858 paper that introduced the *theory of evolution by natural selection*. (Darwin expanded greatly on this idea in a book called *On the Origin of Species*, which was published the following year.) In essence, the **theory of evolution** states that the traits that make an individual organism more successful in competing for survival and in attracting mates are traits that succeeding generations can inherit, because

successful individuals are more likely to have offspring. Individuals whose traits make them less capable of surviving or of attracting a mate are less likely to produce offspring, so their traits do not appear in succeeding generations. This phenomenon, known as *survival of the fittest*, eventually leads to the appearance of new species over the course of thousands of generations. Modern studies using DNA have revealed that natural *mutations* (changes in the structure of the DNA molecule) impact the speed at which evolution takes place, and that evolution also involves the transfer of bits of DNA from one species to another.

Though Darwin's theory of evolution provides a basis for explaining fossil succession, many mysteries still remain surrounding the path of life's evolution on the Earth. Known fossils cannot account for every intermediate step in the evolution of every species. Over the billions of years that life has existed, 5 to 50 billion species may have lived. Paleontologists have discovered only a tiny percentage of these species. Why is the record so incomplete? First, despite all the fossil-collecting efforts of the past two centuries, paleontologists have not even come close to sampling every cubic centimeter of sedimentary rock exposed on the Earth. Just as biologists have not yet identified every living species, paleontologists have not yet identified every fossil species. Second, not all species are represented in the fossil record because not all species have a high preservation potential—only a minuscule fraction of the species that have lived on the Earth have left a fossil record. Finally, as you will learn in the next section of this chapter, the sequence of sedimentary strata that exists on the Earth does not account for every minute of geologic time at every location. Sediments accumulate only in environments whose conditions are appropriate for deposition, so strata accumulate only episodically.

Take-home message . . .

Fossils are the remnants or traces of ancient organisms. Many different kinds of fossils may form in many ways. Not all organisms or their parts have equal preservation potential, so the fossil record is incomplete. Fossils provide a basis for geologists to determine the relative ages of rock layers as well as a record of life's evolution.

Quick Question

How can the principle of fossil succession be used to determine relative ages of rock layers?

9.4 Establishing the Geologic Column

Stratigraphic Formations

Geologists divide strata of a region into distinct units called stratigraphic formations. Formally defined, a **stratigraphic formation** (or simply *formation*) is a sequence of beds that

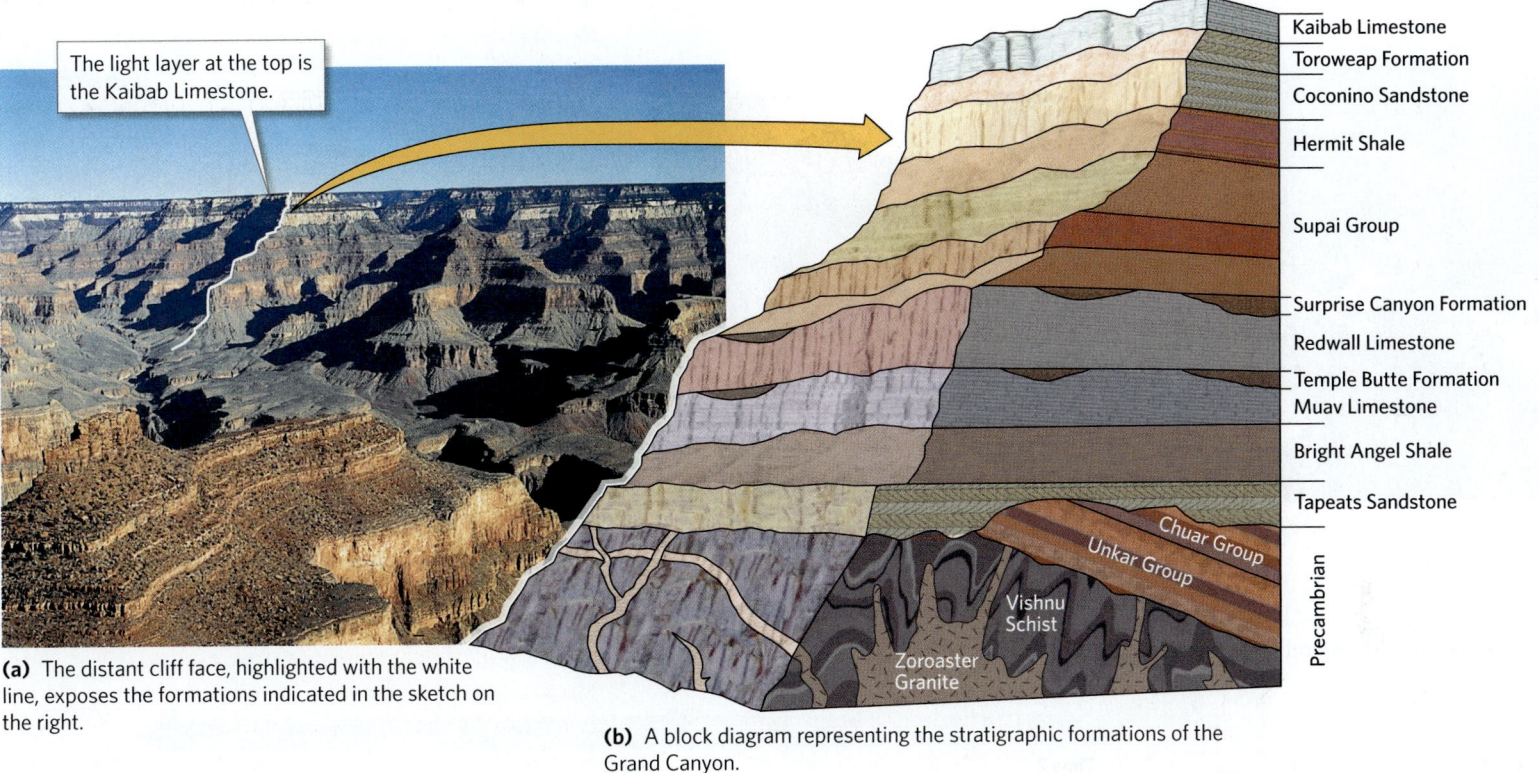

The light layer at the top is the Kaibab Limestone.

Kaibab Limestone
Toroweap Formation
Coconino Sandstone
Hermit Shale
Supai Group
Surprise Canyon Formation
Redwall Limestone
Temple Butte Formation
Muav Limestone
Bright Angel Shale
Tapeats Sandstone
Chuar Group
Unkar Group
Vishnu Schist
Zoroaster Granite
Precambrian

(a) The distant cliff face, highlighted with the white line, exposes the formations indicated in the sketch on the right.

(b) A block diagram representing the stratigraphic formations of the Grand Canyon.

can be traced over a fairly broad region (Fig. 9.10). Some formations include a single rock type, whereas others include interlayered beds of two or more rock types. While most formations consist of sedimentary strata, some also include extrusive igneous rocks. Typically, a formation represents the products of deposition during a definable interval of time. Therefore, a given formation can be assigned a specific age range, and characteristics of rocks in a formation provide clues about the environment at the time of deposition.

Not all formations have the same thickness, and the thickness of a single formation can vary with location. Commonly, geologists name a formation after a locality where it was first identified or first studied. If a formation consists of only one rock type, that rock type may appear as part of the name (for example, the Kaibab Limestone), but if a formation contains more than one rock type, the word *formation* serves as part of the name (for example, the Toroweap Formation). Note that all words in the name are capitalized. Several formations in a succession may be lumped together as a *stratigraphic group*, and several adjacent groups, in turn, may be lumped together as a *supergroup*. The boundary surface between two formations is a *depositional contact*, or simply a **contact**. (A fault surface, or a boundary between an igneous intrusion and its wall rock, is also a type of contact.) We can summarize information about the sequence of formations at a location by

drawing a **stratigraphic column**, a chart representing the order of formations and their relative thicknesses.

Unconformities: Gaps in the Record

To find good exposures of rock, James Hutton sometimes boated along the eastern coast of Scotland, where waves of the stormy North Sea strip away soil and shrubbery. He was particularly puzzled by an outcrop at Siccar Point, which exposes a contact between two distinct sequences of sedimentary rock (Fig. 9.11). In the lower portion of the outcrop, beds of gray sandstone and shale are nearly vertical, whereas in the upper portion, beds of red sandstone display a dip of less than 20°. Further, the gently dipping layers seem to lie across the truncated ends of the vertical layers, like a handkerchief lying across the top of a row of books. We can imagine that, as Hutton was examining this odd relationship, the tide came in and deposited a new layer of sand on top of the rocky shore. With the principle of uniformitarianism in mind, he suddenly realized the significance of what he saw.

Hutton deduced that the sediments that now make up the gray sandstone–shale sequence had been deposited, lithified, tilted, and truncated by erosion before the sediments that now form the red sandstone had accumulated. Therefore, the contact between the gray

FIGURE 9.11 James Hutton deduced that the red sandstone beds above the unconformity at Siccar Point were deposited after the gray siltstone and shale beds below had been lithified, tilted, and eroded.

(a) Siccar Point juts out into the North Sea.

(b) The beds of reddish sandstone above the unconformity dip (are tilted) gently. The beds of gray sandstone and shale below are vertical.

What an Earth Scientist Sees

Devonian "Old Red sandstone"

Unconformity

Silurian sandstone and shale

FIGURE 9.12 The three kinds of unconformities and their formation.

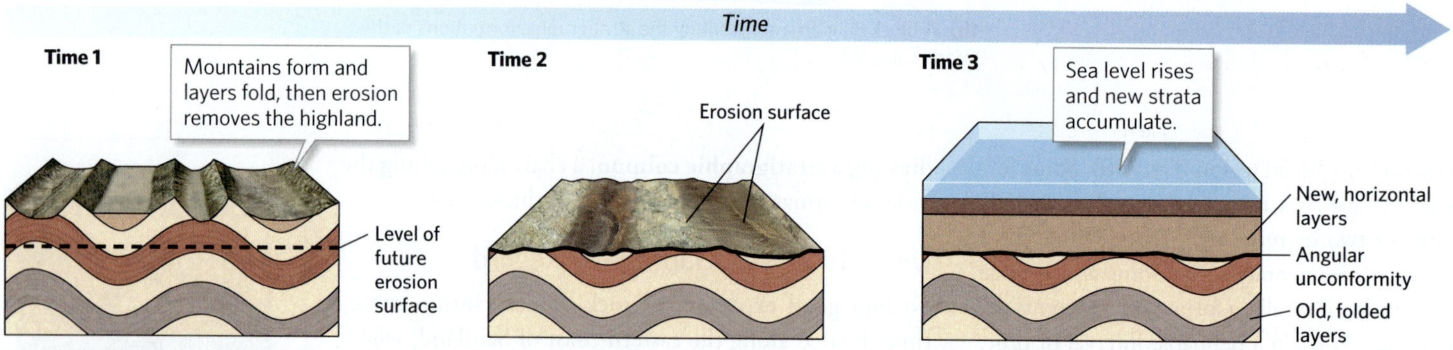

Time

Time 1
Mountains form and layers fold, then erosion removes the highland.

Level of future erosion surface

Time 2
Erosion surface

Time 3
Sea level rises and new strata accumulate.

New, horizontal layers

Angular unconformity

Old, folded layers

(a) Angular unconformity: (1) layers undergo folding; (2) erosion produces a flat surface; (3) sea level rises, and new layers of sediment accumulate.

Time 1
Granite

Future erosion surface

Time 2
Erosion removes cover, so basement lies exposed at the Earth's surface.

Erosion surface

Time 3
Sea level rises and new strata accumulate.

Nonconformity

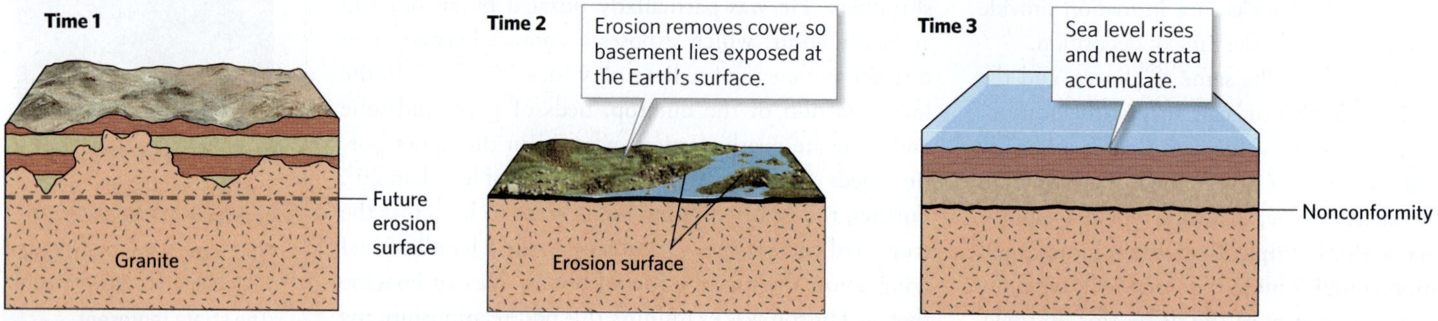

(b) Nonconformity: (1) a pluton intrudes; (2) erosion cuts down into the crystalline rock; (3) new sediment layers accumulate above the erosion surface.

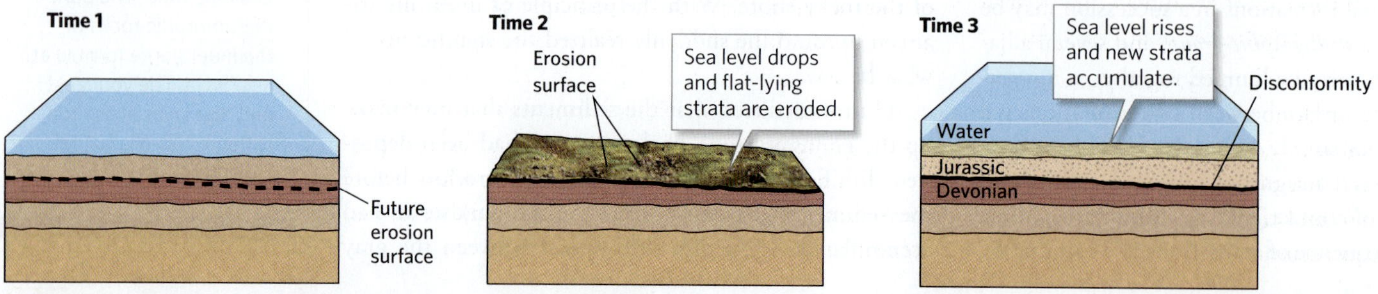

Time 1
Future erosion surface

Time 2
Erosion surface

Sea level drops and flat-lying strata are eroded.

Time 3
Sea level rises and new strata accumulate.

Disconformity

Water

Jurassic

Devonian

(c) Disconformity: (1) layers of sediment accumulate; (2) sea level drops and an erosion surface forms; (3) sea level rises and new sediment layers accumulate.

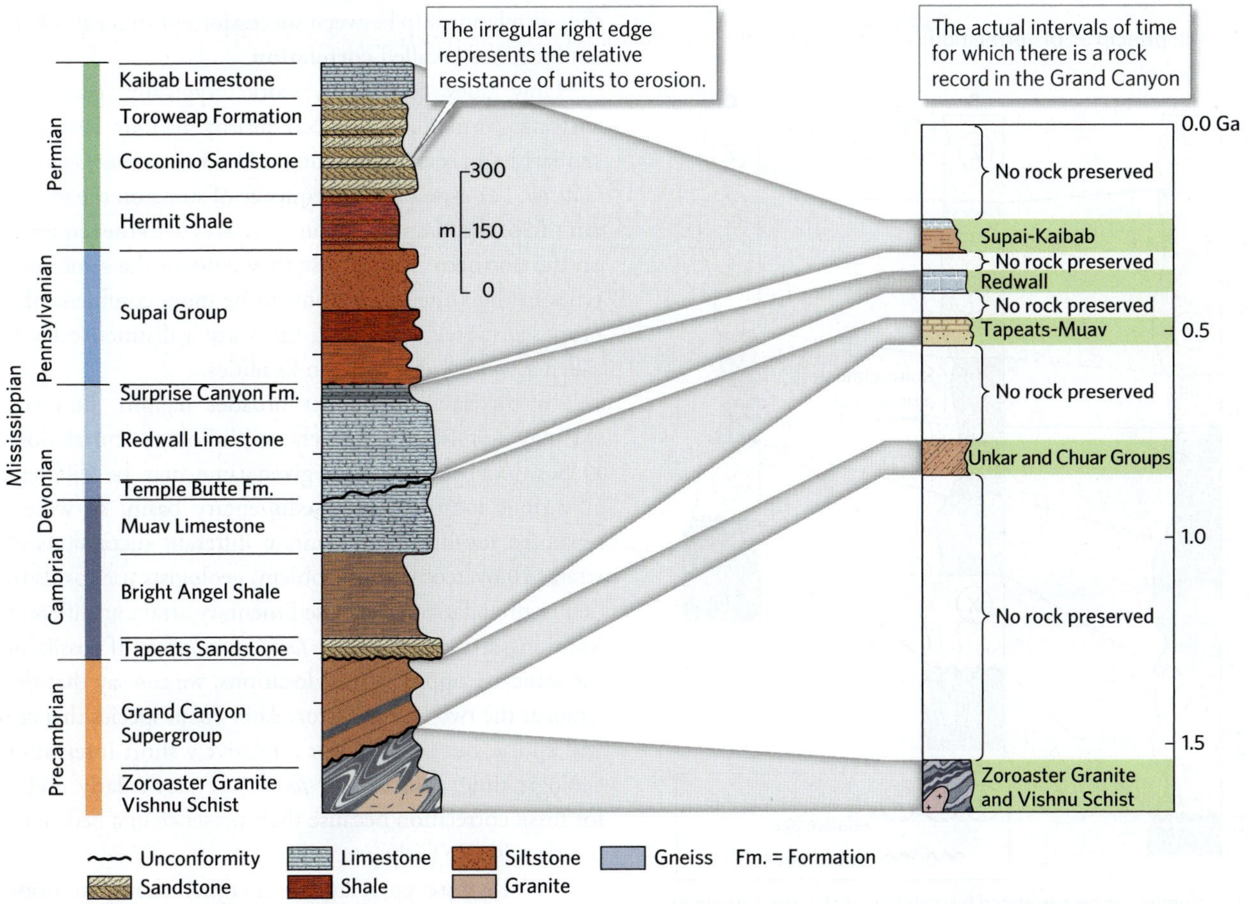

FIGURE 9.13 Interpreting the stratigraphic column of the Grand Canyon.

The irregular right edge represents the relative resistance of units to erosion.

The actual intervals of time for which there is a rock record in the Grand Canyon

(a) On a stratigraphic column, the vertical scale represents thickness, and patterns represent different rock types. The Grand Canyon Supergroup includes the Unkar Group.

(b) Because of unconformities, Grand Canyon strata provide only a partial record of geologic time. This chart, with a time scale as the vertical axis, indicates gaps in the record.

and red stratigraphic formations represented a time interval during which no new strata had been deposited at Siccar Point and older strata had been eroded away. Geologists now refer to such a contact, representing a time period of nondeposition and possibly erosion, as an **unconformity**. The gap in the geologic record represented by an unconformity, meaning the period of time not represented by strata, is called a *hiatus*. There are three common types of unconformities:

- *Angular unconformity:* Rocks below an angular unconformity were tilted or folded before the unconformity developed **(Fig. 9.12a)**. Thus, an **angular unconformity** cuts across the underlying layers, and the orientation of the layers below the unconformity differs from that of the layers above. Siccar Point serves as an example.

- *Nonconformity:* A **nonconformity** is a type of unconformity on which sedimentary rocks overlie a *basement* of older intrusive igneous rocks or metamorphic rocks **(Fig. 9.12b)**. These older rocks underwent cooling, uplift, and erosion before becoming the substrate on which sediment accumulated.

- *Disconformity:* Imagine that a sequence of sedimentary beds has been deposited beneath a shallow sea. Then sea level drops, exposing the beds, so no new sediment accumulates for a time, and some of the pre-existing sediment erodes away. Later, sea level rises again, and a new sequence of sediment accumulates over the old. The boundary between the two sequences is a type of unconformity called a **disconformity** **(Fig. 9.12c)**. The beds above and below the disconformity are parallel, but the contact between them represents a hiatus.

The succession of strata at any location provides a record of Earth history there. But because of unconformities, the geologic record preserved in the rock layers at any particular location is incomplete. For example, the stratigraphic column of the Grand Canyon represents only part of geologic time **(Fig. 9.13)**. It's as if geologic history is being chronicled by a data recorder that turns on only intermittently—when it's on (times of deposition), a geologic record accumulates, but when it's off (times of nondeposition and possibly erosion), an unconformity develops.

Animation
Unconformities: Angular Unconformity

Animation
Unconformities: Nonconformity

Animation
Unconformities: Disconformity

Did you ever wonder . . .
whether the strata exposed in the Grand Canyon represent all of Earth's history?

FIGURE 9.14 The principles of correlation.

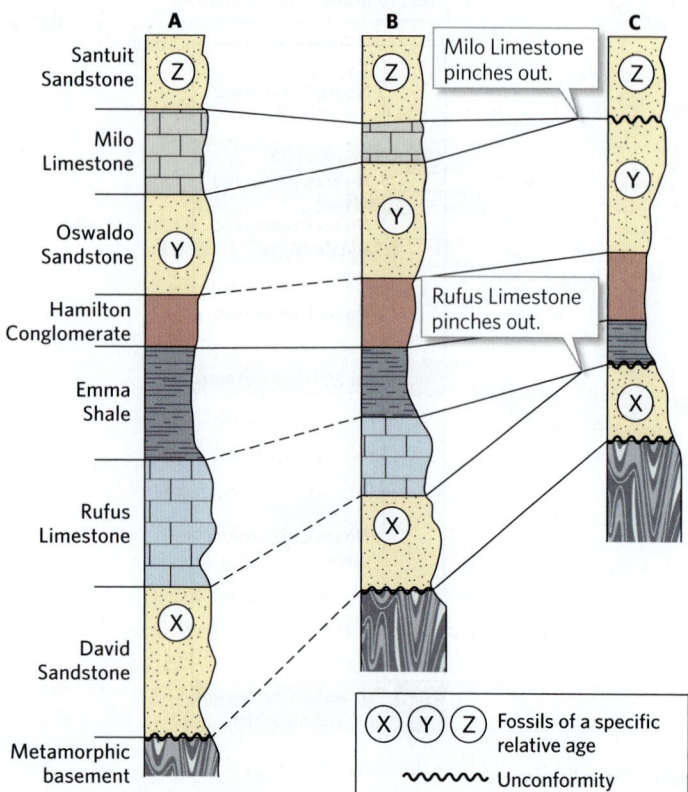

Milo Limestone pinches out.

Rufus Limestone pinches out.

Ⓧ Ⓨ Ⓩ Fossils of a specific relative age

〰〰 Unconformity

(a) Stratigraphic columns can be correlated by matching rock types (lithologic correlation) or by matching fossils (fossil correlation). In this theoretical sequence, the Hamilton Conglomerate is a marker bed. Because some strata pinch out, Column C contains unconformities. Fossil correlation indicates that the youngest formation in C is the Santuit Sandstone.

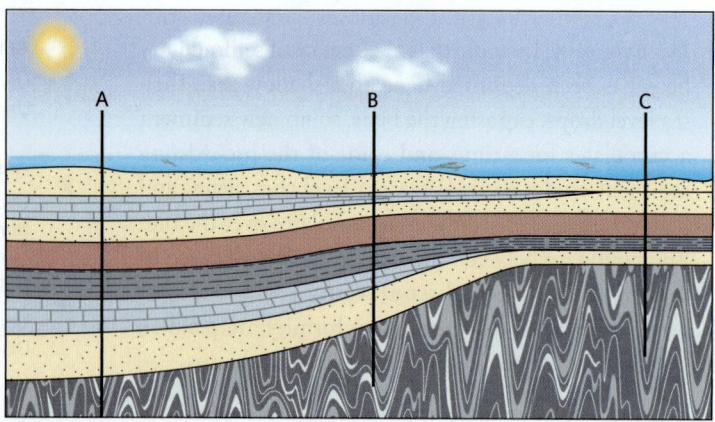

(b) At the time of deposition, locations A, B, and C (which correlate with the columns in part a) were in different parts of a basin. The basin floor was subsiding fastest at A.

Correlation of Stratigraphic Formations

With the concept of stratigraphic formations in mind, we can explore how a succession of beds exposed in one location relates, in terms of relative age, to a succession exposed in another location. This process of determining the age relationship between successions of strata at different locations is called **correlation**.

How does correlation work? Typically, geologists correlate formations between nearby regions based on similarities in rock type, a method known as *lithologic correlation*. For example, the sequence of strata on the southern rim of the Grand Canyon correlates with the sequence on the northern rim because they contain the same rock types in the same order. You can be more confident of a proposed correlation if you can trace a distinctive layer, called a *marker bed*, between localities.

To correlate strata over broader regions, however, we may not be able to rely on lithologic correlation. Depositional settings at a given time may be different at various locations in a sedimentary basin, so widely separated localities may contain different successions of strata. To overcome this problem, geologists use fossils to define the relative ages of sedimentary strata at different locations, a method called *fossil correlation*. If fossils of the same age appear at two locations, we can say that the strata at the two locations correlate. Fossil species that are widespread but survived for a relatively short interval of geologic time, called *index fossils*, are particularly useful for fossil correlation because their presence in a bed characterizes its age fairly precisely.

To illustrate correlation, imagine that you draw stratigraphic columns for three locations (**Fig. 9.14**). You divide the succession of strata in column A into several distinct formations to which you assign names. The successions in columns B and C contain some formations with rock types similar to those in column A, but the formations are not as thick, and not all the formations in column A can occur in the other columns. Lithologic correlation allows you to match the formations of column A with those of column B because you see the same rock types in the same succession. To confirm your correlation, you can compare fossil assemblages—you'll find that the formations in both columns contain fossils of the same ages. Correlation with strata in column C presents more of a challenge because some formations found in the other columns are not present. Fossil correlation, however, allows you to match the sandstone units and locate unconformities.

Dividing Time: From Eons to Epochs

No outcrop on the Earth provides a complete record of our planet's history, for two reasons: first, any outcrop exposes only part of the stratigraphic succession in a given region (older strata remain hidden underground, and younger strata have been eroded away), and second, stratigraphic successions contain unconformities. But by

FIGURE 9.15 Global correlation of rock units led to the development of the geologic column.

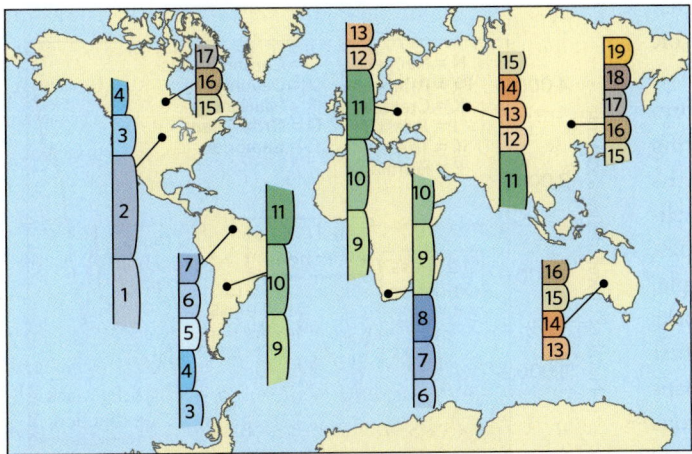

(a) Each of the small columns on this map represents the stratigraphy at a given location. By correlating these columns, geologists determined the relative ages of rock units and filled in the gaps in the record.

Eon	Era	Period	Epoch
Phanerozoic	Cenozoic	Quaternary	Holocene
			Pleistocene
		"Tertiary" Neogene	Pliocene
			Miocene
		"Tertiary" Paleogene	Oligocene
			Eocene
			Paleocene
	Mesozoic	Cretaceous	
		Jurassic	
		Triassic	
	Paleozoic	Permian	
		Carboniferous (Pennsylvanian / Mississippian)	
		Devonian	
		Silurian	
		Ordovician	
		Cambrian	
Precambrian	Proterozoic		
	Archean		
	Hadean		

(b) The geologic column. Geologists divide the column into intervals and assign names to them, but since the column was built without knowledge of numerical ages, it does not show the actual durations of these intervals. Note that the term "Tertiary" is no longer used, but it appears in older literature.

correlating rocks from locality to locality at millions of places around the world, geologists have pieced together a composite stratigraphic column, called the **geologic column**, that symbolizes the entirety of Earth history **(Fig. 9.15)**.

The geologic column can be divided into segments, each of which represents a specific interval of time. The largest subdivisions of Earth history are **eons**. These eons are named, from oldest to youngest, the Hadean, Archean, Proterozoic, and Phanerozoic. Geologists commonly refer to the Hadean, Archean, and Proterozoic together as the **Precambrian**. The suffix *–zoic* means life, so *Phanerozoic* means visible life, and *Proterozoic* means first life. (This 19th-century terminology can be somewhat confusing because in recent decades, geologists have shown that the earliest life—Bacteria and Archaea—appeared in the Archean.) The Phanerozoic Eon is subdivided into **eras**, named in order from oldest to youngest, the Paleozoic (ancient life), Mesozoic (middle life), and Cenozoic (recent life). We further divide each era into **periods** and each period into **epochs**.

Where do the names of the periods come from? They refer either to localities where a fairly complete succession of strata representing that interval was first identified (for example, rocks representing the Devonian Period crop out near Devon, England), or to a characteristic of the time (rocks from the Carboniferous Period contain a lot of coal). The terminology was not organized in a planned fashion that would make it easy to learn. Instead, it grew haphazardly in the years between 1760 and 1845 as geologists began to refine their understanding of geologic history and fossil succession.

Notably, it now appears that the dates that early geologists identified as the boundaries between eras, and between some periods, coincide with **mass-extinction events**, times of short duration when **biodiversity**—the number of different types of organisms living at the same time—decreased substantially **(Fig. 9.16)**. The causes of these mass-extinction events remain a subject of research. They may be the consequence of asteroid impacts, intense volcanic activity, severe climate change, or a combination of these factors. Life diversified after mass extinctions as new organisms appeared, so fossil assemblages in strata deposited before a mass-extinction event differ significantly from those in strata younger than the event.

Vermilion Cliffs, Arizona

Latitude: 36°49′4.81″ N
Longitude: 111°37′56.59″ W

Zoom to an elevation of 2 km (~1.2 mi) and look obliquely north.

Marble Canyon (foreground) is the entry to the Grand Canyon. Outcrops in the distance are the Vermilion Cliffs, exposing reddish-brown sandstone and shale of the Moenkopi Formation. The canyon walls consist of underlying Kaibab Limestone.

With the concept of the geologic column in mind, let's see how geologists have correlated strata across the Colorado Plateau of the southwestern United States, a desert region that contains the Grand Canyon and other spectacular national parks **(Fig. 9.17)**. The oldest Paleozoic sedimentary rocks of the region crop out near the base of the Grand Canyon, while the youngest rocks form the cliffs of Cedar Breaks and Bryce Canyon. Walking through these parks is like walking through the Earth's history: each rock layer gives an indication of the climate and topography of the region at a time in the past **(Earth Science at a Glance**, pp. 274–275). For example, when the Precambrian metamorphic and igneous rocks exposed in the inner gorge of the Grand Canyon first formed, the region was a high mountain range, perhaps as dramatic as today's Himalayas. When the fossiliferous beds of the Kaibab Limestone accumulated where we now see the rim of the canyon, the region was a warm, shallow sea. And when the rocks making up the towering red cliffs of sandstone in Zion Canyon were deposited, the region was a Sahara-like desert, blanketed with huge sand dunes. The spatial relationships among such stratigraphic formations can be portrayed by geologic maps **(Box 9.2)**.

Take-home message . . .

A stratigraphic formation is a sequence of beds that can be mapped over a broad region. Due to unconformities, a stratigraphic sequence at any location is incomplete. Geologists correlate formations based on rock type and fossil content, and can portray the configuration on a geologic map. Correlation led to the creation of the geologic column that shows the entirety of Earth history. It is subdivided into eons, eras, periods, and epochs.

Quick Question _____

Is there any place on Earth where an exposed stratigraphic sequence represents all of geologic time?

9.5 Determining Numerical Ages

Geologists since the time of Hutton could determine the relative age of a rock, but they had no way to specify the rock's **numerical age** (called *absolute age* in older literature), meaning age specified in years. Therefore, they could not define a timeline for Earth history, nor could they determine the duration of geologic events in years. This situation changed with the discovery of **radioactive elements**. Atoms of radioactive elements spontaneously decay to form different elements. The rate of decay is a constant that can be measured in the laboratory and

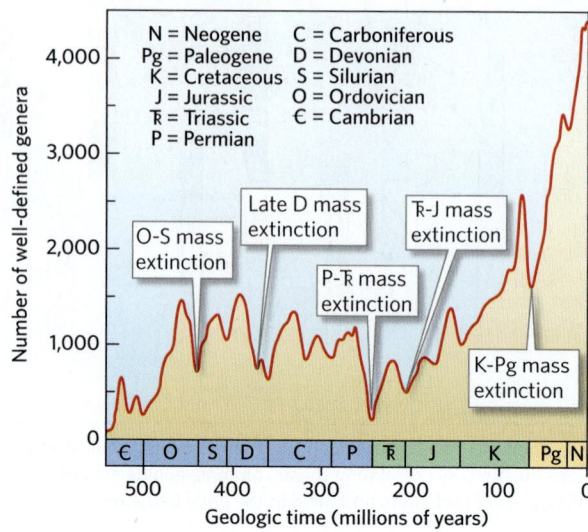

FIGURE 9.16 Changes in the diversity of life over time. Sudden drops in the number of genera indicate mass-extinction events.

can be specified in years. In the 1950s, geologists developed **isotopic dating** (or *radiometric dating*) techniques, which use the ratios of radioactive elements to their decay products as a basis for calculating numerical ages of rocks. The overall study of numerical ages of rocks is now called **geochronology**. Since the 1950s, isotopic dating techniques have steadily improved, and geologists can now make very accurate measurements from very small samples. But the basis of the technique remains the same, and to explain it, we must first review the process of radioactive decay.

Radioactive Decay

All atoms of a given element have the same number of protons in their nucleus—we call this number the *atomic number* (see Chapter 1). However, not all atoms of a given element have the same number of neutrons in their nucleus, which means that not all atoms of an element have the same *atomic mass* (roughly, the number of protons plus neutrons). Different versions, or **isotopes**, of an element have the same atomic number but different atomic masses. For example, all uranium atoms have 92 protons, but the uranium-238 isotope (abbreviated ^{238}U) has an atomic mass of 238, whereas the ^{235}U isotope has an atomic mass of 235; the first isotope has 3 more neutrons than the second.

Some isotopes of some elements are *stable*, meaning that they last essentially forever. Radioactive isotopes, however, are *unstable*. Eventually, they undergo a process called **radioactive decay**, which changes the atomic number, and thus changes an atom of one

FIGURE 9.17 Correlation of strata among the national parks of Arizona and Utah.

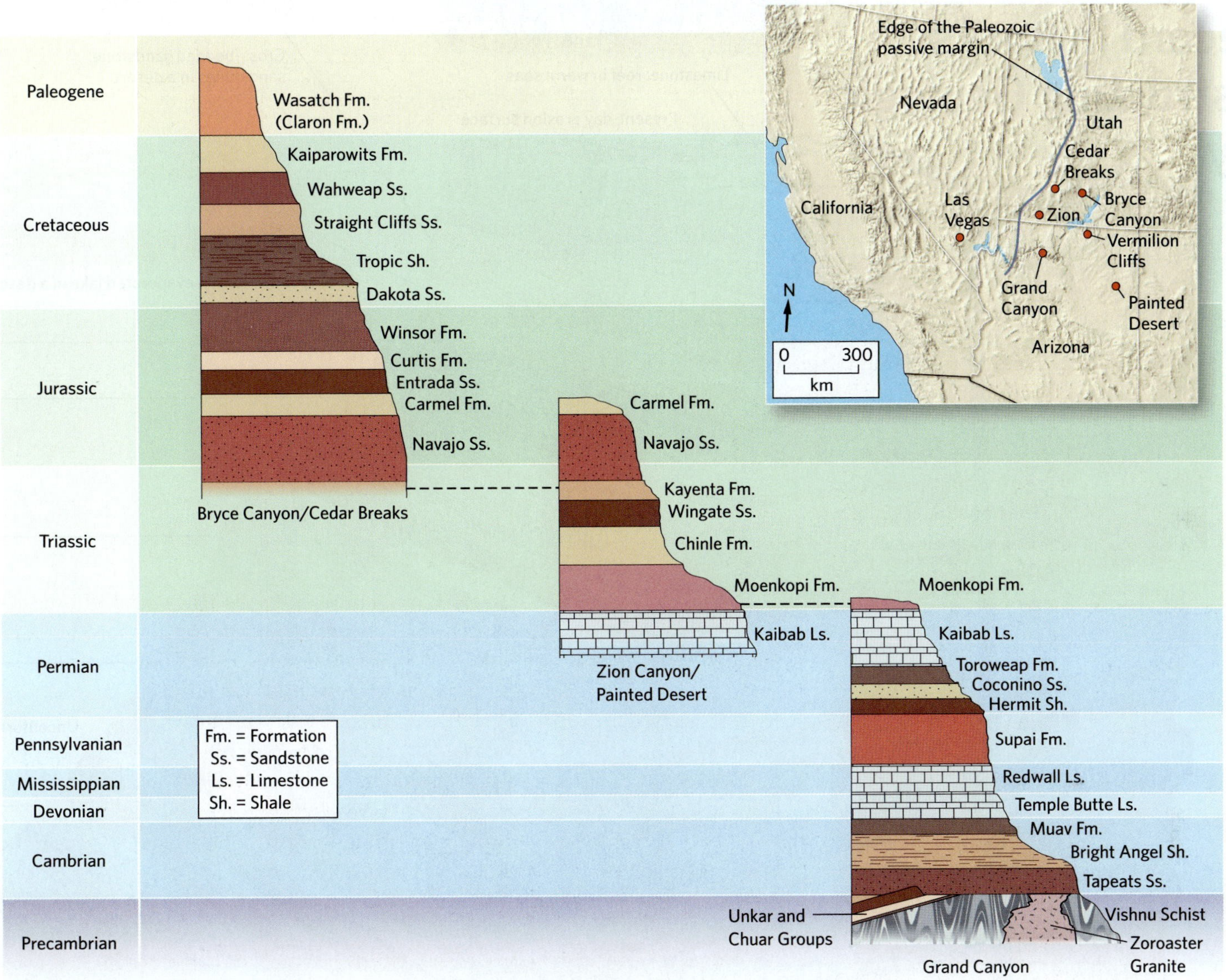

Paleogene
Wasatch Fm.
(Claron Fm.)

Cretaceous
Kaiparowits Fm.
Wahweap Ss.
Straight Cliffs Ss.
Tropic Sh.
Dakota Ss.

Jurassic
Winsor Fm.
Curtis Fm.
Entrada Ss.
Carmel Fm.
Navajo Ss.

Bryce Canyon/Cedar Breaks

Carmel Fm.
Navajo Ss.

Triassic
Kayenta Fm.
Wingate Ss.
Chinle Fm.
Moenkopi Fm.
Kaibab Ls.

Zion Canyon/
Painted Desert

Moenkopi Fm.
Kaibab Ls.

Permian
Toroweap Fm.
Coconino Ss.
Hermit Sh.

Pennsylvanian
Supai Fm.

Mississippian
Redwall Ls.

Devonian
Temple Butte Ls.
Muav Fm.

Cambrian
Bright Angel Sh.
Tapeats Ss.

Fm. = Formation
Ss. = Sandstone
Ls. = Limestone
Sh. = Shale

Precambrian
Unkar and
Chuar Groups
Vishnu Schist
Zoroaster
Granite
Grand Canyon

Edge of the Paleozoic
passive margin

Nevada
Utah
Cedar Breaks
Bryce Canyon
Vermilion Cliffs
California
Las Vegas
Zion
Grand Canyon
Arizona
Painted Desert

N

0   300
km

(a) Different intervals of geologic time are represented by the strata exposed in different parks.

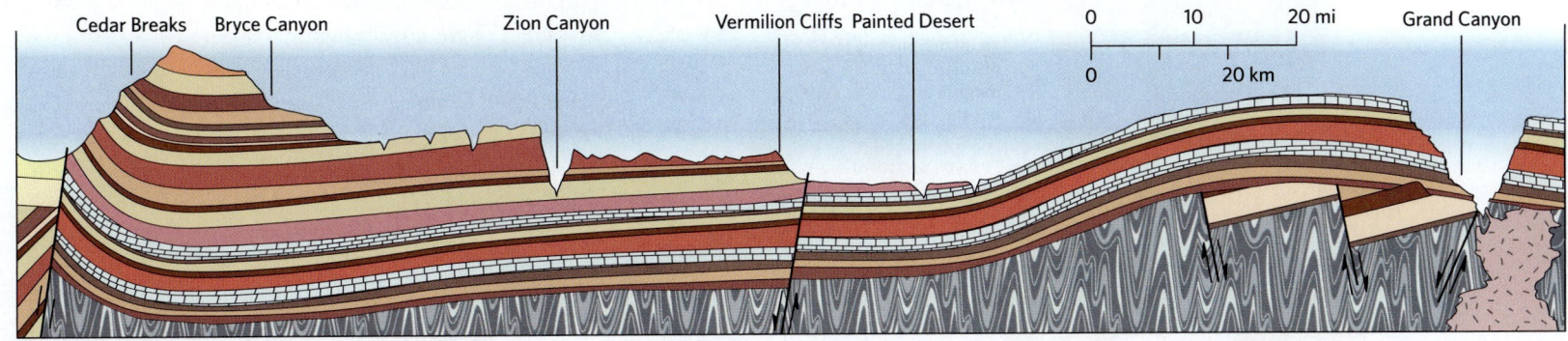

Cedar Breaks   Bryce Canyon   Zion Canyon   Vermilion Cliffs Painted Desert   0   10   20 mi   Grand Canyon

0   20 km

(b) This cross-sectional sketch (vertically exaggerated) shows how the Phanerozoic strata cover the basement of the Colorado Plateau region. Faults cut and warp these strata, and canyons cut into them.

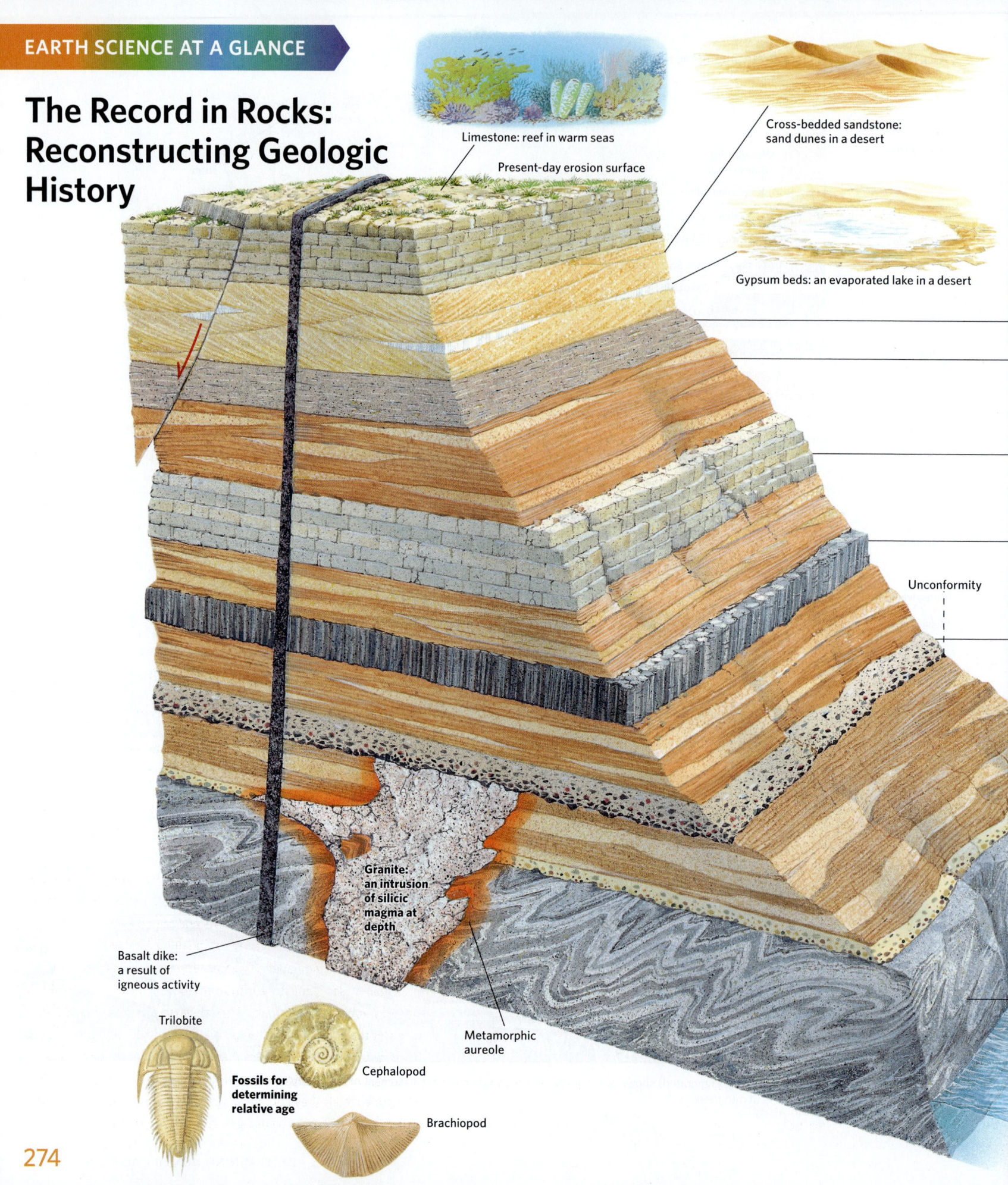

The Record in Rocks: Reconstructing Geologic History

Limestone: reef in warm seas

Present-day erosion surface

Cross-bedded sandstone: sand dunes in a desert

Gypsum beds: an evaporated lake in a desert

Unconformity

Granite: an intrusion of silicic magma at depth

Basalt dike: a result of igneous activity

Metamorphic aureole

Trilobite

Cephalopod

Fossils for determining relative age

Brachiopod

274

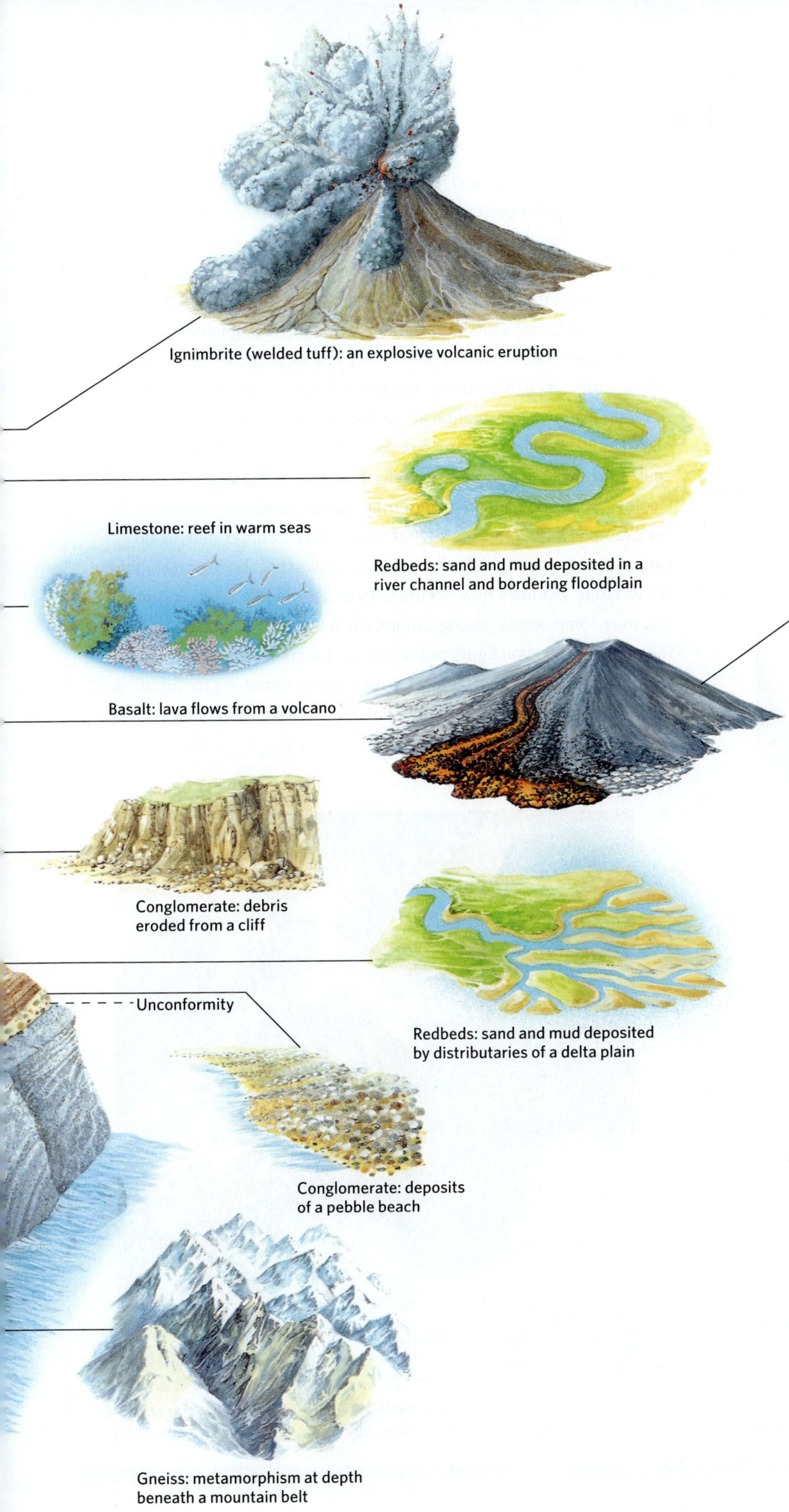

Ignimbrite (welded tuff): an explosive volcanic eruption

Limestone: reef in warm seas

Redbeds: sand and mud deposited in a river channel and bordering floodplain

Basalt: lava flows from a volcano

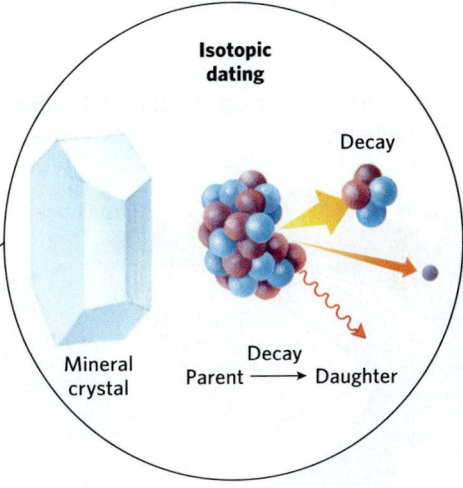

Isotopic dating

Decay

Mineral crystal

Decay

Parent ⟶ Daughter

Conglomerate: debris eroded from a cliff

- - - Unconformity

Redbeds: sand and mud deposited by distributaries of a delta plain

Conglomerate: deposits of a pebble beach

Gneiss: metamorphism at depth beneath a mountain belt

When geologists examine a sequence of rocks exposed on a cliff, they see a record of Earth history that they interpret by applying the basic principles of geology, by searching for fossils, and by using isotopic dating. On this cliff, we see evidence for many geologic events. The layers of sediment (and the sedimentary structures they contain), the igneous intrusions, and the geologic structures tell us about past climates and past tectonic activity.

The insets show the way the region looked in the past, based on the record in the rocks. For example, the presence of gneiss at the base of the canyon indicates that at one time the region was a mountain belt where deeply buried crust underwent metamorphism and deformation. Unconformities indicate that the region later underwent uplift and erosion. Sedimentary successions record transgressions and regressions of the sea. The land surface portrayed in this painting was sometimes a river floodplain or a delta (indicated by redbeds), sometimes a shallow sea (limestone), and sometimes a desert dune field (cross-bedded sandstone). The presence of igneous rocks indicates that, at times, the region was volcanically active, and the presence of faults indicates that it was seismically active. We can gain insight into the age of the sedimentary rocks by studying the fossils they contain, and into the age of the igneous and metamorphic rocks by using isotopic dating methods.

Putting Earth Science to Use

Geologic maps

A **geologic map** portrays the distribution of bedrock units that would be exposed at the Earth's surface if all soil, unconsolidated sediment, vegetation, and human-built structures were removed **(Fig Bx9.2a, b)**. To construct such a map, geologists study *outcrops* (places where bedrock is exposed) and, if available, examine drill cores. As more data becomes available, a map can be updated and refined. On a geologic map, the contacts between stratigraphic formations or between igneous intrusions and wall rock appear as lines. With practice, you can learn to read a geologic map and recognize the patterns of contacts that portray features such as flat-lying strata, unconformities, igneous intrusions, folds, and faults. Specific symbols on a map indicate the orientation of layers and the location of geologic structures.

Geologic maps are important tools for conveying information about the geologic character and history of a location. As such, not only are they used in academic studies, but they have important practical applications. Information about the distribution and orientation of rock units may help energy companies find reserves of oil, gas, or coal, and may help mining companies find reserves of valuable ores. Geologic maps help planners determine where geologic hazards exist. For example, if the map shows that much of the bedrock in a region consists of very weak rock, of if the map shows that the region contains active faults, planners may decide to look elsewhere for building sites. You may come across geologic maps when you visit a national park—the map will help you figure out which stratigraphic formations are exposed in the park **(Fig. Bx9.2c)**. And if you're searching for fossils, a geologic map can show you where outcrops of fossiliferous rock might be exposed.

FIGURE Bx9.2 A geologic map depicts the distribution of stratigraphic formations.

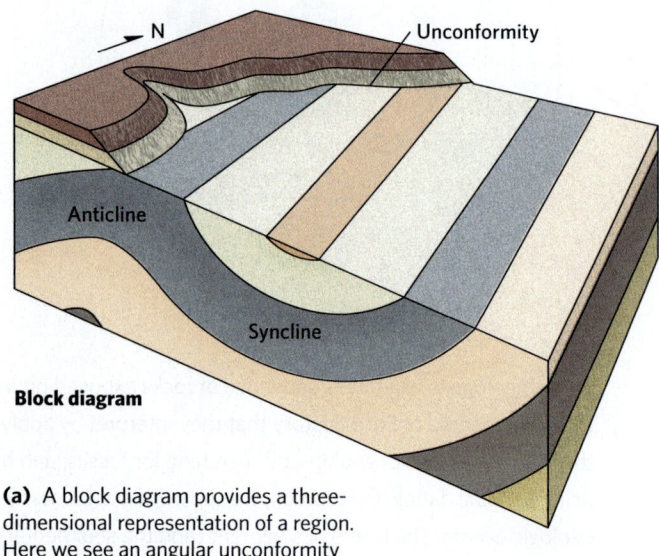

(a) A block diagram provides a three-dimensional representation of a region. Here we see an angular unconformity over folded strata.

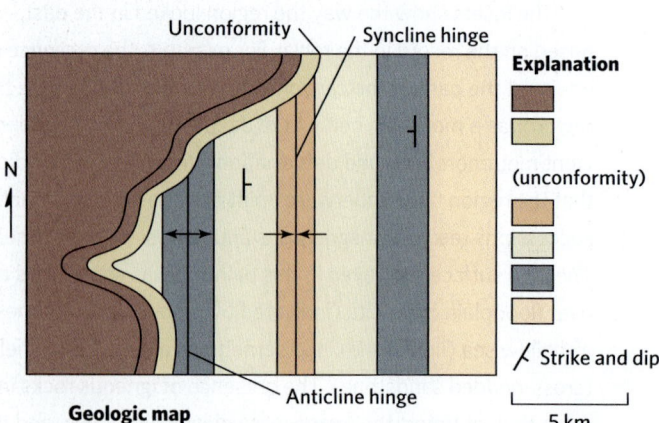

(b) A geologic map portrays the top surface of the block diagram. It shows the distribution of rock units. The black lines represent contacts between units.

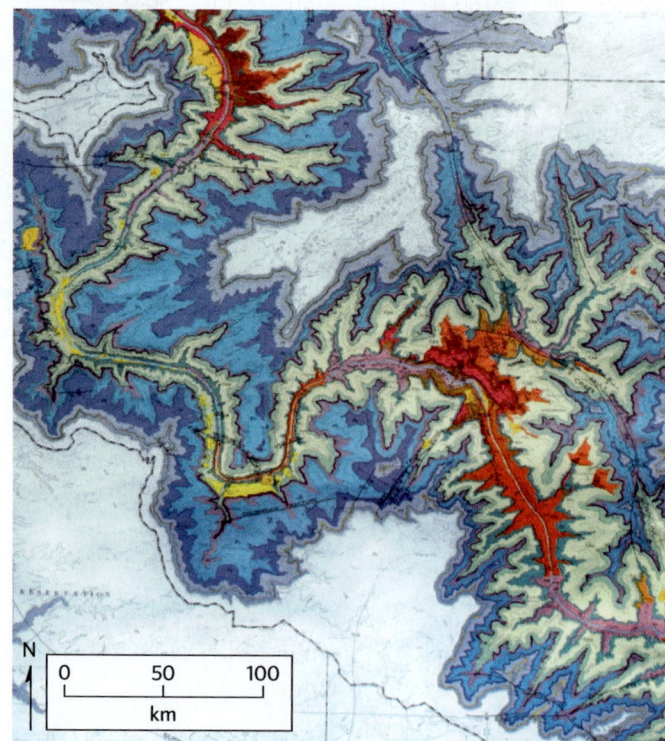

(c) This geologic map portrays the distribution of stratigraphic formations in a portion of the Grand Canyon. Each color represents a different formation.

element into an atom of a different element. We refer to the atom that undergoes decay as the **parent isotope** and the decay product as the **daughter isotope**. For example, the parent isotope ^{238}U ultimately decays to become a daughter isotope, lead-206 (^{206}Pb). In some cases, the decay that takes place to form the final, stable daughter has many steps, and the process involves the formation and decay of other atoms.

Physicists cannot specify how long an individual atom of a radioactive isotope survives before it decays, but they can measure the **half-life**, the time it takes for half of a group of atoms to decay. Because the half-life for a given decay reaction is constant, the ratio of parent to daughter isotopes can serve as a natural clock (Fig. 9.18). Box 9.3 helps you visualize the concept of a half-life.

Isotopic Dating Techniques

Because radioactive decay proceeds at a known rate, it provides a basis for telling time. In other words, because an element's half-life is a constant, we can calculate the age of a mineral by measuring the ratio of parent isotope to daughter isotope in the mineral. To put this concept into practice, geologists have identified isotopes that have long half-lives and occur within common minerals. Table 9.1 lists some particularly useful isotopes, and the minerals that contain them. Note that each radioactive isotope has a unique half-life. We do not list carbon in Table 9.1 because it is not useful for dating rocks. Radioactive carbon (^{14}C) has a very short half-life (about 5,000 years) and accumulates only in living material, so we can use *radiocarbon dating* only to date organic material that is less than about 70,000 years old.

If you find a rock containing dateable minerals, you can obtain its numerical age by following these steps:

- *Collecting appropriate samples:* For isotopic dating methods to work, you need to obtain unweathered rock samples. Chemical weathering reactions may cause the loss of parent or daughter isotopes, so working with weathered rocks might produce incorrect results.

- *Separating dateable minerals:* Once collected, crush the rocks so that you can pick out and isolate the dateable minerals.

- *Extracting parent and daughter atoms:* To separate parent and daughter isotopes from the minerals, you can either dissolve the minerals in acid or evaporate portions of the minerals with a laser.

- *Analyzing the parent-daughter ratio:* To determine the ratio of parent to daughter, you can pass the atoms through a *mass spectrometer*. The magnet in this instrument separates atoms from one another based on

FIGURE 9.18 The concept of a half-life in the context of radioactive decay.

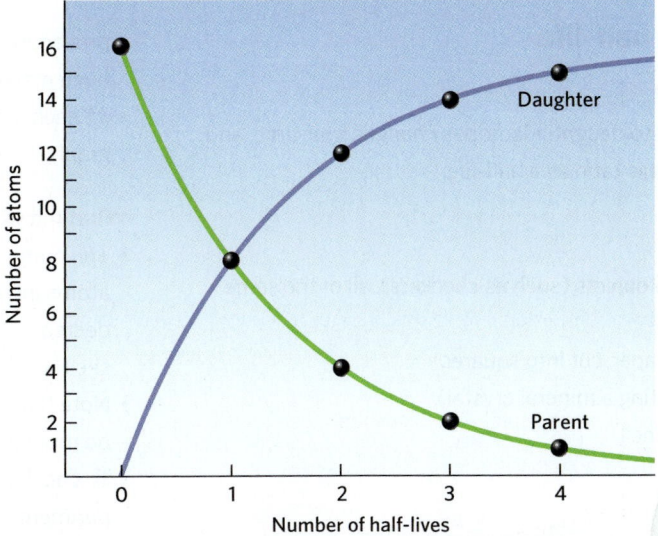

(a) This graph shows how the number of atoms of the parent isotope decreases and the number of atoms of the daughter isotope increases as time passes. (Note that in a real mineral crystal, the number of atoms would be immensely larger.) The rate of change decreases with time.

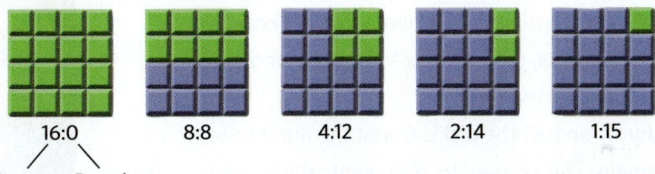

Parent   Daughter

(b) The ratio of parent isotope to daughter isotope changes with the passage of each successive half-life.

(c) In a cluster of atoms undergoing decay, there is no way to predict which atom will decay next.

Did you ever wonder . . .

how geologists can specify the ages of rocks in years?

TABLE 9.1 Isotopes Used in the Isotopic Dating of Rock

Parent → Daughter	Half-Life (years)	Minerals Containing the Isotopes
$^{147}Sm \rightarrow {}^{143}Nd$	106.0 billion	Garnets, micas
$^{87}Rb \rightarrow {}^{87}Sr$	48.8 billion	Potassium-bearing minerals (mica, feldspar, hornblende)
$^{238}U \rightarrow {}^{206}Pb$	4.5 billion	Uranium-bearing minerals (zircon, uraninite)
$^{40}K \rightarrow {}^{40}Ar$	1.3 billion	Potassium-bearing minerals (mica, feldspar, hornblende)
$^{235}U \rightarrow {}^{207}Pb$	713.0 million	Uranium-bearing minerals (zircon, uraninite)

BOX 9.3

BOX 9.3 ▶ How can I explain . . .

The concept of a half-life

What are we learning?

How the ratio of parent to daughter isotopes changes over time, and how we can represent this ratio as a half-life.

What you need:

- 32 small, disk-shaped objects (such as checkers), all of the same color
- Colored sticky-note paper, cut into squares
- A cake pan (representing a mineral crystal)
- Graph paper and a pencil

Instructions:

- Spread the disks around the cake pan. These disks represent the distribution of parent atoms.
- Create a graph that plots number of disks on the vertical y-axis (from 0 to 32) and number of half-lives, from 0 to x, on the horizontal x-axis.
- Plot a point on the graph representing the number of unmarked disks at time 0. Since there are 32 disks, $y = 32$ and $x = 0$. These unmarked disks represent parent atoms.
- Place sticky notes randomly on half the disks. Count the number of unmarked disks that remain. The answer, 16, represents the number of parents remaining after the first half-life has passed. Plot the numbers of parents and daughters at 1 half-life on the graph.
- To represent the number of parent atoms remaining after the second half-life, mark half the remaining unmarked disks (8 disks), then record the numbers of marked and unmarked disks. Repeat the procedure for the remaining half-lives. Continue until all parents have been marked.
- At each stage, record the numbers of parents and daughters on the graph.

What did we see?

- The number of parent atoms decreases and the number of daughter atoms increases as time passes, and the ratio of parents to daughters decreases progressively. If you connect the points on your graph, you see they lie along a curve, not a straight line.
- Note that after enough time passes, the method no longer works, as no more parents remain to decay.
- To check your understanding, imagine that the half-life for this experiment is 10 million years. How old is a mineral after 2.5 half-lives? What is the ratio of parent to daughter?

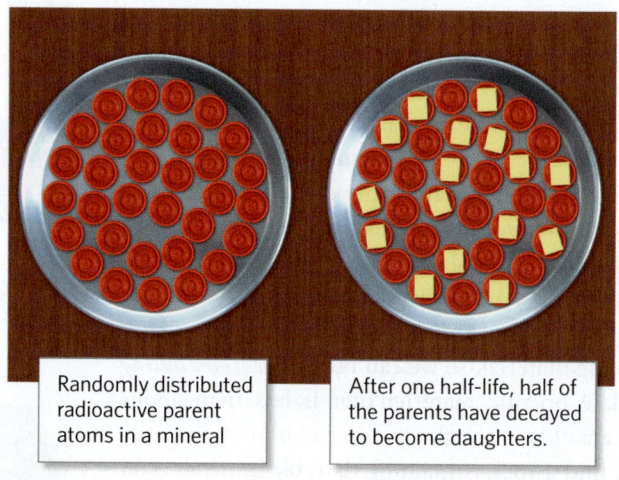

Randomly distributed radioactive parent atoms in a mineral

After one half-life, half of the parents have decayed to become daughters.

FIGURE 9.19 In an isotopic dating laboratory, samples are analyzed using a mass spectrometer.

atomic mass, allowing the number of atoms of a specific isotope to be counted (Fig. 9.19).

At the end of the laboratory process, you know the ratio of parent to daughter isotope in a mineral, so from this ratio you can calculate the age of the mineral. Needless to say, this is a simplified description of the procedure—in reality, obtaining an isotopic date is time-consuming and expensive and requires complex calculations.

What Does an Isotopic Date Mean?

At high temperatures, atoms in a crystal vibrate so rapidly that chemical bonds can break and re-form relatively easily.

As a consequence, isotopes can escape from or move into crystals. Because isotopic dating is based on the parent-daughter ratio, the isotopic clock starts only when crystals become cool enough for both types of isotopes to be locked into the crystal. The temperature below which isotopes can no longer freely move into and out of a crystal is called the **closure temperature** of a mineral. When we specify an isotopic date for a mineral, we are defining the time at which the mineral cooled below its closure temperature.

The concept of closure temperature provides a basis for interpreting the meaning of isotopic dates. In the case of igneous rocks, isotopic dating tells us when a magma or lava cooled to form a solid, cool igneous rock. In the case of metamorphic rocks, isotopic dating tells us when a rock cooled from a metamorphic temperature above the closure temperature to a temperature below it. Not all minerals have the same closure temperature, so in a rock that cools very slowly, different minerals yield different dates.

Can we use isotopes to date a clastic sedimentary rock directly? No. If we date individual minerals in a sedimentary rock, we determine only when these minerals first crystallized as part of an igneous or metamorphic rock, but not when the minerals were deposited as sediment or when the sediment lithified to form a sedimentary rock. So how can we determine the numerical age of a sedimentary rock? We need to answer this question if we want to add numerical ages to the geologic column. Geologists obtain dates for sedimentary rocks by studying cross-cutting relations between sedimentary rocks and dateable igneous or metamorphic rocks. For example, if we find a sequence of sedimentary strata deposited unconformably on dateable granite, the strata must be younger than the granite **(Fig. 9.20)**. If a dateable basalt dike cuts the strata, the strata must be older than the dike. And if dateable volcanic ash buried the strata, then the strata must be older than the ash.

Take-home message . . .

Isotopic dating of rocks specifies numerical ages in years. To obtain an isotopic date, we measure the ratio of radioactive parent isotope to stable daughter isotope in a mineral. An isotopic date gives the time at which a mineral cooled below its closure temperature. We can directly date igneous and metamorphic rocks, but not clastic sedimentary rocks.

Quick Question

How can you obtain a numerical age for a sedimentary rock?

FIGURE 9.20 Using cross-cutting relations to date sedimentary rocks.

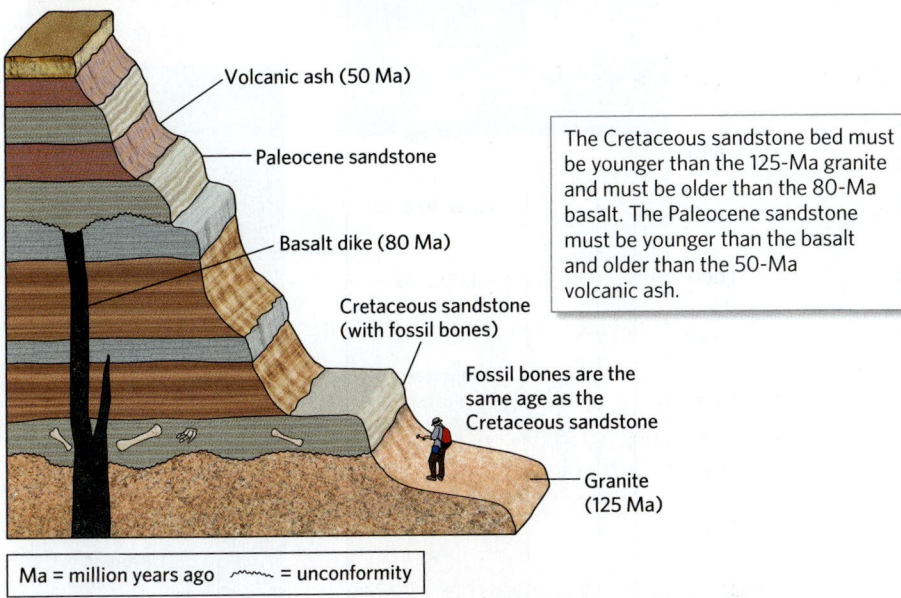

Volcanic ash (50 Ma)

Paleocene sandstone

Basalt dike (80 Ma)

Cretaceous sandstone (with fossil bones)

Fossil bones are the same age as the Cretaceous sandstone

Granite (125 Ma)

The Cretaceous sandstone bed must be younger than the 125-Ma granite and must be older than the 80-Ma basalt. The Paleocene sandstone must be younger than the basalt and older than the 50-Ma volcanic ash.

Ma = million years ago ⁓⁓ = unconformity

9.6 The Geologic Time Scale

Adding Numerical Ages to the Geologic Column

In Section 9.4, we pointed out that the geologic column was established a century before researchers had developed isotopic dating methods, so numerical ages could not initially be assigned to the boundaries between intervals on the geologic column. Once good dating methods became available, geologists searched the world for localities where they could recognize cross-cutting relations between sedimentary rocks and dateable igneous rocks at period boundaries. By isotopically dating the igneous rocks, geologists have been able to determine numerical ages for the boundaries between geologic periods. For example, cross-cutting relations indicate that the Cretaceous Period began 145 million years ago and ended 66 million years ago. (Therefore, the Cretaceous sandstone bed in Figure 9.20 was deposited during the middle part of the Cretaceous, not at the beginning or end of that period.)

As new dates become available, the specific numbers defining the boundaries of geologic periods may shift. For example, around 1995, new studies on ash layers above and below the Precambrian-Cambrian boundary placed this boundary at 542 Ma, in contrast to previous, less definitive studies that had placed the boundary at 570 Ma. Further studies moved the date to 541 Ma. A chart showing the currently favored numerical ages of eras and

If the Eiffel Tower were now representing the world's age, the skin of paint on the pinnacle-knob at its summit would represent man's share of that age; and anybody would perceive that that skin was what the tower was built for. I reckon they would, I dunno.

—MARK TWAIN (AMERICAN WRITER, 1835–1910)

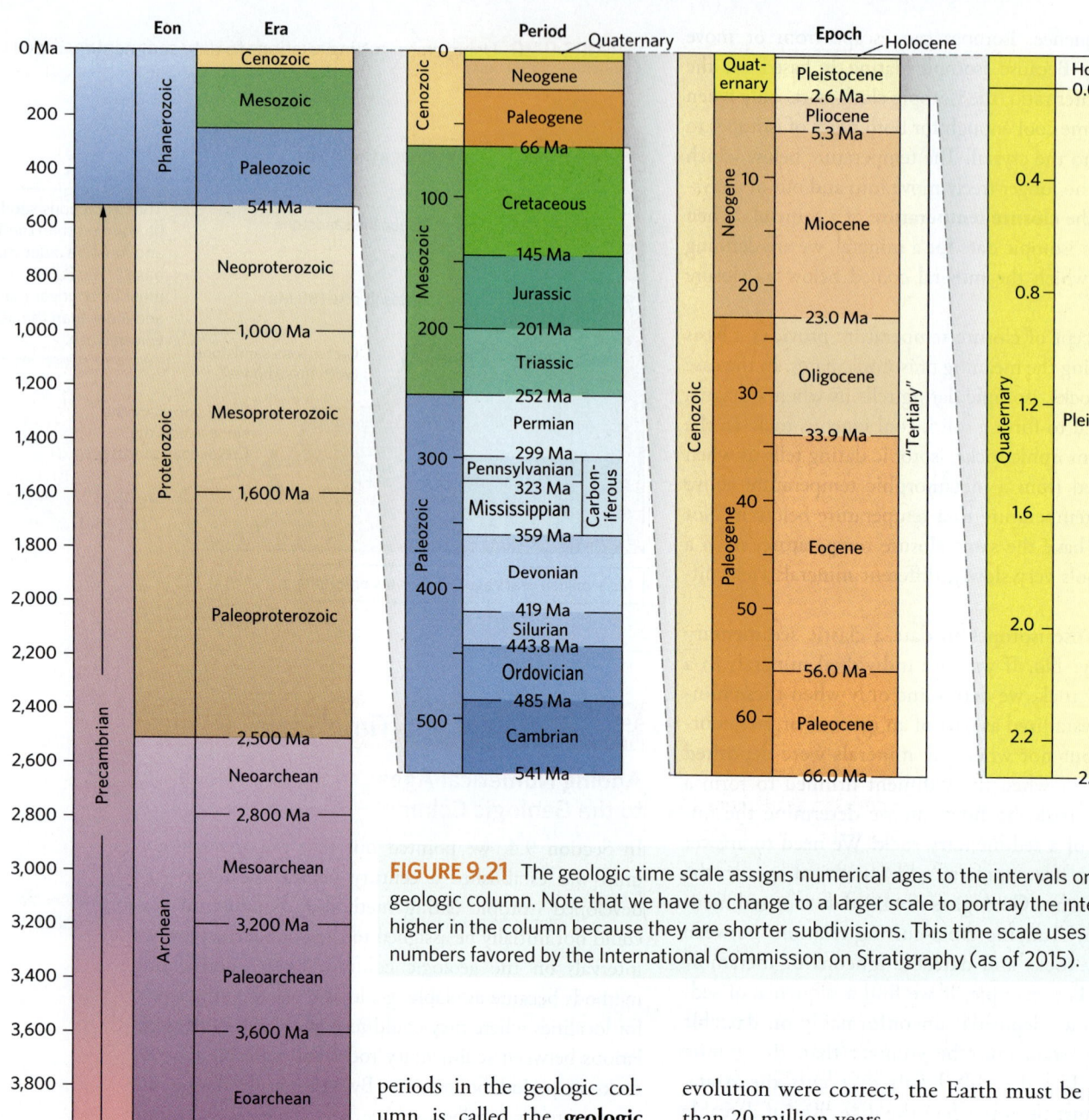

FIGURE 9.21 The geologic time scale assigns numerical ages to the intervals on the geologic column. Note that we have to change to a larger scale to portray the intervals higher in the column because they are shorter subdivisions. This time scale uses the numbers favored by the International Commission on Stratigraphy (as of 2015).

periods in the geologic column is called the **geologic time scale** (Fig. 9.21).

How Old Is the Earth?

During the 18th and 19th centuries, before the discovery of isotopic dating, scientists came up with a great variety of clever answers to the question, "How old is the Earth?"—all of which have since been proved wrong. For example, Lord Kelvin, a renowned 19th-century physicist, estimated that the Earth is 20 million years old by calculating how long it would take a planet the size of the Earth to cool from the temperature of the Sun. Kelvin's estimate contrasted with the longer time estimates promoted by followers of Hutton, who argued that if the concepts of uniformitarianism and

evolution were correct, the Earth must be much older than 20 million years.

Debate continued until isotopic dating became reliable, at which time geologists determined that the Earth is vastly older than Kelvin predicted. (Kelvin did not know that radioactive heat generation within the Earth made the planet cool much more slowly than he estimated.) Geologists have scoured the continents to locate their oldest rocks. Rocks that formed more recently than 3.85 Ga are fairly common. (Recall that Ga means billion years ago.) A few rock samples from several localities (Wyoming, Canada, Greenland, and China) have yielded dates as old as 4.03 Ga. Individual grains of the mineral zircon have yielded dates as old as 4.4 Ga, indicating that rock of this age did once exist. To obtain older dates, geologists have analyzed rocks from elsewhere in the Solar System. Meteorites thought to have come from undifferentiated planetesimals have yielded dates as old as 4.57 Ga, so geologists

FIGURE 9.22 Many people drive by this road cut on a highway in eastern New York State without realizing that it provides amazing insight into the past history of the Earth. We can specify the ages of the events in years. The unconformity was tilted after the rocks above it were deposited. These relations tell us that two different mountain-building events happened here. The mountains themselves have eroded away.

Lower Devonian limestone: 415 Ma (hidden by trees)

Latest Silurian dolostone and limestone: ~420 Ma

Middle Ordovician shale and sandstone: ~470 Ma

Unconformity: a gap of 50 m.y.

What an Earth Scientist Sees

identify 4.57 Ga as the time of planetesimal growth. The oldest meteorites from differentiated planetesimals formed at 4.54 Ga, so if we consider the time of internal differentiation as the time when the Earth formally became a planet, then we can consider geologic time to begin between 4.57 and 4.54 Ga.

The number 4.54 Ga is so staggeringly large that it's hard to understand. One way to grasp the immensity of geologic time is to equate the Earth's history to a single calendar year. On this scale, the oldest rocks preserved on the Earth appear in early February, the first bacteria arrive on February 21, the first shelly invertebrates on October 25, the first amphibians crawl out onto land on November 20, and the continents coalesce into Pangaea on December 7. Dinosaurs appear on December 15 and go extinct on December 25. The last week of December represents the Cenozoic, the last 66 million years of Earth history. The first human-like ancestor appears on December 31 at 3 P.M. Our species, *Homo sapiens*, shows up an hour before midnight, and all of recorded human history takes place in the last 30 seconds. To put it another way, human history occupies the last 0.000001% of Earth history. The Earth is so old that it has had more than enough time for its rocks and life forms to have formed and evolved (Fig. 9.22), for mountain ranges to rise and erode away, and for supercontinents to coalesce and disperse.

Take-home message . . .

Numerical ages for sedimentary rocks can be deduced from isotopic dating of cross-cutting dateable rocks. Such work led to the geologic time scale. The oldest rock of the Earth's crust is about 4 billion years old. Dating of meteorites indicates that planetesimals formed at 4.57 Ga, and that the Earth had differentiated by 4.56 to 4.54 Ga.

Quick Question
Why have the assignments of numerical ages to periods on the geologic time scale changed?

Did you ever wonder . . .

how old the Earth's oldest rock is?

9 CHAPTER REVIEW

- Geologic time refers to the time span since the Earth's formation.
- Relative age specifies whether one geologic feature is older or younger than another, whereas numerical age is the age of a geologic feature in years.
- Using such principles as uniformitarianism, original horizontality, superposition, and cross-cutting relations, we can construct the geologic history of a region.
- Fossils form when organisms or traces of organisms are buried and preserved in rock. Typically, hard parts of organisms are most likely to be preserved.
- The principle of fossil succession states that the assemblage of fossils in younger strata differs from that in older strata. Once a species becomes extinct, it never reappears.
- Fossils provide a record of life's evolution and a way of determining the ages of rock layers relative to one another.
- Darwin's theory of evolution by natural selection states that the fittest organisms survive and pass on their traits to their offspring.
- The fossil record is incomplete because paleontologists have found only a tiny fraction of fossil species, not all organisms are preserved, and preserved strata do not record all of geologic time.
- A stratigraphic column shows the succession of strata in a region. A given sequence of beds that can be traced over a fairly broad region is a stratigraphic formation.
- Strata are not necessarily deposited continuously at a location. An interval of nondeposition or erosion is called an unconformity. Geologists distinguish among angular unconformities, nonconformities, and disconformities.
- Correlation defines the age relationship between formations at one location and formations at another.
- The geologic column represents the entirety of geologic time. Its largest subdivisions are eons. Eons are subdivided into eras, eras into periods, and periods into epochs.
- At several times during the Earth's history, a high percentage of species went extinct. Such mass-extinction events may be a consequence of catastrophic events.
- A geologic map shows the distribution of formations and geologic structures.
- The numerical ages of rocks can be determined by isotopic (radiometric) dating. This method is based on the observation that radioactive isotopes decay at a rate characterized by a known half-life.
- The isotopic age of a mineral specifies the time at which the mineral cooled below a closure temperature. We can use isotopic dating to determine when an igneous rock solidified and when a metamorphic rock cooled.
- To date sedimentary strata, we must examine their cross-cutting relations with dateable igneous or metamorphic rock.
- Isotopic dating indicates that planetesimals started to form at about 4.57 Ga, and that the Earth differentiated at 4.54 Ga.

angular unconformity (p. 269)
baked contact (p. 258)
biodiversity (p. 271)
closure temperature (p. 279)
contact (p. 267)
correlation (p. 270)
cross-cutting relations (p. 258)
daughter isotope (p. 277)
disconformity (p. 269)
eon (p. 271)
epoch (p. 271)

era (p. 271)
extinct (p. 265)
fossil (p. 260)
fossil assemblage (p. 265)
fossilization (p. 260)
fossil succession (p. 265)
geochronology (p. 272)
geologic column (p. 271)
geologic history (p. 258)
geologic map (p. 276)
geologic time (p. 257)
geologic time scale (p. 280)

half-life (p. 277)
inclusion (p. 258)
isotope (p. 272)
isotopic dating (p. 272)
mass-extinction event (p. 271)
nonconformity (p. 269)
numerical age (p. 272)
original horizontality (p. 258)
paleontology (p. 260)
parent isotope (p. 277)
period (p. 271)
Precambrian (p. 271)

preservation potential (p. 261)
radioactive decay (p. 272)
radioactive element (p. 272)
relative age (p. 257)
stratigraphic column (p. 267)
stratigraphic formation (p. 266)
superposition (p. 258)
taxonomy (p. 264)
theory of evolution (p. 266)
unconformity (p. 269)
uniformitarianism (p. 257)

Letters in parentheses correspond to the chapter's learning objectives.

1. Explain the concept of uniformitarianism. **(B)**

2. Compare the meaning of a relative age and a numerical age. **(A)**

3. Describe the principles that allow us to determine the relative ages of geologic events. **(B)**

4. Describe the various processes that can produce fossils. Which type of fossil organism does the sketch show? **(C)**

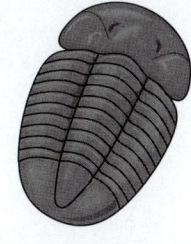

5. Why did geologists determine the change from the Precambrian to the Cambrian to be an eon boundary? **(D)**

6. How does an unconformity develop? Describe the differences among the three kinds of unconformities. **(E)**

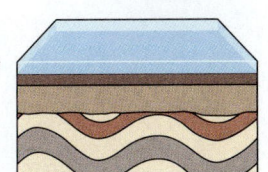

7. Which type of unconformity does this drawing show? **(E)**

8. What is a stratigraphic formation? How are stratigraphic formations portrayed on geologic maps? **(F)**

9. Describe two different methods of correlating rock units. How was correlation used to develop the geologic column? Which of the fossils (X, Y, or Z) in this stratigraphic column is younger? **(F)**

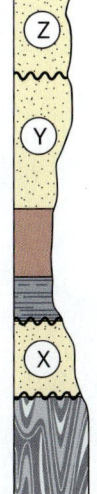

10. What does the process of radioactive decay entail? **(G)**

11. How do geologists obtain an isotopic date? What are some of the pitfalls in obtaining a reliable one? **(G)**

12. What do the numerical ages of igneous and metamorphic rocks mean?

13. Why can't we date sedimentary rocks directly? How do we obtain numerical ages of periods on the geologic column? **(G)**

14. What is the age of the oldest rock yet found on Earth? What is the age of the oldest meteorite? Why is there a difference? **(H)**

15. Imagine an outcrop exposing a formation consisting of alternating sandstone and conglomerate beds. A geologist studying the outcrop notes the following:

    - The sandstone beds contain fossils.
    - A layer of volcanic ash, dated at 300 Ma, overlies the sandstone-conglomerate formation.
    - A paleosol (ancient soil layer) occurs at the base of the ash layer.
    - A 100-Ma basalt dike cuts the ash and the sandstone-conglomerate formation.
    - Pebbles of granite in the conglomerate yield radiometric dates of 400 Ma.

    On the basis of these observations, how old is the sandstone and conglomerate formation? (Specify both the numerical age range and the period or periods of the geologic column during which it formed.) **(F)**

16. Look again at Figure 9.22. The rocks below the unconformity are Middle Ordovician turbidites deposited in deep water, and the rocks above the unconformity are Lower Devonian limestones deposited in shallow water. What type of unconformity does this outcrop expose? Provide a brief geologic history of the outcrop, keeping in mind that the unconformity has been tilted. **(E)**

ANOTHER VIEW Even this humble cliff of the Entrada Sandstone, near Toadstool Hoodoos in southern Utah, have a tale to tell of the Earth's history. The layers of Jurassic strata record a time when what is now a desert region hosted beaches and tidal mudflats along a shallow sea.

10 A BIOGRAPHY OF THE EARTH

By the end of the chapter you should be able to . . .

A. characterize a scientific model for how and when the Earth formed, and define the eons of the Precambrian.

B. describe a model for the formation of the first oceans and continents.

C. explain the concept and history of supercontinents.

D. describe the earliest life and how life evolution is linked to atmospheric evolution.

E. identify stages of life evolution through the Phanerozoic.

F. discuss when and why continental interiors sometimes host shallow seas or glaciers, and how the Earth's climate has changed over time.

G. discuss when and why mountain belts formed at various times and places in the past.

H. define global change, and provide examples of various types of change.

10.1 Introduction

In 1868, Thomas Henry Huxley, a British biologist, presented a public lecture to an audience in Norwich, England. Seeking a way to convey his fascination with the Earth's history, he focused the audience's attention on the piece of chalk he'd been drawing with. And what a tale the chalk had to tell! Chalk consists of microscopic marine plankton shells. The specific chalk that Huxley held came from 90-Ma beds now exposed in the dramatic white cliffs of England's southeastern coast. Geologists in Huxley's day knew that similar chalk beds cropped out throughout much of Europe and that some of these beds contained fossils of bizarre swimming reptiles that were very different from those of modern times. Clues in his humble piece of chalk allowed Huxley to demonstrate that the geography and inhabitants of the Earth in the past differed markedly from those of today. In other words . . . the Earth has a history!

In this chapter, we explore the Earth's history by offering a concise geologic biography of our home planet, from its birth 4.56 to 4.54 billion years ago to the present. We proceed chronologically, beginning with our planet's formation and finishing with the present day. We see how, over billions of years, the Earth's surface, oceans, and atmosphere changed dramatically as continents drifted, mountain belts grew, climates alternated between warmer and cooler, and sea level rose and fell. In addition, we examine how life evolved and diversified. The chapter ends by introducing the concept of global change and the ways in which humanity may be leaving its mark on Earth history.

10.2 The Hadean Eon: Before the Rock Record Began

When James Hutton, the 18th-century Scottish geologist, pondered the implications of his principle of uniformitarianism (see Chapter 9), he realized that, since most geologic processes happen slowly, our planet must be old indeed. Without access to isotopic dating, however, he had no way to measure the actual duration of the Earth's history. Isotopic dating of meteorites indicates that the Earth began to grow by coalescence of planetesimals at about 4.57 Ga (Fig. 10.1a; see also Chapter 1). Formation of the proto-Earth from planetesimals may have happened relatively rapidly, and by some estimates, was substantially complete by 4.56 Ga. Differentiation of the interior into a core and mantle, and formation of a solid, or at least partially solid, crust had happened by 4.54 Ga (Fig. 10.1b). Therefore, the birthday of the Earth depends on which phenomenon one uses to define the completion of Earth formation—substantial accretion, or attainment of its full size and differentiation. We'll use the age range here, and therefore use 4.56 to 4.54 Ga as the age of the Earth.

Significantly, the oldest mineral grain to be formed on the Earth dates to about 4.36 Ga, the oldest whole rock yet found is "only" about 4.03 Ga, and the rock record does not become substantial until about 3.85 Ga. Geologists refer to the mysterious, half-billion-year time interval between 4.54 Ga and about 4.0 Ga as the **Hadean Eon**, named for Hades, the Greek god of the underworld.

Many major changes happened on the Earth during the Hadean. As we saw in Chapter 1, the eon began with differentiation of the Earth's interior, during which gravity pulled molten iron down to the center of the Earth, where it accumulated to form the metallic core, leaving behind a mantle of ultramafic rock. Researchers suggest that about 4.53 Ga, soon after differentiation was complete, a protoplanet collided with the Earth. This impact blasted away a significant fraction of the Earth's mantle to yield a ring of silicate debris that coalesced to form our Moon (Fig. 10.1c).

In the wake of differentiation and Moon formation, the Earth may have been so hot that much of its surface was covered with lava. Rafts of solid rock may temporarily have formed on the surface of the lava ocean, but they sank and remelted. It's this image of a hell-like environment that led to the name Hadean (Fig. 10.1d). How long the Earth's surface stayed molten remains a subject of debate. The young planet contained relatively large quantities of radioactive elements. Elements with short half-lives were decaying and releasing heat, so geologists once thought it remained hot through much of the

> The man who should know the true history of the bit of chalk which every carpenter carries about in his breeches pocket, though ignorant of all other history, is likely . . . to have a truer and therefore a better conception of this wonderful universe and of man's relation to it than the most learned student who [has] deep-read the records of humanity [but is] ignorant of those of nature.
>
> —THOMAS HENRY HUXLEY
> (1825–1895)

<< At first glance, this large roadcut through a ridge of the Appalachian Mountains in Maryland seems like just a cliff of rock. But closer examination tells the story of a portion of the Earth's history. The black layers consist of coal, formed from plants that grew when this location, now a temperate forest, was a tropical swamp. The other layers are sandstones and shales, deposited in rivers and on floodplains. If you try to trace the layers across the road cut (ignoring the terraces that were made to stabilize the cliff face), you can see that they have been bent into a large syncline, a consequence of the collision between Africa and North America a few hundred million of years ago.

FIGURE 10.1 Scenes from the Hadean Eon.

(a) The Earth grew from the accumulation of planetesimals.

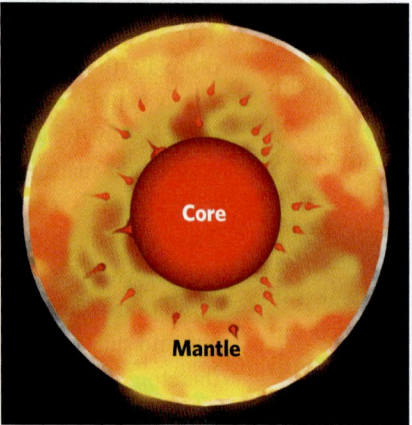

(b) During differentiation, iron sank to the center of the Earth to form its core, leaving a rocky mantle behind.

(c) The Moon formed from material blasted from the Earth by a collision with a protoplanet.

(d) Early on, the surface of the Earth was probably a lava ocean.

(e) Some crust and an early ocean may have developed during the Hadean.

FIGURE 10.2 The late heavy bombardment may have pulverized and remelted the Earth's crust between 4.0 and 3.9 Ga. Few rocks older than this event remain.

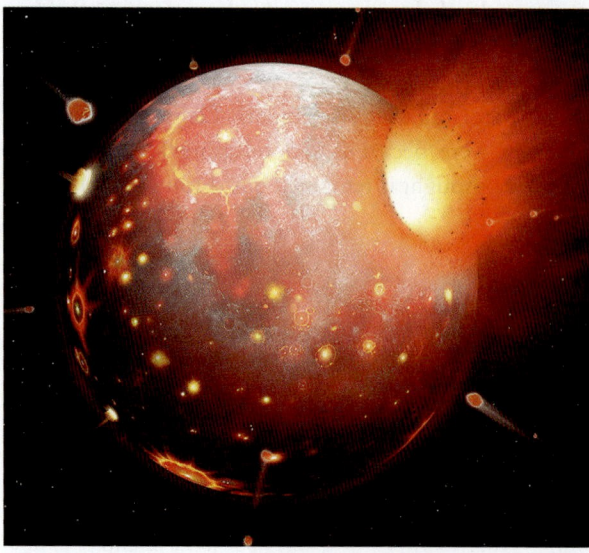

Hadean. But recent calculations hint that it may have taken only a few tens of millions of years after Moon formation for the Earth's surface to have cooled enough to have hosted a crust of solid rock, locally covered by a water ocean **(Fig. 10.1e)**.

During the Hadean Eon, volatile (gassy) elements and compounds that had originally been incorporated into mantle minerals bubbled out of volcanic vents. The products of this *outgassing* accumulated to form the Earth's first atmosphere, a murky mixture consisting mostly of carbon dioxide (CO_2), water vapor (H_2O), nitrogen (N_2), methane (CH_4), ammonia (NH_3), hydrogen (H_2), hydrogen sulfide (H_2S), and sulfur dioxide (SO_2). Some researchers speculate that comets colliding with the Earth may have contributed gas molecules, and perhaps even organic matter, to this early atmosphere.

FIGURE 10.3 A model for crust formation during the Archean Eon.

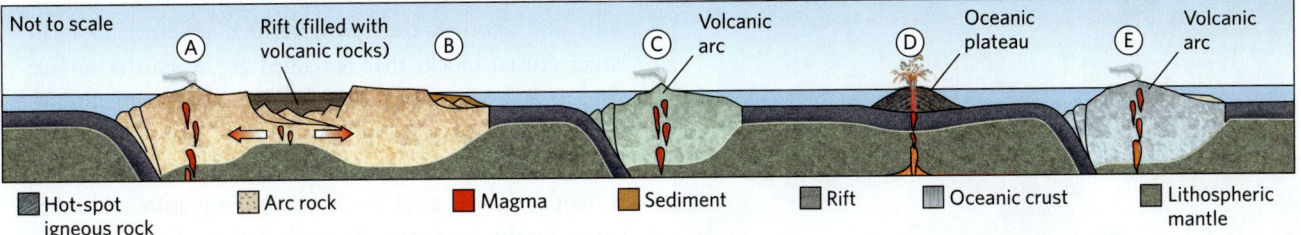

Hot-spot igneous rock | Arc rock | Magma | Sediment | Rift | Oceanic crust | Lithospheric mantle

(a) In the Archean, convergent boundaries and hot spots built small blocks of relatively buoyant crust. Rifting of these blocks may have produced flood basalts, and erosion of the blocks produced sediment.

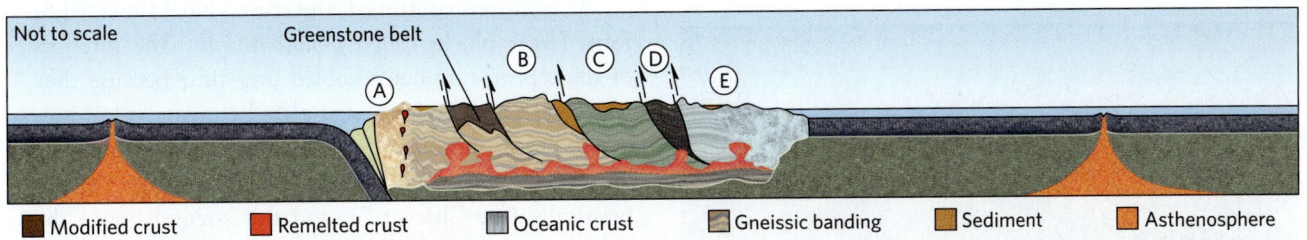

Modified crust | Remelted crust | Oceanic crust | Gneissic banding | Sediment | Asthenosphere

(b) Buoyant blocks collided and were sutured together, forming protocontinents, which were intruded by granite. Eventually, regions of crust cooled, stabilized, and became continents.

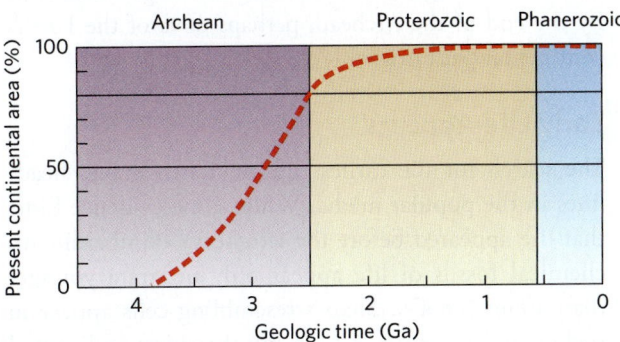

(c) As time progressed, the area of the Earth covered by continental crust increased. Most of the crust that exists today had formed by the end of the Archean.

Take-home message . . .

The Hadean Eon began when the Earth differentiated. Shortly after this, a protoplanet collided with the Earth to form the Moon. Initially molten, the Earth's surface eventually cooled and solidified.

Quick Question
Where did the gases of the Earth's early atmosphere come from, and what did this atmosphere consist of?

10.3 The Archean Eon: Birth of Continents and Life

Geologists now place the start of the **Archean Eon** (from the Greek word *arché*, meaning beginning) at 4.0 Ga, about the age of the oldest known rock. Why are rocks formed before the Archean so rare? Some researchers argue that rocks older than this date have passed through the rock cycle (see Chapter 6) to become components of younger rocks. Others note that studies of the craters visible on the Moon indicate that meteorites battered the inner planets between 4.0 and 3.9 Ga. This **late heavy bombardment** may have pulverized and melted any crust that existed at the time, so that only after these impacts ceased could long-lasting crust, atmosphere, and oceans survive **(Fig. 10.2)**.

By 3.85 Ga, the near-surface realm of the Earth was cool and stable enough for the rock record to become more complete. The cooling Earth System allowed water in the atmosphere to condense, and rains filled permanent oceans. (The evidence for oceans at that time comes from the discovery of 3.85-billion-year-old marine sedimentary rocks.) Most atmospheric CO_2 dissolved into the new oceans. Removal of H_2O and CO_2 from the atmosphere left N_2 gas behind, for nitrogen does not react chemically with or dissolve in other substances. So at the dawn of the Archean, the Earth had a transparent atmosphere composed of 80% N_2 and 20% CO_2 (see Chapter 17).

Continental Crust Appears

The earliest land probably formed at hot-spot volcanoes, which extruded thick plateaus of mafic lava on the floor of the early oceans. So a visitor to the Earth's surface at the beginning of the Archean would probably have seen a water ocean from which only volcanic islands protruded. When plate tectonics began to operate, mafic igneous activity also began to take place at convergent boundaries **(Fig. 10.3a)**. Some of the mafic rock of early Archean island arcs and

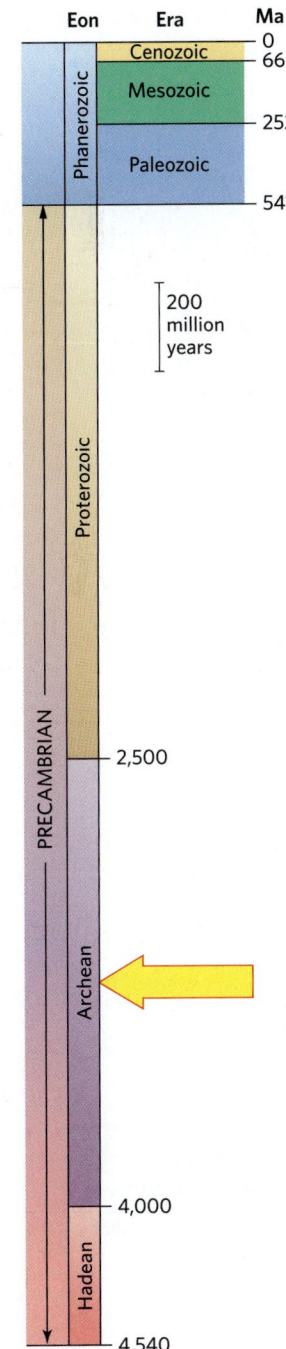

FIGURE 10.4 Examples of Archean life.

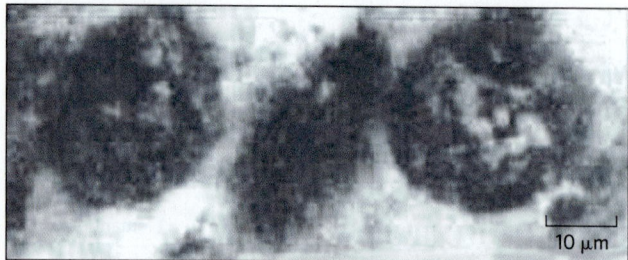

(a) These shapes in 3.2-Ga chert from South America are thought to be fossil bacteria or archaea.

(b) Stromatolites form when sediment sticks to layers of cells. As one layer dies and compacts, a new one grows above.

(c) This weathered outcrop of 1.85-Ga dolostone near Marquette, Michigan, reveals the layer-like structure of stromatolites. The delicate ridges represent the fossilized remnants of cyanobacterial mats. Similar stromatolites occur in exposures of Archean rocks.

(d) Modern stromatolites in Shark Bay, Western Australia.

Did you ever wonder . . .

what the oldest relict of life is?

oceanic plateaus was not subducted, however, because it was less dense than the ultramafic rock of the underlying mantle. When these relatively buoyant blocks collided with one another, they were sutured together to form larger crustal blocks that remained at the Earth's surface **(Fig. 10.3b)**. Metamorphic rocks formed during collision. Sometime after collision, rifting took place and flood basalts extruded at various locations. Eventually, melting at depth in the crustal blocks produced granite and other lower-density igneous rocks, and erosion of exposed crust yielded sediments. As a result of all these processes, the new crustal blocks contained a variety of rock types.

As collisions continued, the crustal blocks merged to form larger blocks called *protocontinents*. The interiors of these protocontinents cooled over time because they were isolated from the heat of plate-boundary volcanism. When protocontinent interiors cooled, the rock they contained became stronger because it could no longer deform plastically (see Chapter 7), and this strength made the protocontinents durable. By 3.2 Ga, several durable protocontinents existed, and by 2.7 Ga, some of them had been sutured together to form the first large continents. By the end of the Archean, perhaps 80% of the Earth's continental crust existed **(Fig. 10.3c)**.

Early Life Appears

The search for the earliest life on Earth makes headlines in the popular media. While some evidence hints that life appeared before the late heavy bombardment, chemical fossils of life appear only in strata younger than about 3.8 Ga. Shapes resembling cells appear in rocks dated as early as 3.5 Ga, but the oldest undisputed fossil cells of bacteria and archaea, the most primitive living organisms, occur in strata dating from 3.2 Ga **(Fig. 10.4a)**. These rocks contain *stromatolites*, distinct layered mounds interpreted to be made up of fossilized mats of *cyanobacteria* (photosynthetic bacteria). The mounds form because sediment settling out of ocean water sticks to a mucus-like substance secreted by the mats of cyanobacteria. As the mat becomes buried, new cyanobacteria colonize the top of the sediment, building the mound upward **(Fig. 10.4b–d)**.

What specific environment on the Archean Earth served as the cradle of life? Laboratory experiments of the 1950s led researchers to speculate that life began in warm pools of surface water beneath a methane- and ammonia-rich atmosphere shocked by bolts of lightning. More recent studies, however, suggest that deep-sea hydrothermal vents may have hosted the first organisms. These vents emit clouds of ion-charged solutions from which sulfide minerals precipitate. Alternatively, hydrothermal craters on land may have hosted the first life. The earliest life in the Archean may well have been heat-loving bacteria or archaea that dined on sulfides emitted from these vents in the dark depths of the ocean. Later in the Archean, when organisms evolved the ability to carry out photosynthesis (between 3.5 and 3.2 Ga), did life move into shallow, brightly lit ocean water. The oxygen produced by these organisms began to change the composition of the atmosphere.

As the Archean Eon came to a close, the first continents had formed, plate tectonics was under way, continental drift was taking place, collisional mountain belts were forming, and erosion was occurring. Life had colonized not only the depths of the sea, but also the shallow marine realm. The atmosphere had transformed from an H_2O- and CO_2-rich one into an N_2-rich one containing traces of O_2. The stage was set, by about 2.5 Ga, for another major change in the Earth System.

> **Take-home message . . .**
>
> During the Archean (4.0–2.5 Ga), the first continental crust formed from colliding volcanic arcs and oceanic plateaus. As oceans filled with water, the atmosphere lost its H_2O and CO_2 and became nitrogen rich. Early life appeared in the sea.
>
> **Quick Question**
> What environment may have hosted the earliest life?

10.4 The Proterozoic Eon: The Earth in Transition

The **Proterozoic Eon**, the last interval of the Precambrian, spans roughly 2 billion years—almost half of the Earth's history—from 2.5 Ga to the beginning of the Cambrian Period at 541 Ma. During the Proterozoic, the Earth's surface environment changed from an unfamiliar world of small continents and an oxygen-poor atmosphere into a world of large continents and an oxygen-rich atmosphere much more like what we see today.

Large Continents Form

New continental crust continued to form during the Proterozoic, but at progressively slower rates. By the middle of the eon, over 90% of the Earth's present continental crust had formed (see Fig. 10.3c). But a glance at a simplified crustal province map of the Earth today indicates that Precambrian rocks currently underlie much less than 90% of the continental crust (Fig. 10.5a). That's because during the Phanerozoic, Precambrian rocks went through various stages of the rock cycle (weathering, melting, metamorphism) and formed new Phanerozoic-age rocks.

Recall that a block of crust that has remained relatively unchanged since at least about 1 Ga is called a **craton**. Phanerozoic continents consist of one or more cratons, surrounded by Phanerozoic orogens or stretched (rifted) crust. Each craton may consist of many Precambrian blocks, but these blocks have remained connected, and have remained relatively cool and strong, for at least the last billion years—we can see this pattern by examining a crustal province map of North America in more detail (Fig. 10.5b, c).

Because of continental drift, the map of the Earth constantly changes. Using a variety of data sources, including correlation of rock units and studies of paleomagnetism, geologists have been able to define the relative positions of continents in the past and to portray their locations on maps. To emphasize that these maps represent times in the past, they are called **paleogeographic maps**. Successive collisions ultimately brought together most continental crust on the Earth into a single supercontinent, named **Rodinia**, by around 1 Ga. On a paleogeographic map of Rodinia (Fig. 10.6a), we can identify the regions that eventually became the familiar continents of today. The last major collision during the formation of Rodinia, an event called the *Grenville orogeny*, affected eastern North America. Several studies suggest that later, between 800 and 600 Ma, Antarctica, India, and Australia broke away from western Rodinia, swung around, and recollided with the larger landmass to form a short-lived supercontinent known as **Pannotia** (Fig. 10.6b).

Life and Climate in the Proterozoic World

Fossil evidence suggests that the Proterozoic saw important steps in the evolution of life. When this eon began, most life was *prokaryotic*, meaning that it consisted of single-celled organisms—archaea and bacteria—that do not contain a cell nucleus (an internal, membrane-surrounded region containing the cell's DNA). Though studies of chemical fossils hint that *eukaryotic* life, consisting of cells that do have nuclei, originated as early as 2.7 Ga, the first possible body fossil of a eukaryotic organism occurs in rocks dated at 2.1 Ga, and abundant body fossils of eukaryotic organisms can be found only in rocks younger than about 1.2 Ga. So the diversification of eukaryotic life, the foundation from which complex organisms, including humans, eventually evolved, took place during the Proterozoic.

The last half billion years of the Proterozoic Eon saw the remarkable transition from simple organisms to complex ones. Sediments deposited perhaps as early as 620 Ma, and certainly by 565 Ma, contain several types of multicellular organisms that together constitute the **Ediacaran fauna**, named for a region in southern Australia. Some of

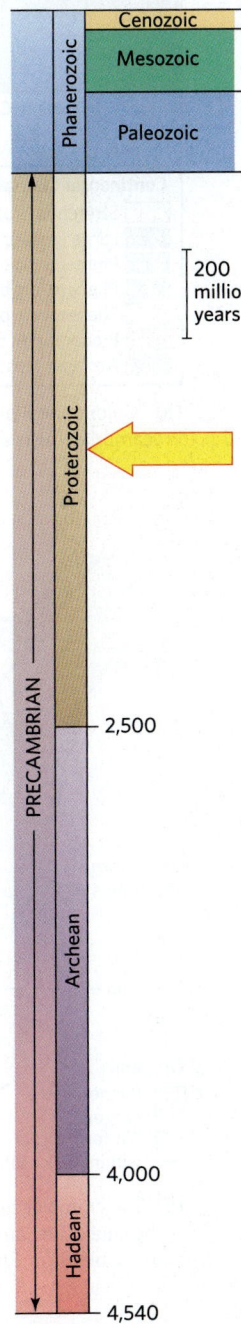

FIGURE 10.5 Different parts of the continental crust are of different ages. The old, stable parts are the cratons.

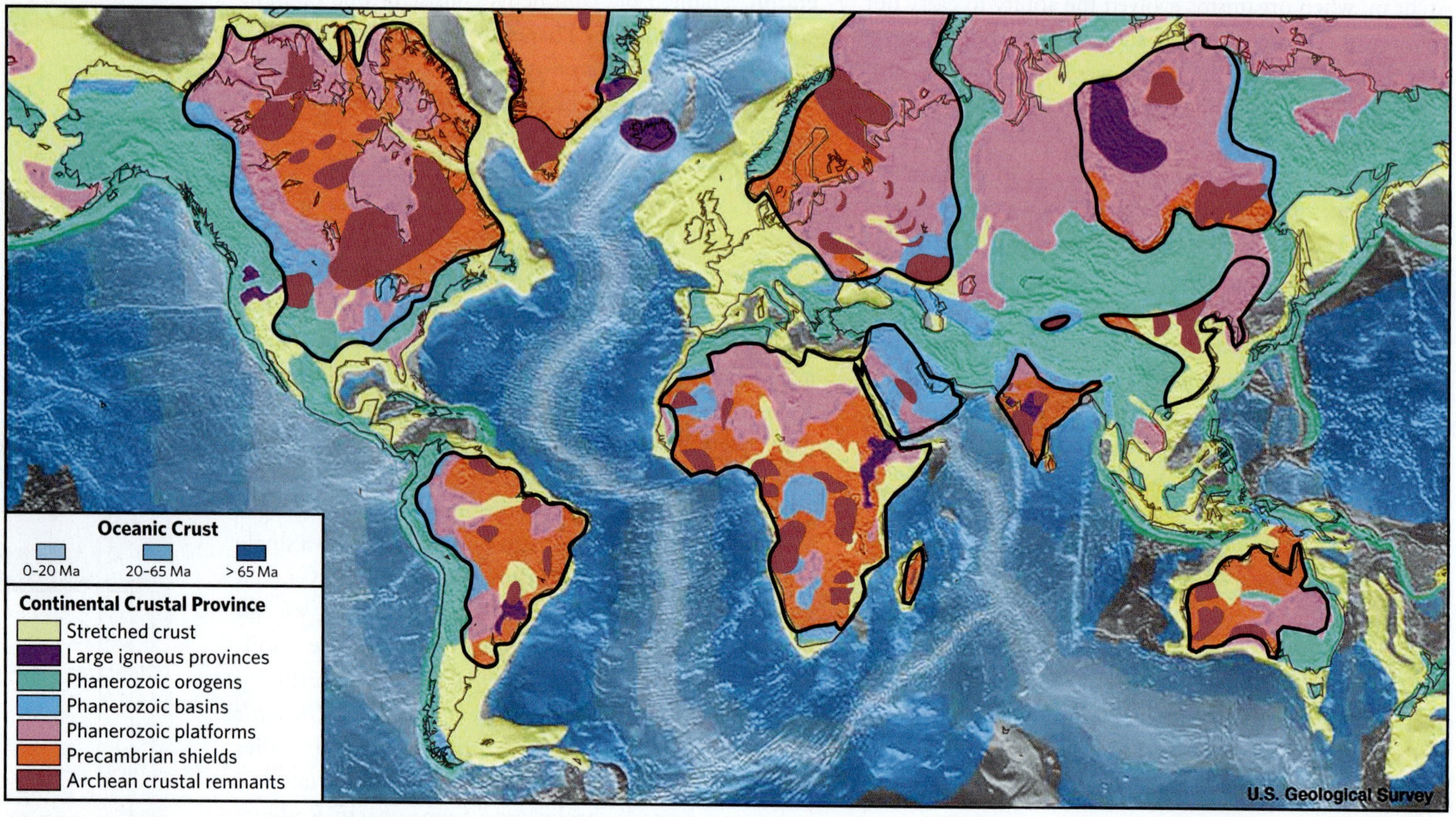

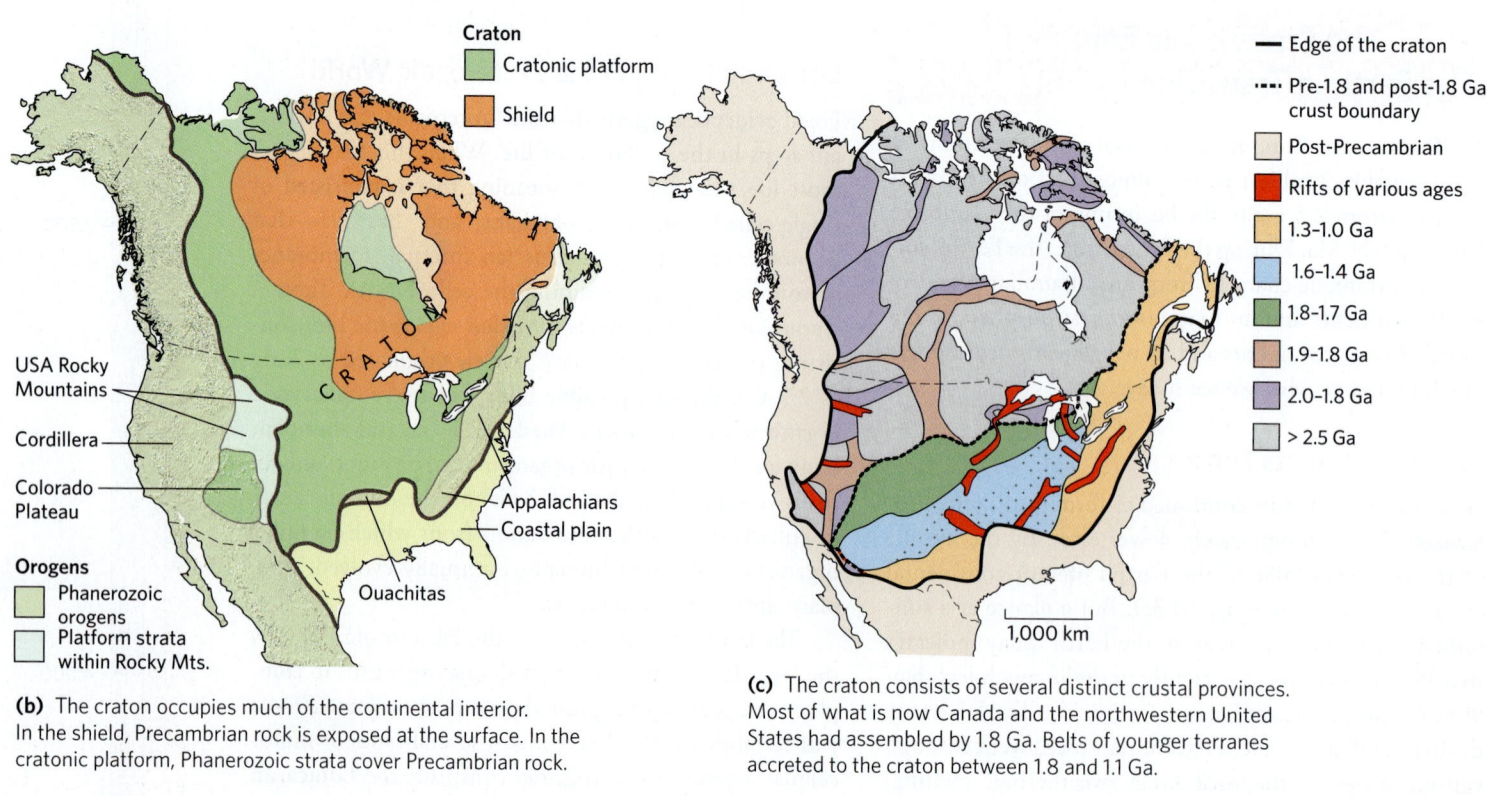

Oceanic Crust
0–20 Ma   20–65 Ma   > 65 Ma

Continental Crustal Province
- Stretched crust
- Large igneous provinces
- Phanerozoic orogens
- Phanerozoic basins
- Phanerozoic platforms
- Precambrian shields
- Archean crustal remnants

U.S. Geological Survey

(a) Major crustal provinces of the Earth. The black lines indicate the border of regions underlain by Precambrian crust. Shields are regions where broad areas of Precambrian rocks are exposed.

Craton
- Cratonic platform
- Shield

USA Rocky Mountains
Cordillera
Colorado Plateau
Appalachians
Coastal plain
Ouachitas

Orogens
- Phanerozoic orogens
- Platform strata within Rocky Mts.

(b) The craton occupies much of the continental interior. In the shield, Precambrian rock is exposed at the surface. In the cratonic platform, Phanerozoic strata cover Precambrian rock.

— Edge of the craton
···· Pre-1.8 and post-1.8 Ga crust boundary
- Post-Precambrian
- Rifts of various ages
- 1.3–1.0 Ga
- 1.6–1.4 Ga
- 1.8–1.7 Ga
- 1.9–1.8 Ga
- 2.0–1.8 Ga
- > 2.5 Ga

1,000 km

(c) The craton consists of several distinct crustal provinces. Most of what is now Canada and the northwestern United States had assembled by 1.8 Ga. Belts of younger terranes accreted to the craton between 1.8 and 1.1 Ga.

FIGURE 10.6 Supercontinents in the Proterozoic.

Rodinia
(at about 750 Ma)

(a) Rodinia formed around 1 Ga and lasted until about 700 Ma. Note that Laurentia consists of the landmasses that are now North America and Greenland.

Pannotia
(at about 570 Ma)

(b) According to one model, by 570 Ma, Rodinia had broken apart; continents that once lay to the west of Laurentia ended up to the east of Africa. The resulting supercontinent, Pannotia, broke up soon after it formed.

these soft-bodied invertebrate organisms resembled jellyfish, and others resembled worms **(Fig. 10.7a)**.

The evolution of life played a key role in changing the composition of Earth's atmosphere. Before life appeared, the atmosphere contained hardly any free oxygen (O_2). With the appearance of photosynthetic organisms, trace amounts of oxygen accumulated in the atmosphere. But it was not until about 2.4 Ga that the concentration of oxygen reached more than a few percent. That increase, called the **great oxygenation event**, happened when surface rocks and ocean water could no longer absorb or dissolve all the oxygen produced by organisms, so oxygen accumulated as a gas in the atmosphere. When the oceans became saturated with oxygen, the iron that had

FIGURE 10.7 Late Proterozoic Ediacaran fauna, and BIF.

(a) An artist's reconstruction of Ediacaran fauna.

(b) An outcrop of banded iron formation (BIF) from the Upper Peninsula of Michigan. The gray stripes are hematite and magnetite and the red stripes are red chert (jasper).

FIGURE 10.8 Snowball Earth.

Strata contain large clasts surrounded by mudstone, for glaciers can carry clasts of all sizes.

Bedding

(a) Layers of Proterozoic glacial till occur in regions that were at low elevation and near the equator in the late Proterozoic.

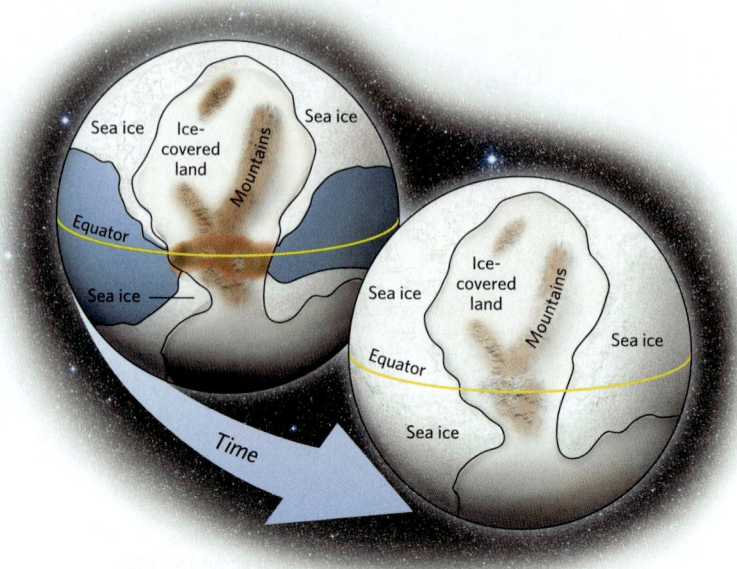

Sea ice · Ice-covered land · Sea ice · Mountains · Equator · Sea ice

Time

Ice-covered land · Mountains · Sea ice · Equator · Sea ice

(b) The planet may have frozen over completely to form "snowball Earth." The ice shell prevented volcanic CO_2 from dissolving in the oceans. Accumulation of CO_2 in the atmosphere eventually caused greenhouse warming that melted the ice.

Did you ever wonder...

whether the oceans have ever frozen over entirely?

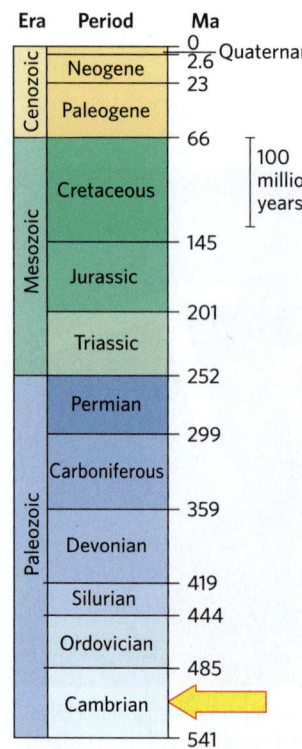

Era	Period	Ma
Cenozoic		0
	Neogene	2.6 — Quaternary
		23
	Paleogene	
		66
Mesozoic	Cretaceous	} 100 million years
		145
	Jurassic	
		201
	Triassic	
		252
Paleozoic	Permian	
		299
	Carboniferous	
		359
	Devonian	
		419
	Silurian	
		444
	Ordovician	
		485
	Cambrian	
		541

been dissolved in seawater bonded to oxygen atoms and formed iron oxide minerals that precipitated from the water and settled on the seafloor. In fact, between 2.4 Ga and 1.8 Ga, so much iron oxide settled out of the ocean that thick, colorful sequences of sedimentary rock, called **banded iron formation** (**BIF**), built up. BIF, which consists of layers of gray iron oxide minerals alternating with layers of red chert, provides most of the iron used by society today (**Fig. 10.7b**).

The Earth's climate cooled at the end of the Proterozoic Eon, and large ice sheets grew on continents. As a result, glacial till occurs worldwide in late Proterozoic sedimentary sequences. What's strange about the occurrence of this till is that it can be found even in regions that were located at the equator (**Fig. 10.8a**), implying that the entire planet was cold enough for glaciers to form. Geologists speculate that when continents became entirely glaciated, the entire ocean surface froze as well, and they refer to the resulting ice-encrusted planet as **snowball Earth** (**Fig. 10.8b**).

Take-home message...

During the Proterozoic (2.5 Ga–541 Ma), the interiors of large continents became cratons. Multicellular organisms appeared, and the atmosphere began to accumulate oxygen. At the end of the Proterozoic, a large supercontinent, Pannotia, existed.

Quick Question

What was snowball Earth?

10.5 The Paleozoic Era: Continents Reassemble and Life Diversifies

The end of the Proterozoic Eon defines the end of the Precambrian and the start of the **Phanerozoic Eon**. Geologists studying the fossil record recognized the significance of this boundary long before they could assign it a numerical age (currently, 541 Ma) because it marks the appearance of creatures with shells, which may have evolved as a means of protection against predators. The atmosphere also changed, as photosynthetic organisms prospered. By the beginning of the Paleozoic, O_2 accounted for about 17% of atmospheric gas. The Phanerozoic Eon consists of three eras: *Paleozoic* (Greek for ancient life), *Mesozoic* (middle life), and *Cenozoic* (recent life) (**Earth Science at a Glance**, pp. 294–295).

The Early Paleozoic Era (Cambrian and Ordovician Periods)

PALEOGEOGRAPHY. At the beginning of the Paleozoic Era, rifting broke apart the supercontinent Pannotia, yielding several smaller continents: **Laurentia** (composed of modern-day North America and Greenland), **Gondwana** (South America, Africa, Arabia, Antarctica, India, and Australia), *Baltica* (Europe), and *Siberia* (**Fig. 10.9a**). The coasts of these continents hosted new passive-margin basins. In addition, sea level rose as the climate warmed. As a result, large regions of continental interiors became flooded with shallow seas, called **epicontinental seas**

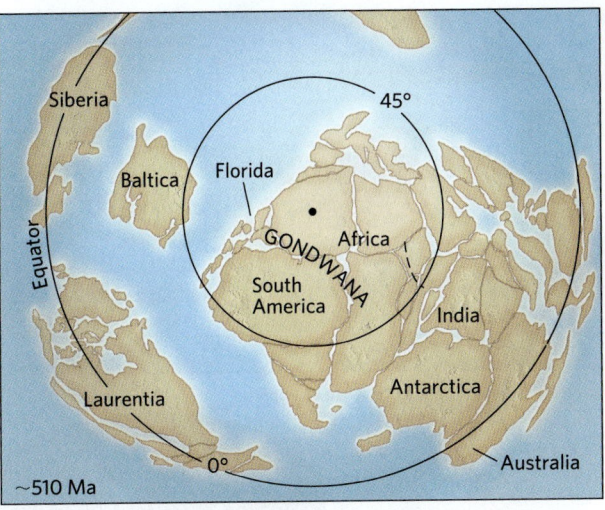

(a) The distribution of continents in the Cambrian Period (510 Ma), as viewed looking down on the South Pole.

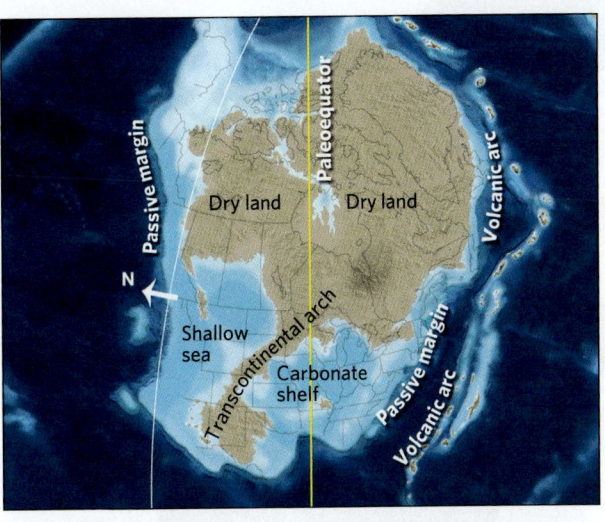

FIGURE 10.9 Land and sea in the early Paleozoic Era.

(b) A paleogeographic map of present-day North America shows the regions of dry land and epicontinental seas in the Late Cambrian.

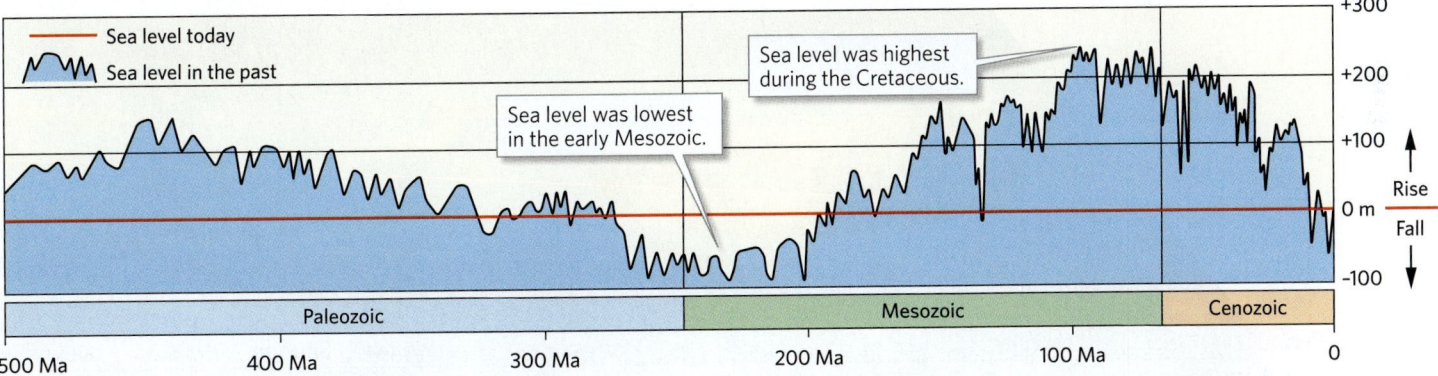

(c) This chart of transgressions and regressions in the stratigraphic record may indicate relative sea-level change during the past half billion years.

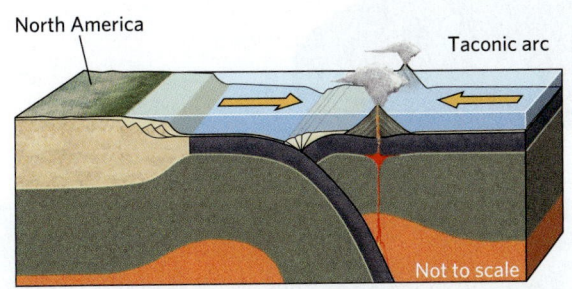

(d) During the Middle Ordovician, epicontinental seas covered much of North America. An island arc off the east coast collided with the continent.

(Fig. 10.9b). The existence of transgressions and regressions in the stratigraphic record of continental interiors indicates that sea level rises and falls, relative to the land surface, over geologic time. Geologists have developed a chart illustrating such sea-level changes (Fig. 10.9c). Charts depicting global sea-level change remain controversial, however, for many factors besides sea-level change can affect the stratigraphic record. Further, because the average radius of the Earth must remain constant, the magnitude of change depicted on the chart may be unrealistic.

Geologic phenomena that might affect sea level (Box 10.1) include the following:

- *Growth or melting of large glaciers:* Formation of a glacier removes water from the sea and traps it on land, causing sea level to fall. The melting of glaciers causes sea level to rise.

- *A change in the capacity of ocean basins:* If the volume of ocean basins decreases, sea level rises and the land floods. This phenomenon may be due to a change in the width and number of mid-ocean ridges, for the crust of mid-ocean ridges is shallower than that of abyssal plains.

- *A change in the area or elevation of continental surfaces:* Mountain building, erosion and deposition, or vertical displacements of a continent caused by processes taking place beneath the lithosphere can cause the surface of the continent to move with respect to sea level. If a continent's surface sinks, the continent may flood.

During the early part of the Paleozoic, passive-margin basins fringed all sides of Laurentia. These basins collected immense amounts of sediment eroded from the continent, and their surfaces became broad continental shelves. In the Middle Ordovician Period, the geologic tranquility of

The Earth Has a History

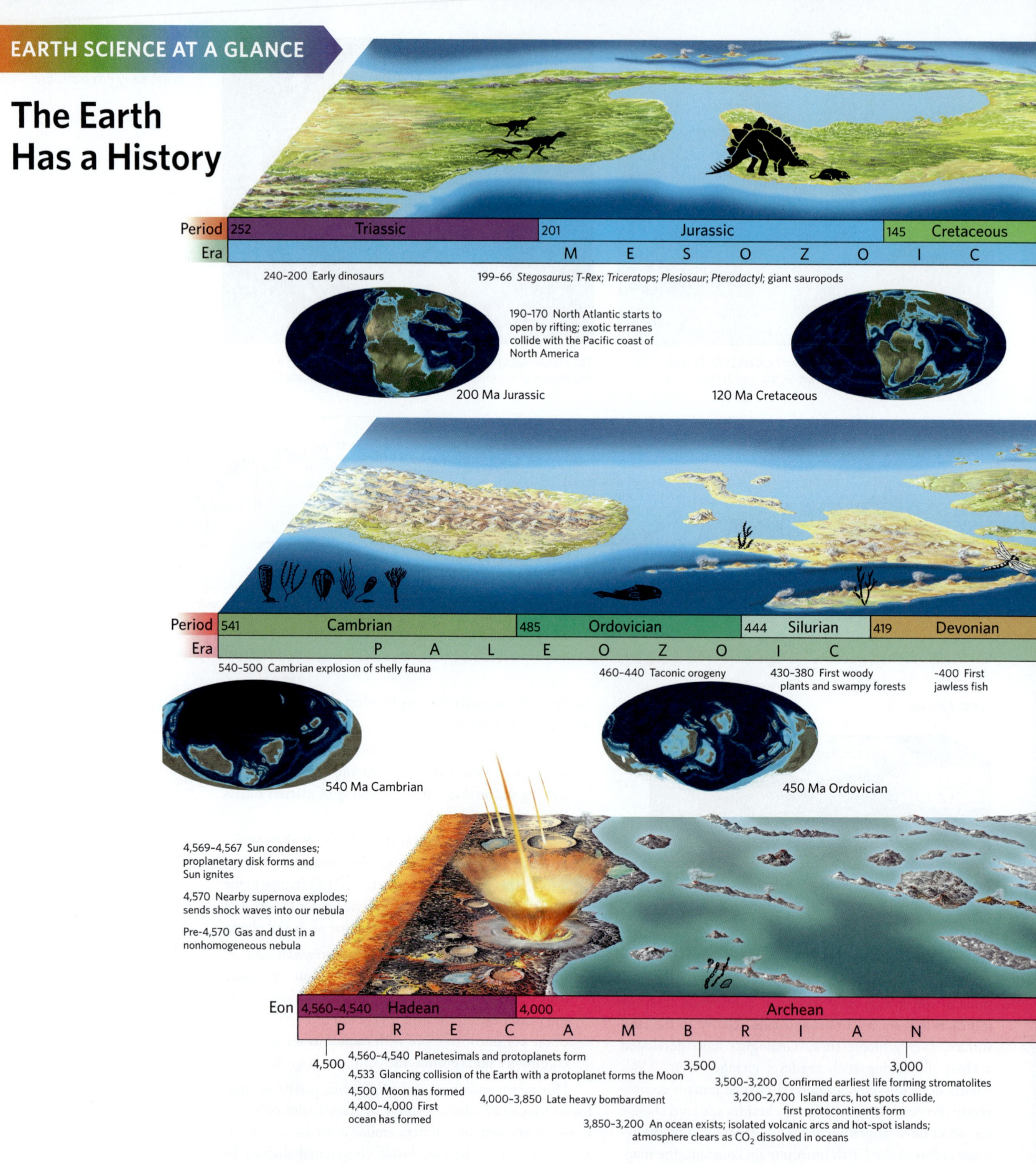

Period	252	Triassic	201	Jurassic	145	Cretaceous
Era		M E S O		Z O		I C

240–200 Early dinosaurs

199–66 *Stegosaurus; T-Rex; Triceratops; Plesiosaur; Pterodactyl; giant sauropods*

190–170 North Atlantic starts to open by rifting; exotic terranes collide with the Pacific coast of North America

200 Ma Jurassic

120 Ma Cretaceous

Period	541	Cambrian	485	Ordovician	444	Silurian	419	Devonian
Era		P A L E O		Z O		I C		

540–500 Cambrian explosion of shelly fauna

460–440 Taconic orogeny

430–380 First woody plants and swampy forests

~400 First jawless fish

540 Ma Cambrian

450 Ma Ordovician

4,569–4,567 Sun condenses; proplanetary disk forms and Sun ignites

4,570 Nearby supernova explodes; sends shock waves into our nebula

Pre-4,570 Gas and dust in a nonhomogeneous nebula

Eon	4,560–4,540 Hadean	4,000	Archean
	P R E C A M	B R	I A N

4,500

4,560–4,540 Planetesimals and protoplanets form

4,533 Glancing collision of the Earth with a protoplanet forms the Moon

4,500 Moon has formed

4,400–4,000 First ocean has formed

4,000–3,850 Late heavy bombardment

3,500

3,000

3,500–3,200 Confirmed earliest life forming stromatolites

3,200–2,700 Island arcs, hot spots collide, first protocontinents form

3,850–3,200 An ocean exists; isolated volcanic arcs and hot-spot islands; atmosphere clears as CO_2 dissolved in oceans

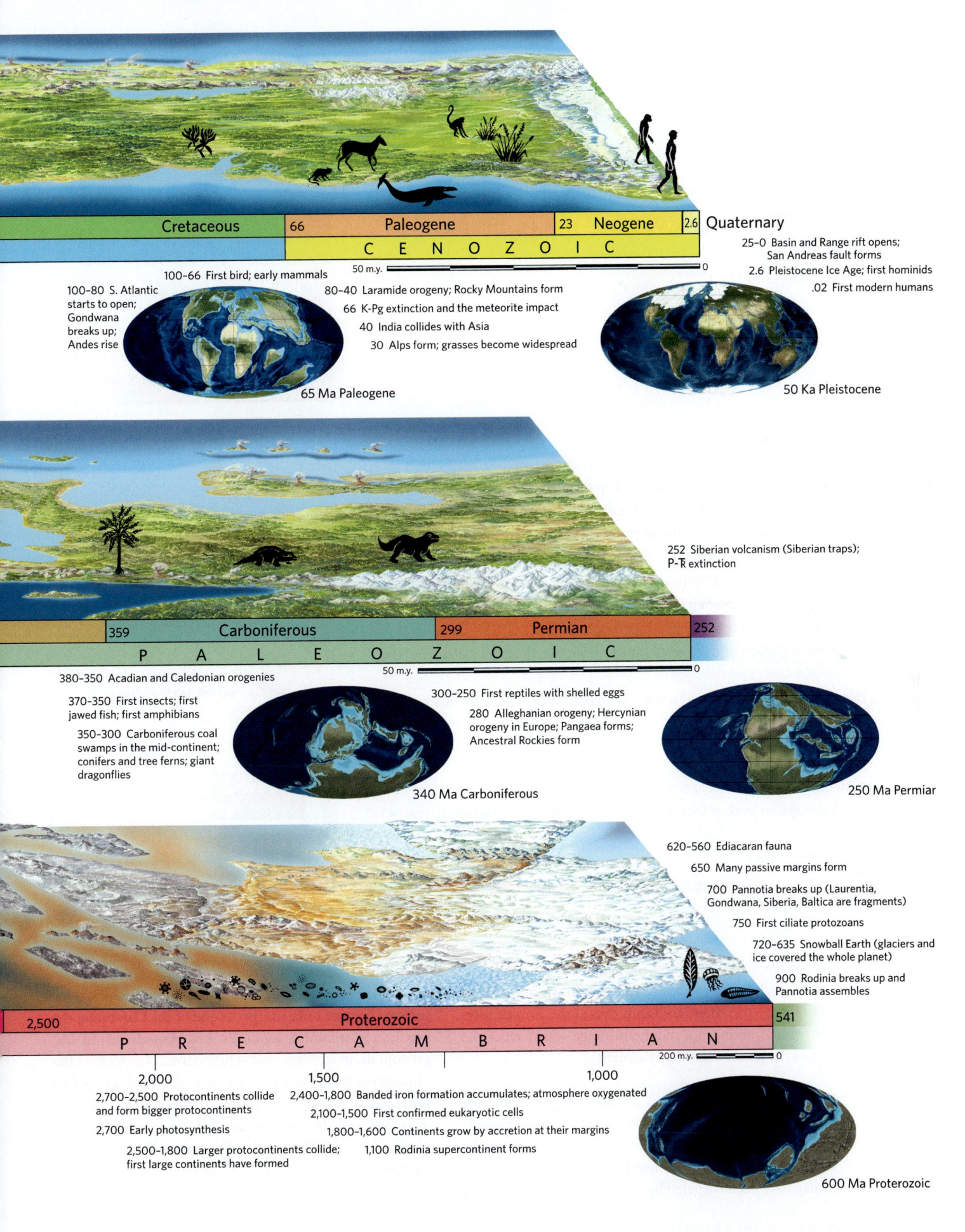

Cretaceous | 66 | Paleogene | 23 | Neogene | 2.6 | Quaternary

C E N O Z O I C

50 m.y. ⊢━━━━━┤ 0

100–66 First bird; early mammals

100–80 S. Atlantic starts to open; Gondwana breaks up; Andes rise

80–40 Laramide orogeny; Rocky Mountains form
66 K-Pg extinction and the meteorite impact
40 India collides with Asia
30 Alps form; grasses become widespread

25–0 Basin and Range rift opens; San Andreas fault forms
2.6 Pleistocene Ice Age; first hominids
.02 First modern humans

65 Ma Paleogene

50 Ka Pleistocene

| 359 | Carboniferous | 299 | Permian | 252 |

P A L E O Z O I C

50 m.y. ⊢━━━━━┤ 0

252 Siberian volcanism (Siberian traps); P-T extinction

380–350 Acadian and Caledonian orogenies

370–350 First insects; first jawed fish; first amphibians

350–300 Carboniferous coal swamps in the mid-continent; conifers and tree ferns; giant dragonflies

300–250 First reptiles with shelled eggs

280 Alleghanian orogeny; Hercynian orogeny in Europe; Pangaea forms; Ancestral Rockies form

340 Ma Carboniferous

250 Ma Permian

620–560 Ediacaran fauna

650 Many passive margins form

700 Pannotia breaks up (Laurentia, Gondwana, Siberia, Baltica are fragments)

750 First ciliate protozoans

720–635 Snowball Earth (glaciers and ice covered the whole planet)

900 Rodinia breaks up and Pannotia assembles

| 2,500 | Proterozoic | 541 |

P R E C A M B R I A N

200 m.y. ⊢━━━━━┤ 0

2,000   1,500   1,000

2,700–2,500 Protocontinents collide and form bigger protocontinents

2,700 Early photosynthesis

2,500–1,800 Larger protocontinents collide; first large continents have formed

2,400–1,800 Banded iron formation accumulates; atmosphere oxygenated

2,100–1,500 First confirmed eukaryotic cells

1,800–1,600 Continents grow by accretion at their margins

1,100 Rodinia supercontinent forms

600 Ma Proterozoic

295

BOX 10.1 ▶ **How can I explain . . .**

Sea-level change

What are we learning?

The concept that sea level can rise and fall relative to the land surface of continents over time, and some of the reasons for this change.

What you need:

- A plastic basin (about 30 × 40 cm, or 1 ft²)
- A standard brick
- A small measuring cup and a ladle
- Five stones (each about 6–10 cm across)
- Two pencils

Instructions:

- Place the brick in the middle of the basin.
- Fill the basin with water to just slightly below the top surface of the brick.
- To simulate a rise in sea level due to growth of the volume of a mid-ocean ridge, line up the stones in the basin. As the "ridge" grows, the water surface rises. Eventually, the water submerges the brick.
- To simulate sea-level fall, remove the stones. Water takes the place of the stones, and the top of the brick re-emerges.
- To simulate growth of continental glaciers and the resulting sea-level fall, place the cup on the brick, then ladle some water out of the basin and place it in the cup. Glaciers store water on land, so the amount of water in the sea decreases, and the water surface drops.
- To simulate melting of continental glaciers and the resulting sea-level rise, pour the water from the cup back into the basin.
- To simulate a tectonic rise of the continent relative to sea level, place the pencils underneath the brick so that its surface rises.

What did we see?

Sea-level changes may be due to changes in the capacity of the ocean basin to hold water, removal of water from the oceans and its storage on land, or local tectonic events that cause the land to rise or fall.

Brick = continent   Water = oceans

Brick becomes submerged   Rocks = formation of the mid-ocean ridge

Cup represents the glacial reservoir

When water is stored in glaciers as land, sea level drops.

---

the passive-margin basin on the eastern side of Laurentia came to a close, for the basin rammed into volcanic island arcs (Fig. 10.9d). These collisions caused the *Taconic orogeny*, which produced a mountain range whose relicts lie within the present-day Appalachians.

EVOLUTION OF LIFE. The Cambrian began with the appearance of shelled organisms. Soon after, biodiversity increased dramatically. This increase, known as the **Cambrian explosion**, continued over several million years. It may have been triggered by the breakup of Pannotia, for when smaller continents formed and drifted apart, many new ecological niches were formed, and populations of organisms became isolated. By the end of the Cambrian, the seas hosted trilobites, mollusks, brachiopods, nautiloids, gastropods, graptolites, and echinoderms (Fig. 10.10). The Ordovician Period saw the first crinoids as well as the first vertebrate animals (animals with backbones) in the form of jawless fish. In the Middle Ordovician, the first land plants, tiny moss-like liverworts, appeared. At the end of the Ordovician, a mass-extinction event took place, and animal life on Earth changed significantly.

FIGURE 10.10 A museum diorama illustrates what early Paleozoic marine organisms may have looked like.

The Middle Paleozoic Era (Silurian and Devonian Periods)

PALEOGEOGRAPHY. When the Silurian began, land was divided among two very large continents, Laurentia and Gondwana, and several small ones, including Baltica (Scandinavia and western Russia) and Siberia. During the Silurian, these continents began to be sutured together, and each collision produced an orogeny. Specifically, Baltica collided with eastern Laurentia, causing the *Caledonian orogeny*. Today, you can find geologic structures and metamorphic

rocks that formed during this event in western Scandinavia, eastern Greenland, and Scotland **(Fig. 10.11a)**. Soon after this, one or more **microcontinents** (continental blocks less than 1,000 km, 600 mi, across), including one called Avalonia, slammed into what is now the eastern United States, causing the *Acadian orogeny* **(Fig. 10.11b)**.

Through most of the middle Paleozoic, the western margin of Laurentia remained a quiet passive-margin basin, even as mountains grew on the eastern side of the continent. But in the Late Devonian, the west-coast passive-margin basin collided with an island arc. This event, the *Antler orogeny*, began the long history of mountain building that has dominated the geologic history of western North America ever since.

EVOLUTION OF LIFE. Middle Paleozoic seas not only welcomed new species of marine invertebrates, which replaced species that disappeared during the mass-extinction event at the end of the Ordovician Period, but also hosted jawed fish such as sharks. Even more dramatic changes took place on land. Vascular plants, which have woody tissues, seeds, and veins for transporting water and food, rooted on land for the first time in the Silurian, and by the Late Devonian, swampy forests made up of tree-sized relatives of club mosses and ferns covered large areas **(Fig. 10.12a)**. The first land animal, a millipede-like organism, appeared in the Silurian. By the Late Devonian, spiders, insects, and

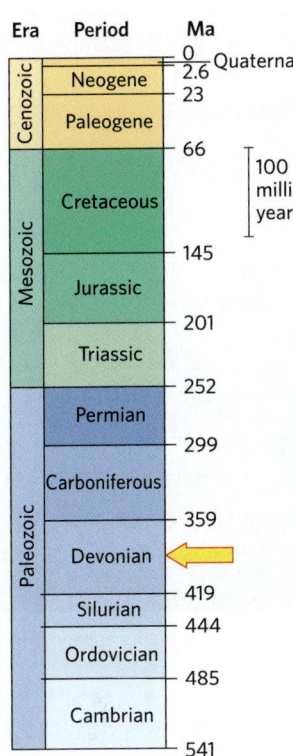

FIGURE 10.11 Paleogeography of middle Paleozoic time.

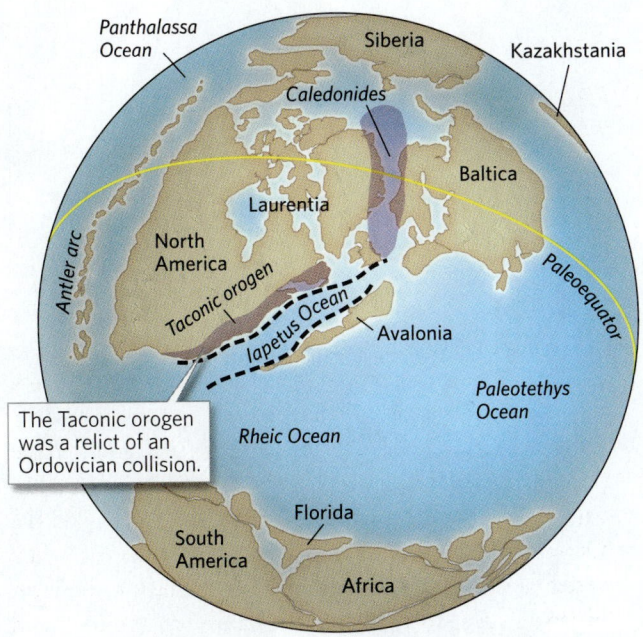

The Taconic orogen was a relict of an Ordovician collision.

(a) During the Caledonian and Acadian orogenies, Laurentia collided with Baltica and Avalonia. Meanwhile, the Antler island arc formed off the west coast.

(b) During the Devonian, shallow seas covered parts of North America's interior. The Acadian orogen shed sediment westward to form the Catskill Delta.

FIGURE 10.12 Examples of middle Paleozoic life.

(a) Vascular plants appeared in the Devonian, allowing the first forests to take root.

(b) A Late Devonian fossil of *Tiktaalik*. This lobe-finned fish was one of the first vertebrates to walk on land.

crustaceans had exploited both dry-land and freshwater habitats, and the first four-legged vertebrate had crawled out of the sea and inhaled air **(Fig. 10.12b)**.

The Late Paleozoic Era (Carboniferous and Permian Periods)

PALEOGEOGRAPHY. The late Paleozoic Era saw another succession of continental collisions, culminating in the formation of Alfred Wegener's supercontinent, **Pangaea (Fig. 10.13a)**. The largest of these collisions occurred when Gondwana rammed into Laurentia and Baltica. This event caused the *Alleghanian orogeny* and built a vast mountain belt whose eroded remnants crop out in the Appalachian Mountains of the eastern United States **(Fig. 10.13b)**. On the continental side of this mountain belt, compression generated the *Appalachian fold-thrust belt*, in which the

FIGURE 10.13 Paleogeography of the late Paleozoic.

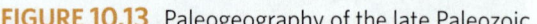

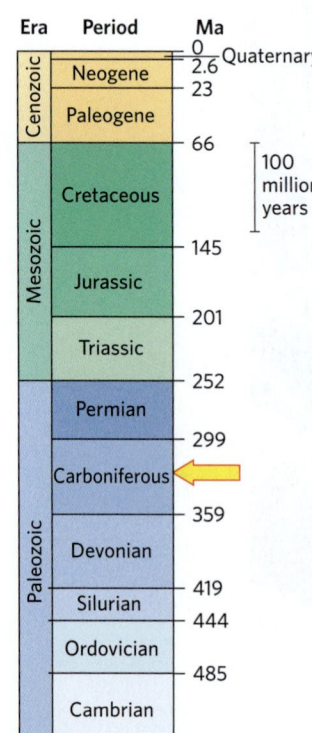

(a) At the end of the Paleozoic, almost all land had combined into a single supercontinent, called Pangaea.

(b) During the Alleghanian and Hercynian orogenies, a huge mountain belt formed. Coal swamps bordered epicontinental seas. The Ancestral Rockies rose in the western part of the continent.

sedimentary layer of crust shortened horizontally by as much as 50% (Fig. 10.14). Collision at this time also produced the Ouachita orogen in the southern U. S., and the Hercynian orogen in Europe. Compression generated during the Alleghanian orogeny was so strong that it caused old, pre-existing faults all across North America to become active again. Movement on these faults produced small mountain ranges and adjacent sediment-filled basins. The largest of these ranges rose, in the late Paleozoic, in the region presently occupied by the Rocky Mountains. Because of their location, this set of late Paleozoic ranges is called the *Ancestral Rockies* (see Fig. 10.13b).

EVOLUTION OF LIFE. The fossil record indicates that during the late Paleozoic Era, plants and animals continued to evolve toward forms that are familiar to us today. The vegetation in huge Carboniferous swamps produced so much O_2 that, for a while, this gas accounted for about a third of the gas in the atmosphere. This growth also left thick piles of plant debris that were eventually transformed into coal after burial. In these swamps, insects with fixed wings, including huge dragonflies, flew through a tangle of ferns, club mosses, and scouring rushes (Fig. 10.15a). Forests containing *gymnosperms* ("naked-seeded" plants such as conifers) and cycads (trees with palm-like stalks and fern-like fronds) became widespread in the Permian. Amphibians populated the land, and reptiles followed (Fig. 10.15b). The success of reptiles on land reflected a radically new component in animal reproduction: eggs with a watertight protective covering.

The Paleozoic Era came to a close with the devastating *Permian-Triassic mass-extinction event* (known as the *P-T extinction*) about 252 Ma, during which over 96% of marine species and 70% of terrestrial species disappeared, making it the largest mass extinction in geologic history. According to one hypothesis, the event followed the

eruption of immense flood basalts in Siberia. This eruption could have led to the emission of gases that clouded the atmosphere and acidified the oceans, making many environments inhospitable for life.

FIGURE 10.14 Features of the Appalachian Mountains in the eastern United States.

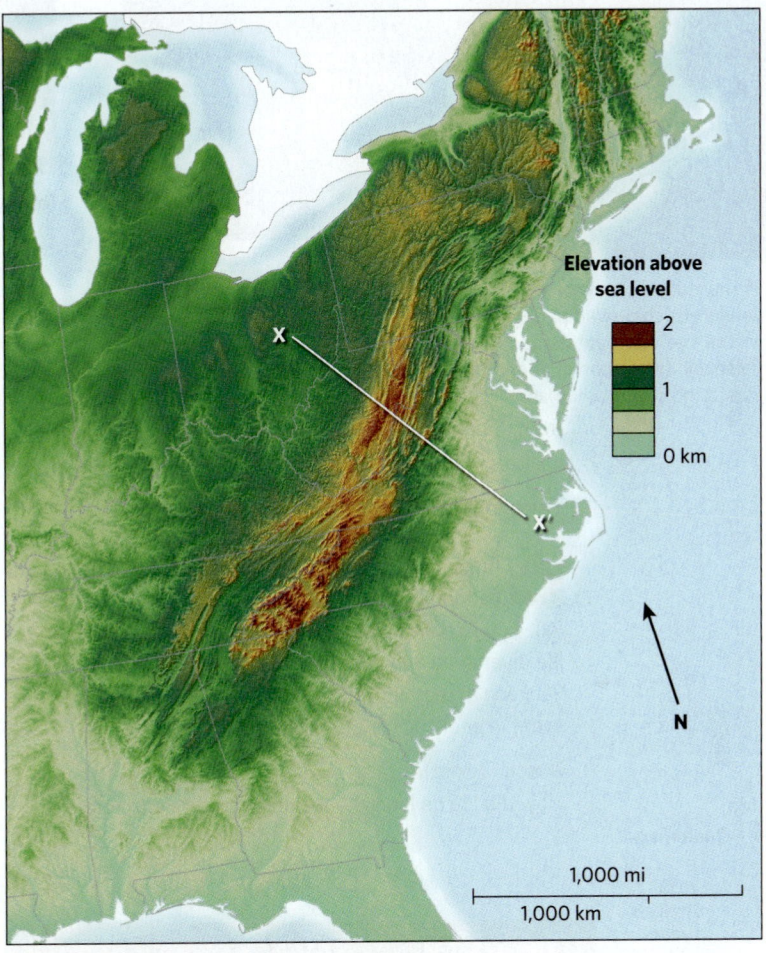

Elevation above sea level

2

1

0 km

1,000 mi

1,000 km

(a) The eroded remnants of the Appalachian orogen stand out in this representation of the topography of the eastern United States. Line X–X′ shows the position of the cross section.

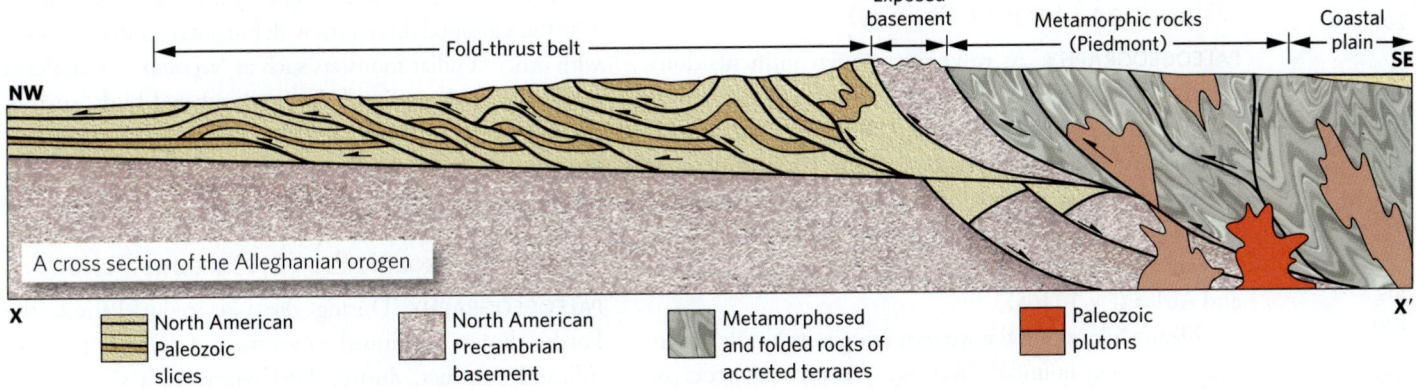

Exposed basement

Metamorphic rocks (Piedmont)

Coastal plain

Fold-thrust belt

SE

NW

A cross section of the Alleghanian orogen

X

X′

North American Paleozoic slices	North American Precambrian basement	Metamorphosed and folded rocks of accreted terranes
Paleozoic plutons		

(b) A cross section of the crust in the Appalachians shows deformation due to Paleozoic orogenies. In the Appalachian fold-thrust belt, strata have been pushed westward in overlapping thrust slices.

FIGURE 10.15 Life in the late Paleozoic.

(a) A museum diorama of a Carboniferous coal swamp. The inset shows the size of a giant dragonfly from the Paleozoic (wingspan of about 1 m) relative to the size of a human.

(b) By the Permian, reptiles such as *Dimetrodon* inhabited the land.

Take-home message . . .

Breakup of the late Precambrian supercontinent Pannotia ushered in the Paleozoic Era. During the Paleozoic, shallow seas covered continents at times, and life diversified and moved onto land. As the era ended, Pangaea assembled, and the largest mass-extinction event in geologic history occurred.

Quick Question

How did life on land change during the Paleozoic?

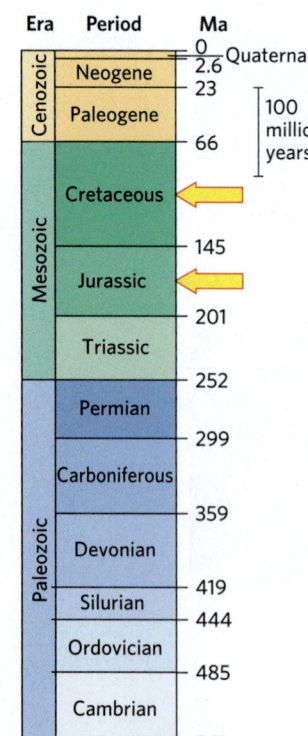

10.6 The Mesozoic Era: When Dinosaurs Ruled

The Early and Middle Mesozoic Era (Triassic and Jurassic Periods)

PALEOGEOGRAPHY. As we've seen, supercontinents don't last forever. Pangaea, which had assembled at the end of the Paleozoic, existed for about 100 million years until, in the Late Triassic and Early Jurassic, rifting began in the region that is now the east coast of North America. By the end of the Jurassic, the North Atlantic Ocean had started to grow, separating North America from Europe and Africa (Fig. 10.16a).

Meanwhile, along the western margin of North America, convergent-boundary tectonics became the order of the day. Beginning in the Late Permian and continuing through the Mesozoic, island arcs and oceanic plateaus

that had developed offshore collided with western North America once the oceanic lithosphere between them and the continent had been consumed by subduction. As these exotic terranes became attached, the continent grew westward. Then, starting at the end of the Jurassic, a large continental volcanic arc, now known as the *Sierran arc*, began to form along the western margin of North America itself.

During the Triassic and Early Jurassic, the Earth had a relatively warm climate, so glaciers melted away entirely in polar regions, and sandy deserts covered portions of western North America. The large dunes of these deserts lithified to become the colorful red sandstones exposed in Zion National Park (Fig. 10.16b). In the Middle Jurassic, sea level began to rise, and epicontinental seas once again flooded portions of the continent.

EVOLUTION OF LIFE. During the early Mesozoic Era, a variety of new plant and animal species appeared. Reptiles swam in the oceans, and colonial corals built extensive reefs. On land, gymnosperms and reptiles diversified, and the Earth saw its first turtles and flying reptiles. At the end of the Triassic, the first true dinosaurs appeared (Fig. 10.17). *Dinosaurs* differed from other reptiles in that their legs extended under their bodies rather than off to the sides. By the end of the Jurassic, 100-ton sauropod dinosaurs, with long necks and tails, along with other familiar monsters such as *Stegosaurus*, thundered across the landscape, and the first feathered birds, such as *Archaeopteryx*, took to the skies (see Fig. 9.6d). Small rat-like creatures, the earliest ancestors of mammals, scurried through Late Triassic underbrush.

The Late Mesozoic Era (Cretaceous Period)

PALEOGEOGRAPHY. During the Cretaceous Period, the Earth's climate continued to warm, and sea level rose significantly. In fact, during the Cretaceous, a shark could have swum across North America from the Gulf of Mexico to the Arctic Ocean.

FIGURE 10.16 Paleogeography of the early Mesozoic.

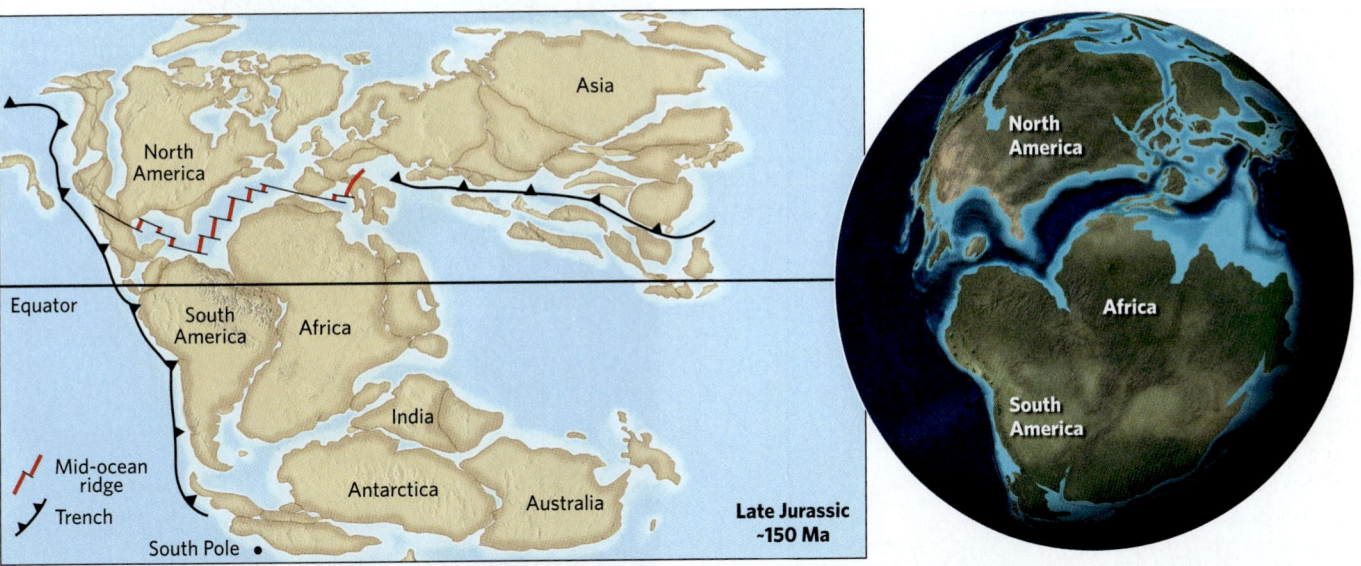

Mid-ocean ridge

Trench

Equator

South Pole •

Late Jurassic ~150 Ma

(a) Pangaea began to break up in the Triassic. By Jurassic time, a narrow North Atlantic Ocean existed.

Location map

(b) During the Jurassic, immense sand dunes blanketed the southwestern United States. These sandstone beds in Zion National Park are the relics of those dunes.

At the dawn of the Cretaceous, the North Atlantic Ocean had grown to a width of 1,500 km (930 mi), and Pangaea continued its breakup. South America finally broke away from Africa, and the South Atlantic started to grow, by the Middle Cretaceous. By the Late Cretaceous, Antarctica and Australia had separated, and India started drifting rapidly northward toward Asia (Fig. 10.18a).

The Sierran arc along the west coast of North America remained active during the Cretaceous. Its volcanoes have long since eroded away, but the granite plutons that intruded beneath it now form the great cliffs of the Sierra

FIGURE 10.17 During the Jurassic, giant dinosaurs roamed the land. This painting shows several species.

FIGURE 10.18 Paleogeography of the late Mesozoic.

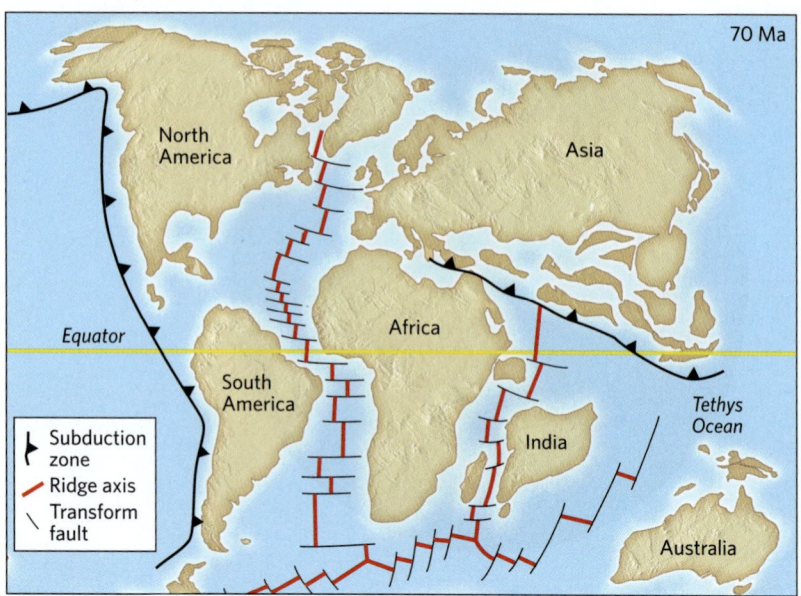

(a) By the Late Cretaceous Period (70 Ma), the Atlantic Ocean had formed, and India was moving rapidly northward toward Asia.

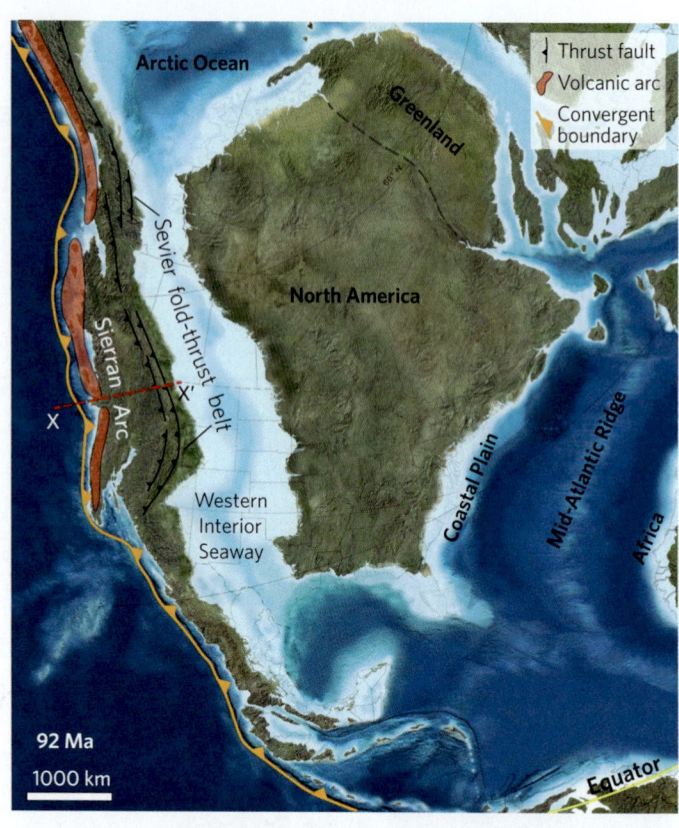

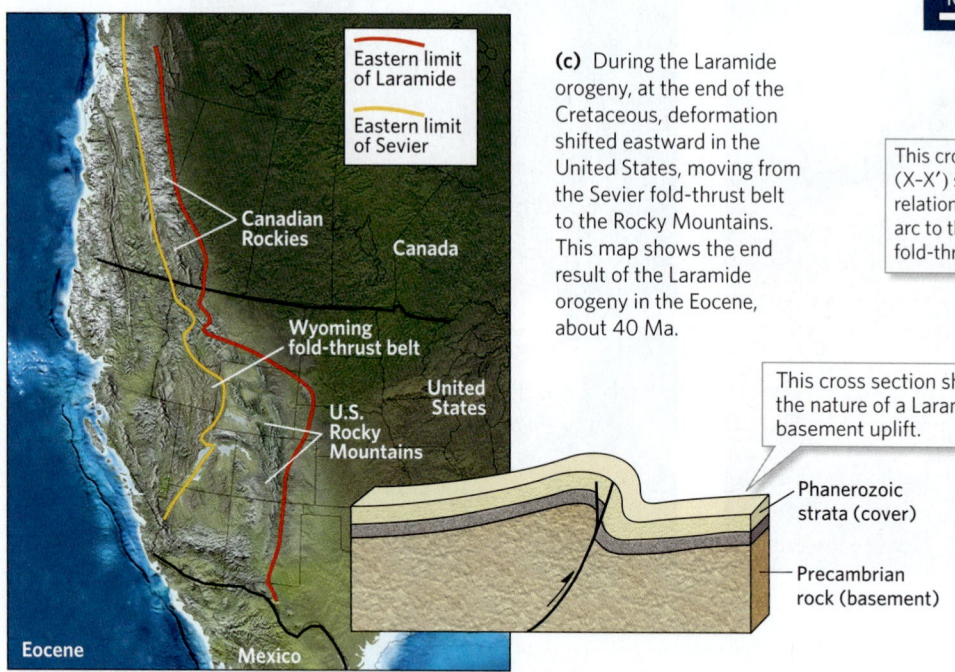

(c) During the Laramide orogeny, at the end of the Cretaceous, deformation shifted eastward in the United States, moving from the Sevier fold-thrust belt to the Rocky Mountains. This map shows the end result of the Laramide orogeny in the Eocene, about 40 Ma.

This cross section shows the nature of a Laramide basement uplift.

This cross section (X–X') shows the relation of the Sierran arc to the Sevier fold-thrust belt.

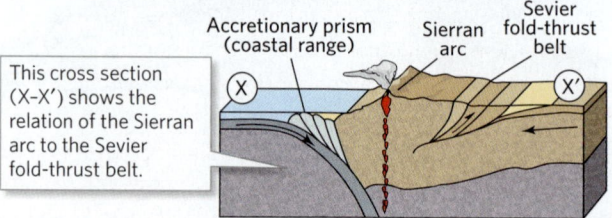

(b) In the Cretaceous, a long seaway flooded the western interior of North America, a large continental volcanic arc (the Sierran arc) grew on its west coast, and the Sevier fold-thrust belt formed to the east of the arc.

moved eastward, and large reverse faults became active in Wyoming, Colorado, eastern Utah, and northern Arizona. These faults penetrated deep into the Precambrian basement, so movement along the faults uplifted basement rocks. This uplift, in turn, caused layers of overlying Paleozoic strata to warp into large monoclines (see Chapter 7). This series of events, which geologists call the *Laramide orogeny*, resulted in the growth of the present-day Rocky Mountains in the United States **(Fig. 10.18c)**.

EVOLUTION OF LIFE. In the seas of the late Mesozoic world, modern fish appeared and became dominant. They served as prey for huge swimming reptiles and gigantic turtles. On land, angiosperms (flowering plants), including hardwood trees, populated the forest **(Fig. 10.19a)**, and dinosaurs inhabited almost all environments. Mammals also diversified and developed larger brains and more specialized teeth **(Fig. 10.19b)**.

Nevada in California. At the same time, a fold-thrust belt, whose remnants crop out in the Canadian Rockies and western Wyoming, developed to the east **(Fig. 10.18b)**. Geologists refer to the deformation that produced this fold-thrust belt as the *Sevier orogeny*. Overall, the western United States of this time would have resembled the Andes of today.

At the end of the Cretaceous, the dip of the oceanic plate subducting beneath the western United States decreased, and the plate started shearing against the base of the continent. As a result, mountain-building activity

Then suddenly, at 66 Ma, the dinosaurs disappeared entirely, along with most other species, during a mass-extinction event now known as the **K-Pg extinction**. This event, which probably resulted from a catastrophic meteorite impact **(Box 10.2)**, brought the era to a close.

Take-home message . . .

During the Mesozoic Era, rifting broke Pangaea apart and produced the Atlantic Ocean. In North America, convergent-boundary tectonics along the west coast produced the granites of the Sierra Nevada and eventually led to the uplift of the Rocky Mountains. Dinosaurs roamed all continents, but they died off, along with many other species, at the end of the era.

Quick Question

Did all of today's continents break off from Pangaea at the same time?

10.7 The Cenozoic Era: The Modern World Comes to Be

PALEOGEOGRAPHY. During the last 66 million years, the map of the Earth's surface has continued to change, gradually producing the configuration of continents and plate boundaries that we see today. The continents that once constituted Gondwana drifted northward as subduction consumed the *Tethys Ocean*, which separated Gondwana from Europe and Asia **(Fig. 10.20a)**. Eventually, the southern margins of Europe and Asia collided with various small crustal blocks and then finally with India and Africa, resulting in the growth of the largest orogen on Earth today, the **Alpine-Himalayan chain**. Continued northward movement of India led to uplift of the Tibetan Plateau.

As the Atlantic Ocean grew and the Americas moved westward, convergent-boundary activity occurred along their western margins. In South America, this activity produced the *Andes*, still an active mountain-building region. In North America, the Laramide orogeny continued until about 40 Ma (the Eocene, in the Middle Paleogene). At this time, the configuration of plates along the western boundary of North America changed, and by 25 Ma, the convergent boundary along the coast was evolving into a transform boundary, defined by the trace of the San Andreas fault. Today, the Pacific Plate moves northward with respect to North America at a rate of about 6 cm (3 in) per year along this strike-slip fault. In the western United States, convergent-boundary activity continues only in Washington, Oregon, and northern California, where subduction of the Juan de Fuca Plate (a small remnant of the Farallon Plate) generates the volcanism of the Cascade Range.

FIGURE 10.19 Life in the Cretaceous.

(a) Flowering plants first appeared in the Cretaceous, and were the favored diet of herbivore dinosaurs, such as *Leptoceratops* (shown here).

(b) This Cretaceous rat-like mammal was carnivorous.

At about the same time that the San Andreas fault became active, the region that had been affected by the Sevier and Laramide orogenies began to undergo rifting (extension) in a roughly northwest-southeast direction. The result was the formation of the **Basin and Range Province**, a broad continental rift **(Fig. 10.20b)** whose name reflects its topography. The region contains long, narrow mountain ranges separated from each other by flat, sediment-filled basins. This topography formed because the crust of the region broke up when normal faults caused blocks of crust above the faults to slip downward and tilt (see Chapter 7). The crests of the tilted blocks formed the ranges, and the depressions between them, which rapidly filled with sediment eroded from the ranges, became basins. Following the initial opening of the Basin and Range, the Colorado Plateau underwent uplift **(Box 10.3)**.

The Basin and Range Province terminates just north of the Snake River Plain, a feature that marks the track of the

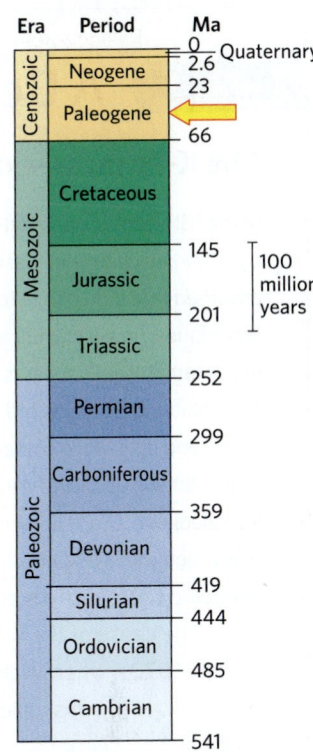

Era	Period	Ma
Cenozoic		0 — Quaternary
	Neogene	2.6
		23
	Paleogene	
		66
Mesozoic	Cretaceous	
		145
	Jurassic	
		201
	Triassic	
		252
Paleozoic	Permian	
		299
	Carboniferous	
		359
	Devonian	
		419
	Silurian	444
	Ordovician	
		485
	Cambrian	
		541

100 million years

See for yourself

Rocky Mountain front, Colorado

Latitude: 39°46′2.32″ N
Longitude: 105°13′45.35″ W

Zoom to 8 km (~5 mi) and look obliquely.

We see the steep face of the Rocky Mountains in Colorado. These mountains were uplifted during the Laramide orogeny. During the event, reactivation of large faults thrust Precambrian rocks up and caused overlying Paleozoic strata to fold.

BOX 10.2 A Deeper Look

The K-Pg mass-extinction event

In the 18th century, paleontologists defined the end of the Cretaceous by a marked change in fossil species. Until the 1980s, most researchers assumed that the change took place over millions of years. Modern dating techniques, however, indicated that this change actually happened almost instantaneously. Dinosaurs, which had dominated the planet for more than 150 million years, vanished, along with 90% of plankton species in the ocean and up to 75% of plant species. This abrupt mass-extinction event is now known as the *K-Pg extinction*; K stands for Cretaceous and Pg stands for Paleogene. (Prior to recent changes in the geologic time scale, the event was called the *K-T extinction*; T stands for Tertiary.) What kind of catastrophe could cause such a sudden and extensive mass extinction?

The cause of the K-Pg extinction remained a mystery until the late 1970s, when Walter Alvarez, an American geologist, and his colleagues examined a shale layer deposited exactly at the K-Pg boundary. They found that this shale contained relatively high concentrations of iridium, an element that comes primarily from meteorites. Further study showed that the clays of this age contained other unusual materials, such as tiny glass spheres that form when a spray of molten rock freezes, grains

The Earth at 66 Ma

of coesite (a mineral that forms when intense shock waves pass through quartz), and even carbon from burned vegetation. All these features pointed to the occurrence of a huge meteorite impact at the time of the K-Pg event **(Fig. Bx10.2a)**. Subsequently, geologists found a meteorite crater, 100 km (62 mi) in diameter and 16 km (10 mi) deep, buried beneath younger strata near the Yucatán Peninsula in Mexico **(Fig. Bx10.2b)**. Isotopic dating indicated that the crater was formed 66 ± 0.4 Ma, the time of the K-Pg event. Because of its age and size, this crater, known as the Chicxulub crater, may be the imprint of the deadly object whose impact with the Earth eliminated so much life.

The K-Pg impact was catastrophic because it not only formed a crater, blasting huge quantities of debris into the sky, but probably also generated 2-km (1.2-mi)-high tsunamis that inundated the shores of continents and generated a blast of hot air that set forests on fire worldwide. The blast and the blaze together could have ejected enough debris into the atmosphere to cause months of perpetual night and winter-like cold. In addition, chemicals ejected into the air could have combined with water to produce acid rain. These conditions could have broken the food chain and triggered extinctions.

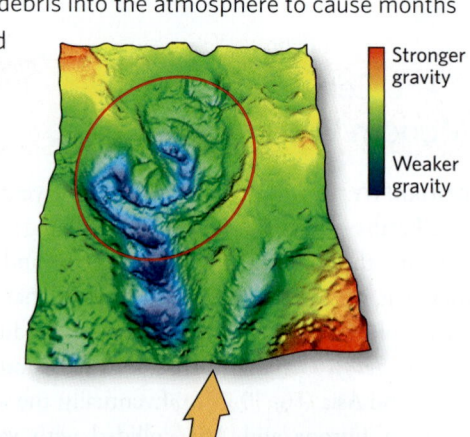

Stronger gravity

Weaker gravity

(a) An artist's image of the 13-km-wide object as it hit. The inset shows how the Earth looked at the time of impact.

(b) The location of the Chicxulub crater today. Gravity anomalies, as shown on the inset, reveal the shape of the crater.

FIGURE Bx10.2 The K-Pg meteorite impact.

FIGURE 10.20 Paleogeography of the Cenozoic.

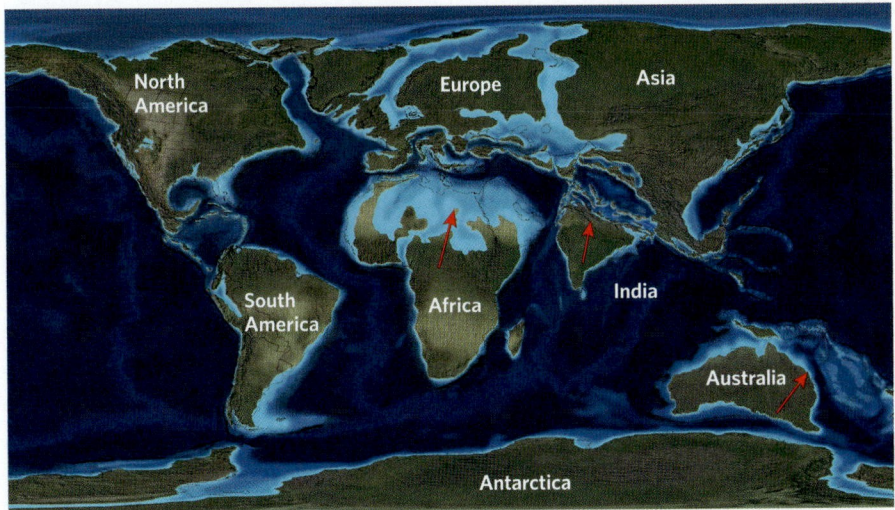

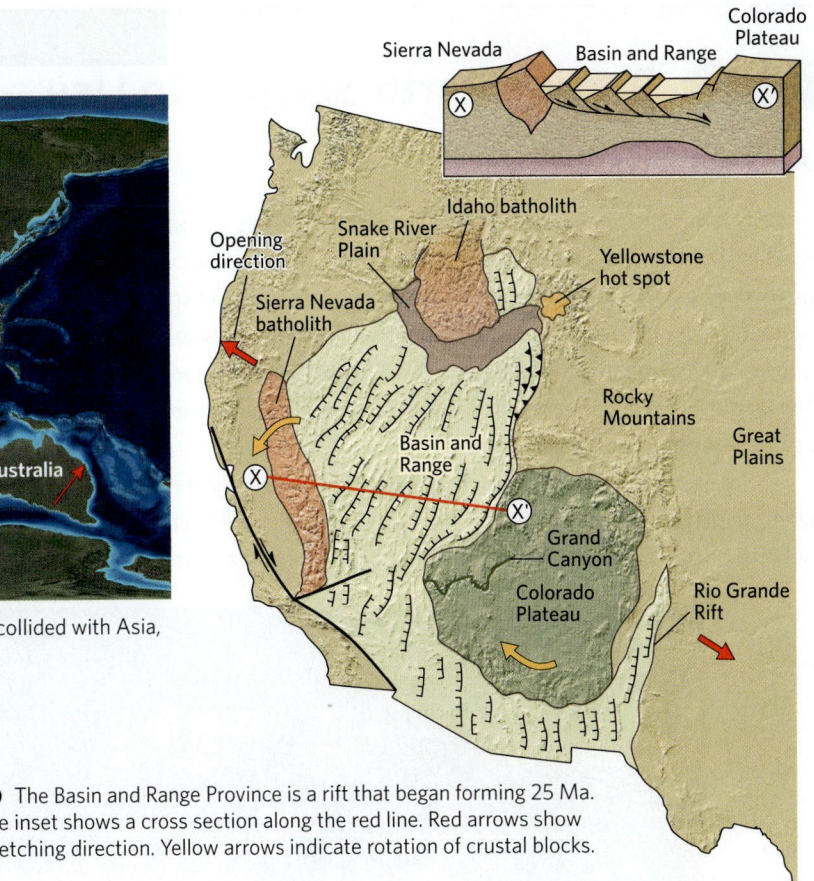

(a) In this Eocene (50 Ma) paleogeographic reconstruction, India has not yet collided with Asia, and Europe and Asia have not yet coalesced into a single landmass.

hot spot that now lies beneath Yellowstone National Park. As the North American Plate drifted westward, volcanic calderas formed along this track. Yellowstone National Park straddles the most recent caldera (see Chapter 4).

By about 15 Ma, the mountain belts and plate motions we see today had been established (Fig. 10.21). Around the world, plate interactions continue to drive seismicity, volcanism, and mountain building. During the Neogene,

(b) The Basin and Range Province is a rift that began forming 25 Ma. The inset shows a cross section along the red line. Red arrows show stretching direction. Yellow arrows indicate rotation of crustal blocks.

FIGURE 10.21 The Alpine-Himalayan orogen formed when Africa and India collided with Asia. The Cordilleran and Andean orogens reflect the consequences of continuing convergent-boundary tectonics along the eastern coast of the Pacific Ocean.

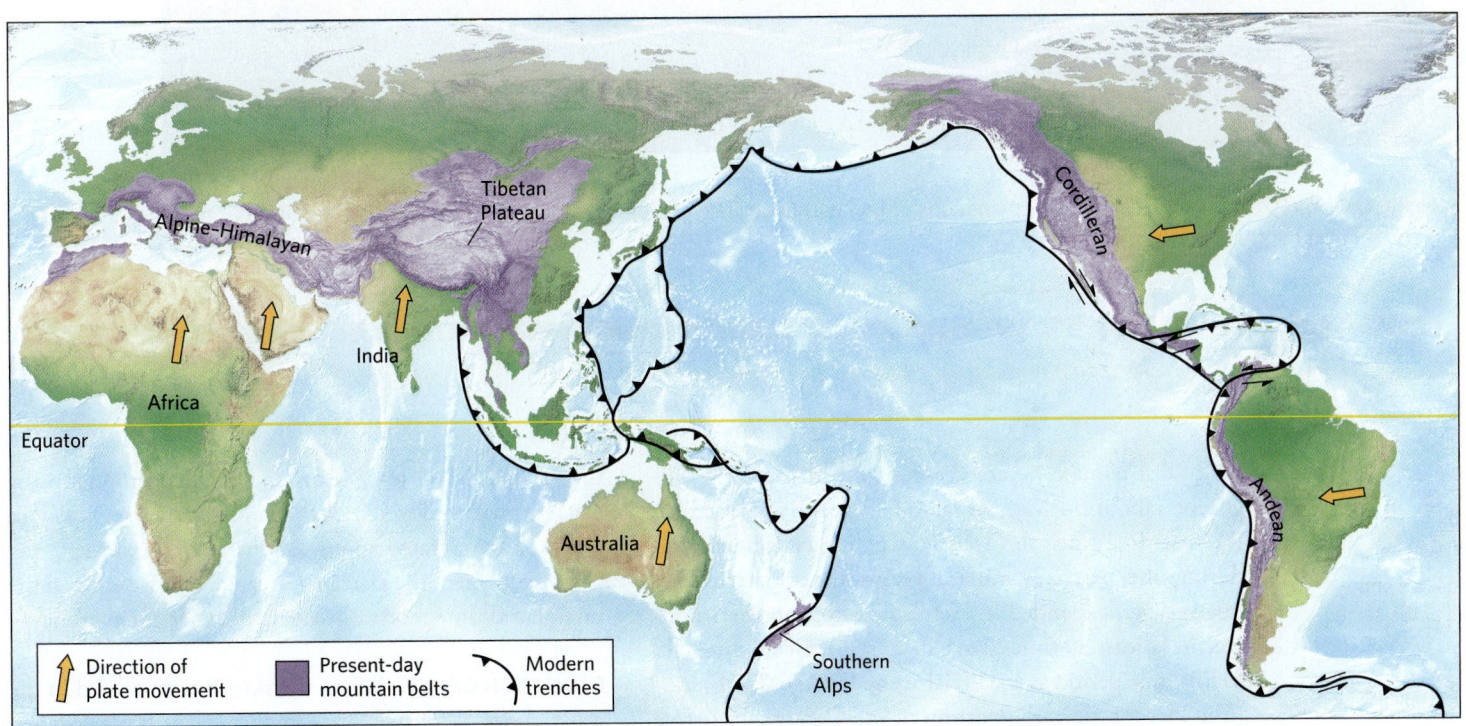

BOX 10.3 ▶ **Putting Earth Science to Use**

Adding context to a view

The site of nearly every national park in the United States was selected because it offers spectacular views of the natural world. Some parks provide access to a forest, some to a seashore or swamp, and some to dramatic rocky cliffs. If you have the opportunity to visit these places of beauty, keep in mind that the Earth has a history, and what you are seeing most likely represents the product of geologic phenomena and processes acting over geologic time.

For example, look at the photo of Bryce Canyon National Park, Utah, displayed in **Figure Bx10.3**. It's not just a pretty cliff. It's a record of Earth history from the Cretaceous through the Eocene, a time when environment and topography changed dramatically in western North America. The oldest rocks in the vicinity, hidden below the base of the cliffs, are lithified sand dunes. They are overlain by shallow marine deposits, and finally by the deposits of lakes and streams. So, the succession of rock layers tells a tale of transgression and regression, and changing climate. The top of Bryce Canyon now lies at an elevation of about 2.7 km (1.7 mi), due to the uplift of the Colorado Plateau. Erosion has attacked the uplifted rocks—frost wedging, flowing water, and gravity have whittled away the bedrock to form the dramatic hoodoos (columns of weathered rock) that decorate the park today.

FIGURE Bx10.3 The intricately carved hoodoos of Bryce Canyon. The geologic history of the western United States provides a context for this view.

Did you ever wonder . . .

whether dinosaurs and humans lived at the same time?

the global climate became cooler, and in the early Oligocene (about 34 Ma), Antarctic glaciers reappeared for the first time since the Triassic. Cooling continued, and during the past 2.6 million years—the Quaternary—huge glaciers expanded and retreated across northern continents at least 20 times (see Chapter 14). Geologists refer to this time period as the **Pleistocene Ice Age (Fig. 10.22)**. (The Pleistocene is the portion of the Quaternary before the last glacial retreat.) Erosion and deposition by the glaciers produced much of the landscape we see today in northern temperate regions. About 11,000 years ago, the climate warmed, the glaciers retreated, and we entered the interglacial time interval we are still experiencing today.

EVOLUTION OF LIFE. When the skies finally cleared in the wake of the K-Pg meteorite impact, plant life recovered,

FIGURE 10.22 The maximum advance of continental glaciers during the Pleistocene Ice Age in North America.

Sea ice surrounded Iceland.

The Bering Strait was a land bridge.

Continental glacier

Sea ice

Unglaciated land

Large lakes formed in the Basin and Range Province.

New York City was under ice.

500 mi   500 km

and the last Denisovans went extinct 25,000 years ago, leaving *Homo sapiens* as the only remaining human species on the Earth.

Take-home message . . .

During the Cenozoic, the mountain belts of today rose, and modern plate boundaries became established. In North America, the San Andreas fault and the Basin and Range Province formed. With the dinosaurs gone, mammals diversified. During the Pleistocene, glaciers covered large areas, and humans appeared.

Quick Question —————
Did *Homo sapiens* live during the Pleistocene Ice Age?

10.8 The Concept of Global Change

Did the Earth's surface look the same in the Jurassic as it does today? Definitely not! As we've seen, in Jurassic time, Africa and South America were part of the same continent, and the call of the wild rumbled from the throats of dinosaurs, whereas today the broad South Atlantic Ocean separates the two landmasses **(Fig. 10.24)**, and the largest animals are mammals **(Fig. 10.25)**. What we see of the Earth today is just a snapshot, an instant in the life story of a constantly changing planet with a long and complex history—an idea that arguably stands as geology's greatest philosophical contribution to humanity's understanding of our Universe.

To finish our consideration of the Earth's biography, let's turn our attention to the concept of change. In the Prelude, we introduced the concept of the *Earth System*: all the physical and biological realms of the Earth along with the complex ways in which they interact with one another. In this context, we can define a **global change** as any modification of the Earth System over time. Researchers distinguish among different types of global change, based first on the rate of change: *gradual change* takes place slowly, over long intervals of geologic time (millions to billions of years), whereas *catastrophic change* takes place relatively rapidly (seconds to millennia). But we can also distinguish among types of change based on the way in which the

and soon forests of both angiosperms and gymnosperms proliferated. By the middle of the Cenozoic Era, grasses had spread across the plains at mid-latitudes. With the dinosaurs gone, birds and mammals diversified. Most of the groups of mammals that live today originated at the beginning of the Cenozoic, giving this era the nickname *Age of Mammals*. During the latter part of the era (Late Neogene and Early Quaternary), huge mammals, such as mammoths and saber-toothed lions, appeared **(Fig. 10.23a)**. But these animals became extinct during the past 10,000 years, probably due to hunting by humans.

According to the fossil record, ape-like primates diversified in the Miocene Epoch (about 20 Ma), and the first human-like primates, *australopithecines*, appeared about 4 Ma **(Fig. 10.23b)**, followed by the first members of the human genus, *Homo*, at 2.4 Ma. *Homo erectus*, a species capable of making stone axes, appeared in Africa about 1.6 Ma, and our species, *Homo sapiens*, has existed since about 0.5 Ma. When modern people first walked the Earth, they shared the planet with two other species of the genus *Homo*—the Neanderthals and the Denisovans. The last Neanderthals died off about 40,000 years ago,

See for yourself

Basin and Range rift, Utah

Latitude: 39°15′1.83″ N
Longitude: 114°38′32.10″ W

Zoom to 250 km (~155 mi) and look down.

In this region of the Cenozoic Basin and Range rift, darker bands are fault-block mountains, whereas lighter areas are sediment-filled basins. White areas are evaporites, formed where lakes dried up.

FIGURE 10.23 Life in the Cenozoic.

(a) Grasslands first appeared in the Middle Cenozoic. Giant mammals ruled the land, for dinosaurs had gone extinct tens of millions of years earlier.

FIGURE 10.24 The map of the Earth's surface changes over time because of plate motions. Here we see the change that has taken place during the past 200 million years.

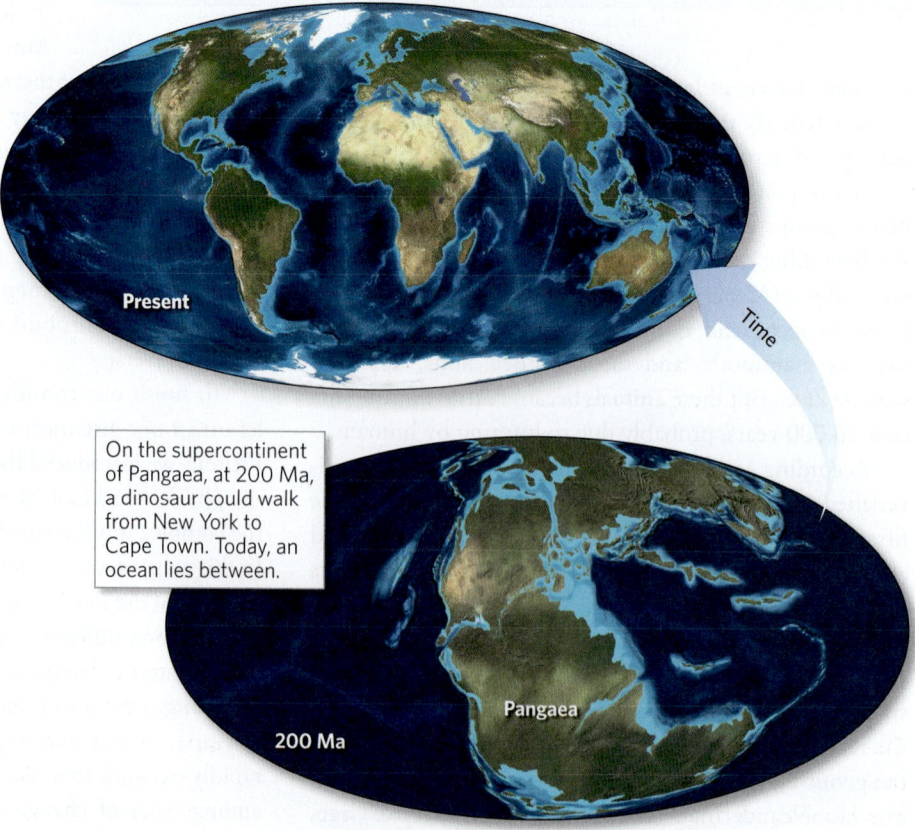

Present

Time

On the supercontinent of Pangaea, at 200 Ma, a dinosaur could walk from New York to Cape Town. Today, an ocean lies between.

200 Ma

Pangaea

(b) Our own human lineage evolved in the Cenozoic. Human-like primates such as *Australopithecus*, shown here, first appeared about 4 Ma.

The largest Mesozoic land animal was a dinosaur. The inset shows an elephant at the same scale.

The largest land animal today is the elephant.

change progresses: *unidirectional change* involves transformations that never repeat; *cyclic change* repeats the same steps over and over, though not necessarily with the same results or at the same rate; and *periodic change* repeats steps with a definable frequency.

Why has the Earth changed so much over geologic time, and how can it continue to change? Ultimately, change can happen both because the Earth's internal heat makes the asthenosphere weak enough to flow and because the Sun's radiation can keep most of the Earth's surface at temperatures above the freezing point of water. Flow in the asthenosphere permits plate tectonics, which in turn leads to continental drift, volcanism, and mountain building. Solar radiation keeps streams, glaciers, waves, and wind in motion, thereby causing erosion and deposition, and it also fuels photosynthesis. If the Earth did not have just the right mix of plate tectonics and solar heat, it would be a frozen dust bowl like Mars, a crater-pocked wasteland like the Moon, or a cloud-choked oven like Venus, and could not host life as we know it.

Let's now briefly re-examine some aspects of Earth history that we've discussed in this chapter in the context of global change. We'll see that some aspects of this history illustrate the different types of change. And we'll see that humans have become an important agent of change that may be recorded in the stratigraphic record.

Unidirectional Changes during Earth History

The Earth System has changed in many irreversible ways over the course of geologic time. For example, differentiation of the Earth's interior into a core and mantle will never

happen in the future, because once a core has formed, it can't form again. The filling of the oceans, and the effect that this process has had on atmospheric composition, also represents a unidirectional change in that once it took place, liquid water persisted on our planet's surface. The existence of liquid water set the stage for the appearance of life. The fossil record, as we have seen, indicates that evolution is a unidirectional change, punctuated by mass extinction **(Fig. 10.26)** over the course of geologic time.

Cyclic Changes in the Earth System

During cyclic change, a sequence of stages repeats over time. Some cyclic changes are periodic, in that cycles happen with a definable frequency, but many are not. Some cycles involve only movements of physical components of the Earth System, while others involve transfer of materials among both living and nonliving components of the Earth System. We have seen examples of all of these types of cyclic changes in this chapter:

- *The rock cycle:* In effect, rocks serve as reservoirs of atoms, and movement of atoms from reservoir to reservoir over time constitutes the rock cycle (see Chapter 6).

- *The supercontinent cycle:* On a time scale of hundreds of millions of years, smaller continental blocks collide and collect into large supercontinents. A supercontinent survives for tens to hundreds of millions of years until it undergoes rifting to form smaller continents, which drift apart due to seafloor spreading. Eventually, the continents reassemble into a new supercontinent, only to break apart again into different blocks **(Fig. 10.27)**.

FIGURE 10.26 Evolution of life in the context of the geologic column. The Earth formed at the beginning of the Hadean Eon. (Not to scale.)

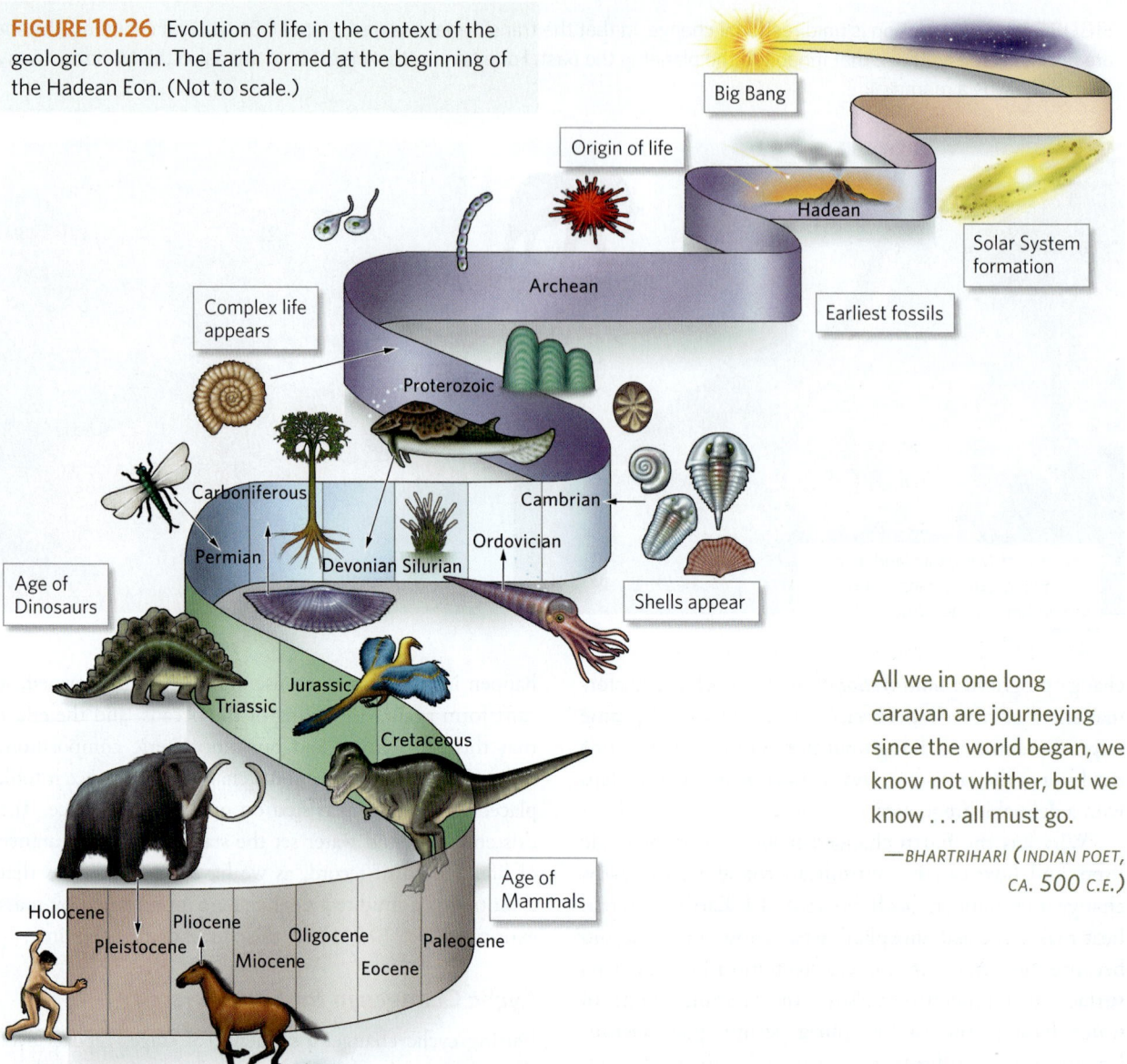

All we in one long caravan are journeying since the world began, we know not whither, but we know . . . all must go.

—*BHARTRIHARI (INDIAN POET, CA. 500 C.E.)*

- *The sea-level cycle:* Relative sea level has gone up and down over Earth history. When sea level rises, continental interiors may become shallow seas.

- *Biogeochemical cycles:* A **biogeochemical cycle** involves the passage of chemicals among nonliving and living reservoirs within the Earth System. Nonliving reservoirs include the atmosphere, the crust, and the oceans, whereas living reservoirs include plants, animals, and microbes. Some stages in a biogeochemical cycle may take only hours, while others may take millions of years. For a time, biogeochemical cycles can attain a **steady-state condition**, meaning that the proportions of a chemical in different reservoirs remain fairly constant even though flow among reservoirs continues. A global change in a biogeochemical cycle modifies the proportions of chemicals in different reservoirs, causing a change in the steady-state condition. We'll be discussing examples of biogeochemical cycles, such as the hydrologic cycle and the carbon cycle, later in the book (see Chapters 12 and 20).

The Geologic Record of Our Time: The Anthropocene?

By the dawn of civilization, at 4000 B.C.E., the human population was at most a few tens of millions. But by the beginning of the 19th century, revolutions in industry and hygiene had substantially lowered death rates and raised living standards, so the population grew at accelerating rates and reached 1 billion in 1850. It then took only 80 years for the population to double, reaching 2 billion in 1930. Today, the doubling time is only 44 years, so the population passed the 6 billion mark just before the year 2000 and surpassed 7.7 billion in 2019 **(Fig. 10.28)**.

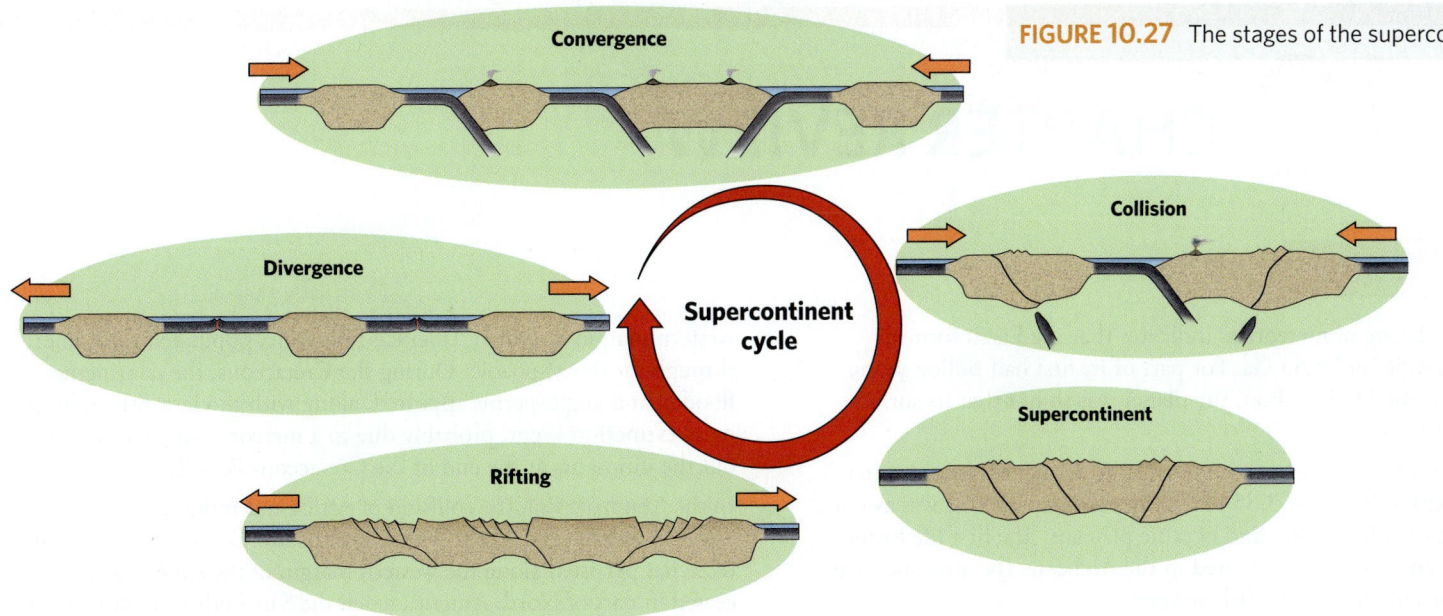

FIGURE 10.27 The stages of the supercontinent cycle.

As the human population grows and our standard of living continues to improve, our use of the Earth's resources increases. We use soil for agriculture and grazing; wood, rock, and gravel for construction; oil and coal for energy and plastics; and ores for metals (see Chapter 11). Every time we move a pile of rock, plow a field, drain a wetland, or pave a road, we change a portion of the Earth's crust. Clearly, our use of resources affects the Earth System profoundly, so humanity has become a significant agent of global change. Furthermore, deforestation, overgrazing, agriculture, and urbanization have led to a marked decrease in biodiversity. This decrease may prove to be so severe that it will appear in the geologic record as a mass-extinction event.

Because of all the human-caused changes that have taken place in the last few centuries, some researchers have suggested that this time period marks the start of a new geologic interval, which they informally refer to as the **Anthropocene**. The geologic record of this time is likely to look distinctly different from the earlier part of the Holocene. (The Holocene is the portion of time from the last glacial retreat to the present.) Human-caused changes in the atmosphere and the landscape, the production of durable trace fossils (such as concrete and glass), the decrease in biodiversity, and the accumulation of long-lived *pollutants* (contaminants that cannot be absorbed or destroyed by natural Earth System processes) may be preserved as a distinct marker bed, discernible by geologists living 100 million years in the future. Use of this term, of course, remains rather controversial. But it does convey the impact that our species will have had on the biography of the Earth as read by observers of the future.

Take-home message . . .

Since our planet first formed, the Earth System has been undergoing major changes. Some of these changes are unidirectional, whereas others are cyclic. Cyclic change can be periodic, but often is not. In the present day, human activities are a powerful cause of global change.

Quick Question

What does the Anthropocene refer to ?

FIGURE 10.28 The human population now doubles about every 44 years. The Black Death pandemic caused an abrupt drop that lasted for a few decades.

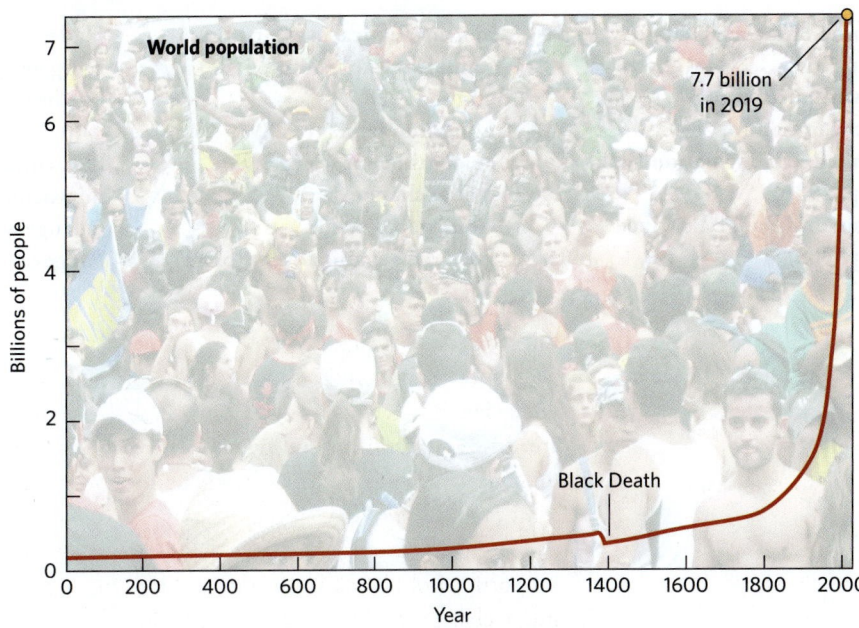

10 CHAPTER REVIEW

Chapter Summary

- Isotopic dating of meteorites indicates that the Earth formed between 4.56 and 4.54 Ga. For part of its first half billion years, known as the Hadean Eon, the planet was so hot that its surface was a lava ocean.

- The Archean Eon began about 4.0 Ga. A permanent liquid-water ocean existed by about 3.85 Ga. Early continental crust was assembled out of volcanic arcs and oceanic plateaus. The first life forms—bacteria and archaea—appeared in the Archean. The atmosphere of that time contained very little oxygen.

- In the Proterozoic Eon, which began at 2.5 Ga, Archean crustal blocks were sutured together to form large continents. Photosynthesis added oxygen to the atmosphere. By the end of the Proterozoic, soft-bodied marine invertebrates populated the planet, and continental crust had accumulated to form a supercontinent.

- As the Paleozoic Era began, rifting yielded several separate continents. Sea level rose and fell. Various orogenies took place during the Paleozoic, and by the end of the era, continents combined to form Pangaea. Early Paleozoic life included invertebrates with shells, and jawless fish. Land plants and insects appeared in the middle Paleozoic, and by the end of the era, there were land reptiles and gymnosperm trees.

- In the Mesozoic Era, Pangaea broke apart and the Atlantic Ocean formed. Convergent-boundary tectonics dominated along the western margin of North America. Dinosaurs populated the planet throughout the Mesozoic. During the Cretaceous, the continents flooded and angiosperms appeared, along with modern fish. A huge mass-extinction event, probably due to a meteorite impact, wiped out the dinosaurs at the end of the Cretaceous Period.

- In the Cenozoic Era, the collision of Africa and India with Europe and Asia formed the Alpine-Himalayan chain. Convergent-boundary tectonics persisted along the western margin of the Americas. It ceased in part of North America when the San Andreas fault formed. Rifting in the western United States produced the Basin and Range Province. Various kinds of mammals took the place of the dinosaurs, and the human genus, *Homo*, appeared. During the Pleistocene, glaciers covered large areas.

- Earth history reflects the process of global change, modification of physical and biological components of the Earth System through time. Unidirectional change results in transformations that never repeat, whereas cyclic change involves repetition of the same steps over and over.

- Humans have changed landscapes, caused a decrease in biodiversity, produced durable trace fossils, and added pollutants to the environment. The record of human activity during the present time interval, which has been called the Anthropocene, may be preserved in the stratigraphic record.

Key Terms

Alpine-Himalayan chain (p. 303)
Anthropocene (p. 311)
Archean Eon (p. 287)
banded iron formation (BIF) (p. 292)
Basin and Range Province (p. 303)

biogeochemical cycle (p. 310)
Cambrian explosion (p. 296)
craton (p. 289)
Ediacaran fauna (p. 289)
epicontinental sea (p. 292)
global change (p. 307)
Gondwana (p. 292)
great oxygenation event (p. 291)

Hadean Eon (p. 285)
K-Pg extinction (p. 303)
late heavy bombardment (p. 287)
Laurentia (p. 292)
microcontinent (p. 297)
paleogeographic map (p. 289)
Pangaea (p. 298)

Pannotia (p. 289)
Phanerozoic Eon (p. 292)
Pleistocene Ice Age (p. 306)
Proterozoic Eon (p. 289)
Rodinia (p. 289)
snowball Earth (p. 292)
steady-state condition (p. 310)

Letters in parentheses correspond to the chapter's learning objectives

1. Approximately when did the Earth form and differentiate? Why are there no whole rocks on Earth that yield isotopic dates older than about 4 billion years? **(A)**

2. Describe the condition of the Earth's crust, atmosphere, and oceans during the Hadean Eon. What defines the Hadean? **(B)**

3. When did the Archean occur? How might the first continental crust have formed? When did cratons appear? **(A, B)**

4. How did the atmosphere and plate tectonics change during the Proterozoic Eon? When did this eon begin? **(A, B)**

5. When did life, as recorded by the first fossils, appear? What environment might have served as the cradle of the earliest life? **(D)**

6. What evidence suggests that the Earth nearly froze over during the Proterozoic Eon? **(A)**

7. According to the graph, when did most continental crust form? Did supercontinents exist in the Proterozoic? If so, when did they break apart? **(B, C)**

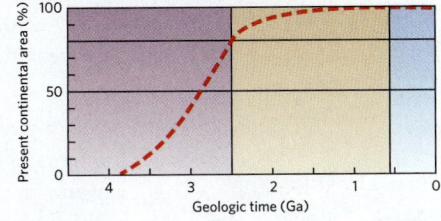

8. Were there multicellular organisms before the Cambrian? How did the Cambrian explosion change the nature of the living world? **(E)**

9. When did continents collide during the Paleozoic? What orogenies mark these collisional events? **(G)**

10. What are the major types of organisms that appeared during the Paleozoic, and in what sequence? **(E)**

11. Would the land surface at the beginning of the Ordovician have looked different from the land surface at the end of the period? If so, how? **(H)**

12. Why did broad areas of continental interiors become covered with layers of marine strata during the Paleozoic? **(F)**

13. What supercontinent existed at the end of the Paleozoic Era, and what ocean formed when it broke apart? What event defines the end of the Paleozoic? **(C, E)**

14. What might have caused the mass extinction at the end of the Paleozoic? **(D, G)**

15. Describe the plate interactions that led to the formation of the Sierran arc, the Sevier orogeny, and the Laramide orogeny. **(G)**

16. What life forms appeared during the Mesozoic, and what caused a mass-extinction event at the end of the Mesozoic? **(E)**

17. What continents formed as a result of the breakup of Pangaea, and what events eventually produced the Himalayas and the Alps? **(C, G)**

18. Which ocean is growing on this paleogeographic map? **(C)**

19. What major geologic features formed in the western United States during the Cenozoic? **(G)**

20. What major climate changes and biological events happened during the Pleistocene? **(F)**

21. What is global change, and how do unidirectional changes differ from cyclic changes? Do all changes happen at the same rate? **(H)**

22. How have humans become agents of global change? What might the evidence of this change look like to a geologist millions of years in the future? **(H)**

23. Geologists have concluded that 80% to 90% of Earth's continental crust had formed by 2.5 Ga. But if you look at a geologic map of the world, you find that only about 10% of the Earth's continental crustal surface is labeled "Precambrian." Why? **(B)**

24. Explain how changes in atmospheric composition may relate to the evolution of life. Keep in mind that organisms that can use oxygen for metabolism can produce much more energy than can organisms that use sulfide minerals for metabolism. **(D)**

11 RICHES IN ROCK
Energy and Mineral Resources

By the end of the chapter you should be able to ...

A. describe how oil and gas form, explain how they can be extracted, and distinguish conventional from unconventional hydrocarbon reserves.

B. explain how coal forms and is mined, and discuss challenges associated with coal use.

C. discuss where nuclear fuel comes from, how a nuclear power plant operates, and why handling nuclear waste can be problematic.

D. describe alternative energy sources, such as hydroelectric, geothermal, wind, and solar energy, and the issues associated with their use.

E. describe where the metals that society uses come from and the challenges we face in obtaining and using them.

F. relate nonmetallic mineral resources such as stone, gravel, and evaporites to their sources, and discuss how people use these resources.

G. identify sustainability issues pertaining to the use of energy and mineral resources.

11.1 Introduction

Picture a wolf living in a forest. To survive, it needs only the food and water that it can find in its immediate surroundings. Humans need more. Even a Late Pleistocene hunter-gatherer used *energy resources* (wood for heat) and *mineral resources* (stone for tools, weapons, and construction (Fig. 11.1). Industrialization in the 19th century greatly increased demands for **resources**, defined broadly as sources of material that sustain and enhance life. In fact, a person living in a modern industrialized society uses 100 times more resources than did a prehistoric hunter-gatherer. Furthermore, some resources essential for technological applications today consist of exotic substances that hadn't even been discovered before the last century (Table 11.1).

Where do resources come from? Some have biological sources (forests, fields, or herds). Others come from Earth materials or from geologic, atmospheric, and oceanographic processes—these are **Earth resources**, so that's why we discuss them in an Earth Science book. This chapter begins by introducing *fossil fuels* (oil, gas, and coal), today's most widely used energy resources. We then consider other energy sources of growing importance in modern society. Finally, we introduce metallic and nonmetallic mineral resources. For each resource type, we review both its geologic origin and issues associated with its extraction, production, and use.

> << The rock that once occupied the volume of this 300-m (1000-ft)-deep open-pit mine in Sonora, Mexico, contained trace amounts of gold. Workers crushed the rock and treated it with chemicals to extract the gold. Currently the mine produces gold worth almost $60,000,000. Clearly, rocks contain riches . . . but they are hard to extract, and obtaining them can cause significant changes to the landscape.

TABLE 11.1 Yearly Per Capita Use of Earth Resources in the United States

Resource	Amount (kg)	Amount (lb)
Stone	4,100	9,000
Sand and gravel	3,800	8,400
Oil	3,600	7,900
Coal	3,300	7,300
Iron and steel	550	1,200
Cement	360	800
Clay	220	480
Salt	200	440
Phosphate	140	300
Aluminum	25	55
Copper	10	22
Lead	6	3
Zinc	5	11
Rare earth elements	0.6	1.3

Note: All amounts are approximate.

11.2 Introducing Fossil Fuels

Fuel is any material that can supply energy when burned. For centuries, people have used wood, dung, grass, and peat as fuel. The energy in such **biomass** (living or recently living organisms) originally came to the Earth as radiation from the Sun. **Photosynthesis**, a series of chemical reactions that take place in cyanobacteria, green algae, and plants, uses this radiation to build organic molecules. Chemical bonds in these molecules store energy. During *combustion* (burning), oxygen reacts chemically with biomass, a process that yields carbon dioxide and water and transforms some of the stored energy into heat and light. Beginning with the 19th-century *industrial revolution*, society's use of engines and furnaces required more fuel

FIGURE 11.1 Standing stones were set in place over 5,300 years ago by inhabitants of what is now Brittany, France. Ancestors of these people made stone tools, such as this ax head.

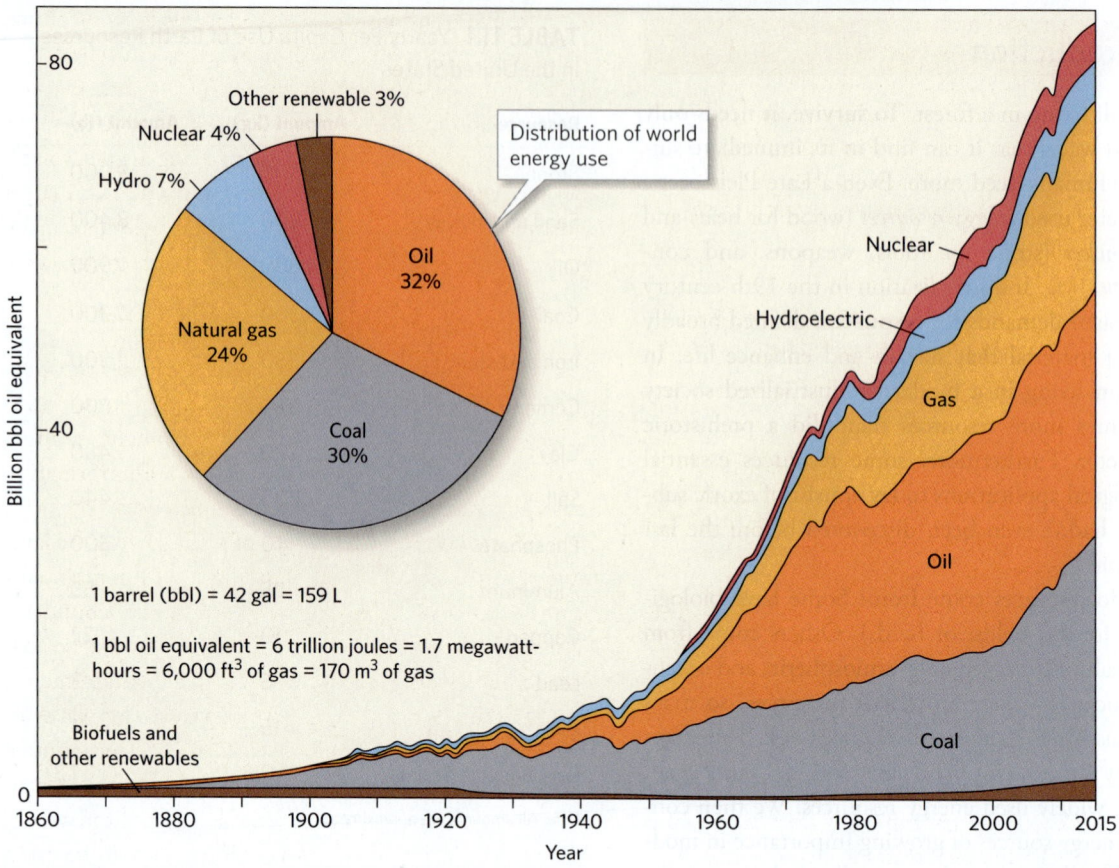

FIGURE 11.2 The proportions of different sources of energy used by human society have changed over time, but the amount of energy used has increased almost continuously.

Distribution of world energy use

Other renewable 3%
Nuclear 4%
Hydro 7%
Oil 32%
Natural gas 24%
Coal 30%

1 barrel (bbl) = 42 gal = 159 L

1 bbl oil equivalent = 6 trillion joules = 1.7 megawatt-hours = 6,000 ft^3 of gas = 170 m^3 of gas

Billion bbl oil equivalent

Nuclear
Hydroelectric
Gas
Oil
Coal

Biofuels and other renewables

Year

than biomass could supply. So, people began to use **fossil fuels**, biomass that has been buried and preserved—"fossilized"—underground. Fossil fuels, in effect, store solar energy that came to the Earth long ago. Burning these fuels releases this energy.

Fossil fuels remain the world's dominant energy source to this day **(Fig. 11.2)**. Fossil fuels can be burned directly to make heat and light, to power engines, or to produce steam to spin turbines that drive electrical generators **(Fig. 11.3)**. These fuels are popular because of their high *energy density* (energy stored per kilogram) and

transportability **(Table 11.2)**. For example, a jet plane can carry enough fuel to fly 500 people halfway around the world, but a plane powered by lead-acid batteries wouldn't even get off the ground. And it takes only 5 minutes to fill an automobile with enough gas to drive 480 km (300 mi), whereas an electric car would need to charge for almost 10 hours to drive that far.

Geologists distinguish between two general categories of fossil fuels: hydrocarbons (oil and natural gas) and coal. These substances differ from each other in their composition and origin, as we'll now see.

FIGURE 11.3 A simplified schematic diagram showing the configuration of a fossil fuel power plant that generates electricity.

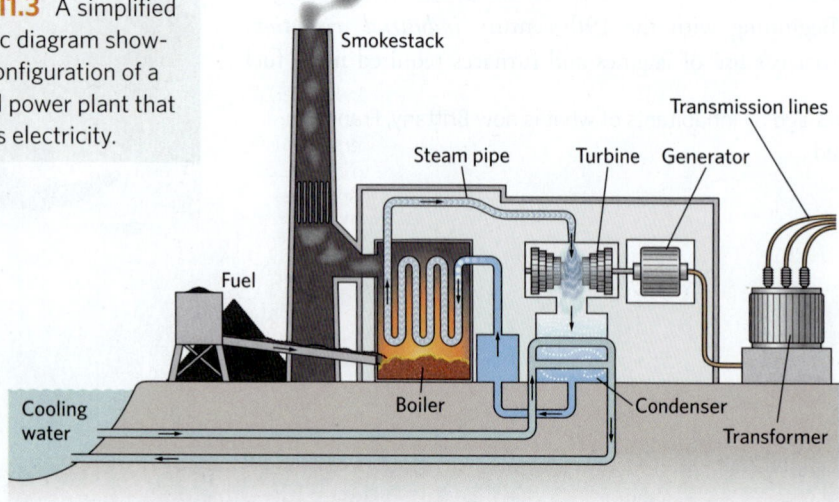

Smokestack
Steam pipe
Turbine
Generator
Transmission lines
Fuel
Cooling water
Boiler
Condenser
Transformer

TABLE 11.2 Energy Density of Common Energy Sources

Source	Energy Density (MJ/kg)*
Uranium-235	79,500,000
Gasoline	46
Propane	46
Coal	24
Wood	16
Gunpowder	3
Lithium battery	1.8
Lead-acid battery	0.2

*MJ = megajoules, a unit of energy.

Take-home message ...

Photosynthesis stores solar energy in biomass. Burning biomass releases this energy as heat and light. Before the industrial revolution, people burned recently harvested biomass directly. Because of increased demand for energy, people now rely on fossil fuels (oil, gas, and coal).

Quick Question

How does the energy density of fossil fuels compare with that of wood, or with that of a lead-acid car battery?

11.3 Hydrocarbons: Oil and Natural Gas

What Is a Hydrocarbon?

Gasoline, diesel fuel, cooking gas, lubricating oil, heating oil, tar, kerosene, and jet fuel are all types of **hydrocarbons**, chain-like or ring-like molecules made of carbon and hydrogen atoms **(Fig. 11.4)**. We distinguish among three general types of hydrocarbons based on their *viscosity* (ability to flow) and their *volatility* (ability to evaporate). **Natural gas** refers to hydrocarbons that can easily exist in a gaseous state at room temperature; **oil** refers to hydrocarbons that generally exist in liquid form at room temperature; and **tar**

refers to hydrocarbons that are solid at room temperature. The viscosity and volatility of a hydrocarbon reflects the size of its molecules, meaning the number of carbon atoms in each molecule. Hydrocarbons with small molecules have lower viscosity (they flow more easily) and higher volatility (they evaporate more easily) than do those with larger molecules. Natural gas consists of small molecules, oil of medium-sized molecules, and tar of very large molecules.

Where Do Hydrocarbons Come From?

Most oil and gas forms from the remains of *plankton*, tiny floating aquatic organisms, including some forms of algae—they are not, as commonly thought, residues from trees or dinosaur carcasses! Hydrocarbon formation requires a succession of special circumstances **(Fig. 11.5)**. The process begins in the water of a shallow sea or lake, which must receive plenty of sunlight and contain sufficient nutrients so that plankton become abundant. When plankton dies, it must settle in calm water so that it, together with *clay* (very fine-grained clastic sediment), can accumulate before being washed away by waves or currents. The water in which dead plankton settles must not contain a high concentration of dissolved oxygen or abundant microbes. If it did, the dead plankton would either react with oxygen to form CO_2 or would be eaten before it could mix with clay. In places where all these

FIGURE 11.4 Motor oil is a liquid composed of hydrocarbons.

FIGURE 11.5 The process of hydrocarbon formation begins when organic debris settles with clay. As these materials are buried by additional sediment, heat and pressure transform them into organic shale, in which the organic matter becomes kerogen. At still higher temperatures, kerogen transforms into oil and gas, which can seep upward.

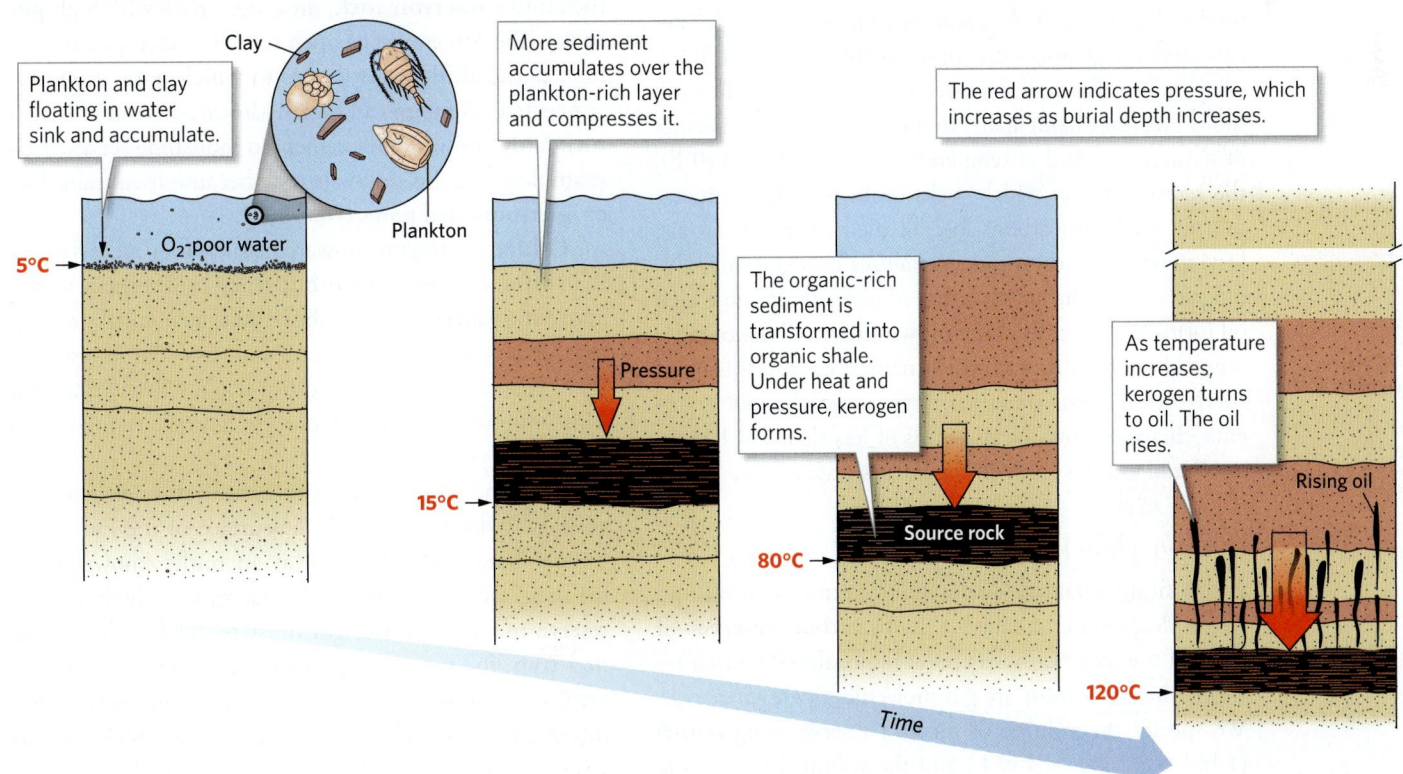

11.3 HYDROCARBONS: OIL AND NATURAL GAS 317

FIGURE 11.6 Porosity and permeability in sedimentary rocks, as seen through a microscope. Hydrocarbons occupy the pore space.

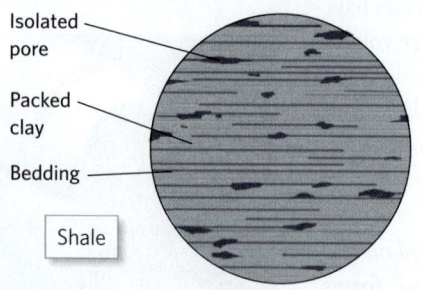

Isolated pore

Packed clay

Bedding

Shale

(a) This rock has low porosity and low permeability.

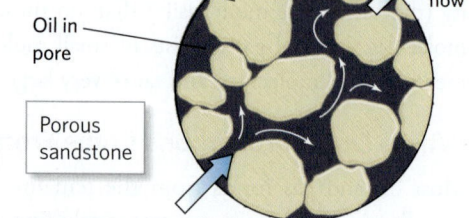

Clast

Oil in pore

Fluid flow

Porous sandstone

(b) This rock has high porosity and high permeability.

conditions have been met, a layer of black *organic ooze* accumulates on the floor of the lake or sea. If this ooze becomes buried deeply enough to undergo lithification, it becomes *organic shale*, a black, fine-grained clastic sedimentary rock containing 25%–75% organic matter. This organic shale (informally known as *black shale*) contains the raw materials from which hydrocarbons can form, so geologists refer to it as **source rock**.

Transformation of the organic material of a source rock into hydrocarbons involves chemical reactions that happen only in a relatively narrow temperature range. At a temperature of 50°C–90°C (122°F–194°F), the organic chemicals (fats, sugars, and carbohydrates) in dead plankton transform into waxy molecules of **kerogen**. If the temperature rises to 90°C–160°C (190°F–320°F), kerogen molecules break down into oil and gas molecules, a process known as *hydrocarbon generation*. Oil molecules can survive at temperatures of up to 160°C (320°F). If the temperature of a source rock becomes warmer than that, oil molecules *crack* (split apart) to form the smaller molecules of natural gas. And at temperatures over 250°C (480°F), the temperature at which rocks begin to undergo metamorphism, hydrocarbons lose all their hydrogen atoms, leaving behind carbon that crystallizes into graphite. Geologists refer to the limited range of temperatures in which oil forms and survives as the **oil window**, and the broader range of temperatures at which natural gas can form and survive as the *gas window*. At typical geothermal gradients, these windows lie at depths of less than 10–15 km, so oil and gas exist only in the Earth's upper crust.

Hydrocarbon Reserves

A significant quantity of potentially extractable oil and gas underground is called a **hydrocarbon reserve**—if most of the hydrocarbons are oil molecules, it's an *oil reserve*, whereas if most are gas molecules, it's a *gas reserve*. We specify the volume of oil in a reserve using *barrels* (1 bbl = 42 gal = 159 L) and the volume of gas using

Did you ever wonder . . .

whether there are actually pools of oil underground?

cubic meters or cubic feet (as measured at 15°C [59°F] and 1.01 bar of pressure). Hydrocarbons underground occupy *pores* (open spaces or cracks) between grains of solid rock—they do not exist in underground lakes or pools.

Geologists distinguish between two categories of hydrocarbon reserves based on how easily the hydrocarbons can be pumped up from underground. Hydrocarbons can be pumped out of the ground relatively easily from a **conventional reserve**. In such reserves, the hydrocarbons have relatively low viscosity and the rock is both *porous* (contains many pores) and *permeable* (the pores connect to one another). Because of these characteristics, the hydrocarbons can readily move through the rock to a pump. Hydrocarbons cannot be pumped out of the ground easily from an **unconventional reserve**. Pumping from such reserves is difficult, either because the hydrocarbons are too viscous to flow or because the rock containing the hydrocarbons is *impermeable* (the pores do not connect to one another). We use the adjectives *conventional* and *unconventional* simply to emphasize that energy companies have been extracting hydrocarbons from conventional reserves since the late 19th century, but have tapped unconventional reserves only for about the last 20 years.

CONVENTIONAL HYDROCARBON RESERVES. Development of a conventional hydrocarbon reserve requires two steps after a source rock has passed through the oil window. First, because source rocks are a type of shale, which consists of tightly packed clay flakes, they are not very porous or permeable **(Fig. 11.6a)**. To form a conventional reserve, the hydrocarbons must migrate out of the source rock into a **reservoir rock**, meaning a rock with high **porosity** (the percentage of open space between grains) and high **permeability** (the degree to which pores are interconnected). A pump can suck hydrocarbons out of such a rock fairly easily. Poorly cemented sandstone serves as an example of a good reservoir rock because it contains lots of interconnected pores **(Fig. 11.6b)**.

Oil and gas migrate upward because they are less dense than groundwater, so their buoyancy causes them to rise above the water. (Oil rises above vinegar in salad dressing for the same reason.) If both oil and gas occur in reservoir rock, gas rises above oil because it's less dense than oil. This *hydrocarbon migration* process can take thousands to millions of years **(Fig. 11.7)**.

If oil and gas move relatively easily through reservoir rock, why don't they reach the surface of the Earth and escape? Sometimes they do, at an outlet known as an **oil seep**. But in order for an underground hydrocarbon reserve to exist, oil and gas must be held underground in a **trap**, the second step in the development of a conventional reserve. Formation of a trap requires two conditions. First, a *seal rock*, an impermeable rock such as shale or salt, must lie above the reservoir rock to prevent hydrocarbons from rising farther. Second, the seal rock and reservoir rock must be configured in a way that confines the hydrocarbons to a relatively restricted area. Geologists recognize several types of trap geometries (Fig. 11.8).

UNCONVENTIONAL HYDROCARBON RESERVES. Formation of an unconventional hydrocarbon reserve does not require migration of hydrocarbons into a reservoir rock, nor does it require confinement within a trap. All it requires is rock containing a large amount of organic matter. Geologists recognize three types of unconventional reserves: (1) **Oil shale** reserves consist of organic-rich source rock that became warm enough for kerogen to form, but not for oil or gas to form. (2) **Tar sand** reserves consist of sandstone whose pores contain hydrocarbons that are too viscous to flow. (3) **Tight oil** and **tight gas** reserves consist of source rock in which kerogen warmed enough to transform into oil and gas, but could not migrate away.

Finding, Extracting, and Refining Hydrocarbons

Prior to the mid-19th century, people used "rock oil," later known as *petroleum* (from the Latin *petra*, meaning rock, and *oleum*, meaning oil), only for greasing wagon axles. Such oil was rare, as it was found only at oil seeps. Then, in 1854, a group of investors speculated that petroleum could replace increasingly scarce whale oil as a lamp fuel if it could be found in reasonable quantities. They hired Edwin Drake to drill for oil at a seep near Titusville, Pennsylvania. On August 27, 1859, Drake and his crew succeeded in finding oil underground. Within a few years, thousands of oil wells had been drilled, and by the turn of the 20th century, civilization was addicted to oil. Initially, most oil was used to produce kerosene for lamps, but as electricity took over as the main power source for lighting, people began using oil as a fuel to generate that electricity, as well as for powering cars and trucks.

The earliest *oil fields* and *gas fields*—areas where numerous wells can be drilled—were found either by searching for seeps or through blind luck. Eventually, energy companies hired geologists to identify source rocks, reservoir rocks, and traps. Most modern exploration relies on **seismic-reflection profiles** to identify hydrocarbon-bearing strata underground (Fig. 11.9). To produce such a profile, a special vibrating truck sends seismic waves into the ground. The waves reflect off contacts between rock layers and return to the ground surface. The time between the generation of each wave and its return indicates the depth of contacts, from which geologists can provide a cross section of the strata.

To extract oil or gas, workers penetrate layers of rock using a **rotary drill**, a metal pipe tipped by a *drill bit*, a metal bulb studded with hard prongs (Fig. 11.10a). As the

FIGURE 11.7 Initially, oil resides in source rock. Because it is buoyant relative to groundwater, the oil migrates into the overlying reservoir rock. The oil accumulates beneath seal rock in a trap. In this example, the oil migrates along a fault into a fold.

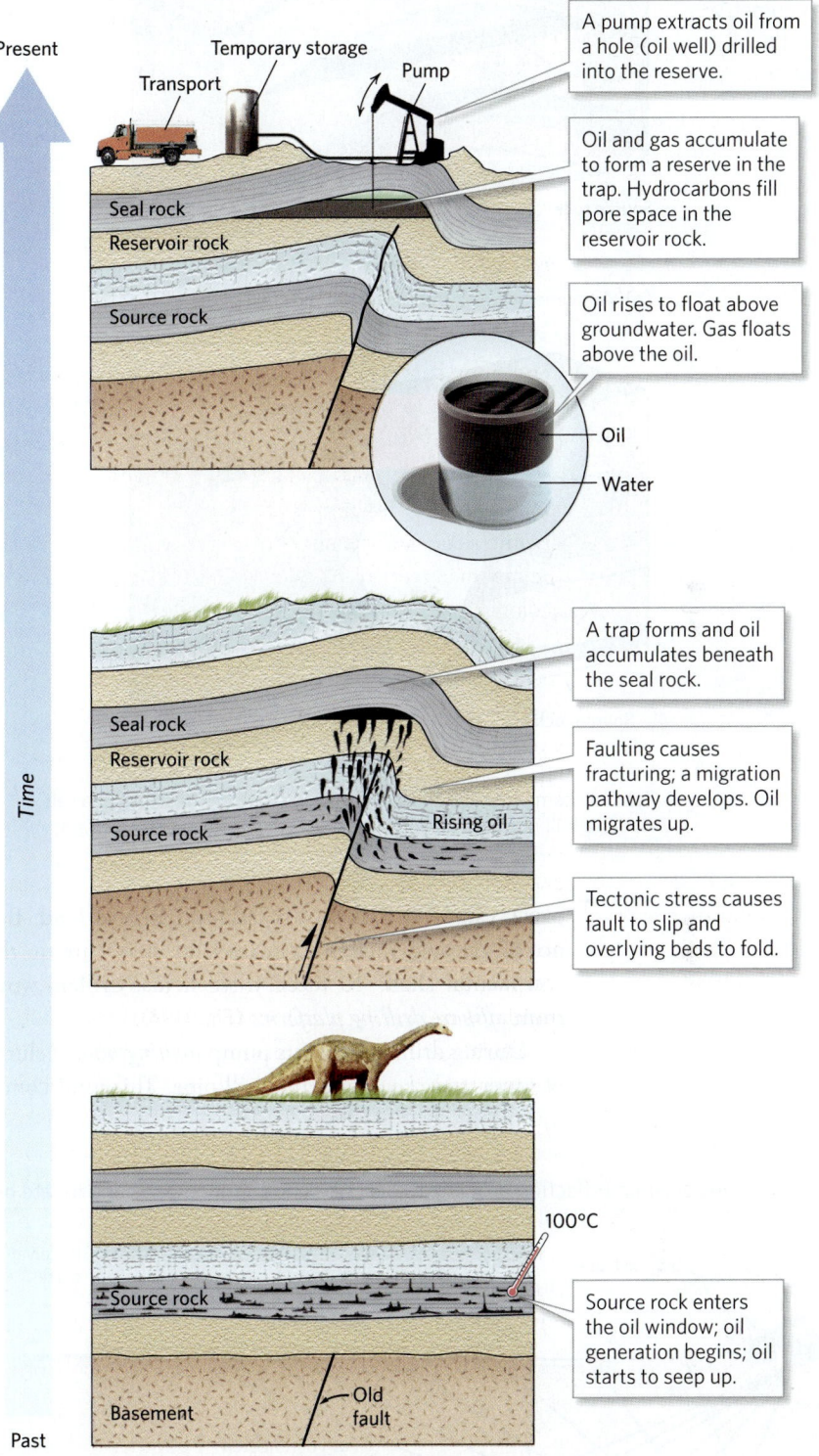

A pump extracts oil from a hole (oil well) drilled into the reserve.

Oil and gas accumulate to form a reserve in the trap. Hydrocarbons fill pore space in the reservoir rock.

Oil rises to float above groundwater. Gas floats above the oil.

A trap forms and oil accumulates beneath the seal rock.

Faulting causes fracturing; a migration pathway develops. Oil migrates up.

Tectonic stress causes fault to slip and overlying beds to fold.

Source rock enters the oil window; oil generation begins; oil starts to seep up.

rotating bit grinds, it transforms the rock into powder and chips, called *cuttings*. Early drilling methods could produce only vertical drillholes, but today, **directional drilling** allows drillholes to be angled (Fig. 11.10b). On land, drilling derricks, which hoist heavy drill pipe into

FIGURE 11.8

A trap is a configuration of seal rock over reservoir rock in a geometry that keeps the oil underground.

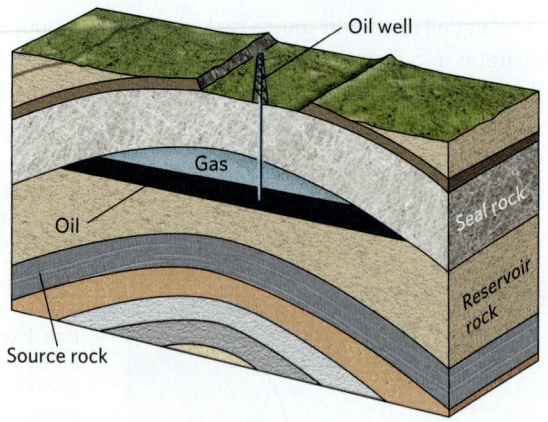

(a) Anticline trap. Oil and gas rise to the crest of the fold.

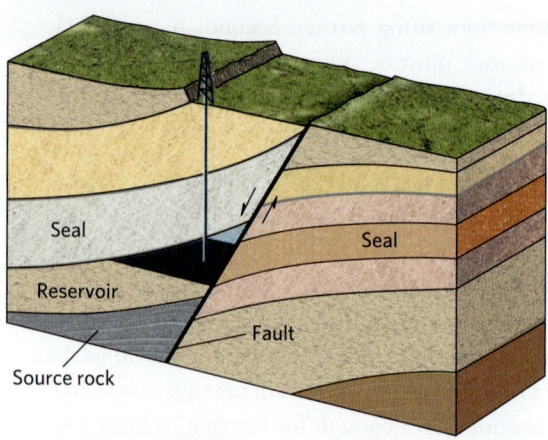

(b) Fault trap. Oil and gas collect in tilted strata adjacent to the fault.

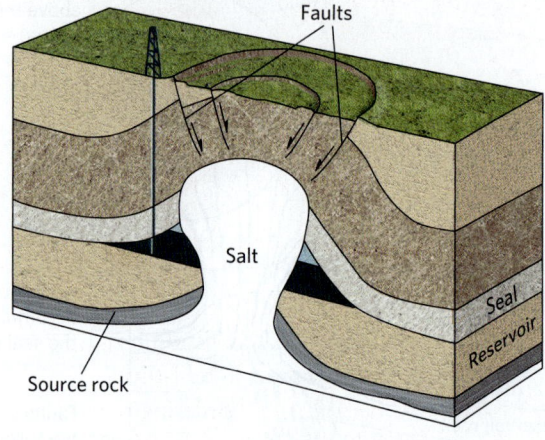

(c) Salt-dome trap. Oil and gas collect in strata on the flanks of a salt dome, a bulbous intrusion of salt that has risen into overlying strata.

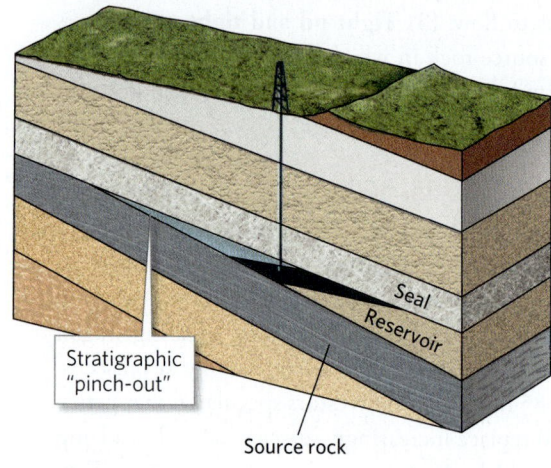

(d) Stratigraphic trap. Oil and gas collect where the reservoir-rock layer pinches out.

place, can be installed on a cleared patch of land. But not all oil or gas fields occur on land; many are on the continental shelf. To reach these fields, drillers work from *offshore drilling platforms* (**Fig. 11.11**).

During drilling, workers pump *drilling mud*, a slurry of water and clay, down the drill pipe. This mud comes out of holes at the end of the bit, then flows back up to the surface in the space between the pipe and the hole. This mud cools the bit, flushes cuttings out of the hole, and prevents subsurface hydrocarbons from entering the hole. Once drilling has been completed, workers set up pumps to suck oil and gas out of rock (**Fig. 11.12a**).

FIGURE 11.9 Using seismic-reflection profiling to describe underground beds and locate hydrocarbon reserves.

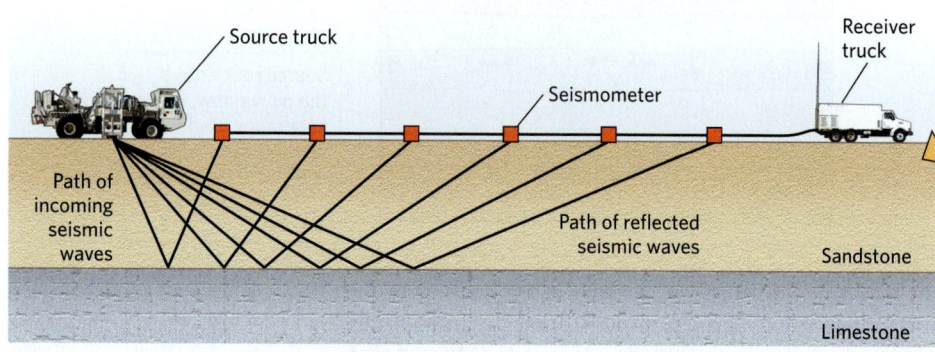

(a) A source truck sends seismic waves into the Earth. The waves are reflected by contacts between rock layers. At the ground surface, they are picked up by seismometers. The waves' travel times indicate the depths of the contacts.

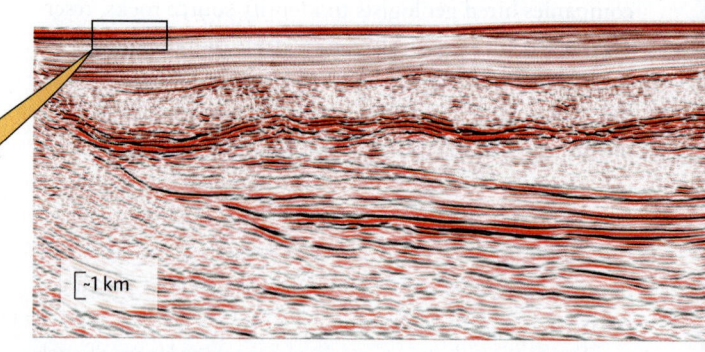

(b) A seismic profile can reveal the presence of geologic structures underground. The colored bands represent layers of strata.

FIGURE 11.10 Drilling for oil.

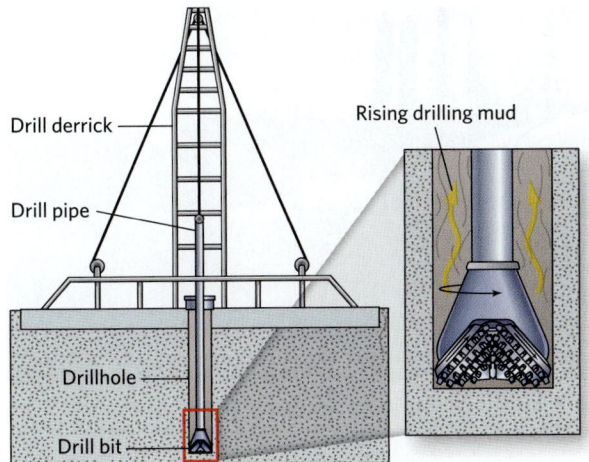

(a) A rotary drill—a rotating pipe tipped by a drill bit—grinds a hole into the ground. Drilling mud, pumped down through the pipe, comes out through holes in the bit, flushing cuttings out of the hole and keeping the hydrocarbons underground.

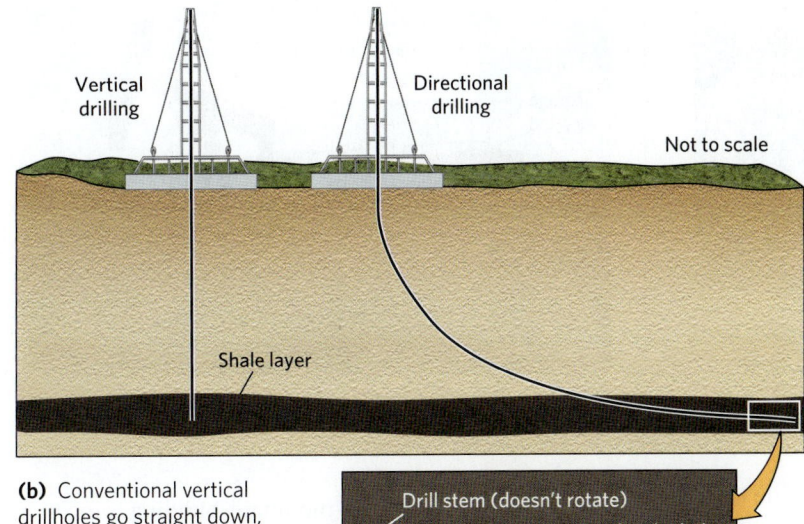

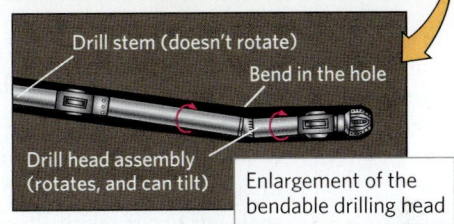

(b) Conventional vertical drillholes go straight down, but directional drilling allows drillers to bend drillholes and hit specific targets. As the inset shows, the head of the drill can bend at an elbow.

Oil taken directly from the ground is called *crude oil*. Once it has been extracted, it can be transported by truck, tanker ship, or pipeline to a *refinery* (Fig. 11.12b, c). At the refinery, crude oil is heated and then placed in a *distillation column*. The smaller (lighter) molecules rise in the column while larger (heavier) molecules sink. Consequently, natural gas flows from outlets at the top of the column, gasoline and motor oil flow from outlets in the middle, and tar (used for the production of plastics) collects at the bottom.

Simple pumping brings only about 30% of the oil in a conventional oil reserve out of the ground. Energy companies use *secondary recovery techniques* to coax out as much as 20% more. One such technique, **hydrofracturing** (commonly known as *hydrofracking* or simply *fracking*), produces new fractures in rock and opens up pre-existing fractures, thereby increasing the permeability of the rock. To "frack" a well, drillers pump *fracking fluid*, a liquid containing 90% water, 9.5% sand, and 0.5% other materials (detergents, lubricants, bactericides, and thickeners) into a sealed-off portion of the drillhole, then increase the pressure in the fluid until the rock around the hole breaks (Fig. 11.13a, b). When they pump out the fluid, some sand remains and prevents cracks from closing tightly, so that oil or gas can flow from the rock into the cracks and up the drillhole (Fig. 11.13c).

Where Do Oil and Gas Occur?

Conventional oil and gas reserves are not randomly distributed around the planet. Some of these reserves

occur along passive margins (onshore and offshore), in rift basins, in intracontinental basins, and in fold-thrust belts (Fig. 11.14). Currently, the countries bordering the Persian Gulf contain the world's largest reserves. Why? Much of the Earth's crust that now lies in the Middle East was situated in tropical areas during the Cretaceous Period (145–66 Ma). Biological productivity in the brightly lit waters of these regions was high, so the mud that accumulated there, when lithified, became excellent source rock. Thick layers of sand buried the source rock and eventually became porous sandstone that serves

▶ **Video**
Hydrofracking

FIGURE 11.11 An offshore drilling platform.

FIGURE 11.12 Pumping, transporting, and refining oil.

See for yourself

Oil Field near Lamesa, Texas

Latitude: 32°33'18.42" N
Longitude: 101°46'55.39" W

Zoom to 8 km (~5 mi) and look down.

Within a grid of farm fields, 1.6 km or 1 mi wide, small roads lead to patches of dirt. Each patch hosts or hosted a pump for extracting oil. These wells are tapping an oil reserve in Permian sandstone reservoirs.

(a) When drilling is complete, the derrick is replaced by a pump, which sucks oil out of the ground.

(b) The Trans Alaska Pipeline transports oil from fields on the Arctic coast to a tanker port on the southern coast of Alaska.

(c) This oil tanker is capable of carrying a million barrels (enough to supply the entire United States for a few hours).

as reservoir rock. Later, mountain-building processes uplifted and folded these layers, producing traps.

In the past 20 years, energy companies have begun to increase their focus on extracting hydrocarbons from unconventional reserves. Major reserves of tight oil and shale gas occur relatively close to highly populated areas **(Fig. 11.15a)**. Hydrocarbons can be extracted from these reserves by using directional drilling and hydrofracturing **(Fig. 11.15b)**. Directional drilling allows a driller to bore into a horizontal bed of organic shale, then extend the drillhole for kilometers within the bed, providing access to large volumes of the shale. Workers can then hydrofracture the entire length of the horizontal drillhole to create permeability through which the hydrocarbons can reach the well. The boom in production of hydrocarbons from unconventional reserves, however, has led to significant concerns about contamination of

FIGURE 11.13 Hydrofracturing technology is able to increase the permeability of rock by producing new fractures and opening existing fractures.

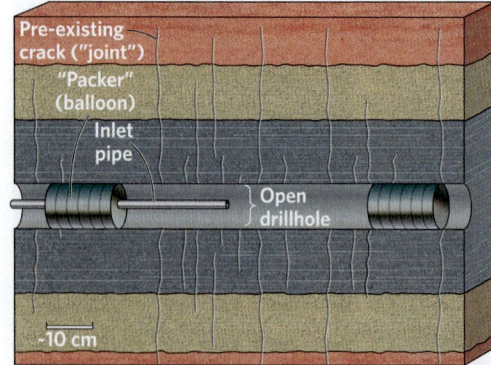

(a) After a hole has been drilled, a portion of the hole is sealed off with expandable cylinders called packers. A pipe is inserted through one of the packers.

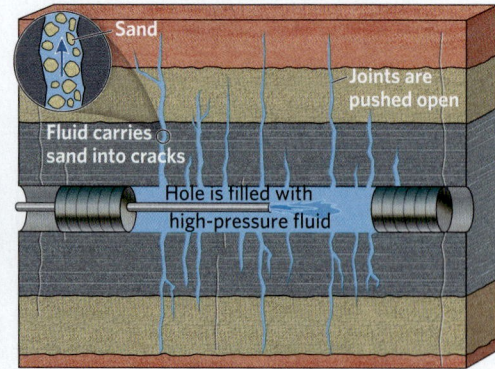

(b) Fluid is pumped under high pressure into the isolated segment of the hole. The pressure pushes open existing cracks and forms new cracks.

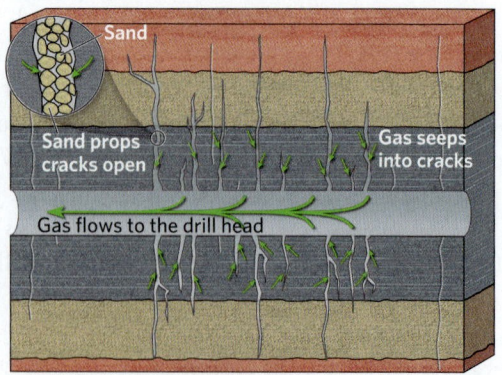

(c) After the packers and the fluid are removed, sand remains behind to prop open the cracks. Oil or gas can seep into the pipe and flow up to the ground.

FIGURE 11.14 Conventional oil reserves.

(a) The distribution of conventional oil reserves around the world. Some are onshore and some are offshore.

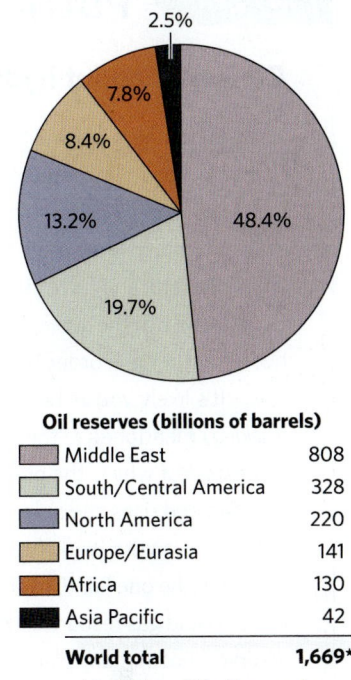

Oil reserves (billions of barrels)

	Middle East	808
	South/Central America	328
	North America	220
	Europe/Eurasia	141
	Africa	130
	Asia Pacific	42
	World total	**1,669***

* includes ~414 of tar sands

(b) Distribution of oil reserves among regions.

freshwater supplies, leakage of methane (a greenhouse gas) into the atmosphere, and triggering small earthquakes (Box 11.1).

Major reserves of tar sand occur in western Canada and in Venezuela. To obtain hydrocarbons from near-surface tar sand deposits, producers dig open-pit mines

FIGURE 11.15 Accessing tight oil and shale gas reserves.

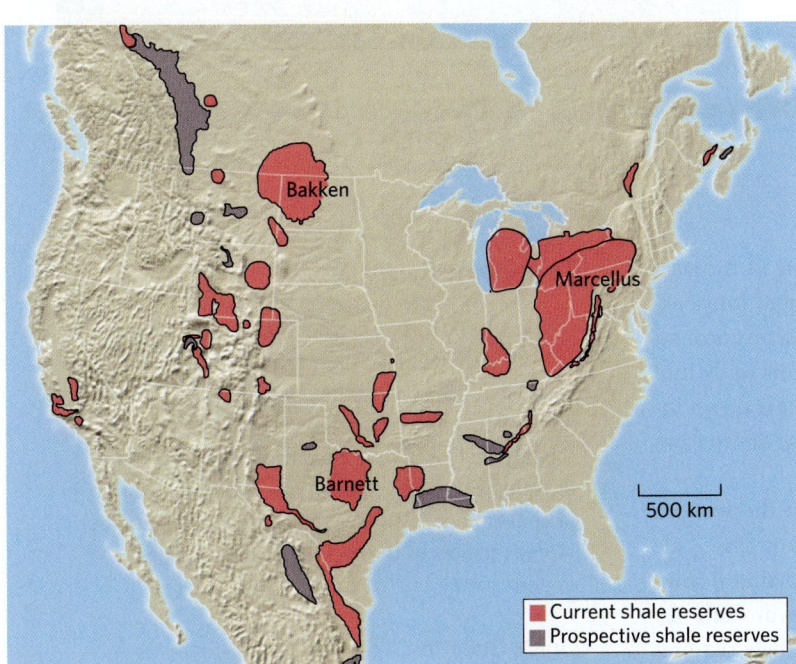

(a) A map of the major unconventional shale gas and oil fields in North America. The three largest sources are labeled.

(b) A drilling site. The trucks and the holding pond are used during hydrofracturing.

BOX 11.1 ▶ **Putting Earth Science to Use**

Concerns about hydrofracturing

The development of directional drilling and hydrofracturing has led to a boom of drilling into horizontal shale beds to extract tight oil and shale gas. Energy companies have spent billions of dollars to obtain drilling rights in large areas of Pennsylvania, Texas, Oklahoma, and North Dakota, and have installed many new wells. Because of the low permeability of the shale, almost all of these wells need to undergo hydrofracturing in order to produce hydrocarbons.

It's likely you've heard hydrofracturing (often shortened to *fracking*) mentioned in the news as a subject of political debates and protests. But why is the process so controversial? Understanding its benefits and risks can help inform your own participation in these debates as a voter and citizen.

On the one hand, hydrofracturing allows access to notably more oil and gas resources than could be produced from conventional reserves. But in many places, a local population boom follows the discovery of a new unconventional field, raising concerns about the strain on local services and the crowding of once-quiet rural areas with trucks carrying water, chemicals, and sand. Also, residents worry that surface-water supplies may be overused for fracking fluid, or that the chemicals in fracking fluid may spill into surface water or leak into groundwater.

Which aspects of fracking pose the greatest risk? Typically, the long horizontal segments of drillholes in source beds lie at depths of over 1.5 km (1 mi), where groundwater tends to be saline and not good for drinking or irrigation, so leakage from the fracked part of the hole at depth might not have immediate effects on water supplies. Fresh, usable groundwater, on the other hand, lies at shallower depths, so

this region has greater vulnerability **(Fig. Bx11.1)**. Therefore, it is critical that the vertical part of each drillhole be sealed to prevent leakage and that operators take great care not to spill fracking fluid on the ground surface or into surface water.

In some cases, oil and gas extraction brings up large volumes of saline groundwater. Saline water can't be dumped into rivers or used for irrigation, so to dispose of it, drillers pump it underground into deep wells called injection wells. Researchers have shown that pumping at very high pressure and at great depth may trigger small earthquakes. Therefore, injection pressures need to be closely regulated.

FIGURE Bx11.1 The potential for groundwater contamination by hydrofracturing.

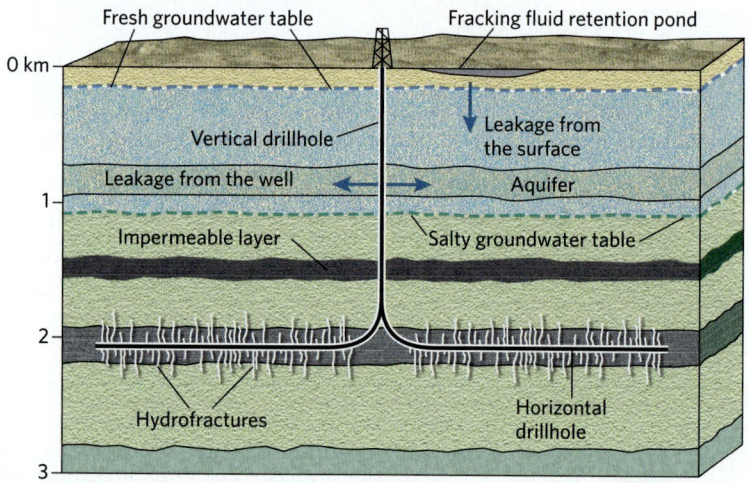

(Fig. 11.16), then heat the extracted sandstone in furnaces until the hydrocarbons become less viscous and can separate from the sand. To extract oil from deep deposits of tar sand, producers drill injection and extraction wells into the deposits. Then they force steam or solvents down injection wells to liquefy the oil, then pump the oil up from extraction wells.

Significant reserves of oil shale occur in the western United States. Producing hydrocarbons from this shale requires processes similar to those used to obtain oil from tar sand. Heating can convert kerogen into hydrocarbons. Notably, producing oil from tar sand or oil shale requires a lot of energy, so it's more expensive, and its production has the potential to damage the landscape.

Did you ever wonder . . .

how coal differs from oil?

Take-home message . . .

Oil and gas form from the remains of plankton that, together with clay, has lithified into a source rock. Heat can convert the organic molecules into oil or gas. In conventional reserves, hydrocarbons have migrated into a porous and permeable reservoir rock and are held underground by an impermeable trap. In unconventional reserves, the hydrocarbons are too viscous to flow or the rock containing them is impermeable.

Quick Question

What steps are involved in finding and producing oil?

FIGURE 11.16 An open-pit mine for tar sand in Alberta, Canada.

11.4 Coal: Energy from the Swamps of the Past

Coal Formation

Imagine a swamp in a tropical climate. Trees, ferns, shrubs, and vines grow in abundance, producing an immense amount of woody biomass (Fig. 11.17a). Over time, the plants die and fall into the stagnant water of the swamp. This water lacks oxygen, so the dead biomass does not rot away entirely, and new plants grow on top of it. Eventually, a layer of compacted organic material, known as **peat**, accumulates. Dried peat can be burned directly for heating.

If a peat layer becomes deeply buried by overlying sediment, it can be preserved over geologic time (Fig. 11.17b). When the overlying sediment becomes thick enough, the organic matter in peat undergoes compaction and heating. Slowly, chemical reactions that release volatile compounds (such as water, methane, and ammonia) transform peat into a dark-brown to black, brittle sedimentary rock, called **coal**, that consists of more than 60% carbon. The carbon in coal occurs in huge organic molecules, called *coal macerals*, each of which contains hundreds of carbon atoms. We consider coal to be a type of fossil fuel because it stores solar energy that reached the Earth long ago—burning coal releases the energy stored in the chemical bonds of coal macerals.

Note that oil and coal have different origins. As we have seen, oil consists of hydrocarbons derived from the remains of buried plankton, whereas coal is derived from woody plants that grew in *coal swamps*, regions that resembled modern rainforests and wetlands in tropical and subtropical coastal areas. Because coal forms in layers within a sedimentary succession, a coal bed, or **coal**

FIGURE 11.17 Coal forms when plant debris becomes deeply buried.

(a) A museum diorama depicting a Carboniferous coal swamp.

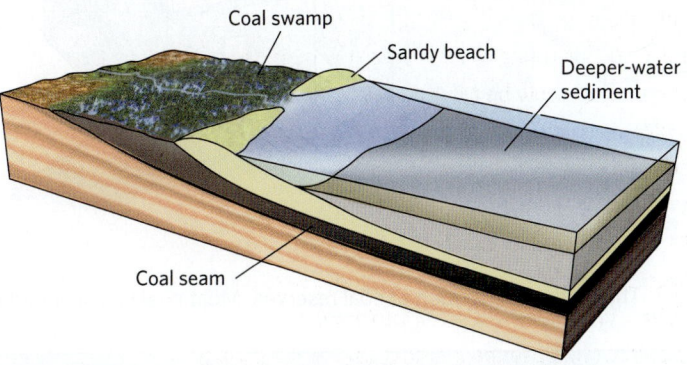

(b) As sea level rises, peat formed in a coal swamp can be buried and preserved.

(c) Coal beds interlayered with beds of sandstone and shale.

seam, appears in an outcrop as a black band between other kinds of sedimentary rocks (Fig. 11.17c). Individual seams may be centimeters to meters thick and may underlie a broad region.

Because the most extensive deposits of coal in the world occur in Carboniferous strata, geologists named

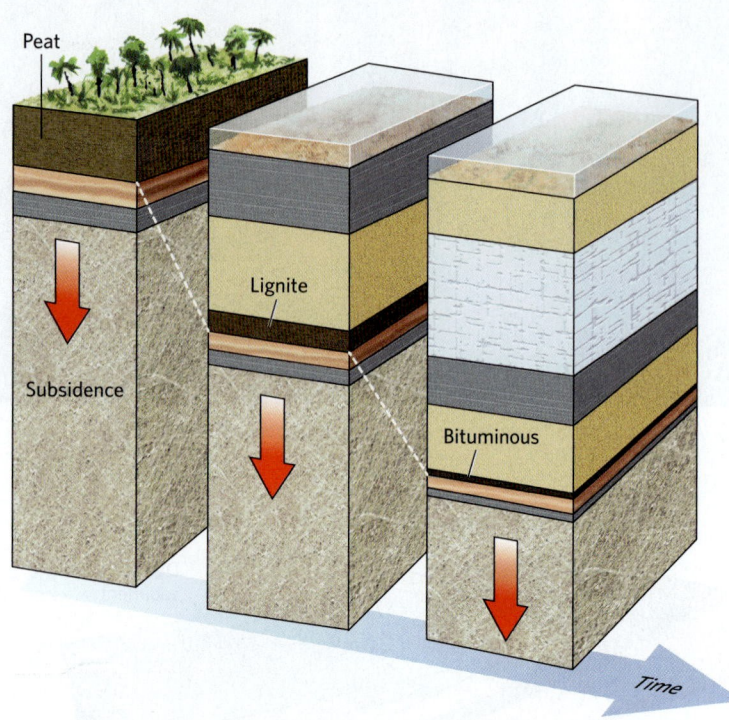

FIGURE 11.18 Burial of peat leads to the formation of coal of progressively higher carbon content.

Peat

Lignite

Subsidence

Bituminous

Time

the time period *Carboniferous*. At this time, the supercontinent Pangaea had assembled, and much of it straddled warm tropical regions with high rainfall. In these regions, land plants flourished in swampy conditions.

Classifying Coal

Coal is formed by the progressive burial of peat **(Fig. 11.18)**. As burial takes place, chemical reactions produce volatiles, which escape, causing the concentration of carbon in the peat to increase. **Coal rank** indicates the extent to which coal has become enriched in carbon. Once the carbon concentration exceeds 60%, the deposit can formally be called coal—specifically, low-rank coal called *lignite* or *brown coal* **(Table 11.3)**. With still deeper burial, lignite warms up and additional volatiles escape, yielding even higher concentrations of carbon, so lignite transforms into dull black intermediate-rank *bituminous coal*. At even higher temperatures, bituminous coal transforms into shiny black high-rank *anthracite*. Notably, the energy density of coal increases with increasing rank (see Table 11.3). Lignite and bituminous coal can be formed by burial in a sedimentary basin. The temperatures needed to form anthracite are so high, however, that such coal

FIGURE 11.19 The global distribution of coal reserves. Most coal has accumulated in mid-continental basins.

■ Anthracite and bituminous coal

■ Lignite

FIGURE 11.20 Coal mining.

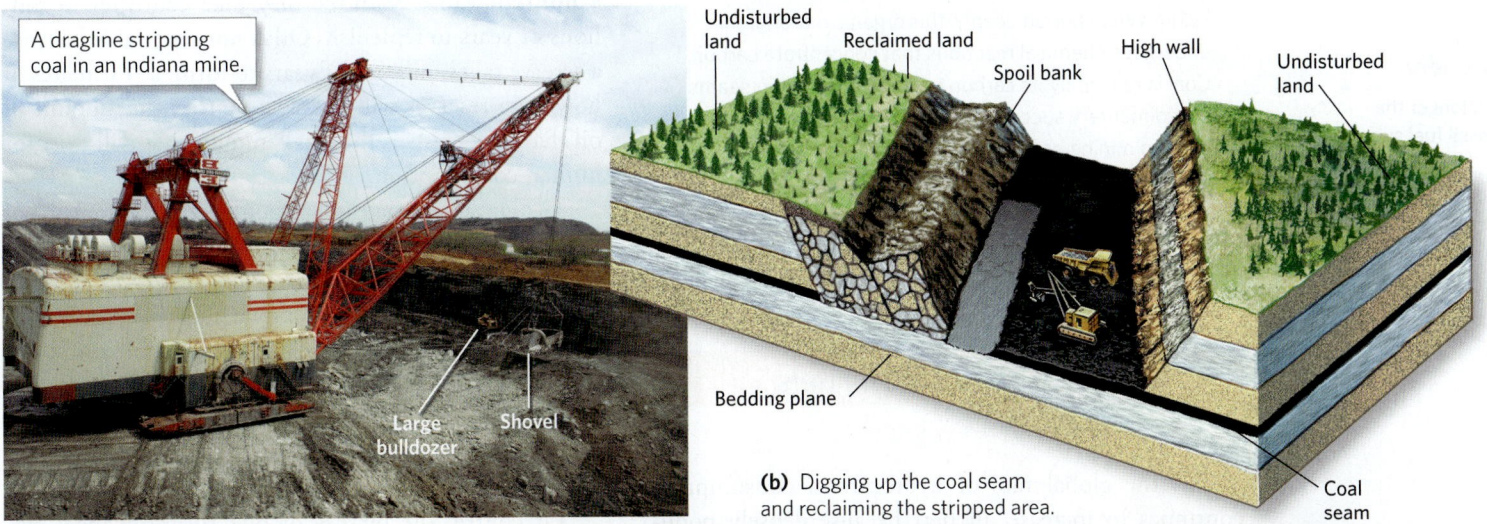

A dragline stripping coal in an Indiana mine.

Large bulldozer

Shovel

(a) A dragline stripping away the soil and sedimentary rock that overlie a coal seam.

Undistributed land

Reclaimed land

Spoil bank

High wall

Undisturbed land

Bedding plane

Coal seam

(b) Digging up the coal seam and reclaiming the stripped area.

Mining Coal

Sedimentary strata of continents contain huge accumulations of coal, known as *coal reserves* **(Fig. 11.19)**. The methods by which energy companies extract coal from these reserves depend on the depth of the coal. If coal seams lie within about 100 m (330 ft) of the ground surface, *strip mining* can be the least expensive method. Here, miners use a giant shovel, or *dragline*, to scrape off the soil and the layers of sedimentary rock that lie above the coal seam **(Fig. 11.20a)**. Miners then use other equipment to dig out the coal and dump it into trucks or onto a conveyor belt **(Fig. 11.20b)**. In hilly areas, miners may employ *mountaintop removal mining*—during this process, they remove all land above the coal and dump it into an adjacent valley in order to expose the coal. In some mines, dragline operators separate out and preserve soil and underlying rock. When the coal has been scraped out, the operator refills the pit.

Deeply buried coal can be obtained economically only by **underground mining**. To develop an underground mine, miners dig a shaft down to the depth of the coal seam and then excavate a maze of tunnels, using huge grinding machines that chew their way into the coal **(Fig. 11.20c)**. Underground coal mining can be dangerous, not only because the sedimentary rocks forming the roof of the mine are weak and may collapse, but also because methane gas released by chemical

(c) Underground coal mining.

reactions in the coal can accumulate in the mine, and if not removed by ventilation fans, can explode. All coal mining produces dust, so unless they breathe through filters, miners risk contracting black-lung disease by inhaling coal dust.

TABLE 11.3 Coal Rank and Associated Energy Density

Rank	Formation Temperature	Carbon Concentration	Energy Density
Lignite	<100°C	60%–70%	18 MJ/kg
Bituminous	100°C–200°C	70%–87%	27 MJ/kg
Anthracite	200°C–300°C	87%–95%	33 MJ/kg

usually can develop only beneath thrust sheets along the margins of mountain belts.

Take-home message . . .

Coal forms from the remains of plant material. When buried deeply, this organic material undergoes chemical reactions that concentrate carbon. Coal is ranked by its carbon content. It occurs in seams in sedimentary successions, from which it is extracted by strip mining or by underground mining.

Quick Question

Why does more deeply buried coal have a higher rank?

11.5 The Future of Fossil Fuels

The Age of Oil

Today, the global rate of hydrocarbon consumption continues to increase, in part because densely populated countries such as China and India have undergone rapid industrialization. Future historians may refer to the present time as the **Oil Age** because so much of the world's economy depends on oil. How long will the Oil Age last? To answer this question, we must keep in mind the distinction between a renewable and a nonrenewable resource: a **renewable resource** can be replaced by nature within months to decades, whereas a **nonrenewable resource** may take centuries to millions of years to replenish. Oil is nonrenewable because a reserve takes millions of years to form, so estimates of how long reserves will last assume that the amount of oil that now exists is the total amount that will exist for human consumption.

Geologists estimate that about 1,700 billion bbl of oil resides in proven conventional reserves. An additional 2,000 billion bbl of oil in conventional reserves may exist, but has not yet been found. In total, therefore, the Earth's sedimentary strata may hold up to 3,700 billion bbl of conventional oil reserves. Presently, humanity guzzles this oil at a rate of about 34 billion bbl per year, which means that conventional oil supplies may last for only about another century.

Of course, the picture of oil's future changes significantly if estimates include unconventional reserves. All told, perhaps 500 billion bbl of tight oil exist, along with 1.5 trillion bbl of oil in tar sand and 4.5 trillion bbl in oil shale. At current rates of consumption, supplies might last for another few hundred years, if oil continues to be consumed that long.

The Future of Coal Reserves

Society still burns a lot of coal. Worldwide coal reserves are estimated at about 850 trillion tons, which could supply approximately the same amount of energy as 11 trillion barrels of oil. But such an estimate of coal reserves does not distinguish clearly between accessible (minable) coal and inaccessible coal, which is too deep to mine. So estimating how long coal reserves can last is a challenge. At current rates of use, many centuries of accessible supplies remain.

Environmental Consequences of Fossil Fuel Use

The extraction, processing, and use of fossil fuels have significant consequences for the Earth's environment and climate. Oil drilling requires large drilling rigs and drilling pads to support the rigs, all of which can damage the land. And, as demonstrated by the 2010 *Deepwater Horizon* blowout in the Gulf of Mexico, offshore oil drilling can lead to tragic losses of life and disastrous marine oil spills **(Box 11.2)**. Oil spills from ships and tankers can also create oil slicks that spread over the sea surface and foul the shoreline **(Fig. 11.21)**. On land, oil spills from pipelines or trucks may sink into the subsurface and contaminate groundwater. And as we've seen, if not handled properly, fracking fluid may contaminate water supplies.

FIGURE 11.21 Marine oil spills can come from drilling rigs or from tankers.

(a) An oil tanker leaking oil on the sea surface.

(b) Oil spills can contaminate the shore and can be very difficult to clean up.

BOX 11.2

A Deeper Look

The *Deepwater Horizon* disaster

A substantial proportion of the world's conventional oil reserves resides in sedimentary basins that underlie the continental shelves of passive continental margins, accessible only by offshore drilling platforms. During both onshore and offshore exploration, drillers worry about the possibility of a **blowout**, when hydrocarbons spurt out of a drillhole under their own pressure. A catastrophic blowout occurred on April 20, 2010, when drillers on the *Deepwater Horizon*, a huge pontoon platform, were completing a 5.5-km (18,000-ft)-long well in 1.5-km (5,000-ft)-deep water southeast of the Mississippi Delta. Due to a series of errors, gassy oil under high pressure in reservoir rock punctured by the well rushed up the drillhole. A backup safety device, called a blowout preventer, was supposed to clamp the wellhead (the outlet of the well) shut, but it failed, so the gassy oil reached the platform and sprayed 100 m (330 ft) into the sky. Sparks from electronic gear triggered an explosion, and the platform became a fountain of flame and smoke that killed 11 workers. An armada of fireboats could not douse the conflagration **(Fig. Bx11.2)**, and after 36 hours, the still-burning platform tipped over and sank.

Robot submersibles sent to the seafloor to investigate found oil and gas billowing from the twisted mess of bent and ruptured pipes at the well's outlet. On the order of 50,000–62,000 bbl of oil entered the Gulf's water from the well each day. Stopping this underwater gusher proved to be an immense challenge, and initial efforts to block the well or to cover the wellhead with a containment dome failed. Not until July 15 was the flow finally stopped. Meanwhile, workers drilled another

FIGURE Bx11.2 Fireboats doused the burning *Deepwater Horizon* platform in vain before it sank.

well from a platform a few kilometers away. Using directional drilling, they managed to intersect the 15-cm (6-in)-diameter *Deepwater Horizon* drillhole, and by September 19, the concrete they pumped into the failed well sealed it permanently. All told, about 4.2 million bbl of hydrocarbons from the *Deepwater Horizon* blowout contaminated the Gulf. The spill devastated wetlands, wildlife, and the fishing and tourism industries. Fortunately, bacteria can digest the oil, so the region's environment is beginning to recover.

Coal mining, in addition to changing the landscape, may produce **acid mine runoff**, a dilute solution of sulfuric acid. Acid mine runoff develops when sulfur-bearing minerals in coal react with rainwater when exposed by mining. If the runoff enters streams, it can kill fish and plants. Another serious problem arises when, due to arson, accident, or spontaneous processes, unmined coal beds ignite. Once started, a *coal-bed fire* may burn for years and may be difficult or impossible to extinguish. Such fires produce toxic fumes that rise through joints to the surface to make the overlying land uninhabitable.

Numerous air pollution issues arise from burning fossil fuels. Emissions from smokestacks or exhaust pipes may introduce soot, carbon monoxide, sulfur dioxide, nitrous oxide, and unburned hydrocarbons into the air. These pollutants can cause deadly smog. In some cases, sulfur-rich emissions combine with moisture in the air to form dilute sulfuric acid that falls as **acid rain**, causing vegetation and fish to die off in regions downwind of smokestacks. *Coal ash*, the residue from burning coal, has become an increasing environmental problem. In some cases, ash reservoirs have ruptured, spreading the ash over a landscape. Even if ash remains isolated, dangerous chemicals that it contains may leach into groundwater and end up in streams. Some researchers have suggested that some of the pollution problems associated with burning coal could be overcome by transforming solid coal into a mixture of burnable gases. This process, called coal gasification, leaves pollutants as a solid residue that never enters the smokestack.

Even if new technologies can reduce the sulfur, soot, and poisonous chemicals emitted by burning fossil fuels, the process still releases carbon dioxide into the atmosphere. As we will see in Chapter 20, CO_2 is a greenhouse gas whose increasing concentration in the atmosphere contributes to global warming. Because of concern about the **carbon footprint** of fossil fuels, meaning the amount of CO_2 and other greenhouse gases emitted by their production and burning, most governments are encouraging a reduction in fossil fuel burning with regulations and by exploring alternative energy sources.

Clearly, society faces difficult choices about where to obtain energy. Some researchers estimate that by 2050, most energy will come from alternatives to fossil fuels. Next, we consider some of these alternatives.

Take-home message ...

Fossil fuels are nonrenewable resources. In fact, at current rates of consumption, oil supplies could run out within a century or two. Production and use of fossil fuels have environmental consequences of concern, so societies are exploring ways to find alternative energy sources.

Quick Question ————————————
Why do acid rain and acid mine runoff develop?

11.6 Other Energy Sources

Nuclear Power

When you watch a fossil fuel burn, you're seeing a *chemical reaction* that releases energy by breaking the chemical bonds that hold atoms together in a molecule. In a nuclear power plant, however, energy comes from *nuclear reactions* that involve the breaking, or *fission*, of nuclear bonds holding protons and neutrons together in an atom's nucleus. Fission takes place when a neutron strikes a radioactive parent atom, causing it to split into smaller daughter atoms. Neutrons released during the fission of one atom strike other atoms, triggering fission of those atoms in turn, in a self-perpetuating process called a **chain reaction**. An extremely rapid, uncontrolled chain reaction is what causes the blast of an atomic bomb of the type used in World War II. Controlled fission of concentrated uranium or plutonium is what produces energy in a nuclear power plant.

HOW DOES A NUCLEAR POWER PLANT WORK? The heart of a nuclear power plant is a **nuclear reactor**, a container holding *fuel rods*, metal tubes filled with concentrated radioactive uranium or plutonium. Fission in the fuel rods produces

heat **(Fig. 11.22a)**. Nuclear reactors also contain *control rods*, composed of substances that can moderate the rate of overall energy production. Because radioactive materials pose a danger to living organisms, engineers place reactors within a containment structure of reinforced concrete.

A nuclear power plant produces electricity in much the same way a fossil fuel plant does. Heat produced by fission transforms water into steam. The steam then pushes the blades of a turbine, which in turn rotates the shaft of a generator.

Where does the uranium used in nuclear power plants come from? Rising granite magma brings uranium atoms into the upper crust, where various geologic processes, such as groundwater flow, concentrates uranium in localized deposits. Uranium extracted from these deposits can't be used in a reactor directly, because the fissionable isotope of uranium that serves as the fuel in reactors accounts for only about 0.7% of naturally occurring uranium. To make a fuel for use in a reactor, the fissionable isotope must be increased to 3%–5%, an expensive process called **enrichment**.

CHALLENGES OF USING NUCLEAR POWER The first nuclear power plants were built in the 1950s, and about 450 nuclear power plants currently operate. To date, two major disasters have occurred at these plants, meaning events during which containment structures were damaged and significant quantities of radiation escaped. The first occurred in 1986 at the Chernobyl power plant in Ukraine, and the second in 2011 at the Fukushima Daiichi power plant in Japan. In both cases, the fuel in the reactor got so hot that it melted (an event called a *meltdown*). The intense heat released caused water molecules in the cooling system to separate into a mixture of hydrogen and oxygen gas, which exploded, causing a rupture of the containment structure and a release of radioactive debris into the environment. The Chernobyl disaster, which caused many deaths, was worse. An accident also happened at the Three Mile Island plant in Pennsylvania in 1979, in which a partial meltdown of the reactor took place. Fortunately, a breach of the containment structure did not happen at Three Mile Island, so less radiation escaped.

Operation of a reactor produces **nuclear waste**, including very radioactive *high-level waste*, such as fuel rods, as well as less radioactive *low-level waste*, such as radiation-contaminated water, piping, and concrete. Some radioactivity in nuclear waste decays relatively quickly (in decades to centuries), but some will remain dangerous for thousands of years. High-level waste needs to be kept cool by being submerged in water. But even low-level waste cannot simply be buried in a landfill because it might leak into water supplies. Instead, it should be isolated in durable, sealed containers. Finding appropriate long-term storage for nuclear waste isn't easy—to date, most such waste remains on the property of the power plants that produced it.

Biofuels

Can we transform modern-day biomass into fuels that work like fossil fuels through the use of laboratory chemistry? The answer is yes, and the resulting materials are called **biofuels**. The most commonly used biofuel is *ethanol* (CH_3CH_2OH), a type of alcohol that can substitute for gasoline in car engines. It can be produced commercially either from corn or from sugarcane. More recently, researchers have been developing processes that yield ethanol from cellulose (the fibrous material in plants), permitting perennial grasses to become a source of biofuel. Another promising method uses algae, which naturally synthesizes fatty organic chemicals from which hydrocarbons can be produced. Recent technologies also include the commercial production of **biodiesel**, a fuel that comes from chemical modification of vegetable oils.

Geothermal Energy

As the name suggests, **geothermal energy** comes from the Earth's internal heat. Significant sources of geothermal energy are available in areas of igneous activity, where high temperatures at relatively shallow depths can make accessible groundwater so hot that, when pumped from the subsurface and run through pipes, it can heat houses and buildings directly. The hottest groundwater turns to steam when it rises and undergoes decompression, and this can drive turbines and generate electricity **(Fig. 11.22b)**.

Hydroelectric and Wind Power

For centuries, people have used water wheels to power mills and factories directly. Modern hydroelectric power plants generate electricity through a similar technology—flowing water turns the blades of a turbine, and the turbine drives an electrical generator. Most hydroelectric plants rely on water held in a reservoir by a dam **(Fig. 11.23a)**. In effect, **hydroelectric power** comes from the potential energy in the reservoir's elevated water, which converts into kinetic energy when the water flows to a lower elevation at the foot of the dam.

Hydroelectric power does not release chemical or radioactive pollutants or greenhouse gases, and it does not consume nonrenewable resources. In addition, the reservoirs of a hydroelectric system may provide irrigation water, flood control, and recreational opportunities. But the construction of dams and reservoirs can also cause problems. For example, damming a river may submerge spectacular scenery, displace towns, and destroy ecosystems. And reservoirs trap sediment and nutrients, not only decreasing reservoir capacity over time, but also preventing these materials from

FIGURE 11.22 Nuclear and geothermal energy.

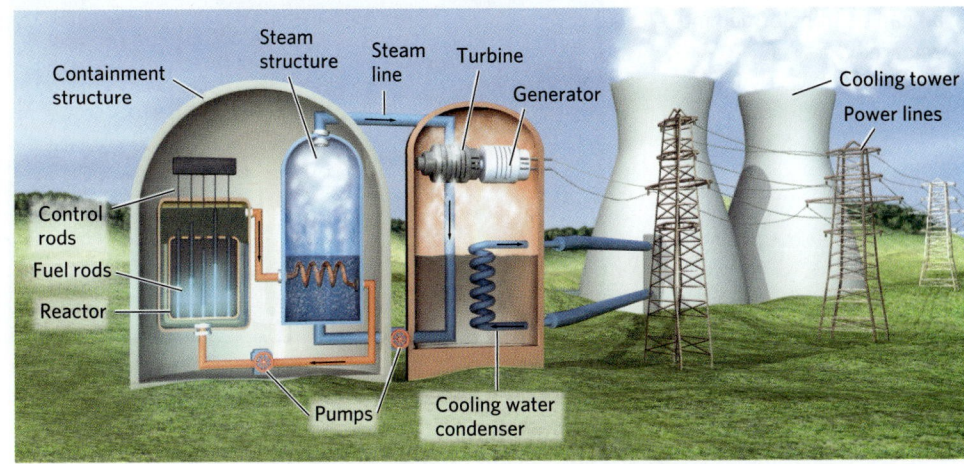

(a) A nuclear reactor heats water to produce high-pressure steam that drives a turbine.

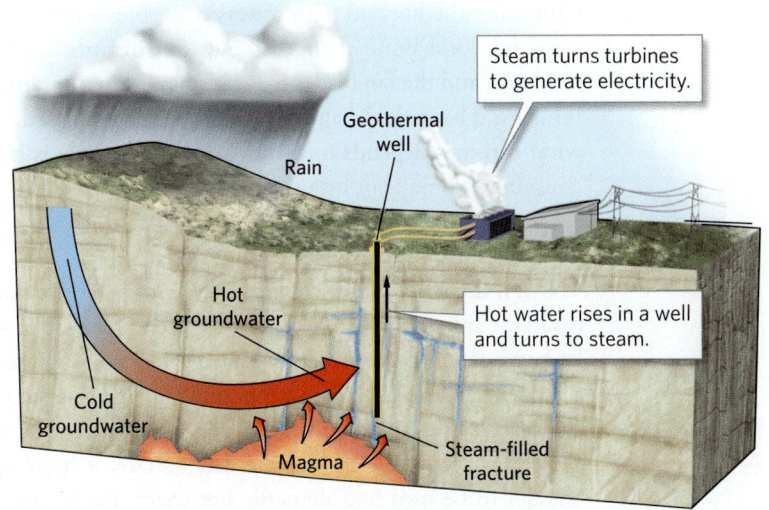

(b) Geothermal power plants use hot groundwater. This groundwater turns to steam when it reaches the surface and undergoes decompression.

reaching floodplains or deltas downstream, adversely affecting agriculture and delta growth.

Not all hydroelectric power generation uses flowing river water. Engineers have been developing new means to tap **tidal energy**, the energy associated with the daily rise and fall of tides. One approach involves building a dam, called a *tidal barrage*, across the entrance to a bay. When the tide rises, water flows into the enclosed area through openings in the barrage. When the tide drops, water trapped behind the barrage flows back to the sea in pipes that carry it through power-generating turbines.

Modern efforts to harness **wind power** are developing rapidly and on a large scale. To produce wind power, meteorologists identify regions that have steady breezes. In these regions, engineers build *wind farms* with numerous

FIGURE 11.23 The kinetic energy of flowing water and air can be transformed into electricity.

(a) The water held back by the Three Gorges Dam, on the Yangtze River in China, flows through turbines to generate electricity.

(b) A wind farm in southwestern England. The towers are about 50 m high.

towers, each of which holds a turbine powered by a set of giant fan blades (Fig. 11.23b). Like a hydroelectric turbine, a wind turbine drives an electrical generator as it turns. Wind power is clean, but it has some drawbacks. Cluttering the horizon with towers may spoil a beautiful view, the loud hum of the turbines can disturb nearby residents, and the fan blades may be a hazard to migrating birds. And because the amount of electricity produced by wind turbines depends on wind speed, the supply is not constant and may not be available when needed.

Solar Energy

The Sun drenches the Earth with energy in quantities that dwarf the amounts stored in fossil fuels. Such *solar energy* can be harnessed in two ways. In a *solar collector*, a dark surface placed beneath a glass plate absorbs sunlight. The glass keeps the heat from escaping, so when water passes through pipes placed between the glass and the dark surface, it heats up enough to be used as a domestic hot water supply. Alternatively, solar energy can be used to produce electricity by

means of **photovoltaic cells** (*solar cells*), shiny panels with two wafers of silicon that each contain specific impurities. When sunlight strikes the cell, electrons flow from one wafer to the other, producing an electric current (Fig. 11.24).

Take-home message …

Society increasingly uses alternative energy sources that do not rely on fossil fuels. Examples include energy produced by biofuels, nuclear, geothermal, hydroelectric, wind, and solar power.

Quick Question

What is the difference between a chemical reaction and a nuclear reaction?

11.7 Metallic Mineral Resources

In January 1848, James Marshall and a crew of workers were finishing the construction of a new sawmill in the foothills of the Sierra Nevada. As Marshall stood admiring the building, he noticed a glimmer of metal in the gravel that littered the bed of the adjacent stream. He picked up the metal, banged it between two rocks to test its hardness, and shouted, "Boys, by God, I believe I have found a gold mine!" Word of the discovery soon spread, and within weeks, all the mill's workers had disappeared into the mountains to seek their own fortunes. Gold fever spread throughout the country, and 1849 saw 40,000 prospectors heading to California.

Gold is just one of the many metals that people rely on for many applications. A **metal** is an opaque, shiny material that can conduct electricity. Most metals can be bent, drawn into wire, or hammered into a thin sheet. The first metals that people learned to use—gold, copper, and silver—occur in rock as *native metal*, meaning that they look and behave like metal in their natural, unprocessed states (Fig. 11.25). Beginning about

FIGURE 11.24 Arrays of photovoltaic cells produce electricity directly from solar radiation.

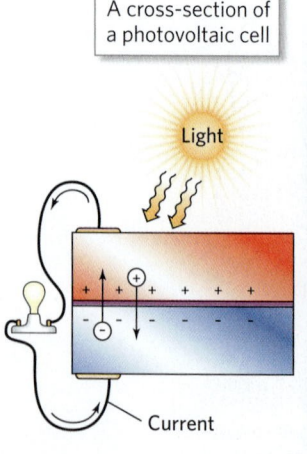

A cross-section of a photovoltaic cell

Light

Current

A commercial solar array

5000 B.C.E., people discovered **smelting**, the process of heating certain rocks to a temperature high enough that the rocks decompose to yield metal plus a nonmetallic residue called *slag*. Only because of smelting can we obtain the variety of metals, and the quantities of metals, that modern society uses. In this section, we explore the nature of metallic mineral resources, the Earth materials from which metals can be obtained.

Ores and Ore Minerals

Geologists use the term **ore** for a rock containing sufficient useful metal to be worth mining. The concentration of a useful metal in an ore determines the **grade** of the ore—the higher the concentration, the higher the grade. Whether or not a mining company will choose to excavate an ore of a given grade, at a given time, depends on the metal's market price (Box 11.3).

As we've noted, a few metals can occur as native metals. More typically, the metal atoms in ores occur in specific minerals, known as **ore minerals**, that contain a high proportion of extractable metal. For example, galena (PbS) contains 50% lead, so it is an ore mineral for lead (Fig. 11.26a). Hematite (Fe_2O_3) and magnetite (Fe_3O_4) are ore minerals for iron. Copper comes from a variety of ore minerals, including chalcopyrite ($CuFeS_2$) and malachite [$Cu_2CO_3(OH)_2$] (Fig. 11.26b). The chemical formulas of these examples show that ore minerals include *sulfides*, *oxides*, and *carbonates* (see Chapter 3).

Formation of Ore Deposits

Ore minerals do not occur uniformly through the rocks of the Earth's upper crust. Fortunately for society, geologic processes concentrate ore minerals in **ore deposits**. Geologists distinguish among different types of ore deposits based on the processes by which the deposits form. Here are a few examples:

- *Magmatic deposits:* In some magmas, ore minerals crystallize and accumulate to form lenses of ore, called *magmatic deposits*, as the magma solidifies into igneous rock (Fig. 11.27a).

- *Hydrothermal deposits:* Hot groundwater circulating through an igneous intrusion and its wall rock can dissolve metal ions. When the resulting solution enters cooler rock, or encounters water containing certain chemical characteristics, the metals precipitate as ore minerals—either in *veins* (mineral-filled cracks) or in pores—to produce a *hydrothermal deposit* (Fig. 11.27b).

- *Seafloor massive sulfide deposits:* Along mid-ocean ridges, black smokers erupt hydrothermal solutions (see Chapter 4). When the solutions mix with cool seawater, the dissolved components precipitate as tiny crystals of sulfide minerals (Fig. 11.27c). These minerals accumulate around the vent to form a *seafloor massive sulfide deposit*.

- *Secondary-enrichment deposits:* In the upper crust, groundwater can dissolve ore minerals. When the resulting solution flows into a different environment, it may precipitate new ore minerals in a high concentration, generating a *secondary-enrichment deposit*.

- *Sedimentary deposits:* A *sedimentary deposit* accumulates from ore minerals that have settled or precipitated out of water as sediment. Examples include banded iron formation (see Fig. 10.7b).

- *Placer deposits:* When rocks containing native metals erode, the resulting sediment contains rock clasts, metal flakes, and *nuggets* (pebble-sized metal clasts). Moving water carries away lighter clasts, but can't move the heavier metal flakes or nuggets as easily, so they concentrate in gravel. Such concentrations constitute *placer deposits* (Fig. 11.28).

FIGURE 11.25 Native metals, such as gold, look much the same in their unprocessed form as they do in familiar objects, like gold bracelets from a Kuwaiti jewelry market.

Gold embedded in quartz

FIGURE 11.26 Examples of ore minerals.

(a) This lead ore from Missouri contains galena (PbS) crystals that grew in dolostone.

(b) Most copper comes from ore minerals that look nothing like metallic copper. This rock has a coating of malachite, a copper carbonate mineral.

BOX 11.3

How can I explain . . .

The economics of mining

What are we learning?

That a lot of waste rock must be produced to obtain a relatively small amount of ore, and that mining a higher-grade ore deposit can be more profitable than a lower-grade deposit.

What you need:

- 4 cups of dry rice
- A tablespoon full of M&M's candies
- 2 large bowls
- 1 small bowl
- A 1-cup measuring cup

Instructions:

- Pour the rice into one of the large bowls. Add the candy and mix thoroughly.
- Scoop out the rice-plus-candy mixture, one cup at a time.
- Sort through each cup and find the candies. Place the candies in the small bowl and the candy-free rice in the other large bowl. Time how long it takes to recover the candy from each cup.
- Repeat the experiment, but this time place the candy in the center of the rice-filled bowl.
- Then, knowing where the candy is, scoop a cup or two from that part of the bowl, and time how long it takes to recover the candy.

What did we see?

- When the candy was dispersed throughout the rice, you could recover very little in a given time. Therefore, you would have to sift through lots of rice to get the candy. If the time you take scooping and sorting represents the time it takes to mine valuable ore minerals, you are spending a lot of time (which equals money) to get the ore minerals.
- Of course, it's much easier to extract the candy if it's concentrated and if you know where it is. This situation represents the discovery of a high-grade ore deposit. Mining such a deposit is much more efficient.

 Video

Formation of Ore Deposits

All the gold which is under or upon the Earth is not enough to give in exchange for virtue.

—PLATO (GREEK PHILOSOPHER, CA. 428–347 B.C.E.)

Residual deposits: Heavy rains in tropical climates leach soluble minerals from the soils, leaving a concentration of insoluble minerals (see Chapter 5). For example, if the soil formed from a rock that originally contained aluminum, insoluble aluminum oxide minerals remain in the zone of leaching to form a *residual deposit.* An aluminum-bearing residual deposit is called *bauxite.*

Where Are Ore Deposits Found?

The Inca Empire of 15th-century Peru boasted elaborate temples decorated with statues and masks of gold. Then, around 1532, Spanish conquistadors arrived and defeated the Incas. Spanish ships began transporting golden treasure back to Spain. Why did the Incas possess so much gold? Or, to ask the broader question, what geologic factors control the distribution of ore?

Several of the ore-deposit types mentioned above occur in association with igneous rocks. As we learned earlier in this book, igneous activity takes place primarily in the volcanic arcs of convergent boundaries, along mid-ocean ridges, in continental rifts, and at hot spots. Therefore, magmatic and hydrothermal deposits, as well as secondary-enrichment deposits and placer deposits

derived from them, develop in these geologic settings. Inca gold, for example, came from placer deposits eroded from hydrothermal ores formed in the Andean continental arc. Copper ore and many other metal ores also form in association with hydrothermal activity. Most iron ore, however, comes from banded iron formation deposited between about 2.4 and 1.8 Ga (see Chapter 10). During this time interval, oxygen accumulated in the atmosphere and dissolved in the sea. Dissolved iron in seawater bonded to the oxygen to form iron oxide minerals, which then settled to the seafloor.

Ore Exploration and Production

Imagine prospectors of days past clanking through the wilderness with worn-out donkeys, searching for ore. What, specifically, were those prospectors hoping to see? In some cases, they scanned hillsides for outcrops of milky-white quartz veins that might contain native metals, or for brightly colored stains caused by oxidation (rusting) of ore minerals. Sometimes they panned streambed gravels, hoping to find placer gold. On finding a possible ore deposit, a prospector would take a sample back to town for an *assay,* a test to determine how much extractable metal the deposit contained. If the

FIGURE 11.27 Some processes that form ore deposits.

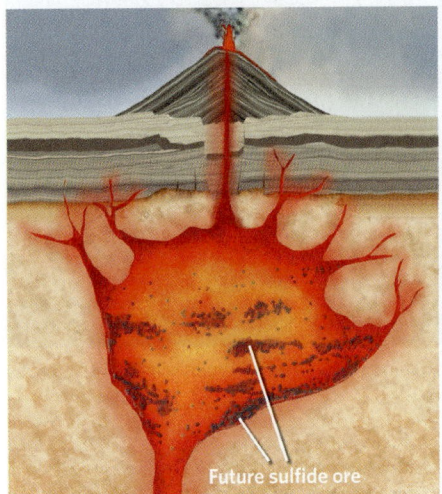

(a) Magmatic ore deposits can form in a magma chamber when ore minerals crystallize.

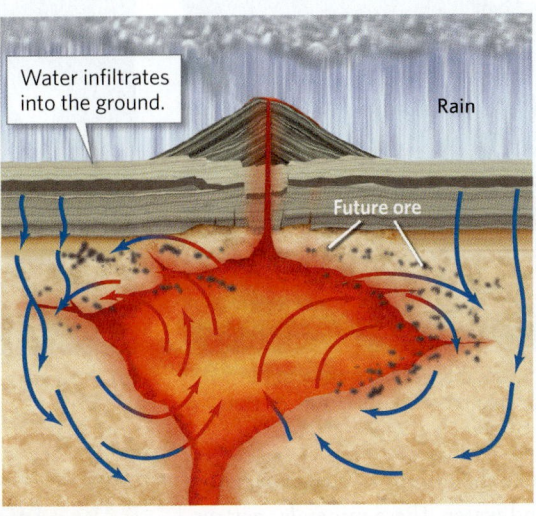

(b) Hydrothermal deposits form when water circulating around and through magma dissolves metals, which then precipitate out of solution elsewhere. (Arrows indicate flowing water.)

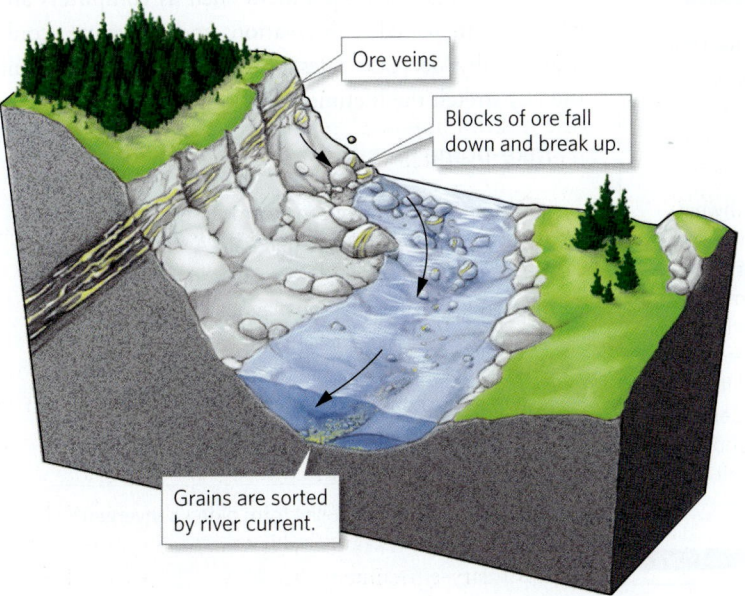

(c) Seafloor massive sulfide deposits form when sulfides precipitate around hydrothermal vents (black smokers) along a mid-ocean ridge.

assay indicated a significant concentration of metal, the prospector might "stake a claim" by literally marking off an area of land with stakes.

These days, commercial mining companies employ geologists to survey potential ore-bearing regions. During surveys, geologists may measure the strength of the local gravitational or magnetic field, because ore minerals tend to be denser and more magnetic than average rocks. They may also sample rocks, soils, and plants to test for metal concentrations. If surface observations hint at ore below, geologists then drill to sample and analyze subsurface rock.

If calculations indicate that mining an ore deposit can yield a profit while accommodating environmental concerns, a company builds a mine. To develop an **open-pit mine (Fig. 11.29a)**, workers use explosives to shatter rock. Front-end loaders dump the rock into giant trucks that can carry up to 27,000 kg (300 tons) of rock in a single load. The trucks dump waste rock into a *tailings pile*, an artificial hill composed of waste rock, and place ore in a crusher that smashes the ore into small fragments. Smelting these fragments separates metal atoms from slag.

To reach ore deposits that lie more than about 100 m (about 330 ft) below the Earth's surface, miners construct an underground mine **(Fig. 11.29b)**. To do so, they either bore a tunnel into the side of a mountain or sink a vertical shaft. At the level in the crust where the ore deposit appears, they excavate a network of tunnels into the *ore body* (the volume of rock containing an economically minable quantity of ore).

Mining and the Environment

Mining activity can leave a big environmental footprint (**Earth Science at a Glance**, pp. 338–339). Some of the

gaping basins that result from open-pit mining are so big that astronauts can see them from space. And both open-pit and underground mining yield immense tailings piles which, without soil, may remain unvegetated for decades (see chapter opening photo). In some places, mining companies douse tailings with acidic solutions to leach out additional metal ions. If these acids escape into the environment, they can damage vegetation and groundwater.

FIGURE 11.28 Placer deposits form where erosion produces clasts of native metals. Sorting by flowing water concentrates the metals.

FIGURE 11.29 Mining ore deposits.

(a) Open-pit mining extracts ore that lies fairly close to the ground surface. The steps cut into the wall help ensure its stability.

(b) An underground mine chiseled into a mountainside in Colorado.

Mining metal ores, like the mining of coal, can expose sulfide minerals to air and water. These minerals may dissolve to produce acid mine runoff, which can kill vegetation downstream (Fig. 11.30). Similarly, the smoke from ore smelting may contain harmful chemicals, including sulfur, which dissolves in water to produce acid rain. Our need for metals continues to grow, so environmental issues associated with their extraction and use will probably continue to challenge future generations.

How Long Will Metal Resources Last?

Most metallic mineral resources are nonrenewable. Geologists have calculated reserves for various minerals, just as they have for fossil fuels, and they are concerned that supplies of some metals may run out in only decades to centuries. Further, because worldwide distribution of metallic mineral reserves varies, not all countries have equal access to mineral supplies. This issue has gained great importance for national security because some minerals—known as *strategic minerals*—are essential for the production of high-tech equipment such as computers and batteries. Increased conservation and recycling could dramatically decrease rates of metal consumption and thereby stretch the lifetime of existing reserves.

> **Take-home** message ...
>
> Ores—rocks that can be processed to produce metals—contain ore minerals. A variety of geologic processes produce ores. Geologists can find ores by measuring gravitational and magnetic fields, studying outcrops, and drilling for samples. Mining takes place either in open-pit mines or underground. Most mineral resources are nonrenewable and are not distributed uniformly around the planet. But some are recyclable.
>
> **Quick Question** ⎯⎯⎯⎯⎯⎯⎯⎯⎯
>
> Why did many ore deposits form along convergent boundaries?

11.8 Nonmetallic Mineral Resources

Consider the materials you can see in a typical house or apartment—concrete, brick, glass, wallboard. Society uses many such **nonmetallic mineral resources**. Where do they come from?

Dimension Stone

The Parthenon, a colossal stone temple, has stood atop a hill in Athens, Greece, for almost 2,500 years. No wonder: **dimension stone** (or just *stone*—an architect's word for rock) outlasts nearly all other construction materials. The names that contractors give to various types of stone may differ from the formal rock names that geologists use. For example, contractors generally refer to any polished carbonate rock as "marble," whether or not it has been metamorphosed, and to any crystalline rock containing feldspar or quartz as "granite," regardless of whether the rock has an igneous or metamorphic texture, or a felsic or mafic composition. To obtain intact slabs or blocks of stone from a quarry, workers either split the stone from bedrock by hammering a series of wedges into it, causing a crack to propagate, or they slice it off the bedrock wall by using various power tools (Fig. 11.31a). Once workers remove blocks of dimension stone from a quarry, the blocks can be cut into smaller pieces (Fig. 11.31b). Rubbing the blocks with abrasive and water creates a shiny polish.

Crushed Stone and Concrete

Crushed stone forms the foundation of highways and railroads and serves as the raw material for manufacturing cement, concrete, and asphalt. In crushed-stone quarries (Fig. 11.32), operators use explosives to break up bedrock into small chunks, which they then transport by truck to a crusher.

Many buildings constructed in the past century have concrete floors, columns, or walls. To make concrete, workers mix cement with *aggregate* (sand and/or gravel)

Did you ever wonder ...

how concrete differs from rock?

See for yourself

Bingham Copper Mine, Utah

Latitude: 40°31′14.66″ N
Longitude: 112°9′1.97″ W

Zoom to 20 km (~12 mi) and look straight down.

The gray patch southwest of Salt Lake City is the largest open-pit mine in the world. Over 17 million tons of copper have been extracted from ore formed when hydrothermal fluids circulated through Cenozoic igneous rock. Zoom in closer to see the pit and tailings pile.

FIGURE 11.30 Acid mine runoff. The orange color is due to iron oxide staining and to sulfide-eating bacteria in the water.

FIGURE 11.31 Stone production in quarries.

(a) An active quarrying operation in Missouri that produces large blocks of cut dimension stone.

(b) Sheets of cut dimension stone being measured for further cutting to become a kitchen countertop.

and water to produce a slurry. When workers pour this wet concrete into molds and let it *set*, it hardens into a solid rock-like material as the chemicals in the cement react and precipitate to produce an interlocking assemblage of new mineral crystals. These crystals bind the grains of the aggregate together. The cement used in concrete consists mostly of lime (CaO), with lesser amounts of silica, aluminum oxide, and iron oxide. In the 18th and early 19th centuries, workers produced cement simply by placing chunks of a special type of limestone—one that contained some quartz and clay in addition to calcite—in a kiln. When heated to a temperature of about 1,450°C (2,640°F), the calcite in the limestone breaks down to form lime and carbon dioxide gas; the clay and quartz provide the other oxides for the cement. The specific type of limestone needed to make such *natural cement* is fairly rare, so today most concrete structures employ *Portland cement*, a mixture of limestone, sandstone, and shale in just the right proportions to provide the proper mix of chemicals to make cement.

Nonmetallic Minerals in Your Home

You'll find a great variety of nonmetallic Earth resources in a building. As we've seen, the concrete in a building's foundation comes from baked limestone, mixed with sand or gravel and water. The *bricks* used in walls are made from clay produced by the chemical weathering of silicate minerals. To make bricks, workers mold wet clay into blocks and then bake them in order to drive out water and cause metamorphic reactions that lead to the growth of stronger minerals. Window glass is made from molten

quartz sand—freezing the melt quickly solidifies it without forming crystals. *Drywall*, the solid sheets used for interior walls, can be made by mixing crushed gypsum, an evaporite mineral, with water to form a slurry, which is then spread into a thin layer. New crystals grow from the slurry, making the layer into a solid board. (Evaporite deposits serve as the source of many chemicals used in daily life, including the lithium used in rechargeable batteries.)

Take-home message ...

Society uses a great variety of nonmetallic mineral resources. These materials include dimension stone, crushed stone, cement (made from baked limestone), evaporites (including gypsum), and clay (used to make bricks).

Quick Question
What's the difference between natural cement and Portland cement?

FIGURE 11.32 A large crushed-stone quarry of Silurian limestone in Illinois. Drillers are working on the shelf in the distance.

Forming and Processing Earth's Mineral Resources

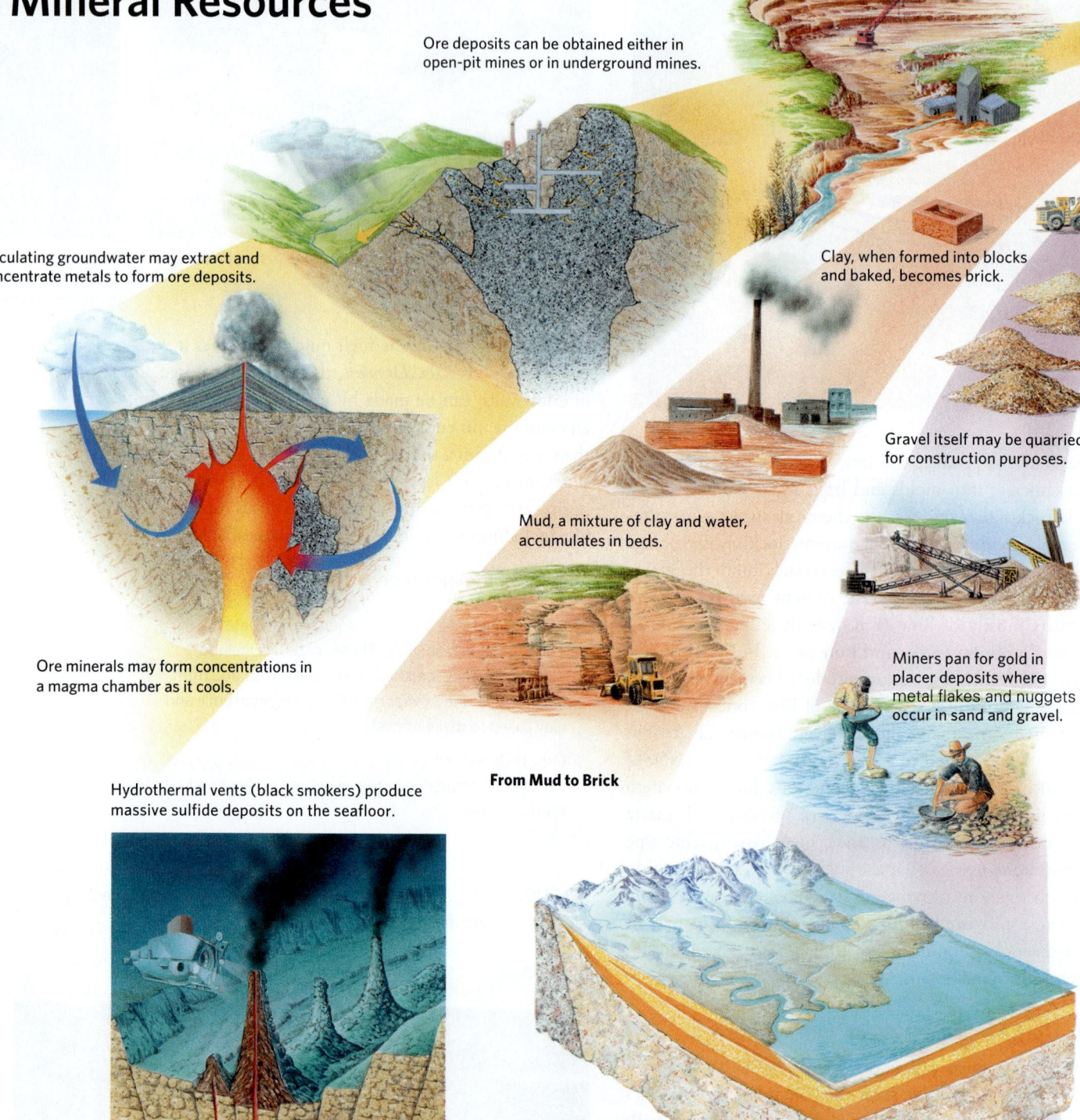

Mining and processing ore has environmental consequences, including acid runoff, acid rain, and groundwater contamination.

Ore deposits can be obtained either in open-pit mines or in underground mines.

Circulating groundwater may extract and concentrate metals to form ore deposits.

Clay, when formed into blocks and baked, becomes brick.

Gravel itself may be quarried for construction purposes.

Ore minerals may form concentrations in a magma chamber as it cools.

Mud, a mixture of clay and water, accumulates in beds.

Miners pan for gold in placer deposits where metal flakes and nuggets occur in sand and gravel.

Hydrothermal vents (black smokers) produce massive sulfide deposits on the seafloor.

From Mud to Brick

Erosion tears down mountains and produces gravel and sand.

From Magma to Metal

From Stream Channel to Roadbed

Earth materials are the substances from which cities grow, but their use has environmental consequences.

A mixture of lime, other elements, sand, and water, when allowed to harden, becomes concrete.

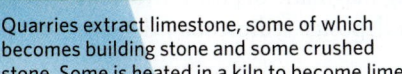

Mixed with water, spread into sheets, and wrapped in paper, gypsum makes drywall.

In quarries, operators dig up gypsum, crush it to powder, and ship it to factories.

Quarries extract limestone, some of which becomes building stone and some crushed stone. Some is heated in a kiln to become lime.

Gypsum is a salt that precipitates when saline lakes evaporate. It grows as white or clear crystals.

From Lake Bed to Drywall

Over millions of years, shells and shell fragments collect and eventually form beds of limestone.

The raw materials from which we manufacture the buildings, roads, wires, and coins of modern society were produced by geologic processes. For example, ore deposits—the concentrations of minerals that are a source of metal—formed during a variety of magmatic or sedimentary processes. Limestone, a rock used for buildings and for making concrete, began as an accumulation of seashells. Brick began as clay, a by-product of chemical weathering. And the gypsum of drywall began as an accumulation of salt in a desert lake. Metal, gravel, lime, and gypsum are all examples of the Earth's mineral resources. We can use some mineral resources right from the Earth, simply by digging them out. But most become usable only after expensive processing.

Organisms extract calcium and carbonate ions from seawater to construct calcite shells.

From Seafloor to Sidewalk

11 CHAPTER REVIEW

Chapter Summary

- Photosynthesis stores energy from the Sun in the bonds within organic chemicals of biomass. Combustion releases this energy. Fossil fuels are biomass that was buried and preserved underground.

- Oil and gas are hydrocarbons formed from the remains of plankton, which become incorporated in organic shale, a source rock. Chemical reactions at elevated temperatures underground convert organic matter to kerogen and then to oil or gas.

- To form a conventional oil reserve, oil must migrate from a source rock into a reservoir rock and must then be confined underground by a trap.

- Substantial volumes of hydrocarbons exist in unconventional reserves (tight oil, shale gas, oil shale, and tar sand). These hydrocarbons are difficult to extract, either because they have high viscosity or because the rock holding them has low permeability.

- Obtaining hydrocarbons from tight oil and shale gas reserves typically involves directional drilling and hydrofracturing.

- For coal to form, abundant plant debris must be deposited in an oxygen-poor environment. Compaction of plant debris produces peat, which, when buried deeply and heated, transforms into coal. Coal occurs in beds and can be mined by either strip mining or underground mining.

- Geologists classify coal as lignite, bituminous, or anthracite based on its carbon content. Higher-rank coal forms at higher temperatures and contains more carbon.

- We now live in the Oil Age. Conventional oil reserves may last for only another century, but unconventional reserves, and coal, could last longer. Our use of fossil fuels has significant environmental consequences.

- Nuclear power plants generate electricity using the energy released by the fission of radioactive uranium or plutonium. The rate of fission reactions must be carefully controlled to avoid overheating or meltdown. Radioactive nuclear waste is difficult to store.

- Geothermal power plants extract the Earth's internal heat from groundwater heated by igneous activity. Hydroelectric and wind power use the energy of flowing water and air, respectively. Photovoltaic cells convert sunlight to electricity.

- Some metals occur in native form, but most occur in ore minerals. An ore is a rock containing native metals or ore minerals in sufficient quantities to be worth mining.

- Ore deposits form in a variety of ways. Geologists distinguish among types, including magmatic deposits, hydrothermal deposits, secondary-enrichment deposits, sedimentary deposits, and placer deposits.

- Metallic mineral resources are nonrenewable. Many are now or may soon be in short supply.

- Nonmetallic mineral resources include dimension stone, crushed stone, clay, sand, and many other materials. A large proportion of the materials in your home (such as concrete, brick, glass, and wallboard) have a geologic ancestry.

- To produce concrete, workers mix aggregate (sand and gravel) with cement. Cement is made from limestone, mixed with lesser amounts of shale and sandstone, heated in a kiln. When dissolved in water and left to set, it solidifies due to precipitation of new minerals.

Key Terms

acid mine runoff (p. 329)
acid rain (p. 329)
biofuel (p. 331)
biodiesel (p. 331)
biomass (p. 315)
blowout (p. 329)
carbon footprint (p. 330)
chain reaction (p. 330)
coal (p. 325)
coal rank (p. 326)
coal seam (p. 325)
conventional reserve (p. 318)

dimension stone (p. 336)
directional drilling (p. 319)
Earth resource (p. 315)
enrichment (p. 331)
fossil fuel (p. 316)
fuel (p. 315)
geothermal energy (p. 331)
grade (p. 333)
hydrocarbon (p. 317)
hydrocarbon reserve (p. 318)
hydroelectric power (p. 331)
hydrofracturing (p. 321)

kerogen (p. 318)
metal (p. 332)
natural gas (p. 317)
nonmetallic mineral resource (p. 336)
nonrenewable resource (p. 328)
nuclear reactor (p. 330)
nuclear waste (p. 331)
oil (p. 317)
Oil Age (p. 328)
oil seep (p. 318)
oil shale (p. 319)

oil window (p. 318)
open-pit mine (p. 335)
ore (p. 333)
ore deposit (p. 333)
ore mineral (p. 333)
peat (p. 325)
permeability (p. 318)
photosynthesis (p. 315)
photovoltaic cell (p. 332)
porosity (p. 318)
renewable resource (p. 328)
reservoir rock (p. 318)

Review Questions

Letters in parentheses correspond to the chapter's learning objectives.

1. What is the source of the organic material in oil, and how is it transformed into oil? **(A)**

2. What is the oil window, and what happens to oil at temperatures higher than the oil window? **(A)**

3. Explain the difference between a conventional and an unconventional hydrocarbon reserve. **(A)**

4. How are hydrocarbons trapped to yield a conventional reserve? Which type of trap does the drawing show? **(A)**

5. What are tight oil and shale gas reserves, and how can hydrocarbons be extracted from them? **(A)**

6. What are tar sand and oil shale reserves, and how can oil be extracted from them? **(A)**

7. Where are most of the world's conventional oil reserves found? Are most unconventional oil reserves found in the same places? **(A)**

8. How is coal formed, and in what class of rocks is coal considered to be? **(B)**

9. What is the difference between a high-rank and a low-rank coal? What are the names of the types of coal in the different ranks? **(B)**

10. Is oil a renewable or a nonrenewable resource? What is the likely future of fossil fuel production and use in the 21st century? **(A, H)**

11. Describe how a nuclear power plant produces electricity. What is a nuclear meltdown? **(C)**

12. What is geothermal energy? What geologic factors limit its use? **(D)**

13. Are there any drawbacks involved in using hydroelectric power? **(D)**

14. Why don't we use ordinary granite as a source for metals? **(E)**

15. Describe some types of ore deposits and how they form. Which type of ore deposit does the drawing show? **(E)**

16. What procedures are used to locate and mine metallic mineral resources today? **(E)**

17. What are some environmental hazards of mining for metals? **(E)**

18. Name some materials in your home that come from nonmetallic mineral resources. **(F)**

19. How is stone cut from a quarry? **(F)**

20. What are the ingredients of concrete, and how are these substances produced? **(F)**

21. Will the supply of mineral resources run out? **(G)**

On Further Thought

22. Do you think it would make sense for an energy company to drill for oil in a locality where beds of anthracite occur in the stratigraphic sequence? Explain your answer. **(A, B)**

23. A body of rock in Arizona has the following characteristics: One portion of the rock is intrusive igneous rock in which tiny grains of copper sulfide minerals are dispersed among the other minerals of the rock. Another nearby portion of the rock consists of limestone in which malachite (a copper carbonate mineral that grows from the interaction of rock with groundwater) fills cavities and pores in the rock. What types of ore deposits are these? Describe the geologic history that led to the formation of these deposits. **(E)**

12 SHAPING THE EARTH'S SURFACE
Landscapes, the Hydrologic Cycle, and Mass Wasting

By the end of this chapter you should be able to . . .

A. explain why the Earth's land surface is not static, like the Moon's, but rather changes constantly over time.

B. differentiate between uplift and subsidence, internal and external energy, and erosion and deposition, and describe their effects on landscapes.

C. sketch a diagram showing processes involved in the hydrologic cycle of the Earth System.

D. describe the characteristics and consequences of different types of mass wasting.

E. explain the concept of a failure surface, and identify factors affecting slope stability.

F. use your knowledge of conditions that lead to mass wasting to assess an area's susceptibility to mass-wasting hazards and to evaluate preventive measures.

12.1 Introduction

A dinosaur, gazing at the Moon, would see the same view that we see today. With the exception of a few new craters here and there, much of the Moon's solid surface has remained nearly unchanged for a few billion years. An observer looking at the Earth, however, would see features of the land surface change radically over time. Some of these changes happen over thousands to millions of years, whereas others happen in seconds to days.

Why do landscapes on the Moon persist while those on the Earth evolve? The Moon's surface remains static because the Moon's outer layer does not consist of moving plates, and because the Moon has no atmosphere, hydrosphere, or biosphere. Therefore, only meteorite impacts or *space weathering* (the breakdown of minerals due to the impact of cosmic rays) alter the Moon's surface. The Earth's surface, in contrast, remains dynamic because the Earth's outer layer does consist of moving plates, and because our planet hosts an atmosphere, hydrosphere, and biosphere. Interactions of lithosphere plates with one another and with the asthenosphere below cause earthquakes, volcanism, mountain building, and basin formation. Movements associated with these phenomena generate slopes down which rocks and sediments can tumble or slide. Meanwhile, chemical and physical weathering, the flow of streams and glaciers, waves, wind, and the activities of life in the Earth System constantly break down and redistribute surface materials.

Because our planet's land surface continues to be dynamic, it hosts a great variety of landscapes. By **landscape**, we mean the character and shape of the land surface in a region. Artists and writers across the centuries have gazed at landscapes for inspiration, for landscapes spark the full range of human emotion (**Fig. 12.1**). Earth scientists similarly feel inspired when they see a landscape, but they can't help wondering, "How did this landscape come to be? How will it change in the future? Do features of this landscape pose a threat to life and property?" This chapter begins to address these questions by describing the nature of changes in land-surface elevation and the energy sources that drive these changes; the *hydrologic cycle*, during which water molecules move from ocean to air to land and back to ocean; and *mass wasting*, during which gravity transports rock and sediment down slopes.

FIGURE 12.1 Examples of the great variety of landscapes on Earth.

(a) Rounded mountains border the beach in Rio de Janeiro, Brazil.

(b) Glaciers carved these rugged peaks of the Alps in France.

(c) A rainforest hides the Amazon River in Peru.

(d) Buttes of sandstone tower above Monument Valley, Arizona.

(e) Cliffs rise from the forest in the Blue Mountains, Australia.

(f) The flat plains of the midwestern United States show a checkerboard of farm fields.

<< Layers of sedimentary rock, once buried deeply underground, now lie exposed on a cliff in Utah due to erosion, which continues to this day. Huge blocks of tan sandstone have tumbled downslope during rockfalls. Landscapes change over geologic time.

BOX 12.1 ▶ Science Toolbox

Characterizing topography on a map

For many applications, it's useful to portray the shape of the land surface on a **topographic map**, which uses contour lines to represent variations in elevation **(Fig. Bx12.1a)**. A **contour line** is an imaginary line on the land surface along which all points have the same elevation. You can picture a contour line as the intersection between the land surface and an imaginary horizontal plane **(Fig. Bx12.1b)**. For example, all points along the 100-m contour line lie at an elevation of 100 m above sea level. The elevation difference between two adjacent contour lines on a topographic map is called the *contour interval*. On a given topographic map, the contour interval is constant. So, for example, if the contour interval is 50 m, the next contour line above the 100-m contour is the 150-m contour, and the one above that is the 200-m contour, and so on. If you walk parallel to a contour line, you stay at the same elevation, but if you walk perpendicular to a contour line, you go upslope (if the next contour you cross is a higher number) or downslope (if the next contour you cross is a lower number). You can picture a slope's angle from the spacing between contour lines. Specifically, closely spaced contour lines represent a steep slope (where you would cross several contour lines while moving a short horizontal distance on the map), whereas widely spaced contour lines represent a gentle slope.

FIGURE Bx12.1 Topographic maps and profiles.

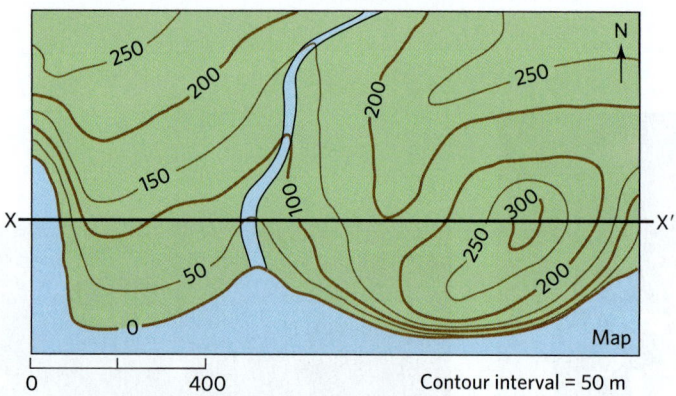

(a) A topographic map depicts the shape of the land surface through the use of contour lines. The difference in elevation between two adjacent lines is the contour interval.

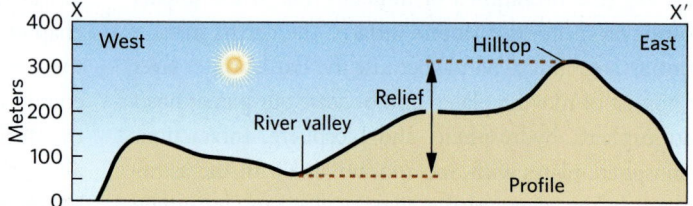

(c) A topographic profile (along section line X–X') shows the shape of the land surface as seen in a vertical slice.

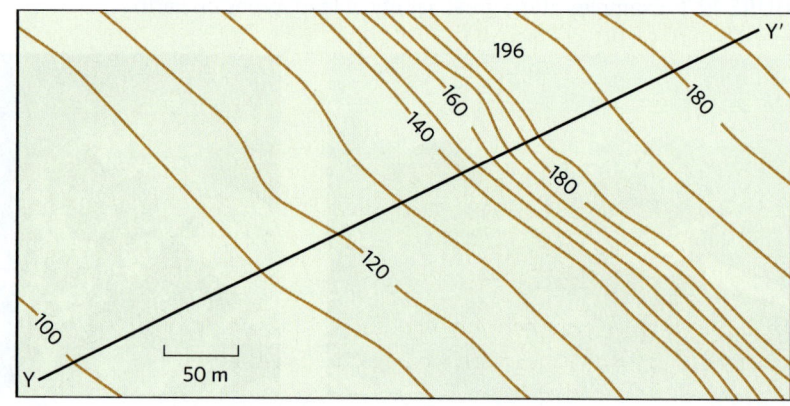

(d) This topographic map shows a distinct cliff.

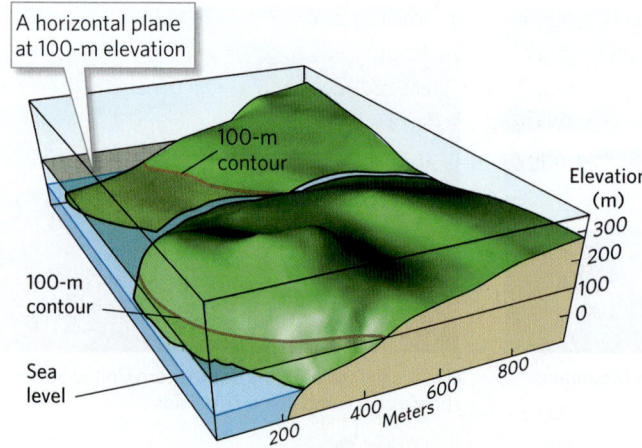

(b) A contour line represents the intersection of a horizontal plane with the land surface. This block diagram shows the area mapped in part (a).

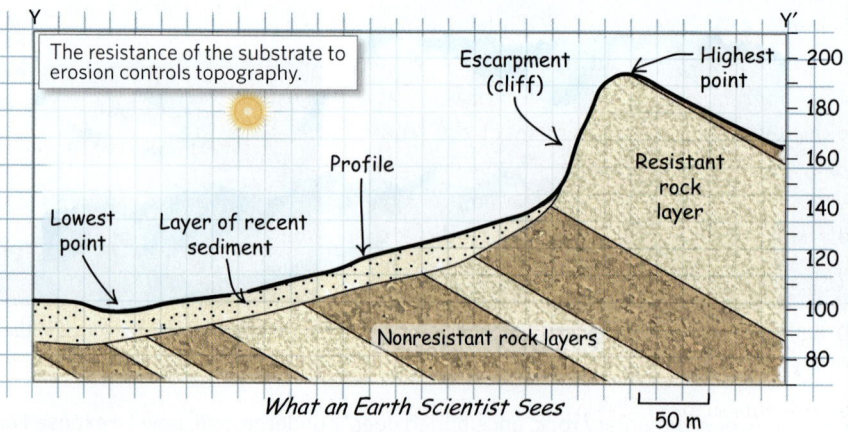

What an Earth Scientist Sees

(e) A geologic cross section depicts an Earth scientist's interpretation of the subsurface along the section line Y–Y' from part (d). The cliff is the edge of a resistant rock layer.

To represent variations in elevation in a given direction, you can sketch a **topographic profile**, the trace of the ground surface as it would appear on a vertical plane that slices into the ground **(Fig. Bx12.1c)**. Put another way, a topographic profile represents the shape of the ground surface as viewed from the side. A topographic profile between two points on a topographic map can help you visualize how the land goes up and/or down between the points.

By combining a topographic profile with a representation of geologic features under the ground, geologists produce a **geologic cross section**. In some cases, geologists can gain insight into subsurface geology by looking at the shape of a landform **(Fig. Bx12.1d, e)**. For example, a steep cliff in a region of sedimentary strata may indicate the presence of beds that are resistant to weathering, whereas low areas or gently sloping areas may be underlain by nonresistant beds.

In recent years, Earth scientists have developed methods for representing a landscape by using a **digital elevation model (DEM)**. Computers construct a DEM from a set of data in which each location on a map has

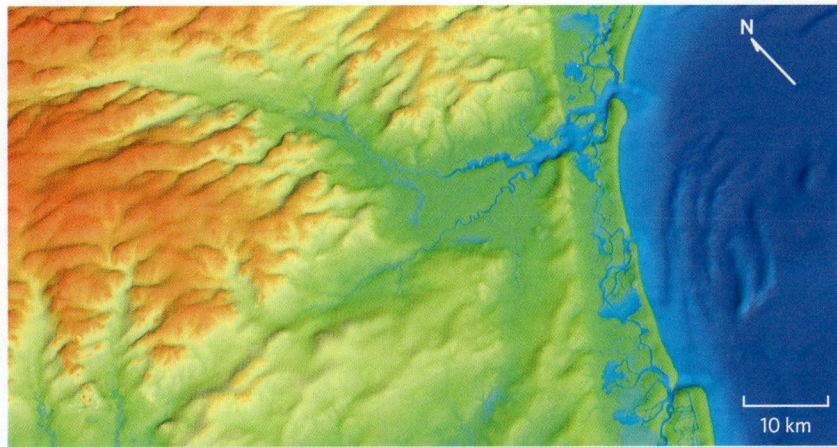

(f) A digital elevation model of the region southwest of Atlantic City, NJ, reveals details of the landscape. Blue represents water; green, lower elevations; orange, higher elevations. All the land shown lies within 20 m of sea level, so the topography is subtle.

three coordinates: latitude, longitude, and elevation. The shape of the land on a DEM, or on a topographic map, can be highlighted by adding shadows to simulate the appearance of the land if it were lit by the Sun when it's low in the sky **(Fig. Bx12.1f)**. The resulting image can also be called a **shaded-relief map**.

12.2 The Earth's Ever-Changing Surface

Generation of Relief

If the Earth's surface were perfectly flat, the great diversity of landscapes that we can see would not exist. But the land surface isn't flat, because moving lithosphere plates interact with one another and with the underlying asthenosphere. These interactions lead to subduction, continental collisions, rifting, and volcanism, and these phenomena cause portions of the land surface to move up or down relative to adjacent regions. We refer to the upward movement of the land surface as **uplift** and to the sinking or downward movement of the land surface as **subsidence (Fig. 12.2)**. Uplift or subsidence of one location relative to a neighboring location generates **relief**—a difference in elevation—and where relief exists, slopes form. Variations in elevation within a region define the shape of the region's landscape. Geologists use the term *topography* to refer to such variations **(Box 12.1)**.

When uplift or subsidence generates relief, other components of the Earth System kick into action. Over time, bedrock undergoes physical and chemical weathering, which causes it to disintegrate into pieces that

undergo mass wasting. During and after weathering and mass wasting, moving water, ice, and air can cause **erosion**, the grinding away and removal of material at the Earth's surface. As we'll see in succeeding chapters, erosion by rivers and glaciers can carve into the land surface and generate steep local relief. Materials or processes that cause erosion are known as *agents of erosion*. Flowing water, ice, or air can transport eroded materials to locations where **deposition**—the settling of sediment—takes place. Overall, mass wasting, erosion, and deposition, acting together, redistribute rock and sediment, ultimately stripping it from higher areas and collecting it in lower areas. Mass wasting and erosion can lower a landscape's elevation, while deposition can cause a landscape's surface to rise.

Water flows humbly to the lowest level. In the world, nothing is more submissive or weak than water, yet for attacking what is hard and strong, nothing can surpass it.

—LAO-TZU (*CHINESE PHILOSOPHER, 604–531 B.C.E.*)

Did you ever wonder . . .

how fast the land surface rises or sinks, on average?

FIGURE 12.2 Uplift raises hills, and subsidence forms basins. These processes generate slopes.

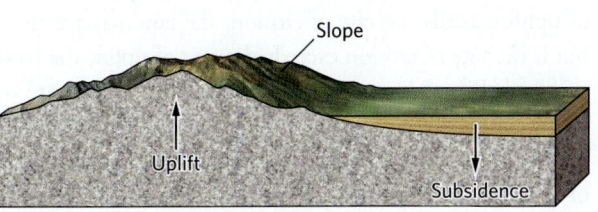

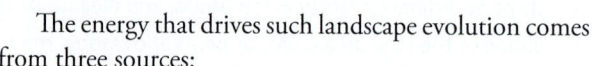

(a) Uplifted beach terraces form where the coast is rising relative to sea level. Here, present-day wave erosion is forming a new terrace and cutting a cliff on the edge of the old one.

(b) So much erosion can take place during a single hurricane that houses built along the beach become undermined.

The energy that drives such landscape evolution comes from three sources:

1. **Internal energy**, the heat from within the Earth, keeps the mantle soft enough to flow, which allows plate movements to take place. These movements, along with the activity of mantle plumes, drive uplift and subsidence.

2. **External energy**, the radiation that comes to the Earth from the Sun, causes air and water near the Earth's surface to become warmer. In the Earth's gravitational field, air circulation produces the wind. Wind can erode and transport sediment, and it can generate water waves. External energy also causes water on the Earth's surface to evaporate. This water, when carried over the land by wind, precipitates as rain or snow, filling rivers and building glaciers.

3. **Gravitational energy**, the potential energy of an object due to the downward pull of gravity, drives mass wasting and, as we have seen, works in concert with other energy sources to drive convective movement in the mantle, oceans, and atmosphere.

Overall, we can think of landscape evolution as a "battle" between what geologists call *tectonic processes* (collision, convergence, rifting, and basin formation) driven by internal energy, which build relief by moving the land surface up or down, and *surface processes* (mass wasting, erosion, and deposition), driven by external energy and gravity, which destroy relief by removing material from high areas and depositing it in low areas. If, in a particular region, the rate of uplift exceeds the rate of erosion, the land surface rises, but if the rate of erosion exceeds the rate of uplift, the land surface becomes lower with time. Similarly, if the rate of subsidence exceeds the rate of deposition, the land surface becomes lower, but if the rate of deposition exceeds the rate of subsidence, the land surface rises.

How rapidly do vertical movements of the Earth's surface take place? Though the Earth's surface can rise or sink by as much as 3 m (10 ft) during a single major earthquake, rates of vertical surface movement, when averaged over time, vary between 0.1 and 10 mm (0.004 and 0.4 in) per year **(Fig. 12.3a)**. Similarly, erosion during a single storm or mass-wasting event can carve tens of meters from the land **(Fig. 12.3b)**, and deposition by a single mass-wasting event can produce a layer of debris tens of meters thick in a matter of minutes to days. However, when averaged over time, erosion and deposition change the land-surface elevation at rates between 0.1 and 10 mm per year. Although these rates seem slow, a change in surface elevation of just 0.5 mm (0.02 in, the thickness of a fingernail) per year can yield a net change of 5 km (3 mi) in 10 million years. Uplift can build a mountain range, and erosion can whittle one down to near sea level—it just takes time!

Controls on Landscape Evolution

Imagine traveling across a continent. On your journey, you see a variety of **landforms** (natural, recognizable features of a landscape), including plains, swamps, hills, valleys, mesas, and mountains. Some of these are **erosional landforms** that result from the breakdown and removal of rock or sediment and develop where agents of erosion carve into the land. Others are **depositional landforms** that result from the deposition of sediment. The specific landforms that develop at a given locality reflect several factors:

- *Eroding or transporting agent:* Water, wind, and ice all cause erosion and all transport sediment. But the shapes of the landforms formed by each are different because of differences in the abilities of each to carve into the land and to carry away debris. Of the three agents, water has the greatest effect on a global basis.

FIGURE 12.4 Human influences on a geologic scale.

(a) The pyramids of Egypt are human-made hills that rise above the desert sands. They have lasted for thousands of years.

(b) In the process of highway construction, deep valleys are cut through high ridges. This example borders a highway near Denver.

(c) This stone dam holds back a reservoir in Colorado.

- *Relief:* The elevation difference, or relief, between adjacent places in a landscape determines the height and steepness of a slope. Steepness, in turn, controls the velocity of ice or water flow down the slope, and it determines whether rock or soil will stay in place on the slope or slip downslope.

- *Climate:* The average temperature, seasonal temperature variation, the overall volume of precipitation in a year, and the distribution of precipitation over time define the *climate* of a region. Climate determines whether running water, flowing ice, or wind serves as the main agent of erosion or deposition, and it affects the lushness of vegetation, which can influence the behavior of slopes.

- *Substrate composition:* The material that makes up the land determines how the **substrate**, the material at and just below the Earth's surface, responds to erosion. For example, strong rocks can stand up to form steep cliffs, whereas soft sediment collapses to generate gentle slopes.

- *Living organisms:* Plants and animals can weaken the substrate (by burrowing, wedging, or digesting) or hold it together (by binding it with roots).

- *Time:* Landscapes evolve through time in response to uplift, subsidence, erosion, and deposition. For instance, a gully that has just started to form in response to the flow of a stream does not look the same as the deep canyon that develops after the same stream has existed for a long time.

For most of geologic time, water, wind, and ice generated most landscapes. During the past few centuries, however, human activities have had an increasingly important impact on the Earth's surface. We have replaced mountains with deep basins (open-pit mines), have built hills (tailings piles and landfills) where once there were valleys, and have made steep slopes gentle and gentle slopes steep **(Fig. 12.4)**. By constructing concrete walls and by dumping piles of debris, we modify the shapes of coastlines, change the courses of rivers, and fill new lakes (reservoirs). In cities, buildings and pavement completely seal the ground, forcing water that might have infiltrated the ground to spill instead into a stream, increasing the stream's flow. In rural areas, agriculture, grazing, water use, and deforestation substantially alter the rates at which natural erosion and deposition take place. Humanity has become a major agent of erosion and deposition, and globally, humans have modified over half of our planet's land area.

Take-home message . . .

Land can undergo uplift or subsidence to yield relief. The energy driving landscape evolution comes from three sources: the Earth's internal energy, external energy from the Sun, and gravitational energy. The nature of a landscape depends on eroding or transporting agents, climate, time, relief, the activity of organisms, and substrate composition.

Quick Question ————————————
What happens to the elevation of the land surface if the rate of uplift exceeds the rate of erosion?

12.3 The Hydrologic Cycle

Did you ever wonder . . .
how long a molecule resides in the ocean, on average, before evaporating?

Because water in its various states—liquid, gas, and solid—plays such an important role in landscape evolution, any study of surface processes requires that we first consider how and why water moves around in the Earth System. Our planet's surface and near-surface water occupies several distinct *reservoirs*, a word used here to mean any realm or part of a realm in the Earth System **(Table 12.1)**.

Atmospheric water occurs as vapor, as tiny droplets or ice crystals in clouds, and as rain, snow, or hail that falls to Earth's surface (see Chapter 17). Surface water collects mostly in oceans, but also in lakes, streams, puddles, and swamps on land. Frozen water collects in snowfields and glaciers. Below the ground surface, some water dampens soil and rock near the surface, and some sinks deeper and fills underground *pores* (small open spaces) and cracks as **groundwater**. *Permafrost* exists where soil water and groundwater, extending down to depths of tens to hundreds of meters underground, remains frozen all year. As discussed in the Prelude, glaciers and permafrost make up the cryosphere. A significant amount of water also resides in the living organisms of the *biosphere*. In fact, about 50% to 65% of your body consists of water. Water constantly flows from reservoir to reservoir, driven by gravity and solar radiation. This

never-ending migration is called the **hydrologic cycle** (**Earth Science at a Glance**, pp. 350–351).

To get a clearer sense of how the hydrologic cycle operates, let's follow the fate of a water molecule that has just reached the surface of the ocean. Solar radiation heats the seawater, and the increased thermal energy of the vibrating water molecules allows them to evaporate and drift upward in a gaseous state to become part of the atmosphere. Atmospheric water vapor moves with the wind to higher altitudes, where it cools, undergoes condensation, and *precipitates* (falls out of the air) as rain or snow. About three-quarters of this water falls directly into the ocean. Of the remainder, which falls on the land surface, most becomes trapped temporarily in the soil or in living organisms and soon returns directly to the atmosphere by *evapotranspiration* (the sum of evaporation from bodies of surface water, evaporation from the ground surface, and release of water by plants and animals). Rainwater that does not become trapped in the soil or in living organisms enters lakes or rivers and ultimately flows back to the sea as surface water, or becomes trapped in glaciers, or sinks deeper into the ground to become groundwater. Groundwater flows through the substrate and ultimately returns to the Earth's surface reservoirs.

The average length of time that water stays in a particular reservoir during the hydrologic cycle is called the **residence time**. Water in different reservoirs has different residence times. For example, a typical molecule of water remains in the oceans for 4,000 years or less, in lakes and ponds for 10 years or less, in rivers for 2 weeks or less, and in the atmosphere for 10 days or less. Groundwater residence times depend in part on the path that the groundwater

TABLE 12.1 Major Water Reservoirs of the Earth

Reservoir	Volume (km³)	% of Total Water	% of Freshwater
Oceans and seas	1,338,000,000	96.5	—
Glaciers, ice caps, snowfields	24,064,000	2.05	68.7
Saline groundwater	12,870,000	0.76	—
Fresh groundwater	10,500,000	0.94	30.1
Permafrost	300,000	0.022	0.86
Freshwater lakes	91,000	0.007	0.26
Salt lakes	85,400	0.006	—
Soil moisture	16,500	0.001	0.05
Atmosphere	12,900	0.001	0.04
Wetlands	11,470	0.0008	0.03
Rivers and streams	2,120	0.0002	0.006
Living organisms	1,120	0.0001	0.003

Source: Data from P. H. Gleick, *Encyclopedia of Climate and Weather* (New York: Oxford University Press, 1996).

follows between the point where it enters the ground and the point where it exits (see Chapter 13). Water can stay underground for anywhere from 2 weeks to 10,000 years or more before it moves into another reservoir.

Take-home message . . .

Water, which plays a major role in landscape evolution on the Earth, moves among various reservoirs (the ocean, the atmosphere, the land surface, the subsurface, and living organisms) during the hydrologic cycle.

Quick Question ————————————————
What does the residence time of water in a reservoir refer to?

12.4 Introducing Mass Wasting

What Is Mass Wasting?

It was Sunday, May 31, 1970, a market day, and thousands of people had crammed into the Andean town of Yungay, Peru, to shop. Suddenly they felt the jolt of an earthquake, strong enough to topple some masonry houses. Unfortunately, worse was yet to come. The ground shaking had also broken a huge ice slab off the end of a glacier at the top of Nevado Huascarán, a nearby 6.6-km (4.1-mi) mountain peak. As it tumbled down the mountainside, the ice disintegrated into a turbulent mass of chunks. Near the base of the mountain, most of the debris channeled into a valley and thickened into a churning cloud as high as a 10-story building, ripping up rocks and soil along the way. Frictional heating transformed the ice into water, which mixed with loose rock and dust to produce a muddy slurry capable of lifting boulders larger than houses.

Near the mouth of the valley, some of the debris came to rest, but part of it rocketed up the sides of the valley and became airborne above a ridge bordering Yungay. As the town's 18,000 inhabitants stumbled out of earthquake-damaged buildings, they heard a deafening roar and looked up to see a wall of debris descending from above. The debris buried the town, along with everyone in it. Today, only a grassy meadow overlies the site where Yungay once bustled **(Fig. 12.5)**.

We commonly assume that the ground beneath us is terra firma, a solid foundation on which we can build our lives. But a catastrophe like the one at Yungay shouts otherwise. Much of the Earth's surface hosts *unstable slopes*, meaning that the materials of the slope—including rock, **regolith** (a general term for soil, unconsolidated sediment, and weathered fragments of rock that lie above coherent bedrock), snow, or ice—might start moving downslope if disturbed. Geologists refer to the gravity-driven tumbling, flowing, or sliding of these materials downslope from higher elevations to lower ones as **mass wasting**, or *mass movement*. Mass wasting plays a

FIGURE 12.5 The May 1970 landslide disaster in Yungay, Peru.

(a) Before the landslide, the town of Yungay perched on a hill near the ice-covered mountain Nevado Huascarán.

(b) The landslide completely buried the town beneath debris. A landslide scar is visible on the mountain in the distance.

critical role in the rock cycle as the first step in the transportation of sediment, and it serves as the most rapid means to modify slope shapes. Like earthquakes, volcanic eruptions, storms, and floods, mass wasting is a type of *natural hazard*, a feature of the environment that can cause damage to landscapes and to human society. Unfortunately, mass wasting becomes more of a threat to society every year because, as the world's population grows, cities have expanded to include areas of unstable slopes.

Most people refer to any mass-wasting event as a **landslide**. Geologists and engineers, however, find it useful to distinguish among types of mass-wasting events based on four features: (1) the type of material involved

The Hydrologic Cycle

Water circulates through a number of reservoirs in the Earth System. The largest reservoir by far is the ocean, which covers 71% of the Earth's surface. Water evaporates from the ocean and enters the atmosphere. Thus, the atmosphere serves as another reservoir. Atmospheric water gradually condenses and forms clouds, which drop rain or snow onto the oceans or land.

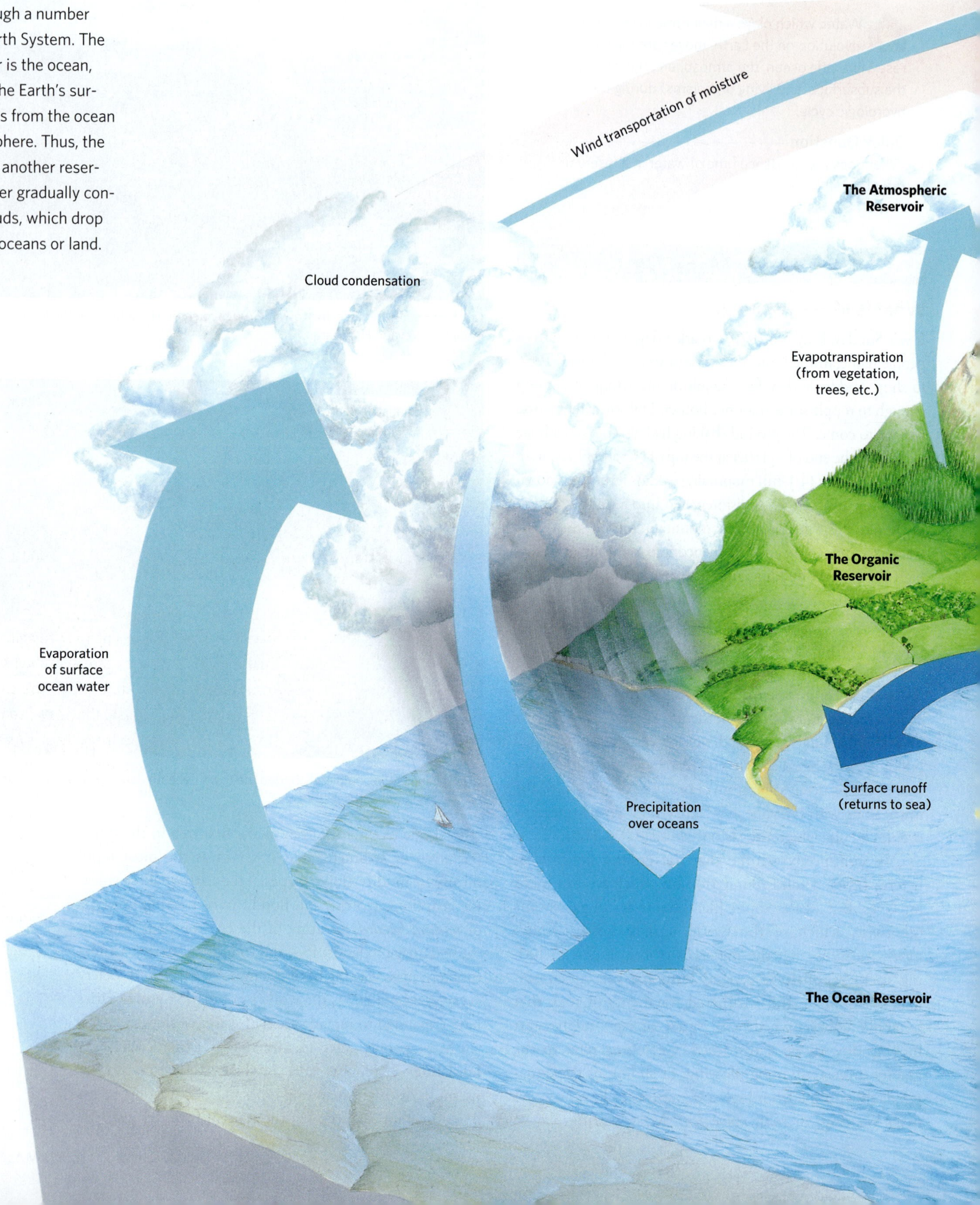

Wind transportation of moisture

Cloud condensation

The Atmospheric Reservoir

Evapotranspiration (from vegetation, trees, etc.)

The Organic Reservoir

Evaporation of surface ocean water

Precipitation over oceans

Surface runoff (returns to sea)

The Ocean Reservoir

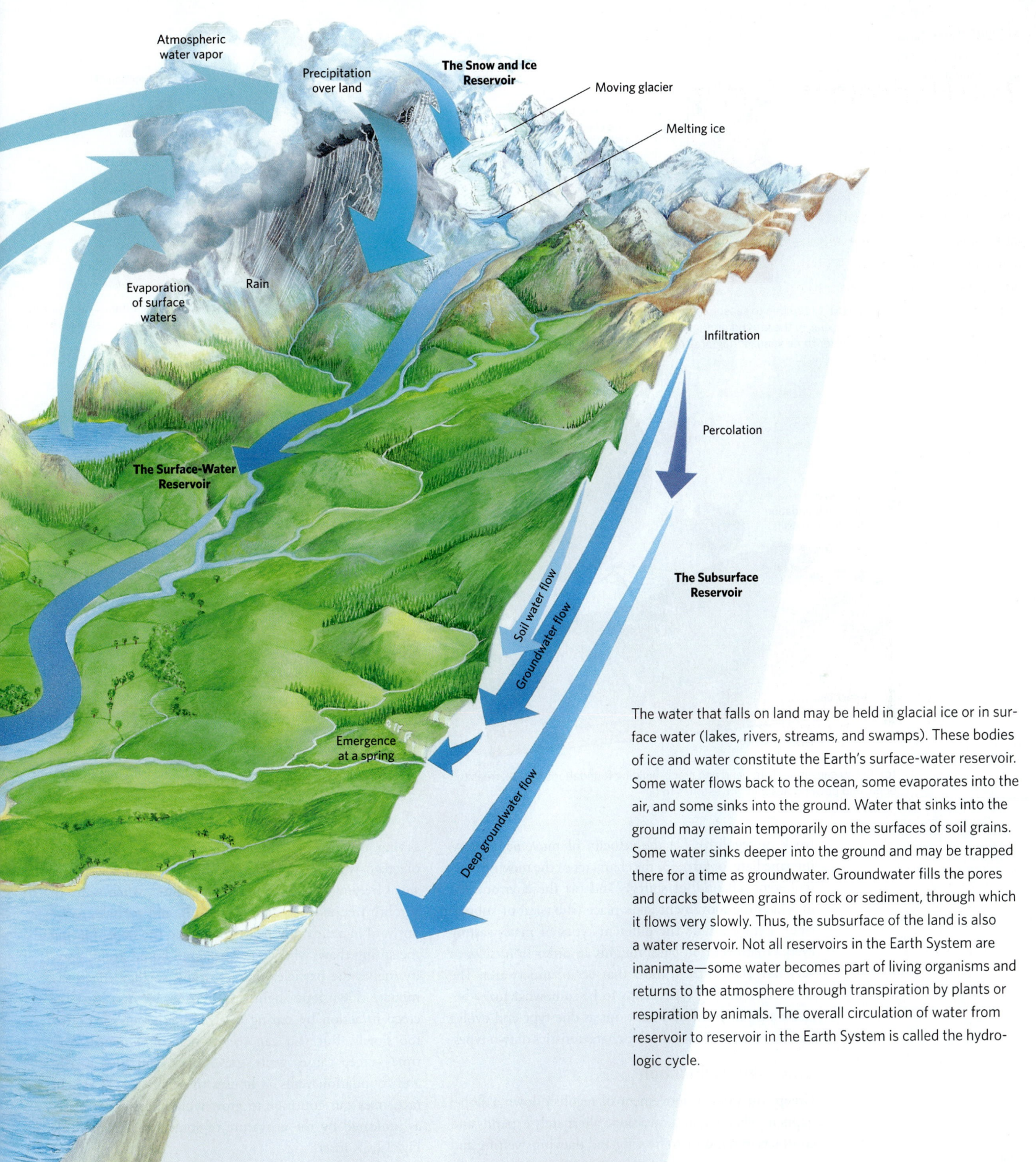

Atmospheric water vapor

Precipitation over land

The Snow and Ice Reservoir

Moving glacier

Melting ice

Evaporation of surface waters

Rain

Infiltration

Percolation

The Surface-Water Reservoir

The Subsurface Reservoir

Soil water flow

Groundwater flow

Emergence at a spring

Deep groundwater flow

The water that falls on land may be held in glacial ice or in surface water (lakes, rivers, streams, and swamps). These bodies of ice and water constitute the Earth's surface-water reservoir. Some water flows back to the ocean, some evaporates into the air, and some sinks into the ground. Water that sinks into the ground may remain temporarily on the surfaces of soil grains. Some water sinks deeper into the ground and may be trapped there for a time as groundwater. Groundwater fills the pores and cracks between grains of rock or sediment, through which it flows very slowly. Thus, the subsurface of the land is also a water reservoir. Not all reservoirs in the Earth System are inanimate—some water becomes part of living organisms and returns to the atmosphere through transpiration by plants or respiration by animals. The overall circulation of water from reservoir to reservoir in the Earth System is called the hydrologic cycle.

351

FIGURE 12.6 Slow mass wasting: creep and solifluction.

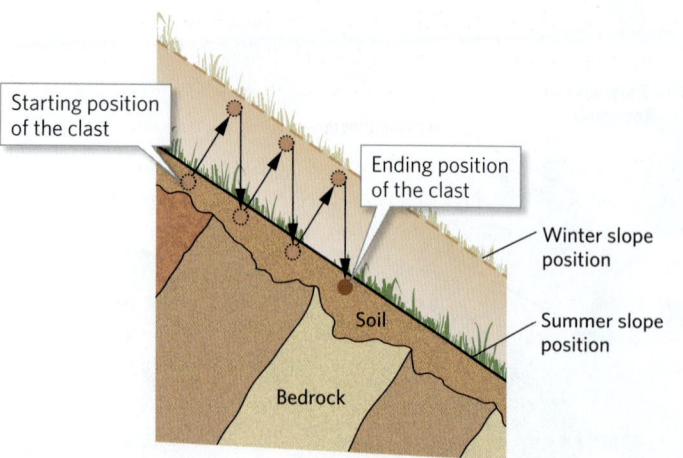

Starting position of the clast

Ending position of the clast

Winter slope position

Summer slope position

Soil

Bedrock

(a) Creep due to seasonal freezing and thawing: A clast rises perpendicular to the ground during freezing, but sinks vertically during thawing. After three years, it migrates to the position shown.

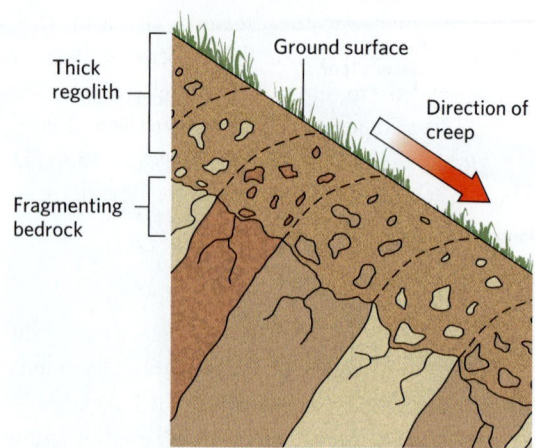

Ground surface

Thick regolith

Direction of creep

Fragmenting bedrock

(b) As rock layers weather and break up, the resulting debris creeps downslope.

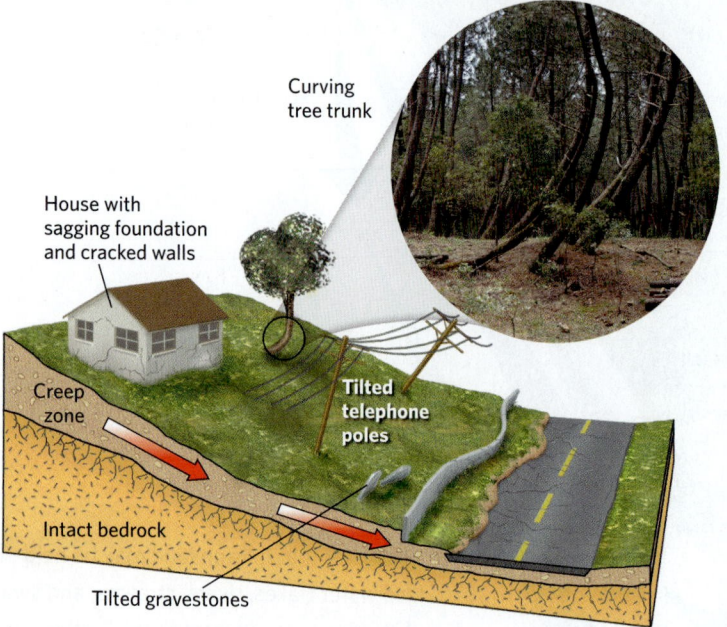

Curving tree trunk

House with sagging foundation and cracked walls

Creep zone

Tilted telephone poles

Intact bedrock

Tilted gravestones

(c) Creep causes walls to bend and crack, building foundations to sink, trees to bend, and power poles and gravestones to tilt.

(d) Solifluction on a hillslope in the tundra.

(rock or regolith); (2) the velocity of movement (slow, intermediate, or fast); (3) the character of the moving mass (coherent, chaotic, or slurry); and (4) the environment in which the movement takes place (subaerial or submarine). Let's examine the different types of mass-wasting events that occur on land, roughly in order from slow to very fast, then turn to those that occur underwater. The distinction among types tends to be somewhat fuzzy because a given event may start out as one type and evolve into another, or it may display characteristics of two types.

Creep and Solifluction

Creep, the gradual movement of regolith down a slope, happens when regolith on a slope alternately expands and contracts in response to freezing and thawing, wetting and drying, or warming and cooling. To see how the process of creep works, let's focus on the consequences of seasonal freezing and thawing. In the winter, when water in regolith freezes, the regolith expands, and particles move outward in a direction perpendicular to the slope. During the spring thaw, when water becomes liquid again, gravity makes the particles sink vertically, so they effectively migrate downslope slightly **(Fig. 12.6a, b)**. You can't see creep in action by staring at a hillside because it occurs too slowly. But you can see its net consequences over time—creep causes walls, gravestones, and trees to tilt, and foundation walls of houses to crack **(Fig. 12.6c)**. In fact, trees can continue to grow while creep takes place, as indicated by the curvature of some trees' trunks (see Fig. 12.6a, inset).

In Arctic or high-elevation regions, a warm spell during the summer may cause the uppermost 1–3 m (3–10 ft) of permafrost to thaw. Because meltwater cannot sink into the still solid permafrost below, the melted layer becomes soggy and weak and flows slowly downslope in overlapping sheets. Geologists refer to this kind of creep as **solifluction** (Fig. 12.6d).

Slumping

The majestic Holbeck Hall Hotel had perched on a cliff along the eastern coast of England since 1879. Its guests could enjoy a spectacular view out across the North Sea—until June 5, 1993. On that day, a block of land bordering the cliff slipped down and rotated seaward, taking half the hotel with it. Fortunately, telltale cracks in the ground weeks before, and a smaller collapse the day before, had led officials to evacuate the hotel, so no one was hurt. But the landmark, along with an immense amount of

the substrate beneath, slid out onto the beach, where the pounding surf of the North Sea eventually carried it away.

Geologists refer to a relatively slow mass-wasting event during which moving rock or regolith stays somewhat coherent (meaning that it doesn't completely break apart) and slides down along a concave-up spoon-shaped surface as a **slump**. The moving mass itself is called a *slump block*, the surface on which it moves is a **failure surface**, and the process of movement is *slumping*. The exposed upslope edge of a failure surface forms a *head scarp*—a new cliff face—and the downslope end of a slump block forms the block's *toe* (Fig. 12.7). In some cases, the upslope ends of a slump block break into a series of discrete slices, each separated from its neighbor by a small failure surface. These blocks tend to rotate around a horizontal axis, so that their surfaces tilt toward the head scarp. The downslope ends may divide into discrete overlapping slices or may break up to form a chaotic jumble. Slumps come in all sizes, from only a few

Video

Developing and Detecting Slumps

FIGURE 12.7 The process of slumping on a hillslope.

(a) A head scarp on a hillslope.

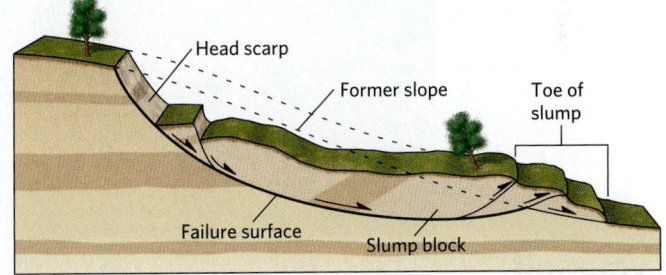

(b) Cross section of a slump.

(c) Slumping has dumped sediment into this river in Costa Rica.

(d) A slump beginning to form along a highway in Utah.

FIGURE 12.8 Examples of mudflows and debris flows.

(a) A 2011 mudflow destroyed houses along this hillside in Brazil.

(b) These mudflows of 2011 stripped away forests on hillslopes in Brazil.

(c) A recent debris flow in Utah. Note the chaotic mixture of rock chunks and mud.

(d) A mudflow down a rain-soaked hillslope buried a highway in Taiwan in 2010.

See for yourself

La Conchita Mudflow, California

Latitude: 34°21′50.29′ N
Longitude: 119°26′46.85′ W

Look NE from 250 m (~820 ft).

Tectonic activity has up-lifted the coast of California to form a terrace bordered on the ocean side by a steep escarpment. Below, you can see a mudflow that overran houses built at the base of the escarpment. Note the head scarp.

meters to tens of kilometers across, and they move at speeds from millimeters per day to tens of meters per minute.

Mudflows, Debris Flows, and Lahars

In the tropical climate of Brazil, bedrock undergoes intense chemical weathering and slowly transforms into a thick, clay-rich regolith that becomes saturated with water when it rains heavily. Where slopes are steep, the saturated regolith can transform into a slurry of mud containing up to 30% water, with the consistency of wet concrete, that flows downslope. Geologists refer to such a moving slurry of mud as a **mudflow**, or *mudslide*, if it consists mostly of clay and water. If it also carries cobbles and boulders suspended in the mud, geologists refer to it as a **debris flow**. Mudflows or debris flows may start out as slumps, but because of their high water content, they cannot stay coherent. Their movement strips away forests and can knock down buildings and overwhelm towns **(Fig. 12.8)**.

Mudflows or debris flows travel at speeds of 20–80 km per hour (10–50 mph). The speed at which a given flow moves depends on both the water content and the slope angle: wetter ones move faster than dryer ones, and flows move faster down steep slopes than they do down gentle slopes. Because mudflows and debris flows have much greater viscosity than clear water, they can carry large rock chunks and huge trees as well as houses and cars. They typically follow channels downslope, and at the base of the slope they may spread out into a broad fan.

The downslope movement of mud and debris can be deadly. Twenty-one people were killed in January 2018 by debris flows that struck the California city of Montecito and the surrounding areas of Santa Barbara County. The flows were caused by a storm in which half an inch of rain fell in just 5 minutes, causing mud and boulders to flow down the Santa Ynez Mountains and into the towns **(Fig. 12.9)**.

When mudflows or debris flows spill down the river valleys bordering volcanoes, they consist of a mixture of volcanic ash, coarser pyroclastic debris, and water. The water may come from snow or ice melted by the volcano's heat or from heavy rains. As we saw in Chapter 4, such volcanic mudflows are known as *lahars* (**Fig. 12.10a**). Lahars generally follow river valleys and typically travel for distances of tens to a few hundred kilometers away from the volcano.

When lahars happen in the wilderness, they can tear away forests and bury the landscape with muck. When they happen near populated areas, they spell disaster. And because lahars can be activated by rainwater, not just meltwater, they can happen years after a large eruption of pyroclastic debris. For example, devastating lahars carrying ash from the 1991 eruption of Mt. Pinatubo spilled down valleys into the surrounding lowlands in the Philippines for each of the next four rainy seasons after the eruption (**Fig. 12.10b**).

Rockslides and Rockfalls

In the early 1960s, engineers built a huge new dam across a river on the northern side of Monte Toc, in the Italian Alps, to create a reservoir for generating electricity. The Vaiont Dam was an engineering marvel, a concrete wall rising as high as an 85-story skyscraper above the valley floor. Unfortunately, the dam's builders did not recognize the hazard posed by Monte Toc. The side of Monte Toc facing the reservoir is underlain by limestone beds interlayered with weak shale beds, all of which dip parallel to the surface of the mountain (**Fig. 12.11a**). The shale beds serve as failure surfaces. As the reservoir filled, the mountain flank began to crack and rumble, and chunks slid down into the reservoir. Local residents began to refer to Monte Toc as *la montagna che cammina* (the mountain that walks).

On October 9, 1963, after several days of rain, more cracks formed, and Monte Toc started to rumble so much that engineers lowered the water level in the reservoir. They thought the wet ground might slump a little into the reservoir, with minor consequences, so authorities did not order an evacuation of the town of Longarone, a few kilometers down the valley from the dam. Unfortunately, the engineers underestimated the problem. At 10:30 that evening,

FIGURE 12.9 A debris flow buried a home in California.

a huge chunk of Monte Toc detached from the mountain and slid down its side, along a weak shale bed. Some of the debris rocketed up the opposite wall of the reservoir to a height of 260 m (850 ft) above the original reservoir level, and some remained in the reservoir, displacing 50 million m³ (1.8 billion ft³) of water. This water splashed over the top of the dam, then rushed down the valley below (**Fig. 12.11b**). When the wall of water had passed, nothing remained of Longarone and its 1,500 inhabitants. Though the dam still stands today, it holds back only debris, and it has never provided electricity.

Geologists refer to such a sudden movement of rock down a non-vertical slope as a **rockslide**, and if the mass consists mostly of regolith, then it's a *debris slide*. If the rock or debris free-falls vertically during part of its journey, the movement can also be referred to as a **rockfall** or *debris fall*. Once one of these mass-wasting events has taken place, it leaves a scar on the slope and forms an accumulation of rock fragments at the base of the slope (**Fig. 12.12**).

See for yourself

Debris Flow, Yungay, Peru
Latitude: 9°7′21.42′ S
Longitude: 77°39′44.86′ W
Looking NE from 6.4 km (~4.3 mi).

Here you can see the steep, glaciated face of Nevado Huascarán. In 1970, a debris flow from the mountain rushed down the valley in the foreground and buried the landscape. More recent, smaller debris flows have accumulated on top of the larger one.

FIGURE 12.10 Examples of lahars.

(a) A lahar that rushed down the side of Mt. St. Helens, Washington.

(b) Damage to a town in the Philippines, caused by a lahar.

FIGURE 12.11 The Vaiont Dam disaster.

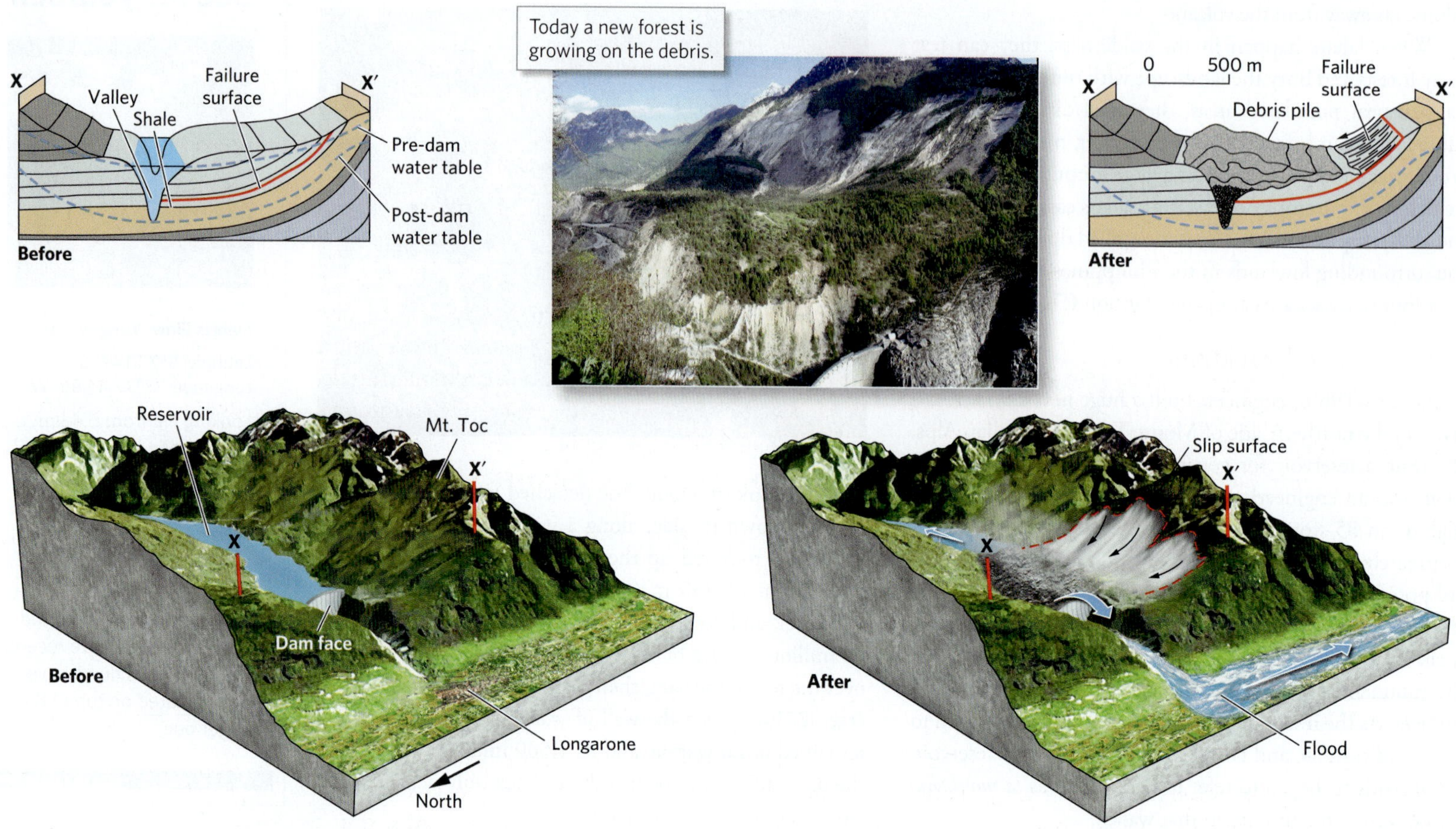

Today a new forest is growing on the debris.

Before — Valley, Shale, Failure surface, Pre-dam water table, Post-dam water table, X, X'

After — Debris pile, Failure surface, 0 500 m, X, X'

Before — Reservoir, Mt. Toc, X, X', Dam face, Longarone, North

After — Slip surface, X, X', Flood

(a) Before the landslide, the north flank of Monte Toc was forested. When the reservoir was filled with water, the slope became unstable. The cross section shows the shale bed that became the failure surface.

(b) 260 million cubic meters (9.2 billion cubin ft) of rock and debris slid down the mountainside and displaced water in the reservoir. The water surged over the dam and swept away a town in the valley below.

FIGURE 12.12 Examples of rockfalls.

(a) Successive rockfalls have littered the base of this sandstone cliff with boulders. Note the talus at the base of the cliff.

(b) Debris from a rockfall in the Andes of Peru has accumulated downslope. The fresh rock exposed by the rockfall has a lighter color.

FIGURE 12.13 The angle of repose is the steepest slope at which a pile of unconsolidated sediment can remain stable.

(a) This talus pile at the base of a cliff in the Uinta Mountains, Utah, has a relatively steep angle of repose.

Well-rounded sand has a small angle.

Irregularly shaped gravel has a large angle.

(b) The angle of repose depends on the shape and size of the grains in the pile.

Rockslides happen when bedrock or regolith detaches from a slope, then slips rapidly downslope on a failure surface. Rockfalls happen when rocks on a cliff break off along a joint and topple out away from the cliff. When they strike other rocks downslope, they shatter into smaller fragments. If the fragments accumulate over time at the base of the cliff, they can build an apron or pile of **talus** (Fig. 12.13a). Talus, like all granular debris, tends to pile up, producing the steepest slope it can without collapsing. The angle of this slope, called the **angle of repose**, has a value between 30° and 37° for most dry, unconsolidated (loose) materials, such as dry sand. Steeper angles of repose, up to 45°, can form where debris consists of irregularly shaped grains that can interlock (Fig. 12.13b).

Rockslides and rockfalls may move at speeds of up to 300 km/h (185 mph), depending on the steepness of the slope and the distance that they move. They become particularly fast when a cushion of air gets trapped beneath them, for in such cases, there is virtually no friction between the slide and its substrate, so the mass moves like a hovercraft. Large, fast rockslides or rockfalls push the air in front of them, generating a short blast of hurricane-like wind. The wind in front of a 1996 rockfall in Yosemite National Park, for example, flattened over 2,000 trees.

Rockslides and rockfalls, like slumps and other forms of mass wasting (Fig. 12.14), happen at a variety of scales, depending on the amount of material that detaches and on whether the energy of tumbling material moves debris in its path. The largest can involve a whole mountainside, as we saw in the example of the Monte Toc catastrophe. Even smaller falls, involving just a few cobbles, can be a hazard, particularly along highways, where the debris can tumble onto the road surface or onto cars. Hence, highway crews erect "Falling Rock" signs along stretches of highways that border cliffs.

Avalanches

In the winter of 1999, an unusual weather system passed over the Austrian Alps. First it snowed. Then the temperature warmed and the snow began to melt. But then the weather turned cold again, and the melted snow froze into a hard, icy crust. This cold snap ushered in a blizzard that blanketed the icy crust with tens of centimeters (1–2 ft) of new snow. When the new snow became heavy enough, the icy crust acted as a failure surface, and the snow began to slip. As it accelerated, the mass transformed into a **snow avalanche**, a turbulent cloud of snow mixed with air, that surged downslope at 290 km per hour (180 mph). At the bottom of the slope, the avalanche overran a ski resort, crushing and carrying away buildings, cars, and trees and killing over 30 people. It took searchers and their specially trained dogs many days to find buried survivors and victims under the 5- to 20-m (16- to 66-ft)-thick pile of snow that the avalanche deposited (Fig. 12.15a).

What triggers snow avalanches? As we've seen, some occur when a broad slab of snow on a slope detaches from its substrate along an icy failure surface. Others take place when a *cornice*, a large drift of snow that builds up on the side of a windy mountain summit, suddenly gives way and falls onto the slopes below, where it knocks additional snow free. Avalanches behave differently depending on the temperature. *Wet avalanches*, which occur at warmer temperatures, move relatively slowly as a slurry of solid and liquid water, whereas *dry avalanches*, which occur at colder temperatures, tumble rapidly downslope as a cloud of powder (Fig. 12.15b).

Did you ever wonder . . .

why highway engineers erect "Falling Rock" signs?

See for yourself

Rockfalls in Canyonlands, Utah

Latitude: 38°29′50.07′ N
Longitude: 110°0′15.63′ W

Zoom to 5 km (3 mi) and look down.

The Green River has carved a deep valley through horizontal layers of strata. Shale forms the slopes, and sandstone forms the top ledge. Erosion of the shale undercuts the sandstone, which breaks away at joints and tumbles downslope to build a talus pile.

FIGURE 12.14 As we have seen, different kinds of mass wasting can modify the landscape.

Solifluction and creep

Slumping

Lahars and mudflows

Rockfalls and rockslides

Debris flows

Geologists sometimes use the word *avalanche* in a broad sense to refer to any mass-wasting event during which solid fragments become suspended in a fluid, so that the flowing mixture behaves like a turbulent cloud. In this context, a *debris avalanche* is a turbulent cloud containing rock or sediment, suspended in air, that surges down a slope. Below, we will see that avalanche-like flows can also occur underwater.

FIGURE 12.15 Snow avalanches.

(a) The aftermath of a 1999 avalanche in the Austrian Alps. Masses of snow buried homes, crushed and ripped through cars, and killed 31 people.

Origin of the avalanche

Toe of the avalanche

(b) This dry snow avalanche in Alaska is a turbulent cloud. The arrow indicates the flow direction.

FIGURE 12.16 Submarine mass wasting.

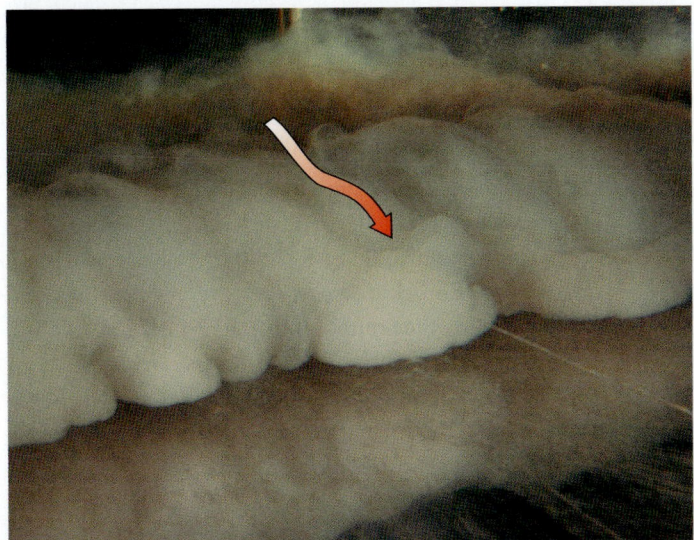

(a) A laboratory model of a turbidity current. The cloud is entirely underwater and consists of fine clay suspended in water.

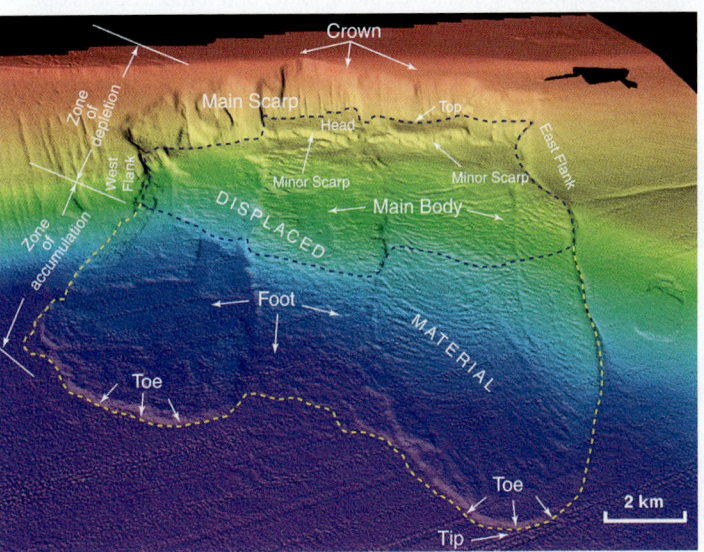

(b) A digital bathymetric model of a submarine slump along the coast of California. The parts of the slump are labeled. Blue is deeper, orange is shallower.

(c) A digital bathymetric model showing the slumps bordering the island of Oahu, Hawaii.

Submarine Mass Wasting

So far, we've focused on mass wasting that occurs on the land surface—we refer to such events as *subaerial*, to emphasize that they occur beneath the air. But mass wasting also happens underwater. Geologists distinguish among three types of submarine mass wasting based on whether the mass remains coherent or disintegrates as it moves. In a **submarine slump**, semicoherent blocks slip downslope along a failure surface. In a **submarine debris flow**, the moving mass breaks apart

to form a slurry containing pebbles, cobbles, and boulders suspended in a mud matrix. And in a **turbidity current**, the moving sediment disperses in water to yield a turbulent cloud of suspended sediment that rushes downslope like an underwater avalanche **(Fig. 12.16a)**.

As revealed by shaded-relief maps of the seafloor, submarine slopes bordering hotspot volcanoes and active plate boundaries are scalloped by many immense slumps because earthquakes frequently jar these areas and set masses of material in motion **(Fig. 12.16b)**. Since a submarine slump can quickly displace a large area of the seafloor, it can trigger a tsunami as devastating as one caused by an earthquake (see Chapter 8).

Debris from countless slumps over millions of years has substantially modified the flanks of the Hawaiian Islands **(Fig. 12.16c)**. In fact, the cliff that follows the axis of the island of Oahu represents the head scarp of a giant slump. Debris from this slump spread eastward for a distance of 180 km (110 mi) on the floor of the Pacific Ocean. Huge slumps have also been mapped along the coasts bordering the Atlantic Ocean, indicating that passive continental margins are not immune to slumping. About 8,000 years ago, for example, the 100-km (60-mi)-wide Storegga Slide slipped down the face of the Atlantic passive margin west of Norway to produce a tsunami that washed away Stone Age villages around the coast of the North Sea.

FIGURE 12.17 Jointing has broken up this thick sandstone bed along a cliff in Utah. Blocks of sandstone break free along joints and tumble downslope.

Take-home message . . .

Mass-wasting events differ from one another based on the speed, content, and character of the moving material. Creep, slumping, and solifluction are slow. Mudflows and debris flows move faster, and avalanches and rockfalls move the fastest. Mass wasting occurs both on land and underwater.

Quick Question

In what way is a dry snow avalanche like a turbidity current?

12.5 Why Does Mass Wasting Occur?

Mass wasting takes place when the rock or regolith underlying a slope has become weak enough, or the strength of attachment across a failure surface beneath the rock or

FIGURE 12.18 Downslope movement is determined by the balance between two forces.

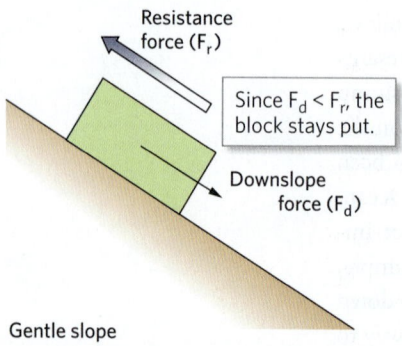

(a) If the resistance force is greater than the downslope force, the block does not move.

Resistance force (F_r)

Since $F_d < F_r$, the block stays put.

Downslope force (F_d)

Gentle slope

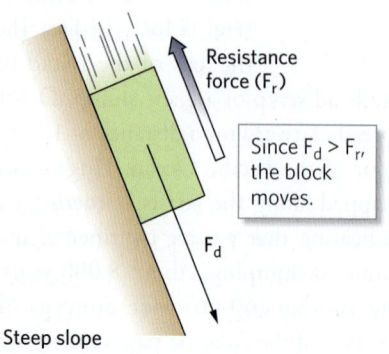

(b) If the slope angle increases, the downslope force increases.

Resistance force (F_r)

Since $F_d > F_r$, the block moves.

F_d

Steep slope

regolith has diminished enough, that the pull of gravity can cause the material to start moving. If material on a slope becomes susceptible to movement, we say that the slope has become *unstable*. Let's look at various conditions that can lead to instability.

Weakening of the Substrate

If the Earth's surface exposed only fresh, intact bedrock, mass wasting would be of little concern, for rock has great strength and could easily hold up stalwart mountain cliffs. Fresh bedrock doesn't occur everywhere, however; in many places, the ground's substrate consists of regolith. Furthermore, intact rock is rare because in most places, joints or faults break rock into pieces **(Fig. 12.17)**. Regolith and fractured rock are much weaker than intact rock and can collapse in response to gravity. In other words, weathering, jointing, and faulting ultimately make mass wasting possible.

Why are regolith and fractured rock weaker than intact bedrock? The answer comes from looking at the strength of the attachments holding materials together. Intact bedrock is relatively strong because the chemical bonds within its interlocking grains, or within the cements between grains, can't be broken easily. A mass of regolith, in contrast, is relatively weak because the grains are held together by friction, by weak bonds between electrically charged surfaces of grains, or by bonds between water molecules, and none of these attachments can be as strong as the bonds holding atoms together in minerals. To picture this contrast, think about how much easier it is to destroy a sand castle than it is to destroy a sandstone sculpture.

Slope Failure

When material starts moving on an unstable slope, we say that **slope failure** has taken place. Whether or not a slope fails depends on the balance between two forces: the *downslope force*, caused by gravity, and the *resistance force*, which inhibits sliding. The resistance force comes from bonds holding the grains of the slope material together, or from friction across the failure surface. If the downslope force exceeds the resistance force, the slope fails, and mass wasting results **(Fig. 12.18)**. Note that for a given mass, downslope force depends on slope steepness **(Box 12.2)**.

In some locations, downslope movement begins on a weak failure surface beneath the ground surface. Geologists recognize several different kinds of weak surfaces that are likely to become failure surfaces. These include wet clay layers, unconsolidated sand layers, exfoliation joints, weak beds of shale, and metamorphic foliation planes **(Fig.12.19)**. Weak surfaces that are parallel to the land surface of a slope are particularly likely to fail because the downslope force parallels the weak surface.

FIGURE 12.19 Several kinds of weak surfaces can become failure surfaces.

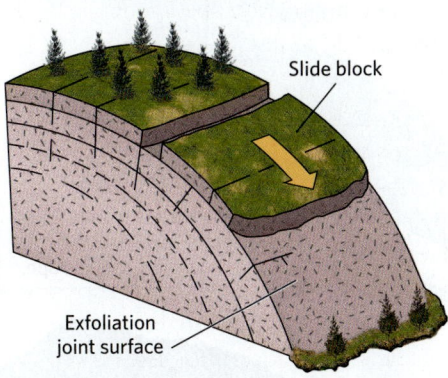

(a) Exfoliation joints, which form parallel to slope surfaces in granite, may become failure surfaces.

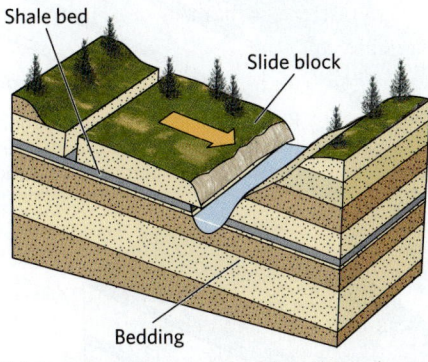

(b) In sedimentary rock, bedding planes (particularly in weak shale) may become failure surfaces.

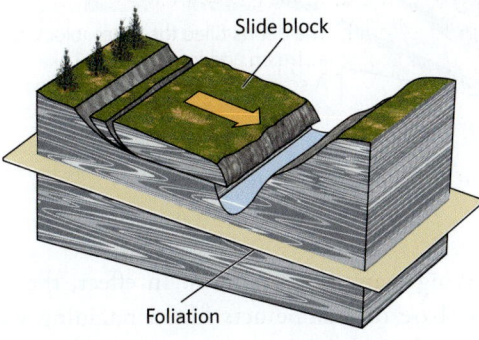

(c) In metamorphic rock, foliation planes (particularly in mica-rich schist) may become failure surfaces.

| BOX 12.2 | How can I explain . . . |

The concept of slope stability

What are we learning?

How slope angle and friction can affect slope stability.

What you need:

- A brick and a smooth cutting board
- A protractor, sandpaper, and tape

Instructions:

- Place the brick on the surface of the cutting board.
- Slowly lift one end of the board until the brick becomes unstable and slips. Measure the angle of the slope.
- Repeat the tilting experiment, but this time, tilt the board to an angle that is less than the previous angle at which the brick slipped. At this angle, shake the board slightly. Can you get the brick to move?
- Tape the sandpaper to the board, place the brick on it, and press down.
- Repeat the tilting experiment. Does the angle at which the brick becomes unstable change?

What did we see?

- The surface of the board represents a failure surface. Tilting the board represents steepening of the slope. The brick may be stable on gentle slopes, but it becomes unstable on steeper slopes.
- Shaking the board (which represents ground shaking) decreases the slope angle at which the brick can remain stable.
- Increasing friction across the surface by adding the sandpaper increases the slope angle at which the brick remains stable.

What causes the balance of forces to change so that the downslope force exceeds the resistance force and causes a slope to suddenly fail? Here, we look at various phenomena that can trigger slope failure.

GROUND SHAKING. An earthquake, a storm, a large truck driving by, or a blast at a construction site may cause a mass on the verge of moving to begin moving. Ground shaking can trigger sliding by breaking bonds that hold rock or regolith in place. It can also cause slip to begin on a failure surface by causing the rock above the failure surface to move upward slightly, thereby decreasing friction across the failure surface. If the decrease in friction causes a sufficient decrease in the resistance force, then the downslope force sets the mass above the failure surface in motion. Finally, ground shaking can cause liquefaction of wet sediment, either by increasing water pressure in the pores between grains, so that the grains are pushed apart, or by breaking the cohesion between the grains (see Chapter 8). When liquefaction takes place, the resistance force effectively disappears, so that material above the liquefied layer begins to move on even the slightest slope.

CHANGES IN SLOPE LOAD AND SUPPORT. Slope loads change when the weight of the material above a potential

FIGURE 12.20 Events leading to the 1925 Gros Ventre Slide in Wyoming.

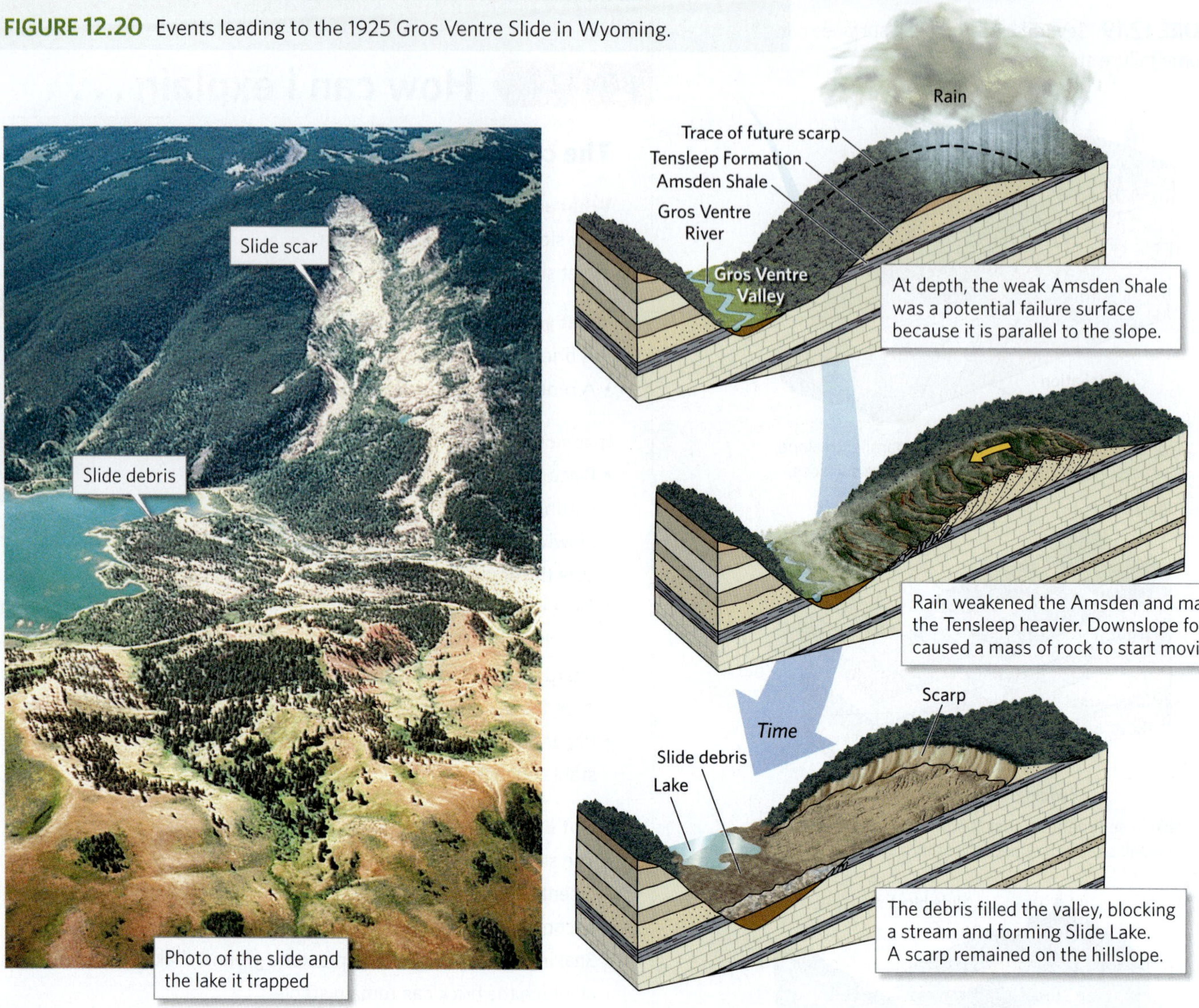

Rain

Trace of future scarp

Tensleep Formation

Amsden Shale

Gros Ventre River

Gros Ventre Valley

At depth, the weak Amsden Shale was a potential failure surface because it is parallel to the slope.

Rain weakened the Amsden and made the Tensleep heavier. Downslope force caused a mass of rock to start moving.

Time

Scarp

Slide debris

Lake

The debris filled the valley, blocking a stream and forming Slide Lake. A scarp remained on the hillslope.

Slide scar

Slide debris

Photo of the slide and the lake it trapped

failure surface changes. If the load increases—for example, due to construction of buildings on top of a slope or due to saturation of regolith by heavy rains—the downslope force increases and may exceed the resistance force. Seepage of water into the ground may also weaken underground failure surfaces, further decreasing resistance force. Water seepage triggered the largest observed landslide in American history, the Gros Ventre Slide, which took place in 1925 on the flank of Sheep Mountain, near Jackson Hole, Wyoming **(Fig. 12.20)**. Almost 40 million m³ (1.4 billion ft³) of rock, as well as the overlying soil and forest, detached from the side of the mountain and slid 600 m (2,000 ft) downslope, filling a valley and forming a natural dam across the Gros Ventre River.

Removal of support at the base of a slope by erosion or by human excavation plays a major role in triggering many slope failures. In effect, the material at the base of a slope acts like a retaining wall. Removing this "wall" by undercutting allows the material farther up the slope to start sliding down. Undercutting contributed to making slopes bordering a river near Oso, Washington, susceptible to slumping. A disastrous slump happened there in 2014, burying a housing development and highway **(Fig. 12.21)**. In some cases, erosion by a river or by waves eats into the base of a cliff and produces an overhang. When such undercutting has occurred, the rock making up the overhang eventually breaks away from the slope and falls **(Fig. 12.22)**.

CHANGES IN SLOPE STRENGTH. The stability of a slope depends on the strength of the material constituting it. If the material weakens with time, the slope becomes

FIGURE 12.21 The Oso, Washington, slump of 2014.

(a) The aftermath of the slump. Note the head scarp.

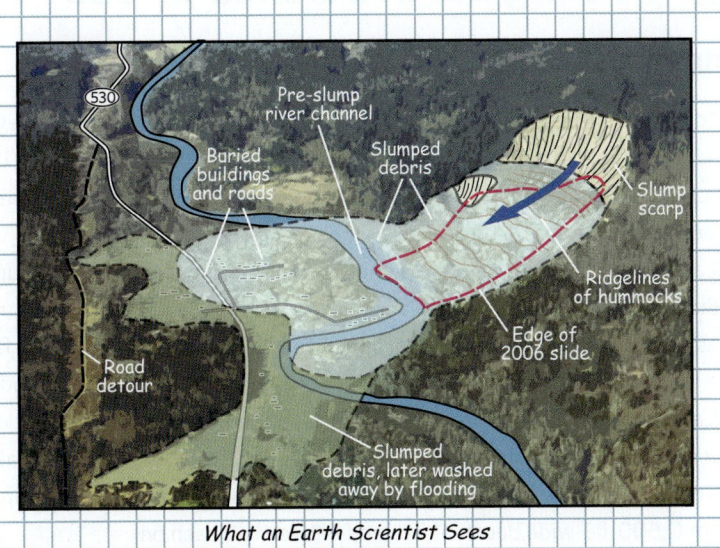

What an Earth Scientist Sees

(b) A geologist's sketch indicates where buildings were overwhelmed.

weaker and eventually collapses. Four factors influence the strength of slopes:

- *Weathering:* With time, chemical weathering produces weaker minerals, and physical weathering breaks rocks apart. When this happens, a formerly strong rock transforms into weak regolith.

- *Vegetation cover:* In the case of slopes underlain by regolith, vegetation tends to strengthen the slope because the roots hold otherwise unconsolidated grains together. Furthermore, plants absorb water from the ground, keeping it from turning into slippery mud. The removal of vegetation therefore has the net result of making slopes more susceptible to downslope movement. Deforestation, due to either logging or fire, can lead to catastrophic mass wasting of the forest's substrate.

- *Water content:* Water affects slope stability in many ways. If the water content of the regolith increases too much, the regolith may turn into a muddy slurry that can be susceptible to flow. Also, if the regolith contains *expanding clay*—certain types of clay whose grains absorb water and swell up—the regolith can expand and, as a consequence, break up. Finally, water infiltration may make failure surfaces underground more slippery by forming weak films on mineral surfaces, and the weight of added water can increase the downslope force.

- *Water pressure:* Addition of water at the ground surface can change the pressure of water in the rock of the failure surface. The increase in pressure effectively pushes grains apart along the failure surface, decreasing friction and, therefore, the resistance force along the failure surface.

Take-home message . . .

Weathering and fragmentation weaken slope materials and make them more susceptible to mass wasting. Slope failure occurs when downslope force exceeds resistance force due to ground shaking, changing slope angles and strength, changing water content, and changing slope support and loads.

Quick Question

Can a slope's strength change over time?

FIGURE 12.22 Undercutting and collapse of a sea cliff.

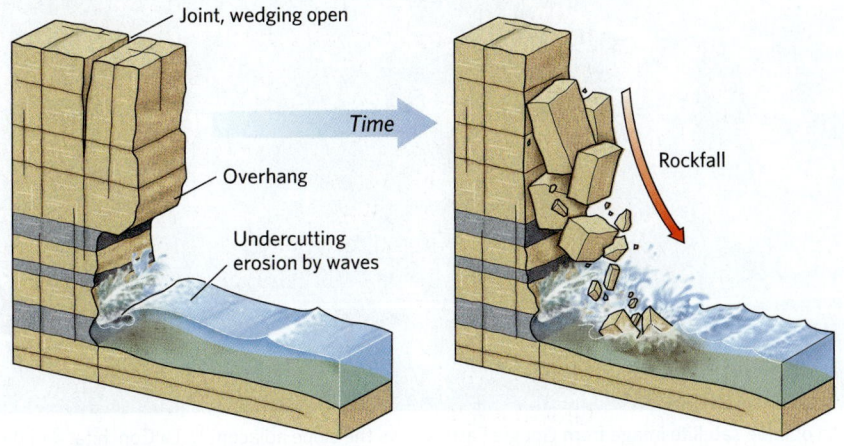

(a) Undercutting by waves removes the support beneath an overhang.

(b) Eventually, the overhang breaks off along joints, and a rockfall takes place.

BOX 12.3

Putting Earth Science to Use

Thinking about slope stability at La Conchita, California

Many factors go into the decision of whether or not to buy a home: cost; access to schools, shopping, jobs, and recreation; construction quality; and beauty come to mind. In areas on or near a slope, stability of the slope should be considered as well. In most cases, it's important to consult a qualified engineer before making a decision about slope stability. But as a potential home buyer, you can start by looking for some clues to slope stability on your own. Specifically, evidence of past slope movements hint that future movements might occur. The region of La Conchita, California, about 100 km (62 mi) northwest of Los Angeles, provides an illustration.

La Conchita, a development of about 200 houses, was built on a 500-m (1,600-ft)-wide bench of flat land between a beach on the southwest, and a steep slope 180 m (600 ft) high on the northeast. Above the slope, a broad terrace hosts irrigated orchards. Due to intense weathering, the terrace and bluff are underlain by weak, clay-rich regolith. The terrace originally formed by wave action at sea level; compression caused by the push of the Pacific Plate against the North American Plate subsequently caused it to undergo uplift to its present height.

Look closely at the slope **(Fig. Bx12.3a)**. Notice that the face of the slope isn't a smooth, planar surface, but rather is hummocky, and several curving, unvegetated steps of various sizes have developed on this face. These steps look like head scarps of slumps that formed recently enough that they have not had time to erode away or be revegetated. The overall impression of the slope shape suggests that it has undergone mass wasting in the not-too-distant past, implying that it has the potential to be unstable.

Of course, a statement that a slope that has the potential to be unstable generally doesn't imply an exact time frame for the next failure. Perhaps the slope will fail within a year, or perhaps it might not fail again for decades. Like any other natural hazard (such as flooding, earthquakes, fire, and severe weather), a statement about risk represents a possibility, not a certainty. At La Conchita, unfortunately, concern about instability was justified, for slope failure did happen again.

If the area were uninhabited, such mass wasting would simply be part of the natural process of landscape evolution. But mass-wasting events at La Conchita have made headlines, because they caused loss of life and property damage. In 1995, following heavy rains, which weakened the regolith and made it heavier, slumping took place on part of the bluff. The wet slump block turned into a mudflow which moved rapidly downslope and overwhelmed 9 houses. An even more devastating slump and mudflow happened in 2005 **(Fig. Bx 12.3b)**. This event killed 10 people, buried 13 houses, and damaged 23 houses.

FIGURE Bx12.3 The 2005 La Conchita mudflow.

Edge of the plateau
Possible slump

Terrace

Slope

Pacific Ocean

(a) An oblique satellite image from Google Earth shows the slope adjacent to La Conchita, and the terrace above. Note the highlighted head scarps.

Plateau

Head scarp

Road

Buried houses

(b) During heavy rains, the bluff gave way, and heavy mud flowed downslope, burying houses and taking several lives.

FIGURE 12.23 Surface features warn that a large slump is beginning to develop. Cracks that appear at the head scarp may drain water and kill trees. Power poles tilt, and the lines become tight. Fences, roads, and houses on the slump begin to crack.

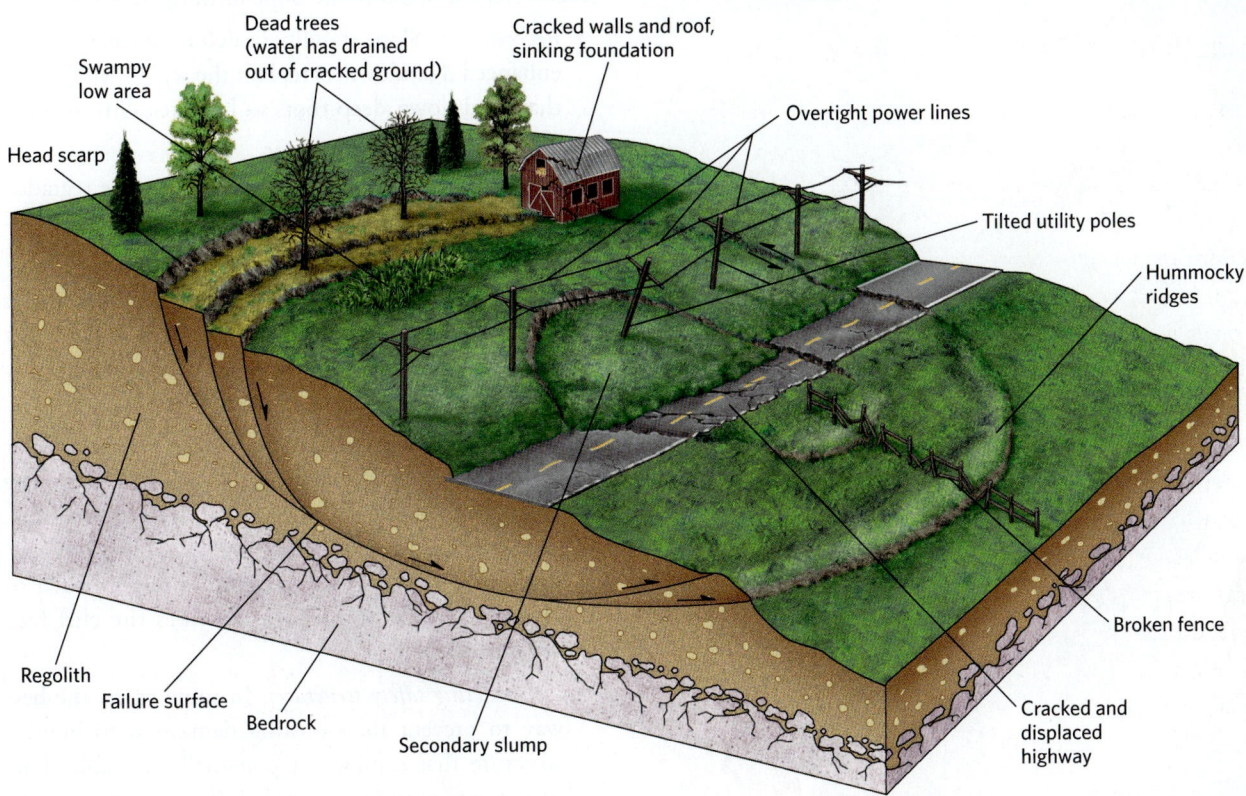

12.6 Can We Prevent Mass-Wasting Disasters?

Clearly, landslides, mudflows, and slumps are natural hazards we cannot ignore. Too many of us live in regions where mass wasting has the potential to kill people and destroy property. In many cases, the best solution is avoidance—don't build, live, or work in an area where mass wasting can take place. But avoidance is possible only if we know where the hazards are and how they might be triggered.

Identifying Regions at Risk

To pinpoint locations susceptible to mass wasting, geologists look for landforms known to result from mass wasting, for where downslope movement has happened in the past, it might happen again in the future. Features such as head scarps, swaths of forest in which trees have been tilted, piles of loose debris at the bases of hills, and hummocky land surfaces all indicate recent mass wasting **(Box 12.3)**.

In some cases, geologists may also be able to detect slopes that are beginning to move **(Fig. 12.23)**. For example, roads, buildings, and pipes begin to crack over unstable ground. Power lines may be too tight or too loose because their support poles have moved together or apart, and trees may tilt. Visible cracks may form on the ground at the potential head of a slump, and the ground may bulge up at the toe of the slump. Locally, subsurface cracks may drain water from an area and kill off vegetation, or subsidence may cause an area to become swampy.

In recent years, new, extremely precise laser-based surveying methods have permitted surveyors to detect the beginnings of ground movements that have not yet visibly affected the land surface. Geologists have also begun to identify hazardous areas by using computer programs to evaluate factors that trigger mass wasting, including slope steepness, substrate strength, degree of water saturation, orientation of potential failure surfaces relative to the slope, nature of vegetation cover, potential for heavy rains, potential for undercutting, and likelihood of earthquakes. Such hazard assessment studies permit compilation of **landslide-potential maps**, which rank regions according to the likelihood that mass wasting will occur **(Fig. 12.24a)**. Detailed mapping of landslides that have already happened in an area can provide additional insights into hazards by characterizing the probable dimensions of landslides **(Fig. 12.24b)**.

FIGURE 12.24
Mapping helps
to characterize
landslide potential.

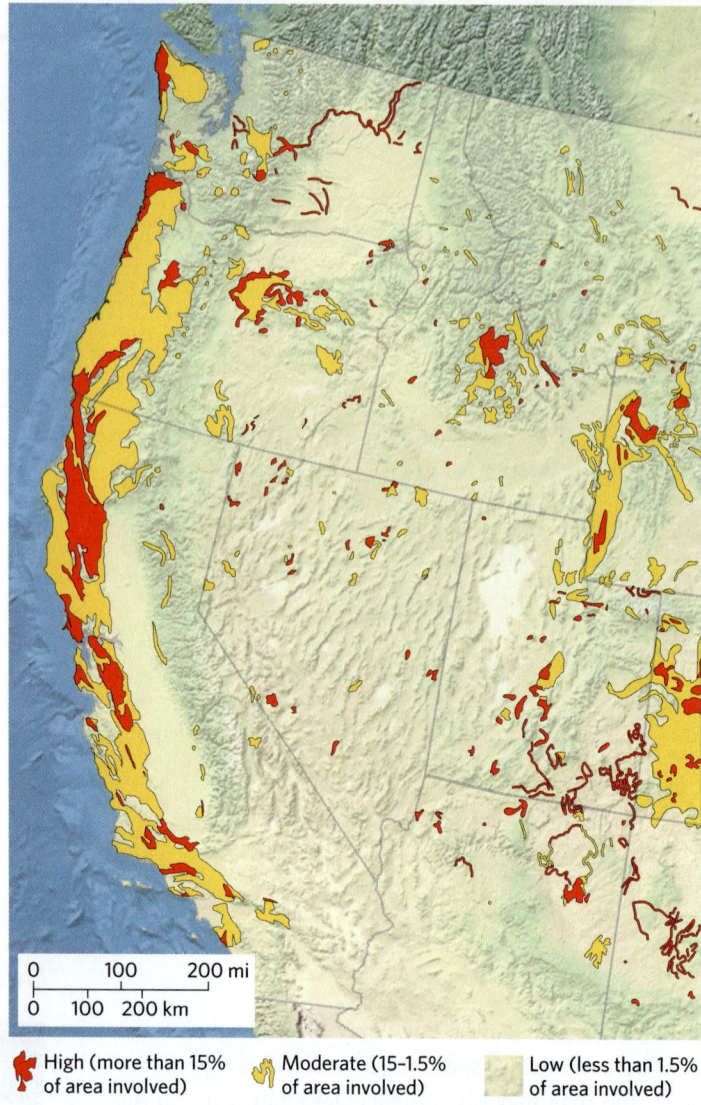

High (more than 15% of area involved)

Moderate (15–1.5% of area involved)

Low (less than 1.5% of area involved)

(a) A landslide-potential map of the western United States.

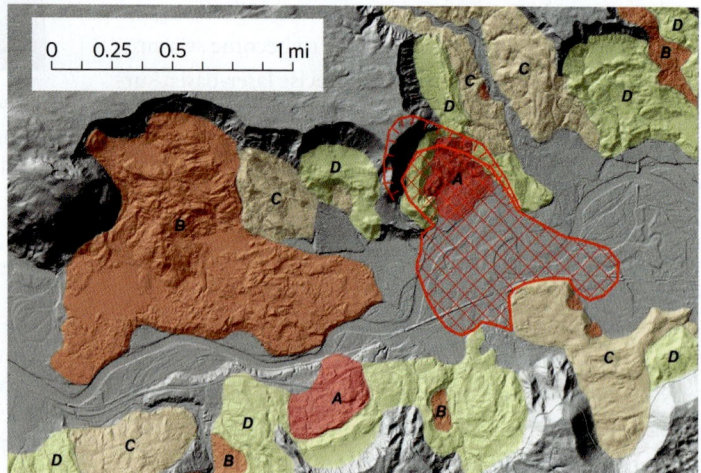

(b) A digital elevation model showing the extent of landslides that have happened during the past several thousand years along the Stillaguamish River, near the Oso landslide (shown by the red lines). Colored areas are places that have slumped. D slumps are the oldest and A slumps are the youngest.

Preventing Mass Wasting

In areas where landslide potential exists, people can take certain steps to stabilize the slope or mitigate the hazard.

- *Revegetation:* Slope stability in deforested areas can be enhanced if landowners replant the region with plants that send down deep roots to bind regolith together **(Fig. 12.25a)**.

- *Regrading:* A dangerously steep slope can be regraded or terraced so that it does not exceed the angle of repose **(Fig. 12.25b)**.

- *Reducing subsurface water:* Slope instability may be decreased either by improving drainage, so that water does not enter the subsurface in the first place, or by extracting water from the ground **(Fig. 12.25c)**.

- *Preventing undercutting:* In places where a river undercuts a cliff face, engineers can divert the river **(Fig. 12.25d)**. Similarly, in coastal regions, they may build an offshore breakwater or pile riprap (loose boulders or concrete) along a beach to absorb wave energy before it strikes the cliff face **(Fig. 12.25e)**.

- *Constructing safety structures:* In some cases, the best way to prevent mass-wasting damage is to build a structure that stabilizes a potentially unstable slope or protects a region downslope from debris if mass wasting does occur. For example, engineers can build retaining walls or bolt loose slabs of rock to more coherent masses in the substrate in order to stabilize highway embankments **(Fig. 12.25f, g)**. The danger from rockfalls can be decreased by covering a road cut with chain link fencing or by spraying it with concrete. Highways at the base of an avalanche-prone slope can be covered by an avalanche shed whose roof keeps debris off the road **(Fig. 12.25h)**.

- *Controlled blasting of unstable slopes:* When it is clear that unstable ground or snow threatens a particular region, the best solution may be to blast the unstable ground or snow loose at a time when its movement can do no harm.

Take-home message . . .

Various features of the landscape may help geologists to identify unstable slopes and estimate the hazard they pose. Systematic study allows production of landslide-potential maps. Engineers use a variety of techniques to stabilize slopes.

Quick Question

What clues indicate that a slump may be starting to form?

FIGURE 12.25 A variety of remedial steps can stabilize unstable slopes or mitigate landslide hazards.

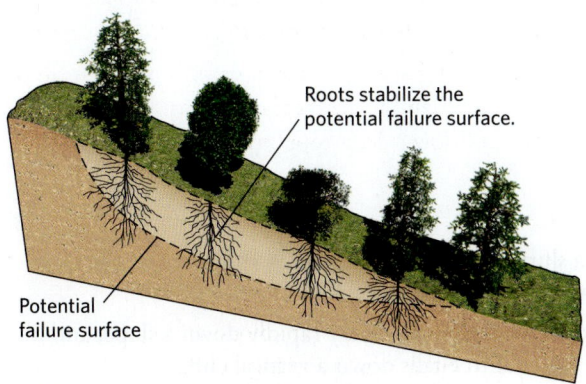

Roots stabilize the potential failure surface.

Potential failure surface

(a) Revegetating a slope results in the growth of roots that can hold the slope together.

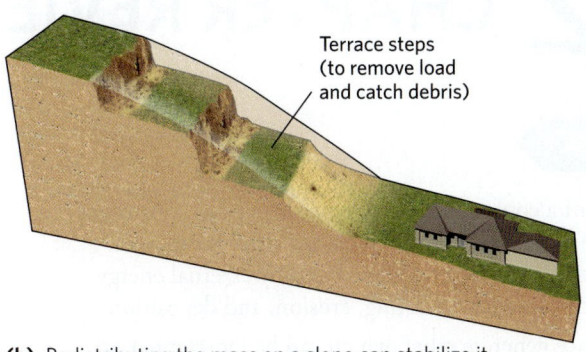

Terrace steps (to remove load and catch debris)

(b) Redistributing the mass on a slope can stabilize it. Terracing can help catch debris.

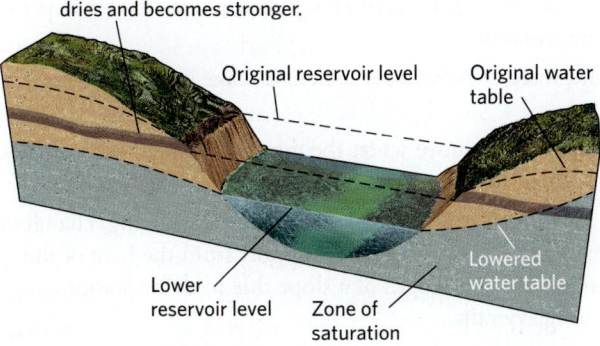

Potential failure surface dries and becomes stronger.

Original reservoir level

Original water table

Lower reservoir level

Zone of saturation

Lowered water table

(c) Lowering the water table can strengthen a potential failure surface.

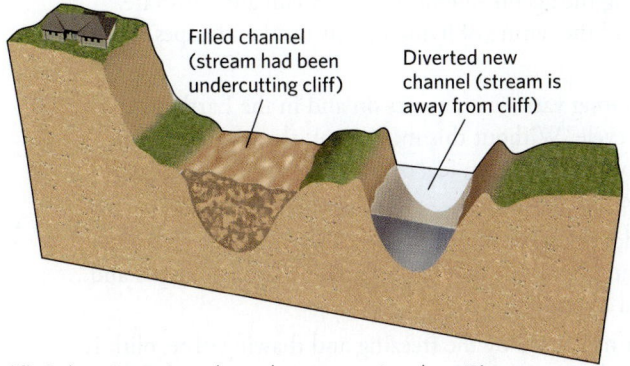

Filled channel (stream had been undercutting cliff)

Diverted new channel (stream is away from cliff)

(d) Relocating a river channel can prevent undercutting.

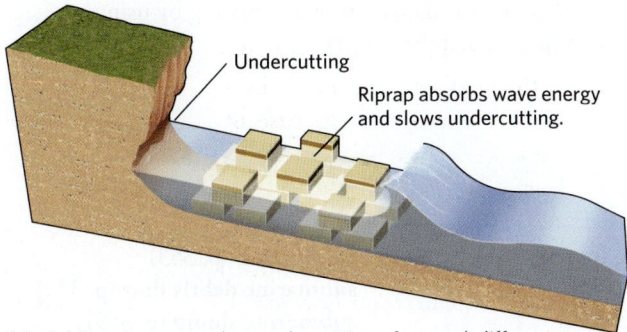

Undercutting

Riprap absorbs wave energy and slows undercutting.

(e) Adding riprap can slow undercutting of coastal cliffs.

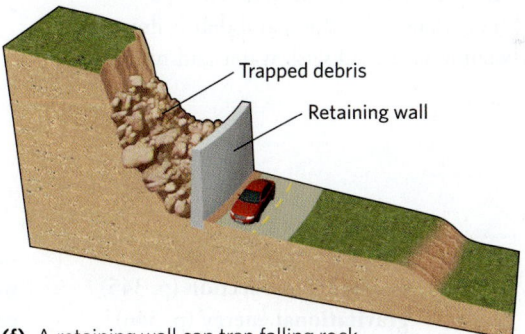

Trapped debris

Retaining wall

(f) A retaining wall can trap falling rock.

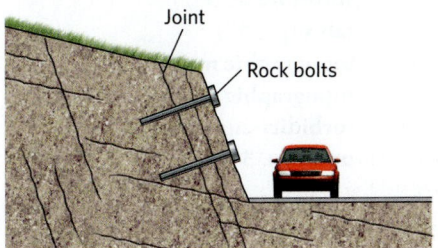

Joint

Rock bolts

(g) Bolting or screening a cliff face can hold loose rocks in place.

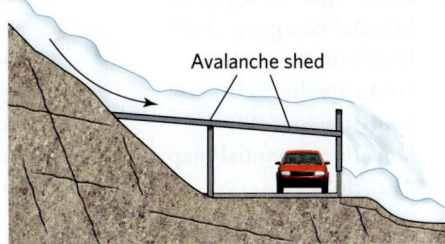

Avalanche shed

(h) An avalanche shed diverts debris or snow over a roadway.

⊙12 CHAPTER REVIEW

- Landscapes represent a consequence of the battle between processes driven by the Earth's internal energy, which cause uplift or subsidence, and processes driven by gravity and by external energy from the Sun, which cause mass wasting, erosion, and deposition.

- Uplift and subsidence generate relief, which can be represented on topographic maps or digital elevation models.

- The types of landforms that develop at a location depend on many factors, including the agents of erosion, relief, climate, substrate composition, and the nature of living organisms. Landscapes change over time.

- Water moves among various reservoirs on and in the Earth during the hydrologic cycle. Without this movement, there would be no erosion or deposition by water, ice, or wind.

- Rock or regolith on unstable slopes has the potential to move downslope under the influence of gravity. This process, called mass wasting, plays an important role in the evolution of landscapes and can be a natural hazard.

- Slow mass wasting caused by the freezing and thawing of regolith is called creep. In places where slopes are underlain with permafrost, solifluction causes a melted layer of regolith to flow downslope.

- During slumping, a relatively coherent mass of material slips down a spoon-shaped failure surface. Mudflows and debris flows occur where regolith has become saturated with water and moves

downslope as a slurry. If the mud consists of volcanic ash, the flow is called a lahar.

- Rockslides and debris slides move very rapidly down a slope, and in a rockfall, the material free-falls down a vertical cliff.

- During snow avalanches, snow mixes with air and moves downslope as a turbulent cloud.

- Large mass-wasting events—submarine slumps, submarine debris flows, and turbidity currents—can take place on underwater slopes. Some generate tsunami.

- Fracturing and weathering weaken rock and make it more susceptible to mass wasting.

- Unstable slopes start to move when the downslope force exceeds the resistance force that holds material in place.

- Downslope movement can be triggered by ground shaking, changes in the steepness of a slope, removal of support from the base of the slope, or changes in the strength of a slope due to deforestation, weathering, or heavy rain.

- Geologists can sometimes detect unstable ground before it begins to move, and they can produce landslide-potential maps to identify areas susceptible to mass wasting.

- Engineers can help prevent dangerous mass wasting by using a variety of techniques to stabilize slopes.

Key Terms

angle of repose (p. 357)
contour line (p. 344)
creep (p. 352)
debris flow (p. 354)
deposition (p. 345)
depositional landform (p. 346)
digital elevation model (DEM) (p. 345)
erosion (p. 345)
erosional landform (p. 346)
external energy (p. 346)

failure surface (p. 353)
geologic cross section (p. 345)
gravitational energy (p. 346)
groundwater (p. 348)
hydrologic cycle (p. 348)
internal energy (p. 346)
landform (p. 346)
landscape (p. 343)
landslide (p. 349)
landslide-potential map (p. 365)

mass wasting (p. 349)
mudflow (p. 354)
regolith (p. 349)
relief (p. 345)
residence time (p. 348)
rockfall (p. 355)
rockslide (p. 355)
shaded-relief map (p. 345)
slope failure (p. 360)
slump (p. 353)
snow avalanche (p. 357)

solifluction (p. 353)
submarine debris flow (p. 359)
submarine slump (p. 359)
subsidence (p. 345)
substrate (p. 347)
talus (p. 357)
topographic map (p. 344)
topographic profile (p. 345)
turbidity current (p. 359)
uplift (p. 345)

Letters in parentheses correspond to the chapter's learning objectives.

1. Why does the Earth's surface constantly undergo change? **(A)**

2. Distinguish between internal and external sources of energy in the Earth System. **(B)**

3. What is the hydrologic cycle and what drives it? **(C)**

4. Explain how soil creep operates. Explain the meaning of the arrows in the sketch. **(D)**

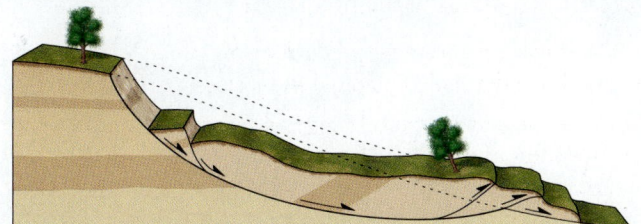

5. What factors do geologists use to distinguish among various types of mass-wasting events? **(D)**

6. Identify the key differences between a slump, a debris flow, a lahar, an avalanche, a rockslide, and a rockfall. Which type of mass wasting does the sketch show? Identify parts of the sketch. **(D)**

7. Why is intact bedrock stronger than fractured bedrock? Why is it stronger than regolith? **(F)**

8. Explain the difference between a stable and an unstable slope. What factors determine the angle of repose of a material? What features are likely to serve as failure surfaces? **(E)**

9. Discuss the variety of phenomena that can cause a stable slope to become so unstable that it fails. Identify the forces in the sketch. **(E)**

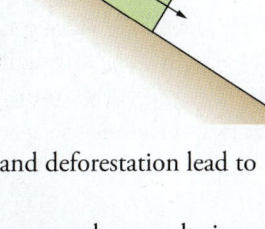

10. How can ground shaking cause fairly solid layers of sand or mud to become weak slurries capable of flowing? **(E)**

11. Discuss the role of vegetation and water in slope stability. Why can fires and deforestation lead to slope failure? **(E)**

12. What factors do geologists take into account when producing a landslide-potential map, and how can they detect the beginning of mass wasting in an area? **(F)**

13. What steps can people take to avoid landslide disasters? What is the purpose of the rock bolts shown in the sketch? **(F)**

14. Imagine that you have been asked by a large bank to determine whether it makes sense to build a dam in a steep-sided, east-west-trending valley in a small central Asian nation. The local government has lobbied for the dam because the country's climate has gradually been getting drier, and the area's farms are running out of water. The bank is considering making a loan to finance dam construction, a process that would employ thousands of now-jobless people. Initial investigation shows that the rock of the valley wall consists of schist containing strong foliation that dips toward the valley and is parallel to the slope of the valley wall. Outcrop studies reveal abundant fractures in the schist along the valley floor. Moderate earthquakes have rattled the region. What would you advise the bank to do? Explain the hazards and what might happen if the reservoir were filled. **(F)**

ANOTHER VIEW The floor of this canyon has been littered with the products of rockfalls. On the far wall, above the people, blocks of sandstone recently detached from bedrock along joints and bedding planes.

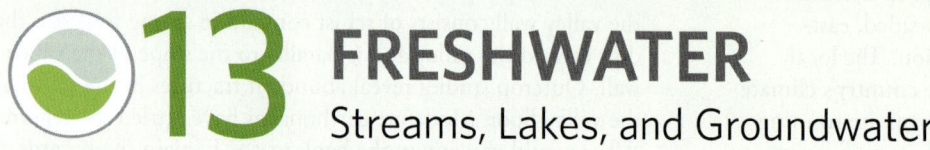

13 FRESHWATER
Streams, Lakes, and Groundwater

By the end of the chapter you should be able to . . .

A. characterize drainage networks, lakes, and wetlands, and describe how they form and evolve.

B. describe the nature of stream discharge and factors that control it.

C. explain how streams produce distinct landscapes by both erosion and deposition.

D. discuss the nature and causes of flooding and how society can protect against flood damage.

E. define groundwater, explain where it comes from and where it occurs, and describe factors that control its flow.

F. distinguish aquitards from aquifers, define the water table, and discuss factors that affect water-table depth.

G. contrast the ways in which groundwater reaches the surface through wells and through springs.

H. discuss the origin of cave networks and related karst landscapes and their relationship to groundwater.

I. identify sustainability and environmental issues that pertain to freshwater resources.

13.1 Introduction

In the 1880s, developers built a clay-and-gravel dam across Pennsylvania's Conemaugh River, trapping a reservoir of cool water to provide a pleasant setting for summer homes for wealthy residents of Pittsburgh. Unfortunately, when torrential rains drenched the state on May 31, 1889, the water level in this reservoir rose so much that water flowed over the dam. Eventually, the soggy structure collapsed, and a wall of water roared downstream and slammed into the city of Johnstown. The infamous "Johnstown flood" killed 2,300 residents and transformed bridges and buildings into twisted wreckage.

Sadly, as residents of Johnstown learned, the Earth's freshwater can occasionally cause catastrophes. But **freshwater**—defined as water that contains less than about 500 parts per million, or ppm (0.05%) dissolved salts—also plays key roles in the Earth System, both as an essential component of life on land and as a sculptor and transporter of Earth materials. Furthermore, freshwater provides food, transportation, and power to human society. Surprisingly, freshwater accounts for just 2.5% of the total volume of water in the Earth's surface and near-surface realms **(Fig. 13.1)**. Of this water, about 69% resides as ice in glaciers and permafrost, and about 30% lies hidden beneath the surface as groundwater, so fresh

FIGURE 13.1 The amount and distribution of the Earth's freshwater.

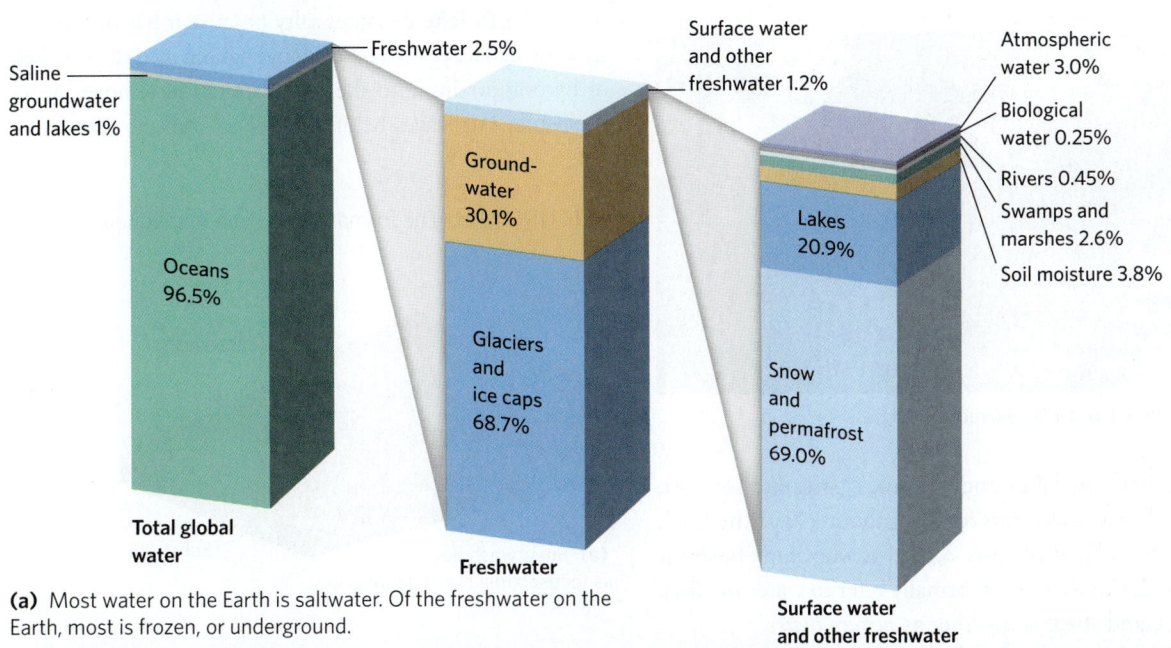

(a) Most water on the Earth is saltwater. Of the freshwater on the Earth, most is frozen, or underground.

<< After a heavy rain, surface water spills over a waterfall near Hilo, Hawaii.

(b) All the water in the world could be collected in a sphere with a diameter of 1,300 km (860 mi). The freshwater portion could be collected in a sphere with a diameter of only 275 km (170 mi).

FIGURE 13.2 Surface water in lakes and wetlands.

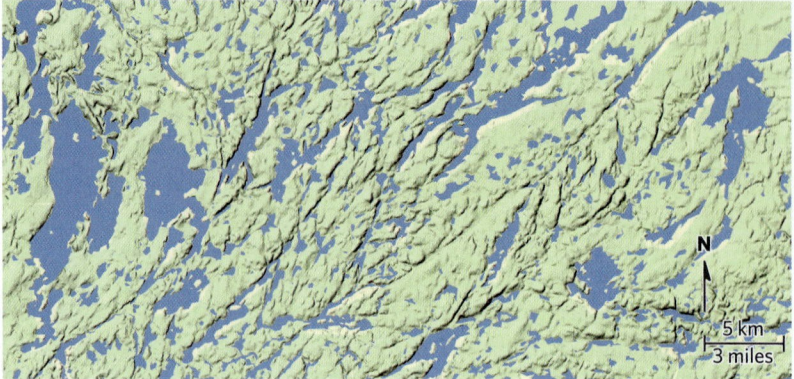

(a) Countless lakes cover the land about 80 km (50 mi) north of the boundary between western Ontario (Canada) and Minnesota. The water table in this cool, rainy region is high, so depressions in the land fill with water.

(b) A swamp in Illinois. The stagnant water is covered with algae.

liquid water—in lakes and wetlands, streams, soil, air, and organisms—accounts for only about 1% of the total. Geologists and hydrologists, scientists who study freshwater in the Earth System, informally refer to water in lakes, wetlands, and streams together as *surface water*.

In this chapter, we first discuss the Earth's surface water. We begin by briefly introducing lakes and wetlands, the "standing" bodies of water. Then we focus on streams to see how they drain the land, how they modify landscapes, and why they sometimes cause devastating floods. We then consider subsurface water, or *groundwater*, to see where it comes from, how it flows, and why it sometimes produces cave networks. This chapter concludes with a look at the challenges of maintaining safe, sustainable freshwater resources.

13.2 Lakes and Wetlands

As we'll discuss in more detail below, streams are flowing bodies of water. In contrast, a **lake** is a standing body of water that occurs on land. By "on land," we mean that lakes do not overlie oceanic crust, and by "standing body," we mean that, to an observer standing on the shore, the

water of a lake seems to stay in place, and not flow over the landscape like water in a stream.

Most lakes contain freshwater. As new water enters a lake, carried by rain, inflowing streams, or springs (seeps from underground), some existing water leaves the lake via an outlet stream. In effect, the water in a freshwater lake does flow, but on a much longer time scale than it does in a stream. The residence time of water in a lake generally ranges from days (for smaller lakes) to years (for larger lakes). A lake's water level can vary over time, depending on the balance between inflow and outflow. Permanent lakes can have enough inflow to keep the bed of the lake submerged all year, whereas in ephemeral lakes, water may dry up during dry seasons.

Some lakes, such as the Great Salt Lake of Utah, contain saltwater. In fact, such lakes may be saltier than the ocean. A **salt lake** becomes salty because it has no outlets, so water can leave the lake only by soaking into the lake bed or by evaporating into the air. Evaporation removes water molecules but leaves behind dissolved ions, so over time,

FIGURE 13.3 The formation of stream channels.

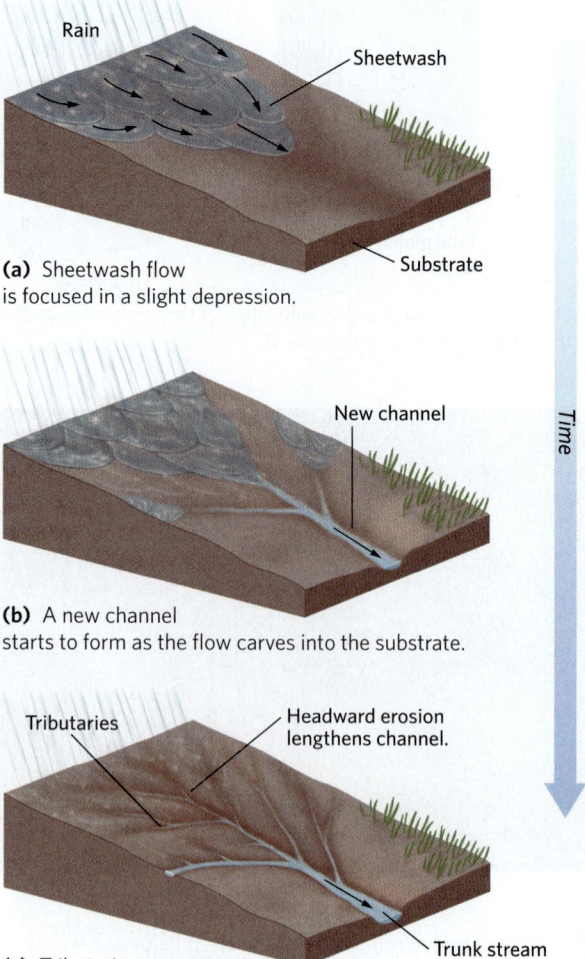

(a) Sheetwash flow is focused in a slight depression.

(b) A new channel starts to form as the flow carves into the substrate.

(c) Tributaries start to develop and feed into the trunk stream.

salts become progressively concentrated in the lake water.

Many lakes exist simply because they are low areas that receive inputs from streams, rain, meltwater, or spring water (Fig. 13.2a). Some lakes are a consequence of specific geologic phenomena, such as the collapse of ground over a cave; the isolation of a segment of a river; glacial erosion or deposition; collapse of a volcanic caldera; blockage of a stream by a landslide; or subsidence due to rifting. Small lakes (called ponds) may be only tens to hundreds of meters across. The largest lakes are hundreds of kilometers across, so wide that you can't see the other side. The Earth's deepest lake, Lake Baikal in Siberia (1.6 km, or 1 mi, deep), and its longest, Lake Tanganyika in Africa (660 km, or 410 mi, long), both formed over narrow rifts. Lake Superior, one of the Great Lakes along the border between Canada and the United States, has the largest surface area—84,000 km² (32,000 mi²).

In a *wetland*, the water is shallow enough that vegetation growing in it rises above the surface (Fig. 13.2b). Earth scientists distinguish among different types of wetlands based on the type of vegetation they host. Specifically, woody plants grow in a *swamp*, grasses grow in a *marsh*, and decaying vegetation and moss fill a *bog*. Wetlands serve as important breeding grounds for birds, fish, insects, and amphibians.

Take-home message . . .

Though lakes are defined as standing bodies of water on land, in fact, freshwater lakes have an outlet, so water in the lake has a residence time measured in days to years. Lakes without outlets become salty.

Quick Question
How does a wetland differ from a lake?

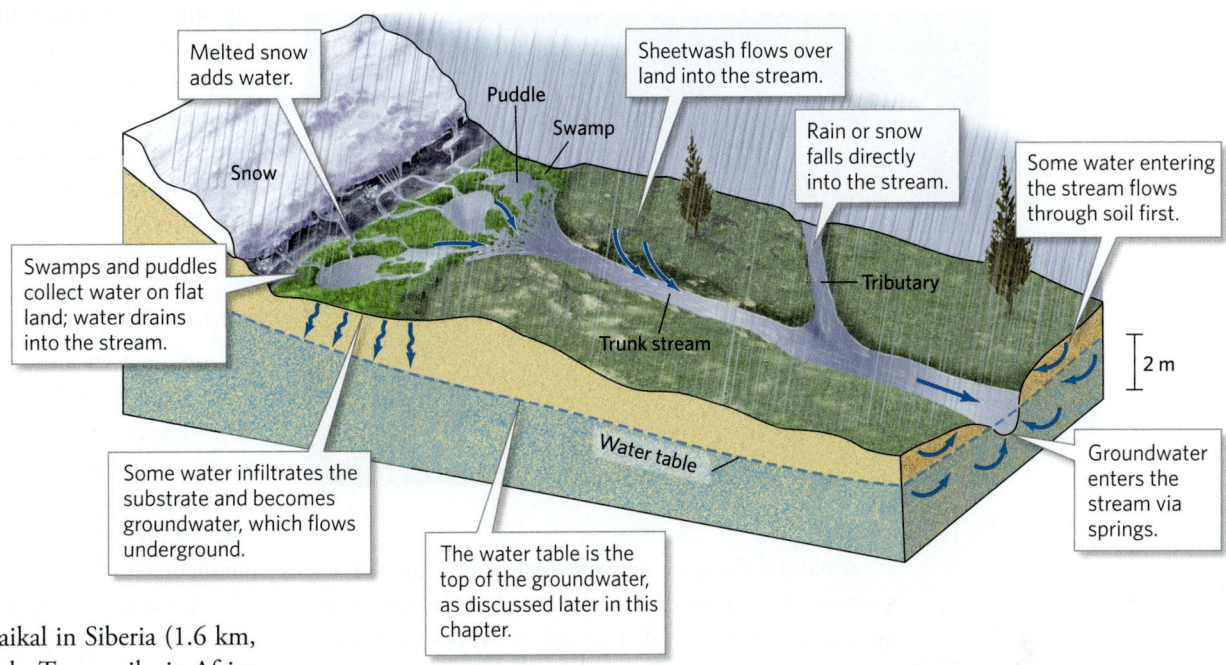

FIGURE 13.4 Runoff comes from rain, from melting ice and snow, and from springs. On flat ground, water accumulates in puddles or swamps, but on slopes it flows downslope in streams.

13.3 Draining the Land

Formation of Streams and Drainage Networks

During the hydrologic cycle, water that entered the atmosphere by evaporation condenses and falls back to the Earth's surface as rain or snow. Some of this *precipitation* accumulates on land in surface water (puddles, lakes, wetlands, or snowfields), some sinks into the ground, and some gets absorbed by plants. Gravity causes excess surface water to flow downslope as **runoff** (Fig. 13.3a). When runoff starts flowing, it does so as a thin film known as *sheetwash*. Where the land surface happens to be a little weaker, or flow happens to be a bit faster, the flow digs down into its *substrate* (the material at and just below the surface) and produces a trough-like depression, or **channel** (Fig. 13.3b, c).

Geologists and hydrologists commonly use the term **stream** for any water flowing along a channel. In everyday English, we refer to large streams as *rivers* and to medium-sized ones as creeks or brooks. Streams receive water from many sources in addition to sheetwash (Fig. 13.4). Inputs come from rain and snow, soil, and *springs* (outlets through which groundwater returns to the surface).

Over time, continued erosion causes a stream channel to deepen, a process called **downcutting**, as well as to lengthen at its origin (*headwaters*), a process known as **headward erosion** (Fig. 13.5). As downcutting progresses along a stream, the surrounding land surface starts to slope toward the stream's channel. Eventually, new side channels, or **tributaries**, form on this land surface and flow into the initial, dominant channel, or *trunk stream* (see Fig. 13.3c). This process establishes a **drainage network**, a linked association of channels that remove surface water from a **drainage basin** or *watershed*.

See for yourself

Headward Erosion, Utah
Latitude: 38°17′58.16″ N
Longitude: 109°50′18.76″ W

Zoom to 8 km (5 mi) and look down.

Intermittent streams flow down tributary canyons into the Colorado River in Canyonlands National Park, Utah. The tributaries cut horizontal strata. A steep escarpment forms at the head of each tributary canyon, where headward erosion takes place.

FIGURE 13.5 An example of headward erosion, as seen from an airplane.

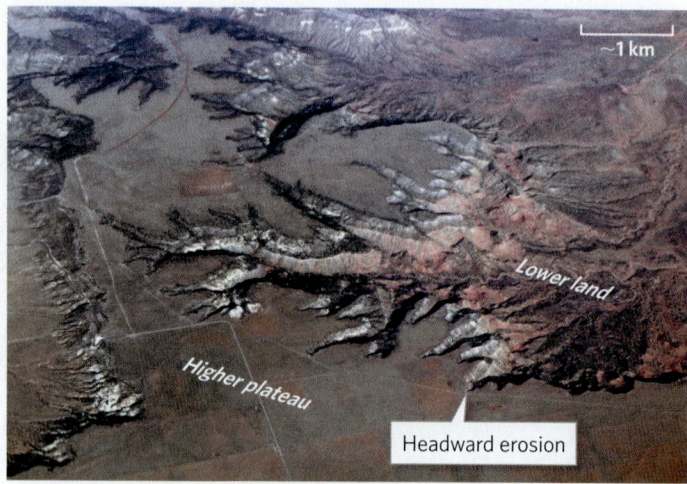

A natural drainage network drains the land the way a network of gutters drains a parking lot. Geologists recognize several forms of drainage networks—dendritic, radial, rectangular, trellis, and parallel—on the basis of the network's map pattern **(Fig. 13.6)**. North America's largest drainage basin, which spans the entire mid-continent region, feeds the Mississippi River. The largest drainage basin in the world covers much of northern South America and feeds the Amazon River **(Fig. 13.7a)**. A **drainage divide** separates one drainage basin from another **(Fig. 13.7b)**. Some divides only separate the streams on opposite sides of a hill or ridge. The largest—*continental divides*—separate drainage networks flowing into one ocean from those flowing into another. The crest of the Andes defines the continental divide of South America (see Fig 13.7a). In North America, "the Continental Divide," whose name you may see on highway signs, separates streams that flow into the Pacific Ocean from streams that flow into the Atlantic Ocean **(Fig. 13.7c)**.

Stream Discharge

Imagine two streams, a larger one in which water flows slowly and a smaller one in which water flows rapidly. Which stream carries more water? To answer this question, researchers compare the streams' average **discharge**—the volume of water that passes through a cross section of the stream (meaning a vertical plane perpendicular to the stream channel) in a given time. We calculate discharge by a simple formula: $D = A \times v$. In this formula, A is the cross-sectional area of the stream, and v is the average velocity at which water moves in the downstream direction. Note that we specify discharge (D) in units of volume (cubic meters or cubic feet) per unit of time.

The average velocity of stream water (v) can be difficult to calculate because not all water in a stream travels at the same velocity. Why? Friction slows water flow,

FIGURE 13.6 Five types of drainage networks.

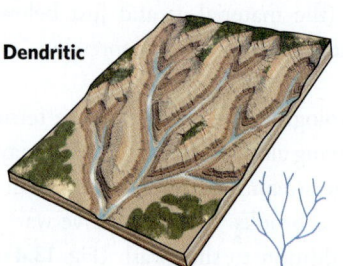

Dendritic

(a) Streams erode a uniform substrate, and connect like the branches of a tree.

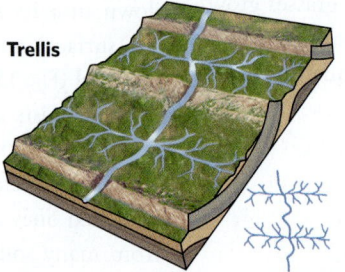

Trellis

(b) Tributaries in valleys intersect a trunk stream that cuts across the ridges.

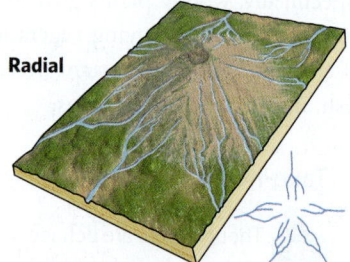

Radial

(c) Streams radiate from a central peak, like the spokes of a wheel.

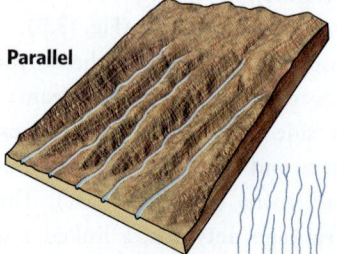

Parallel

(d) Streams form on a uniform slope, and all trend parallel to one another.

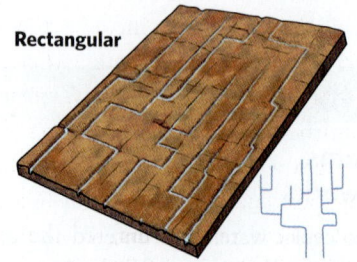

Rectangular

(e) Streams intersect at right angles because they follow pre-existing cracks.

FIGURE 13.7 Drainage divides and basins.

(a) The Amazon watershed of South America stretches from the crest of the Andes to the Atlantic Ocean.

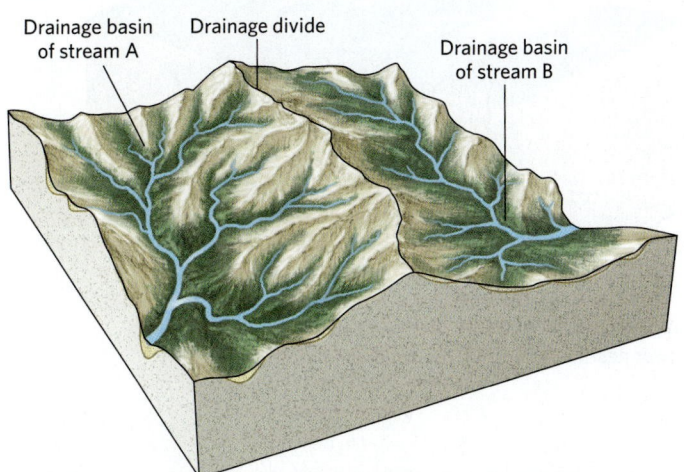

(b) A drainage divide is a ridge that separates two drainage basins.

(c) The major drainage basins of North America. The Great Basin is a region of interior drainage with no outlet to the sea.

so water near the channel walls or near the *streambed* (the floor of the channel) generally moves more slowly than does water in the middle of the flow **(Fig. 13.8a)**. In detail, **turbulence**—the twisting, swirling motion of a moving fluid—causes water to follow paths that greatly exceed the channel's length. In fact, water may flow upstream in part of an *eddy*, and within *whirlpools* it flows in a spiral **(Fig. 13.8b)**.

A stream's average discharge reflects the size of its drainage basin and the climate where the stream flows. For example, the Amazon River, which drains a huge rainforest, has the largest average discharge in the world—about 210,000 m³ (7.4 million ft³) per second. The Mississippi River's drainage basin has half the area of the Amazon's, but its average discharge equals only about 8% of the Amazon's because the central United States receives much less rain than does the Amazon rainforest. Discharge varies with the seasons—it tends to be greatest when winter snows are melting rapidly, or during a rainy season. Discharge may also vary along a stream's length. Specifically, in a wet region, discharge tends to increase downstream because each tributary adds more water, whereas in a dry region, discharge tends to decrease downstream because water seeps into the streambed or evaporates. In fact, not all streams contain water all the time, so geologists distinguish between **permanent streams**, which flow all year **(Fig. 13.9a)**, and **ephemeral streams**, which flow only part of the year. When flow in an ephemeral stream vanishes, its channel becomes a *dry wash* **(Fig. 13.9b)**.

Take-home message . . .

Stream channels form by downcutting and lengthen by headward erosion. Drainage networks remove surface water from a drainage basin and can have a variety of different configurations. Stream discharge depends on the area of a drainage basin and on climate, and can vary along a stream. Some streams are ephemeral.

Quick Question

Why isn't the velocity of stream flow the same everywhere in a stream?

FIGURE 13.8 Flow velocity and its measurement in streams.

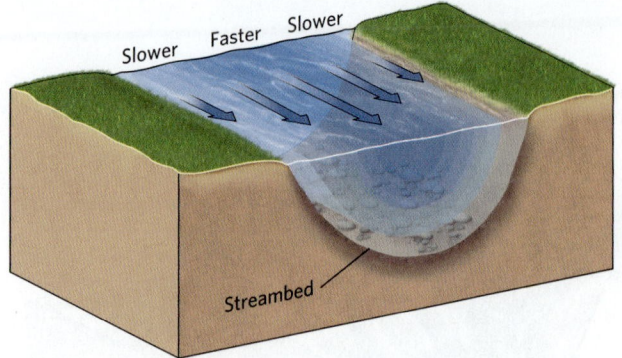

Slower   Faster   Slower

Streambed

(a) Overall, velocity varies with location in a stream. Velocity is slower near the banks and the streambed.

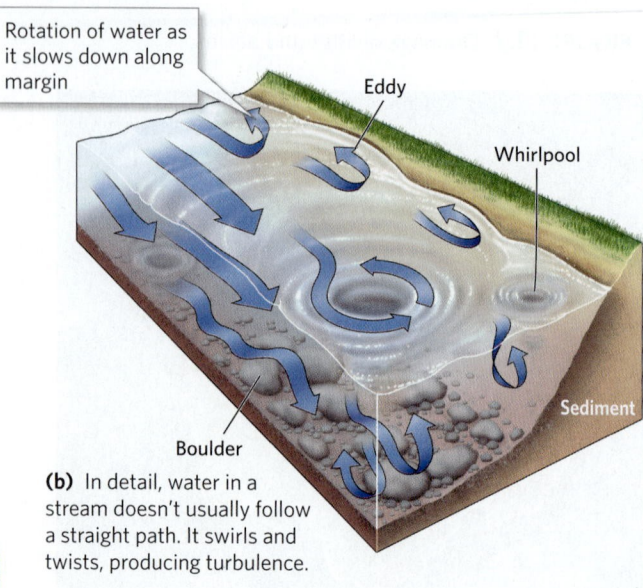

Rotation of water as it slows down along margin

Eddy

Whirlpool

Boulder

Sediment

(b) In detail, water in a stream doesn't usually follow a straight path. It swirls and twists, producing turbulence.

13.4 The Work of Running Water

How Do Streams Erode the Earth's Surface?

If you spray the ground with a hose and watch as the water digs into the soil and carries it away, you are seeing erosion by running water. In the case of a natural stream, gravity causes water to move downslope, and the resulting flow erodes the Earth's surface in four ways: (1) *Scouring* happens when running water picks up and carries away loose sediment. (2) *Breaking and lifting* takes place when running water separates and lifts chunks of rock from the channel.

FIGURE 13.9 The contrast between permanent and ephemeral streams.

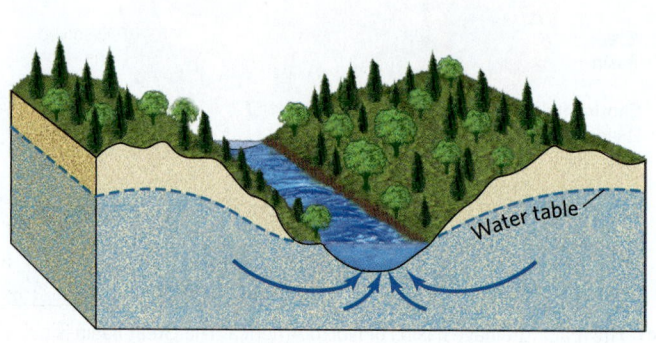

Water table

Permanent stream in Wyoming

(a) The channel of a permanent stream in a temperate climate lies below the water table. Springs add water from below, so the stream contains water even between rains.

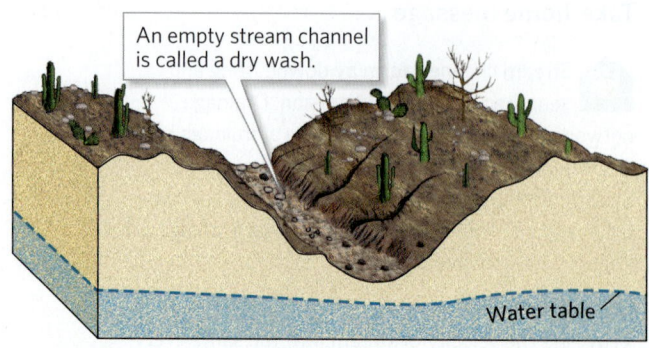

An empty stream channel is called a dry wash.

Water table

Dry wash in Arizona

(b) The channel of an ephemeral stream lies above the water table, so the stream flows only when water enters the stream faster than it can seep into the ground.

FIGURE 13.10 Erosion by streams.

(a) This slot canyon in Arizona has been polished by abrasion.

(b) During floods, sandy water polished this rock outcrop and carved potholes.

(3) *Dissolution* separates ions from minerals in the channel wall, and the stream carries these ions away as solutes in solution. (4) *Abrasion* happens when sediment-laden water acts like sandpaper and rasps away at the channel **(Fig. 13.10a)**. In places where turbulence produces long-lived whirlpools, abrasion can carve a bowl-shaped depression, called a **pothole**, into the streambed **(Fig. 13.10b)**.

The efficiency of stream erosion (or *fluvial erosion*, from the Latin *fluvius*, meaning river) depends on the discharge of the stream and on the stream's sediment content. Therefore, a large volume of fast-moving, turbulent, sandy water causes much more erosion than a trickle of quiet, clear water, so most stream erosion takes place during floods. Notably, the supply of sediment in a stream comes not only from erosion of the stream's channel, but also from landslides that carry debris down slopes bordering the stream and dump it into the stream. As discussed in Chapter 12, stream erosion at the foot of a slope can trigger landslides.

How Do Streams Transport Sediment?

The Mississippi River earned its nickname "Big Muddy" for a reason: the water in it tends to be brown because of the clay and silt it carries. A stream's **sediment load**, meaning the total volume of sediment carried by a stream, includes three components **(Fig. 13.11a)**: (1) *suspended load*, consisting of silt- or clay-sized grains that swirl along with the water without settling to the streambed **(Fig. 13.11b)**; (3) *bed load*, consisting of relatively large clasts that bounce or roll along the streambed; and (3) *dissolved load*, consisting of ions in solution.

When describing a stream's ability to carry sediment, geologists distinguish between competence and capacity.

Stream **competence** refers to the maximum clast size the stream can carry—a stream with more competence can carry larger clasts, while one with less competence can carry only small ones. Competence depends on both water velocity and viscosity, so a fast-moving, turbulent stream

FIGURE 13.11 Streams transport sediment in many forms.

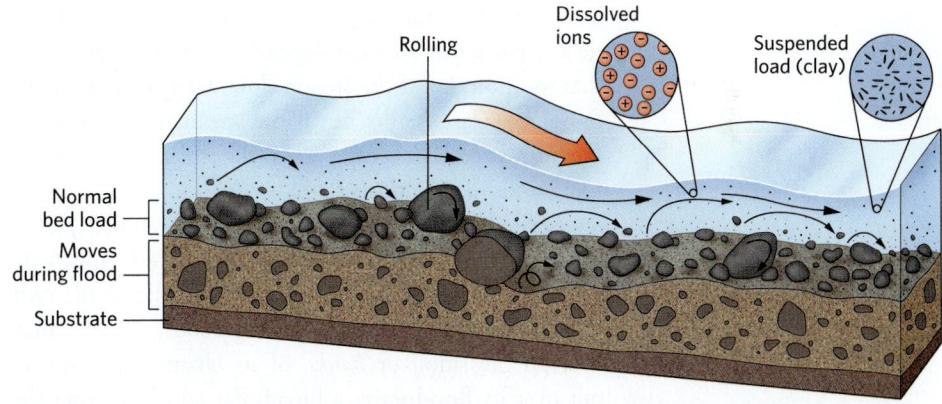

(a) Dissolved ions are carried in solution; tiny suspended grains are distributed throughout the water; and larger clasts slide or roll along the streambed.

(b) The brown color of this turbulent stream water is due to suspended clay and silt.

FIGURE 13.12 Examples of alluvium deposited by streams.

(a) Gravel in the bed of a mountain stream in Denali National Park, Alaska. The large clasts were carried during floods.

(b) Bars of sediment deposited by a stream in the Canadian Rockies.

containing suspended sediment has greater competence than does a slow-moving, clear stream. Stream **capacity** refers to the total quantity of sediment the stream can carry. A stream's capacity depends on both competence and discharge, so a large, fast-flowing river has more capacity than does a small, slowly flowing creek.

How Do Streams Deposit Sediment?

If the flow velocity of a stream decreases, then the competence of the stream also decreases, and its sediment load starts to settle out. The sizes of the clasts that settle at a particular location depend on the flow velocity at that location. So, if the stream slows by a small amount, only large clasts settle, but if the stream slows by a lot, medium-sized clasts settle, and if the stream slows almost to a standstill, fine clasts settle.

Sediments deposited by a stream are called *fluvial deposits* or **alluvium** (**Fig. 13.12a**). Coarser alluvium may accumulate along the streambed in elongate mounds known as **bars** (**Fig. 13.12b**). During floods, a stream may overtop the sides, or *banks*, of its channel and spread out over its **floodplain**, a broad, flat area bordering the stream. Friction slows flow on the floodplain, so fine-grained alluvium settles out to form *floodplain deposits*. Where a stream empties into a standing body of water, the slowdown of flow causes a wedge of sediment called a *delta* to accumulate. (We discuss floodplains and deltas further in Section 13.5.)

How Do Streams Change along Their Length?

In 1803, the United States bought the Louisiana Territory, a vast tract of land encompassing the western half of the Mississippi drainage basin. President Jefferson asked Meriwether Lewis and William Clark to lead a voyage of exploration from St. Louis to the Pacific and to make a map of what they saw along the way. Lewis and Clark, together with about 40 men, began their expedition at the outlet, or *mouth*, of the Missouri River, where it joins the Mississippi. At this juncture, the Missouri is a wide stream of muddy water that can be navigated fairly easily. The farther upstream they went, however, the more difficult their journey became, for the **stream gradient**, meaning the slope of the stream's surface, became progressively steeper, and the stream's discharge diminished. When Lewis and Clark reached North Dakota, they had to abandon their original boats and haul smaller vessels up intervals of the stream where fast, turbulent water swirled over a rocky bed. When they reached Montana, they abandoned boats entirely, and instead trudged along the stream's banks on foot, climbing steep gradients until they reached the Continental Divide.

Geologists can represent the change in stream gradient that Lewis and Clark experienced on a graph that plots elevation on the vertical axis and distance from the mouth on the horizontal axis (**Fig. 13.13**); a curve on this graph depicts the stream's *longitudinal profile*. An idealized longitudinal profile has a concave-up shape, emphasizing that streams have steep gradients near their headwaters and gentle gradients near their mouths. Longitudinal profiles of real streams are not perfectly smooth curves, but rather display local steps.

The lowest elevation on a longitudinal profile, meaning the lowest elevation to which a stream can downcut, defines the **base level** of the stream. At places where a tributary joins a larger stream, the surface of the larger stream acts as the base level for the tributary. A lake, reservoir,

FIGURE 13.13 The character of a stream changes along its length.

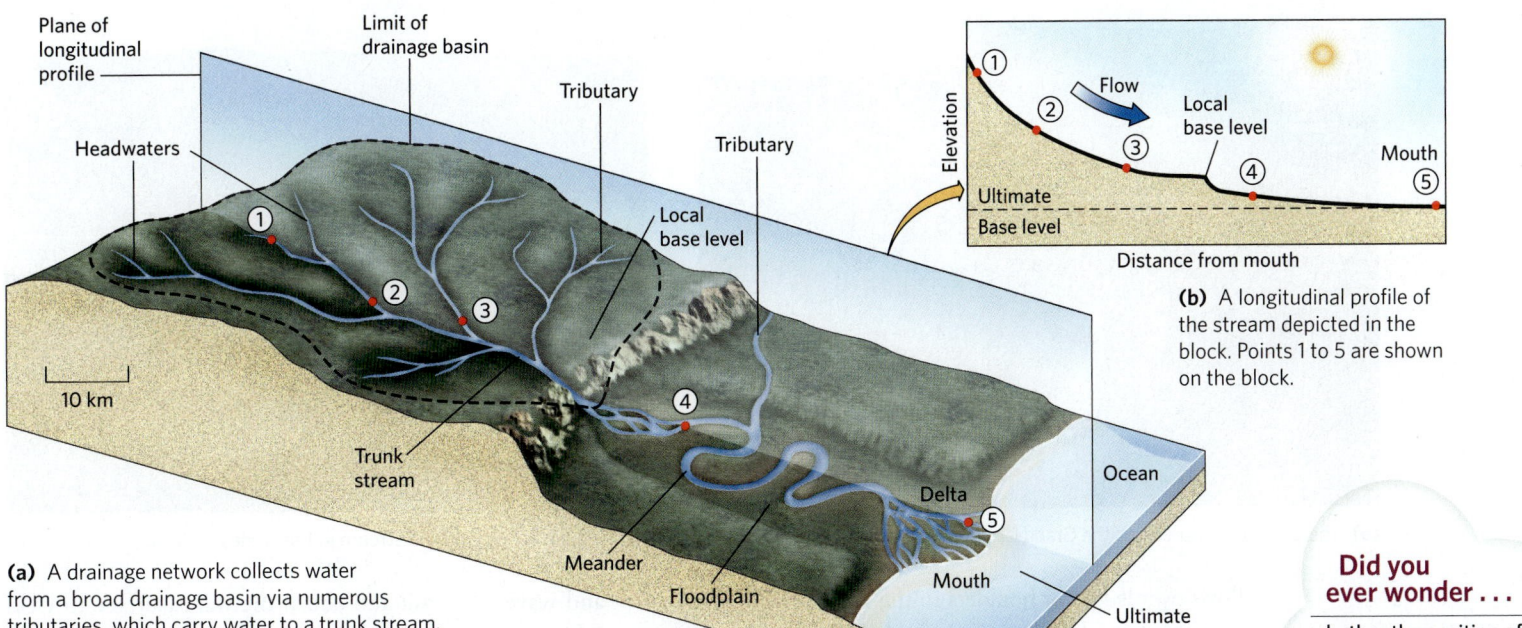

(a) A drainage network collects water from a broad drainage basin via numerous tributaries, which carry water to a trunk stream.

(b) A longitudinal profile of the stream depicted in the block. Points 1 to 5 are shown on the block.

or resistant rock cliff can act as a *local base level* along a stream—these local base levels are the local steps, mentioned above, that appear along a longitudinal profile. A standing body of water at the mouth of a trunk stream serves as the *ultimate base level* of a drainage network, the lowest level of its longitudinal profile. For streams that flow into the ocean, sea level represents the ultimate base level—the surface of a stream can't be lower than sea level, for if it were, the stream would have to flow upslope to enter the sea.

Take-home message . . .

Streams erode the Earth's surface and transport sediment. The size of the clasts a stream carries depends on its flow velocity and water viscosity. Where velocity decreases, sediment settles out. Typically, the gradient of a stream decreases downstream to a base level; sea level serves as the ultimate base level of drainage networks that flow into the ocean.

Quick Question

What's the difference between the competence and the capacity of a stream?

13.5 Fluvial Landscapes

As streams flow, they both carve into and remove their substrate and leave behind sedimentary deposits, producing a variety of distinctive *fluvial landscapes*. The character of these landscapes depends on location along the stream's longitudinal profile, on the nature of the substrate, on

vegetation, and on climate. Let's examine a few of the landforms found in fluvial landscapes.

Valleys and Canyons

About 17 million years ago, a block of crust in the region of what is now the southwestern United States began to rise—its surface became the Colorado Plateau. As the plateau rose, downcutting by the Colorado River produced the Grand Canyon, whose floor now lies 1.6 km (1 mi) below the plateau's surface **(Fig. 13.14a)**. The formation of the Grand Canyon emphasizes that in regions where the land surface lies well above the base level, a stream can carve a trough much deeper than the channel itself. We refer to the trough as a *canyon* if its walls slope steeply, and as a *valley* if they slope gently **(Fig. 13.14b)**.

Whether stream erosion produces a canyon or a valley depends on the rate at which downcutting takes place relative to the rate at which mass wasting causes the walls on either side of the stream to collapse. In places where a stream downcuts faster than the walls collapse, a canyon develops **(Fig. 13.15a)**, whereas in places where the walls collapse as fast as the stream downcuts, a valley develops **(Fig. 13.15b)**. If a stream cuts through alternating layers of resistant and nonresistant rock, the resulting canyon has a stair-step profile **(Fig. 13.15c)**.

Rapids and Waterfalls

Rafters and kayakers seek out **rapids**, intervals of a stream where the water surface is particularly turbulent and rough **(Fig. 13.16a)**. Rapids can develop where water

Did you ever wonder . . .

whether the position of a waterfall can change over time?

See for **yourself**

River-cut gorge in the Himalayas

Latitude: 28°9'41.82" N
Longitude: 85°23'1.04" E

Zoom to 6 km (~4 mi) and look obliquely to the NE.

A deep gorge, about 50 km (31 mi) north of Katmandu, was cut by a river flowing out of the Himalayas. This upper reach of the river has a steep gradient. If you look straight upstream, you'll note the valley's V-shaped profile.

FIGURE 13.14 Examples of landscapes cut by streams.

(a) The Colorado River carved the Grand Canyon, here seen at its western end.

(b) A deep valley cut by a stream in the Andes of Peru.

flows over ledges or boulders in the streambed, where the channel abruptly narrows, or where the channel's gradient abruptly changes. Turbulence in rapids generates eddies

FIGURE 13.15 The shape of a canyon or valley formed by downcutting depends on the resistance of its walls to erosion.

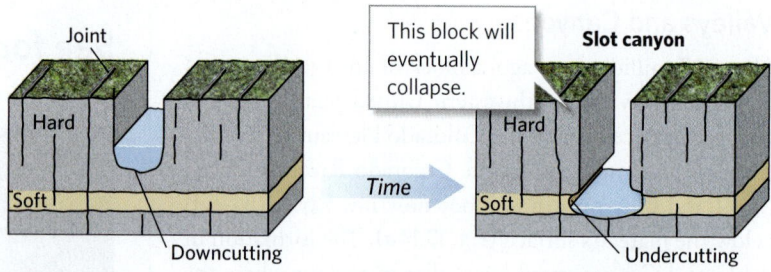

(a) If downcutting by the stream happens faster than mass wasting on the walls, a slot canyon forms. The canyon widens as the stream undercuts the walls.

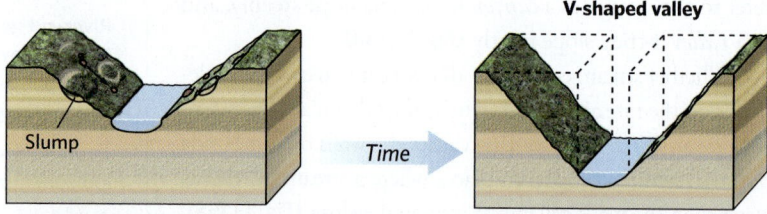

(b) If mass wasting takes place as fast as downcutting occurs, a V-shaped valley develops.

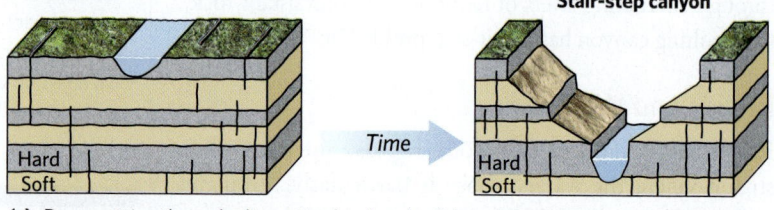

(c) Downcutting through alternating hard and soft layers produces a stair-step canyon.

and waves that roil and churn the water surface to yield *whitewater*, a mixture of bubbles and water.

A **waterfall** exists where the stream gradient becomes so steep that the stream's water free-falls above the streambed **(Fig. 13.16b)**. The energy of falling water may excavate a *plunge pool* at the base of the waterfall. Waterfalls evolve as headward erosion eats away the resistant ledge that underlies them. You can see an example of this process at Niagara Falls, where water flowing from Lake Erie to Lake Ontario drops over a 55-m (180-ft)-high escarpment of resistant dolostone overlying weak shale **(Fig. 13.16c)**. Over time, erosion of the shale undercuts the dolostone, leaving an overhang that eventually collapses, causing the position of the waterfall to migrate upstream **(Fig. 13.16d)**.

Braided and Meandering Streams

In places where runoff flows over a gently sloping land surface, two distinctive types of streams evolve: braided streams and meandering streams. The type you'll find at a locality depends on the character and volume of the stream's sediment load and on the *cohesion* of the stream's banks (meaning the degree to which they can hold together).

A **braided stream** develops where flowing water carries abundant coarse sediment (because the sediment source lies relatively nearby) and where the stream's banks have low cohesion. The adjective *braided* emphasizes that the stream consists of multiple channels that weave around elongate sediment bars and therefore entwine like hair in a braid **(Fig. 13.17)**. How does a braided stream form? During floods, water in the stream rises and picks up large volumes of sediment, which it transports downstream. When flooding stops and the water slows, sediment settles out in elongated bars, and the remaining water divides to flow in the low areas between the bars.

FIGURE 13.16 Rapids and waterfalls.

(a) These rapids in the Grand Canyon formed when a flood from a side canyon dumped debris into the channel of the Colorado River.

(b) Horseshoe Falls, a part of Niagara Falls.

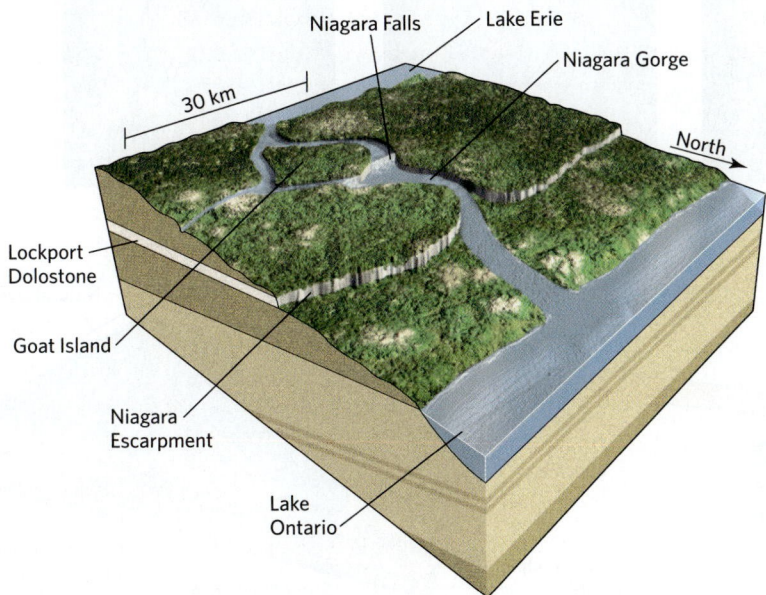

(c) The falls developed where water spills over the Lockport Dolostone, a resistant rock layer that forms the Niagara Escarpment.

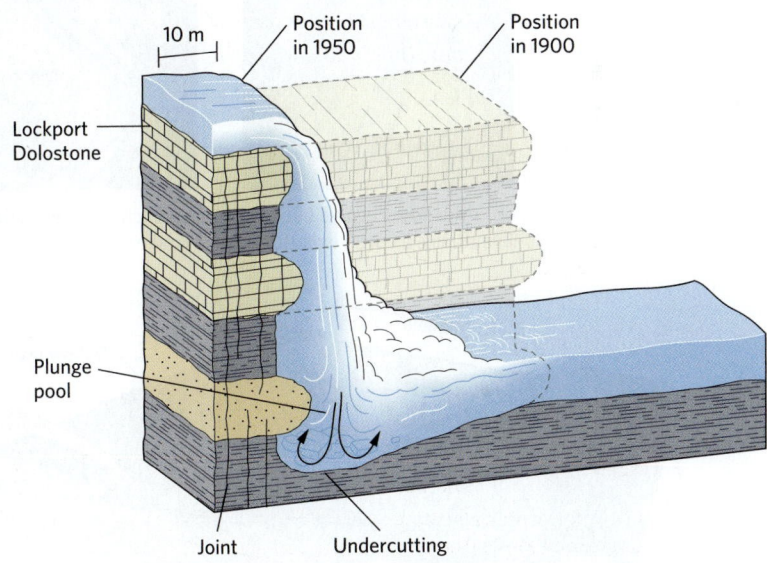

(d) Over time, as the erosion of shale undercuts the dolostone, the position of the falls migrates upstream.

A **meandering stream** develops where sediment doesn't choke the stream and where the stream's banks have high cohesion either because their materials stick together or because they are bound by plant roots. The adjective *meandering* emphasizes that the stream's channel winds back and forth in a succession of snake-like curves called **meanders (Fig. 13.18a)**. How do meanders form and evolve? Even if a stream starts out with a straight channel, the location of the strongest current in the stream tends to wander, so that it sometimes lies nearer the center of the channel and sometimes nearer the banks **(Fig. 13.18b)**. Water erodes more rapidly where

FIGURE 13.17 Strands of this braided stream carry meltwater from a glacier near Denali, Alaska. The sediment bars were deposited at times when the stream was in flood and had greater competence.

FIGURE 13.18
Development
of meandering
streams.

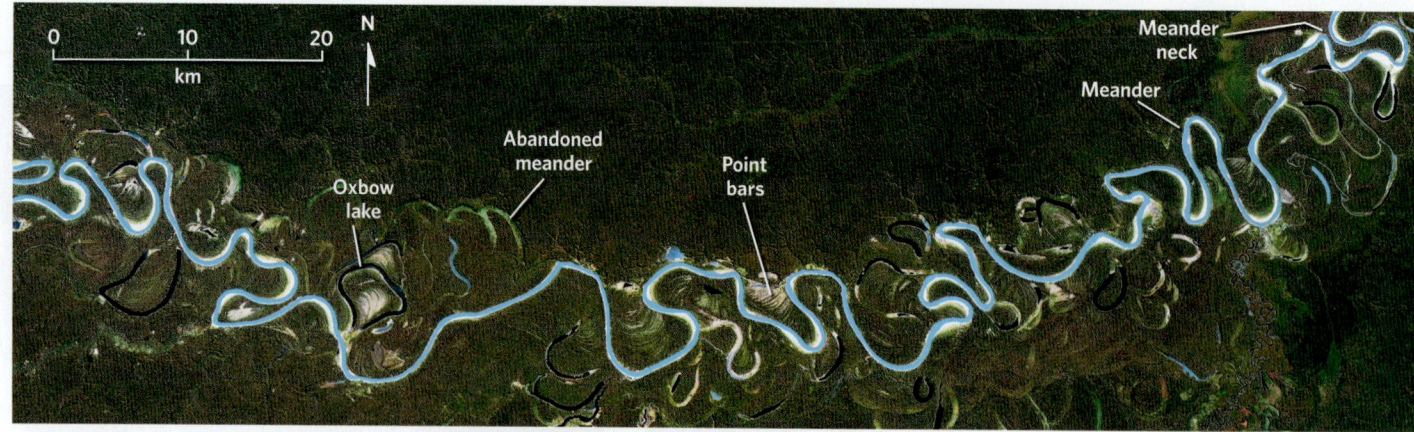

(a) A meandering stream in Brazil, as viewed from space. Note the various landscape features.

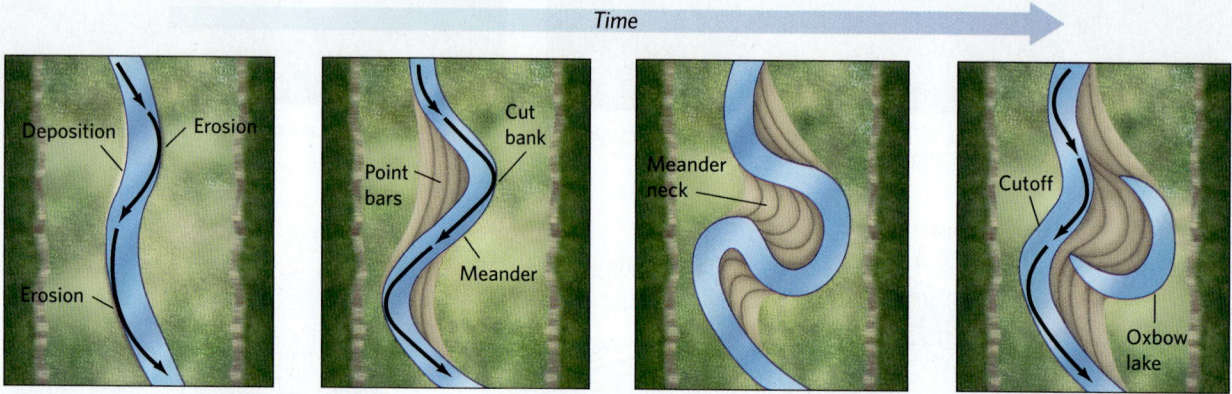

(b) Meanders evolve because erosion occurs faster on the outer bank of a curve, while deposition takes place on the inner bank. Eventually, a cutoff isolates an oxbow lake.

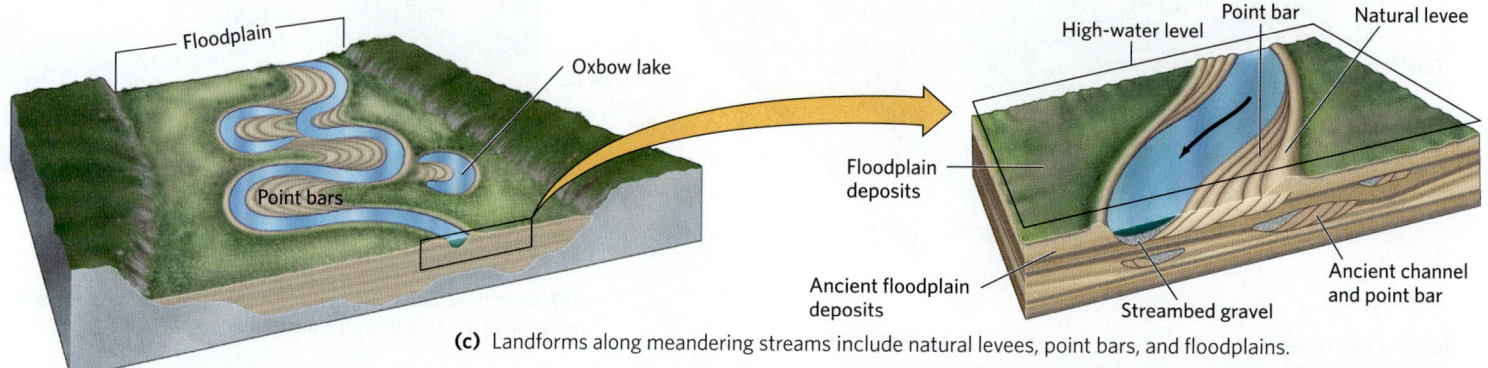

(c) Landforms along meandering streams include natural levees, point bars, and floodplains.

(d) Two oxbow lakes are visible in this aerial photo of a meandering stream.

(e) The flat land of this floodplain hosts farm fields. The dark area is the stream channel, bordered by trees.

it flows faster, so when the fastest current runs along the bank, the stream digs into the bank, carving into it and producing a small escarpment (a steep slope) called a *cut bank*. As the process continues, the deepest part of the channel (known as the *thalweg*) stays put along the outer arc of the curve. Meanwhile, on the inside edge of the curve, water slows down, its competence decreases, and sediment accumulates in a wedge-shaped deposit called a **point bar (Fig. 13.18c)**. As time passes, the cut bank moves outward, the point bar grows wider, and the curvature of the meander becomes more pronounced. Eventually, a meander may curve through more than 180°, and two neighboring meanders may approach each other until only a narrow isthmus, known as a *meander neck*, separates them. When erosion finally eats through a meander neck, a *cutoff* develops. The meander that has been cut off then becomes a curving **oxbow lake (Fig. 13.18d)**. If sediment fills in the lake, the curving depression remains visible as an *abandoned meander*. Production of new meanders and point bars and abandonment of old ones can cause the course of a meandering stream to change markedly on a time scale of years to centuries.

Most meandering stream channels cover only a relatively small portion of a broad, nearly flat *floodplain* **(Fig. 13.18e)**. As we noted earlier, water overtops the banks of the stream channel and spreads out over the floodplain during a flood. Notably, the initial slowdown of water as it overtops the banks causes some sediment to settle out directly along channel's banks. This sediment builds up to produce a pair of low ridges, called **natural levees**, on either side of the channel. Commonly, a floodplain terminates along an escarpment bounding the edge of higher land that doesn't flood (see Fig. 13.18c).

The End of the Line for Sediment: Alluvial Fans and Deltas

We've seen that fluvial sediment gets deposited along streams, in bars between or along channels, or out on the floodplain. Some sediment, however, makes it all the way to the mouth of the stream and accumulates there. Where a stream exits a narrow canyon onto a plain in a dry environment, this sediment forms a wedge called an **alluvial fan (Fig. 13.19a)**, and where a stream enters a standing body of water, such as a lake or the sea, it forms a **delta (Fig. 13.19b)**. The name *delta* comes from the observation by an ancient Greek historian, Herodotus, that the surface of the wedge of sediment deposited by the Nile River where it enters the Mediterranean Sea resembles the shape of the Greek letter delta (Δ) **(Fig. 13.20a)**. Relatively few deltas have the distinct triangular shape of the Nile Delta—some are more rounded, and some look like the scrawny feet of birds **(Fig. 13.20b)**. The particular shape of a delta depends on the interplay between the rate at which the river supplies new sediment and the rate at which waves or currents remove sediment. A delta deposited by a major river along the seacoast may become very thick and wide, so its surface can become a broad *delta plain*. On alluvial fans and deltas, streams divide into several branches known as *distributaries*.

Evolution of Stream-Eroded Landscapes

Imagine a place where geologic processes cause the land surface of a region to rise. When this happens, the landscape evolves **(Fig. 13.21)**. Stream downcutting initially produces deep valleys or canyons with steep gradients. Over time, mass wasting and erosion transform the

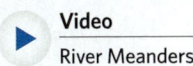
Video
River Meanders

FIGURE 13.19 Examples of depositional landforms formed at the mouths of streams.

(a) This alluvial fan in Death Valley, California, consists of sand, gravel, and debris flows. The curving black line is a road.

(b) A small delta in Costa Rica.

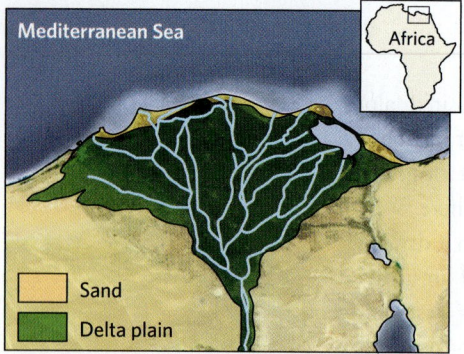

Mediterranean Sea

Sand

Delta plain

(a) The Nile Delta is shaped like the Greek letter Δ.

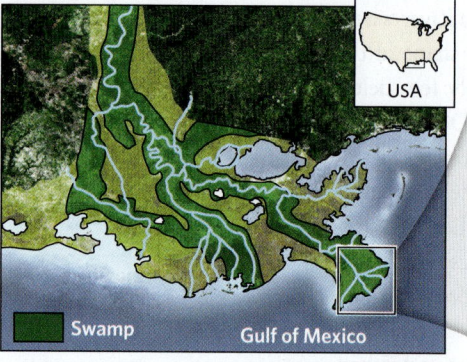

USA

Swamp

Gulf of Mexico

(b) The Mississippi Delta is a bird's-foot delta.

FIGURE 13.20 Not all deltas look the same.

Mouth of the Mississippi as seen from space.

Natural levee

13.6 Raging Waters: River Flooding and Flood Control

During a **flood**, the volume of water flowing down a stream exceeds the volume of the stream channel, so water overtops the banks and spreads out over a flood-plain or delta plain, or fills a canyon to a greater depth than normal. When news reports say that a stream has risen above **flood stage**, they mean that the water level in the stream lies above the elevation of the stream's floodplain, has overtopped the stream's banks, or has submerged property of value. Floods can be classified in different ways: location-based classifications distinguish among river floods, coastal floods, and urban floods, whereas time-based classifications distinguish between *slow-onset floods* and *flash floods*. We'll focus on the latter approach here.

Slow-Onset Floods

A flood that takes days to develop, lasts for days to weeks, and involves the trunk stream of a drainage network can be called a **slow-onset flood (Fig. 13.24)**. Slow-onset floods may happen when the ground becomes saturated and cannot absorb any more water, so runoff increases. They may occur during the spring when snowmelt happens quickly, during sustained rains in a wet season (see Chapter 18), or when a storm system remains stationary over a region for a long time. Note that because of the link between the timing of slow-onset floods and the seasons, some slow-onset floods can also be called *seasonal floods*.

Examples of slow-onset floods make headlines every year **(Box 13.1)**. For example, during 2010, particularly intense monsoonal rains in Pakistan caused the Indus River and its tributaries to rise slowly until water covered about one-fifth of the country's land area. The flooding affected about 20 million people and led to about 2,000 fatalities. In 1931, a slow-onset flood of the Yellow and Yangtze Rivers in China may have led to the deaths of over 3 million people. This terrible event happened when melting of the winter's heavy snowfall was followed by

rugged landscape into low, rounded hills, and valleys broaden into wide floodplains with gentle gradients. Eventually, even the low hills get eroded down, leaving a nearly flat land surface at an elevation close to that of the drainage network's base level. On our dynamic planet, however, changes in land elevation or sea level commonly prevent landscapes from ever getting shaved all the way down to the base level. If the land surface under a drainage network rises or the base level of the network falls, **stream rejuvenation** takes place, meaning that streams start to downcut again. Rejuvenation of a meandering stream may produce dramatic *incised meanders* **(Fig. 13.22)**. If regional uplift causes the tilt of the land to change, the flow of streams in a drainage network may eventually change direction, a phenomenon known as a *drainage reversal*.

The details of fluvial landscape evolution can also be complicated by local uplift or subsidence. For example, if uplift of a mountain range in the path of a stream happens faster than the stream can erode the range, the uplift diverts the path of the stream, but if erosion can keep pace with uplift, the stream may cut across the range. If erosion takes place on one side of a drainage divide faster than it does on the other side, the position of the drainage divide may move over time. And if headward erosion causes the channel of one stream to intersect the channel of another, the water of one stream may start to flow down the channel of the other, a phenomenon known as *stream piracy* **(Fig. 13.23)**.

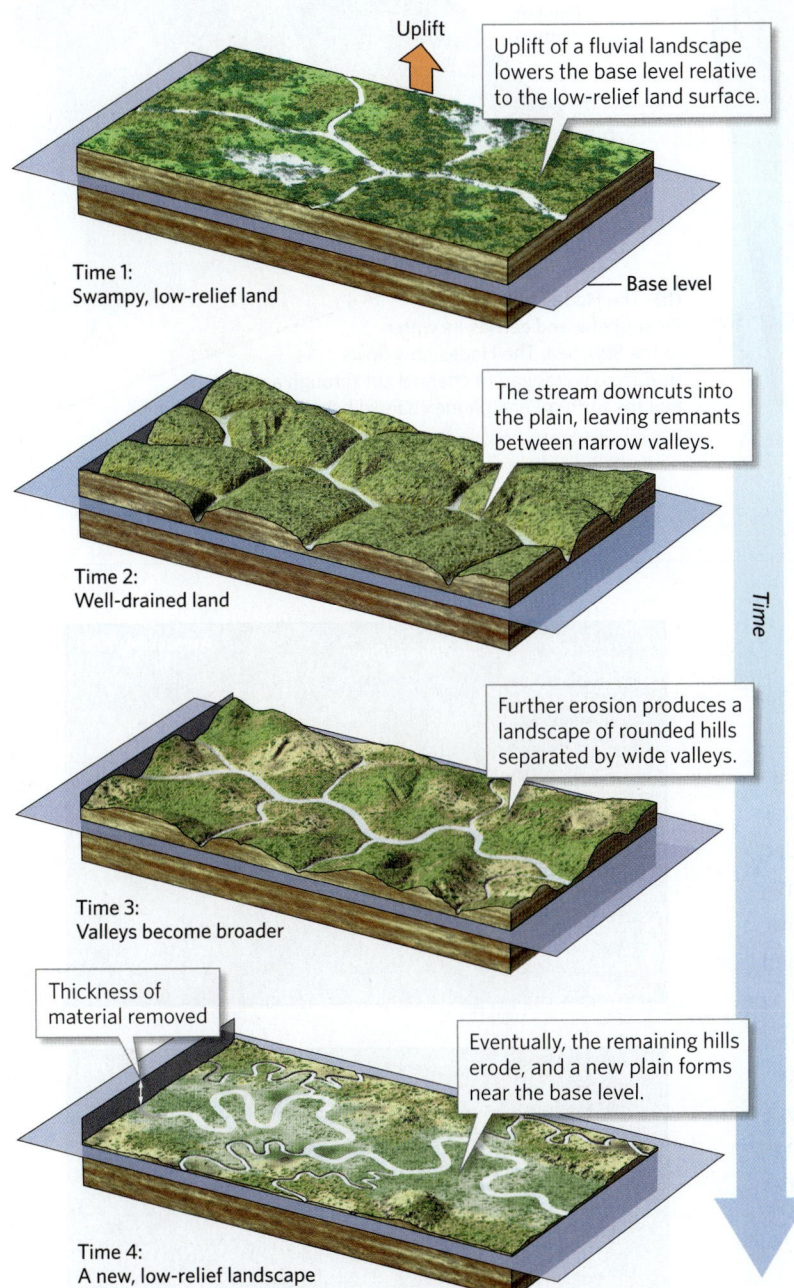

FIGURE 13.21 Fluvial landscapes evolve over time. When the base level drops, streams downcut, and a hilly landscape develops. The resulting relief, however, eventually erodes away.

Uplift

Uplift of a fluvial landscape lowers the base level relative to the low-relief land surface.

Time 1:
Swampy, low-relief land

Base level

The stream downcuts into the plain, leaving remnants between narrow valleys.

Time 2:
Well-drained land

Further erosion produces a landscape of rounded hills separated by wide valleys.

Time 3:
Valleys become broader

Thickness of material removed

Eventually, the remaining hills erode, and a new plain forms near the base level.

Time 4:
A new, low-relief landscape

Time

heavy monsoon rains and a succession of typhoons (hurricanes). At its peak, water in the Yangtze River rose to 16 m (53 ft) above flood stage. In 2017, disastrous floods due to persistent storms affected Texas and California. The California floods were associated with *atmospheric rivers*, streams of very moist air that can drop torrential rains over land.

Flash Floods

An event during which the discharge of a stream increases so fast that it may be difficult to escape from the path of the water is a **flash flood (Fig. 13.25)**. The floodwaters may subside in minutes to hours and generally affect relatively small areas. Flash floods may happen during particularly intense rainfall—when so much water drenches an area that there isn't time for it to sink into the ground, so it becomes runoff—or when dams collapse (as in the 1889 Johnstown flood). During a flash flood, a narrow canyon or valley may fill to a level many meters above the channel banks in a matter of minutes, and the leading edge of the flood may rush downstream as a wall of water. Flash floods can become so turbulent and fast, and carry so much sediment, that they carry away everything in their path—boulders, trees, bridges, cars, and houses—and leave devastation in their wake. Flash floods can take place in any climate, but are especially dramatic in arid regions, where a thunderstorm can suddenly trigger a muddy, turbulent flow in a dry wash.

FIGURE 13.22 The goosenecks of the San Juan River, Utah, are incised meanders.

FIGURE 13.23 Stream piracy.

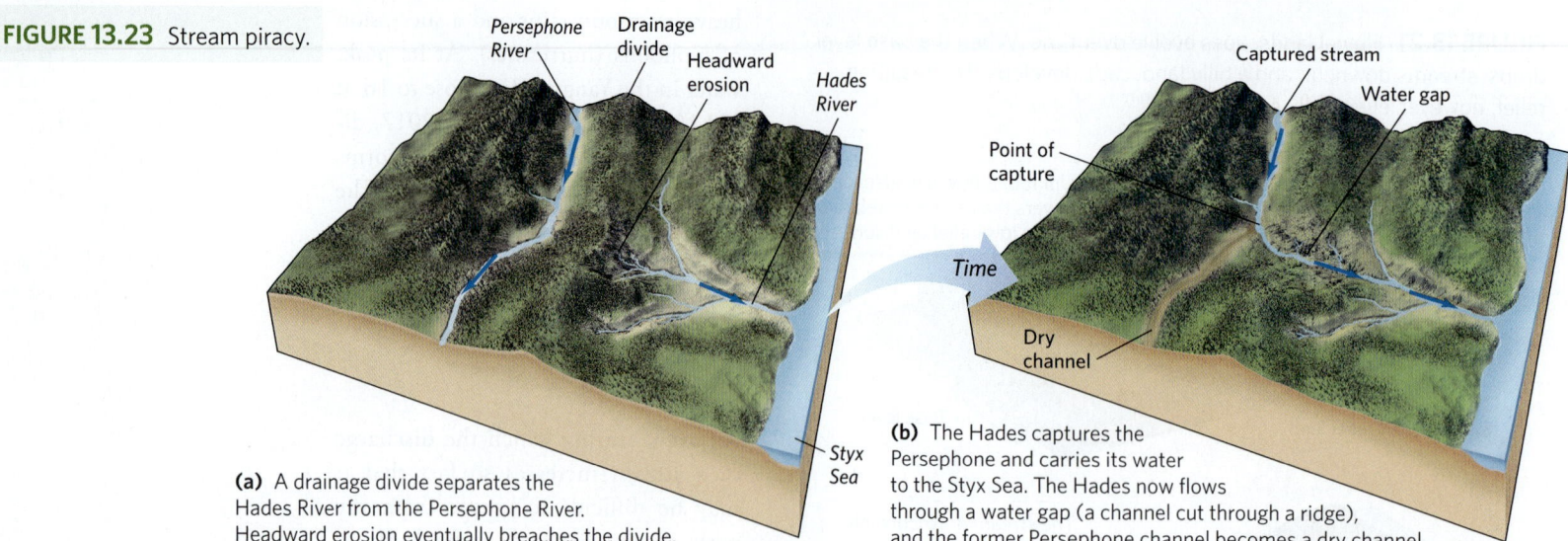

(a) A drainage divide separates the Hades River from the Persephone River. Headward erosion eventually breaches the divide.

(b) The Hades captures the Persephone and carries its water to the Styx Sea. The Hades now flows through a water gap (a channel cut through a ridge), and the former Persephone channel becomes a dry channel.

BOX 13.1 ▸ **A Deeper Look**

The 1993 Mississippi River flood

The Mississippi River drainage network has had its share of slow-onset floods. The worst in recent decades took place in the summer of 1993, when unusual weather patterns caused a whole year's supply of rain to fall across the upper Midwest of the United States during a period of a few weeks—some regions received 400% more rain than usual. Immense volumes of runoff flowed into the Mississippi drainage network. Eventually, the water of the Mississippi and Missouri Rivers spread out over the rivers' floodplains, and by July, parts of nine states were underwater (Fig. Bx13.1). Bridges and roads were washed away and towns were submerged. Rowboats replaced cars as the favored mode of transportation in towns where only the rooftops remained visible. In St. Louis, Missouri, the river reached a maximum level, or *crested*, at 14 m (47 ft) above flood stage. For 79 days, the flooding continued. When the water finally subsided, it left behind a thick layer of sediment, filling living rooms and kitchens in towns and burying crops in fields. In the end, more than 40,000 km² (15,400 mi²) of the floodplain had been submerged, 50 people had died, and at least 55,000 homes had been destroyed. Comparable flooding struck the Mississippi River and its tributaries again in 2011 and in 2019. The 2019 floods were caused because heavy winter snow melted suddenly, making the ground soggier than had ever been recorded over the previous 124 years. When spring rains began, and were heavier than usual, water couldn't soak into the ground because it was already saturated. This meant that rivers overflowed and floodplains became submerged. Many states declared states of emergency.

FIGURE Bx13.1 Satellite photos show the extent of inundation just before and during the 1993 Mississippi River flood.

FIGURE 13.24 Examples of slow-onset flooding.

(a) This flooding in Dhaka, the capital city of Bangladesh, was caused by monsoon rains and submerged city streets, making travel hazardous.

(b) Flooding in the Sacramento Valley of California in February 2017 due to rain from an "atmospheric river."

Living with Floods

Mark Twain once wrote of the Mississippi that we "cannot tame that lawless stream, cannot curb it or confine it, cannot say to it, 'go here or go there,' and make it obey." Was Twain right? Since ancient times, people have attempted to control the courses of rivers to prevent flooding. In the 20th century, such *flood control* intensified as populations living along rivers increased. For example, following a disastrous flood in 1927, the US Army Corps of Engineers began a program to control Mississippi flooding. First, they built about 300 dams along the river's tributaries to store excess runoff until it could be released safely. Second, they built *artificial levees* of sand and mud, as well as *floodwalls* of concrete (Fig. 13.26a, b). These structures are designed to increase the channel's volume and to isolate portions of the floodplain from flooding.

Although the Corps' strategy worked for floods up to a certain size, it was insufficient to handle the 1993, 2011, and 2019 Mississippi floods, when reservoirs filled to capacity and additional runoff headed downstream. The river rose until it spilled over the tops of some levees and undermined others. *Undermining* occurs when rising water levels increase the water pressure at the base of the river channel, forcing water through sand under the levee (Fig. 13.26c). Water then spurts out of the ground on the floodplain side of the levee, thereby washing away the levee's support until the levee collapses (Fig. 13.26d). Newer flood control approaches involve both restoration of wetland areas along rivers—for wetlands can absorb significant quantities of floodwater—and designation of *floodways*, regions in which construction has been banned so that floodwaters have room to spread out without causing expensive damage.

Because flooding is a danger to society, it's important to be able to estimate how susceptible an area might be to floods of a given size. To do this, researchers keep records of the variation in a stream's discharge, as well as when and by how much a stream rises above flood stage. Such data helps them calculate the average frequency of floods of a given size (Box 13.2).

Take-home message . . .

Slow-onset floods submerge broad areas for days or weeks at a time, whereas flash floods are sudden and short-lived. Engineers can build reservoirs, levees, and floodwalls to limit casualties and damage due to flooding.

Quick Question
Does a 100-year flood only happen every 100 years? Why or why not? What's its annual probability?

Did you ever wonder . . .
what newscasters mean by a "100-year flood"?

13.7 Introducing Groundwater

When it rains, some of the water that falls on the land sinks or percolates down into the ground, a process called **infiltration**. Of the water that infiltrates, some descends only into the soil. This water, *soil moisture*, may later evaporate, be absorbed by plant roots, rise back to the ground surface, or seep into streams. The remainder sinks deeper into sediment or rock, where, along with water trapped in rock at the time the rock

FIGURE 13.25 Examples of flash flooding.

(a) A flash flood in a desert region of Israel has washed over a highway, forcing the evacuation of truckers.

(b) A 2013 flash flood in India washed away bridges and stranded 40,000 people.

(c) A 2019 flash flood in western China filled a dry wash with muddy water and submerged a road.

(d) During the 1976 Big Thompson River flash flood, this house was carried off its foundation and dropped on a bridge.

formed, it makes up **groundwater**. Significantly, groundwater exists only in the Earth's upper crust, for below depths of 20–30 km (12–19 mi), metamorphic reactions incorporate water, plastic flow of rock destroys space for water, or water no longer can exist in liquid form.

Porosity and Permeability

Geologists refer to a small space within a volume of sediment or within a body of rock as a **pore**, and to the total volume of open space within a solid material, specified as a percentage, as **porosity** (Fig. 13.27a). For example, if we say that a block of rock has 30% porosity, then 30% of the volume of the block consists of open space. Porosity includes both pores between solid mineral grains in sediment or rock and cracks of various sizes **(Fig. 13.27b)**.

For groundwater to flow, pores or cracks in rock or sediment must be linked by *conduits* (openings). The resulting interconnectivity determines a material's **permeability** **(Fig. 13.27c)**. Groundwater flows easily through a permeable material such as loose gravel, flows slowly through a low-permeability material, and can't flow at all through an impermeable material. The permeability of a material depends on three factors:

1. *Number of conduits:* As the number of conduits increases, permeability increases.

2. *Conduit size:* Fluid travels faster through wider conduits than through narrower ones.

FIGURE 13.26 Flood control strategies.

(a) Artificial levees have been built to protect the town of Galena, Illinois.

(b) A concrete floodwall at Cape Girardeau, Missouri. When floods threaten, a crane drops a gate into the slot to keep out the Mississippi River.

3. *Conduit straightness:* Water flows faster through straight conduits than it does through crooked ones.

Note that the factors controlling permeability resemble those controlling the ease with which traffic moves through a city. Traffic can flow quickly through cities with many straight multilane boulevards, whereas it flows slowly through cities with only a few narrow, crooked streets. Note, too, that a high-porosity rock or sediment does not necessarily have high permeability—a material whose pores are isolated from one another can have high porosity but low permeability **(Box 13.3)**.

Aquifers, Aquitards, and the Water Table

With the concepts of porosity and permeability in mind, geologists distinguish between an **aquifer**, rock or sediment with high permeability and porosity, and an **aquitard**, rock or sediment with low permeability regardless of porosity. Geologists further distinguish between an *unconfined aquifer*, which starts at the ground surface and extends downward, and a *confined aquifer*, which is separated from the ground surface by an aquitard **(Fig. 13.28a)**. Not all rock or sediment underground contains water-filled pores, so geologists also distinguish between the *unsaturated zone*, in which water only partially fills pores so that some air-filled pore space remains, and the *saturated zone*, in which water completely fills pore spaces. The underground boundary between the unsaturated zone (above) and the saturated zone (below) is called the **water table (Fig. 13.28b)**. In effect, the water table forms the top boundary of groundwater. Typically, some groundwater

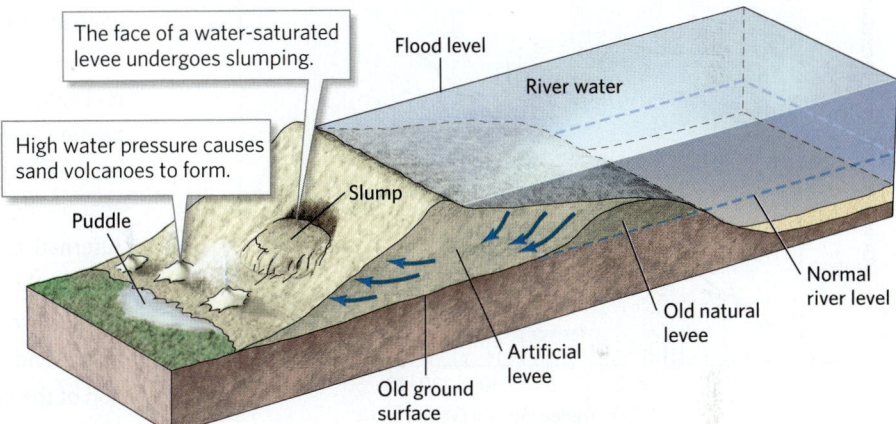

(c) Artificial levees may be undermined when infiltrated by water.

(d) At a levee breach, water flows from the river to the floodplain.

BOX 13.2 ▶ **Putting Earth Science to Use**

Evaluating flooding hazards

When making decisions about investing in flood-control measures, mortgages, or insurance, you need a basis for defining the hazard or risk posed by flooding. Hydrologists characterize the risk of flooding in two ways. The **annual probability** of flooding indicates the likelihood that a flood of a given size will happen along a specified portion of a stream during any given year. For example, if we say that a flood with a given discharge has an annual probability of 1%, then we mean there

is a 1-in-100 chance that a flood of this size will happen in any given year. The **recurrence interval** of a flood with a given discharge indicates the average number of years between successive floods of that size **(Fig. Bx13.2a)**. For example, if a flood with a given discharge happens once in 100 years on average, then it has a recurrence interval of 100 years and can be called a *100-year-flood*. Annual probability and recurrence interval are related by the following equation: Annual probability = 1 ÷ recurrence interval. For example, the annual probability of a 50-year flood is 1/50, which can also be written as 0.02 (or 2%). The recurrence interval for a flood with a larger discharge is longer than that for a smaller flood because large floods happen less frequently than small ones.

Unfortunately, people commonly misunderstand the meaning of a recurrence interval, believing that they do not face a future flooding hazard if they buy a home within an area where a 100-year flood has just occurred. Their confidence comes from making the incorrect assumption that because such flooding just happened, it won't happen again until another 100 years have passed. It's important to keep in mind that flooding does not take place periodically. Two 100-year floods can occur in consecutive years, or even in the same year. Alternatively, the interval between such floods could be, say, 210 years. Clearly, the recurrence interval only indicates average timing.

By knowing the discharge for a flood of a specified annual probability, and by knowing the shape of the stream channel and the elevation of the land bordering the stream, geologists can predict the extent of land that will be submerged by a flood of a given size. Such data, in turn, permit geologists to produce *flood-hazard maps* **(Fig. Bx13.2b, c)**. It's wise to avoid buying a home in a location that has a high probability of flooding!

FIGURE Bx13.2 The relationship between flood size and probability.

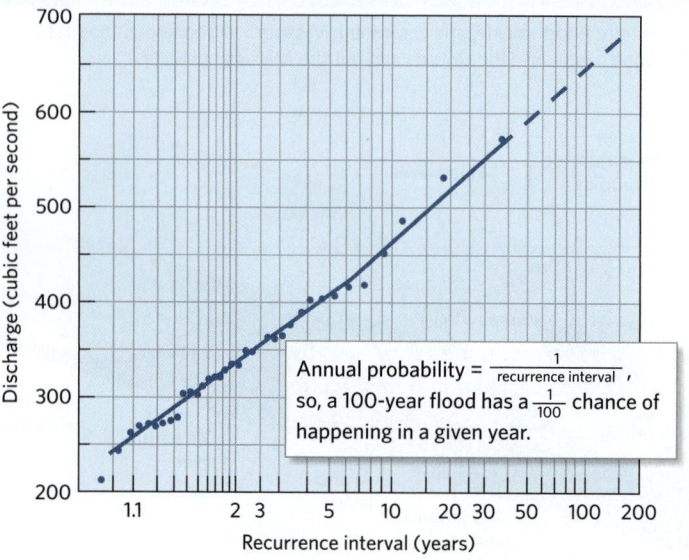

Annual probability = $\frac{1}{\text{recurrence interval}}$, so, a 100-year flood has a $\frac{1}{100}$ chance of happening in a given year.

(a) A flood-frequency graph shows the relationship between the recurrence interval and the discharge for an idealized river.

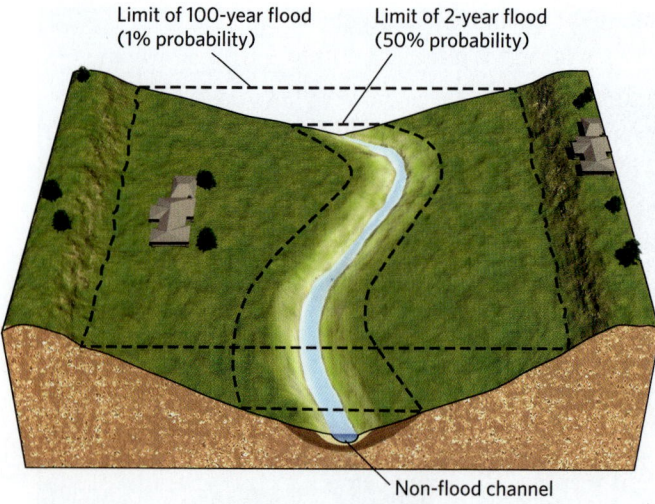

(b) A 100-year flood covers a larger area than a 2-year flood and occurs less frequently.

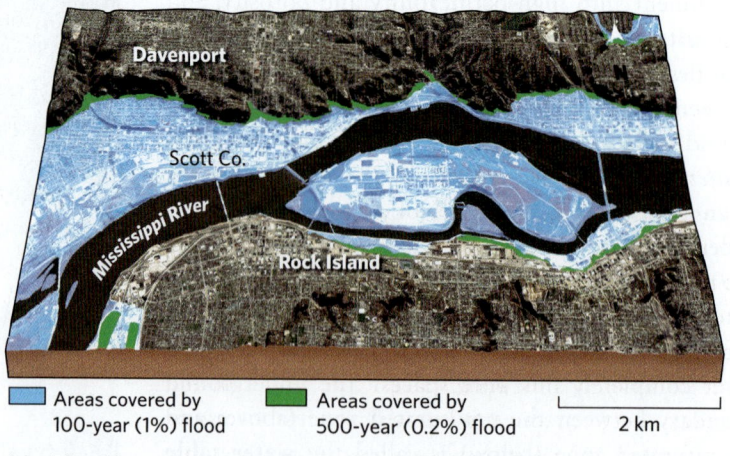

(c) A flood-hazard map shows areas likely to be flooded. Here, near Rock Island, Illinois, even large floods are confined to the floodplain.

FIGURE 13.27 Porosity and permeability.

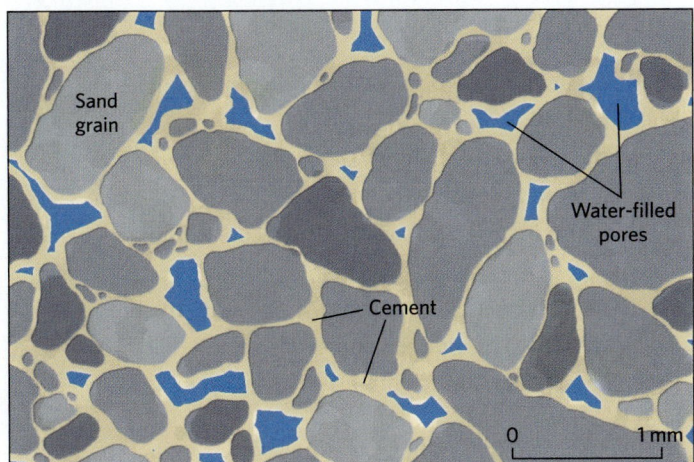

(a) Isolated pores occur in the spaces between grains in a sandstone. Water or air can fill these pores.

These fractures have been enlarged by dissolution.

(b) This limestone outcrop on the coast of Ireland contains abundant fractures that provide porosity.

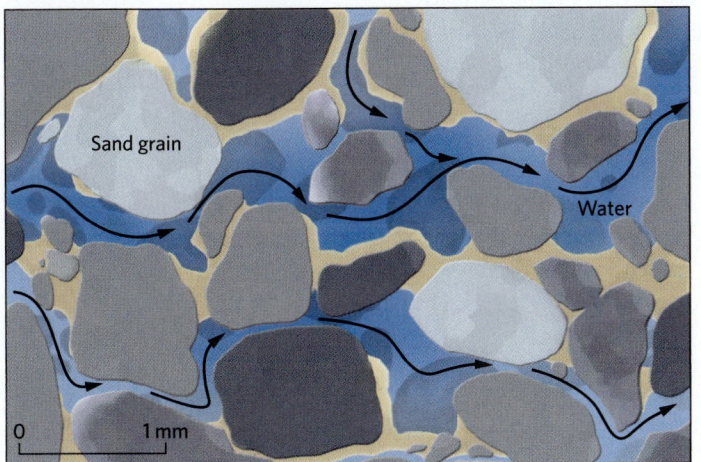

(c) Permeability is the degree to which pores are linked and water can move from pore to pore in rock or sediment.

seeps up from the water table due to capillary action caused by the attraction of water molecules to one another and to grain surfaces, producing a wet *capillary fringe* just above the water table.

The depth of the regional water table varies with location. In places where it lies at the ground surface, no unsaturated zone exists, and the ground is wet and soggy. Where the water table lies hidden below the ground surface, the ground surface can be dry. But in depressions or channels where the ground surface lies below the water table, the low area fills with water and becomes a lake or stream—in such places, the surface of the lake or stream represents the water table (Fig. 13.28c). In humid regions, the water table lies within a few meters or tens of meters of the surface, whereas in arid regions, it may lie hundreds to thousands of meters below the surface. Rainfall changes the water-table depth at a given

locality, so the water table drops during the dry season and rises during the wet season. If the water table drops below the bed of a lake or stream, the water body may dry up (Fig. 13.28d). In some locations, lens-shaped horizons of an aquitard (such as shale) lie within a thick aquifer. If a mound of groundwater accumulates above an aquitard lens, its top surface will lie above the regional water table. Geologists refer to the top surface of such a mound as a **perched water table** (Fig. 13.28e).

In hilly regions, the shape of the regional water table tends to mimic, in a subdued way, the shape of overlying topography (Fig. 13.29). This means that the water table lies at a higher elevation beneath hills than it does beneath valleys. It may seem surprising that the elevation of the water table varies as a consequence of ground-surface topography. (After all, when you pour a bucket of water into a pond, the surface of the pond immediately adjusts to remain horizontal.) The elevation of the water table varies because groundwater moves so slowly through rock and sediment that it cannot quickly assume a horizontal surface. For example, when rain falls on a hill and water infiltrates to the water table, the water table rises. During a dry spell, the water table below the hill sinks so slowly that by the time it rains, causing the water table to rise again, the water table hasn't had time to sink very far.

Groundwater Flow

In the unsaturated zone, water simply percolates downward, like the water passing through a drip coffee maker, for this water moves only in response to the downward

How can I explain . . .

Porosity and permeability

What are we learning?
- The difference between porosity and permeability.
- Factors that affect porosity and permeability.

What you need:
- Two large glass beakers and a screen made of soft plastic mesh
- A supply of pebbles and sand, a measuring cup, two strong rubber bands, and a stopwatch

Instructions:
- Determine the volume of the beakers (by filling them with water from a measuring cup).
- Drain the beakers, then fill one with pebbles and the other with sand.
- Using the measuring cup, add water to each beaker until the water table in the beaker reaches the top. Compare the volume of water

needed to fill the pebble-filled beaker with the volume needed to fill the sand-filled beaker.
- Calculate the porosity of each material. Determine which material has higher porosity, and suggest why.
- Place a screen over the top of each beaker and secure it with a rubber band.
- Tilt the sand-filled beaker so it's horizontal, and record the time it takes for the water to spill out. Repeat for the pebble-filled beaker. Based on the result, which material has greater permeability? Why?

What did we see?
- In the ground, water can fill pores between solid grains, but not all materials have the same porosity.
- Not all materials have the same permeability. Permeability can be affected by the size and continuity of conduits between grains.

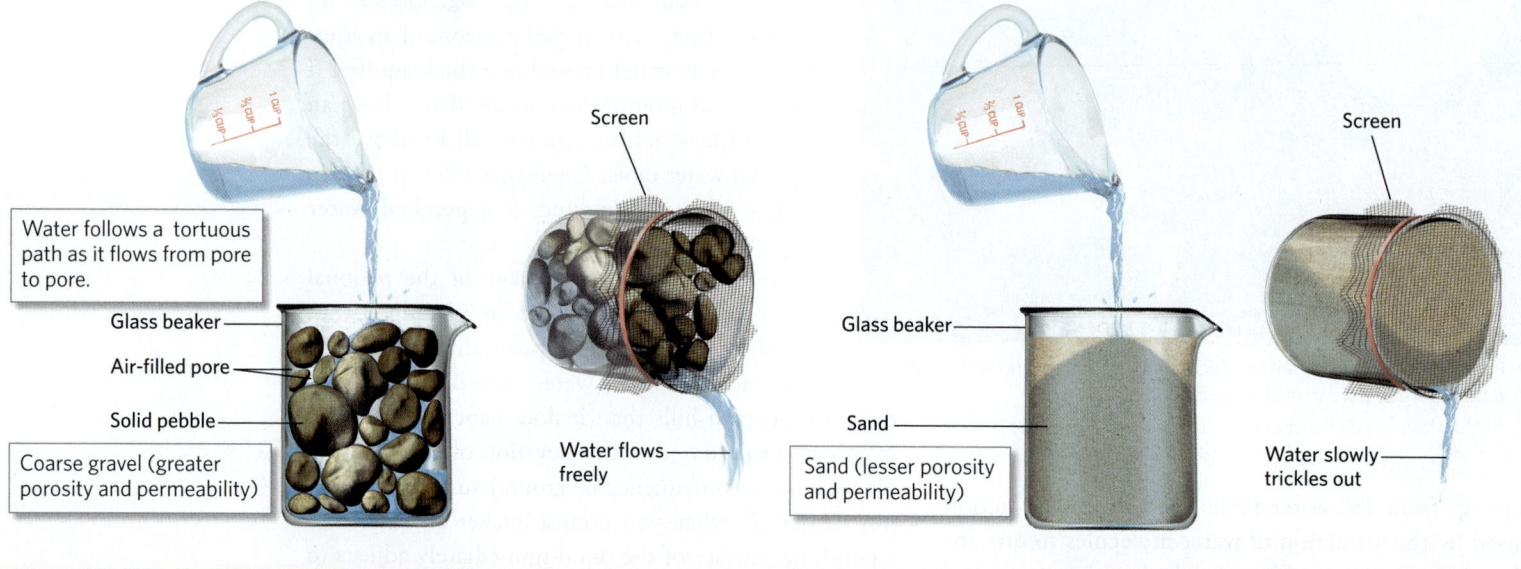

Water follows a tortuous path as it flows from pore to pore.

Screen

Glass beaker

Air-filled pore

Solid pebble

Coarse gravel (greater porosity and permeability)

Water flows freely

Glass beaker

Sand

Sand (lesser porosity and permeability)

Screen

Water slowly trickles out

pull of gravity. Below the water table, however, groundwater flow can be complex. Here, groundwater moves not only in response to gravity, but also in response to differences in pressure. In fact, pressure can cause groundwater to flow sideways or even upward. Therefore, to understand the nature of groundwater flow, we must first understand the origin of pressure in groundwater. For simplicity, we'll consider only the case of groundwater in an unconfined aquifer.

The pressure in groundwater at any specific point underground comes from the weight of all overlying water from that point up to the water table. (The weight of overlying rock does not contribute to the pressure exerted on groundwater because contact points between mineral grains bear the rock's weight.) As a result, a point at a greater depth below the water table feels more pressure than does a point at lesser depth. If the water table is horizontal, an imaginary plane at a given

FIGURE 13.28 Water underground: aquifers, aquitards, and the water table.

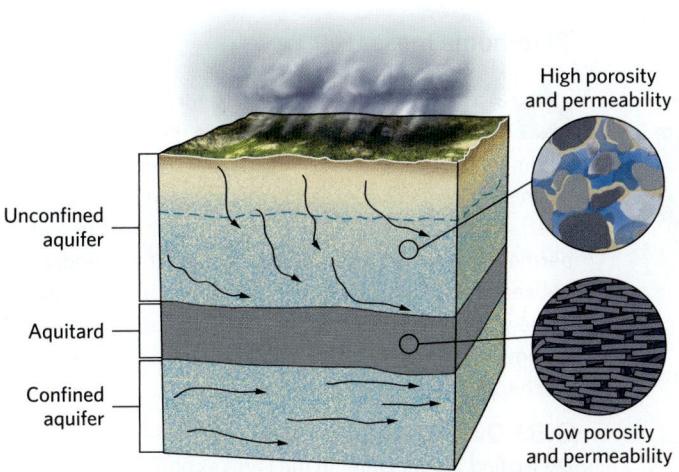

(a) An aquifer consists of high-porosity, high-permeability rock. An aquitard consists of low-permeability rock. Some aquifers are unconfined, and some are confined.

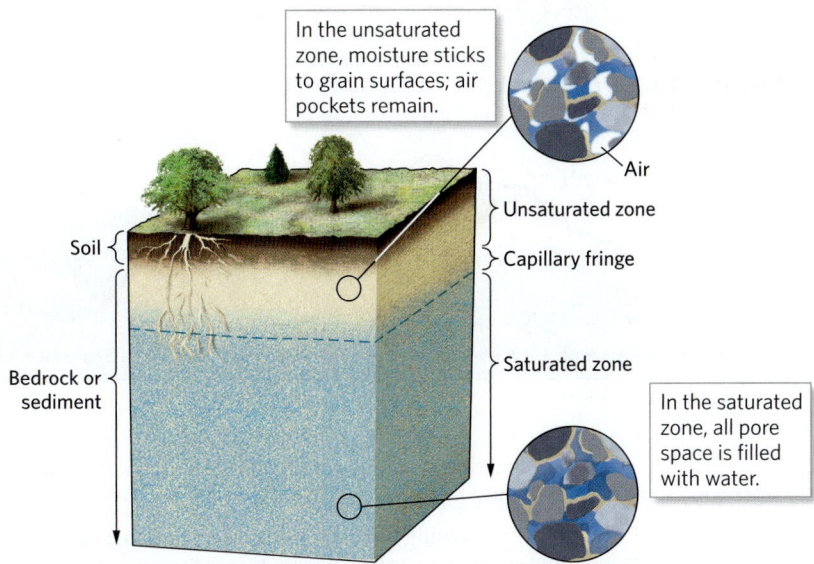

(b) The water table is the top of the groundwater reservoir. It separates the unsaturated zone above from the saturated zone below. The capillary fringe lies just above the water table.

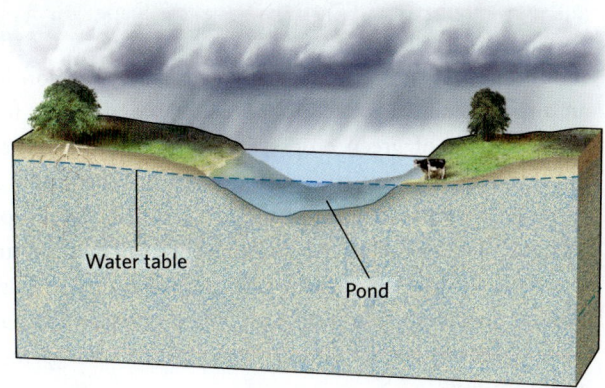

(c) Where the water table lies close to the ground surface, ponds remain filled—the surface of the pond represents the water table.

(d) In dry regions and during dry seasons, the water table sinks deep below the surface. Water that collects temporarily in low areas infiltrates the subsurface.

depth, feels the same pressure everywhere. But if the water table isn't horizontal, the pressure exerted at the imaginary horizontal plane changes from point to point on the plane as the elevation of the water table changes (see Fig. 13.29).

Both the elevation of a volume of groundwater and the pressure within the water provide potential energy that, if given a chance, drives groundwater flow. Hydrogeologists have determined that regional groundwater flow typically follows concave-up curved paths, as viewed in cross section (**Fig. 13.30a**). These curved paths eventually take groundwater from regions where the water table is high (such as under a hill) to regions where the water table is low (such as below a valley). A location where water enters the ground, meaning a place where flow has a downward trajectory, is a **recharge area**, whereas a

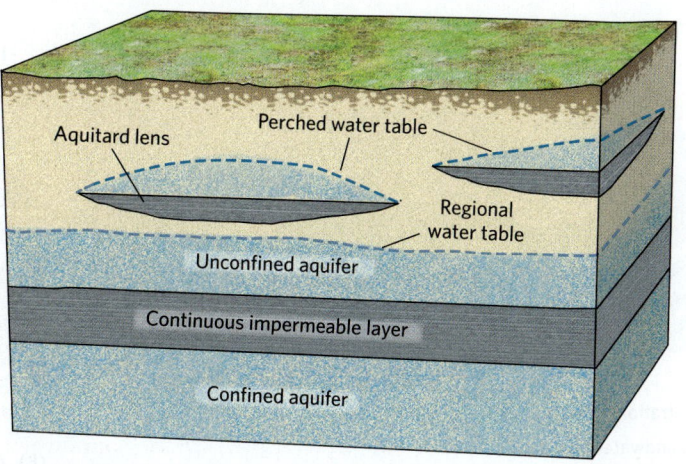

(e) A perched water table occurs where a mound of groundwater becomes trapped above an aquitard lens that lies above the regional water table.

FIGURE 13.29 The shape of a water table beneath hilly topography. Its shape sets up differences in water pressure that influence groundwater flow.

Point h_1 on the water table is higher than Point h_2 relative to a reference elevation (sea level). The pressure at p_1 is, therefore, higher than the pressure at p_2.

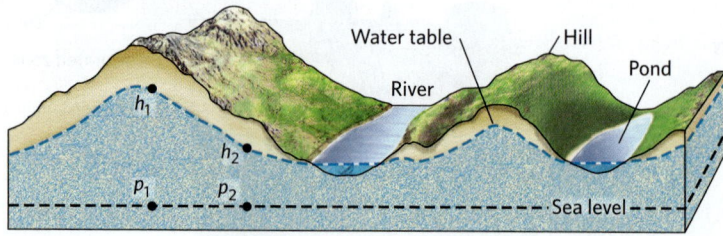

location where groundwater flows back up to the ground surface is a **discharge area**.

Flowing water in a steep river channel can reach speeds of 30 km (20 mi) per hour. In contrast, groundwater moves at rates between 5 and 500 m (15 to 1,500 ft) per year, so it may remain underground from a few months up to tens of thousands or even millions of years before returning to the surface **(Fig. 13.30b)**. Groundwater moves so slowly because it follows a crooked network of tiny conduits and must travel a much greater distance than it would if it followed a straight path, and friction between groundwater and conduit walls slows its flow. Basically, the velocity of groundwater flow at a location depends on the slope of the water table and on the permeability of the material through which the groundwater flows. Groundwater moves faster through high-permeability rocks than it does through low-permeability rocks, and it moves faster in regions where the water table has a steep slope than it does in regions where the water table has a gentle slope.

> Thousands have lived without love, not one without water.
>
> —W. H. AUDEN (ENGLISH-AMERICAN POET, 1907–1973)

A French researcher, Henry Darcy, developed an equation (*Darcy's Law*) that describes controls on groundwater flow more completely, as discussed in more advanced books on the topic.

Take-home message . . .

Most underground water fills pores and cracks in rock or sediment. Porosity is the total volume of open space within a material, whereas permeability is the degree to which that material's open spaces connect. Aquifers have high porosity and permeability, whereas aquitards do not. The boundary between the unsaturated zone and the saturated zone is the water table. Gravity and pressure cause groundwater to flow slowly from recharge areas to discharge areas.

Quick Question ——————————
Do we find groundwater in the Earth's core?

13.8 Tapping Groundwater Supplies

If you have an opportunity to fly over an arid region in an airplane, take a look out the window and watch for circular fields of green crops **(Fig. 13.31)**. In the center of each field, a pump sucks up groundwater to supply a sprinkler that pivots around the field to keep the crops irrigated. Groundwater may be the only source of clean freshwater in arid regions where surface water doesn't exist or in temperate or humid regions where surface water has been polluted. People obtain groundwater from *springs* and from *wells*.

FIGURE 13.30 The flow of groundwater.

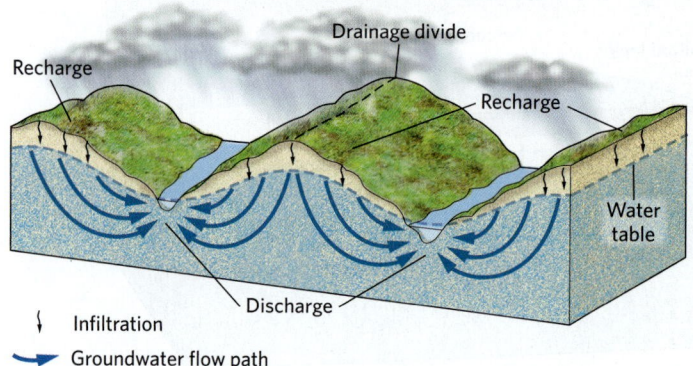

↓ Infiltration

→ Groundwater flow path

(a) Groundwater flows from recharge areas to discharge areas. Typically, the flow follows curving concave-up paths.

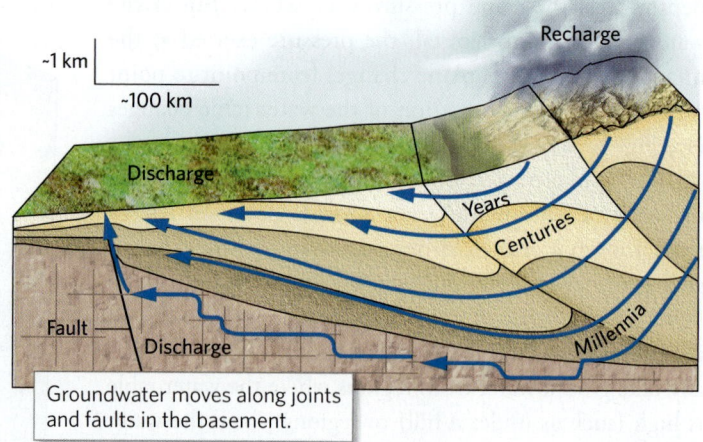

Groundwater moves along joints and faults in the basement.

(b) A large elevation difference in the water table may drive groundwater hundreds of kilometers, across regional sedimentary basins. Groundwater that follows deeper flow paths stays underground longer.

Springs

Imagine an oasis of green palm trees surrounded by the dry sands of a desert. Where does the water that these plants need to survive come from? It flows from a **spring**, a natural outlet from which groundwater spills or seeps onto the ground surface. Springs can provide freshwater for drinking or irrigation without the expense of drilling or digging, so many villages and towns worldwide have grown up adjacent to springs. Springs form in a variety of geologic settings **(Fig. 13.32)**. Notably, while we can see springs that spill water onto dry land, many springs lie submerged beneath streams or lakes.

Wells

Millennia ago, farmers learned that they could get groundwater, even where no springs exist, by digging a **well**, a hole that reaches down to the water table. Today, people still construct wells by digging holes, but it's usually easier to bore a well by drilling **(Fig. 13.33a)**. Hydrogeologists distinguish between two types of wells based on the pressure of water in the well.

In an *ordinary well*, the base of the well lies below the water table, so water simply seeps from the aquifer into the well and fills it to the level of the water table. Drilling into an aquitard, or into rock or sediment in the unsaturated zone, will not provide water, and yields a *dry well*.

FIGURE 13.31 Irrigated circular fields in eastern Colorado, as seen from a satellite.

1 km

FIGURE 13.32 Springs may form in a variety of geologic settings.

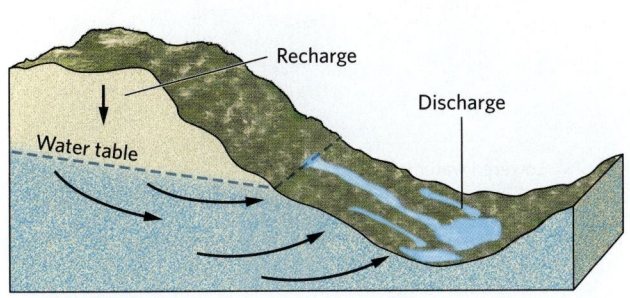

(a) Groundwater reaches the ground surface in a discharge area.

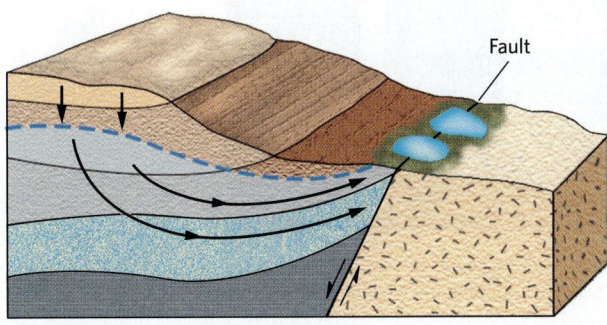

(b) Where groundwater reaches an impermeable barrier, it rises.

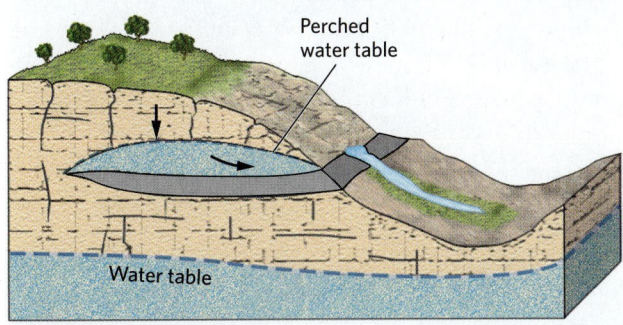

(c) Groundwater seeps from the ground where a perched water table intersects a slope.

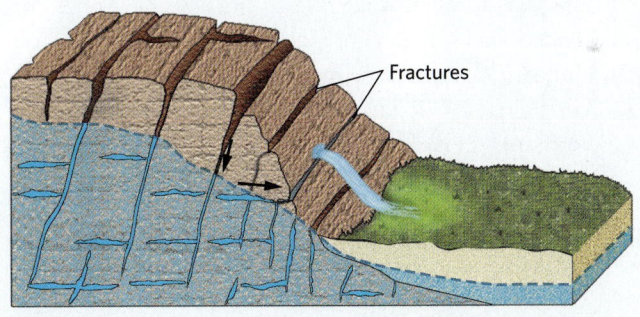

(d) A network of interconnected fractures channels water to the surface of a hill.

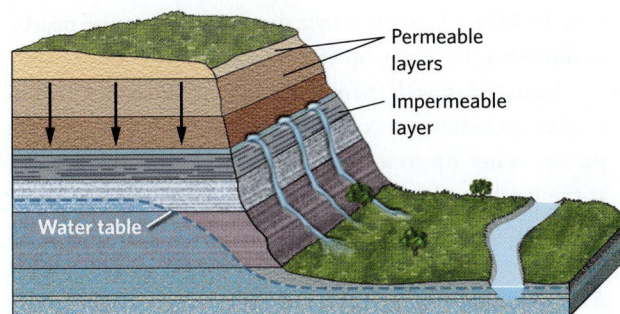

(e) Groundwater seeps out of a cliff face at the top of an impermeable bed.

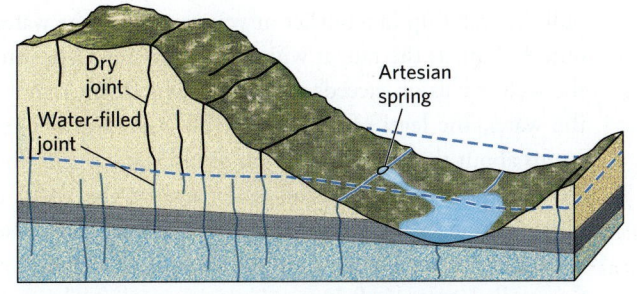

(f) Where water under pressure lies below an aquitard, a crack may provide a pathway for an artesian spring to form.

FIGURE 13.33 The nature of ordinary wells.

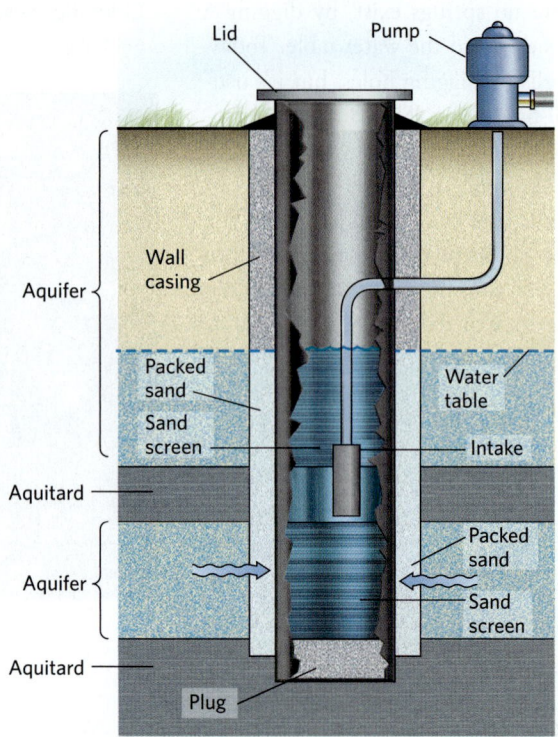

(a) A modern ordinary well sucks up water with an electric pump. The packed sand surrounding the well filters the water.

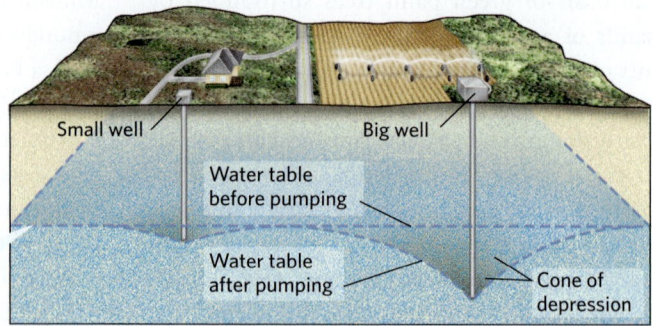

(b) Pumping water from a well faster than groundwater flow can refill it forms a cone of depression in the water table.

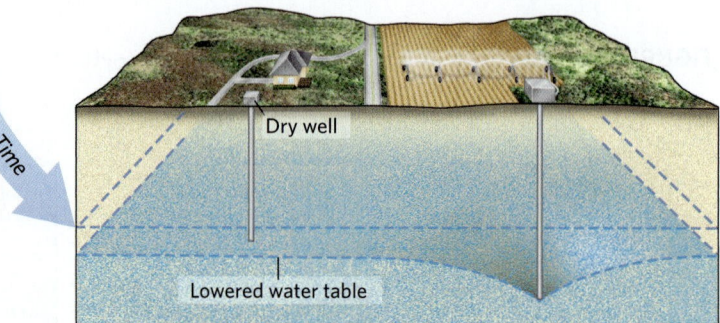

(c) Pumping by the big well may be enough to make the small well run dry.

(d) Irrigation of fields in Utah uses large volumes of groundwater and causes drawdown.

To extract water from an ordinary well, you can either pull the water up in a bucket or you can pump the water out. As long as the rate at which groundwater seeps into the well equals or exceeds the rate at which you remove the water, the level of the water table near the well remains about the same. But if you pump water out of the well too fast, the water table sinks around the well—a phenomenon called *drawdown*. When this happens, the water table forms a downward-pointing, cone-shaped surface, called a **cone of depression**, surrounding the well (Fig. 13.33b). Drawdown by a deep well may cause nearby shallower wells to dry up (Fig. 13.33c, d).

An **artesian well**, named for the province of Artois in France, penetrates a confined aquifer in which pressure pushes water up to a level above the top of the aquifer. If the level to which the water table would rise lies below the ground surface, the well is a *nonflowing artesian well*. But if the level lies above the ground surface, the well is a *flowing artesian well*—the water from such a well actively fountains out of the ground. Artesian wells exist where the recharge area for the confined aquifer lies at a higher elevation than the site of the well, so that if the aquifer were not confined, the water table would be higher (Fig. 13.34).

Hot Springs

Hot springs, defined as natural springs whose water has a temperature above about 30°C (86°F), can be found in two types of geologic settings. Some occur where groundwater can rise relatively rapidly from several kilometers below the surface, depths at which rock is warmer even with a normal geothermal gradient. The rate at which the groundwater moves upward must be fast enough so that the water doesn't lose all its heat before bubbling

FIGURE 13.34 The nature of artesian wells.

(a) A flowing artesian well in Wisconsin. Groundwater rises from a confined aquifer into the corrugated pipe without the need of pumping.

out of a spring. Such flow can happen in places where fractures and faults provide a high-permeability pathway. Hot springs may also develop in *geothermal areas*, regions of igneous activity where magma-heated rock lies relatively close to the Earth's surface and warms up water that has infiltrated only a short distance downward **(Fig. 13.35a)**. Some hot springs become popular spas **(Fig. 13.35b)**.

Numerous distinctive features form in association with hot springs. For example, when hot water that contains a high concentration of dissolved minerals spills out of hot springs and then cools, the minerals precipitate, forming colorful mounds or terraces of travertine and other chemical sedimentary rocks **(Fig. 13.35c)**. Natural pools fed by mineralized hot springs may harbor brightly colored thermophilic (heat-loving) bacteria and archaea that thrive in the hot water **(Fig. 13.35d)**. In places where the hot water rises into volcanic ash, a viscous slurry forms and fills bubbling *mud pots*.

In some geothermal areas, scalding water spurts out of hot springs to form **geysers**, fountains of steam and hot water that erupt episodically **(Fig. 13.35e)**. Geysers form when groundwater seeps into irregular fractures in hot rock. Under the elevated pressure underground, the groundwater can become *superheated*, meaning that its temperature exceeds the temperature at which water boils at the Earth's surface. When superheated groundwater moves upward through a fracture to shallower depths, the pressure acting on it decreases until, eventually, some of it transforms into steam. The resulting expansion pushes some overlying water out of the geyser's vent, causing the pressure in the superheated groundwater to decrease even more. This new, rapid drop in pressure, in turn, causes more superheated groundwater to vaporize instantly, forming steam

Did you ever wonder . . .

why huge underground caves form?

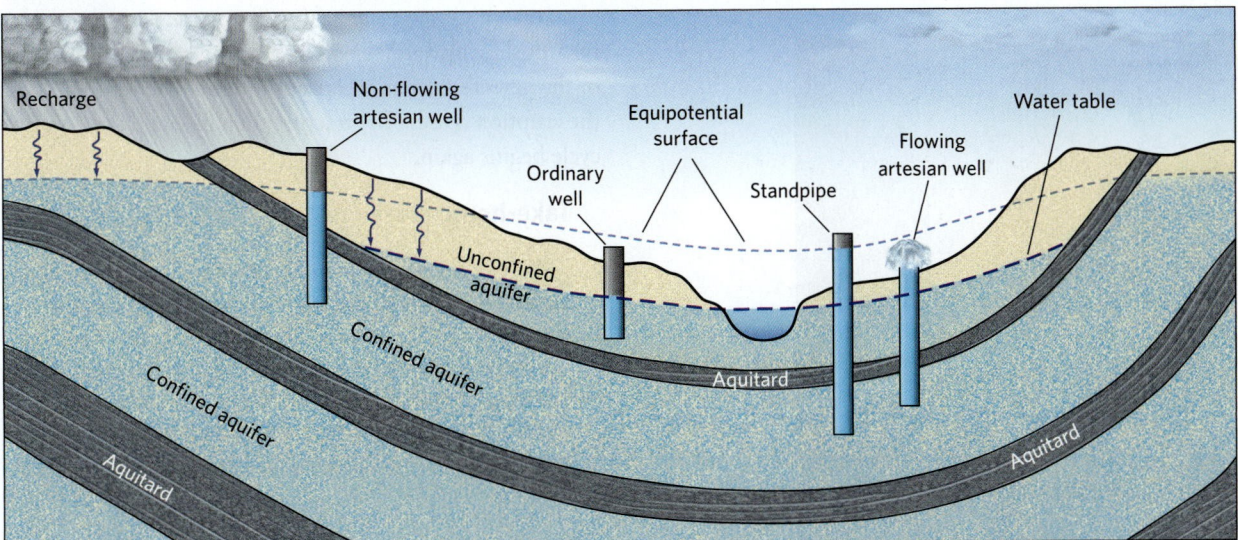

(b) The configuration of a regional artesian system. Water rises above the confined aquifer because the water is under pressure.

FIGURE 13.35 Hot springs in areas of igneous activity.

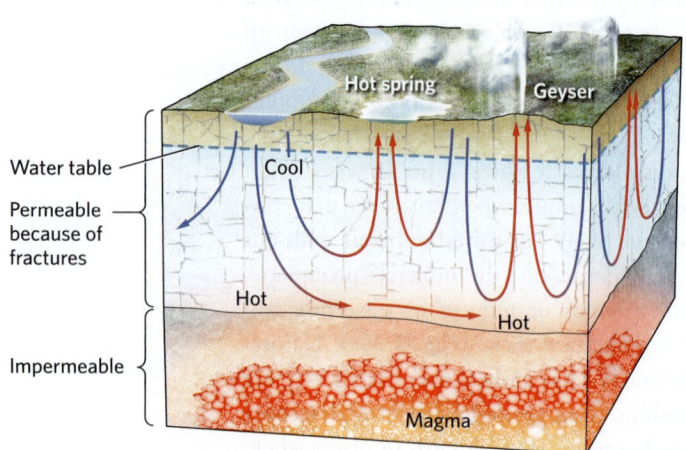

(a) Hot springs occur where groundwater, heated at depth, rises to the ground surface.

(b) Hot springs in Iceland, warmed by magma below, attract tourists from around the world.

(c) Terraces of minerals precipitated at Pamukkale, Turkey.

(d) Colorful bacteria- and archaea-laden pools at Yellowstone National Park.

(e) The Old Faithful geyser in Yellowstone National Park erupts predictably.

that pushes upward, ejecting all the water above it out of the geyser's vent as an eruption. Once the vent empties, the eruption ceases. The fracture then fills, and the eruptive cycle begins again.

Take-home message . . .

Groundwater can be obtained from springs and wells. In ordinary wells, water must be lifted to the surface, but in artesian wells and springs, it rises due to pressure within a confined aquifer. Pumping groundwater faster than it can be replaced lowers the water table and may produce a cone of depression around a well. Springs form in a variety of geologic settings.

Quick Question

In what settings do hot springs develop?

13.9 Caves and Karst Landscapes

Cave Networks

In 1799, as legend has it, a hunter by the name of Houchins was tracking a bear through the woods of Kentucky when he suddenly felt a draft of surprisingly cool air. Curious, Houchins searched for the source of the air, and he found that it emanated from a dark portal in the hillslope above him. Houchins had discovered an entrance to Mammoth Cave, an immense network of interconnected *chambers* or *caverns*—large underground open spaces—and *passages*—tunnel-shaped or slot-shaped underground openings (**Earth Science at a Glance**, pp. 400–401). In some locations, chambers host underground lakes, and some passages serve as conduits for underground streams.

Most large cave networks develop in limestone bedrock because limestone dissolves relatively easily in slightly acidic groundwater. Cave formation primarily takes place just below the water table, for in this region groundwater acidity remains high, the mixture of groundwater and downward-percolating rainwater has not become saturated with dissolved ions, and groundwater flows relatively fast. Typically, groundwater becomes acidic when it infiltrates downward through organic matter in overlying soil. In some cases, however, acidity results from the interaction of groundwater with sulfur-bearing hydrocarbons.

The shape of a cave reflects variations in the composition and in the permeability of the rock from which it formed. Specifically, larger open spaces develop where the limestone is most soluble or where groundwater flow is fastest. Passages typically follow pre-existing joints, for the joints provide conduits along which groundwater flows more easily **(Fig. 13.36)**.

Cave Deposits (Speleothems)

When the water table drops below the level of a cave, the cave becomes an open space filled with air. In places where downward-percolating groundwater containing dissolved calcite drips from the ceiling, the surface of the cave gradually changes, for as soon as groundwater enters the air, it releases some of its dissolved carbon dioxide and evaporates a little. As a result, calcite precipitates out of the water, producing travertine. The travertine builds into intricately shaped cave deposits, or **speleothems**.

Spelunkers (cave explorers) and geologists have developed a detailed nomenclature for different kinds of speleothems **(Fig. 13.37a)**. Where water drips from the ceiling of the cave, the precipitated limestone builds an icicle-like cone called a **stalactite**. Where the drips hit the floor, the resulting precipitate builds an upward-pointing cone called a **stalagmite**. If this process continues long enough,

FIGURE 13.36 The locations of caves are controlled by jointing and bedding in limestone. Passages follow joints, and chambers preferentially form in more soluble beds.

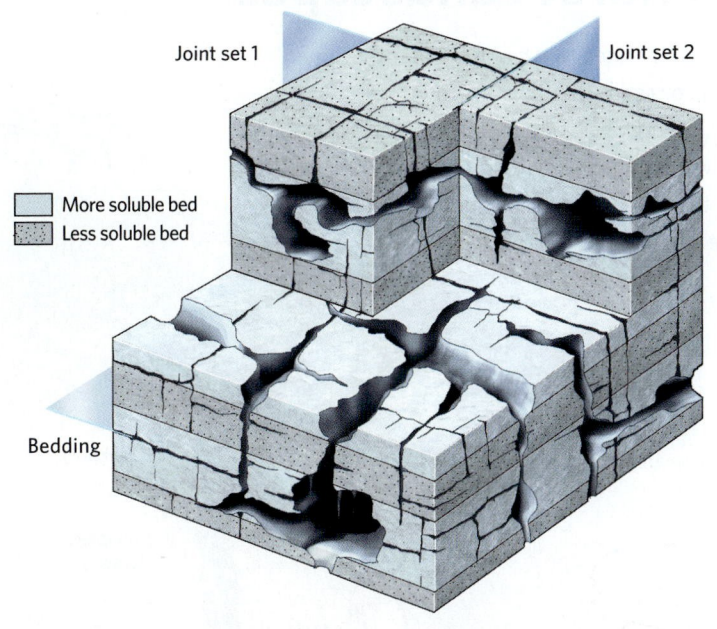

More soluble bed
Less soluble bed

FIGURE 13.37 Speleothems in caves.

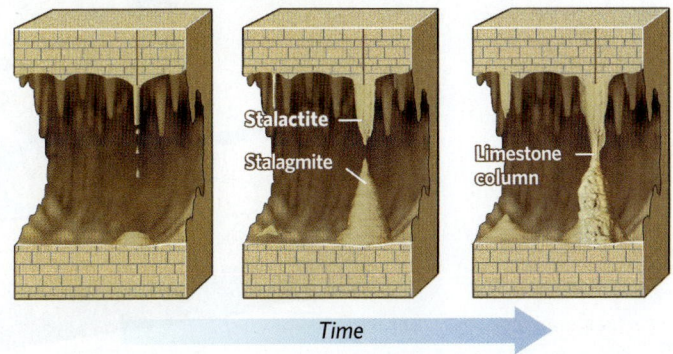

Stalactite
Stalagmite
Limestone column

Time

(a) A drip from the ceiling evolves into a stalactite. Precipitation from water that reaches the floor can form a stalagmite.

1 m

(b) Flowstone on the wall of a cave in Vietnam.

Caves and Karst Landscapes

Limestone is soluble in acidic water. Much of the water that falls to the ground as rain or seeps through the ground as groundwater is acidic, so in regions of the Earth where bedrock consists of limestone, we find signs of dissolution. Networks of underground openings that develop by dissolution are called caves or caverns. Some parts of these may be large, open

Limestone pavement, Ireland

Disappearing stream

Sinkhole

Collapsed breccia

Stalagmite

Stalactite

Soda-straw stalactite

Dissolved joint

Flowstone

Chamber

Stalactite

Limestone column

Underground stream

Underground pool

Passage

Sinkholes

Underground pool, Mexico

Emerging spring

Natural Bridge, Virginia

Spelunker crawling in a cave

chambers, whereas others are long, narrow passages. Underground lakes and streams may cover the floor. A cave's location depends on the orientation of bedding and joints, for these features localize the flow of groundwater.

In many locations, groundwater drips from the ceiling of a cave or flows along its walls. As the water evaporates, calcite precipitates. Over time, this calcite builds into cave formations, or speleothems, such as stalactites, stalagmites, columns, and flowstone. Distinctive landscapes, called karst landscapes, develop at the Earth's surface over limestone bedrock that has undergone dissolution.

Soda-straw stalactites, Utah

5cm

See for yourself

Sinkholes in Central Florida

Latitude: 28°37'50.59" N
Longitude: 81°23'13.60" W

Zoom to 20 km (~12.5 mi) and look down.

You can see several sink-holes, ranging from about 100–800 m (330–2,600 ft) across, that lie within sub-urban developments. The sinkholes have filled with water and are now lakes.

FIGURE 13.38 Development of sinkholes in central Florida.

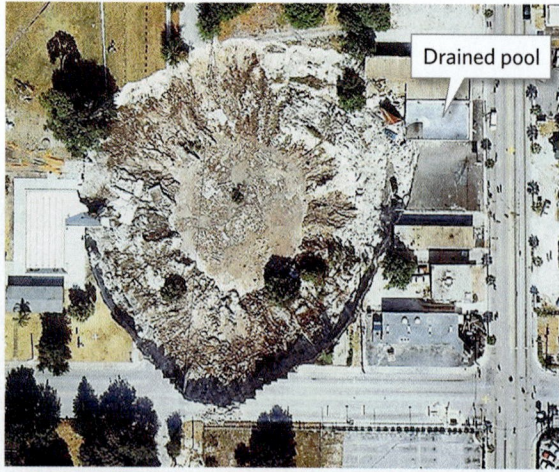

(a) The Winter Park sinkhole, as seen from a helicopter.

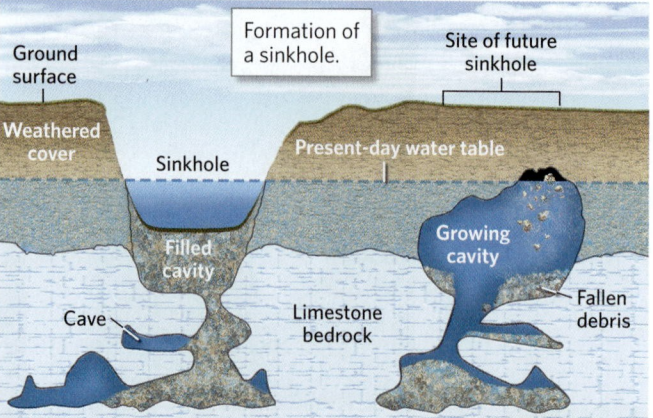

(b) The weathered cover that forms the roof of a cave slowly washes into the cave, forming a cavity that grows ever closer to the ground surface. When the roof of this cavity collapses, a sinkhole forms. (Not to scale.)

(c) An aerial view of Florida sinkholes that have become lakes.

FIGURE 13.39 Features of karst landscapes.

(a) The rough, rocky surface of the Kras Plateau, for which karst landscapes were named, is formed by numerous sinkholes.

(b) A small disappearing stream in the Hudson Valley of New York; the water is spilling into a subsurface cave.

FIGURE 13.40 Tower karst in the Guilin region of China.

(a) The landscape is treeless today, a consequence of industrialization policies in the 1950s.

(b) Chinese artists painted scrolls depicting forested towers of karst.

a stalagmite can merge with an overlying stalactite to form a *column*. In some cases, groundwater flows along the surface of a wall and precipitates to produce drape-like sheets of travertine called *flowstone* (**Fig. 13.37b**).

Formation of Karst Landscapes

When Mae Owens looked out her window on May 8, 1981, she discovered that the large tree in the backyard of her Winter Park, Florida, home had suddenly disappeared. She went outside to investigate, and found that more than the tree had vanished—her whole backyard had become a deep, gaping pit! The pit continued to grow for a few days until it finally swallowed Owens's house and six other buildings as well as a municipal swimming pool, part of a road, and several Porsches in a car dealer's lot (**Fig. 13.38a**).

What had happened? The bedrock under Winter Park consists of limestone. Water had gradually dissolved the limestone to produce caves underground. On May 8, the roof of the cave underneath Owens's backyard began to collapse, forming a circular depression called a **sinkhole** (**Fig. 13.38b**). Eventually, the sinkhole filled with water, and now it's a circular lake. Similar sinkhole lakes have developed throughout central Florida (**Fig. 13.38c**). Sinkhole formation can take place without warning and, sadly, has caused fatalities.

Geologists refer to regions such as central Florida, where surface landforms develop as caves collapse, as **karst landscapes** or *karst terrains*, named for the Kras Plateau on the eastern side of the Adriatic Sea (**Fig. 13.39a**).

Karst landscapes are characterized not only by sinkholes, but also by several other distinctive landforms. For example, surface streams in karst landscapes may flow into cracks or holes that link to caves below (**Fig. 13.39b**). Such *disappearing streams* re-emerge from a cave entrance downstream. In places where most of the ground has collapsed into sinkholes, the landscape consists of narrow ridges with bowl-like depressions in between. Over time, the lower parts of a ridge may erode away, leaving a natural bridge. When these bridges collapse, all that remains are tall spires or pinnacles of limestone (**Fig. 13.40a**). One of the world's most spectacular examples of such *tower karst* decorates the Guilin region of China and has inspired generations of artists (**Fig. 13.40b**). Karst landscapes evolve over time, and as the water table drops, different levels of caves may form (**Fig. 13.41**).

Take-home message . . .

Reaction with acidic groundwater dissolves limestone underground to form cave networks of chambers and passages. Most of this dissolution takes place just below the water table. If the water table sinks, water dripping in caves can produce speleothems. Collapse of a cave produces a karst landscape.

Quick Question
How do sinkholes form?

See for yourself

Karst Landscape in Puerto Rico
Latitude: 18°23′53.04″ N
Longitude: 66°25′49.43″ W

Zoom to 7 km (~4.3 mi) and look down.

We see a karst terrain in central Puerto Rico. Each of the rounded depressions is a sinkhole. Ridges of limestone separate adjacent sinkholes.

FIGURE 13.41 The progressive formation of karst landscapes.

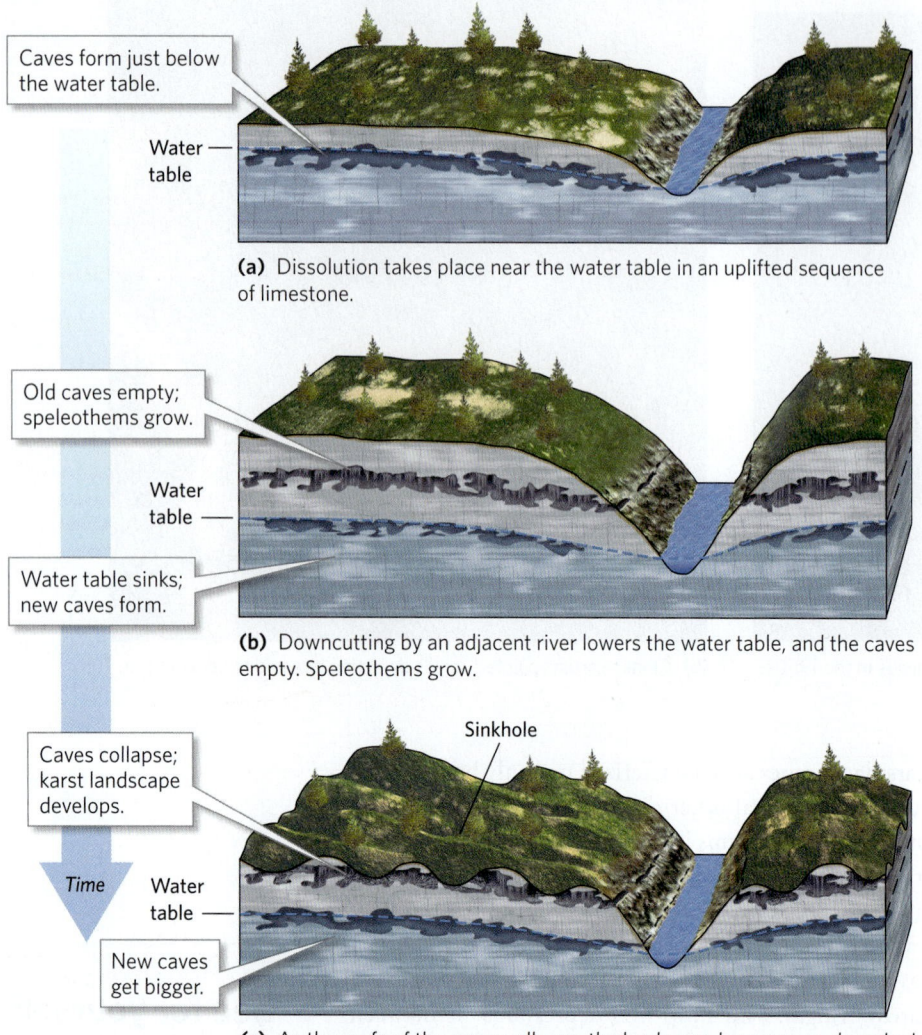

Caves form just below the water table.

Water table

(a) Dissolution takes place near the water table in an uplifted sequence of limestone.

Old caves empty; speleothems grow.

Water table

Water table sinks; new caves form.

(b) Downcutting by an adjacent river lowers the water table, and the caves empty. Speleothems grow.

Caves collapse; karst landscape develops.

Sinkhole

Time

Water table

New caves get bigger.

(c) As the roofs of the caves collapse, the landscape becomes pockmarked with sinkholes.

13.10 Threats to Freshwater Supplies

Surface-Water Challenges

The lands bordering streams and lakes have long been desirable places for people to settle, for these water bodies provide drinking and irrigation water, food, avenues for transportation, sources of power, water for industry, and sites for recreational activities. In recent centuries, unfortunately, the sustainability of clean surface-water supplies has come under threat, for many reasons:

• *Pollution:* People dump sewage, garbage, runoff from streets, spilled hydrocarbons, toxic chemicals from industrial sites, fertilizer, and animal waste into streams **(Fig. 13.42a)**. These pollutants poison aquatic life and make water unsafe to drink.

FIGURE 13.42 Sustainability challenges for surface-water supplies.

(a) Large quantities of garbage and various pollutants end up in rivers.

(b) An algal bloom produces green slime on the surface of a lake that has undergone eutrophication.

• *Eutrophication:* Clear, well-oxygenated freshwater hosts an amazing variety of life. But an influx of nutrients—from sewage, fertilizer-rich runoff from fields or lawns, and animal waste—can disrupt freshwater ecosystems. An oversupply of nutrients, or

eutrophication, can trigger algal blooms that not only turn a water body bright green but can also make it toxic. In addition, the decay of dead algae can remove so much oxygen from the water that fish and other organisms living in it die off **(Fig. 13.42b)**.

- *Water depletion:* In many localities, society consumes so much of a stream's water (for irrigation, industry, and civic use) that only a saline trickle reaches the stream's mouth **(Fig. 13.43)**. In locations where streams supply water to a lake, such *water depletion* can cause the lake to dry up.

- *Dam construction:* Dam construction impounds reservoirs. Filling a reservoir may flood upstream towns and forests as well as scenic canyons and rapids. Also, reservoirs trap sediments and nutrients carried by the stream, preventing them from reaching downstream floodplains and deltas that host farmland.

- *Changes in infiltration and sediment supply:* Urbanization transforms fields and forests into parking lots, roads, and buildings made of concrete and asphalt. Such impermeable materials decrease rainfall infiltration and therefore increase runoff, a change that can contribute to flooding. Conversion of forests and fields into farmland increases the amount of sediment in streams because farmland hosts relatively little plant cover and may be barren for much of the year, so sheetwash flowing across farmland carries sediment into streams.

Groundwater Challenges

MINERALIZED AND SALINE GROUNDWATER. Commonly, you can drink groundwater directly from a spring or well, for rocks and sediments serve as natural filters capable of trapping suspended solids, and most groundwater contains low concentrations of dissolved chemicals and salts. In fact, commercial distribution of bottled groundwater (labeled "spring water") has become a major business worldwide. However, dissolved salts and minerals make some natural groundwater unusable. For example, in many locations, fresh groundwater overlies older, saline groundwater. If wells or springs tap into the saline groundwater, they may yield water that cannot be used for drinking or irrigation. In addition, groundwater that has passed through limestone or dolomite contains calcium, magnesium, and bicarbonate ions; such *hard water* causes problems for consumers because carbonate minerals precipitate from it to form *scale* that clogs pipes, and because soap won't develop a lather in it. Groundwater that has passed through iron-bearing rocks may contain dissolved iron oxide that precipitates to form rust stains. And arsenic, a highly toxic chemical, can enter groundwater by the dissolution of

FIGURE 13.43 The Central Arizona Project canal shunts water from the Colorado River to urban and agricultural areas of the desert in south-central Arizona.

arsenic-bearing minerals. In Southeast Asia, a significant percentage of wells, drilled so that people could avoid drinking polluted surface water, contain dangerous levels of arsenic.

GROUNDWATER CONTAMINATION. In recent decades, increasing amounts of contaminants have been introduced into aquifers by human activities **(Fig. 13.44a)**. The underground cloud of contaminated groundwater that flows away from a contaminant source constitutes a **contaminant plume (Fig. 13.44b)**. Fortunately, in some cases, natural processes can clean up groundwater contamination, for clay binds to some contaminants, oxygen in the water reacts with and destroys others, and natural bacteria in the water metabolize still others. But where natural processes can't remove all the contamination from groundwater, several approaches can be taken to maintain safe groundwater supplies. First, the direction of plume flow can be tracked, so that wells in the path of the contaminant plume can be shut down to prevent consumption of contaminated water. Second, if resources permit, the groundwater can be extracted and purified. If extraction isn't feasible, engineers may install an underground wall of chemicals that will react with the contaminants and cause them to precipitate into solids that won't migrate with groundwater, or they may inject oxygen and nutrients into the plume to foster growth of bacteria that can react with and break down the

Did you ever wonder . . .

where bottled "spring water" comes from?

FIGURE 13.44 Contamination of groundwater and its remediation.

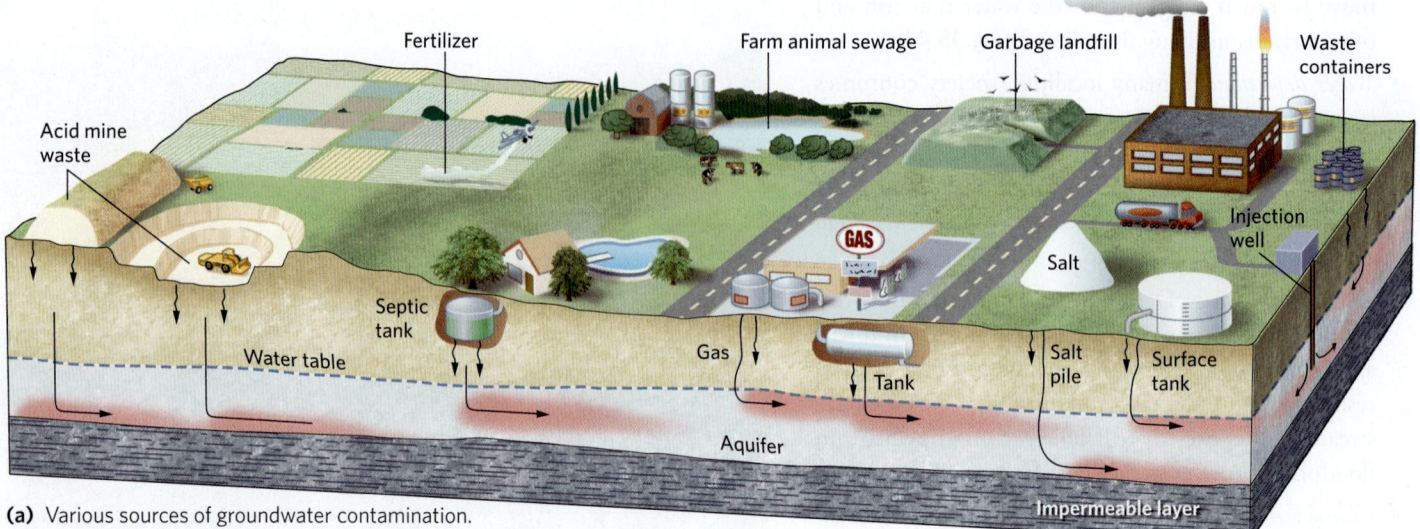

(a) Various sources of groundwater contamination.

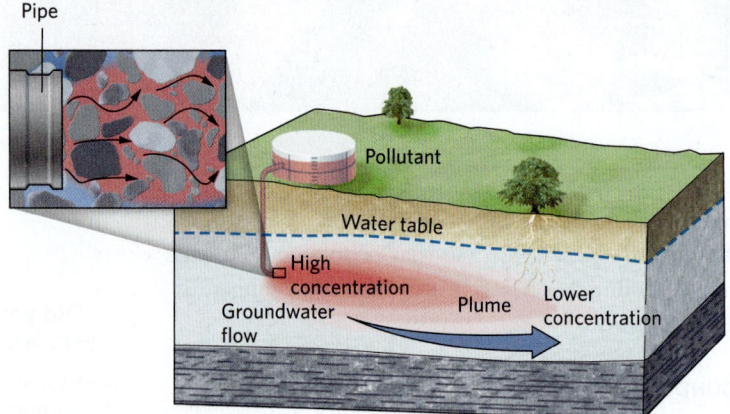

(b) A contaminant plume as seen in cross section. The darker the color, the greater the concentration of the contaminant.

▶ **Video**

Groundwater Removal

contaminants. In some cases, sadly, it's just not possible to clean contaminated groundwater.

GROUNDWATER DEPLETION. Is groundwater a renewable resource? In a time frame of 10,000 years, the answer is yes, for the hydrologic cycle will eventually replenish depleted reserves. But in a time frame of 100 to 1,000 years, groundwater in nontropical regions should be viewed as a nonrenewable resource (see Chapter 11). In effect, by pumping water out of the ground at a rate faster than nature replaces it, people are "mining" groundwater reserves. A number of problems accompany such *groundwater depletion*:

- *Lowering of the water table:* When we extract groundwater from wells at a rate faster than it can be resupplied by infiltration, the water table drops, and springs, streams, lakes, swamps, and nearby wells dry up **(Fig. 13.45a)**.

- *Reversal of groundwater flow direction:* The cone of depression around a well creates a local slope in the water table large enough to reverse the flow direction of nearby groundwater **(Fig. 13.45b)**. This change may cause contaminant plumes to head toward wells that were once clean.

- *Saltwater intrusion:* In coastal areas, fresh groundwater lies in a layer above denser saltwater that has entered the aquifer from the adjacent ocean **(Fig. 13.45c)**. Pumping water from a well too quickly can suck saltwater up into the well, causing *saltwater intrusion*. The same phenomenon can happen inland, if lowering of the water table causes people to drill wells down far enough to tap deeper, saltier groundwater.

- *Pore collapse and land subsidence:* Water can't be compressed, so pore water holds sediment grains apart in the subsurface. Air, however, can be compressed, so removal of groundwater allows grains around pores to pack together more closely. Such *pore collapse* decreases the porosity and permeability of an aquifer. It also decreases the volume of the aquifer, causing land above the aquifer to undergo subsidence. Commonly, subsidence results in fissuring of the ground surface **(Fig. 13.45d)**.

Take-home message . . .

Freshwater is not a renewable resource in many places. Unfortunately, society has diminished and damaged both surface water and groundwater supplies through overuse and contamination.

Quick Question

What are the consequences of lowering the water table?

FIGURE 13.45 Effects of groundwater depletion.

Before

After

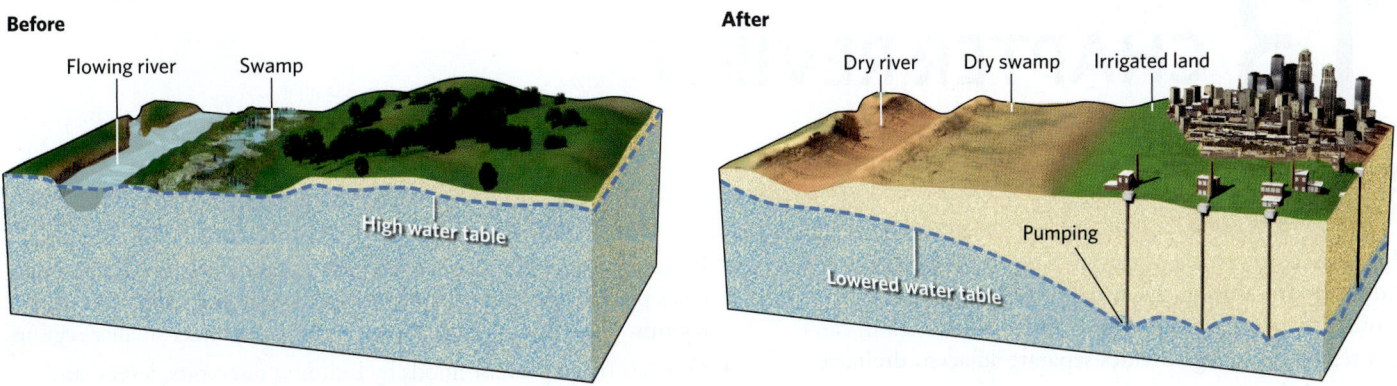

(a) Lowering of the water table. (Left) Before humans start pumping groundwater, the water table is high. A swamp and permanent stream exist. (Right) Pumping for consumers in a nearby city causes the water table to sink, so the swamp and stream dry up.

Before

After

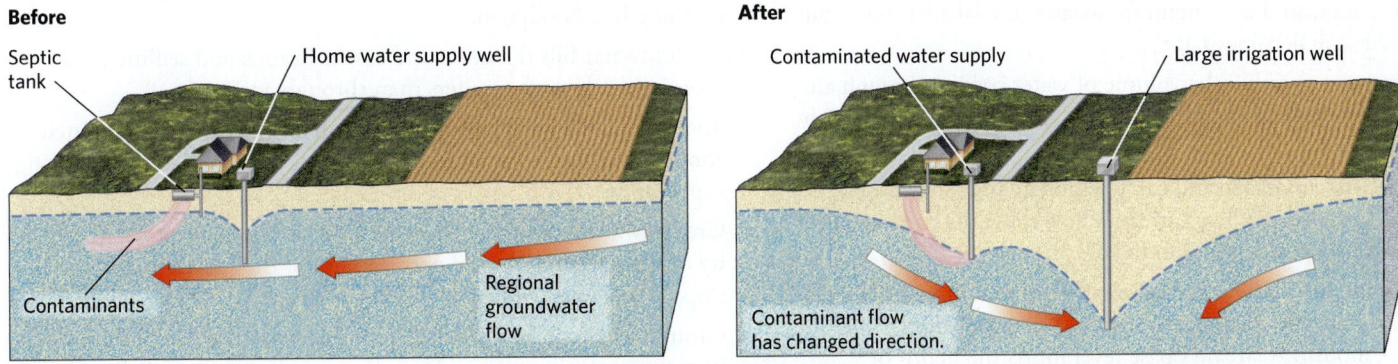

(b) Contamination of wells. (Left) Before pumping, effluent from a septic tank drifts with the regional groundwater flow, and the home well pumps clean water. (Right) After pumping by a nearby irrigation well forms a cone of depression, effluent flows into the home well in response to the new local slope of the water table.

Before

After

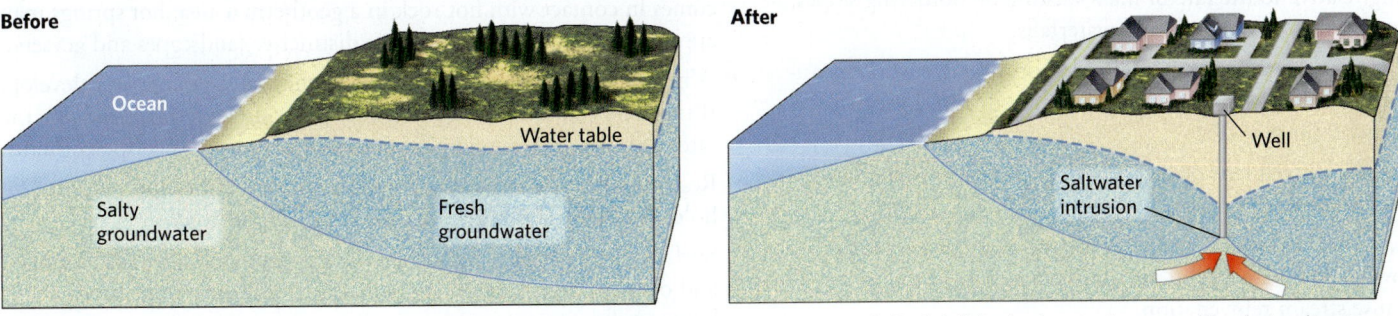

(c) Saltwater intrusion. (Left) Before pumping, fresh groundwater forms a lens below the ground. (Right) If the freshwater is pumped too fast, saltwater from below is sucked up into the well.

Before

After

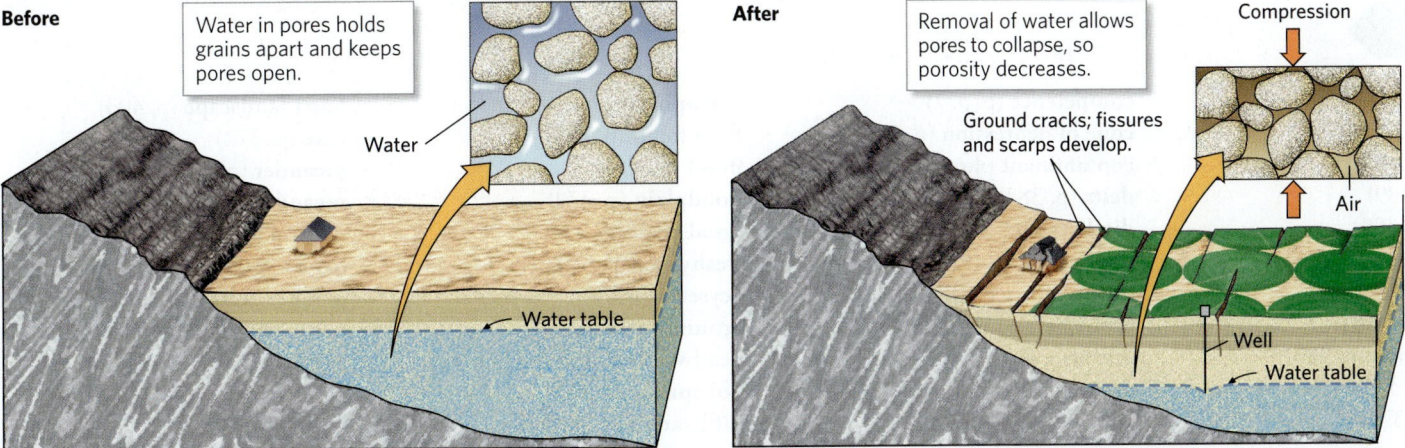

(d) Pore collapse and land subsidence. (Left) Before pumping, groundwater in pores holds sediment grains apart. (Right) When intensive irrigation removes groundwater, pore space in an aquifer collapses. As a result, the land surface sinks, leading to the formation of ground fissures and causing houses to crack.

13 CHAPTER REVIEW

Chapter Summary

- Streams drain the land surface of runoff. They grow by down-cutting and headward erosion. Eventually, a drainage network, consisting of many tributaries that flow into a trunk stream, can develop in a region. Drainage divides separate adjacent drainage networks.

- Lakes are bodies of standing water on land. Most lakes have both an inlet and an outlet, so they contain freshwater. If a lake has inlets but no outlets, the lake becomes salty.

- The discharge of a stream—the volume of water passing through an area in a given time—varies with location along the stream and with climate and season.

- Streams erode the landscape and carry sediment as dissolved, suspended, and bed loads. When a stream's flow slows, its competence decreases, and it deposits alluvium in bars.

- A stream's gradient can be depicted in a longitudinal profile. Typically, a stream has a steeper gradient at its headwaters than near its mouth. A stream's base level limits the depth of downcutting.

- Whether a stream cuts a valley or a canyon depends on the rate of downcutting relative to the rate of mass wasting on bordering slopes. Locally, streams flow over rapids and waterfalls.

- On gentle gradients, streams can be braided or meandering. A meandering stream winds back and forth across a floodplain. Eventually, a meander may be cut off.

- Streams deposit alluvial fans or deltas at their mouths, where flow slows and competence decreases.

- Over time, fluvial landscapes evolve as the regional landscape gets beveled down to the base level. Land uplift or a drop in the base level can cause stream rejuvenation.

- If more water enters a stream than the channel can hold, flooding takes place. Slow-onset floods develop slowly and submerge large regions. Flash floods happen very rapidly and affect smaller regions.

- Officials try to prevent floods by building reservoirs, levees, and floodwalls. Nevertheless, it's important to pay attention to flooding hazard (defined by annual probability or recurrence interval) when building in a floodplain.

- Groundwater fills the pores and cracks in rock and sediment. It flows more easily through aquifers than through aquitards.

- The water table separates the unsaturated zone from the saturated zone. The shape of the water table reflects the shape of the overlying topography.

- Groundwater flows from recharge areas to discharge areas. The velocity of flow depends on the permeability of the substrate and on the slope of the water table.

- Groundwater can be obtained from springs and wells. Springs form in many geologic settings. An ordinary well penetrates below the water table, but in an artesian well, water rises on its own.

- If deep groundwater rises quickly to the surface, or groundwater comes in contact with hot rock in a geothermal area, hot springs may appear. These springs can produce distinctive landscapes and geysers.

- Where limestone dissolves just below the water table, caves develop. If the water table drops, the caves empty out, and travertine precipitates out of water dripping from cave roofs to produce speleothems.

- Regions where caves have collapsed to form features such as sinkholes and natural bridges are called karst landscapes.

- Surface water supplies have been damaged by pollution, damming, and overuse. Some regions have lost their groundwater supplies because of overuse or contamination.

Key Terms

alluvial fan (p. 383)
alluvium (p. 378)
annual probability (p. 390)
aquifer (p. 389)
aquitard (p. 389)
artesian well (p. 396)
bar (p. 378)
base level (p. 378)
braided stream (p. 380)
capacity (p. 378)
channel (p. 373)

competence (p. 377)
cone of depression (p. 396)
contaminant plume (p. 405)
delta (p. 383)
discharge (p. 374)
discharge area (p. 394)
downcutting (p. 373)
drainage basin (p. 373)
drainage divide (p. 374)
drainage network (p. 373)
ephemeral stream (p. 375)

eutrophication (p. 405)
flash flood (p. 385)
flood (p. 384)
floodplain (p. 378)
flood stage (p. 384)
freshwater (p. 371)
geyser (p. 397)
groundwater (p. 388)
headward erosion (p. 373)
hot spring (p. 397)
infiltration (p. 387)

karst landscape (p. 403)
lake (p. 372)
meander (p. 381)
meandering stream (p. 381)
natural levee (p. 383)
oxbow lake (p. 383)
perched water table (p. 391)
permanent stream (p. 375)
permeability (p. 388)
point bar (p. 383)
pore (p. 388)

porosity (p. 388)
pothole (p. 377)
rapid (p. 379)
recharge area (p. 393)
recurrence interval (p. 390)
runoff (p. 373)

salt lake (p. 372)
sediment load (p. 377)
sinkhole (p. 403)
slow-onset flood (p. 384)
speleothem (p. 399)
spring (p. 395)

stalactite (p. 399)
stalagmite (p. 399)
stream (p. 373)
stream gradient (p. 378)
stream rejuvenation (p. 384)
tributary (p. 373)

turbulence (p. 375)
waterfall (p. 380)
water table (p. 389)
well (p. 395)

Review Questions

Letters in parentheses correspond to the chapter's learning objectives.

1. Define runoff and explain how runoff can initiate formation of a stream. **(A)**

2. What factors determine whether a lake is fresh or salty? What is the difference between a lake and a wetland? Distinguish among swamps, marshes, and bogs. **(A)**

3. What is a drainage network? Describe the different patterns of drainage networks recognized by geologists. Which type does the figure show? **(A)**

4. Define stream discharge and explain how it can vary along a stream's length. How can discharge be affected by climate? Why are some streams permanent and some ephemeral? **(B)**

5. Describe how streams erode and carry sediment. Distinguish between competence and capacity. **(C)**

6. How does a stream's gradient vary along its length? What's the difference between a local and an ultimate base level? What factors control the locations of rapids and waterfalls? **(C)**

7. How does a braided stream differ from a meandering stream, and what factors determine which forms at a location? Describe how meanders form and evolve. **(C)**

8. How does a drainage network evolve over time, and how does it change when stream rejuvenation takes place? What is stream piracy? What causes a drainage reversal? **(C)**

9. How do deltas grow, and how do they differ from alluvial fans? **(C)**

10. What is the difference between a slow-onset flood and a flash flood? What phenomena can cause flooding, and how do people try to prevent flooding? **(D)**

11. What is the annual probability of a flood, and how is it related to the recurrence interval? **(D)**

12. What is groundwater, and where does it reside? How do porosity and permeability differ? Using these terms, contrast an aquifer with an aquitard. **(E, F)**

13. What is the water table, and what factors affect its level? What factors affect the rate and flow direction of the groundwater? Why did the stream in the diagram dry up? **(F)**

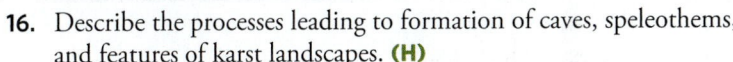

14. How is an artesian well different from an ordinary well? **(G)**

15. Why do natural springs form? Describe where hot springs form and why geysers erupt. **(G)**

16. Describe the processes leading to formation of caves, speleothems, and features of karst landscapes. **(H)**

17. How have humans overused and abused surface-water supplies? **(I)**

18. Is groundwater a renewable or a nonrenewable resource? Describe ways in which human activities affect groundwater. **(I)**

On Further Thought

19. Records indicate that the height of the Mississippi River above flood stage, for a given amount of discharge, has been rising since 1927, when a system of levees began to block off portions of the floodplain. Why? **(D)**

20. The population of a desert town in the southwestern United States has been doubling every 10 years. The town has been growing on a flat, gravel-filled basin between two small mountain ranges. Where does the water supply for the town come from? What do you predict will happen to the water table of the area in coming years? **(F, I)**

14 EXTREME REALMS
Desert and Glacial Landscapes

By the end of the chapter you should be able to . . .

A. describe the conditions that lead to the designation of a region as a desert, and explain why such conditions develop.

B. explain how weathering, erosion, and deposition lead to the formation of desert landscapes.

C. analyze the potential of human activities to change regions into deserts.

D. explain how glacial ice forms and flows, and distinguish among various kinds of glaciers.

E. discuss how glaciers advance and retreat, and how they erode the land surface.

F. distinguish between sediments and depositional landforms left by glaciers and those formed in other ways.

G. define what an ice age is, and provide the evidence used to determine when ice ages took place.

H. evaluate the ideas proposed to explain why ice ages happen and why glaciations during an ice age are periodic.

14.1 Introduction

By the end of the 18th century, European navigators had compiled maps of coastlines worldwide. But large inland regions—the extreme realms with inhospitable climates (our planet's rainforests, deserts, frozen terrains, and high mountains)—remained unknown. Europeans didn't map these realms in earnest until the 19th century, and it wasn't until the 20th century that images obtained from airplanes and satellites allowed researchers to develop a scientific understanding of them. Now, in the 21st century, you can tour even the Earth's most isolated localities on a cell-phone screen.

In this chapter, we introduce two of our planet's extreme realms: deserts and glacial landscapes. First, we learn why deserts develop, how erosion and deposition shape their surfaces, and how nondeserts may become deserts. Then we turn our attention to landscapes that are, or were, covered by glaciers. We see how glaciers form and move, and how they sculpt the ground and deposit sediment. This chapter concludes by describing the consequences and causes of *ice ages*, times when glaciers covered large areas of the continents.

<< Because of climate extremes, we see vast differences in landscapes on our planet. You can see these differences by comparing a vista of sand dunes and rocky ridges from the hot, parched Gobi Desert of Western China with one of snow-dusted, glaciated peaks in the chilly French Alps.

14.2 The Nature and Locations of Deserts

What Is a Desert?

Camels can walk for up to 3 weeks without drinking or eating because they barely sweat, they can use their own body fat as a water source, and they can withstand severe dehydration. Such adaptations are essential for animals that survive in deserts because, by definition, a **desert** receives average rainfall (or snowfall equivalent) of less than 25 cm (10 in) per year, and is therefore so *arid* (dry) that it supports vegetation on no more than 15% of its surface. Because of their lack of water, deserts host no permanent streams, except for ones that bring in water from elsewhere. Note that the definition of a desert depends on a region's aridity, not on its temperature, so geologists distinguish between *cold deserts*, where temperatures generally stay below about 20°C (68°F), and *hot deserts*, where temperatures often exceed 35°C (95°F). In some hot deserts, temperatures have reached 58°C (136°F). Deserts of all types together occupy about 25% of our planet's land surface (**Fig. 14.1**).

Where Do Deserts Exist?

Geologists classify deserts by the reason why the desert formed. Types of deserts include the following:

- *Subtropical deserts:* At equatorial latitudes, intense sunlight strikes the ground. Heat rising from the ground warms the base of the atmosphere. As its temperature increases, near-surface air becomes less dense than overlying cooler air, so it rises—like a

Did you ever wonder . . .

how hot it can get in a desert?

FIGURE 14.1 The global distribution of deserts.

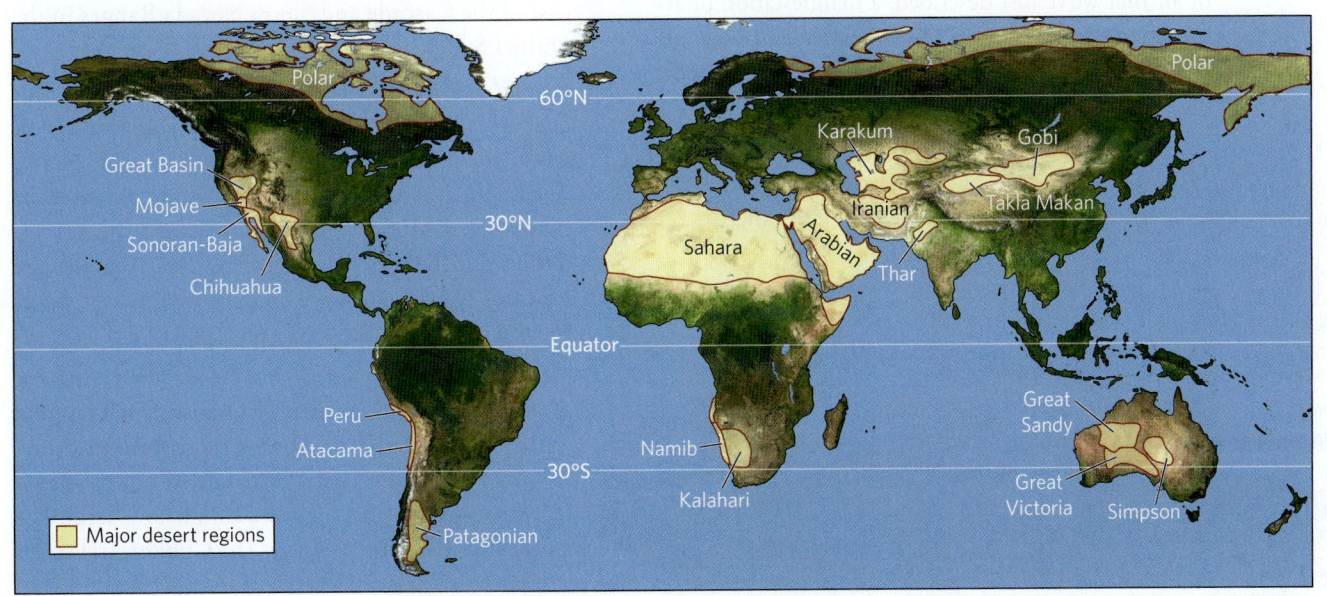

Major desert regions

FIGURE 14.2 Formation of subtropical deserts. Subtropical deserts form because they lie beneath the portion of an atmospheric circulation pattern, the Hadley cell, where dry air descends and warms. As a result, rain clouds rarely form.

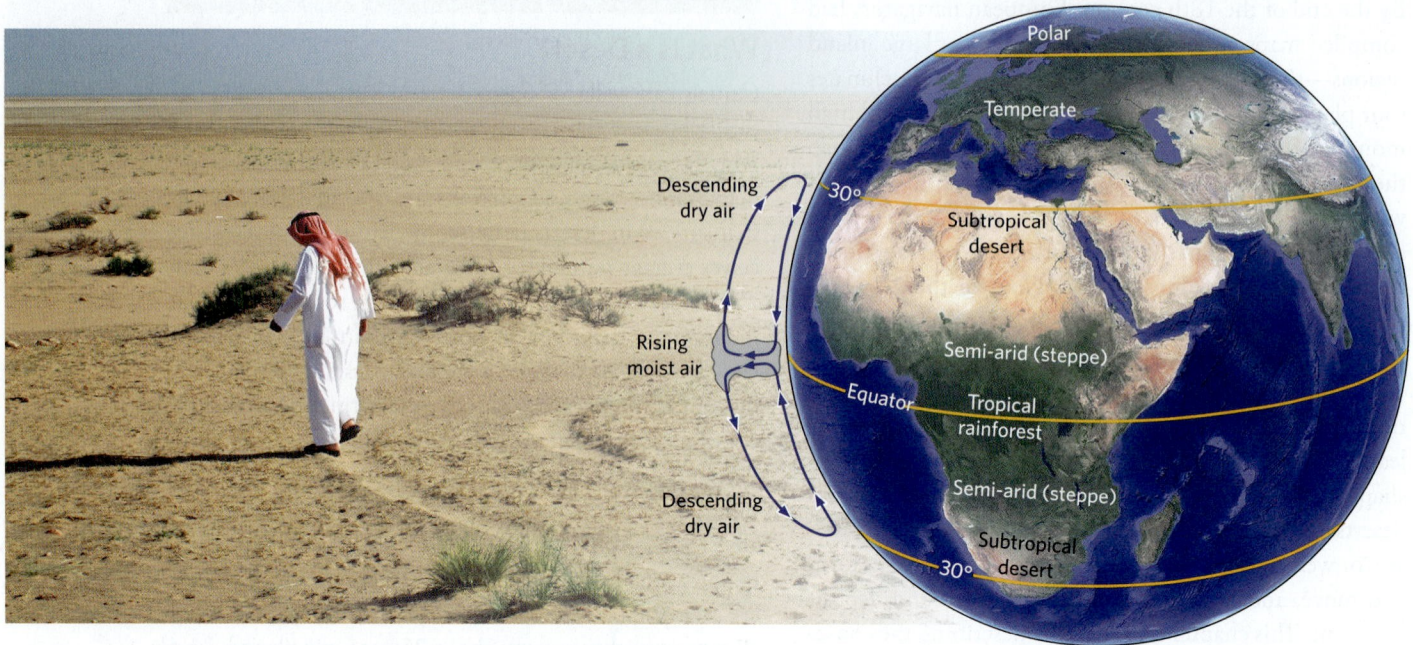

hot-air balloon—to high elevation, where it expands and cools. Cooling of the rising air causes its moisture to condense and fall as rain, which drenches the tropical rainforests below. At high elevation, the now-drier air flows away from the equator to subtropical latitudes, where it sinks. As it sinks, this air undergoes compression, which causes it to heat up. Rain-producing clouds can't form in this dry descending air, so the land below becomes a *subtropical desert*. Most of the Earth's largest deserts, including the Sahara, the Arabian, the Thar, the Kalahari, and the deserts of Australia, are subtropical deserts. The flow pattern of air that we've just described, a manifestation of atmospheric convection, is known as a *Hadley cell*, after the researcher who proposed it (see Chapter 18);

one Hadley cell extends north of the equator, and one south **(Fig. 14.2)**.

- *Rain-shadow deserts:* Air flowing toward a mountain range must rise when it encounters the range's slopes. As it does, it expands and cools, so water vapor in it condenses and falls as rain on the windward flank of the mountains, nurturing coastal rainforests. When the air moves farther inland and descends on the leeward side of the mountains, it has lost its moisture and can no longer provide rain. As a consequence, the land on the leeward side becomes a *rain-shadow desert* **(Fig. 14.3)**. Examples lie east of the Cascade and Sierra Nevada Ranges in the United States.

- *Coastal deserts:* Air above cold ocean water cools. This cool air is too dense to rise into overlying warmer air where the water in it could condense to form clouds. Therefore, mid-latitude and low-latitude coastal lands bordering cold ocean currents receive little precipitation and become *coastal deserts*. An example, the Atacama Desert of western South America, is probably the driest place on the Earth **(Fig. 14.4)**.

- *Continental-interior deserts:* As air flows from the ocean across a continent, its moisture falls as rain. By the time the air reaches the continent's interior, far from a coast, it has dried out so that it can't produce rain, and the land below becomes a *continental-interior desert*. The Gobi of Asia serves as an example.

FIGURE 14.3 Formation of rain-shadow deserts. Moist air rises and drops rain on the windward side of a mountain range. By the time it reaches the leeward side of the mountains, the air is dry.

FIGURE 14.4 Formation of coastal deserts.

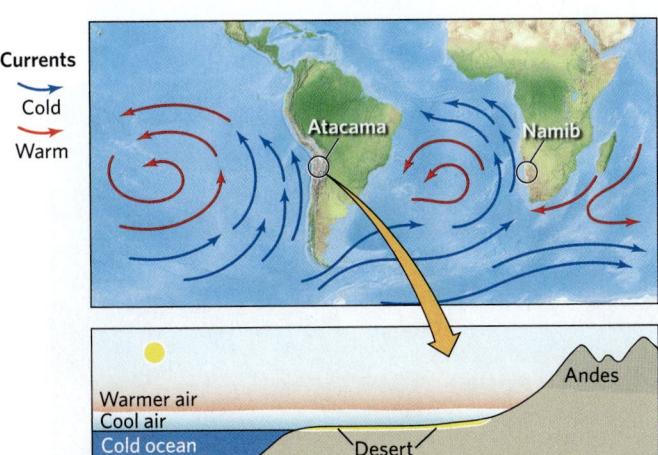

Currents
Cold
Warm

(a) Cold ocean currents cool and moisten the air along the coast. This air is too dense to rise and produce rain clouds.

(b) The Atacama Desert is the driest place on the Earth. While clouds form there, they don't produce rain.

- *Polar deserts:* The very cold, dense air above polar regions holds very little moisture, so minimal precipitation falls in polar regions, making these regions actually very dry. What snow does fall, however, survives for a long time, so polar deserts can retain substantial snow cover.

Take-home message . . .

Deserts receive an average of less than 25 cm (10 in) of rain a year, which makes them so arid that they host only sparse vegetation. Deserts develop in several settings: subtropical dry climates, rain shadows, coasts bordered by cold currents, continental interiors, and polar regions.

Quick Question

Why does the world's largest desert, the Sahara, exist?

14.3 Weathering, Erosion, and Deposition in Deserts

Desert Weathering and Soil Formation

Physical weathering in deserts begins, as it does elsewhere, when joints split bedrock into pieces that can tumble down slopes and fragment into smaller pieces. Chemical weathering also happens in deserts when moisture from rain or dew seeps into rock and dissolves calcite, quartz, and various salts there, and breaks down silicate minerals as well. Over time, this process causes fresh rock to crumble and become sediment. However, because there's less water available to react with rock, chemical weathering in deserts happens more slowly than in moister climates.

Because rain happens so rarely in deserts, little vegetation grows there, and infiltrating water cannot carry dissolved ions completely away. Therefore, soils of arid regions differ from those formed in wetter climates (see Chapter 5). Specifically, a desert soil does not have an organic-rich O-horizon, and calcite can precipitate beneath the ground surface to form *caliche*. Notably, the lack of plant cover in deserts allows color variations in bedrock and soil to stand out, so slight variations in the concentration of iron, or in the degree of iron oxidation, produce spectacular color banding on the land surface. The Painted Desert of northern Arizona earned its name from the brilliant and varied hues of oxidized iron in the region's shale bedrock **(Fig. 14.5a)**.

Exposed rock surfaces in deserts commonly display **desert varnish**, a dark, shiny, rusty-brown coating **(Box 14.1)**. Recent studies suggest that desert varnish forms when windblown clay settles on the surfaces of rocks in the presence of a small amount of moisture. Over the course of centuries to millennia, chemical reactions and the activity of microbes extract iron and manganese ions from the dust, incorporating them into shiny oxide minerals that bind to the rock. Traditional cultures create artistic figures, called *petroglyphs*, by chipping away the varnish to reveal the underlying lighter-colored rock **(Fig. 14.5b)**.

BOX 14.1 ▶ Putting Earth Science to Use

Recognizing young landslides in desert regions

As we discussed in Chapter 12, landslides carry rock and debris from higher up on steep slopes down to lower elevations. How can you recognize where a relatively recent landslide has occurred if the landslide was not observed directly by people? In temperate or tropical areas, the character of vegetation on a hillslope can provide a clue to the timing of landslides because mass wasting disrupts vegetation. Trees and shrubs on a new landslide are tilted chaotically. A somewhat older landslide may host a patch of dead vegetation because mass wasting disrupts plant roots, so after weeks to months, the plants die. A still older landslide may be covered by regrowth that looks younger than vegetation on slopes that haven't slipped.

Can you interpret the relative ages of landslides in desert regions, where vegetation doesn't grow? Yes. The visual cue comes from looking at the color of desert varnish. The boulders of very old landslides have a very dark coating of varnish, whereas many boulders of recent landslides don't have a coating of varnish at all (because the downslope movement of the boulders has turned some of them over) **(Fig. Bx14.1)**. Therefore, the surface of a young landslide, overall, has a lighter color. Intermediate-aged landslides will be darker than new ones, but lighter than very old ones. For those looking to live or build in desert regions—or just wanting to be able to interpret the history of a desert region during a hike—color is a useful clue for reading the landscape.

FIGURE Bx14.1 On the wall of the Grand Canyon, rocks develop desert varnish over time. Fresh rocks, exposed by a landslide, do not have varnish yet.

FIGURE 14.5 Desert colors.

(a) The red hues of the Painted Desert in Arizona are due to the oxidation (rusting) of iron in the rock.

(b) By chipping away desert varnish to reveal the lighter rock beneath, Native Americans produced art and symbols.

FIGURE 14.6 Evidence of erosion by water in deserts.

(a) These hills in the desert near Las Vegas, Nevada, are bone dry, but their shape indicates erosion by water. Note the numerous stream channels.

(b) Badlands topography occurs when water erodes unvegetated weak sediment.

(c) Gravel and sand have been left behind on the floor of a dry wash after a flash flood in Death Valley. Erosion by water is cutting a channel.

Erosion by Water

Although heavy rains rarely fall in deserts, when they do, they can radically alter a landscape in a matter of minutes, for vegetation doesn't protect the land surface. In fact, in most deserts, water causes more erosion than does wind **(Fig. 14.6a)**. Erosion by water begins with the impacts of raindrops, which knock sediment from the ground into the air. On a hill, this sediment lands downslope. When water starts flowing across the ground surface, it carries away loose sediment. On hillslopes composed of soft substrate, water erosion may produce *badlands topography*, a network of parallel channels that merge downslope **(Fig. 14.6b)**.

During intense rainstorms, channels in valley floors fill with a turbulent mixture of water and sediment, which rushes downstream as a flash flood, picking up sediment and carrying it along. When the rain stops, the water sinks into the streambed and disappears, so such streams are ephemeral. The dry channels that exist between floods are called **dry washes**, *arroyos*, or *wadis* **(Fig. 14.6c)**.

Erosion by Wind

Because vegetation doesn't provide protection in deserts, winds have direct access to the ground and can pick up and transport large quantities of sediment. Geologists distinguish between two components of the sediment load carried by wind: suspended loads and surface loads.

Suspended loads consist of fine-grained sediment—clay and silt—that can move with the wind and can stay aloft for a long time. In fact, this sediment can be carried so high (up to several kilometers above the Earth's surface) and so far downwind (tens to thousands of kilometers)

that it may move completely out of its source region. In fact, dust blown from the Sahara crosses the Atlantic Ocean **(Fig. 14.7a)**. Particularly strong winds can generate dramatic **dust storms** up to 100 km (60 mi) long and 1.5 km (1 mi) high. An approaching dust storm resembles a roiling, opaque wave. Large dust storms occasionally engulf cities **(Fig. 14.7b)**.

Surface loads, or *bed loads*, start moving only when the wind becomes strong enough to roll and bounce sand grains along the ground. This motion, known as **saltation**, begins when air turbulence lifts sand grains **(Fig. 14.7c)**. The grains move downwind, following an asymmetrical, arc-like trajectory. When they return to the ground, they strike other sand grains, causing those grains to bounce up and drift or roll downwind. Saltating grains generally rise no more than 0.5–2 m (1.5–6.5 ft) above the ground.

Just as sandblasting cleans the grime off the surface of a building, windblown sand and dust grind away at surfaces in the desert. Over long periods, such wind abrasion can sculpt bedrock, producing streamlined outcrops called *yardangs* **(Fig. 14.8a)**. At a smaller scale, saltating grains carve facets on pebbles and cobbles, producing *ventifacts* **(Fig. 14.8b, c)**. Locally, wind may carry away so much sand and dust that remaining pebbles and cobbles become concentrated at the ground surface as a **lag deposit** **(Fig. 14.8d)**.

Deposition in Deserts

Recall from Chapter 12 that rocky debris falling from cliffs tumbles downslope and may accumulate in a **talus pile**, a steep, sloping apron of debris at the cliff base **(Fig. 14.9a)**. In deserts, talus piles remain unvegetated for a long time, and their rocks may become

FIGURE 14.7 Wind transport of sediment in deserts.

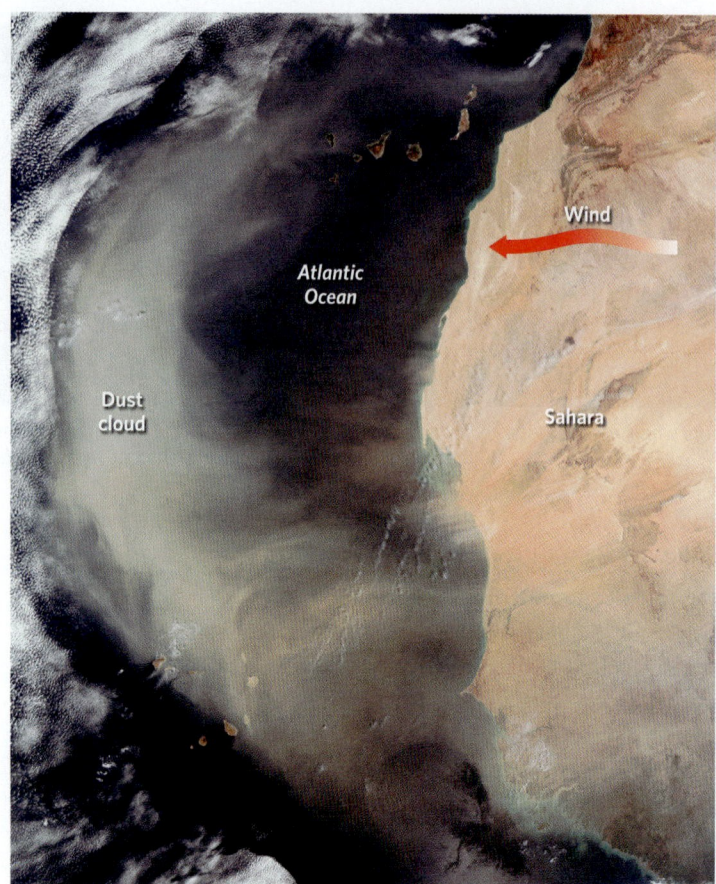

(b) A huge dust storm approaching Phoenix, Arizona.

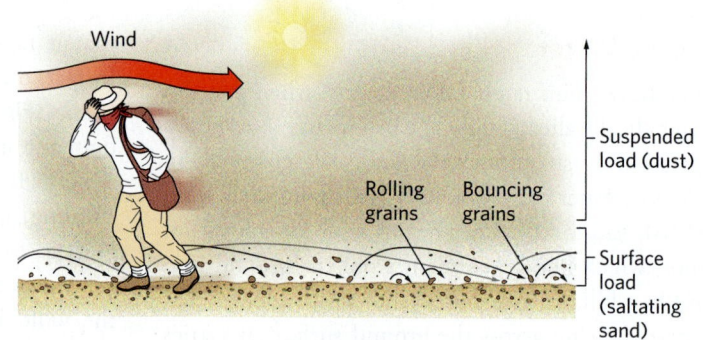

(a) In this satellite image, a huge dust cloud that originated in the Sahara blows across the Atlantic Ocean.

(c) Wind transports desert sediment both in suspension and in a saltating layer.

coated with desert varnish. When sediment eventually makes it into a stream channel, it will be transported downstream by flash floods. As floodwaters subside, this sediment accumulates to form bars that remain as elongate mounds of gravel in a dry wash. Where a stream emerges from a canyon mouth and onto a plain, its channel typically subdivides into a number of smaller braided channels, or *distributaries*, that diverge outward and deposit debris in an **alluvial fan**, a broad wedge- or apron-shaped pile of sediment **(Fig. 14.9b)**. Not all distributaries of an alluvial fan flow at the same time. Rather, flows carry sediment along one part of the fan for a while until the sediment piles up. The next flow finds an easier path, to the side of the previously deposited mound of sediment. Repetition of this process over time causes deposition to move from one part of the fan to another, so overall, the fan maintains its symmetrical shape. Alluvial fans emerging from adjacent valleys may eventually merge and overlap along the front of a mountain range, producing an elongate wedge of sediment bordering the range, called a *bajada*.

During particularly wet seasons, lowlands or basins between mountain ranges in a desert may fill with water to form shallow lakes. During drier times, when a desert lake evaporates entirely, salts in the lake precipitate, and the lake bed becomes exposed to form a flat, salt-and-clay-coated surface known as a **playa (Fig. 14.10a)**. Over multiple cycles of flooding and drying, the salty sediment layer, composed of halite and gypsum mixed with clay, can become quite thick **(Fig. 14.10b)**.

Take-home message . . .

Weathering breaks down rocks in deserts, as in other places, but because of the scarcity of water, soils are thin, and desert varnish may coat rock surfaces. Wind transports sediment and can erode rock. Heavy rains cause significant erosion and sediment transport in deserts, but since streams rarely flow, their channels are usually dry washes. Debris formed by desert erosion accumulates in talus piles and alluvial fans.

Quick Question
How do playas form?

FIGURE 14.8 Examples of wind erosion in deserts.

(a) Yardangs in western China.

Time 1

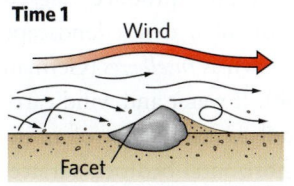

Windblown sand abrades the face of a rock, forming a facet.

Time 2

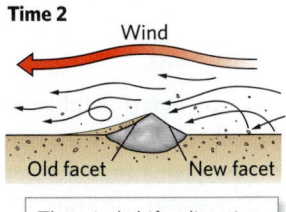

The wind shifts direction, and a new facet forms.

(b) Ventifacts form when windblown sediment erodes the surface of a rock.

(c) An example of a multifaceted ventifact in the Gobi Desert. The shoe provides scale.

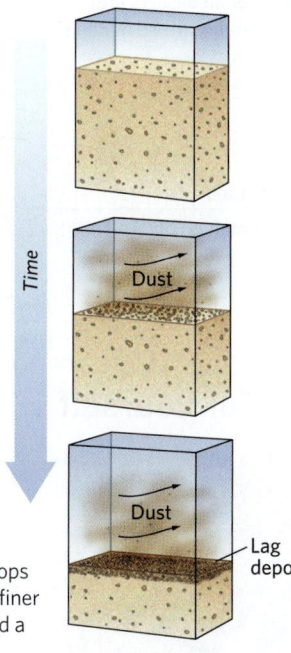

(d) A lag deposit develops when wind blows away finer sediment, leaving behind a layer of coarser grains.

FIGURE 14.9 Accumulation of debris in deserts.

(a) This talus pile along the base of a desert cliff formed from rocks that broke off and tumbled down the cliff.

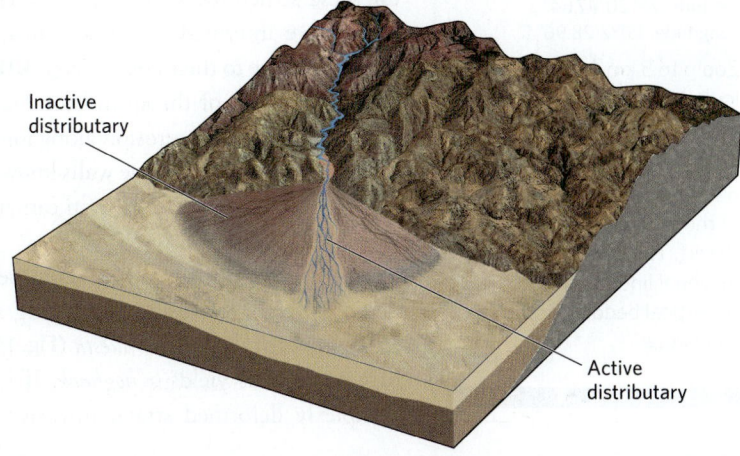

(b) Distributaries carry water and sediment to different parts of an alluvial fan at different times, thereby maintaining the fan shape.

FIGURE 14.10 The formation of playas.

A close-up of salt crystals.

(a) This playa in California formed at the base of a bajada, as seen in this aerial, oblique view.

(b) White salt crystals encrust the floor of a playa in Death Valley.

Did you ever wonder . . .

whether all deserts are completely covered by sand?

 Video

Evolution of Deserts

See for **yourself**

Uluru, Australia

Latitude: 25°20′47.64″ S
Longitude: 131°2′28.96″ E

Zoom to 5 km (~3 mi) and look down.

The red sandstone of Uluru (Ayers Rock), an inselberg, rises above the stony plains of the central Australian desert. The NW-trending diagonal lines are traces of vertical bedding in the sandstone.

14.4 Desert Landscapes

Popular media commonly portray deserts as endless seas of sand. In reality, desert landscapes display much more variation. In addition to sand, deserts can host stony plains and steep rocky cliffs. Let's examine these contrasting landscapes.

Rocky Cliffs, Mesas, and Arches

Because thick soil cover doesn't accumulate on steep slopes in deserts, these slopes become rocky ridges and cliffs. In places underlain by horizontal sedimentary strata, or by horizontal layers of volcanic rock, cliff faces tend to form when rock separates along vertical joints, and topples. By breaking off at successive joints, a cliff face steps back into the land while retaining roughly the same profile, a phenomenon known as **cliff retreat**. Where beds have different resistance to erosion, cliffs develop a step-like shape—the resistant layers become cliffs, whereas the nonresistant layers become rubble-covered slopes **(Fig. 14.11a)**. Due to cliff retreat, erosion of a flat-topped plateau underlain by horizontal layering produces flat-topped hills **(Fig. 14.11b)**. Of these structures, *mesas* have the largest surface area, *buttes* have intermediate surface area, and *chimneys* are narrow relative to their height **(Fig. 14.11c)**. In places where bedrock consists of thick sandstone layers cut by widely spaced vertical joints, erosion along joints can transform a layer into a set of sandstone walls known as *fins*. Localized erosion through the base of a fin can yield a **natural arch** **(Fig. 14.12)**.

Of course, horizontally layered bedrock doesn't underlie all deserts. Erosion of dipping strata can produce an asymmetrical ridge, or *cuesta* **(Fig. 14.13a)**, and erosion of vertical beds yields a *hogback*. If bedrock consists of complexly deformed strata, intrusive igneous rocks, or metamorphic rocks, erosion typically produces jagged rocky ridges. Eventually, erosion of a desert landscape leaves isolated ridges of rock, known as *inselbergs* (German for island mountain; **Fig. 14.13b**). If water and wind carry away the sediment that accumulates at the base of an inselberg, then a broad, nearly horizontal bedrock surface, called a *pediment*, surrounds the inselberg **(Fig. 14.13c)**. Generally, sediment-filled plains lie between neighboring inselbergs.

Stony Plains and Desert Pavement

The surfaces of low-relief, sand-free desert landscapes typically host a layer of pebbles and cobbles, so they are called *stony plains*. Portions of these plains may develop **desert pavement**, a surface that consists of countless separate tabular (broad and flat) fragments of rock that fit together like pieces of a tile mosaic **(Fig. 14.14a, b)**. Researchers suggest that development of desert pavement involves several processes. The tabular fragments may form when differential heating of rock blocks during the Sun's daily cycle causes the blocks to split along parallel vertical cracks. The resulting pieces eventually fall over, and windblown dust settles between them. During rains, clay in the dust absorbs water and swells, and during dry times, the clay shrinks. This repeated swelling and shrinking causes the tabular fragments to settle into an overall flat surface. As this process continues, additional clay settles between the fragments, and soil builds up beneath them **(Fig. 14.14c)**. Gradually, desert varnish coats the surface of the pavement.

Seas of Sand: The Nature of Dunes

Places where wind deposits thick accumulations of sand can become sand seas, known as *ergs* in Arabic, that range from just a few kilometers to a few hundred

FIGURE 14.11 Consequences of cliff retreat.

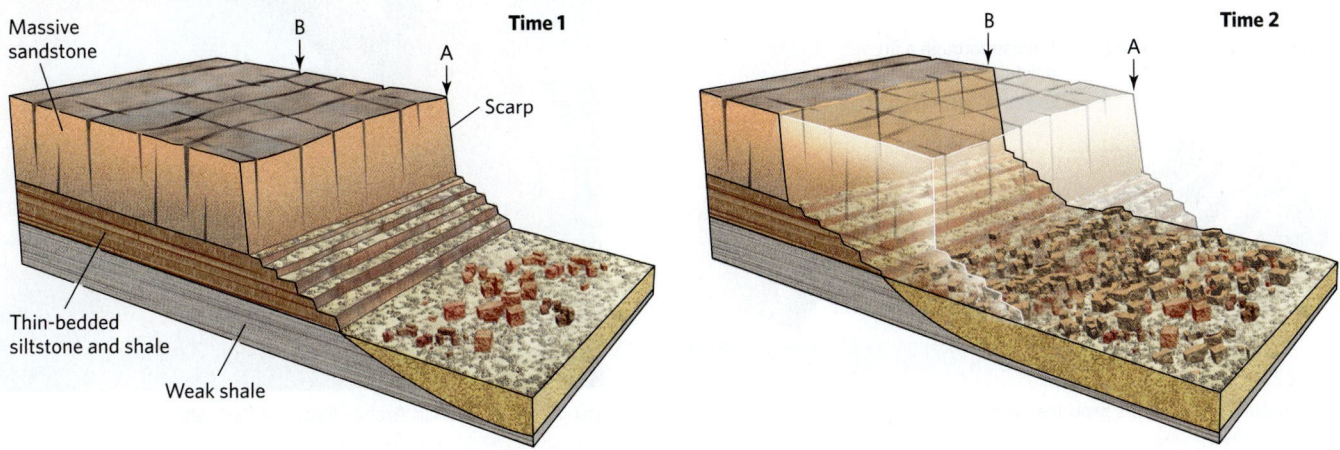

Time 1 — Massive sandstone, B, A, Scarp; Thin-bedded siltstone and shale; Weak shale

Time 2 — B, A

(a) Cliff retreat happens when rock breaks off along vertical joints parallel to the cliff face.

Mesa

Butte — Chimney

Time (increasing amount of erosion)

(b) A once-continuous layer of rock evolves into a series of isolated mesas, buttes, and chimneys. If the bedding is horizontal, the resulting landforms have flat tops.

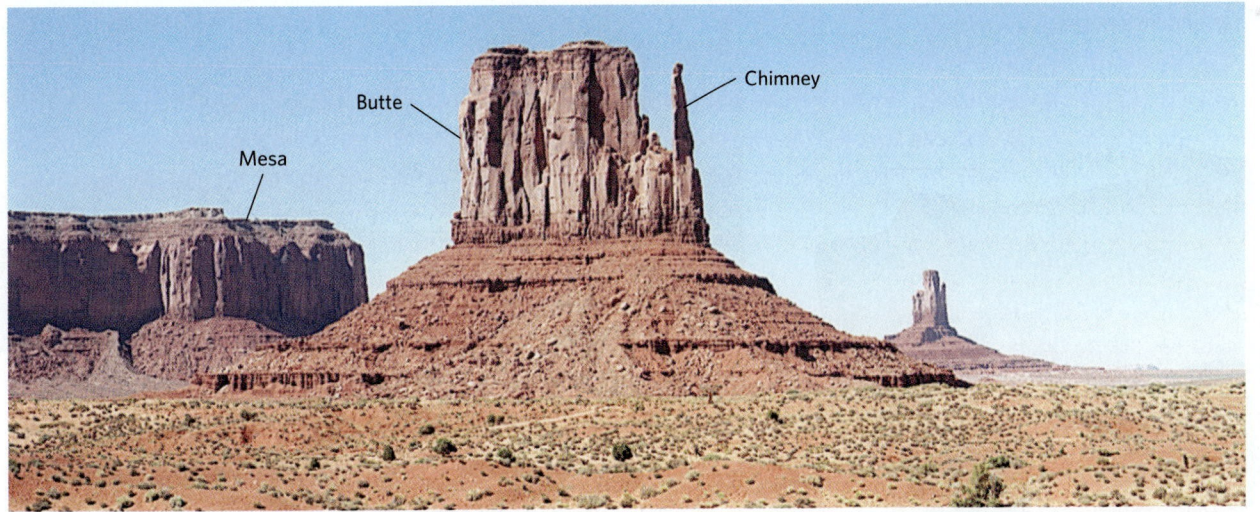

Mesa — Butte — Chimney

(c) Buttes and mesas tower above the floor of Monument Valley, Arizona.

kilometers across. Within these regions, the wind builds **sand dunes**, localized elongate hills of sand **(Fig. 14.15a)**. Dune formation begins when sand becomes trapped on the windward side of an obstacle. Gradually, the sand builds downwind and buries the obstacle. Sand saltates up the windward side of the dune, blows over the crest of the dune, and then settles on the dune's steeper, leeward face. When this face attains the angle of repose, the maximum slope of a freestanding pile of sand, it becomes unstable, and sand slides down the slope, so

FIGURE 14.12 Formation of natural arches.

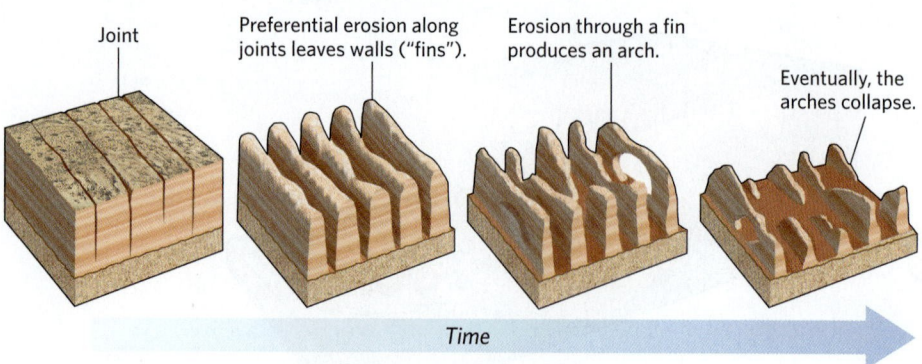

Joint

Preferential erosion along joints leaves walls ("fins").

Erosion through a fin produces an arch.

Eventually, the arches collapse.

Time

(a) Erosion that occurs preferentially along joints produces wall-like fins of rock. When the lower part of a fin erodes, a natural arch remains.

(b) A natural arch in Arches National Park, Utah.

geologists refer to the lee side of a dune as the *slip face*. As more and more sand accumulates on the slip face, the crest of the dune migrates downwind, and former slip faces are preserved as cross beds inside the dune **(Fig. 14.15b)**.

Dunes display a variety of shapes and sizes, depending on the character of the wind and the sand supply **(Fig. 14.15c)**. In regions of scarce sand and steady wind, beautiful crescents called *barchan dunes* develop, with the tips of the crescents pointing downwind. If the wind direction shifts frequently, a group of crescents pointing in different directions overlap one another,

resulting in the formation of *star dunes*. Where enough sand accumulates to bury the ground surface completely, and only moderate winds blow, sand piles into simple, wave-like shapes called *transverse dunes*, with their crests trending perpendicular to the wind direction. Strong winds may break through transverse dunes and transform them into *parabolic dunes*, whose ends point in the upwind direction. Finally, if there is abundant sand and a strong, steady wind, the sand streams into *longitudinal dunes*, whose crests lie parallel to the wind direction. Active sand dunes constantly move and change as the wind blows.

FIGURE 14.13 Formation of cuestas and inselbergs.

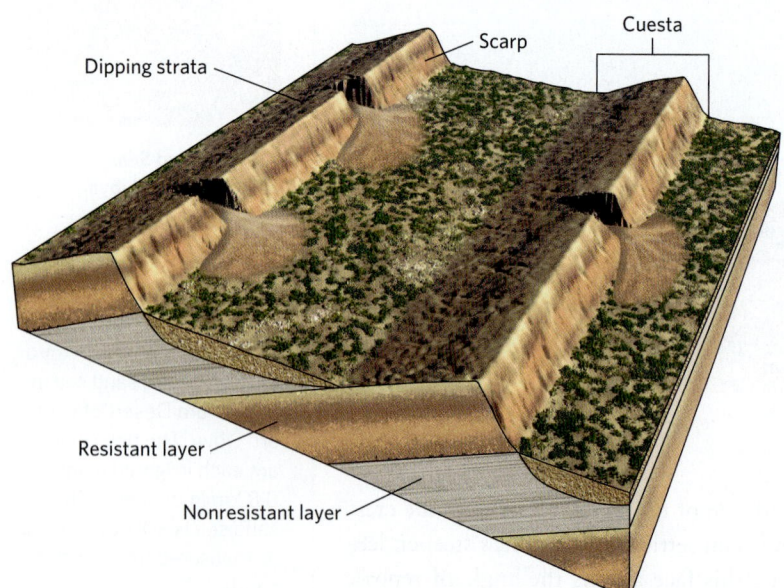

Dipping strata

Scarp

Cuesta

Resistant layer

Nonresistant layer

(a) Asymmetrical ridges called cuestas develop where strata are not horizontal.

(b) Aerial view of an inselberg surrounded by alluvium in China.

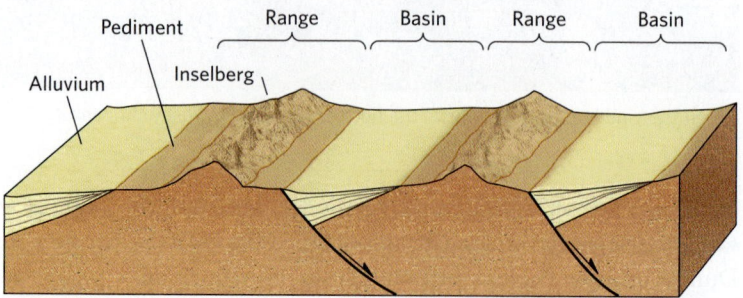

Pediment

Range   Basin   Range   Basin

Alluvium

Inselberg

(c) In the Basin and Range Province of the southwestern United States, tilted fault-block ranges evolve into inselbergs, bordered by pediments and basins.

FIGURE 14.14 Desert pavement and a hypothesis for its formation.

(a) A well-developed desert pavement in the Sonoran Desert, Arizona. The inset shows a close-up of the pavement.

(b) Students standing at the edge of a trench cut into desert pavement. Note the soil between the pavement and the underlying alluvium.

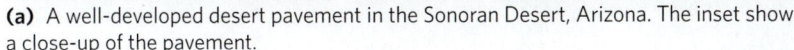

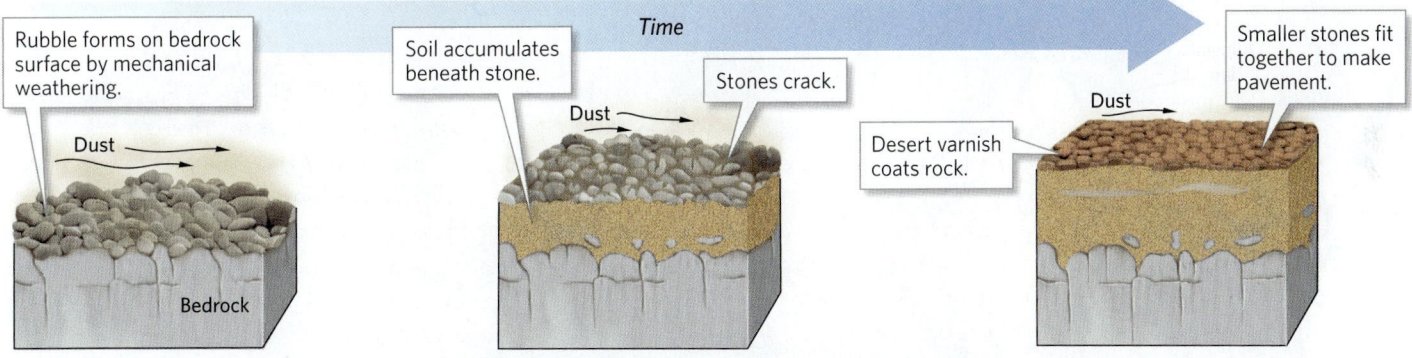

Time

Rubble forms on bedrock surface by mechanical weathering.

Dust

Bedrock

Soil accumulates beneath stone.

Dust

Stones crack.

Smaller stones fit together to make pavement.

Dust

Desert varnish coats rock.

(c) Desert pavement forms in stages. First, loose pebbles and cobbles collect at the surface. Dust settles among the stones and builds up a soil layer below. The stones eventually crack into smaller pieces and settle to form a mosaic-like pavement.

FIGURE 14.15
Sand dunes in deserts.

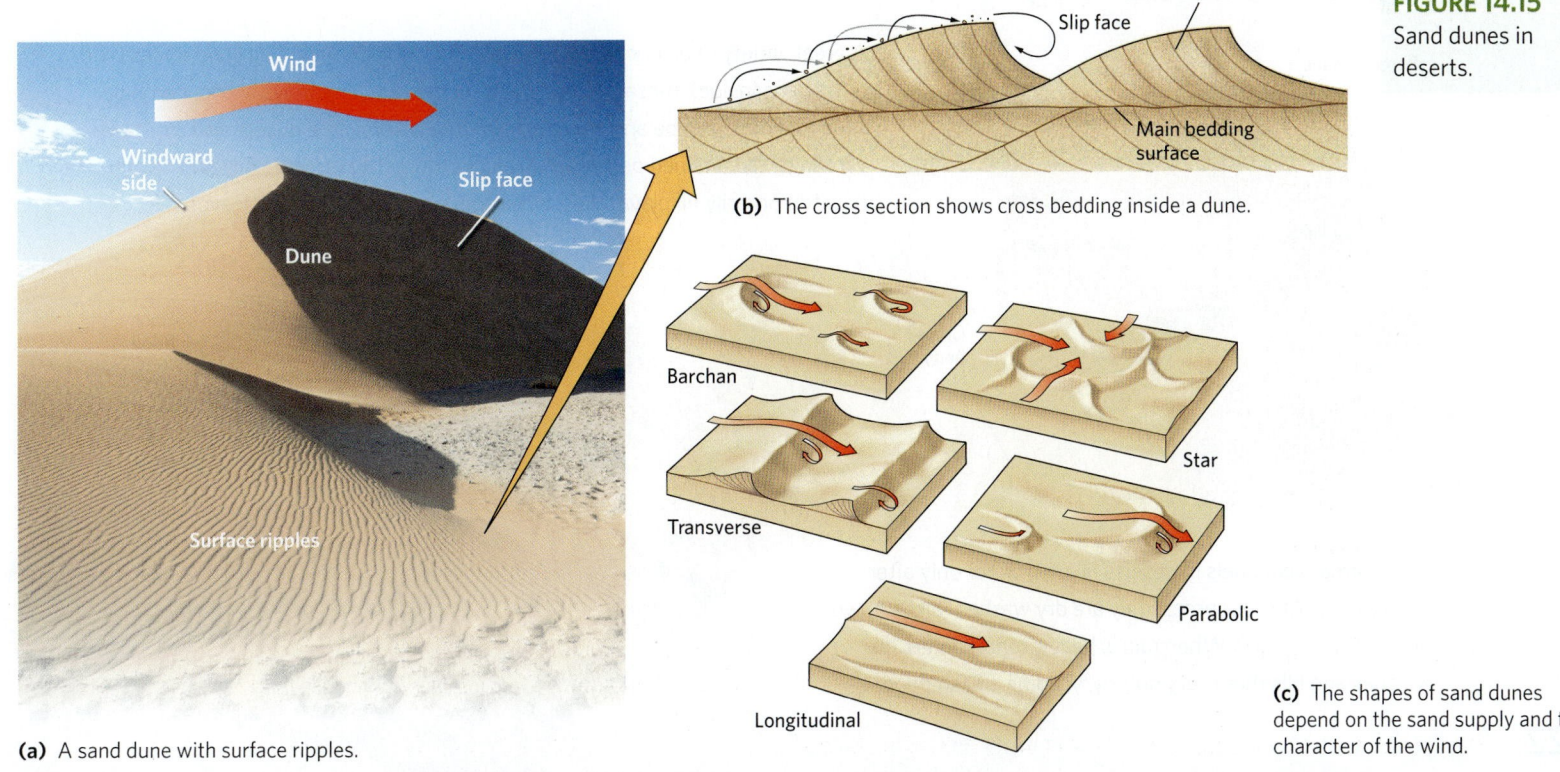

Wind

Windward side

Dune

Slip face

Surface ripples

(a) A sand dune with surface ripples.

Cross-bed surface

Slip face

Main bedding surface

(b) The cross section shows cross bedding inside a dune.

Barchan

Star

Transverse

Parabolic

Longitudinal

(c) The shapes of sand dunes depend on the sand supply and the character of the wind.

The Desert Realm

The desert of the Basin and Range Province in Utah, Nevada, and Arizona consists of alternating basins (grabens or half-grabens) separated by narrow ranges (tilted fault blocks), depicted schematically here. The Sierra Nevada, underlain largely by granite, borders the western edge of the province, while the Colorado Plateau, underlain by flat-lying sedimentary strata, borders the eastern edge. The overall climate of the region is dry.

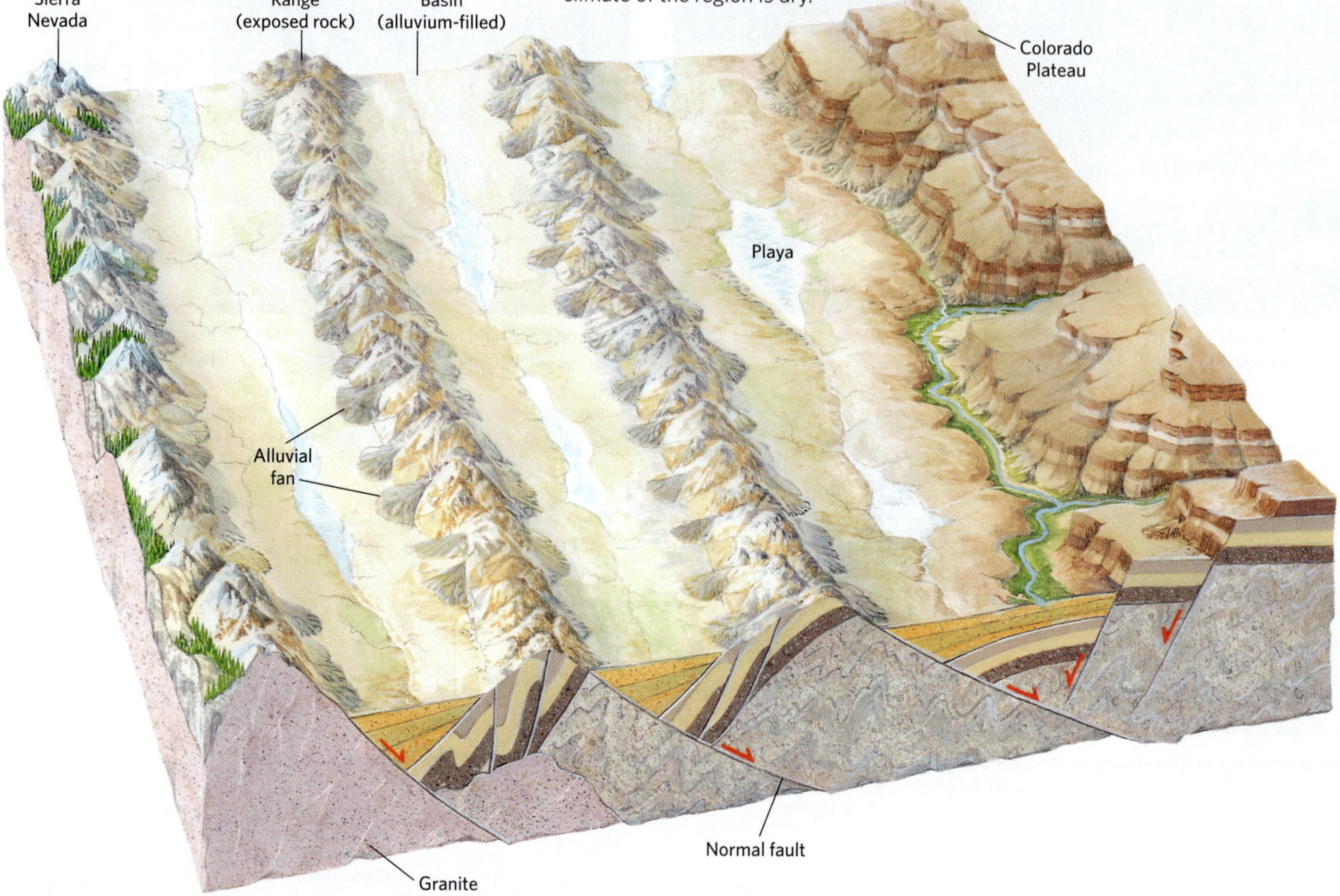

Sierra Nevada

Range (exposed rock)

Basin (alluvium-filled)

Colorado Plateau

Playa

Alluvial fan

Normal fault

Granite

Barchan dune

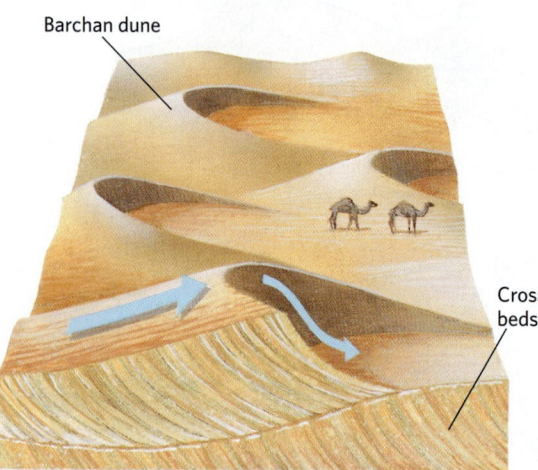

Cross beds

A great variety of distinct landforms occur in deserts, including sand dunes (with cross beds inside), mesas, alluvial fans, arches, and chimneys. Wind, carrying sand and dust, can be an effective agent of erosion in the desert, and can carve yardangs, streamlined pedestals. In places, desert pavements develop. Water may temporarily fill playa lakes; when this water evaporates, it leaves salt.

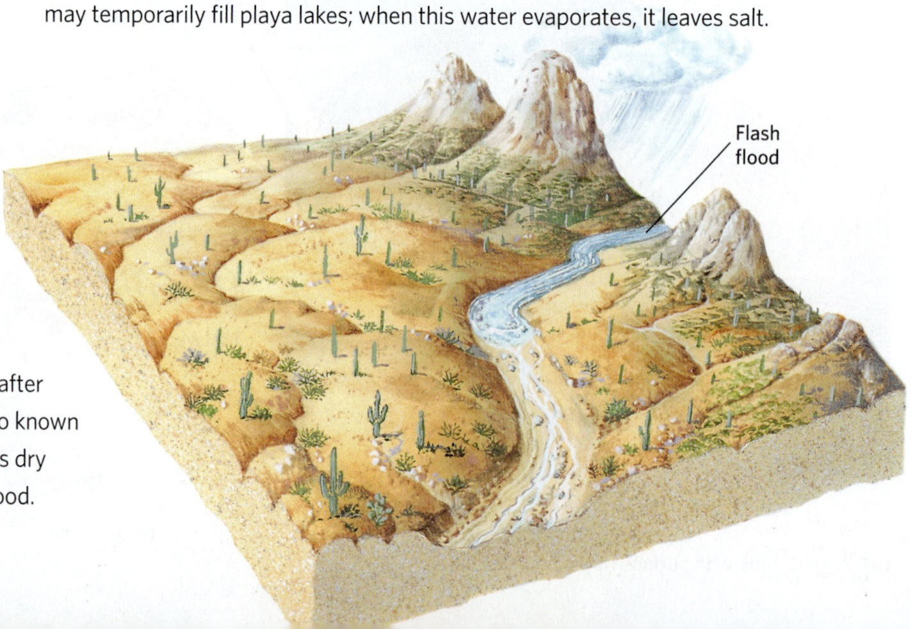

Flash flood

Most stream channels in deserts fill with water only after heavy rains. At other times, they are dry washes (also known as *arroyos* or *wadis*). When rain is heavy, runoff enters dry washes and fills them very quickly, yielding a flash flood.

422

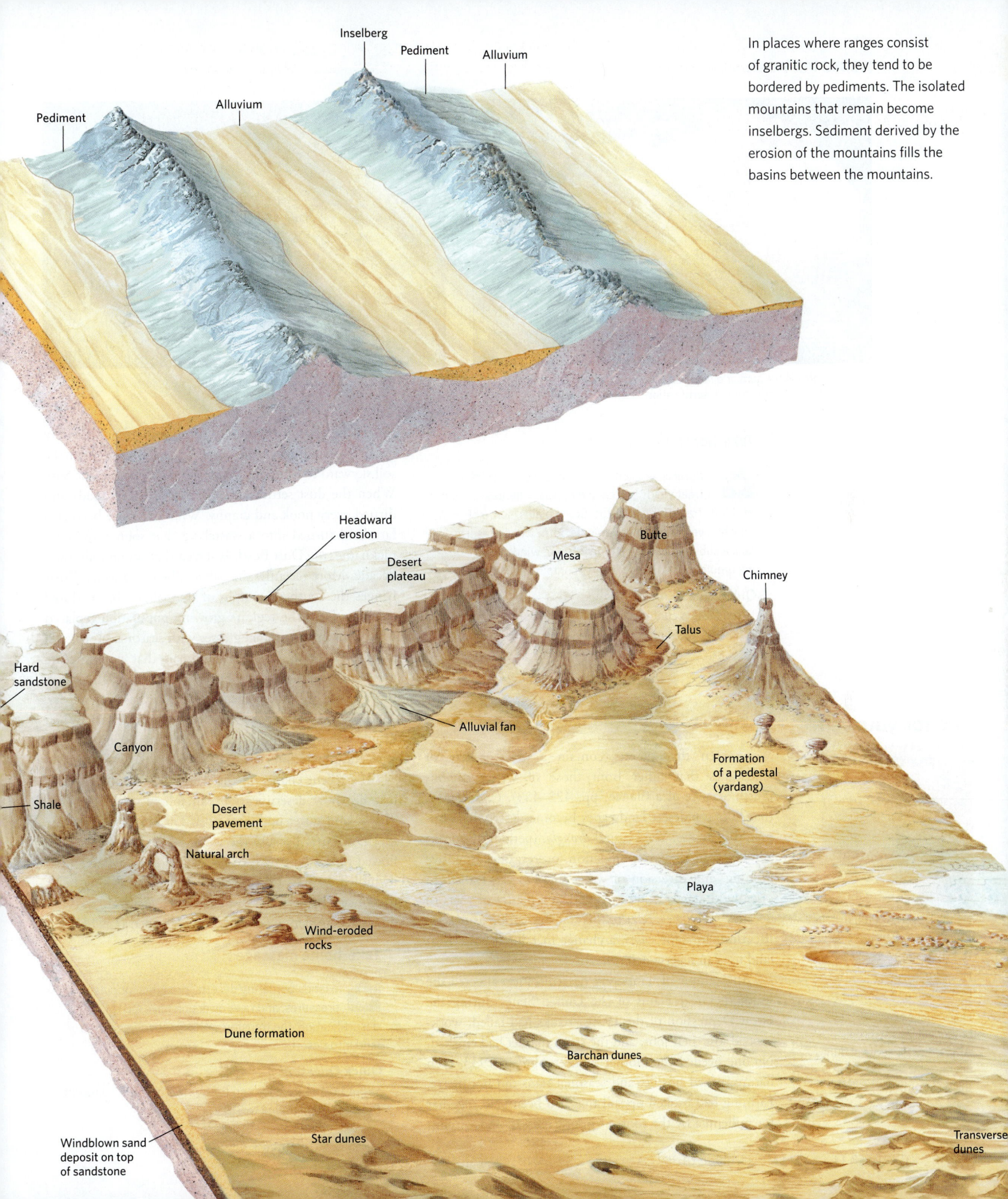

Pediment

Inselberg

Pediment

Alluvium

Alluvium

Pediment

In places where ranges consist of granitic rock, they tend to be bordered by pediments. The isolated mountains that remain become inselbergs. Sediment derived by the erosion of the mountains fills the basins between the mountains.

Headward erosion

Desert plateau

Mesa

Butte

Chimney

Talus

Hard sandstone

Alluvial fan

Canyon

Formation of a pedestal (yardang)

Shale

Desert pavement

Natural arch

Playa

Wind-eroded rocks

Dune formation

Barchan dunes

Windblown sand deposit on top of sandstone

Star dunes

Transverse dunes

FIGURE 14.16 Desertification in the Sahel.

(a) The Sahel is the semiarid land along the southern edge of the Sahara. Large parts have undergone desertification.

(b) In parts of the Sahel, the soil has dried up.

Take-home message . . .

In deserts, erosion of horizontal beds by cliff retreat yields buttes and mesas, whereas erosion of tilted strata forms cuestas. Stony plains, some of which develop desert pavements, form where deserts lack a substantial sand supply. Where winds build accumulations of sand, deserts contain sand dunes.

Quick Question

What factors control the shape, dimensions, and orientation of sand dunes?

14.5 Desert Problems

We've seen that deserts contain a great variety of landscapes that differ from those of wetter regions (**Earth Science at a Glance**, pp. 422–423). Can regions that host nondesert landscapes become deserts in a human time frame? Yes. Natural *droughts* (periods of unusually low rainfall), overpopulation, overgrazing, intensive farming, and diversion of water supplies can transform a semiarid grassland into a desert in a matter of years to decades. Geologists refer to this transformation as **desertification**. In some cases, a change in rainfall pattern or land use can reverse desertification.

An example of temporary desertification affected the southern Great Plains of the United States during the Great Depression of the 1930s. Banks had failed, workers had lost their jobs, and the stock market had crashed. No one needed yet another disaster—but in 1933, even nature turned hostile, and seasonal rains did not arrive. Farming had broken up soil that had been held together for millennia by the roots of prairie grasses, and without water, the topsoil of the croplands turned to powdery dust. Strong storms blew across the plains and sent the soil skyward to form dust storms that blotted out the Sun. When the dust settled, it buried houses and roads and dirtied every nook and cranny. What had once been rich farmland turned into a wasteland that soon acquired a nickname, the Dust Bowl. It stayed that way for almost a decade. More recently, the Sahel, a belt of grassland that fringes the southern margin of the Sahara, has endured a similar fate. The region's growing population replaced soil-preserving perennial grasses with seasonal crops, and introduced herds of grazing animals that not only ate grass down to the ground but also compacted the soil as they walked. Without vegetation, the air has grown drier, and portions of the Sahel have turned into desert. When droughts have hit the changed landscape, its inhabitants have endured famine (**Fig. 14.16**).

The recent change in the Aral Sea of central Asia illustrates how desertification can follow river diversion. This inland water body was over 250 km (155 mi) wide in 1990. Now, because the water in the rivers that once flowed into it is being used for irrigation elsewhere, it has dwindled to only 10 km (6 mi) wide. The fishing boats that once plied its waters lie as rusting hulks amid the sand dunes accumulating on the former lake bed (**Fig. 14.17**).

Take-home message . . .

In lands bordering deserts, droughts, population pressures, and river diversion have led to desertification, the transformation of once-vegetated land into desert.

Quick Question

What factors transformed the southern Great Plains into the Dust Bowl during the 1930s?

See for yourself

Namib Desert, Namibia

Latitude: 24°44′27.80″ S
Longitude: 15°27′58.92″ E

From 20 km (~12.5 mi), look down and zoom in.

Huge dunes of orange sand border a dry wash in western Africa. The slip faces of dunes are recognizable. A road in the wash provides scale.

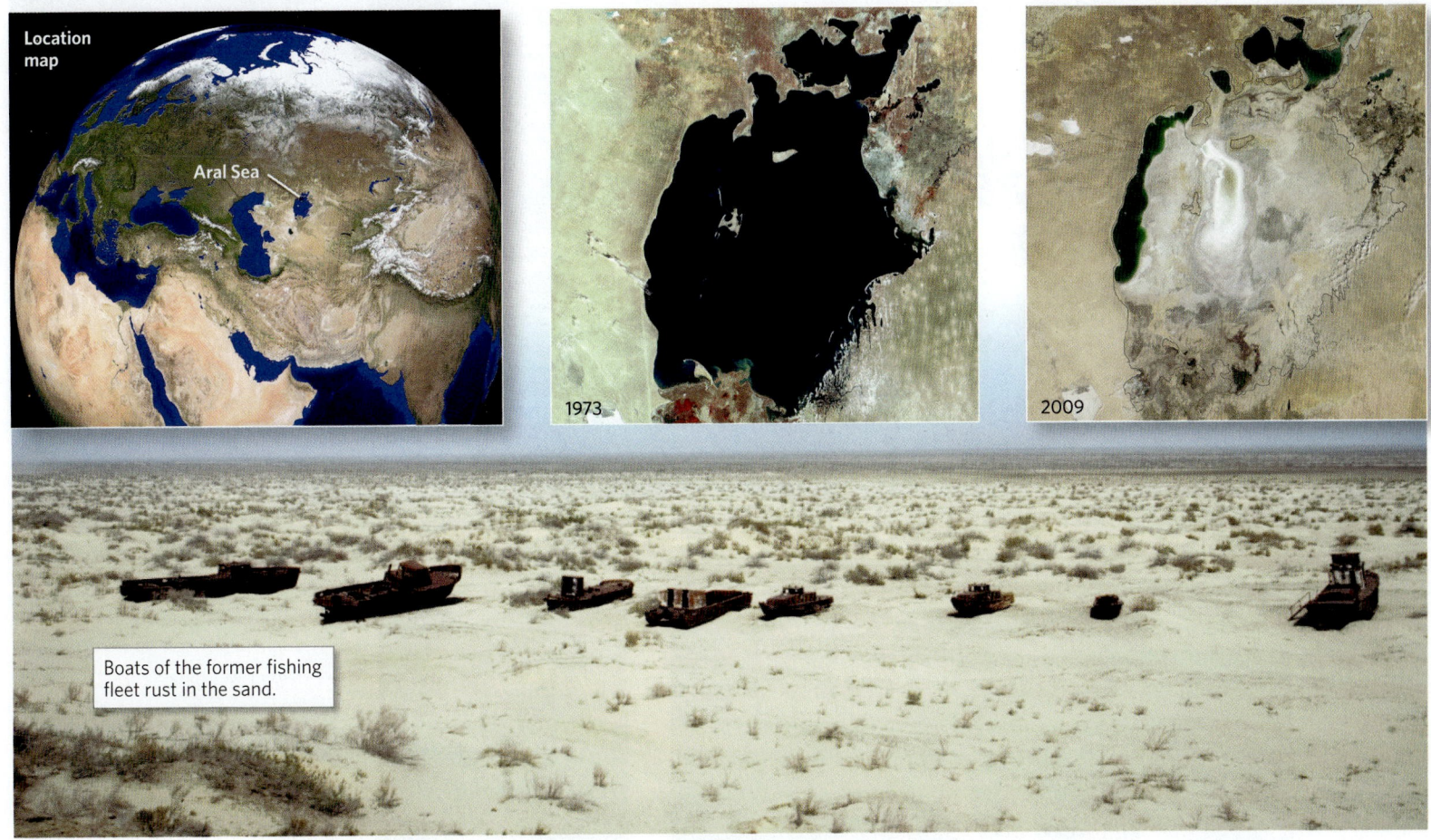

FIGURE 14.17 Diversion of rivers that flowed into the Aral Sea caused the sea to shrink.

Location map

Aral Sea

1973

2009

Boats of the former fishing fleet rust in the sand.

14.6 Ice and the Nature of Glaciers

We've just discussed some of the hottest places on our planet. Now we shift our attention to consider the coldest places. In cold realms, water exists in solid form—as *ice*—for most, if not all, of the year and can build into **glaciers**, very slowly flowing sheets or streams of ice. To begin our discussion of glaciers and their consequences, we must first understand what they consist of, how they form, and how they move.

Forming Glaciers

Ice consists of water in its solid state, formed when liquid water cools below its freezing point. To understand natural occurrences of ice in a geologic context, let's apply the concepts introduced in the chapters on minerals and rocks. We can think of a single ice crystal as a mineral specimen—a naturally occurring, inorganic solid with a definite chemical composition (H_2O) and a regular crystal structure **(Fig. 14.18a)**. A layer of snow, in this context, represents a sediment layer, and a layer of snow that has been compacted so that the grains stick together resembles a sedimentary rock layer **(Fig. 14.18b)**. We can think of the ice on the surface of a pond as an igneous rock, for it forms when liquid water ("molten ice") solidifies. What about glacial ice? It's comparable to a metamorphic rock, in that it develops when pre-existing ice recrystallizes in the solid state **(Fig. 14.18c)**.

Glaciers can develop in polar regions, even though relatively little snow falls there, because temperatures remain so cold, even in summer, that ice survives all year. Glaciers also develop in mountains, even at low latitudes, because temperature decreases with elevation, so that at high elevations, the average annual temperature stays cold enough for ice to survive all year.

How does freshly fallen snow, which consists of about 10% delicate snowflakes and about 90% air, transform into glacial ice? The process begins as snow ages and the points of the snowflakes either *sublimate* (evaporate directly into water vapor) or *melt* (transform into liquid water), so the flakes pack together more tightly. When packed snow becomes buried deeply enough, the weight of the overlying snow generates sufficient pressure to cause remaining points of contact between snowflakes to melt

FIGURE 14.18 The nature of ice and the formation of glaciers.

(a) The shapes of snowflakes reflect the hexagonal crystal structure of ice.

A boundary between layers

(b) Accumulating layers of snow are essentially layers of sediment.

The layers in the photo at left are part of this glacier in the Alps.

The wall of a tunnel bored into a glacier

(c) Glacial ice is blue. As revealed by a microscope, it has coarse grains and contains air bubbles.

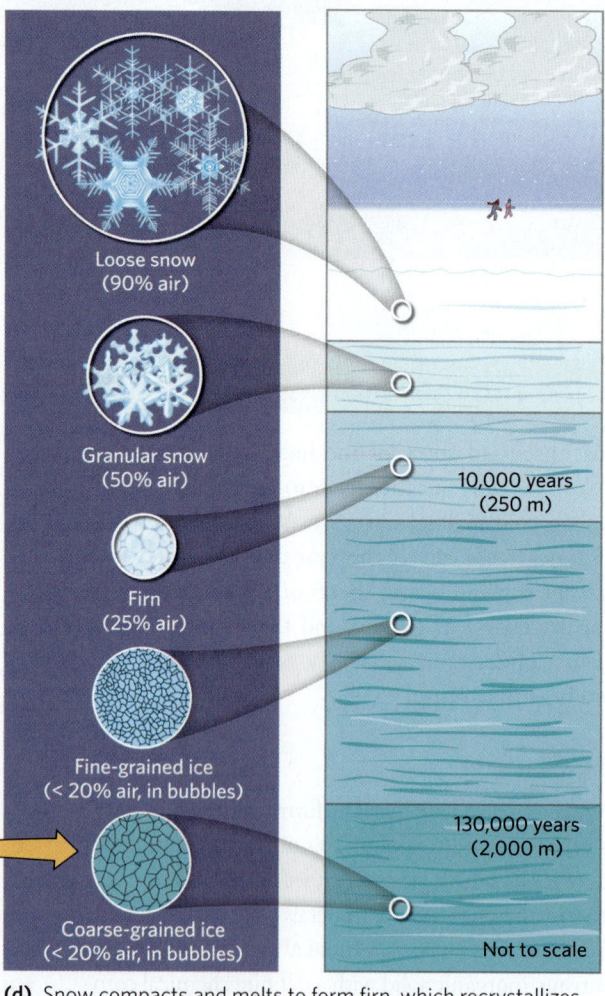

Loose snow (90% air)

Granular snow (50% air)

Firn (25% air)

Fine-grained ice (< 20% air, in bubbles)

Coarse-grained ice (< 20% air, in bubbles)

10,000 years (250 m)

130,000 years (2,000 m)

Not to scale

(d) Snow compacts and melts to form firn, which recrystallizes into glacial ice in a process comparable to metamorphism. Crystal size increases with depth.

FIGURE 14.19 A variety of glacier types form in mountainous areas.

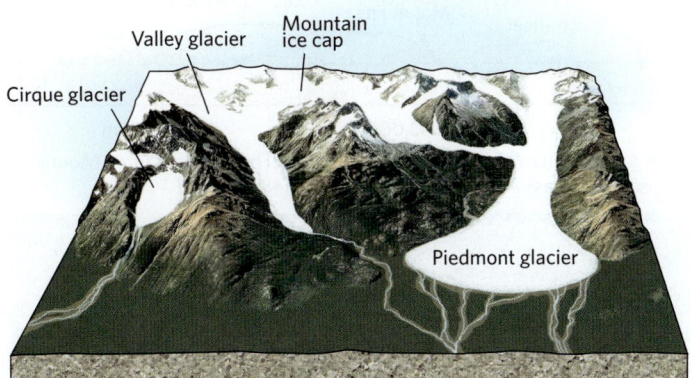

(a) Mountain glaciers are classified by their shape and position.

(b) A valley glacier and cirque glaciers in Switzerland.

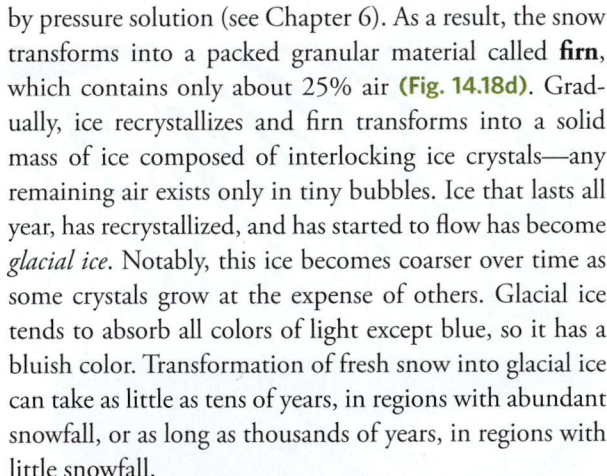

(c) Valley glaciers draining a mountain ice cap in Alaska.

(d) A piedmont glacier near the coast of Alaska.

by pressure solution (see Chapter 6). As a result, the snow transforms into a packed granular material called **firn**, which contains only about 25% air (**Fig. 14.18d**). Gradually, ice recrystallizes and firn transforms into a solid mass of ice composed of interlocking ice crystals—any remaining air exists only in tiny bubbles. Ice that lasts all year, has recrystallized, and has started to flow has become *glacial ice*. Notably, this ice becomes coarser over time as some crystals grow at the expense of others. Glacial ice tends to absorb all colors of light except blue, so it has a bluish color. Transformation of fresh snow into glacial ice can take as little as tens of years, in regions with abundant snowfall, or as long as thousands of years, in regions with little snowfall.

Categories of Glaciers

Geologists distinguish between two general categories of glaciers. **Mountain glaciers**, or *alpine glaciers*, grow in or adjacent to mountainous regions (**Fig. 14.19a**). This category includes *cirque glaciers*, which fill a bowl-shaped depression, known as a **cirque**, on the flank of a mountain (see Fig. 14.27c); *valley glaciers*, rivers of ice that flow down valleys; mountain *ice caps*, mounds of ice that submerge peaks at the crest of a mountain range; and *piedmont glaciers*, fans or lobes of ice that form where a valley glacier emerges from a valley and spreads out over the nearby plain (**Fig. 14.19b–d**). Mountain glaciers range in size from a few hundred meters to a few hundred kilometers long, and from tens of meters to 1.5 km (1 mi) thick, and flow overall from higher elevations to lower elevations. **Continental glaciers**, or *continental ice sheets*, are vast sheets of ice that form at high latitudes and spread over thousands of square kilometers of continental crust. They now exist only on Antarctica and Greenland (**Fig. 14.20**), but during ice ages, they have covered large portions of other continents. Continental glaciers flow outward from their thickest point, which may be up to 4 km (2.5 mi) thick, and thin toward their margins where they may be only tens to hundreds of meters thick.

The Movement of Glacial Ice

The manner in which ice moves within a glacier typically varies with depth and with temperature. Under the

Did you ever wonder . . .

how a glacier moves?

 **Animation**
Glacial Dynamics: Base of Glacier (Interactive)

FIGURE 14.20 The Antarctic ice sheet is one of the two continental glaciers that exist today.

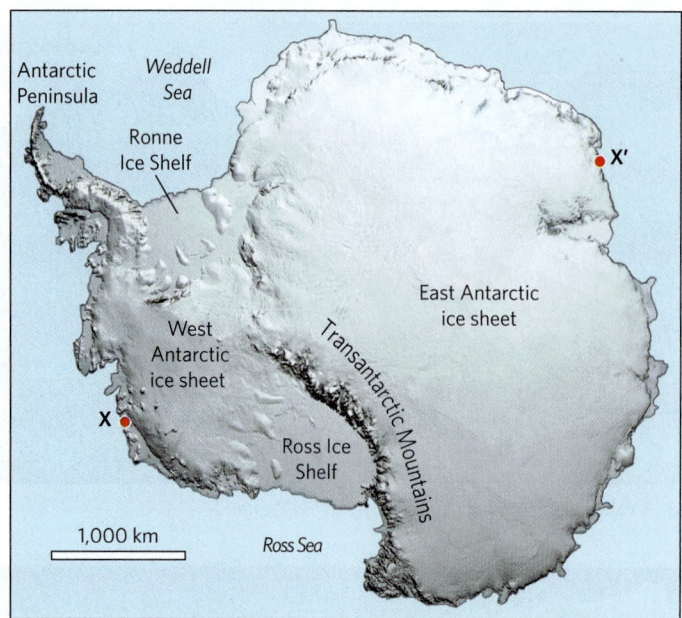

(a) A digital elevation model of the Antarctic ice sheet in map view. Valley glaciers carry ice from the East Antarctic portion of the ice sheet down to the Ross Ice Shelf.

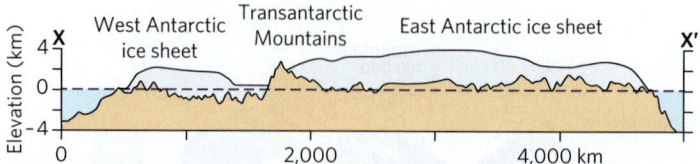

(b) A cross section of the Antarctic ice sheet. The Transantarctic Mountains separate East Antarctica from West Antarctica.

(c) A view across the Antarctic ice sheet. Wind has blown away most loose snow, and the ice surface itself has become rippled.

pressures found at depths of greater than about 60 m (200 ft) below the glacier's surface, ice can undergo *plastic deformation*, meaning that the grains within it change shape very slowly without breaking, and new grains grow while old ones disappear (see Chapter 7). In *polar glaciers*, in which ice remains completely frozen all year, plastic deformation involves only the rearrangement of water molecules within ice grains. In *temperate glaciers*, in which ice can be at its melting temperature for part of the year, water films develop along grain boundaries, so plastic deformation may also involve the slip of ice crystals past their neighbors.

FIGURE 14.21 Crevasses form in the upper layer of a glacier, in which the ice is brittle. Commonly, cracking takes place where the glacier bends while flowing over steps or ridges in its substrate.

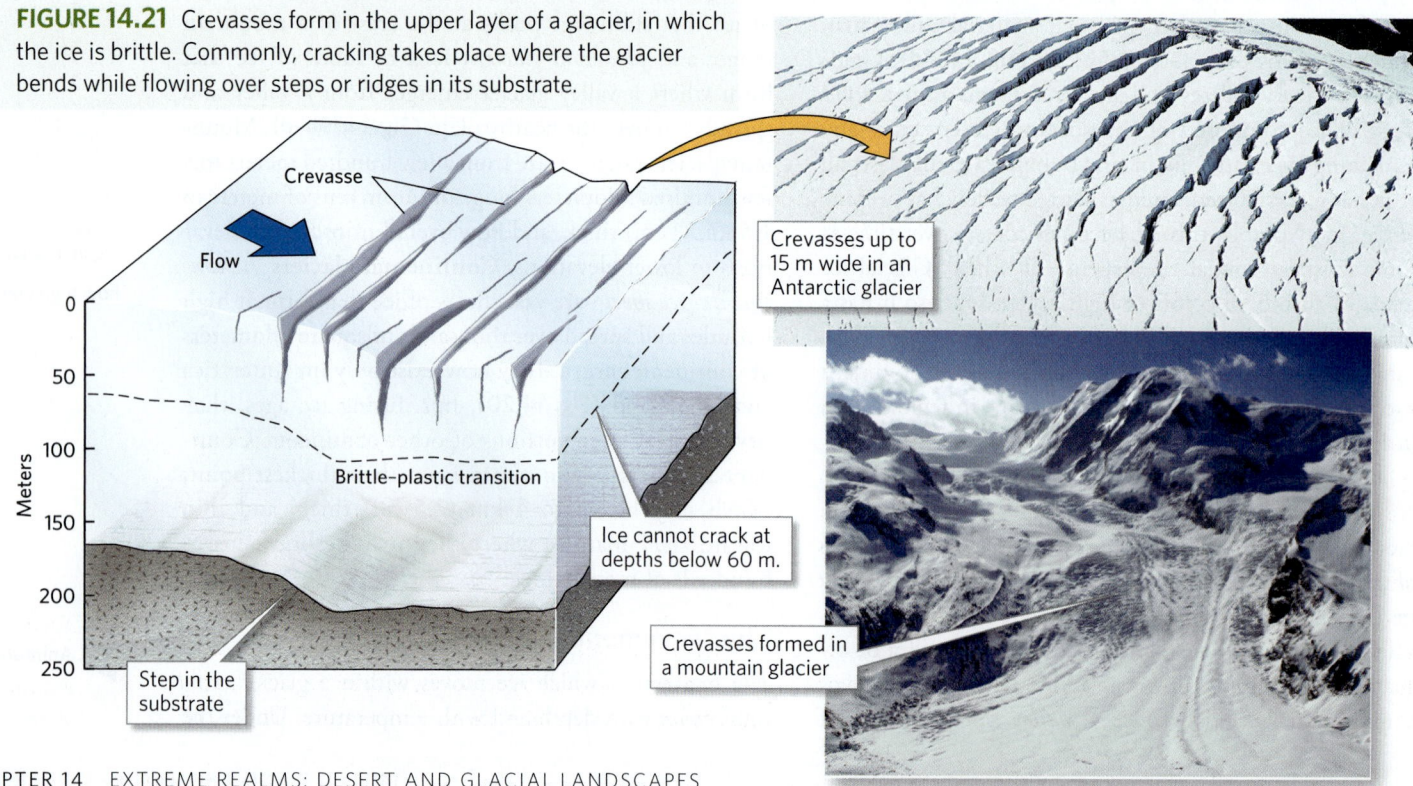

Crevasse

Flow

Brittle–plastic transition

Ice cannot crack at depths below 60 m.

Step in the substrate

Crevasses up to 15 m wide in an Antarctic glacier

Crevasses formed in a mountain glacier

FIGURE 14.22 Forces that drive the movement of glaciers.

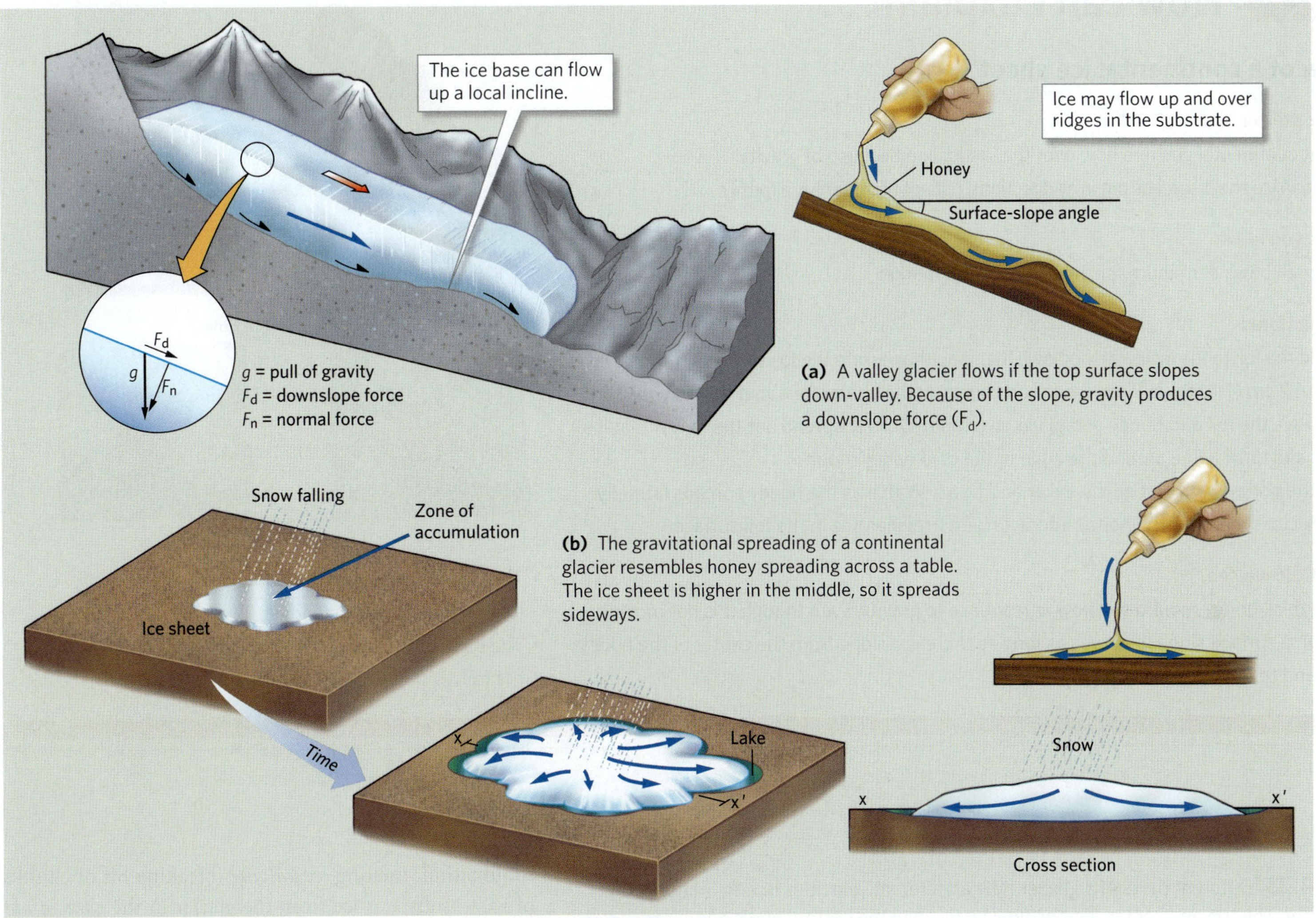

The ice base can flow up a local incline.

Ice may flow up and over ridges in the substrate.

Honey

Surface-slope angle

g = pull of gravity
F_d = downslope force
F_n = normal force

(a) A valley glacier flows if the top surface slopes down-valley. Because of the slope, gravity produces a downslope force (F_d).

Snow falling

Zone of accumulation

Ice sheet

(b) The gravitational spreading of a continental glacier resembles honey spreading across a table. The ice sheet is higher in the middle, so it spreads sideways.

Time

Lake

Snow

Cross section

At depths of less than about 60 m within a glacier, above a boundary known as the *brittle-plastic transition,* ice can't flow. This brittle ice gets carried along as the ice beneath it flows, and it adjusts to changes in the shape of the glacier, as may happen when the glacier flows over a ledge or around a curve, by cracking. A crack can open into a fissure called a **crevasse (Fig. 14.21)**. In large glaciers, crevasses can be hundreds of meters long, and can open to be up to 15 m (50 ft) across.

Why do glaciers move? Ultimately, they move because ice is relatively weak, so the pull of gravity alone can cause it to deform. A glacier flows in the direction in which the top surface of the ice slopes **(Box 14.2)**. Therefore, a mountain glacier flows from its *head* (origin) to its *toe* or *terminus* (downslope edge). As long as the glacier's surface overall slopes down-valley, the base of a glacier can move up and over obstacles **(Fig. 14.22a)**. Continental glaciers spread outward from their thickest point, behaving much like a mound of honey poured on a plate **(Fig. 14.22b)**.

Glaciers generally flow at rates between 10 and 300 m (30–1,000 ft) per year. Notably, not all parts of a glacier move at the same rate, because friction between the glacier and rock, where they come in contact, can slow the movement of the ice. Therefore, the center of a valley glacier moves faster than its edges, and its top moves faster than its base **(Fig. 14.23)**. The movement of temperate glaciers may be accelerated by a process called *basal sliding*, in which the whole mass of the glacier moves relative to its substrate by sliding on a layer of liquid water or water-saturated sediment. The water or watery sediment holds the glacier above bedrock and decreases friction. Locally, basal sliding may trigger a *glacial surge*, during which ice accelerates to speeds of 10–110 m (30–360 ft) per day!

Glacial Advance and Retreat

Glaciers resemble bank accounts in that ice can be added or removed. Snowfall adds to the glacier in the *zone of accumulation*, while **ablation**—the removal of ice by

Animation

Glacial Dynamics: Zone of Accumulation (Interactive)

How can I explain . . .

Flow of a continental ice sheet

What are we learning?

- That continental glaciers flow over the landscape because of gravity.
- That an ice sheet spreads out in all directions, but not necessarily uniformly.

What you need:

- A small bowl; aluminum foil; honey; a felt-tip marker

Instructions:

- Mold the foil to the inside of the bowl; then pour honey into the foil so that it makes a layer about 2 cm (1 in) thick. Place the foil and honey in the freezer for about half an hour.
- Remove the foil and honey and place on a table. Quickly spread out the foil so that it's flat, and draw a line around the edge of the cold honey mound.
- As the honey warms, it gets weaker and flows. Watch as the honey spreads laterally. Look closely at how the honey interacts with irregularities in the foil surface.

What did we see?

- The thick honey mound will flow in response to gravity, even though it is not moving down a surface slope. It moves as long as there is a slope from the center of the honey mound to the edge.

Time 1   Time 2

- The honey travels farther in low areas, so distinct lobes develop along its edges.
- The base of the honey can ride up over bumps in the foil.

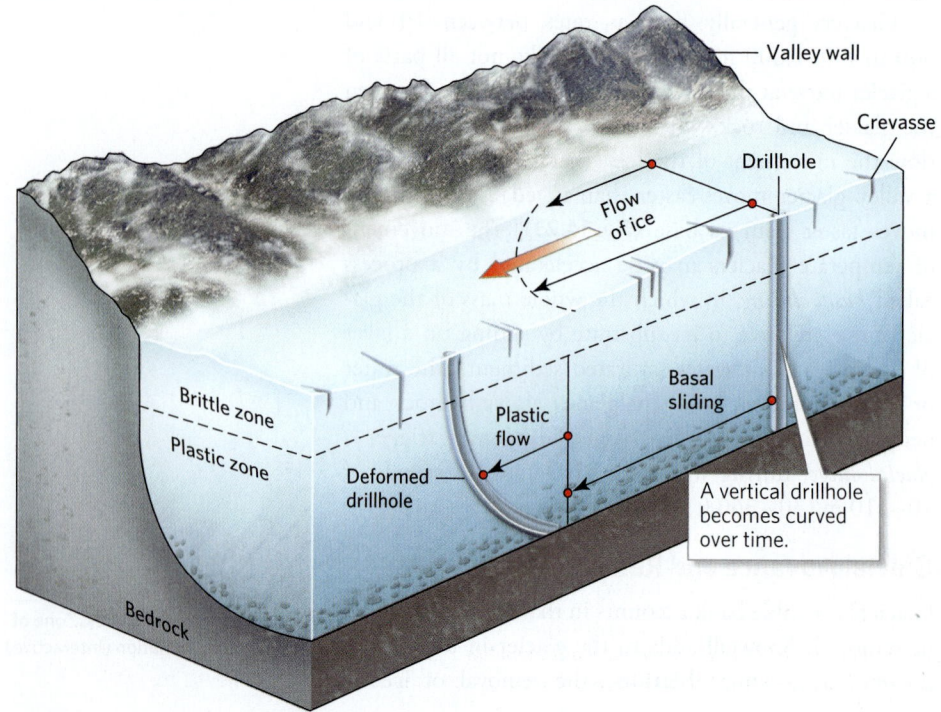

FIGURE 14.23 Different parts of a glacier flow at different velocities due to friction with the substrate. The top and center regions flow faster than the bottom and sides.

Valley wall

Crevasse

Drillhole

Flow of ice

Brittle zone

Plastic zone

Plastic flow

Basal sliding

Deformed drillhole

A vertical drillhole becomes curved over time.

Bedrock

sublimation, melting, or *calving* (breaking off of chunks of ice)—subtracts ice from the glacier in the *zone of ablation*. The **equilibrium line** defines the boundary between these zones of the glacier **(Fig. 14.24)**. During flow, ice follows a curved path, sinking down toward the base of the glacier in the zone of accumulation as new ice accumulates above it, and rising up toward the surface of the glacier in the zone of ablation as overlying ice melts or sublimates.

The balance between accumulation and ablation affects the position of a glacier's toe. Specifically, if the rate of accumulation equals the rate of ablation, then the position of the toe remains fixed, even though ice within the glacier continues to flow toward the toe **(Fig. 14.25a)**. If, however, the rate of accumulation exceeds the rate of ablation, then the toe moves forward into previously unglaciated regions, a change known as a **glacial advance (Fig. 14.25b)**. In mountain glaciers, the position of the toe moves downslope during an advance, and in continental glaciers, the toe moves outward, away from the glacier's origin, so the glacier may move toward lower latitudes. If the rate of ablation exceeds the rate of accumulation, then the position of the toe moves back toward the glacier's origin, a change known as a **glacial retreat (Fig. 14.25c)**. During a mountain glacier's retreat, the position of the toe moves upslope, and during a continental glacier's retreat, lower latitudes undergo *deglaciation*. Note that when a glacier retreats, it's only the position of the toe that moves back toward the origin; ice itself continues to flow toward the toe, for it cannot move against gravity.

Ice in the Sea

On the moonless night of April 14, 1912, the ocean liner RMS *Titanic* struck an **iceberg**, a large block of floating ice, in the frigid North Atlantic. Lookouts had seen the ghostly mass only minutes earlier and had alerted the ship's pilot, but the ship had been unable to turn fast enough to avoid disaster. The force of the blow split its hull, allowing water to gush in, and less than 3 hours later, the *Titanic* disappeared beneath the surface.

Where do icebergs, such as the one responsible for the *Titanic*'s fate, originate? At high latitudes, mountain glaciers and continental glaciers flow down to and, in some cases, out into coastal waters to become **tidewater glaciers**. A tidewater glacier that forms where a mountain glacier enters the sea may protrude a few kilometers out

FIGURE 14.24 The equilibrium line separates the zone of accumulation from the zone of ablation. As indicated by arrows, ice sinks while flowing downhill in the zone of accumulation and rises while flowing downhill in the zone of ablation. Overall, it follows a curving path.

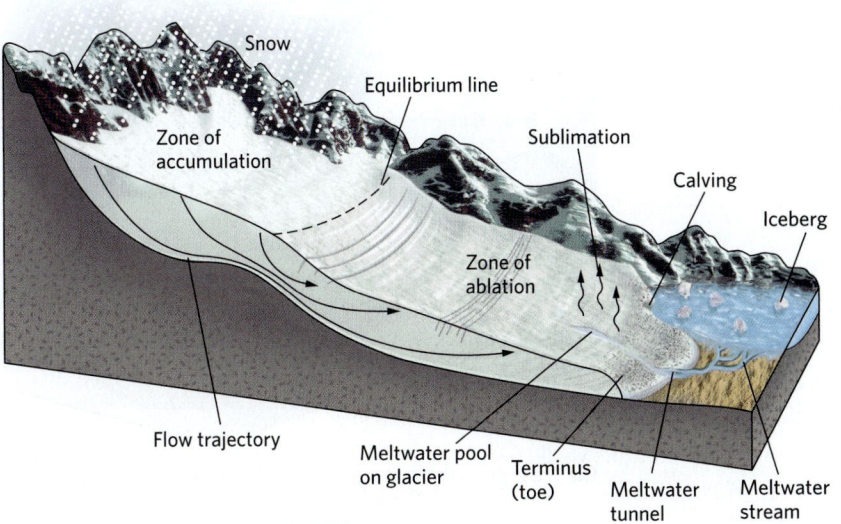

 Animation
Glacial Dynamics: Glacial Terminus (Interactive)

as an *ice tongue*, whereas a continental glacier entering the sea becomes a broad, flat **ice shelf**. In shallow water, glacial ice remains grounded, but where the water becomes deep enough, the ice floats **(Fig. 14.26a)**. At the toe of a tidewater glacier, blocks of ice calve off and tumble into the water with an impressive splash, producing blocks of various sizes. Those that rise at least 6 m (20 ft) above the water and are at least 15 m (50 ft) long can formally be called icebergs. Since four-fifths of a floating body of ice lies below the surface of the sea, the base of a very large iceberg may actually extend a few hundred meters below sea level **(Fig. 14.26b)**.

Not all ice floating in the sea originates as glaciers on land. In polar climates, the surface of the ocean itself freezes, forming **sea ice (Fig. 14.26c, d)**. The north polar ice cap of the Earth consists of a 2–6-m

FIGURE 14.25 Glacial advance and retreat.

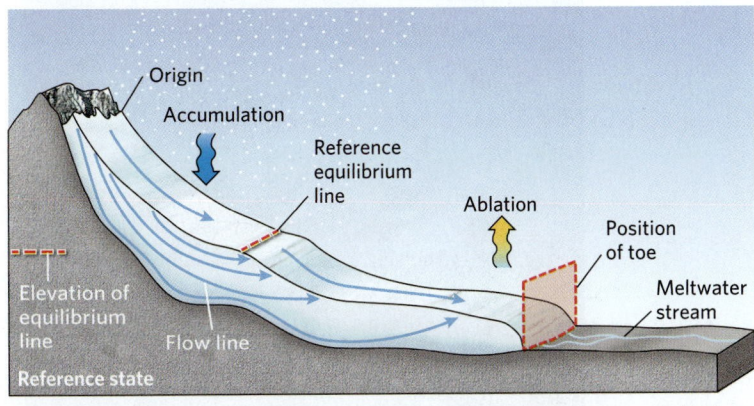

(a) The position of the toe represents a balance between accumulation and ablation.

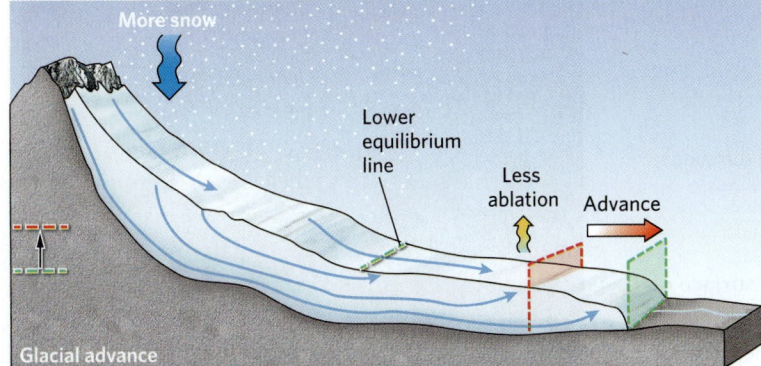

(b) If accumulation exceeds ablation, the glacier advances, the toe moves farther from the origin, and the ice thickens.

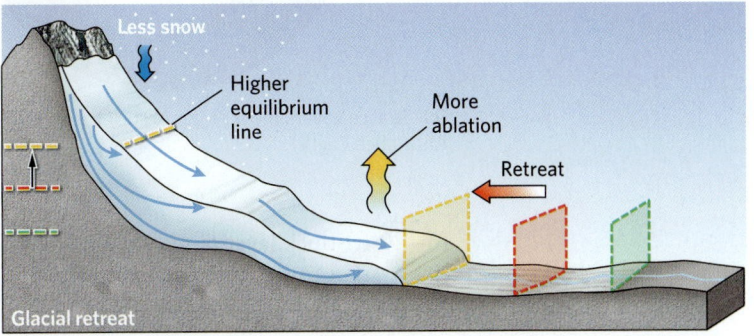

(c) If ablation exceeds accumulation, the glacier retreats and thins. The position of the toe moves back, even though ice continues to flow toward the toe.

FIGURE 14.26 Ice in the sea.

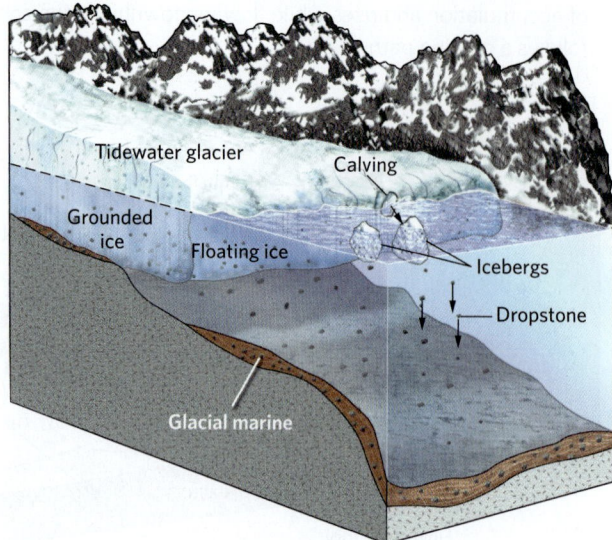

(a) Ice is grounded in shallow water, but floats in deep water.

(c) Sea ice covers most of the Arctic Ocean (left) and surrounds Antarctica (right).

(6–20-ft)-thick layer of sea ice formed on the surface of the Arctic Ocean. This sea ice slowly moves with ocean currents at a rate of about 2 km (1.2 mi) per day. A wide belt of sea ice encircles Antarctica during much of the year.

(b) This artist's rendition of an iceberg emphasizes that most of the ice is underwater.

(d) In summer, some of the sea ice of Antarctica breaks up to form icebergs.

Take-home message . . .

🏠 Glaciers form when snow persists all year, gets buried deeply, and turns to ice. Mountain glaciers form at high elevations, whereas continental glaciers form at high latitudes and spread over continents. Ice flows by plastic deformation at depth. The upper, brittle portion of a glacier fractures to form crevasses. The balance between accumulation and ablation controls glacial advance or retreat. In polar regions, sea ice covers large areas.

Quick Question ———————————
Does ice actually flow uphill during a retreat of a valley glacier?

14.7 Carving and Carrying: Erosion by Ice

Glaciers serve as very powerful agents of erosion, capable of carving deep valleys and knife-edged ridges in mountains, and of beveling and polishing broad areas of continents. Here we look at how glacial erosion takes place and at specific landforms it produces.

Glacial Erosion and Its Products

Glacial ice plucks up fragments of its substrate as it moves. Pure ice is too soft to erode rock, but hard clasts embedded in moving ice act like the teeth of a giant rasp and grind away the substrate. Abrasion by incorporated sand-sized grains can polish bedrock at the base or sides of the glacier **(Fig. 14.27a)**. Larger clasts, or trains of clasts, embedded in ice gouge out grooves or scratches, called **glacial striations**, that range from 1 mm–1 m (0.04–40 in) across and may be tens of centimeters to tens of meters long **(Fig. 14.27b)**. As you might expect, glacial striations trend parallel to the flow direction of the ice.

Mountain glaciers sculpt a variety of erosional landforms. Frost wedging due to freezing and thawing breaks up bedrock bordering the head of the glacier high in the mountains. This fractured rock falls on the ice, or gets picked up at the base of the ice, and moves downslope with the glacier. As a consequence, a bowl-shaped depression, or cirque, develops on the side of the

FIGURE 14.27 Glaciers are powerful agents of erosion.

(a) A glacially polished outcrop in Central Park, New York City.

(b) Glacial striations in Victoria, British Columbia.

(c) Examples of a cirque and an arête in the Alps.

(d) The Matterhorn in Switzerland is bordered by cirques.

FIGURE 14.28 Stages during the development of a glacially carved mountainous landscape.

Time

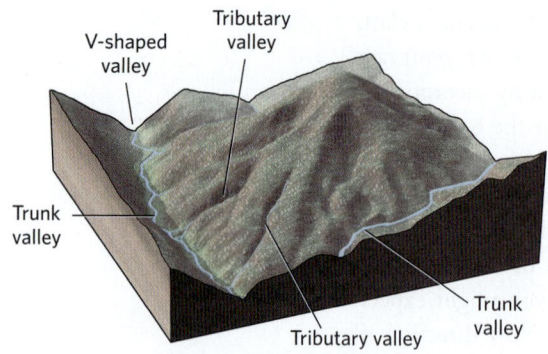

(a) Before glaciation, valleys are V-shaped, and tributary mouths are at the same elevation as the trunk stream.

(b) During glaciation, ice erodes both the floor and sides of valleys.

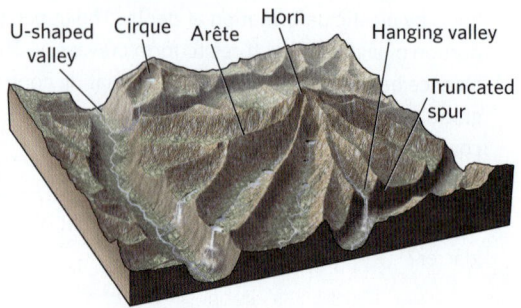

(c) After glaciation, valleys are U-shaped, and hanging valleys intersect the trunk valley. Some peaks become horns.

FIGURE 14.29 Examples of glacially eroded valleys.

(a) A U-shaped valley bordered by Half Dome, in Yosemite National Park, California.

See for yourself

Glaciated Peaks, Montana

Latitude: 48°56′33.66″ N
Longitude: 113°49′54.59″ W

Zoom to 8 km (~5 mi) and look down.

Three cirques bound a horn in the Rocky Mountains, north of Glacier National Park. Note the knife-edge arêtes between the cirques. The glaciers that carved the cirques have melted away. The thin stripes on the mountain face are traces of sedimentary beds that make up the bedrock.

(b) A U-shaped glacial valley in the Tongass National Forest, Alaska.

(c) A waterfall spilling out of a U-shaped hanging valley in the Sierra Nevada.

mountain (Fig. 14.27c). An **arête** (French for ridge) separates adjacent cirques, and when cirques form on more than one side of a mountain, the mountain becomes a pointed peak called a **horn (Fig. 14.27d)**.

Glacial erosion dramatically modifies the shape of a valley **(Fig. 14.28)**. To see how, compare a river-eroded valley with a glacially eroded valley. If you look along the length of a river in unglaciated mountains, you'll see that it typically flows down a *V-shaped valley*, with the river channel forming the point of the V. The V shape develops because river erosion occurs only in the channel, and mass wasting causes the valley slopes to approach the angle of repose (see Chapter 13). But if you look down the length of a glacially eroded valley, you'll see that it resembles a U, with steep walls. Such a **U-shaped valley (Fig. 14.29a, b)** forms because glacial erosion not only lowers the floor of the valley but also bevels its sides.

Glacial erosion in mountains also modifies the intersections between tributary valleys and the trunk valley. Recall that the trunk stream of a drainage network transporting water serves as the base level for its tributaries, so a tributary stream and a trunk stream intersect at the same elevation. Tributary glaciers, however, do not erode their valleys to the same depth as the valley carved by a trunk glacier at the same elevation. As a consequence, when tributary glaciers melt away, they leave behind

FIGURE 14.30 The sculpting of a roche moutonnée.

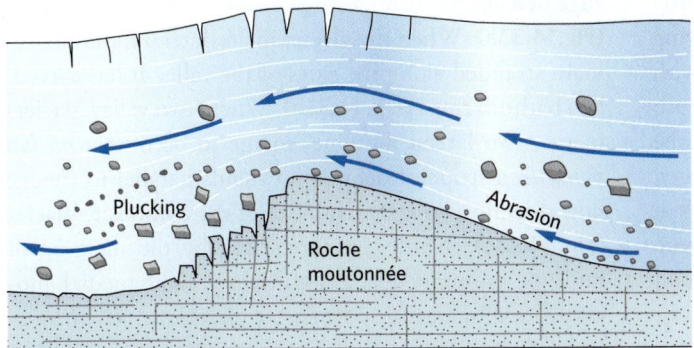

(a) Glacial abrasion rasps the upstream side, and glacial plucking carries away fracture-bounded blocks on the downstream side.

(b) An example of a roche moutonnée in the Sierra Nevada. The glacier that shaped this hill flowed from right to left.

hanging valleys, whose mouths typically perch well above the floor of the trunk valley. A present-day tributary stream cascades over a waterfall where it emerges from a hanging valley **(Fig. 14.29c)**.

The erosional features produced by continental glaciers depend, to a large extent, on the nature of the pre-glacial landscape. Where an ice sheet spreads over a region of low relief, glacial erosion produces a vast area of polished, striated surfaces. Where an ice sheet spreads over a hilly area, it deepens valleys and smooths hills. Glacially eroded hills may be elongate in the direction of ice flow and asymmetrical because glacial abrasion smooths the upstream part of the hill, yielding a gentle slope, whereas glacial plucking eats away at the downstream part, producing a steep slope. Geologists refer to such a hill as a **roche moutonnée**, French for sheep rock, because its profile resembles that of a sheep resting in a meadow **(Fig. 14.30)**.

Fjords: Submerged Glacial Valleys

Glacially carved valleys that later fill with water produce deep, elongate bodies of water known as **fjords**. Some fjords, such as the Finger Lakes of New York State, are inland freshwater lakes **(Fig. 14.31a)**. Where a glacially

FIGURE 14.31 Glacial fjords.

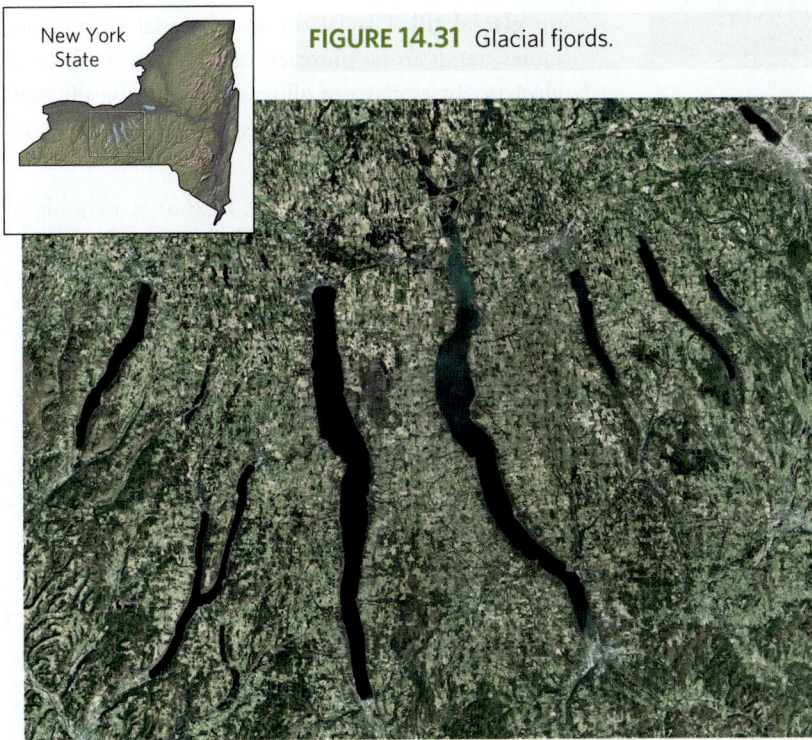

New York State

(a) The Finger Lakes in New York State are freshwater glacial fjords.

(b) One of the many spectacular fjords of Norway. The water is an arm of the sea that fills a U-shaped valley. Tourists are standing on Pulpit Rock (Preikestolen).

Baffin Island, Canada

Latitude: 67°8′27.56″ N
Longitude: 64°49′49.31″ W

Zoom to 40 km (~25 mi)
and look down.

Two valley glaciers draining
the Baffin Island ice cap
merge into a trunk glacier
that flows NE and then
into a fjord, partly filling a
U-shaped valley. Note the
lateral and medial moraines.
In some nearby fjords,
meltwater has deposited a
delta of outwash at the end
of the glacier.

carved valley meets the sea, however, it becomes a marine fjord, an elongate bay filled with seawater, when the ice melts. The floor of a marine fjord may lie significantly below sea level, because at the time the glacier carving the fjord existed, its base could remain in contact with the substrate until it reached a point where the sea's depth exceeded four-fifths of the glacier's thickness (as is the case for icebergs). And, as we'll see later in this chapter, sea level was significantly lower during the ice age when the valleys were cut. Spectacular marine fjords with steep walls occur along the coasts of Norway, New Zealand, Chile, Maine, and Alaska **(Fig. 14.31b)**.

Take-home message . . .

A glacier plucks and scrapes up rock from its substrate and carries it, along with debris that falls on its surface, in the direction of ice flow. Abrasion by incorporated clasts erodes the glacier's substrate, producing polished rock, striations, and distinctive landforms such as U-shaped valleys and cirques. U-shaped valleys that fill with water become fjords.

Quick Question

Why do we find hanging valleys in mountains that have been eroded by glaciers?

14.8 Deposition Associated with Glaciation

The Glacial Conveyor

Glaciers can carry sediment of any size and, like a conveyor belt, transport it in the direction of ice flow **(Fig. 14.32)**. Some of the sediment moving in this *glacial conveyer* comes from the substrate under the ice. The rest tumbles onto the glacier during rockfalls or slides down slopes bordering the glacier.

Geologists refer to a pile of sediment carried or left by a glacier as a **moraine**. Sediment tumbling onto the edge of a glacier forms a stripe known as a *lateral moraine* **(Fig.14.33a)**. When the glacier melts, lateral moraines remain stranded along the sides of the valley it has carved, like bathtub rings **(Fig. 14.33b)**. Where two valley glaciers merge, two lateral moraines merge to become a *medial moraine*, a stripe of debris that trends parallel to the ice flow direction down the interior of the composite glacier **(Fig. 14.33c)**. Trunk glaciers created by the merging of multiple tributary glaciers contain several medial moraines. Sediment transported to a glacier's toe by the glacial conveyor accumulates in a pile at the toe and builds up to form an *end moraine* **(Fig. 14.34a, b)**. Geologists refer to the end moraine at the farthest limit of ice flow as the glacier's **terminal moraine**. A large terminal moraine, formed during the last ice age, built the ridge of sediment that now forms Long Island and Cape Cod in the eastern United States **(Fig. 14.34c)**. As a glacier recedes, it may pause temporarily and build up a *recessional moraine*. Sediment left when a glacier recedes without pausing, so that a distinct ridge doesn't form, makes up *ground moraine*.

Types of Glacial Sedimentary Deposits

Geologists distinguish among several different types of sediment that accumulate in glacial environments:

- *Till:* Sediment transported by ice and deposited beneath, at the side of, or at the toe of a glacier makes up **glacial till**. Glacial till, the material within moraines, tends to be unsorted because the solid ice of glaciers carries clasts of all sizes. So till typically consists of cobbles or boulders suspended in clay and silt **(Fig. 14.35a)**.

- *Erratics:* A glacial **erratic (Fig. 14.35b)** is a cobble or boulder that has been transported a distance from its source and dropped by a glacier. The word comes from the Latin *errare*, which means to wander.

FIGURE 14.32 The concept of the glacial conveyor.

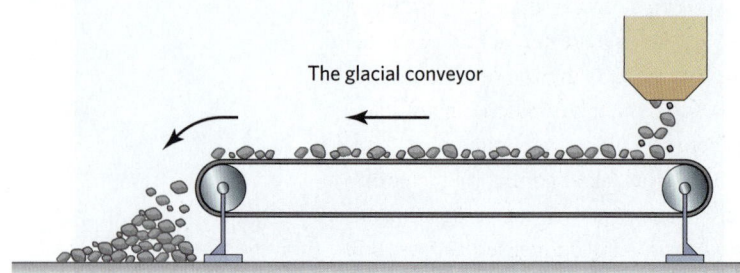

The glacial conveyor

(a) In effect, a glacier acts like a conveyor belt, moving sediment in the direction of flow.

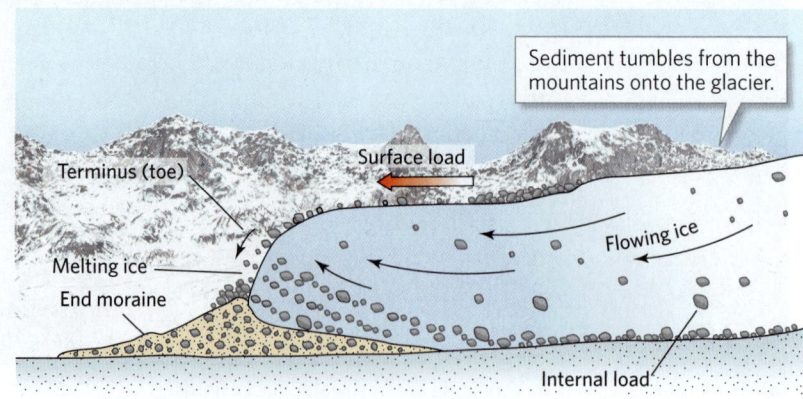

Sediment tumbles from the mountains onto the glacier.

Surface load

Terminus (toe)

Flowing ice

Melting ice

End moraine

Internal load

(b) Sediment in and on a glacier gets carried to the toe.

FIGURE 14.33 Medial moraines.

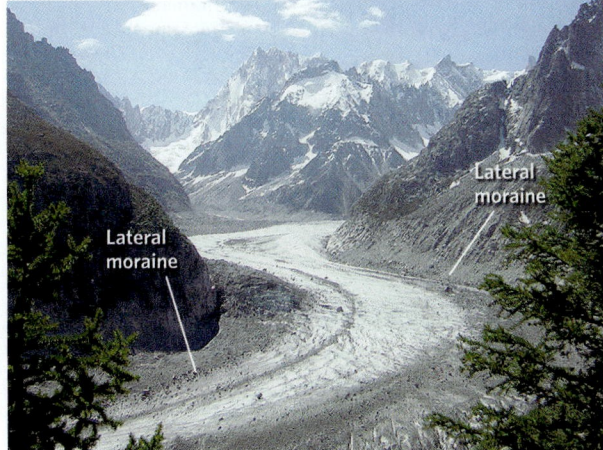

(a) This glacier in the French Alps has a medial moraine and two lateral moraines.

(b) The large ridge of sediment is a lateral moraine left when the glacier that once filled the valley in the foreground melted away.

(c) These medial moraines formed where lateral moraines of two valley glaciers merged in Alaska.

- *Glacial marine sediment:* Where a sediment-laden glacier flows into the sea, icebergs that calve off the toe float clasts out to sea. As the icebergs melt, the clasts they carry sink to the seafloor. These *dropstones* mix with marine sediment to form *glacial marine* (see Fig. 14.26a).

- *Glacial outwash:* Till deposited by a glacier at its toe may be picked up and transported by meltwater streams. Such water-transported sediment, known as **glacial outwash**, tends to be sorted by water to form bars of gravel and sand **(Fig. 14.35c)**.

- *Loess:* Strong winds in glacial environments pick up fine clay and silt produced by glacial erosion and transport it away from the glacier. This wind-transported sediment eventually settles out of the air to form deposits of **loess (Fig. 14.35d)**.

- *Glacial lake-bed sediment:* Some of the sediment yielded by glaciers accumulates on the floors of **glacial lakes**,

bodies of liquid freshwater that form due to the growth or recession of glaciers. These lakes contain meltwater from the glacier and from winter snows, as well as water that falls as rain during the summer. Glacial lakes may form along the toe of a glacier, in depressions between end moraines, in the bowl of a cirque, and in mountain valleys whose outlets have been blocked by the toe of a glacier. (A lake formed at the toe of a glacier can also be called an *ice-margin lake.*) When a glacial lake dries up, it leaves behind a plain underlain by thin beds of sediment. Notably, such lake-bed sediments tend to consist of alternate layers of silt (brought into the lake during spring floods) and clay (which settles from still water when the lake freezes in the winter). A pair of layers, representing deposition during an entire year, is called a *varve* **(Fig. 14.35e)**.

FIGURE 14.34 Formation of end moraines.

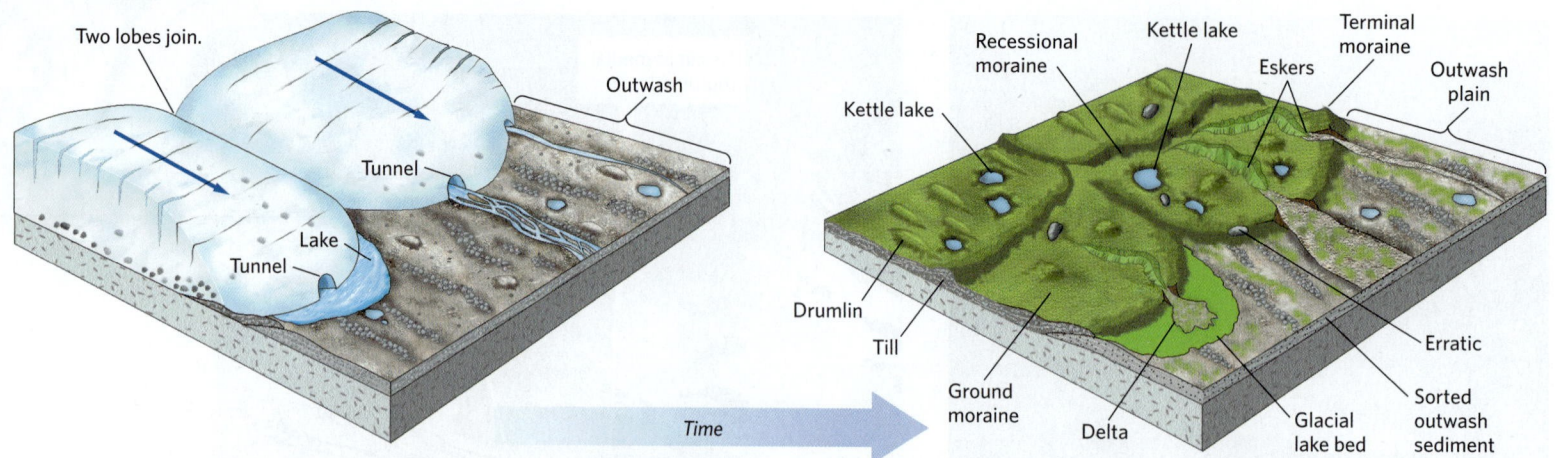

(a) When a glacier begins to recede, it drops sediment and releases meltwater.

(b) A number of distinct depositional landforms, including several types of moraines, form as a consequence of glaciation.

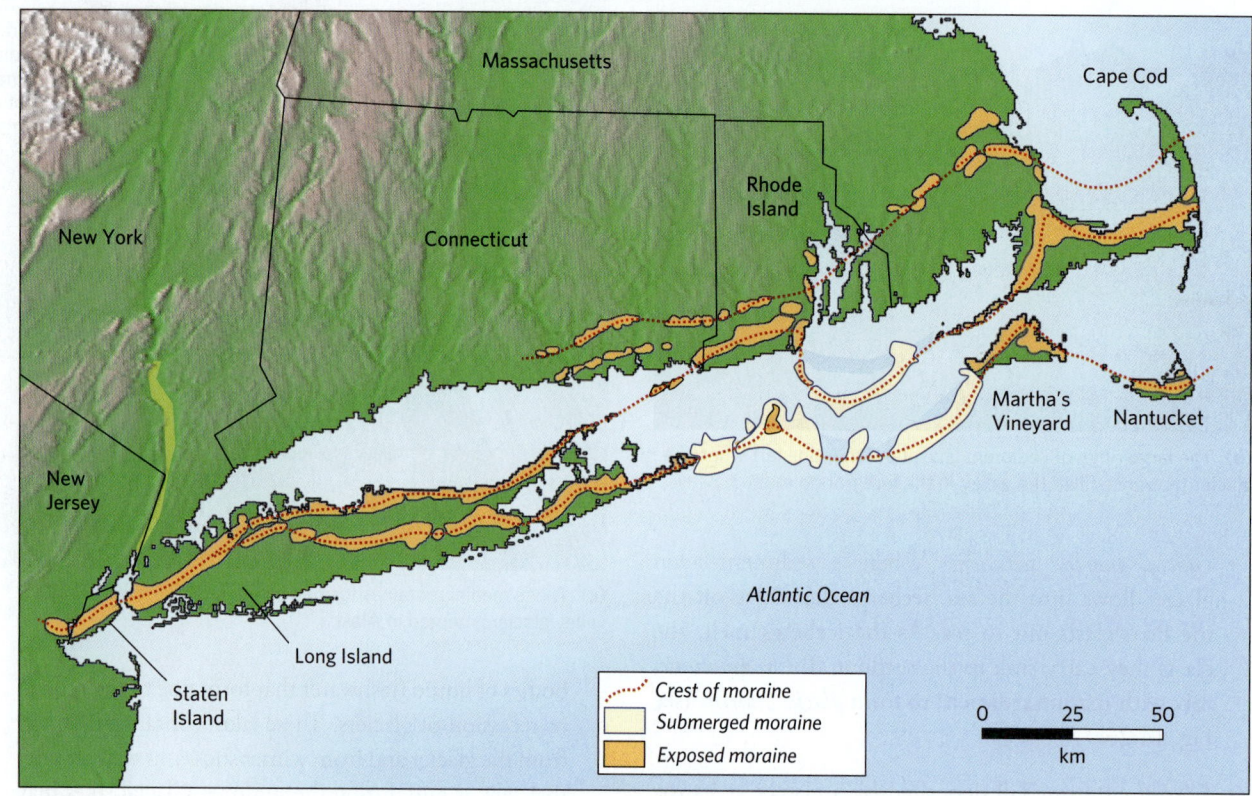

(c) The terminal and recessional moraines of a continental glacier underlie Cape Cod and Long Island in the northeastern United States.

Depositional Landforms of Glacial Environments

Imagine what you would see if you stood on the toe of a retreating glacier. Looking in one direction, you would see nothing but ice. Looking in the other direction, you would see a variety of landscape features produced by glacial erosion and deposition (**Earth Science at a Glance**, pp. 446–447).

For example, from your perch, you might see a few curving end moraines. Moraines are *hummocky*,

meaning that they are bumpy topographic surfaces with countless small hills and depressions. This kind of landscape develops both because of variations in the amount of sediment supplied by the ice and because of the development of **kettle holes**, depressions formed when sediment-covered ice blocks, left behind as the glacier retreats, melt away (**Fig. 14.36**). Beyond an end moraine, the land surface may host braided streams of sediment-choked meltwater. These streams

FIGURE 14.35 Sediment types associated with glaciation.

(a) This glacial till in Ireland is unsorted because ice can carry sediment of all sizes.

(b) Glacial erratics resting on a glacially polished surface in Wyoming.

(c) Braided streams choked with glacial outwash in Alaska. The streams carry away finer sediment and leave the gravel behind.

(d) Thick loess deposits underlie parts of the prairie in Illinois.

(e) (Left) In the quiet water of an Alaskan glacial lake, fine-grained sediments accumulate. (Right) Alternating layers (varves) in these lake-bed sediments, now exposed in an outcrop near Puget Sound, Washington, reflect seasonal changes.

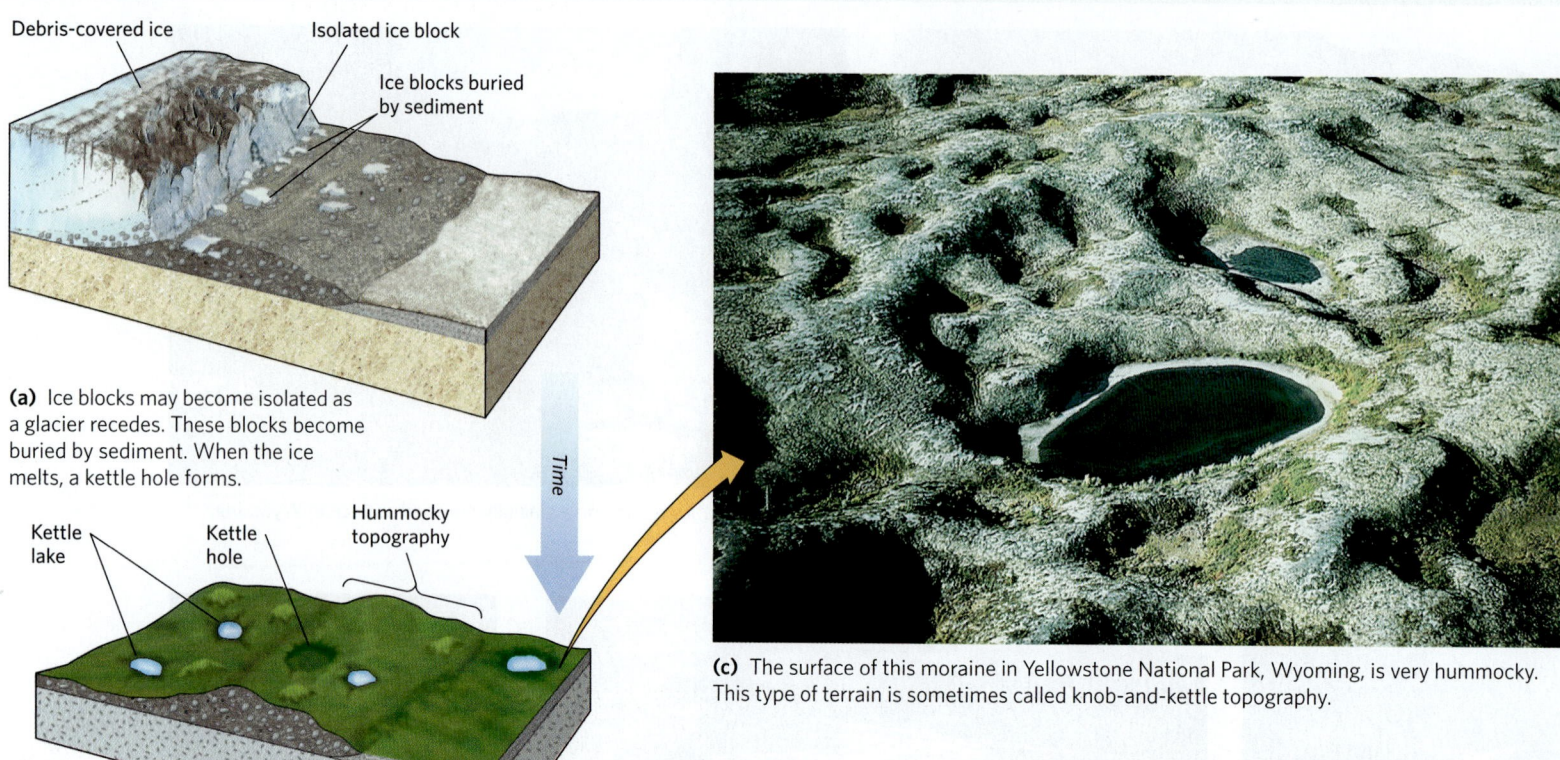

Debris-covered ice

Isolated ice block

Ice blocks buried by sediment

(a) Ice blocks may become isolated as a glacier recedes. These blocks become buried by sediment. When the ice melts, a kettle hole forms.

Time

Kettle lake

Kettle hole

Hummocky topography

(b) If the water table is high, kettle holes fill with water and turn into roughly circular lakes.

(c) The surface of this moraine in Yellowstone National Park, Wyoming, is very hummocky. This type of terrain is sometimes called knob-and-kettle topography.

deposit bars of gravel and sand, which build a broad *glacial outwash plain*. You may also see numerous glacial lakes.

Retreating glaciers also leave behind landforms that develop at the base of the glacier. These landforms include **drumlins (Fig. 14.37a, b)**, asymmetrical elongate hills of sediment molded by ice flow, and **eskers**, sinuous ridges composed of sediment deposited by meltwater in tunnels at the base of a glacier **(Fig. 14.37c, d)**.

Take-home message . . .

Flowing ice does not sort sediment, so when a glacier melts, it deposits unsorted till. Meltwater streams and wind sort and transport glacial sediment to form glacial outwash plains and loess deposits, respectively. Deposition by glaciers produces distinctive landforms such as moraines, kettle holes, drumlins, and eskers.

Quick Question

What does the sediment deposited in glacial lakes look like?

14.9 Additional Consequences of Continental Glaciation

Glacial Subsidence and Glacial Rebound

When a large ice sheet (more than 50 km, or 30 mi, in diameter) grows on a continent, its weight causes the surface of the lithosphere to sink. In other words, an ice load causes **glacial subsidence**. The lithosphere, the Earth's relatively rigid outer shell, can sink because the underlying asthenosphere is soft enough to flow slowly out of the way **(Fig. 14.38a)**. Very large ice-margin lakes develop at the toes of continental glaciers, due to glacial subsidence. The largest ice-margin lake, known as Glacial Lake Agassiz, covered portions of south-central Canada and the north-central United States between 11,700 and 9,000 years ago **(Fig. 14.39)**. At times, this lake submerged an area greater than that of all the present Great Lakes combined. When the ice sheet melts away, the surface of the underlying continent rises back up, a process called **glacial rebound**, and the asthenosphere flows back underneath to fill the space **(Fig. 14.38b)**. Glacial rebound takes many thousands of years because asthenosphere flows so slowly. Because of glacial rebound, beaches formed at sea level thousands of years ago now lie well above sea level.

Effects of Glaciers on Drainage

A continental glacier disrupts drainage networks, for the glacier may completely cover a drainage network, causing

FIGURE 14.37 Development of drumlins and eskers.

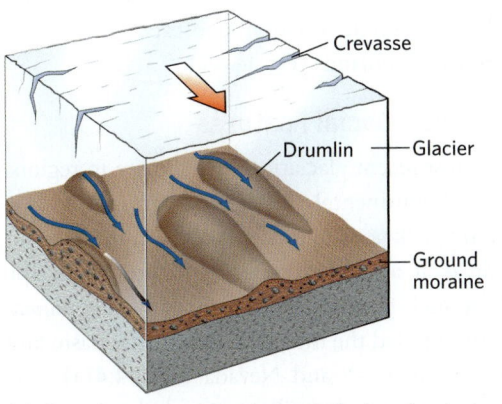

(a) Drumlins are sculpted beneath the ice of a glacier and are aligned with its flow direction.

View looking southeast

(b) Drumlins form a set of hills south of Lake Ontario in central New York State.

Shaded relief map of the drumlins in central New York. Their SSE angle gives the direction of glacial flow.

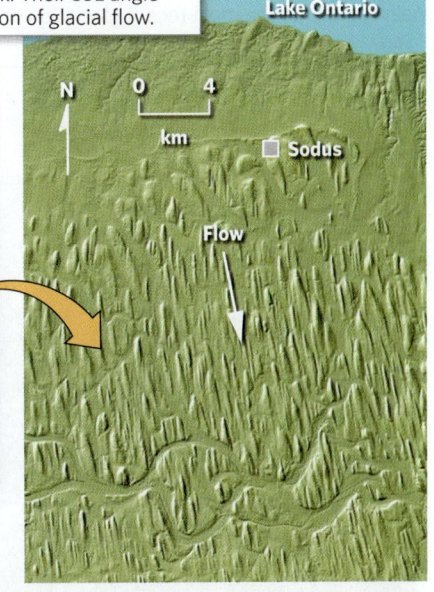

Lake Ontario

N   0   4
        km
        Sodus

Flow

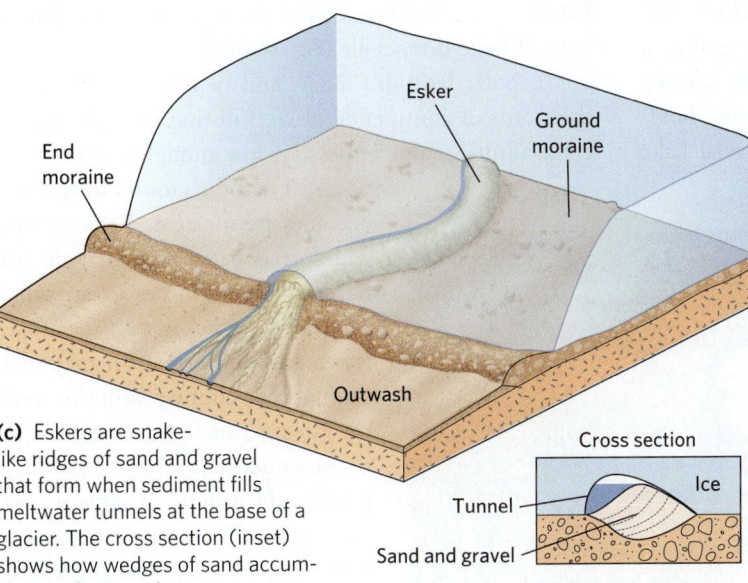

(c) Eskers are snake-like ridges of sand and gravel that form when sediment fills meltwater tunnels at the base of a glacier. The cross section (inset) shows how wedges of sand accumulate in the tunnel.

Cross section

Ice

Tunnel

Sand and gravel

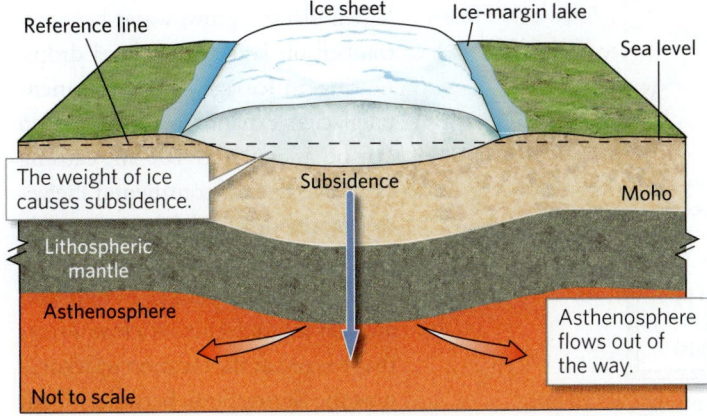

Esker

(d) An example of an esker in an area that was once glaciated.

FIGURE 14.38 The concept of glacial subsidence and rebound.

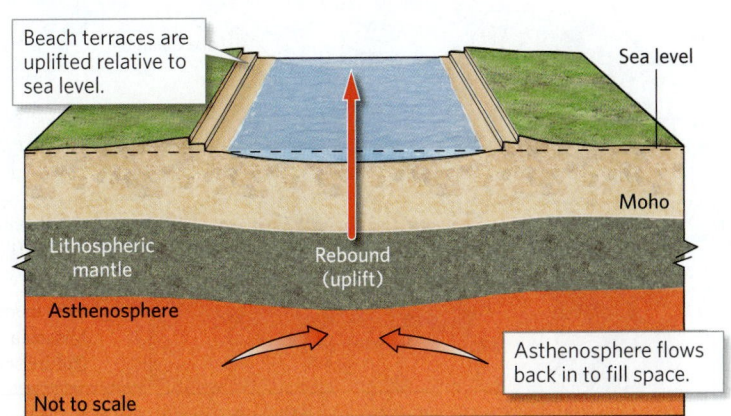

Reference line   Ice sheet   Ice-margin lake   Sea level

The weight of ice causes subsidence.

Subsidence

Moho

Lithospheric mantle

Asthenosphere

Asthenosphere flows out of the way.

Not to scale

(a) The weight of an ice sheet causes the surface of the lithosphere to sink (subside).

Beach terraces are uplifted relative to sea level.   Sea level

Lithospheric mantle

Rebound (uplift)

Moho

Asthenosphere

Asthenosphere flows back in to fill space.

Not to scale

(b) When the glacier melts, the land surface rises (rebounds).

FIGURE 14.39 Glacial Lake Agassiz was a glacial lake that formed near the end of the most recent glaciation of the Pleistocene Ice Age.

pre-existing streams to find different routes. By the time the glacier melts away, these new streams may have become so well established in valleys that old river courses may remain abandoned.

As noted earlier, glaciation can lead to the development of glacial lakes. Short-lived *glacial torrents*, floods caused by the sudden release of water from glacial lakes, can modify downstream landscapes significantly. Several such torrents were released when ice dams holding back Glacial Lake Missoula in Montana broke during times of glacial recession, and the lake drained in a matter of hours to days. The catastrophic floods stripped away soil in eastern Washington State, leaving behind a barren landscape now known as the Channeled Scablands (Fig. 14.40).

Pluvial and Periglacial Features

During the most recent glaciation, the climate in regions to the south of continental glaciers was wetter than it is today. Fed by enhanced rainfall, lakes accumulated in low-lying areas at a great distance from the toe of the glacier. Many such **pluvial lakes** (from the Latin *pluvia*, meaning rain) flooded the interior basins of the Basin and Range Province in Utah and Nevada (Fig. 14.41a). The largest of these, Lake Bonneville, covered almost a third of western Utah. When this lake suddenly drained after a natural dam holding it back broke, it left a bathtub ring of shoreline sediments rimming the mountains near Salt Lake City. Today's Great Salt Lake represents a small remnant of Lake Bonneville (Fig. 14.41b).

In polar latitudes today, and in regions adjacent to the fronts of continental glaciers during ice ages, the average annual temperature stays low enough (below −5°C, or 23°F) that the ground freezes and becomes **permafrost**. Regions with widespread permafrost that do not have a cover of snow or ice are called *periglacial environments*. The upper few meters of permafrost may melt during the summer months, only to refreeze when winter comes. As a consequence of the freeze-thaw process, the ground of some permafrost areas splits into pentagonal or hexagonal shapes, producing a landscape known as **patterned ground** (Fig. 14.42).

Sea-Level Changes Related to Glaciation

When glaciers grow, water becomes trapped on land, so sea level drops. During an ice age, when continental glaciers expand greatly, this drop can be as much as 100 m, causing extensive areas of continental shelves to become dry land (Fig. 14.43a, b). During times of low sea level during the most recent ice age, people and animals migrated across a *land bridge* that existed at the location of what is now the Bering Strait between North America and northeastern Asia.

FIGURE 14.40 Consequences of glacial torrents

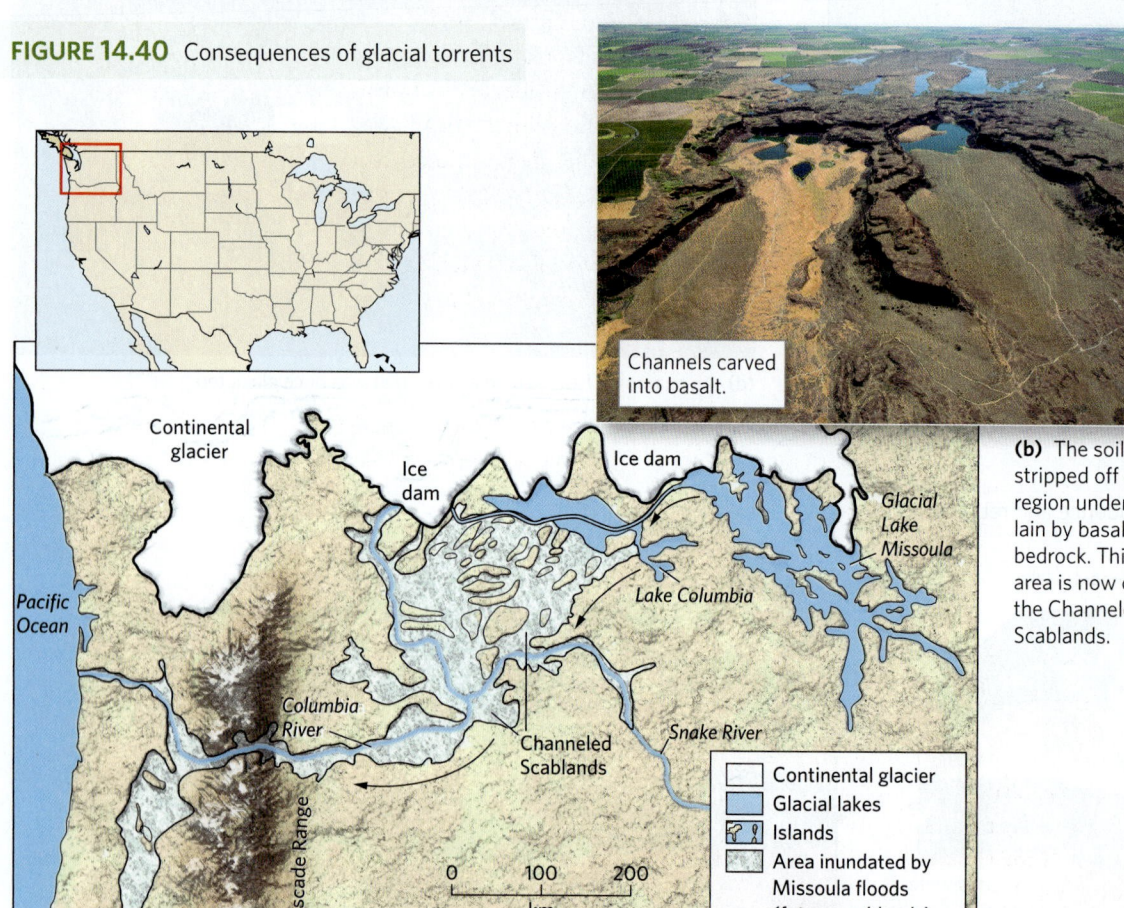

Channels carved into basalt.

(b) The soil was stripped off of a region underlain by basalt bedrock. This area is now called the Channeled Scablands.

⬜	Continental glacier
🟦	Glacial lakes
🏝	Islands
▨	Area inundated by Missoula floods (future scablands)

0   100   200
km

(a) An ice dam blocked the outlet of Glacial Lake Missoula. When the dam broke, floodwaters swept westward across central Washington State and down the Columbia River valley.

FIGURE 14.41 Pluvial lakes in the western United States.

The Great Salt Lake is a remnant of Lake Bonneville.

(a) During the last ice age, a wetter climate produced pluvial lakes throughout the Basin and Range Province of Nevada and Utah.

When continental glaciers melt, sea level rises again **(Fig. 14.43c)**.

Take-home message ...

The weight of a continental glacier can cause the lithosphere to subside, and melting of the glacier can cause it to rebound. The land beyond a continental glacier may be covered with permafrost or pluvial lakes. Continental glaciers store large amounts of water, so glacial advance or retreat affects sea level.

Quick Question

How can glaciation affect drainage in a region?

14.10 Ice Ages

The Discovery of the Pleistocene Ice Age

When 19th-century farmers in northern Europe prepared their land for spring planting, they occasionally broke their plows on large boulders scattered through the soil of their fields. Such glacial erratics fascinated geologists of the day. Most assumed that the erratics had been transported by immense floods. Then, in 1837, a Swiss geologist named Louis Agassiz proposed an alternative solution to the mystery of

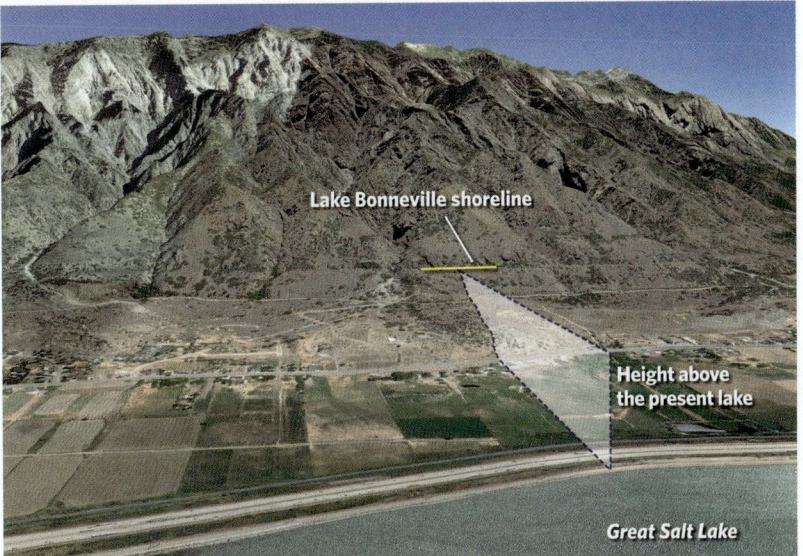

Lake Bonneville shoreline

Height above the present lake

Great Salt Lake

(b) The Great Salt Lake of Utah is a small remnant of Lake Bonneville. Subtle horizontal terraces, the remnants of beaches, now lie about 100 m above the present lake level.

Europe's wandering boulders. Agassiz lived near the Alps and knew that glaciers could carry boulders as well as sand and mud. He proposed that the erratics had been left by glaciers, implying that Europe had once been in the grip of an **ice age**, a time when climate was significantly colder and vast glaciers covered much of the continent.

Agassiz's ice age had clearly taken place fairly recently in Earth history—otherwise, the landscapes and sediment deposits it produced would have been either eroded away or buried. Twentieth-century studies demonstrated that the deposits were left as glaciers advanced and retreated many times during the Pleistocene, the epoch that spans the interval between 2.6 million and 11,700 years ago. So geologists now refer to Agassiz's ice age as the **Pleistocene Ice Age**.

FIGURE 14.42 An example of patterned ground in a periglacial environment in Manitoba, Canada. The polygons are tens of meters across.

FIGURE 14.43 The link between sea level and global glaciation.

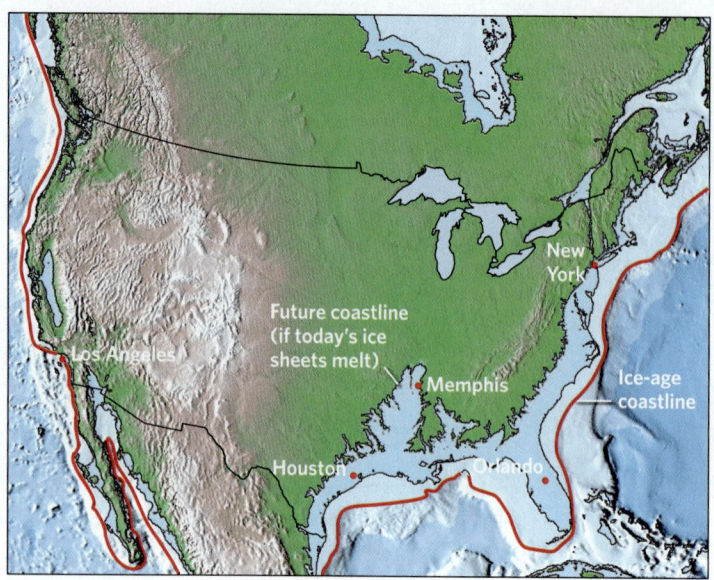

(a) The red line shows the coastline of North America during the most recent continental glaciation, when much of the continental shelf was dry. If the present-day ice sheets melt, areas that are coastal lands today will be flooded.

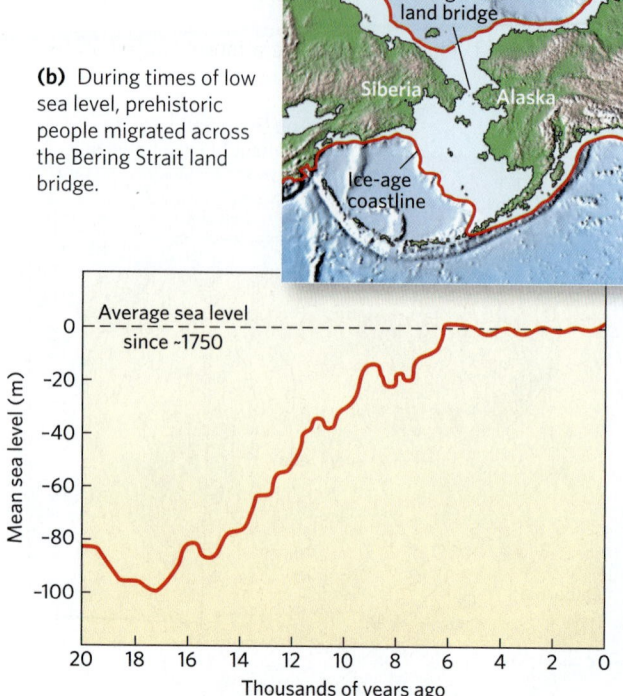

(b) During times of low sea level, prehistoric people migrated across the Bering Strait land bridge.

(c) Sea level rose between 17,000 and 7,000 B.C.E. due to the melting of continental glaciers.

The Extent of Pleistocene Glaciers

Today, most of the land surface in New York City lies hidden beneath concrete, but in Central Park, it's still possible to see land in a semi-natural state. If you stroll through the park, you'll find that the top surfaces of outcrops are smooth and polished (see Fig. 14.27a) and have been grooved and scratched, and that here and there, glacial erratics rest on the bedrock. You are seeing evidence that an ice sheet once scraped along this now-urban ground.

FIGURE 14.44 Pleistocene glaciers

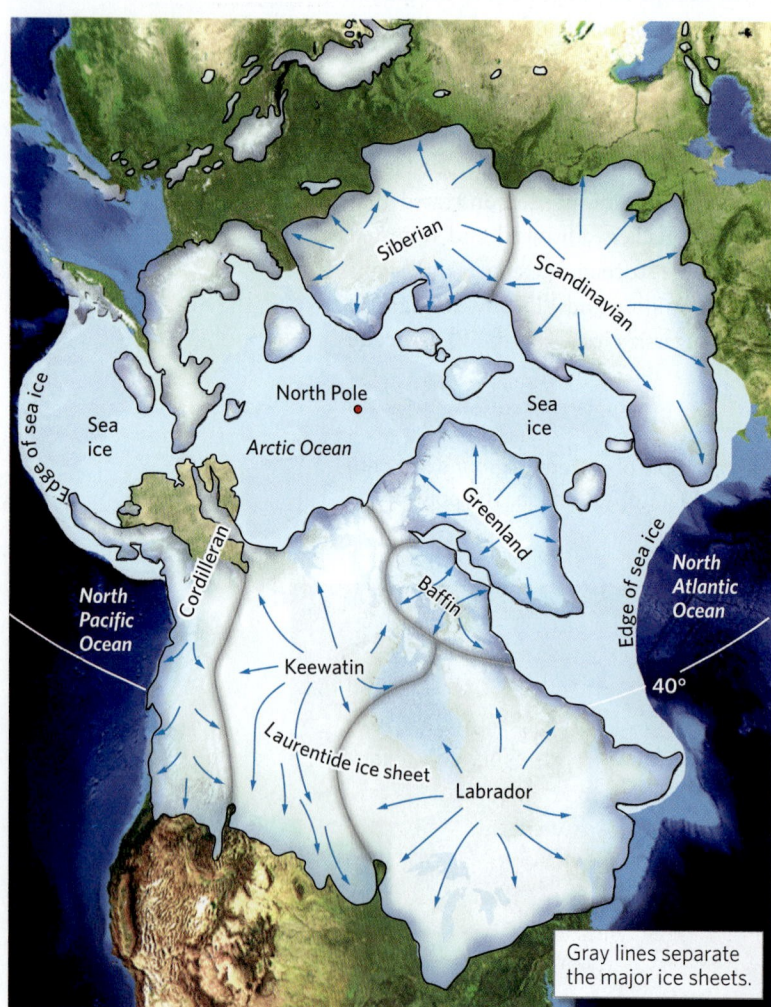

(a) Ice sheets covered much of the northern hemisphere.

(b) The southward shift of climate belts allowed cold-adapted large mammals, now extinct, to roam now-temperate regions of eastern North America.

FIGURE 14.45 The timing of glaciations.

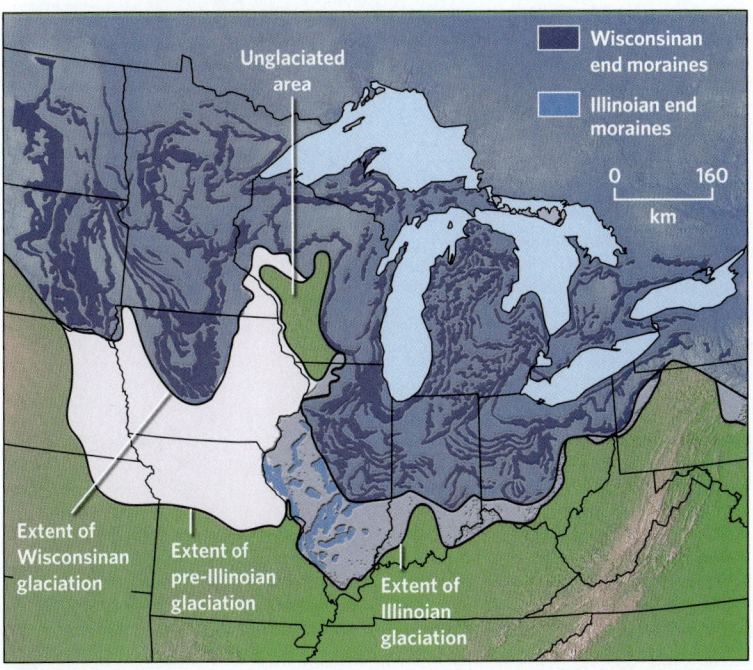

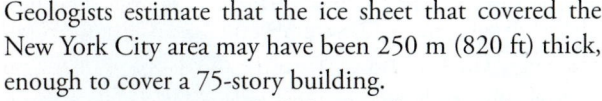

(a) Pleistocene glacial deposits in the north-central United States. The record shows that glaciers advanced more than once. Curving moraines reflect the shape of a glacier's toe.

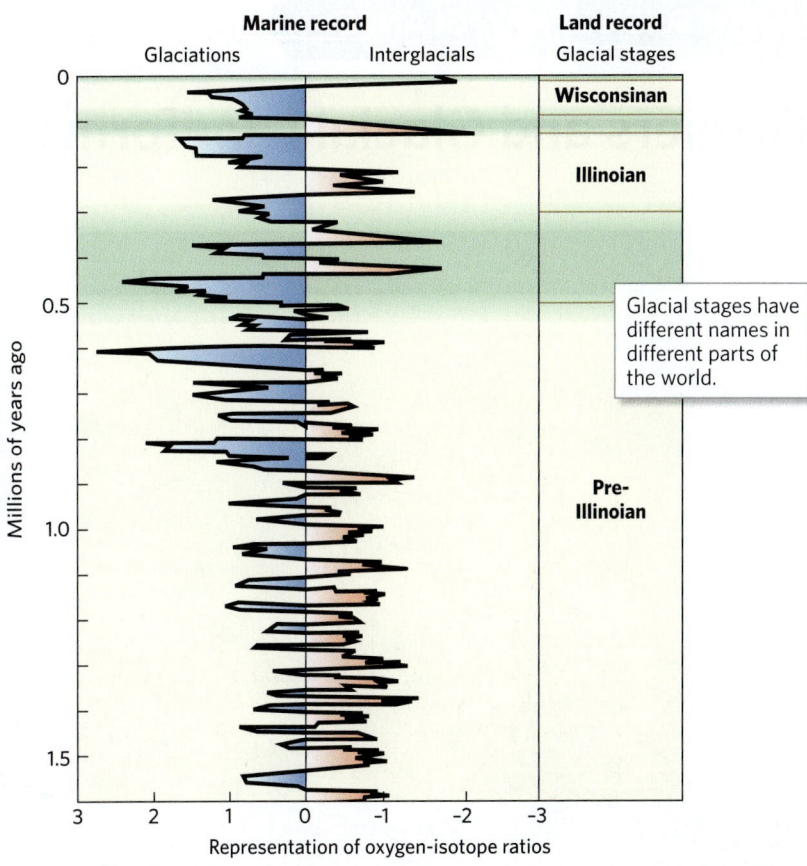

Glacial stages have different names in different parts of the world.

(b) Oxygen-isotope ratios from marine sediment define 20 to 30 glaciations in the Pleistocene. The tan bands represent the traditionally recognized glaciations of the north-central United States.

Geologists estimate that the ice sheet that covered the New York City area may have been 250 m (820 ft) thick, enough to cover a 75-story building.

Geologists have studied the distribution and orientation of such glacial erosion features, as well as the distribution of moraines and the sources of erratics, to map out not only the extent of the Pleistocene glaciers, but also their origins and flow directions. In North America, major ice sheets originated in at least three locations **(Fig. 14.44a)**. These sheets, together with one or more smaller ones, merged to form the giant *Laurentide ice sheet*, which at times covered all of Canada east of the Rocky Mountains and spread southward over the northern portion of the United States. A separate ice sheet, the *Cordilleran ice sheet*, originated in the mountains of western Canada, then spread westward to the Pacific coast and eastward until it merged with the Laurentide ice sheet. Other ice sheets formed in Greenland, Scandinavia, and Russia. During the ice age, climate belts shifted southward, and sea ice covered all of the Arctic Ocean and part of the North Atlantic **(Fig. 14.44b)**.

Glacial Advances and Retreats during the Pleistocene Ice Age

Louis Agassiz assumed that only one ice age had affected the planet. But close examination of glacial deposits on land revealed that *paleosols* (ancient soils preserved in the stratigraphic record), as well as beds containing fossils of warmer-weather animals and plants, occur in sediment deposits between layers of glacial till. This observation suggested that between episodes of glaciation, glaciers receded and warmer climates prevailed. In the second half of the 20th century, when modern methods for dating geologic materials became available, the age differences between layers of glacial sediment could be confirmed. It's now clear that glaciers advanced and retreated more than once during the Pleistocene. Times during which glaciers covered large areas are called **glaciations**, and times between glaciations are called **interglacials**.

Using the on-land sedimentary record, geologists initially recognized 5 Pleistocene glaciations in Europe and 4 in the north-central United States **(Fig. 14.45a)**. This interpretation was revised in the 1960s, when researchers found a much more complete record of glaciations preserved in successions of Pleistocene marine sediments. By studying the ratio of heavier to lighter oxygen isotopes, a number that represents global temperature (see Chapter 20), this record suggested that there were as many as 30 distinct glaciations during the Pleistocene **(Fig. 14.45b)**. The on-land record may preserve evidence of only the largest ones.

Glaciers and Glacial Landforms

Continental ice sheet

Crevasses

Ice shelf

Higher sea level

Lower sea level

Dropstones

Iceberg

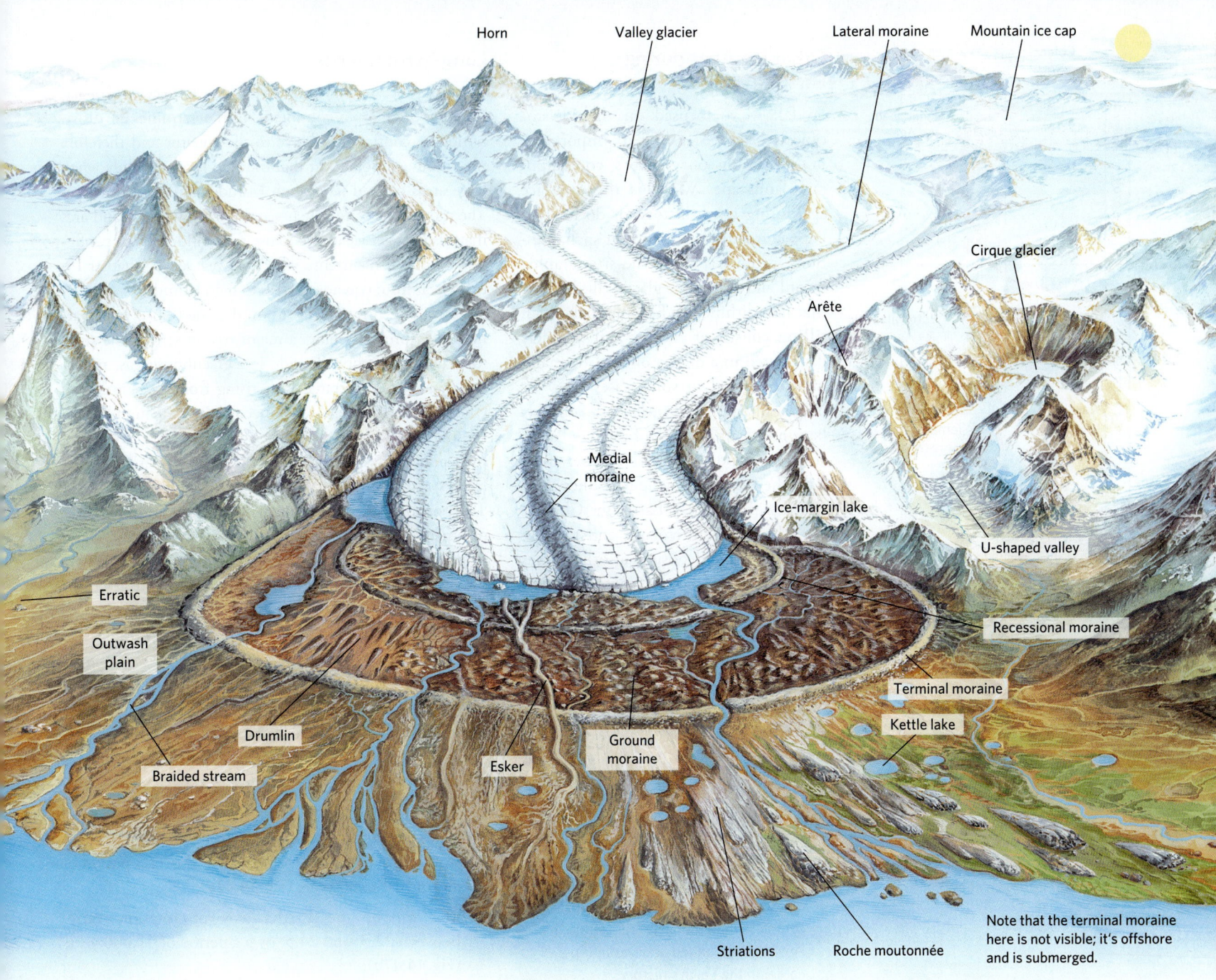

Horn Valley glacier Lateral moraine Mountain ice cap

Cirque glacier

Arête

Medial moraine

Ice-margin lake

U-shaped valley

Erratic

Outwash plain

Drumlin

Braided stream

Esker

Ground moraine

Striations

Roche moutonnée

Recessional moraine

Terminal moraine

Kettle lake

Note that the terminal moraine here is not visible; it's offshore and is submerged.

Continental glaciers, vast sheets of ice up to a few kilometers thick, covered extensive areas of land during times when the Earth had a colder climate. They form from snow that accumulates at high latitudes. When buried deeply enough, the snow packs together and recrystallizes into glacial ice. Although it is solid, ice is weak, so ice sheets spread over the landscape like syrup over a pancake, though much more slowly. When a continental glacier reaches the sea, it becomes an ice shelf. At the edge of the shelf, icebergs calve off and float away, and sediment carried by the ice falls to the seafloor. A second

category of glaciers, called mountain or alpine glaciers, grow in mountainous areas because snow can last all year at high elevations. Valley glaciers are confined to valleys, and ice caps cover the peaks of mountains. These glaciers carve a landscape containing U-shaped valleys, cirques, horns, and hanging valleys. Lateral moraines form along their sides.

During an ice age, a mountain glacier advances and may eventually flow out onto the land surface beyond the mountain front. When climate warms, the glacier starts to recede. The landscape in front of the glacier retains

erosional features, such as striations and roches moutonée—formed when the area was covered by flowing ice.

When the glacier pauses, till (unsorted glacial sediment) accumulates to form an end moraine. Glacial lakes filled with meltwater gather at the toe, and streams pick up till, sort it, and redistribute it as glacial outwash. Sediment that accumulates in ice tunnels, exposed when the glacier melts, make up sinuous ridges called eskers. Flowing ice can mold underlying sediment into drumlins. Ice blocks buried in till melt away and leave behind kettle holes.

447

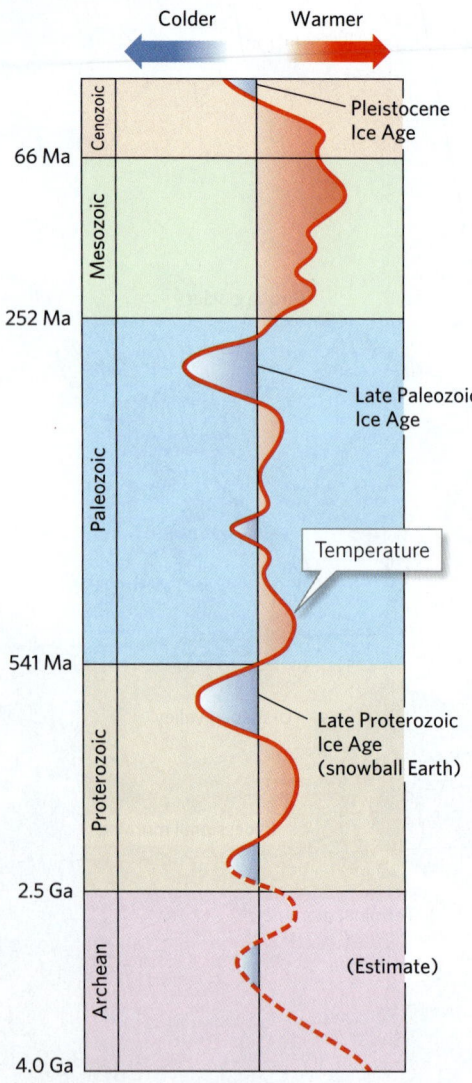

FIGURE 14.46 The Pleistocene is not the only ice age in Earth's history. Glaciations happened in colder intervals of earlier eras, too.

Other Ice Ages during Earth History

So far, we've focused on the Pleistocene Ice Age because of its importance in developing our planet's contemporary landscape. Was this the only ice age during Earth history, or do ice ages happen frequently? To answer such questions, geologists study the stratigraphic record and search for glacial till deposits that have lithified to become rock. Such **tillites** consist of cobbles and boulders distributed in a matrix of sandstone and mudstone, and in some cases, they lie on glacially polished surfaces. Geologists have now found tillites of the late Paleozoic (the deposits Alfred Wegener studied), the late Proterozoic (the ones associated with snowball Earth; see Chapter 10), the early Proterozoic, and the late Archean. Strata deposited at other times in Earth history do not contain tillites, so it appears that glacial advances and retreats do not occur regularly throughout Earth history, but only during specific ice ages, of which there have been four or five **(Fig. 14.46)**.

Take-home message . . .

During the Pleistocene Ice Age, continental glaciers advanced and retreated many times, causing many glaciations and interglacials. The record of these events on land is less complete than that in marine strata. Ice ages also happened at discrete times earlier in the Earth's history.

Quick Question
What is the evidence for multiple Pleistocene glaciations?

14.11 The Causes of Ice Ages

As we've just seen, ice ages occur only during restricted intervals of Earth history, hundreds of millions of years apart. But within an ice age, glaciers advance and retreat with a frequency measured in tens of thousands to hundreds of thousands of years. Geologists, therefore, have searched for both long-term and short-term controls on glaciation to explain this timing.

Long-Term Causes

Tectonic processes play a role in the long-term control of glaciation. For example, continental drift determines the distribution of continents relative to the equator and therefore the amount of solar heat that the land receives. Only when large areas of land lie at high latitudes can ice ages begin. The movement of continents and the growth of island arcs may also trigger ice ages because they influence the configuration of ocean currents, which, as we'll see in Chapter 15, carry heat from equatorial areas to high latitudes.

The concentration of carbon dioxide (CO_2) in the atmosphere probably plays a major role in setting the stage for ice ages. Carbon dioxide is a greenhouse gas, meaning that it traps infrared heat rising from the Earth (see Chapter 20), so when CO_2 concentrations are high, the atmosphere becomes warmer. Ice sheets cannot form during such periods, even if other factors favor glaciation. What factors cause long-term changes in CO_2 concentrations? The possibilities include changes in the population sizes of organisms that extract dissolved carbonate from water to make shells, or of organisms that extract CO_2 during photosynthesis; changes in the amount of chemical weathering on land (a process that absorbs the gas); and changes in the amount of volcanic activity (which emits the gas).

Short-Term Causes

Why do glaciers advance and retreat periodically during an ice age? In 1920, Milutin Milanković, a Serbian researcher, came up with an explanation. He evaluated the Earth's movement through space over time and characterized three types of periodic changes, as follows:

1. The shape of the Earth's orbit (its *eccentricity*) changes from more circular to more elliptical over a period of around 100,000 years **(Fig. 14.47a)**.
2. The *tilt* of the Earth's axis of rotation varies between 22.5° and 24.5° over a period of 41,000 years **(Fig. 14.47b)**.
3. The Earth's axis wobbles (a movement called *precession*) over a period of 23,000 years **(Fig. 14.47c)**.

These cycles are now called **Milankovitch cycles** (using an English spelling of the researcher's name). All three Milankovitch cycles can affect both the total annual **insolation** (the amount of solar energy arriving at the top of the Earth's atmosphere) at mid- to high latitudes and the seasonal distribution of that insolation. In fact, Milanković found that the cycles could cause insolation to change by as much as 25% **(Fig. 14.47d)**! Geologists have found that the times of Pleistocene glaciations correlate with the times of minimum insolation in high latitudes predicted by Milanković. This correlation implies that glaciers advance during times when high latitudes have cooler summers, during which ice from the previous winter can't melt completely.

Subsequent work suggests that the cooling predicted by Milanković might not be enough to trigger a glaciation, so other factors may come into play during times of minimum insolation to cool the Earth even more and trigger a glacial advance. For example, the presence of snow on the ground, or of high-level clouds in the sky, may help cool the Earth by increasing the *albedo* (reflectivity) of the planet—the reflected sunlight doesn't reach the Earth's surface, so it can't warm the atmosphere.

Taking into account the various inputs that could trigger cooling of the Earth, let's look at a possible case history of a single advance and retreat of the Laurentide ice sheet. The process starts when the Earth reaches a point in the Milankovitch cycles when summer temperature at the latitude of northern Canada becomes cold enough that winter snow does not entirely melt away. Because the snow reflects sunlight, its presence makes the region grow still colder, so more snow accumulates. Eventually, the base of the snow cover turns to ice, and when the newborn glacier gets thick enough, it begins flowing laterally across the land. As the glacier broadens, it reflects more sunlight, so the climate cools even more, and because of the cold temperatures, broad areas of high-latitude oceans freeze over.

A glaciation, of course, doesn't last forever, for several reasons: as insolation increases during a Milankovitch cycle, air temperatures rise; as glacial subsidence progresses, the glacier's surface drops down to lower, warmer elevations; and as sea ice cuts off the supply of moisture to the air, snowfall decreases. Taken together, these changes eventually cause the rate of ablation to exceed the rate of accumulation, so the glacier begins to retreat. The shrinking of the glacier decreases the area of high albedo, so the climate warms even more, and snowfalls turn to rain. Rainfall increases the rate of melting, so retreat accelerates until the glacier vanishes.

Will another glacial advance occur? Considering the periodicity of glaciations predicted by the Milankovitch cycles, we may well be set for a new glaciation to begin. But, as we will discuss in Chapter 20, global temperatures have been warming in the past two centuries, and glaciers worldwide have been retreating. So whether ice can return in the next few thousand years remains unknown.

Take-home message . . .

Ice ages occur when the distribution of continents, the configuration of ocean currents, and the atmospheric concentration of CO_2 are conducive to ice-sheet formation. Glacial advances and retreats during an ice age are triggered by Milankovitch cycles, defined by variations in the Earth's orbit and axis of rotation.

Quick Question

What factors, other than Milankovitch cycles, may cause a glaciation to occur?

FIGURE 14.47 Milankovitch cycles cause the amount of insolation received at high latitudes to vary over time. (Not to scale.)

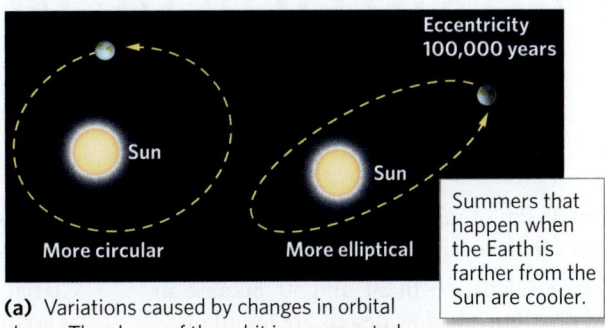

Eccentricity 100,000 years

Sun | Sun

More circular | More elliptical

Summers that happen when the Earth is farther from the Sun are cooler.

(a) Variations caused by changes in orbital shape. The shape of the orbit is exaggerated.

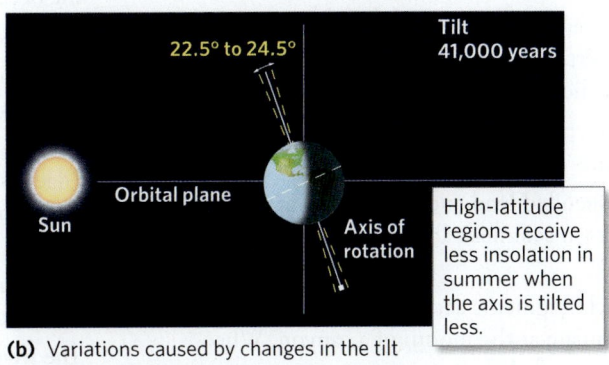

Tilt 41,000 years

22.5° to 24.5°

Orbital plane

Sun

Axis of rotation

High-latitude regions receive less insolation in summer when the axis is tilted less.

(b) Variations caused by changes in the tilt of the Earth's axis.

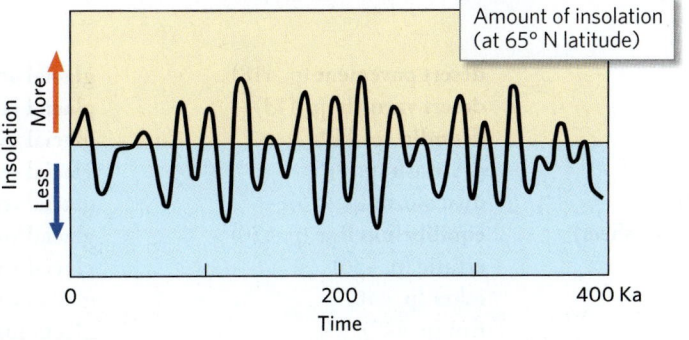

Precession 23,000 years

Wobble of axis

Orbital plane

Sun

Axis of rotation

23.5°

(c) Variations caused by the precession of the Earth's axis.

Amount of insolation (at 65° N latitude)

Insolation — More / Less

0 | 200 | 400 Ka

Time

(d) Combining the effects of eccentricity, tilt, and precession produces varying amounts of insolation during the summer at mid- to high latitudes.

14 CHAPTER REVIEW

Chapter Summary

- Deserts receive an average of 25 cm of rain per year or less, so vegetation covers less than 15% of their surfaces.

- Subtropical deserts form between latitudes of 20° and 30° because air at these latitudes is hot and dry. Rain-shadow deserts are found on the leeward sides of mountain ranges, coastal deserts develop on land adjacent to cold ocean currents, continental-interior deserts form far from the ocean, and polar deserts form at high latitudes.

- Physical weathering in deserts produces rocky debris. Without constant availability of water, chemical weathering happens slowly.

- Water causes erosion in deserts during downpours. Flash floods carry large quantities of sediments down ephemeral streams.

- Wind picks up dust as suspended load and causes sand at the ground surface to saltate. Windblown sediment abrades desert surfaces and sculpts ventifacts.

- Talus piles form where rock fragments accumulate at the base of a cliff. Alluvial fans accumulate at the mouth of a canyon. When temporary desert lakes dry up, a playa remains.

- In some desert landscapes, erosion causes cliff retreat, eventually resulting in the formation of mesas, buttes, and inselbergs. Desert pavements are mosaics of varnished stones.

- Where sand is abundant, desert winds build it into sand dunes of various shapes, including barchan, star, transverse, parabolic, and longitudinal dunes.

- Drought and careless agricultural practices may cause desertification.

- Glaciers are streams or sheets of ice that flow slowly in response to gravity. They form where snow persists year-round.

- Mountain glaciers exist at high elevations and fill cirques and valleys. Continental glaciers spread over substantial areas of continents.

- Glaciers move by plastic deformation of ice grains or by basal sliding at rates of tens of meters per year. Gravity drives glacial flow.

- Glacial advances or retreats depend on the balance between the rate of accumulation and the rate of ablation.

- Icebergs break off glaciers that flow into the sea. Sea ice forms where the ocean surface freezes.

- Glaciers erode by abrasion and plucking.

- Mountain glaciers carve numerous landforms, including cirques and U-shaped valleys. Fjords are water-filled valleys carved by glaciers.

- Glaciers transport sediment of all sizes. Lateral moraines accumulate along the sides of valley glaciers, medial moraines form down the middle, and end moraines form at a glacier's toe.

- Glacial depositional landforms include moraines, kettle holes, drumlins, eskers, ice-margin lakes, and glacial outwash plains.

- The weight of a large ice sheet causes the continental lithosphere to subside. When the glacier melts away, the land surface rebounds.

- Permafrost develops where unglaciated land remains frozen all year.

- When water is stored in continental glaciers, sea level drops. When those glaciers melt, sea level rises.

- During the Pleistocene Ice Age, large continental glaciers periodically covered much of North America, Europe, and Asia.

- The stratigraphy of Pleistocene glacial deposits indicates that glaciers advanced and retreated many times during the Pleistocene Ice Age.

- Long-term causes of ice ages include continental drift and changes in the concentration of CO_2 in the atmosphere. Short-term causes include the Milankovitch cycles.

Key Terms

ablation (p. 429)
alluvial fan (p. 416)
arête (p. 434)
cirque (p. 427)
cliff retreat (p. 418)
continental glacier (ice sheet) (p. 427)
crevasse (p. 429)
desert (p. 411)
desertification (p. 424)

desert pavement (p. 418)
desert varnish (p. 413)
drumlin (p. 440)
dry wash (p. 415)
dust storm (p. 415)
equilibrium line (p. 430)
erratic (p. 436)
esker (p. 440)
firn (p. 427)
fjord (p. 435)

glacial advance (p. 430)
glacial lake (p. 437)
glacial outwash (p. 437)
glacial rebound (p. 440)
glacial retreat (p. 431)
glacial striation (p. 433)
glacial subsidence (p. 440)
glacial till (p. 436)
glaciation (p. 445)
glacier (p. 425)

hanging valley (p. 435)
horn (p. 434)
ice age (p. 443)
iceberg (p. 431)
ice shelf (p. 431)
insolation (p. 448)
interglacial (p. 445)
kettle hole (p. 438)
lag deposit (p. 415)
loess (p. 437)

Review Questions

Letters in parentheses correspond to the chapter's learning objectives.

1. What factors determine whether a region can be classified as a desert? Why do deserts form? **(A)**

2. How do weathering and soil-forming processes in deserts differ from those in temperate or humid climates? **(B)**

3. Describe how water modifies the landscape of a desert. Be sure to discuss both erosional and depositional landforms. **(B)**

4. Explain the ways in which desert winds transport sediment. **(B)**

5. Explain how desert varnish, desert pavement, and ventifacts form. **(B)**

6. Describe the process of cliff retreat and the resulting landforms. **(B)**

7. What type of sand dune is pictured in the figure? **(B)**

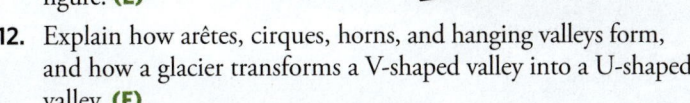

8. What is the process of desertification, and what causes it? **(A, C)**

9. How do mountain glaciers and continental glaciers differ? **(D)**

10. Describe the mechanisms that enable glaciers to move, and explain why glaciers move. **(D)**

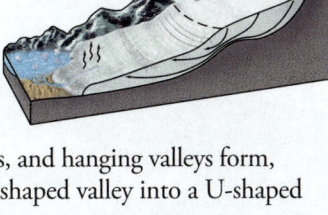

11. Identify the zones of ablation and accumulation in the figure. **(E)**

12. Explain how arêtes, cirques, horns, and hanging valleys form, and how a glacier transforms a V-shaped valley into a U-shaped valley. **(F)**

13. Describe the various kinds of glacial deposits. Be sure to note the materials from which the deposits are made and the landforms that result from their deposition. **(F)**

14. How many Pleistocene glaciations were identified using the on-land record? How was this number modified with the study of marine sediment? Were there ice ages before the Pleistocene? **(G)**

15. What are some of the long-term causes that lead to ice ages? What are the short-term causes? **(H)**

On Further Thought

16. The Namib Desert lies to the north and west of the Kalahari Desert, in Africa. The reason that the former region is a desert differs from the reason that the latter is a desert. Explain. **(A)**

17. Have today's deserts always been deserts? (Hint: Keep in mind the consequences of tectonic processes.) **(A)**

18. If you fly over the barren fields of Illinois in early spring, slight differences in soil color outline polygonal shapes tens of meters across. How and when did these patterns form? **(E, F)**

ANOTHER VIEW As populations grow, more people are moving into the desert realm. In northwestern China, high-speed trains and super highways now cross the dunes and stony plains. The train tracks must be lifted above the shifting sands to be stable.

15 OCEAN WATERS
The Blue of the Blue Marble

By the end of the chapter you should be able to . . .

A. place oceanographic discovery in the context of human history.

B. explain where the ocean's water and salt come from and how sea level varies over time.

C. discuss how the salinity, temperature, and density of ocean water changes with location and depth.

D. explain what surface currents are and how their configuration depends on wind, the Coriolis force, and pressure variations.

E. describe why upwelling and downwelling take place, relate variations in ocean-water density to the formation of deep currents, and describe the importance of these currents.

F. discuss why waves form, how they behave, and how big they can get.

G. distinguish among oceanic zones by water depth and proximity to land, and explain how these zones affect life in the ocean.

15.1 Introduction

During a storm in 1992, a container holding 29,000 plastic bath toys washed off a cargo ship and into the middle of the Pacific Ocean. Ten months later, little yellow duckies started appearing on beaches all around the Pacific. A small number floated through the Bering Strait, moved with sea ice around the Arctic Ocean to the Atlantic, and years later, ended up on the shores of Great Britain (Fig. 15.1). The amazing voyage of these toys emphasizes that, though we divide it into several geographic regions with different names, there is really only one global **ocean**, a layer of saltwater with an average depth of 4–5 km (2.5–3 mi) that covers about 70.8% of our planet's surface. This layer contains 1.3 billion km³ (300 million mi³) of water.

The Earth's ocean is a dynamic environment in constant motion. Its waters flow over vast distances in *currents* (stream-like flows), and its surface elevation changes due to both global-scale *tides* (the daily rise and fall caused by the gravitational attraction of the Moon, Sun, and other forces) and the development of *waves* (local surface undulations). Despite all this motion, ocean water remains nonhomogeneous, for characteristics such as salt content and temperature vary regionally and with depth.

In this chapter, we explore the waters of the world's ocean—the part of our planet's surface that looks blue from space (Fig. 15.2). We begin with a quick survey of ocean exploration. Next, we describe the physical and chemical characteristics of seawater and show how they vary. We then turn our attention to the movement of seawater by describing currents, upwelling and downwelling, and waves. (Our discussion of tides appears in Chapter 16, as the consequences of tides primarily affect coasts.) We conclude this chapter by focusing on how zones of the ocean, distinguished from one another based on depth and proximity to shore, provide varying environments for different assemblages of living organisms.

15.2 The Blue of the Blue Planet

Discovering the Sea

No one knows when the first person sailed over the horizon, but archaeological evidence suggests that as early as 3000 B.C.E., Polynesians built large canoes to travel among isolated islands in the western Pacific. By the time Egyptians constructed the Great Pyramid (2570 B.C.E.), vessels routinely navigated inland seas, and by 1000 C.E., Vikings from Norway had crossed the Atlantic. The end of the Middle Ages saw the beginnings of modern ocean

The Sea, once it casts its spell, holds one in its net of wonder forever.

— JACQUES-YVES COUSTEAU (FRENCH EXPLORER, 1910–1997)

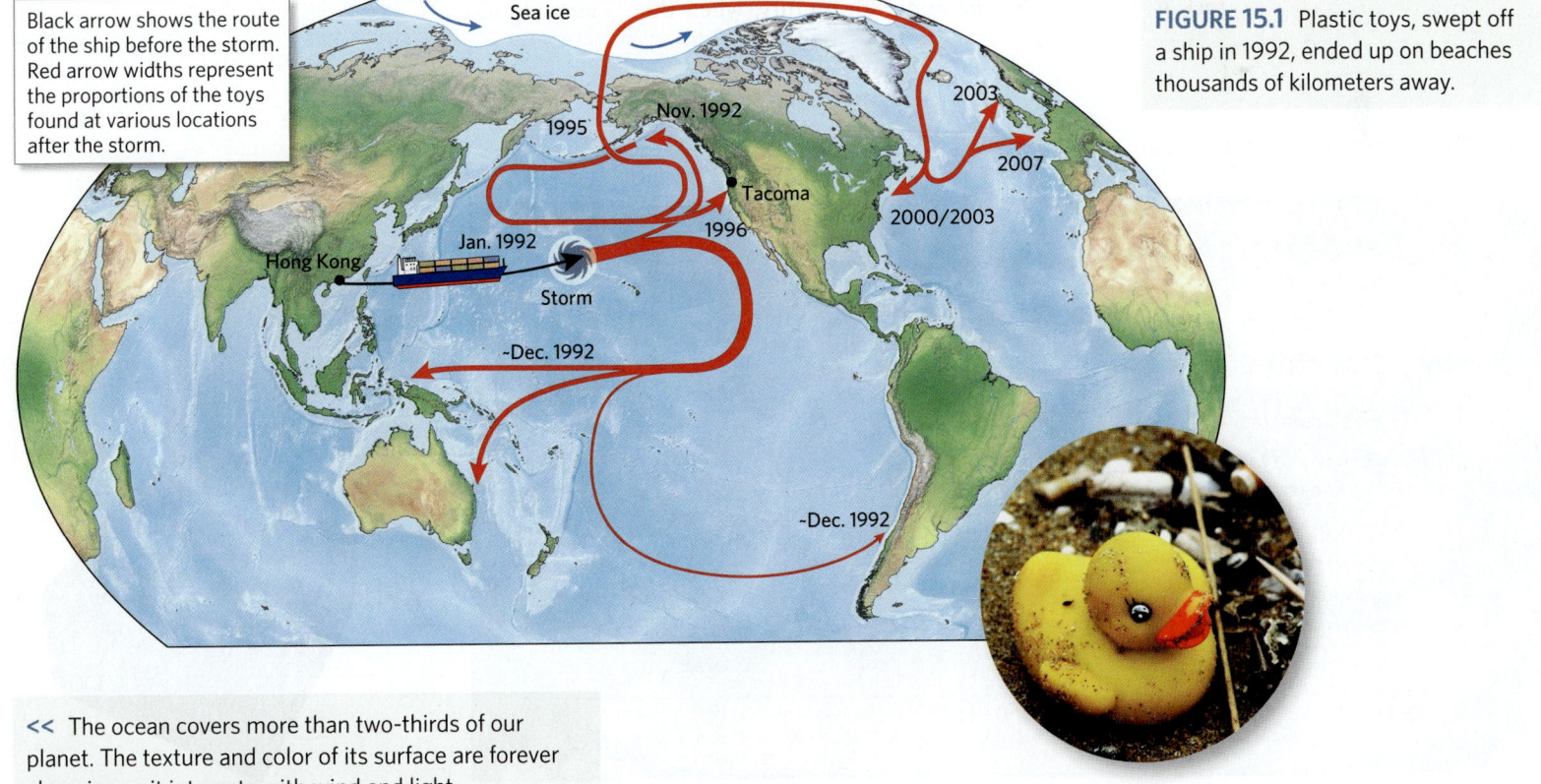

Black arrow shows the route of the ship before the storm. Red arrow widths represent the proportions of the toys found at various locations after the storm.

FIGURE 15.1 Plastic toys, swept off a ship in 1992, ended up on beaches thousands of kilometers away.

<< The ocean covers more than two-thirds of our planet. The texture and color of its surface are forever changing as it interacts with wind and light.

FIGURE 15.2 A view of the Earth from space emphasizes that blue ocean water covers most of its surface.

exploration. Admiral Cheng Ho of China explored the Pacific and Indian Oceans between 1405 and 1433 C.E. ships up to 120 m (400 ft) long. The focus of global oceanic exploration then shifted to Europe.

Christopher Columbus's journey across the Atlantic, in 1492, followed by the discovery of sea routes to China in 1498, opened the European *Age of Discovery*. A quarter century later, the sole surviving ship of Ferdinand Magellan's fleet became the first known vessel to journey entirely around the globe. Each voyage of this era brought back a little more information about what the edges, or **shorelines**, of continents looked like, so by the beginning of the 17th century, the basic layout of continents and the oceans between them had been established **(Fig. 15.3a)**. These early European maps, however, had major inaccuracies, for although explorers had long known how to measure the angle of the Sun above the horizon at noon to calculate *latitude* (representing the distance north or south of the equator), it was not until the mid-18th century, when the

invention of a mechanical clock permitted navigators to keep accurate time on a moving ship **(Fig. 15.3b)**, that it became possible to measure *longitude* (the distance east or west). Measurement of both latitude and longitude then became routine, so explorers could add detail to their maps and could measure the true dimensions of the seas. Improvements in navigation technology, along with aerial photos, satellite images, and GPS satellites, have now characterized every detail of every shoreline.

The early explorers who ventured across the sea undertook their voyages primarily in search of gold and spices. Starting in the latter half of the 18th century, however, some European explorers also collected data about the sea. Between 1768 and 1780, for example, Captain James Cook of England documented currents, collected biological specimens, and measured water temperatures in addition to mapping coastlines. But the modern science of **oceanography**, the study of the sea, did not begin until 1872, when Britain's HMS *Challenger* begin a 4-year cruise that collected enough new information about ocean water and the seafloor to fill 50 books **(Fig. 15.4a)**. By using research ships, submersibles, and satellites **(Fig. 15.4b–f)**, modern *oceanographers* (who study the physical and chemical characteristics and movements of ocean water), along with *marine geologists* (who study the seafloor and coasts) and *marine biologists* (who study marine life), continue to enhance our knowledge of our planet's amazing seas.

FIGURE 15.3 Early mapping of the ocean.

(a) A 1598 map by a Dutch cartographer showing the coastlines as they were known at the time.

(b) A 19th-century chronometer.

FIGURE 15.4 Many types of research vessels study the oceans.

(a) HMS *Challenger*, the first dedicated oceanographic research vessel, set sail in 1872.

(b) A modern oceanographic research vessel.

(c) Submersibles use spotlights to see at depth.

(d) Robotic submersibles provide information without risking lives.

(e) A single-pilot submersible can navigate along the seafloor.

(f) SEASAT is a NASA satellite designed to monitor the world's oceans.

The Map of the Modern Ocean

Geographers divide the global ocean into five separate bodies—the Atlantic, Pacific, Indian, Arctic, and Southern Oceans—outlined by continents. The equator divides the Atlantic and Pacific into northern and southern portions. In addition, geographers recognize several *seas* (such as the Mediterranean, Black, Red, and Caribbean Seas), *bays* (such as Hudson Bay), and *gulfs* (such as the Gulf of Mexico), which are smaller regions of seawater that are partially surrounded by land **(Fig. 15.5)**. Of course, the shapes and positions of today's oceans and seas differ markedly from the way they appeared in the past, and the way they will appear in the future, because of plate motions.

Sea Level and Sea-Level Change

Sea level refers to the elevation of the boundary between ocean water and the air above. When we use the term *sea level*, we are actually talking about *mean sea level*, the average height of this boundary over the course of a year **(Fig. 15.6a)**. At any given moment, the sea surface at a given locality may be higher or lower than mean sea level because of passing waves or the rise or fall of tides.

The geologic record indicates that liquid water has filled the oceans at least since the beginning of the Archean—except, perhaps, for "snowball Earth" episodes in the Proterozoic, when the sea surface was frozen—and that the overall volume of surface water on the Earth has stayed roughly constant (see Chapter 10). However, *relative sea level*, the position of sea level with respect to the land surface, does vary significantly over geologic time, due to a variety of geologic phenomena **(Fig. 15.6b)**. Some examples are: (1) when water gets trapped in large glaciers on land during an ice age, sea level falls relative to land everywhere, whereas when large glaciers melt at the end of an ice age, sea level rises; (2) when a region of land undergoes uplift, sea level falls relative to the region, whereas when the region undergoes subsidence, sea level rises; and (3) when the volume of the ocean basins

FIGURE 15.5 Geographers distinguish many oceans and seas. Some are separated by land and some by latitude.

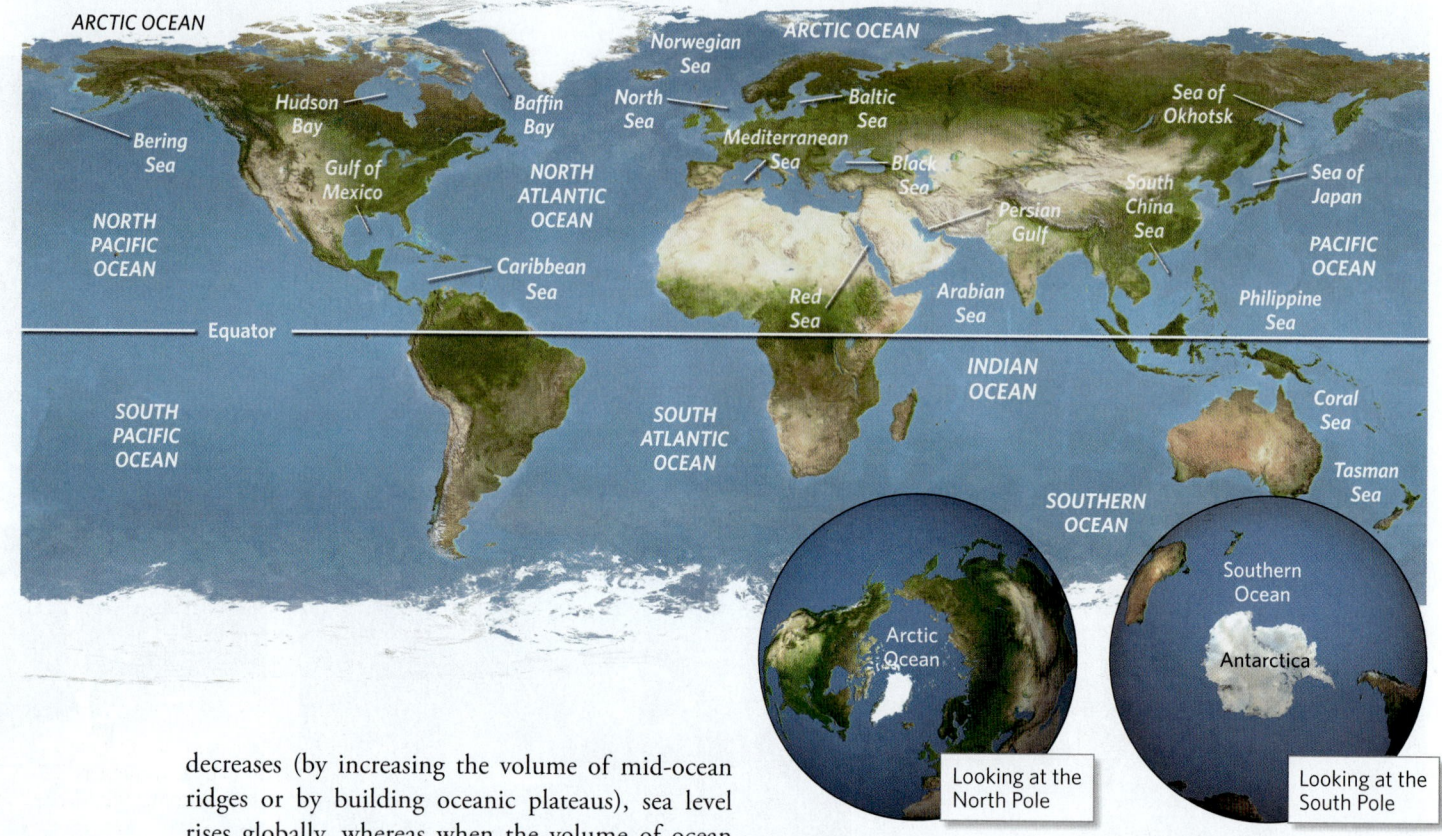

decreases (by increasing the volume of mid-ocean ridges or by building oceanic plateaus), sea level rises globally, whereas when the volume of ocean basins increases, sea level falls. When relative sea level rises significantly, ocean water may submerge large areas of continents to form *epicontinental seas* (Fig. 15.6c), and when sea level falls, portions of *continental shelves*, the relatively shallow part of the ocean adjacent to continents, may be exposed.

Sea level changed significantly during the Pleistocene in association with glaciations and interglacials (see Chapter 14). During the Holocene, as the Laurentide ice sheet melted away, sea level rose by about 120 m (400 ft), and in the past 150 years, it has risen by about 23 cm (9 in). Researchers suggest that this recent rise primarily reflects an increase in seawater temperature, for liquid water expands when heated, but it also reflects the continued melting of glaciers (see Chapter 20).

Take-home message . . .

A layer of saltwater forms a single global ocean. The continents and the equator divide the ocean into distinct geographic regions, and partial enclosure of oceanic regions by land defines seas and bays. Mean sea level, the average elevation of the sea surface, has varied over geologic time.

Quick Question
What factors can cause a change in mean sea level?

15.3 Characteristics of Ocean Water

Where Does the Water in the Oceans Come From?

When the Earth first formed, most of the hydrogen and oxygen atoms that would later compose molecular water (H_2O) existed instead as ions bonded to solid minerals in rock of the crust and mantle. Melting released these ions, which then bonded to yield water molecules that bubbled out of volcanoes as vapor. During the Earth's very early history, the planet remained so hot that this water vapor stayed in the atmosphere. Oceans formed when the planet cooled enough for water vapor to condense and fall as liquid water that could collect in ocean basins.

While the volume of ocean water has stayed fairly constant for most of geologic time, individual molecules of water do not remain in the ocean for very long, for the ocean serves as but one of several reservoirs in the hydrologic cycle (see Chapter 12). In fact, on average, a given molecule of water remains in liquid form in the ocean for less than 4,000 years before it evaporates to become vapor in the air. Then, within hours to days, the vapor condenses or crystallizes and falls back to the Earth's surface.

FIGURE 15.6 The concept of sea level and its variation over time.

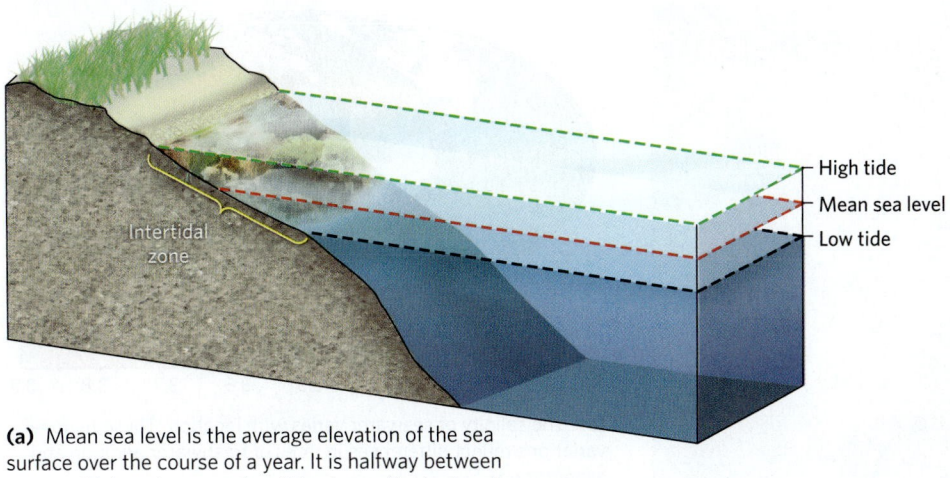

(a) Mean sea level is the average elevation of the sea surface over the course of a year. It is halfway between average high and average low tide at a particular location.

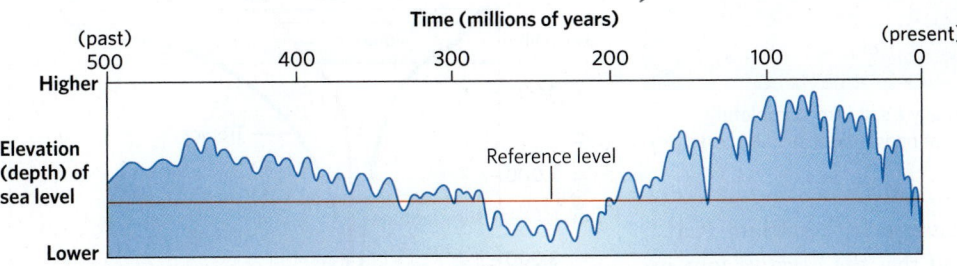

(b) A simplified estimate of sea-level change over the last half billion years.

(c) At times, sea level rises so high, relative to land, that water floods large areas of continental interiors. During the Mississippian, for example, much of the interior of the United States lay beneath an epicontinental sea.

The Salt of the Sea

If you swim in the ocean, you'll immediately notice its salty taste and observe that you float more easily in ocean water than you do in freshwater **(Fig. 15.7a)**. That's because ocean water is a solution containing dissolved salts. In a solution, dissolved ions fit between water molecules without changing the overall volume of the water, so adding salt to water increases the water's density. You float higher in a denser liquid (seawater) than in a less dense liquid (freshwater) because, according to *Archimedes' principle*, an object sinks in water only to a depth at which the weight of the water displaced equals the weight of the object.

How salty does the sea get? It has about the same saltiness as a solution you can prepare by stirring a teaspoon of salt into a cup of warm water. More precisely, typical seawater contains about 96.5% water (H_2O) and 3.5% salt by weight. In contrast, typical freshwater contains less than 0.02% salt. Put another way, seawater contains about 220 times the amount of salt in freshwater. There's so much salt in the sea that if all of the sea's water molecules evaporated, the layer of salt left behind would be about 67 m (220 ft) thick, the height of a 25-story building. Oceanographers refer to the concentration of salt in water as **salinity**.

Ocean salinity depends, in part, on water temperature, because warm water can hold more salt in solution than can cold water. It also depends on the balance between inputs of freshwater and removal of freshwater by evaporation. For example, in **estuaries**, places where seawater floods the mouth of a river, salty seawater mixes with fresh river water to produce *brackish water* that has less salinity than seawater. Brackish water also develops where glacial meltwater enters the sea or where heavy rain falls on the sea surface. In contrast, along the margins of restricted seas in hot, arid climates, so much water evaporates that the remaining water becomes *brine*, meaning that its salinity becomes greater than that of seawater. Therefore, while ocean salinity averages 3.5%, it varies with location, ranging from about 1.0% to about 4.1% **(Fig. 15.7b)**. Ocean salinity also varies with depth **(Fig. 15.7c)**. The largest variations in salinity occur in the upper 1 km of the sea, the portion most affected by evaporation or by freshwater inputs. Salinity in deeper water tends to be more homogeneous. Oceanographers refer to the gradational boundary between surface-water salinities and deep-water salinities as the **halocline**.

What does the salt in the ocean consist of? Chemical analyses indicate that the positive ions in seawater

FIGURE 15.7 Salinity varies within the ocean.

(a) The Dead Sea (which is actually a salt lake, completely surrounded by land) is so salty that people float in it like corks.

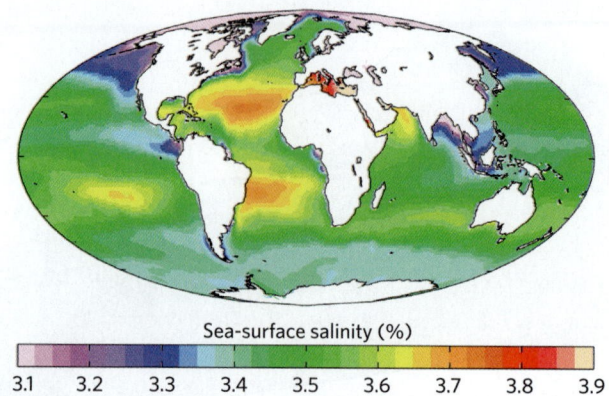

Sea-surface salinity (%)

3.1   3.2   3.3   3.4   3.5   3.6   3.7   3.8   3.9

(b) The salinity of seawater varies with location. These regional variations reflect differences in rates of freshwater addition and evaporation.

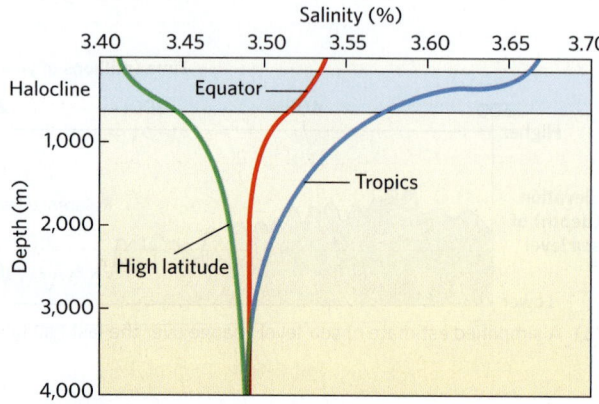

(c) Salinity varies with depth as well as with climate. Variation in salinity is greatest in the uppermost 1 km of ocean water.

include sodium (Na⁺), potassium (K⁺), calcium (Ca²⁺), and magnesium (Mg²⁺), and that the negative ions include chloride (Cl⁻) and sulfate (SO₄²⁻). Most of the ions are sodium and chloride **(Fig. 15.8a)**. Therefore, if seawater evaporates entirely, about 85% of the salt that precipitates consists of the mineral halite (NaCl). The remainder consists of other mineral salts, including gypsum (CaSO₄·2H₂O), anhydrite (CaSO₄), magnesium sulfate (MgSO₄), magnesium chloride (MgCl₂), and potassium chloride (KCl). Seawater doesn't need to evaporate entirely to precipitate salt—salt can precipitate directly

from brines **(Fig. 15.8b)**. Different minerals precipitate from a brine depending on how concentrated the brine has become. As seawater starts to evaporate, NaCl precipitates first. Most common table salt, the kind in your salt

FIGURE 15.8 Seawater contains a variety of salts.

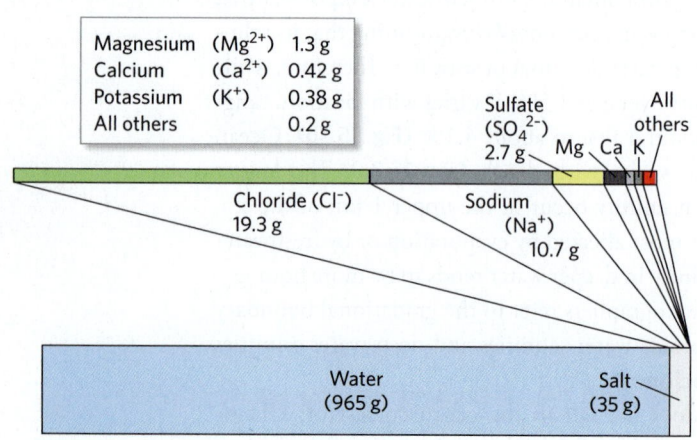

Magnesium	(Mg²⁺)	1.3 g
Calcium	(Ca²⁺)	0.42 g
Potassium	(K⁺)	0.38 g
All others		0.2 g

Sulfate (SO₄²⁻) 2.7 g   Mg   Ca   K   All others

Chloride (Cl⁻) 19.3 g

Sodium (Na⁺) 10.7 g

Water (965 g)   Salt (35 g)

(a) The average chemical composition of seawater, in grams per liter.

Close-up of salt crystals

Salt accumulation

(b) Sea salt precipitates in puddles that fill with spray from waves.

shaker, comes from purification of salt deposits preserved in successions of sedimentary strata. It contains almost entirely NaCl. The "sea salt" that has become popular in gourmet stores contains 95% to 98% NaCl; the remainder consists of traces of other salts. Workers produce sea salt by evaporating ocean water under controlled conditions. Evaporation can also be used to extract freshwater from saltwater (Box 15.1).

Where does the salt in the sea come from? In 1715, Sir Edmond Halley suggested that the rivers flowing on the Earth's surface transported salt to the sea, for he realized that even though river water is fresh enough to drink, it contains tiny amounts of dissolved salts. Halley proved to be correct: most dissolved ions in the sea are produced by the chemical weathering of rock and are carried to the sea by flowing groundwater and river water. In fact, rivers deliver over 2.5 billion tons of salt to the sea every year; lesser amounts come from submarine hydrothermal vents (see Chapter 4).

Notably, the proportions of ions that enter the ocean from rivers differ from those of ions dissolved in the sea. In fact, more than half of the negative ions in river water consist of HCO_3^-, traditionally known as bicarbonate. Ocean water loses its HCO_3^- because organisms extract the ion to build calcite shells, and when the organisms die, the calcite sinks to the seafloor, where it eventually becomes buried. Na^+ and Cl^- do not get extracted to make shells, so these ions stay behind in seawater and become relatively concentrated.

Temperature Variations in the Ocean

When RMS *Titanic* sank after striking an iceberg in the North Atlantic, passengers and crew who jumped or fell into the sea died of cold within minutes. The seawater temperature at the site of the tragedy hovered just below the freezing temperature of freshwater, and such cold water removes heat from a body very rapidly. Yet swimmers can play for hours in the Caribbean, where sea-surface temperatures routinely reach 28°C (83°F). Though the global annual sea-surface temperature averages around 17°C (63°F), it ranges from almost 35°C (95°F) in restricted tropical seas to −2°C (28°F), the freezing temperature of saltwater, near the poles (Fig. 15.9a).

The correlation of average annual sea-surface temperature with latitude exists because the total amount of solar radiation reaching the Earth's surface varies with latitude. The amount of solar radiation also varies with the seasons, so sea-surface temperature varies with the seasons as well, but not by as much as air temperature does, because water has a high *heat capacity*, meaning that it can absorb or release large amounts of heat without changing its temperature very much. For example, in mid-latitudes, the difference between winter and summer temperatures in

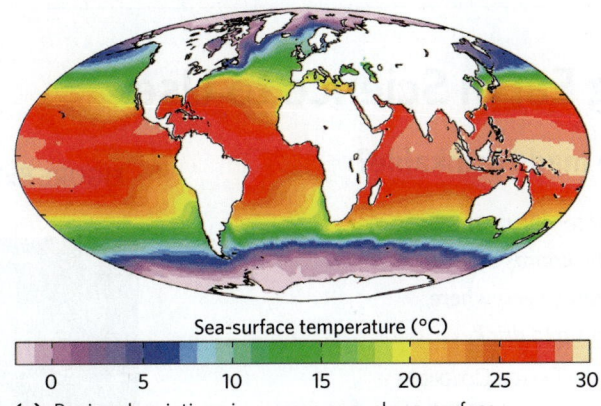

FIGURE 15.9 Temperature varies within the ocean.

Sea-surface temperature (°C)

0   5   10   15   20   25   30

(a) Regional variations in average annual sea-surface temperature correlate roughly with latitude.

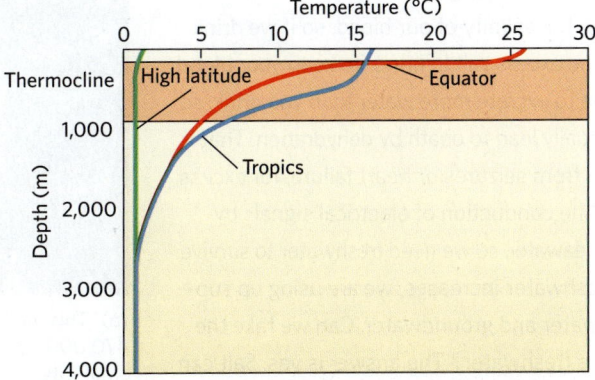

(b) The temperature of seawater changes with depth. Oceanic bottom waters are close to the freezing point of freshwater.

the ocean averages about 8°C (14°F). In contrast, average air temperature in the interior of North America changes by 43°C (75°F) over the course of a year.

Water temperature in the ocean varies markedly with depth in tropical and temperate latitudes, for water heated by the Sun expands and becomes less dense than colder water. Therefore, warmer water "floats" above denser, colder water; the boundary between warmer water above and colder water below defines the **thermocline** (Fig. 15.9b). In the tropics, the top of the thermocline lies at about 200 m (660 ft) below the ocean surface and the bottom at about 900 m (3,000 ft); below the thermocline, seawater temperatures remain fairly constant. A pronounced thermocline doesn't develop in polar seas because surface waters are already so cold that there's hardly any temperature contrast between sea surface and seafloor.

Ocean-Water Density and Water Masses

Both temperature and salinity control the density of seawater, so ocean-water density varies laterally and vertically. Water density at the ocean surface ranges between 1.02 and 1.03 g/cm³, so it's about 2% to 3% denser than freshwater. Water at the seafloor, due to a combination of the pressure caused by the weight of overlying water, low temperature, and high salinity, can reach a

BOX 15.1

Putting Earth Science to Use

How can seawater be made drinkable?

> *Water, water, everywhere,*
> *And all the boards did shrink;*
> *Water, water, everywhere,*
> *Nor any drop to drink.*
>
> —SAMUEL TAYLOR COLERIDGE
> (ENGLISH POET, 1772–1834)

Human blood has a salinity of 0.9% by weight, much less than that of seawater. Our kidneys regulate the salinity of our blood, so if we drink seawater, our bodies excrete the excess salt in urine. To keep your blood salinity normal, you would have to excrete more water than you drink, so drinking seawater would eventually lead to death by dehydration. That is, of course, if you didn't die first from seizures or heart failure, for excess salt in your blood can disrupt the conduction of electrical signals by nerves. Clearly, we can't drink seawater, so we need freshwater to survive.

As society's need for freshwater increases, we are using up supplies of natural fresh surface water and groundwater. Can we take the salt out of seawater to produce freshwater? The answer is yes. Salt can be removed from water by **desalination**. How? Desalination involves either distillation or reverse osmosis.

During *distillation,* workers pump seawater into a tank and heat the water to accelerate evaporation. When seawater evaporates, it leaves its salt behind, so the resulting vapor consists of freshwater—and condensation of the vapor produces liquid freshwater. Most commercial distillation operations use *multi-stage flash distillation* **(Fig. Bx15.1a)**. This process takes advantage of the fact that water's boiling temperature depends on atmospheric pressure—under lower pressures, water boils at lower temperatures. During multi-stage flash distillation, seawater that has been warmed in a furnace flows into a chamber from which air has been extracted. At the lower pressure in this chamber, some of the water boils, producing vapor that can be collected and removed. Evaporation also causes the remaining water to cool (see Chapter 18). This slightly cooler water then flows into another chamber with still lower pressure, where it boils and evaporates. By repeating the process in a series of chambers, each with a pressure lower than the previous one, workers can remove about 85% of the water molecules from seawater.

Reverse osmosis purifies seawater without heating it. During this process, workers apply pressure to push water through a semipermeable membrane that allows water molecules, but not salt molecules, to pass through it **(Fig. Bx15.1b)**. Why is this process called "reverse" osmosis? During normal osmosis, water from a solution with a lower concentration of salt flows through a membrane into a region with a higher concentration of salt. Reverse osmosis requires the application

FIGURE Bx15.1 Desalination of seawater.

(a) This multi-stage flash distillation plant on the coast of a desert produces 470,000,000 L (125,000,000 gal) of freshwater a day, enough to supply a small city.

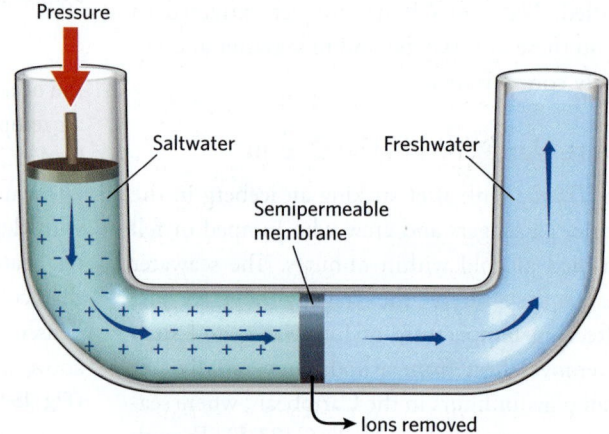

(b) During reverse osmosis, pressure is applied to saltwater to push it through a semipermeable membrane, which filters out salt.

of pressure to push water from the higher-salt-concentration side to the lower-salt-concentration side of the membrane.

Both distillation and reverse osmosis take energy. By one estimate, the United States would have to increase its energy consumption by 10% if it were to obtain all its water by desalination. Both processes also produce brine, which can harm the environment if not disposed of properly. Understanding the benefits and costs of desalination can help you become more informed about this issue, which affects an essential element of human health.

FIGURE 15.10 Water density varies within the ocean.

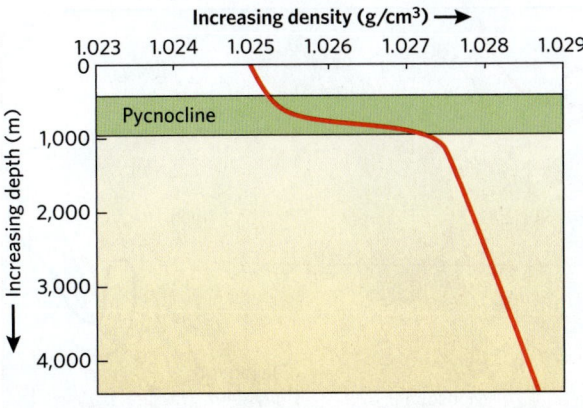

(a) Seawater density varies with depth and changes most rapidly over an interval called the pycnocline. The depth and thickness of the pycnocline vary with latitude.

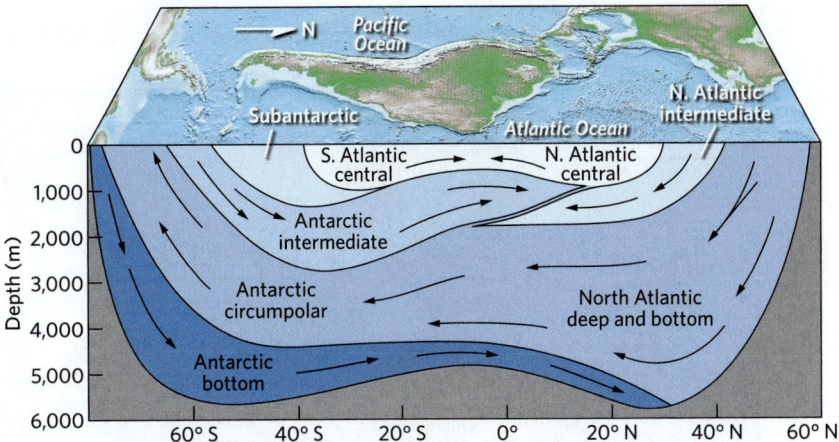

(b) Due to variations in density, the oceans are stratified into distinct water masses.

density of 1.05 g/cm³. In some parts of the world, ocean-water density changes relatively rapidly across a boundary, called the **pycnocline**, at a depth of about 1 km **(Fig. 15.10a)**. A less dense *shallow layer* extends down from the surface to the pycnocline, and a denser *deep layer* continues from the pycnocline down to the seafloor.

Over time, prevailing conditions at a location may cause a large volume of ocean water to attain a characteristic temperature and salinity and, therefore, a characteristic density; oceanographers refer to such a volume as a **water mass**. Oceanic water masses remain fairly distinct for long periods as they move relative to their neighbors because they do not mix easily at their margins. For example, cold water from the coast of Antarctica sinks to the floor of the ocean, forming a layer on the Atlantic Ocean seafloor known as *Antarctic bottom water*. Similarly, cold water from the Arctic region sinks to form the *North Atlantic bottom water*. The former overlaps the latter, and in equatorial regions, other water masses form above these bottom waters. Interleaving of distinct water masses causes an overall layering, or stratification, of the ocean **(Fig. 15.10b)**.

Take-home message . . .

The ocean's water originated as vapor released during volcanic activity; salts are carried into the sea by rivers. Ocean salinity, temperature, and density vary with depth and location. Differences in these characteristics cause ocean water to form water masses that do not mix readily with others, forming distinct layers.

Quick Question

What factors affect the salinity of ocean water?

15.4 Currents: Rivers in the Sea

The ocean's waters do not sit still, but rather flow or circulate at speeds between 1 and 10 km per hour (0.6–6 mph). Oceanographers refer to a band of flowing water that moves distinctly faster than adjacent water as a **current**. Unlike a stream on land, an ocean current does not move within a distinct physical channel. But because currents transport water from one environment to another, temperature and salinity may change across the boundary of a current, making it visible in satellite images **(Fig. 15.11)**. Currents flow within two layers in the ocean: **surface currents** affect only the upper 100–400 m (330–1,300 ft) of water, whereas **deep-sea currents** affect water at great depth and may even move water along the seafloor. Let's now consider the factors that drive currents, starting with surface currents.

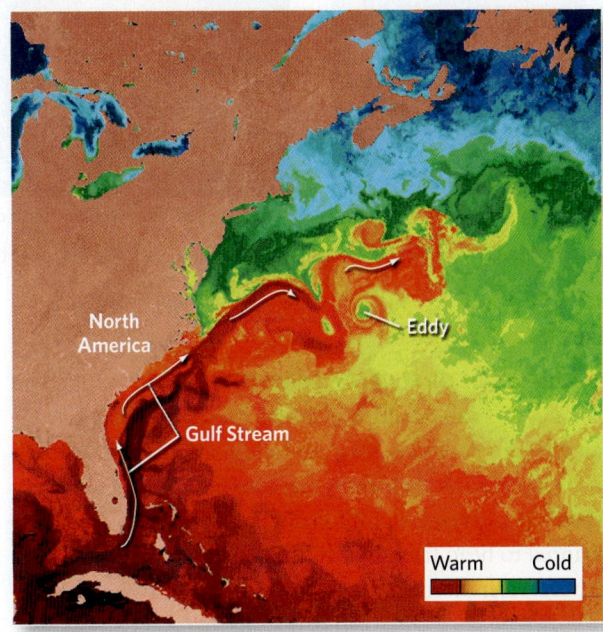

FIGURE 15.11 The Gulf Stream is a current that carries warm water northward along the southeastern coast of North America and then across the North Atlantic. On this map, the colors represent surface-water temperatures.

FIGURE 15.12 Surface currents flow through the upper 400 m (1,300 ft) of ocean water.

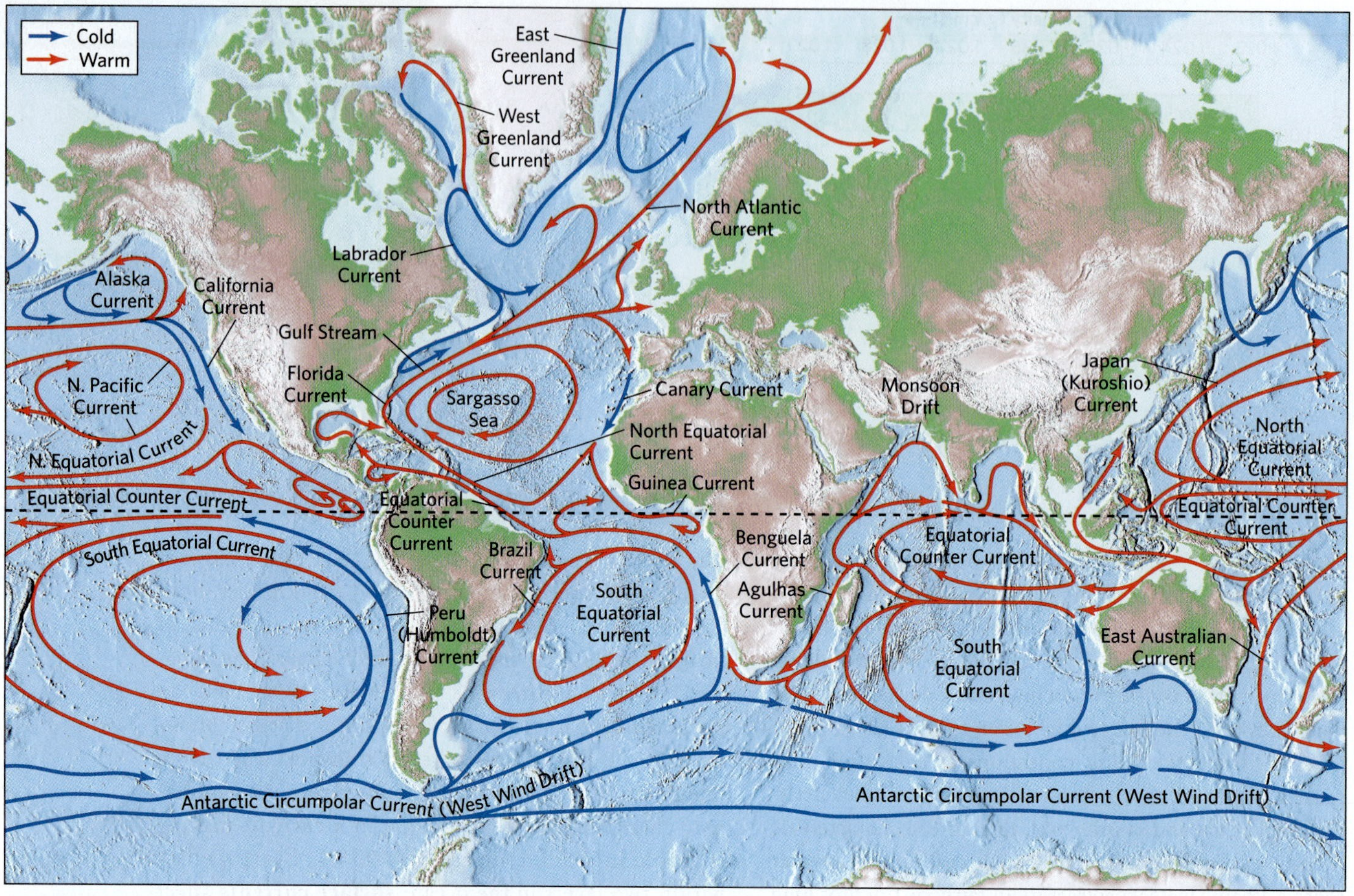

(a) In each major ocean basin, these currents carry water around large circular paths known as gyres.

Western Indian Ocean   Southern Ocean, south of Africa   Western North Atlantic and Caribbean

(b) These images from NASA show the presence of many eddies, swirling currents that are smaller than gyres. The white lines represent flow.

Surface Currents

THE PATTERN OF SURFACE CURRENTS. When skippers in the days of sailing ships planned routes between Europe and North America, they paid close attention to the directions of surface currents, because sailing with a current sped up a voyage significantly. For example, when they wanted to head from North America toward Europe, they would steer north so as to stay in an eastward-flowing surface current, the Gulf Stream. The major surface currents display circle-like paths, called **gyres**, each of which follows the margins of the ocean basin in which it has formed **(Fig. 15.12a)**. Generally, gyres flow clockwise in the northern hemisphere and counterclockwise in the southern hemisphere. Locally, currents define smaller swirls, or **eddies**, whose circumference can range from a few hundred to over a thousand kilometers **(Fig. 15.12b)**.

FIGURE 15.13 The Great Pacific Garbage Patch.

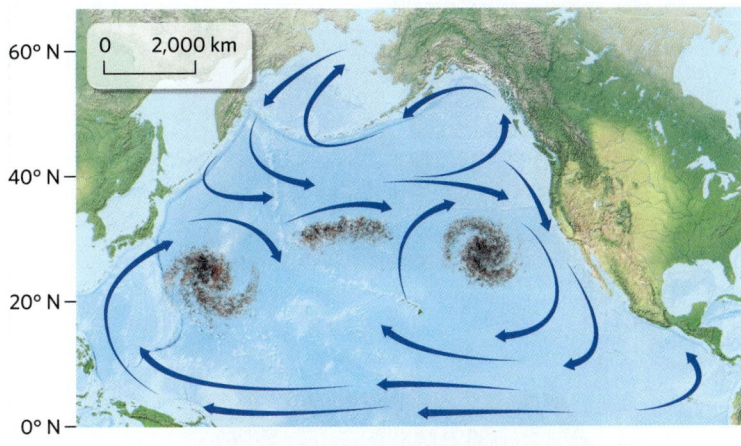

(a) Floating garbage (dark areas) collects within the North Pacific Gyre.

(b) An example of the floating garbage, mostly plastic, as seen from below.

Notably, floating debris gets trapped in the relatively still water that lies in the center of a gyre. The center of the North Atlantic Gyre, for example, collects a type of floating seaweed called sargassum, so sailors refer to the region as the *Sargasso Sea*. The interior of the North Pacific Gyre has collected vast quantities of floating plastic garbage and chemical sludge, so the region has come to be known as the *Great Pacific Garbage Patch* (Fig. 15.13).

Why do surface currents form? Why do they flow in the directions that they do? Surface currents begin because of the frictional interaction between moving air and the surface of the sea. But the specific path that a current follows depends on the influence of two other forces—the Coriolis force and the pressure-gradient force—and on the shapes of ocean basins. Let's examine each of these influences.

WIND PUSHING WATER: FRICTIONAL DRAG. You're already familiar with *friction* as resistance to sliding. It's the force

that slows down a book when you push it across a table. This force exists because protruding molecules on the table surface push against protruding molecules at the base of the book. In the familiar experiment we've just described, the table stays still while the book slows. If you repeat the experiment by placing a sheet of paper between the book and the table, you'll see that the sliding book "drags" the paper along (Box 15.2). The force applied by a moving material to another material across an interface is **frictional drag**. Air and water are both composed of molecules, so the movement of air against the surface of water applies frictional drag to the water and causes the water to move. The *wind*—horizontally flowing air—applies frictional drag to water at the sea surface, causing the water to start moving. This movement initiates a current in the same direction as the wind.

In a given region, winds tends to blow in about the same direction, the *prevailing wind direction*, during most of a season (Fig. 15.14; see Chapter 18). If prevailing winds didn't

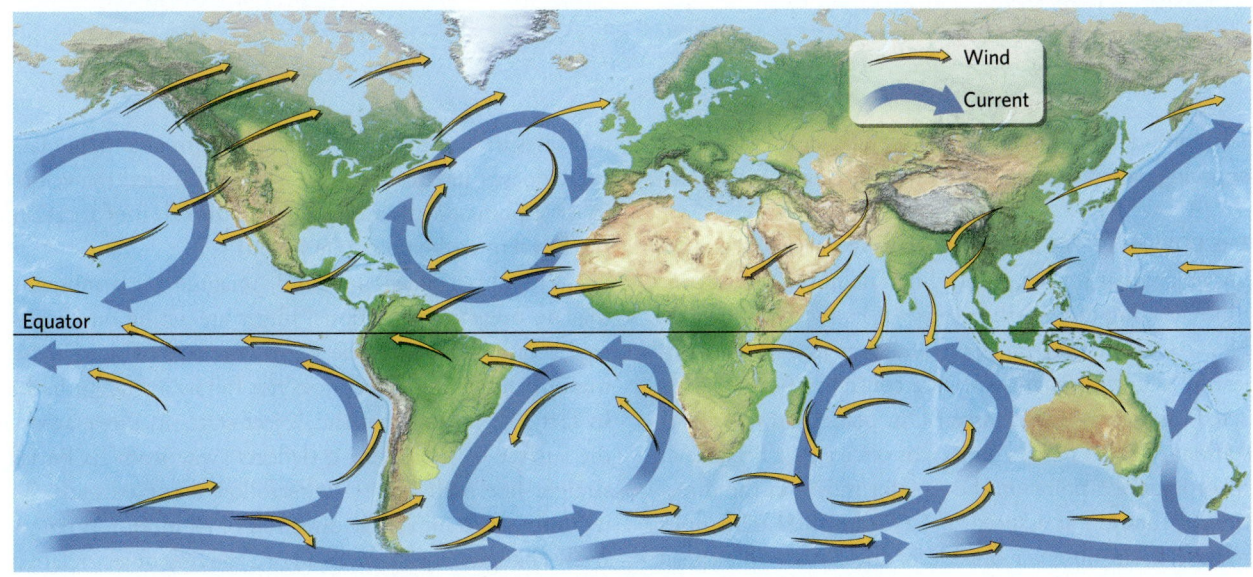

FIGURE 15.14 Global prevailing winds control the overall pattern of ocean currents.

How can I explain . . .

Frictional drag

What are we learning?
Friction caused by a moving object or fluid can carry underlying material along with it.

What you need:
- A textbook, masking tape, a sheet of paper, and a slippery tabletop
- A shallow pan filled with soapy water, with a few small scraps of paper floating in it

Instructions:
- Mark out a reference line on the table with masking tape.
- Align the sheet of paper so that its edge is flush with the line.
- Place the book on top of the paper, then push it forward by about 15 cm (6 in). Measure the distance that the paper moves.
- Place the pan of soapy water on the tabletop.
- Blow hard at a shallow angle to the water surface.

What did we see?
- Friction slows the movement of the book. But frictional drag at the base of the book can move a layer of material beneath it. Materials moved by frictional drag do not move as fast as the materials that drive the motion.
- The same process happens when wind blows on water. You can see this by watching the paper scraps and bubbles move as you blow on the water in the basin. If you add a drop of food coloring, you'll see that the velocity of the water decreases with depth.

exist, well-defined surface currents would not develop, for opposing movements of water would cancel one another out.

THE CORIOLIS EFFECT. If the Earth were stationary, a moving object that started off heading due north would continue heading due north because of its inertia. (Sir Isaac Newton defined *inertia* as the tendency of a moving object to follow a straight path at a constant velocity unless acted on by an outside force.) Our planet, however, isn't stationary, for it rotates on its axis once a day. Because of this rotation, a point on the equator moves in the direction of rotation at a speed of 1,670 km/h (1,037 mph) relative to an observer outside of the Earth. This speed

decreases toward the poles because a point at a higher latitude doesn't have as far to go in a day. In fact, the velocity of a point on the surface in the direction of rotation reaches zero at the poles (Fig 15.15a).

With this concept in mind, imagine for a moment that no pressure or frictional forces are acting on an object as it moves across the Earth. Because of the Earth's rotation, a moving object does not follow a straight path. In fact, the instant that the object starts to move across the surface of the Earth, it deflects away from its initial straight-line path and starts to follow a curved path. As it continues to move, the direction of its motion progressively changes. Such deflection of a moving object relative

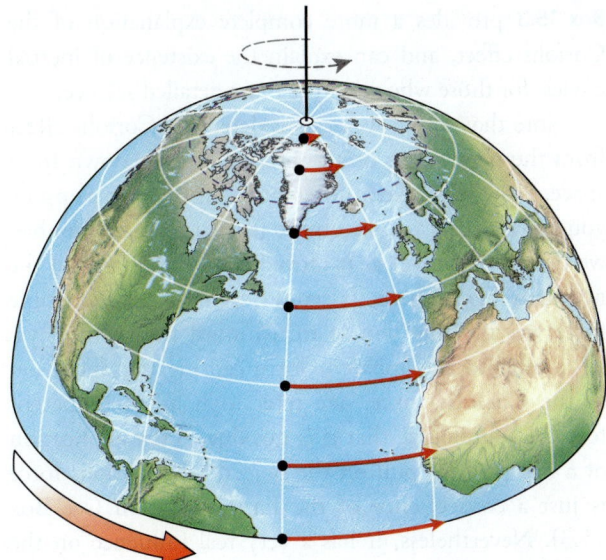

(a) The linear velocity of a point on the Earth's surface varies with latitude, because all points must make one revolution per day, but points nearer the equator have farther to travel than do points near the poles.

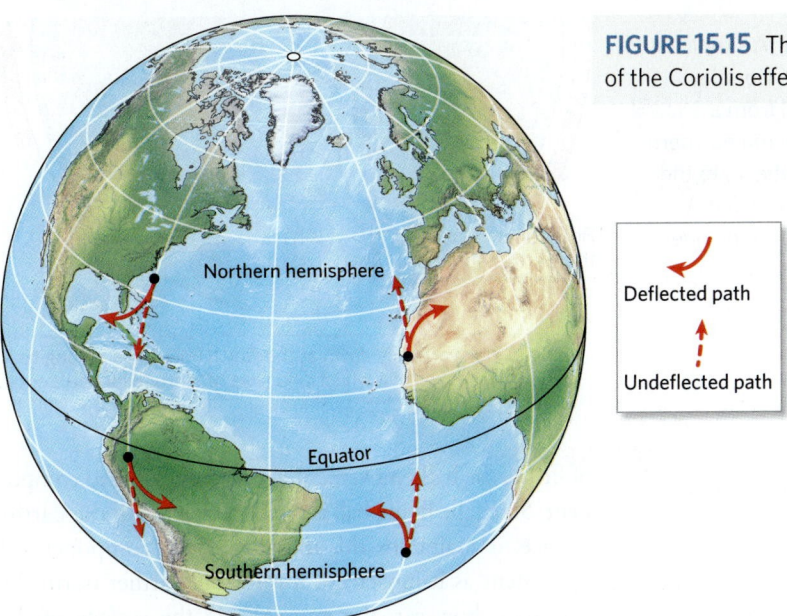

FIGURE 15.15 The concept of the Coriolis effect.

Deflected path

Undeflected path

(b) Moving objects in the northern hemisphere deflect to the right, whereas moving objects in the southern hemisphere deflect to the left.

to the surface of the rotating Earth beneath it is called the **Coriolis effect**, named for the French physicist who explained the phenomenon in 1835.

Because of the Coriolis effect, any moving object undergoes a deflection regardless of the initial direction in which it was moving—unless the object happens to lie exactly on the equator and happens to start moving exactly parallel to the equator. The sense of the deflection depends on the hemisphere in which the object moves **(Fig 15.15b)**. Objects that start moving in the northern hemisphere deflect to their right and follow a clockwise path, whereas objects in the southern hemisphere deflect to their left and follow a counterclockwise path. Specifically, an object in the northern hemisphere that starts moving due north deflects to the east, one that starts moving due south deflects to the west, one that starts moving west deflects to the north, and one that starts moving east deflects to the south. In contrast, an object in the southern hemisphere that starts moving north deflects to the west, one that starts moving south deflects to the east, one that starts moving east deflects to the north, and one that starts moving west deflects to the south. If the object keeps moving long enough, without friction or other forces acting on it, it follows a circular path, called an *inertial circle*, and ends up back where it started **(Fig. 15.15c)**. The diameter of this inertial circle depends on the initial speed of the object and on its latitude: faster objects follow larger inertial circles at a given latitude, and inertial circles get smaller closer to the poles. Figure 15.15c shows inertial circles for air moving at 140 km/h, a speed characteristic of winds in the upper atmosphere. Inertial circles are much smaller for ocean currents because they flow much more slowly.

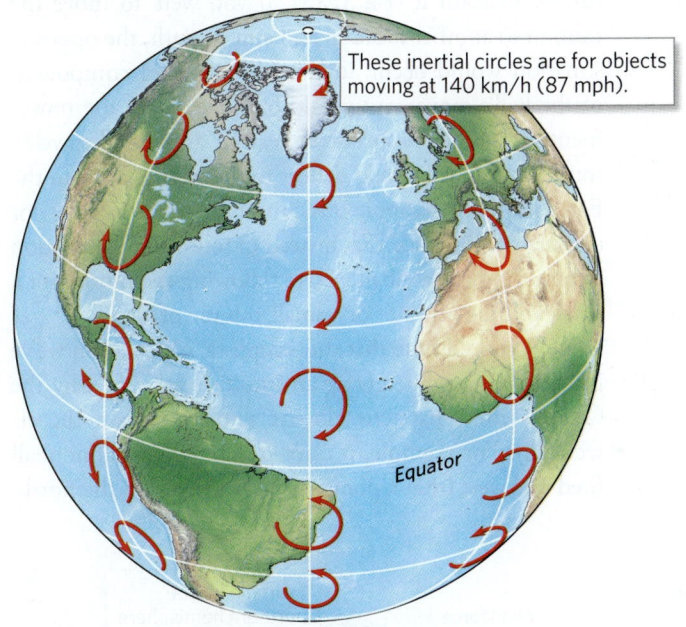

These inertial circles are for objects moving at 140 km/h (87 mph).

(c) Due to the Coriolis effect, an object that keeps moving on the Earth's surface follows a circular path called an inertial circle. The diameter of the circle depends on the object's latitude and speed.

For example, at 40° N, the diameter of an inertial circle traced by a current flowing at 5 km/h, or 3 mph, would be only about 14 km, or 9 mi.

The motion of ocean currents is strongly influenced by the Coriolis effect. But why does the Coriolis effect exist? To develop a simplistic model of its origin, imagine that you place a cannon at a latitude of 45° N, aim due north, and fire. Assuming (unrealistically) that no friction acts on the cannonball, the instant that the ball comes out of the cannon's muzzle, it's moving north. But because the cannon itself was moving with the Earth's surface at the time

FIGURE 15.16 A simplistic analogy for the Coriolis effect. The ball from a cannon fired in the northern hemisphere deflects to the right. So, the ball from A deflects east, and the one from B deflects west.

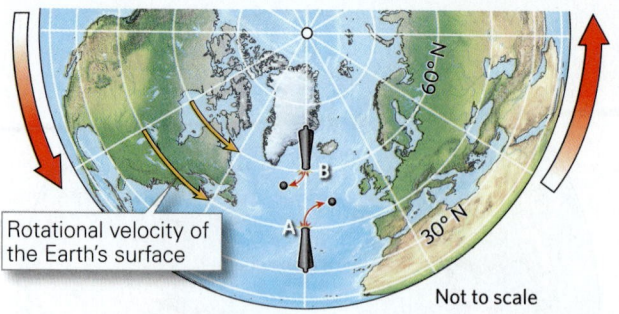

Rotational velocity of the Earth's surface

Not to scale

If you look vertically down at the North Pole, lines of longitude look like spokes of a wheel, and lines of latitude look like circles.

of firing, the ball's movement also has an eastward component. Since the flying ball is not in contact with the Earth's surface, it maintains this initial eastward component of movement as it moves northward. The farther north the ball goes, however, the more slowly the surface of the Earth beneath it is moving in the direction of rotation. Therefore, the ball deflects to the east relative to the Earth's surface beneath it **(Fig. 15.16)**. If you were to move the cannon to another location and aim it south, the opposite situation would occur, in that the eastward component of the ball's movement would be slower than the movement of the Earth's surface beneath it as the ball traveled south, so the ball would deflect to the west relative to the Earth's surface. To apply this concept to ocean currents or air currents, imagine that instead of following the motion of a moving cannonball, you're following the motion of a small volume of water or a small volume of air.

(We should note that this simplistic explanation we've just provided, while giving an intuitive sense of why the Coriolis effect happens, doesn't actually explain the effect fully. For example, it can't explain why a cannonball fired due west from a point at 30° N deflects to the north.)

Box 15.3 provides a more complete explanation of the Coriolis effect, and can explain the existence of inertial circles, for those who want the more detailed science.

Note that so far, we've considered the Coriolis effect from the perspective of an observer looking down from space. Now let's take a different point of view. Suppose you were sitting on an object (such as the cannonball we described above) as it moved across the Earth in the northern hemisphere. It would seem like an invisible force was pushing you to turn toward the right. Physicists refer to the apparent force causing the deflection as the **Coriolis force**. We refer to it as an "apparent force" to emphasize that it is not due to the application of a real push or pull on the object—the Coriolis force is just a consequence of the Earth's rotation (see Box 15.3). Nevertheless, it has a very real influence on the movement of both ocean currents and air currents. The Coriolis force has the following properties: (1) it causes a deflection to the right in the northern hemisphere and a deflection to the left in the southern hemisphere; (2) it affects the direction in which an object moves across the Earth's surface, but not the object's overall speed; (3) the magnitude of the force increases as the speed of the object increases; and (4) its value is zero at the equator and a maximum at the poles.

INFLUENCE OF THE CORIOLIS EFFECT DEEPER DOWN: EKMAN TRANSPORT. As we've seen, moving air molecules of the wind exert frictional drag on water molecules at the surface layer of the ocean, causing these water molecules to start to move. What happens to water below the surface? To answer this question, let's picture a volume of ocean water at a location in the northern hemisphere as a stack of thin water layers (labeled Layers 1, 2, and 3, with Layer 1 at the ocean surface) **(Fig. 15.17a)**. As the water in Layer 1 starts to

FIGURE 15.17 The concept of Ekman transport.

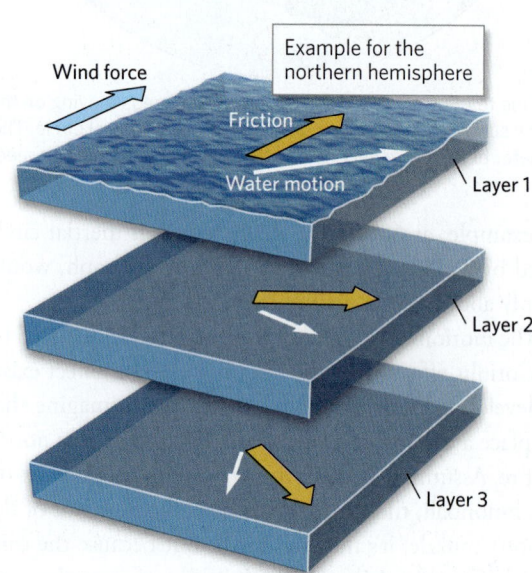

Wind force

Example for the northern hemisphere

Friction

Water motion

Layer 1

Layer 2

Layer 3

Wind direction

Surface current

Net flow (Ekman transport)

(a) The near-surface realm of the ocean can be pictured as a series of layers. When wind shears the top layer, this layer shears the layer below, and so on. Due to the Coriolis effect, the motion of each layer is deflected relative to the layer above.

(b) The resulting pattern of flow is called an Ekman spiral. As a consequence, the overall flow, known as Ekman transport, trends at 90° to the wind direction.

move in response to the wind, the Coriolis effect deflects the flow to the right. The water of Layer 1, in turn, shears Layer 2 beneath it. Because of the Coriolis effect, the water in Layer 2 experiences an additional deflection to the right of the water in Layer 1. The same effect occurs in Layer 3, and in successive layers below: the flow in each layer turns farther to the right than the flow in the layer above it. In each layer, the flow moves more slowly than in the layer above it, because the effects of frictional drag dissipate until finally, at some critical depth, the wind no longer has an effect, and water no longer moves.

The progressive turning of water flow with depth defines the *Ekman spiral* (Fig. 15.17b), named for its discoverer. The existence of an Ekman spiral in a column of water causes the water's overall flow to deflect at an angle of about 90° to the wind shearing the surface of the water. Oceanographers refer to this overall flow as **Ekman transport**, and as we see next, it plays a major role in governing the geometry of regional currents.

A BALANCING ACT: GEOSTROPHIC FLOW AND THE DEVELOPMENT OF GYRES. By this point, you might still be puzzled by the overall flow pattern of surface currents in the ocean. It might seem at first that ocean currents, due to the Coriolis effect, should follow relatively small inertial circles. A map of currents, however, shows that surface currents follow gyres that are vastly bigger than inertial circles, and that current directions are actually rather similar to prevailing wind directions. (see Fig. 15.14). What's going on?

The answer comes from considering a third force acting on ocean water: the **pressure-gradient force**. This force develops because Ekman transport causes a slow drift of water toward the center of an ocean. As a result of this converging flow, sea level in the center of an ocean becomes slightly higher (about 1 m, or 3 ft, higher) than at the ocean's margins—it's as if there is a flat-topped mound of water in the middle of the ocean. (The peak of the mound actually lies to the west of the ocean's center, for reasons beyond the scope of our discussion.) The mound is so broad, and its height so slight, that you don't notice it when you sail across the ocean (Fig. 15.18a), but it is real. If you picture a horizontal surface at a depth beneath this mound, the pressure acting on that surface is greater under the peak of the mound than at its edges. In other words, the slight variation in sea-level height caused by Ekman transport produces a *horizontal pressure gradient* in the ocean between the center and periphery of an ocean basin. The resulting outward-directed pressure-gradient force pushes back against the inward-directed Coriolis force responsible for Ekman transport. When the inward and outward forces are equal and opposite, we can say that *geostrophic balance* exists. When geostrophic balance

exists, water must circulate around the circumference of the mound (Fig. 15.18b). The roughly circular currents that result are the gyres, which, because they develop when geostrophic balance has been established, are also known as **geostrophic currents**. Because in reality, the mound rises higher on the western side of an ocean, the current on the western side of an ocean is stronger than on the eastern side (Fig. 15.18c).

THE EFFECT OF OCEAN-BASIN SHAPE. As we have seen, continents and the equator separate the global saltwater layer into separate oceans. Because of these divisions, a distinct gyre forms in each of the oceans. Different parts of gyres have been given different names (see Fig. 15.12a). Gyres don't cross the equator because the direction of the Coriolis force changes there. The southern edges of gyres in the northern hemisphere and the northern edges of gyres in the southern hemisphere produce westward-flowing equatorial currents. (A weaker, eastward-flowing countercurrent generally exists between the North Equatorial Current and the South Equatorial Current.)

Land acts as an obstacle to ocean currents and can block their flow, causing them to bend and change direction or even to split into two branches flowing in directions opposite to each other. South of 55° S to the shores of Antarctica, however, no land blocks ocean flow, so the Antarctic Circumpolar Current carries water in a circle completely around Antarctica (Fig. 15.19). This current, which ranges between 800 and 2,000 km (500–1,200 mi) wide, extends deeper, and carries more water, than any other current. On average, it transports about 140 times the amount of water in all the world's rivers combined. The existence of the Antarctic Circumpolar Current prevents warm waters from getting close to Antarctica, thereby keeping the continent cold enough to remain covered by ice and fringed by sea ice year-round.

In places where currents flow through relatively narrow spaces between landmasses, such as between Africa and Madagascar, frictional interaction with shallow water bordering the land causes larger currents to break up into smaller eddies (see Fig. 15.12b; Fig. 15.18c). Eddies also form at the boundary between a major gyre and the slower water outside the gyre due to the shear between the moving water masses. We can see this phenomenon clearly in the satellite image of the Gulf Stream in Figure 15.11.

Upwelling, Downwelling, and Deep-Sea Currents

Surface currents are not the only movements of water in the ocean. Water also circulates in the vertical direction: oceanographers identify **downwelling** zones as places where near-surface water sinks and **upwelling** zones as places where subsurface water rises.

BOX 15.3 ▶ **Science Toolbox**

What really causes the Coriolis effect?

To streamline our discussion of why the Coriolis effect exists, we must begin by introducing a few key terms from physics:

1. *Forces and vectors:* A force, simplistically, is a push or pull in a given direction. Since the consequences of applying a force depend on both its magnitude (strength) and the direction in which it has been applied, we cannot represent a force on a drawing with just a number, but have to represent it with an arrow called a vector. By definition, a *vector* is a quantity that contains information about both magnitude and direction. The length of the vector represents the magnitude of the force, and the orientation of the vector represents the direction of the force. A vector can be divided into components. A *component* of a vector is the proportion of the vector in a specified direction.

2. The *Earth's axis of rotation*: The Earth's axis of rotation is an imaginary line passing through the center of the Earth around which our planet spins. Because the Earth is a sphere, not all points on its surface are equidistant from its axis of rotation. Specifically, points on the equator lie farthest from the axis of rotation, while points at the poles lie nearest. This change in distance to the axis of rotation with latitude affects how centrifugal force, discussed below, acts on objects moving on the Earth's surface.

3. *Centrifugal force:* The apparent outward-directed force that an object resting on the surface of a spinning disk or sphere experiences is called *centrifugal force*. We refer to it as an "apparent force" because it exists only from the perspective of the object—it is actually just a manifestation of inertia. The magnitude of the centrifugal force acting on an object at a point on the spinning Earth depends on the square of the object's velocity at that point divided by the distance of the point from the axis of rotation. Therefore, an object moving east at 100 km/h at a latitude of 60° N feels more centrifugal force than does an object moving east at 100 km/h at the equator.

4. *Components of gravity:* Gravity is the attractive force between two objects. Because of gravity, centrifugal force doesn't cause objects to fly off the spinning surface of the Earth. The force of gravity acting on an object at the surface of our planet pulls toward the center of the Earth. The red arrows in **Figure Bx15.3a** represent the direction and magnitude of gravitational force at different points on the Earth's surface. Note that at the equator, the direction of this force is perpendicular to the Earth's axis of rotation, whereas at the poles, it is parallel to the axis. The magnitude of the force stays the same. As an object on the Earth's surface moves from the equator toward the poles, the magnitude of the gravitational force acting on the object in the direction pointing toward the axis of rotation (represented by the blue arrows in Fig. Bx15.3a) progressively decreases.

5. *Conservation of angular momentum:* Imagine swinging a weight attached to a string in a circle around your head **(Fig. Bx15.3b)**. An object moving around a circle has *angular momentum*, defined as the product of its mass (m), its speed in the direction of rotation (w), and its distance from the axis of rotation (r). Written as an equation, angular momentum = $m \times w \times r$. Newton showed that if no twisting force (called a torque) acts on an object in rotation, there must be *conservation of angular momentum*. In other words, if the mass of the object doesn't change, a decrease in its the distance from the axis (r) means that its rotational velocity (w) must increase in order for the angular momentum to stay the same.

Using the above terms, we can develop a clearer sense of the cause of the Coriolis effect **(Fig. Bx15.3c)**.

Let's start by considering the forces acting on a stationary object sitting on the Earth's surface at a latitude in the northern hemisphere (Time 1). The object is rotating around the Earth's axis with the surface of the Earth, so it has a velocity in the direction of rotation appropriate for its latitude. Gravity applies a downward force, pulling the object toward the center of the Earth, and the

FIGURE Bx15.3 The physical origin of the Coriolis effect.

90° N

Total force of gravity

Component of gravity pointing toward the Earth's axis

60° N

30° N

Center of the Earth

0° N

(a) Gravity (red arrows) pulls toward the center of the Earth at every point on the Earth. The proportion of gravitational force pointing toward the Earth's rotational axis varies with latitude (as shown by the lengths of the blue arrows). It is strongest at the equator and weakest at the poles.

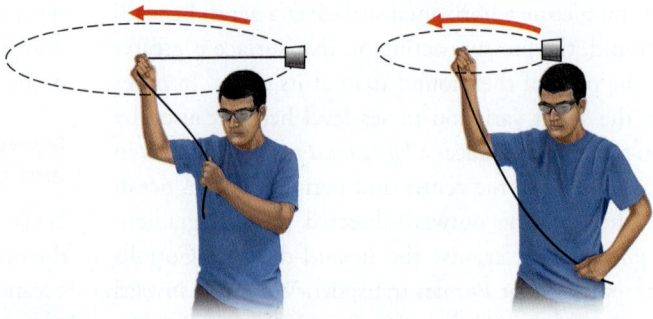

(b) When swinging a weight overhead, shortening the string increases the weight's rotational velocity in order to conserve its angular momentum.

Earth's surface provides an upward force to prevent the object from sinking into the Earth's interior. Meanwhile, this centrifugal force also acts on the object due to the Earth's rotation.

Now, imagine that the object starts moving to the north (Time 2). (For simplicity, we'll ignore friction.) In profile, we see that as the object moves north, it gets closer to the Earth's axis of rotation, so the object's speed in the direction of rotation must increase to conserve angular momentum. Now the object has a greater speed than the Earth below, so it deflects to the right (east). Note that as the object turns eastward, it is still moving at the same overall speed—it has simply changed direction from moving straight north to turning east.

As the object moves north, the forces acting on it change. Specifically, since the object speeds up, the outward centrifugal force increases. At the same time, the component of the gravitational force pointing toward the Earth's axis of rotation decreases, because the total pull of gravity acts at a smaller angle to the Earth's axis. The changes in these two forces result in a net force outward, away from the Earth's axis. The net force pulls the object outward, perpendicular to the Earth's surface, and southward, parallel to the Earth's surface. The outward component is not sufficient to lift the object, but the southward component of the net force decreases the object's northward movement until the object is moving due east, and then deflects the object southward, back toward the equator. (It seems like an external force, the Coriolis force, has been applied to the object, but in fact, all that's happened is that the relative magnitudes of other forces have changed.) Eventually, the object crosses the latitude where it first started to move, this time heading south (Time 3). It continues moving because of its inertia, much as a pendulum moves past its equilibrium point as it swings.

What happens next? As the object continues southward, it moves farther away from the Earth's axis, so its speed in the direction of rotation decreases (Time 4). Now, the object is moving more slowly relative to the Earth's surface below, so it turns right (west). How did the southward change in its position affect the balance of forces? With the object farther south, the centrifugal force decreases because the object is farther from the axis of rotation and has a slower speed in the direction of rotation, even though its overall speed relative to the Earth below remains unchanged. Meanwhile, the component of the gravitational force pointing toward the Earth's axis increases because the total gravitational force is now at a greater angle to the Earth's axis. The net force pulls the object inward, perpendicular to the Earth's surface, and northward, parallel to the Earth's surface. It can't move inward, but it does move northward. As a result, the object continues to turn right and heads back north (Time 5). Eventually, it reaches the point where it started its journey, so it has traced out an inertial circle across the Earth's surface.

Time 1: Object starts moving northward from its starting point.

- Gravity toward axis
- Centrifugal force
- Total force
- Direction of acceleration

Starting position

The component of gravity toward the axis and centrifugal force initially balance.

Time 2: Object moving north turns east. At its highest latitude, it is moving eastward.

Net southward-directed force

Component of gravity toward the axis decreases; centrifugal force increases.

Time 3: Object moving eastward turns south, and passes across its starting latitude.

Forces balance, but object has momentum, so it keeps moving south.

Time 4: Object moving south turns west. It reaches its lowest latitude.

Component of gravity toward axis increases; centrifugal force decreases.

Small net northward-directed force

Time 5: Object moving west starts moving north. It returns to its starting point.

Forces balance, but the object has momentum, so it keeps moving north past the original latitude.

(c) As an object moves across the surface of the rotating Earth, the relative magnitudes of centrifugal and gravitational forces acting on it change, so in order to conserve angular momentum, the object changes direction. The blue line on the globe represents the longitude of the object at the time indicated.

FIGURE 15.18 Understanding geostrophic currents.

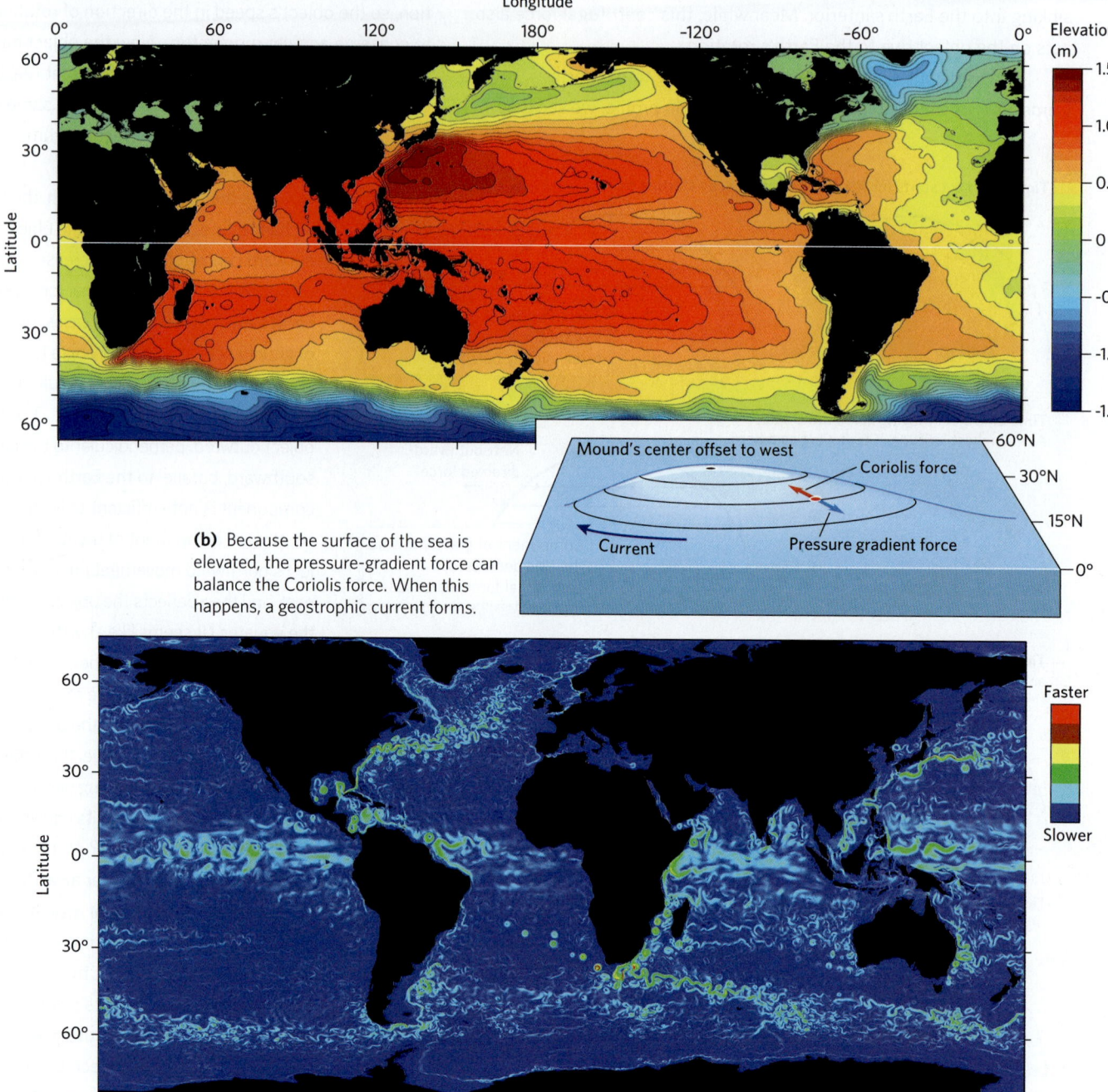

(a) A map showing the subtle topography of ocean surfaces. The elevated areas rise because of Ekman transport. These elevations do not include the effects of tides and waves.

(b) Because the surface of the sea is elevated, the pressure-gradient force can balance the Coriolis force. When this happens, a geostrophic current forms.

(c) A map showing variations in current velocity reveals that geostrophic currents are faster on the western side of oceans, where the sea surface is highest. Note the many eddies that develop in association with regional currents. The colors indicate current velocity.

What causes downwelling and upwelling? Several factors come into play. Coastal downwelling occurs where Ekman transport causes water to move toward the coast **(Fig. 15.20a)**. Since the excess water can't move onto the land, it piles up and sinks. Coastal upwelling occurs near coasts where Ekman transport causes surface water to flow away from the coast. As the surface water moves oceanward, deep water rises, or upwells, to replace it **(Fig. 15.20b)**. Upwelling continually brings cold, nutrient-rich water from the depths to the surface, as happens along the west coast of North America **(Fig. 15.20c)**.

Upwelling of subsurface water also occurs along the equator because the winds there blow steadily from east to west. Due to the Coriolis effect, water just to the north of the equator deflects to the right, and water just to the south of the equator deflects to the left. These deflections remove surface water, so upwelling along the equator must take place to replace this water from below.

FIGURE 15.19 The Antarctic Circumpolar Current. Note the many eddies along the overall current.

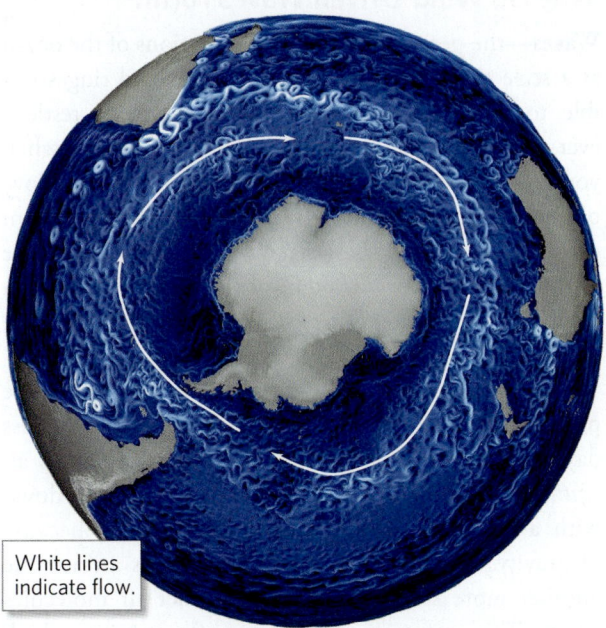

White lines indicate flow.

FIGURE 15.20 Coastal upwelling and downwelling in the northern hemisphere.

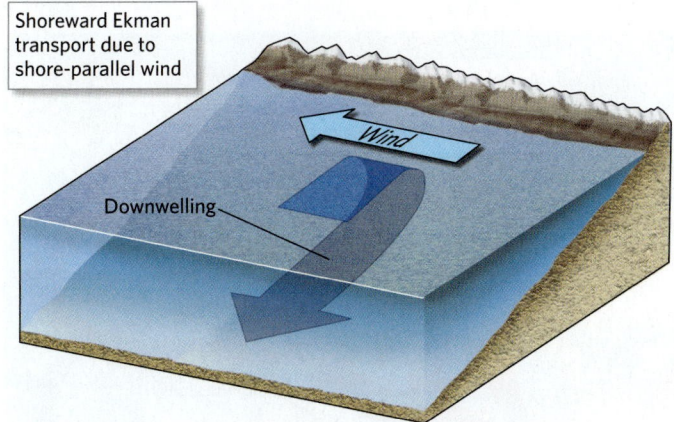

Shoreward Ekman transport due to shore-parallel wind

Wind

Downwelling

(a) A surface flow that moves toward the shore causes downwelling.

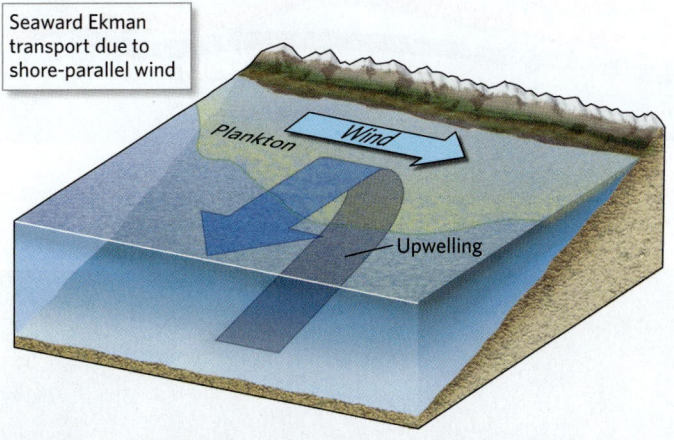

Seaward Ekman transport due to shore-parallel wind

Plankton

Wind

Upwelling

(b) A surface flow that moves water away from the shore causes upwelling. Upwelling brings up nutrients and fosters the growth of algae near the coast.

So far, we've discussed upwelling and downwelling driven ultimately by winds. These processes can also be driven by differences in seawater temperature and salinity because these differences produce contrasts in water density, and in a fluid, density contrasts cause buoyancy differences: denser water sinks, and less dense water buoyantly rises. We refer to the rising and sinking of water driven by density contrasts as a **thermohaline circulation (Fig. 15.21)**. As a result of thermohaline circulation, colder or saltier water sinks because it is denser, where as warmer or less salty water rises because it is less dense. Therefore, cold water in polar regions sinks and flows along the bottom of the ocean as a deep-sea current, flowing back toward the equator at the ocean bottom. The combination of surface currents and thermohaline circulation, like a conveyor belt, moves water and heat throughout the various ocean basins.

Take-home message . . .

Ocean water is in constant motion due to wind-driven surface currents, upwelling and downwelling, and thermohaline circulation. The paths that surface currents follow depend on a combination of factors: wind directions, the Coriolis effect and associated Ekman transport, the pressure-gradient force, and the shapes of ocean basins. Overall, the largest surface currents are gyres, which carry water around the periphery of the ocean basins.

Quick Question

Why does downwelling of ocean water occur at polar latitudes and upwelling at the equator?

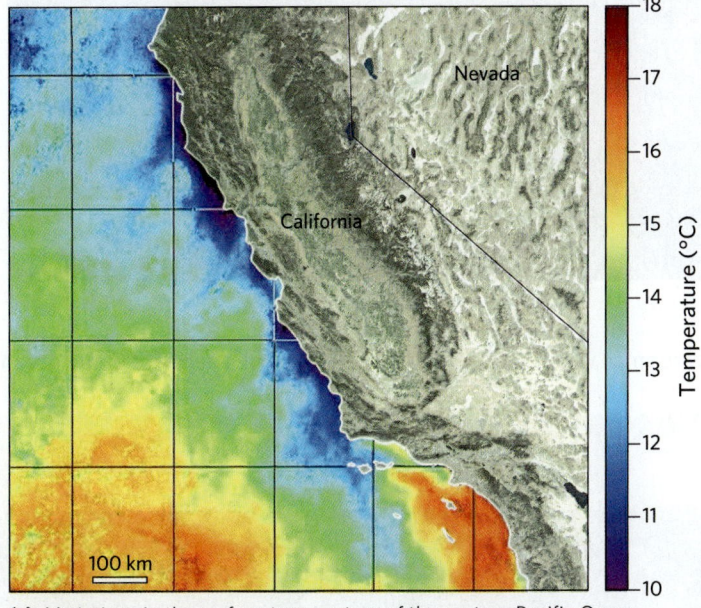

Nevada

California

100 km

Temperature (°C)

(c) Variations in the surface temperature of the eastern Pacific Ocean show upwelling of cold water along the west coast of the United States.

FIGURE 15.21 Thermohaline circulation results in a global-scale conveyor belt that circulates water throughout the ocean system. Because of this circulation, the ocean mixes entirely in a 1,500-year period.

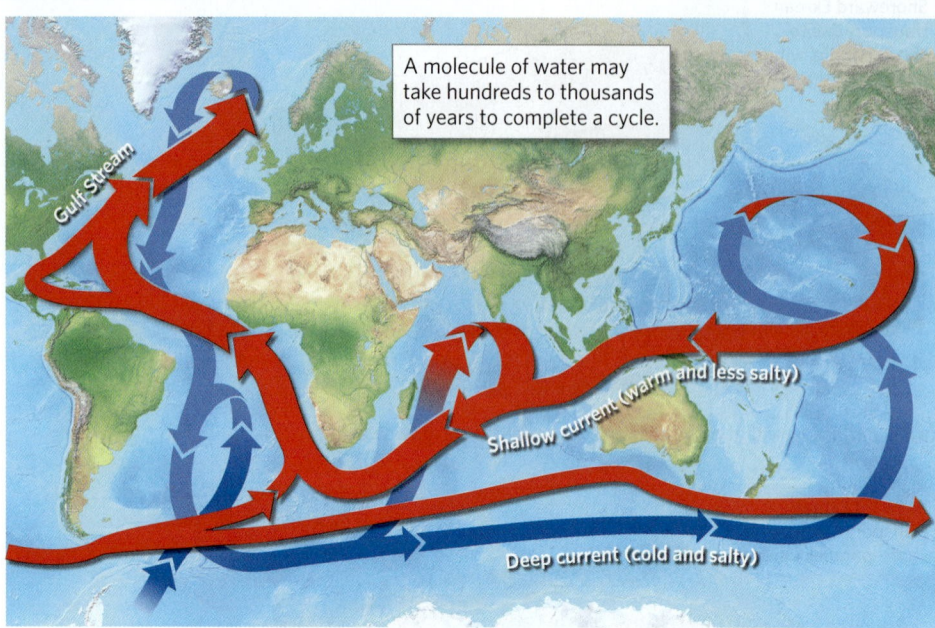

A molecule of water may take hundreds to thousands of years to complete a cycle.

Gulf Stream

Shallow current (warm and less salty)

Deep current (cold and salty)

FIGURE 15.22 Wind-driven waves provide a playground for surfers off Hawaii.

FIGURE 15.23 Ripples are small waves that develop due to the elastic behavior of the ocean surface.

15.5 Wave Action

Why Do Wind-Driven Waves Form?

Waves—the periodic up-and-down motions of the ocean at a scale that generates distinct troughs and ridges visible to your eye—make the ocean surface a restless, ever-changing vista **(Fig. 15.22)**. The waves you see when you look out at the sea from the shore, a plane window, or a ship's deck are **wind-driven waves**. These waves form due to the interaction between moving air and the surface of the ocean. (*Tsunamis*, a type of wave caused by sudden displacement of water due to slip on a fault or a landslide, are relatively rare.)

How do wind-driven waves begin? To picture the process, imagine still water in a pond on a windless day. The horizontal surface of the water represents an *equilibrium level*—if you pushed the surface up or down with a paddle, it would return to that level because of gravity. At the surface, water molecules attract one another more strongly than they attract air molecules above. This attraction produces *surface tension*, which makes the water surface behave somewhat like an elastic sheet. When a breeze starts to blow, frictional drag causes this elastic surface to stretch. As soon as it stretches, elastic rebound causes it to twang back like a rubber band, and as a consequence, it wrinkles into small waves called **ripples (Fig. 15.23)**.

Once ripples exist, wind can push against their sides, causing them to build still higher. A ripple can evolve into a wave that lifts water well above the equilibrium level. When this happens, gravity acts on the elevated water in the wave, causing it to sink. Inertia causes the sinking water surface to descend below the equilibrium level **(Fig.15.24a)**. Next, the water surface bounces back up and rises above the equilibrium level, and the process repeats, like the up-and-down motion of a weight suspended from a vertical spring. You can see this phenomenon happening when you drop a pebble into a pond **(Fig. 15.24b)**. A *wave train* propagates outward, across the water surface, and can move a long distance from the location of the disturbance. Wave trains in the ocean can move far from the location where they were generated by the wind.

FIGURE 15.24 Development of wave trains.

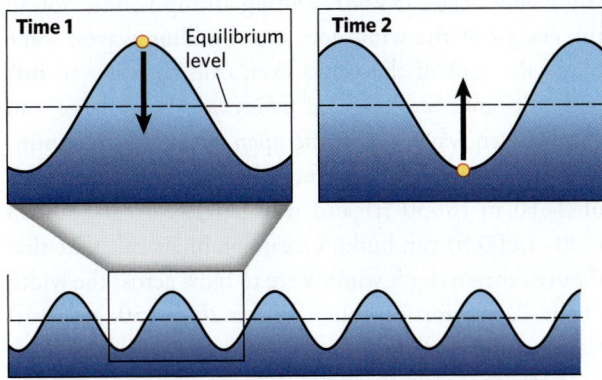

(a) Once a wave builds to a sufficient size, gravity acts as a restoring force that causes its elevated surface to sink below the equilibrium level, as shown here in cross section. Then the water surface rebounds like a weight hanging from a spring.

(b) An example of a wave train formed when a pebble displaces the surface of a pond.

Describing Waves and Wave Motion

We can use the same basic terminology for water waves that we use for any waves. The top of a wave is its *crest* and the base is its *trough* (Fig. 15.25a). The vertical distance between the crest and the trough is the *wave height*, and half the wave height represents the wave's *amplitude*. The horizontal distance between two successive troughs or two successive crests defines the *wavelength*. The number of wave crests (or troughs) that pass a point in a given time defines the *wave frequency*, and the time between the passage of two successive crests (or troughs) is the *wave period*. The horizontal velocity at which a crest or trough moves gives the *wave speed*.

When you watch a wave travel, you may get the impression that the whole mass of water constituting the wave moves with the wave. But drop a cork into the water, and you'll see that it bobs up and down and back and forth as a wave passes—it does not move along with the wave. That's because a particle of water within an ideal wave moves in a circle (see Fig. 15.25a). The diameter of the circle is greatest at the ocean's surface, where it equals the amplitude of the wave. With increasing depth, the diameter of the circle decreases until, at a depth equal to about half the wavelength—called the **wave base**—no wave movement takes place at all. Submarines traveling below the wave base enjoy smooth water while ships toss about on the sea surface above.

FIGURE 15.25 The motion in ocean waves.

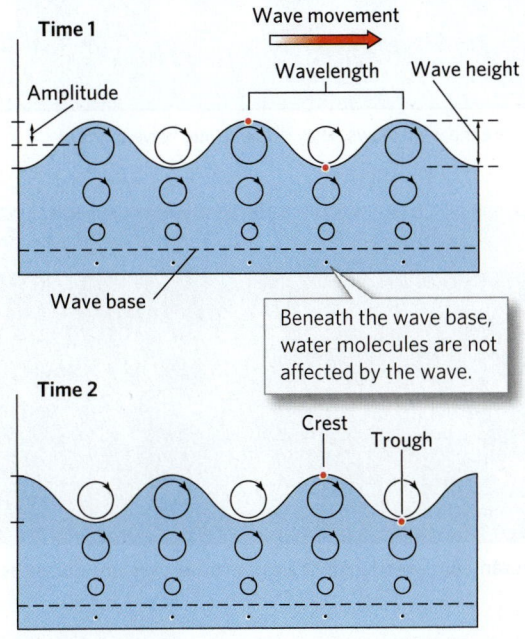

(a) The terminology used for ocean waves is the same as that used for seismic waves. Within a passing wave, a particle of water follows a circular path. The diameter of these circles decreases with depth. The wave base occurs at a depth of one-half the wavelength.

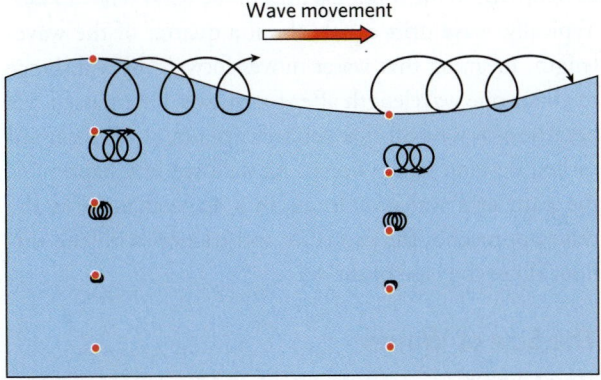

(b) Near the surface of the ocean, wind can cause wave drift. When this happens, particles of water follow a looping path.

FIGURE 15.26 Waves build progressively as the wind continues to blow.

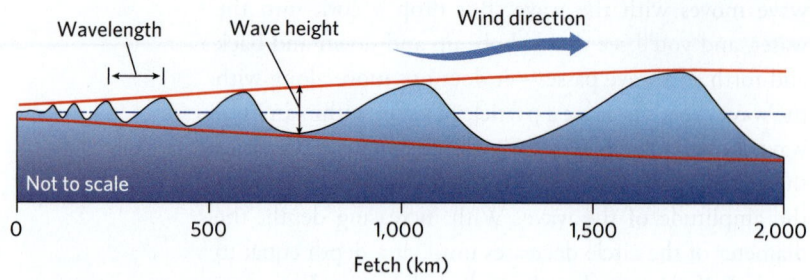

(a) The size of a wave depends in part on the fetch of the wind.

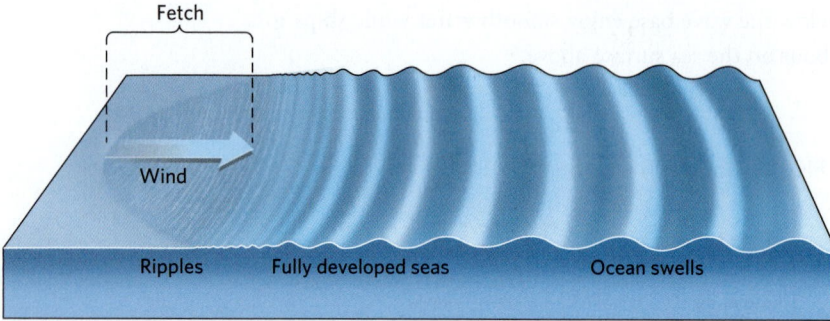

(b) Many waves are generated where the wind blows. They disperse into wave trains of different wavelengths.

(c) An example of swells in the ocean. Long-wavelength waves can travel over long distances as swells.

Although the motion of a water molecule in an ideal wave defines a circle, *wave drift*, an overall lateral motion accompanying the passage of a wave, occurs where waves develop due to the frictional drag of the wind (Fig. 15.25b). Typically, wave drift equals about a quarter of the wavelength, meaning that water moves forward by a distance of about one wavelength after four waves have passed. It's because of wave drift that surface currents, as we discussed earlier, develop in the ocean. As we'll see in Chapter 16, the path of a water molecule in a wave changes as the wave approaches the shoreline and friction with the seafloor slows the base of the wave.

Did you ever wonder . . .

how big an ocean wave can get?

The Size of Waves

The size of waves generated by wind in the open ocean depends on the strength of the wind (how fast the air

moves), on the **fetch** of the wind (the distance over which it blows), and on the length of time during which the wind blows (Fig. 15.26a). During strong winds, not all the energy of the wind goes into building waves; some blows the tops of the waves over, causing water to mix with air and resulting in *whitecaps*. How large can wind-driven waves get in the open ocean? With continued blowing over a long fetch, waves with amplitudes of 2–10 m (6–30 ft) and wavelengths of 40–500 m (130–1,600 ft) can build. Oceanographers calculate that if hurricane-strength winds were to blow across the width of the Pacific for 24 hours, 15- to 20-m (50- to 70-ft) waves would develop.

Even though waves always move more slowly than the wind that generates them, wave speed depends on wind speed: faster wind causes faster waves. In the open ocean, waves typically travel between 10 and 50 km/h (6–30 mph). Physicists have shown that the speed of waves depends on their wavelength: waves with longer wavelengths travel faster than waves with shorter wavelengths. When a storm blows over a region, it generates many different wave trains, each characterized by a different wavelength. Because the wave trains travel outward from their source at different speeds, when they are far from the source, they separate from one another, a phenomenon known as *wave dispersion* (Fig. 15.26b). As a result, a ship in a storm gets tossed about on a chaotic sea surface as it rides over wave trains with different wavelengths, whereas a ship that is far from the source may intersect regularly spaced waves. Waves that have traveled away from their source, and are not being driven by the local wind, are called **swells**; they may travel for thousands of kilometers across the ocean (Fig. 15.26c).

Particularly large waves may form by *constructive interference*, which takes place when two wind-driven waves moving from different directions come together in such a way that the wave crests overlap to form a single crest that is higher than that of either wave. During the 1979 Fastnet yacht race off Ireland, constructive interference produced waves large enough to capsize 23 out of the 300 participating sailboats. The waves developed when swells from an east-blowing gale collided with those generated by a west-blowing gale. Interactions of wind-driven waves with strong currents, or the focusing of waves into a small area by the shape of the coastline or seafloor, can also form very large waves.

Occasionally, these processes lead to the formation of a **rogue wave**, defined as an isolated wave that rises more than twice as high as most large waves passing a locality during a specified time interval. Long thought to exist only in the imaginations of sailors, rogue waves have now been documented numerous times. For example, during the 1990s, workers on oil platforms in the North Sea

FIGURE 15.27 Rogue waves.

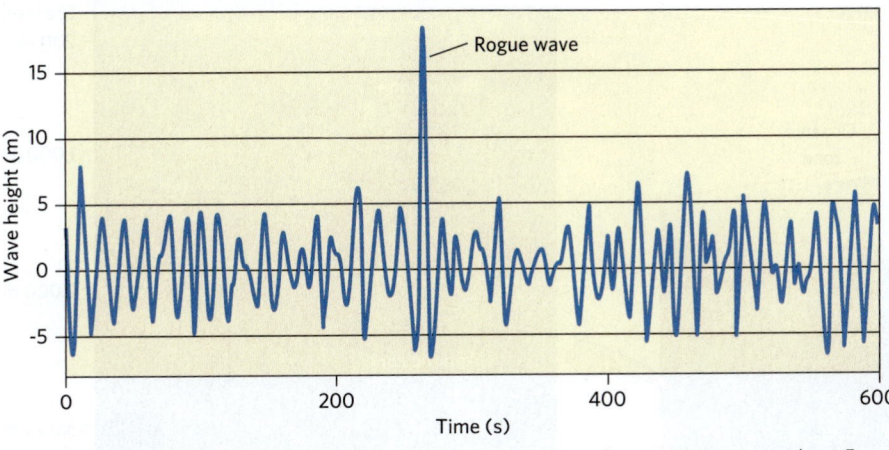

(a) A recording of waves in the North Sea over 10 minutes. Note that most waves are about 5 m (16 ft) high. The rogue wave is 18 m (60 ft) high.

(b) A rogue wave washing over the deck of a large ship.

recorded almost 500 encounters with rogue waves—some of which were three to five times higher than other waves present at the time (Fig. 15.27a). The decks of large ships have been swamped by rogue waves in the open ocean (Fig. 15.27b). The largest wave known to have encountered a ship reached a height of 34 m (112 ft). Mysterious losses of ships in the open ocean may be a consequence of rogue waves. If a rogue wave reaches the shore, it can wash unsuspecting bystanders off piers or beaches.

Take-home message . . .

Friction between wind and the sea surface produces waves. Within a wave, water moves approximately in a circle; the amount of motion decreases with depth. Constructive interference and other factors can generate huge waves.

Quick Question
What is a rogue wave?

15.6 Life in the Sea

Zones of the Ocean and Their Effect on Life

The character of the ocean changes with distance from the shore. When discussing these differences, oceanographers divide the ocean into three zones (Fig. 15.28a; Earth Science at a Glance, pp. 478–479), starting from the shore and moving outward:

- The **littoral zone** consists of the intertidal zone (the area that is above water during low tide and under water during high tide) and the adjacent areas of coastal land that become submerged or drenched in

spray during unusually high tides or fierce storms. Specialized organisms have evolved to live in these conditions.

- The **neritic zone** is the relatively nearshore, shallow realm of the ocean. Water in the neritic zone receives inputs of chemicals and sediments from the land, and within the neritic zone, sunlight penetrates to the seafloor. Light undergoes *scattering* (absorption by atoms, followed immediately by reemission in a different direction) when it passes through water. Most of the scattered light heads upward or sideways, so the intensity of light decreases progressively with depth (Box 15.4). Even in clear water, then, only 1% of sunlight penetrates to a depth of 200 m (660 ft). Oceanographers consider this depth to be the maximum base of the neritic zone. By this definition, most of the continental shelf lies within the neritic zone.

- The **oceanic zone** consists of the open-water portion of the ocean that does not interact with the coast. The water layer in the oceanic zone is deep enough to absorb all sunlight, so the seafloor beneath this zone remains in perpetual darkness. The oceanic zone includes regions above abyssal plains, mid-ocean ridges, and deep-sea trenches. Oceanographers use different names for different depths in the oceanic zone. Specifically, *bathyal depths* lie between 1,000 m (3,300 ft) and 4,000 m (13,000 ft), whereas *abyssal depths* lie between 4,000 m and 6,000 m (20,000 ft).

Oceanographers also describe depth zones in the ocean based on position relative to the seafloor. In this context, the **pelagic zone** includes all water above the seafloor, whereas the **benthic zone** includes the seafloor itself, as well as the water immediately above the seafloor and the sediment immediately below.

FIGURE 15.28 Zones of the ocean.

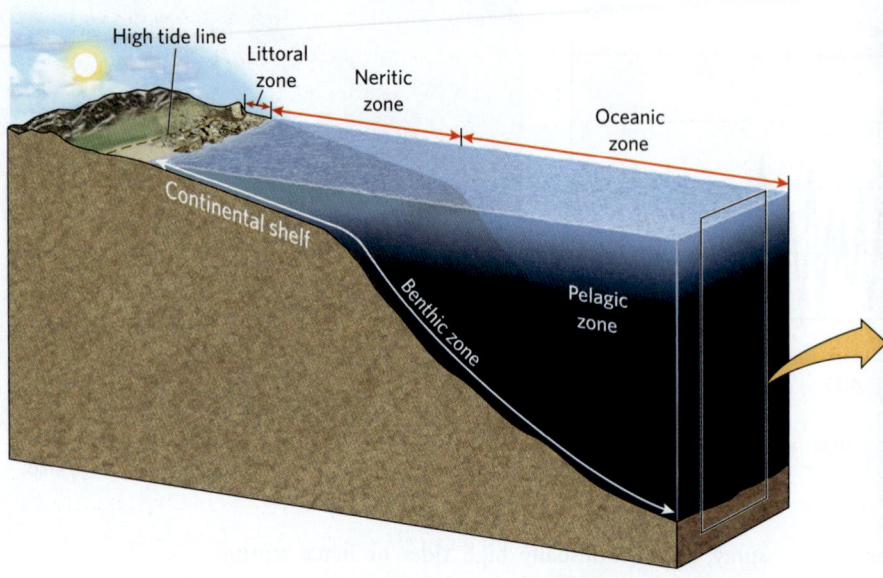

(a) Oceanographers distinguish among littoral, neritic, and oceanic zones, based on distance from shore, and between pelagic and benthic zones, based on depth.

(b) Marine biologists distinguish between photic and aphotic zones, based on light penetration.

Environmental Controls on Living in the Sea

The ocean hosts an incredibly diverse assemblage of life because not all of its water has exactly the same properties. Temperature, salinity, and light penetration all vary from place to place, and different assemblages of organisms have adapted to different conditions. Light penetration serves a particularly important role in determining what types of living organisms can thrive because the availability of light determines whether or not photosynthesis can take place. Marine biologists, therefore, distinguish the **photic zone** (from the Greek *phot*, meaning light), in which sunlight penetrates water, from the **aphotic zone**, in which sunlight does not **(Fig. 15.28b)**. The boundary between the two zones lies at the depth where light has less than 1% of its intensity at the surface. This depth varies depending on the clarity of the water. In very clear water, enough light penetrates down to about 200 m (660 ft) that photosynthesis can take place. A tiny bit of light may reach a depth of 600–1,000 m (2,000–3,300 ft) at noon in the clearest water, so this depth represents the maximum base of the photic zone. If water contains suspended particles, the base of the photic zone may lie at depths much shallower than 200 m.

By the definitions given earlier, all of the neritic zone lies within the photic zone. The photic zone also includes the shallower portions of the oceanic zone. Notably, about 90% of marine life lives in the upper 70 m (230 ft) of the photic zone, where light intensity can support photosynthesis—oceanographers sometimes refer to this upper, brightly lit part of the photic zone as the *euphotic zone*. Organisms that navigate below the euphotic zone live in a dimly lit environment, and those that live in the aphotic zone survive in perpetual darkness.

The Categories of Ocean Life

On land, some organisms live on or are rooted in the ground, others fly through the air, and some survive underground. The sea hosts such variation as well. Some organisms float on the surface of the sea, some remain suspended in the water at various depths, some swim actively through the water, some root in or crawl about on the seafloor, and some exist in sediments beneath the seafloor. Marine biologists distinguish among three distinct categories of marine organisms to emphasize these differences.

PLANKTON. All organisms that drift or are suspended in the water column, but cannot move against a moderate current, are plankton (from the Greek *planktos*, meaning drifting) **(Fig. 15.29a–c)**. Plankton include archaea, bacteria, single-celled microbes, multicellular protists (protozoans and most algae), and some larger multicellular organisms. Larger plankton include tiny shrimp known as krill, some species of **seaweed** (non-microscopic multicellular algae), and jellyfish **(Fig. 15.29d)**. Taken together, plankton account for most of the ocean's living organic matter, or *biomass*.

Most plankton species have no means of moving and consist of immobile living masses that simply float in or on the water. But some can move, though not quickly

▶ **A Deeper Look**

The color of ocean water

If you've looked at photos of the ocean, had the opportunity to walk along its shores, or sailed across it, you may have noticed that the ocean's color can vary greatly. It can be deep blue, turquoise, dark green, or even gray (Fig. Bx15.4a–c). And when churned to a froth by wind or waves, it becomes white. What gives water its color, and why does its color vary so much?

Clean water in a clear glass looks colorless—you can see through it almost as easily as you can see through the glass itself. That's because you're looking through only about 8 cm (3 in) of water. When you look at the ocean, you are looking at water that can be meters to kilometers deep. Under a sunny sky, this water looks blue, and the deeper the water, the bluer it looks. You see this color for two reasons. First, water molecules absorb red, green, and yellow light but scatter blue light. Second, when you look at the sea beneath a blue sky, you are seeing the reflection of the blue sky by the water surface. On an overcast day, the water might look greenish gray.

Where water is shallow enough for substantial light to penetrate to the seafloor, the color of the water depends in part on the color of the seafloor: water over white sand will appear lighter than water over dark seaweed. The presence of tiny suspended particles in water also affects its color. For example, the presence of abundant phytoplankton, which contain green chlorophyll, can make the water greenish; the presence of abundant plankton with white calcite shells tends to make water turquoise, because the shells reflect light and brighten the water; and the presence of suspended clay can make water a brownish color.

Because water scatters different wavelengths of light by different amounts, the colors that you see reflected from objects in the water change with depth. For example, the color of a fish as seen by a scuba diver at a depth of 20 m (60 ft) isn't the same as the color of the same fish at the surface.

FIGURE Bx15.4 Colors of the ocean.

(a) On a sunny day, water off a Caribbean island looks blue. The darker-colored water is deeper.

(b) On an overcast day, the ocean water off Cuba looks gray.

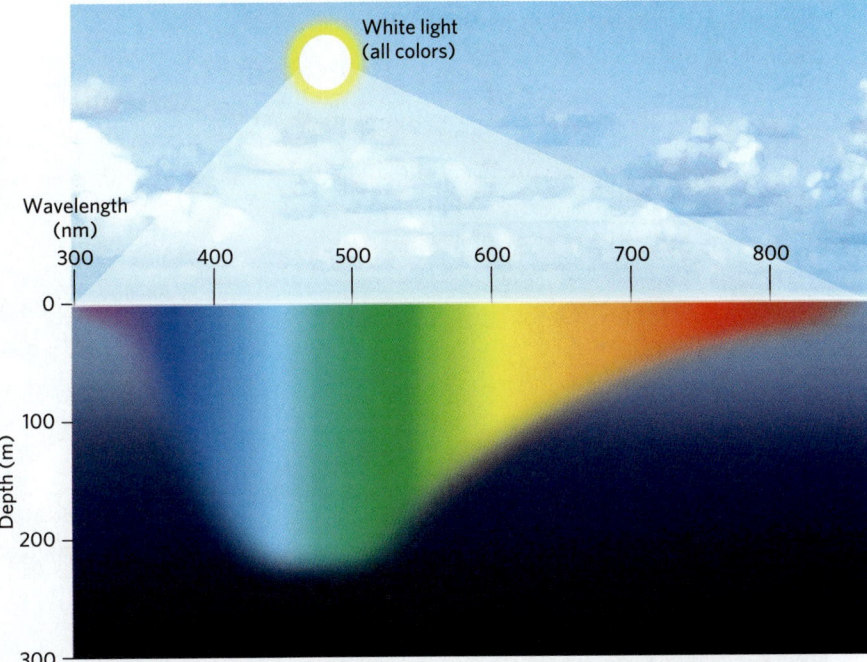

(c) Water absorbs most other colors of light more efficiently than it absorbs blue light. Blue light, therefore, penetrates deeper. In clear water, light can penetrate to a depth of about 200 m (650 ft). At noon in extremely clear water, it may penetrate deeper.

The Water of the Sea

Our planet's oceans host a variety of realms that differ in temperature, salinity, and brightness. The water stays in motion constantly, at a range of rates.

Water masses and currents

Oceanographers divide the ocean, at the scale of a whole ocean basin, into different layers, or water masses. In the Atlantic Ocean, for example, the Antarctic bottom water, a cold and saline mass, spreads northward from the southern continent. North Atlantic deep water sinks to depth in the polar regions of the north and spreads southward to form a layer above the Antarctic bottom water. North Atlantic intermediate water, in turn, overlies the North Atlantic deep water. Relatively warm water forms the top layer. The near-surface portion of the top layer circulates in a huge gyre whose western margin, the Gulf Stream, flows northeast along the coast of the United States.

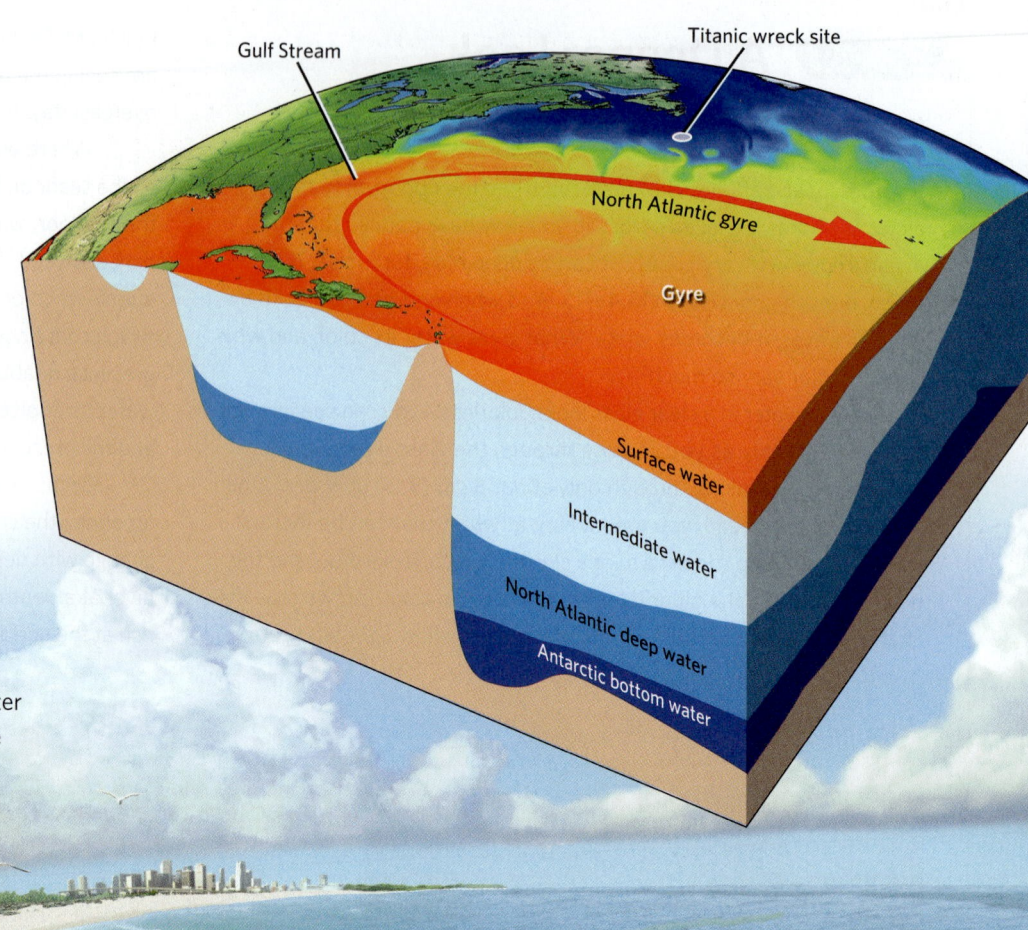

Gulf Stream

Titanic wreck site

North Atlantic gyre

Gyre

Surface water

Intermediate water

North Atlantic deep water

Antarctic bottom water

Seaweed

Coral reef

Benthos

The nearshore neritic zone

The neritic zone of the ocean, the portion where the seafloor receives at least some light, encompasses most of the continental shelf. In its nearshore portion, it hosts a diversity of organisms, including coral reefs, seaweed, and a variety of nekton, plankton, and benthos.

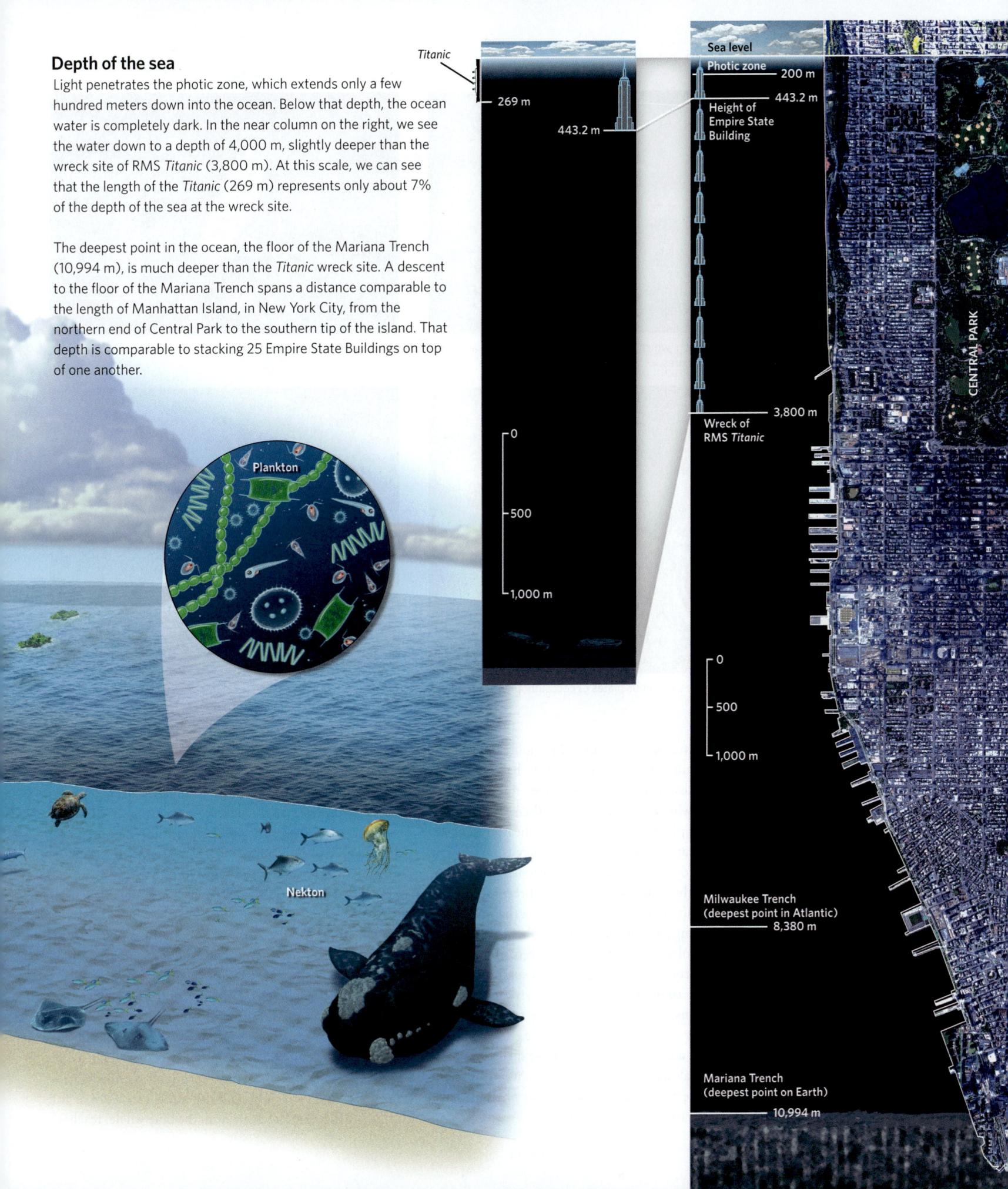

Depth of the sea

Light penetrates the photic zone, which extends only a few hundred meters down into the ocean. Below that depth, the ocean water is completely dark. In the near column on the right, we see the water down to a depth of 4,000 m, slightly deeper than the wreck site of RMS *Titanic* (3,800 m). At this scale, we can see that the length of the *Titanic* (269 m) represents only about 7% of the depth of the sea at the wreck site.

The deepest point in the ocean, the floor of the Mariana Trench (10,994 m), is much deeper than the *Titanic* wreck site. A descent to the floor of the Mariana Trench spans a distance comparable to the length of Manhattan Island, in New York City, from the northern end of Central Park to the southern tip of the island. That depth is comparable to stacking 25 Empire State Buildings on top of one another.

Titanic

269 m

443.2 m

0

500

1,000 m

Plankton

Nekton

Sea level

Photic zone — 200 m

443.2 m

Height of Empire State Building

3,800 m

Wreck of RMS *Titanic*

CENTRAL PARK

0

500

1,000 m

Milwaukee Trench (deepest point in Atlantic) — 8,380 m

Mariana Trench (deepest point on Earth) — 10,994 m

FIGURE 15.29 Examples of marine plankton.

(a) Phytoplankton, which contain chlorophyll, include diatoms and coccolithophores.

(b) Zooplankton include tiny shrimp, forams, and radiolarians.

(c) The intricate calcareous plates of coccolithophore shells look like wheels.

(d) Jellyfish drift in the water column. Some can move, but not quickly enough to swim against a current.

enough to overcome currents. Examples of moving plankton include *dinoflagellates*, which have whip-like tails; krill, whose tiny legs propel them through water; and jellyfish, which move by making their whole bodies pulsate. Many planktonic organisms are heavier than water, so if the water were perfectly still, they would sink to the bottom. Fortunately for them, currents or local turbulence keep them suspended, just as a breeze keeps dust suspended in air. Some plankton maintain their buoyancy by incorporating gas into their bodies.

Marine biologists distinguish between **phytoplankton**, which carry out photosynthesis to sustain their life processes, and **zooplankton**, which survive by ingesting other organisms, living or dead. Many types of phytoplankton have tiny shells. For example, *diatoms* make shells of silica, and *coccolithophores* make shells composed of intricate calcareous (containing calcium carbonate) plates. Because they are photosynthetic, phytoplankton are considered to be a type of algae.

NEKTON. Organisms that actively swim in the open water of the ocean and can move against a current are known as **nekton** (from the Greek *nektos*, meaning swimming). Some nekton migrate vast distances across the pelagic zone, while others move about a local coastal area of only a few square meters, such as a patch of reef. Most nekton can survive only within limited ranges of such physical conditions as temperature, pressure, and salinity, so they don't stray beyond a relatively small region. Nekton survival also depends on the availability of food, so they often migrate with their food supplies.

Nekton include a great variety of organisms whose sizes range from nearly microscopic to gigantic. Examples of nekton include invertebrates such as squid and octopuses **(Fig. 15.30a)**; swimming reptiles such as sea turtles and sea snakes; marine mammals such as whales, porpoises, and seals **(Fig. 15.30b)**; and all varieties of fish **(Fig. 15.30c)**. The largest nekton, the blue whale, reaches a length of 32 m (110 ft), making it the largest animal ever to inhabit our planet.

The largest category of nekton, fish, includes two principal groups. *Bony fish* have a skeleton with a skull, jaws, spine, and many bones protruding from the spine. Examples include familiar species such as tuna, swordfish, cod, and salmon (Fig. 15.31a). Most bony fish have gas-filled swim bladders that allow them to regulate their buoyancy and remain at a given depth without swimming. Bony fish lay eggs that are fertilized and hatch outside the mother's body. The *non-bony fish*, such as sharks and rays, have cartilaginous skeletons, meaning that their skeletons consist of strong but flexible connective tissue (Fig. 15.31b). Non-bony fish do not have swim bladders, so they must actively swim in order to maintain their depth.

BENTHOS. Organisms that live on, just above, or just below the seafloor are known as **benthos** (from the Greek *benthos*, meaning bottom). The seafloor, as we have seen, ranges from fairly shallow, in the littoral and neritic zones, to moderate depths (between the depth of the continental shelf and the depth of the abyssal plain), to great depths, on abyssal plains or the floor of deep-sea trenches. The nature of benthos varies radically among these depths. Coastal benthic communities include a variety of animals such as corals, sponges, and bryozoans, which build spectacular reefs (Fig. 15.32a). Even in the great depths of the oceans, where no light penetrates, organisms such as snails, crabs, starfish, and mollusks creep through the mud (Fig. 15.32b). Benthos also include a great variety of seaweed types that anchor to the seafloor. The largest, giant kelp, grows stalks as long as 200 m (600 ft) (Fig. 15.32c). Some seaweed can survive in the intertidal zone (Fig. 15.32d).

Food Webs in the Oceans

Who eats whom in the ocean? The answer to this question defines the ocean's **food web**, the group of

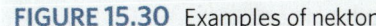

FIGURE 15.30 Examples of nekton.

(a) A squid, a type of marine invertebrate, propels itself by ejecting jets of water.

(b) A whale, a type of marine mammal, can grow to become larger than the largest dinosaur.

(c) A school of fish swimming above a reef.

(a) The bluefin tuna is a type of bony fish.

(b) A shark is an example of a non-bony fish.

FIGURE 15.31 Two groups of fish.

FIGURE 15.32 Examples of benthos.

(a) In the neritic zone, benthos include many varieties of coral and other reef-building organisms.

10 cm

(b) Some organisms, such as this sea cucumber, move through the clay in the aphotic zone, leaving tracks across the ocean floor.

(c) Kelp, a type of seaweed, can grow to be hundreds of meters long.

(d) Intertidal seaweed can survive exposure to air.

organisms associated by feeding relationships. In the food web, organisms that survive by extracting nutrients from nonliving energy sources, and thus bring energy into the web, are known as **primary producers**, or *autotrophs*. In the euphotic zone, phytoplankton serve as the primary producers—these organisms obtain their energy from sunlight. Undersea exploration in the past few decades has revealed complex benthic food webs surrounding deep-sea hydrothermal vents (black smokers). In these aphotic food webs, the primary producers are microbes that use the chemical bonds in sulfide minerals as their energy source. **Consumers**, or *heterotrophs*, in the food web include a vast variety of organisms, ranging from tiny zooplankton to fearsome killer whales.

Nutrients and Dissolved Oxygen

In order to survive, primary producers ingest **nutrients**, elements that an organism uses to construct chemicals and carry out metabolic reactions. Nutrients of particular importance to marine organisms include phosphate (PO_4^{3-}), nitrate (NO^{3-}), and iron. The presence or absence of

nutrients determines the biomass that a volume of water can support at a given latitude.

Where do nutrients come from? Organisms accumulate nutrients during their lifetimes, and when they die and sink, they carry those nutrients with them to the seafloor, where decomposition releases them back into the water. So, unless new nutrients flow to the sea from rivers or coastal runoff, or rise from the depths back to the ocean surface by upwelling, the concentration of nutrients in the photic zone may become too low for life to thrive. Therefore, if we map the distribution of photosynthetic organisms by means of satellite detection of chlorophyll concentrations at the ocean surface **(Fig. 15.33)**, we see that photosynthetic organisms increase in abundance near the shore, where nutrient sources enter the sea from land, and in offshore localities where upwelling brings nutrients from the aphotic zone back to the photic zone.

Consumers such as zooplankton and marine animals cannot survive without molecular oxygen, which they use for respiration. On land, organisms inhale oxygen in gaseous form from the air. Marine reptiles, birds, and

Did you ever wonder . . .

how fish can breathe?

mammals must surface to inhale air, but have adaptations that allow them to hold their breath and stay underwater for relatively long periods—in fact, a sperm whale can stay submerged for an amazing 90 minutes, and during this time it can descend to depths of as much as 3 km (9,800 ft). Fish and marine invertebrates never need to surface, although they do require oxygen. They survive by extracting *dissolved oxygen* from water passing through their gills.

Where does the dissolved oxygen in ocean water come from? Some enters the ocean at the water surface and then diffuses downward, and some comes from algae as a by-product of photosynthesis. Animals remove some dissolved oxygen from the water by breathing, but most removal takes place as a result of decomposition.

Because decomposition reactions remove oxygen from seawater, an *algal bloom* (a rapid growth of lots of algae) can remove so much oxygen that it makes the water *anoxic* (oxygen free). This may seem counterintuitive, but decomposition of the algae, along with respiration by zooplankton feeding on the algae, removes more oxygen than the algae can produce by photosynthesis. Such anoxic conditions yield a **dead zone** in the water,

where fish and shellfish die from lack of oxygen. A large dead zone has developed at the mouth of the Mississippi, where the river dumps vast amounts of nutrients into the Gulf of Mexico. These nutrients, which come from runoff from Midwestern farm fields, feed algal blooms that strip the oxygen from the water.

Clearly, interactions between the sea and the land affect oceans significantly. In the next chapter, we turn our focus to the coast, the boundary where these interactions take place.

Take-home message . . .

Most marine life lives in the euphotic zone, where sunlight intensity is sufficient to support photosynthesis. Marine organisms can be categorized as plankton, nekton, or benthos, depending on where in the ocean they live and what type of movement they are capable of. Marine life plays an active role in cycling nutrients through the Earth System.

Quick Question

Does classification of an organism as plankton depend on its size?

FIGURE 15.33 The colors on this map represent concentrations of chlorophyll in near-surface seawater. Chlorophyll serves as an indicator of primary production in the ocean's food web. Green represents the highest concentrations of chlorophyll.

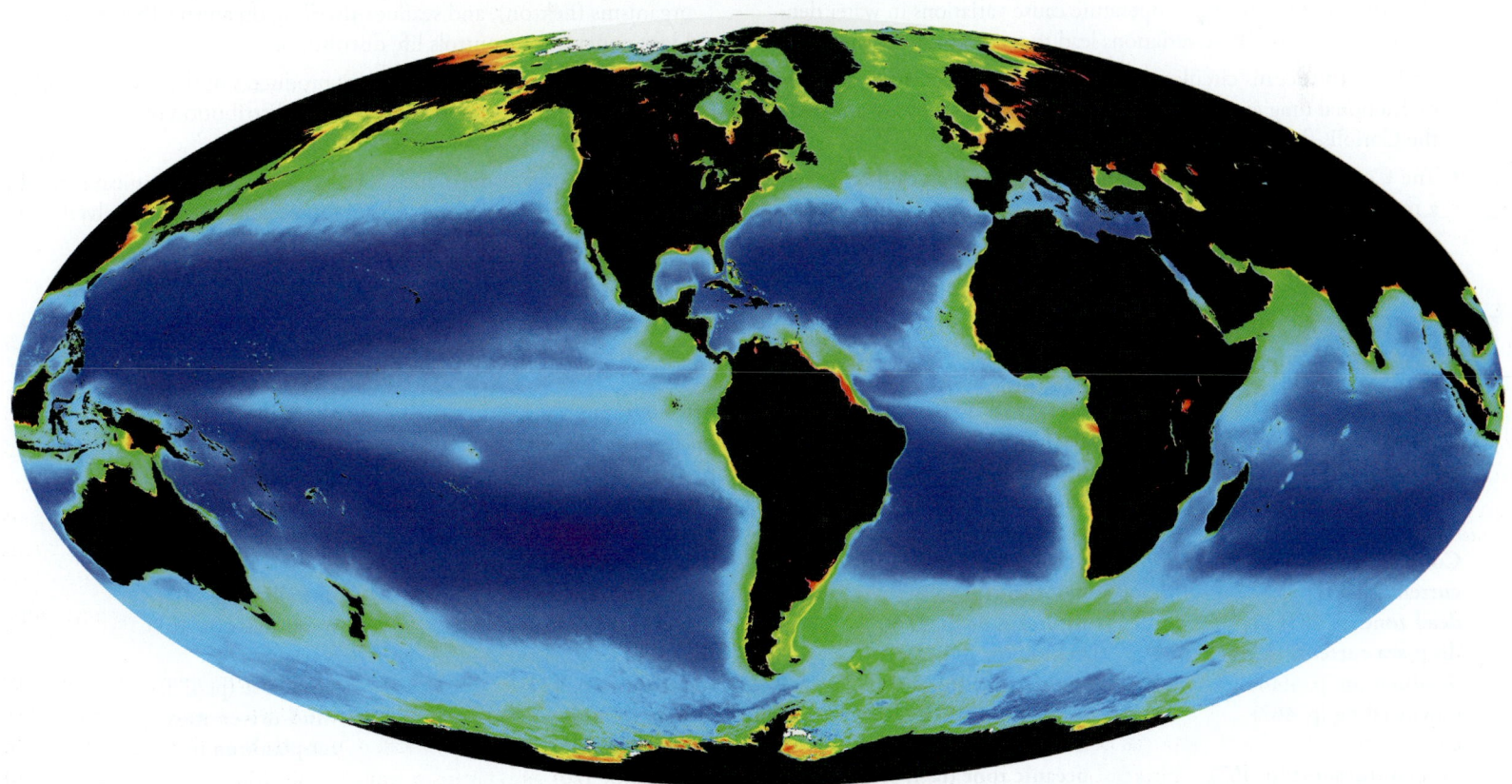

15 CHAPTER REVIEW

Chapter Summary

- Mapping of the oceans began in the 15th and 16th centuries, but maps could not be accurate until longitude could be measured. There are several distinct ocean basins, but water flows among them.

- Mean sea level represents the elevation of the boundary between ocean water and air averaged over a year. It changes over geologic time by as much as a few hundred meters. Changes in sea level reflect changes in the volume of continental glaciers and mid-ocean ridges as well as vertical land movement.

- Ocean water originally came from water vapor that bubbled out of volcanoes.

- Most salt in the sea consists of NaCl, but seawater contains other salts as well. The salinity of seawater averages about 3.5%, but varies with location and depth.

- Most of the salt in seawater is carried to the sea by rivers, although some comes from hydrothermal vents. Evaporation of seawater removes water and leaves behind salt.

- The ocean surface is warmer at the equator than at the poles. Because warm water is less dense than cold water, it stays above warmer water, so at low latitudes, a thermocline separates warm surface water from cool deep water.

- Variations in salinity and temperature cause variations in water density in the oceans. These variations lead to stratification of the oceans.

- Water in the oceans circulates in currents. Surface currents are driven by frictional drag applied by the wind, but their paths are affected by the Coriolis force and the pressure-gradient force.

- The Coriolis effect causes Ekman transport, which builds a mound of slightly elevated water in the middle of an ocean. Geostrophic balance develops between the resulting outward-directed pressure-gradient force and the inward-directed Coriolis force. As a result, large gyres (geostrophic currents) circle these mounds.

- Smaller currents are, effectively, eddies that form where a larger current shears against nonflowing water or against land.

- Upwelling and downwelling due to surface-water flow relative to the coast, or to thermohaline circulation, contribute to the formation of deep currents.

- Waves are initiated by frictional drag where the wind shears across the surface of the ocean. Once waves become large enough, gravity causes the up-and-down movement of the water surface.

- Ideally, water particles move in a circle as a wave passes. The circle gets smaller with depth, so that below the wave base, there is no water movement. In reality, the upper layer of water undergoes wave drift.

- Oceanographers divide the ocean into littoral, neritic, and oceanic zones based on proximity to shore. They also distinguish among pelagic and benthic zones, based on proximity to the seafloor, and photic and aphotic zones, based on light penetration.

- Life in the sea includes floating organisms (plankton), swimming organisms (nekton), and seafloor-dwelling organisms (benthos). Light penetration controls life distribution.

- Phytoplankton serve as the primary producers in the oceanic food web and account for most biomass. The distribution of nutrients affects the abundance of phytoplankton.

- Consumers, organisms that survive by eating other organisms, need dissolved oxygen to survive. Algal blooms can remove dissolved oxygen and produce a dead zone in the ocean.

Key Terms

aphotic zone (p. 476)
benthic zone (p. 475)
benthos (p. 481)
consumers (p. 482)
Coriolis effect (p. 465)
Coriolis force (p. 466)
current (p. 461)
dead zone (p. 483)
deep-sea currents (p. 461)
desalination (p. 460)
downwelling (p. 467)
eddy (p. 462)
Ekman transport (p. 467)

estuary (p. 457)
fetch (p. 474)
food web (p. 481)
frictional drag (p. 463)
geostrophic current (p. 467)
gyre (p. 462)
halocline (p. 457)
littoral zone (p. 475)
nekton (p. 480)
neritic zone (p. 475)
nutrient (p. 482)
ocean (p. 453)
oceanic zone (p. 475)

oceanography (p. 454)
pelagic zone (p. 475)
photic zone (p. 476)
phytoplankton (p. 480)
plankton (p. 476)
pressure-gradient force (p. 467)
primary producer (p. 482)
pycnocline (p. 461)
ripple (p. 472)
rogue wave (p. 474)
salinity (p. 457)
sea level (p. 455)
seaweed (p. 476)

shoreline (p. 454)
surface current (p. 461)
swell (p. 474)
thermocline (p. 459)
thermohaline circulation (p. 471)
upwelling (p. 467)
water mass (p. 461)
wave (p. 472)
wave base (p. 473)
wind-driven wave (p. 472)
zooplankton (p. 480)

Letters in parentheses correspond to the chapter's learning objectives.

1. How much of the Earth's surface do the oceans cover? When were the oceans mapped? **(A)**

2. What do we mean when we refer to sea level, and how has sea level changed over Earth history? **(B)**

3. Where does the salt in the ocean come from? How does ocean salinity vary with location, and why? Which line in the graph represents the tropics? **(B, C)**

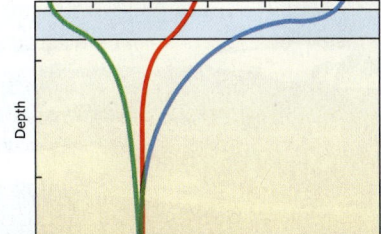

Salinity (%)

Depth

4. How does the temperature of ocean water vary with latitude and depth, and why? Why does a thermocline develop at the equator, but not at the poles? **(C)**

5. What factors determine ocean-water density? Is the ocean homogeneous or stratified? Explain. **(E)**

6. What forces play a role in establishing surface currents in the ocean? **(D)**

7. Why do gyres and eddies form? Why does garbage accumulate in the center of the Pacific Ocean? **(D)**

8. What causes upwelling and downwelling to take place along a coast? **(E)**

9. Explain the thermohaline circulation and its role in producing deep-sea currents. **(E)**

10. What forces contribute to the formation of ocean waves? Describe the motion of water molecules in a wave. **(F)**

11. What's the difference between a ripple and a swell? How large can waves become? What is a rogue wave? **(F)**

12. Explain the differences between the littoral, neritic, and oceanic zones. What is the relationship between the neritic and the photic zones? **(G)**

13. Explain the differences among plankton, nekton, and benthos. Which is depicted in the figure? **(G)**

14. Describe the basic components of the oceanic food web and how light penetration affects it. **(G)**

15. Columbus didn't actually reach continental North America when he sailed west in 1492. Rather, he reached the Bahamas, to the southeast of Florida, about 70 days after leaving Spain, after traveling about 7,000 km (4,300 mi). Why did he get to the Bahamas? What was his average speed? How long might the journey have taken if he had just drifted, without sails? **(D)**

16. Compare the water of the Persian Gulf, a confined sea at low latitude, with the water of the northernmost Pacific Ocean, an open ocean at high latitude. **(C)**

17. If wind patterns were to change so that upwelling no longer occurred along the west coast of the Americas, what would happen to the fish population? **(E)**

ANOTHER VIEW At the end of the day, the local fishing fleet has anchored in Portsmouth's harbor, along the south coast of England. The life of the sea has always been an important food source for humanity.

16 MARINE GEOLOGY
The Study of Ocean Basins and Coasts

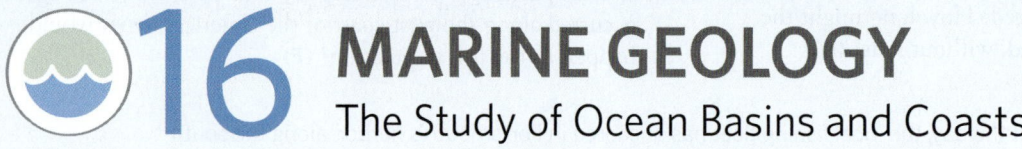

By the end of the chapter you should be able to . . .

A. distinguish among the various bathymetric provinces of ocean basins, and explain how they form in the context of plate tectonics theory.

B. discuss ocean tides and their causes, and interpret why some regions have large tidal ranges while others do not.

C. describe the processes that shift sand on beaches and produce observed beach profiles.

D. develop a model explaining the evolution of rocky coasts.

E. contrast estuaries with fjords and wetlands with reefs, and explain their origins.

F. explain the variation in the character of coastlines, and the role of relative sea-level change in coastline evolution.

G. describe how wave erosion, and human responses, can modify coasts.

16.1 Introduction

A thousand kilometers from the nearest land, two scientists and a pilot wriggle through the entry hatch of the research submersible *Alvin*, ready for a cruise to the floor of the ocean and, hopefully, back. *Alvin* consists of a super-strong metal sphere embedded in a cigar-shaped tube **(Fig. 16.1)**. The sphere protects its crew from the immense water pressures of the deep ocean, and the tube holds motors and oxygen tanks. A cruise begins when, after sealing the hatch, the pilot releases cables tethering *Alvin* to its mother ship, and the submersible sinks like a stone. Most of this journey takes place in utter darkness, for the photic zone extends down only a couple of hundred meters (see Section 15.6). On reaching the seafloor, at a depth of 4.5 km (2.8 mi), the cramped explorers turn on spotlights and gaze at a stark vista of loose sediment, black rock, and an occasional sea creature. For the next 5 hours they take photographs and use a robotic arm to collect samples. When finished, they release some ballast, and *Alvin* rises like a bubble, reaching the surface about 2 hours later.

In the 21st century, human-piloted submersibles like *Alvin* are but one of many tools employed to collect data about the 70.8% of the Earth's solid surface that lies hidden beneath seawater. Orbiting satellites can characterize regional-scale **bathymetry** (the shape of the seafloor) in general **(Fig. 16.2)**, and shipborne sonar

<< This beach, along the coast of Martha's Vineyard in Massachusetts, is a place of beauty and peace on a sunny afternoon. During a storm, breakers wash over the beach and erode into the adjoining cliff, emphasizing that such coastal landscapes can change on a human time scale.

FIGURE 16.2 Bathymetry of the seafloor.

(a) Regional bathymetry of the North Atlantic. The data used to make this Google Earth image came from satellite measurements of variations in the strength of the Earth's gravitational pull.

(b) Detailed bathymetry can be produced by sonar. This oblique view shows the Turnif Seamount, northwest of Hawaii. The colors indicate depth. Red areas are shallower; blue areas are deeper.

FIGURE 16.1 *Alvin* is a workhorse submersible that researchers use to explore the seafloor.

(a) *Alvin* is lowered into the sea from its mother ship, RV *Atlantis*.

(b) The small porthole provides the only view out.

487

FIGURE 16.3 Seismic-reflection profiling helps characterize the layering of sediments and rocks beneath the seafloor. Seismic signals are produced by an air gun and are recorded by hydrophones. The resulting profile reveals layering.

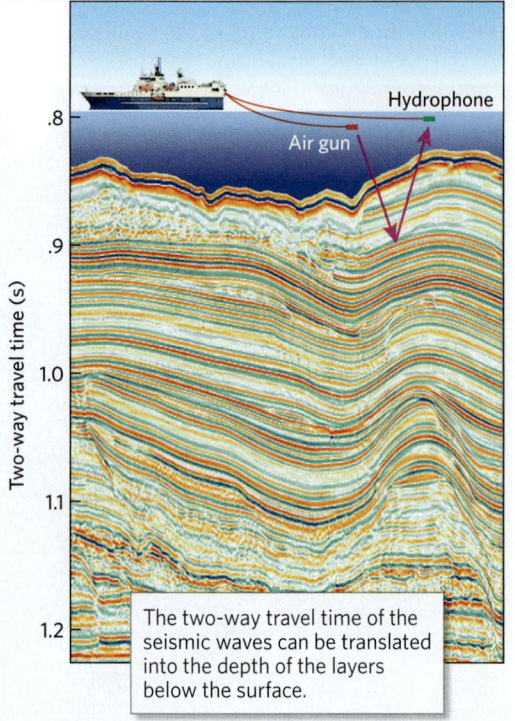

Hydrophone

Air gun

Two-way travel time (s)

.8
.9
1.0
1.1
1.2

The two-way travel time of the seismic waves can be translated into the depth of the layers below the surface.

can define local-scale bathymetry in detail. Instruments towed by ships, and robotic submersibles that can cruise at great depths without risking lives, can study physical characteristics of the seafloor, such as variations in its temperature or magnetic field strength. Shipborne tools can also yield *seismic-reflection profiles*, cross-sectional images that reveal the configuration of layering under the seafloor **(Fig. 16.3)**. By dredging (pulling a chain net along the seafloor) and by coring (plunging hollow tubes into the seafloor), researchers can collect samples of seafloor rocks and sediment. And by using *drilling ships* that are capable of boring holes, researchers have recovered oceanic crust from depths of up to 4 km (2.5 mi) below the seafloor **(Fig. 16.4)**. Geologists use many of the same tools to study the **coast** (the shore and the adjacent areas of land and sea). But on coasts, they can also make direct field observations. The results of all this research—from land, sea, and sky—provide the knowledge base of **marine geology**, the study of the ocean's floor and margins.

In this chapter, we explore the results of modern marine geology research. We first survey the fundamental features of ocean basins and discuss how these features form. Then we focus on the landforms and habitats that develop along the coast, where over 60% of the global human population lives today. We'll see how coasts interact with tides and waves, and how they evolve over time in response to geologic phenomena and human activities.

FIGURE 16.4 Deep-sea drilling and coring bring up sediment and rock from below the seafloor.

(a) The *JOIDES Resolution* drills holes into the seafloor.

(b) Cores can be preserved for future analysis.

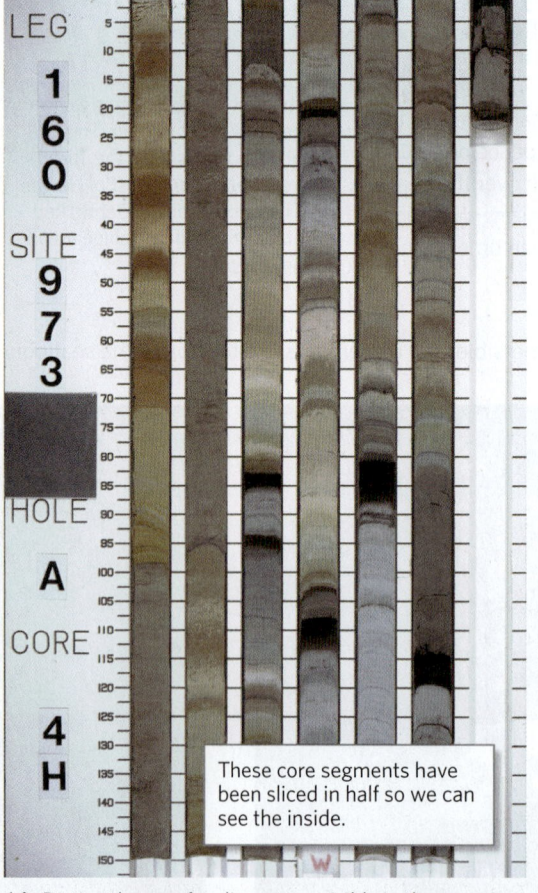

LEG 160 SITE 973 HOLE A CORE 4 H

These core segments have been sliced in half so we can see the inside.

(c) Distinct layers of sediment are visible in these core segments. Numbers on the scale indicate length (in cm).

16.2 What Controls the Depth of the Sea?

If the Earth's crust were at the same elevation everywhere, the water currently held in the oceans would cover the surface of the Earth uniformly to a depth of about 2.5 km (1.5 mi). But the elevation of the Earth's surface is not uniform; rather, it consists of higher areas (the continents) and lower areas (the **ocean basins**) that differ in elevation by an average of 4.5 km (2.8 mi) **(Fig. 16.5a)**. Due to gravity, water flows downslope, so it drains from continents into

FIGURE 16.5 Contrasts between continental lithosphere and oceanic lithosphere.

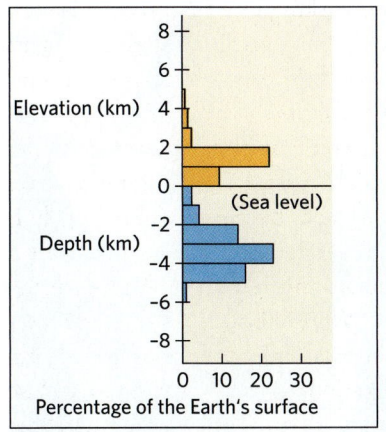

(a) A graph showing the percentage of the Earth's surface at different elevations above sea level or depths below the sea.

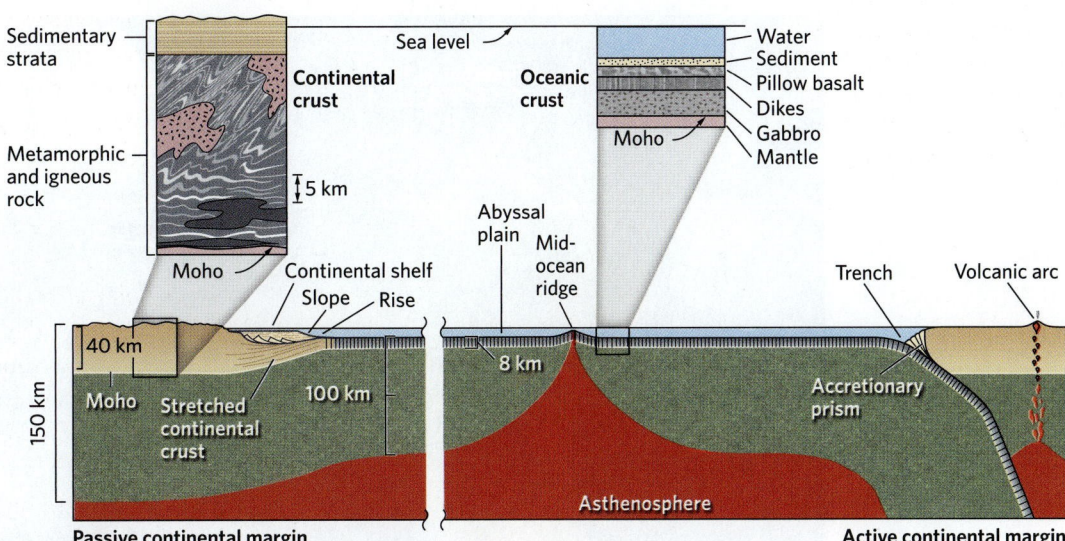

(b) The crustal portion of continental lithosphere differs markedly from that of oceanic lithosphere.

(c) A simple model showing that the surface of a thick pine block sits higher than the surface of a thin, dense oak block when both are placed in water.

the ocean basins to fill the oceans, leaving the surface of continents as dry land.

Why do distinct ocean basins exist? Recall from Chapter 2 that oceanic lithosphere and continental lithosphere differ markedly in terms of their composition and thickness **(Fig. 16.5b)**. Oceanic lithosphere has a maximum thickness of 100 km (60 mi) and includes a 7- to 10-km (4- to 6-mi)-thick crust of relatively dense rock, whereas continental lithosphere reaches a thickness of about 150 km (90 mi) and includes a 25- to 70-km (15- to 45-mi)-thick crust of relatively less dense rock. Because of these differences, the surface of oceanic lithosphere sits deeper than the surface of the continental lithosphere—the low areas, underlain by oceanic lithosphere, are the ocean basins. To picture this contrast, imagine two blocks of wood **(Fig. 16.5c)**: a thicker one composed of less dense pine, which represents a continent, and a thinner one composed of denser oak, which represents the ocean floor. If you place both blocks in a basin of water, the surface of the oak block sits lower than the surface of the pine block, once both blocks have attained equilibrium (see Box 7.3). In this model, water flows out of the way so the blocks can attain their equilibrium positions. In the case of the Earth, the plastic asthenosphere flows out of the way to let the different types of lithosphere attain their equilibrium positions.

Have you ever wondered what the ocean floor would look like if all the water evaporated? Studies of bathymetry indicate that the ocean floor can be divided into *bathymetric provinces*, distinguished from one another by their water depth **(Fig. 16.6)**. We'll now revisit these provinces, first mentioned in Chapter 2, to characterize their surfaces—the landscape of the seafloor—in a bit more detail.

The Shallower Realms: Continental Shelves

Imagine you're cruising in a submersible, heading east from the coast of North America into the North Atlantic. Moving out from the shore for a distance of 150–500 km (100–300 mi), you will be above the **continental shelf**, a relatively shallow portion of the ocean that fringes the continent. Over the continental shelf, water depth does not exceed about 200 m (650 ft), and across the width of the shelf, the ocean floor slopes seaward at an angle of only 0.3° **(Fig. 16.7a)**. At its seaward edge, the continental shelf terminates at the *continental slope*, a surface that descends at an angle of about 2° from a depth of 200 m to nearly 4 km (2.5 mi). From about 4 km down to about 4.5 km (2.8 mi), the slope angle decreases, defining a region called the *continental rise*. At a depth of 4.5 km, you find yourself above a vast, nearly horizontal surface known as the **abyssal plain**.

Broad continental shelves, like that of eastern North America, form along *passive continental margins*. These margins are not plate boundaries, and they lack seismicity (see Chapter 2). Passive margins originate after rifting succeeds in breaking a continent in two, so that a new mid-ocean ridge forms and seafloor spreading begins. At this time, the stretched continental lithosphere of what

FIGURE 16.6 Bathymetric provinces of the South Atlantic Ocean.

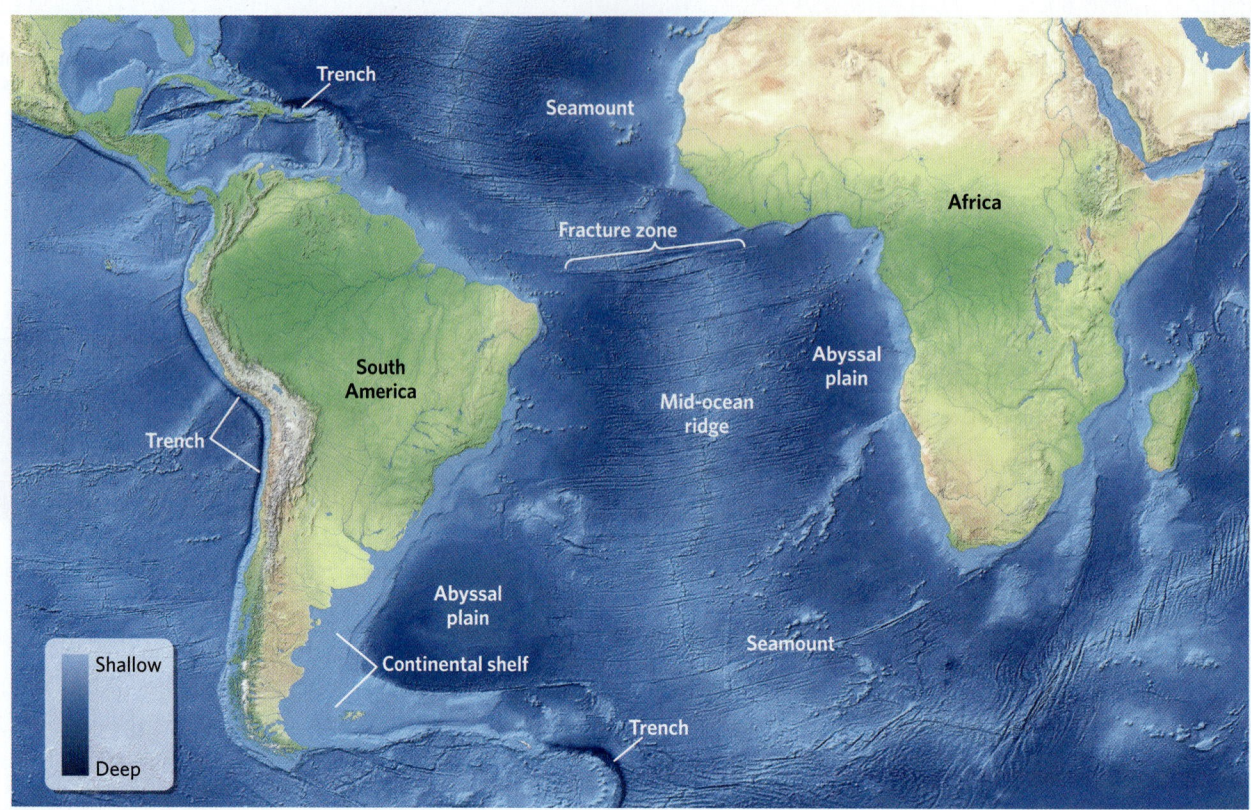

(a) A bathymetric map of the South Atlantic and the eastern South Pacific shows the complexity of the seafloor landscape.

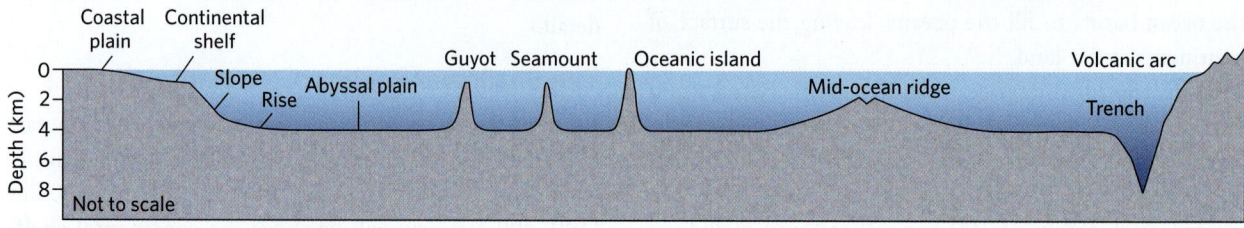

(b) A generalized schematic profile of an ocean floor shows key geologic features.

had been the rift forms a transitional region between newly formed oceanic lithosphere on one side and un-stretched continental lithosphere on the other. Over time, the stretched lithosphere cools and becomes denser and thicker. As a result, its surface slowly sinks, or sub-sides, just as the deck of a cargo ship moves down, closer to the sea surface, when heavy cargo has been placed in the ship's hold (Fig. 16.7b, c). As it sinks, the surface of the former, now inactive, rift becomes a low area that collects sand and mud washed off the continent as well as shells of plankton and other marine creatures. Deposition of sediment keeps pace with the rate of subsidence, so after tens of millions of years, a very thick layer of sediment covers the region of stretched continental crust. This sediment-filled depression is a **passive-margin basin**, and the flat surface of this sediment layer constitutes the continental shelf. In some cases, the sediment layer in a

passive-margin basin reaches a thickness of 15–20 km (9–12 mi).

If you were to take your submersible to the western coast of South America and cruise out into the Pacific, you would find an *active continental margin*, a seismically active plate boundary where the Pacific Ocean floor subducts beneath the continent (Fig. 16.8a). This margin looks very different from a passive margin. After crossing a relatively narrow continental shelf, you would find yourself over a continental slope that descends at a relatively steep angle of 3.5° into a deep-sea trench whose floor lies at a depth of over 8 km (5 mi). In this location, the continental slope is the face of an accretionary prism, a wedge of sediment that has been scraped off the seafloor, like snow in front of a plow, during subduction. The narrow continental shelf consists of a sediment apron that has spread out over the top of the accretionary prism (Fig. 16.8b).

FIGURE 16.7 Continental shelf of a passive margin.

(a) An oblique view, looking northwest, of the continental shelf along part of North America's east coast. It is much shallower than the adjacent abyssal plain.

1: Continental shelf
2: Continental slope
3: Continental rise
4: Abyssal plain

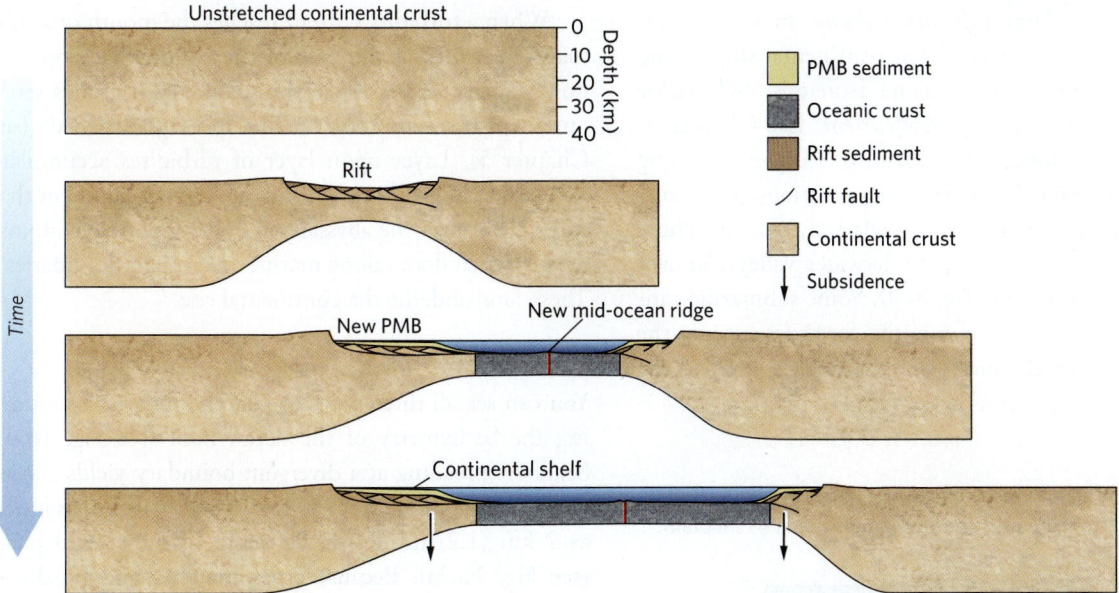

(b) The formation of a passive-margin basin (PMB). The top surface of the PMB is the continental shelf.

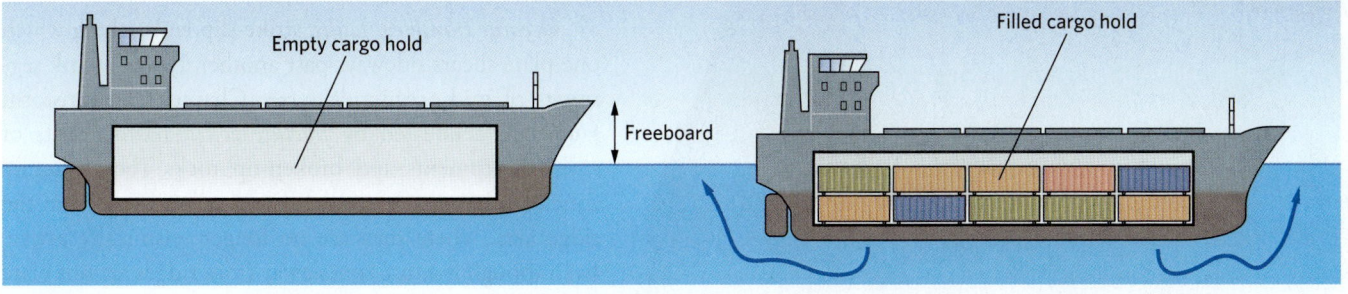

(c) The subsidence of a passive-margin basin happens for the same reason that a heavier ship moves downward relative to a lighter one. As the keel sinks down, water flows out of the way (blue arrows).

FIGURE 16.8 Continental shelf of an active margin.

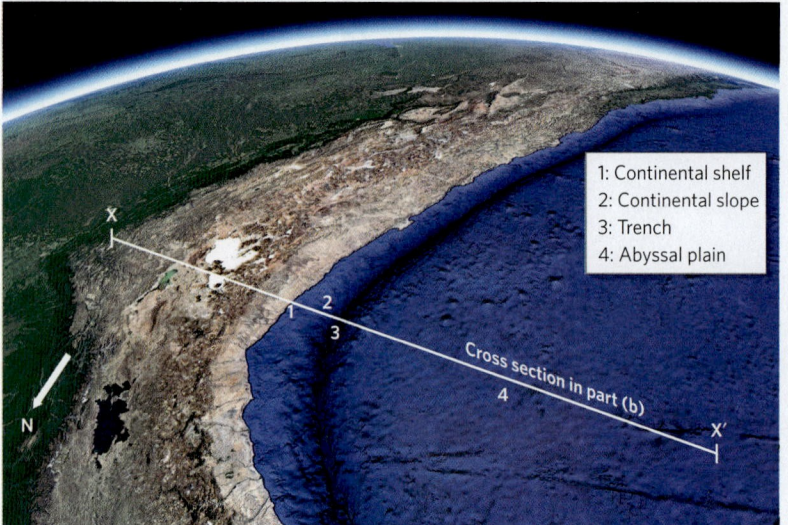

(a) Looking southeast, obliquely, at the western coast of South America, we see a narrow continental shelf, built on top of an accretionary prism. The tan area represents the Andes.

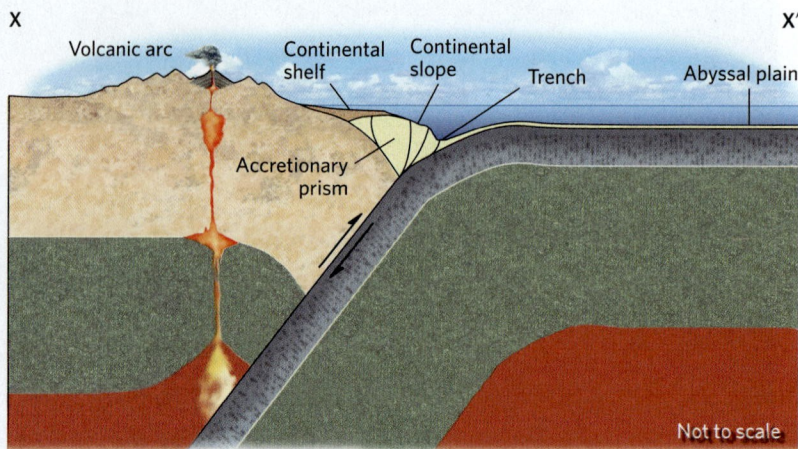

(b) A schematic cross section shows the development of an accretionary prism next to a trench at a convergent boundary.

Between continental shelves and the abyssal plain or the trench floor are slopes where submarine mass wasting events occasionally take place. As we mentioned in Chapter 12, large *submarine slumps* may carry immense volumes of debris down continental slopes, and these movements may generate tsunamis. Submarine avalanches known as *turbidity currents* (see Chapter 5) send abrasive clouds of suspended sediment rushing down continental slopes. In many locations, these turbidity currents focus down a relatively narrow chute and, over time, carve deep underwater valleys, known as **submarine canyons (Fig. 16.9)**. Some submarine canyons start offshore of river mouths, for rivers cut into the continental shelf at times when sea level is low and the shelf is exposed, producing a trough that can focus turbidity currents to the head of a submarine canyon when sea level rises.

When a turbidity current reaches the mouth of a submarine canyon, at the base of the continental slope, its flow velocity decreases, so the sediments it carries settle out to produce *turbidites* made up of graded beds (see Chapter 5). Layer upon layer of turbidites accumulate and build out a **submarine fan**, a wedge of sediment that spreads out over the abyssal plain (along passive margins) or the trench floor (along marine convergent boundaries). These fans underlie the continental rise.

Plate Boundaries on the Seafloor

You can see all three types of plate boundaries by studying the bathymetry of the ocean floor (see Fig. 16.6). Seafloor spreading at a divergent boundary yields a *mid-ocean ridge*, a submarine mountain belt that rises as much as 2 km (1.2 mi) above the depth of the abyssal plain (see Fig. 16.2a). Because crust stretches and breaks as seafloor spreading takes place, the ridge axis may be bordered by escarpments (cliffs), a result of normal faulting **(Fig. 16.10a)**.

Oceanic transform faults, strike-slip faults along which one plate shears sideways past another, typically link segments of mid-ocean ridges (see Chapter 2). Transform faults are delineated by *fracture zones*, narrow belts of steep escarpments and broken-up rock. These fracture zones can be traced into the oceanic plate away from the ridge axis, where they are no longer seismically active. Even though fracture zones away from ridges are not plate

FIGURE 16.9 A submarine canyon off the coast of New Jersey. Turbidity currents flow down the canyon and carry sediment to a submarine fan on the abyssal plain.

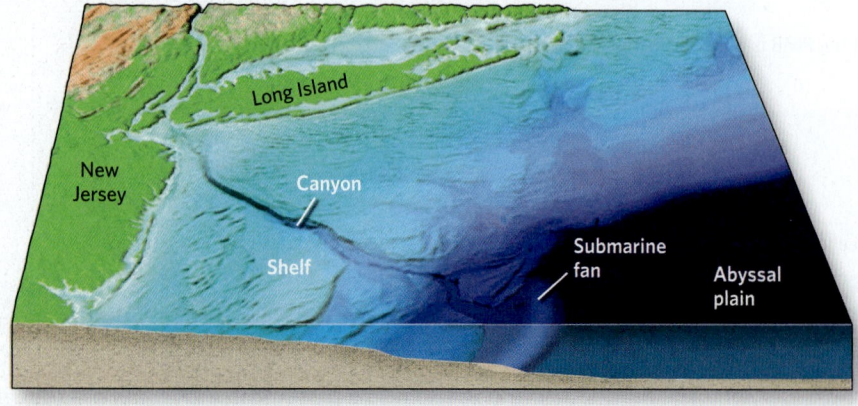

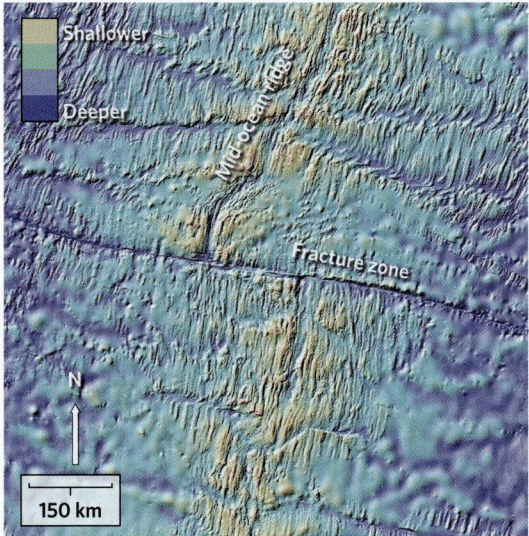

(a) Close-up showing the bathymetry of the Mid-Atlantic Ridge at 46° W, 42° N.

FIGURE 16.10 Plate boundaries on the seafloor.

(b) Oblique view of the Mariana Trench, in the western Pacific.

boundaries, they delineate the boundary between oceanic lithosphere of different ages (see Fig. 2.22b).

Subduction at convergent boundaries yields a *deep-sea trench*, a deep, elongate trough bordering a volcanic arc (see Fig. 16.8). As we noted, trenches typically reach depths of over 8 km (5 mi). The deepest, the Mariana Trench in the western Pacific, extends to a depth of 11 km (6.8 mi). Some trenches border continents, lying seaward of an active continental volcanic arc, as is the case along the western coast of South America. Others border *island arcs*, curving chains of active volcanic islands such as Alaska's Aleutian Islands (Fig. 16.10b).

Abyssal Plains and Seamounts

As oceanic crust moves away from the axis of a mid-ocean ridge and gets progressively older, it cools and thickens, and as it does so, its surface sinks and its slope decreases (see Chapter 2). As a result, the surface of the seafloor becomes a nearly horizontal abyssal plain. A blanket of *pelagic sediment* gradually accumulates and covers the basalt of the oceanic crust (Fig. 16.11a). This blanket consists mostly of microscopic plankton shells and fine flakes of clay (from volcanic ash or windblown dust), which slowly fall like snow from the ocean water and settle on the seafloor. The older the seafloor, the longer the time during which pelagic sediment has been accumulating, so over older oceanic lithosphere of the abyssal plain, it becomes thick enough (up to 0.6 km, or 0.4 mi) to bury irregularities in the basaltic crust of the seafloor. The top of this sediment layer forms the nearly featureless surface of the abyssal plain (Fig. 16.11b).

Oceanic islands (whose peaks protrude above sea level) and **seamounts** (whose peaks are submerged) rise above abyssal plains or mid-ocean ridges (Fig. 16.12a). These features result from hot-spot volcanic activity. Oceanic islands or seamounts that currently lie over hot spots are active volcanoes, whereas those that have moved off the hot spot are extinct (see Chapter 2). Some seamounts are hot-spot volcanoes that did not become tall enough to protrude above sea level (Fig. 16.12b), but many started out as oceanic islands that later sank below sea level.

FIGURE 16.11 The origin and nature of abyssal plains.

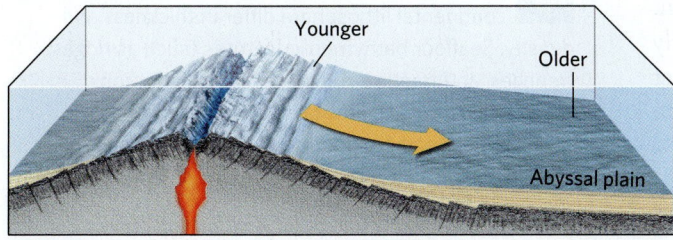

(a) The seafloor sinks progressively as it moves away from a ridge axis. It gets buried by sediment, which thickens away from the ridge axis.

(b) The surface of the abyssal plain has a covering of fine, loose pelagic sediment.

FIGURE 16.12 Oceanic islands and seamounts.

(a) An oceanic island protruding from the Pacific Ocean.

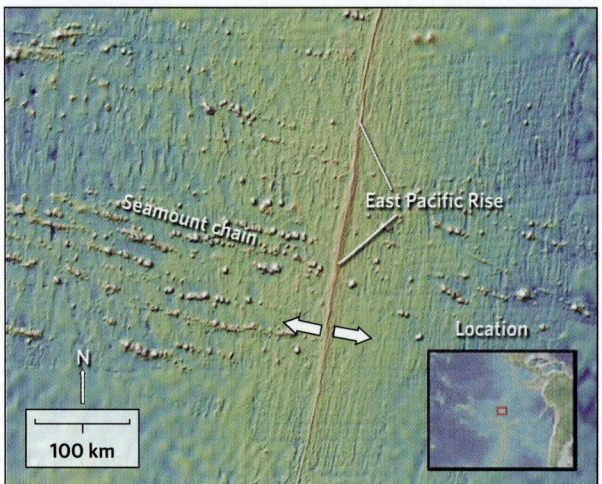

(b) Seamounts adjacent to the East Pacific Rise (112° W, 18° S), a mid-ocean ridge. Note that the seamount chains are perpendicular to the ridge axis and, therefore, are parallel to the direction of plate motion.

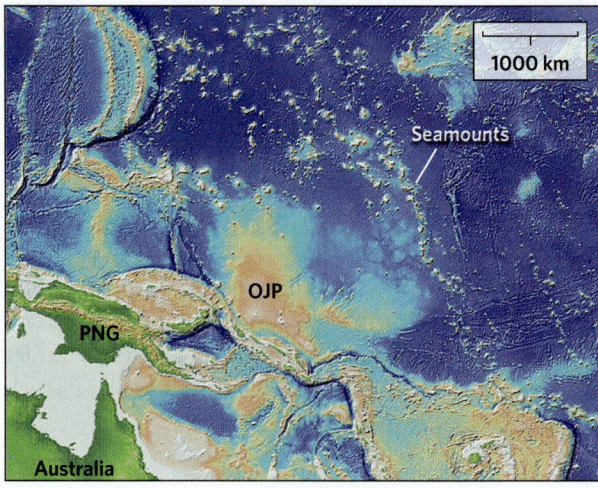

(c) The Ontong-Java Plateau (OJP) lies to the north of Papua New Guinea (PNG). Orange represents shallower water; blue represents deeper water.

Why do they sink? During and after the end of volcanic activity on an oceanic island, the island erodes and undergoes slumping and, over time, the seafloor beneath it ages and subsides. Oceanic islands and seamounts align in a chain parallel to the motion of the oceanic plate over the hot spot—as we saw in Chapter 4, these chains, called hot-spot tracks, define the direction of plate motion. Where hot-spot igneous activity becomes particularly voluminous, a broad *oceanic plateau*, composed of a layer of basalt up to 3 km (2 mi) thick, develops **(Fig. 16.12c)**. Oceanic plateaus represent submarine large igneous provinces (LIPs; see Chapter 4).

In some cases, the top of a seamount may be beveled flat by erosion as its summit reaches sea level. Or the island may be overgrown by a coral reef as it sinks. When such islands sink below sea level, they become flat-topped seamounts, known as **guyots**.

Take-home message . . .

Ocean basins exist because oceanic and continental lithosphere differ in thickness and density. Seafloor bathymetric features (such as ridges, trenches, and fracture zones) reflect plate tectonic processes. Abyssal plains overlie old oceanic lithosphere, and broad continental shelves overlie passive-margin basins.

Quick Question ⎯⎯⎯⎯⎯⎯⎯⎯⎯⎯
How do submarine canyons form?

16.3 The Tides Come In . . . the Tides Go Out . . .

Fishermen hoping to sail from a shallow port must pay attention to the **tide**, the generally twice-daily rise and fall of sea level. At *low tide*, when sea level sinks to its lowest elevation, boats might run aground, whereas at *high tide*, when sea level rises to its highest elevation, they can make it to open water easily (Fig. 16.13a, b). We've already mentioned tides in Chapter 15, and their existence is common knowledge. Here, we look more closely at how they are manifested and why they form. Tides affect the entire ocean, so the transition from high tide to low tide, or vice versa, involves the movement of a huge volume of water on a global scale. We discuss tides in this chapter because this water movement primarily affects coastal areas.

Tidal range, the difference between sea level at high tide and at low tide (Fig. 16.13c), varies greatly with location. Some regions experience a small tidal range, under 0.5 m (20 in), whereas others experience a large one. The largest tidal range on the planet occurs in the Bay of Fundy (Fig.16.13d), on the eastern coast of Canada, where the tidal range can be as much as 16.3 m (53.5 ft). Large tides also occur along the coast of western Europe. In the open ocean, tidal range averages about 0.6 m (2.0 ft). At several locations on our planet, the tidal range is zero.

Tides affect the coast because during a rising tide, or *flood tide*, the **shoreline** (the boundary between water and land) moves inland, and during a falling tide, or *ebb tide*, the shoreline moves seaward. The surface of the seafloor that lies between the shoreline at high tide and the shoreline at low tide constitutes the **intertidal zone** (Fig. 16.14a). The horizontal distance over which the shoreline migrates between high and low tides—and, therefore, the width of the intertidal zone—depends on both the tidal range and the slope of the seafloor surface (Box 16.1). Where the tidal range is large and the slope is

FIGURE 16.13 Evidence of ocean tides.

(a) Low tide at Perranporth, in Cornwall, England.

(b) High tide from the same viewpoint as in the previous photo.

(c) The tidal range is the vertical distance between low tide and high tide. This photo of a French harbor was taken at low tide.

(d) At low tide, the muddy floor of this small cove along the coast of the Bay of Fundy is exposed.

BOX 16.1 ▶ How can I explain . . .

Variations in the width of the intertidal zone

What are we learning?

That the slope of the shore determines the width of the intertidal zone at a location on the coast.

What you need:

- A small plastic cutting board
- A rectangular glass pan, large enough to hold the cutting board, deep enough to hold several centimeters (a couple of inches) of water (An aluminum pan can work, too.)
- Water that has been dyed blue with food coloring
- A protractor, ruler, and graph paper
- A ladle, a pitcher, and a measuring cup

Instructions:

- Fill the pan with blue water. Place the cutting board and ruler as shown.
- Ladle out enough water to lower the water level by half; reserve the water in a measuring cup.
- Repeat the experiment for different slope angles of the cutting board.
- Plot a graph of slope angle (measured with the protractor) against width of the intertidal zone (measured with a ruler). How are the quantities related?

What did we see?

- Simple geometry requires that the gentler the slope, the wider the intertidal zone. The smallest intertidal zones occur where the shore is a vertical cliff.
- The intertidal zone can be very wide along very gently sloping coasts.

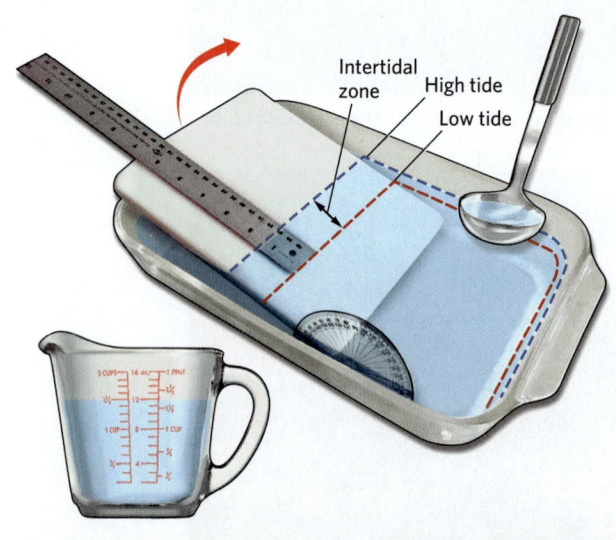

FIGURE 16.14 The intertidal zone and tidal bores.

(a) The intertidal zone on a beach on Cape Cod. Loose seaweed was deposited at high tide; growing weed is exposed at low tide.

(b) A tidal flat on the coast of France.

(c) A tidal bore moves up an estuary along the Bay of Fundy.

gentle, the position of the shoreline can move by kilometers during a tidal cycle, so at low tide, the intertidal zone becomes a broad *tidal flat* exposed to the air **(Fig. 16.14b)**. Large tidal flats can be dangerous places—shellfish hunters digging for clams in the mud at the seaward edge of a tidal flat have drowned when caught by the rising tide because they couldn't make it back to dry land before the water became too deep.

Arrival of a flood tide in the mouth of an estuary, where the tide moves against a river current, can produce a **tidal bore**, a visible wall of water, ranging from a few centimeters to a few meters high. Tidal bores can move at speeds of up to 35 km/h (22 mph), faster than a person can run (Fig. 16.14c). In a few places, tidal bores become large enough for surfers to ride. A skilled surfer once rode a tidal bore for 1 hour and 10 minutes, during which he traveled 17 km (11 mi)!

In nontechnical discussions, tides are attributed simply to the gravitational pull of the Moon and Sun. Indeed, gravitational pull does play a key role in driving tides; in fact, the Moon contributes most of the force that causes tides. (The Sun, even though it is vastly larger, lies so far away that its contribution to tides is only 46% that of the Moon.) But the gravitational pull of these objects doesn't completely explain why tides happen. In detail, tides take place because the water of the ocean moves in response to two forces: (1) the gravitational attraction of the Moon and the Sun, and (2) centrifugal force caused by the revolution of the Earth-Moon system. Oceanographers refer to the combination of these forces as the **tide-generating force** (Box 16.2).

The tide-generating force creates two tidal bulges in the global ocean, forming an envelope of water that is more oval shaped than the nearly spherical solid Earth (Fig. 16.15a). One bulge, the *sublunar bulge*, lies on the side of the Earth that is closer to the Moon, because the

Moon's gravitational attraction (the strongest contributor to tides) is greatest at this point. The other, the *secondary bulge*, lies on the opposite side of the Earth, pushed outward by the centrifugal force (see Box 16.2). A depression in the global ocean surface separates the two bulges. When a coastal location passes under a tidal bulge as the Earth spins, it experiences a high tide, and when it passes under a depression, it experiences a low tide. Tides are higher underneath the sublunar bulge than underneath the secondary bulge.

If the Earth's solid surface were smooth and completely submerged beneath the ocean, so that there were no continents or islands, the timing and height of tides would be simple to understand—each location would experience two tides a day, and the tidal range would depend on the position of the location relative to the highest part of the bulge. But the story isn't quite that simple, for many other factors affect the timing and magnitude of tides:

- *Tilt of the Earth's axis:* Because the Earth's axis of rotation is not perpendicular to the orbital plane of the Earth-Moon system, any given point on the Earth passes between a high part of one bulge during one part of the day and a lower part of the other bulge during another part of the day (Fig. 16.15b).

FIGURE 16.15 Tides.

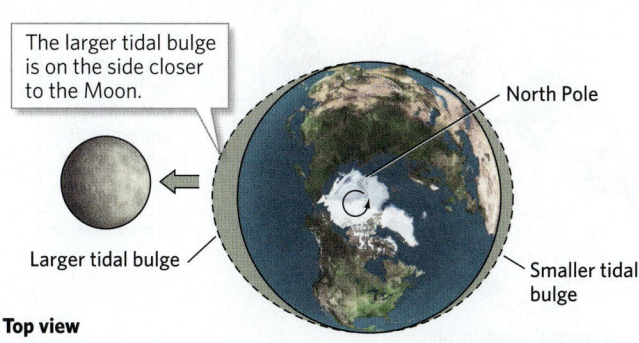

Top view

(a) The larger (sublunar) tidal bulge always faces the Moon, and the smaller (secondary) tidal bulge is always on the opposite side of the Earth.

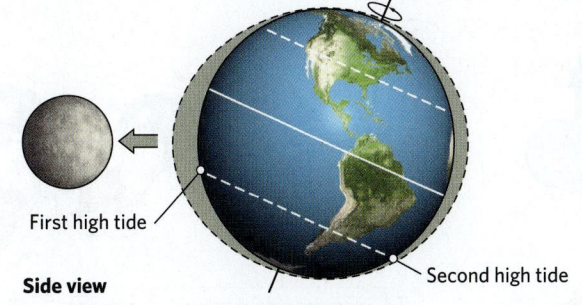

Side view

(b) Viewed from the side, the sublunar bulge does not align with the equator.

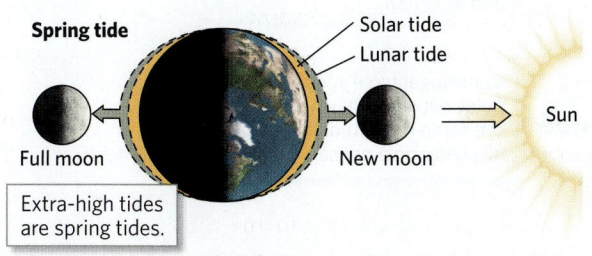

(c) When the Sun is aligned with the Moon, stronger, higher tides, called spring tides, result.

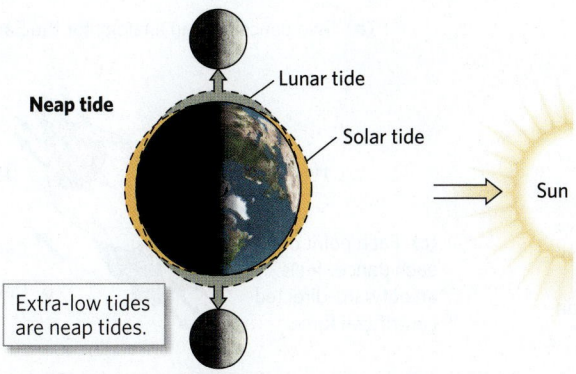

(d) When the Sun is at right angles to the Moon, weaker, lower tides, called neap tides, result.

BOX 16.2

Science Toolbox

The tide-generating force

Tides form in response to two forces acting at the same time: gravitational pull exerted by the Moon and Sun on the Earth, and centrifugal force caused by the revolution of the Earth around the center of mass of the Earth-Moon system. To explain this statement, we must first review a few physics terms:

• *Gravitational pull* is the attractive force that one mass exerts on another. Its magnitude depends on the amount of mass in each object and on the distance between the two masses. (The pull increases as mass increases, and the pull decreases as distance between masses increases.)

• *Centrifugal force*, a manifestation of inertia, is the apparent outward-directed (center-fleeing) force that a body on or in an object feels when the object spins or moves in orbit around a point (see Box 15.3). It is an "apparent force" in that it exists only from the perspective of the moving object.

• The *Earth-Moon system* refers to our planet and the Moon, viewed as a gravitationally linked pair as they move together through space.

• The *center of mass* is the point within an object, between objects, or within a group of objects, about which mass is evenly distributed.

Because the Earth is 81 times more massive than the Moon, the center of mass of the Earth-Moon system lies 1,700 km (1,060 mi) below the Earth's surface.

With these terms on hand, let's begin by considering the origin and consequence of centrifugal force in the Earth-Moon system. To do this, we must observe the way in which the system moves. Although we usually picture the center of the Earth as following a simple orbit around the Sun, it's actually the center of mass of the Earth-Moon system that follows this orbit **(Fig. Bx16.2a)**. To see why, picture the Earth-Moon system as a pair of dancers, one much heavier than the other. The dancers face each other, hold hands, and whirl as they cross the dance floor **(Fig. Bx16.2b)**. The center of mass of the pair defines their overall path—each dancer crosses back and forth across the overall path as the dance progresses.

With this image in mind, let's turn our attention back to the Earth-Moon system. As the Earth revolves around the system's center of mass, centrifugal forces develop on both the Earth and the Moon that would cause the two bodies to fly away from each other, were it not for the gravitational attraction holding them together. We can understand this statement by thinking again of our dancer analogy—the centrifugal force acting on each dancer points outward, away from his or her partner, and is the same for all points on each dancer **(Fig. Bx16.2c)**. We can represent the direction and magnitude of

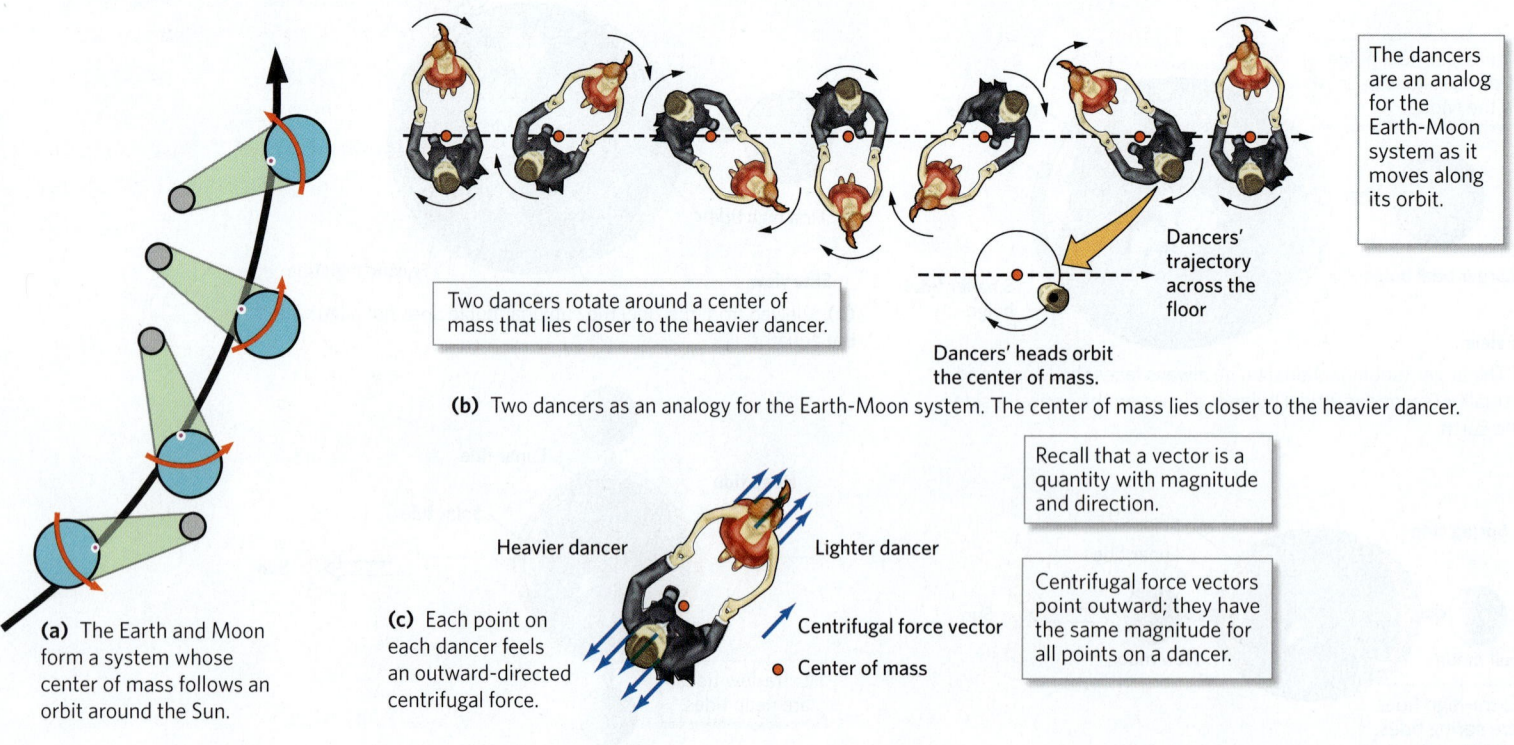

FIGURE Bx16.2 The concepts of the center of mass of the Earth-Moon system and the tide-generating force.

The dancers are an analog for the Earth-Moon system as it moves along its orbit.

Two dancers rotate around a center of mass that lies closer to the heavier dancer.

Dancers' trajectory across the floor

Dancers' heads orbit the center of mass.

(b) Two dancers as an analogy for the Earth-Moon system. The center of mass lies closer to the heavier dancer.

(a) The Earth and Moon form a system whose center of mass follows an orbit around the Sun.

(c) Each point on each dancer feels an outward-directed centrifugal force.

Heavier dancer

Lighter dancer

Recall that a vector is a quantity with magnitude and direction.

Centrifugal force vector

Centrifugal force vectors point outward; they have the same magnitude for all points on a dancer.

• Center of mass

centrifugal force with an arrow, or *vector*, whose length represents the size of the force and whose orientation indicates the direction of the force. Centrifugal force vectors at all points all around the surface of the Earth point away from the Moon.

Now, let's consider how the force of gravity comes into play in causing tides. To simplify this discussion, we examine only the effect of the Moon's gravity on Earth. Vectors representing the magnitude and direction of the Moon's gravitational pull, at any point on the surface of the Earth, point toward the center of the Moon. Because the magnitude of gravity depends on distance, the Moon exerts more attraction on the side of the Earth closer to the Moon than on the side of the Earth farther from the Moon.

How do centrifugal force and gravity work together to produce tides? If we draw vectors representing both centrifugal force and gravitational force at various points on or in the Earth, we see that the arrows representing centrifugal force do not have the same length as those representing gravitational attraction, except at the Earth's center **(Fig. Bx16.2d)**. Moreover, the vectors representing centrifugal force do not point in the same direction as the vectors representing gravitational attraction. The force that ocean water feels is the sum of the two forces acting on the water. You can determine the sum of two vectors by drawing the vectors to touch head to tail—the sum is the vector that completes the triangle. This sum of the gravitational vector and the centrifugal force vector at a point on the Earth's surface is the *tide-generating force* **(Fig. Bx16.2e)**.

The magnitude and direction of the tide-generating force vary with location on the Earth. For example, on the side of the Earth closer to the Moon, gravitational force is greater than centrifugal force, so adding the two gives a net tide-generating force that causes the sea surface to bulge toward the Moon. On the side of the Earth farther from the Moon, centrifugal force is the larger vector and causes the surface of the sea to bulge away from the Moon. Therefore, the ocean has two tidal bulges—one on the side of the Earth close to the Moon, and one on the opposite side. The bulge closer to the Moon is always larger.

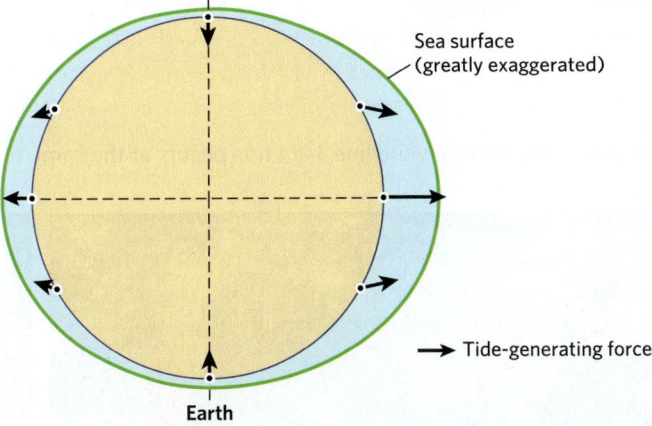

(d) Each point on the Earth's surface feels the same centrifugal force, but also feels a gravitational pull from the Moon. A vector is a quantity with both magnitude and direction.

(e) The tide-generating force is the sum of the centrifugal force and the gravitational force vectors. The bulge of the sea surface is exaggerated.

- *The Moon's orbit:* The Moon progresses in its 28-day orbit around the Earth in the same direction as the Earth rotates. High tides arrive 50 minutes later each day because of the difference between the time it takes for the Earth to spin on its axis and the time it takes for the Moon to orbit the Earth.

- *The Sun's gravity:* When the Sun lies on the same side of the Earth as the Moon (a new Moon) or on the side opposite the Moon (a full Moon), we experience particularly high flood tides, called *spring tides*, because the Sun's gravitational attraction adds to that of the Moon **(Fig. 16.15c)**. When the Moon and Sun are 90° apart relative to the Earth (quarter Moon), we experience lower flood tides, called *neap tides*, because the Sun's gravitational attraction counteracts that of the Moon (Fig. 16.15d).

- *Ocean-basin shape:* The shape of an ocean basin influences the sloshing of water back and forth as tides rise and fall. Depending on its timing and magnitude, this sloshing can add to or subtract from the tidal bulge on a local scale. In some locations, sloshing entirely cancels out one of the daily tides.

- *Focusing effect of bays:* On a still more local scale, the shape of the shoreline influences the tidal range. In a bay that narrows to a point, the flood tide brings a large volume of water into a small area, so the point at the end of the bay experiences an especially high flood tide.

Because of the complexity of these contributing factors, the timing and magnitude of tides vary significantly around the ocean and along the coast (Fig. 16.16). Nevertheless, at any given location, the tides are periodic and can be predicted (Box 16.3). For early civilizations, tides provided a rudimentary way to tell time. In fact, in some languages, the word for tide is the same as the word for time.

Have tides on the Earth been the same through Earth history? Probably not. When the Moon first formed, it was much closer to the Earth than it is now, so the tides in early oceans were larger than they are now. Friction between ocean water and the ocean floor causes the movement of the tidal bulge to lag slightly behind the movement of the Moon around the Earth. The Moon, therefore, exerts a slight pull on the side of the bulge. This pull acts like a brake and slows the Earth's spin, so that days are growing longer at a rate of about 0.002 seconds per century. Over geologic time, these seconds add up—a day was only 21.9 hours long in the Middle Devonian Period (390 Ma). The process also slows the Moon's orbit, causing the distance to the Moon to increase at a rate of about 4 cm (2 in) per year. (This change in distance happens to conserve angular momentum, a concept discussed in Box 15.3.)

Take-home message . . .

Tides are the generally twice-daily rise and fall of the sea surface. They are caused by the tide-generating force, a consequence of gravitational pull by the Moon and Sun and of centrifugal force due to the revolution of the Earth-Moon system. Many factors affect the specific tidal range—the difference between the elevation of high tide and that of low tide—at a given location.

Quick Question
What factors determine the width of the intertidal zone in a region?

FIGURE 16.16 A map showing the variation of tidal range around the world. On a given white line, high tide occurs at the same time.

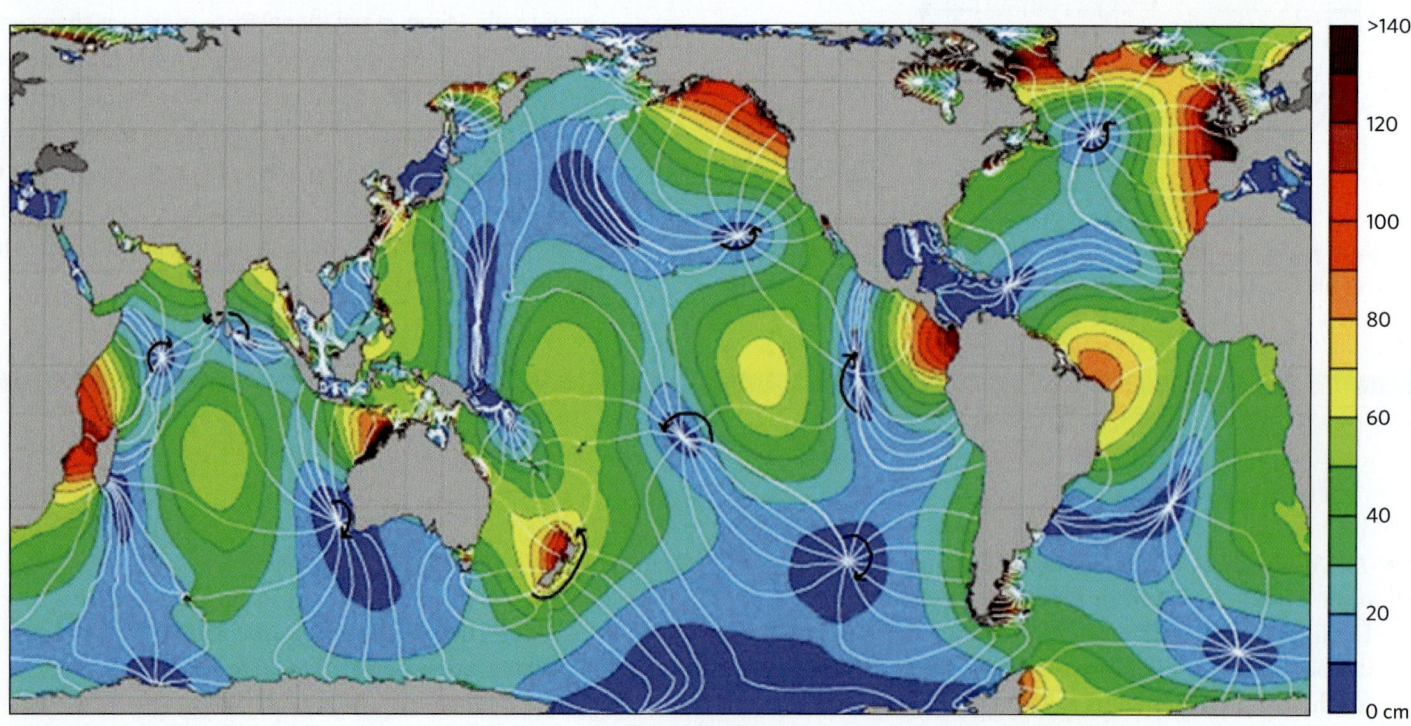

FIGURE 16.17 Spectacular coastal scenery from localities around the world.

(a) Rocky cliffs form the Amalfi Coast of western Italy.

(b) Steep green slopes form the Na Pali coast of Kauai, Hawaii.

(c) A sandy beach along the coast of St. John, US Virgin Islands.

(d) The abrupt edge of the Nullarbor Plain forms the southern coast of Australia.

16.4 Where Land Meets Sea: Coastal Landforms

Tourists along the Amalfi Coast of Italy thrill to the sound of waves crashing on rocky shores, but in the Virgin Islands, sunbathers can find seemingly endless white sand beaches. A fisherman plying the channels of the Mississippi Delta will find vast swamps along the seacoast, but a sailor approaching Rio de Janeiro, Brazil, will come face-to-face with granite domes rising directly from the sea (see Figure 12.1), and a sailor landing at the southern shore of Australia will have to climb a 100-m (330-ft) vertical cliff (Fig. 16.17). As these examples illustrate, coasts vary dramatically in terms of their topography and associated landforms (Earth Science at a Glance, pp. 504–505). The shape of a coast depends on many factors, but first and foremost, it reflects deposition or erosion by waves. So we start our discussion of coasts by examining how waves change when they approach the shore.

Waves Approaching the Shore

In Chapter 15, we discussed the motion of waves in the open ocean and saw that in an ocean wave, water

molecules follow circular paths. The diameter of these paths decreases with depth until, at a distance below the surface equal to half a wavelength, water remains stationary (see Fig. 15.25). In the open ocean, the *wave base*, the lowest level at which water moves in a wave, for wind-driven waves, lie far above the seafloor, so wave motion has no effect on the ocean floor. In the *shoaling zone*, nearer shore, however, the wave base touches the ocean floor. At the seaward edge of the shoaling zone, a passing wave causes a slight back-and-forth motion of sediment. Still closer to shore, as the water gets shallower, friction between the wave and the seafloor slows the deeper part of the wave, and the circular motion in the wave becomes more elliptical (Fig. 16.18a). Eventually, water at the top of the wave curves over its base, and the wave becomes a **breaker**, ready for surfers to ride (Fig. 16.18b). The toppling water at the crest of a breaker mixes with air to form a white froth. Breakers crash onto the shore in the *surf zone*, sending a surge of water up the beach. This upward surge, or **swash**, continues until friction and gravity bring water motion to a halt. Then gravity draws the water back down the beach as **backwash** (Fig. 16.18c).

The crests of waves that make an angle to the shoreline bend as they approach the shore, a phenomenon called

The three great elemental sounds in nature are the sound of rain, the sound of wind in a primeval wood, and the sound of the outer ocean on a beach.

—HENRY BESTON (AMERICAN NATURALIST, 1888–1968)

Did you ever wonder . . .

why the breakers that surfers love form near the shore?

FIGURE 16.18 The interaction of waves with the shore.

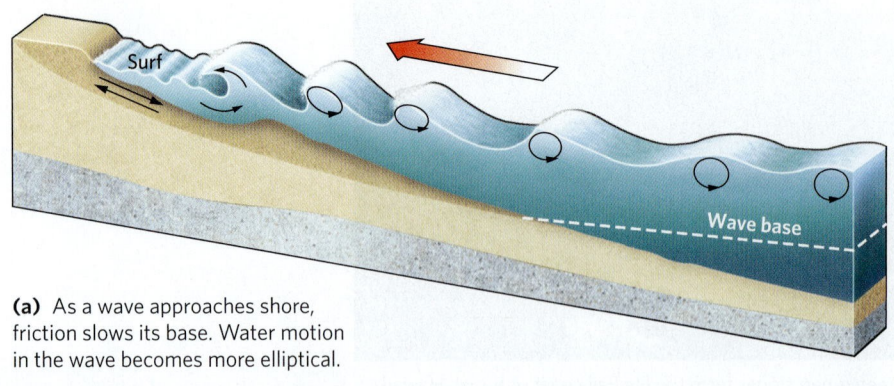

(a) As a wave approaches shore, friction slows its base. Water motion in the wave becomes more elliptical.

(b) Waves becoming breakers as they approach a beach.

(c) Swash and backwash on a beach. The water from the previous wave covered the wet area, but it has since flowed down the beach as backwash. The swash of the next wave is moving up the beach.

wave refraction. Typically, by the time the wave reaches the shore, the angle between the crest and the shoreline decreases to less than 5° **(Fig. 16.19)**. To understand why wave refraction happens, imagine a wave approaching the shore so that its crest initially makes an angle of 45° with the shoreline. The end of the wave closer to the shore touches bottom first and slows down because of friction, whereas the end farther offshore continues to move at its original velocity. This difference swings the whole wave around so that it becomes more parallel with the shoreline.

Although wave refraction decreases the angle at which waves move close to shore, it does not necessarily eliminate that angle. Where waves do arrive at the shore obliquely, water in the nearshore region has a component of motion that trends parallel to the shore. This **longshore current** causes swimmers floating in the water just offshore to drift gradually in a direction parallel to the shoreline (see Fig. 16.19).

Waves push water toward the shore incessantly. As the backwash moves back to the sea, it may concentrate

FIGURE 16.19 Wave refraction and its consequences along the shore.

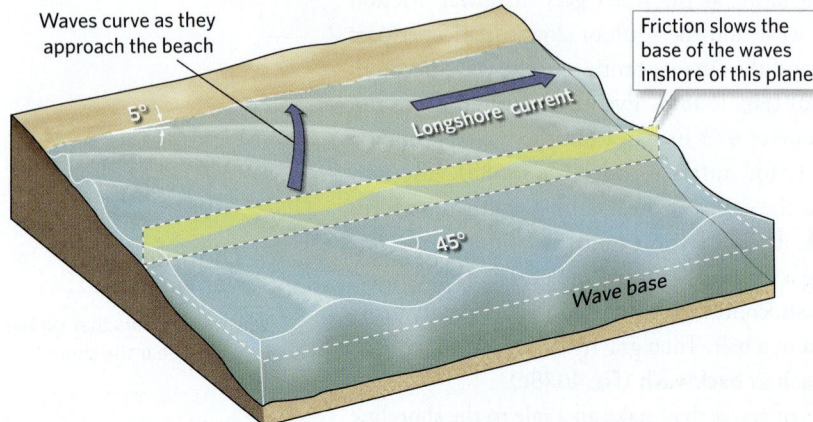

(a) Wave refraction occurs when waves approach the shore at an angle. Waves arriving at the shore obliquely can cause a longshore current.

(b) An example of waves refracting along a beach.

BOX 16.3 **Putting Earth Science to Use**

Planning for clamming: reading tidal charts

Imagine that you are an independent shell fisherman, or a tourist taking a seaside holiday, and you want to harvest clams from tidal mudflats (Fig. Bx16.3a). If the mudflats occur in an area with a significant tidal range, you'll need to pay attention to when highs and lows occur. Clams lie buried in the mud, so it makes sense to dig for them at low tide, when you can walk further offshore to find clams without being submerged. But pay close attention to when the tide starts coming in. If you get caught by a rising (flood) tide, you may have to wade or swim through deep, cold water to get back to shore, and risk exhaustion or hypothermia, even death. If you anchor a boat nearshore at high tide, you may find your boat stuck in the mud during a falling (ebb) tide, and you'll have to wait many hours before getting underway again.

In the United States, NOAA uses computers that take into account the many variables that affect tides, in order to predict when highs and lows will take place. You can find NOAA's data and charts online at https://tidesandcurrents.noaa.gov. Graphs depict water level on the vertical axis, and time on the horizontal axis, allowing you to read off the water level at a specific time, and to see when tides rise or fall (Fig. Bx16.3b). A graph for tides over the course of a month shows how tidal range varies, due primarily to the relative positions of the Sun and the Moon (Fig. Bx16.3c).

FIGURE Bx16.3 Reading tidal charts.

(a) Digging for clams at low tide.

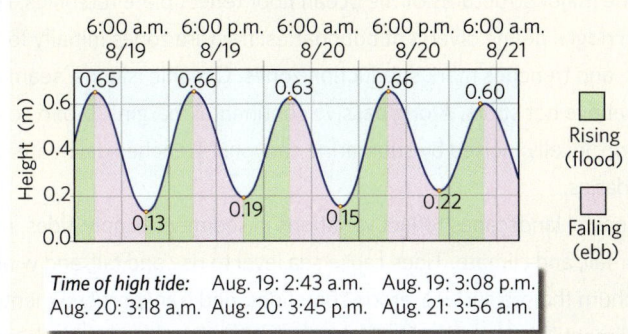

(b) The NOAA tidal chart for Schooner Bay, Virginia, for August 19-21, 2019. Note that high tides arrive a little later each day.

Time of high tide: Aug. 19: 2:43 a.m. Aug. 19: 3:08 p.m.
Aug. 20: 3:18 a.m. Aug. 20: 3:45 p.m. Aug. 21: 3:56 a.m.

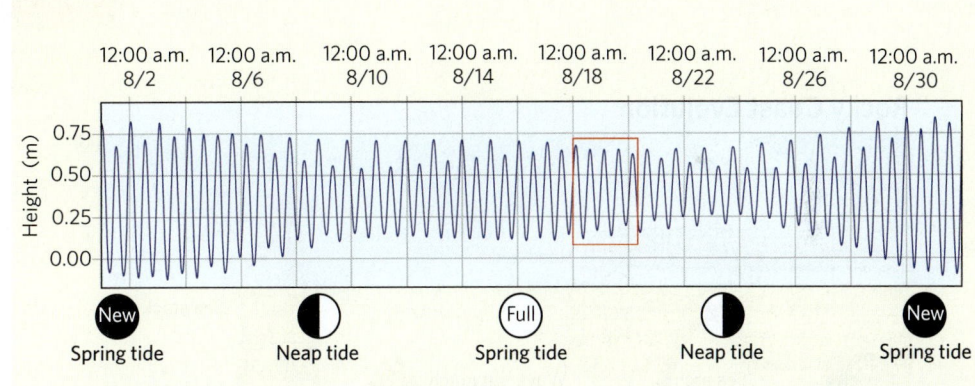

(c) The tidal range varies with the phases of the Moon. Note how the tidal range during neap tide is less than that at spring tides, and that tides are particularly high when the Moon is new. The red rectangle shows the days depicted in (b).

---

into a strong seaward flow, called a *rip current*, that moves perpendicular to the shoreline (Fig. 16.20). Rip currents cause many drownings along beaches every year because they can carry unsuspecting swimmers out into deeper water.

Beaches and Tidal Flats

For millions of vacationers, the ideal holiday includes a trip to a **beach**, a gently sloping fringe of sediment along the shore. Some beaches consist of pebbles or boulders,

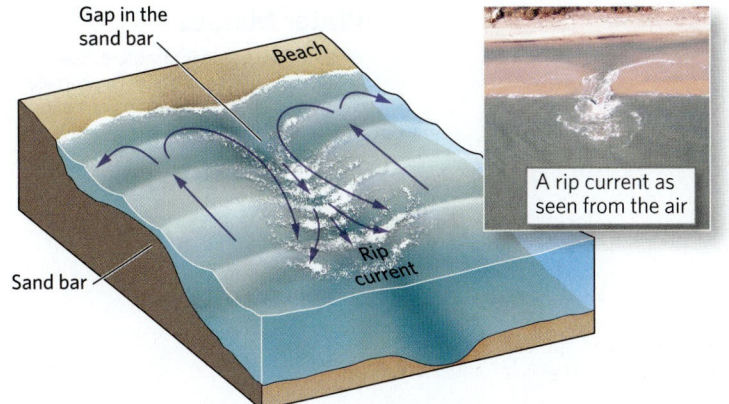

FIGURE 16.20
A rip current forms when backwash concentrates into a narrow, local current that carries water farther offshore.

A rip current as seen from the air

The Seafloor and Coasts

The oceans of the world host a diverse array of environments and landscapes, illustrating the complexity of the Earth System. Tectonic processes and surface processes, working alone or in tandem, generate unique features beneath the sea and along its coasts.

The major structures of the ocean floor reflect plate tectonics. For example, mid-ocean ridges define divergent boundaries, fracture zones initially form along transform faults, and trenches mark subduction zones. Oceanic islands, seamounts, and plateaus build above hot spots. Along passive continental margins, broad continental shelves develop, locally incised by submarine canyons. Trenches delineate convergent boundaries.

Coastal landscapes reflect variations in sediment supply, tides, relative sea-level rise or fall, and climate. Tides cause sea level to rise and fall, and wind builds waves that churn the sea surface, erode shorelines, and transport sediment. Where the supply of sediment is low and the landscape is rising relative to sea level, rocky shores with dramatic cliffs and sea stacks may evolve. Where sediment is abundant, sandy beaches and bars develop. Regions where glaciers carved deep valleys now feature spectacular fjords. Protected coastal areas, especially those in warm climates, host unique coastal ecosystems. Corals may contribute to growth of broad reefs along the shore.

Temperate and Tropical Coastal Landforms

Along sandy shores, sand builds beaches, sand spits, and bars.

Turbidity currents carve submarine canyons and produce submarine fans.

In tropical environments, mangroves live along the shore and coral reefs grow offshore.

Along rocky coasts, sea cliffs, sea arches, and sea stacks evolve.

Offshore, reefs grow

At a passive margin, a broad continental shelf develops. Submarine slumping may occur along the shelf.

The ocean teems with life.

Rocky Coast Evolution

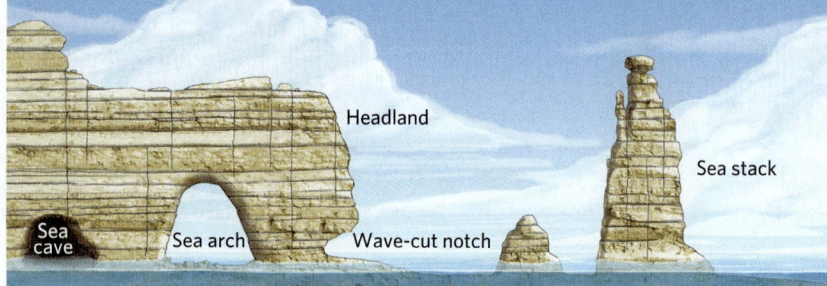

Headland

Sea stack

Sea cave

Sea arch

Wave-cut notch

As waves erode rocky coasts, bedrock breaks away at joints and along bedding planes. As a result, several distinct landforms evolve over time.

Water Masses

Water of the ocean is stratified, due to thermohaline circulation. The surface layer flows in ocean-spanning currents, and locally in eddies.

Gulf Stream

North Atlantic gyre

Surface water

Intermediate water

North Atlantic deep water

Antarctic bottom water

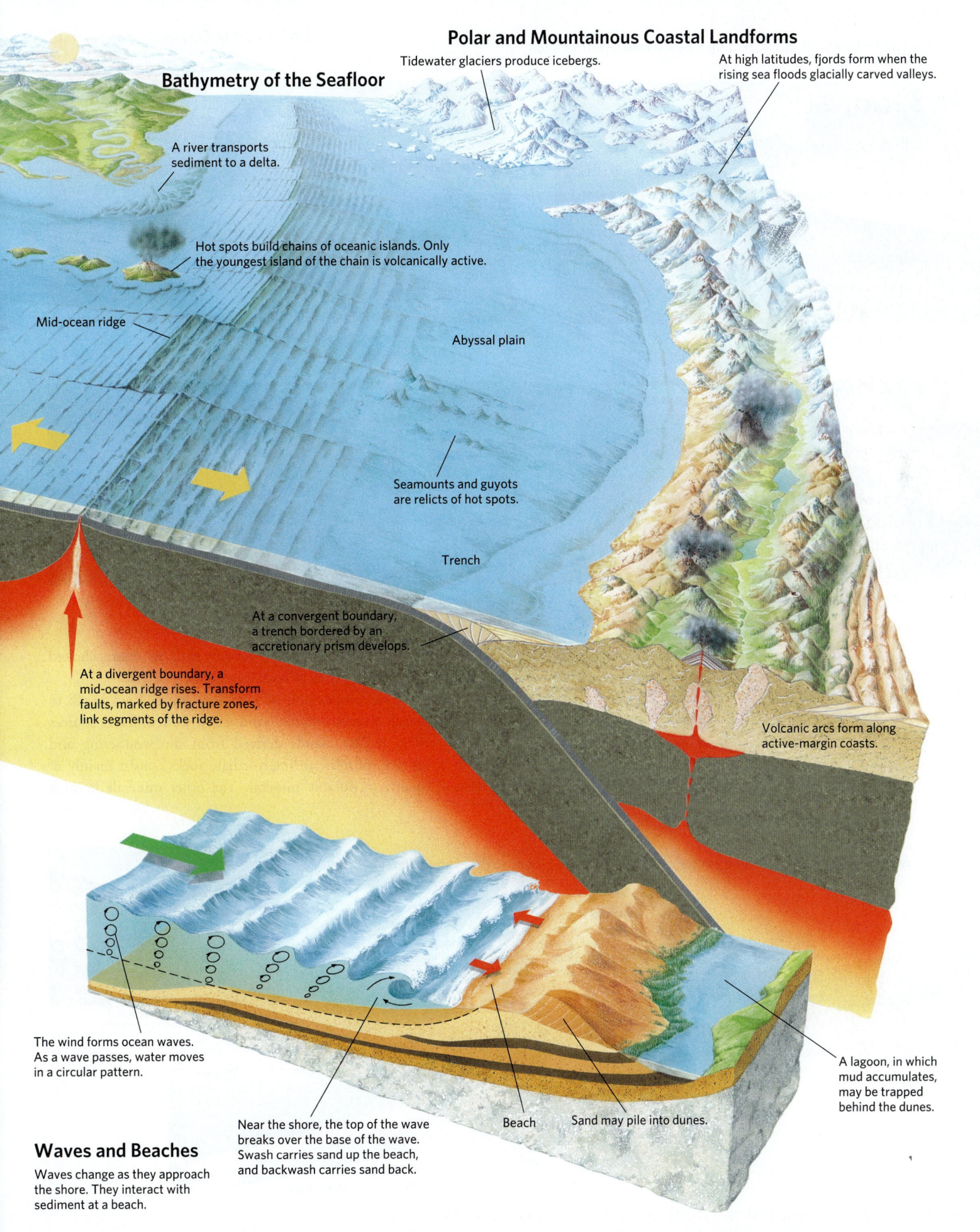

Bathymetry of the Seafloor

Polar and Mountainous Coastal Landforms

Tidewater glaciers produce icebergs.

At high latitudes, fjords form when the rising sea floods glacially carved valleys.

A river transports sediment to a delta.

Hot spots build chains of oceanic islands. Only the youngest island of the chain is volcanically active.

Mid-ocean ridge

Abyssal plain

Seamounts and guyots are relicts of hot spots.

Trench

At a convergent boundary, a trench bordered by an accretionary prism develops.

At a divergent boundary, a mid-ocean ridge rises. Transform faults, marked by fracture zones, link segments of the ridge.

Volcanic arcs form along active-margin coasts.

The wind forms ocean waves. As a wave passes, water moves in a circular pattern.

A lagoon, in which mud accumulates, may be trapped behind the dunes.

Near the shore, the top of the wave breaks over the base of the wave. Swash carries sand up the beach, and backwash carries sand back.

Beach

Sand may pile into dunes.

Waves and Beaches

Waves change as they approach the shore. They interact with sediment at a beach.

505

FIGURE 16.21 The type of sediment that makes up a beach varies with location.

(a) Cobble-sized clasts make up a beach in Oregon.

(b) A beach in Puerto Rico consists of quartz sand.

Quartz sand

Carbonate sand

Basalt sand

(c) Different kinds of sand make up different sandy beaches. Quartz sand comes from erosion of felsic rock, carbonate sand from fragmentation of shells, and black sand from basalt.

Did you ever wonder . . .

why beautiful sandy beaches don't form along all coasts?

whereas others consist of sand grains **(Fig. 16.21a, b)**. This is no accident, for waves winnow out finer sediment such as silt and mud and carry it to quieter water offshore, where it settles. Storm waves can smash cobbles against one another with enough force to shatter them, but they have little effect on sand, for sand grains can't collide with enough energy to crack. So cobble, or "shingle," beaches may persist only where nearby cliffs or debris-choked streams supply large rock fragments.

The composition of sand itself varies from beach to beach because different sands come from different sources **(Fig. 16.21c)**. Sands derived from the weathering and erosion of felsic to intermediate rocks consist mainly of quartz, a durable mineral. The other minerals in such

FIGURE 16.22 Interaction with waves, over time, produces a distinct beach profile, with several different zones.

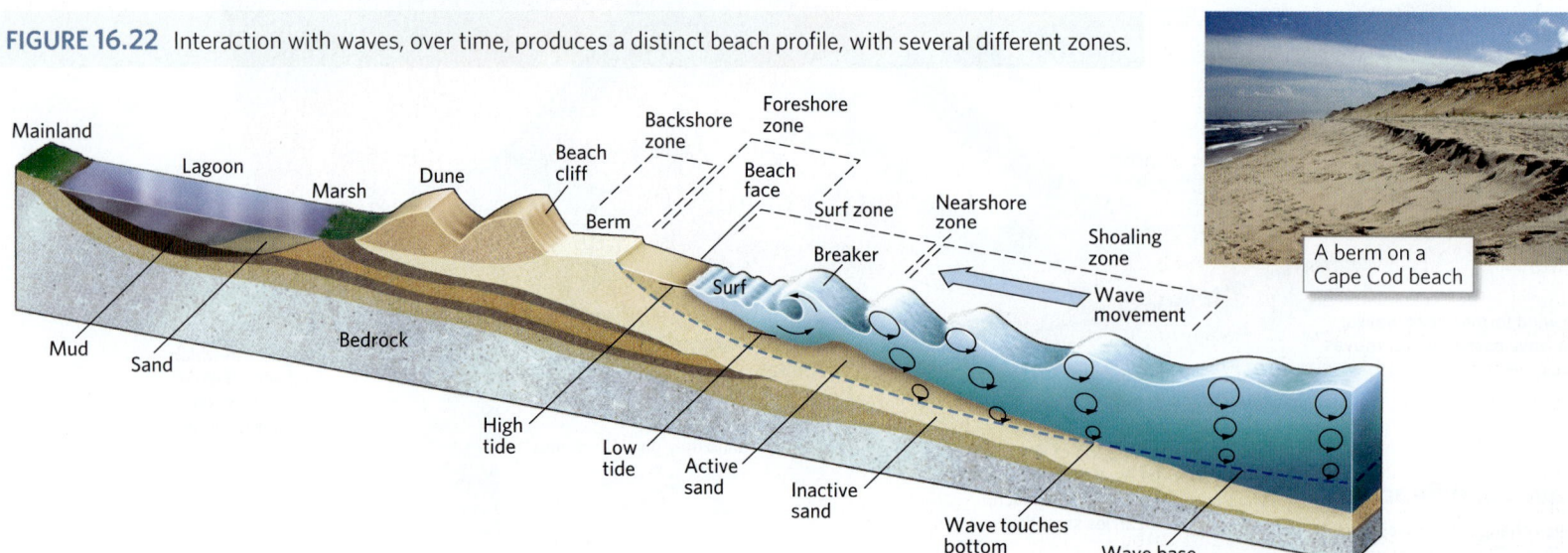

A berm on a Cape Cod beach

FIGURE 16.23 The width of a beach can change seasonally.

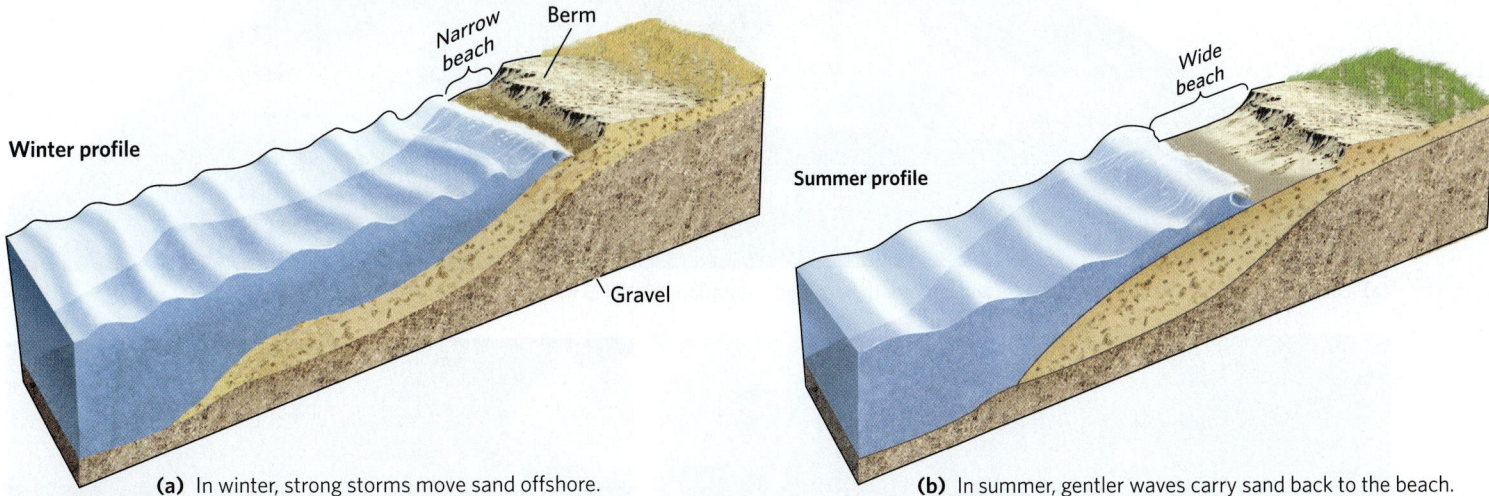

(a) In winter, strong storms move sand offshore.

(b) In summer, gentler waves carry sand back to the beach.

rocks weather chemically to form clay, which washes away in the waves. Beaches made from the erosion of limestone, coral reefs, or shells consist of carbonate sand, including tiny shell chips. And beaches derived from the erosion of basalt boast black sand, made of tiny basalt grains.

Beaches consist of distinct zones, as illustrated by a **beach profile**, a cross section drawn perpendicular to the shore **(Fig. 16.22a)**. Starting from the sea and moving landward, the *foreshore zone* includes the intertidal zone, across which the tide rises and falls, as well as the *beach face*, a steeper, concave-up area that forms where the swash of the waves actively scours the sand. The *backshore zone* extends from a small step, cut by high-tide swash, to the far edge of the beach farther inshore. The backshore zone includes one or more *berms*, horizontal to slightly landward-sloping terraces made up of sediment deposited during storms **(Fig. 16.22b)**. The sand on the beach, if exposed to persistent winds, may build into sand dunes, similar to those of deserts, along the landward edge of the beach. The inshore edges of some beaches, however, border cliffs or vegetated land not affected by waves.

Beach sand doesn't sit still, for typical wave action moves an *active sand layer* back and forth. (The *inactive sand layer* below moves only during severe storms or not at all.) The width of a beach ultimately reflects the width of the area subjected to frequent wave action. Beach profiles may vary seasonally depending on the types of storms in the area during different seasons. For example, in mid-latitudes, winter storms tend to be

stronger and more frequent than summer ones. The larger, shorter-wavelength waves of winter storms wash beach sand into deeper water, making the beach narrower. But smaller summer waves with longer wavelengths bring sand in from offshore and deposit it on the beach **(Fig. 16.23)**.

Where waves roll onto the shore at an angle, grains in the active sand layer follow a sawtooth pattern of movement that results in a gradual net transport of the sediment parallel to the beach; such movement is called **longshore drift** (or *beach drift*). This sawtooth pattern

FIGURE 16.24 The concept of longshore drift. Swash carries sand obliquely up the beach, whereas backwash carries it straight downslope. So sand grains follow a sawtooth pattern, yielding longshore drift (yellow arrow).

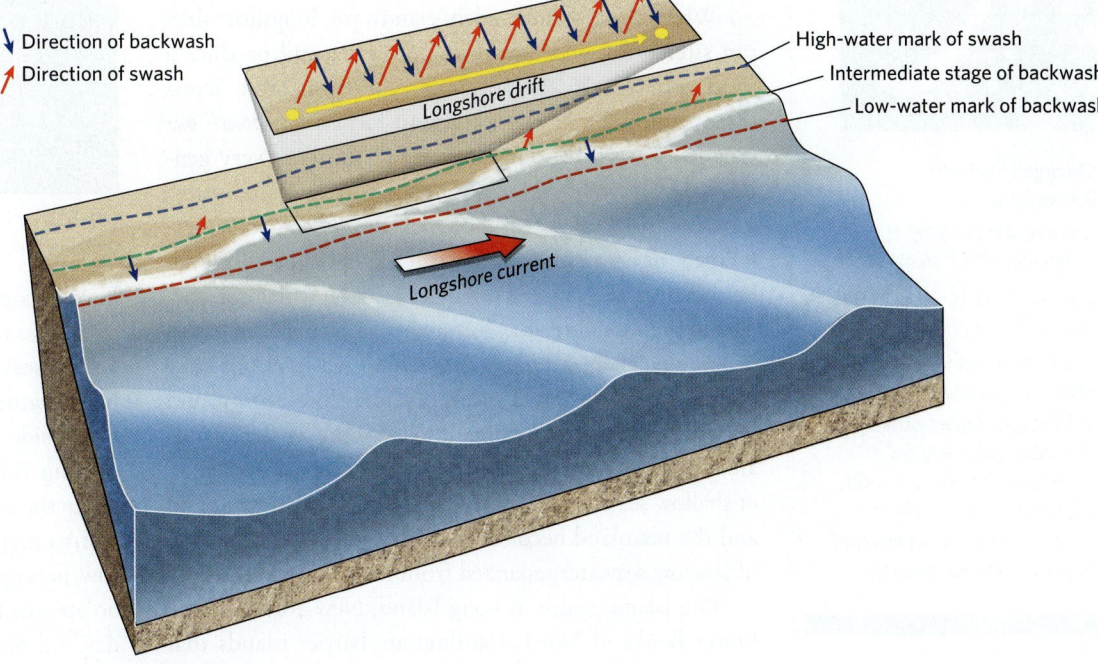

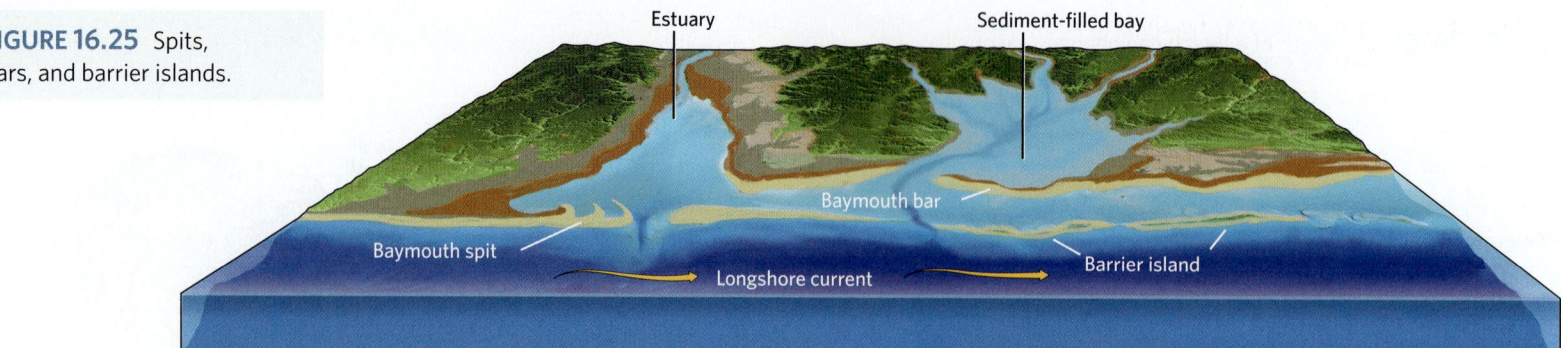

FIGURE 16.25 Spits, bars, and barrier islands.

(a) Longshore drift can generate sand spits and baymouth bars. Sedimentation may fill in the region behind the bar.

(b) An example of a sand spit.

(c) An example of a baymouth bar.

happens because the swash of a wave moves perpendicular to the wave crest, so an oblique wave carries sediment diagonally up the beach. Backwash, however, flows straight down the slope of the beach due to gravity. Over time, longshore drift may move sand tens to hundreds of kilometers along a coast (Fig. 16.24). For this reason, geologists sometimes describe beaches as "rivers of sand."

Where the coastline indents landward, longshore drift can stretch a beach out into open water and produce a **sand spit** (Fig. 16.25a, b). Some sand spits grow across the opening of a bay or estuary to form a *baymouth bar* (Fig. 16.25c). In regions where the coast has a very gentle slope and there is an abundant supply of sediment, a narrow ridge of sand, known as an *offshore bar* if submerged, or a **barrier island** if its crest lies above sea level, lies offshore (Fig. 16.25d). Barrier islands, which are parallel to the shoreline, may form when ancient beach dunes become partially submerged by rising sea level. Or they may form where waves break offshore, lose energy, and deposit some of their sediment load. Some barrier islands are simply very long sand spits that built out into a region of shallow sea offshore. The water between a barrier island and the mainland becomes a quiet-water **lagoon**, a body of shallow seawater separated from the open ocean.

Fire Island, south of Long Island, New York, and the Outer Banks of North Carolina are barrier islands that

(d) Barrier islands may be separated from the mainland by a lagoon.

have become popular with developers wanting to build expensive resorts. Unfortunately, such barrier islands, on a time scale of decades to millennia, are temporary geologic features. Wind and waves pick up sand from the ocean side of the island and drop it on the lagoon side, causing the island to migrate landward; longshore drift shifts the sand of the island and modifies its shape; and storms may breach the island to produce an *inlet* (a narrow passage of water between a lagoon and the ocean). So an area might seem like a prime construction site one day, but may be gone in the near future.

Tidal flats, which are broad, nearly horizontal areas of mud and silt that are exposed, or nearly exposed, at low tide but submerged at high tide, develop in regions protected from strong wave action (Fig. 16.26). They are typically found along the margins of lagoons or on shores protected by barrier islands, where, in relatively quiet water, mud and silt can accumulate in thick, sticky layers. Tidal flats provide a home for burrowing organisms such as clams and worms, so *bioturbation* (which means stirring by life) constantly mixes tidal-flat sediments and disrupts bedding.

Because of the movement of sediment on beaches and in adjacent areas, the **sediment budget**—the balance between sediment supplied and sediment removed—plays an important role in determining the long-term evolution of a coastal area. Let's look at how the sediment budget works for a small segment of beach (Fig. 16.27). Sand may be added to the segment by local rivers. It may also be brought from just offshore by waves, or from far away by longshore drift. (In fact, the large quantity of sand along the beaches of the southeastern United States may have been carried there by longshore drift from the outwash plains of Pleistocene Ice-Age glaciers in coastal New England.) Conversely, some of the sand may be removed from the segment by longshore drift, be carried offshore by backwash or rip currents, or be blown inland by wind. The offshore sediment either settles in deeper water or tumbles down a submarine canyon into the deep sea. If new sand does not replace the removed sand, the beach segment grows narrower. If, however, the supply of sand exceeds the amount that washes away, the beach becomes wider.

Rocky Coasts

More than one ship has met its end, smashed and splintered in the spray and thunderous surf, on a **rocky coast**, where bedrock cliffs rise directly from the sea (Fig. 16.28). Lacking the protection of a beach, rocky coasts feel the full impact of ocean breakers. Water pressure generated during the impact of a breaker can pick up boulders and smash them together until they shatter, and it can squeeze air into cracks, creating enough force to pluck chunks free from bedrock. Further, because of its turbulence, the water hitting a cliff face carries suspended sand that can abrade the cliff.

The combined effects of shattering, wedging, and abrading, together called **wave erosion**, can gradually undercut a cliff face by making a *wave-cut notch* (Fig. 16.29a). Undercutting continues until the overhanging rock becomes unstable and breaks away at a joint, toppling down to form a pile of rubble at the base of the cliff. Then, over time, wave action breaks up the rock until fragments become small enough to be transported offshore by moving water. In this

FIGURE 16.26 Mont-Saint-Michel, on the coast of France, was built on a tidal flat.

way, wave erosion continually cuts away at a rocky coast, so that the cliff gradually migrates inland. Such *cliff retreat* leaves behind a **wave-cut platform**, or *wave-cut bench*, that becomes visible at low tide (Fig. 16.29b).

Other processes besides wave erosion break up rocks along coasts. Salt wedging, for example, occurs as salt spray coats the cliff face above the waves and infiltrates

FIGURE 16.27 The sediment budget along a segment of coast. Sediment enters the system from rivers and by erosion of the land along the shore. Some sediment is moved along the coast by longshore drift, and some washes out to sea or avalanches down submarine canyons. Wind may blow beach sand inland.

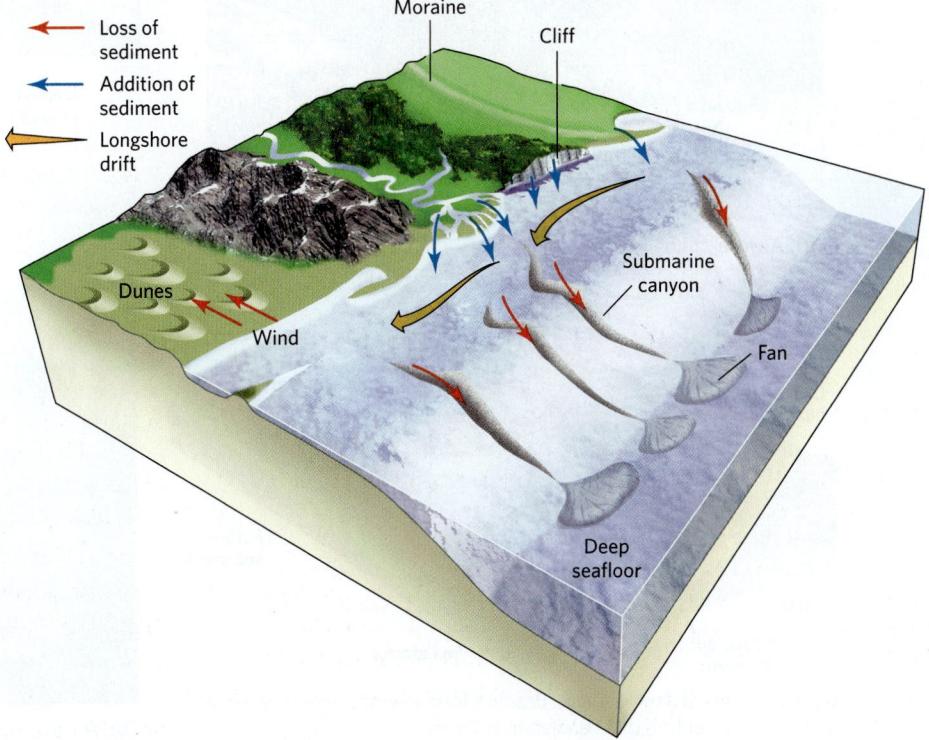

FIGURE 16.28 Exposed bedrock along rocky coasts is subjected to powerful, concentrated wave action. Rocky coasts develop unique landforms as a result.

(a) Cliffs rise from the sea along the coast of Wales.

(b) There's no beach to protect these rocky headlands in France from wave erosion.

FIGURE 16.29 Wave erosion of a rocky coast.

(a) A wave-cut notch.

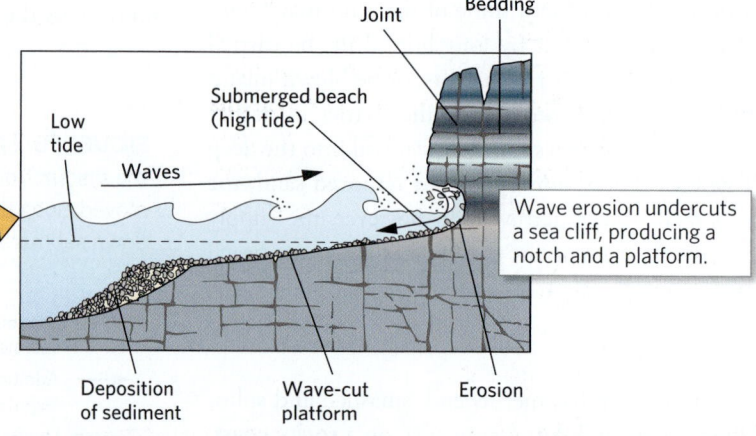

Joint · Bedding · Submerged beach (high tide) · Low tide · Waves · Wave erosion undercuts a sea cliff, producing a notch and a platform. · Deposition of sediment · Wave-cut platform · Erosion

Wave-cut platform exposed at low tide

Headland · Embayment · Tombolo · Sea cave · Wave-cut notch · Gravel beach · Pillar · Wave-cut platform · Sea stacks · Sea arch · Future sea stack

(c) Landforms of a rocky shore. Beaches form in embayments, whereas erosion is concentrated at headlands.

(b) A wave-cut platform at the foot of the cliffs at Êtretat, France.

FIGURE 16.30 Wave refraction at headlands.

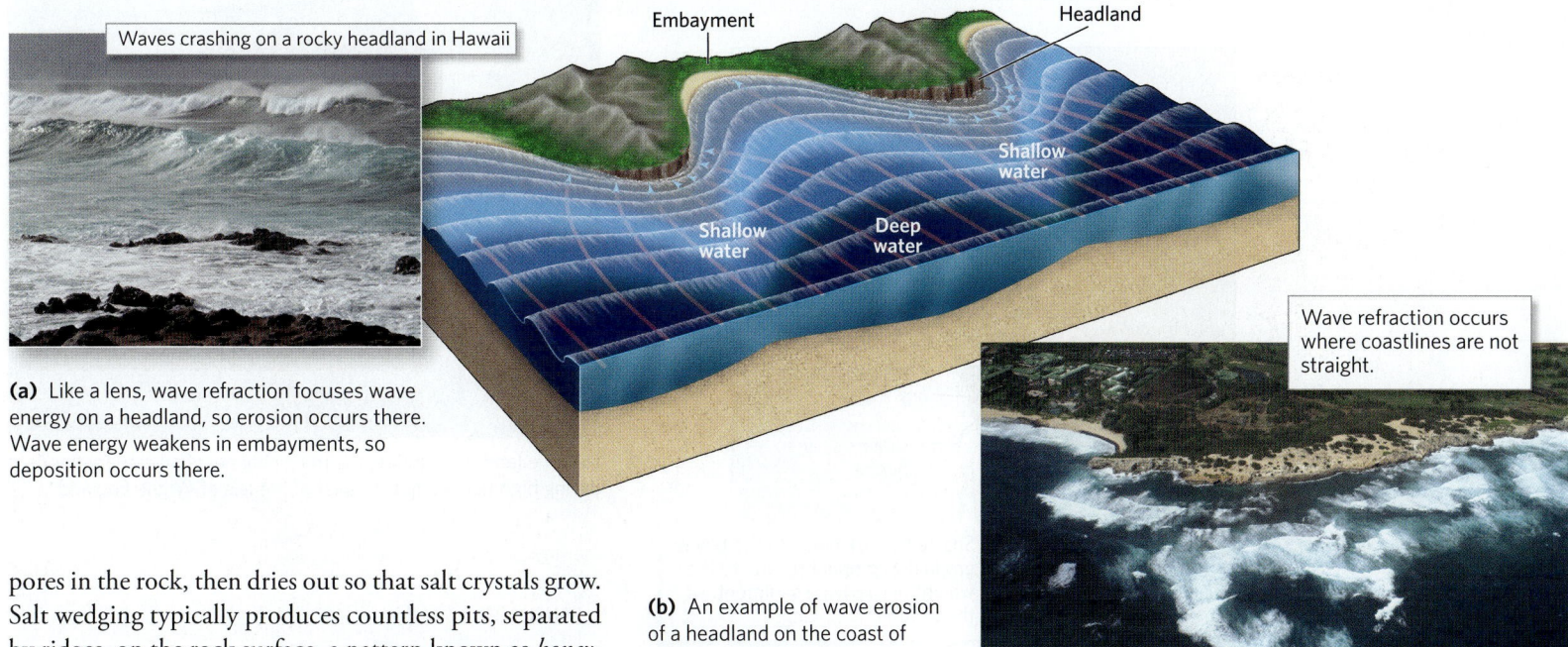

Waves crashing on a rocky headland in Hawaii

Embayment

Headland

Shallow water

Shallow water

Deep water

Wave refraction occurs where coastlines are not straight.

(a) Like a lens, wave refraction focuses wave energy on a headland, so erosion occurs there. Wave energy weakens in embayments, so deposition occurs there.

(b) An example of wave erosion of a headland on the coast of Hawaii.

pores in the rock, then dries out so that salt crystals grow. Salt wedging typically produces countless pits, separated by ridges, on the rock surface, a pattern known as *honey-comb weathering*. Biological processes also contribute to erosion as plants and animals in the intertidal zone bore into the rocks and gradually break them up.

Many rocky coasts start out with an irregular coastline, with **headlands** protruding into the sea and **embayments** set back from the sea **(Fig. 16.29c)**. Such irregular coastlines tend to be temporary features in the context of geologic time. As a result of wave refraction, wave energy focuses

on headlands and disperses in embayments. The resulting pattern of wave erosion removes debris at headlands, and sediment accumulates in embayments **(Fig. 16.30)**, so over time, the shoreline becomes less irregular.

A headland erodes in stages **(Fig. 16.31a)**. Because of refraction, waves curve and attack the sides of a headland, slowly eating through it to create a *sea arch* connected

FIGURE 16.31 Stages in the erosion of a headland.

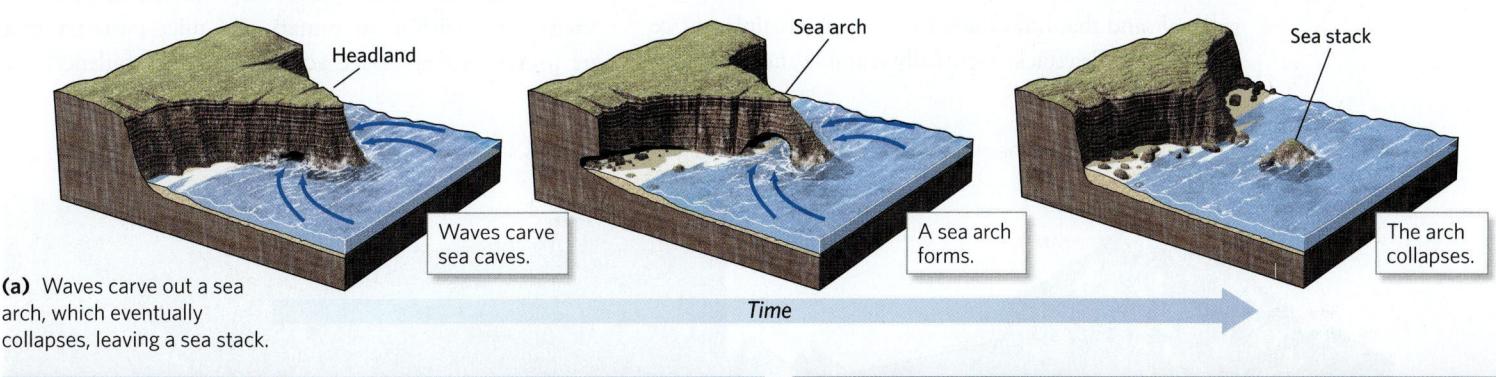

Headland

Sea arch

Sea stack

Waves carve sea caves.

A sea arch forms.

The arch collapses.

Time

(a) Waves carve out a sea arch, which eventually collapses, leaving a sea stack.

(b) An example of a sea arch (left) and sea stacks (right) on the southern coast of Australia. The sea stacks are part of a group known locally as the Twelve Apostles.

FIGURE 16.32 River valleys that are flooded by sea-level rise are called estuaries.

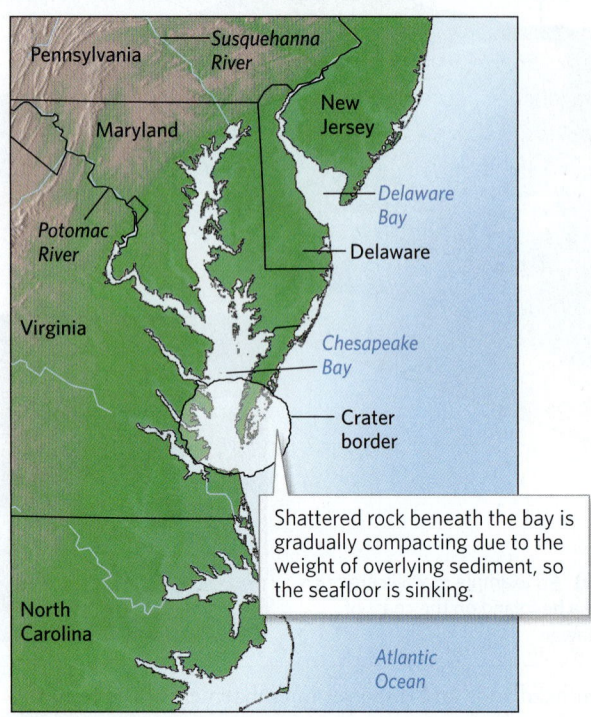

(a) Chesapeake Bay is an estuary at the mouth of the Susquehanna and Potomac Rivers. Recent research suggests that the mouth of these rivers subsided because a meteorite crater lies underneath.

Shattered rock beneath the bay is gradually compacting due to the weight of overlying sediment, so the seafloor is sinking.

(b) The land surrounding the main channel of an estuary floods during high tide along the coast of the Isle of Wight, England.

to the mainland by a narrow bridge. Eventually, the arch collapses, leaving isolated **sea stacks** just offshore **(Fig. 16.31b)**. Once formed, a sea stack protects the adjacent shore from waves. Therefore, sand collects in the lee of the stack, slowly building a *tombolo*, a narrow ridge of sand that links the sea stack to the mainland (see Fig. 16.29c). Sea stacks eventually crumble and disappear.

Estuaries

Along some coasts, a relative rise in sea level causes the sea to flood river valleys that merge with the coast, resulting in **estuaries**, where seawater and river water interact. Estuaries can develop where, when sea level was lower, a river cut a valley below present sea level, or where the region containing the river valley is slowly subsiding (sinking). You may be able to recognize an estuary on a map by the branching pattern of its river-carved coastline, inherited from the shape of dendritic drainage networks **(Fig. 16.32a)**. Estuaries are commonly bordered by marshes that flood at high tide **(Fig. 16.32b)**.

Seawater and river water may interact in two different ways within an estuary. In quiet estuaries that are protected from wave action or river turbulence, the

FIGURE 16.33 Coastal fjords form where sea-level rise drowns glacially carved valleys.

Glacial valleys have a U shape, so fjords have steep sides.

Fjord

A fjord in Norway

water becomes stratified, and the denser seawater flows upstream as a wedge beneath the less dense freshwater during flood tide. Such a *saltwater wedge* migrates about 100 km (62 mi) up the Hudson River in New York, and another moves about 40 km (25 mi) up the Columbia River in Oregon. In turbulent estuaries, such as the Chesapeake Bay, seawater and river water mix to produce nutrient-rich brackish water with a salinity between those of oceans and rivers. Estuaries typically host complex ecosystems inhabited by unique species of shrimp, clams, oysters, worms, and fish that can tolerate large changes in salinity.

Coastal Fjords

During the Pleistocene Ice Age, glaciers carved deep valleys in coastal mountain ranges. When the last ice age came to an end, the glaciers melted away, leaving deep, U-shaped valleys (see Section 14.7). When the water stored in the glaciers returned to the sea and caused sea level to rise, it flooded glacial valleys to produce coastal **fjords**, narrow fingers of the sea surrounded by hills or mountains. Because of their deep-blue water and steep walls of polished rock, fjords are distinctively beautiful (**Fig. 16.33**).

Take-home message . . .

A great variety of landscapes form along coasts. In regions with ample sediment supply, beaches build up, each with a distinctive profile. Longshore drift leads to the development of sand spits. In areas protected from wave action, muddy tidal flats develop. Rocky coasts feel the brunt of wave erosion and evolve over time. Where the sea floods the mouth of a river valley, an estuary develops, and where the sea floods a glacially carved valley, a fjord results.

Quick Question
How do lagoons and barrier islands develop?

16.5 Organic Coasts

So far, we've discussed coasts where the nature and distribution of sediments and rocks primarily determine the character of the shore. Along many coasts, however, living organisms control the landforms of shore and nearshore regions. Such **organic coasts** include coastal wetlands, where salt-resistant plants thrive, and coral reefs, where tiny marine animals build mounds of biochemical limestone. The nature of an organic coast depends on the types of organisms that live there, which in turn depends on climate.

Coastal Wetlands

It would be hard to find a more dramatic contrast to the crashing waves of rocky coasts than the gentlest type of nearshore environment, the coastal wetland. A **coastal wetland** is a vegetated, flat-lying stretch of coast that floods at high tide, becomes partially exposed at low tide, and does not feel the impact of strong waves. Some wetlands grow directly along the shore, whereas others occupy lagoons separated from the sea by a beach. Vast numbers of marine species spawn in wetlands—in fact, wetlands account for 10% to 30% of all marine organic productivity, even though they constitute just a tiny portion of ocean regions.

In mid-latitude climates, coastal wetlands include *swamps* (wetlands dominated by trees), *marshes* (wetlands dominated by grasses; **Fig. 16.34a**), and *bogs* (wetlands dominated by mosses and shrubs). In tropical or subtropical climates (between 30° N and 30° S), *mangrove swamps* thrive along the shore (**Fig. 16.34b**). Mangroves are trees that have evolved roots that can filter salt out of water, so they can survive in either freshwater or saltwater. Some mangrove species form a broad network of roots above the water surface, making the tree look like an octopus standing on its tentacles, and some send up small protrusions from the roots that rise above the water and

(a) A marsh along the coast of Cape Cod.

(b) A mangrove swamp growing along the shore of southern Florida.

FIGURE 16.34 Coastal wetlands.

FIGURE 16.35 The character and evolution of coral reefs.

(a) A reef off the shore of Honolulu, Hawaii.

(b) A reef as seen underwater; corals have a great variety of colors and shapes.

(c) A close-up of living coral polyps.

allow the plant to breathe. Dense stands of mangroves absorb the impact of waves and, by doing so, prevent coastal erosion.

Coral Reefs

Along the azure coasts of Hawaii, colorful mounds of living coral form shallow reefs just offshore (Fig. 16.35a). Snorkelers swimming over a reef will see a great variety of different coral species. Some species build mounds that look like brains, others like elk antlers, and still others like delicate fans (Fig. 16.35b). Sea anemones, sponges, clams, and many other organisms grow on and around the coral. Though at first glance, coral looks like a plant, it is actually a colony of tiny invertebrates related to jellyfish. An individual coral animal, or polyp, has a tube-like body with a head of tentacles (Fig. 16.35c). Corals obtain some of their nutrition by filtering nutrients from seawater; the

remainder comes from algae that live within the corals' tissues. Corals have a *symbiotic* (mutually beneficial) relationship with these algae, in that the photosynthetic algae provide nutrients and oxygen to the corals while the corals provide carbon dioxide and other nutrients to the algae.

Coral polyps secrete calcite shells, which gradually build into a mound of solid limestone. At any given time, only the surface of the mound is alive; the mound's interior consists of shells from previous generations of coral (Fig. 16.36). The realm of shallow water underlain by coral mounds, associated organisms, and debris constitutes a **coral reef**. Living corals must remain submerged, so the tops of the shallowest reefs lie just below the level of low tide, and they must receive abundant sunlight, so most growing reefs lie at depths of less than 60 m (200 ft). Coral reefs absorb wave energy and thus serve as a living buffer that protects coasts from erosion.

Corals need clear, warm (18°C–30°C, or 64°F–86°F) water with normal ocean salinity, so coral reefs grow only along unpolluted coasts at latitudes below about 30° (Fig. 16.37a). The largest reef in the world, the Great Barrier Reef, extends for a distance of close to 2,000 km (1,200 mi) along the northeastern coast of Australia and reaches a width of 120 km (75 mi) (Fig. 16.37b). Most reefs are much smaller.

Marine geologists distinguish three different shapes of coral reefs (Fig. 16.38). A *fringing reef* forms directly along the coast, a *barrier reef* lies offshore (separated from the coast by a lagoon), and an *atoll* forms a circular ring surrounding a lagoon. As Charles Darwin first recognized back in 1859, coral reefs associated with oceanic islands in the Pacific start out as fringing reefs, then later become barrier reefs and, finally, atolls. Darwin suggested, correctly, that this progression reflects the continued growth

FIGURE 16.36 What's inside a reef?

(a) A quarry in Florida cuts into coral that formed thousands of years ago.

(b) A close-up shows the internal skeleton of a long-dead coral.

of the reef as the island around which it formed gradually subsides. Eventually, the reef itself sinks too far below sea level to remain alive and becomes the cap of a guyot.

Take-home message . . .

Organic coasts are places where living organisms control landforms along the shore. Some organic coasts consist of wetlands in the intertidal zone, which host grasses, mosses and shrubs, or mangrove trees, depending on climate. Coral reefs, which develop primarily in warm, shallow, clear water, consist of generations of shells produced by corals and other organisms.

Quick Question

Why do fringing reefs evolve into atolls around oceanic islands of the Pacific?

16.6 Causes of Coastal Variation

Emergent versus Submergent Coasts

Changes in *relative sea level*, meaning the position of sea level relative to the land surface at a given location, can result either from *global sea-level changes* (the rise or fall of sea level worldwide; see Section 10.5) or from local vertical movement (uplift or subsidence) of coastal land

FIGURE 16.37 Reefs of the world.

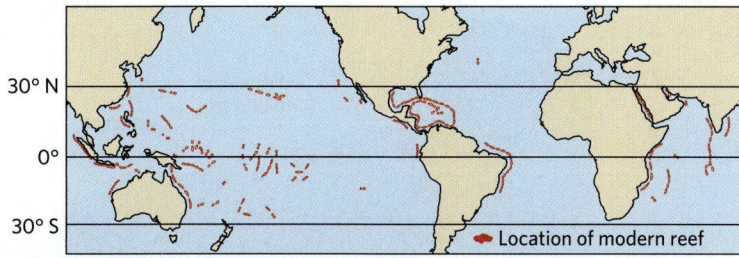

(a) Most coral reefs lie between 30° N and 30° S latitude.

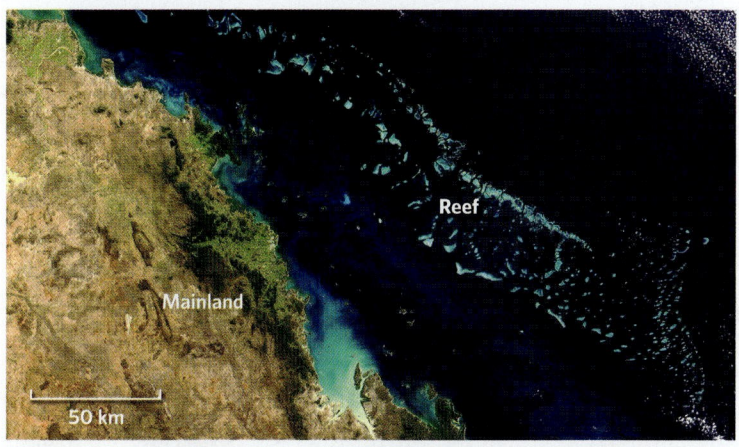

(b) The Great Barrier Reef forms shoals (submerged ridges) off the coast of northeastern Australia.

FIGURE 16.38 Evolution of reefs around oceanic islands in the Pacific.

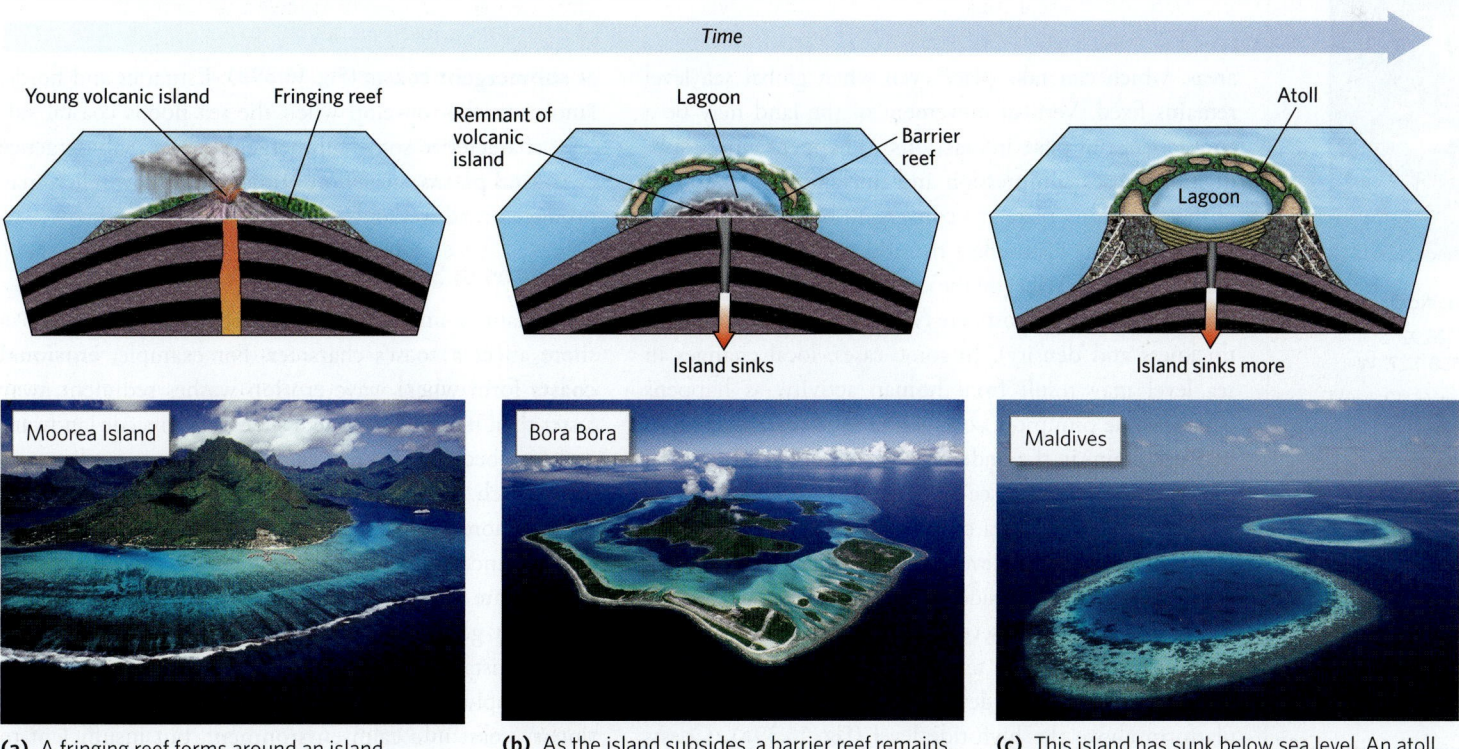

(a) A fringing reef forms around an island.

(b) As the island subsides, a barrier reef remains.

(c) This island has sunk below sea level. An atoll remains.

FIGURE 16.39 Features of emergent and submergent coasts.

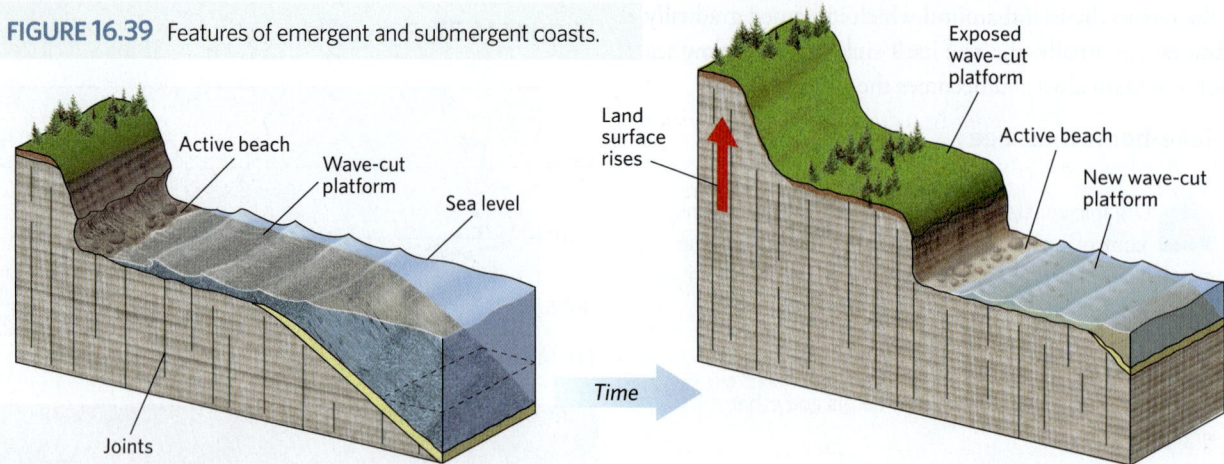

(a) Wave erosion produces a wave-cut platform along an emergent coast. As the land rises, the platform becomes a terrace, and a new wave-cut platform forms.

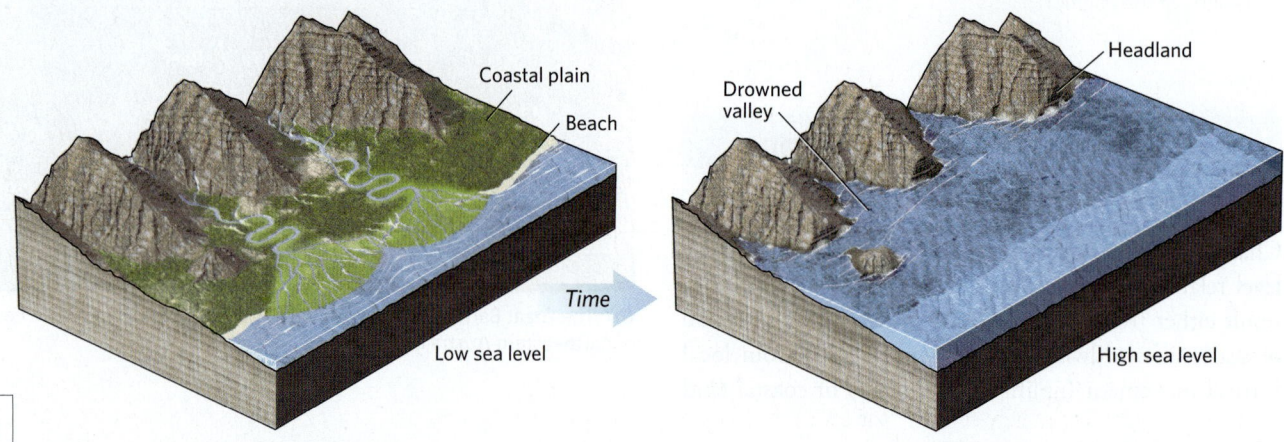

(b) Rivers drain valleys and deposit sediment on a coastal plain along a submergent coast. As the land sinks, sea level rises and floods the valleys, and waves erode the headlands.

areas, which can take place even when global sea level remains fixed. Vertical movement of the land may be a consequence of plate interactions, as happens where subduction causes compression and thickening in the crust of the overriding plate at a convergent boundary. Vertical movement may also reflect the addition or removal of a load (such as a glacier) on the crust's surface or the cooling and/or heating of lithosphere (which changes lithospheric thickness and density). In some cases, local changes in sea level may result from human activity, as happens when people pump out so much groundwater that pores between grains in the underground sediment collapse and the land surface sinks (see Section 13.9).

Geologists refer to a coastal location where the land is rising relative to sea level as an **emergent coast**. Along emergent coasts, steep-sided cliffs or hills may border the shore. In some cases, several step-like terraces delineate emergent coasts; these terraces form where a wave-cut platform has time to develop before uplift brings the platform above the high-tide level (**Fig. 16.39a**). Coasts where the land is sinking relative to sea level are known

as **submergent coasts** (**Fig. 16.39b**). Estuaries and fjords, landforms that develop when the sea floods coastal valleys, characterize some submergent coasts. Submergence of **coastal plains**, nearshore landscapes of low relief, may produce broad wetlands and lagoons.

Changes in Sediment Supply and Climate

The quantity and character of sediment supplied to a shore affect a coast's character. For example, **erosional coasts** form where wave erosion washes sediment away faster than it can be supplied; such coasts recede landward and may become rocky if there's insufficient sand to supply a beach. In contrast, **accretionary coasts**, those that receive more sediment than they lose by erosion, grow seaward and develop broad beaches.

Climate also affects the character of a coast. Shores that enjoy generally calm weather erode less rapidly than those constantly subjected to ravaging storms. A sediment supply may be large enough to generate an accretionary coast in a calm environment, but insufficient to prevent the development of an erosional coast in a stormy environment. The climate also affects biological activity along coasts. For example, in the warm water of tropical climates, mangrove swamps flourish along the shore and coral reefs form offshore. In cooler climates, marshes develop, and in polar regions, the coast may be a stark environment of lichen-covered rock and barren sediment.

Take-home message . . .

At emergent coasts, the land is rising relative to sea level, and terraces may develop. In contrast, at submergent coasts, sinking land allows the sea to flood valleys or to form broad wetlands. Whether coasts are emergent or submergent depends not only on global sea-level change, but also on local uplift or subsidence. The sediment supply relative to the erosion rate determines whether beaches build outward or erode away.

Quick Question ───────────────────
How does climate affect the character of the coast?

16.7 Challenges to Living on the Coast

People tend to view a shoreline as a permanent entity. But, as we have seen in this chapter, shorelines, like many geologic features, are ephemeral, in that they can change on a time scale of hours to millennia. Let's look at some of the changes now taking place along shorelines and how people respond to them.

Coastal Storms

When the winds of large storms, such as hurricanes, blow across the surface of the sea, they can generate immense waves. When these waves reach the shore, they become giant breakers that cause intense erosion and can destroy buildings, ships, and ports. During some types of storms, the wind, as well as changes in atmospheric pressure, causes sea level to rise to form a mound beneath the storm. This **storm surge** is independent of waves and tides. When storm surge reaches the coast at the same time as high tide and is accompanied by high waves, sea level temporarily becomes so high that ocean water can inundate land far inland of a beach. In a matter of hours, a storm can radically alter a beach that took centuries or millennia to build. The backwash of storm waves sweeps vast quantities of sand seaward, leaving the beach a skeleton of its former self. Surf during storms can submerge and shift barrier islands and can cut new inlets. Waves and wind together can rip out mangrove swamps and marshes and break up coral reefs, thereby destroying the organic buffer that normally protects the coast and leaving it

FIGURE 16.40 Beach erosion due to a hurricane.

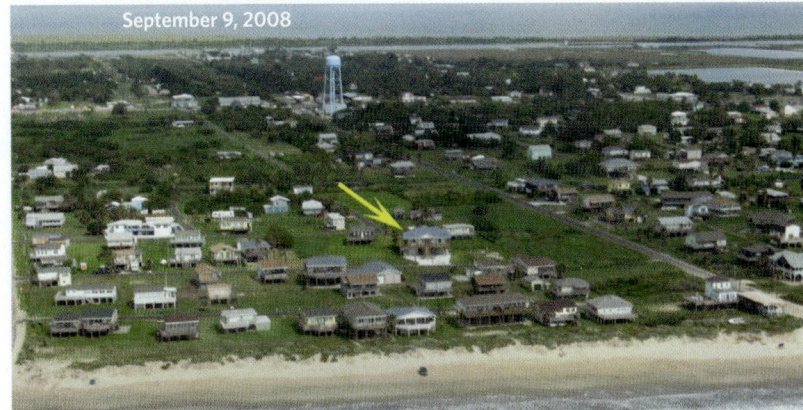

(a) A community in Galveston, Texas, before Hurricane Ike.

(b) The same location, after destruction by storm surge and storm waves.

vulnerable to accelerated erosion for years to come. Surf can also undermine coastal cliffs and trigger landslides.

Of course, major storms also destroy human constructions: erosion undermines seaside buildings, causing them to collapse into the sea; wave impacts smash buildings to bits; and storm surge floats buildings off their foundations **(Fig. 16.40)**. Because storms can be immensely destructive, coastal population centers need safety and evacuation plans, which should ensure that residents know what to do when storms approach.

Ongoing Sea-Level Rise

Global sea level is rising today at an estimated rate of about 3.2 mm/year (0.13 in/year), as we discuss further in Chapter 20. At this rate, sea level will be a meter or two higher on a time frame of centuries. Many of the world's large cities have been built on coastal land less than a few meters above sea level, so sea-level rise will be disruptive in those communities. Already, some coastal cities are enduring flooding during times of particularly high tides, known as *nuisance tides*, or when storms drive storm surge onto the land. By one estimate, 15% of the oceanic

FIGURE 16.41 Areas of the United States coast that may be submerged if sea level rises by 1.0–1.5 m.

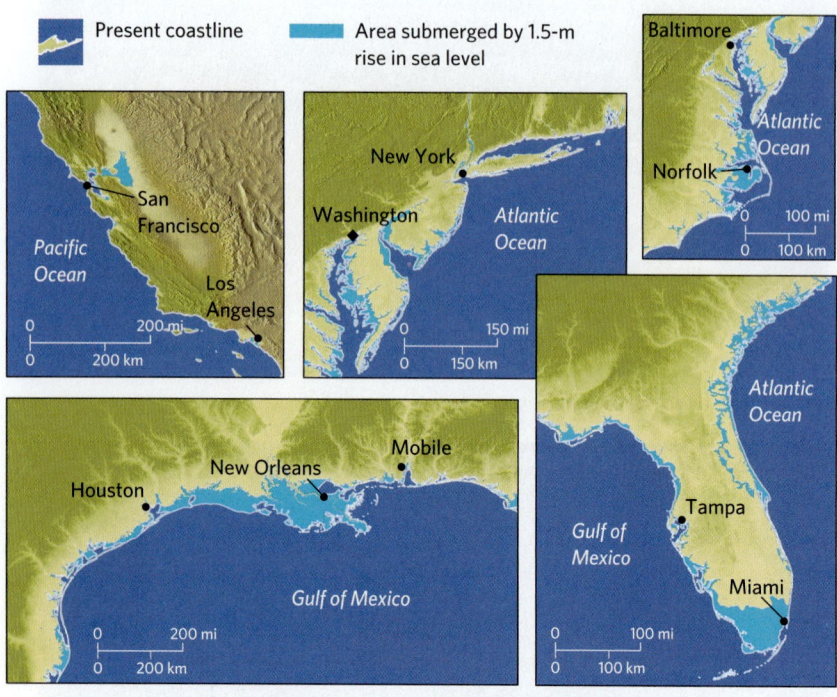

islands in the Pacific will eventually become submerged; some low-lying oceanic islands have all but disappeared already. In the United States, flooding will affect a greater area of the east and Gulf coasts than of the Pacific coast because slopes along the Pacific coast are much steeper (Fig. 16.41). As saltwater encroaches on the land, groundwater may become salty, and as a result, plants without salt tolerance will die off.

How will society deal with coastal flooding due to this ongoing sea-level rise? Solutions remain a work in progress. Some options, all of which would come at great

cost, include moving towns inland, adding fill to make the land surface higher, constructing barriers against the sea, and installing pumps and drainage canals to keep rising waters out.

Beach Destruction—Beach Protection?

As we've seen, storms can cause major changes to beaches overnight. But even less dramatic events, such as the loss of river sediment, a gradual rise in sea level, a change in the shape of a shoreline, or the destruction of coastal vegetation, can alter the sediment budget of a beach. If the sediment supply decreases, erosion eventually removes so much sediment that the beach becomes much narrower; sometimes beach erosion exposes underlying bedrock (Fig. 16.42). Cliff retreat at the back edge of a beach may cause the whole beach to migrate landward, a process known as *beach retreat*. Surveys indicate that beach retreat at some locations takes place at rates of 1–2 m (3–6 ft) per year. Some homeowners react by picking up and moving their houses. Even large lighthouses have been moved to keep them from washing away or tumbling down eroded headlands.

In many parts of the world, beachfront property has great value. Property owners may construct artificial barriers to protect their stretch of coast or to shelter the mouth of a harbor from waves. These barriers alter the natural movement of sand and can change the shape of the beach, sometimes with undesirable results. For example, people may build *groins*—concrete or stone walls perpendicular to the shore—to prevent longshore drift from removing sand from a beach (Fig. 16.43a, b). Sand accumulates on the updrift side of the groin, forming a triangular wedge, but sand erodes away on the downdrift side. Needless to say, the property owner on the downdrift side does not appreciate this process. At the entrance to a harbor, engineers may construct a pair of walls called *jetties* (Fig. 16.43c). Jetties, however, effectively extend the river channel into deeper water, which may lead to the deposition of an offshore sandbar. Engineers may also build an offshore wall called a *breakwater*, parallel or at an angle to the beach, to prevent the full force of waves from reaching the beach (Fig. 16.43d, e). With time sand builds up in the lee (landward side) of the breakwater, and the beach grows seaward. To protect expensive seaside homes, people build *seawalls* out of riprap (large stone or concrete blocks) or reinforced concrete on the landward side of the backshore zone (Fig. 16.44). But seawalls reflect wave energy, which crosses the beach back to sea, so this process increases the rate of erosion at the foot of the seawall. During a large storm, the seawall may be undermined so much that it collapses.

FIGURE 16.42 Examples of beach erosion.

(a) On the coast of Cape Cod, wave erosion has removed the entire beach right up to the coastal dune.

(b) Along the shore at Palm Beach, Florida, all the sand has been removed, and a ledge of rock has been exposed.

FIGURE 16.43 Techniques used to preserve beaches.

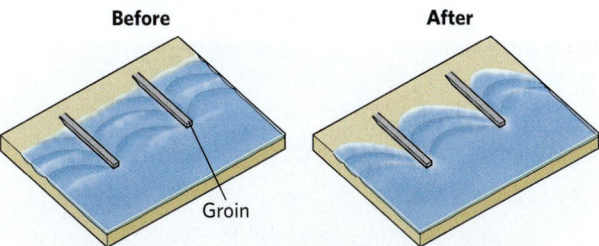

(a) The construction of groins may produce a sawtooth beach.

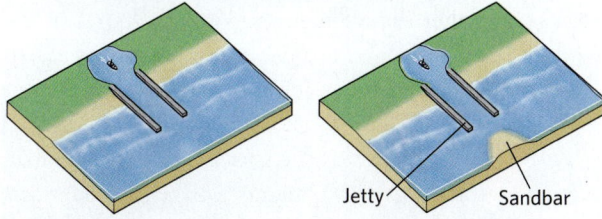

(c) Jetties extend a river channel farther into the sea, but may cause a sandbar to form at the end of the channel.

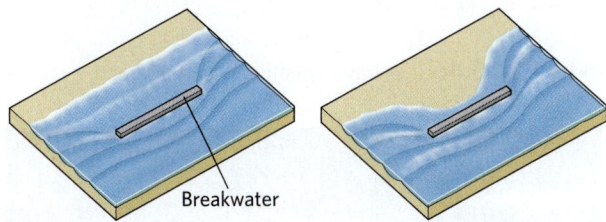

(d) A beach may grow seaward behind a breakwater.

(b) Groins along the shore in southern England.

(e) A breakwater along a beach near Boston, Massachusetts.

In some places, people have given up trying to decrease the rate of beach erosion and instead have worked to increase the sediment supply. To do this, they pump sand from farther offshore or bring in sand from elsewhere by truck or barge to replenish a beach **(Fig. 16.45)**. This procedure, called **beach nourishment** or *beach replenishment*, can be hugely expensive and at best may provide a temporary fix. The backwash and longshore drift that formerly removed the sand continue unabated as long as the wind blows and the waves break.

Pollution Problems

Bad cases of beach pollution create headlines. Because of longshore drift, garbage dumped in the sea in an urban area may drift along the shore and be deposited on a beach far from its point of introduction **(Fig. 16.46)**. For example, hospital waste from New York City has washed up on beaches tens of kilometers to the south. Oil spills, which may come from ships that flush their bilges with seawater, from tankers that have run aground or foundered in stormy seas, or from blowouts at offshore wells (see Box 11.2), have contaminated shorelines around the world.

Threats to Organic Coasts

Organic coasts, a manifestation of interaction between the physical and biological components of the Earth System, are particularly susceptible to changes in the environment. The loss of such landforms can increase a coast's vulnerability to erosion and, because organic coasts provide spawning grounds for marine organisms, can upset the food web of the global ocean.

In wetlands and estuaries, sewage, chemical pollutants, and agricultural runoff cause havoc. Toxins, along with clay, settle and concentrate in sediment, where they contaminate burrowing marine life. Fertilizers and sewage that enter the sea with runoff increase the nutrient content of seawater, stimulating algal blooms that absorb dissolved oxygen from the water and therefore kill animal and plant life. Coastal wetlands face destruction by development: many wetlands have been filled or drained for use as farmland or suburbs, and many have been used as garbage dumps. Throughout the world, between 20% and 70% of coastal wetlands have been destroyed in the last century.

Coral reefs, which depend on the health of delicate coral polyps, can be devastated by even slight changes in

FIGURE 16.44 A seawall can protect the sea cliff under most conditions, but during a severe storm, wave energy reflected by the seawall helps scour the beach. As a result, the wall may be undermined and will collapse.

Animation
Living with the Coast

Seawall

Beach

Reflected wave energy

Wall is undermined. Scouring

Time

Eroded cliff face

Rubble from seawall

Beach has disappeared.

A seawall built to protect expensive homes

FIGURE 16.45 Beach nourishment by piping in sand from offshore.

FIGURE 16.46 Garbage washing up on a beach.

the environment. Pollutants, particularly hydrocarbons, will poison the polyps. Sewage fosters algal blooms that rob ocean water of dissolved oxygen and suffocate the corals. And suspended sediment introduced to coastal waters, either by terrestrial runoff or by beach-nourishment projects, reduces light levels, killing the algae that live in the coral polyps, and clogs the pores that the polyps use to filter water. Changes in water temperature, salinity, and acidity caused by global warming of the atmosphere also destroy reefs, for reef-building organisms are very sensitive to temperature changes. People can destroy reefs directly by dragging anchors across reef surfaces, setting off explosives to kill fish, touching reef organisms, or quarrying reefs to obtain construction materials.

Sadly, in the last two decades, on the order of 50% of reefs worldwide have lost their color and died (Fig. 16.47). This process, called **reef bleaching**, may be due to the removal or death of symbiotic algae in response to the warming of seawater, or it may be a result of the dust carried by winds from desert or agricultural areas. Reef die-off has become a global crisis.

Clearly, managing the fragile realm of the coast has become a challenge for society. Understanding the processes we have discussed in this chapter may help us to define a way forward in addressing these problems. We'll return to this topic in Chapter 20, in the context of discussing global climate change.

Take-home message . . .

Coastal landscapes can be affected by storms, human activities, and sea-level change. Beaches naturally erode, and people try to protect them by constructing barriers or replenishing sand, but these actions can lead to other problems. Pollution and warming of seawater are destroying wetlands and coral reefs.

Quick Question
Why can a beach become polluted even if it is far from a source of pollution?

FIGURE 16.47 Reef bleaching in the Caribbean. The coral here is now colorless and mostly dead.

16 CHAPTER REVIEW

Chapter Summary

- The landscape of the seafloor depends on the character of the underlying lithosphere. Wide continental shelves form over passive-margin basins. Continental shelves may be cut locally by submarine canyons. Abyssal plains develop on old, cool oceanic lithosphere. Oceanic islands form above hot spots, and trenches, fracture zones, and mid-ocean ridges define plate boundaries.

- The largest tidal range is about 16 m (52 ft). Where coastlines have a gentle slope, the position of the shoreline can change by kilometers between high and low tide. Tides form as the Earth spins under two tidal bulges. Because of variations in ocean-basin and coastal shape, tides vary significantly around the globe.

- Tides are caused by the tide-generating force, a combination of forces applied by the gravity of the Moon and Sun, and by centrifugal force caused by rotation of the Earth-Moon system around its center of mass.

- The motion of waves changes as they approach the coast, where they can transform into breakers. Where waves intersect the shore at an angle, wave refraction and longshore drift of beach sediment take place.

- The swash and backwash of waves move sand on a beach and yield a distinctive beach profile. In some locations, barrier islands develop offshore.

- On rocky coasts, waves grind away at rocks, yielding such features as wave-cut notches and platforms, and sea stacks. Wave energy focuses on rocky headlands.

- Coastal wetlands, including marshes and mangrove swamps, host salt-resistant plants. Coral reefs grow offshore in warm, clear water. Reefs around oceanic islands evolve as the islands slowly sink.

- Differences in coastal landscapes depend on whether relative sea level is rising or falling (which may reflect global sea-level change or local uplift or subsidence) and on whether the coast is undergoing erosion or deposition.

- Coastal areas face many threats, both from nature (in the form of large storms and ongoing sea-level rise) and from human activities. To protect beach property, people build groins, jetties, breakwaters, and seawalls. These structures change the distribution and movement of sediment on beaches.

- Human activities have led to the pollution of coasts. Changes in oceanic environments have killed about half of the world's coral reefs.

Key Terms

abyssal plain (p. 489)
accretionary coast (p. 516)
backwash (p. 501)
barrier island (p. 508)
bathymetry (p. 487)
beach (p. 503)
beach nourishment (p. 519)
beach profile (p. 507)
breaker (p. 501)
coast (p. 488)
coastal plain (p. 516)
coastal wetland (p. 513)
continental shelf (p. 489)

coral reef (p. 514)
embayment (p. 511)
emergent coast (p. 516)
erosional coast (p. 516)
estuary (p. 512)
fjord (p. 513)
guyot (p. 494)
headland (p. 511)
intertidal zone (p. 495)
lagoon (p. 508)
longshore current (p. 502)
longshore drift (p. 507)
marine geology (p. 488)

ocean basin (p. 488)
oceanic island (p. 493)
organic coast (p. 513)
passive-margin basin (p. 490)
reef bleaching (p. 521)
rocky coast (p. 509)
sand spit (p. 508)
seamount (p. 493)
sea stack (p. 512)
sediment budget (p. 509)
shoreline (p. 495)
storm surge (p. 517)
submarine canyon (p. 492)

submarine fan (p. 492)
submergent coast (p. 516)
swash (p. 501)
tidal bore (p. 496)
tidal flat (p. 509)
tidal range (p. 495)
tide (p. 495)
tide-generating force (p. 497)
wave-cut platform (p. 509)
wave erosion (p. 509)
wave refraction (p. 502)

Letters in parentheses correspond to the chapter's learning objectives.

1. How does the lithosphere beneath a continent differ from that beneath an abyssal plain? **(A)**

2. How do the shelf and slope of an active continental margin differ from those of a passive margin? Why do passive-margin basins exist? **(A)**

3. Why is the sediment on the right side of the figure thicker than that near the ridge axis? **(A)**

4. What are tides, and why do they exist? Is the tidal range the same everywhere? Explain your answer. **(B)**

5. Which tidal bulge on the diagram is closer to the moon? **(B)**

6. How do waves change as they approach the shore? **(C)**

7. Why does longshore drift take place along some beaches? What landforms can result from longshore drift? **(C)**

8. Describe the components of a beach profile. What is a barrier island? **(C)**

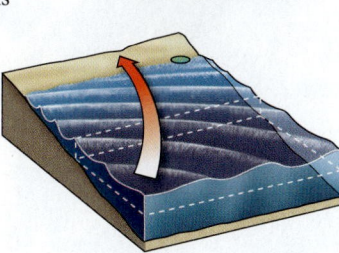

9. Identify the direction of longshore drift on the diagram. **(C)**

10. Describe how rocky coasts evolve. Why do headlands erode more rapidly than embayments? **(D)**

11. What is the difference between an estuary and a fjord? How do seawater and river water interact in an estuary? **(E)**

12. Discuss the different types of coastal wetlands and the factors that determine which type occurs in a given locality. **(E)**

13. How does a coral reef form, and how does a coral reef surrounding an oceanic island change over time? **(F)**

14. Which type of reef does the diagram show? **(F)**

15. What factors affect the local shape and evolution of a coastline? How does the sediment supply affect a beach? **(F)**

16. Explain the difference between an emergent and a submergent coast. **(F)**

17. Explain how storms and rising sea level affect coastal areas. **(F)**

18. In what ways do people try to modify or stabilize coasts? How do those efforts threaten coastal areas? **(G)**

On Further Thought

19. A hotel chain would like to build a new beachfront hotel along a north-south-trending stretch of beach where a strong longshore current flows from south to north. The neighbor to the south has constructed an east-west-trending groin on the property line. Will the groin pose a problem for the hotel beach? If so, what solutions could the hotel try? **(C, G)**

20. Much of southern Florida lies at elevations of less than 6 m (20 ft) above sea level. In recent years, as sea level has risen, parts of Miami Beach are flooding. The bedrock beneath the city consists of porous limestone, the remnants of ancient reefs. What protective measures might help mitigate this situation, if any? **(G)**

ANOTHER VIEW Coasts can be hazards for ships, so for centuries, people have marked their presence with lighthouses, such as this one in Maine.

17 THE AIR WE BREATHE
Introducing the Earth's Atmosphere

By the end of the chapter you should be able to . . .

A. explain how the atmosphere has evolved and how photosynthesis contributed to the modern atmosphere, which contains oxygen.

B. define the quantities that meteorologists use to describe the atmosphere, and list the instruments they use to monitor the atmosphere.

C. interpret a drawing that displays atmospheric layers.

D. evaluate the importance of atmospheric aerosols and the role that human activities play in generating them.

E. describe what clouds are, interpret the clouds you see in the sky, and explain how clouds produce precipitation.

F. create a model of the interactions between radiation and the atmosphere that control the temperature of the Earth's atmosphere and surface.

FIGURE 17.1 Atmospheric phenomena can be both beautiful and dangerous.

(a) A storm in the midwestern United States pours rain from the base of a swirling cloud.

(b) Arcs and halos at sunrise caused by interaction of sunlight with ice crystals in the air.

(c) A blizzard can bring traffic in a city to a standstill.

(d) Fog at the entrance to San Francisco Bay, California, hides part of the Golden Gate Bridge.

Introduction

Breathe deeply. You've just inhaled **air**, the unique mixture of nitrogen, oxygen, water vapor, carbon dioxide, and other gas molecules that makes up the Earth's **atmosphere**, the envelope or blanket of gas that surrounds our planet. Amazingly, some of the nitrogen that just entered your lungs was exhaled by Madame Curie, George Washington, Julius Caesar, Confucius, and Cleopatra; some of the water molecules once flowed down the Congo River, rained from Hurricane Katrina, and watered the gardens of Versailles; and some of the

>> In this view from an airplane window, we can see various types of clouds and the reflection of sunlight on the ocean below. The Sun's radiation drives the evaporation of water to provide the moisture in air, and it provides energy that keeps the air in motion.

oxygen molecules came from plants on your windowsill, algae in the ocean, trees of the rainforest, and grasses of the prairie. That's because the atmosphere flows and stirs constantly on local to global scales and interacts with many other components of the Earth System.

Without the atmosphere, the Earth would be as barren as the Moon. The atmosphere provides plants with carbon dioxide for photosynthesis, provides animals with oxygen for metabolism, shields our planet's surface from the dangerous ultraviolet radiation in sunlight, and carries water from where it evaporates to where it falls as *precipitation* (rain, snow, or hail). The atmosphere also serves as the medium in which birds, insects, and airliners fly and through which our voices carry. In human culture, the atmosphere's visual splendor—its auroras, rainbows, sunsets, and clouds—have inspired artists and sparked romances. The atmosphere can also unleash violent *storms*, intense episodes of locally strong winds and precipitation **(Fig. 17.1)**.

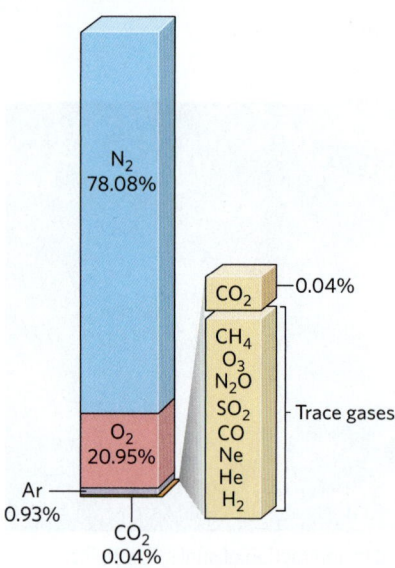

FIGURE 17.2 Gases of the Earth's atmosphere. The proportions shown here are those for an average sample of dry air; water vapor is not included.

Though we are merely bottom dwellers in this shared sea of air, we have become its caretakers, for our activities have changed, and will continue to change, the composition of the atmosphere. Furthermore, our vulnerability to storms and other atmospheric events increases as cities and farms spread into lands susceptible to flooding and drought. To prevent calamities due to atmospheric phenomena, and to ensure that the atmosphere can continue hosting our planet's rich biosphere into the future, it is essential that we strive to understand our atmosphere's character and behavior, and that we learn how to predict its short-term and long-term future. This chapter provides a foundation for addressing these challenges by discussing several basic questions: What is the atmosphere? How did it form and attain its present composition? What quantities can we use to describe atmospheric conditions? What are clouds, and why are there so many different types? Where does the energy that drives atmospheric phenomena come from? And how does the atmosphere interact with light?

17.2 Atmospheric Composition

The Recipe for Air

What is air made of? An average sample of dry air, meaning air from which all water has been removed, consists mainly of two gases: 78% (by volume) molecular nitrogen (N_2) and 21% molecular oxygen (O_2) **(Fig. 17.2)**. The remaining 1% includes argon (Ar) and carbon dioxide (CO_2), and trace gases such as ozone (O_3) and sulfur dioxide (SO_2). Although these other gases occur in such minuscule quantities, some, as we'll see, play key roles in controlling atmospheric temperature.

Note that when we describe the composition of the Earth's atmosphere, we consider only dry air. That's because in nature, the amount of water (H_2O) in air varies greatly from place to place at the same time and from

FIGURE 17.3 Water in the atmosphere occurs in three forms.

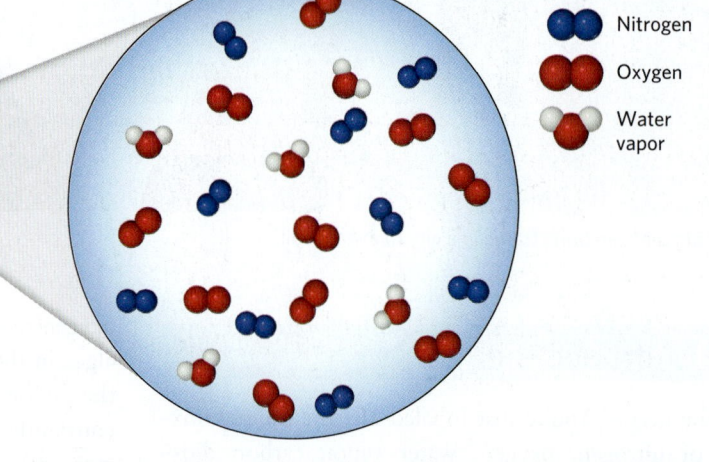

(a) Air contains invisible water vapor, even on this cloudless day on Cape Cod. Water molecules in vapor are mixed in with other gas molecules.

(b) Clouds in the warm sky over Brazil consist of liquid water droplets.

(c) High clouds over the University of Illinois consist of tiny ice crystals.

time to time in the same place, so we can't specify a single percentage that characterizes the average proportion of H₂O in the whole atmosphere. Water in the atmosphere occurs in three forms: as an invisible gas called **water vapor** (Fig. 17.3a), as a liquid (in tiny *droplets* or larger *drops*) (Fig. 17.3b), and as a solid (in the form of ice, snow, or hail) (Fig. 17.3c). Looking up, we often see **clouds**, collections of countless water droplets or tiny ice crystals suspended in air. Clouds normally appear to be floating above the Earth's surface, but when they come in contact with the ground, they are called *fog*.

Atmospheric Aerosols

Press the button on a can of air freshener, and countless extremely tiny particles spray into the air, particles so small that they can be wafted away by even gentle air currents. Researchers refer to such tiny particles—small enough to remain suspended in air—as **aerosols**. Air always contains many types of aerosols (Fig. 17.4a). Inorganic aerosols include specks of mineral dust lofted by winds blowing over soil, salts and sulfates carried into the air by sea spray, fine ash injected into the atmosphere by volcanic eruptions, and soot (carbon) particles billowing from forest fires (Fig. 17.4b). Organic aerosols include pollen, bacteria, molds, and viruses, as well as detritus from decaying organisms. Many aerosols are produced by human activities such as burning fossil fuels, spraying pesticides on crops, or using household products such as oven cleaner or deodorant.

Atmospheric aerosol concentrations vary with location. They tend to be greater over land surfaces than over the sea because there are more aerosol sources on land. Aerosols from land sources differ in composition from their oceanic counterparts. Most aerosols found over the oceans are salts and sulfates, whereas over land, most arise from

FIGURE 17.4 Atmospheric aerosols.

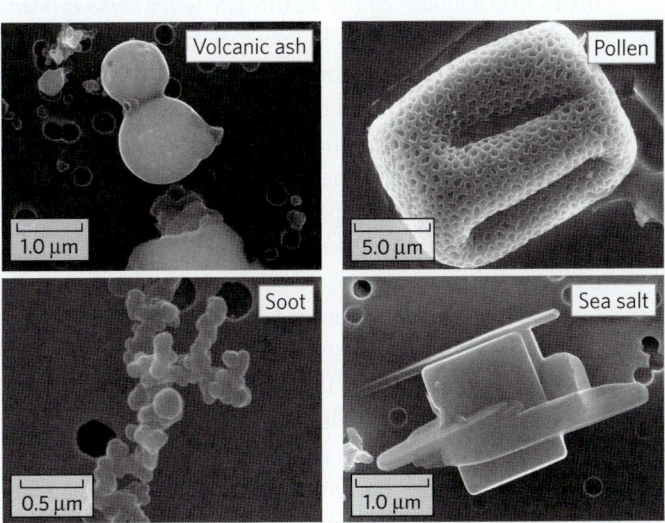

(a) Aerosols viewed through a scanning electron microscope (1 μm = 0.001 mm).

(b) Smoke rising from a forest fire.

(c) Dust blowing from the Sahara out over the Atlantic, as viewed from space.

(d) Haze over the Blue Ridge Mountains, part of the Appalachians.

(e) Photochemical smog at the Beijing airport includes particles from power plants, factories, and car exhausts.

windblown dust, fires, and human activities. Aerosols can travel long distances. Saharan dust blown from Africa, for example, reaches the Caribbean (Fig. 17.4c). Aerosol particles absorb light, so an abundance of aerosols in the air limits our ability to see into the distance. Under high-humidity conditions, some aerosols capture water molecules in the air and dissolve in the water to form microscopic liquid droplets, which produce **haze**. This haze causes the decrease in visibility that you see on sultry days (Fig. 17.4d).

Aerosols, along with certain gases such as ozone, contribute to air pollution. In discussions of pollution, aerosols may be referred to as *fine particulate matter*. The dank, dark *smog* that engulfed industrial cities of the 19th century formed when factories and heating stoves produced coal smoke that mixed with fog. In modern cities, *photochemical smog* develops when exhaust from cars and trucks reacts with air in the presence of sunlight to produce an ozone-rich brown haze (Fig. 17.4e). Government agencies monitor the aerosol concentrations in air to characterize air quality—high concentrations lead agencies to post health-hazard warnings.

Take-home message . . .

Dry air consists of nitrogen (78%), oxygen (21%), and other gases (1%). The amount of water in the atmosphere varies considerably over time and space. Water can occur as a gas, liquid, or solid. The atmosphere also contains tiny particles called aerosols, the presence of which can cause haze or smog.

Quick Question
What does smog consist of?

17.3 The Earth's Atmosphere in the Past

The Earth has not always hosted the oxygen-rich atmosphere that we inhale today. In fact, if you stepped out of a time machine into the air of the Precambrian, you would suffocate instantly. Let's look briefly at how our atmosphere has evolved over time, from the Earth's birth to today.

The First Atmosphere

When the Earth first formed, between 4.56 and 4.54 Ga, molecules of hydrogen (H_2) and atoms of helium (He), the most common gases of interstellar nebulae, along with molecules of ammonia (NH_3) and methane (CH_4), surrounded our planet (Fig. 17.5). This primordial atmosphere didn't last long. The gases were blown away by particles streaming from the newly formed Sun, and any that remained were blasted away during the Moon-forming collision. Afterwards, volcanic eruptions, as well as the arrival of comets, added molecules of many other gases (such as H_2O, CO_2, N_2, and SO_2) to the air. By about 4.0 Ga, the Earth had an atmosphere whose assemblage of gases differed markedly from that of a nebula. Specifically, the atmosphere consisted of about 55% H_2O, 15% CO_2, 15% N_2, 15% ammonia (NH_3), and traces of methane (CH_4), gases that came from *volcanic outgassing*—the emission, from volcanoes, of gases once bonded to minerals inside the Earth—but some probably came from impacting comets.

In sum, the Earth's *first atmosphere* began as a mix of nebular gases, but it quickly evolved into one composed of various volcanic gases. The atmosphere may have been modified further by reaction of CO_2 with the solid crust, perhaps during times when Hadean oceans formed. A rock record doesn't exist for this time period, and interpretations remain controversial.

The Second Atmosphere

Between about 4.0 and 3.9 Ga, so many meteorites pummeled the Earth that its surface may have temporarily become molten, and water could not remain liquid. Any ocean that had existed before this *late heavy bombardment* would have evaporated. The geologic record shows that by 3.85 Ga, after heavy bombardment ceased, our planet's surface once again cooled below the temperature at which water vapor condenses into liquid, and it has stayed that way ever since (except, perhaps, for short time intervals after occasional huge impacts). Liquid water collected as surface water (oceans, lakes, and rivers), as groundwater, and in solid form as ice.

Transfer of H_2O from the air into the liquid water reservoirs of the Earth caused the atmosphere's H_2O concentration to plummet. The appearance of liquid water, in turn, led to a major reduction in the atmosphere's CO_2 concentration, for large amounts of CO_2 dissolved in seawater and then precipitated to form solid carbonate sediment. Chemical weathering of rocks on land also absorbed CO_2. These processes locked much of the CO_2 that had been in the atmosphere into solid rock of the crust, so by about 3.8 Ga, the Earth had its *second atmosphere*, an atmosphere composed of volcanic gas, from which most water and CO_2 has been extracted. This atmosphere was composed predominantly of N_2, along with about 20% CO_2, lesser amounts of other volcanic gases, and variable amounts of water (see Fig. 17.5). Why did N_2 gas become predominant in the atmosphere? N_2 is an *inert gas*, meaning that it doesn't react with other chemicals. Therefore, when other gases became bonded to minerals or dissolved in the sea, N_2 remained behind in the atmosphere.

The Third Atmosphere

If the Earth were devoid of life, the atmosphere would contain hardly any *free oxygen* (O_2 molecules) because volcanoes do not emit O_2 gas. Fortunately for us, the Earth

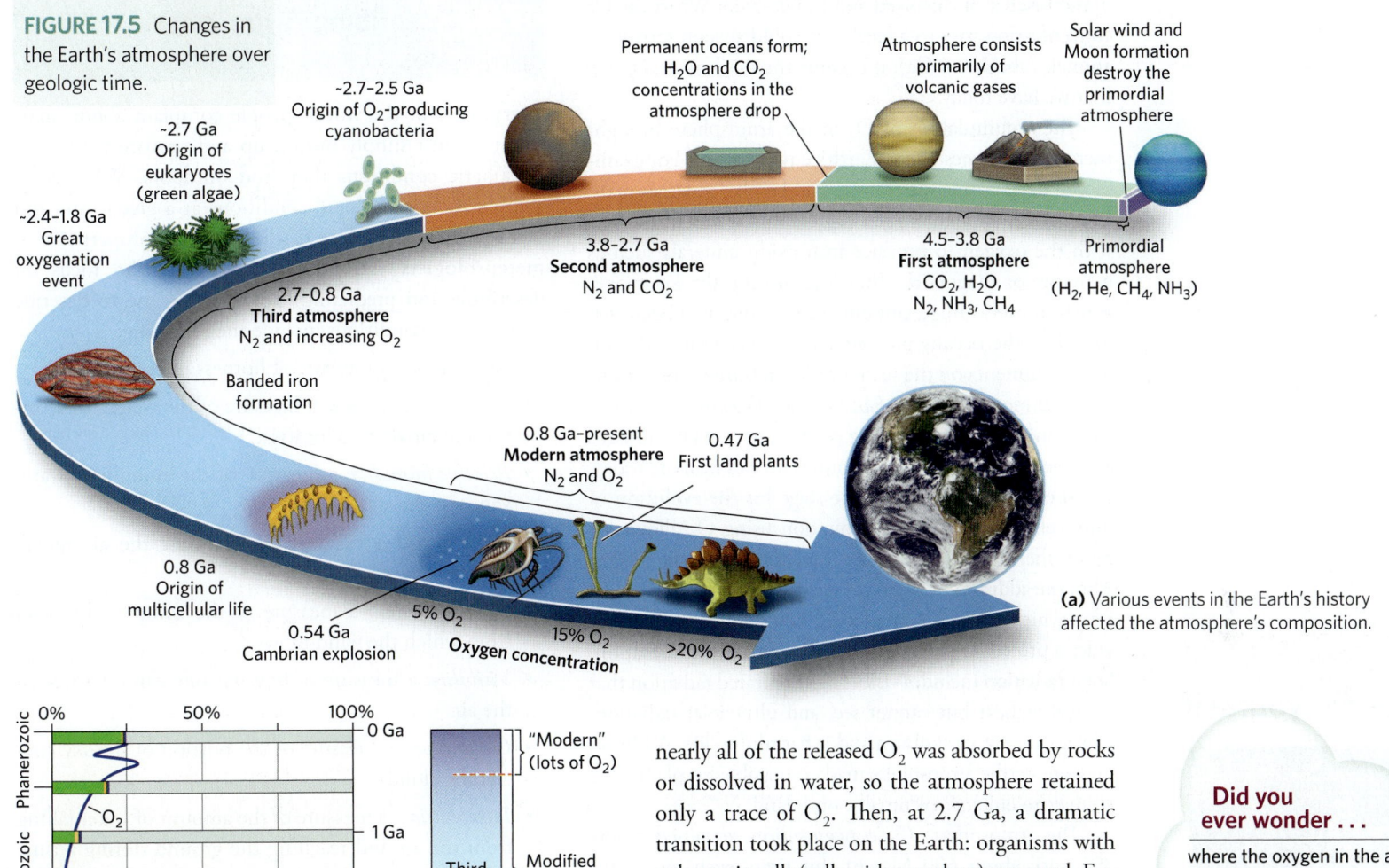

FIGURE 17.5 Changes in the Earth's atmosphere over geologic time.

~2.7–2.5 Ga
Origin of O_2-producing cyanobacteria

~2.7 Ga
Origin of eukaryotes (green algae)

Permanent oceans form; H_2O and CO_2 concentrations in the atmosphere drop

Atmosphere consists primarily of volcanic gases

Solar wind and Moon formation destroy the primordial atmosphere

~2.4–1.8 Ga
Great oxygenation event

3.8–2.7 Ga
Second atmosphere
N_2 and CO_2

4.5–3.8 Ga
First atmosphere
CO_2, H_2O, N_2, NH_3, CH_4

Primordial atmosphere (H_2, He, CH_4, NH_3)

2.7–0.8 Ga
Third atmosphere
N_2 and increasing O_2

Banded iron formation

0.8 Ga–present
Modern atmosphere
N_2 and O_2

0.47 Ga
First land plants

0.8 Ga
Origin of multicellular life

0.54 Ga
Cambrian explosion

5% O_2 15% O_2 >20% O_2
Oxygen concentration

(a) Various events in the Earth's history affected the atmosphere's composition.

0%   50%   100%

Phanerozoic · Proterozoic · Archean · Hadean

0 Ga · 1 Ga · 2 Ga · 3 Ga · 4 Ga

O_2

"Modern" (lots of O_2)

Third — Modified by life

Second — Volcanic gases (after ocean formation)

First — Volcanic gases (before ocean formation)

"Primordial" (gases from nebula; before Moon-forming collision)

O_2 | CO_2 | H_2O | N_2 | H_2 & He | NH_3 & CH_4

(b) A model for the compositional change of the atmosphere over geologic time. The first atmosphere may have changed many times, due to temporary episodes of ocean formation, and due to collision with comets. These changes aren't shown.

abounds with life. The first organisms appeared at about 3.8 Ga, and by 2.7 Ga, cyanobacteria, tiny blue-green single-celled organisms that carry out photosynthesis, were thriving in the ocean. *Photosynthesis*, a chemical process that takes place within certain types of living cells, extracts CO_2 from the atmosphere or ocean, converts it into organic chemicals, and releases O_2 as a by-product. At first,

nearly all of the released O_2 was absorbed by rocks or dissolved in water, so the atmosphere retained only a trace of O_2. Then, at 2.7 Ga, a dramatic transition took place on the Earth: organisms with eukaryotic cells (cells with a nucleus) appeared. Eukaryotic cells are more efficient than bacterial cells at carrying out photosynthesis, so the global rate of O_2 production started to increase, and between 2.4 Ga and 1.8 Ga, a time known as the **great oxygenation event**, O_2 began to accumulate to yield more than trace quantities in the atmosphere.

Photosynthetic organisms not only release oxygen, but also absorb CO_2. If all *biomass* (the material in organisms) decayed when the organisms died, the atmosphere's CO_2 concentration would have remained at about 20%, because decay returns CO_2 to the atmosphere. In the Earth System, however, not all biomass decays. Instead, some gets deposited, along with sediment, on the seafloor or on lake beds. If this organic-rich sediment becomes buried deeply enough to turn into rock, the biomass it contains effectively becomes locked underground. The trapping of organic matter in sedimentary rocks led to a further decrease of the CO_2 concentration in the atmosphere, from about 20% to a trace amount. As a result, the proportion of O_2 in air increased, and the Earth developed its *third atmosphere*, one influenced by life, and composed predominantly of N_2 and O_2. This atmosphere became progressively more

Did you ever wonder . . .

where the oxygen in the air comes from?

oxygen-rich as life evolved and proliferated. When the O_2 concentration rose to a level that could sustain terrestrial animals (about 0.5 Ga), it became the *modern atmosphere* that we have today.

The accumulation of O_2 in the atmosphere brought incredible changes to the Earth's environment. For example, in oxygen-free water, iron can dissolve in large quantities. But if enough oxygen is present in water, iron reacts with the oxygen to produce iron oxide minerals such as hematite or magnetite. Therefore, during the great oxygenation event, huge amounts of iron that had been dissolved in the oceans precipitated and accumulated with other sediments on the seafloor. When buried, these iron-rich sediments became *banded iron formation* (BIF), a rock consisting of alternating layers of iron oxide minerals and chert. BIF serves as our main source of iron ore today.

The presence of O_2 set the stage for the evolution of multicellular life because respiration using O_2 allows for more efficient metabolism and, therefore, larger organisms. Also, the addition of O_2 to the atmosphere provided atoms from which ozone (O_3) molecules could form. Ozone provides a protective shield by absorbing ultraviolet radiation. Solar radiation includes visible light, infrared radiation that we feel as heat but cannot see, and ultraviolet radiation, which can give us sunburn and is harmful to life. Addition of ozone to the atmosphere made it possible for plants and animals to emerge, eventually, onto land.

The atmospheric O_2 concentration remained below 5% until about 600 Ma. At this time, even more efficient oxygen-producing multicellular organisms evolved. During the Phanerozoic, the atmospheric concentration of O_2 has fluctuated. The highest concentrations occurred when the vast, O_2-producing forests of the late Paleozoic grew. (The buried debris from these forests became coal.) Notably, these high concentrations of O_2 allowed very large insects, such as 1-m (3-ft)-wide dragonflies, to survive in late Paleozoic forests. The atmosphere's O_2 concentration cannot rise higher than about 35%, however. If it did, immense wildfires would consume O_2-producing land plants, for burning takes place rapidly in O_2-rich air.

Take-home message . . .

The composition of the Earth's atmosphere has changed radically since the Earth first formed. A primordial atmosphere of nebular gases was quickly replaced by one dominated by volcanic gases. When the oceans formed, H_2O and CO_2 were removed from the atmosphere. Billions of years later, photosynthesis added O_2 to the atmosphere, and today, air consists almost entirely of N_2 and O_2.

Quick Question ——————————
Explain why the atmosphere could never contain 100% O_2.

See for yourself

Evidence of the Great Oxygenation Event

Latitude: 46°26′54.63″ N
Longitude: 87°37′31.97″ W

Look down from an altitude of 16 km (10 mi) and you can see a large open-pit mine near Ishpeming, Michigan. Miners here are digging up banded iron formation (BIF), formed from sediment deposited about 2 billion years ago, when ocean water began to contain significant quantities of dissolved oxygen. Iron oxide minerals give the BIF its dark, gray-red color. The rusty orange color in the settling pool comes from the iron oxide in the water.

17.4 Describing Atmospheric Properties

Everywhere in the world, people complain about, marvel at, or just simply button up and prepare for the atmospheric conditions they find outdoors. We refer to the specific atmospheric conditions at a given time and place as **weather**. Some of the common properties that **meteorologists**—atmospheric scientists who focus on describing and predicting the weather—use to describe atmospheric conditions include the following:

- *Temperature:* a measure of hotness or coldness
- *Atmospheric pressure:* a measure of the weight of a column of air above a location
- *Relative humidity:* a measure of the amount of water vapor in the air
- *Wind speed:* a measure of how fast the air moves horizontally
- *Wind direction:* a measure of the compass direction from which the wind blows
- *Visibility:* a measure of how far one can see through the air
- *Cloud cover:* a measure of the portion of the sky covered by clouds
- *Precipitation:* a measure of the amount of water falling from the air and reaching the ground during a time interval

Let's look at each of these properties in more detail and learn how they can be measured.

Air Temperature

We all pay attention to the air temperature—if it's hot, you might stroll about in shorts, but if it's cold, you need to put on a coat. But what exactly does *temperature* measure? In the air, gas molecules move rapidly in random directions, vibrate, and occasionally collide with one another. **Air temperature** represents the average speed at which these molecules move. The faster the average speed of the molecules, the higher the temperature. Note that *temperature* is not synonymous with *heat*. *Heat* refers to the total thermal energy contained in a material, meaning all the energy due to vibrations and movements of all atoms or molecules in the material. A volume of air near the top of the atmosphere contains few, but very fast-moving, gas molecules. It has a very high temperature, but it does not contain as much heat as the same volume of air at the same temperature at sea level, for the sea-level air contains a great many more molecules.

We measure temperature with a **thermometer**, calibrated in degrees according to standard scales (**Fig. 17.6**). Originally, thermometers consisted of a thin column of

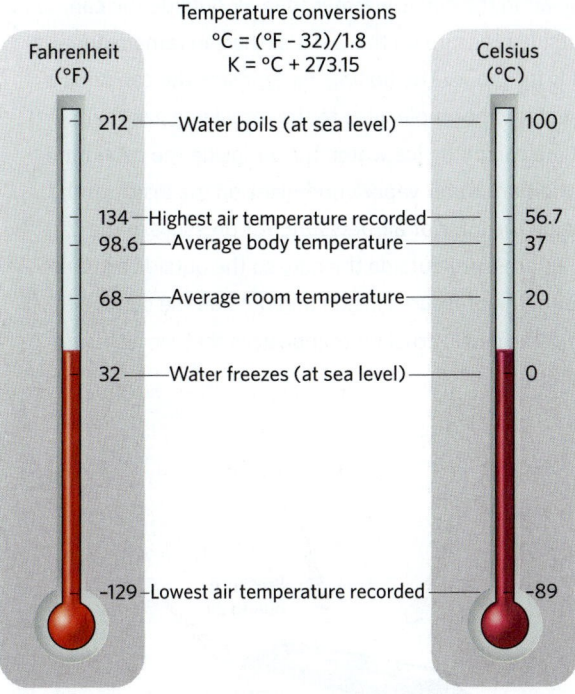

lowest air temperature was recorded at Vostok Station, Antarctica, where the thermometer dropped to −89.2°C (−128.6°F).

Atmospheric Pressure and the Height of the Atmosphere

Atmospheric pressure refers to the force applied by the overlying air on a surface of a specified area, such as a square meter, a square inch, or a square centimeter. You can picture atmospheric pressure as the weight of a column of air over a unit area at the base of the column. For example, using English units, imagine a column of air that is 1 in by 1 in at its base and extends from the ground at sea level all the way through the atmosphere to the edge of space. On average, this column weighs 14.7 lb, so we can say that the average atmospheric pressure at sea level is 14.7 lb/in^2. Atmospheric pressure exists because air, like any substance, has mass, so in the Earth's gravitational field, it has weight.

How is pressure different from the *directed force* that you feel when your friend pushes you? Pressure, by definition, acts in all directions equally. You don't get knocked over by air pressure in still air because the air pressure pushing on your front is the same as that pushing on your back. Similarly, an empty can doesn't collapse because air pressure pushing on its inside walls equals that pushing on its outside walls (Box 17.1).

Note that pressure affects the density of a gas. As pressure increases, air molecules get pushed closer together, and as pressure decreases, air molecules move farther apart. The density of a gas also depends on temperature. If you heat a given volume of air, it expands, because the molecules start moving faster. If the air is confined in a container, the pressure in the container increases. Similarly, if you cool air, it contracts, and if the air is confined in a container, its pressure decreases.

As Table 17.1 shows, we can use several different units to specify atmospheric pressure. These days, meteorologists prefer either *millibars* (mb) or the numerically

mercury in a calibrated glass tube. Today, most thermometers use alcohol because mercury is toxic. When the alcohol warms, it expands, so it rises in the tube. Meteorologists can also measure temperature with a *thermistor*, a sensor containing a conductor in which the resistance to the flow of electric current depends on temperature. Most thermometers in the United States still use the *Fahrenheit scale*, whereas those in the rest of the world use the *centigrade scale*, also known as the *Celsius scale*. At sea level, water boils at 212°F and freezes at 32°F on the Fahrenheit scale, and it boils at 100°C and freezes at 0°C on the centigrade scale. An increment of temperature, a *degree*, represents a greater temperature change on the centigrade scale than on the Fahrenheit scale. Scientists commonly use another scale, the Kelvin scale. *Absolute zero*, the lowest temperature possible, is 0 K (−273.15°C)—at 0 K, molecules stop moving or vibrating. (Note that when we write a temperature using the Kelvin scale, we do not use the word *degree* or the degree symbol.)

At any particular location, air temperature fluctuates daily as the Sun rises and sets. It also changes seasonally because the number of daylight hours and the Sun's position in the sky vary with the time of year. Temperature at a location can also change when volumes of cooler or warmer air move in from elsewhere. The highest air temperature recorded on land was measured in Death Valley, California, where the thermometer reached 56.7°C (134°F). The

TABLE 17.1 Common Units Used for Measuring Atmospheric Pressure

Unit	System	Average Atmospheric Pressure at Sea Level
Pounds per square inch	English	14.7 lb/in^2
Pascals (Pa)	Metric	101,325 (= 1,013.25 hPa)
Atmospheres (atm)	By convention*	1.000 (= 1.013 bar)
Bars (= 1,000 Pa)	By convention*	1.01325 (= 1 atm)
Millibars (mb) (= 0.001 bar)	By convention*	1,013.25 (= 1,013.25 hPa)
Inches of mercury	Traditional	29.92 in.

*"By convention" means that the number was defined by an international agency.

BOX 17.1

How can I explain . . .

Atmospheric pressure

What are we learning?
- That the force exerted by atmospheric pressure is real.
- That atmospheric pressure is related to the density of air.

What you need:
- An open, empty aluminum soda can
- A bowl filled with water and ice
- A small hot plate and a pair of tongs

Instructions:
- Put a small amount of water in the can. Set the hot plate on low heat and place the can, open top up, on the heating element. Wait for the air in the can to heat and the water to boil.
- Using the tongs so you don't burn yourself, grasp the can, invert it, and place it open-side-down into the ice water.
- The can will immediately crumple.

What did we see?
- Before the can is heated, the atmospheric pressure exerted on the inside of the can wall and the outside of the can wall is the same. When you put the can on the hot plate, the gas in the can begins to expand, so it escapes from the top opening.
- As air escapes, the amount of air, and therefore the air pressure inside the can, decreases.

- However, because air in the can is warmer than air outside the can, the inside and outside pressure on the walls of the can remain equal, and the can retains its shape. The boiling water inside the can adds a large number of water vapor molecules to the remaining gas there.
- When you invert the can in the ice water, the air inside the can immediately cools. In addition, water vapor condenses on the inside of the can, further reducing the interior air pressure. Air pressure in the can no longer equals air pressure outside the can, so the outside pressure crushes the can until the pressures inside and outside the can are the same. The fact that the can is crushed emphasizes that air, although invisible, exerts pressure.

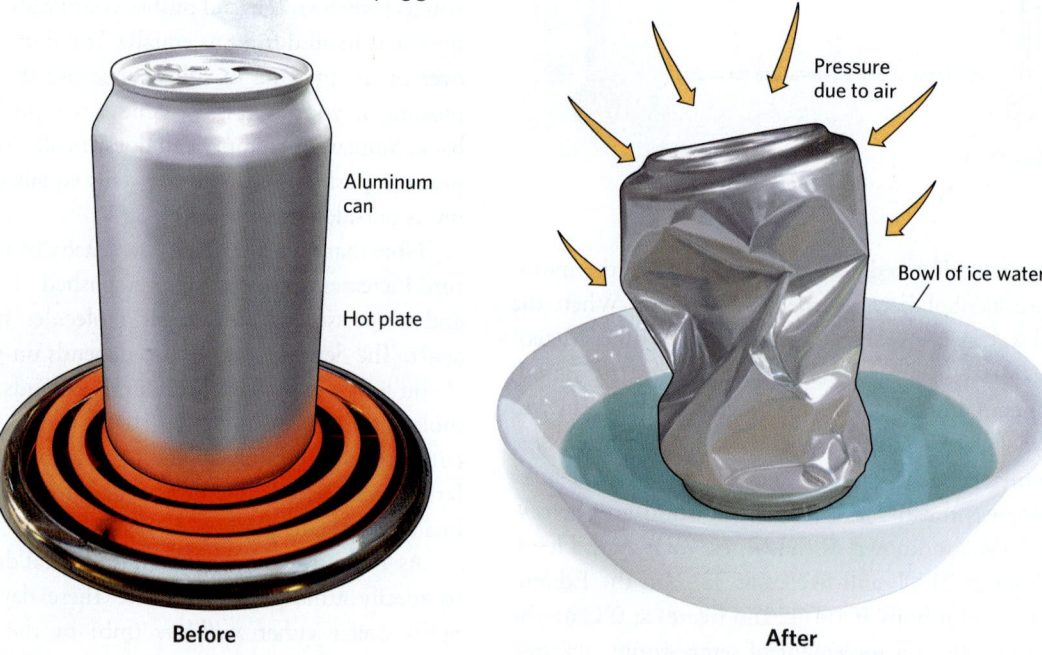

Escaping gas

Aluminum can

Hot plate

Pressure due to air

Bowl of ice water

Before

After

---

equivalent *hectopascals* (hPa). In this book, we'll use the more common unit, millibars. The numbers given for "average atmospheric pressure at sea level" in Table 17.1 are just that—averages. The specific value of atmospheric pressure varies with time and location at a given elevation, but it always decreases as elevation increases because the higher you go, the less gas lies above you.

Meteorologists measure atmospheric pressure with a **barometer**. A traditional mercury barometer consists of a dish of mercury into which a vertical glass tube has been inserted **(Fig. 17.7a)**. The tube is open on the bottom and closed at the top, and the air has been pumped out of the tube, so it contains a vacuum. As the atmosphere pushes down on the surface of the mercury, the mercury rises into the tube. The higher the air pressure, the greater the weight of air pushing on the mercury, so the higher the mercury rises in the tube. When the pressure equals the average (mean) atmospheric pressure at sea level (1,013.25 mb), the column attains a height of 29.92 in (see Table 17.1). Most television meteorologists

FIGURE 17.7 How a barometer works.

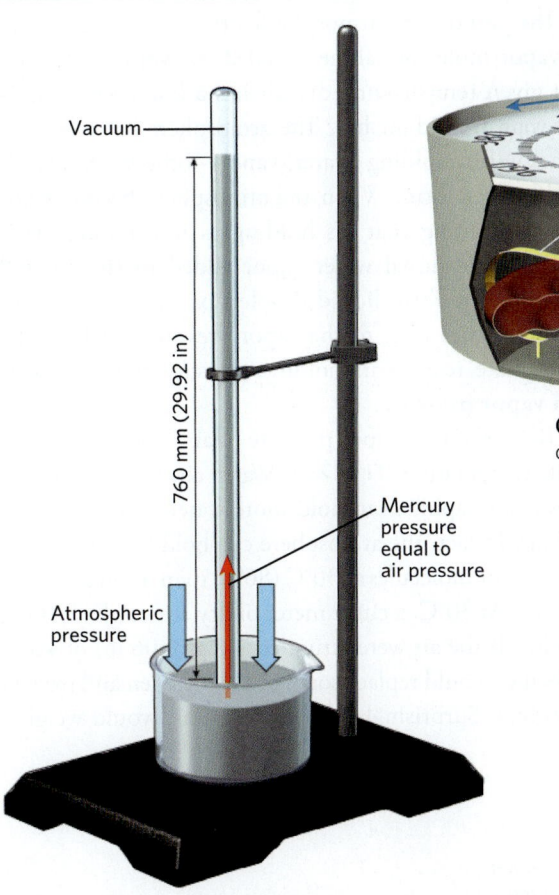

As air pressure decreases, the diaphragm expands; levers cause the dial to rotate counterclockwise.

Pointer

Spring   Lever

Diaphragm

(b) In an aneroid barometer, compression or expansion of a diaphragm moves a dial.

Vacuum

760 mm (29.92 in)

Mercury pressure equal to air pressure

Atmospheric pressure

(a) In a traditional mercury barometer, air pressure pushes mercury up a vacuum tube. The greater the pressure, the higher the mercury rises.

and newspaper weather pages in the United States still report pressure in "inches of mercury," the height of the mercury column in a mercury barometer. If atmospheric pressure is decreasing, meteorologists say that "the barometer is falling" because the height of the mercury column decreases. Conversely, if air pressure is increasing, they say that "the barometer is rising." Mercury barometers were actually replaced long ago by *aneroid barometers*, which consist of a flexible metal diaphragm containing a vacuum (Fig. 17.7b). As pressure increases, the walls of the diaphragm move closer together, and if pressure decreases, the walls move farther apart. This movement drives a lever, which in turn moves a dial calibrated in pressure units.

Atmospheric pressure, and therefore atmospheric density, decreases by about half for every 5.6 km (3.5 mi) of elevation gain (Fig. 17.8). As a result, about 50% of the atmosphere's gas lies below an elevation of 5.6 km, and 75% lies below an elevation of 11.2 km (7 mi), the elevation at which commercial jets fly. People can't survive long at elevations above about 6 km (3.7 mi) because air doesn't contain enough oxygen.

Continuing upward, we find that 99% of gas in the atmosphere lies below 50 km (31 mi). Nevertheless, there's enough gas above that elevation that meteors begin to heat up and vaporize when they pass below 70–120 km (43–75 mi). The "top" of the atmosphere, where the density of the atmosphere equals the density of interplanetary space, varies between 350 and 800 km (217 and 497 mi) in elevation.

Sea-level air pressure varies across the globe at any given time—rarely will a direct measurement of atmospheric pressure at sea level at a particular location yield a number exactly equal to average sea-level pressure. Corrected for elevation, the highest sea-level pressure ever recorded was 1,085 mb in Siberia, whereas the lowest ever recorded occurred in the center of a typhoon (a Pacific hurricane), where the pressure dropped to 870 mb. Regions of higher pressure develop beneath colder, and therefore denser, air, whereas regions of lower pressure occur beneath warmer, less dense air. As we'll see in Chapter 18, horizontal variations in atmospheric pressure control the speed and direction of the wind.

Moisture and Relative Humidity

Water vapor, an invisible gas, consists of freely moving water molecules. These molecules enter the atmosphere by evaporation from the Earth's oceans, lakes, and rivers and from plants by transpiration.

FIGURE 17.8 Atmospheric pressure and air density decrease with increasing elevation.

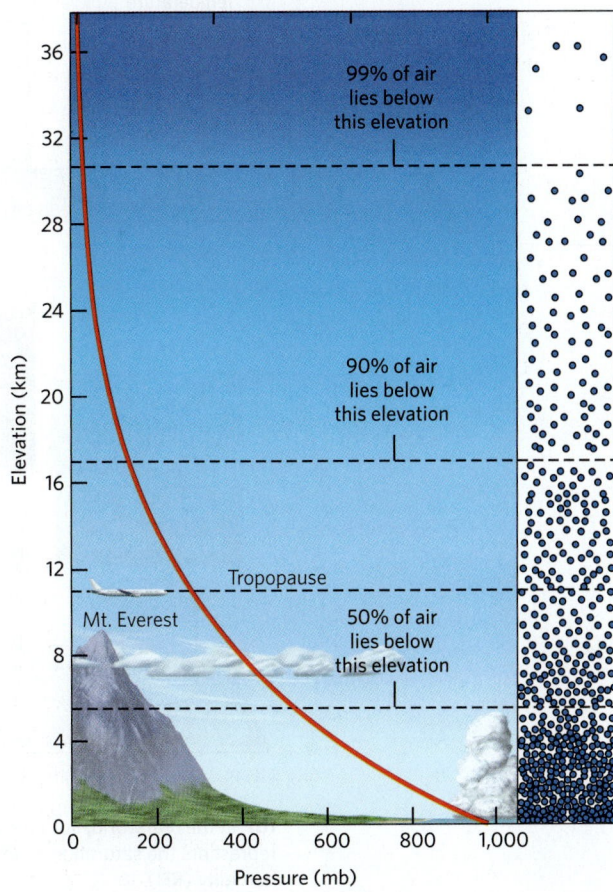

99% of air lies below this elevation

90% of air lies below this elevation

Tropopause

Mt. Everest

50% of air lies below this elevation

Elevation (km)

Pressure (mb)

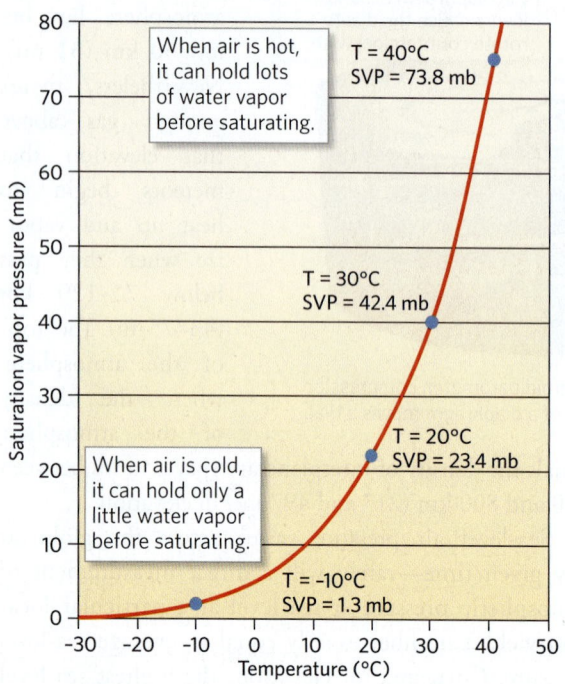

FIGURE 17.9 Saturation vapor pressure (SVP) decreases rapidly as temperature (T) decreases.

When air is hot, it can hold lots of water vapor before saturating.

T = 40°C
SVP = 73.8 mb

T = 30°C
SVP = 42.4 mb

T = 20°C
SVP = 23.4 mb

When air is cold, it can hold only a little water vapor before saturating.

T = −10°C
SVP = 1.3 mb

Saturation vapor pressure (mb)

Temperature (°C)

Meteorologists use several quantities to characterize the water vapor content of the atmosphere.

The part of total atmospheric pressure exerted by water vapor molecules alone is called the **vapor pressure**. At a given temperature, dry air has a lower vapor pressure than humid air has. The atmosphere has a limited capacity for holding water vapor—otherwise, clouds would never form. When the atmosphere becomes *saturated*, meaning that it's holding as much water as it can, any additional water vapor added to the air will condense and form liquid droplets or solid ice crystals. Meteorologists refer to the vapor pressure at which the atmosphere reaches its holding capacity as the **saturation vapor pressure**.

The saturation vapor pressure of air depends strongly on its temperature (Fig. 17.9). Warm air, in which molecules jostle rapidly, can hold more water vapor than can cold air. In fact, the atmosphere can hold 84 times more water vapor molecules at 30°C (86°F) than it can at −30°C (−22°F). At 30°C, a cubic meter of dry air weighs 1,160 g (2.6 lb). If the air were saturated, 28 g (0.06 lb) of water molecules would replace some of the nitrogen and oxygen molecules. Surprisingly, the saturated air would weigh a

FIGURE 17.10 Relative humidity.

High RH   Temperature   Low RH

(a) Because cold air can hold less water vapor than warm air, the same volume of water vapor in the air makes cold air more humid.

Hot, dry day

Warm, humid day

Cool, dry day

Foggy, chilly day

| 10°C | 20°C | 30°C | 40°C |
| RH = 100% | RH = 20% | RH = 80% | RH = 5% |

(b) In the real world, both the amount of water vapor in the air and the temperature vary from place to place. The size of each beaker represents the saturation vapor pressure. The ratio of the amount of water in the beaker to the beaker's capacity represents the relative humidity (RH).

(a) The air where this spider built its web has cooled to the dew point, and water has condensed on the web.

(b) On a cold day, frost forms on leaves.

FIGURE 17.11 The dew point is the temperature at which air becomes saturated with water vapor.

bit less than the dry air, because water molecules (H_2O) weigh less than the replaced nitrogen (N_2) and oxygen (O_2) molecules.

Humans can sense the air's moisture content because our bodies are cooled by evaporating perspiration. When hot air is humid, our perspiration can't evaporate quickly, so we feel hot and sticky. In dry air at the same temperature, our perspiration evaporates quickly, so we feel cool and comfortable. Significantly, it's not the absolute amount of water in the air (as indicated by the vapor pressure) that makes it feel humid, but rather how closely the air approaches saturation. To convey a sense of how damp air feels, meteorologists characterize the air's moisture content by specifying its **relative humidity** (**RH**), defined as the amount of water vapor in the air (the measured vapor pressure) divided by the air's capacity for holding water vapor (the saturation vapor pressure). We can express this quotient as a percentage:

$$RH = \frac{vapor\ pressure}{saturation\ vapor\ pressure} \times 100\%$$

Clearly, relative humidity depends on both the amount of water vapor in the air and the air's holding capacity. For example, if two volumes of air contain the same number of water vapor molecules, the warmer one will have a lower relative humidity (Fig. 17.10a). Therefore, a given location may feel cool and be foggy because the air there has a high relative humidity. Another location may feel hot and dry because the air has a low relative humidity, even if it contains the same number of water vapor molecules as the cool, foggy air does (Fig. 17.10b). When the relative humidity is 90%, an actual temperature of 84°F might feel like 98°F to you because your perspiration won't evaporate. The relationship between human comfort, temperature, and relative humidity can be expressed by the *heat index*, a calibration of how hot air feels, given its humidity.

While vapor pressure provides a good way to describe the actual amount of moisture in the air, it's very difficult to measure. Instead, meteorologists rely on another variable, the dew point, to characterize the absolute, as opposed to the relative, amount of water vapor in the air. The **dew point** is the lowest temperature to which air can be cooled, at constant pressure, before becoming saturated. To measure the dew point, we simply cool air at constant pressure and measure the temperature at which liquid droplets first appear. When temperatures cool overnight, air may cool to the dew point and become saturated, so that tiny liquid drops, called *dew*, form on exposed surfaces (Fig. 17.11a). If the process happens when the temperature is below freezing, *frost*, a thin ice rind on solid surfaces, will form. The temperature at which frost appears is called the *frost point* (Fig. 17.11b). We can estimate relative humidity by comparing the dew point with the air temperature. When the dew point approaches the air temperature, the relative humidity is high, as happens on muggy days. In contrast, when the two temperatures are far apart, the relative humidity is low, as happens on crisp, dry days.

Wind Speed and Direction

Everyone knows the feel of flowing air. It can cool your skin, rustle the leaves, or—if stronger—rip off your roof. Atmospheric scientists define **wind** specifically as the horizontal movement of air. When the wind blows against your face, you feel more air pressure on your face than you do on the back of your head. Of course, air in nature doesn't just move horizontally—it can also flow up as an *updraft* or down as a *downdraft*. While up-and-down air movements have real consequences, such as triggering storms or causing discomfort for airline passengers, vertical air movements are not represented by a standard "wind" measurement.

To describe wind, we use two familiar quantities: wind direction and wind speed. By convention, the **wind direction** refers to the direction from which the wind blows. For example, in the northern hemisphere, a

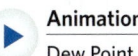

Animation
Dew Point

Did you ever wonder . . .

why you feel hotter on a steamy, humid day in Florida than on a hot, bone-dry day in the Arizona desert?

FIGURE 17.12 Measuring wind speed and direction.

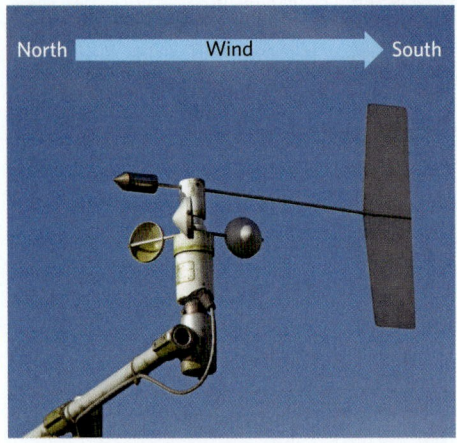

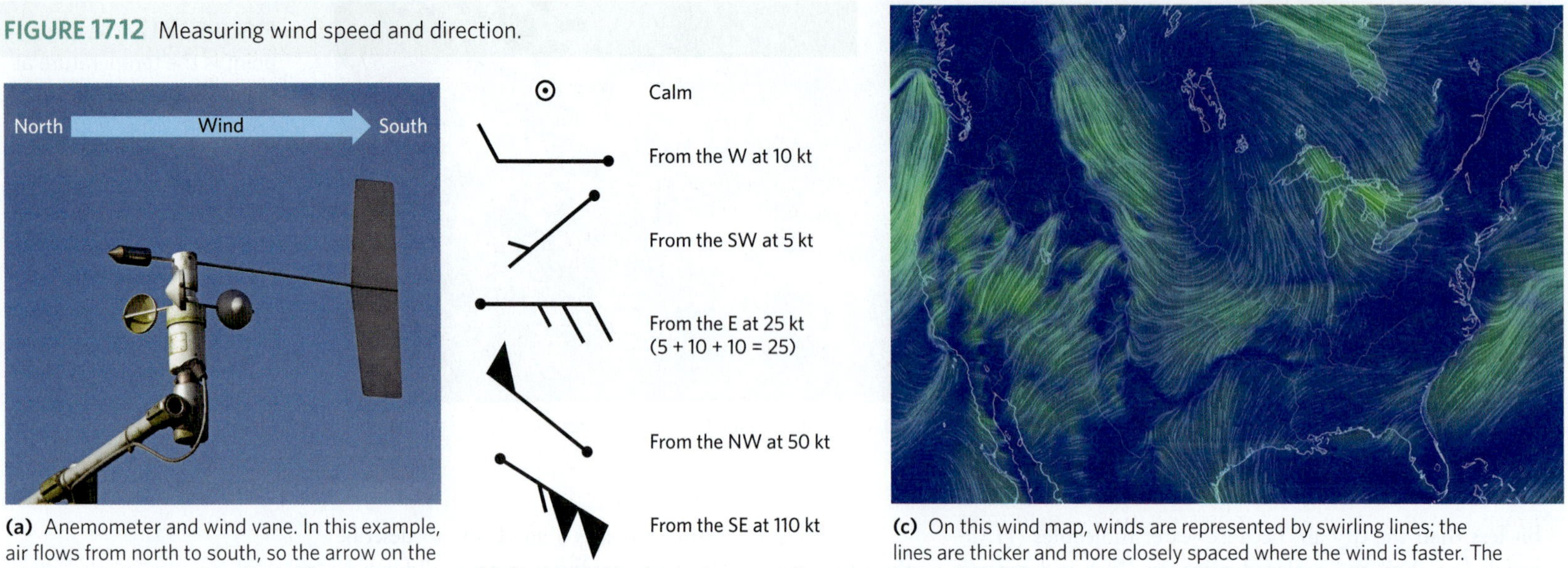

North   Wind   South

(a) Anemometer and wind vane. In this example, air flows from north to south, so the arrow on the wind vane indicates a "north wind."

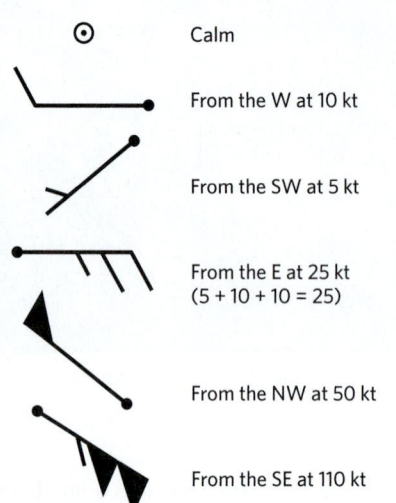

⊙   Calm

From the W at 10 kt

From the SW at 5 kt

From the E at 25 kt (5 + 10 + 10 = 25)

From the NW at 50 kt

From the SE at 110 kt

(b) Key to wind barbs used on a weather map.

(c) On this wind map, winds are represented by swirling lines; the lines are thicker and more closely spaced where the wind is faster. The orientation and taper of the colored lines represent wind direction.

north wind brings colder air from the north toward the south. We can measure wind direction with a *wind vane*, mounted so that it swings about a vertical axis; the push of the wind reorients an arrow so that the arrowhead points in the direction the wind comes from **(Fig. 17.12a)**. **Wind speed** is the horizontal rate of air movement. We can measure wind speed with an *anemometer*, a device consisting of three cups that catch the wind and, therefore, spin around a vertical axis (see Fig. 17.12a). Meteorologists typically measure wind speed in *knots* (kt), where 1 kt = 1 nautical

mile per hour. Since a nautical mile is longer than a statute mile, 1 kt = 1.15 mph. Converting to the metric system, 1 kt = 1.85 km/h. Meteorologists display winds by placing *wind barbs* on a weather map **(Fig. 17.12b)**. They may also display winds on a map by drawing lines parallel to the wind direction **(Fig. 17.12c)**.

Visibility and Cloud Cover

The old phrase "On a clear day, you can see forever" may be an exaggeration, but it does draw attention to the fact

FIGURE 17.13 Composite satellite image showing the Earth's cloud cover.

that air doesn't always have the same transparency. Meteorologists describe air transparency by specifying **visibility**, the distance from which a person with normal vision can identify objects while looking through the air. In the past, visibility measurements were made by a human observer looking at objects at known distances. Today, meteorologists measure visibility automatically with a sensor that detects air transparency. What factors affect visibility? Pure air is virtually transparent, so visibility decreases when air contains aerosols, water droplets, or ice particles. A fog can reduce visibility to near zero, making driving hazardous, whereas on a clear day, visibility may be "unlimited" in that you can see to the horizon and beyond.

Meteorologists define **cloud cover** as an estimate of how much of the sky is obscured by clouds at a given time. In the past, human observers would determine cloud cover visually by estimating what fraction of the sky contained clouds. Today, laser technology allows us to determine what fraction of the time clouds lie directly overhead. When describing cloud cover, we typically use familiar descriptive terms (clear skies, few clouds, scattered clouds, broken clouds, or overcast skies) to indicate progressively more cloud cover. Cloud cover varies radically with location and time of year. At any given time, clouds hide approximately 70% of the Earth's surface (Fig. 17.13).

Precipitation

In the geology chapters of this book, we generally use the word *precipitation* to refer to the formation of a solid when a liquid solution becomes supersaturated. In atmospheric science, **precipitation** generally refers to water that condenses to a liquid or solid state and falls from the sky. Precipitation includes brief showers, heavy downpours, snow flurries, hailstorms, and blizzards.

To characterize precipitation, meteorologists specify both the *precipitation rate*, a measure of how fast rain or snow is falling in inches per hour or millimeters per hour, and the *total precipitation*, the cumulative amount that falls during a specified time period, such as a single storm, a single month, a season, or a whole year. We can report total rainfall in inches or millimeters. For total snowfall, the measurement indicates either an *accumulated depth*, the thickness of the snow layer, or its *water equivalent*, the depth of the water that would be formed if the snow melted completely. Meteorologists report snowfall differently than rainfall because snow density can vary significantly depending on temperature: warmer snow tends to be dense, whereas colder snow tends to be light and fluffy. Meteorologists traditionally measure precipitation using a *precipitation gauge*, a container that collects either rain or snow and measures its weight or volume (Fig. 17.14). Today, we can also use information from *weather radar* to estimate precipitation over large regions (Box 17.2).

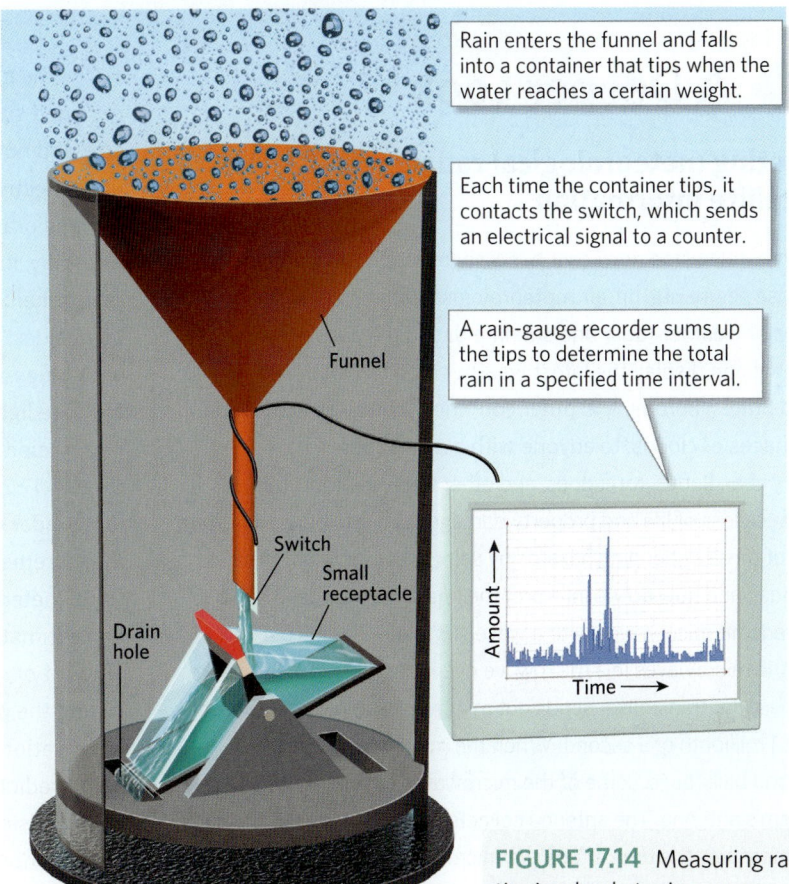

Rain enters the funnel and falls into a container that tips when the water reaches a certain weight.

Each time the container tips, it contacts the switch, which sends an electrical signal to a counter.

A rain-gauge recorder sums up the tips to determine the total rain in a specified time interval.

Funnel

Switch

Small receptacle

Drain hole

Amount

Time

FIGURE 17.14 Measuring rainfall using a tipping-bucket rain gauge.

Weather Stations

When you see weather maps in a newspaper or on television, you're seeing a compilation of data collected at hundreds of *weather stations*, sites at which meteorologists have set up instruments that automatically measure temperature, dew point, wind speed and direction, visibility, cloud type and coverage, and precipitation type and intensity (Fig. 17.15). In the United States, the National Weather Service operates the *Automated Surface Observing System*, while the Federal Aviation Administration and Department of Defense operate the *Automated Weather Observing System*. The instruments in these systems update observations 24 hours a day and send electronic records of their measurements to data centers (Box 17.3). Weather stations on ships or on buoys provide data for oceanic regions.

Take-home message . . .

To characterize weather, meteorologists measure temperature, atmospheric pressure, dew point, wind speed and direction, cloud cover, visibility, and precipitation. These characteristics vary with location at a given time and with time and elevation at a given location.

Quick Question

How can meteorologists detect a hurricane over the ocean?

BOX 17.2

A Deeper Look

Interpreting meteorological radar and satellite information

Almost every television newscast has a segment on the weather. During those segments, on-air meteorologists almost always show animations of weather radar and satellite data. Free apps are also available that can display the latest weather radar animations from the United States, Europe, and other countries, as well as worldwide satellite images of clouds, to anyone with a smartphone.

Every year, floods, tornadoes, and other types of hazardous weather cause loss of life and property damage. Meteorologists can warn the public of these impending hazards by using weather radar. What is weather radar, and how do we interpret the images it provides? Radar systems transmit microwaves with a wavelength of 1–10 cm (0.4–4 in), much like the microwaves in a microwave oven **(Fig. Bx17.2a)**. But unlike your oven, a radar transmitter sends out microwaves in pulses that last only about 1 millionth of a second. When the microwaves encounter raindrops and hailstones, some of the microwaves scatter back to the radar system's antenna. The antenna collects this energy, called the *radar echo,* and measures the time it took the microwave pulse to travel out to the precipitation and back. Because microwaves travel at the speed of light, we can calculate the distance to the precipitation from the following equation: distance = velocity × time. By knowing the angle at which the antenna points to the sky, we can precisely determine the elevation of the precipitation within the atmosphere.

The character of precipitation affects the strength of the returned signal, or the *radar reflectivity*. For example, the greater the size and number of particles the pulse intercepts, the greater the radar reflectivity will be. High radar reflectivity values, depicted with red and orange colors on a display, indicate heavy rain or hail. In contrast, low values, depicted with blues and greens, indicate light drizzle or nonprecipitating clouds **(Fig. Bx17.2b)**. By examining radar reflectivity measurements over time, meteorologists can estimate the total amount of rain that fell during the period of observation. These data help predict flash floods. On television, meteorologists normally show the radar reflectivity to give viewers an idea of where rain is falling and the intensity of the rainfall.

Visible channel

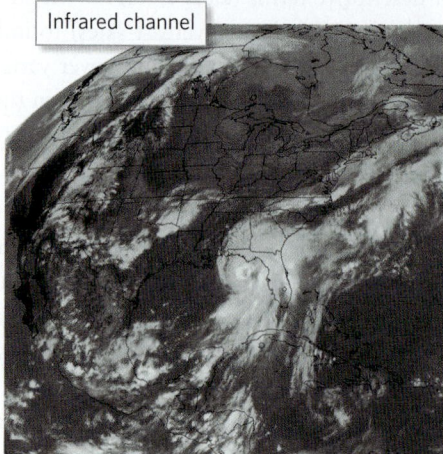

Infrared channel

FIGURE Bx17.2 Weather radar and satellites.

(a) A radar antenna used in research.

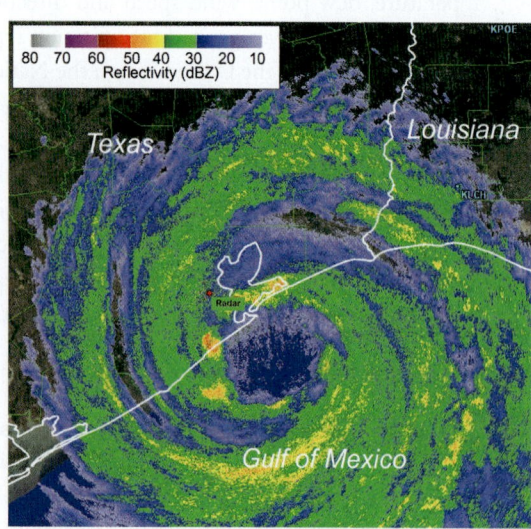

(b) Radar reflectivity from Hurricane Ike as it made landfall near Galveston, Texas, in 2008.

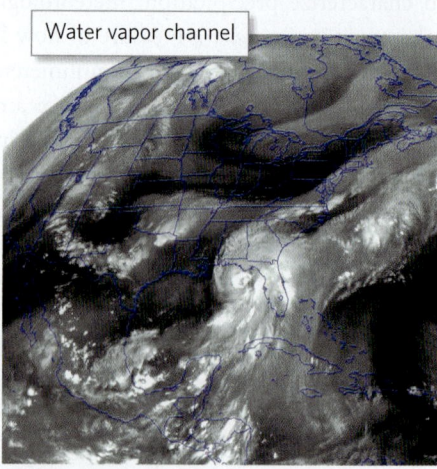

Water vapor channel

(c) The same image from a weather satellite, viewed in different channels. The area shown by the map covers the eastern United States.

Radar systems transmit microwave energy at specific frequencies. When the radar echo returns to the antenna, the frequency of the returned energy has typically shifted slightly as a result of the movement of the raindrops or ice particles along the direction of the radar beam. **Doppler radar** systems measure this shift and use it to determine the speed of the wind in the direction parallel to the beam. Doppler radar unfortunately cannot measure air motion across the beam. Nevertheless, by mapping the wind in the direction of the beam, Doppler radar can identify strong straight-line winds as well as rotation in the air flow, which may be evidence of a tornado. Therefore, meteorologists who monitor severe weather commonly use Doppler radar to detect these hazards and provide early warning to the public. The development of *polarization Doppler radar*, which employs beams in which the microwaves move only in a given plane, can now even help meteorologists identify tornadoes that have lofted airborne debris.

Meteorologists supplement radar measurements from ground-based locations with data collected by orbiting weather satellites, which can acquire images of the atmosphere over a wide region, or even the entire world. Animated satellite images allow meteorologists to watch storms develop and to monitor the progress of a storm as it moves across the Earth. Weather satellites use different wavelengths of radiation to gather different kinds of information **(Fig. Bx17.2c)**. The *visible channel* uses reflected sunlight to show clouds as you would see them from space, in black and white or even in color. The *infrared channel* shows the temperatures of cloud tops or the ground, with cold temperatures indicated by bright tones and warm temperatures by dark tones. The *water vapor channel* indicates regions of high moisture content with bright tones and dry areas with dark tones. Meteorologists on television often show satellite data to inform viewers about clouds moving into or out of the regions they serve.

17.5 Vertical Structure of the Atmosphere

Thermally Defined Layers

Imagine that you're sitting outside reading this book on a warm, calm summer day. Right now, only 10 km (6 mi) directly above you, the wind may be blowing at over 160 km/h (100 mph), the temperature may be hovering at a frigid −50°C (−58°F), and the gas in the air has such low density that you couldn't survive by breathing it. If you're a passenger in an airliner that has climbed to cruising altitude, these conditions exist right outside your window. Clearly, atmospheric conditions change rapidly with elevation.

FIGURE 17.15 Surface weather stations in North America.

(a) Each white dot on this map represents a weather station.

(b) Equipment set up at a weather station.

BOX 17.3 — **Putting Earth Science to Use**

Interpreting a weather forecast

How do meteorologists forecast the weather? And how do they come up with statements like "Tomorrow there is a 60% chance of rain" or "Tuesday will be partly cloudy"?

Weather forecasting is a complicated process that begins with the collection of weather data worldwide from airports and cities, weather balloons, satellites, radar systems, and other instruments. The data flow continuously into national data centers, such as the National Center for Environmental Prediction in Washington, DC. These centers process the data using supercomputers, arranging the information on orderly grids that cover all or parts of the planet. Numerical models, sophisticated computer codes consisting of mathematical equations that describe the forces that govern atmospheric motion and other factors that control the behavior and evolution of the atmosphere, use these data to calculate the future state of the atmosphere as far out as 14 days. The accuracy of these predictions is strongest within a 3-day range, but models can help estimate future states at longer intervals.

So how does all that relate to the chance of rain in Cincinnati at noon on Tuesday? To pinpoint that type of information, meteorologists compare, using statistics, data from the current model forecast with thousands of past model forecasts and historical data about rainfall in Cincinnati (and all other weather data relevant to people living at thousands of other locations!). These model output statistics provide the information you actually receive on your television and internet-based forecasts. For example, when a meteorologist states that there is a 60% chance of rain tomorrow afternoon in Cincinnati, the forecaster means that, historically, rain occurred 60% of the time that the weather conditions forecasted by the model were present in the afternoon in the Cincinnati area. A meteorologist might forecast clear, partly cloudy, mostly cloudy, or cloudy weather (sometimes meteorologists use sunny and partly sunny for daytime hours—these terms have the same meaning as clear and partly cloudy). These predictions, again, result from model output statistics, which provide the forecaster with the percentage of the time clouds are expected at a specific location. Forecasters always have the option of "tweaking" the model predictions based on their local knowledge, but their forecasts always start with the computer-based predictions and model output statistics. The data are also compiled to create the weather maps and forecasts that you see on television, on the internet, or in newspapers (Fig Bx 17.3).

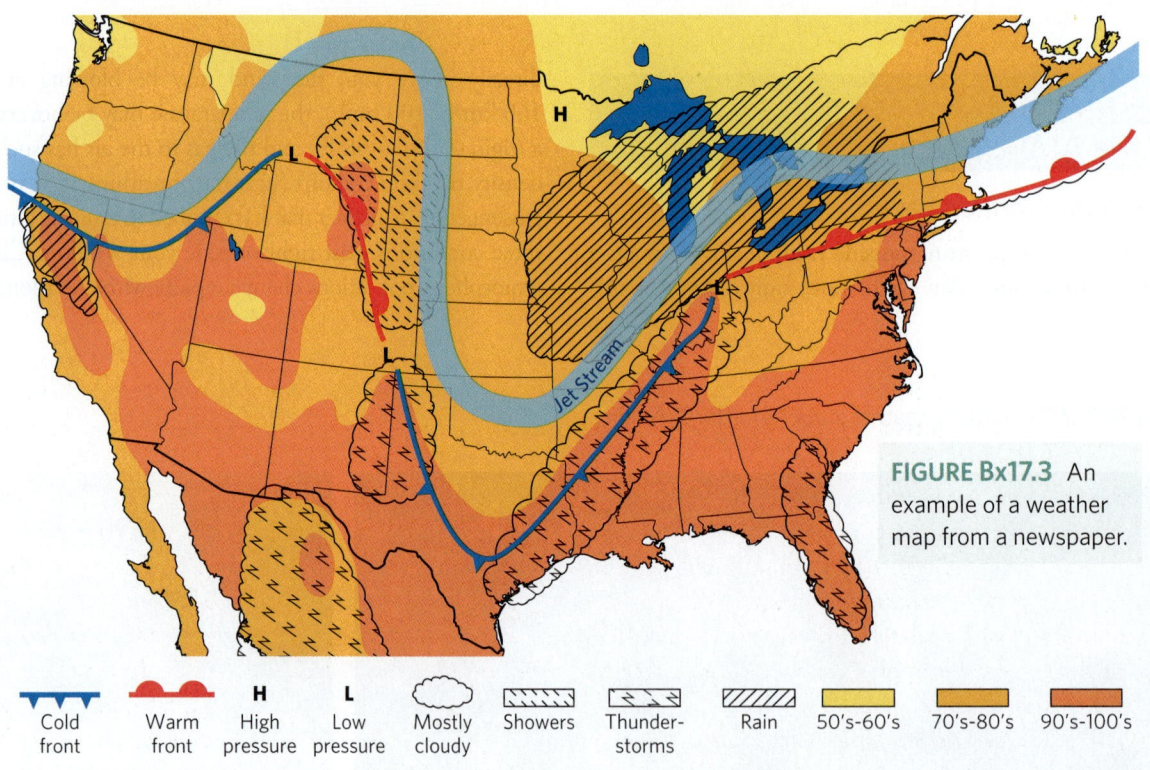

FIGURE Bx17.3 An example of a weather map from a newspaper.

| Cold front | Warm front | H High pressure | L Low pressure | Mostly cloudy | Showers | Thunderstorms | Rain | 50's-60's | 70's-80's | 90's-100's |

Atmospheric scientists use balloon-lofted instrument packages called *rawinsondes* to measure temperatures and other properties of the atmosphere at high elevations (Fig. 17.16a). A rawinsonde radios back information about atmospheric pressure, temperature, dew point, and wind direction and speed continuously as it rises, producing a "sounding" that depicts the vertical structure of the atmosphere (Fig. 17.16b). By international agreement, rawinsondes are launched simultaneously twice a day at hundreds of localities worldwide. On average, rawinsonde launch sites

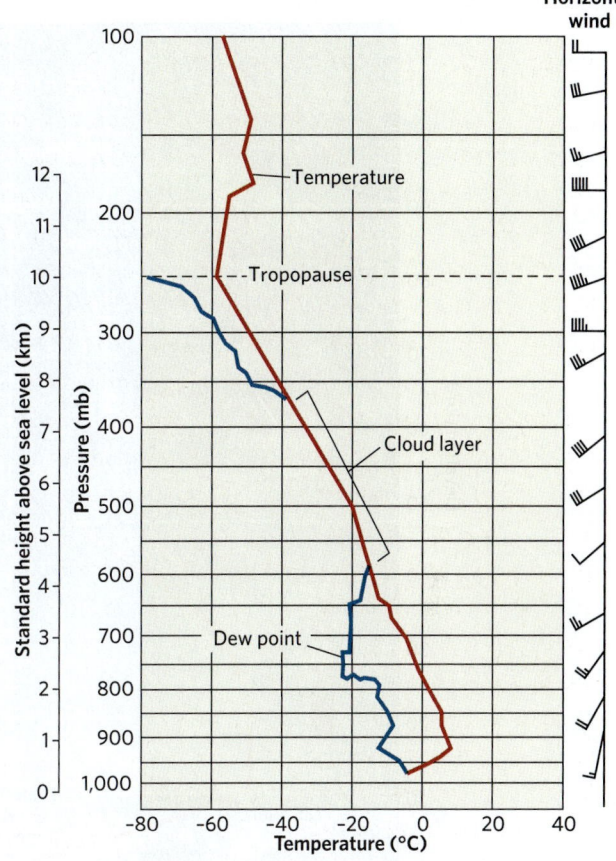

(a) Launching a weather balloon with a rawinsonde attached. The inset shows a rawinsonde before it was launched, as well as a parachute.

(b) A vertical profile of atmospheric temperature, dew point, and wind from a rawinsonde. See figure 17.12 for an explanation of the wind symbols at the right.

on land are located about 500 km (200 mi) apart, so their spacing is much greater than that of ground-based weather stations. As a result, high-elevation data are not as detailed as ground-based data. Because atmospheric pressure decreases with elevation, weather balloons expand as they rise. When a balloon reaches about 20 km (12.5 mi), it breaks, and the rawinsonde it carried parachutes back to the ground.

Temperature also changes with elevation. Research shows that in some layers of the atmosphere, temperature increases as elevation increases, whereas in others, temperature decreases as elevation increases. Elevations at which the direction of change reverses (from decreasing to increasing, or vice versa) delineate boundaries that separate four distinct atmospheric layers (Fig. 17.17). Let's examine these layers in sequence, beginning at the Earth's surface.

Since solar energy reaches the atmosphere from above, it's a common misperception that the air around you becomes warm because it's "broiled from above." In fact, air is transparent to most incoming radiation (such as visible light)—that's why you can see through air. This energy reaches the Earth's surface, which absorbs it. The surface then radiates the energy back into the air as infrared radiation, radiation that you can't see but can feel as heat, some of which gets trapped by the air and warms it. (We'll discuss this *greenhouse effect* in more detail later in this chapter.) Put another way, the Earth's surface acts as a heater that warms the air at the base of the atmosphere—in effect, the air is "baked from below."

As we move upward, away from the Earth's surface, the distance from that heat source decreases, and the air gets cooler. So, if you climb a tall mountain, you'll find that as your elevation increases, air temperature decreases. This decrease in temperature with increasing elevation continues up to about the cruising altitude of commercial airliners, where air temperature ranges between −40°C and −60°C (−40°F and −76°F). Meteorologists refer to this lowest layer of the atmosphere, in which temperature decreases with elevation, as the **troposphere**. The rate of change of temperature with elevation, a quantity known as the **environmental lapse rate**, varies substantially from place to place and from time to time, but on average has a value of 6.5°C/km

FIGURE 17.17 Major layers in the lower 110 km (70 mi) of the atmosphere.

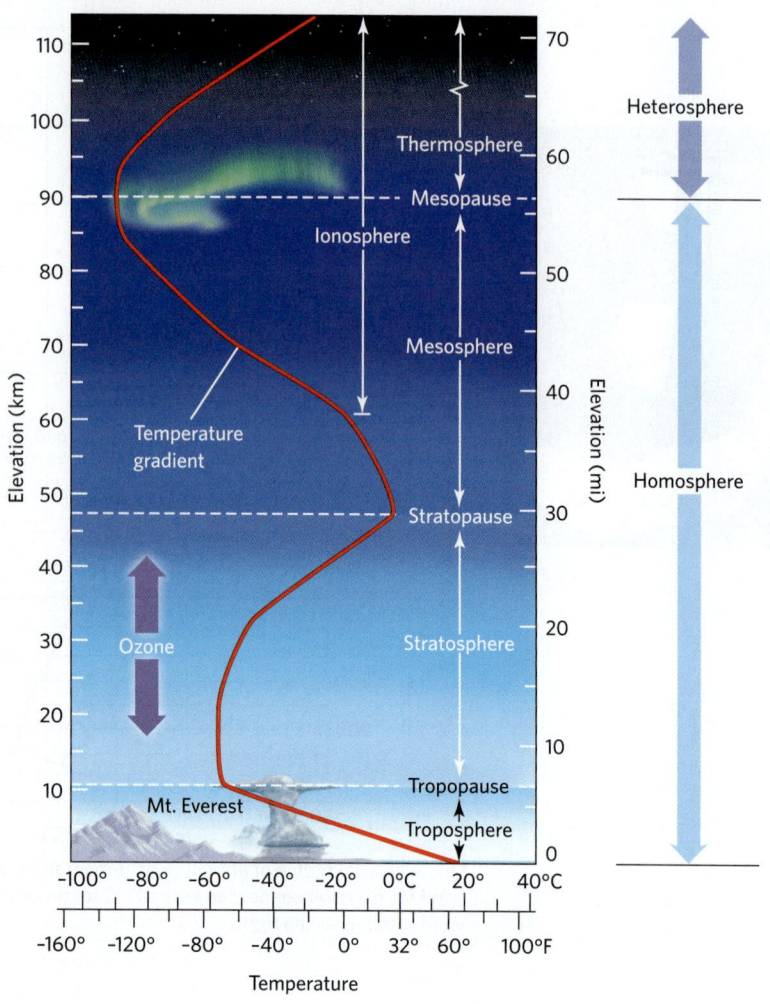

(3.5°F/1,000 ft). The thickness of the troposphere varies with latitude and time of year: it decreases from about 20 km (12.5 mi) above the equator to about 7 km (4.3 mi) above the frigid poles. The Earth's storms occur entirely within the troposphere.

Above the *tropopause*, the top of the troposphere, air temperature increases with increasing elevation. The layer in which this increase takes place, the **stratosphere**, reaches up to a height of about 50 km (31 mi). The increase in temperature with elevation occurs because ozone (O_3) molecules in the stratosphere absorb ultraviolet radiation from the Sun. This absorbed radiation causes air molecules to move about faster, so their average kinetic energy—their temperature—increases.

Ultraviolet radiation can harm living organisms, so its absorption by ozone makes life on land possible. Unfortunately, human activities have caused the concentration of ozone in the stratosphere to decrease in recent decades. When emitted into the atmosphere, chemicals such as chlorofluorocarbons (CFCs) react with and break up ozone molecules. This reaction happens most rapidly on the surfaces of tiny ice crystals in polar stratospheric clouds, so an **ozone hole**, a region of diminished ozone concentration, forms over the South Pole during the southern hemisphere spring (Fig. 17.18a). Because of the importance of stratospheric ozone, a 1987 international summit in Montreal led to an agreement to reduce CFC emissions globally.

Above about 50 km, an elevation called the *stratopause*, there's not enough ozone absorption of solar radiation to heat the air, so air temperature again begins to decrease with increasing elevation. This elevation marks the bottom of the **mesosphere**, which continues upward

FIGURE 17.18 High-altitude zones of the atmosphere.

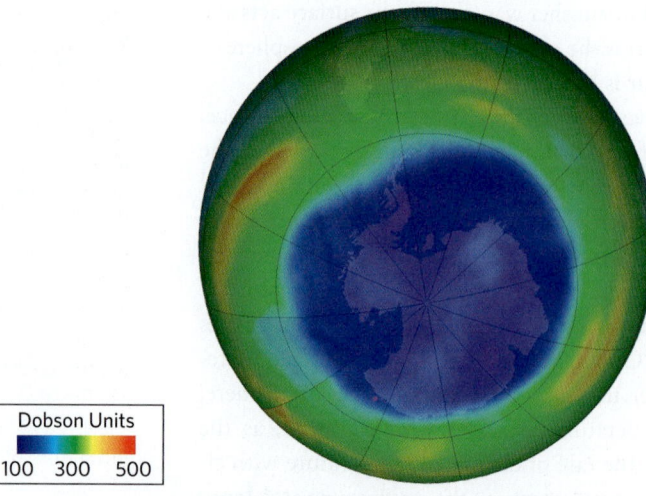

(a) The ozone hole over Antarctica in August 2018. A Dobson unit measures the concentration of ozone in the atmosphere; larger numbers mean more ozone.

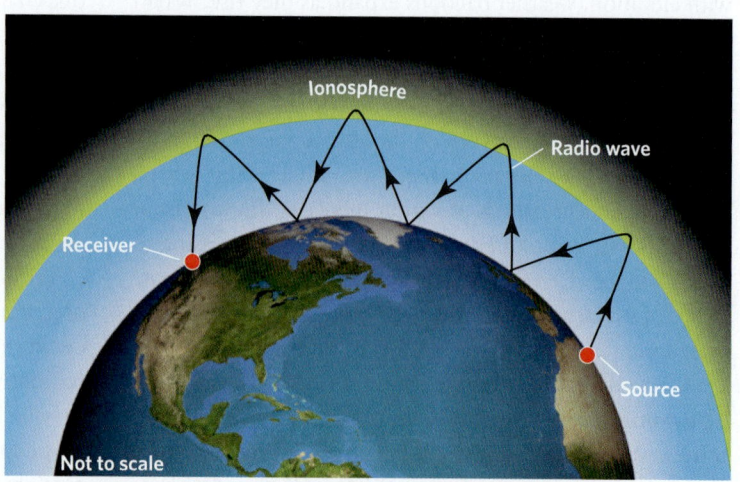

(b) The ionosphere, a layer containing charged ions, causes certain radio waves to reflect back to the Earth, from which they bounce up again. These waves can be detected beyond the horizon.

to the *mesopause*, at an elevation of about 90 km (56 mi). Above the mesopause lies the outermost layer of the atmosphere. In this layer, the **thermosphere**, temperature increases with increasing elevation, reaching a maximum of 1,500°C (2,700°F), as the few remaining air molecules absorb high-energy radiation from the Sun. These high temperatures are deceptive, however, for an astronaut entering the thermosphere without first donning a space suit would freeze instantly. That's because the air density in the thermosphere is so low that a given volume of air holds hardly any heat. All the air molecules colliding with the astronaut, though traveling very fast, could not provide enough heat to counteract the loss of heat by radiation from the astronaut's body. The density of air in the thermosphere decreases upward until, at an elevation that ranges from 350–800 km (217–500 mi), it equals that of interplanetary space. This elevation, which marks the top of the atmosphere, varies depending on the quantity of incoming atoms brought to the Earth in the solar wind.

Layers Defined by Composition and Electric Charge

Below an altitude of about 90 km, air circulation mixes the atmosphere so thoroughly that it has a composition that is mostly the same from place to place. Researchers refer to this well-mixed layer as the **homosphere** (see Fig. 17.17). We use the qualifier "mostly" because the concentration of certain gases varies. For example, ozone forms and occurs primarily in the stratosphere, and gases that enter the atmosphere from localized sources at the Earth's surface tend to decrease in concentration with distance from the source. Also, water vapor, which has a short *residence time* in the atmosphere (meaning that it enters and leaves the atmosphere relatively quickly), varies substantially with location and time. Note that the homosphere includes all of the troposphere, stratosphere, and mesosphere. Above the homosphere, gases sort gravitationally according to their molecular weight, with lighter gases rising and heavier ones sinking. This stratified layer, the **heterosphere**, coincides with the thermosphere.

Meteorologists also recognize another layer, not based on temperature or composition, but rather on electrical characteristics. In this layer, known as the **ionosphere**, the atmosphere contains a relatively high concentration of *ions*, molecules that have gained or lost electrons and thus have a negative or positive electric charge, respectively. The ionosphere, which occurs above elevations of 60 km (37 mi) during the day and 100 km (62 mi) at night (see Fig. 17.17), serves an important role in human communications because it reflects certain frequencies of radio transmissions. These transmissions, notably AM radio, bounce back and forth between the ionosphere and the Earth's surface, which allows them to travel over long distances **(Fig. 17.18b)**.

FIGURE 17.19 The aurora borealis, as seen over Alaska.

Incoming high-energy particles from the Sun are channeled along magnetic field lines to the poles of the Earth. When they interact with certain molecules in the ionosphere, these molecules emit light. We see this light as a shimmering **aurora**, a gauzy curtain of undulating green and red color in the sky **(Fig. 17.19)**. We refer to the aurora above northern polar regions as the *aurora borealis*, or northern lights, and that over southern polar regions as the *aurora australis*, or southern lights.

Take-home message . . .

Atmospheric scientists divide the atmosphere into layers based on changes in its temperature, composition, and electric charge. The layers defined by changes in temperature are the troposphere, stratosphere, mesosphere, and thermosphere. The layers defined by consistency of composition are the homosphere and heterosphere, and the layer defined by electric charge is the ionosphere.

Quick Question
What role does the ionosphere play in radio transmission?

17.6 Clouds and Precipitation

From the vantage point of an astronaut in the International Space Station, the pattern of clouds stands out as one of the most striking features of the Earth. Some clouds look like small cotton puffs, others like broad white blankets, and still

See for yourself

The Atmosphere

Latitude: 38°57'33.82" N
Longitude: 95°15'55.74" W

Look down from an altitude of 11,000 km (6,835 mi). You will see the entire Earth below, centered on North America. On the View tab, turn on Atmosphere. The atmosphere will appear as a very thin haze above the Earth's surface along the Earth's boundary with space. Notice how thin the atmosphere really is: 99% of the atmosphere's mass is contained in a shell only 31 km (19 mi) thick. For reference, the island of Puerto Rico (18°12'10" N, 66°18'45" W) is 160 km (100 mi) long.

others like elegant swirls (see Fig. 17.13). What do they consist of? Clouds are not composed of pure water vapor, which, as gaseous water, is transparent. Rather, clouds are collections of countless water droplets and/or tiny ice crystals suspended in the sky. These particles reflect, scatter (emit in random directions), and absorb light, so thin clouds appear whitish and thick clouds look gray and ominous. Let's examine how clouds form, how clouds differ from one another, and why some produce rain or snow while others do not.

Cloud Formation

As we have seen, air can hold only a certain amount of water vapor before it becomes saturated, and the amount it can hold depends on its temperature. If saturated air cools, it becomes *supersaturated*, and when this happens, water

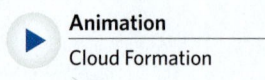

Animation
Cloud Formation

molecules begin to attach to the surfaces of aerosols. In effect, a tiny aerosol can serve as a *condensation nucleus*. If the air temperature is above freezing, a droplet of liquid water grows around the nucleus as more and more water molecules attach. If, however, the air temperature is below freezing, an aerosol can serve as a *deposition nucleus*, to which water molecules attach to produce a crystal of solid ice. Water vapor is invisible, but water droplets and ice particles are visible—so, simply put, clouds form when water molecules in supersaturated air collect on aerosols to form liquid droplets and ice particles. Every raindrop and snowflake you see initially formed on an aerosol, so if there were no aerosols, there would be no rain, and life on land would never have evolved.

How does air become locally supersaturated? Generally, supersaturation happens where air rises. Air rises when it undergoes *lifting* by any of several processes that force air upward. Lifting can happen, for example, when a mass of less dense, warmer air encounters and flows up and over a mass of denser, cooler air (Fig. 17.20a). The boundary where two such contrasting air masses meet is called a **front**. A similar phenomenon happens along a shoreline when cool air moves onshore and lifts warm air. Air can also rise as it encounters and flows over a mountain range (Fig. 17.20b). And air can rise when it becomes *buoyant*, meaning that it becomes less dense than its surroundings (Fig. 17.20c). Buoyancy can develop, for example, when heat rising from the ground below warms air near the ground surface. You've seen this process happening if you've ever seen a hot-air balloon head skyward. We'll discuss additional lifting mechanisms in Chapter 19.

Atmospheric pressure decreases with altitude, so rising air expands as it moves upward. In order to expand, the air must push against surrounding air molecules. But pushing takes energy, so the air loses heat energy and becomes cooler (as we'll see in Section 19.3). Cooling, in turn, reduces the

FIGURE 17.20 Several mechanisms can cause air to rise.

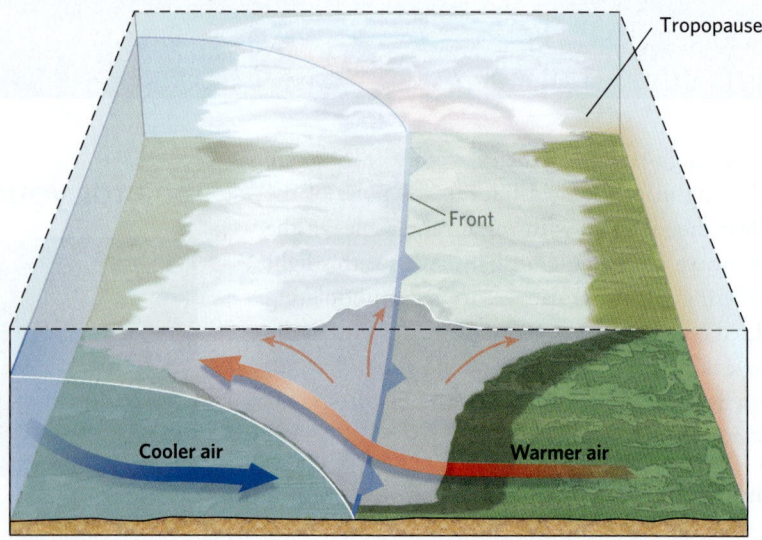

(a) When a mass of warm air meets an advancing mass of cold air, the warm air is forced to rise. The leading edge of the cold air mass defines the location of the front.

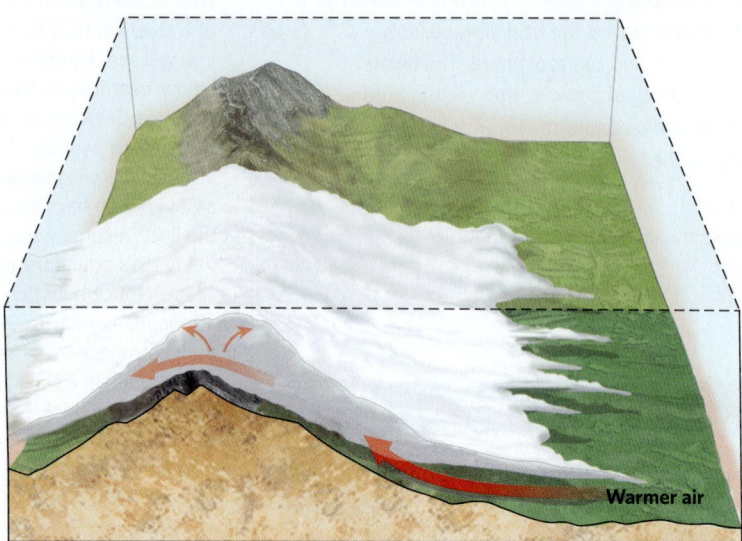

(b) When flowing air reaches a mountain range, it is forced to rise.

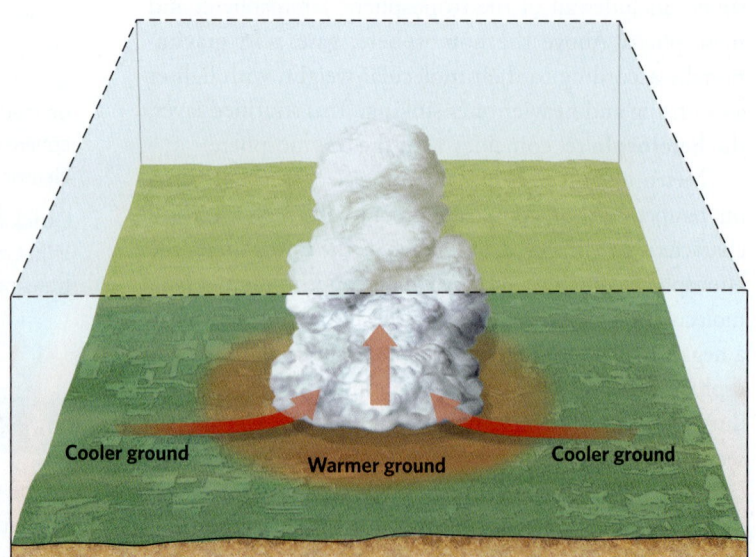

(c) Air becomes buoyant and rises over warmer ground. Cooler air then flows in toward the column of rising air.

air's capacity to hold water vapor, so if its temperature decreases enough, the air reaches its dew point. At this point, any further rising, which leads to further cooling, causes supersaturation and, therefore, cloud production.

Types of Clouds

Clouds fascinate people because of their beauty and the way they constantly change. They come in many shapes and sizes and form at many altitudes within the troposphere. Meteorologists classify clouds by their altitude and shape and by whether or not they produce precipitation, using terminology first proposed in 1802 by a British pharmacist named Luke Howard (Fig. 17.21). Howard, adapting Latin vocabulary, coined the terms *cirrus* for high, wispy clouds, *stratus* for sheets of clouds that cover vast areas, and *cumulus* for tall, puffy clouds with a cauliflower-like appearance. Combining these terms with others provides additional detail about clouds. For example, adding the word *nimbus* to a cloud name indicates that the cloud produces rain, so a *cumulonimbus cloud* is a towering puffy cloud that produces rain. Today, we recognize many more types of clouds, but still use many of the names Howard first proposed in 1802.

Why are there so many types of clouds? The character of a cloud depends on the amount of water vapor

available, the nature of the lifting mechanisms that drive cloud formation, the temperature in the troposphere, and the velocity of winds at various altitudes.

HIGH-ALTITUDE CLOUDS. As we've noted, wispy, elongate clouds that develop very high in the troposphere (above 6 km, or 20,000 ft) are known as **cirrus clouds (Fig. 17.22a)**. The temperature is so cold where cirrus clouds form that these clouds consist entirely of tiny ice crystals. As the ice crystals fall through the atmosphere, winds at different altitudes cause them to line up in wisp-like strands. Cirrus clouds are too high and thin to produce precipitation. If they do accumulate into layers thick enough to dim the Sun, they become *cirrostratus clouds*, and if they grow into puffy billows, they become *cirrocumulus clouds*.

MID-ALTITUDE CLOUDS. Mid-altitude clouds form between 3 and 6 km (10,000–20,000 ft) above the Earth's surface (Fig. 17.22b). Meteorologists refer to them as *altostratus clouds* if the clouds are layered, uniformly gray, and thick enough to hide the Sun. If the clouds resemble roll-like patches or puffs, they are *altocumulus clouds*. Altostratus and altocumulus clouds, like cirrus clouds, are too high and too thin to produce precipitation.

Did you ever wonder . . .

why clouds have different shapes?

FIGURE 17.21 The most common cloud types. Clouds come in a variety of sizes and shapes.

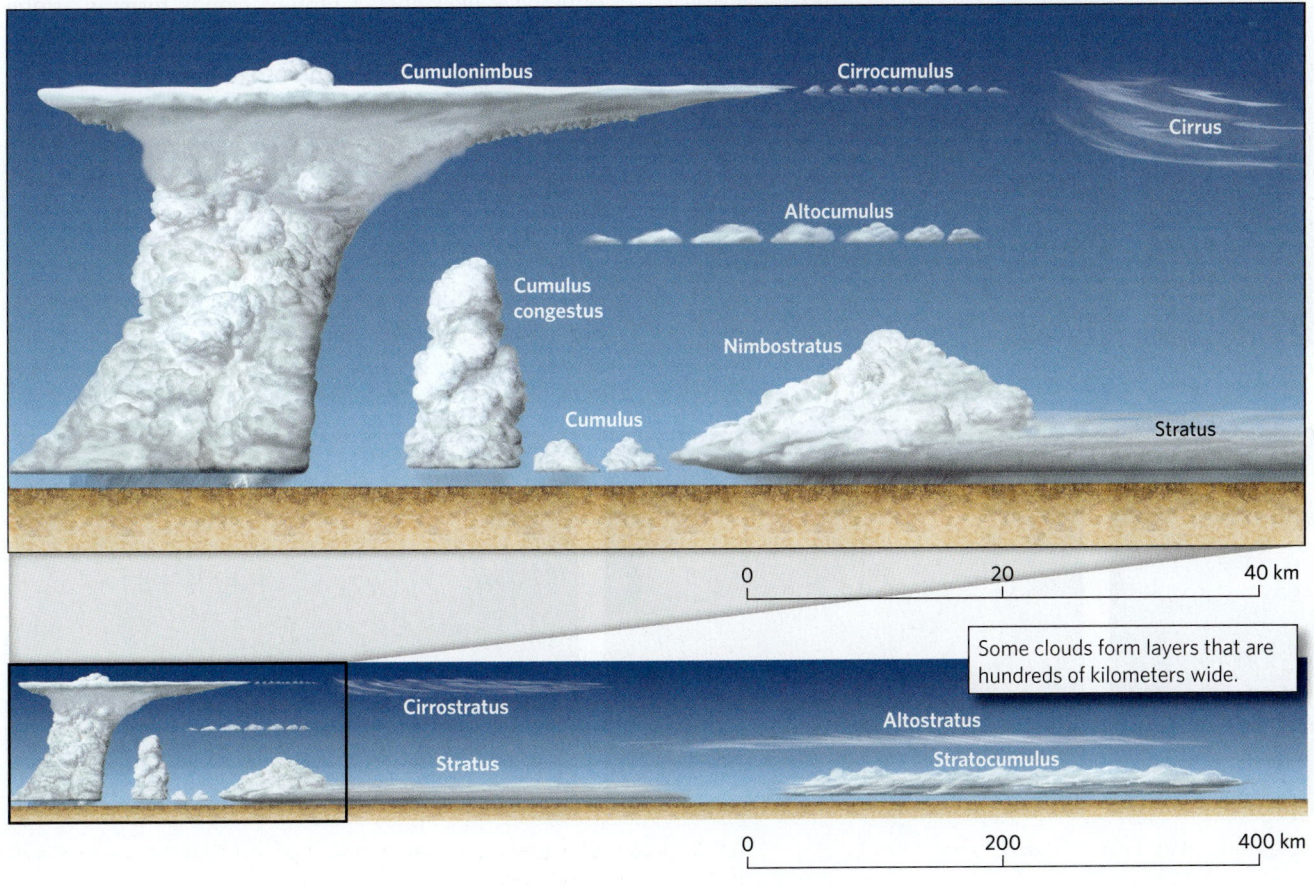

FIGURE 17.22 Photos of some common cloud types.

Cirrus

Cirrostratus

Cirrocumulus

(a) High-altitude clouds.

Altostratus

Altocumulus

(b) Mid-altitude clouds.

Stratocumulus

Stratus

(c) Low-altitude clouds.

LOW-ALTITUDE CLOUDS. Widespread, layered clouds that spread out like great blankets at relatively low altitudes (below 3 km, or 10,000 ft) are called **stratus clouds (Fig. 17.22c)**. The width of stratus clouds greatly exceeds their thickness. If portions of stratus clouds billow into puffy shapes, they become *stratocumulus clouds*, and if stratus clouds thicken sufficiently that precipitation falls from them, they become *nimbostratus clouds*.

TOWERING CLOUDS. Unlike the altitude-defined clouds described above, *towering clouds* rise across a range of altitudes **(Fig. 17.22d)**. The largest extend from low altitudes to the tropopause. Small towering clouds are called **cumulus clouds** if they are growing vertically and have cauliflower-like lobes. Cumulus clouds form in strong updrafts and, in general, have comparable horizontal and vertical dimensions. If cumulus clouds grow very large, they become *cumulus congestus clouds*, and if they produce precipitation, they become *cumulonimbus clouds*. Some cumulonimbus clouds rise into the base of the stratosphere, where strong winds blow their tops into a broad sheet; these clouds are known as *anvil clouds*.

UNUSUAL CLOUDS. Some clouds take on strange shapes. For example, under special conditions, lens-shaped clouds, known as *altocumulus lenticularis clouds*, may

Cumulus

Cumulus congestus

Cumulonimbus

(d) Towering clouds.

FIGURE 17.24 Three types of fog.

(a) Radiation fog.

develop; these clouds have been mistaken for flying saucers (Fig. 17.23). Thunderstorms can produce ominous cloud formations with names such as wall clouds, shelf clouds, mammatus clouds, and roll clouds, which we'll examine in our discussion of thunderstorms in Chapter 19.

Fog

On a foggy day, visibility can become so poor that you can't see more than a few meters in front of you. **Fog** exists when the base of a cloud lies at the Earth's surface and cloaks the ground. Three types of fog develop under different conditions. The first type, known as *radiation fog*, tends to form at night when, in the absence of sunlight, the ground cools by radiating heat upward (Fig. 17.24a). The cooled ground absorbs heat from the air above it, causing the air to reach saturation. The second type, *advection fog*, forms when warm, humid air flows over a cooler surface, as can happen when warm air moves over cold ocean water or a field of snow (Fig. 17.24b). The third type, *evaporation fog*, forms where enough surface water evaporates into air that the air becomes saturated (Fig. 17.24c). Such fog might develop over a warm lake or marsh.

(b) Advection fog.

Precipitation and Its Causes

Unless you live in a bone-dry desert, you've probably experienced *precipitation* (rain, hail, or snow) many times, and you're familiar with the common terms (drizzle, downpour, flurries, heavy snow) used to describe precipitation. What conditions lead to precipitation, and why do some clouds pass overhead without dropping any?

WARM-CLOUD PRECIPITATION. Meteorologists define a *warm cloud* as one in which the temperature stays above 0°C throughout, so that the cloud consists only of liquid water droplets. In such a cloud, the individual *cloud droplets* are so small that they remain suspended in air. It's only when the droplets grow sufficiently large that they become **raindrops**,

(c) Evaporation fog.

spheres of water heavy enough to overcome air resistance and fall toward the ground. *Rain* consists of a vast collection of raindrops—during a downpour, trillions of raindrops fall on a square kilometer of the Earth's surface.

To understand how raindrops form in warm clouds, we must first see how tiny cloud droplets, which range from 0.01–0.1 mm (0.0004–0.004 in) in diameter, grow big enough to fall as raindrops (Fig. 17.25a). The formation of raindrops begins in a cloud if the air remains supersaturated. In such a cloud, random motions of water molecules cause additional water molecules to collide with and bond to those at the surfaces of the existing droplets. Such migration of atoms or molecules due to their random motions is called *vapor diffusion*. Vapor diffusion operates very slowly, however, so that process alone can't produce raindrops. For raindrops to form, the cloud droplets themselves must come in contact with one another and merge, a process called **collision-coalescence**. As a somewhat larger droplet falls faster than its smaller neighbors, it collides with them, collects these additional droplets, and continues to grow. Small raindrops range from 0.5–1.0 mm (0.02–0.04 in) in diameter, large ones range from 1.0–4.0 mm (0.04–0.16 in) across, and very large ones are 4.0–6.0 mm (0.16–0.24 in) across. For a single large (4-mm) raindrop to form, it must collect 8 million 0.01-mm cloud droplets during its fall!

Popular media typically depict raindrops as having a teardrop shape—round on the bottom, tapering to a point at the top. Real raindrops don't look like teardrops at all. In fact, cloud droplets and small raindrops have a spherical shape because of *surface tension*. Surface tension exists because the water molecules within a drop bond to other water molecules in all directions around them, but can't bond to air molecules around the drop. Because the bonds on the surface of the droplet are all oriented inward, the water drop assumes the shape of a sphere. Due to air resistance, large raindrops flatten into a hamburger shape as they fall. Raindrops larger than about 6.0 mm rarely develop because when such large drops collide with their neighbors, they break apart (Fig. 17.25b).

COLD-CLOUD PRECIPITATION. Many clouds extend to altitudes where atmospheric temperatures fall below 0°C. In these *cold clouds*, ice particles grow by one of two paths, depending on whether supercooled droplets are present (Fig. 17.26). *Supercooled droplets* are droplets that remain in a

FIGURE 17.25 Raindrop growth and breakup.

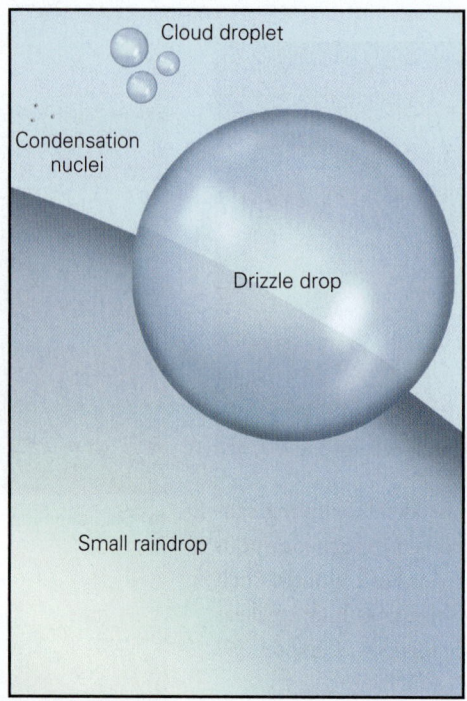

(a) Cloud droplets form when water vapor condenses around a condensation nucleus. Rain forms as droplets collide with one another and grow.

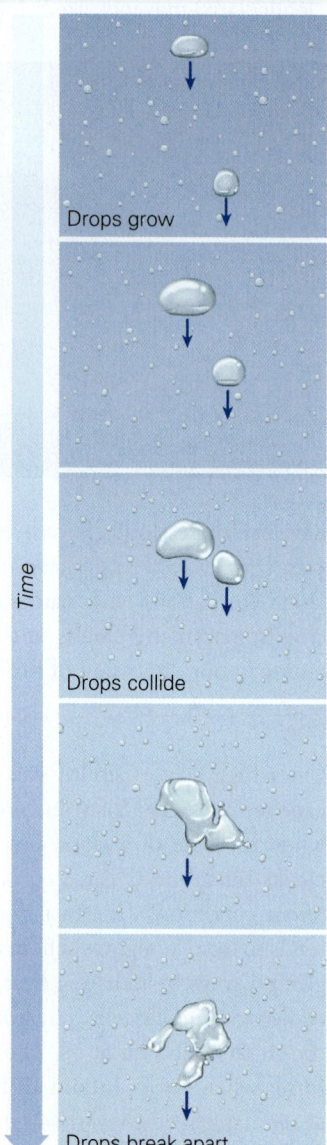

(b) When large raindrops collide, they can break up into many more, smaller raindrops.

liquid state even though their temperature is well below 0°C (32°F).

First, tiny ice crystals grow on special types of aerosols called *ice nuclei*. Ice nuclei are mineral grains whose crystal structure resembles that of ice. Because of this similarity, water molecules can find favorable attachment sites. Once these tiny crystals have formed, more water vapor molecules attach to their solid surfaces, eventually building beautiful hexagonal crystals. These crystals grow into a wide variety of shapes, depending on the temperature and humidity. Some crystals have branched arms, so that when they collide with neighboring ice crystals, they latch onto one another and clump together. The resulting composite of numerous ice crystals becomes a **snowflake** (see Fig. 17.26).

Not all ice particles grow entirely by deposition of water vapor directly on ice nuclei, however. If sufficient ice nuclei do not exist in a cloud, water droplets remain in a liquid state, but they become supercooled. When ice crystals falling through clouds encounter these droplets, the droplets instantly freeze onto the ice and form *rime* on the crystals. Ice particles growing by collection of supercooled droplets eventually grow into small, soft spheres of ice called **graupel**. A graupel ball that continues to grow and harden becomes a *hailstone* (see Fig. 17.26). In some cases, raindrops become supercooled in a layer of cold air near the ground. These drops can reach the ground as *freezing rain*, which freezes to surfaces on contact to form an ice crust on the ground surface, power lines, and trees.

PRECIPITATION INVOLVING BOTH WARM- AND COLD-CLOUD PROCESSES. As we learned earlier in this chapter, temperature decreases with altitude in the troposphere, so even if warm temperatures exist at ground level, the upper parts of towering clouds have temperatures well below freezing. Therefore, water vapor may attach to ice nuclei in the upper part of a tall cumulus congestus cloud. The resulting ice particles collect supercooled water droplets as they fall, forming graupel. But when the graupel particles fall to lower altitudes, where the temperatures rise above freezing, they melt to form raindrops. If strong updrafts exist in the cloud, graupel balls may remain aloft long enough to grow into hailstones. If the hailstones become large enough to fall through the updraft, or if the updraft weakens, they can reach the ground before completely melting and land as hail.

In winter, a warm, above-freezing layer of air may exist aloft while the air beneath it is below freezing. If this happens, raindrops falling from the warm layer aloft sometimes freeze into tiny pellets called **sleet** when they enter the cold air near the ground. At other times, they arrive as **freezing rain**, supercooled water droplets that freeze on contact with the ground, trees, and wires, to produce an ice-covered landscape.

Clouds and Energy

Recall that atmospheric water can exist in three states: gas (vapor), liquid, and solid (ice). A **phase change** happens when water in one state converts to another, such as when ice *melts* (solid → liquid) or *sublimates* (solid → gas), or a water droplet *freezes* (liquid → solid) or *evaporates* (liquid → gas). Phase changes can happen when air containing water cools or warms **(Fig. 17.27)**. Because phase changes don't take place instantly, and because temperature varies with altitude, a cloud can contain all three phases at once.

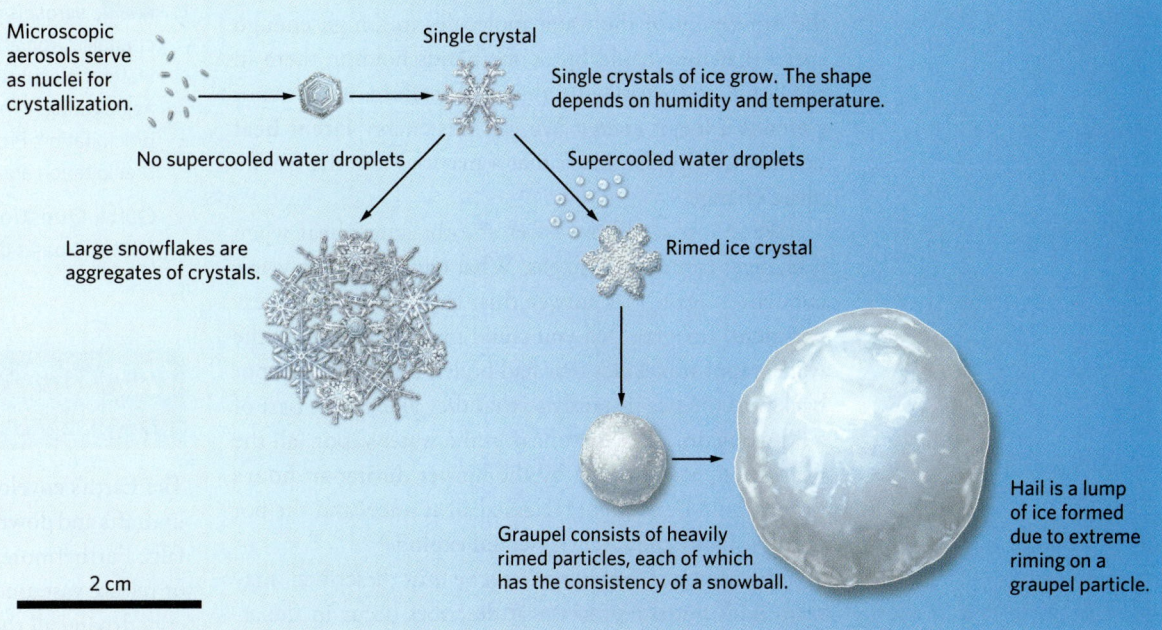

FIGURE 17.26 Ice particle growth in a cold cloud.

Microscopic aerosols serve as nuclei for crystallization.

Single crystal

Single crystals of ice grow. The shape depends on humidity and temperature.

No supercooled water droplets

Supercooled water droplets

Large snowflakes are aggregates of crystals.

Rimed ice crystal

Graupel consists of heavily rimed particles, each of which has the consistency of a snowball.

Hail is a lump of ice formed due to extreme riming on a graupel particle.

2 cm

Phase changes absorb or release energy. To understand the relationship between a phase change and energy, imagine a simple experiment. Take a large, full pot of water out of the refrigerator and place it on the red-hot burner of a stove. Heat the water for, say, 10 minutes—time for enough heat to transfer from the burner to the water to bring the water to a boil. Once the water has come to a boil, continue heating it for another hour, until nearly all the liquid water in the pot has become water vapor. If you measure the temperature of the water during boiling, you'll see that it remains unchanged, meaning that all the energy from the burner goes into driving the phase change. This experiment tells us that it

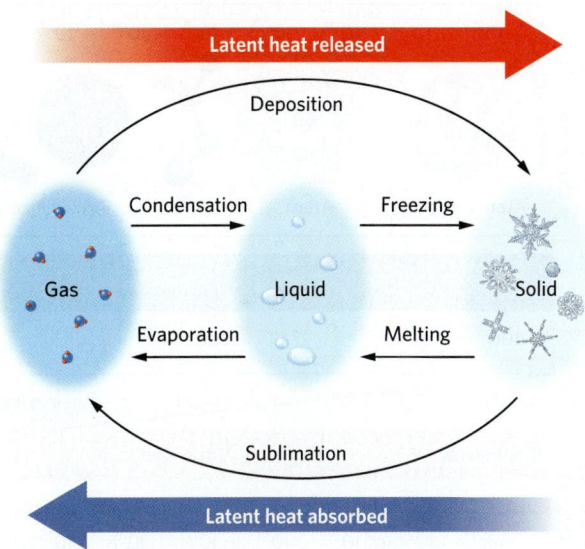

FIGURE 17.27 Latent heat is either released or absorbed in the atmosphere when water undergoes phase changes.

Latent heat released

Deposition

Condensation

Freezing

Gas

Liquid

Solid

Evaporation

Melting

Sublimation

Latent heat absorbed

took an hour's worth of energy from the red-hot burner to convert the pot's water into vapor—that's a lot of energy! Where did all that energy go? It went into accelerating the movement of the water molecules to a high enough speed that they could break the bonds holding them in the liquid and escape into the air. Therefore, water vapor contains a lot of energy. We call this energy **latent heat** because it's "hidden heat" that a material inherits from a phase change.

We've just seen how water absorbs latent heat when it changes from liquid to gas. What happens when water condenses again? To answer this, let's continue our experiment. Imagine that you could magically force all the water vapor molecules that had boiled out of the hot pot back into the pot instantly so that they all became part of a liquid again. The latent heat in the water vapor (all the heat added to the water by the burner during an hour's time) would be suddenly released all at once, and the pot (and probably the kitchen) would explode!

The thought experiment we've just described may seem odd, but the processes it describes occur in the atmosphere all the time. Water in the oceans absorbs solar energy, evaporates, and enters the atmosphere as vapor. This vapor stores the solar energy as latent heat. When clouds form, water vapor condenses to liquid, thereby releasing its latent heat. This release, in turn, warms the air. Similarly, since it's necessary to add heat to ice to get it to melt, liquid water stores latent heat that will be released when the liquid water freezes. We'll see in Chapter 19 that the energy from phase changes contributes to the development of storms.

Take-home message . . .

Clouds form when water molecules attach to aerosols and form either liquid droplets or ice crystals. Droplets and crystals grow both by absorbing water molecules from the air and by collecting and aggregating. When they become large enough, they fall as precipitation. Processes involved in cloud formation and precipitation absorb or release large amounts of energy.

Quick Question
On what basis do meteorologists classify clouds?

17.7 The Atmosphere's Power Source: Radiation

The Earth's envelope of gas does not sit still—winds blow, updrafts and downdrafts form, and precipitation forms and falls. Furthermore, phase changes of water in the air absorb or release vast amounts of latent heat. Where does the energy driving all these processes come from? In a word, the Sun. In this section, we examine how solar energy moves through and interacts with the Earth's atmosphere.

The Nature of Electromagnetic Radiation

All solar energy arrives at the top of the Earth's atmosphere as radiation. In a general sense, **radiation** is energy that travels away from its source either through a material or through a vacuum. Radiation comes in two general forms: *electromagnetic radiation* (such as visible light) and

FIGURE 17.28 The electromagnetic spectrum.

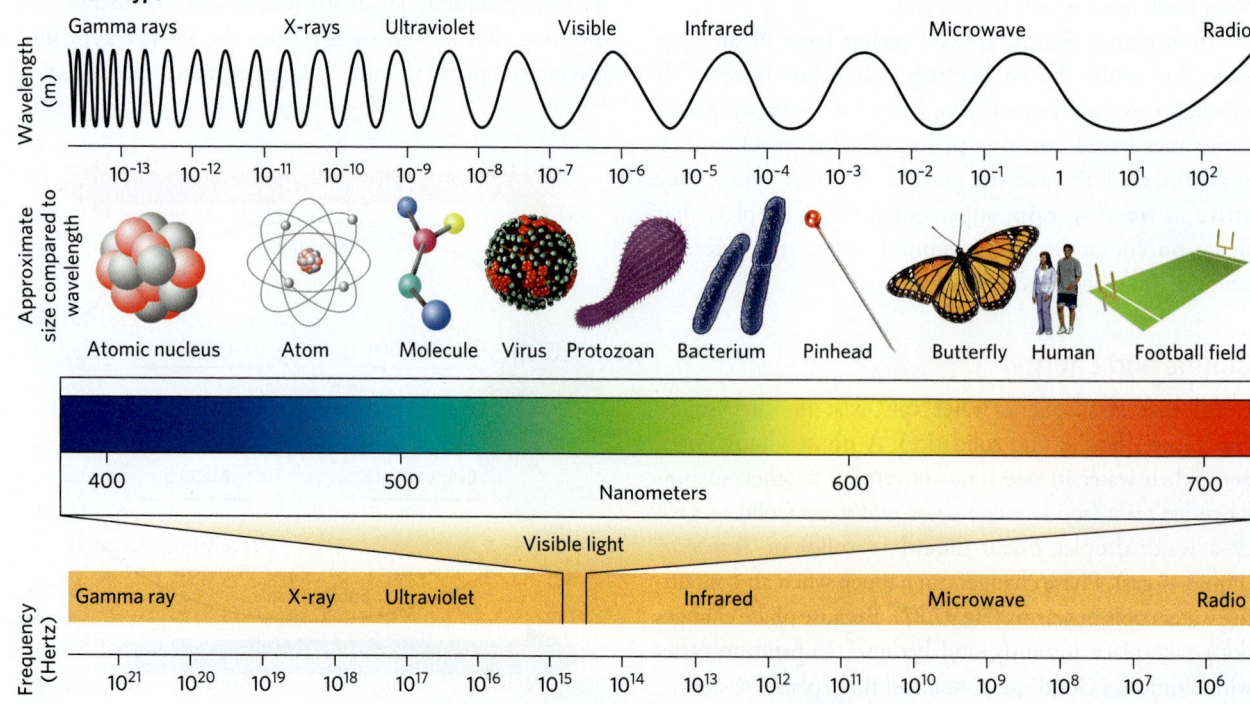

1 meter (m) = 100 centimeters (cm)
1 cm = 10 millimeters (mm)
1 mm = 1,000 micrometers (μm)
1 μm = 1,000 nanometers

particulate radiation (such as particles emitted from radioactive elements). Here, we focus our attention on electromagnetic radiation—specifically, the solar energy you feel warming (or burning) you on a bright sunny afternoon.

All objects—the Sun, you, the ground beneath you—emit electromagnetic radiation as long as they have a temperature above absolute zero (0 K). Physically, **electromagnetic radiation** is a form of energy that travels in the form of electric and magnetic waves through space. In the early 20th century, Albert Einstein proposed that the smallest unit of energy in radiation is a massless particle, which later came to be known as a *photon*. Groups of photons moving through space constitute *electromagnetic waves*, and groups of electromagnetic waves constitute *beams*. Physicists generally portray electromagnetic waves as a set of periodic up-and-down curves that, on a drawing, resemble the shape of ocean waves. But physically, electromagnetic waves differ greatly from ocean waves—electromagnetic waves can transmit energy through a vacuum and are themselves invisible, though they can be detected by instruments and by our senses. (Ocean waves, in contrast, exist only if water moves; see Section 15.5.)

We can characterize electromagnetic waves by specifying their **wavelength**, the distance between wave crests, and their **frequency**, the number of wave crests passing a point in space in 1 second (Fig. 17.28). Electromagnetic waves come in a huge range of wavelengths and, by implication, frequencies. Because all electromagnetic waves travel at the same velocity in a vacuum—the speed of light (300,000 km, or 186,000 mi, per second)—frequency increases as wavelength decreases. The full range of possible wavelengths constitutes the **electromagnetic spectrum**. Physicists use different names for different parts of the spectrum. *Shortwave radiation* includes gamma rays, X-rays, and ultraviolet (UV) radiation, as well as visible light and some infrared radiation. Infrared is divided into (a) shorter-wavelength near-infrared and, (b) longer-wavelength far-infrared. *Longwave radiation* includes far-infrared radiation as well as microwaves (used in radar) and radio waves. The energy carried by electromagnetic radiation depends on its wavelength and frequency. Shortwave (high-frequency) radiation contains more energy than longwave (low-frequency) radiation.

Energy Balance in the Earth-Sun System

THE CONCEPT OF BLACKBODY RADIATION. To understand why the Earth's atmosphere has the temperature that it does, we have to dig into a few background concepts from physics. When discussing the behavior of electromagnetic radiation, physicists picture an imaginary object, called a **blackbody**, that absorbs all radiation that shines on it. Any radiation absorbed by a blackbody must eventually be re-radiated back into space as *blackbody radiation*—otherwise, the object would become infinitely hot. The Earth,

the Sun, and, for that matter, all objects in space, act like blackbodies. Three laws govern the behavior of blackbodies:

- A warmer object emits more radiation, at all wavelengths, than a colder object.
- The total amount of radiation an object emits increases rapidly as its temperature increases.
- As the temperature of an object increases, the wavelength of maximum radiation emission shifts toward shorter wavelengths.

A physics book can explain why these laws operate as they do, but for our purposes, we'll simply keep them in mind as we explore how energy is added to the Earth's atmosphere by sunlight.

THE CONCEPT OF RADIATIVE EQUILIBRIUM. The Sun's surface reaches a temperature of about 5,600°C (10,000°F), hot enough to vaporize the Earth (see Chapter 23). But because the Sun sends energy in all directions and lies so far from the Earth, our planet intercepts only a tiny fraction (0.00000015%) of the Sun's total radiation. Nevertheless, if the Earth were to absorb all of the energy that it receives from the Sun, without releasing any back into space, it would eventually vaporize. This, of course, doesn't happen, because the Earth, like any blackbody, constantly re-radiates energy into space. Over time, the total energy arriving at the Earth from the Sun exactly balances the amount of energy re-radiated from the Earth into space. On the time scale of a decade, the rate at which solar energy reaches the Earth, a quantity called the *solar energy flux*, remains nearly constant. If the Earth had no atmosphere, the solar energy flux would cause the Earth to stay at a constant −18°C (0°F), a temperature known as the Earth's *radiative equilibrium temperature*.

Because the Sun is so hot and the Earth is so cool, the spectrum of the solar radiation arriving at the Earth differs greatly from the spectrum of the radiation that the Earth re-radiates into space, as the laws of blackbody radiation predict (Fig. 17.29). Specifically, solar energy reaching the Earth arrives predominantly as shortwave radiation—ultraviolet light (the radiation that can give us sunburns), visible light (the radiation our eyes detect), and near-infrared radiation (the radiation we can feel as heat but cannot see). In contrast, the radiation emitted by the Earth consists of longwave radiation (far-infrared radiation).

THE ATMOSPHERIC GREENHOUSE EFFECT. If the Earth's surface were at its radiative equilibrium temperature, you wouldn't be reading this book, for life would not exist. Fortunately, the actual average surface temperature of our planet, 15°C (59°F), is much warmer, so liquid water can exist and life can flourish. This difference is due entirely to the *atmospheric greenhouse effect*—also called, simply, the **greenhouse effect**. We commonly hear about the

Incoming solar shortwave radiation
=
Outgoing terrestrial longwave radiation
Shortwave radiation
Longwave radiation
Relative amount of energy
Ultraviolet
Visible light
Near-infrared
Far-infrared
0.1
0.15
0.2
0.3
0.5
1.0
1.5
2
3
4
5
10
15
20
30
50
100
Wavelength (μm)
This curve indicates the amount of energy at a given wavelength received by the Earth. The total energy is the area under the curve.
This curve indicates the amount of energy at a given wavelength emitted by the Earth to space. The total energy is the area under the curve.

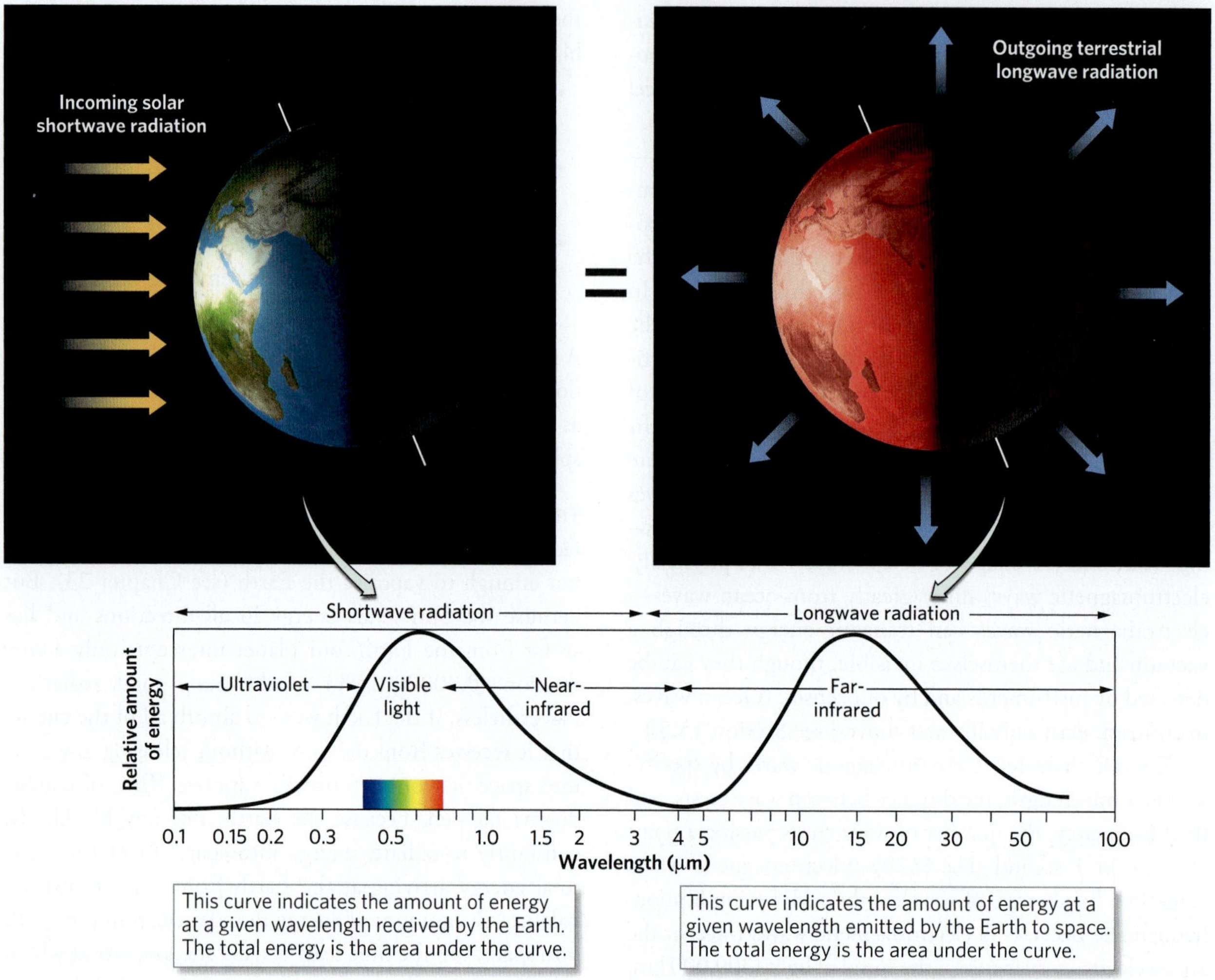

FIGURE 17.29 In order for the Earth to be in equilibrium, the shortwave radiation that the Earth receives from the Sun must equal the longwave radiation that the Earth emits to space.

greenhouse effect in news concerning the Earth's changing climate (see Chapter 20). Here, let's explore how it works and why it is important to our survival.

While the Earth itself acts like a blackbody, its atmosphere does not. Rather, each atmospheric gas absorbs and emits radiation selectively, meaning that it takes in only certain wavelengths and sends out only certain wavelengths. Why does the atmosphere exhibit this behavior? It's because of the nature of the molecules that air contains. Molecules such as O_2 and N_2, which consist of only two atoms apiece, do not absorb or emit very much solar radiation or terrestrial radiation. In contrast, molecules with more than two atoms (such as H_2O, CO_2, O_3, CH_4, and N_2O) are excellent absorbers and emitters of radiation *at certain wavelengths*. This difference in behavior can be traced to how the molecules vibrate when exposed to radiation.

Because of the different absorption and emission characteristics of the various gases in the atmosphere, the atmosphere is transparent to two ranges of solar and terrestrial wavelengths **(Fig. 17.30)**. Put another way, the atmosphere

Shortwave radiation
Longwave radiation
Visible window
Infrared window
0.2 μm
0.5
1.0
5
10
50
100
100
0
% absorption by atmospheric gases
O_2 O_3
Blue
Green
Red
Visible
H_2O
H_2O
CO_2
H_2O
O_3
CO_2
H_2O

FIGURE 17.30 Different atmospheric gases absorb radiation of different wavelengths. Low-absorption regions of the spectrum are called atmospheric windows.

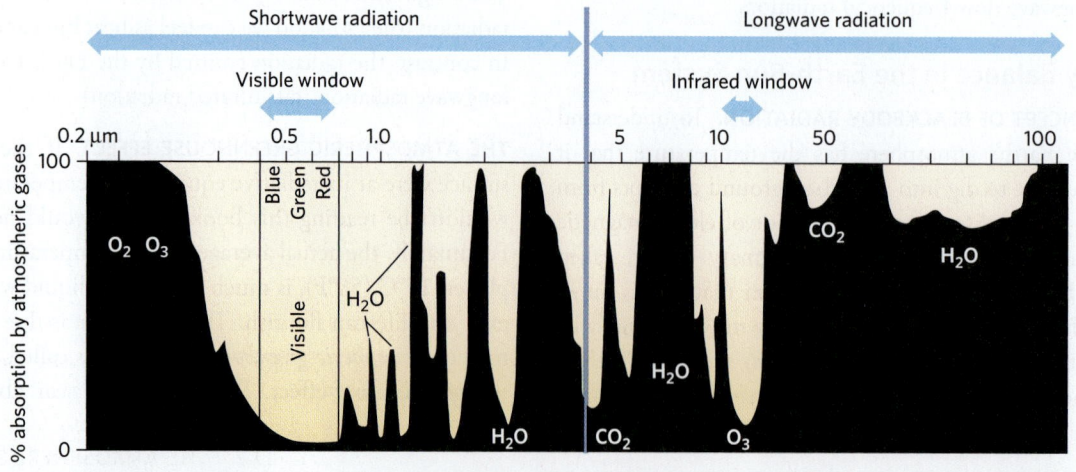

FIGURE 17.31 A simple depiction of the atmospheric greenhouse effect.

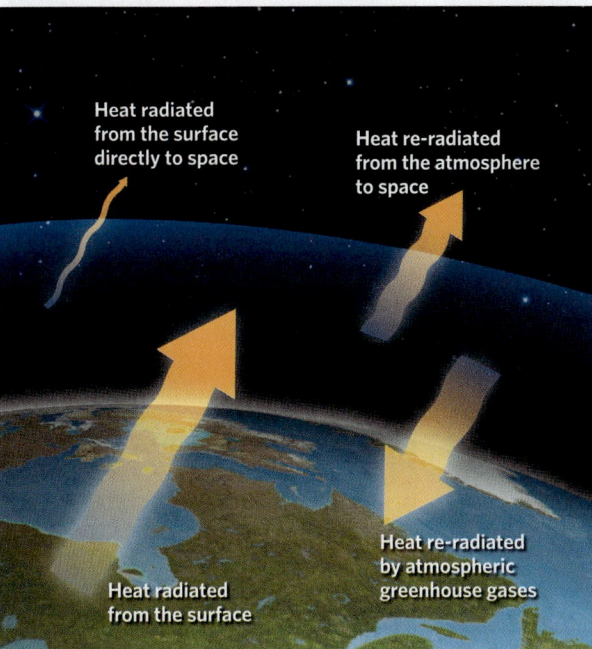

Heat radiated from the surface directly to space

Heat re-radiated from the atmosphere to space

Heat re-radiated by atmospheric greenhouse gases

Heat radiated from the surface

Take-home message . . .

The Sun provides energy to the Earth and its atmosphere in the form of electromagnetic radiation. Visible light and some ultraviolet radiation that pass through the atmosphere are absorbed by the Earth's surface, which in turn re-radiates this energy as infrared radiation. Greenhouse gases, which trap some of this re-radiated energy, keep the atmosphere warm enough for liquid water—and life—to exist on the Earth.

Quick Question

If there were no atmosphere, what would the temperature of the Earth's surface be?

17.8 Optical Effects in the Atmosphere

Why Is the Sky Blue?

When astronauts walked on the Moon and looked upward, they saw a black sky, even though they were bathed in sunlight. Similarly, when they looked into the shadows behind boulders, they saw only black. In contrast, when you look up from the surface of the Earth on a sunny day, you see a blue sky, and if you look into the shadow behind a boulder, you can still see the ground. Why?

Blue skies exist because air molecules cause **light scattering**, a phenomenon that takes place when molecules absorb visible light and then immediately re-emit it in random directions, as we mentioned in Chapter 15. The degree of scattering depends on the sizes of the molecules relative to the wavelengths of the light (**Fig. 17.32**). Because air molecules are small relative to the wavelengths of visible light, short-wavelength light (purple and blue) scatters more effectively than long-wavelength light (orange and red). When a beam of sunlight passes through the atmosphere from directly overhead, blue wavelengths within the beam preferentially scatter again and again. After these many scattering events, blue wavelengths arrive at our eyes from somewhere in the sky, making the sky appear blue (see Fig. 17.32a). Also because of light scattering, some light energy makes it into the shadows behind objects, so shadows on the Earth aren't completely black.

When the Sun lies low in the sky, the distance through the atmosphere that a beam of sunlight must travel to reach our eyes increases. As a result, nearly all blue, green, and yellow wavelengths scatter back into space, leaving only reds and oranges, the color of the rising or setting Sun (see Fig. 17.32c).

Mirages

Desert travelers sometimes see what looks like a lake of shimmering water on bone-dry land in the distance.

has two **radiation windows** through which certain types of solar and terrestrial radiation can pass between the Earth's surface and space without interference. The first window is transparent to the visible wavelengths (allowing incoming sunlight to reach the Earth's surface and allowing us to see), and the second window is transparent to certain infrared wavelengths (allowing radiation from the Earth to escape back into space). For all other wavelengths, the atmosphere behaves as a strong absorber. It re-radiates the absorbed energy in random directions. Some of the re-radiated energy heads back to the Earth's surface, heating the planet, and some gets absorbed by other atmospheric molecules, increasing the average kinetic energy (that is, the temperature) of the air. Thus, the atmosphere acts like a blanket over the Earth, trapping radiation and preventing it from escaping to space (**Fig. 17.31**).

The absorption and re-radiation of energy by the atmosphere has come to be known as the *atmospheric greenhouse effect* because the way the atmosphere retains heat makes people think of the way in which a glass greenhouse retains heat. The analogy is not perfect, however, because the heat in a real greenhouse gets trapped simply because the warm air inside cannot escape to the outside. In contrast, the atmospheric greenhouse effect works because, although solar energy can pass through the atmosphere to the ground, certain atmospheric molecules prevent the Earth's infrared radiation from escaping from the atmosphere back into space. The atmospheric gases that absorb and re-emit infrared radiation are known as **greenhouse gases**.

FIGURE 17.32 The many colors of the sky.

(a) At midday, the sky appears blue.

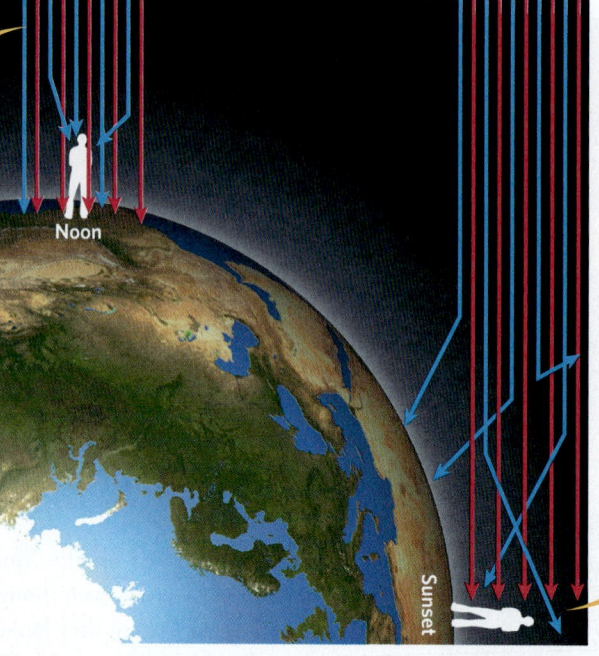

More blue light than red light reaches the observer's eyes.

More red light than blue light reaches the observer's eyes.

Noon

Sunset

(b) When sunlight passes through the atmosphere, short wavelengths (primarily blue) are scattered out of the main solar beam. At noon, the observer sees blue light coming from many directions. At sunrise or sunset, only the red light reaches the observer.

(c) At sunrise or sunset, the sky appears orange or red.

FIGURE 17.33 Mirages.

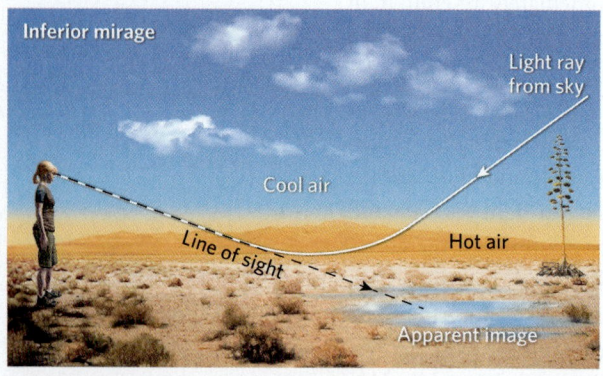

Inferior mirage

Light ray from sky

Cool air

Line of sight

Hot air

Apparent image

A mirage on a desert highway

(a) An inferior mirage forms when hot air is present near the ground and bends light upward. The mirage looks like shimmering water because the hot air at ground level is not still. The shimmering water is actually refracted blue light from the sky.

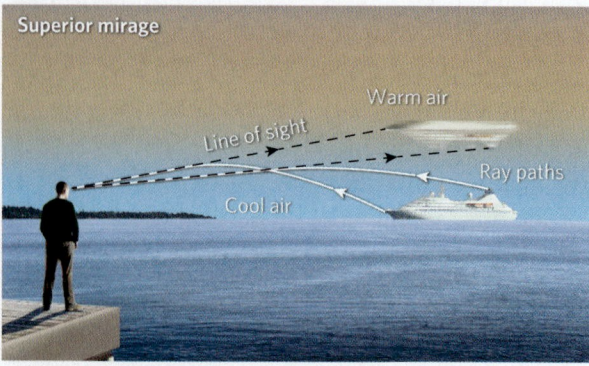

Superior mirage

Warm air

Line of sight

Ray paths

Cool air

A mirage along the seashore

(b) A superior mirage forms when warm air overlies cooler air and bends light downward. Generally, but not always, the mirage is upside down.

Sailors might see ships floating upside down in the sky, while polar explorers look twice when they see ice castles floating above the horizon. Such images are mirages, and they disappear when you approach them. Simply put, a **mirage** is an image formed when the light reflected from an object bends, or *refracts*, though the atmosphere so that the object seems to appear where it doesn't actually exist.

Mirages develop when the air near the ground surface is either exceptionally warm or exceptionally cold relative to the air above. As light passes from less dense warm air into denser cold air (or vice versa), it undergoes refraction. The presence of a hot layer of air beneath a cool layer produces an *inferior mirage* (meaning that the image you see lies below the real object) **(Fig. 17.33a)**, and the presence of a cold layer produces a *superior mirage* (meaning that the image you see lies above the real object) (Fig. 17.33b). In deserts, where a layer of very hot air lies at ground level, you'll see inferior mirages of the sky on the ground surface. Because the hot air moves, the light of the mirage shimmers, so it looks like water. Superior mirages develop mostly in the Arctic and Antarctic, where a layer of cold air develops at the Earth's surface, so that light bends downward as it travels from colder air upward into warmer air.

The Atmosphere Showing Off: Colorful Optical Effects

When sunlight passes through air containing water droplets or ice particles, a variety of colorful optical phenomena may develop. For example, when the Sun lies close to the horizon and its rays pass through clear air behind you into rain in front of you, you'll see a **rainbow**, an arc displaying all the colors of visible light in sequence. If you look closely, sometimes you'll see two rainbows, a lower one called the *primary bow* and an upper one called the *secondary bow* (Fig. 17.34a). Note that the order of colors in the secondary bow is the opposite of that in the primary bow.

Rainbows develop when light reflects within and refracts through raindrops. Specifically, when an incoming beam of sunlight strikes the spherical surface of a raindrop, the drop acts as a prism and splits the light into colors. The light then reflects off the interior of the drop's back surface and refracts again as it exits the drop (Fig. 17.34b). Light in a secondary bow undergoes an additional reflection.

Many other, less frequently observed optical phenomena develop locally in the atmosphere (Earth Science at a Glance, pp. 556–557). Examples include *cloud iridescence*, bands of shimmering colors seen when light passes through thin clouds; *glory*, a series of colored rings that surround the shadows of aircraft on the tops of clouds when viewed from above; and *halos* and *arcs*, circles and arches of light and color that develop when light passes through a thin cloud or fog of ice crystals (see Fig. 17.1b). All of these phenomena are associated with the manner in which light passes around and through water droplets and ice crystals in the air.

Take-home message . . .

Radiation interacting with the atmosphere produces an array of optical phenomena. Mirages develop when the ground-level layer of air has a different temperature than the air above, so that light bends while passing through the air. Rainbows, halos, and other brilliant phenomena form when light interacts with water droplets and ice crystals.

Quick Question

Where is the Sun, relative to you, when you see a rainbow?

FIGURE 17.34 Rainbows delight the eye.

(a) If you're at the right position, with the Sun behind you and rain in front of you, a rainbow will appear. Sometimes you'll see a double rainbow.

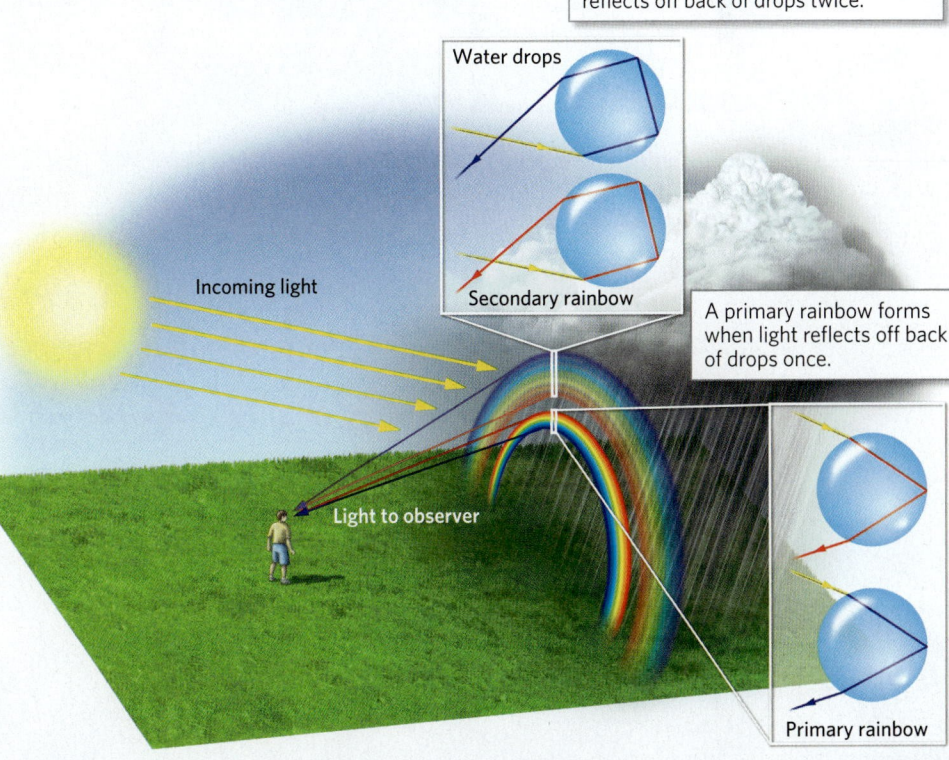

A secondary rainbow forms when light reflects off back of drops twice.

A primary rainbow forms when light reflects off back of drops once.

(b) A rainbow forms because light refracts and reflects as it interacts with raindrops.

Atmospheric Optics

Light, as it passes through the atmosphere, interacts with aerosols, water droplets, and ice crystals to produce a bewildering array of optical phenomena that delight the eyes, warm the heart, and astonish those lucky enough to witness them. From a scientific perspective, these effects arise as a result of six different ways light interacts with the material in the atmosphere. These are *absorption*, where light is removed from a beam; *reflection*, where light bounces off a surface such as a mirror; *refraction*, where light bends as it passes between denser and less dense substances, such as water and air; *diffraction*, where light waves bend around objects such as small raindrops; *scattering*, where light is absorbed and then immediately emitted in a different direction; and *emission*, where light is produced when high-energy particles from the Sun crash into oxygen and nitrogen molecules high in the atmosphere.

Crepuscular rays form when clouds absorb sunlight, shadowing some, but not all solar beams.

Irridescence and glory are diffraction effects caused by light waves bending around tiny water droplets. Glory can appear when looking down from an aircraft passing over water clouds.

Water

Shadow

Sun pillar (caused by reflection off bottom of ice crystals)

Auroral colors are caused when light is emitted during collisions between solar particles and molecules in the Earth's upper atmosphere at different altitudes.

Oxygen; higher altitudes

Oxygen; lower altitudes

Nitrogen; much lower altitudes

Arcs, halos, and sundogs form as light refracts through columnar and plate-like ice crystals falling with different orientations.

Arcs

Halos

Sundogs

Ice

Upper atmosphere

⊚17 CHAPTER REVIEW

- Dry air consists of nitrogen (78%), oxygen (21%), argon (1%), carbon dioxide (0.04%), and trace gases (1%). The amount of water vapor in the atmosphere varies considerably over time and space.

- In addition to gases, the atmosphere contains tiny particles called aerosols.

- The Earth's atmosphere evolved over time. Before the appearance of oceans, it consisted of volcanic gases. Formation of the oceans removed most of the atmosphere's water vapor and CO_2, leaving N_2 behind. When photosynthetic organisms evolved, they absorbed CO_2, which became buried, and released O_2.

- Addition of O_2 to the atmosphere also led to the formation of ozone, which absorbs harmful components of sunlight, allowing life on land.

- Atmospheric pressure decreases rapidly with elevation. Half of the atmosphere's mass lies below an elevation of 5.6 km (3.5 mi).

- The atmosphere's capacity to hold water vapor increases as its temperature increases. The relative humidity represents the amount of water vapor in air relative to its holding capacity. Another quantity, the dew point, represents the actual amount of water vapor in air.

- We measure properties of the atmosphere (such as temperature, pressure, humidity, and wind direction and speed) at ground level using a variety of instruments at a weather station. To measure conditions high in the atmosphere, meteorologists launch rawinsondes on weather balloons.

- Four atmospheric layers are distinguished by patterns of temperature change with elevation: the troposphere, stratosphere, mesosphere, and thermosphere. The atmosphere can also be divided into layers that are well mixed (homosphere) and unmixed (heterosphere). Another layer, the ionosphere, is recognized by its high concentration of ions.

- Clouds form when air rises, expands, and cools, as happens when air is lifted or when it becomes buoyant. We classify clouds by their altitude, shape, and whether or not they produce precipitation.

- Cloud droplets grow into raindrops by first absorbing water molecules from water vapor, and then by coalescing with neighboring droplets. Ice particles grow to precipitation size by first incorporating water molecules, and then by either linking together to form snowflakes or collecting supercooled water droplets to form rimed ice particles, graupel, or hail.

- Water in the atmosphere constantly undergoes phase changes, releasing or absorbing latent heat in the process.

- The Sun provides energy to the Earth in the form of electromagnetic radiation, which includes many different wavelengths. An object that absorbs all radiation it receives is called a blackbody. A blackbody re-emits this radiation at wavelengths that depend on its temperature.

- Solar energy arrives at the Earth's surface as shortwave radiation, but the much cooler Earth re-radiates this energy as longwave radiation. Certain gases in the atmosphere trap some of this energy, causing the greenhouse effect, without which the Earth's surface would be below freezing.

- Visible light interacts with air molecules and scatters, making the sky appear blue. Mirages form when sunlight bends due to variations in air temperature with elevation. Rainbows, halos, and other brilliant phenomena form when light interacts with water droplets and ice crystals in the atmosphere.

aerosol (p. 527)
air (p. 525)
air temperature (p. 530)
atmosphere (p. 525)
atmospheric pressure (p. 531)
aurora (p. 543)
barometer (p. 532)
blackbody (p. 551)
cirrus cloud (p. 545)
cloud (p. 527)
cloud cover (p. 537)
collision-coalescence (p. 548)
cumulus cloud (p. 546)
dew point (p. 535)
Doppler radar (p. 539)

electromagnetic radiation (p. 551)
electromagnetic spectrum (p. 551)
environmental lapse rate (p. 541)
fog (p. 547)
frequency (p. 551)
freezing rain (p. 549)
front (p. 544)
graupel (p. 549)
great oxygenation event (p. 529)
greenhouse effect (p. 551)
greenhouse gas (p. 553)
haze (p. 528)
heterosphere (p. 543)
homosphere (p. 543)

ionosphere (p. 543)
latent heat (p. 550)
light scattering (p. 553)
mesosphere (p. 542)
meteorologist (p. 530)
mirage (p. 554)
ozone hole (p. 542)
phase change (p. 549)
precipitation (p. 537)
radiation (p. 550)
radiation window (p. 553)
rainbow (p. 555)
raindrop (p. 547)
relative humidity (RH) (p. 535)
saturation vapor pressure (p. 534)

sleet (p. 549)
snowflake (p. 548)
stratosphere (p. 542)
stratus cloud (p. 546)
thermometer (p. 530)
thermosphere (p. 543)
troposphere (p. 541)
vapor pressure (p. 534)
visibility (p. 537)
water vapor (p. 527)
wavelength (p. 551)
weather (p. 530)
wind (p. 535)
wind direction (p. 535)
wind speed (p. 536)

Letters in parentheses correspond to the chapter's learning objectives.

1. What is the present composition of the atmosphere? Which gases have concentrations that vary little over time, and which vary rapidly over time and from location to location? **(A)**

2. How has the atmosphere evolved over time? At what point did it become hospitable to life on land? **(A)**

3. You decide to climb a mountain and take along a thermometer and a barometer. During the climb, you take readings. At the end, you plot graphs of the results. What will the graphs look like? **(B)**

4. Explain why the absolute amount of moisture in the atmosphere over a hot desert can be larger than that over the Arctic tundra, even though the relative humidity over the tundra may be 90%, while over the desert it may be 10%. **(B)**

5. A southwest wind blows from where to where? **(B)**

6. Describe the primary layers in the atmosphere as characterized by temperature, atmospheric composition, and electric charge. What are the boundaries between the temperature layers called? **(C)**

7. Explain why aerosols are a critical component of the Earth's hydrologic cycle. **(D)**

8. Is latent heat released to the atmosphere or absorbed from the atmosphere when raindrops evaporate? How about when snowflakes melt? How about when a cloud forms? **(E)**

9. Are you more likely to find clouds on the upwind or downwind side of a mountain range? Why? **(E)**

10. Summarize the primary cloud types. What type of cloud is pictured in the figure? **(E)**

11. Is the surface of the Sun hotter or colder than the electric coils on your kitchen stove when they're fully heated? Explain. (Hint: Recall Figure 17.27 and the laws governing the behavior of blackbody radiation.) **(F)**

12. What is meant by shortwave radiation and longwave radiation? **(F)**

13. Which type of radiation is emitted by the Sun? By the Earth? (Hint: Consider the figures.) **(F)**

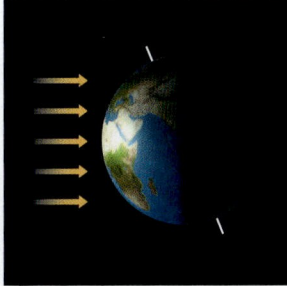

14. Why can we consider radiation arriving at the top of the Earth's atmosphere from the Sun as distinct from radiation emitted by the Earth-atmosphere system? **(F)**

15. Which gases are important selective absorbers of radiation? What relevance does this property of gases have to the atmosphere? **(F)**

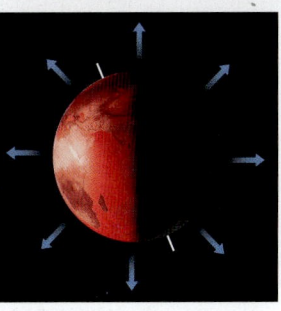

16. What is the atmospheric greenhouse effect? Which atmospheric gases are most responsible for the greenhouse effect? **(F)**

17. From weather reports you can obtain on the internet, determine the current temperature, dew point, and relative humidity in a number of cities. Rank the cities according to their relative humidity, placing the city with the highest relative humidity at the top of the rank. Now subtract the dew point (D) from the temperature (T), and rank the cities by the difference in these two temperatures ($T - D$), but place the city with the smallest difference at the top. Compare the two rankings. Summarize your findings. **(B)**

18. Clouds generally move too slowly for the human eye to appreciate the lifting mechanisms that produce them. High-speed video circumvents this problem. Examine YouTube videos of clouds by searching for "high-speed clouds" or "time-lapse clouds." Summarize what you observe for the different cloud types described in the chapter. **(E)**

18 WINDS OF THE WORLD
Global and Local Wind Systems

By the end of the chapter you should be able to . . .

A. explain why our planet's atmosphere constantly flows on a global scale.

B. contrast how air flows around high-pressure and low-pressure centers in both hemispheres.

C. discuss why tropical rainforests and subtropical deserts develop where they do.

D. describe monsoons, and explain why these meteorological phenomena are so important to society.

E. explain why the polar front exists and how it is related to the polar-front jet stream.

F. differentiate between the polar-front jet stream and the subtropical jet stream in terms of altitude and reason for existence.

G. discuss how local weather systems develop near land-water boundaries and in association with mountain ranges.

18.1 Introduction

When Sir Francis Drake set sail with the *Golden Hind* and its sister ships in 1577 (Fig. 18.1) to begin what would become the second circumnavigation of the globe, he and his crews were literally at the mercy of the wind, and during their three-year voyage, they experienced most of the wild variations in wind and weather that the Earth System can conjure. Sailing south, they blew right into the clutches of a swirling storm large enough to cover a wide swath of an ocean or continent. Winds and waves battered the ships so badly that Drake had to return to England for repairs.

Drake's second effort proved more successful, and his ships reached the belt of *trade winds*—westward-blowing winds that were so named because of their importance to transatlantic commerce. These winds filled the explorers' sails and blew them steadily southwestward. When the ships reached the equator, fortunately missing a fierce hurricane over the Caribbean to their west, they entered the sultry *doldrums*, locations near the equator where the wind goes calm. After bobbing helplessly with limp sails for endless days, they crossed the equator and entered another belt of trade winds. These winds blew toward the northwest, so it took all the crew's skill to maintain an overall southward route. When the ships reached the southern tip of South America, they encountered perpetually strong winds blowing from west to east, the *prevailing westerlies*. Sea spray doused the ships, coated their decks and sails with ice, and thereby disabled part of the fleet. The luckier ships made it into the Pacific and headed north along the west coast of the Americas. Drake and his remaining crew then turned west. As they crossed the Pacific and Indian Oceans, they encountered many challenging weather conditions, such as winds associated with *typhoons* (Pacific hurricanes) and *monsoons* (seasonally changing tropical winds driven by temperature differences between land and sea). Finally, after rounding the southern tip of Africa, the *Golden Hind*, now the only remaining ship, headed home, its hold filled with plundered gold and its sailors full of tales about the winds of the world.

Atmospheric scientists refer to the flow pattern of air around the globe as the atmosphere's **general circulation**. Drake had only a vague sense of this pattern. Today, you can see it on a computer screen by looking at successive satellite images that show the distributions of clouds and moisture moving with the winds (Fig. 18.2). By compiling measurements from weather stations worldwide,

<< Strong winds on the Solar de Ulyuni, a wide salt flat on the Altiplano, a high plateau in Bolivia, cause flags to flap violently. Wind, the horizontal movement of air, circulates and mixes the atmosphere.

FIGURE 18.1 The *Golden Hind*, the second ship to circumnavigate the world, was driven by wind power.

atmospheric scientists can generate maps that display variations in wind direction and speed around the world (**Earth Science at a Glance**, pp. 562–563). In this chapter, we focus on our atmosphere's currents of air—both large and small—and how those winds provide the environment for storms, the subject of our next chapter.

FIGURE 18.2 The pattern of atmospheric water vapor, as measured by satellites, shows that the general circulation of the atmosphere is complex, with many currents and eddies.

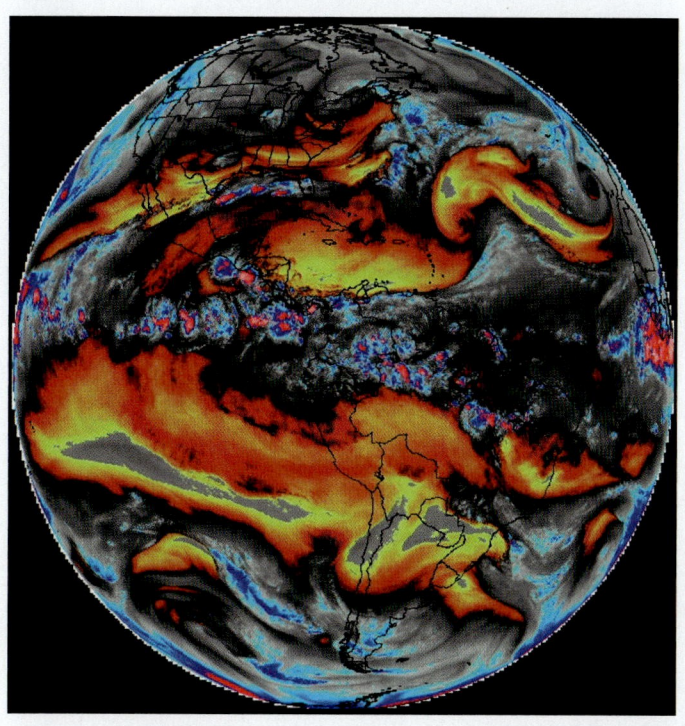

Winds of the World:
A Resource for Power

Wind maps

The four global wind maps illustrate the directions and speeds of surface
winds at different locations, at the same time, and therefore emphasize that
wind direction and speed vary dramatically around the world. The
orientation and taper of the colored lines represent wind direction, and their
brightness represents speed—brighter lines indicate faster speeds. Areas where
the lines have a reddish tint are regions where winds are particularly fast (> 70 km/h).
Note that surface air flows in a loop around Antarctica, but that locally, it curves into eddies.
In the northern hemisphere, several distinct cyclones stand out, as do the trade winds and the doldrums.
Overall, surface winds over land are much slower than those over the ocean, due to frictional interaction between air and land.

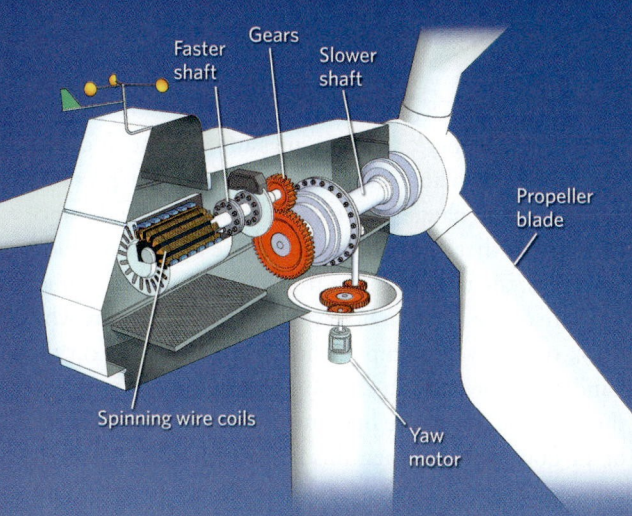

Faster
shaft
Gears
Slower
shaft

Propeller
blade

Spinning wire coils

Yaw
motor

Map of potential wind power

Because of the variation in the strength and steadiness of winds, some regions have the potential to generate more wind power than do others. On the map below, darker colors represent more potential power. The strongest steady winds are over the oceans, but the technology for building wind farms beyond the continental shelf does not yet exist.

The inside of a wind turbine

The propellers of modern wind turbines translate the flow of air in the wind into the rotation of a shaft. Gears shift the relatively slow rotation of the shaft connected to the spinning blades into the relatively fast rotation of a shaft that enters a generator. In the generator, magnets cause an iron core to rotate within a coil of conducting wire. The movement of the iron core induces an electrical current in the wire.

We begin this chapter by investigating the nature of wind itself. We discover why the wind blows and examine factors that control its direction and intensity. With this background, we then show how the characteristics of the general circulation vary with latitude. We start with the *tropics*, the region that straddles the equator, where intense solar heating drives the planet's largest distinct air circulations: *Hadley cells*, the *Southern Oscillation*, and *monsoons*. We then shift to the Earth's mid-latitudes, a region where *jet streams* play a major role in controlling the weather. Finally, we wrap up with an examination of some notable local wind systems around the globe.

18.2 Why Do the Winds Blow?

Our world would be a very different place without **wind**, the horizontal component of air flow. You can feel the wind, for when it blows, air molecules bombard your skin, and your nerves sense their combined impact, even though each molecule alone is too tiny to be visible. The faster the molecules move, the more force they exert, so during storms, winds can rip roofs from buildings or blow trains off their tracks. In *steady winds*, air flow maintains the same direction and speed, whereas in *turbulent winds*, the flow of air twists and turns, and the speed at a locality changes constantly. Winds form at all scales, from small gusts that rustle a few leaves on a summer day to vast currents that circle the planet or swirl into gigantic storms.

Have you wondered why the wind blows? Three forces play a key role in determining the direction and speed of the wind: (1) the pressure-gradient force, (2) the Coriolis force, and (3) the frictional force. Let's look at these forces and their consequences in turn. You'll recall that we introduced some of these forces in Chapter 15, in our discussion of ocean currents.

The Pressure-Gradient Force

In the atmosphere, the air pressure at one location may differ from that at another. We define a **pressure gradient** as a change in pressure divided by the distance over which that change occurs. Where a pressure gradient exists, air tries to move from the location of higher pressure to the location of lower pressure. To picture why, think of a bicycle pump—when you push down on the handle, the pressure in the pump's cylinder increases, so the air flows out of the open nozzle. Because the velocity of air flow changes when this movement takes place, we can say that a pressure gradient causes air to accelerate, and we refer to the force causing this acceleration as the *pressure-gradient force*.

To visualize how a pressure-gradient force can exist in the atmosphere, think of a sailing ship on the ocean. If the air pressure behind the ship exceeds the air pressure ahead of it, then the air density behind the ship exceeds the air density in front. Air molecules are in constant motion, so more air molecules strike the back side of the sails than the front side, and the ship moves forward **(Fig. 18.3)**. The same concept applies even if the "ship" is the size of a single air molecule: more collisions happen on the side of the air molecule where the pressure is greater, so the molecule, on average, moves in the direction of lower pressure. When this phenomenon affects a vast number of air molecules, the whole air volume moves in the direction of lower pressure, and wind blows.

The wind's speed—measured in meters per second, miles per hour, or *knots* (nautical miles per hour)—depends on the magnitude of the pressure gradient, in that the wind blows faster where the air pressure changes more quickly over a given horizontal distance than where it changes more slowly. To picture why, imagine the difference between stepping on an air-filled plastic bag with all your weight and squeezing the bag gently with your hands. If there is an opening in the bag, the air escapes faster when you step on the bag than when you squeeze it gently.

We can represent pressure gradients on a map of constant elevation by plotting contour lines, or **isobars**, along which all points have the same air pressure. Where isobars lie closer together, there's a larger (or steeper) pressure gradient, so stronger winds blow **(Fig. 18.4a)**, and where isobars lie farther apart, there's a smaller (or gentler) pressure gradient, so weaker winds blow **(Fig. 18.4b)**.

Why does air pressure differ across locations in the atmosphere? One primary reason involves heating and cooling of air. As we saw in Chapter 17, the Earth's surface absorbs incoming solar energy, then re-radiates it as infrared energy, which heats air at the base of the atmosphere.

FIGURE 18.3 The relationship between wind and a gradient in air pressure. There are more air molecules to the left of the ship, so the air pressure is greater on the left, and the wind blows from left to right.

FIGURE 18.4 The relationship between wind and the spacing of isobars.

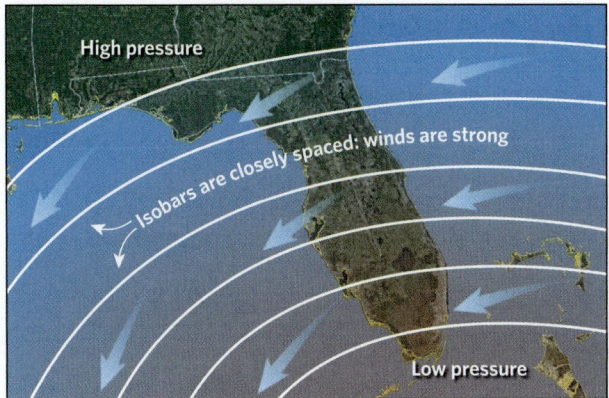

(a) Closely-spaced isobars depict a strong pressure gradient and fast winds.

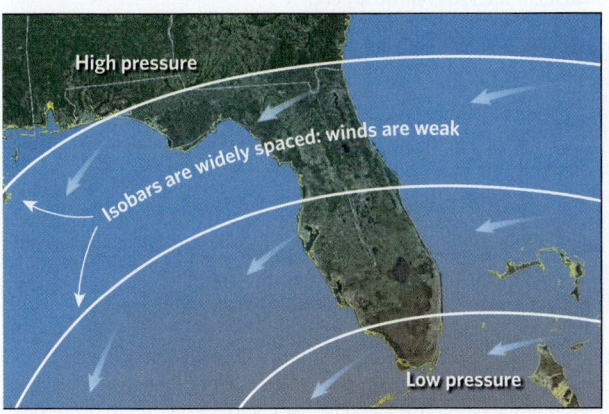

(b) Widely-spaced isobars depict a weak pressure gradient and slow winds.

Consequently, the near-surface air expands and decreases in density. This change can create low pressure at the Earth's surface if it occurs over a localized region. To understand why, we first introduce the concept of an isobaric surface. Imagine that we ascend in the atmosphere to a level where the pressure has a certain value, say 500 mb. We place a floating marker there. We move to different locations at 500 mb and do the same, again and again. Imagine now that we connect all the markers to form an **isobaric surface**, a level above the Earth where the pressure is the same everywhere. Now let's imagine an atmosphere, at Time 1, where temperature at any elevation is uniform and the isobaric surfaces are horizontal **(Fig. 18.5a)**. In this situation, pressure decreases uniformly with elevation. Let's now imagine, at Time 2, that a region of the atmosphere is heated **(Fig. 18.5b)**. Expansion of air in the heated region causes the pressure in a column of air above it to become higher than in surrounding regions. In the upper atmosphere, this situation results in an outward-directed pressure-gradient force, and air starts to flow out of the column. As a result, the total mass of air within the column has decreased by Time 3. Because air pressure at the Earth's surface represents the weight of the overlying column of air,

the expansion associated with heating ends up decreasing the air pressure at the surface.

The opposite effect happens when air cools **(Fig. 18.5c)**. Once again, imagine air at Time 1 with uniform temperature at a given elevation. Large-scale cooling—for example, over Canada or Siberia in winter—causes near-surface air to contract and increase in density, which makes the pressure in a column of air above it lower than that in surrounding regions. In the upper atmosphere, this results in an inward-directed pressure-gradient force so that air flows into the column aloft (meaning above ground) (Time 2). As a result, the total mass of the air column increases, and high pressure develops at the Earth's surface (Time 3). Development of such pressure gradients at the Earth's surface due to heating or cooling can cause winds to blow.

The Coriolis Force

In Chapter 15, we introduced the Coriolis force in the context of ocean currents (see Box 15.3). The Coriolis force also influences wind. To see why, imagine a small volume of air, which we'll call an *air parcel*, that is moving parallel to the Earth's surface at a constant speed. (To simplify our discussion, let's ignore the frictional force and

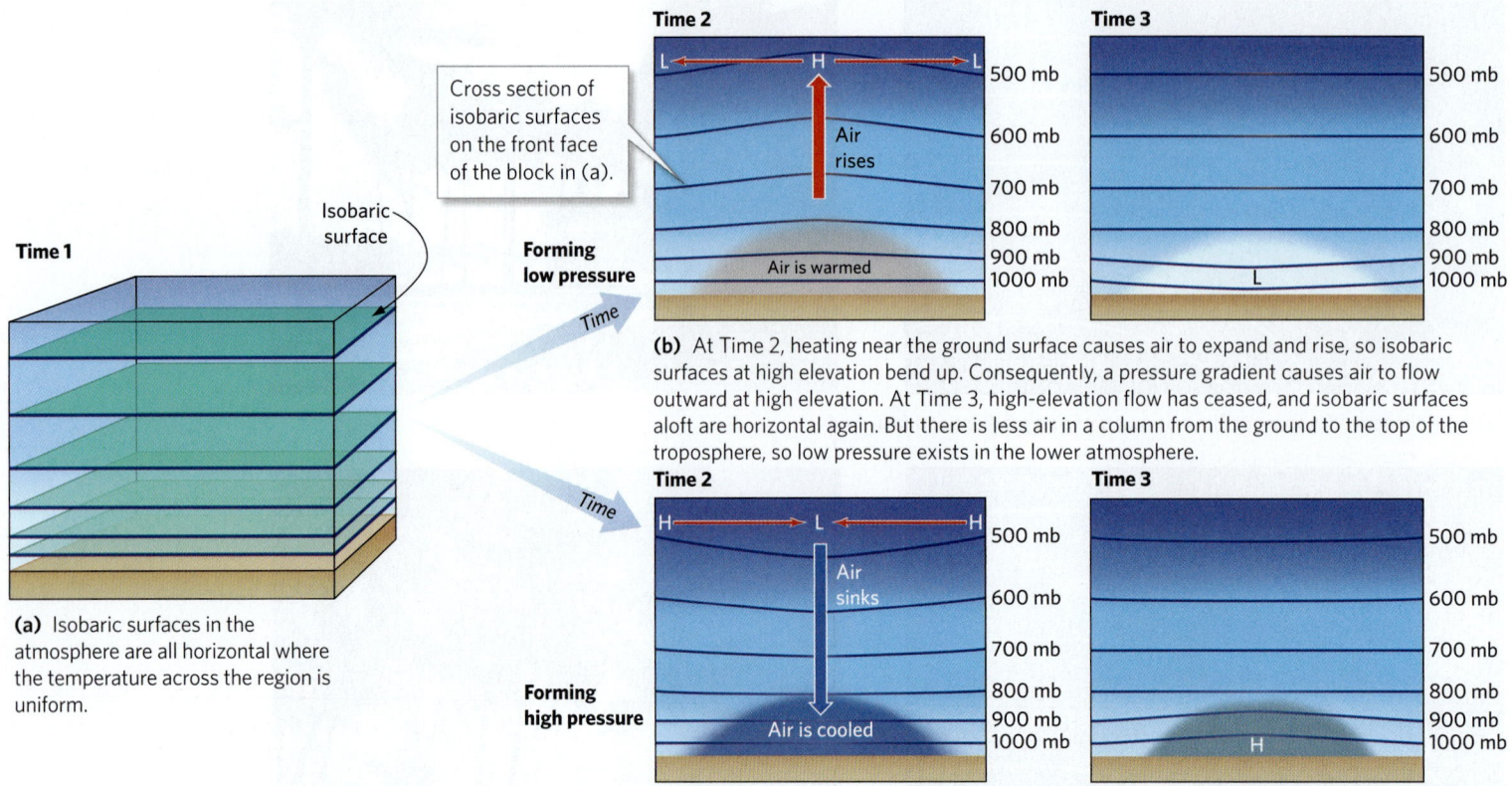

(a) Isobaric surfaces in the atmosphere are all horizontal where the temperature across the region is uniform.

Cross section of isobaric surfaces on the front face of the block in (a).

(b) At Time 2, heating near the ground surface causes air to expand and rise, so isobaric surfaces at high elevation bend up. Consequently, a pressure gradient causes air to flow outward at high elevation. At Time 3, high-elevation flow has ceased, and isobaric surfaces aloft are horizontal again. But there is less air in a column from the ground to the top of the troposphere, so low pressure exists in the lower atmosphere.

(c) At Time 2, cooling near the ground causes air to contract and sink, so isobaric surfaces at high elevation bend down. Consequently, a pressure gradient causes air to flow inward at high elevation. At Time 3, high-elevation flow has ceased, and isobaric surfaces aloft are horizontal again. But there is more air in a column from the ground to the top of the troposphere, so high pressure exists in the lower atmosphere.

pressure-gradient force for now.) What path will the air parcel follow? As we learned in Chapter 15, because the Earth rotates on its axis, a moving fluid doesn't move in a straight line relative to the Earth, but rather follows a curved path.

The Coriolis force has four important properties that affect the movement of air parcels: (1) it causes moving parcels to veer to the right of their direction of motion in the northern hemisphere and to the left in the southern hemisphere; (2) it affects the direction in which a parcel moves across the Earth's surface, but has no effect on its speed; (3) it is strongest for parcels that move fast relative to a point on the surface of the Earth and has a value of zero for stationary parcels; and (4) it has a value of zero on the equator and a maximum value at the poles.

The Frictional Force

When you picture a *frictional force*, you might think of the force that slows a book that you've slid across a table. Friction occurs in the atmosphere, too, either when a volume of faster-moving air molecules shears against a volume of slower-moving air molecules, or when a volume of air molecules shears against the Earth's surface. The frictional force always acts opposite the direction of motion of an object, so friction reduces the speed of air flow. The frictional force increases near the Earth's surface, both because moving air interacts with the surface and because air density increases toward the base of the atmosphere. The rougher the surface, the greater the frictional force, so more friction exists between moving air and a city of tall buildings than between moving air and a calm sea. For this reason, winds tend to be stronger out over the ocean (see Earth Science at a Glance, pp. 562–563).

Atmospheric scientists refer to the layer of air adjacent to the Earth's surface in which friction significantly affects air movement and generates turbulence as the **boundary layer**. The thickness of the boundary layer depends on several factors: the roughness of the underlying surface, surface heating (which causes the turbulent layer to loft to higher elevations), the presence of updrafts and downdrafts, and the wind speed. Clouds may form at the top of the boundary layer in buoyant air. On a cold, calm winter morning at sunrise, the boundary layer over

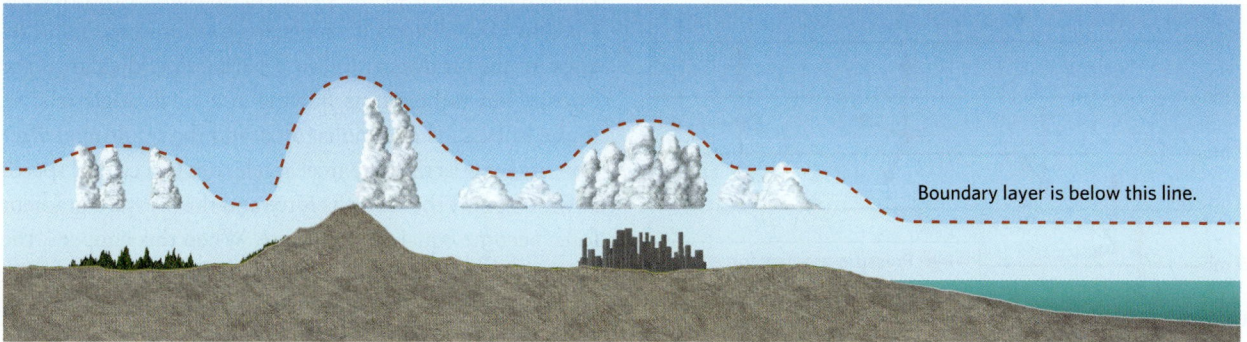

(a) The boundary layer's thickness depends on the roughness and temperature of the surface below; it can vary from a few hundred meters to a few kilometers. Here, warmer air over cities and hillslopes produces updrafts that increase the thickness of the boundary layer and can cause clouds to form.

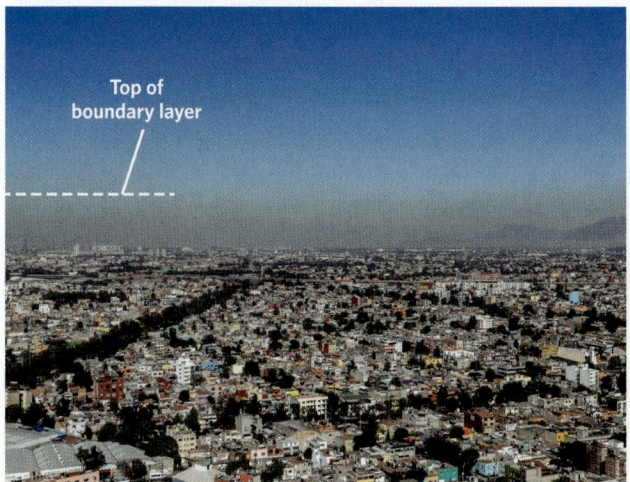

(b) Pollution or low-level clouds can sometimes show where the top of the boundary layer lies.

a large ice-covered lake extends upward only a few hundred meters. In contrast, on a hot, windy summer afternoon over a city or over hills, the boundary layer extends upward a few kilometers (Fig. 18.6a). The frictional force can generate turbulence in the boundary layer. *Eddies* (whirlpool-like swirls) tend to develop, for example, when wind speed changes rapidly with distance, or wind shears against the ground, or air passes between buildings or over mountain peaks. The boundary layer is sometimes visible as a layer of pollution or shallow clouds capped by clear air (Fig. 18.6b).

Wind Direction: A Consequence of Geostrophic Balance

If the temperature of the atmosphere at each altitude across the Earth were the same everywhere, and if the Earth's surface were uniform, atmospheric pressure at any given altitude wouldn't vary from location to location, and the wind would stop blowing. This doesn't happen in reality, because various sources add energy to the atmosphere, and the

amount of added energy varies with location. For example, solar radiation provides more heat at the equator than at the poles, land and sea absorb and release heat at different rates, the strength of the frictional force depends on the roughness of the Earth's surface below, and the density of air changes with elevation at different rates in different locations. Given all these variables, the atmosphere never achieves a balanced state and remains in constant motion.

While a balanced state can't be achieved permanently everywhere, atmospheric conditions do approach a balanced state locally, particularly at elevations above the boundary layer. This condition, called **geostrophic balance**, develops when the horizontal pressure-gradient force and the Coriolis force are equal and opposite and the frictional force is insignificant.

We first discussed geostrophic balance in the context of ocean currents (see Chapter 15). To understand geostrophic balance in the atmosphere, let's examine how air moves across North America when a pressure gradient exists. Picture an initially stationary air parcel within the region of a pressure gradient (Fig. 18.7a). The pressure-gradient force

FIGURE 18.7 Geostrophic wind.

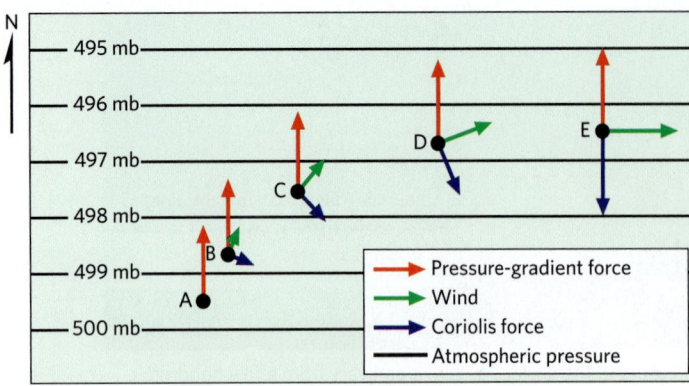

(a) Air at A starts from rest. As it accelerates from A to E, the Coriolis force turns the air's path to the right until it flows parallel to the isobars.

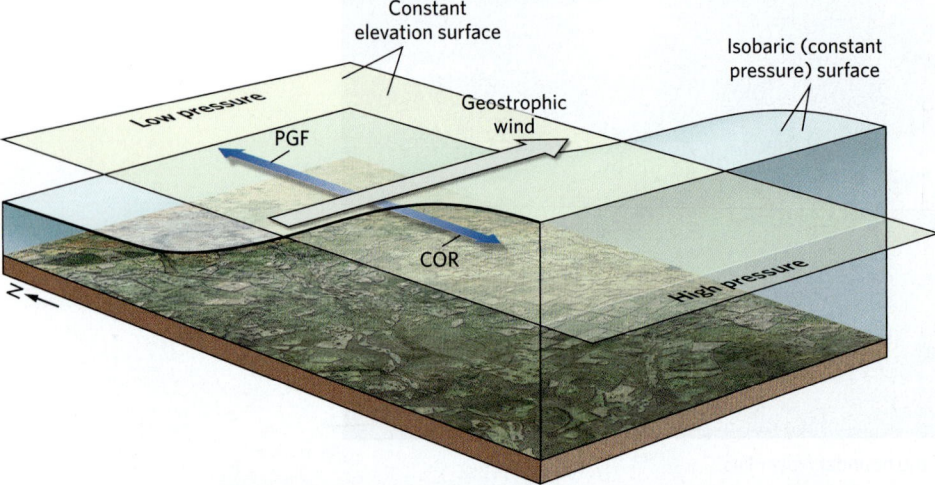

(b) When there is a balance between the pressure-gradient force (PGF) and the Coriolis force (COR), the resulting wind is called the geostrophic wind.

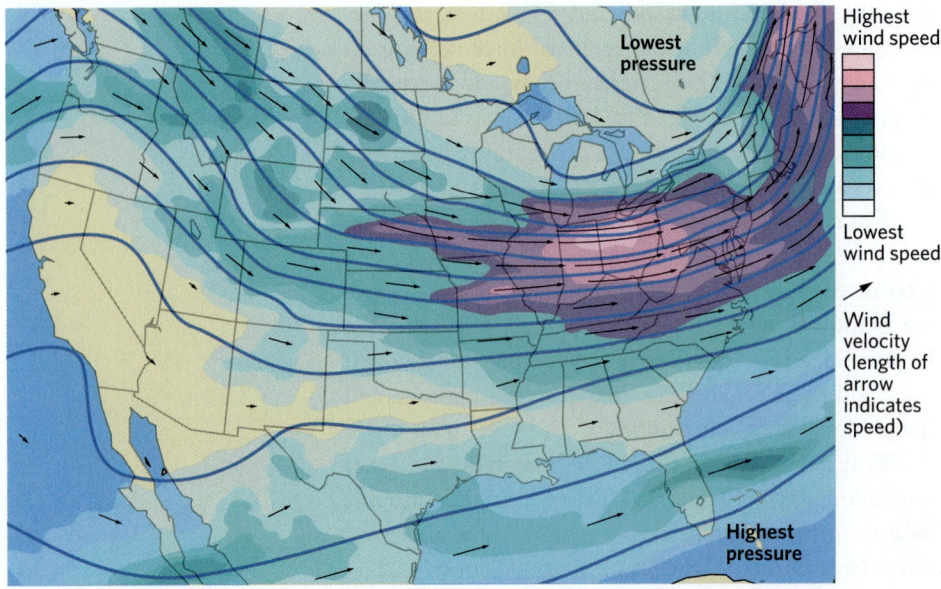

(c) The wind blows roughly parallel to the isobars at 10 km (6 mi) above sea level, the elevation of the jet stream. Winds blow faster where the pressure gradient is greater.

initially accelerates the air and causes it to move from a region of higher pressure toward one of lower pressure. However, as soon as the parcel of air starts to move, the Coriolis force deflects it, so it starts to turn to the right. As a result, the air doesn't follow a path perpendicular to the isobars, but rather starts flowing at a small angle relative to the isobars. The Coriolis force, and the resulting deflection, increase as the air parcel accelerates to a critical speed, at which point the Coriolis force and the pressure-gradient force become equal and opposite. When this happens, the air has attained geostrophic balance (Fig. 18.7b), and the wind that results, the **geostrophic wind**, flows parallel to isobars (Fig. 18.7c). In nature, air is nearly, but not exactly, in geostrophic balance at elevations above the boundary layer over much of the Earth. (We'll see in the next chapter how deviations from geostrophic balance contribute to the formation of large storms.)

High-Pressure and Low-Pressure Systems

As we've just seen, the development of high pressure and low pressure significantly affects the pattern of air flow near the Earth's surface. When a high-pressure center develops, it generates a pressure-gradient force that causes air parcels to flow outward from that center into surrounding regions. Recall that when an air parcel starts to move horizontally on the spinning Earth, the Coriolis force deflects it to the right (in the northern hemisphere). If the pressure-gradient force and the Coriolis force were the only forces affecting air flow, air would circle the high-pressure center in a clockwise direction (in the northern hemisphere). But because friction slows the air, and because speed determines the magnitude of the Coriolis force, the pressure-gradient force exceeds the Coriolis force, so that air spirals outward from the high-pressure center. These outward-spiraling clockwise winds generate *divergence* at the base of the troposphere, which causes air to descend over the low-pressure center (Fig. 18.8a). Meteorologists refer to a high-pressure center, together with the associated three-dimensional pattern of wind circulation, as a **high-pressure system**. The downward vertical air movement that takes place in high-pressure systems affects the relative humidity and, therefore, the cloudiness in the air column. When the air within a high-pressure system sinks and spirals downward, it undergoes compression and warms, so its relative humidity decreases. As a result, high-pressure systems tend to host relatively dry, cloud-free air—in other words, they produce *fair weather*.

The development of a low-pressure center near the Earth's surface also leads to the formation of a pressure gradient. Alone, this gradient would cause air from surrounding regions to flow in toward the center. However, the Coriolis force deflects moving air to the right (in the northern hemisphere). If only the pressure-gradient and

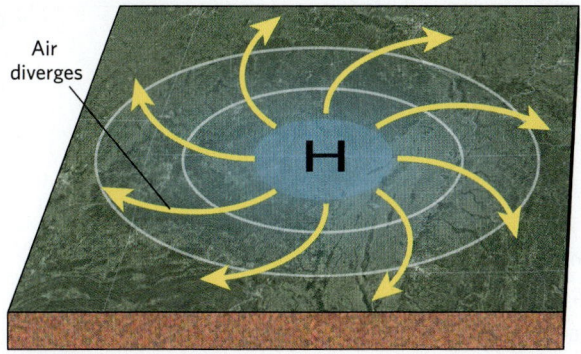

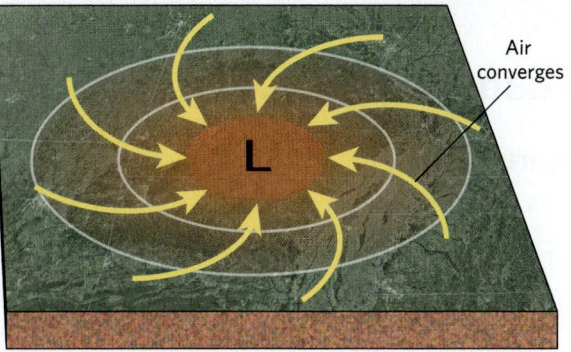

Air
diverges

Air
converges

FIGURE 18.8 Air flow in surface high- and low-pressure systems.

Low-altitude divergence

(a) Air spirals out from a high-pressure system and descends toward the ground.

Low-altitude convergence

(b) Air spirals into a low-pressure system and ascends away from the ground.

Coriolis forces acted on the moving air, the net result would be that air would circle the low-pressure center in a counterclockwise direction (in the northern hemisphere). However, because friction slows the air, the pressure-gradient force exceeds the Coriolis force, so air spirals inward toward the low-pressure center (Fig. 18.8b). This inward-spiraling, counterclockwise flow causes *convergence* at the base of the troposphere, which causes air to rise over the high-pressure center. Air rises in low-pressure systems as a result of air converging toward the low-pressure center (Box 18.1). We'll see in the next chapter that rising air is associated with clouds and storms, including tropical and mid-latitude cyclones. A low-pressure center, together with the associated three-dimensional pattern of air circulation, is called a **low-pressure system**.

The Earth's atmosphere hosts two types of low-pressure systems, migrating and semi-permanent. The great storms of the Earth's atmosphere, tropical and mid-latitude cyclones, are examples of migrating low-pressure systems. We will defer discussion of these storms to Chapter 19. The remaining low-pressure systems, and most high-pressure systems, are semi-permanent, meaning that they form and re-form over the same geographic area in a given season. Here we consider these *semi-permanent* pressure systems and how they affect the Earth's winds and weather.

What happens to the temperature outdoors after the Sun goes down? We know from common experience that air starts cooling and the temperature drops. We often feel a chill in the air. At night, the Sun no longer heats the land, but the Earth's surface and the air above it cool as they radiate infrared energy upward into the higher atmosphere and space. Now imagine what happens in winter in large regions of Canada, Siberia, and Antarctica, regions that are in nearly perpetual darkness because of their high latitude. These vast landmasses continually cool for months at a time during winter. Through the processes described in the previous section, this perpetual cooling leads to the development of a long-lived zone, called a

semi-permanent high, near the Earth's surface over broad regions of these continents (Fig. 18.9). Even if air flows out of these regions to lower latitudes, high pressure will quickly re-form as the new air moving into the regions continues to cool.

Perhaps somewhat surprisingly, cooling also happens over the Atlantic and Pacific Oceans at mid-latitudes during the summer. This cooling happens when air that

FIGURE 18.9 Semi-permanent high- and low-pressure systems.

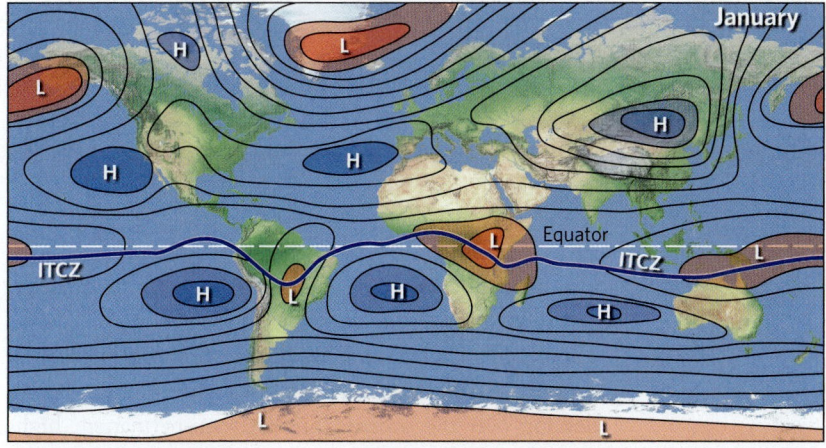

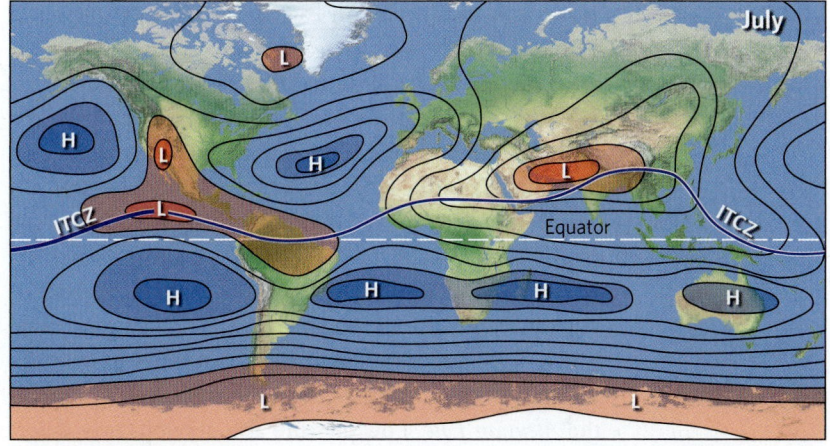

BOX 18.1

How can I explain . . .

Flow in a low-pressure system

What are we learning?

How friction near the ground causes air to converge in the center of a low-pressure system.

What you need:

- A cup filled about three-fourths of the way with transparent tea
- Loose tea leaves in the bottom of the cup
- A spoon to stir the tea

Instructions:

- Let the liquid come to rest so that the tea leaves are spread out and resting at the bottom of the cup.
- Stir the tea vigorously in a continuous direction so that the fluid rotates in the cup.
- Observe the tea leaves.

What did we see?

- This experiment illustrates fluid flow in a low-pressure system. As the tea in the cup rotates, the top surface of the tea becomes funnel-shaped. As a consequence, there's less liquid in the middle of the cup than along the sides, so there's less pressure on the bottom of the cup in the center than along the sides. This distribution of mass must mean that low pressure exists at the bottom of the cup at its center. One might expect the tea leaves to follow a circular path around the edge of the cup due to centrifugal force.
- Meanwhile, however, the fluid experiences a pressure gradient that should drive it toward the low pressure at the center of the cup. If a balance were to be established between centrifugal force and pressure-gradient force, the tea leaves would follow a circular path as you stirred. But toward the bottom of the cup, friction between the tea and the cup slows the flow, so the inward-directed pressure-gradient force (toward the center) exceeds the centrifugal acceleration of the spinning fluid, and the leaves spiral toward the center of the cup. The converging liquid then rises in the middle of the cup.
- A similar process occurs in low-pressure systems in the atmosphere. Air near the Earth's surface converges toward the low-pressure center, and as the air approaches the center, it rises, causing clouds and precipitation.

had warmed over the hot land surface of the adjacent continents moves out over the cold water of the oceans. This cooling causes **semi-permanent highs**, areas of high pressure that form and re-form throughout the season, to develop over the oceans. (Similar semi-permanent highs form in subtropical latitudes during the winter.) In the North Atlantic and North Pacific, the summertime semi-permanent high-pressure systems are called the *Bermuda High* and the *Pacific High*, respectively (see Fig. 18.9).

As you might anticipate, low pressure develops in the lower atmosphere where the Earth's surface is warm and continually heating air above it. This commonly occurs in winter over the North Atlantic south of Iceland and the North Pacific south of the Aleutian Islands. There, the wintertime oceans are much warmer than the surrounding continents, so the oceans warm air moving over the water. The **semi-permanent lows** that develop over the northern oceans in winter are called the *Icelandic Low* and the *Aleutian Low*. Low pressure also develops over desert

regions, such as the southwestern United States, in summer due to the strong solar heating that occurs during the daytime.

Take-home message . . .

Three forces (the pressure-gradient, frictional, and Coriolis forces) determine the speed and direction of the wind. Air accelerates from regions of higher pressure toward regions of lower pressure. As it does so, the Coriolis force increases and eventually balances the pressure-gradient force. Where this happens, geostrophic balance is attained, and the wind flows parallel to isobars. Heating or cooling at the base of the atmosphere lowers or raises air pressure at the Earth's surface, creating semi-permanent high- and low-pressure systems.

Quick Question

Why does the thickness of the boundary layer vary with location?

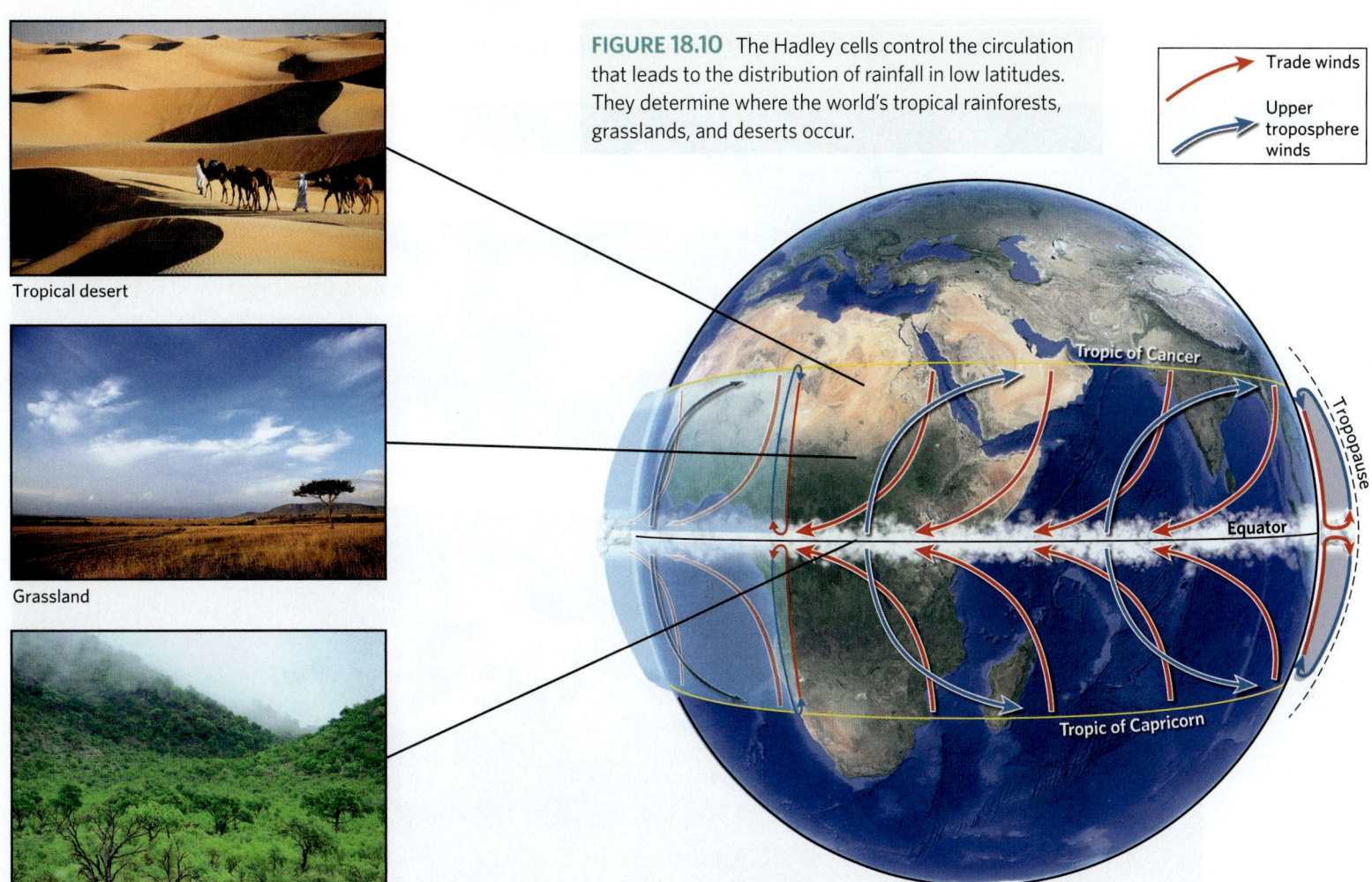

Tropical desert

Grassland

Tropical rainforest

FIGURE 18.10 The Hadley cells control the circulation that leads to the distribution of rainfall in low latitudes. They determine where the world's tropical rainforests, grasslands, and deserts occur.

→ Trade winds

→ Upper troposphere winds

Tropic of Cancer

Tropopause

Equator

Tropic of Capricorn

18.3 Air Circulation in the Tropics

What image comes to mind when you hear the word *tropics*—warm sunny breezes and swaying palm trees? In reality, the **tropics**—which lie between the Tropic of Cancer and the Tropic of Capricorn—cover two-thirds of the Earth's surface and host a great variety of landscapes. They include not only rainforests, but also portions of deserts and broad grasslands in which rain falls only seasonally. Regions in the tropics do not experience large seasonal changes in temperature. North of the Tropic of Cancer and south of the Tropic of Capricorn lie regions known as the *subtropics*. These are zones of transition between the hot tropics and the temperate mid-latitudes. Weather in the tropics and subtropics results from a global air circulation system driven by differential solar heating and by related variations in sea-surface temperatures. Let's explore how this system operates.

Hadley Cells and the Intertropical Convergence Zone

George Hadley, a British lawyer who had a strong interest in meteorology, often wondered why winds blow. In

1735, he published an article in which he suggested that the warming of air at the equator—where the Sun heats the Earth most intensely—would cause air to rise from low elevations to the top of the atmosphere, where, unable to rise higher, it would start to flow poleward. As the poleward-moving air aloft cooled and became denser, Hadley proposed that it would sink back toward the surface and, nearer the ground, would flow back toward the equator. Such a circulation caused by convective flow (see Chapter 2) is called a **convective cell**.

Hadley's model was partially correct, but he did not take the Earth's rotation, or the decrease in atmospheric temperature with altitude, into account. If the Earth did not rotate, the poleward-flowing component of Hadley's proposed convective cell might theoretically extend well into the polar regions. But the Earth does rotate, so the Coriolis force comes into play and deflects the high-elevation poleward flow eastward. As a consequence, air flowing poleward at the top of the troposphere reaches a latitude of only about 25° before it begins flowing roughly parallel to that line of latitude (Fig. 18.10). Furthermore, by 25° latitude, all of the air that rose to the top of the convective cell near the equator is sinking back toward

FIGURE 18.11 The intertropical convergence zone (ITCZ) forms at the junction of the Hadley cells in the two hemispheres.

Less rain — More rain

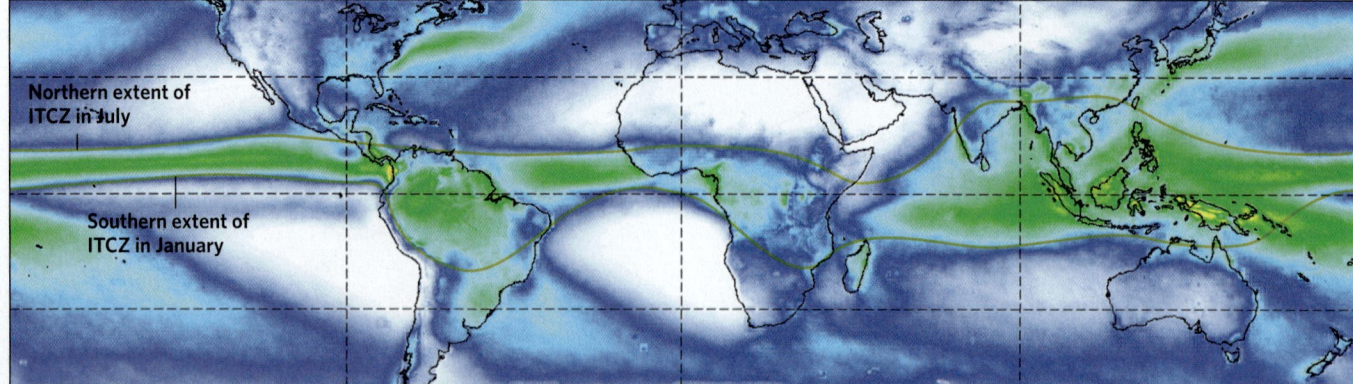

Northern extent of ITCZ in July

Southern extent of ITCZ in January

(a) Global annual rainfall distribution. Heavy rains fall along the ITCZ. Its position varies during the year.

(b) The thunderstorms that occur along the ITCZ are visible in this satellite image.

the Earth's surface. When the sinking air reaches a low elevation, it flows back toward the equator. The Coriolis force deflects this near-surface air westward, so that in the northern hemisphere it flows southwest at a decreasing angle to lines of latitude as it approaches the equator. By the time the air reaches the equator, it's flowing almost parallel to the equator. This tropical surface flow constitutes the **trade winds**, the same winds that sent Sir Francis Drake westward across the Atlantic and Pacific. Atmospheric scientists refer to this convective cell, which extends from the equator to a latitude of about 25°, as a **Hadley cell**, in honor of George Hadley. A Hadley cell exists in each hemisphere.

Near the Earth's surface, the southwest-blowing trade winds of the northern hemisphere and the north-west-blowing trade winds of the southern hemisphere

converge near the equator, producing the **intertropical convergence zone**, or **ITCZ**. The exact position of the ITCZ approximately follows the line along which the Sun's heat is most intense. Because of the Earth's tilt, the position of this line changes over the course of a year. In general, the ITCZ lies north of the equator during the northern hemisphere summer and south of the equator during the northern hemisphere winter. The exact position of the ITCZ also depends on the temperature of the Earth's surface below, so it can be affected by ocean currents, land cover, and topography **(Fig. 18.11a)**.

Air converging in the ITCZ has nowhere to go but up (since it can't sink into the Earth), so it forms the rising part of the Hadley cell. This air, because it's so warm, absorbs water evaporating from the ocean surface below. For reasons we'll discuss in the next chapter, rising moist air produces **thunderstorms**, localized areas of strong winds and heavy precipitation accompanied by lightning. On a satellite image, clusters of thunderstorms can be seen ringing the globe along the ITCZ (Fig. 18.11b). These storms produce heavy rainfall that provides water for tropical rainforests. Because air in the ITCZ primarily moves upward, air at the surface barely moves horizontally. This relatively still air, known to sailors as the **doldrums**, becalms sailing ships.

As we have noted, to the north and south of the ITCZ, at about 25° latitude, the air of the Hadley cell descends from the cold upper troposphere to the middle and lower troposphere. This air contains very little water, for most of its moisture rained out in the ITCZ. As this dry air descends, it compresses and warms, so its relative humidity decreases even more. Skies of the northern and southern tropics and subtropics, therefore, tend to host the clear, hot, dry weather that characterizes the world's largest deserts. The *steppes*, the grasslands between subtropical deserts and tropical rainforests, have a long dry season and a short rainy season. The rainy season occurs during the summer, when the ITCZ moves overhead.

The Subtropical Jet Stream

A strong temperature contrast develops in the upper troposphere above subtropical latitudes, where the poleward-flowing part of the Hadley cell meets the cooler air of the mid-latitudes. As we will see in Section 18.4, this temperature change leads to a pressure gradient that accelerates air movement at the top of the troposphere. The Coriolis force deflects the moving air eastward. As a result, a very fast (>160 km/h, or 100 mph), high-altitude river of air, called the **subtropical jet stream**, encircles the globe. It tends to be relatively narrow, with a width of only 300–500 km (180–300 mi), measured perpendicular to the air-flow direction. This jet stream can affect weather in the mid-latitudes as well as in the tropics. At any given time, two jet streams exist in each hemisphere; we'll discuss the other one, the polar-front jet stream, later in this chapter.

The Southern Oscillation and El Niño

Often, when a major weather event happens, someone in the media blames El Niño or La Niña. What are these phenomena, and how do they affect the world's weather? To address this question, we must first introduce the *Walker circulation*, a cell of flowing air that spans the Pacific Ocean along the equator.

The **Walker circulation**, named after British physicist Gilbert Walker (1868–1958), extends vertically between the Earth's surface and the tropopause and horizontally from South America's western coast to Australia and Indonesia, a distance of more than 16,000 km (9,940 mi). This circulation, which appears as a distinctive feature only within about 5° of the equator, can be thought of as a secondary, east-west-flowing convective cell within the Hadley cells. Due to the Walker circulation, air flows westward in the lower troposphere, rises over the western equatorial Pacific Ocean, returns eastward in the upper troposphere, and sinks over the eastern equatorial Pacific (Fig. 18.12a).

Why does the Walker circulation exist? The western Pacific tends to be warmer than the eastern Pacific because of the configuration of surface currents in the ocean. As a consequence, air over the western Pacific basin becomes warmer than that over the other parts of the Pacific, generating a region of low pressure at the Earth's surface. Meanwhile, a region of high pressure develops at the base of the atmosphere in the eastern Pacific, where the ocean water is cooler, because as air temperature decreases above the cooler water, the air becomes denser. Because the Coriolis force near the equator is very weak (zero at the equator), the pressure-gradient force drives air flow from the high-pressure zone of the eastern Pacific toward the low-pressure zone of the western Pacific.

When the temperature contrast between the eastern and western Pacific becomes large, strong westward-blowing trade winds push surface water westward (see Chapter 15). These winds cause sea level to rise by 10–20 cm (4–8 in) in the western Pacific relative to the eastern Pacific. Because the ocean currents carry surface water away from the eastern Pacific, cool water upwells along the coast of South America to replace the surface water (Fig. 18.12b). This water brings up nutrients that nourish the plankton that fish eat, so fish populations thrive and fishing crews have successful hauls.

Every few years, for reasons that atmospheric scientists cannot fully explain, the intensity of the Walker circulation weakens. This weakening reduces the speed of the trade winds blowing west across the equatorial Pacific. When this happens, the tilted sea surface flattens, and warm ocean water from the western Pacific spreads eastward along the equator all the way to the South American coast (Fig. 18.12c, d). As a result, the temperature of the ocean water becomes more uniform across the Pacific, so the pressure-gradient force decreases. In fact, surface water in the eastern Pacific may become warmer than that in the western Pacific, leading to a reversal in the pressure gradient that, in turn, causes a brief reversal in the direction of the equatorial trade winds. Upwelling along the coast of South America shuts off. Over the course of a year, the Walker circulation slowly strengthens, low pressure again develops in the western Pacific and high pressure in the eastern Pacific, and the trade winds again begin to blow strongly westward.

Did you ever wonder . . .

what an El Niño is and why it may affect your local weather?

FIGURE 18.12 The Walker circulation and the Southern Oscillation.

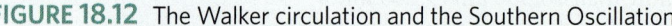

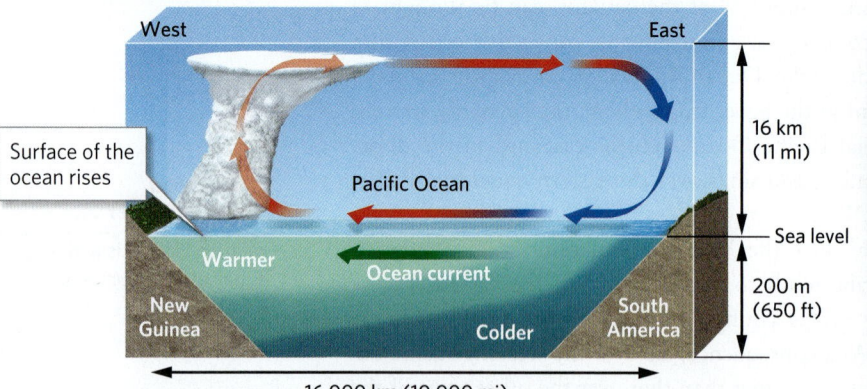

(a) Normal Walker circulation. During La Niña, this circulation strengthens.

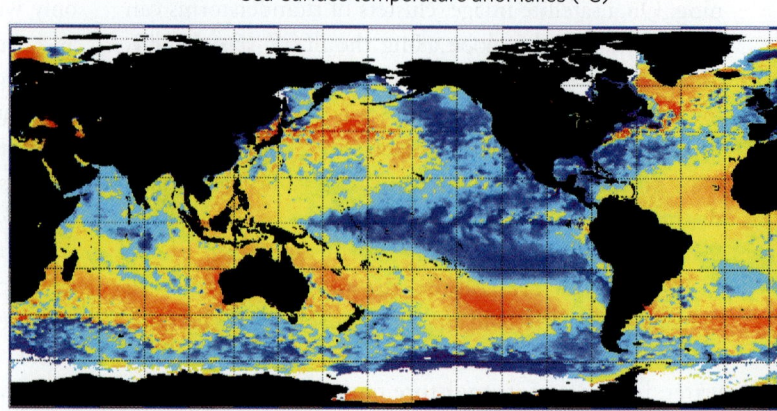

Sea-surface temperature anomalies (°C)

(b) Sea-surface temperature anomalies during La Niña.

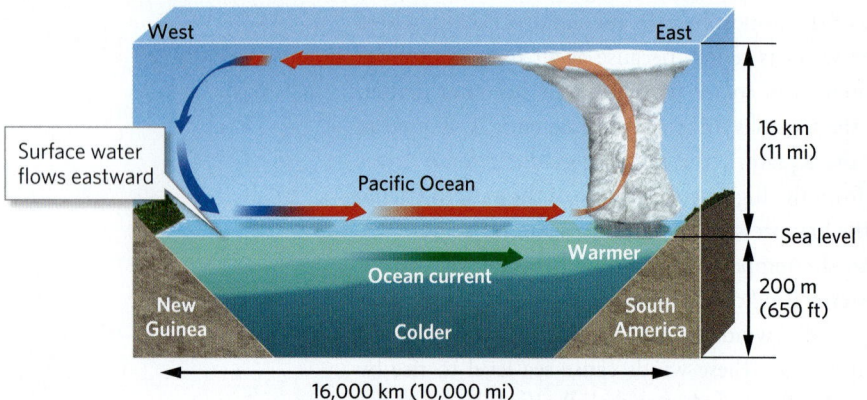

(c) During El Niño, the Walker circulation weakens or reverses.

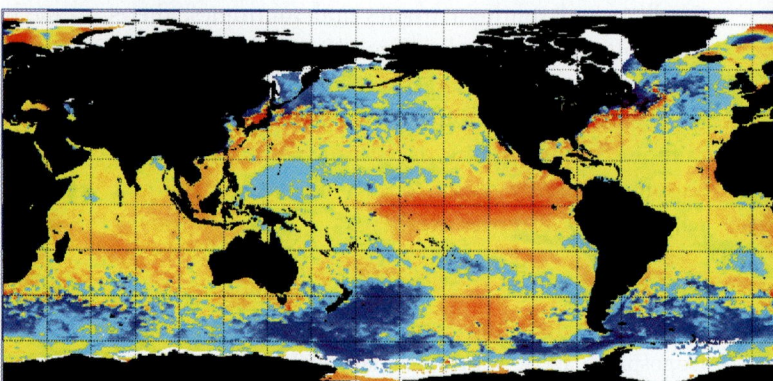

(d) Sea-surface temperature anomalies during El Niño.

FIGURE 18.13 The Southern Oscillation affects global weather.

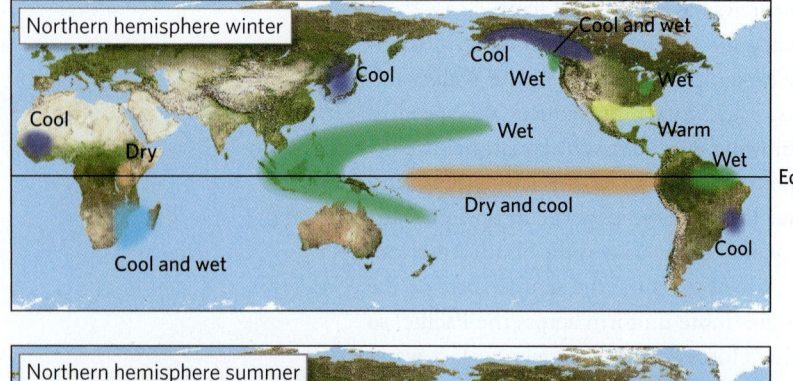

(a) Effects of El Niño.

(b) Effects of La Niña.

Atmospheric scientists refer to the east-west seesaw in surface air pressure across the equatorial Pacific that accompanies these changes in the Walker circulation as the **Southern Oscillation**. The warming of ocean temperatures in the eastern equatorial Pacific has come to be known as **El Niño**, a term originally coined by Peruvian fishermen because the onset of the warming often coincides with the Christmas season (*El Niño* is Spanish for little boy, meaning the Christ Child). El Niño threatens the livelihoods of fishermen because the shutoff of upwelling nutrients decreases fish populations. Atmospheric scientists coined the term **La Niña** (Spanish for little girl) to describe times when the Walker circulation becomes very strong and the upwelling of cold water along coastal South America increases. The acronym **ENSO** (pronounced enn-soh), which stands for El Niño/Southern Oscillation, is used for the overall seesaw pattern and its consequences. During the past century, more than two dozen El Niño events have taken place; one of the strongest happened in 2015.

If it were just a local phenomenon, El Niño would not have become so famous. But the rise of sea-surface temperatures in the eastern Pacific has far-reaching effects. It shifts the center of tropical thunderstorm activity eastward, and these thunderstorms influence the strength of both the Hadley cells and the subtropical jet streams. Therefore, ENSO modifies atmospheric circulation around the globe, even in the mid-latitudes, influencing weather worldwide. For example, El Niño triggers droughts in Australia and Indonesia, particularly rainy weather in the eastern Pacific, and flooding along South America's northwestern coast (Fig. 18.13a). In North America, El Niño can cause storms that normally move from the Pacific into Washington and Oregon to move northward instead, toward the Gulf of Alaska, or can cause storms that normally migrate to the south of the United States to move north instead, triggering floods in California and winter storms farther east. Strong La Niña conditions also affect weather around the globe (Fig. 18.13b).

Monsoons

Think of the word *monsoon* and you probably picture heavy rain and overflowing rivers. In fact, the word has a more general meaning. Technically, a **monsoon** is a seasonally changing tropical wind circulation accompanied by a rainy season and a dry season. In regions that experience monsoons, summer winds blow from the ocean toward the land, and winter winds blow from the land toward the ocean.

Monsoons develop because during the summer months, solar energy heats rock and soil on land more rapidly than it heats water in the ocean (Fig. 18.14a). As a result, air over land becomes warmer than air over the adjacent ocean. The air over land, therefore, expands and

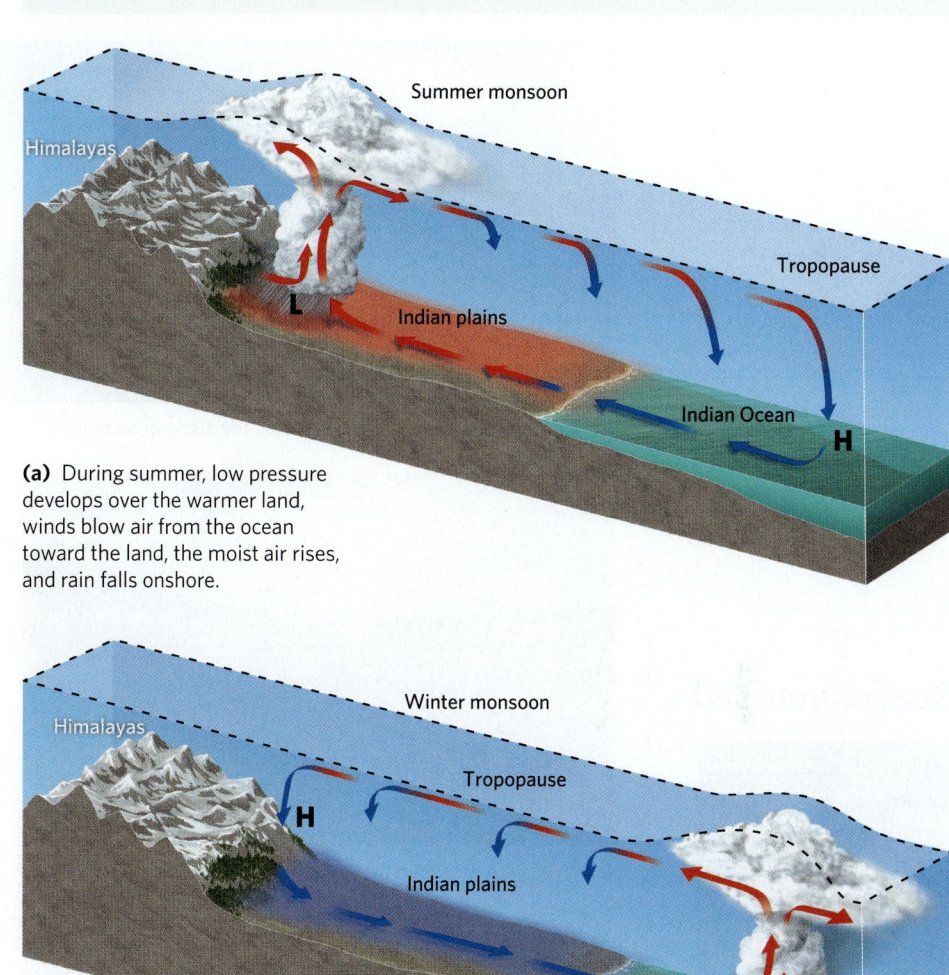

FIGURE 18.14 Formation of monsoons in southern Asia.

(a) During summer, low pressure develops over the warmer land, winds blow air from the ocean toward the land, the moist air rises, and rain falls onshore.

(b) During winter, high pressure develops over the cooler land, winds blow dry air toward the sea, the air moistens over the sea, then rises, and it rains offshore.

rises, so the air pressure over land at the Earth's surface becomes less than that over the ocean. The resulting pressure gradient causes moist air to flow from the ocean over the land. Once over the hot surface of the land, the air warms, becomes buoyant, and rises. The rising moist air produces thunderstorms that drench the land, leading to the common association of the word *monsoon* with flooding. During the winter, the situation reverses (Fig. 18.14b). The land cools faster than the ocean because water cannot lose heat as quickly as rock or soil. Air over the land cools and sinks, forming a region of high pressure. When the air pressure over the land exceeds that over the ocean, the resulting pressure gradient generates a wind that blows seaward; air over the land becomes clear and dry, while rain clouds form over the sea.

FIGURE 18.15 The southern Asian monsoon.

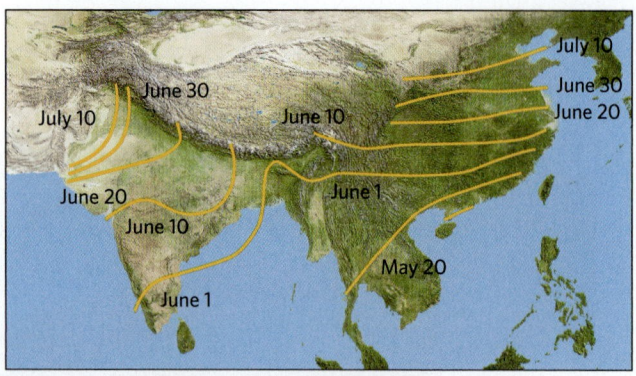

(a) The typical advance of monsoons in the summer season.

(b) Monsoon thunderstorms bring heavy floods.

(c) When the monsoon fails to bring rain, major droughts occur.

The world's largest monsoon occurs over southern Asia, where the seasonal reversals in wind direction cause distinct rainy seasons and dry seasons. Moist air blowing from the Indian Ocean northward over the land rises along the front of the Himalayas and the Tibetan Plateau. This rise spawns strong thunderstorms that produce torrential rainfall. Over a two-month-long period, thunderstorms spread northward across the continent (**Fig. 18.15a**). The water they supply causes rivers to rise and flood large areas (**Fig. 18.15b**). Monsoons determine the success or failure of agriculture in southern Asia—abnormally wet or dry monsoon years can destroy a region's crops and livestock and lead to economic disaster (**Fig. 18.15c**).

Monsoons also occur in other tropical and subtropical regions. For example, the rainy season in South America develops when air warms and rises over the deserts of the Andean region, causing moist winds to blow from the Atlantic westward over the continent. The resulting storms dump rain on the eastern coast of South America and the Amazon rainforest. The summer wet season in the southwestern United States and northwestern Mexico develops when summer sunlight heats the deserts of this region while the waters of the eastern Pacific remain relatively cool. The rains that fall when moist winds blow eastward over the deserts provide nearly 70% of the region's summer rainfall.

Take-home message . . .

Atmospheric circulations in the tropics are associated with the formation of large convective cells in the atmosphere. Air warmed by strong solar heating near the equator rises to the tropopause along the ITCZ and flows poleward, forming two Hadley cells, one north and one south of the equator. The surface flows of these cells are the trade winds. Other circulations caused by differential temperatures at the Earth's surface drive the Southern Oscillation and monsoons.

Quick Question —————
What's the difference between El Niño and La Niña?

18.4 Air Circulation at Mid-Latitudes

People living at tropical and subtropical latitudes typically experience weeks or months of the same weather—dry during the dry season and wet during the wet season—and do not endure any days of cold and ice. This is not so in the *mid-latitudes*, the regions between latitudes of 30° and 60°. Large portions of these regions have *temperate climates* (see Chapter 20), in which weather changes not only as seasons progress, but even during the course of a single day. A typical late summer afternoon in the central United States, for example, may feel hot and humid. In a matter of minutes, skies grow overcast and towering thunderstorms roll by. An hour later, the Sun shines again, and the air feels cool and dry. In winter, a warm southerly breeze bringing mild air can quickly change to a northern blast of frigid air, followed by heavy snow. Why do such radical changes in weather happen?

To develop an answer, we need to step back and think again about how the Sun heats the Earth's surface. During most of the year, much less solar energy reaches the surface in the polar regions than in the tropics. Because of this imbalance, the polar regions remain cool, and in winter, when they experience nearly perpetual night, they become very cold. In the tropics, the Sun rises high in the sky all year, and incoming solar energy exceeds the energy lost as outgoing infrared radiation.

If there were no atmosphere or oceans, the temperature differences between the poles and the equator would be much greater than they are. But the atmosphere and oceans reduce these differences by transporting warm air and water from the tropics toward the poles, and cold air and water from the poles toward the tropics. In other words, winds and ocean currents act like giant conveyor belts that redistribute heat between the hot tropics and the cold polar realms in a never-ending attempt to balance their differences in temperature.

The Earth's mid-latitudes, which lie between the tropics and the poles, serve as the battleground in the war to balance temperature extremes. When cold polar air wins a battle, it pushes back warm tropical air and envelops a region of the mid-latitudes. When tropical air wins, the reverse happens. Clashes between bodies of air generate the large storms of the mid-latitudes, which we will examine in the next chapter. Let's now explore what happens to the wind systems in this complex zone of air circulation.

As we discuss further in Chapter 19, an *air mass* is a regional-scale volume of air that has a characteristic temperature, humidity, and pressure at each elevation. The boundary between two air masses is called a *front*. If we take a global view, we can think of the warm air over the tropics and subtropics and the cold air over the polar regions as two large masses of air in each hemisphere. In fact, the boundary between these two air masses, which meteorologists call the **polar front**, is often quite distinct and can be traced around the globe in the mid-latitudes. Much like the leading edge of cold syrup flowing across a plate, the three-dimensional shape of the cold air mass where it meets warm air forms a dome **(Fig. 18.16)**. The front is the surface of the dome. As cold and warm air move across the Earth, the shapes of the front and of the dome of cold air change over time.

The contrast in temperature across the polar front exerts a major influence on wind velocities at the top of the troposphere. Why? Warm air occupies more volume than cold air, so the vertical distance between isobaric surfaces is greater on the warm side of the front than on the cold side. Isobaric surfaces are therefore lower on the cold side of the front than on the warm side, so at the front, isobaric surfaces slope downward from the warm side toward the cold side. Their slopes become progressively steeper at higher altitudes, and they are steepest near the tropopause (see Fig. 18.16). As a

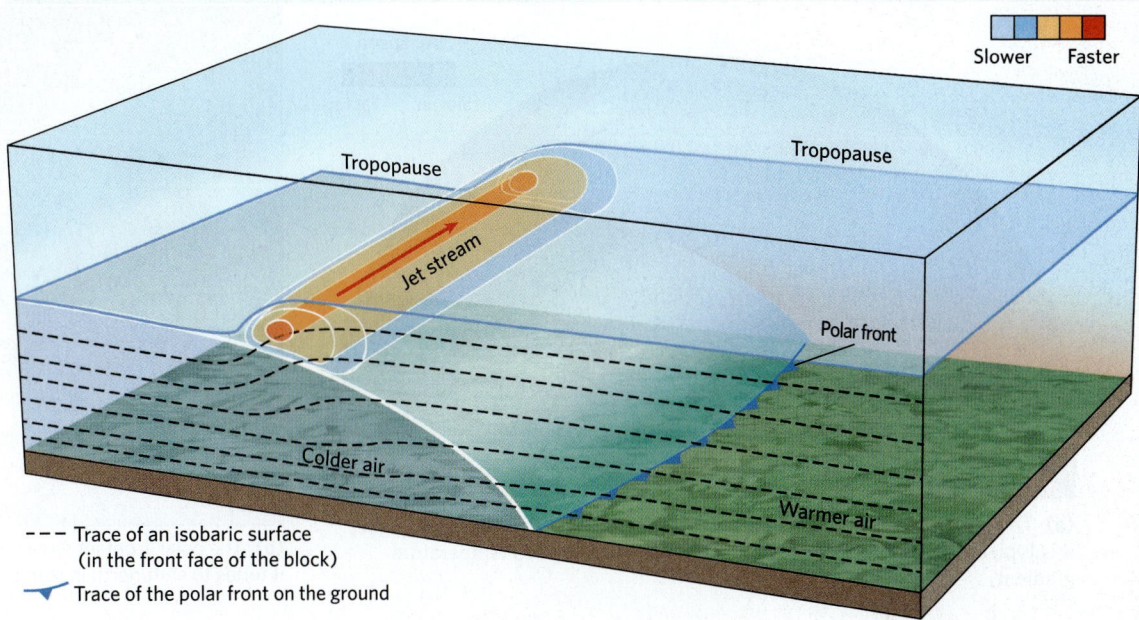

FIGURE 18.16 The relationship between a front and a jet stream. The strongest winds are at the center of the jet stream.

Slower   Faster

Tropopause

Tropopause

Jet stream

Polar front

Colder air

Warmer air

- - - Trace of an isobaric surface (in the front face of the block)
▼ Trace of the polar front on the ground

result, the horizontal pressure-gradient force near the tropopause over the polar front is very strong—much stronger than it is at lower altitudes or far away from the front. Since pressure gradients, as we have seen, drive the wind, this very steep pressure gradient generates very fast winds. The river of fast winds that is present over the polar front is called the **polar-front jet stream (Fig. 18.17a)**. Note in Figure 18.17 that the winds of this jet stream generally flow from west to east. Meteorologists refer to the mid- to upper-tropospheric winds as the *mid-latitude westerlies*.

Note that we have introduced two jet streams in this chapter: the subtropical jet stream and the polar-front jet stream. Both result from strong pressure gradients that develop across a boundary between different air masses. The subtropical jet stream forms at the boundary between the high-latitude end of a Hadley cell and the cooler air of mid-latitudes. This boundary, called the subtropical front, exists only in the upper troposphere. The polar-front jet stream, in contrast, lies over the polar front, an air mass boundary that extends from the Earth's surface upward to the tropopause. Temperature differences across the polar and subtropical fronts increase in winter, so both jet streams strengthen during winter **(Fig. 18.17b)**. Because of the near balance between the pressure-gradient force and the Coriolis force, each jet stream becomes a nearly geostrophic wind that flows more or less parallel to isobars. The polar-front jet stream tends to occur at an altitude of about 10 km (6 mi), lower than the 10–13 km (6–8 mi) altitude of the subtropical jet stream. The difference results from the decrease in the altitude of the tropopause with increasing latitude.

As the atmosphere attempts to balance the results of differential heating in the tropics and in polar regions, cold air

FIGURE 18.17 The polar-front jet stream.

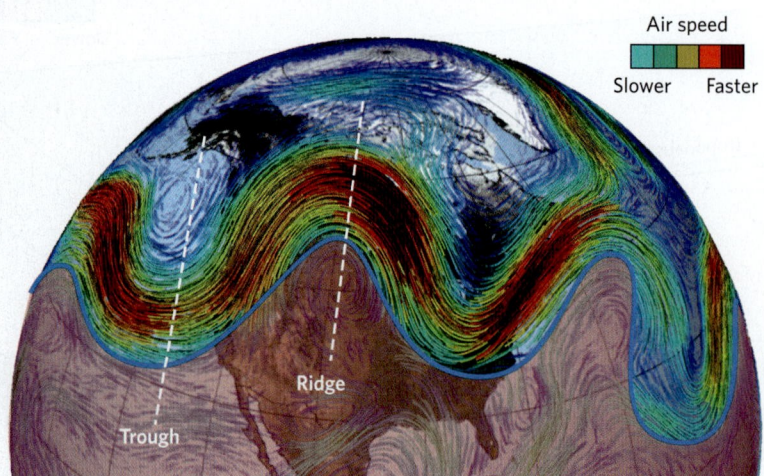

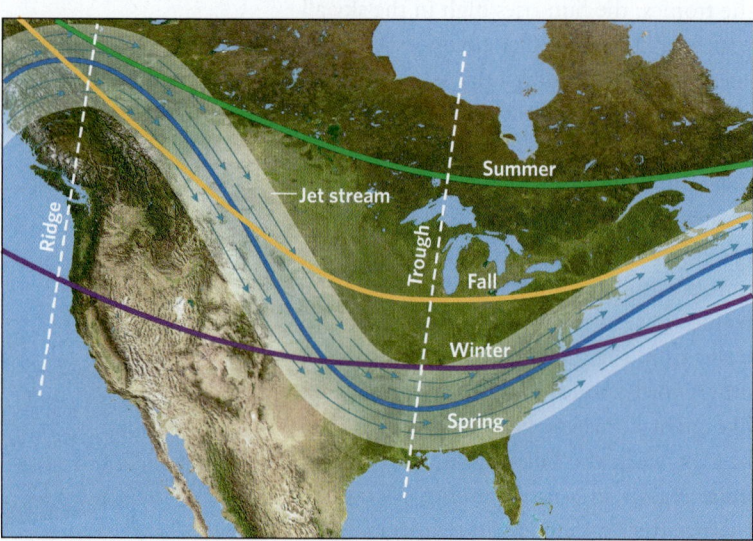

(a) The polar-front jet stream takes on a wavy pattern as it circles the globe. It is typically located over the polar front, which has a steep temperature gradient.

(b) The polar-front jet stream's position changes with time. Over North America, it tends to shift north in summer and south in winter.

masses north of the polar front frequently advance southward and descend in altitude, and warm air masses south of the polar front move northward and ascend in attitude. Viewed at a global scale from above one of the poles, the polar front therefore typically follows an undulating wave-like north-south pattern that marks the warm air–cold air boundary. For example, over North America, the polar-front jet stream may flow northward over the Pacific Northwest, southward along the Rockies into the southeastern United States, turn northeastward, and exit the United

States flowing northward toward the North Atlantic. Meteorologists describe a region where the jet stream bows toward the poles as a **ridge** and one where it bows toward the equator as a **trough** (see Fig. 18.17). Warm air lies beneath a ridge, and cold air lies beneath a trough. Why are they called ridges and troughs? Picture an imaginary surface representing uniform air pressure near the top of the troposphere. Because of variations in air temperature below this isobaric surface, it has ups and downs, and in three dimensions, it has a wave-like shape. A line along the high points in this isobaric surface corresponds to a ridge, and a line along the low points corresponds to a trough (Fig. 18.18).

The location and intensity of the polar-front jet stream changes over time in response to the movements of air masses and the positions of fronts. As we've noted, the path of the jet stream, in map view, displays wave-like undulations as it curves around ridges and troughs. Therefore, the latitude at which the polar-front jet stream flows varies greatly around the planet, and at any given time, it may carry air from polar latitudes into mid-latitudes, or vice versa. As air masses move, the positions of troughs and ridges, and therefore the positions and shapes of the waves in the jet stream, change.

Air traffic must take into account the position and direction of the polar-front jet stream. Specifically, eastbound aircraft flying with the westerlies can take advantage of the significant tailwind that the jet stream provides to save on time and fuel costs. Planes flying westward, however, face into the westerlies and experience a strong headwind. As a consequence, the difference in travel time for westbound and eastbound flights crossing a continent or ocean can be more than an hour!

FIGURE 18.18 The shape of a surface of constant pressure in the upper troposphere. A ridge is a region where this isobaric surface bows upward, and a trough is a region where it forms a valley.

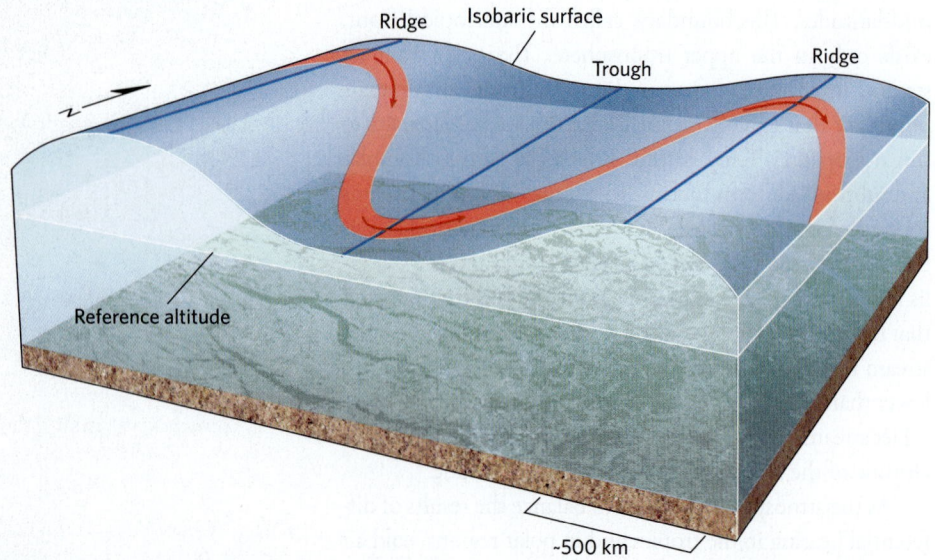

FIGURE 18.19 The sea breeze.

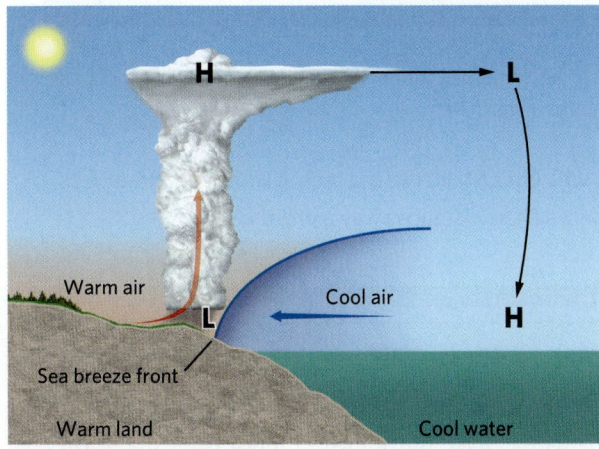

(a) A sea breeze circulation during the daytime along a coastline can trigger thunderstorms.

(b) Storms form along the sea breeze fronts on both Florida coasts.

18.5 Local Wind Systems

Many regions of the world have unique local wind systems that reflect interactions between the atmosphere and specific surface features, as well as specific seasonal conditions. Particularly notable examples occur along land-water boundaries and along and within mountain belts.

Sea Breezes

Land-sea boundaries generate several types of local wind systems. Oceans have a huge *heat capacity*, which means that they can hold a lot of heat relative to the land and that their temperature changes very slowly. Furthermore, land surfaces warm up rapidly during the day, while the ocean does not—you'll burn your feet walking on the beach in the tropics, but your feet will be quite comfortable immersed in the adjacent sea. Why? The same solar rays warm both, but on the opaque, immobile land, the warmth remains at the surface, whereas warming of the ocean affects deeper layers because water is transparent and because it circulates. The contrast between warmer land surfaces and cooler water drives local air circulations along coastlines. Specifically, during the afternoon, air rises over the hot land, flows seaward aloft, descends over the cooler ocean, and flows landward just above the ocean surface (Fig. 18.19a). This circulation, called a **sea breeze**, occurs along many coastlines of the world (Fig. 18.19b). At night, some coastlines experience a *land breeze*, a mirror image of a sea breeze, because the land cools off faster than the sea does, so the direction of the breeze reverses.

Weather associated with a sea breeze depends on the characteristics of the warm and cool air on either side of the *sea breeze front*, the boundary between oceanic cool air and warm air over the land. If the warm air over the land contains sufficient moisture and warms up enough, it rises and triggers clouds and thunderstorms (Box 18.2).

Along the Pacific coast of the United States, ocean water remains cold in summer, and the air over the land is dry. When the land heats up in the early part of the day, the sea breeze causes cool air to blow onshore. In the Los Angeles basin, which is surrounded by mountains, the cool, dense marine air collects at low elevations over the city and remains there for most of the day, particularly during peak traffic hours. This cool marine air is not buoyant and does not rise, so urban air pollutants accumulate within it throughout the day, producing smog (Fig. 18.20).

Winter Winds over the Great Lakes

Because water has a high heat capacity, the water of a large lake may remain liquid in early winter even if the air above it drops to temperatures well below freezing. As a consequence, the lake can provide moisture and heat to

Putting Earth Science to Use

Beach vacations and the sea breeze

Along the east and Gulf coasts of the United States, vacationers flock to the shore to enjoy summer sunshine and warm ocean water. Some choose to reside along the coast during their vacations, while others pitch tents or pull up trailers at campgrounds farther inland to have outdoor fun and reduce their costs. These coastlines host a strong sea breeze circulation on many summer days, particularly when the morning sky is clear and the land heats quickly relative to ocean water. Clouds and thunderstorms often develop along the sea breeze front as cool oceanic air flows onshore and the morning sea breeze intensifies. The storms typically develop several miles inland, and they often drift farther inland as the sea breeze strengthens during the afternoon (Fig. Bx18.2). From a vantage point on the coast, vacationers can gaze inland and watch these storms clatter and bang while enjoying an otherwise sunny day at the beach. Unfortunately, for folks inland, where storms develop overhead, the day appears washed out by the rain, thunder, and lightning. Discouraged, many never head to the beach. Those

FIGURE Bx18.2 Sea breeze clouds forming over central Florida, as seen looking inland from the shore.

few vacationers who understand the sea breeze know that more pleasant weather is only a short distance away along the shore.

---

air flowing across it, and it can therefore produce local weather systems called **lake-effect storms** (Fig. 18.21a). Lake-effect storms commonly occur over and adjacent to the Great Lakes of North America when cold air from northern Canada flows southward over the lakes. As this

FIGURE 18.20 Smog trapped in the marine air over the Los Angeles basin.

air passes, heat and moisture move from the warm lake surface into the boundary layer, while the air above remains cold. The warmed air starts to rise to form cumulus clouds. These clouds become taller as the layer of warm air reaches higher elevations as it moves across the lake. Eventually, the clouds start to precipitate snow.

When the relatively warm, moist air arrives at the downwind shore, friction with trees and buildings reduces the wind speed near the ground. Effectively, the near-surface wind piles up as it reaches the shore and is forced to rise. This additional lifting triggers production of even more clouds and yields heavy snowfalls downwind for several kilometers inland from the shore. As a consequence, regions downwind of the Great Lakes sometimes receive very heavy snow—up to 150 cm (60 in) may fall during a single lake-effect storm (Fig. 18.21b). Such lake-effect snowfall can continue for days, sometimes at a rate of more than 2.5 cm (1 in) per hour, resulting in enormous accumulations in areas within a few tens of kilometers of the shore (Fig. 18.21c).

Mountain-Valley Winds

Winds in mountain valleys during cloud-free days and nights flow in response to the heating of the valley and

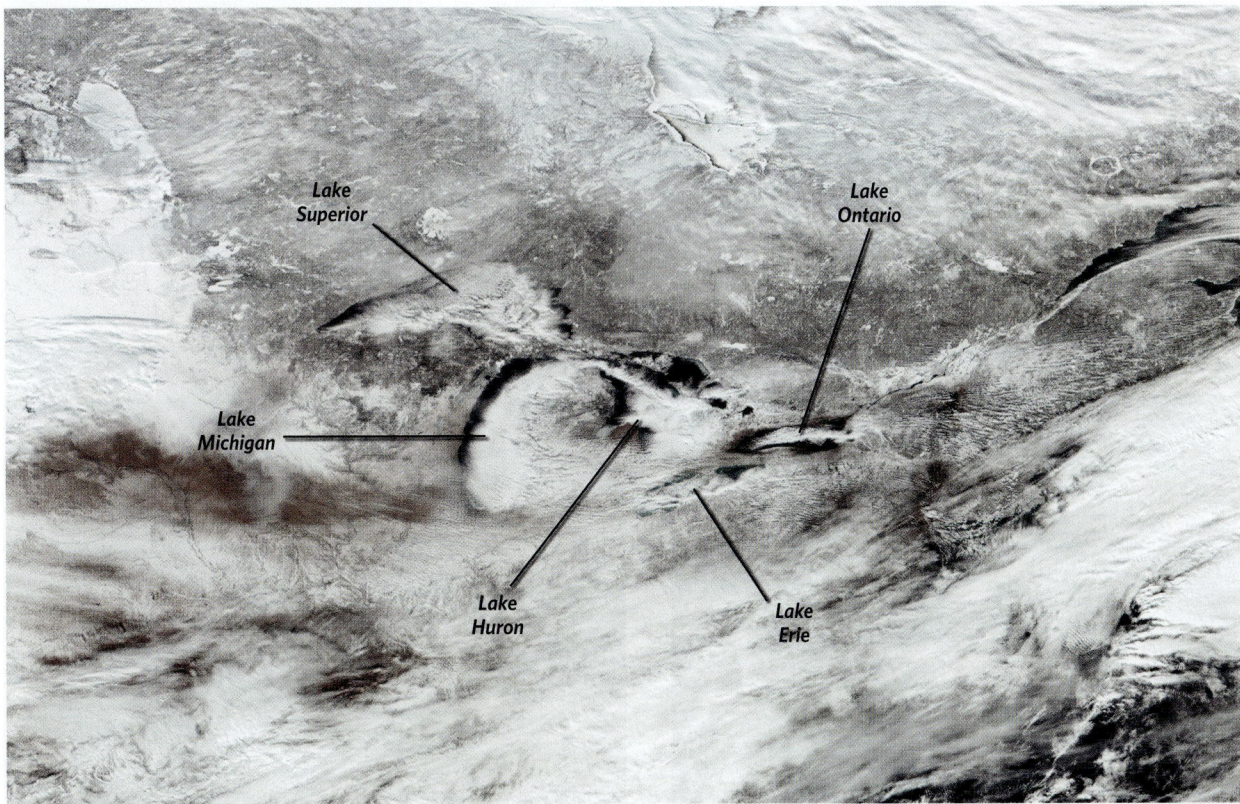

FIGURE 18.21 Lake-effect storms.

(a) Lake-effect clouds develop when cold air flows over the Great Lakes.

(b) Lake-effect storms produce heavy snowfall.

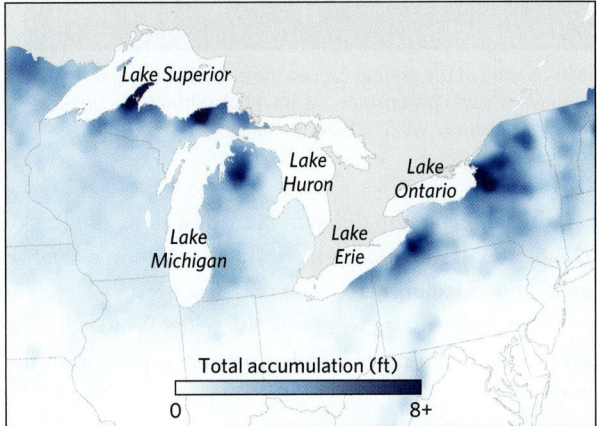

Total accumulation (ft)

0   8+

(c) Lake-effect snowfall accumulation over a winter season.

slopes in the daytime and the cooling that occurs at night (Fig. 18.22). As the Sun heats the slopes just after sunrise, air adjacent to the slopes warms. Because air near the slopes is warmer than air over the central part of the valley, it becomes buoyant and starts to rise along the surface of the slopes. This creates low surface pressure near the top of the valley and a pressure gradient along the valley that causes air to flow up the valley. As the day continues to warm, this flow of air up the valley becomes stronger. However, as the Sun starts to set, the heating stops. The slopes now radiate heat back to space, and the air near the slopes starts to cool. The cool, dense air along the slopes descends into the valley, creating high pressure at the top of the valley and a flow of air down the valley that continues through the night.

The up- and down-valley seesaw of the winds is a common feature of mountain valleys that experience clear skies.

Downslope Windstorms in Mountains

Because air moving over a mountain range loses most of its moisture as rain or snow on the upwind, or *windward*, side of the range, it is relatively dry as it descends on the downwind, or *leeward*, side of the range. In addition, as it descends, the air compresses, becomes denser, and warms. As a consequence, the leeward side of the range may be a rain-shadow desert, even where the windward side is wet and vegetated (see Chapter 14). In addition to dry weather, the leeward side of a mountain range occasionally experiences a **downslope windstorm**. This type of windstorm

FIGURE 18.22 Mountain-valley wind systems.

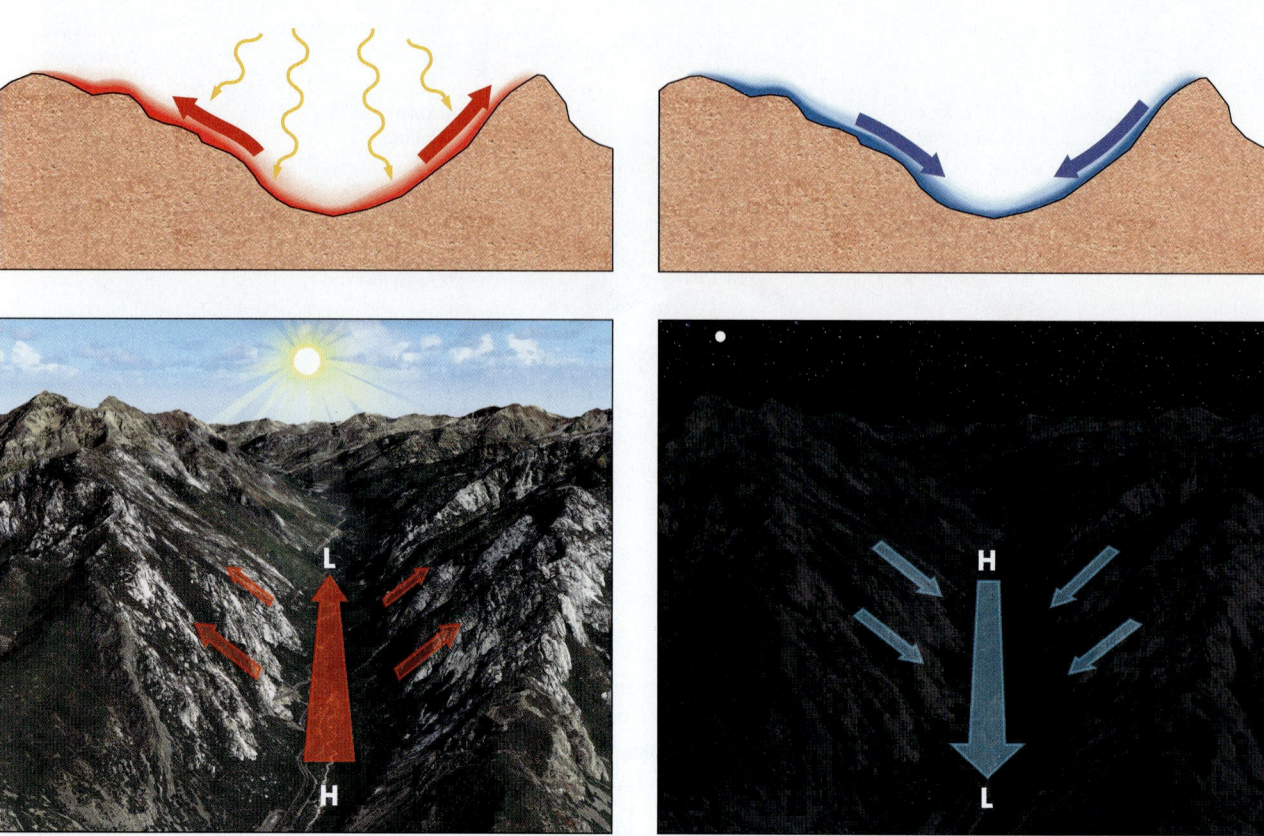

(a) Solar heating of the ground causes the air just above the ground to warm and flow upslope. As the photo shows, the upslope movement also causes wind to blow up-valley

(b) At night, the slopes cool, and cooler, denser air flows downslope. This, in turn, causes wind to blow down the valley axis.

has different names in different locations: along the Rocky Mountains, it's a *chinook wind*; in the Alps, it's a *foehn wind*; and in California, it's a *Santa Ana wind*.

Why do downslope windstorms develop? Wind will blow across a mountain range if higher pressure is present on the windward side than on the leeward side. This situation might occur along the Rockies, for example, if a low-pressure system lies east of the Rockies at the same time a high-pressure system develops west of the Rockies. As we saw earlier, stronger horizontal pressure gradients cause faster winds. When a

FIGURE 18.23 Development of a downslope windstorm. Winds blow faster where the streamlines (blue lines) are closer together. Interaction of the flowing air with the mountains generates a wave in the flow and produces locally faster winds.

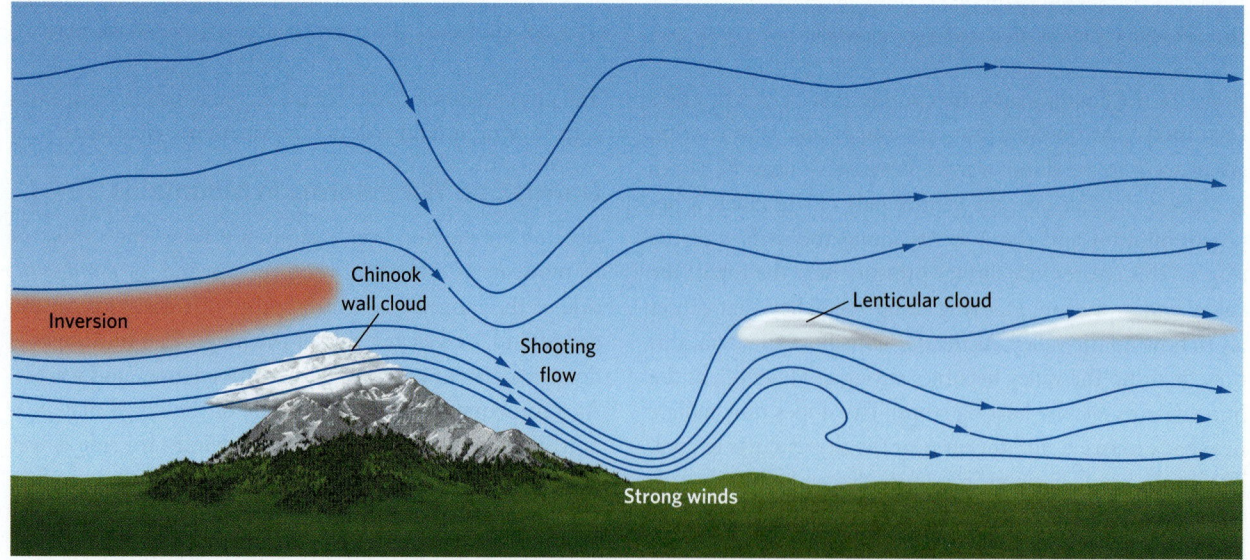

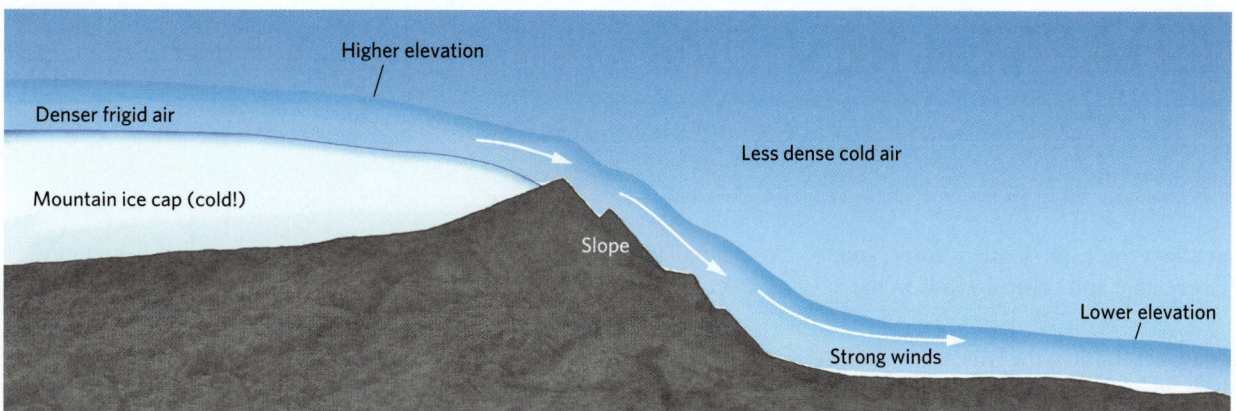

Higher elevation

Denser frigid air

Less dense cold air

Mountain ice cap (cold!)

Slope

Lower elevation

Strong winds

Katabatic winds in Antarctica

particularly strong pressure gradient exists across a mountain range, the stage is set for a downslope windstorm.

In the case of chinook and foehn winds, air flow becomes more rapid as air rises over the crest of the mountains in the shape of a wave, much like the wave that forms when a stream of water flows up and over a large boulder. Clouds form on the upwind side of the mountains as the air rises. Viewed from the downwind side, the clouds appear like a wall, called a *chinook cloud wall*. The background pressure gradient above the mountains accelerates the air passing across the range. This wave of air takes on a special shape when an **inversion**—a layer in the atmosphere in which the temperature increases with altitude—forms upwind of the mountain crest at an elevation just above it (Fig. 18.23). The inversion acts like a lid, confining the wind to a narrow layer above the mountaintop. As a consequence, the wind accelerates even more. You may have seen a similar phenomenon if you've blocked the end of a water hose with your thumb so that water spurts out of the hose more quickly. This strong wind comes down the leeward side of the mountain range in a *shooting flow* and descends to lower elevations, where it accelerates further—the strongest downslope winds can reach 160 km/h (100 mph). Because the air warms as it descends, it can melt snow rapidly.

Santa Ana winds blow when high pressure develops over the desert regions to the east of California's mountain ranges. If a low-pressure zone exists to the west, on the coastal side of California, the pressure gradient causes winds to blow from the desert westward over the mountains. These winds rush down the valleys of southern California, and as with chinook winds, the air warms as it descends. During a dry season, these winds often fan disastrous wildfires.

The downslope windstorms we have just described are driven by strong pressure gradients. A second type of strong windstorm occurs in cold mountainous regions and polar regions where the top surface of a large ice sheet has a higher elevation than its surroundings. These **katabatic winds** develop because air above the mountain range or ice sheet becomes very cold and therefore relatively dense. As a result, it flows downslope, sometimes reaching speeds of up to 160 km/h (100 mph). Katabatic winds can begin abruptly when a dome of extremely cold, dense air has built up over an ice sheet. The winds begin when gravity finally causes the cold air to spill off the ice sheet and rush outward onto the land or sea below (Fig. 18.24). Katabatic winds are common in Antarctica, along the coast of Greenland, and in mountainous regions of Canada and Alaska that have permanent glaciers.

Take-home message . . .

Some local weather systems form only at specific locations. For example, the contrast in heat capacity between land and water can drive sea breezes or lake-effect storms. Mountain topography can lead to the development of strong downslope winds, such as chinook and Santa Ana winds. Cold air pools over ice sheets and flows to lower elevations to produce katabatic winds.

Quick Question

Why does a sea breeze in Florida flow from the ocean toward land?

18 CHAPTER REVIEW

Chapter Summary

- Three forces—the pressure-gradient force, the Coriolis force, and the frictional force—together generate and control winds in the atmosphere.

- The frictional force, which always acts to slow the winds, is most important in the lower atmosphere, in a region called the boundary layer.

- Above the boundary layer, wind speed and direction are largely controlled by the balance between the pressure-gradient and Coriolis forces. Where this balance has been attained, winds are geostrophic and blow parallel to isobars.

- In the northern hemisphere, air flows clockwise around a high-pressure center, while spiraling outward. Air flows counterclockwise around low-pressure centers, while spiraling inward.

- Semi-permanent high-pressure systems develop over large regions where air is cooled, such as Canada in winter. Semi-permanent low-pressure systems develop over regions that are warmed, such as the oceans south of Iceland and the Aleutian Islands in winter.

- Hadley cells consist of air that rises at the ITCZ near the equator, then flows poleward, then descends from the upper troposphere of each hemisphere, and finally returns toward the equator above the Earth's surface. The Hadley cells strongly influence the distribution of tropical rainforests, steppes, and deserts, and their air flow near the surface causes the trade winds.

- The Walker circulation is an east-west-flowing convective cell covering the equatorial Pacific. An El Niño event occurs when the Walker circulation weakens, and a La Niña event occurs when it strengthens. The resulting pattern of changes in surface air pressure, called the Southern Oscillation, can affect weather around the world.

- A monsoon is a seasonally changing air circulation. The largest monsoon is located over southern Asia, where it brings heavy rains in summer and dry weather in winter.

- Jet streams, narrow bands of strong winds that encircle the Earth in the upper troposphere, form at subtropical and mid-latitudes. The subtropical jet stream forms below the tropopause near the poleward extent of each Hadley cell. The polar-front jet stream is found over the polar front, where warm tropical and cold polar air masses meet.

- A sea breeze is a circulation driven by the difference between daytime heating of the shore and of the ocean. It consists of air flowing onshore at low altitudes, rising over land, flowing oceanward aloft, and sinking over the ocean. It may trigger thunderstorms.

- Lake-effect storms develop over large lakes during winter as moisture and heat rise from the lake and moisture condenses to form clouds above the lake and its downwind shores.

- Chinook, foehn, and Santa Ana winds are strong downslope winds that develop occasionally on the leeward side of a mountain range when a strong pressure gradient exists across the range.

- Katabatic winds develop when exceptionally cold air cascades off a high-elevation ice sheet.

Key Terms

boundary layer (p. 566)
convective cell (p. 571)
doldrums (p. 573)
downslope windstorm (p. 581)
El Niño (p. 575)
ENSO (p. 575)
general circulation (p. 561)
geostrophic balance (p. 567)
geostrophic wind (p. 568)

Hadley cell (p. 572)
high-pressure system (p. 568)
intertropical convergence zone (ITCZ) (p. 572)
inversion (p. 583)
isobar (p. 564)
isobaric surface (p. 565)
katabatic wind (p. 583)
lake-effect storm (p. 580)

La Niña (p. 575)
low-pressure system (p. 569)
monsoon (p. 575)
polar front (p. 577)
polar-front jet stream (p. 577)
pressure gradient (p. 564)
ridge (p. 578)
sea breeze (p. 579)
semi-permanent high (p. 570)

semi-permanent low (p. 570)
Southern Oscillation (p. 575)
subtropical jet stream (p. 573)
thunderstorm (p. 573)
trade wind (p. 572)
tropics (p. 571)
trough (p. 578)
Walker circulation (p. 573)
wind (p. 564)

Letters in parentheses correspond to the chapter's learning objectives.

1. What forces act on air and cause it to flow? Which of these forces can cause the wind to increase in speed? **(A)**

2. Can air be in geostrophic balance in the boundary layer? **(A)**

3. In the southern hemisphere winter, would you expect high pressure or low pressure to develop over Antarctica? **(B)**

4. In summer, the east coast of the United States is often warm, while the west coast is cool. How might this difference be related to air flow around the Bermuda High and the Pacific High? **(B)**

5. Identify the Hadley cells on the diagram. Explain their relationship to the locations of the world's tropical rainforests and subtropical deserts. **(C)**

6. Where are you more likely to need an umbrella? Under the south end or the north end of the Hadley cell in the northern hemisphere? Explain. **(C)**

7. Is the sea surface off the shores of Peru warmer or colder than normal during an El Niño event? **(D)**

8. How do sea-surface temperatures change during the Southern Oscillation, and why?

9. What causes monsoons, and why do they occur only at relatively low latitudes? **(D)**

10. The polar front is the global boundary between which air masses? Why does air in it flow so fast? **(E)**

11. Which jet stream occurs at a higher altitude, the polar-front jet stream or the subtropical jet stream? Which occurs at a higher latitude? **(F)**

12. Identify the flow of warm and cold air in the diagram of a sea breeze. **(G)**

13. If you were riding a bicycle along a mountain valley during the daytime, and you wanted a tailwind for your ride, should you start at the top of the valley and ride down, or start at the bottom and ride up? State your reasoning. **(G)**

14. What conditions lead to the development of strong downslope winds on the leeward side of a mountain range? **(G)**

15. The summer-winter transition in weather in the southwestern United States and northwestern Mexico has been described as the "North American monsoon." Based on your understanding of monsoons, can you propose reasons why meteorologists might use this terminology? **(D)**

16. Airplanes flying from Seattle, on the west coast of the United States, to Boston, on the east coast, take about 5 hours to make the trip in winter. When they return to Seattle, they take about 7 hours. Explain why these times are so different. Will the difference change with the season? **(F)**

17. Lake-effect storms typically produce greater snowfall in November and December than in January and February. Why might this be so? (Hint: What changes are likely to occur in a lake as winter progresses?) **(G)**

ANOTHER VIEW Clouds move across the Earth's northern latitudes, carried by the winds that encircle the planet.

⊘19 DANGER IN THE AIR
Stormy Weather

By the end of the chapter you should be able to . . .

A. distinguish among different types of air masses and fronts.

B. explain the basic conditions that lead to the formation of a thunderstorm.

C. describe the difference between an ordinary thunderstorm and other types of thunderstorms, such as supercells.

D. explain how to protect yourself against thunderstorm hazards.

E. describe how tornadoes form, how they are classified, and how they cause destruction.

F. explain how mid-latitude cyclones develop and how they affect weather.

G. describe the relationships among weather, fronts, and jet streams within mid-latitude cyclones.

H. illustrate how forces acting in the atmosphere generate hurricanes.

I. explain why hurricanes are so destructive.

(a) The Parsons manufacturing plant before the storm.

(b) The plant after the storm.

19.1 Introduction

In the late morning of Tuesday, July 13, 2004, *thunderstorms*—billowing clouds that produce rain, lightning, and thunder—began to develop over central Illinois, in the midwestern United States. Meteorologists at the NOAA Storm Prediction Center worried that these storms could eventually produce a *tornado*, a vertical column of intensely spinning air, so they issued a *tornado watch* for the region, meaning that conditions existed that might lead to tornado development. Unfortunately, the thunderstorms indeed worsened, and in the early afternoon, the National Weather Service (NWS) issued a *severe thunderstorm warning* for north-central Illinois, meaning that a storm had been observed that could produce not only lightning, but also particularly strong winds, hail, and possibly a tornado.

At the Parsons manufacturing plant **(Fig. 19.1a)**, the NWS warning interrupted normal broadcasting on the radio in the front office, causing the factory's designated emergency response team leader to look outside. What he saw made him gasp. Not only was a towering dark cloud, backlit by lightning, heading straight toward the factory, but a tornado extended beneath it. He grabbed a microphone and bellowed the words no one wanted to hear: "This is not a drill. Move immediately to designated tornado shelters . . . now!" Thanks to a training exercise three months earlier, everyone knew exactly where to go and moved quickly.

<< Supercell thunderstorms billowed into the skies of Kansas in May 2016. One produced a tornado, visible in the distance. Storms like these can be thrilling to watch and can bring much-needed rain. But they may also bring devastation and represent a hazard that can strike unexpectedly.

The plant's owner had witnessed terrible tornado damage years earlier, so when constructing the plant, he had insisted that the restrooms have walls and ceilings of 20-cm (8-in)-thick reinforced concrete so they could serve as tornado shelters. All 140 employees in the building were already on their way into these shelters when the NWS sent out a *tornado warning*, meaning that a tornado had been sighted. This fearsome funnel of destruction, with winds spinning at nearly 290 km/h (180 mph), was a few hundred meters (less than a half-mile) away when the employees bolted the shelter doors. Two minutes later, it struck! There was a deafening roar, cars in the parking lot tumbled like toys, and the plant's roof detached and shredded as its walls collapsed. Only one terrifying minute later, the howl of winds and clatter of debris stopped, for the tornado had passed. Employees emerged into a chaotic jumble of twisted beams, smashed machines, and crushed vehicles **(Fig. 19.1b)**, but not one person was seriously injured. Warnings and preparations had saved everyone's lives.

Every year, we hear of devastating thunderstorm and tornado damage, often with results much deadlier than what we've just described. Thunderstorms and tornadoes are examples of *hazardous weather*, weather that has the potential to damage property and injure people. In this chapter, we describe hazardous weather and its causes. We begin by investigating smaller hazardous-weather events—thunderstorms and associated phenomena (tornadoes, hail, wind, lightning, and flash floods). We then shift to the Earth's largest storms—cyclones—of which there are two types: *mid-latitude cyclones* and *tropical cyclones*. Mid-latitude cyclones form over land or oceans at latitudes between 30° and 60°, and they can trigger weather phenomena from thunderstorms to blizzards. Tropical cyclones, known as *hurricanes* when winds

587

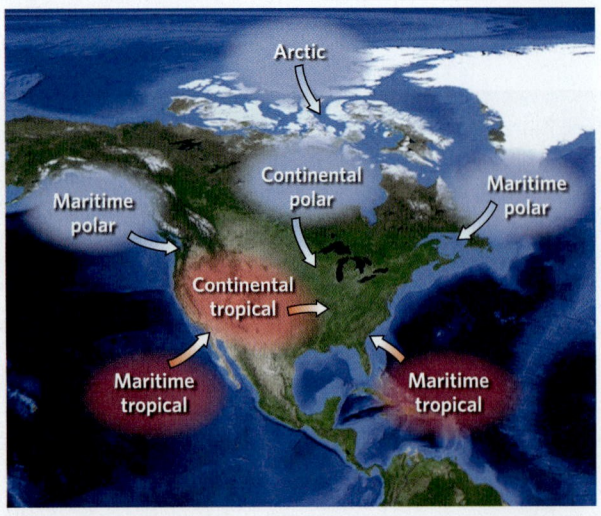

air masses by their source region (continental or maritime) and their latitude of origin (tropical, polar, or arctic).

Seven major air masses typically lie over and around North America and influence its weather (Fig. 19.2). For example, a *continental polar air mass* typically resides over a large part of Canada and the northern United States in winter, and an *arctic air mass* lies over the Arctic Ocean and northern Canada. *Maritime polar air masses*, which are cool and humid, reside over the North Atlantic and North Pacific Oceans. *Maritime tropical air masses*, which are warm and humid, develop over the tropical North Atlantic, the Gulf of Mexico, and the tropical North Pacific.

Because of the global atmospheric circulation, air masses may remain nearly stationary over a period of days, move slowly, or sweep rapidly (at speeds up to 60 km/h, or 40 mph) across a continent or ocean. When an air mass moves out of its source region, a new one soon regenerates there.

What Is a Front?

The boundary between two air masses typically occurs along a narrow zone called a **front**. Because the air masses on either side of a front have different characteristics, fronts delineate an abrupt contrast in temperature and, commonly, relative humidity as well. Most stormy weather in mid-latitudes, particularly in cooler seasons, develops along fronts, because air on the warm side of a front flows up and over air on the cold side, and this movement can produce clouds and precipitation.

Depending on how the air masses in contact at a front are moving, the front may remain stationary or it may move. When a front moves, meteorologists refer to its direction of movement as the direction in which the colder air mass is moving relative to a location on the ground. If the cold air mass is *advancing*, it's moving toward the location, whereas if the cold air mass is *retreating*, it's moving away from the location.

The direction in which the cold air mass moves is used to classify fronts. At a **cold front**, a cold air mass advances under a warm air mass, forcing the warm air to rise (Fig. 19.3a). Typically, the surface of the cold air mass has a dome-like shape, with the boundary between warm and cold air sloping toward the warm air. Meteorologists represent the trace of the front where it intersects the ground as a blue line. Blue triangles are placed along the warm side of this line to indicate the direction in which the front is advancing. A **warm front** exists where a cold air mass retreats and a warm air mass advances (Fig. 19.3b). Generally, the surface of the cold air mass slopes toward the warm air more gradually at a warm front than it does at a cold front. On a map, meteorologists represent the trace of a warm front on the ground with a red line marked by red half circles—the half circles lie on the cold side of the front and indicate

become particularly strong, form only over oceans at tropical latitudes between 5° and 30°, but they can eventually move over land and cause immense devastation.

19.2 Key Features of the Atmosphere in Mid-Latitudes

In mid-latitude regions, which lie between latitudes of 30° to 60°, weather can change not only as seasons progress, but even during a single day. For example, a late summer day in the central United States may be hot and humid until, suddenly, the skies become overcast and a chain of thunderstorms rolls by. A short time later, the Sun shines again and the air feels cool and dry. In winter, a southerly breeze bringing mild air can suddenly be replaced by a northerly blast of frigid air accompanied by heavy snow. Why do such radical changes in weather take place? These changes reflect the movement of large bodies of air known as *air masses*. When a *front*, the boundary between air masses, passes a location, the character of the air above that location changes radically and suddenly. As we'll see, such changes may be accompanied by stormy weather. Because of their importance, let's examine air masses and fronts in more detail. Ultimately, you'll see that the interactions between fronts are skirmishes in a broader battle between cool polar air and warm tropical air, as the atmosphere and oceans work like a giant conveyor to redistribute heat in the atmosphere at a global scale.

What Is an Air Mass?

An **air mass** is a broad body of air within which temperature and humidity are relatively uniform. Air masses are typically several kilometers thick and a thousand or more kilometers wide. A single air mass can cover a quarter to a half of a continent or ocean, so interactions between air masses can affect weather over a broad area. Numerous distinct air masses exist worldwide at any given time. Meteorologists characterize

Did you ever wonder . . .

why weather at a location can change so radically in a matter of minutes?

the direction in which the cold air is retreating. At a warm front, the warm air flows toward and then over the cold air mass. When a front passes over a location, weather conditions can change very rapidly. For example, temperatures often may drop by over 8°C (15°F) when a cold front passes.

In some cases, an advancing cold front overtakes a warm front. When this happens, the cold front lifts the cool air behind the warm front, so the warm front no longer intersects the ground surface, and two cold air masses come into contact at the ground surface. At this time, the warmer air forms a layer over the top of both fronts (Fig. 19.3c). Meteorologists indicate the presence of such an **occluded front** by a purple line with alternating purple triangles and half circles, pointing away from the coldest air. Under other conditions, the position of a front can remain stationary even though air on both sides of the front stays in motion. At such a **stationary front**, air on the cold side of the front flows nearly parallel to the front, but air on the warm side typically rises over the cold air (Fig. 19.3d). Along stationary fronts, the same weather may remain over a location for a long time. A stationary front is indicated by a line of alternating blue and red segments. Blue triangles on the blue line segments point toward the warm air, and red half circles on the red line segments point toward the cold air.

Take-home message . . .

Air masses are broad bodies of air over which temperature and humidity are relatively uniform. The boundaries between air masses, called fronts, are where most stormy weather occurs, for at fronts, warm, moist air flows upward over colder air, and this movement can produce clouds and precipitation.

Quick Question —————————————

What is the difference between a cold front and a warm front?

19.3 Thunderstorms

What Is a Thunderstorm?

Imagine sitting outside on a warm, calm summer day. Suddenly, thunder rumbles in the distance. Soon dark, turbulent clouds hide the Sun. Then gusts of cool wind begin to blow, followed by drenching rain, stronger winds, and possibly hail. As the sky crackles with lightning flashes, you hear the booms of thunder. A thunderstorm has arrived (Fig. 19.4).

Formally defined, a **thunderstorm** consists of towering cumulonimbus clouds that produce lightning and thunder. These clouds typically rise from an elevation of

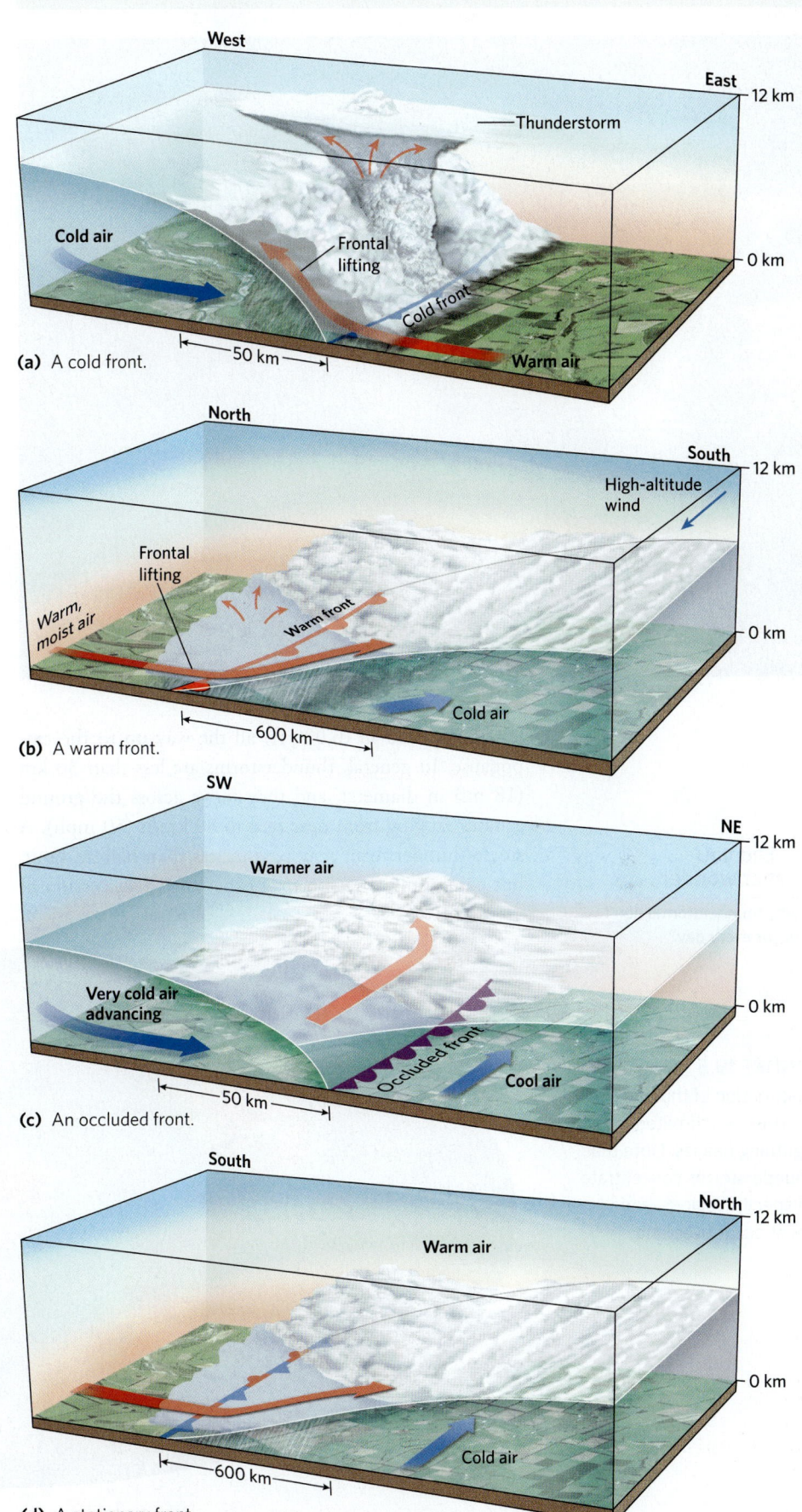

FIGURE 19.3 A front marks the boundary separating two air masses of different temperatures.

(a) A cold front.

(b) A warm front.

(c) An occluded front.

(d) A stationary front.

FIGURE 19.4 Lightning flashes during an intense thunderstorm in Nebraska.

thunderstorms that together last for a few hours or more. Thunderstorms can happen during the day or at night. During daytime thunderstorms, thick clouds may block so much light that the sky becomes quite dark. All thunderstorms pose danger because of lightning, but *severe thunderstorms* are particularly hazardous. A **severe thunderstorm**, by definition, produces hail larger than 2.5 cm (1 in) in diameter, winds exceeding 50 knots (92 km/h, or 56 mph), or a tornado.

Thunderstorms take place frequently in some regions, but rarely or never in other regions **(Fig. 19.5)**. Many develop over the tropics, where they provide the rain that sustains rainforests. During the warm season, they occur fairly frequently in mid-latitudes. They rarely take place in high latitudes, in subtropical deserts, or over large parts of the world's oceans. In the United States, thunderstorms primarily affect regions east of the Rocky Mountains, and they are particularly frequent over the southeastern states. In fact, in the eastern two-thirds of the United States, a given location will experience an average of 30 to 50 *thunderstorm days* (days on which at least one thunderstorm occurs) per year **(Fig. 19.6)**. Note that during a thunderstorm day, only an hour or two of the day may actually host a thunderstorm at a particular location. Florida holds the record for thunderstorm days: parts of the state endure 65 to 80 thunderstorm days per year.

Why don't thunderstorms occur everywhere all the time? They form only when three special conditions exist: (1) the lower atmosphere must hold *moist air*; (2) a *lifting mechanism* must be present to initiate an **updraft** (an upward-moving air flow that can transport the moist air to a higher elevation); and (3) *atmospheric instability* must

less than 2,000 m (6,000 ft) all the way up to the tropopause. In general, thunderstorms are less than 30 km (18 mi) in diameter, and they move across the ground at rates ranging from near zero to 80 km/h (50 mph). A single thunderstorm may pass in less than half an hour, but commonly, a region may experience a succession of

Did you ever wonder . . .

why thunderstorms don't occur every day?

FIGURE 19.5 The global distribution of thunderstorms, as estimated from lightning flashes. Note that thunderstorms concentrate in specific regions, and most occur over land.

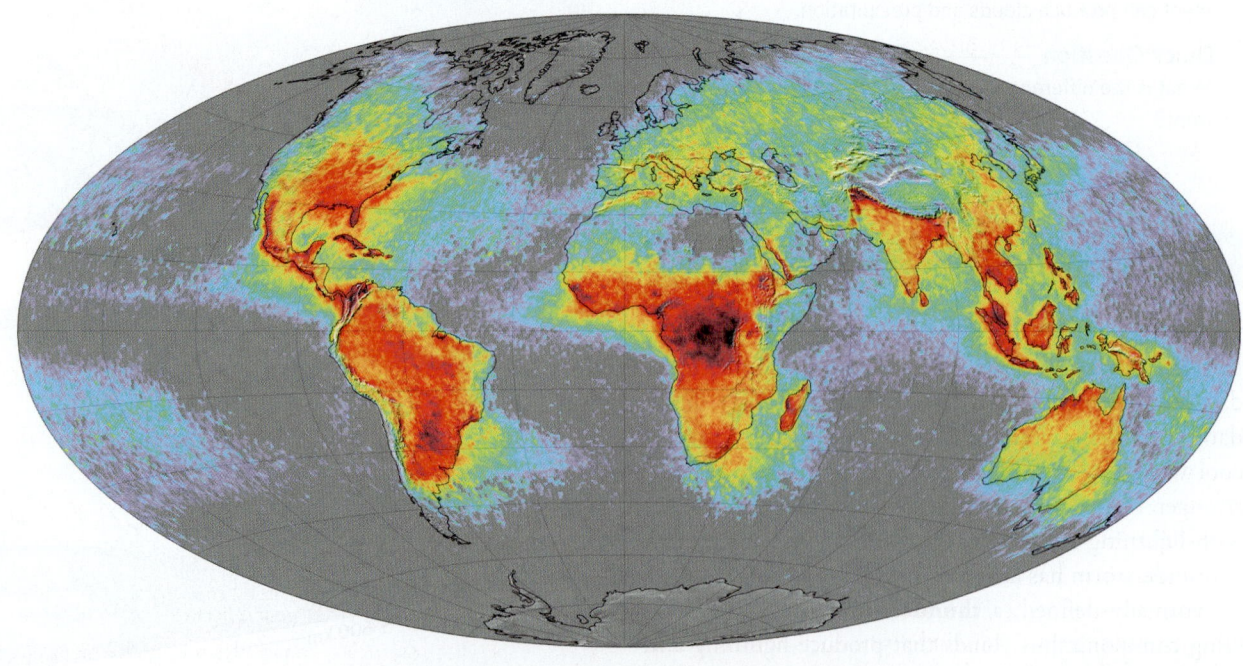

Lightning flashes (per km² per year)

0.1   0.4   1.4   5   20   70

allow the air to keep rising buoyantly to the top of the troposphere once it has started to rise. Let's consider each of these conditions in turn.

The Source of Moist Air

Moisture fuels a thunderstorm, and without it, the storm can't grow. So air in the lower atmosphere must have a high relative humidity in order for a thunderstorm to form. Once formed, a typical thunderstorm contains over 3.7 million tons of water at any given time, and over its lifetime many times that amount passes through the storm.

Where does all the moisture in a thunderstorm come from? Most evaporates from the world's oceans, the largest reservoir of water in the hydrologic cycle. For example, the moisture fueling thunderstorms in the central and eastern United States typically comes from water that evaporated from the Gulf of Mexico and the Atlantic Ocean and was then blown over the land by wind. Water that enters thunderstorms can also come from lakes and wetlands, from soils that have absorbed water from earlier rainfalls, and from transpiration by plants.

Lifting Mechanisms

Development of a thunderstorm begins when moist air starts to rise as an updraft from the lower troposphere. Any process that initiates this upward motion is called a **lifting mechanism**. Atmospheric scientists distinguish among several types of lifting mechanisms including:

1. *Lifting due to movement of a front:* Many thunderstorms develop when the advancing dome of cool air behind a cold front physically lifts warm, moist air ahead of the front (Fig. 19.7a). Lifting by an advancing cold front may produce several thunderstorms at roughly the same time, forming in a line along the front (Fig. 19.7b). Although less common, the lifting of moist air that triggers thunderstorms can also take place along a warm front.

2. *Lifting due to gust fronts:* As we'll see later in this section, when rain starts to fall in a thunderstorm, it produces an intense **downdraft** (a downward flow) of cool air that rushes from the storm to the ground below. When this downdraft reaches the ground, it spreads outward laterally into the regions around the storm. The leading edge of the resulting *cold pool*, an outflow of cool, dense air, becomes a **gust front** of strong wind. An advancing gust front acts like a local cold front in that it lifts warm air ahead of it. At places where gust fronts from different earlier storms collide, or where a gust front and a cold front collide, focused

FIGURE 19.6 Numbers of thunderstorm days per year in the United States.

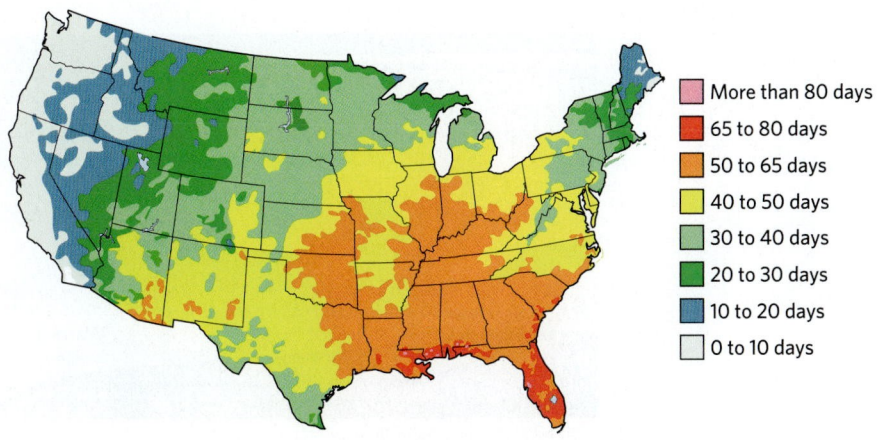

- More than 80 days
- 65 to 80 days
- 50 to 65 days
- 40 to 50 days
- 30 to 40 days
- 20 to 30 days
- 10 to 20 days
- 0 to 10 days

FIGURE 19.7 Thunderstorms may form as a result of frontal lifting.

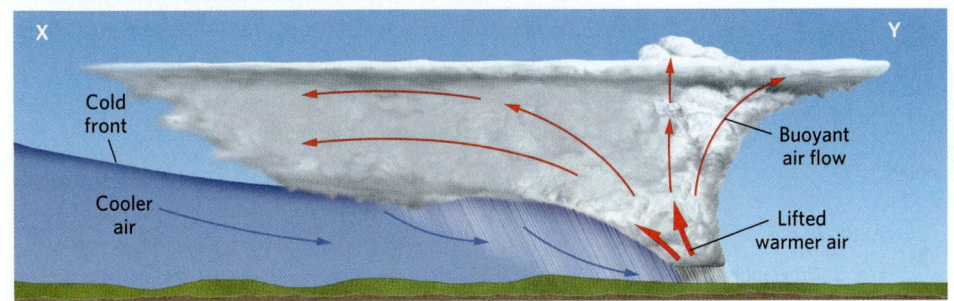

(a) Thunderstorms form as warm air is lifted ahead of an advancing cold front.

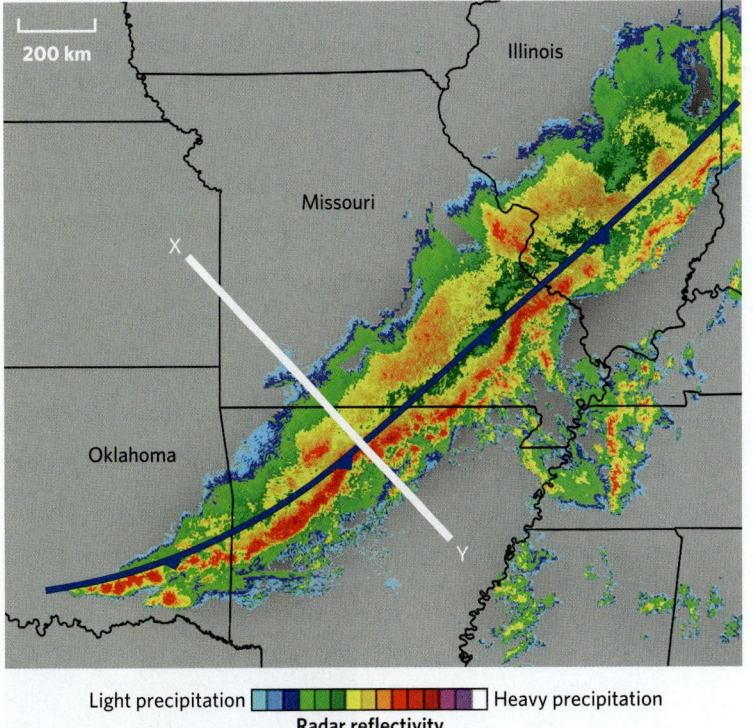

Light precipitation ▬▬▬▬▬▬▬ Heavy precipitation
Radar reflectivity

(b) A radar reflectivity display showing thunderstorms along an advancing cold front in the central United States. The line X-Y shows the location of the cross section in part (a).

FIGURE 19.8 Development of thunderstorms along colliding gust fronts. Each storm sends out a gust front from its base. Where the gust fronts collide, air rises, and a new storm forms.

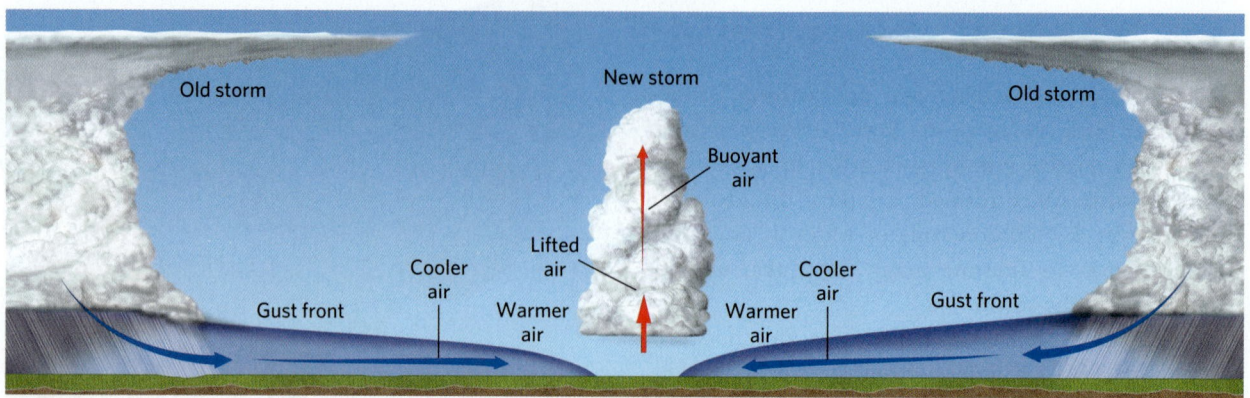

lifting and particularly strong updrafts can evolve into strong thunderstorms (Fig. 19.8).

3. *Lifting where winds interact with high topography:* When wind reaches hills or mountains, the air can no longer travel horizontally and must rise up the slopes. This process, called **orographic lifting**, can produce updrafts that can trigger thunderstorms along the mountain slope.

4. *Lifting due to convergence of air:* In general, any time winds near the ground cause air to converge at a location, air will be forced to rise at that location. Convergence can occur when air flowing from different directions collides, as happens in summer when the sea breezes that flow from the western and eastern coastlines of Florida meet in the central part of the peninsula and trigger large thunderstorms (Fig. 19.9).

5. *Lifting due to warming from below:* Air does not always have to be physically pushed upward for an updraft to begin. In some cases, moist air near the ground starts to rise simply because the hot ground beneath it warms the air enough to make it buoyant relative to surrounding air. This process typically happens in summer, when intense sunlight strikes the ground. This type of warming commonly takes place in mountain ranges when, during a hot afternoon, the slopes of steep mountains warm the adjacent air, and this warm air rises (Fig. 19.10).

Atmospheric Instability

Imagine that we can follow a small volume of air—an *air parcel*—as it moves through the atmosphere. Atmospheric scientists say that a region contains *stable air* when air parcels within the body of air do not continue to rise buoyantly after undergoing initial lifting by one of the mechanisms we described earlier. In contrast, a region contains *unstable air* when air parcels within the body of air do continue to rise buoyantly after undergoing initial lifting. In other words, if air has a tendency to continue rising after it has undergone initial lifting, we can say that **atmospheric instability** exists (Box. 19.1). Thunderstorms develop only where atmospheric instability exists at a level that air, after initial lifting, becomes buoyant enough to rise as an updraft toward the tropopause. Note that the initial lifting of air, alone, does not produce a thunderstorm.

How can air become buoyant as it rises? The answer is a bit complicated. Because the density and temperature of the atmosphere decrease with altitude, a rising air parcel decompresses and expands as it rises. In order to expand, air molecules within the parcel must work to push aside the surrounding air (Fig. 19.11). Since the energy that does this work must come from within the parcel, the air making up the parcel gives up some of its heat energy, so the temperature within the parcel decreases. If the air in the parcel

Animation
Fronts: Frontal Lifting

Animation
Fronts: Gust Front Lifting

Animation
Fronts: Orographic Lifting

Animation
Fronts: Warming from Below

FIGURE 19.9 Convergence of sea breeze flows in mid-afternoon often trigger thunderstorms over central Florida.

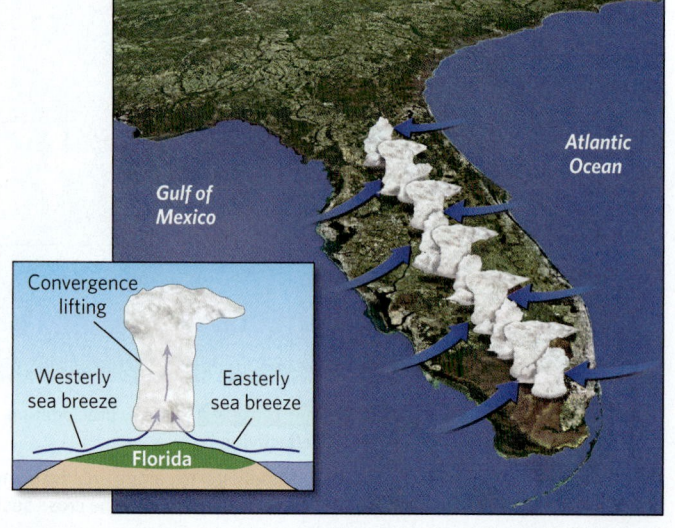

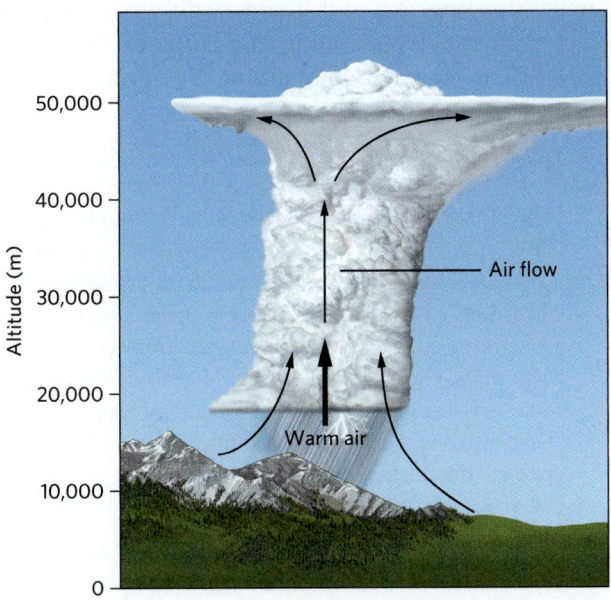

(a) Updrafts can begin over mountain ranges when, during a hot afternoon, the slopes of steep mountains warm the adjacent air.

(b) An example of thunderstorms brewing over mountains.

does not exchange heat or mass with its surroundings as it expands, we call the process **adiabatic expansion**.

Due to adiabatic expansion, a rising air parcel may eventually reach an altitude at which it has become cooler and denser than the surrounding air. If this happens, the parcel's rise ceases. Air composed of such parcels is stable and cannot produce a thunderstorm. Whether air is stable or unstable depends on whether parcels of air in it stop rising, or become buoyant and continues rising, respectively. This contrast depends on two factors. First, stability depends on the **environmental lapse rate**, the rate at which the temperature of the air surrounding the rising air parcels (the parcels' "environment") decreases with altitude. Typically, environmental lapse rates range between 4°C/km and 9°C/km (12°F/mi to 26°F/mi), depending on location. Second, it depends on the moisture content of the rising air parcels. *Moist air*—air that is saturated with water vapor—forms clouds as it rises and cools. *Dry air*—which may contain some water vapor, but is unsaturated—won't form clouds as it rises. The presence or absence of cloud formation plays an important role in determining whether air parcels become buoyant. That's because the rate at which air cools during adiabatic expansion depends on whether or not the water vapor within it condenses into cloud droplets. Let's see why.

When a parcel of unsaturated air rises, cloud droplets do not form. The temperature in the rising parcel decreases by about 10°C for each kilometer (29°F/mi) that it rises, a change known as the *dry adiabatic lapse rate* **(Fig. 19.12a)**. If the dry adiabatic lapse rate exceeds the environmental lapse rate, the rising dry air parcel eventually

FIGURE 19.11 A parcel of air expands as it rises. Note that the number of air molecules in the parcel doesn't change as the parcel rises.

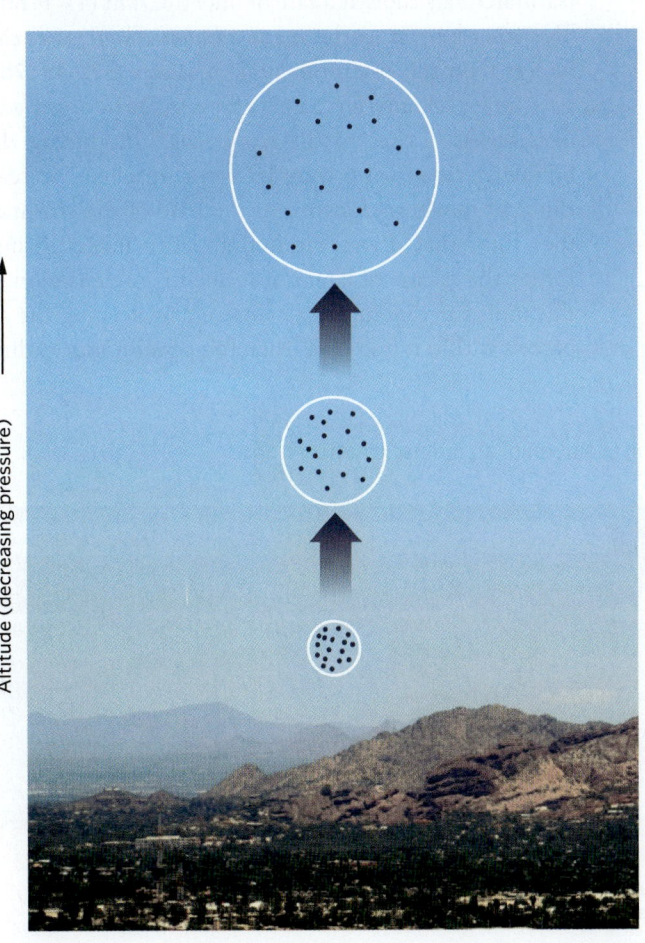

FIGURE 19.12 Stability under different environmental conditions.

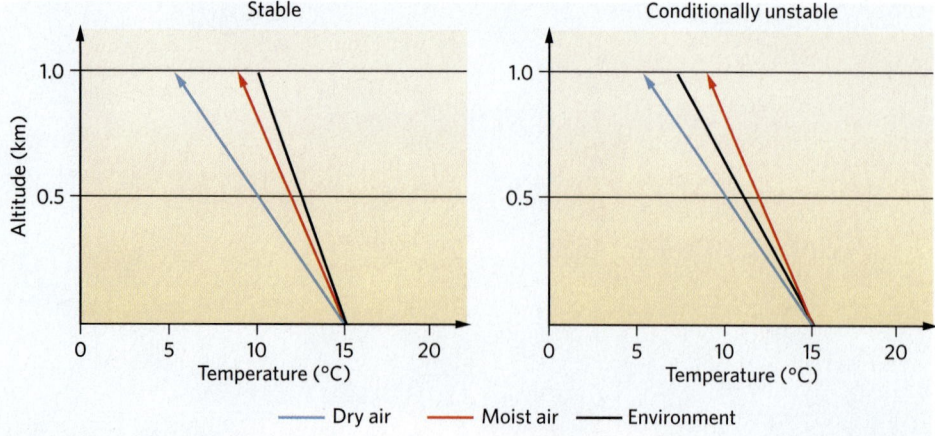

(a) In this environment, both moist and dry air parcels, when lifted, would return to their original altitude.

(b) In this environment, a dry air parcel, when lifted, would return to its original altitude, but a moist air parcel would rise buoyantly.

becomes colder and denser than the surrounding air. As a result, it's no longer buoyant and stops rising.

In contrast, when saturated air rises, water droplets do form. Significantly, condensation of water vapor to liquid water releases latent heat (see Chapter 17), and adding latent heat to a rising air parcel slows the rate of cooling during the parcel's ascent. In fact, rising moist (saturated) air cools at a rate of only 6°C/km (17°F/mi). This rate, known as the *moist adiabatic lapse rate*, can be less than the environmental lapse rate (Fig. 19.12b), so a moist, cloudy air parcel can remain buoyant and unstable as it rises to the tropopause, producing the billowing clouds of a thunderstorm. In effect, we can think of moisture drawn into the base of the storm as the "fuel" that drives the updraft after initial lifting, for it's the release of latent heat due to condensation of this moisture that makes air unstable by keeping the air parcels within it buoyant. Therefore, conditions leading

to thunderstorm formation exist only when air in the lower troposphere is warm and moist while air in the upper troposphere is cool. These conditions can happen all year in the tropics, but in mid-latitudes they generally become common only during late spring through early fall.

Take-home message . . .

Three conditions are necessary for a thunderstorm to form: moist air near the Earth's surface, a lifting mechanism, and local atmospheric instability. Once lifted, a moist air parcel can become buoyant when its moisture condenses into cloud droplets because the release of latent heat makes the parcel cooler and less dense than its surroundings. Such instability can produce an updraft that carries rising air up to the tropopause, producing large billowing clouds that evolve into a thunderstorm.

Quick Question

What mechanisms can lift air to trigger thunderstorm development?

19.4 Types of Thunderstorms

Towering dark clouds, flashes of lightning, claps and rumbles of thunder, torrential rains, and sometimes the clatter of hail—these are the signatures of thunderstorms. But even though all thunderstorms share these basic characteristics, not all are alike. Here, we examine the different ways that thunderstorms develop and evolve.

Ordinary Thunderstorms

Meteorologists refer to an isolated thunderstorm that does not rotate—meaning that the air in the storm does not spin around a vertical axis—as an **ordinary**

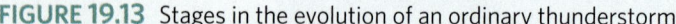

FIGURE 19.13 Stages in the evolution of an ordinary thunderstorm.

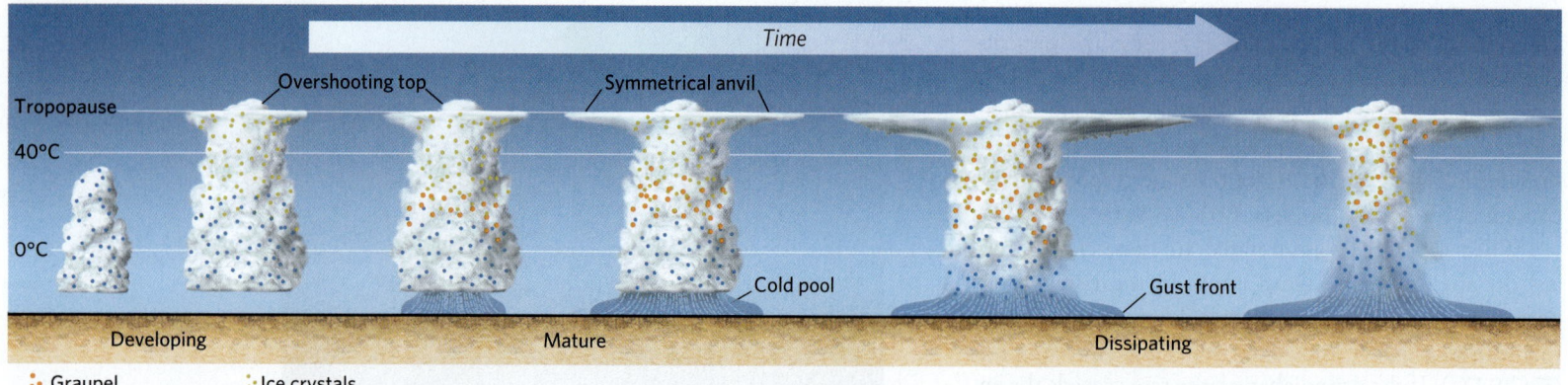

• Graupel   •: Ice crystals
•: Small cloud droplets   // Raindrops

BOX 19.1

How can I explain . . .

Atmospheric instability

What are we learning?
The concept of atmospheric instability.

What you need:
- Vinegar
- Olive oil
- A jar or other glass container
- A spoon

Instructions:
We can simulate the concept of atmospheric instability by doing two experiments using familiar kitchen materials.

- Start by filling a jar a third of the way with vinegar, which has a density of 1.05 g/cm³. Then add olive oil, which has a density of 0.85 g/cm³. Because it's less dense, the olive oil forms a horizontal layer floating above the vinegar. Now, plunge a spoon down below the boundary and lift a blob of vinegar up. What happens to the blob of vinegar?
- Starting again with an empty jar, invert the layering by pouring the oil in before the vinegar. Carefully add vinegar on top of the layer of oil. What happens to the oil?

What did we see?
- In the first experiment, if you remove the spoon, the vinegar sinks back down, so the boundary between oil and vinegar returns to horizontal. We say that the fluid in the jar is stable because if we displace some of the fluid upward, it returns to its original state, even though it has undergone lifting.
- In the second experiment, the fluid in the jar is unstable. If you were careful adding the vinegar, you might have been able to briefly get a layer of vinegar to remain on top of the oil. However, even when left undisturbed, such layering can persist for only a short while. The fluid isn't stable, meaning that a blob of oil beneath the vinegar will rise upward. In fact, the rest of the oil will follow the blob until all the oil ends up on top of the vinegar. You can see this same process in a lava lamp, where buoyant blobs of fluid rise to the top.
- In this context, we are using the term *stable* to mean that a denser fluid lies beneath a less dense fluid. The denser fluid, if lifted, has no buoyancy and simply sinks downward, and the system returns to its initial condition. Similarly, we're using the term *unstable* to mean that a less dense fluid underlies a denser fluid. If the less dense fluid on the bottom get an initial upward push, its buoyancy causes it to continue to rise upward while the denser fluid above sinks to take its place.

Salad dressing

Oil (less dense)

Vinegar (denser)

Stable condition

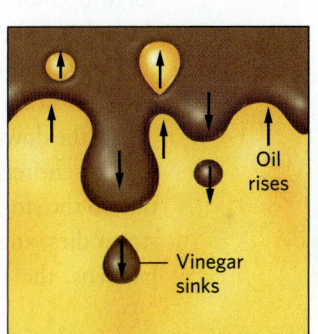

Oil rises

Vinegar sinks

Unstable condition

Lava lamp

---

thunderstorm. Such storms form where winds do not change substantially in direction or speed with altitude. They may develop when air undergoes lifting by any mechanism. In general, ordinary thunderstorms form during hot afternoons, last about an hour before they dissipate, and rarely become severe. At any given time, as many as 2,000 ordinary thunderstorms exist worldwide.

Meteorologists divide the life cycle of an ordinary thunderstorm into three stages (Fig. 19.13). During the *development stage*, warm, moist air undergoes lifting to an altitude at which it becomes buoyant and rises. As the rising air expands and cools, its water vapor condenses into liquid droplets or, at higher altitudes, solidifies into tiny ice crystals, forming cumulus clouds. Strong updrafts carry the rising air to the tropopause. The tropopause acts as a lid on storms, for above this level, rising air cannot remain buoyant. High-altitude winds cause clouds at the top of the thunderstorm to spread laterally into a flat sheet. A storm cloud with this shape is called an **anvil cloud**, so named because of its resemblance to an anvil used by a blacksmith (Fig. 19.14).

With the formation of the anvil cloud, the thunderstorm enters the *mature stage* of its life cycle. When this happens, graupel or hail forms in the upper, cold part of

FIGURE 19.14 When a thunderstorm rises to the tropopause, the top of the cloud spreads out into a flat layer, making the cloud overall resemble a classic blacksmith's anvil.

the cloud. These ice fragments eventually fall to warmer altitudes where they melt into rain. As the falling rain encounters drier air drawn into the storm from the area surrounding the storm's clouds, it evaporates. Such evaporation cools the air within the storm, thereby causing the air to become denser. This denser air sinks, producing a downdraft where there once had been an updraft. When rain reaches the ground within the downdraft, the cloud becomes a *cumulonimbus* cloud.

As the downdraft intensifies, it eventually destroys the updraft entirely, thereby shutting off the supply of moisture to the storm. Without a source of new moisture, the storm dies, and the clouds begin to dissipate. When this happens, the storm enters its *dissipation stage*. Notably,

when the cooled air of the downdraft reaches the ground, it spreads out. The leading edge of this spreading cold pool of air is sometimes marked by strong winds that form a gust front. The gust front may lift warm air ahead of it, triggering new thunderstorms.

Squall-Line Thunderstorms

Meteorologists refer to thunderstorms aligned in a long chain as **squall-line thunderstorms**. These storms produce much of the summer rainfall over the interior plains of the United States. If you look toward squall-line thunderstorms approaching from a distance, you'll probably see a dark and ominous sheet of gray called a *shelf cloud*, which forms over the gust front (**Fig. 19.15**). Typically, only a small amount

FIGURE 19.15 A shelf cloud develops when warm air is lifted over an approaching gust front rushing out from beneath a thunderstorm. The inset shows a closeup of a shelf cloud.

Shelf cloud

Heavy rain

A research radar truck, called the "Doppler on Wheels."

FIGURE 19.16 A squall line develops when ordinary thunderstorms coalesce and a large cold pool forms.

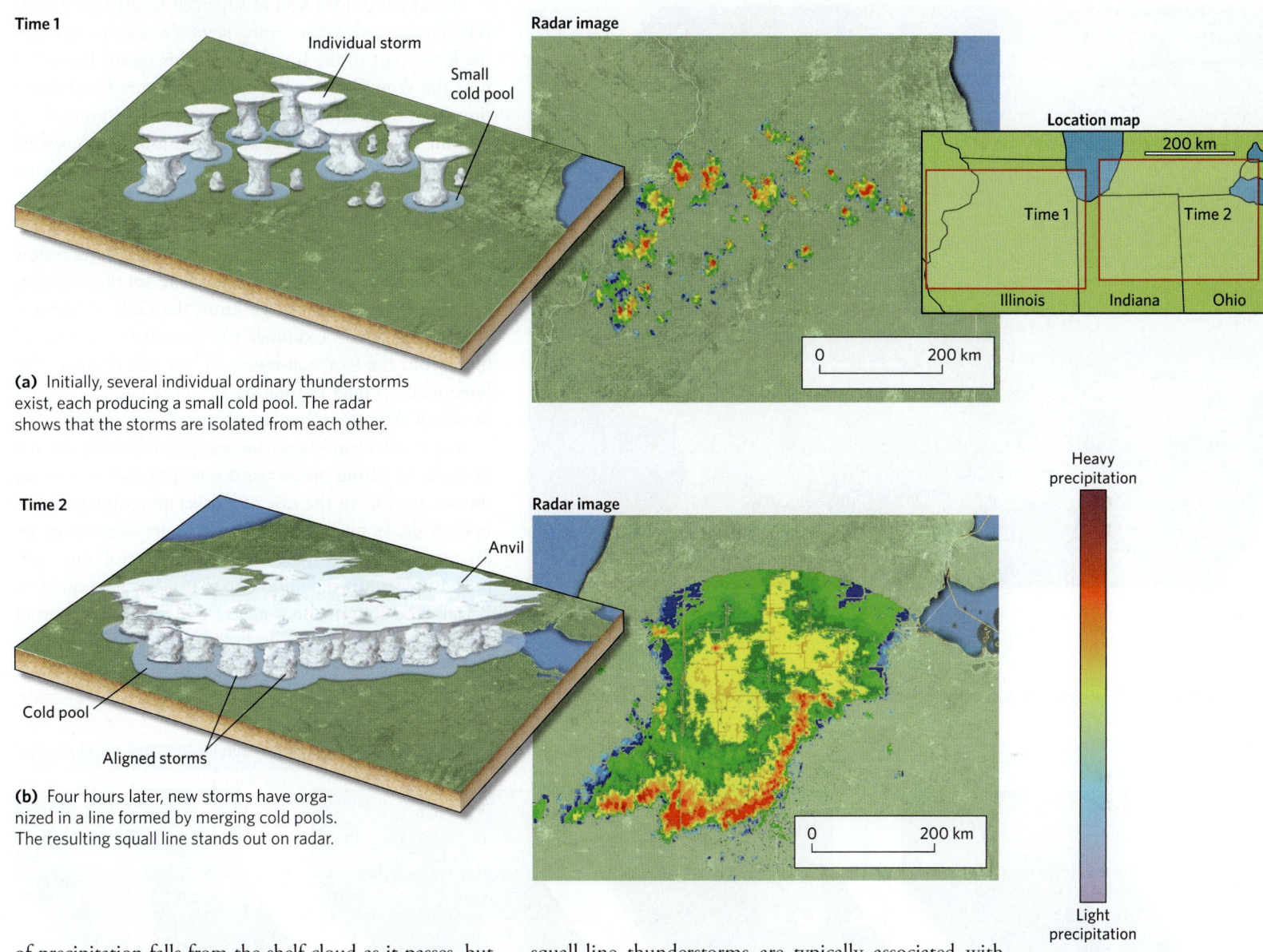

Time 1

Individual storm

Small cold pool

Radar image

Location map

200 km

Time 1

Time 2

Illinois   Indiana   Ohio

0   200 km

(a) Initially, several individual ordinary thunderstorms exist, each producing a small cold pool. The radar shows that the storms are isolated from each other.

Time 2

Anvil

Cold pool

Aligned storms

Radar image

0   200 km

(b) Four hours later, new storms have organized in a line formed by merging cold pools. The resulting squall line stands out on radar.

Heavy precipitation

Light precipitation

of precipitation falls from the shelf cloud as it passes, but the wind that accompanies it can be hazardous. Behind the gust front come **straight-line winds**—winds that blow in a single direction, rather than in a spiral as in a tornado. Strong straight-line winds can damage buildings, topple trees, and pose a hazard for low-flying airplanes.

During warm summer months, squall-line thunderstorms may originate from a disorganized cluster of several ordinary thunderstorms (Fig. 19.16a). As time progresses, downdrafts from the ordinary thunderstorms form a large cold pool at the ground. This cold pool spreads outward, and its leading edge forms a strong gust front. Lifting along this gust front triggers the formation of new thunderstorms defining a squall line (Fig. 19.16b). In cooler seasons, lifting becomes focused along strong cold fronts (Fig. 19.17). Very long *frontal squall lines* can develop along such fronts. As we will see, such frontal

squall-line thunderstorms are typically associated with mid-latitude cyclones.

Supercell Thunderstorms

When a radio or TV station broadcasts a severe thunderstorm warning, most often they are referring to a supercell thunderstorm. Formally defined, a **supercell thunderstorm** is a thunderstorm containing a particularly strong and long-lasting, rotating updraft. (Ordinary thunderstorms are weaker and, as we noted earlier, their updrafts do not rotate.) Because of this distinctive structure, supercell thunderstorms (or, simply, *supercells*) often become severe and can produce tornadoes. Supercells are much less common than ordinary thunderstorms, and they are usually isolated from neighboring thunderstorms, though in some cases, several supercells can form in advance of a squall line. They typically have an overall diameter, at the base, of less

FIGURE 19.17 A frontal squall line along a cold front within a mid-latitude cyclone.

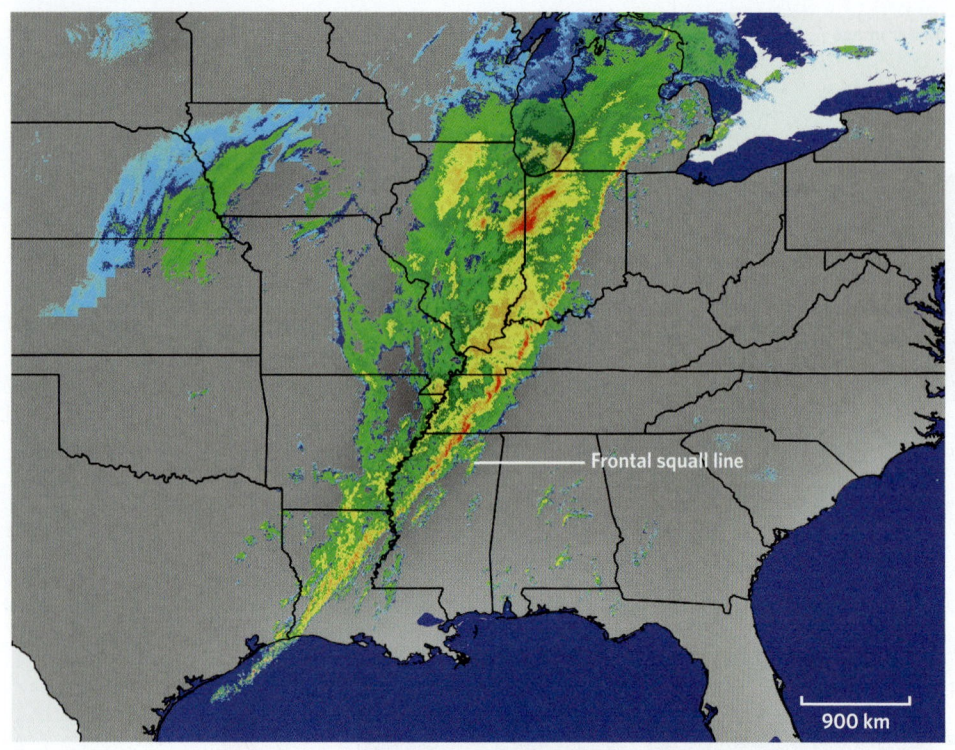

Frontal squall line

900 km

than 50 km (30 mi); the most dangerous region, the storm's updraft, has a diameter of 5–10 km (3–6 mi) or less.

What conditions lead to supercell formation? Supercells develop only when strong *vertical wind shear* exists in the lower part of the troposphere. This means that wind near the ground differs significantly in speed and direction from wind a kilometer or two above the ground. As a result of this difference, air in the lower troposphere tends to swirl around a horizontal axis. You can picture this movement by imagining a horizontal cylinder resembling a cardboard tube (Fig. 19.18a). In the central United States, vertical wind shear typically develops when slow winds near the ground blow from the south-southeast, bringing in warm, moist air from the Gulf of Mexico, while faster winds a kilometer or two above the ground blow from the west-southwest. When this happens, the horizontal cylinder of rotating air in the lower kilometer or two of the troposphere trends nearly due north.

Supercells form where a strong updraft develops by any of the same lifting mechanisms that give rise to ordinary thunderstorms. In the case of a developing supercell, the updraft draws up the horizontal cylinder of rotating air. To picture this, imagine that the cardboard tube gets pulled upward in the middle so that it becomes an arch. Because the updraft also triggers cloud formation, most of

FIGURE 19.18 Stages in the development of a supercell thunderstorm.

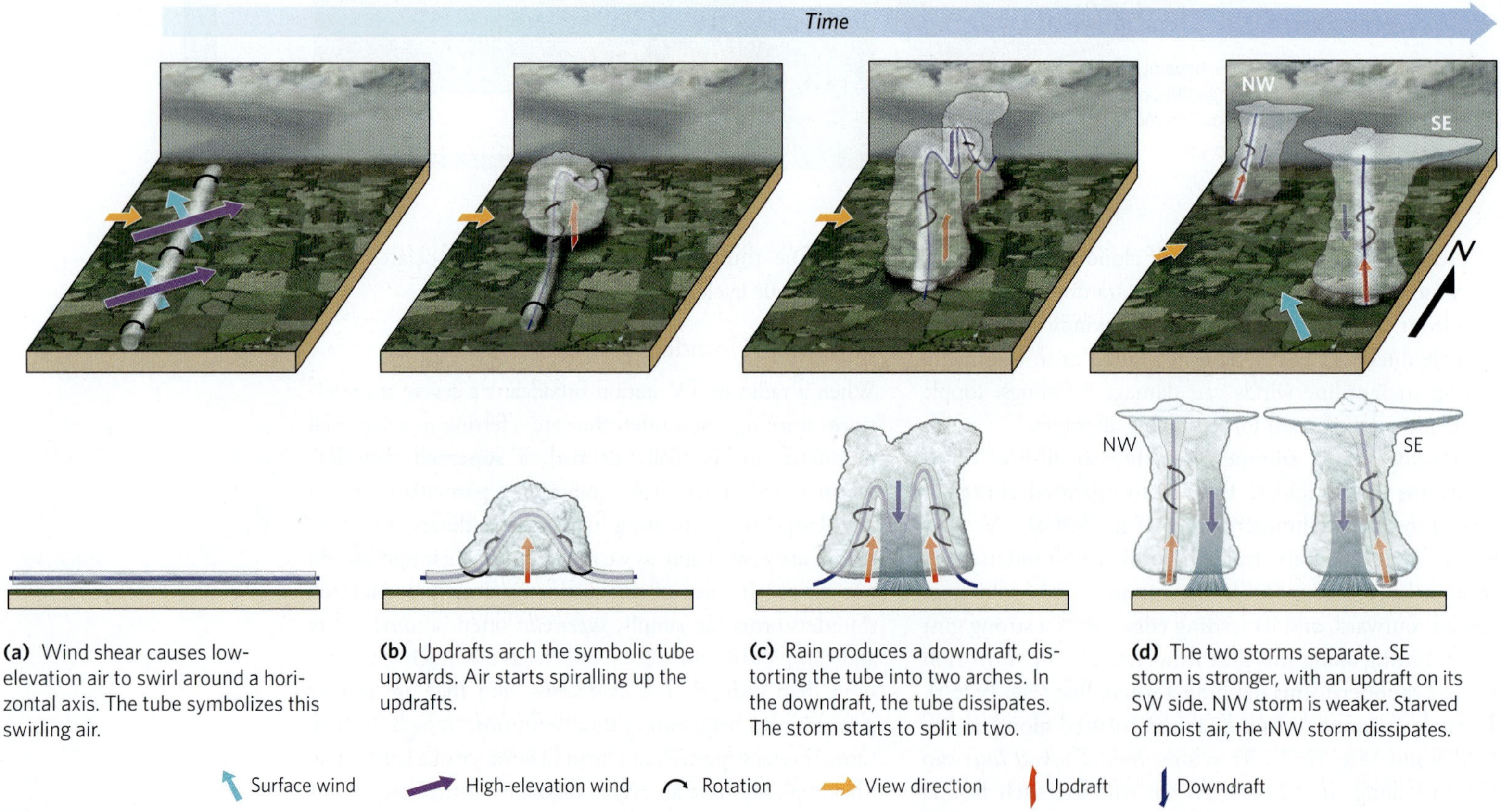

Time

NW   SE

NW   SE

N

(a) Wind shear causes low-elevation air to swirl around a horizontal axis. The tube symbolizes this swirling air.

(b) Updrafts arch the symbolic tube upwards. Air starts spiralling up the updrafts.

(c) Rain produces a downdraft, distorting the tube into two arches. In the downdraft, the tube dissipates. The storm starts to split in two.

(d) The two storms separate. SE storm is stronger, with an updraft on its SW side. NW storm is weaker. Starved of moist air, the NW storm dissipates.

↑ Surface wind   → High-elevation wind   ⌒ Rotation   → View direction   ↑ Updraft   ↓ Downdraft

(a) A composite image showing a supercell thunderstorm, as viewed from the southwest.

(b) This view of a supercell thunderstorm from a high elevation shows the anvil and the overshooting top.

(c) A close-up view of the base of a supercell, showing rotating clouds and the wall cloud.

the cylinder lies hidden within a growing cumulus cloud (Fig. 19.18b). Due to the updraft, air on the steeply tilted sides of the cylinder spirals upward.

As soon as precipitation begins, and the cloud becomes a cumulonimbus cloud, a downdraft forms, driving the center of the arch downward. This process separates the initial arch such that two arches of rotating air form, and the original storm starts to split in two. Very quickly, however, the arms of the arches within the downdraft dissipate into turbulence and disappear (Fig. 19.18c). At this point, two distinct thunderstorms exist—in the northern storm, air rotates clockwise as it spirals upward in the updraft, whereas in the southern storm, air rotates counterclockwise. Since the warm, moist air that feeds the storm comes from the south, the southern storm draws in the moisture and grows, while the northern storm, starved of warm, moist air, dies out (Fig. 19.18d). Therefore, the big, dangerous supercell storms that pose a hazard dangerous supercell storms that pose a hazard in the central United States rotate counterclockwise. Meteorologists refer to the rotating updrafts of these storms as **mesocyclones**, meaning middle-sized (much smaller than a hurricane) and counterclockwise-spinning. Within the updraft that forms the heart of such a storm, vertical wind velocity can reach 160 km/h (100 mph).

When the updraft of a supercell thunderstorm nears the tropopause, the high-altitude winds of the polar-front jet stream blow the top of the storm downwind to produce a large asymmetrical anvil cloud, and they may cause the whole storm to tilt, so that the axis of the updraft is not exactly vertical. Therefore, supercell thunderstorms have a distinctive structure characterized by an asymmetrical anvil at the top and a strong mesocyclone within (Fig. 19.19a, b). Upward flow in the mesocyclone can be so strong that it pushes air up into the base of the stratosphere, forming a bubble of billowing clouds, called an *overshooting top*, on the top of the anvil cloud.

In the central United States, supercell thunderstorms usually move from southwest to northeast and tilt toward the northeast. Meteorologists refer to the northeastern side of the storm as the *forward flank* and to the southwestern side as the *rear flank*. Heavy rain on the forward flank produces an intense downward air flow known as the *forward-flank downdraft* (FFD). A strong downdraft also develops on the rear flank of the storm, producing a *rear-flank downdraft* (RFD). Significantly, because of the storm's northeastward tilt, the FFD lies to the northeast of the mesocyclone, so that precipitation falls outside (to the northeast) of the updraft. Therefore, unlike an ordinary thunderstorm, whose updraft tends to be destroyed by the rain-associated downdraft almost immediately after the storm matures, a supercell thunderstorm can survive for a relatively long time (a few hours). A rotating **wall cloud** typically develops beneath the base of the storm's updraft when cooler, rain-saturated air gets drawn into the mesocyclone from the FFD; the wall cloud outlines the base of the mesocyclone and, as we'll see, may be the site of tornado formation (Fig. 19.19c).

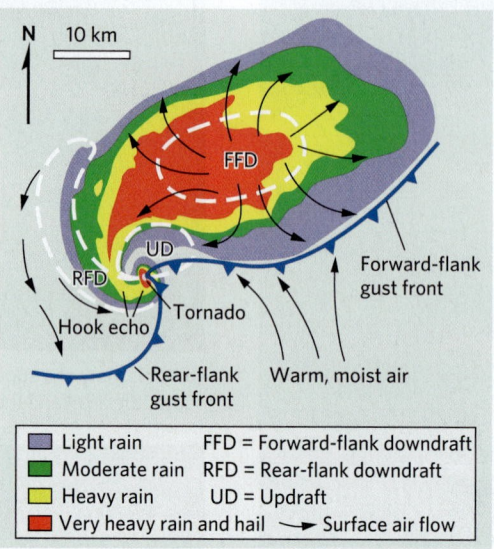

Because of their structure, supercell thunderstorms have a distinctive signature on radar images (Fig. 19.20). Most notably, a *hook echo* appears on the southwestern side of the supercell and wraps around the southwestern side of the updraft. The updraft itself appears as an echo-free region within the hook because no rain falls out of the updraft, and the large radar echo to the northeast of the updraft indicates the presence of heavy rain and the location of the associated downdraft. When meteorologists see a hook echo on their screens, they start worrying about tornado formation.

Take-home message . . .

An ordinary thunderstorm has a vertical, non-rotating updraft and tends to dissipate fairly quickly. Squall-line thunderstorms form in a row along storm-generated cold pools of air or along strong cold fronts. The most violent thunderstorms, supercells, form where vertical wind shear exists. A supercell has a rotating updraft, a particularly broad, asymmetrical anvil cloud, and an overshooting top.

Quick Question

Why do supercell thunderstorms tend to survive for a relatively long time?

19.5 Thunderstorm Hazards

Lightning

A **lightning stroke** (*lightning bolt*) is a short-lived and powerful electric current, and **lightning** is the occurrence of one or more lightning strokes. The process of generating lightning begins when electrons jump between tiny ice crystals and larger ice particles, such as hailstones, when they collide in thunderstorm clouds. As a consequence, the tiny crystals become positively charged, while the larger particles become negatively charged. Updrafts sweep the tiny crystals upward to high elevations in the cloud, while the larger, heavier particles fall toward the lower part of the cloud, producing a *charge separation* within the cloud. This charge separation can be maintained because air is an *electrical insulator*, meaning that it prevents electrons from flowing easily between a negatively charged and a positively charged region. The charge separation produces an *electrostatic potential*, stored energy that could drive electrons from the region of negative charge to the region of positive charge, if given the opportunity. A lightning stroke occurs when the electrostatic potential in a thunderstorm becomes great enough to drive electrons across a large expanse of insulating air.

Most lightning that we see occurs as **cloud-to-cloud lightning**, formed when an electric current jumps from a negatively charged part of a cloud to a positively charged part (Fig. 19.21). Cloud-to-cloud strokes may hit an airplane flying through a storm. Fortunately, electricity can be conducted along the metal skin of a plane without penetrating its interior, so lightning typically causes minimal damage, if any, to airplanes.

Cloud-to-ground lightning, which strikes buildings, trees, people, or other objects, poses a greater hazard. Some cloud-to-ground strokes travel between the positively charged anvil cloud and negatively charged ground. Most cloud-to-ground strokes, however, travel between the base of the storm cloud and the ground. Conditions leading to such lightning start to develop when negative charges accumulate toward the base of the cloud. The negatively charged cloud base drives away negative charges on the ground (because like charges repel), so the ground develops a net positive charge. Because unlike charges attract, electrons surge from the cloud's interior toward the cloud base and then toward the ground in a series of steps, producing a complexly branched *stepped leader* (Fig. 19.22a). When a branch approaches the ground surface, positively charged particles jump upward from an object on the ground, producing a *positive streamer*. When the positive streamer connects with the stepped leader, a channel for electron flow becomes established. (The channel provides an easy pathway for current flow because it contains charged particles—so unlike the air around it, which acts as an insulator, it acts as an electrical conductor.) The instant that the positive streamer connects with a branch of the stepped leader, a powerful *return stroke*—the main discharge of lightning—flashes as vast numbers of electrons stream from the cloud to the ground (Fig. 19.22b). The stroke runs from the cloud to the point where the positive

Did you ever wonder . . .

why lightning strikes tall trees?

streamer started. Additional leaders, known as *dart leaders*, subsequently form and lead to the generation of new return strokes (Fig. 19.22c, d)—these strokes follow the previously established channel, as it has become an electrical conductor. The whole process—from generation of the stepped leader to the formation of several return strokes—happens so fast that you may perceive that you're seeing only a single lightning bolt that flashes like a strobe light.

Needless to say, lightning is dangerous. If lightning strikes a tree, it can cause the sap in the tree to vaporize instantly, so that the tree explodes. If it strikes a house, it can pass into the frame of the house and ignite flammable materials. If lightning strikes power lines, it can blow out transformers and cause a blackout. While lightning strikes cars, it rarely does them harm because their metal skins, like those of airplanes, conduct the electricity and keep it from entering the vehicle. Unfortunately, when

FIGURE 19.21 Different forms of lightning (cloud-to-cloud and cloud-to-ground) can occur during a thunderstorm.

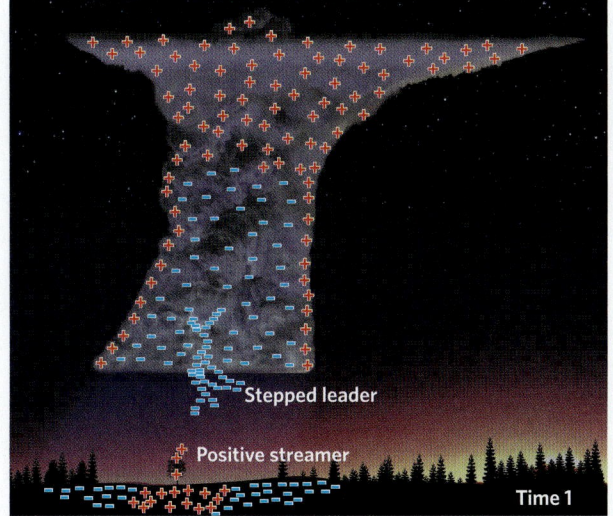

(a) A stepped leader descends from the cloud, while a positive streamer rises from the ground below.

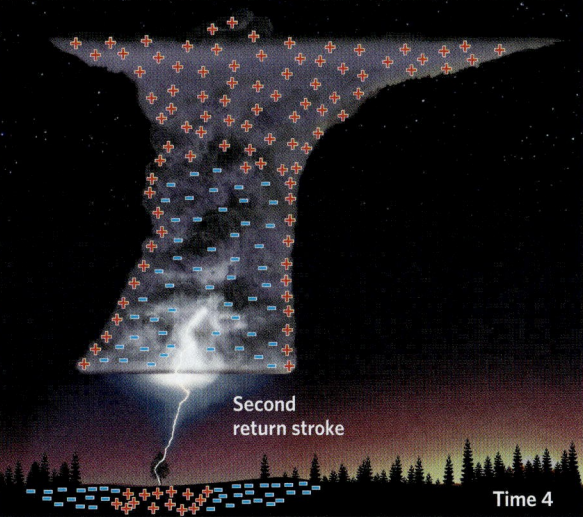

(b) A return stroke makes a brilliant flash.

FIGURE 19.22 The steps leading to the production of a cloud-to-ground lightning bolt.

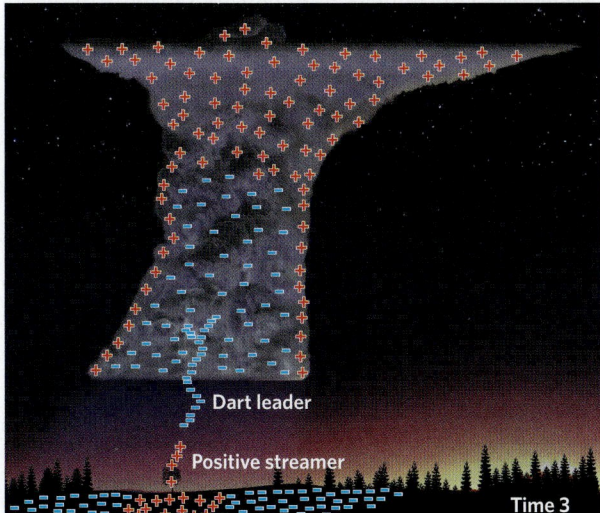

(c) A dart leader descends from the cloud, while a positive streamer flows upward.

(d) The second return stroke flashes.

FIGURE 19.23 Microbursts can be a serious hazard in the vicinity of an airport. If a plane takes off into head-winds from a microburst, it lifts into the sky easily. But when it crosses into the downdraft itself, it gets pushed down, and when it encounters tailwinds in the portion of the microburst blowing in the same direction that it's flying, it loses lift.

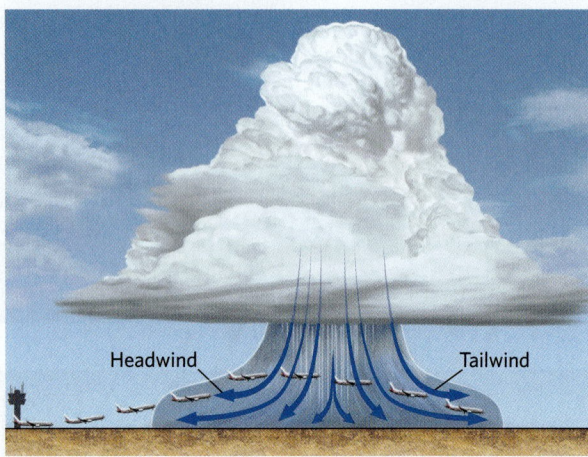

Headwind   Tailwind

FIGURE 19.24 Hail occurs over a relatively small area, but can cause significant damage.

(a) A hailstone that fell in Vivian, South Dakota, on July 23, 2010, holds the US record for diameter (20.3 cm, or 8.0 in) and weight (0.879 kg, or 1.9375 lb).

(b) Hail damaged this car's back window.

lightning strikes a person, it can cause permanent harm or even instant death. Although people struck by lightning can survive, many suffer burns, neurological damage, and other lasting effects. Injury can happen even if a lightning stroke hits the ground or a tree near a person, for the electricity can flow through the ground to the person. In the United States, lightning strokes kill, on average, about 50 people per year. To avoid being a victim, go inside (and stay away from electrical conductors) or hop in a car when a thunderstorm approaches. If you can't take cover, don't stand in the middle of an open area or near isolated tall objects, or be on open water where you would be the highest object. If there's no place to go and you feel your hair standing on end (because it's becoming positively charged), crouch down as close to the ground as possible with your feet together and your heels up to minimize ground contact.

All lightning produces **thunder**, a loud crack or boom. Thunder develops because the intense energy in a lightning stroke instantly heats adjacent air to a temperature five times hotter than the surface of the Sun. This heating causes the air to expand explosively. When the stroke ceases, the air contracts suddenly and generates sound waves, much like the sound produced when you clap your hands. Since sound travels so much more slowly than light, you can estimate the distance to a lightning stroke by counting the seconds that pass between the flash and the thunder—if you divide by three, you get the distance in kilometers, and if you divide by five, you get the distance in miles. Thunder, while frightening, is not hazardous.

Downbursts and Straight-Line Winds

As we've seen, very strong downdrafts develop in association with thunderstorms. When a downdraft reaches the ground, the air must turn and flow outward. The outrushing wind emerging from a downdraft is called a *downburst*. Downbursts flow radially outward from the base of a downdraft, but at a given location, the wind of a downburst flows straight and parallel to the ground, so meteorologists refer to it as a *straight-line wind*. These winds may travel a few kilometers in front of the storm at velocities of 30–130 km/h (20–80 mph).

The area affected by a downburst depends on the size and maturity of a storm. A small area of straight-line wind, known as a *microburst*, affects an area only about 5 km² (2 mi²). Microbursts can be dangerous to aircraft approaching or departing airports **(Fig. 19.23)**. Major airports have installed radar to detect microbursts. When downbursts from several thunderstorms along a squall line merge, they can affect a broad area. In fact, squall lines occasionally produce a swath of destructive straight-line winds extending for hundreds of kilometers along the length of the squall line. Meteorologists refer to these large-scale, destructive windstorms as *derechos*.

Hail and Flooding

In nearly all thunderstorms, **hail**, which consists of spherical or irregularly shaped lumps of solid ice (each of which is a *hailstone*), forms at high altitudes. Most hail is small (pea-sized to marble-sized) and melts before reaching the ground, but some hailstones can grow large enough to fall to the ground. The heaviest hailstone ever measured in the United States, weighing 0.8 kg (1.94 lb), fell during a 2010 storm in South Dakota (Fig. 19.24a). Hail can cause major damage by flattening crops, denting cars, and breaking glass (Fig. 19.24b).

The process of hail formation begins when updrafts carry liquid cloud droplets upward to altitudes where temperatures are well below freezing. These droplets become *supercooled*, meaning that they remain liquid even though their temperature lies below 0°C. When ice particles falling from higher altitudes collide with these supercooled droplets, the droplets bond to the ice particles and immediately freeze, so an ice particle quickly grows into a hailstone (see Chapter 17). In an updraft, upward air flow keeps hailstones floating in the cloud, where they can keep growing. Eventually, however, a hailstone drifts out of the updraft, or its weight overcomes the force of the updraft, and it starts to fall. It collects more supercooled droplets on its way down, so it continues to grow until it encounters above-freezing temperatures in the lower part of the cloud and starts to melt. Hailstones reach the ground only if they grow large enough to avoid melting away as they fall. This happens when a thunderstorm contains particularly strong updrafts, so the most damaging hail tends to develop in supercell thunderstorms.

Thunderstorms can dump a lot of rain in a short time, which can trigger *flash floods*, localized short-duration flooding that happens rapidly with little warning. Flooding can be a particular problem when thunderstorms move slowly, as storms formed by orographic lifting along mountain slopes sometimes do, or when a squall line develops along a stationary front so that several individual thunderstorms move parallel to the front and dump rain on the same location in succession.

Take-home message . . .

Thunderstorms produce several dangerous phenomena. When the electric charge separation in a thunderstorm becomes great enough, a lightning stroke flashes from one part of a cloud to another or from the cloud to the ground. Strong downdrafts in thunderstorms can cause damaging straight-line winds. Strong updrafts can produce large hail, and heavy rains may trigger flash floods.

Quick Question
How does a cloud-to-ground lightning stroke develop?

19.6 Funnels of Destruction: Tornadoes

What Is a Tornado?

Perhaps no other weather phenomenon can be as terrifying as a **tornado** (informally called a *twister*), a violently rotating funnel, or *vortex*, of air that extends from the ground into the lower part of a severe thunderstorm (Fig. 19.25a, b). The

Did you ever wonder . . .

how big a hailstone can grow?

FIGURE 19.25 Examples of tornadoes in the United States.

(a) Small tornado in Kansas.

(b) Giant tornado (1 km- or .6 mi-wide) in Oklahoma.

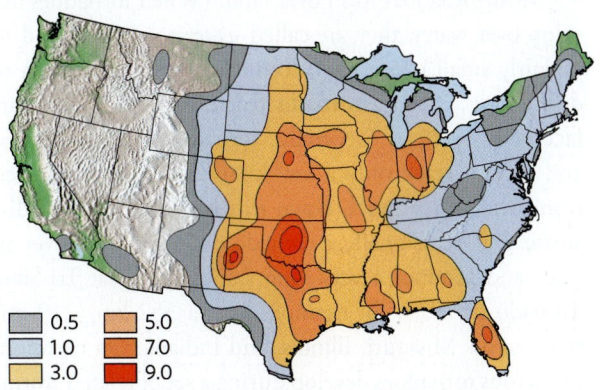

0.5	5.0
1.0	7.0
3.0	9.0

(c) Number of tornadoes per year (per 26,000 km², or 10,000 mi²) in the United States over a 27-year period.

FIGURE 19.26 Tornado tracks.

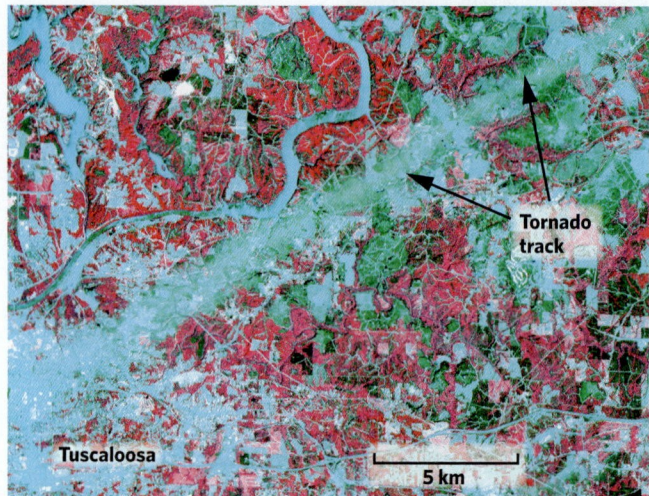

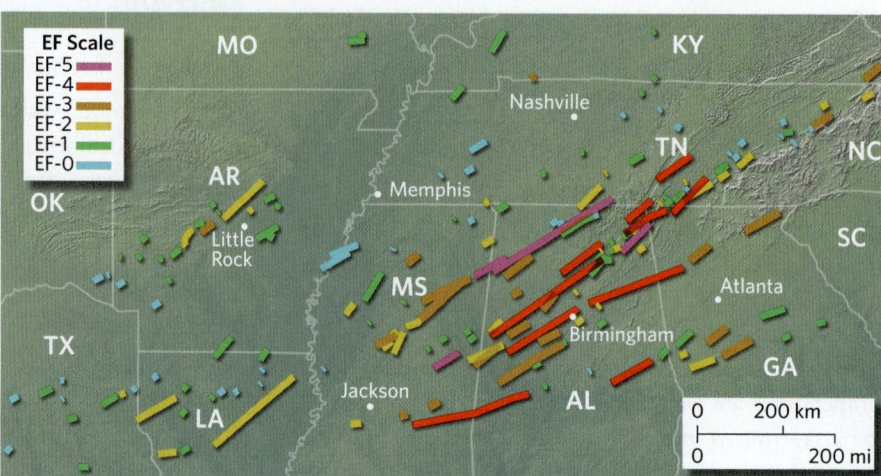

(a) This false-color satellite image shows the track of a tornado near Tuscaloosa, Alabama. Red areas are forested—the winds ripped out the trees along the tornado track.

(b) A series of tornado tracks from an outbreak of tornadoes in April 2011.

strongest tornadoes can damage or destroy steel-reinforced concrete structures, can throw trucks long distances, and can sweep houses completely away. At its base—the end touching the ground—a tornado typically ranges in width from 50 m to 1 km (150 ft–0.6 mi), but some are much larger—the largest tornado ever recorded reached a width of 4.2 km (2.6 mi) at its base. Tornado winds generally blow between 100 and 320 km/h (65–200 mph), but the fastest tornado wind speed ever recorded was 484 km/h (301 mph). A tornado moves across the countryside along with its host storm, so the speed at which its base moves across the ground can vary from near zero to 110 km/h (70 mph). The central and midwestern United States, where warm, moist air from the Gulf of Mexico meets cold air from Canada along strong fronts in spring and early summer, hosts the world's largest number of tornadoes, giving this region the nickname *Tornado Alley* (Fig. 19.25c). Tornadoes are also common across the southeastern United States and in Florida.

Most tornadoes form over land. (When tornadoes develop over water, they are called *waterspouts*, and tend to be fairly small.) As a tornado moves, it leaves a swath of destruction, called a **tornado track**, on the ground surface (Fig. 19.26a). Generally, tornadoes are short-lived and leave a track only a few kilometers long, measured from where they touch down to where they lift and dissipate. But occasionally, a tornado can survive for over an hour and produce a very long track. The great Tri-State Tornado of 1925, for example, tore a 352-km (219-mi) track across Missouri, Illinois, and Indiana. On occasion, numerous tornadoes develop during a set of related storms within a relatively short time. The world's largest tornado

outbreak, which took place in late April 2011, produced 363 tornadoes over the course of 3 days (Fig. 19.26b).

Tornado Formation

Though tornadoes occasionally form along squall lines or within hurricanes, most develop within supercell thunderstorms, so we illustrate the origin of a tornado in a supercell in the central United States (Fig. 19.27a). Recall that a supercell contains a strong rotating updraft, called a mesocyclone. (Note that the mesocyclone itself is not a tornado, but its existence plays a role in tornado formation.) Supercells in the central United States typically move from southwest to northeast, so the forward-flank downdraft forms the storm's northeastern flank, while a strong rear-flank downdraft develops on the storm's southwestern flank (Fig. 19.27b). Production of a tornado begins when the RFD reaches the ground, spreads outward laterally, and produces a horizontal roll of rotating air at its periphery (Fig. 19.28a). When part of this roll moves under the base of the mesocyclone, the rotating air along the roll gets pulled upward into the mesocyclone (Fig. 19.28b). Initially, the roll warps into an arch, but the clockwise-rotating arm soon weakens and dissipates, while the counterclockwise-rotating arm, rotating in the same direction as the mesocyclone, continues to move upward, stretching into the updraft and becoming a nearly vertical vortex (Fig. 19.28c). As the vortex then tilts with the mesocyclone, it also stretches vertically and becomes narrower, so, to conserve angular momentum (see Chapter 15), its winds accelerate dramatically. The resulting tornado extends beneath the wall cloud, which outlines the base of the mesocyclone.

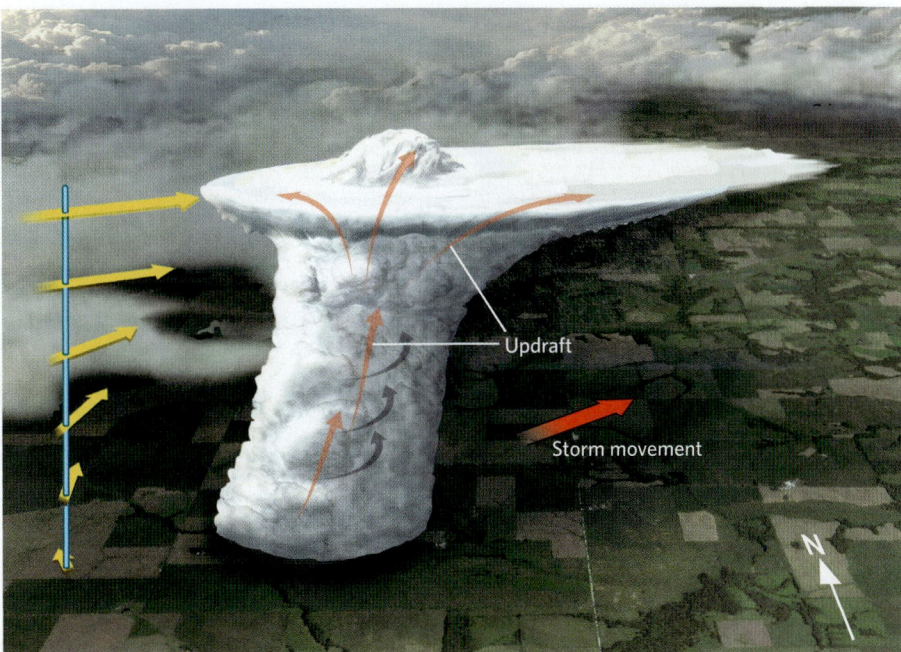

(a) A supercell storm migrates across the plains of the central United States.

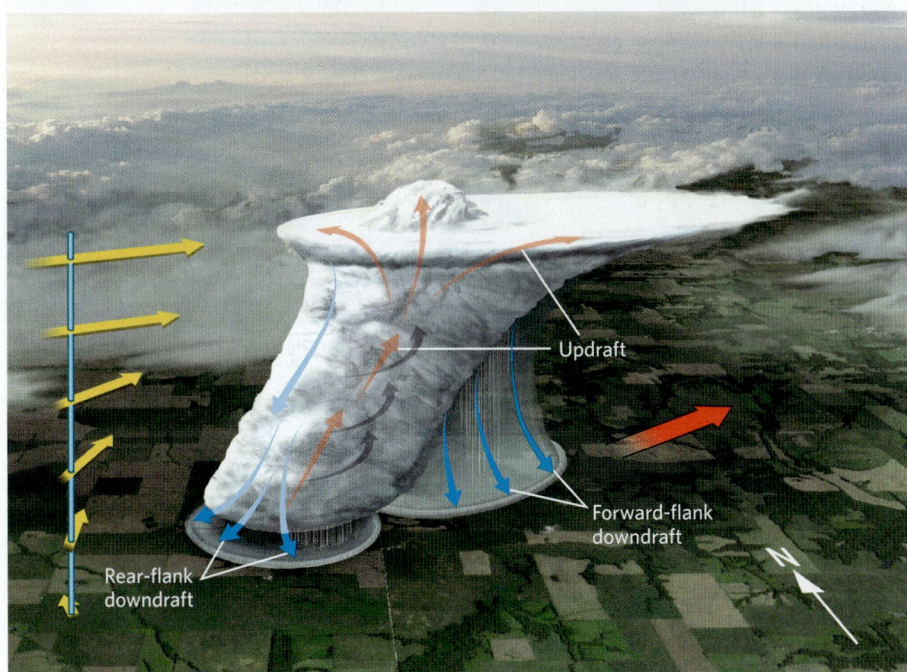

(b) Strong winds aloft cause the updraft (red arrows) to tilt. Both forward-flank and rear-flank downdrafts develop. When the rear-flank downdraft reaches the ground, it spreads outward, with some of the air flowing beneath the mesocyclone.

We can see tornadoes because rapidly rising air produces a zone of extremely low air pressure within the vortex, causing moist air that gets drawn into the tornado to condense into a spinning cloud of water droplets. Also, at the ground, the tornado rips up soil and vegetation and tears apart buildings and even highways. Some of the resulting dust and debris gets carried into the tornado, turning the vortex dark.

In some cases, a particularly large tornado evolves to include several vortices. This happens when the low pressure at the base of the vortex draws air downward within the tornado's core (Fig. 19.29a, b). The resulting downdraft pushes the sides of the tornado outward, so it becomes wider (Fig. 19.29c). When the downdraft reaches the ground, the difference in wind direction between the downdraft outflow and the rotating winds of the tornado causes the main tornado to break down into several smaller tornadoes, called *suction vortices*, moving around the main tornado (Fig. 19.29d). Such *multiple-vortex tornadoes* can be particularly destructive. Their most violent winds have the combined speeds of the suction vortices and the main tornado.

Tornado Destruction and Intensity

Tornadoes destroy property and landscapes and often kill (Fig. 19.30). Most injuries and fatalities happen when people are struck by flying debris. Even a splinter of wood, when carried by a tornado wind, becomes a deadly projectile. Sadly, about 60 people per year, on average, die in tornadoes in the United States. Some years have had particularly high death tolls due to tornadoes—in 2011, for example, 553 people lost their lives, and the 1925 Tri-State Tornado killed 747 people. Fatalities tend to be much lower now because modern warning systems allow people to take precautions (Box 19.2).

The damage caused by a tornado depends, of course, on wind velocity, but it's not possible to measure the wind velocity in every tornado because appropriate instruments

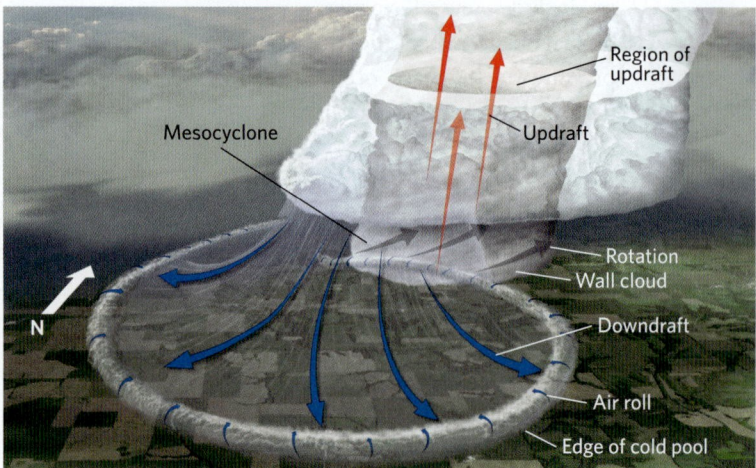

(a) The boundary of the rear-flank downdraft forms a horizontal roll of rotating air close to the ground.

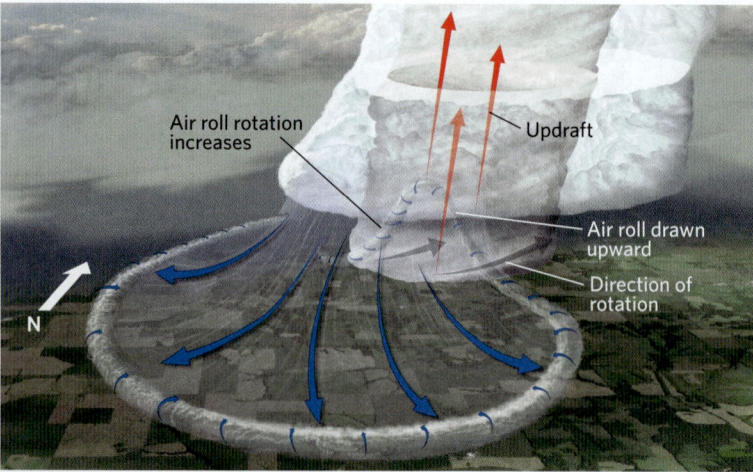

(b) The part of the roll under the updraft gets drawn upward and tilted into the updraft. As it stretches and thins, it spins faster.

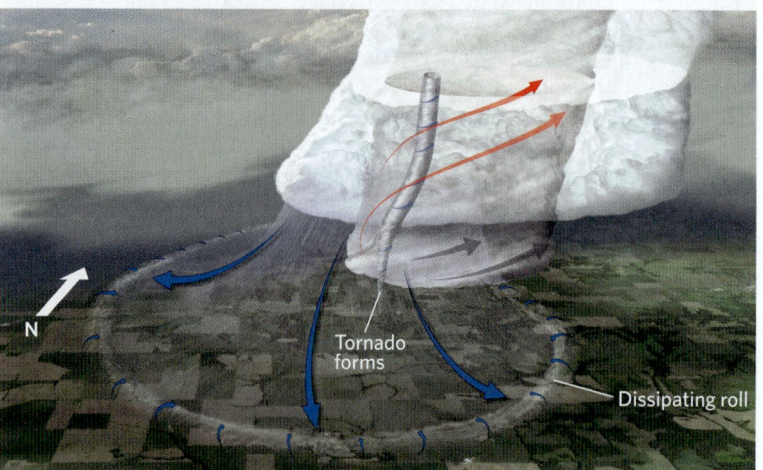

(c) The air roll tightens into a tornado and protrudes from the base of the wall cloud.

are not always nearby. To solve this problem, Ted Fujita, a professor at the University of Chicago, classified tornado intensity on the basis of the damage caused and provided estimates of how damage correlates with wind velocity. His classification system, published in 1971, came to be known as the Fujita scale. In 2007, a more accurate version of the Fujita scale, called the **Enhanced Fujita (EF) scale**, became available, and this is the scale that we use today (Table 19.1). The EF scale divides tornadoes into six categories, labeled EF0 (weak) through EF5 (catastrophic). Note that an EF rating does not depend on the width of the tornado, or on the length of its track. But in general, more intense tornadoes are wider and have longer

TABLE 19.1 The Enhanced Fujita Scale

EF Rating	Wind (km/h)	Wind (mph)	Typical Damage
EF0	105–137	65–85	*Minor damage:* Peels off some shingles; breaks branches; topples weak trees
EF1	138–177	86–110	*Moderate damage:* Severely strips roofs; overturns mobile homes; breaks windows
EF2	178–217	111–135	*Considerable damage:* Tears roofs off; shifts houses off foundations; destroys mobile homes; uproots large trees; flings debris and lifts cars
EF3	218–266	136–165	*Severe damage:* Destroys upper stories of well-built houses; severely damages large buildings; overturns trains; throws cars; debarks trees
EF4	267–322	166–200	*Extreme damage:* Completely levels well-built houses; cars and trucks thrown about
EF5	>322	>200	*Catastrophic damage:* Tall buildings collapse; structures made of reinforced concrete severely damaged; cars and trucks carried for more than a kilometer

FIGURE 19.29 Formation of a multiple-vortex tornado.

Updrafts form along the side.

A downdraft forms in the center, with eddies along its edge.

(a) A small tornado, consisting of a single vortex, forms beneath the wall cloud at the base of a supercell.

(b) A downdraft forms in the middle of the tornado and pushes the sides of the tornado outward.

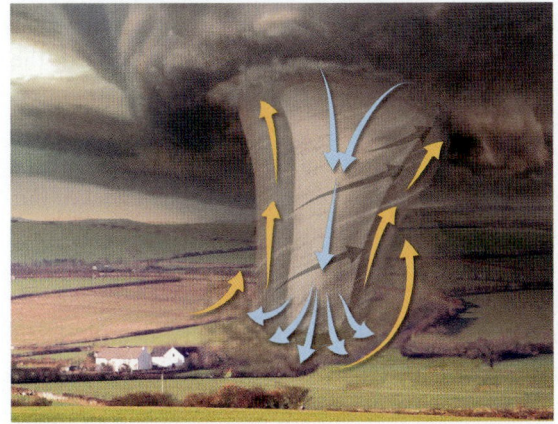

Map view at ground looking down

1 km

(c) The central downdraft reaches the ground, so the whole tornado is now almost as wide as the wall cloud.

(d) Within the overall huge tornado, a number of short-lived but very intense suction vortices form. These vortices contain the most intense winds.

FIGURE 19.30 Tornado damage.

(a) A tornado leaves a swath of destruction.

(b) During a strong tornado, trees snap off and even brick walls fall.

tracks. As a tornado evolves, its EF rating may change. The rating of record represents the worst damage the tornado causes along its track (**Earth Science at a Glance**, pp. 610–611). Notably, because the scale depends on interpreting damage to structures, a tornado that passes across an open field and causes no structural damage will be assigned EF0.

Take-home message . . .

A tornado is a rapidly rotating vortex of air that extends from the ground into the base of a severe thunderstorm. Most tornadoes form when a rotating cylinder of air along the boundary of the rear-flank downdraft of a supercell gets drawn into the updraft of the mesocyclone. As it's pulled up, the cylinder stretches, narrows, and spins faster. The Enhanced Fujita scale classifies tornado intensity on the basis of the damage caused. Doppler radar can detect tornadoes.

Quick Question

What produces the visible funnel of a tornado?

19.7 Mid-latitude Cyclones

Huge, swirling currents of air occasionally develop at latitudes between 30° and 60° north or south of the equator when tropical air masses and polar air masses interact. In the northern hemisphere, these currents are known as *cyclones* if they rotate counterclockwise and *anticyclones* if they rotate clockwise. In the southern hemisphere, cyclones rotate clockwise and anticyclones rotate counterclockwise. Cyclones and anticyclones play a major role in controlling the weather, outside of the tropics, in both hemispheres.

In some cases, cyclones grow into huge storms that span areas as large as half the width of North America. These storms, whose cloud cover resembles a huge comma when viewed from a satellite **(Fig. 19.31a)**, are called **mid-latitude cyclones**. Because they occur outside of tropical latitudes, they are also referred to as *extratropical cyclones*.

The name *mid-latitude cyclone* may be unfamiliar to you because TV meteorologists don't often use the term. But you may have endured their consequences. The fronts that trigger the severe weather events affecting large population centers and agricultural areas in North America are commonly associated with mid-latitude cyclones. Much of the weather that interrupts commerce in the winter is a manifestation of mid-latitude cyclones. Indeed, the infamous "Storm of the Century," three days of thunderstorms, tornadoes, and blizzards in March 1993, affected 40% of the American population—the winds and precipitation of the storm killed over 200 people, knocked out the power to 10 million homes, raked the east coast with storm surge and high waves, brought transportation to a standstill, and caused over $3 billion in damage. To understand such weather, it's important to understand how mid-latitude cyclones form and evolve. This requires

FIGURE 19.31 Examples of mid-latitude cyclones.

(a) A mid-latitude cyclone covers the upper Midwest and the Southeast of the United States in February.

(b) A nor'easter moves up the east coast of North America.

608   CHAPTER 19   DANGER IN THE AIR: STORMY WEATHER

BOX 19.2

Putting Earth Science to Use

Modern tornado detection and your safety

In 1925, people in the path of the deadly Tri-State Tornado had virtually no warning that it was coming, unless they happened to be outside scanning the horizon. Fortunately, the skill of tornado detection has become much more advanced, for Doppler radar (see Box 17.2) can detect rotating winds by distinguishing the part of the vortex moving toward the observer from the part moving away. In some cases, flying debris scatters radar energy and appears as a bright spot called a *debris signature* on a radar screen **(Fig. Bx19.2)**. So when a meteorologist sees a supercell thunderstorm with a hook echo on the southwestern side, rotation within the hook, and a debris signature at the tip of the hook, it's a sure sign that a tornado is on the ground and that a warning should go out. In tornado-prone areas, trained tornado spotters will also be on alert to report visual sightings of tornadoes. Tornado warnings trigger weather alerts on radio, TV, and cell phones. In some communities, emergency managers activate sirens.

The NWS recommends that families, schools, and businesses have safety plans and hold frequent drills for tornadoes. In general, when a tornado warning goes out, you should move to a predetermined shelter, such as a basement or a storm cellar. If an underground location is not available, the safest location is an interior room or a hallway on the lowest floor of a building, especially an interior bathroom, where plumbing provides additional wall support. Put as many walls as possible between you and the tornado, and avoid windows. Abandon mobile homes, which may well be crushed. If caught outdoors, move away from potential sources of airborne debris and lie in the lowest spot available. In an urban or congested area, abandon your car for a sturdy shelter.

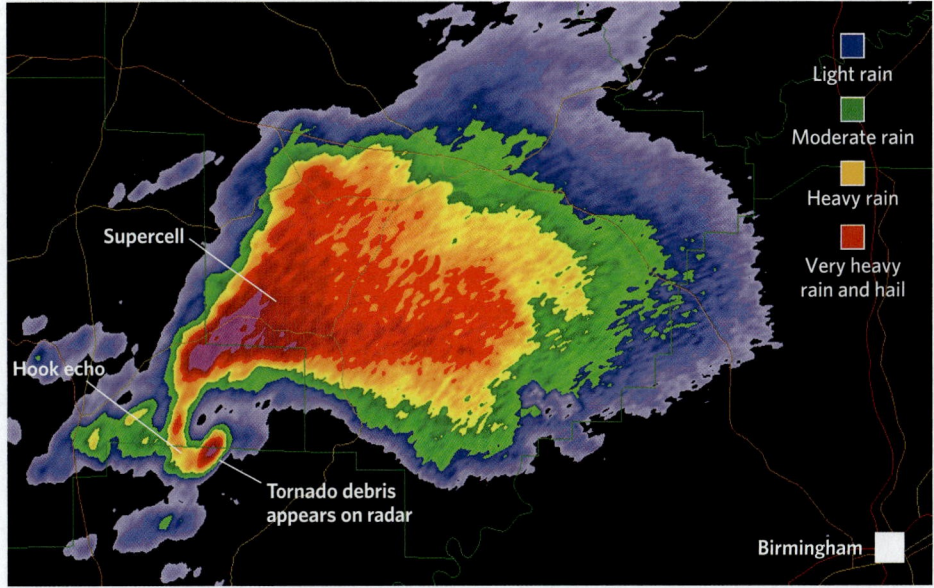

FIGURE Bx19.2 Radar signature of a tornado that struck Alabama in 2013. The bright spot at the tip of the hook echo comes from debris thrown upward by the tornado.

understanding the relationship between air flow and the formation of high- and low-pressure centers.

Mid-latitude cyclones forming along coasts have different local names. For example, a mid-latitude cyclone that develops over the northeastern Pacific Ocean, off the west coast of North America, is called a *Pacific cyclone*. When such storms reach the mountain ranges of the west coast, the moisture-laden air that they carry flows up the slopes of the mountains. This lifting triggers precipitation in the form of torrential rains or heavy snows, depending on the air temperature and altitude. A mid-latitude cyclone formed along the east coast of North America is called a **nor'easter** because its strongest winds come from the northeast, due to the overall counterclockwise flow in the storm **(Fig. 19.31b)**.

Because nor'easters blow water and waves toward the shore, they can cause significant coastal damage.

Development of High- and Low-pressure Centers

In the previous chapter, we learned that heating and cooling of the Earth's atmosphere in specific locations leads to the formation of high-pressure and low-pressure centers at the Earth's surface. The flow of the polar-front jet stream can also contribute to the development of high- and low-pressure centers. Where horizontal air flow in the jet stream slows down, air converges, or "piles up," meaning that the flow carries more air into the locality (in this case, a segment of the jet stream) than is carried

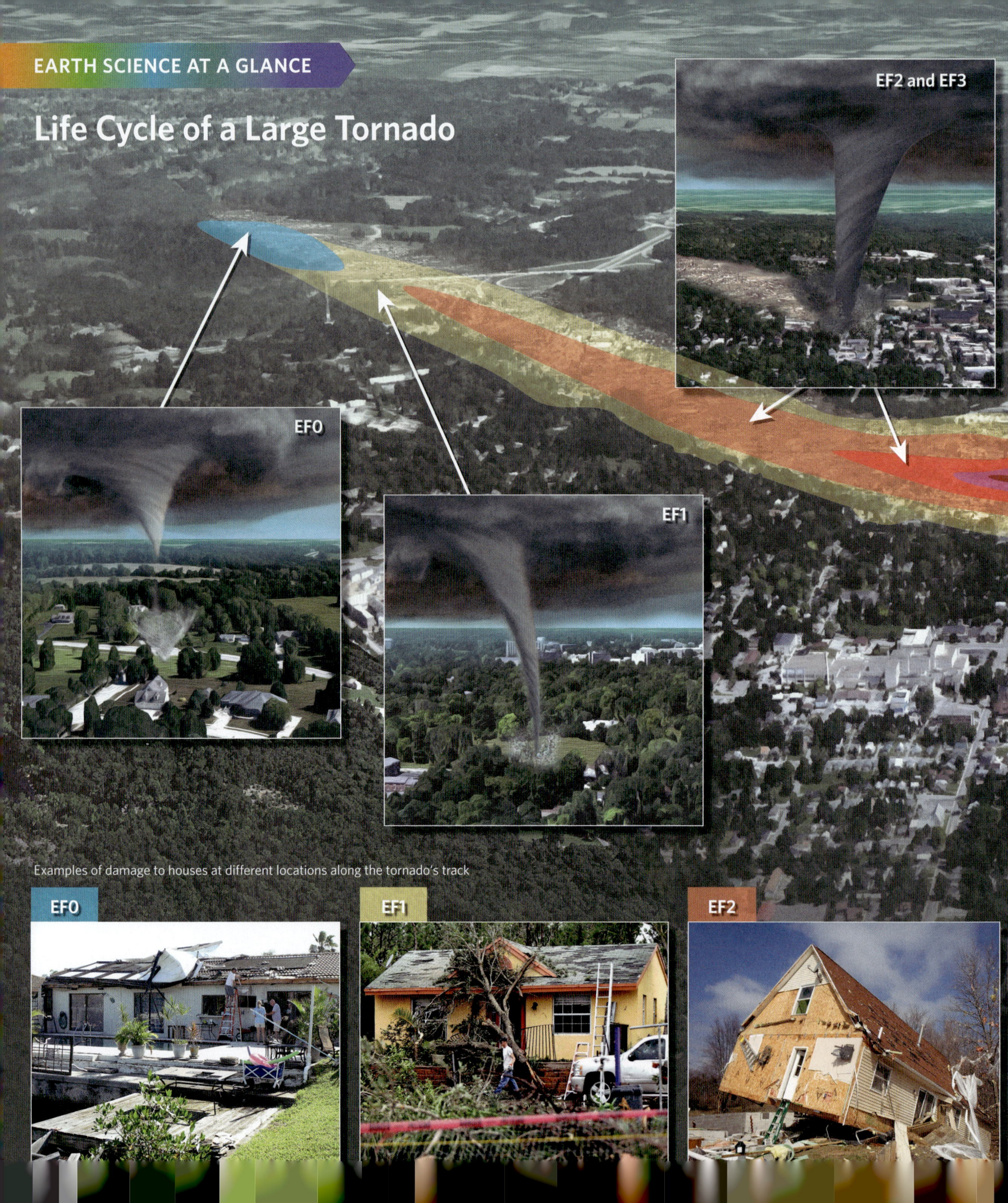

Life Cycle of a Large Tornado

EF2 and EF3

EF0

EF1

Examples of damage to houses at different locations along the tornado's track

EF0

EF1

EF2

EF4 and EF5

EF2 and EF3

EF1

A tornado evolves over time. It starts small, grows larger, and then eventually dissipates. In the record books, the overall rating of a tornado using the Enhanced Fujita (EF) scale reflects the maximum damage that the tornado causes somewhere along a portion of its path. Here, we see the swath of destruction that a tornado produced as it passed through a town, and a representation of the shape and size of the tornado at various times during its life cycle. This example was an EF-5 tornado along only a small part of its track.

EF3

EF4

EF5

FIGURE 19.32 The relationship between convergence and divergence in the jet stream and the development of high- and low-pressure centers at the surface.

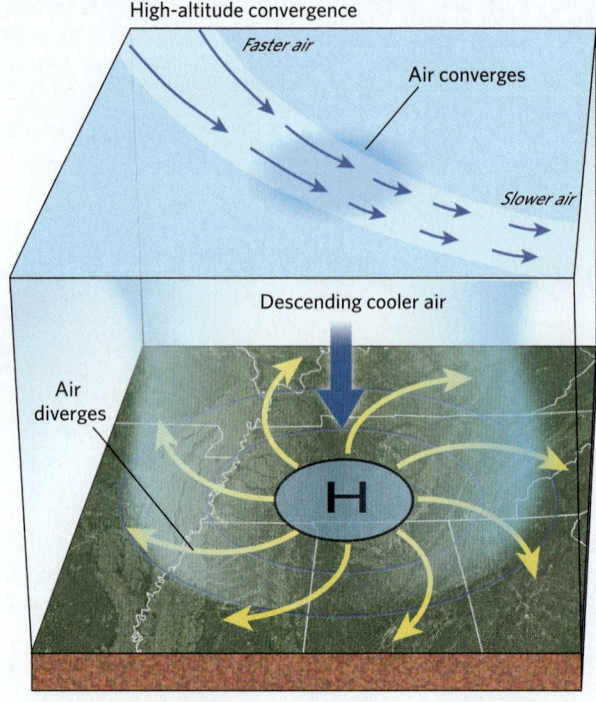

(a) Convergence aloft generates a surface high-pressure center.

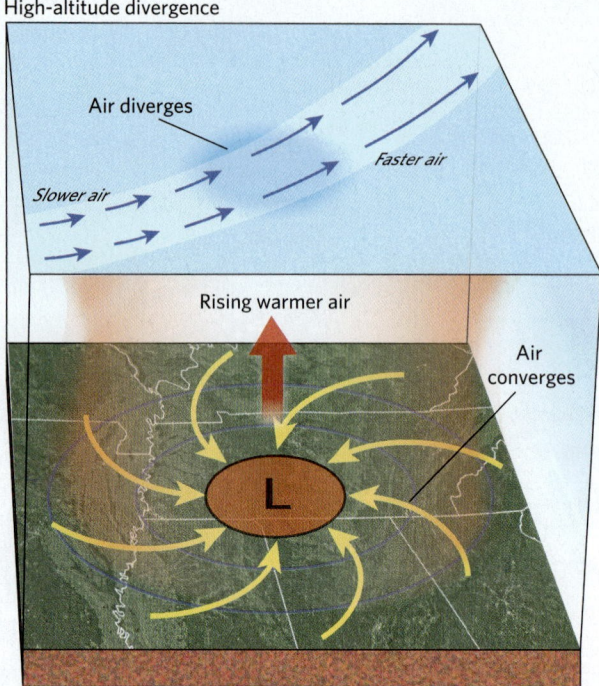

(b) Divergence aloft generates a surface low-pressure center.

away. This **convergence** of air at the top of the troposphere causes the mass of the underlying column of air to increase, so a high-pressure center develops at the base of the troposphere **(Fig. 19.32a)**. Air sinks downward

FIGURE 19.33 The relationship between the polar-front jet stream and the formation of a mid-latitude cyclone. Low pressure develops at the Earth's surface below a zone of divergence, which lies to the east of a trough and just north of the strongest winds in the jet stream.

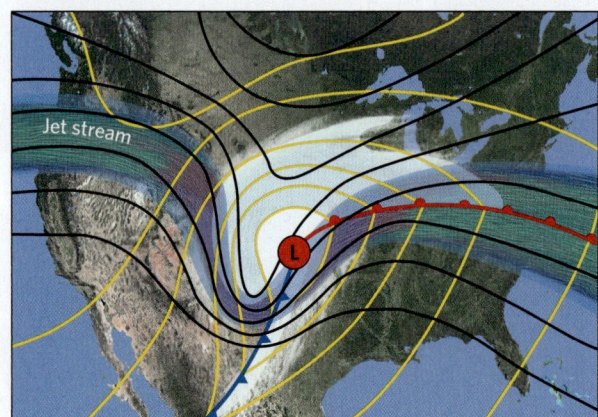

Upper-atmosphere isobars ——————
Air pressure at sea level ——————
Low-pressure center (L)

from the convergence zone aloft toward the high-pressure center at the Earth's surface below. In other segments of the jet stream, where horizontal air flow speeds up, the flow carries more air out of a locality than has entered that locality **(Fig. 19.32b)**. This **divergence** of air, and the resulting air deficit at the top of the troposphere, causes the mass of the underlying column of air to decrease, so a low-pressure center develops at the base of the troposphere. To replace the deficit of air at jet-stream altitude, air from lower in the troposphere rises in and around the low-pressure center. The high-altitude convergence and divergence zones arise from changes in jet-stream velocity where the jet stream curves around ridges and troughs, and where it flows especially fast **(Fig. 19.33)**.

Mid-latitude cyclones are associated with low-pressure systems formed in association with the polar-front jet stream. In a low-pressure system, air preferentially rises along fronts. The lifting of air along fronts within low-pressure systems produces huge comma-shaped storms like those illustrated in Figure 19.31.

Life Cycle of a Mid-latitude Cyclone

Mid-latitude cyclones begin where a strong low-pressure center develops along a front. Due to the counterclockwise convergent flow around the low-pressure center (see Fig. 19.32b), the front on the northeastern side of the

FIGURE 19.34 Evolution of a mid-latitude cyclone.

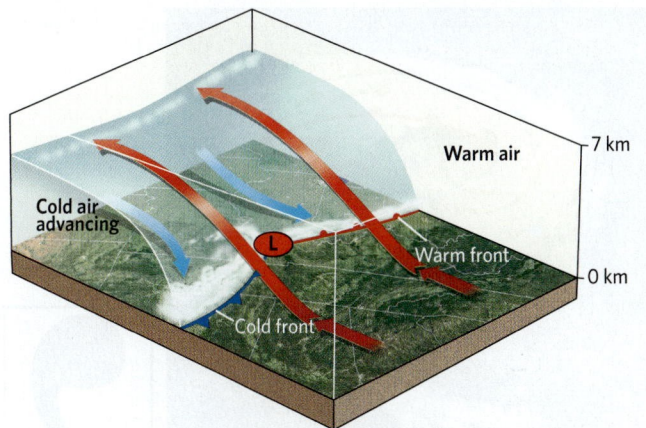

(a) Warm air initially flows upward over the fronts.

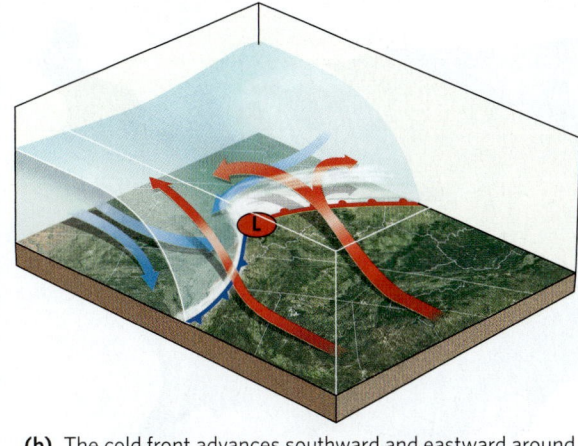

(b) The cold front advances southward and eastward around the center of low pressure.

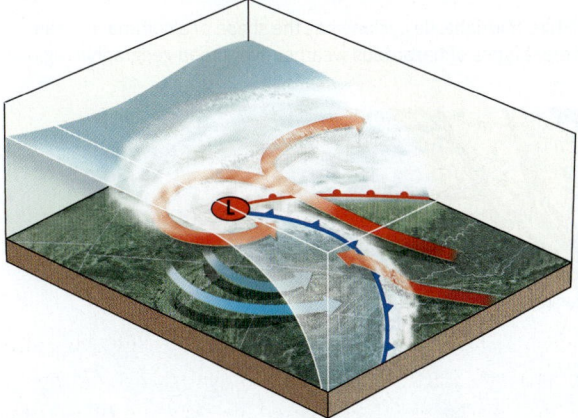

(c) Eventually, the cold front reaches the warm front.

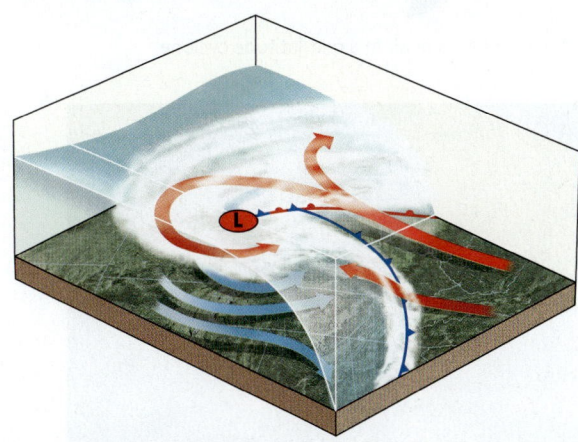

(d) The cold front overtakes the warm front, creating an occluded front.

low-pressure center becomes a warm front that retreats northward, while the front on the southern side of the low-pressure center becomes a cold front that advances southeastward. Initially, warm air ahead of the cold front undergoes lifting as the cold air advances, and warm air south of the warm front flows up and over the warm front as the cold air retreats (Fig. 19.34a). As low-altitude air starts to spiral counterclockwise around the center of the developing cyclone, the cold front advances rapidly eastward, while the warm front migrates slowly northward (Fig. 19.34b). In many mid-latitude cyclones, the cold front eventually catches up with and wraps into the warm front (Fig. 19.34c). When this happens, the cool air north of the warm front undergoes lifting along the face of the cold front, yielding an occluded front (Fig. 19.34d). Once the occluded front forms, cold air from the south side of the cold front comes into contact with cold air from the north side of the warm front, and the warm air that once lay to the south of the warm front becomes a layer that overlies both cold air masses.

A typical mid-latitude cyclone develops and intensifies rapidly, generally reaching its maximum intensity (the lowest air pressure at the low-pressure center) within 36 to 48 hours of initiation. The storm can remain at maximum intensity for another day or two, but then it begins to weaken. This weakening, which may take about a week, happens when low-altitude convergence feeds more air into the air column in the lower troposphere than gets removed by jet-stream divergence in the upper troposphere. The addition of air to the column causes the air pressure to rise at the center of the cyclone and, therefore, causes the surrounding pressure gradient to diminish. When the low-pressure center weakens sufficiently, the storm dissipates. Overall, a mid-latitude cyclone can survive as a distinct circulation for up to 2 weeks, during which it drifts eastward with the overall movement of the trough with which it is associated. As a consequence, a single cyclone can cross a large part of a continent or ocean.

The Characteristic Comma Shape of Mid-latitude Cyclones

The strong pressure gradient that develops in a mid-latitude cyclone generates strong winds that flow

FIGURE 19.35 Features of a large mid-latitude cyclone.

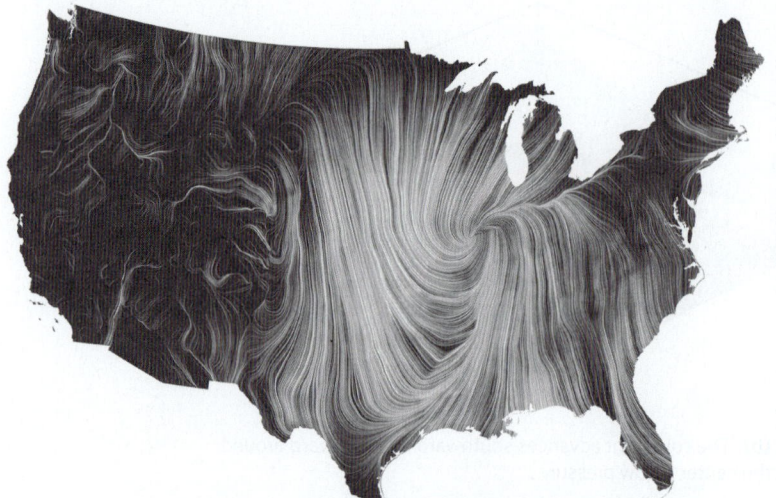

(a) A wind map shows the circulation of air in a mid-latitude cyclone.

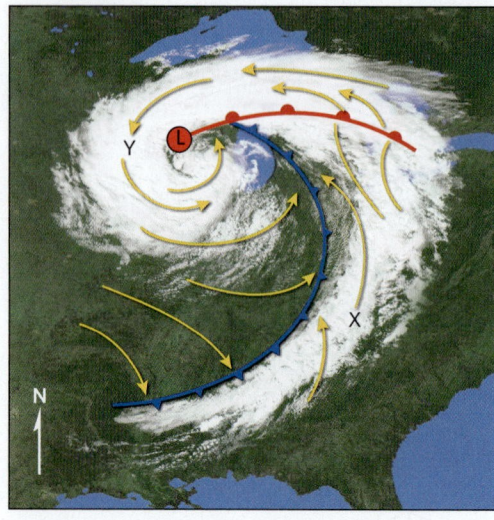

(b) A mature mid-latitude cyclone has the shape of a comma and can bring different types of hazardous weather over broad geographic regions.

(c) A thunderstorm occurs at point X, shown on the map in part (b).

(d) A blizzard occurs at point Y, shown on the map in part (b).

counterclockwise (in the northern hemisphere) at ground level within the cyclone's overall spiral flow (Fig. 19.35a). Convergence at the low-pressure system at the center of the cyclone causes both lifting along the system's fronts and rising of low-elevation moist air, producing broad areas of cloud cover and precipitation in the shape of a comma (Fig. 19.35b). The "tail" of the comma lies over the cold front and typically consists of a line of precipitating clouds or, in some cases, thunderstorms (Fig. 19.35c). The "head" of the comma lies over the low-pressure center as well as the occluded front and the warm front. It, too, produces large amounts of precipitation. In warmer weather, the precipitation falls as rain, but when the weather is cold enough, the precipitation falls as snow (Fig. 19.35d). During fall, winter, and spring, rain may be falling along the comma's tail, while snow falls around the comma's head, due to the poleward decrease in temperature characteristic of mid-latitude regions.

Blizzards and Ice Storms during Mid-latitude Cyclones

A gentle snowfall brings delight, as the shiny flakes that float down from the sky glint in the light and seem to blanket the surroundings with calm and quiet. A blizzard, however, is anything but calm and quiet. Meteorologists formally define a **blizzard** as a snowstorm during which winds exceed 56 km/h (35 mph) for a period of at least 3 hours and visibility decreases to less than 0.40 km (0.25 mi). Blizzards sometimes lead to *white-out conditions*, in which visibility drops to nearly zero (Fig. 19.36). Though the definition of a blizzard does not specify an amount of snowfall, blizzards may drop 30 cm (1 ft) of snow or more. Heavy snowfall in a blizzard can accumulate quickly, and snow blown by the wind—either during or after a snowfall—can build high *snowdrifts* (dunes made of snow). The combination of snowfall, drifting, and decreased visibility can force airports to close and can trap cars on highways.

The extreme winds in a blizzard can also cause people to suffer from *frostbite* (freezing skin) or *hypothermia* (lowering of body temperature), for the heat loss that people experience when it's windy is greater than that at the same air temperature in still air. For example, an air temperature of −7°C (20°F) feels like −16°C (−4°F) when the wind blows at 32 km/h (20 mph). This **wind chill** happens because flowing cold air removes heat more efficiently from the human body than does calm cold air. Weather reports in the media often specify the *wind-chill temperature*, the temperature that you feel when you go outside, so that you can be prepared.

In regions where the temperature is hovering around freezing, a mid-latitude cyclone may produce an **ice storm**, an event in which precipitation lands on the ground (or on trees and houses) as water but instantly freezes to ice. Such *freezing rain* produces an ice glaze that makes roads extremely slippery. Ice storms happen when snow falling from high in the clouds passes downward into a lower atmospheric layer—typically 0.5–1.5 km (0.3–1 mi) above the ground—where the temperature exceeds 0°C (32°F) **(Fig. 19.37a)**. As a consequence, the snowflakes melt and become very cold raindrops. If a *temperature inversion* exists in the lowest part of the atmosphere—meaning that the temperature increases with altitude—the temperature in the layer of air below

FIGURE 19.36 During a blizzard, visibility falls to near zero.

an altitude of about 0.5 km can remain below freezing. If this happens, the raindrops refreeze when they strike the ground. The ice that builds up on trees and power lines can become so heavy that it causes branches to snap off or, in some cases, whole trees to topple, and may cause power lines to collapse, producing blackouts that last for days **(Fig. 19.37b)**.

FIGURE 19.37 Ice storms are another winter hazard associated with mid-latitude cyclones.

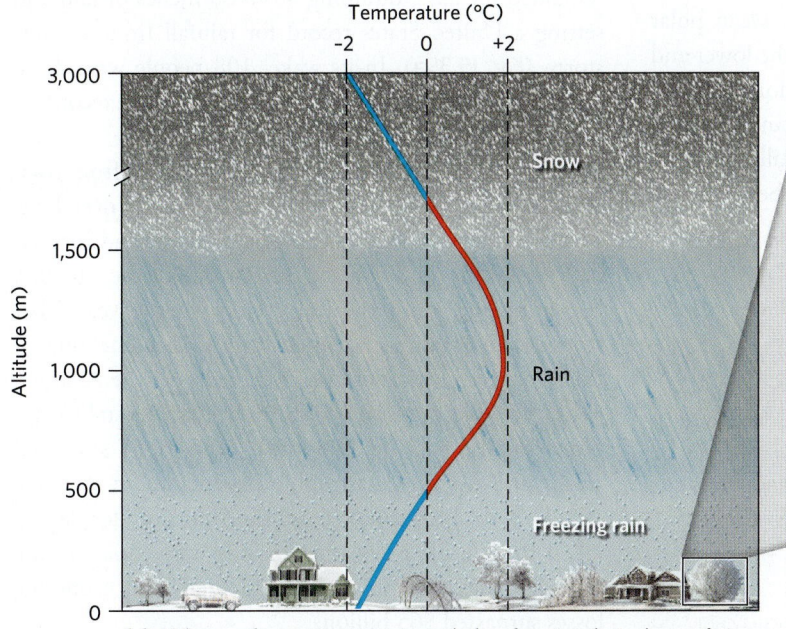

(a) When surface temperatures are below freezing, but a layer of air aloft is above freezing, snow from high in the clouds can melt into rain and then freeze when it hits the ground.

(b) The weight of accumulated ice can cause trees and power lines to collapse.

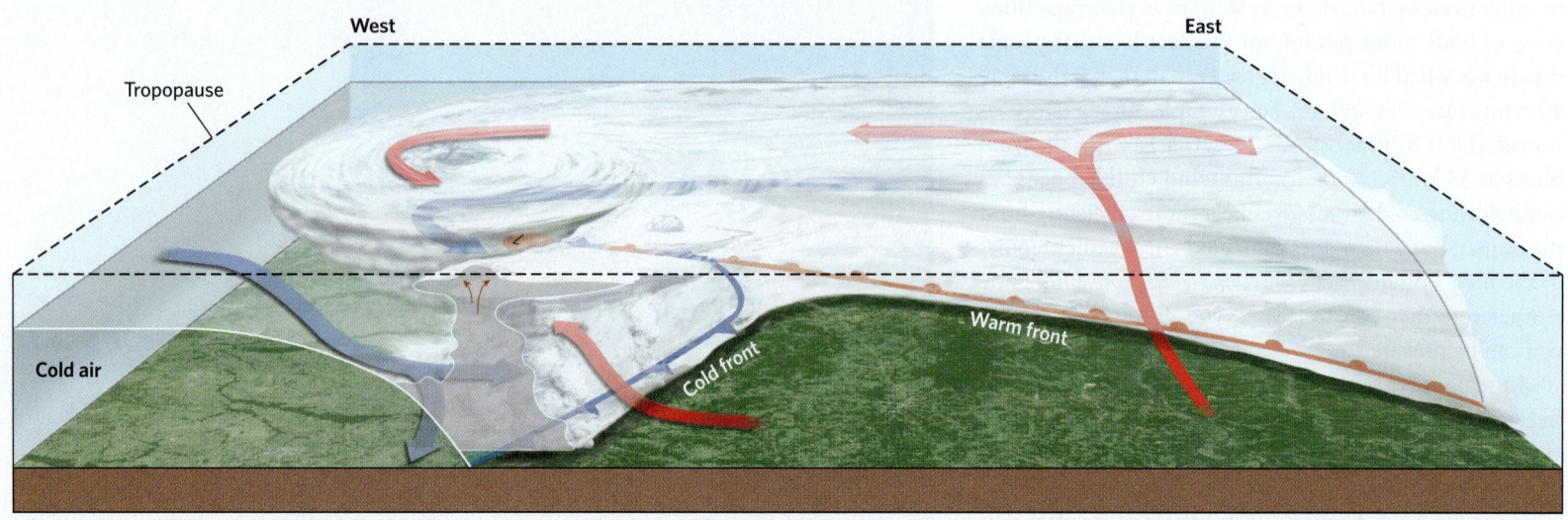

Anyone who says they're not afraid of a hurricane is either a fool or a liar, or a little of both.

—ANDERSON COOPER (AMERICAN NEWSCASTER, 1967–)

Redistribution of Heat by Mid-latitude Cyclones

Before finishing our discussion of mid-latitude cyclones, let's consider the role that these storms play in redistributing atmospheric heat across the Earth **(Fig. 19.38)**. Prior to the formation of a mid-latitude cyclone, warm, humid air lies near the Earth's surface at subtropical latitudes, while cold air lies near the surface at high latitudes. In addition, cold air occurs high in the troposphere. When a mid-latitude cyclone forms, warm air flows northward as well as upward, while cold air flows southward as well as downward. This conveyor-belt-like redistribution of heat reduces the temperature difference between polar and subtropical regions, as well as between the lower and upper troposphere. In this manner, each cyclone contributes to a never-ending process of rebalancing the heat distribution in an atmosphere that continually becomes unbalanced due to uneven solar heating between the polar and tropical regions of the Earth.

Take-home message . . .

Mid-latitude cyclones are large, counterclockwise-rotating systems that can result in hazardous weather across the mid-latitudes. They form in association with low-pressure systems and generally move eastward. Their most dangerous weather occurs along fronts, which move as the result of air flow within the cyclone.

Quick Question

How do mid-latitude cyclones play a role in redistributing atmospheric heat?

19.8 Tropical Cyclones: Hurricanes and Typhoons

The year 2017 was one that would go down in history for hurricane-related calamity. On August 26, Hurricane Harvey, the eighth named tropical cyclone of the 2017 Atlantic season, slammed into the Texas coastline near San Jose Island, causing coastal flooding and widespread wind damage. The storm weakened rapidly, drifting northeastward along the Texas coastline, before finally stalling over the Houston metropolitan area. There it remained for days, dumping 40 to 60 inches of rain and setting a United States record for rainfall from a single storm **(Fig. 19.39a)**. In its wake, 108 people were dead. Economic losses reached $105 billion, tying the record set by Hurricane Katrina in 2005.

But the weather disasters of 2017 were far from over. As September dawned and Harvey dissipated over Louisiana, a mass of clouds swirled, organized, and intensified near Cape Verde on the eastern side of the Atlantic, becoming the ninth tropical cyclone of the year. That storm, Hurricane Irma, intensified to become one of the strongest hurricanes ever observed over the Atlantic. While at peak intensity, the storm overran several Caribbean islands, packing 298-km/h (185-mph) sustained winds and leaving catastrophic destruction in its wake **(Fig. 19.39b)**. Irma then crossed the Straits of Florida and raked the Florida Keys and Florida's west coast. By the time it dissipated,146 people had died, and economic losses surpassed $65 billion.

As Irma was dissipating, yet another storm, Hurricane Maria, was forming. This storm grew to be even

FIGURE 19.39 Destructive hurricanes in 2017.

(a) Flooding in Port Arthur, Texas, near Houston, during Harvey.

(b) Buildings destroyed by Irma in the Virgin Islands.

(c) Power lines downed by Maria along a street in Puerto Rico.

What Makes a Hurricane a Hurricane?

Atmospheric scientists use the term **tropical cyclone** for any rotating spiral-shaped storm that originates over warm tropical ocean waters. In the northern hemisphere, these storms rotate counterclockwise, and in the southern hemisphere, they rotate clockwise. Very weak tropical cyclones, with sustained winds of less than 63 km/h (39 mph), are *tropical depressions*, and those with sustained winds between 63 and 118 km/h (39–73 mph) are **tropical storms**. If a tropical storm in the North Atlantic or eastern Pacific strengthens to the point that it has sustained winds of over 119 km/h (74 mph), it becomes a **hurricane**. The same type of storm is called a **typhoon** if it forms over the northwestern Pacific, and simply a **cyclone** if it develops over the Indian Ocean or anywhere in the southern hemisphere. For simplicity, we'll generally refer to all strong tropical cyclones as hurricanes in this chapter.

In a hurricane, only the central region, with an area of 200 km (120 mi) in diameter or less, hosts hurricane-force winds. But the entire storm, in which strong winds spiral around the center and dense clouds hide the sky, may cover an area between 500 and 1,500 km (300–900 mi) in diameter. Because they affect such a broad area, hurricanes are the most destructive storms on the planet. About a hundred tropical cyclones originate every year over the world's tropical oceans and seas.

stronger than Irma, and when it struck Puerto Rico and Dominica, it destroyed much of the islands' infrastructure, going on record as the worst-ever natural disaster of the region. Parts of Puerto Rico remained without power more than a year later, and the economic losses to Puerto Rico and the nearby US Virgin Islands exceeded $90 billion **(Fig. 19.39c)**.

Clearly, hurricanes are extremely dangerous storms. They threaten ships at sea, coastal communities in tropical and mid-latitudes, and in many cases inland regions as well. Everyone has heard of hurricanes. But what exactly makes a given storm a hurricane? Here, we define this type of storm, and we explain how it forms and why it can cause so much damage.

TABLE 19.2 The Saffir-Simpson Scale for Hurricane Intensity

Category	Sustained Wind Speed	Types of Damage Resulting from Wind
1	74–95 mph 64–82 knots 119–153 km/h	**Dangerous winds produce some damage:** Well-constructed frame homes could have damage to roof shingles, vinyl siding, and gutters. Large branches of trees snap, and shallowly rooted trees may topple. Some power outages occur.
2	96–110 mph 83–95 knots 154–177 km/h	**Extremely dangerous winds cause extensive damage:** Well-constructed frame homes sustain major roof and siding damage. Many shallowly rooted trees are snapped or uprooted, blocking roads. Near-total power outages occur.
3	111–129 mph 96–112 knots 178–208 km/h	**Devastating damage occurs:** Well-built frame homes may incur major damage or removal of roof decking. Large trees snap or are uprooted. Numerous roads become blocked. Electricity and water are unavailable for days to weeks.
4	130–156 mph 113–136 knots 209–251 km/h	**Catastrophic damage occurs:** Well-built homes sustain severe damage. Most trees snap or are uprooted, and most power lines are downed. Debris isolates residential areas, and power outages last weeks to months. The area becomes temporarily uninhabitable.
5	≥157 mph ≥137 knots ≥252 km/h	**Total catastrophic damage occurs:** Most homes are destroyed. Only strongly reinforced buildings remain standing. Debris isolates large areas, and power outages last for weeks to months. Most of the area remains uninhabitable for weeks or months.

Meteorologists at the NWS assign names to Atlantic and eastern Pacific hurricanes. The first hurricane of the season gets a name that starts with the letter A, and each successive hurricane gets one that starts with the next letter of the alphabet, skipping some letters, such as Q, that are uncommon for names. To communicate these storms' severity to the public, meteorologists use the **Saffir-Simpson scale**, which rates hurricanes on the basis of their maximum sustained wind speed (Table 19.2). In this scale, the strongest hurricanes are Category 5.

What's Inside a Hurricane?

Well-developed hurricanes have a distinct internal structure (Fig. 19.40). The top of a hurricane, at the level of the tropopause, looks like a broad sheet of thick clouds. At the center of this cloud sheet is a nearly cloud-free circular area, typically between 10 and 50 km (6–30 mi) across at the top, called the **eye**. In three dimensions, a hurricane's eye resembles a vertical downward-tapering funnel, extending from the top of the storm to an elevation of less than 1 km (0.6 mi) above sea level. Its face, the **eye wall**, consists of dense, rapidly rotating clouds. *Spiral rainbands*, distinct arcs of narrow but tall clouds, some of which contain thunderstorms, extend from the eye wall outward for a distance of several hundred kilometers, and define the spiral shape of the hurricane. These thunderstorms can be detected from space because their overshooting tops bubble up above the high-altitude clouds at the tropopause. Torrential rain falls from the eye wall clouds, and very heavy rain falls from the spiral rainbands. Tornadoes sometimes form at the bases of thunderstorms within the spiral rainbands. In the gaps between rainbands, there's less rain or no rain, and no rain

at all falls right beneath the eye. In fact, someone standing beneath the eye may see blue sky above.

The structure of a hurricane reflects the movement of air within the storm. Air near the Earth's surface spirals inward toward the eye of the hurricane. As it does so, it moves progressively faster, like a spinning skater who moves faster as she pulls her arms inward, to conserve angular momentum (see Box 15.3). At the eye wall, the winds reach their greatest velocity. Air then rises along an ascending spiral in the eye wall until it reaches the tropopause. From the tropopause, nearly all the rising air then spirals outward, eventually sinking back toward the surface well beyond the margins of the hurricane. A small amount of air from the eye wall flows inward toward the eye, and within the eye, this air slowly descends—as it does so, it undergoes compression and warms, so its relative humidity decreases, which is why the eye remains nearly clear. Air also ascends in the rainbands, then generally sinks slowly between them.

Where Does the Energy in a Hurricane Come From?

Energy from the ocean transfers into the base of the atmosphere in two ways. A small portion of the energy in a hurricane comes from the conduction of heat from warm water into the overlying air. The rest of its energy comes from the vast amount of latent heat stored in water vapor that has evaporated from the sea surface and has been carried into the storm. Condensation of water vapor to form water droplets in the hurricane's clouds releases this latent heat, providing the thermal energy that drives the storm. Put another way, evaporated ocean water incorporated into the storm supplies the storm's energy. The rate at which water evaporates from the ocean surface increases substantially as wind speed increases, for the froth and spray generated when wind tears at the water surface and builds huge waves can increase the rate of water vapor transfer to the atmosphere by a factor of 100 to 1,000.

Because evaporating water is so importance in fueling hurricanes, such storms can't form over land. In fact, when a hurricane moves over land, it loses strength and eventually dissipates. And, because warm ocean water provides the energy for the hurricane, the storms grow only over ocean water that has a temperature above 26°C (79°F) for a depth of at least 60 m (200 ft). Because water attains such temperatures only in the tropics, almost all hurricanes form in the tropics (which is why they are called *tropical cyclones*). When they move to higher latitudes over colder water, they weaken and dissipate.

How Do Hurricanes Form?

A tropical depression begins when convergence of air currents at sea level forces moist air upward. As we have seen, condensation in rising moist air forms clouds and releases latent heat. This latent heat warms the air and makes it buoyant, so it continues rising to the tropopause and produces thunderstorms. A cluster of thunderstorms developing over the warm sea serves as the "seed" from which a hurricane can grow.

The convergence of air that forms hurricane-spawning tropical thunderstorms takes place for two reasons. First, convergence occurs along the ITCZ, where trade winds of the northern hemisphere converge with those of the southern hemisphere. This convergence serves as the primary source of Pacific typhoons and Indian Ocean cyclones. Second, convergence occurs where *easterly waves* develop within the trade winds. These air circulations are called "easterly" because their wave-like shape, as seen in map view, moves progressively from east to west, and "waves" because they represent north-south oscillations in the direction of the wind (similar to the ridges and troughs of a jet stream, but in the lower atmosphere). You can see their wave-like shape by looking at a map of variations in wind direction (Fig. 19.41). Easterly waves, which develop when storms originating in Africa move westward into the Atlantic, are the cause of most Atlantic hurricanes. Convergence of air within easterly waves occurs because (in the northern hemisphere) air moves somewhat faster through the southern part of the wave than through the northern part, causing air in the south-to-north part of the flow to pile up and rise, triggering thunderstorms.

What makes a cluster of thunderstorms over warm ocean water grow into a tropical depression and then a tropical cyclone? As the thunderstorms grow, their winds build waves and spray, so more water vapor enters the air. As the water vapor condenses and its latent heat warms

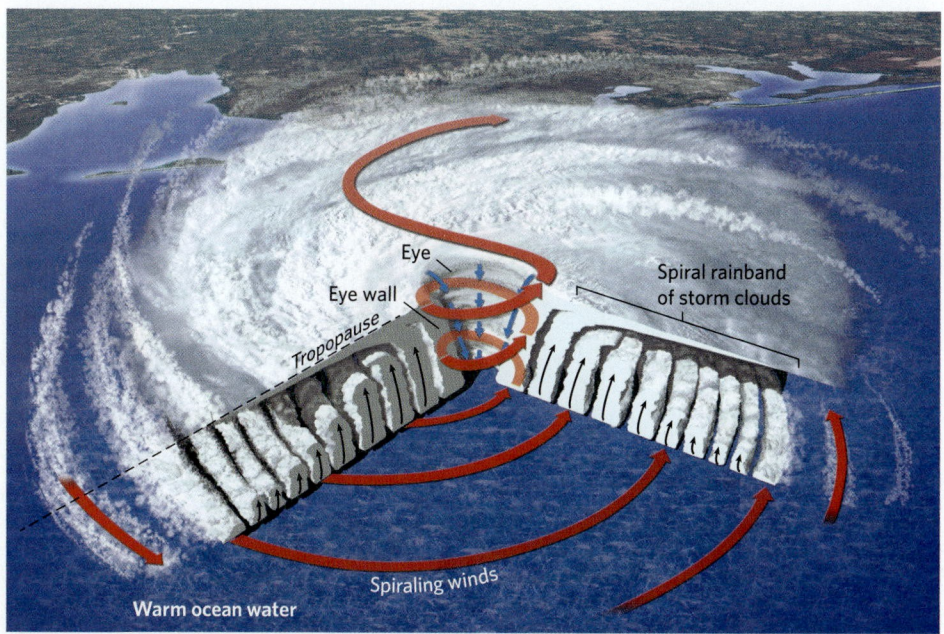

FIGURE 19.40 Internal structure of a hurricane.

(a) Bands of storm clouds spiral toward a central eye. The strongest winds occur within the eye wall.

(b) Hurricane Katrina over the Gulf of Mexico.

A radar map of Katrina

the air, the warmed air expands and rises, and air pressure in the storm decreases. Consequently, the pressure gradient associated with the storm becomes steeper, and steepening of the pressure gradient, in turn, causes winds to become stronger, while the Coriolis force causes the air to rotate around the developing low-pressure center (see Chapter 18). Overall, therefore, storm growth represents a *positive feedback* process, meaning that each step enhances the process and amplifies the consequences. In addition, for a tropical storm to grow into a hurricane, winds at high altitude must be weak, for if strong winds existed aloft, the spiraling circulation of the storm would be torn apart.

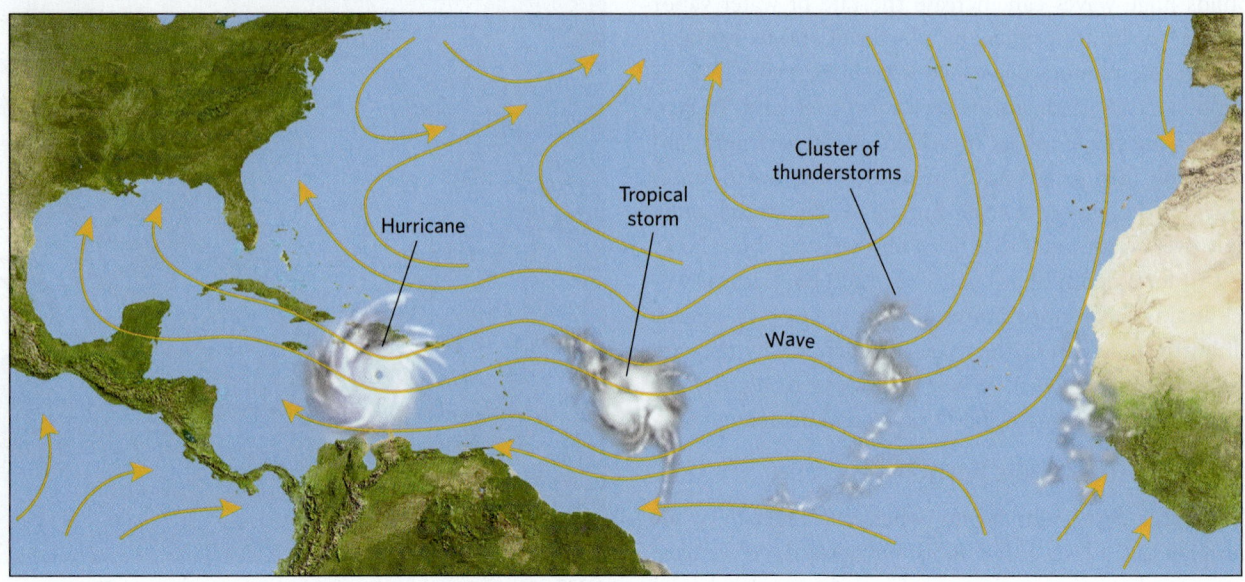

FIGURE 19.41 A typical Atlantic hurricane evolves from a cluster of thunderstorms off Africa into a tropical depression, then a tropical storm, and finally a hurricane. The thunderstorms are associated with convergence within easterly waves in the trade winds.

As a tropical storm begins to rotate faster, air pressure at the base of the storm's center decreases. This decrease starts to develop because, from the viewpoint of a moving air parcel, an outward-directed centrifugal force, due to the rotation of air in the storm, drives air outward, away from the center of the storm, much as a ball on a string moves outward when you start swinging it around your head. This outward air movement, or *divergence*, reduces the weight of the air column at the center of the storm, and therefore reduces the pressure at the base of the storm's center. Once this process has started, the downward pull of gravity overcomes the upward push of the vertical pressure gradient in the storm's center. As a result, air from high elevations begins to sink slowly downward in the storm's center. As this air sinks, it undergoes compression and therefore warms and dries. The warming causes the air to expand, and the expansion moves more air molecules sideways, out of the storm's center, further reducing the air pressure below the center. Meanwhile, the drying causes clouds to dissipate, so the center becomes a visible eye. When air pressure has become low enough for an eye to form, an extreme pressure gradient exists between the storm's center and the region outside. Because winds blow due to pressure gradients, the greater the pressure gradient, the faster the wind—the extreme pressure gradient across an eye wall accelerates air sufficiently to develop hurricane-strength winds from the eye wall outward for as much as 100 km (60 mi) in a strong hurricane.

Significantly, tropical cyclones do not develop directly over the equator, or cross the equator, despite its plentiful supply of warm seawater. In fact, for a strong composite cluster of thunderstorms to start rotating and become a tropical cyclone, it must lie at least 5° north or south of the equator. Why? The rotation of a cyclone is due to the Coriolis force. The Coriolis force is zero at the equator, and it becomes strong enough to cause rotation only at a latitude of 5° (see Chapter 15).

How Do Hurricanes Cause Damage?

WIND DAMAGE. We've already seen that the winds in the eye wall of a hurricane have immense power. During Hurricane Patricia in 2015, winds over the eastern Pacific reached a record 345 km/h (245 mph)—virtually no building can remain standing in such winds (see Table 19.2). The wind speed at a location within a hurricane depends on two factors: first, the speed due to the spiraling of the air around the hurricane, which decreases with distance from the eye wall; and second, the speed of the overall movement of the hurricane as the entire storm drifts along, propelled by the regional atmospheric circulation. Typical forward speeds of hurricanes range between 15 and 90 km/h (10–60 mph), though some hurricanes stall so that the eye remains stationary for a while. As a consequence of a hurricane's forward movement, winds on the right side of a counterclockwise-rotating hurricane (as viewed looking in the direction the hurricane is moving) are stronger than those on the left side. For example, if a hurricane rotates at 150 km/h while moving forward at 40 km/h, the wind speed on the right side will be 150 + 40 = 190 km/h, whereas on the left side the wind speed will be 150 − 40 = 110 km/h (Fig. 19.42). For this reason, the side of a hurricane rotating

620   CHAPTER 19   DANGER IN THE AIR: STORMY WEATHER

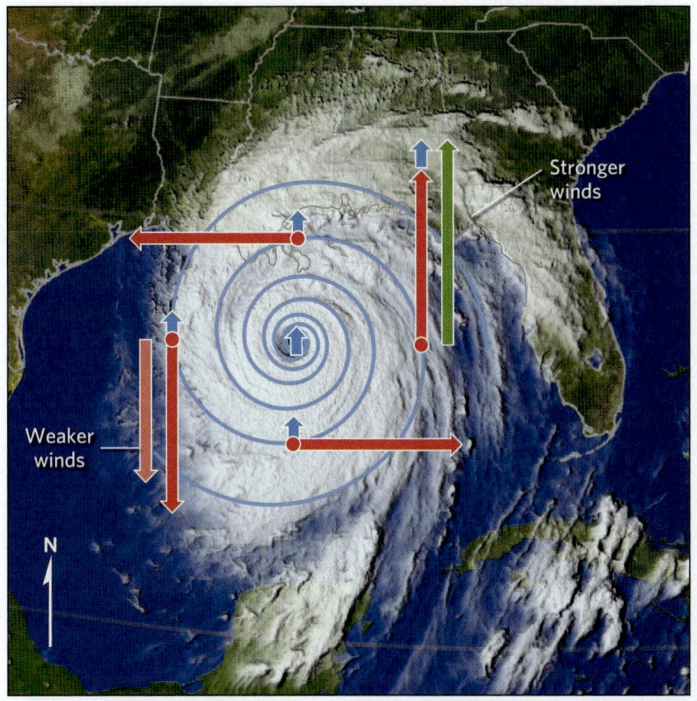

in the direction of its forward motion causes more wind damage than the side rotating in the direction opposite its forward motion.

DAMAGE DUE TO WAVES AND STORM SURGE. Because hurricanes come from the sea, they can cause immense damage to the coast, not only because of towering waves, but also because of **storm surge**, a dome of water that forms in front of a hurricane (Fig. 19.43a). Storm surge develops because the wind of the storm drives water in front of it, effectively building a pile of water. To a lesser extent,

FIGURE 19.43 Storm surge and its consequences.

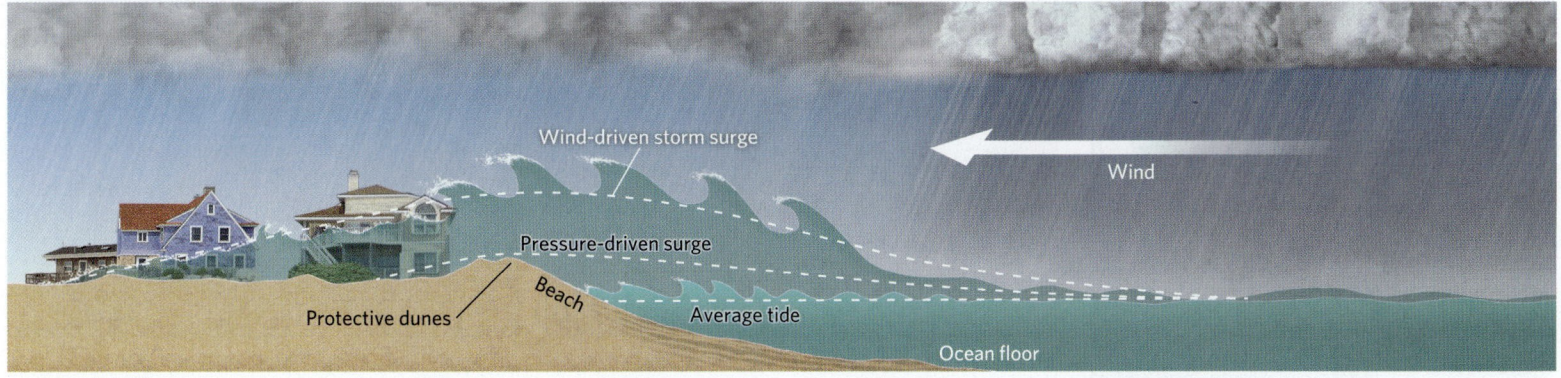

(a) Wind and low atmospheric pressure combine to produce a bulge of water.

(b) Storm surge accompanying a hurricane floods coastal areas and leaves behind devastation.

FIGURE 19.44 Hurricane tracks.

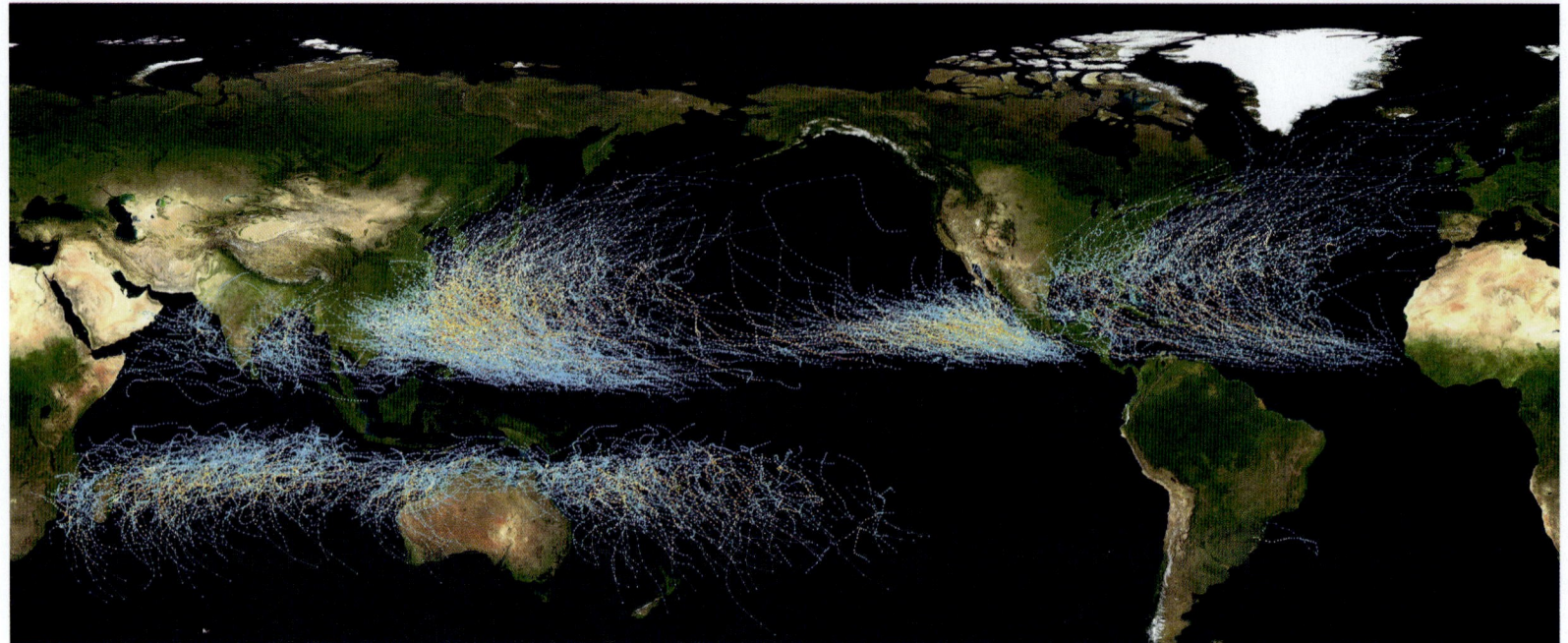

(a) Tracks of all documented hurricanes. Light blue lines are weaker storms, and yellow lines are stronger storms. Note that none of these tracks crosses the equator.

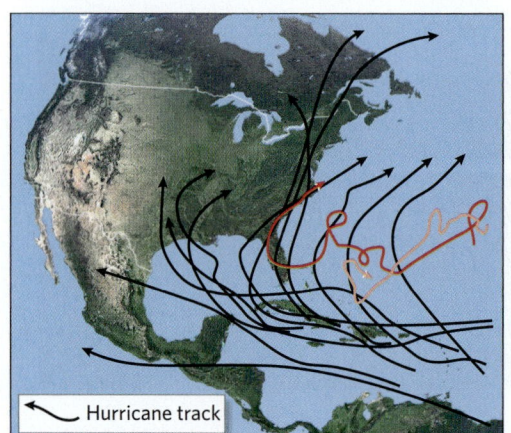

(b) Tracks of representative Atlantic hurricanes. Most head west and then curve north.

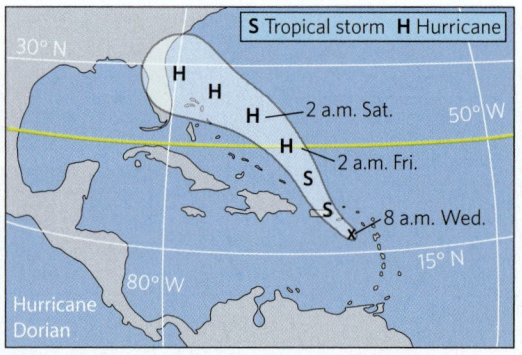

(c) Prediction of the track of Hurricane Dorian in 2019, five days before it reached the U.S. coast. The width of the track indicates uncertainty, not storm size.

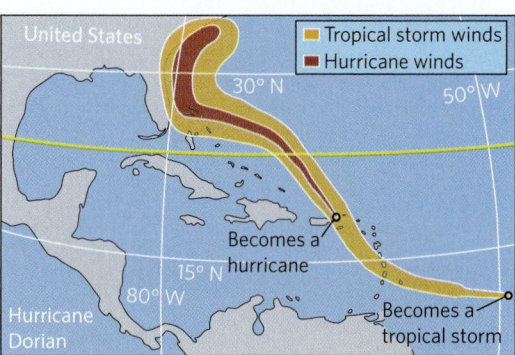

(d) A wind-swath map indicates the actual track of the storm after it has passed. The colors represent the areas affected by winds of a given strength.

it develops because the low atmospheric pressure beneath the storm reduces the weight of the atmosphere at the sea surface, allowing the sea surface to rise. The consequences of storm surge can be catastrophic **(Fig. 19.43b)**. When a cyclone moved over the Ganges River delta, at the head of the Bay of Bengal, in 1970, storm surge lifted sea level by about 6 m (20 ft), and 10-m (33-ft)-high waves built on top of the surge. As a result, water inundated the delta, and 300,000 people drowned. If storm surge arrives at high tide, the water level will become even higher. Storm surge was responsible for the failure of New Orleans's levees during Hurricane Katrina and for the destruction along coastal New Jersey during Hurricane Sandy in 2012.

DAMAGE DUE TO RAINS INLAND. Though hurricane winds slow when the storms head over land, their clouds can still dump torrential rains. For example, when Hurricane Mitch stalled over the mountains of Central America in 1998, the downpours it released caused numerous flash floods. Rainwater also soaked into the ground and saturated the soil, causing unstable slopes to give way. Some of the resulting mudslides buried whole towns.

Hurricane Tracks

Once a hurricane exists, its "engine" will run as long as it has fuel (warm ocean water) and there are no strong winds in the upper troposphere to break apart the storm's spiral circulation. The temperature and supply of warm water, the character of the surrounding winds, the thickness of the troposphere at the latitude where the hurricane resides, and friction with the Earth's surface all control the strength of a hurricane.

Hurricanes move over the course of their lifetime, tracing out a path known as a **hurricane track**. Because hurricanes form in the tropics, within the belt of westward-flowing trade winds, they normally drift westward as they form and mature (Fig. 19.44a). In the northern hemisphere, hurricanes may start to drift northward as they move westward, a consequencs of the Coriolis force (Fig. 19.44b). When they reach a latitude of about 30°, they come under the influence of mid-latitude winds, which generally blow from west to east. As a result, hurricanes in the northern hemisphere typically move eastward after passing a latitude of 30° N. Knowledge of these principles allows meteorologists to predict the hurricane track, and they can even estimate the width of the belt in which hurricane-force winds will occur. Consequently, many countries have set up monitoring centers that use data from satellites, radar, weather stations, and aircraft to produce computer models of hurricane tracks (Fig. 19.44c, d). The reliability of these tracks decreases with time from the prediction, but they provide emergency managers with guidance on where to begin evacuations and deploy resources as these monstrous storms approach the coast.

Take-home message . . .

Tropical cyclones—hurricanes and typhoons—are the most destructive storms on the planet. They originate from clusters of thunderstorms and grow into giant spiraling storms, nourished by latent heat in water evaporating from warm seas. The strongest winds occur along the eye wall. The winds, waves, storm surge, flooding, and mudslides of these storms produce devastation.

Quick Question
Why can't a hurricane form on the equator?

ANOTHER VIEW Snowplows (at top of photo) clear a runway at O'Hare International Airport after the passage of a mid-latitude cyclone across northern Illinois.

⟲19 CHAPTER REVIEW

Chapter Summary

- An air mass is a large body of air with relatively uniform temperature and humidity characteristics that covers a large region of the Earth. The boundaries between air masses are called fronts. We distinguish among cold fronts, warm fronts, occluded fronts, and stationary fronts by the direction in which the colder air mass moves.

- For a thunderstorm to form, three conditions must exist: abundant moisture in the lower atmosphere, a lifting mechanism, and atmospheric instability.

- Lifting occurs when air is forced upward by a front, gust front, convergence of air flow, or high topography, or when it is warmed from below by the Earth's surface.

- Rising moist air, because of the release of latent heat due to condensation, becomes buoyant relative to its surroundings when atmospheric instability is present. Updrafts in unstable air can sometimes extend all the way to the tropopause. In stable air, lifted air sinks back down.

- Ordinary thunderstorms form where winds do not vary substantially with altitude, so these storms do not rotate. They evolve through three stages: the development stage, when warm, moist air is lifted and becomes buoyant; the mature stage, when precipitation produces a downdraft; and the dissipation stage, when the downdraft destroys the updraft and the clouds dissipate.

- In squall-line thunderstorms, several thunderstorms develop and organize into a line along a gust front or cold front. The combined outflow of cool air from such storms can trigger formation of new thunderstorms.

- Supercell thunderstorms develop if wind changes rapidly in speed and direction with elevation, causing the updraft of the storm to rotate. Rain falls downwind from the updraft. As a result, the updraft is not destroyed by the downdraft, and the storm can survive for several hours.

- Supercell updrafts are so strong that an overshooting top rises above the anvil cloud of the storm.

- Lightning is caused by electric charge separation in a thunderstorm: the cloud base accumulates negative charge, and the upper parts of the cloud accumulate positive charge.

- Because of charge separation, a spark—lightning—can jump across the cloud, or between the cloud and the ground.

- Hailstones form when supercooled water droplets freeze onto ice particles in clouds. The largest hailstones form in supercell thunderstorms.

- Flash floods occur when slow-moving thunderstorms, or a succession of thunderstorms, dump a large amount of rain over an area in a short time.

- Tornadoes typically occur on the rear flank of a supercell thunderstorm. They form when part of the rear-flank downdraft gets drawn into the storm's updraft.

- Tornadoes can cause immense damage along a tornado track. Their intensity is rated using the Enhanced Fujita (EF) scale, which is based on the damage the tornado causes.

- Mid-latitude cyclones form between latitudes of 30° and 60° when a large low-pressure system develops at ground level beneath an area of divergence in the polar-front jet stream.

- When viewed from a satellite, the clouds of a mid-latitude cyclone have the shape of a comma. The comma tail follows a cold front, and along it, thunderstorms often form. The comma head overlies a warm front, where rain, snow, or freezing rain may fall. Blizzards and ice storms can occur under the clouds of the comma head in winter.

- Tropical cyclones, also known as hurricanes or typhoons, form over tropical oceans and are the most destructive storms on the planet. They have distinctive features, including a nearly cloud-free eye, a violently rotating eye wall, and spiral rainbands. Hurricane destruction results from high winds, storm surge, and torrential rains.

Key Terms

adiabatic expansion (p. 593)
air mass (p. 588)
anvil cloud (p. 595)
atmospheric instability (p. 592)
blizzard (p. 614)
cloud-to-cloud lightning (p. 600)
cloud-to-ground lightning (p. 600)
cold front (p. 588)
convergence (p. 612)

cyclone (p. 617)
divergence (p. 612)
downdraft (p. 591)
Enhanced Fujita (EF) scale (p. 606)
environmental lapse rate (p. 593)
eye (p. 618)
eye wall (p. 618)
front (p. 588)

gust front (p. 591)
hail (p. 603)
hurricane (p. 617)
hurricane track (p. 623)
ice storm (p. 615)
lifting mechanism (p. 591)
lightning (p. 600)
lightning stroke (p. 600)
mesocyclone (p. 599)

mid-latitude cyclone (p. 608)
nor'easter (p. 609)
occluded front (p. 589)
ordinary thunderstorm (p. 594–595)
orographic lifting (p. 592)
Saffir-Simpson scale (p. 618)
severe thunderstorm (p. 590)
squall-line thunderstorm (p. 596)

stationary front (p. 589)
storm surge (p. 621)
straight-line wind (p. 597)
supercell thunderstorm (p. 597)

thunder (p. 602)
thunderstorm (p. 589)
tornado (p. 603)
tornado track (p. 604)

tropical cyclone (p. 617)
tropical storm (p. 617)
typhoon (p. 617)
updraft (p. 590)

wall cloud (p. 599)
warm front (p. 588)
wind chill (p. 615)

Review Questions

Letters in parentheses correspond to the chapter's learning objectives.

1. What is an air mass, and what are the major sources of air masses for North America? **(A)**

2. What is a front? How can we distinguish among different types of fronts? **(A)**

3. What distinguishes a thunderstorm from other kinds of rainstorms? What characteristics does a thunderstorm have if it is classified as severe? **(B)**

4. On a day when thunderstorms occur, do the thunderstorms last all day? **(B)**

5. List the three conditions that must exist for a thunderstorm to develop. **(B)**

6. Where does the moisture that feeds thunderstorms come from? **(B)**

7. What lifting mechanism is depicted in the figure? **(B)**

8. Explain the difference between stable and unstable atmospheric conditions. **(B)**

9. How does the rate at which temperature changes with altitude influence whether lifted air will become buoyant and trigger a thunderstorm? **(B)**

10. What are squall-line thunderstorms? How can the gust fronts that they create produce new thunderstorms? **(C)**

11. Explain how a supercell thunderstorm forms and how it is different from an ordinary thunderstorm. Why do these storms typically have overshooting tops? What do they look like on a radar screen? **(C)**

12. What is a downburst, and how can it pose a hazard to aircraft? **(D)**

13. Why is hail more common in supercell thunderstorms than in other thunderstorms? **(D)**

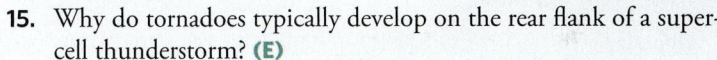

14. Where is the tornado located on the radar image? What features helped you identify it? **(E)**

15. Why do tornadoes typically develop on the rear flank of a supercell thunderstorm? **(E)**

16. On what basis do meteorologists classify tornadoes? **(E)**

17. How is a tornado watch different from a tornado warning? What should you do if a tornado is coming your way? **(E)**

18. What is a mid-latitude cyclone, why does it form, and how is it related to the polar-front jet stream? What kinds of weather form in association with mid-latitude cyclones? **(F, G)**

19. Explain how a hurricane forms and how it gets its energy. **(H)**

20. What are the key structural features of a hurricane? **(H)**

21. What causes destruction when a hurricane strikes a coastline? **(I)**

22. Aside from where they form, describe three fundamental differences between tropical and mid-latitude cyclones. **(F, H)**

On Further Thought

23. Describe the sequence of weather conditions that you would experience as a supercell thunderstorm, moving southwest to northeast, approached and then progressed over your house. **(C)**

24. Outside, it is cool and rainy, and the wind is from the east. Two hours later, the rain has ended, the temperature has increased, and the air now feels humid. The wind is from the south. What type of front passed your location? **(A)**

⟲ 20 CLIMATE AND CLIMATE CHANGE

By the end of the chapter you should be able to . . .

A. explain the difference between climate and weather, and describe the variables that determine the climate of a region.

B. recognize the boundaries among the Earth's major climate zones on a map, and relate their locations to latitude and proximity to the ocean.

C. interpret the factors that control the climate conditions that prevail in a region.

D. describe the techniques that researchers use to reconstruct climates of the past.

E. explain causes of both long-term and short-term climate change.

F. discuss how climate has changed in the recent past, and describe the consequences of this change.

G. evaluate evidence concerning modern-day global warming and its causes.

20.1 Introduction

In 1955, Charles Keeling constructed an instrument that could measure atmospheric carbon dioxide (CO_2) very accurately. This instrument had to be sensitive indeed, because as we've seen, CO_2 is a *trace gas*, one that accounts for far less than 1% of the Earth's atmosphere. Keeling made a set of measurements in California and learned that atmospheric CO_2 concentrations can change measurably over the course of a single day. Significantly, he also found that the average of his measurements was higher than the average of measurements made in the mid-19th century. This finding supported a hypothesis, put forward years earlier, that CO_2 concentrations had increased globally over the previous century.

In 1958, to better understand variations in global CO_2 concentrations, Keeling installed a monitoring station on Mauna Loa, a Hawaiian volcanic peak that rises 3.4 km (11,141 ft) above the Pacific Ocean **(Fig. 20.1a)**. He chose this location—far from cities, factories, forest fires, and other sources of the gas—so that his measurements would represent only CO_2 that had mixed thoroughly within the atmosphere. When he looked over the measurements he had collected during the first year of his study, Keeling found that the CO_2 concentration ranged between 312 and 318 ppm (parts per million by volume), with an average value for the year of 315 ppm (= 0.0315%) **(Fig. 20.1b)**. The CO_2 concentration dropped in the summer, when plants of the northern hemisphere absorbed the gas during photosynthesis, and rose in winter, when those plants shed their leaves, which then decayed and released CO_2.

In effect, Keeling had observed the Earth System inhaling and exhaling CO_2!

Keeling continued making measurements on Mauna Loa several times a year during the remainder of his career, and other researchers have continuously maintained the CO_2-monitoring station at the Mauna Loa Observatory to this day. As a result, we now have a nearly 60-year-long record, known as the **Keeling curve**, proving that the average annual atmospheric concentration of CO_2 is steadily increasing **(Fig. 20.1c)**. In May 2019, air at the station contained 414.7 ppm of CO_2, an increase of 30% since Keeling's first measurements and an increase of over 40% since the mid-19th century.

FIGURE 20.1 The Keeling curve provides a record of atmospheric carbon dioxide concentrations measured at the Mauna Loa Observatory.

(a) The observatory lies at a high elevation in the middle of the Pacific Ocean.

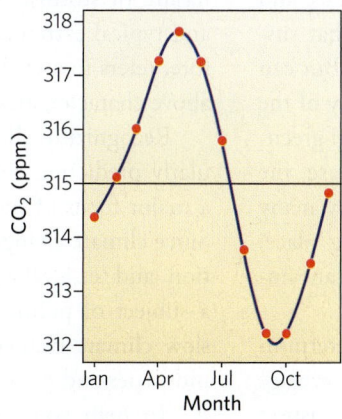

(b) The variation in CO_2 concentration during 1959, the first full year in which Keeling measured it.

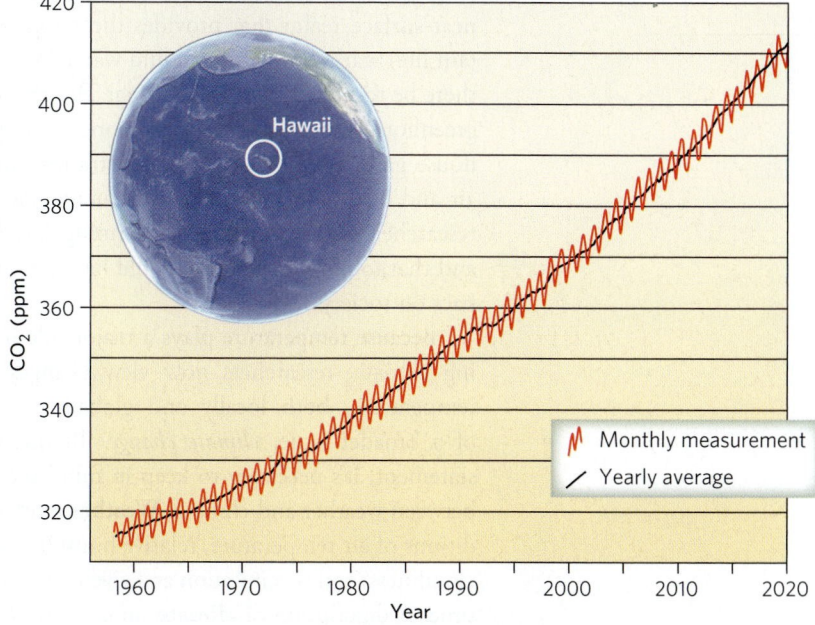

(c) The measured record of CO_2 at Mauna Loa from 1958 to 2020.

<< As a result of climate change, ice sheets along the edges of Greenland and Antarctica have been breaking up.

How can I explain . . .

The difference between weather and climate

What are we learning?

How meteorologists average data to obtain climate statistics.

What you need:

- A calculator
- A computer or smartphone

Instructions:

- Using a standard weather app available on a smartphone, or a local newspaper, record the current temperature for your hometown, and the predicted hourly temperatures for that whole day.
- Next, make a list of the predicted high and low temperatures for each day of the week in the town. Calculate the difference between the high and low for each day. Finally, calculate the average of the daily highs and daily lows for the whole week.
- Using the internet, find a website that lists the monthly high and low temperatures for your hometown. In many locations, a climatology

office keeps a record of these numbers. Calculate the temperature difference between the high and low temperatures for each month.

- Now, compare the high and low temperatures (and the difference between them) during one day, to those of a week, to those of the current month, to those of a month from half a year earlier. Are the numbers the same or different?

What did we see?

Temperature at a given time on a given day provides one aspect of the weather. Weather can change from hour to hour, day to day, and month to month. Averages and variations of temperatures over the course of a year are one aspect of the region's climate. If you were to repeat the measurements that you did above based on another location far from your hometown, you would see that the averages are different, because climate varies with location.

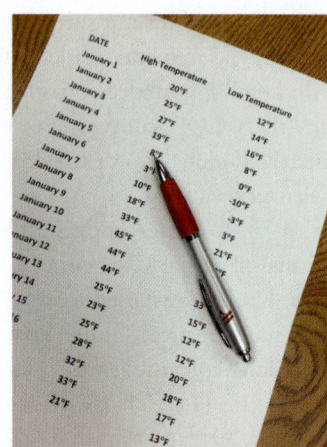

Why should society care about the atmospheric concentration of CO_2? Earlier in this book, we pointed out that CO_2 is a greenhouse gas that traps infrared radiation in the atmosphere. By doing so, it plays a crucial role in keeping the Earth System's **critical zone** (the surface and near-surface realm that provides the resources that sustain life) warm enough for liquid water to exist. But can there be too much of a good thing? The efficiency of the greenhouse effect depends on the concentration of greenhouse gases, so when CO_2 concentrations increase, the air and oceans warm up. Measurements made by many researchers indicate that this warming is taking place, and that too much warming could have a significant impact on society.

Because temperature plays a major role in determining climate, researchers now view changes in average temperature, both locally and globally, as one aspect of a broader issue, *climate change*. To understand this statement, it's necessary to keep in mind the distinction between weather and climate. **Weather** refers to the conditions of air temperature, relative humidity, wind speed, cloudiness, and precipitation at a given location at a given time. A description of **climate**, in contrast, characterizes the average weather conditions and the range of weather

conditions for a region over the course of decades to millennia (Box 20.1). More specifically, a region's climate depends on its average temperature and typical temperature range, the amount and distribution of precipitation, the nature of storms, typical seasonal variation in weather, and typical extremes of weather. **Climate change**, therefore, refers to a shift in the average of one or more of the above characteristics.

Recognizing climate change in the past, and particularly predicting climate change in the future, has been a major focus of Earth science research since the 1970s. Since climate change affects water supplies, food production, and sea level within our lifetimes, it has also become a subject of political discussion, because steps taken to slow climate change might affect the viability of some industries and the sustainability of some lifestyles.

To help you develop an understanding of climate and climate change, this chapter begins by discussing the Earth's present climates and the factors that control them. Then we look back in time to examine changes in climate over the Earth's history and phenomena that may have caused these changes. We conclude by examining how human society is modifying the Earth's climate today and what the impacts of such climate change might be.

Picture a day in the Amazon rainforest. The air feels sultry, nearly everything in sight has a lush covering of green plants, downpours drench the land, and vibrant sounds rise from countless species of insects, monkeys, and birds. Now, picture the same day in the Sahara desert. The Sun beats down relentlessly, and the air is so dry that your lips are chapped. If you look to the horizon, you see mostly rocks and sand in hues of brown and tan. Scrubby plants poke up here and there, but you hear hardly any animals. Finally, picture a day in the northern Canadian tundra, where the ground remains frozen solid almost all year. It's so cold that your breath freezes, and you see few living creatures, for most animals have migrated south or are hibernating.

By comparing your perceptions of a rainforest, a desert, and a tundra, you'll realize that climate affects many of our senses. From a meteorological perspective, however, the major factors that contribute to our experience of climate reflect two variables—surface temperature and precipitation—and their variation over the course of a year. To a large extent, these variables correlate with latitude, but many other factors, such as elevation and proximity to oceans, influence them, and therefore climate, as well.

Global Temperature and Rainfall Patterns

Since temperature serves as a key indicator of climate, atmospheric scientists compile data on temperature at many locations around the world. To determine the *average air temperature* for a day at a particular location, they record the temperature several times during the day, then calculate the average by dividing a sum of the results by the number of measurements. The monthly average temperature for a location, in turn, comes from averaging the daily temperatures during the month. Such information is useful to track how average air temperature changes with location and over time. To be able to compare average temperatures globally, atmospheric scientists rely on measurements recorded either electronically or manually, at weather stations, by a thermistor located 2 m (6 ft) above the ground surface.

Look at a map that shows how average air temperature for the month of January varies with location on the Earth (Fig. 20.2a). You can see a distinct pattern—bands of warm temperatures occur in equatorial latitudes, bands of cold temperatures occur in polar latitudes, and bands of intermediate temperatures lie in between. Now look at the global temperature map for July (Fig. 20.2b). You'll see the same overall relationship between average temperature and latitude, but the boundaries between temperature bands occur at different latitudes. Patterns

FIGURE 20.2 Global map of average surface temperatures.

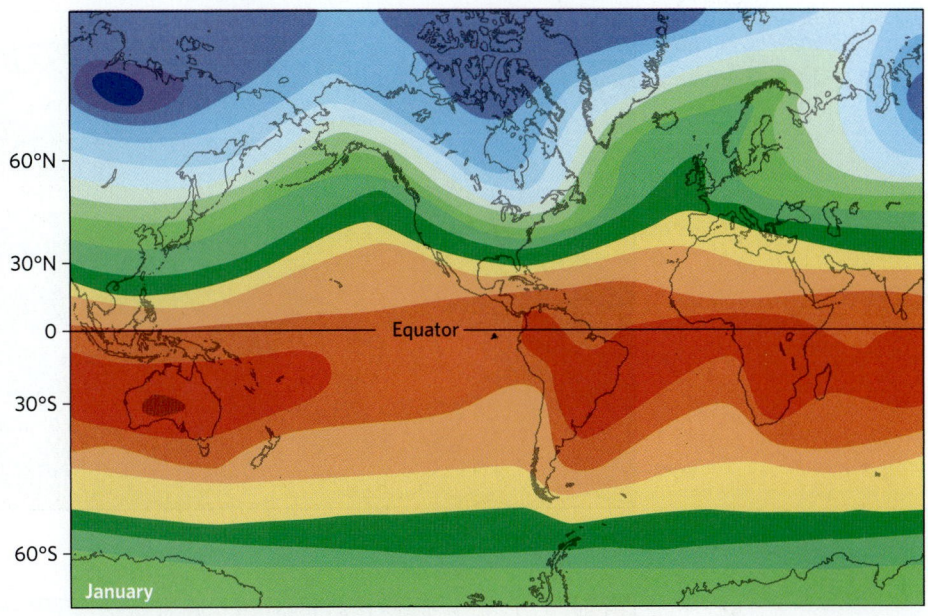

(a) In January, northern mid-latitudes are cool and southern mid-latitudes are warm.

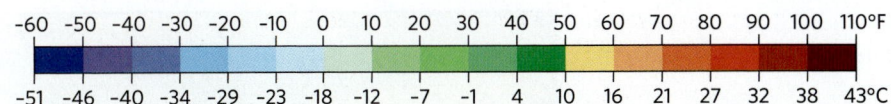

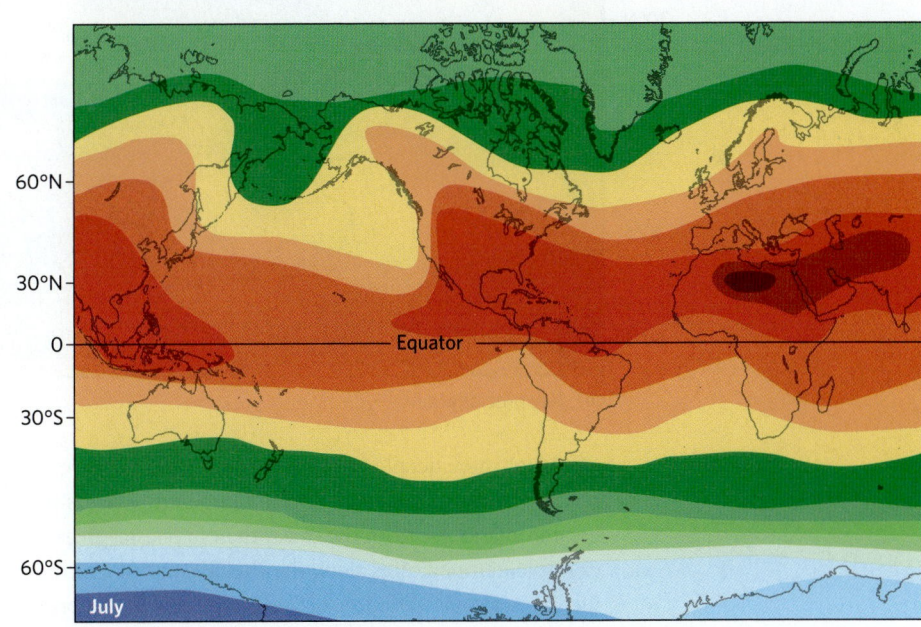

(b) In July, northern mid-latitudes are warm and southern mid-latitudes are cool.

on these maps emphasize that low-latitude regions always receive more solar radiation than high-latitude regions do, but that the temperature at a given latitude varies during the course of a year. This *seasonal change*, as we'll see shortly, is due to the change in the tilt direction of the Earth's axis, relative to the Sun, as our planet orbits the Sun.

FIGURE 20.3 Global map of average precipitation.

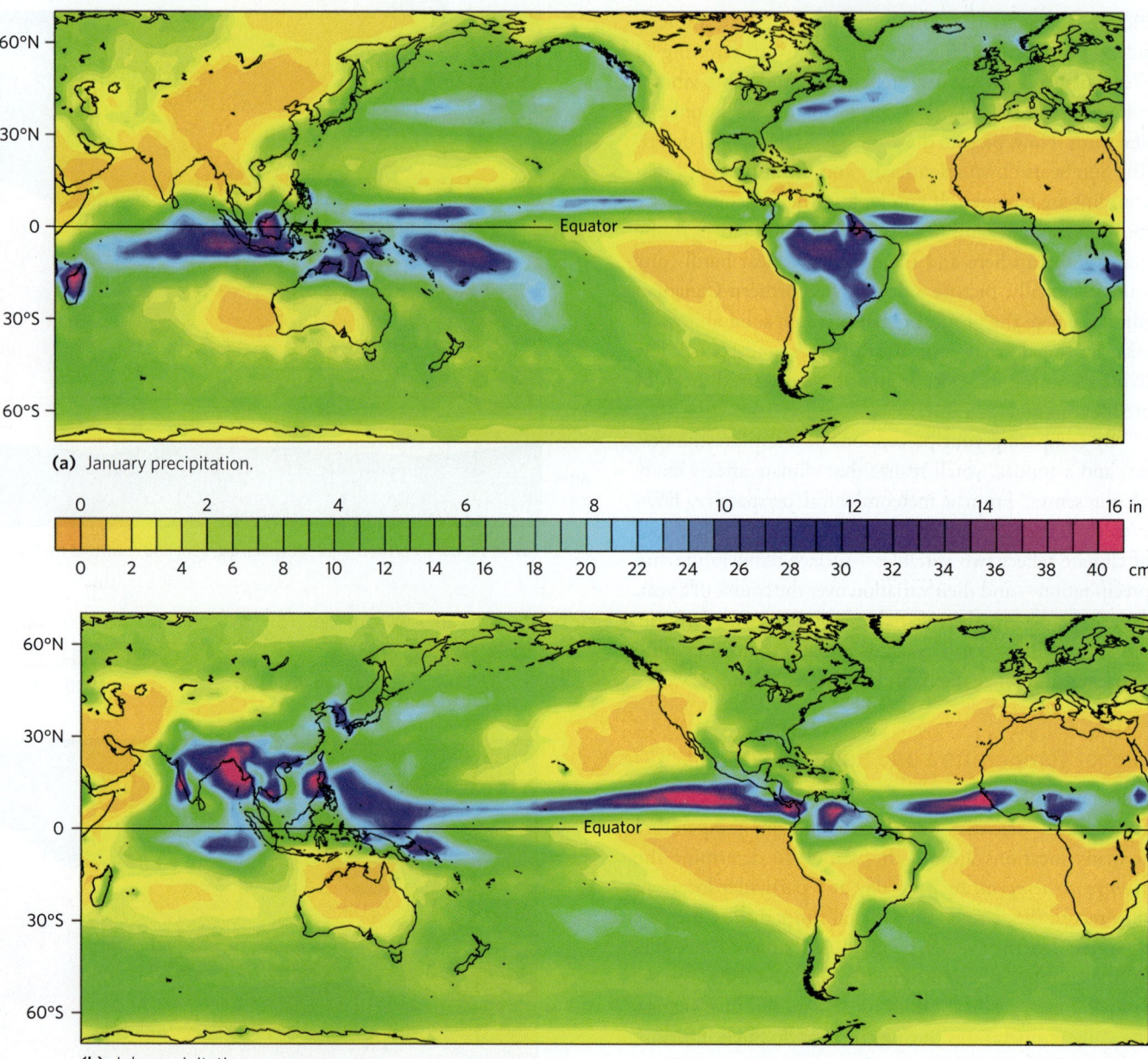

(a) January precipitation.

(b) July precipitation.

Maps more detailed than those provided in Figure 20.2 reveal that the annual variation in temperature tends to be greater inland, far from the coast, than in coastal areas. For example, temperatures along the Pacific coast of California display seasonal variation of only 7°C (13°F), whereas a location at the same latitude in Missouri, in the middle of the United States, experiences seasonal variation of 28°C (50°F). Proximity to the ocean affects seasonal temperature variation because oceans serve as vast heat reservoirs—they slowly absorb a large amount of heat during the summer and slowly release it during the winter—and can moderate temperature. Oceans behave this way for two reasons. First, water has a high *heat capacity*, meaning that it can absorb a large amount of heat without changing temperature by very much. Second, since water is transparent and circulates, and the ocean becomes warm down to the depth of the thermocline, 100–200 m (330–660 ft) below the surface, a great volume of water can absorb heat.

Because of the impact of ocean temperature on the adjacent coast, the proximity of a location to a specific ocean current can also affect temperature at the base of the atmosphere. For example, Ireland and the United Kingdom have relatively mild winters, despite their high-latitude locations, because the Gulf Stream carries warm water to high latitudes in the Atlantic (see Fig. 20.2a). The coast of Labrador, Canada, which lies at the same latitude as Ireland, has a January average temperature that's about 15°C (27°F) cooler than that of Ireland.

The global distribution of precipitation also varies significantly with latitude and season (Fig. 20.3). Large quantities of rain fall along the intertropical convergence zone (ITCZ) at the equatorial edges of the Hadley cells (see Chapter 18), where moist air rises and provides fuel for large thunderstorms. In contrast, lands beneath the descending parts of the Hadley cells host subtropical deserts. Seasonally varying air circulations also affect rainfall. For example, in tropical latitudes subject to monsoons, most rain falls during the summer season.

The Köppen-Geiger Climate Classification

Wladimir Köppen spent most of his adult life in Germany, where he became one of the first *climate scientists* (researchers who focus on understanding climate). In 1884, Köppen published a classification of the Earth's climate zones and portrayed those zones on a map. Later in life, Köppen worked with Rudolf Geiger to modify the original scheme. The resulting **Köppen-Geiger climate classification** (**KGCC**) remains in use today (Table 20.1). Köppen and Geiger used natural vegetation as the primary criterion to characterize the climate of a given area because the assemblages of plant species that thrive in an area reflect the average, range, and seasonal variation of weather conditions over a period of many years.

The KGCC map shows how variables in the Earth System control climate by distinguishing five broad *climate groups*: tropical, arid, temperate, cold, and polar (Fig. 20.4). These climate groups differ from one another in their average temperatures, in the temperature range between their hottest and coldest months, and in their average annual precipitation. Each climate group, in turn, includes a few *climate types* based on other characteristics. For example, the three types of tropical climates (rainforest, monsoonal, and savannah) can be distinguished from one another based on whether or not rain falls throughout the year. A third level of subdivision (*climate subtypes*) distinguishes among climate types, such as deserts, that can be either hot or cold.

Take-home message . . .

The Earth hosts many different climates. A region's climate primarily reflects temperature and precipitation and their variation during the year. Local vegetation serves as the primary basis for defining climate groups and types in the Köppen-Geiger climate classification.

Quick Question —————————————
How does climate vary with latitude and with proximity to the ocean?

20.3 What Factors Control the Climate?

In the alternative universe of science-fiction books, authors sometimes depict planets as disks instead of spheres. If the Earth were a smooth disk with no atmosphere, its climate might be much the same everywhere: cold and dry. But of course the Earth isn't flat—it's a rotating sphere with a tilted axis. Furthermore, it has an atmosphere, its surface consists of both land and ocean, the water in the ocean circulates in currents, the land has topography, and large areas of both land and sea sometimes host ice cover. All these characteristics influence overall global temperature and the variation of climate around the planet. In this section, we explore why.

The Effect of the Atmosphere on the Earth's Energy Balance

The Sun bathes the Earth with an immense amount of energy, mostly as visible light and UV radiation, every second. What happens to this energy? If the Earth had no atmosphere, all the energy would reach its surface. Some of this energy would instantly be reflected back into space, but most would be absorbed by the Earth's surface, warming it. Eventually, however, the Earth's surface re-radiates the energy back into space as infrared radiation. All energy reaching the Earth must ultimately return to space—otherwise, the Earth would become so hot that it would vaporize.

As we discussed in Chapter 17, the existence of our atmosphere significantly changes the fate of both solar energy reaching the Earth and re-radiated energy leaving the Earth (Fig. 20.5). Specifically, some of the incoming solar energy is absorbed by air molecules or by airborne water droplets and ice particles, thereby warming the atmosphere, and some reflects off clouds directly back into space. As a result, the Earth's surface actually absorbs only about half of the Sun's arriving energy. When this absorbed energy re-radiates, a small amount travels directly back into space, but most gets absorbed by clouds or by H_2O, CO_2, and methane (CH_4) molecules in air, causing the atmosphere to warm further. The atmosphere doesn't hold on to this energy permanently, though—some returns to the Earth's surface and contributes to surface warming, while some escapes into space. A small amount of energy from the Earth's surface also rises into the atmosphere by conduction and convection or by the evaporation of water and its re-condensation into clouds. When averaged over time, of course, the amount of energy arriving at the Earth from the Sun must equal the amount of energy emitted into space by the Earth. This relationship is called the Earth's **energy balance**. If an energy balance didn't exist, as we noted before, the Earth would keep getting hotter and hotter.

Did you ever wonder . . .

why tropical rainforests lie near the equator?

TABLE 20.1 The Köppen-Geiger Climate Classification (KGCC)

Group	Type	Subtype	Description
A			**TROPICAL:** Temperature of the coldest month is 18°C or higher.
	f		Rainforest: Precipitation in driest month is at least 6 cm.
	m		Monsoonal: A short dry season, with precipitation in the driest month of < 60 mm, but ≥ [100 − (R/25) mm].[1]
	w		Savannah: Well-defined winter dry season, with precipitation in the driest month of < 60 mm or < [100 − (R/25) mm].
B[2]			**ARID:** Either ≥ 70% of the annual precipitation falls in the summer half of the year and $R < [20T + 280]$ mm, or ≥ 70% of the annual precipitation falls in the winter half of the year and $R < 20T$ mm. *Alternatively*, neither half of the year has ≥ 70% of annual precipitation and $R < [20T + 140]$ mm.
	W		Desert: R is < one-half of the upper limit for classification as a B type.
	S		Steppe: R < the upper limit for classification as a B type, but is > 50% of that amount.
		h	*Hot:* $T \geq 18°C$
		k	*Cold:* $T < 18°C$
C			**TEMPERATE:** Temperature of the warmest month ≥ 10°C, and temperature of the coldest month < 18°C but > −3°C.
	s		Dry summer (Mediterranean): Precipitation in the driest month of the summer half of the year is < 30 mm and is < 33% of the amount in the wettest month of the winter half.
	w		Dry winter: Precipitation in the driest month of the winter half of the year < 10% of the amount in the wettest month of the summer half.
	f		No dry season: Precipitation is more evenly distributed throughout the year; criteria for neither s nor w satisfied.
		a	*Hot summer:* Temperature of the warmest month is ≥ 22°C.
		b	*Warm summer:* Temperature of each of the four warmest months is ≥ 10°C, but the warmest month is < 22°C.
		c	*Cold summer:* Temperature of one to three months is ≥ 10°C, but the warmest month is < 22°C.
D			**COLD:** Temperature of the warmest month is ≥ 10°C, and temperature of the coldest month is ≤ −3°C.
	s		Same as for Group C
	w		Same as for Group C
	f		Same as for Group C
		a	Same as for Group C
		b	Same as for Group C
		c	Same as for Group C
		d	*Very cold winter:* temperature of the coldest month is < −38°C (d designation is then used instead of a, b, or c).
E			**POLAR:** Temperature of the warmest month is < 10°C.
	T		Tundra: Temperature of the warmest month is > 0°C but < 10°C.
	F		Frost (Ice cap): Temperature of the warmest month is ≤ 0°C, and ice covers most land.

[1]In the formulas given, R refers to the annual rainfall in millimeters; T refers to the average annual temperature in degrees centigrade. The summer half of the year is defined as the months April–September for the northern hemisphere and October–March for the southern hemisphere.
[2]Any climate that satisfies criteria for designation as a B type is classified as such, regardless of its other characteristics.

The trapping of re-radiated energy by the atmosphere is a process that atmospheric scientists refer to as the *atmospheric greenhouse effect* (see Chapter 17), and the specific gases in the atmosphere that absorb re-radiated energy from the Earth's surface are called *greenhouse gases*. Because of the greenhouse effect, the Earth's atmosphere acts like a blanket, keeping the Earth's surface warmer than it would be if the Earth had no atmosphere. In fact,

FIGURE 20.4 The Köppen-Geiger climate classification (KGCC).

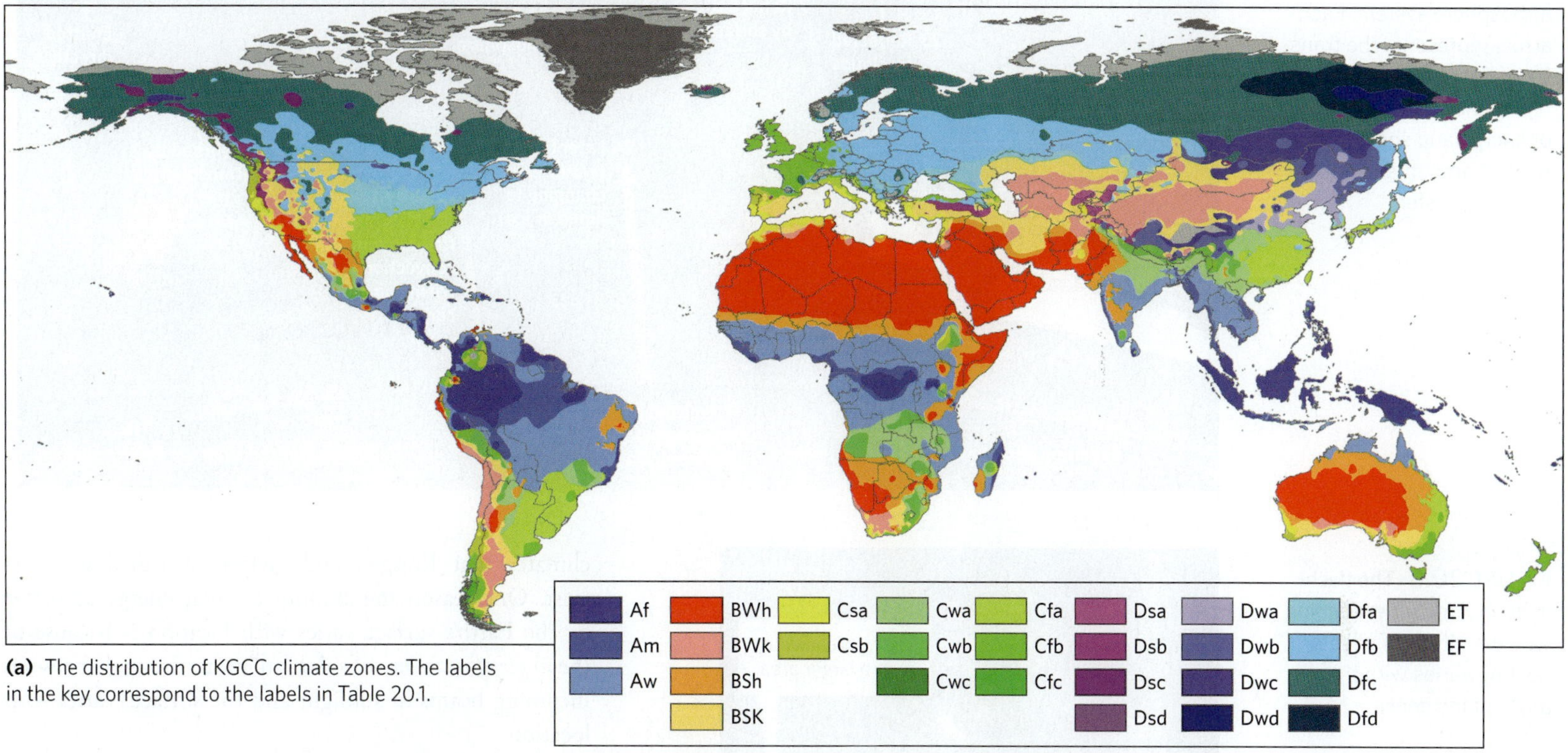

	Af		BWh		Csa		Cwa		Cfa		Dsa		Dwa		Dfa		ET
	Am		BWk		Csb		Cwb		Cfb		Dsb		Dwb		Dfb		EF
	Aw		BSh				Cwc		Cfc		Dsc		Dwc		Dfc		
			BSK								Dsd		Dwd		Dfd		

(a) The distribution of KGCC climate zones. The labels in the key correspond to the labels in Table 20.1.

Tropics
BWh

Af

Mid-latitude
Dfb

CFa

Pole
EF

ET

(b) Examples of climate zones. The inset in each photo indicates the KGCC climate zone it represents.

researchers estimate that without the greenhouse effect, the Earth's average global surface air temperature would be about 33°C (59°F) lower than it is today, and our planet's surface would be a frozen wasteland. Greenhouse gases make the Earth's climate habitable!

The Relationship of Energy Input to Angle of Incidence

If all of the Earth's surface received the same amount of solar energy, the temperature of the lower troposphere would be the same everywhere, and the great variety of

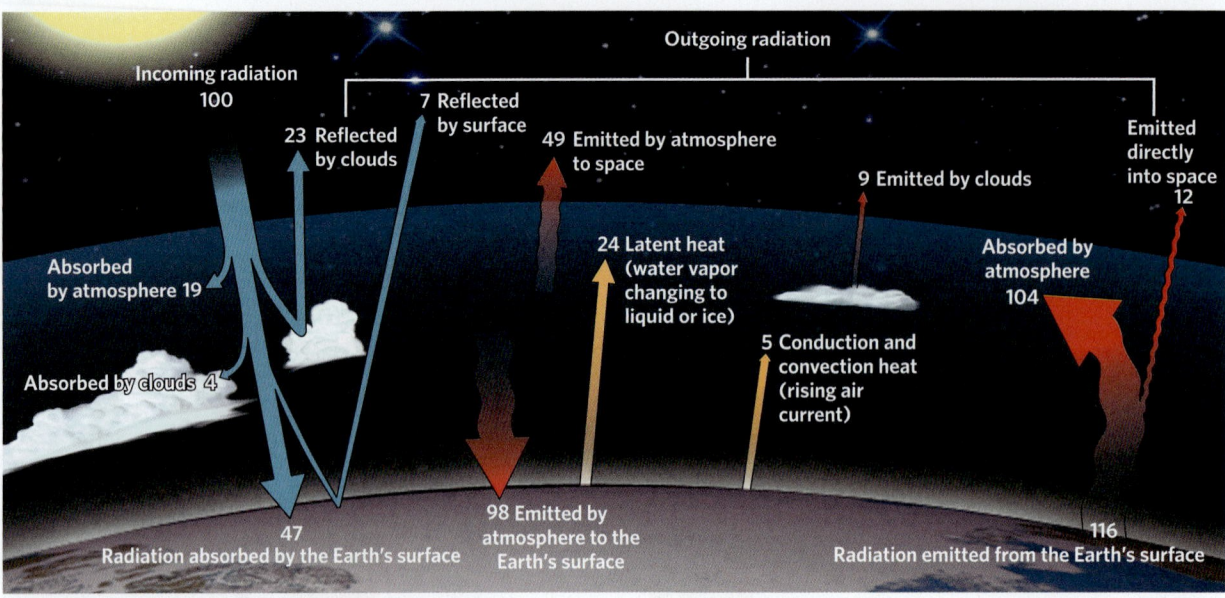

FIGURE 20.5 The energy balance in the Earth-atmosphere system. Each arrow represents the transfer of energy. The numbers represent arbitrary units of energy and show the relative amounts of energy in each transfer.

Incoming radiation 100

Outgoing radiation

23 Reflected by clouds

7 Reflected by surface

49 Emitted by atmosphere to space

9 Emitted by clouds

Emitted directly into space 12

Absorbed by atmosphere 19

24 Latent heat (water vapor changing to liquid or ice)

Absorbed by atmosphere 104

Absorbed by clouds 4

5 Conduction and convection heat (rising air current)

47 Radiation absorbed by the Earth's surface

98 Emitted by atmosphere to the Earth's surface

116 Radiation emitted from the Earth's surface

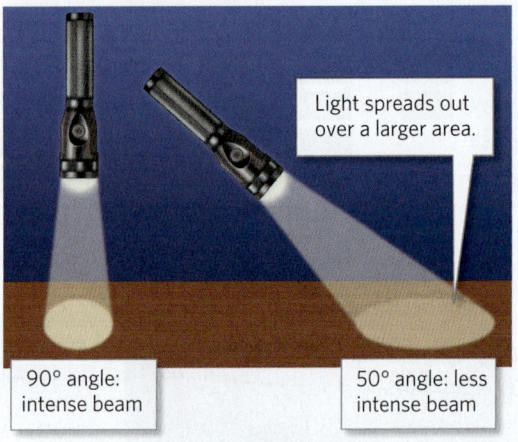

FIGURE 20.6 The flashlight analogy for explaining why the intensity of solar heating varies with the angle of incidence.

Light spreads out over a larger area.

90° angle: intense beam

50° angle: less intense beam

climates that Köppen and Geiger identified wouldn't exist. One reason the amount of solar energy absorbed by the Earth's surface varies with location is because of the *angle of incidence* of solar radiation, the angle between incoming beams of sunlight and the surface, varies with location.

To see why the angle of incidence matters, let's do a simple experiment (Fig. 20.6). Aim a flashlight straight down on a tabletop from a distance of about 10 cm (4 in). The beam intersects the tabletop at an angle of 90° and forms a small circle of bright light. If you then tilt the flashlight and aim it so that it intersects the tabletop at a shallower angle, the beam spreads out over an ellipse. Even though the

FIGURE 20.7 The reason for the seasons. The Earth's axis of rotation tilts at an angle of 23.5° to a line perpendicular to the Earth's orbital plane and stays in that orientation as the Earth orbits the Sun. The seasons are labeled for the northern hemisphere.

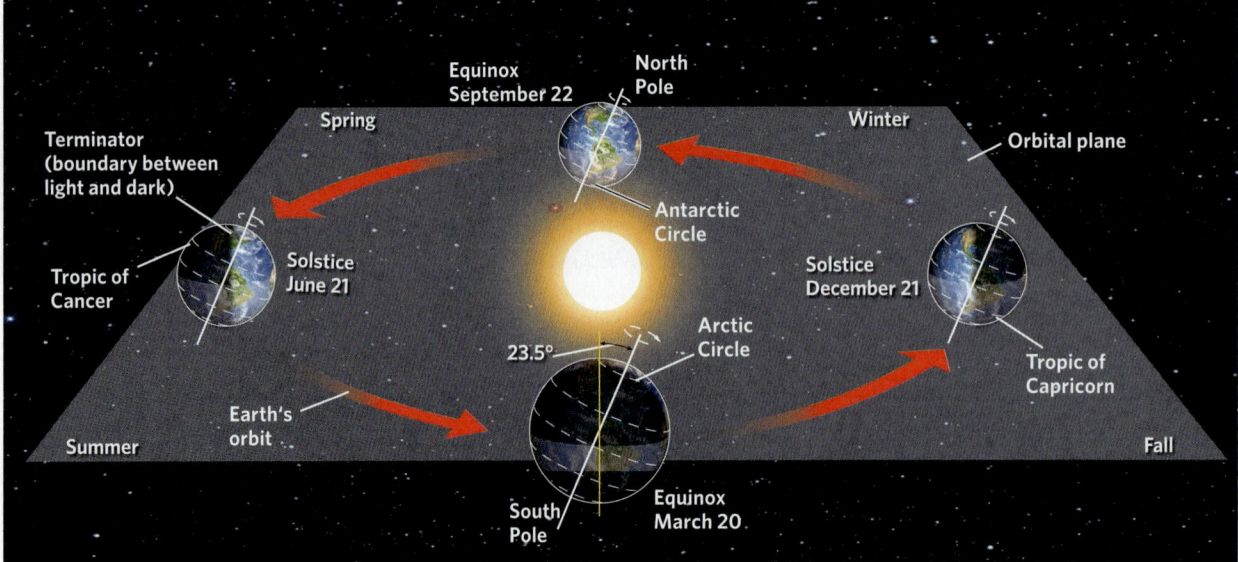

Equinox September 22

North Pole

Spring

Winter

Orbital plane

Terminator (boundary between light and dark)

Antarctic Circle

Tropic of Cancer

Solstice June 21

Solstice December 21

Tropic of Capricorn

Arctic Circle

23.5°

Earth's orbit

Summer

Fall

South Pole

Equinox March 20

amount of light energy emitted by the flashlight remains the same, the light in the ellipse looks dimmer because the light has spread out over. With this concept in mind, note that because the Earth is a sphere, sunlight strikes different latitudes at different angles at any given time. For example, at noon, a square meter at the equator, where the angle of sunlight is steep like that of the straight-down flashlight beam, receives more solar energy than a square meter at a latitude of 40°, where the angle of sunlight is shallow and light spreads over a broader area.

Seasons and Their Consequences

If the Earth's axis of rotation were exactly perpendicular to the plane of the Earth's orbit around the Sun (its *orbital plane*), the variation in temperature from equator to poles that we've just described would be the same all year everywhere on the Earth's surface. It isn't (see Fig. 20.2). Most of the Earth experiences distinct **seasons** during the course of the year, each characterized by different overall weather conditions: days are longer and warmer during the summer and shorter and colder during the winter. Fall and spring represent transitions between summer and winter.

The Earth has seasons because its axis of rotation tilts at an angle of 23.5° from a line drawn perpendicular to the orbital plane **(Fig. 20.7)**. The Earth's axis stays in the same orientation, relative to an observer outside of the Solar System, as the Earth goes around the Sun. So during one half of the year, the southern hemisphere faces the Sun more directly, and during the other half of the year, the northern hemisphere faces the Sun more directly. The resulting change in insolation affects average temperature for three reasons, all of which determine the amount of solar energy that reaches a location on the Earth's surface:

- *Number of daylight hours:* The tilt of the Earth's axis affects the number of daylight hours that a region receives at a given time of year **(Fig. 20.8)**. Since the number of daylight hours is greater in summer than in winter, a region receives more energy during a summer day than during a winter day. In polar regions, no solar radiation at all arrives during winter, but during summer, the Sun shines 24 hours a day.

- *Angle of incidence:* As our flashlight experiment showed, the angle of incoming sunlight relative to the Earth's surface determines how much solar energy reaches a square meter of the surface at a given time. When the northern hemisphere tilts toward the Sun, the angle of incidence of incoming light is steeper, so the hemisphere becomes warmer and experiences summer. When the northern hemisphere tilts away from the Sun, the angle of incidence is less steep, so the hemisphere becomes colder and experiences winter.

FIGURE 20.8 Solar heating of the Earth's surface varies with season because of the angle of the Earth's axis relative to incoming beams of sunlight.

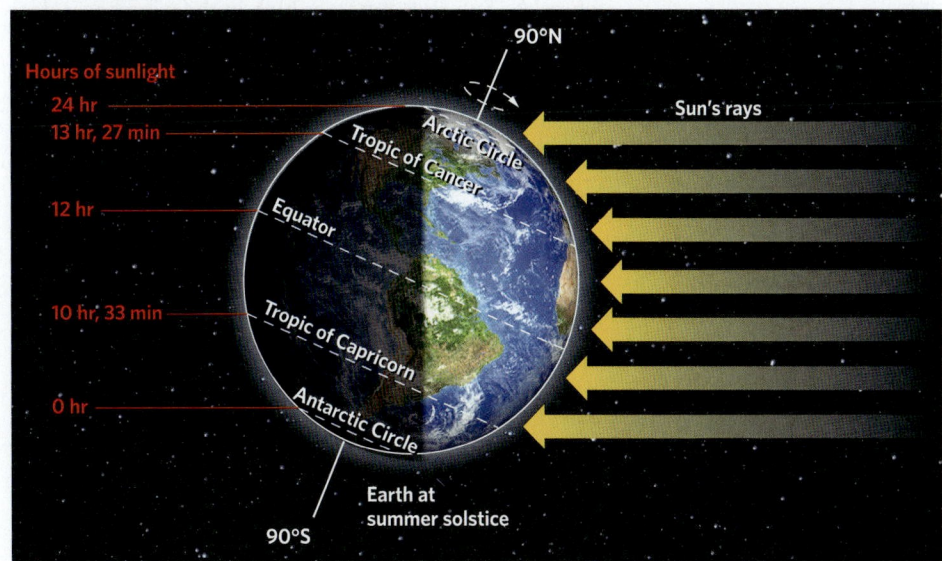

(a) At the northern hemisphere summer solstice, the North Pole receives 24 hours of sunlight, and northern mid-latitudes have long, warm days.

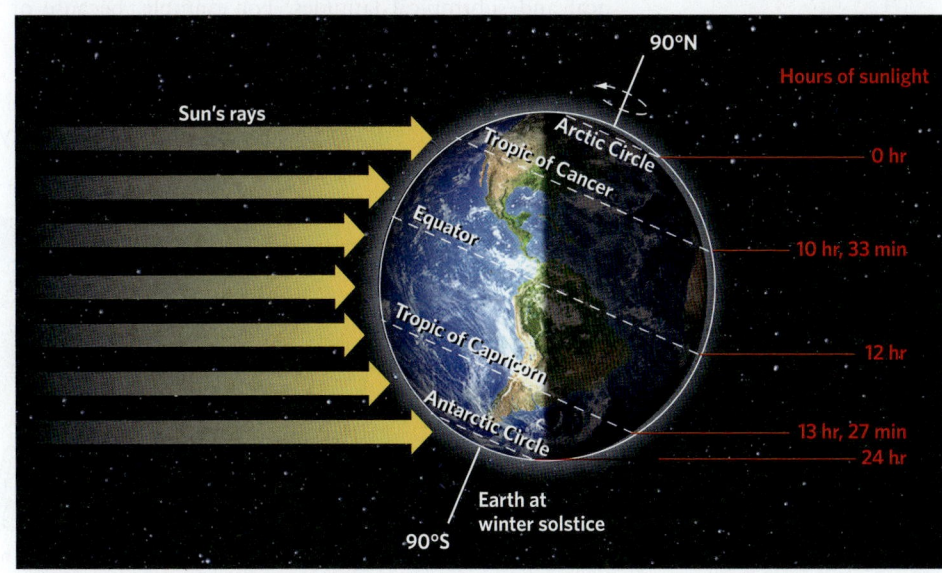

(b) At the northern hemisphere winter solstice, the North Pole is dark all day, and northern mid-latitudes have short, cool days.

- *The amount of air through which sunlight passes:* The energy that reaches the Earth's surface at a given location depends on the amount of atmosphere through which the Sun's rays must pass. Air molecules scatter some radiation back into space before it can reach the Earth's surface. When the Sun's rays intersect the atmosphere at a steep angle, as happens in summer when the Sun rises high above the horizon, they encounter fewer air molecules before reaching the ground than they do during winter, when the Sun stays closer to the horizon and the rays arrive at a shallow angle.

Did you ever wonder . . .

why the Earth has seasons?

As a result of these three factors, the Earth's surface at mid- to high latitudes receives less solar energy overall in the winter than it does in the summer. In the tropics, where the Sun rises almost as high in the winter as it does in the summer, hardly any seasonal change takes place.

Some people mistakenly believe that seasons reflect changes in the distance between the Earth and the Sun over the course of a year. That is not the case. The Earth's orbit is elliptical, but only very slightly so (see Chapter 21), therefore the distance between the Earth and the Sun varies by only 3% during the course of a year, and this variation has almost no effect on climate. In fact, the Earth is closest to the Sun during the northern hemisphere's winter!

The Effect of Atmospheric and Oceanic Circulation on Climate

The tropical atmospheric circulations described in Chapter 18—specifically, the Hadley cells, the Walker circulation, and the monsoons—all redistribute heat and moisture in the atmosphere, so they affect the amount of rainfall and the timing of wet and dry seasons at tropical and subtropical latitudes. For example, because of

the Hadley cells, lots of precipitation falls in the tropics whereas very little falls in the subtropics. This means that we see a transition from rainforest to steppe to desert across the width of each Hadley cell. Because of the Walker circulation, patterns of weather vary in response to ENSO (El Niño and La Niña). And because of monsoons, most rain falls only during certain times of the year across broad regions of some continents.

Atmospheric circulations at mid-latitudes are complicated because, as we saw in Chapter 18, large *air masses*—bodies of air over which temperature and humidity are relatively uniform—form in certain locations. Semi-permanent (long-lasting) regions of high atmospheric pressure develop at the Earth's surface beneath cold air masses, whereas semi-permanent regions of low atmospheric pressure develop beneath warm air masses (Fig. 20.9). These pressure systems have a dramatic effect on local climates at mid- and high latitudes because the pressure gradients between them drive the winds. At the Earth's surface, winds blow counterclockwise around low-pressure systems and clockwise around high-pressure systems in the northern hemisphere. (The opposite is true in the southern hemisphere.) Therefore, regional winds

FIGURE 20.9 The northern hemisphere's average sea-level pressure patterns. Winds flow counterclockwise around low-pressure systems and clockwise around high-pressure systems in the northern hemisphere.

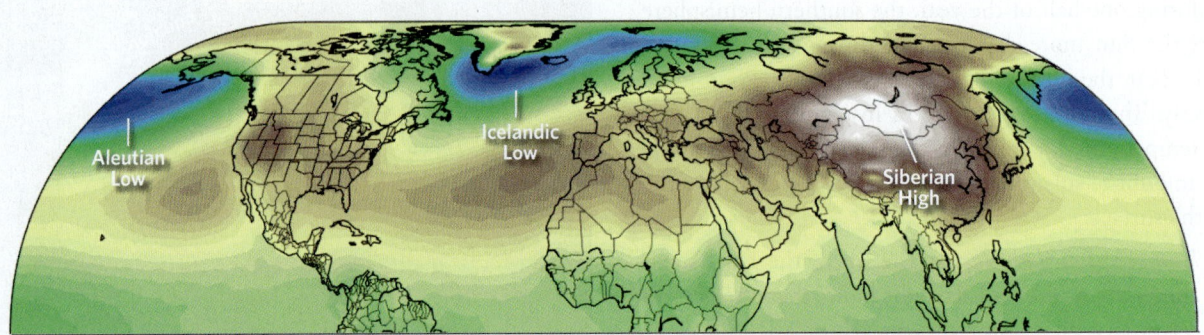

(a) Northern hemisphere winter (December through February). This map indicates names of semi-permanent highs and lows.

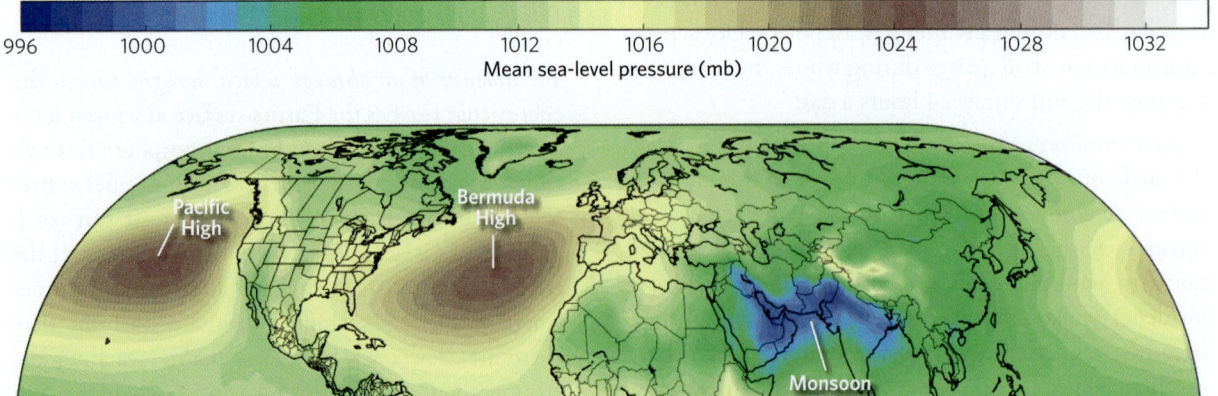

(b) Northern hemisphere summer (June through August). The Bermuda High and the Pacific High become pronounced during the summer.

Stream in the western Sierras

Desert near Reno

associated with the semi-permanent high-pressure systems over the North Atlantic and North Pacific in the northern hemisphere summer transport warm air poleward along the east sides of continents and cold air toward the equator along the west sides of continents. These winds, in turn, drive ocean currents that accelerate heat exchange between the tropics and polar regions (see Chapter 15). Together, the regional winds and associated ocean currents cool the west coasts of continents and warm the east coasts of continents in both hemispheres. In addition, the thermohaline circulation, which acts like a global conveyor belt to transport heat throughout the global ocean, serves to moderate global climates overall.

The Effect of Land Distribution and Elevation on Climate

If the surface of the Earth were all ocean, or all flat land, seasonal variations at a given latitude would be uniform. The wide range of different climates that we can find at a given latitude on the real Earth reflects the distribution of land and sea as well as topography.

We've already seen that continental interiors experience larger seasonal changes in temperature than coastal areas do. The Earth's mountain ranges also affect temperature. Temperature decreases with height in the troposphere, so high-altitude locations remain colder than nearby low-altitude locations. In fact, as you climb a high mountain in the mid-latitudes, you'll pass through the same climate groups of the KGCC that you would if you hiked from central Europe or the United States to the Arctic Ocean. Recall that the presence of a mountain range also controls distribution of rainfall (see Chapter 14), so the climate on the windward side of the range may be very different from that in the rain shadow on the leeward side **(Fig. 20.10)**.

How Snow and Ice Cover Affect Climate

If you try to traverse a snowfield or ice field on a sunny day, you have to wear very dark glasses to protect your eyes from glare, for snow and ice efficiently reflect incoming solar energy back to space. Snow and ice also serve as efficient radiators of infrared energy. For this reason, the temperature over snow-covered ground drops very rapidly at night, compared with that over bare or vegetated ground, so snow-covered land has a cooler climate than dry land at the same latitude and distance from the coast.

Take-home message . . .

Because the atmosphere contains greenhouse gases, the Earth's surface temperature overall is warmer than it would be without those gases. Climate variation depends on many factors, including the effect of latitude on solar radiation, seasons, redistribution of heat and moisture by atmospheric and oceanic circulations, the distribution of land and sea, topography, and snow and ice cover.

Quick Question

How does the atmosphere affect the Earth's energy balance?

20.4 Climate Change

How often have you seen a news report proclaiming, "Record High Temperature Expected Tomorrow"? Does such a headline mean that the climate is changing? Not necessarily—a single hot spell or cold snap may simply represent *natural variation* (random changes that can't be predicted and probably represent the response of a system to input from several subtle causes). But if a new set of conditions—an overall increase in average temperature or an overall change in precipitation—becomes the norm for a region, then climate change has occurred. An increase in average global atmospheric and sea-surface temperatures represents **global warming**, and a decrease represents **global cooling**.

Needless to say, climate change has become a high-profile political issue because of its potential impact on economies and lifestyles. In this section, we provide a context for understanding *contemporary climate change* (the change over the past few and next few centuries) by first examining the kinds of data used to characterize climates of the past, and then by considering the record of climate change through the Earth's history.

Characterizing Paleoclimate

What we know about the Earth's past climate, or **paleoclimate**, comes from studying a variety of *paleoclimate indicators*, meaning clues to paleoclimate preserved in wood, shells, ice, sediment, or rock. By determining the numerical ages of these indicators, researchers can compile a record of how climate has changed over time. *Paleoclimatologists*, researchers who study the prehistoric record of climate change, investigate climate change at two different scales: *long-term climate change*, which takes place over millions to hundreds of millions of years, and *short-term climate change*,

which takes place over decades to hundreds of thousands of years. Notably, the *resolution* (detail) of the paleoclimate record generally permits us to characterize short-term climate changes for only the past few million years or so.

Long-Term Climate Change

The stratigraphic record (see Chapter 5) provides paleoclimatologists with a basic history of global climate over millions to billions of years. That's because the depositional settings in which sediment accumulates, and the assemblages of organisms that live in or on sediment, depend on the climate. For example, beds of organic debris typically accumulate only in warm climates, whereas beds of till accumulate only in glacial climates. Therefore, a succession of strata that includes coal overlain by tillite (rock made from till) records a change from a warm to a cold climate.

Taken as a whole, the stratigraphic record shows that the Earth's surface temperature has stayed between the freezing point of water and the boiling point of water for almost all of geologic time since the beginning of the Archean. The Earth's temperature has not remained uniform, however. Specifically, sometimes the Earth's atmosphere has been significantly warmer than average, and sometimes it has been significantly cooler. Geologists refer to the warmer periods as **hothouse periods** (or *greenhouse periods*) and to the colder ones as **icehouse periods** (Fig. 20.11a). During hothouse periods, even lands at polar latitudes were largely ice free, whereas during icehouse periods, polar regions were ice covered. The more familiar term *ice age* refers to the portions of some icehouse periods when ice sheets advanced and covered substantial areas of continents. During most of the approximately half-dozen ice ages that have occurred during Earth history, ice sheets covered land at mid-latitudes. In the late Proterozoic ice age, however, glaciers appear to have covered all land—even at the equator—and the oceans appear to have frozen over. Researchers refer to this condition as *snowball Earth* (see Chapter 10).

If we focus on the climate record of the past 100 million years, we can get a sense of the range of typical long-term climate changes that happen in the Earth System. The climate of the late Mesozoic was much warmer than that of today. In fact, temperatures may have been 2°C–6°C (36°F–43°F) warmer at the equator and 20°C (68°F) warmer at the poles. As a consequence, dinosaurs lived within 1,500 km (1,000 mi) of the poles. (By comparison, land that could host most large animals today lies at least 4,500 km, or 2,500 mi, from the poles.) It was so warm that there were no polar ice caps, and sea-surface temperatures in some parts of the ocean may have been as much as 17°C (21°F) warmer than today. Starting about 80 Ma, however, the Earth's surface and atmosphere began to cool. The cooling trend has continued overall, except for one 10-million-year-long interval—known as the *Eocene*

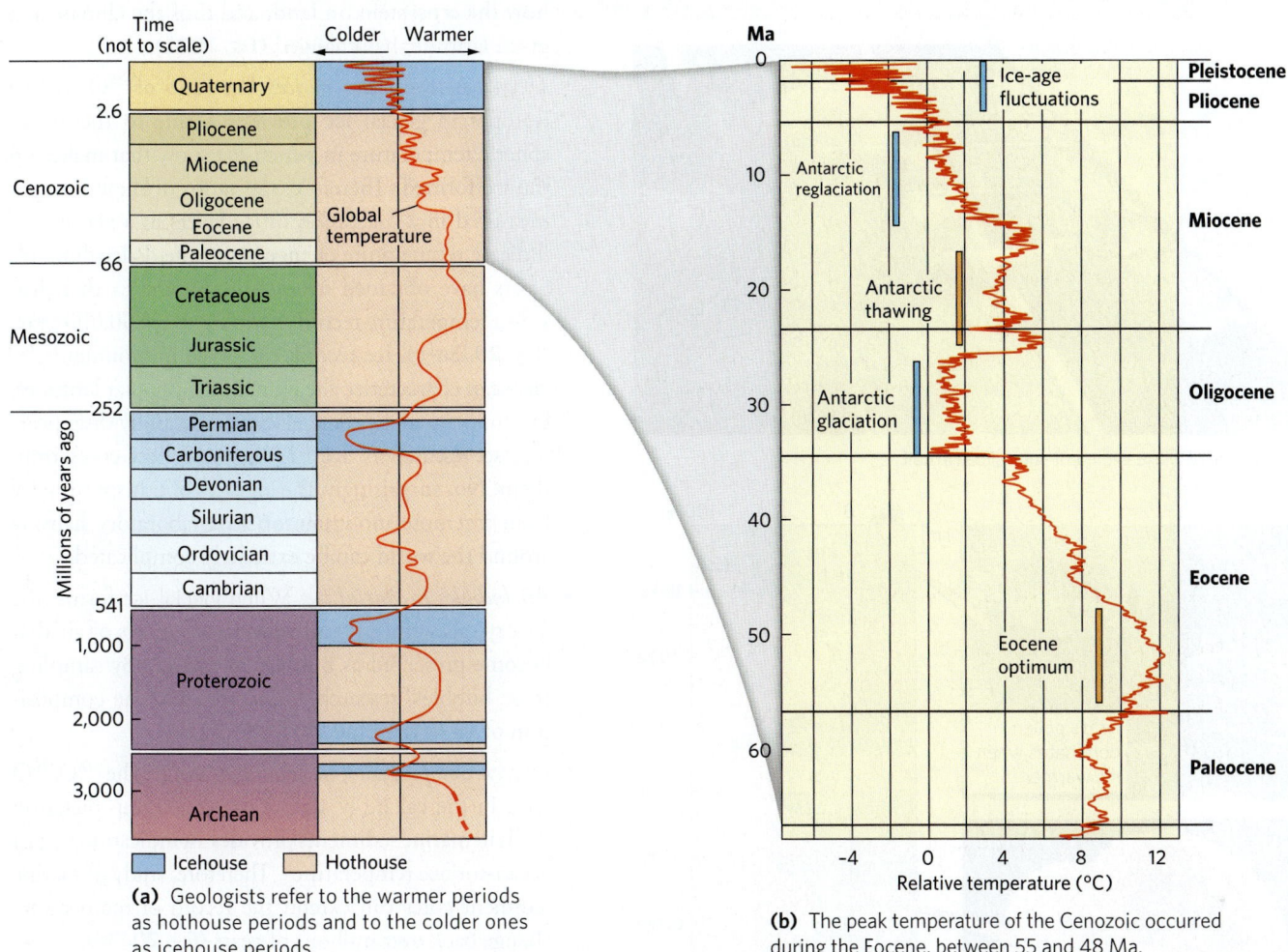

(a) Geologists refer to the warmer periods as hothouse periods and to the colder ones as icehouse periods.

(b) The peak temperature of the Cenozoic occurred during the Eocene, between 55 and 48 Ma.

climatic optimum—when global warming took place **(Fig. 20.11b)**. After that, our planet entered an icehouse period at about 34 Ma, the time when the Antarctic ice sheet started to form. Beginning about 2.6 Ma, the Pleistocene Ice Age began, and glaciers advanced repeatedly over large areas of the continents of the northern hemisphere.

Short-Term Climate Change

About 18,000 years ago, the site of what is now Chicago lay beneath a 1.5-km (0.9-mi)-thick glacier, and sea level was 100 m (300 ft) lower than it is today, so that people could walk across the Bering Strait from northeastern Asia to North America. Today, in contrast, Chicago has snow cover only during the winter season, and stormy seas occupy the Bering Strait. When the first farmers cultivated wheat 4,500 years ago in what is now Iraq, the region enjoyed frequent rains and hosted lush vegetation. Today, it hosts an arid climate with desert landscapes. And when Queen Elizabeth I reigned over England 450 years ago, glaciers were advancing down valleys in the mountains of Europe, and the canals of

the Netherlands froze over every winter. Today, the glaciers have retreated and the canals rarely freeze. Clearly, climate can change noticeably on a scale of centuries to millennia—in other words, on a human time scale.

To characterize short-term climate change over the past thousands to millions of years, paleoclimatologists examine several types of paleoclimate indicators:

- *Historical records:* Because human life depends directly on climate conditions, people tend to remember unusual conditions such as droughts, floods, long cold snaps, or severe storms. Prior to the development of writing, these memories were oral traditions, but since the advent of writing, they have appeared in diaries and histories.

- *Microfossils:* Fossilized marine plankton in Pleistocene and Holocene deposits provide a record of climate change because the assemblage of species living when the deposits formed reflects water temperature at that time, and because these plankton have distinctive shells that allow the identification of species. Similarly, different plant species produce pollen grains

(a) Example of a lake sediment core, sliced in half.

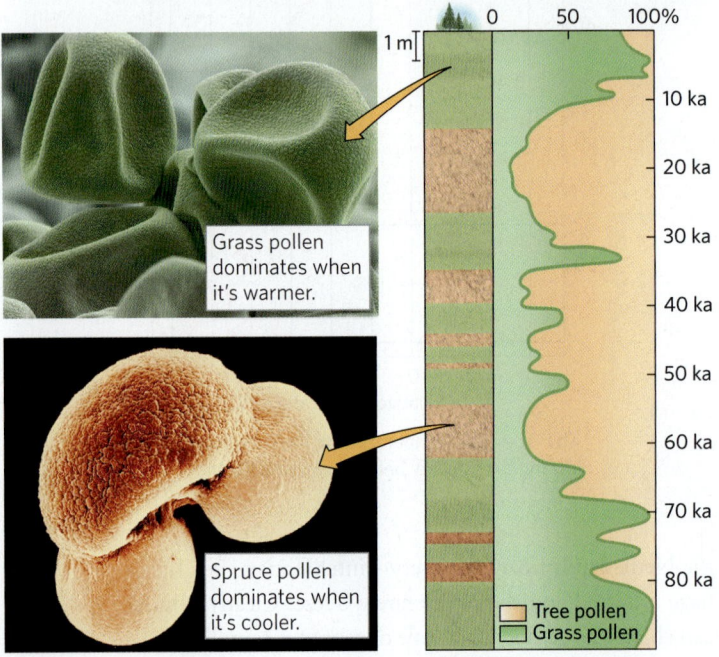

Grass pollen dominates when it's warmer.

Spruce pollen dominates when it's cooler.

Tree pollen
Grass pollen

(b) The pollen preserved in a lake sediment core can be used to determine the atmospheric temperatures at the time the sediment was deposited.

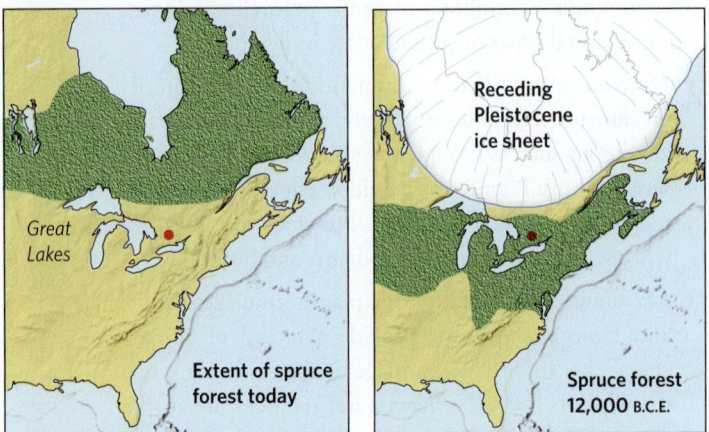

Great Lakes

Extent of spruce forest today

Receding Pleistocene ice sheet

Spruce forest 12,000 B.C.E.

(c) As we see from the pollen analysis, spruce forests (green) grew farther south 12,000 years ago than they do today. The red dots point out the location where the core sample was taken.

with distinctive shapes, so the study of pollen preserved in lake deposits allows researchers to determine how the ecosystem on land, and thus the climate at a given latitude, has changed (Fig. 20.12).

• *Oxygen-isotope ratios in ice:* The ratio of ^{18}O to ^{16}O isotopes in glacial ice provides a clue to the atmospheric temperature in which the snow that makes up the ice formed. Therefore, the ratios of these isotopes measured in a succession of ice layers in a glacier can indicate temperature change over time. Paleoclimatologists have obtained *ice cores* in Antarctica that provide a temperature record spanning over 800,000 years (Fig. 20.13a–c). Ice records preserved in mountain glaciers can characterize the climate at nonpolar latitudes, but unfortunately, these glaciers are rapidly disappearing, so researchers are rushing to collect cores from them. Not surprisingly, the logistics of transporting ice from a remote mountaintop to a laboratory halfway around the world can be extremely complicated.

• *Air bubbles in glacial ice:* When glacial ice forms, the ice crystals within it surround tiny pockets of air that become preserved as bubbles in the ice. By sampling these bubbles, researchers can measure the composition of air at the time the ice formed.

• *Oxygen-isotope ratios in plankton shells:* The $^{18}O/^{16}O$ ratio in the calcite ($CaCO_3$) that makes up plankton shells in marine sediments provides an indication of past ocean-surface temperatures. Therefore, study of marine sediment cores can extend the record of temperature change back over millions of years (Fig. 20.13d).

• *Growth rings:* Trees, corals, clams, and many other organisms develop distinct rings as they grow because their rate of growth varies with the season (Fig. 20.14). These **growth rings** can be used as markers for a given year and, in some cases, to characterize climate conditions during that year. The thickness of growth rings in trees, for example, depends on precipitation and temperature (with more growth in warm, wet years), so variations in rings from year to year serve as an indicator of annual variations in temperature and precipitation. Distinct patterns of growth rings serve as "bar codes" that allow researchers to correlate rings in living trees with rings in dead trees, and even with rings in buried logs. Such information helps paleoclimatologists to extend the record of climate variation far back in time.

By compiling data from the study of paleoclimate indicators, researchers have identified several alternating warming and cooling intervals during the past 800,000 years (Fig. 20.15). And since the retreat of the last Pleistocene ice sheet, trends of cooling or warming have lasted decades to millennia. For example, in the northern hemisphere, the time between 15,000 and 10,500 B.C.E.

FIGURE 20.13 Paleoclimate data can be obtained by studying oxygen-isotope ratios in ice and sediment cores.

(a) Researchers drilling an ice core.

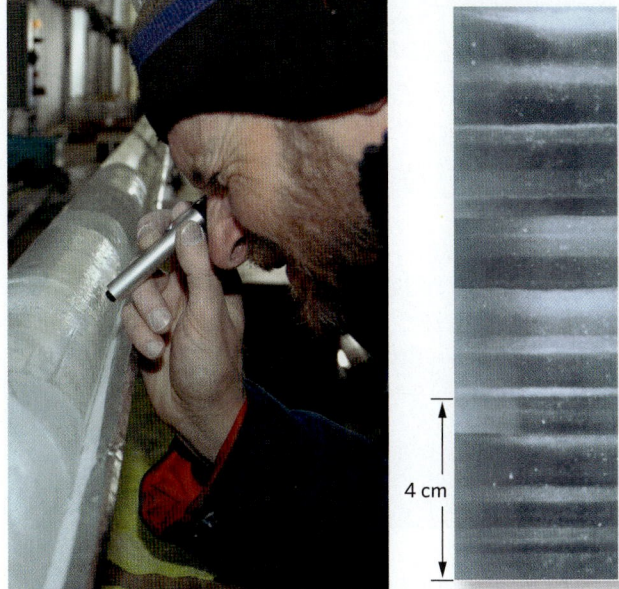

(b) Close examination of an ice core reveals distinct annual layers.

4 cm

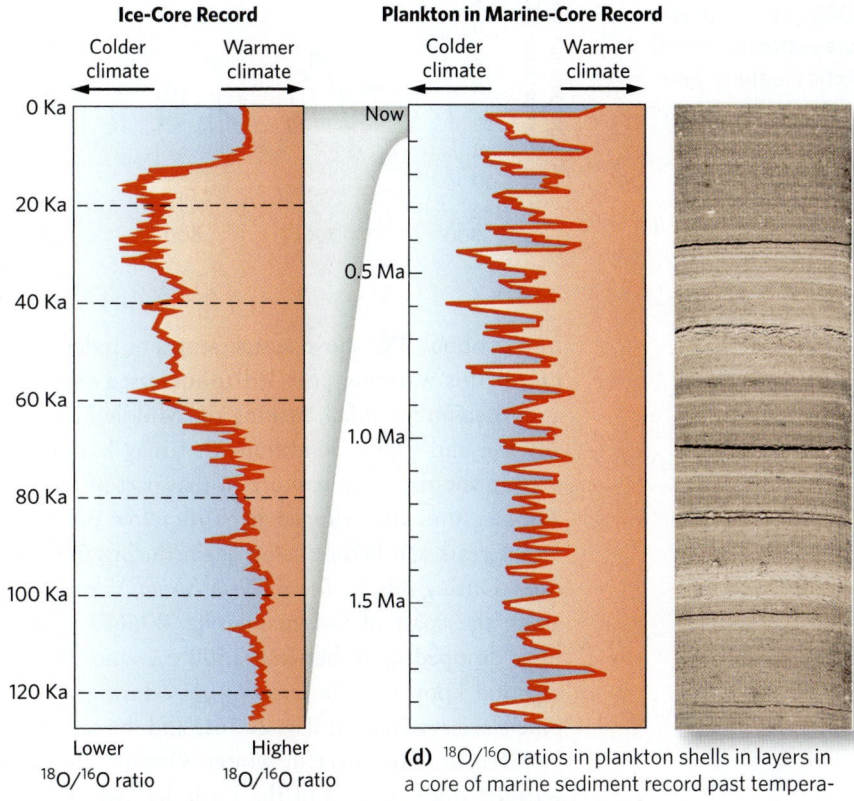

Ice-Core Record

Colder climate ← → Warmer climate

0 Ka
20 Ka
40 Ka
60 Ka
80 Ka
100 Ka
120 Ka

Lower $^{18}O/^{16}O$ ratio   Higher $^{18}O/^{16}O$ ratio

(c) $^{18}O/^{16}O$ ratios in the ice indicate temperatures over the past 120,000 years.

Plankton in Marine-Core Record

Colder climate ← → Warmer climate

Now
0.5 Ma
1.0 Ma
1.5 Ma

(d) $^{18}O/^{16}O$ ratios in plankton shells in layers in a core of marine sediment record past temperatures over longer time scales.

FIGURE 20.14 The spacing of tree rings records a history of temperature and precipitation during the tree's lifetime. More growth happens in warm, wet years than in cool, dry ones.

was a warming period during which the last continental glaciers of the Ice Age melted away entirely **(Fig. 20.16a)**. This warming trend was followed by an interval of cooler temperatures during which much of Europe was treeless tundra. Researchers have named this interval the *Younger Dryas* after an Arctic flower that became widespread during that time. Temperatures then warmed again, so that between 9,000 and 5,000 years ago—an interval called the *Holocene maximum*—average temperatures

Researchers extract cores to see rings in living trees without damaging the tree.

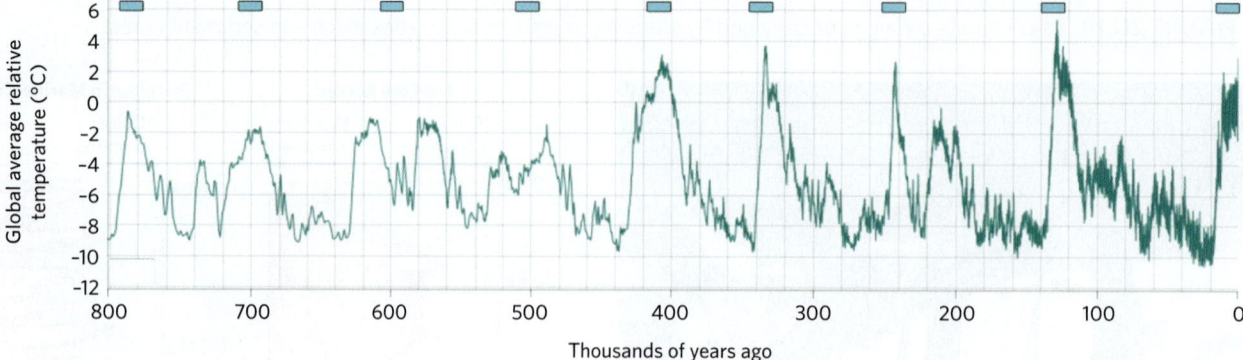

rose to about 2°C above temperatures of today. Significantly, this warming peak led to increased evaporation and, therefore, rainfall, making the Middle East unusually wet and fertile, conditions that may have contributed to the rise of agriculture in that part of the world. Temperatures then dipped back to a low point about 3,000 years ago before rising again during the Middle Ages. During this *Medieval Warm Period*, Vikings settled along the coast of Greenland **(Fig. 20.16b)**. Temperatures dropped again between 1500 c.e. and 1800 c.e., a period known as the *Little Ice Age*, during which Alpine glaciers advanced **(Fig. 20.16c)** and the canals of the Netherlands froze over in winter. Overall, climate has warmed since the end of the Little Ice Age, becoming as warm, or warmer, than it was during the Medieval Warm Period.

At the end of the Little Ice Age, this tributary glacier in France reached the main valley floor.

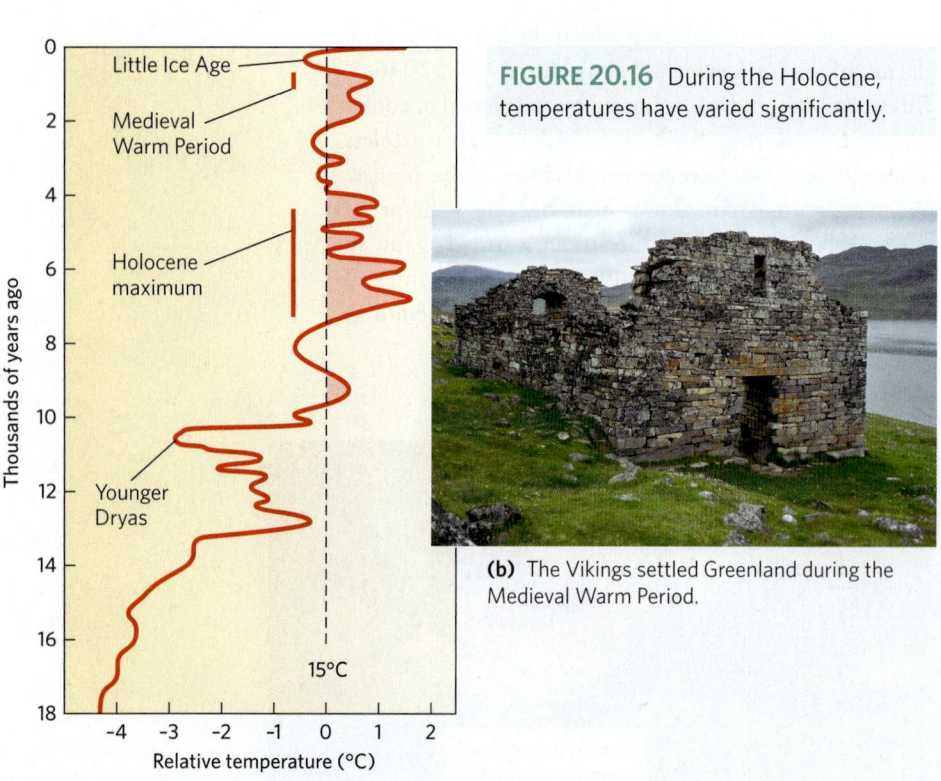

FIGURE 20.16 During the Holocene, temperatures have varied significantly.

(a) There have been several temperature highs and lows since the most recent glaciation.

(b) The Vikings settled Greenland during the Medieval Warm Period.

Today, glaciers extend only partway down the side valleys.

Lateral moraine

Toe today

Toe during Little Ice Age

(c) During the Little Ice Age, glaciers in the Alps advanced. They have since melted away.

Climate change over geologic time includes episodes of both global warming and global cooling. Researchers use the stratigraphic record to characterize long-term changes (taking place over millions to hundreds of millions of years). They use other indicators, such as plankton species, tree rings, pollen, or variations in isotope ratios, to document short-term changes that have happened over centuries to millennia.

Quick Question

How has climate changed in the northern hemisphere since the end of the last glaciation?

20.5 Causes of Climate Change

The Earth has hosted living organisms for at least the last 3.8 billion years. Since life requires liquid water, the presence of fossils throughout the stratigraphic record indicates that at least part of our planet's surface temperature has remained between the freezing and boiling points of water through this time. Researchers refer to this situation as the *Goldilocks effect*, for like Baby Bear's porridge in the tale of *Goldilocks and the Three Bears*, the Earth is not too hot and not too cold . . . it's just right for liquid water, and therefore, for life to survive (Fig. 20.17). The Goldilocks effect exists on the Earth for two reasons. First, the distance that our planet lies from the Sun is just right so that not too much and not too little solar radiation reaches the planet's surface. Put another way, the Earth lies in the *habitable zone* of the Solar System. Second, the concentration of greenhouse gases in the air is enough to keep air warm, but not too much to make it scalding.

Notably, although the Goldilocks effect keeps the Earth habitable, the temperature in the lower atmosphere has changed significantly over time, as we've seen. Why do these changes take place? The answer depends on the duration of the change under discussion. As we'll now see, factors controlling *long-term climate change*, meaning changes that take place over the course of millions of years, are not necessarily the same as factors controlling *short-term climate change*, meaning changes that take place over the course of decades to millennia. One factor playing an important role in both, however, is the *carbon cycle* (Bx. 20.2)

Long-Term Climate Controls

No single factor determines when the Earth's climate, overall, shifts from hothouse to icehouse conditions. We discuss several possible factors below.

SOLAR ENERGY OUTPUT. Astronomers have determined that the intensity of radiation reaching the Earth from the Sun has changed significantly over geologic time. In fact,

the Sun may be 30% brighter today than it was in the early Archean. This change has been steady, because it relates to the Sun's progressive increase in density that happens as nuclear reactions fuse hydrogen atoms together to form helium (see Chapter 23). Therefore, it doesn't correlate with changes from hothouse to icehouse conditions, and changes in solar-energy output do not contribute to long-term climate change.

Notably, if the Sun's intensity were the only factor controlling the Earth's temperature, the planet's surface water should all have frozen during the early Archean, but stratigraphic records prove that this was not so. Researchers refer to the apparent contradiction between the temperature that should have existed under a weaker Sun and the actual temperature of the early Earth as the *faint young Sun paradox*. This paradox can possibly be resolved by keeping in mind that the atmosphere of the early Earth contained much more CO_2 than it does today (see Fig. 17.5). The greenhouse effect caused by the additional CO_2 conceivably counteracted the effect of having a fainter Sun, and kept the Earth's surface temperature above freezing.

CHANGES IN ATMOSPHERIC COMPOSITION. The Earth's early atmosphere consisted mostly of H_2O and CO_2 gas (see Chapter 17). An atmosphere so rich in greenhouse gases would make the Earth's surface realm into an oven, like that of Venus today. Fortunately, the Earth cooled sufficiently for water to precipitate, and once the oceans had formed, CO_2 dissolved in the water and then became incorporated into rock. As a result, the atmospheric concentration of CO_2 decreased substantially, temperatures dropped, and life could exist. Enough CO_2 has remained in the atmosphere, however, to keep the oceans above the freezing temperature of water. When photosynthetic

FIGURE 20.17 The Goldilocks effect, as applied to the Earth. Our planet is just right for liquid water, and therefore for life, to exist.

Venus is too hot.

The Earth is just right.

Mars is too cold (and too small).

BOX 20.2 **A Deeper Look**

The carbon cycle

As we discussed in Chapter 10, climate change is an aspect of global change. Global change includes *unidirectional changes*, those that don't repeat during Earth history, and *cyclic changes*, those that repeat the same steps over and over, though not always with the same outcome. Researchers refer to certain types of cyclic changes as *biogeochemical cycles* because they involve exchanges of chemicals among a variety of living and nonliving reservoirs in the Earth System.

In Chapter 12, we discussed an important example of a biogeochemical cycle, the hydrologic cycle (cycling of H_2O). Another example, the **carbon cycle**, involves the transfer of carbon among various reservoirs **(Fig. Bx20.2)**. This cycle plays a key role in climate change because some carbon occurs in greenhouse gases (such as CO_2 and CH_4). The Earth System contains many reservoirs of carbon. Carbon initially enters the carbon cycle when it bubbles out of volcanoes and into the atmosphere as CO_2. Some of the CO_2 in air dissolves in the oceans to form bicarbonate (HCO_3^-) ions. This carbon later becomes incorporated into calcite ($CaCO_3$) or related minerals in shells such as those in reefs. If

buried deeply, this calcite cements together to form limestone, which, if metamorphosed, forms marble.

Carbon extracted from the air as CO_2 by plants may later be released again as CO_2 when plant biomass burns, or as CH_4 when plant biomass decomposes. Animals that eat plants may later release carbon as CO_2 and CH_4 as a consequence of respiration and flatulence, respectively. Some organic matter, however, is preserved in soil, permafrost, and gas hydrates (an ice-like material containing water and methane), and some undergoes burial and remains preserved underground over geologic time as fossil fuels. When people burn fossil fuels for energy, carbon returns to the atmosphere as CO_2. When permafrost melts and rots, or gas hydrates melt, CH_4 returns to the atmosphere.

FIGURE Bx20.2 The Earth System contains many reservoirs among which carbon is exchanged during the carbon cycle.

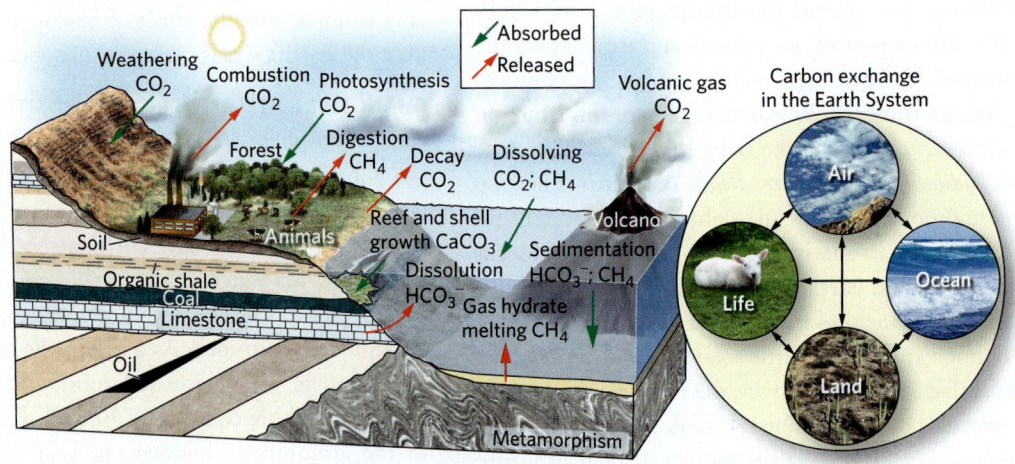

organisms appeared, they extracted additional CO_2 from seawater, and when those organisms died, the extracted carbon was mixed into sediment on the seafloor. Burial of this organic-rich sediment locked even more carbon into strata underground. Overall, the transfer of greenhouse gases from the air into the solid Earth led to a decrease in global temperature. Only after this had happened, by 2.4 Ga, could icehouse conditions even become a possibility. Not surprisingly, the earliest ice age began after 2.4 Ga.

Later changes in atmospheric greenhouse gas concentrations during Earth history caused subsequent climate changes that were recorded in the stratigraphic record as well. For example, some researchers speculate that the growth of immense coal swamps during the Carboniferous decreased atmospheric CO_2 concentrations enough to cause global cooling and the initiation of the Late Carboniferous

ice age. Overall, therefore, changes in greenhouse gas concentration can contribute to long-term climate change.

DISTRIBUTION OF LAND. The late Proterozoic, late Paleozoic, and Pleistocene ice ages all occurred when substantial areas of land lay at high latitudes. Why? When large continents drift to high latitudes, less ocean water can reside in their vicinity. As we have discussed, water has a much higher heat capacity than land. This means that, without oceans at high latitudes, land at those latitudes becomes colder. If broad areas of high-latitude land lie at a great distance from the warmth of ocean waters, conditions on land can remain cold enough all year for continental ice sheets to grow. The distribution of continents also affects the amount of precipitation on land. For example, tropical rainforests can grow only when large areas of land lie at low latitudes, where they receive heavy rainfall.

In addition, the distribution of land affects ocean currents, which control heat distribution around the planet. In some cases, even a small amount of land, when strategically located, can affect currents. For example, when the Central American volcanic arc grew in the Cenozoic, it blocked currents from transporting warm Pacific water into the Atlantic, and this change affected temperatures around the Atlantic.

SEA-LEVEL CHANGE. At times during Earth history, the interiors of continents have been flooded by shallow seas (see Chapter 10). When water covers significant portions of the land, the distribution of heat around the globe changes, and therefore, the climate changes.

VOLCANIC ACTIVITY. Generally, volcanic activity adds only relatively small amounts of new CO_2 to the air, and the Earth System quickly redistributes this gas into other reservoirs. But at certain times in geologic history, the amount of volcanic activity increased dramatically. This change may have been sufficient to cause a substantial increase in the atmospheric concentration of CO_2, which in turn would have caused global warming. For example, widespread rift-related volcanism associated with the breakup of Pangaea, along with the eruption of large igneous provinces, may have contributed to Cretaceous warming.

UPLIFT OF LAND SURFACES. Tectonic events that lead to the long-term uplift of large areas of land to form mountains or plateaus may affect atmospheric CO_2 concentrations, because these events expose rock to chemical weathering reactions that absorb CO_2 (see Fig. Bx20.2). Uplift of the Himalayas and the Tibetan Plateau, for example, may have contributed to the global cooling that occurred after the Eocene climatic optimum. Mountain building also affects global atmospheric circulation patterns and, therefore, the distribution of temperature and rainfall.

Natural Short-Term Climate Controls

During the Pleistocene Ice Age, continental glaciers advanced and retreated perhaps 20 to 30 times. The most recent advance, or *glaciation*, ended only about 11,000 years ago, and since then the planet has been in an *interglacial*. What factors can cause such changes, or comparable ones, that have relatively short durations? Some changes of very short duration (a year or decade) may simply reflect natural variation. But to explain events that last centuries to millennia, researchers point to the following mechanisms:

MILANKOVITCH CYCLES. As Milutin Milanković recognized in 1920, the shape of the Earth's orbit (eccentricity) changes over a period of 100,000 years, the tilt of its axis changes over a period of 41,000 years, and the axis undergoes precession (wobble) over a period of 19,000 to 23,000 years. Such periodic changes are called Milankovitch cycles (see Fig. 14.47), and each affects the amount of solar energy

received at high latitudes during the summer (Fig. 20.18). The temperature difference at high latitudes between times when the cycles reinforce one another and times when they cancel one another strongly influenced the onset and retreat of glaciers during the Pleistocene ice ages.

CHANGES IN OCEAN CURRENTS. Recent studies suggest that the configuration of currents can change quickly, and that these changes could affect climate. The Younger Dryas, for example, may have resulted when a sudden release of water from melting glaciers spread a layer of freshwater over the North Atlantic and temporarily stopped the thermohaline circulation throughout the oceans (see Chapter 15). Rapid growth or loss of sea ice may similarly affect currents.

LARGE ERUPTIONS OF VOLCANIC AEROSOLS. Not all of the sunlight that reaches the Earth penetrates its atmosphere and warms its surface. Some gets reflected by the atmosphere. The degree of reflectivity, or **albedo**, of the atmosphere increases if the concentration of volcanic aerosols in the atmosphere increases. Particularly large eruptions can emit enough aerosols (particularly sulfur dioxide, SO_2) to affect global temperature for months to years.

In 1815, Mt. Tambora in Indonesia exploded violently. This eruption followed other large volcanic

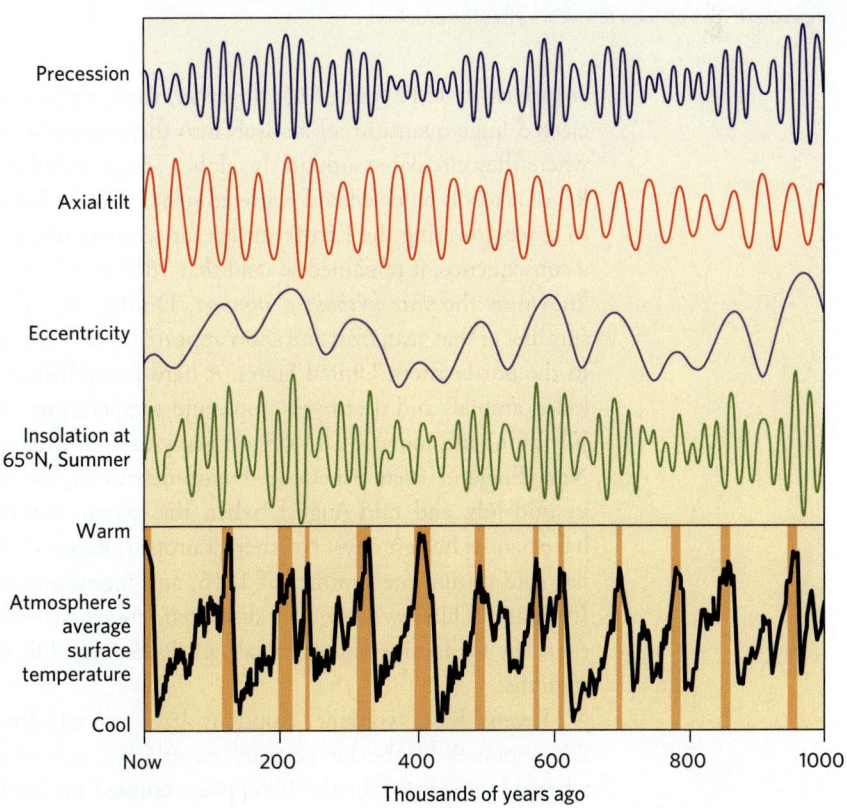

FIGURE 20.18 When the effects of the Milankovitch cycles—precession, axial tilt, and eccentricity—are combined, they cause changes in the daily insolation (the amount of heat added to the atmosphere) at 65°N. The brown stripes are the warmest times of interglacials.

FIGURE 20.19 Effects of the eruption of Mt. Pinatubo in 1991.

(a) A view from the space shuttle orbiting the Earth shows the haze caused by Pinatubo's aerosols in the stratosphere.

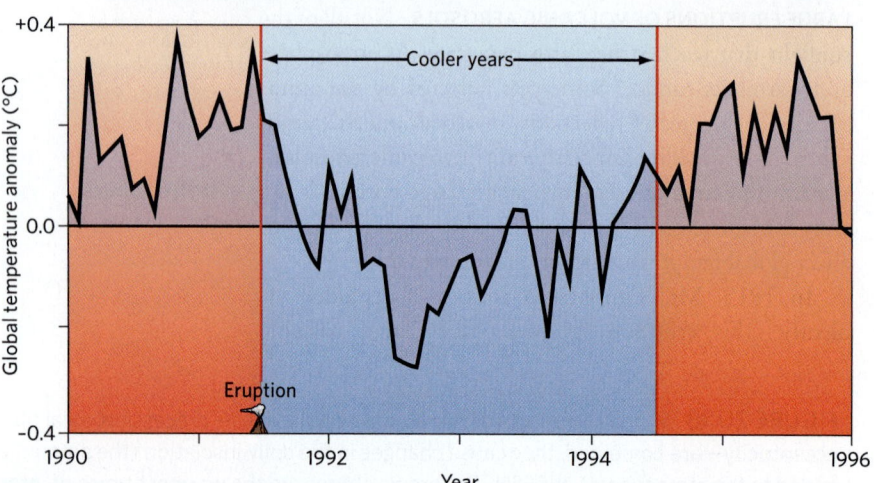

(c) The large injection of dust and aerosols into the atmosphere caused the average global temperature to drop for a few years after the eruption.

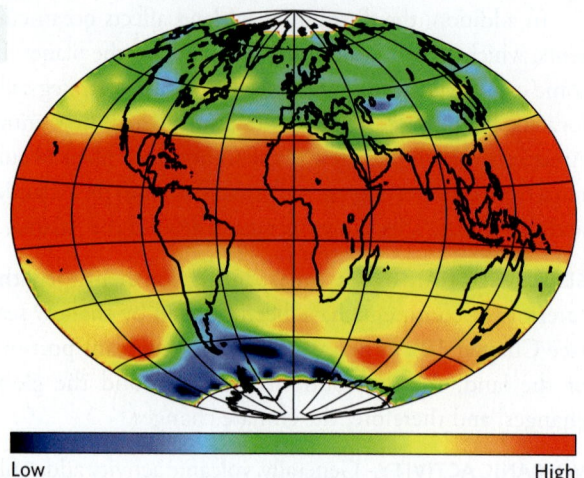

Low   High
Aerosol concentration

(b) Two months after the eruption, volcanic aerosols surrounded the Earth.

(see Fig. Bx4.3c), researchers documented a 3-year-long global decrease in temperature and precipitation (Fig. 20.19).

CHANGES IN SURFACE ALBEDO. Regional-scale changes in the nature of vegetation cover, in the proportion of snow and ice cover, or in the distribution of light-colored volcanic ash can affect the albedo of the Earth's surface. Increased albedo causes cooling, whereas decreased albedo causes warming.

ABRUPT CHANGES IN CONCENTRATIONS OF GREENHOUSE GASES. A relatively sudden change in greenhouse gas concentrations in the atmosphere can affect climate. Such a change might happen in several ways. If ocean temperatures warmed or sea level dropped, some of the gas hydrates (methane-containing ice that forms when CH_4 produced by decaying organic matter combines with water ice) buried in seafloor sediments would melt suddenly. Such melting would release CH_4 to the atmosphere. A similar release of CH_4 might happen if permafrost started to melt and the organic matter within it began to rot. Finally, because growth of photosynthetic organisms removes CO_2 from the atmosphere, algal blooms and changes in forest cover could also conceivably change atmospheric CO_2 concentrations.

Take-home message . . .

Factors causing long-term climate change include changes in atmospheric CO_2 concentrations, changes in the distribution of continents and oceans, and uplift of land surfaces. Those causing short-term climate change include the Milankovitch cycles and large volcanic eruptions.

Quick Question

What factors lead to changes in atmospheric CO_2 concentrations over geologic time?

eruptions in 1812 and 1814. Together, these eruptions ejected huge quantities of aerosols into the stratosphere, where they circulated around the globe. The global blanket of aerosols reflected and scattered solar radiation back to space, lowering the Earth's surface air temperature. As a consequence, it remained so cold that 1816 came to be known as the *year without a summer*. During the early summer of that year, frost and snow appeared several times in the northeastern United States. A hard freeze in June killed animals and destroyed crops, and temperatures in July didn't rise above 7°C (45°F) on many days. Northern New England even experienced crop-destroying frosts in mid-July and mid-August, when the region should have had its hottest days. Northern Europe suffered similar cold during the summer of 1816, and monsoons in India and China were severely disrupted, events that set the stage for famine and outbreaks of disease that killed millions.

Recent lesser volcanic eruptions have caused similar responses in the temperature record. For example, when Mt. Pinatubo in the Philippines erupted in 1991

FIGURE 20.20 Measurements of global warming.

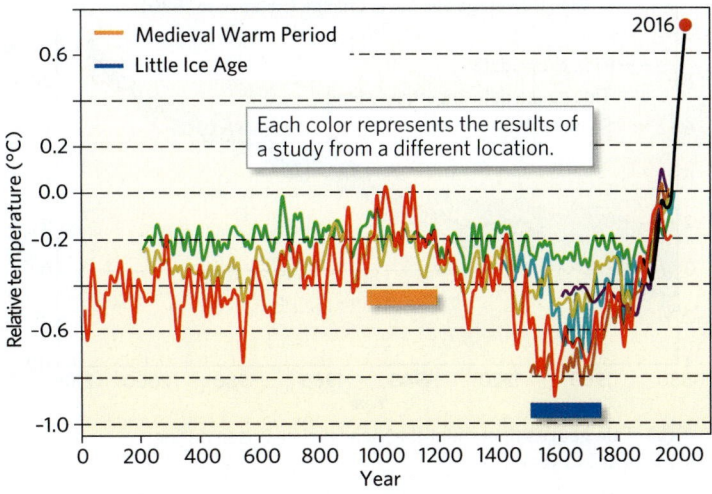

(a) Reconstructions of average global temperatures during the past 2,000 years. The black line represents direct measurements.

(b) Average annual surface air temperatures, 1880–2016. Squares and circles represent annual values, and lines are 5-year averages.

20.6 Evidence for and Interpretation of Recent Climate Change

Let's now turn our focus to the climate of the past few centuries, the time during which human society has had a significant impact on the Earth System. Researchers constantly publish new data and interpretations pertaining to this topic. In fact, the amount of information relevant to documenting and understanding climate change became so overwhelming that in 1988, the World Meteorological Organization, in collaboration with the United Nations Environment Program, founded the **Intergovernmental Panel on Climate Change (IPCC)** to evaluate published climate-related studies and summarize their conclusions in a report for a general audience. This report, a new edition of which comes out every 5 years, has had an important influence on political discussions pertaining to climate change. In this section, we examine evidence for global temperature change and its consequences during the Earth's recent past, as summarized by the IPCC.

Observed Temperature Change in the Recent Past

By combining direct measurements made by thermometers with measurements of paleoclimate indicators, researchers have pieced together a graph showing how *average annual global surface temperature* has changed over the past 2,000 years **(Fig. 20.20a)**. Prior to about 1400, the average global surface temperature rose and fell within a limited range. Over the next four centuries, the Earth experienced the Little Ice Age, in which the average temperature was 0.2°C–0.4°C (0.4°F–0.7°F) colder than

in the previous centuries. Since 1800, the temperature has warmed, with the largest increase after the beginning of the industrial revolution. Because of the shape of the curve depicting this change, the graph has come to be known, informally, as the *hockey-stick diagram*. It remains the subject of research because not all researchers interpret the paleoclimate data it portrays in the same way.

If we zoom in on the recent past (1880–2018), we see that warming hasn't been continuous. We also see that overall, the northern hemisphere has warmed more than the southern hemisphere **(Fig. 20.20b)**: the former has warmed by about 1.2°C (2.1°F) and the latter by about 0.9°C (1.6°F).

FIGURE 20.21 Pattern of temperature change based on measurements taken between 1961 and 2014.

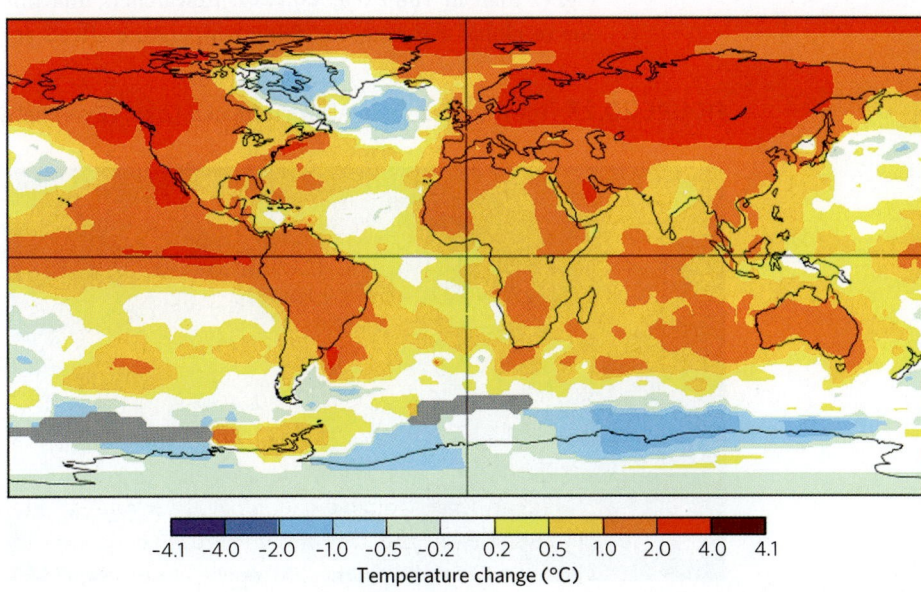

FIGURE 20.22 Changes in sea level since the end of the last glaciation.

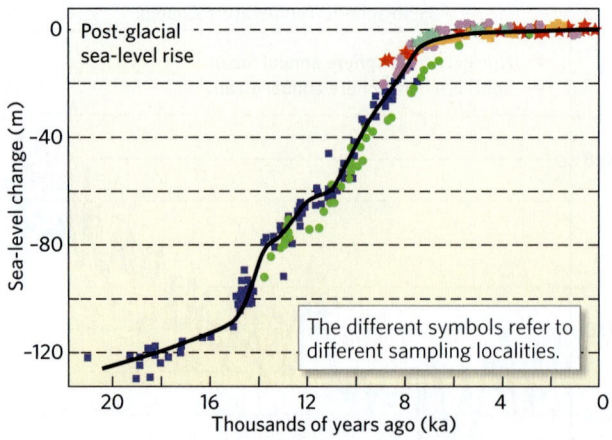

(a) Sea level rose rapidly after the last glaciation due to melting of continental glaciers.

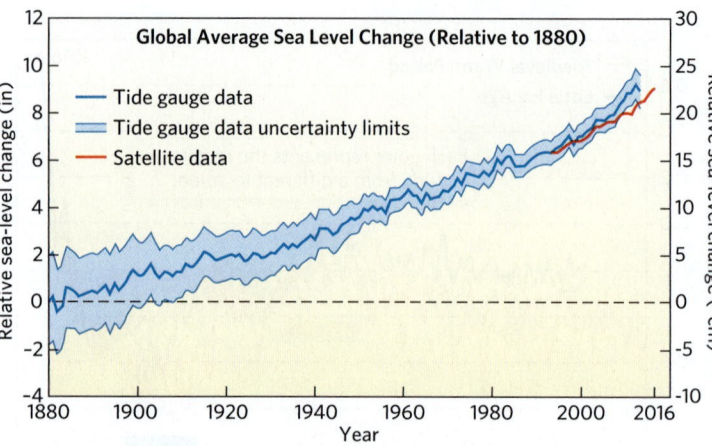

(b) Tide gauges document that sea level has risen during the past 130 years.

Land areas have been affected most, with the greatest warming taking place in the Arctic, northern North America, and northern Asia (Fig. 20.21). Air over the oceans has undergone less warming because the oceans' high heat capacity moderates temperature change.

Observed Recent Global Sea-Level Change

During the last glaciation of the Pleistocene Ice Age, a huge volume of water moved from the ocean reservoir into the glacial reservoir of the hydrologic cycle (see Chapter 12), and sea level dropped. When the glaciers melted, that water returned to the ocean, and sea level rose by about 120 m (about 400 ft) (Fig. 20.22a). This rise tapered off about 8,000 years ago, when the last of the North American and Asian continental glaciers disappeared.

In the past few centuries, sea level has started to rise again. In fact, sea level today lies about 23 cm (9 in) higher than in 1885 (Fig. 20.22b). Researchers link this rise to global warming, for rising temperature affects sea

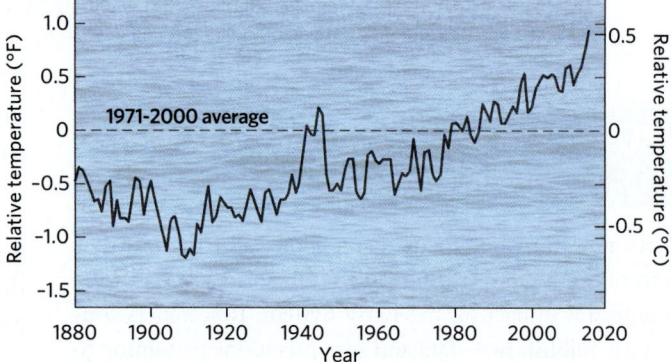

(c) Average global sea-surface temperatures have increased since 1900. This graph shows values relative to the average for 1971–2000.

level in two ways. First, the water in the ocean is warming (Fig. 20.22c), and as liquid water warms, it expands. Because oceans are confined on their bottoms and sides by rock and sediment of the ocean floor, this expansion causes the sea surface to rise. Second, when temperature rises, more glacial ice melts, so water that had been locked in contemporary glaciers moves into the oceans.

FIGURE 20.23 The Muir Glacier in Alaska retreated 12 km (7 mi) between 1941 and 2004.

FIGURE 20.24 Changes in glacial volumes.

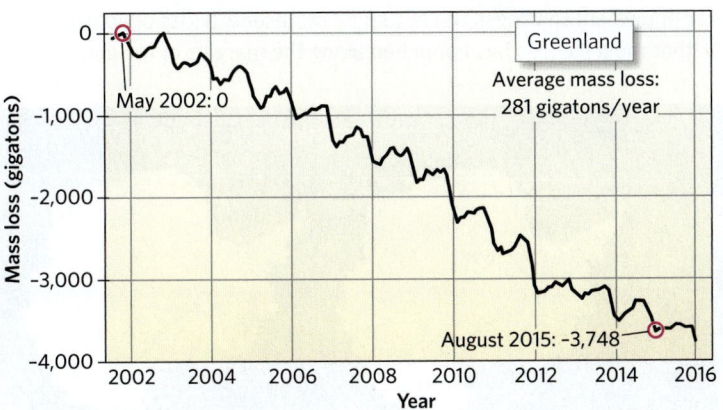

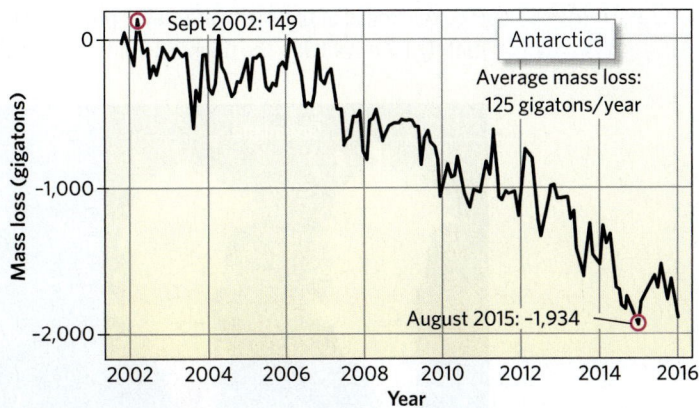

(a) Ice-mass loss in gigatons (G) as measured by the GRACE satellites in Greenland and in Antarctica over the period 2002–2016.

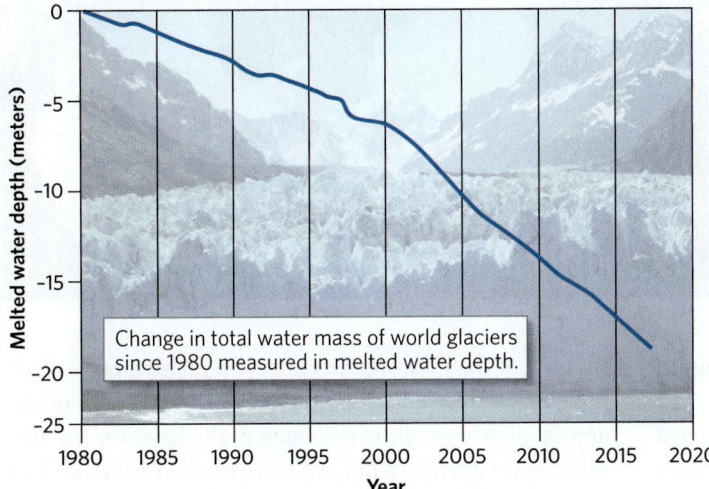

Change in total water mass of world glaciers since 1980 measured in melted water depth.

(b) Global glacial ice volume has been decreasing by 400 km³ (about 100 mi³) per year.

Animation

Global Sea Level Change

Are the ice sheets covering Antarctica and Greenland also shrinking? To answer this question, researchers have used the GRACE (Gravity Recovery and Climate Experiment) satellites to measure the gravitational pull of the Earth very precisely. The resulting measurements indicate that the gravitational pull exerted by the masses of ice in the Antarctic and Greenland ice sheets has decreased noticeably **(Fig. 20.24a)**. Furthermore, their rates of ice loss have accelerated. As a result, the Greenland ice sheet now thins by about 1 m (3 ft) per year. Addition of glacial melting measurements

Other Indicators of Recent Climate Change

In order to further test the idea that the Earth's climate has been warming in recent centuries, researchers have examined several other features of the Earth System. They have focused on features whose characteristics reflect temperature trends averaged over a time frame of decades, rather than on indicators of annual or seasonal natural variation.

GLOBAL GLACIAL RETREAT. Compare an old photograph of a valley glacier with a present-day photograph of the same glacier **(Fig. 20.23)**. The difference is obvious. Most valley glaciers worldwide have retreated and thinned dramatically, leaving their lateral and terminal moraines stranded (see Chapter 14). In many cases, glaciers that once filled valleys have disappeared entirely. For example, Glacier National Park in Montana had 150 glaciers in 1850, but now has only 25. The remaining glaciers in the park continue to lose volume by 2% to 6% per year, so all of the glaciers in the park may be gone in just a few decades.

FIGURE 20.25 Lakes on the Greenland ice sheet often disappear suddenly as they drain through crevasses to the base of the ice sheet.

FIGURE 20.26 Ice loss from the Earth's continental glaciers, as measured by the GRACE satellite. Note that because not all ice has the same density, researchers represent the thickness of the ice layer lost in terms of water-equivalent thickness (the thickness of the meltwater layer produced). Water-equivalent thickness of 1 m (3.3 ft) represents a loss of approximately 1.2 m (4 ft) in the thickness of the glacier. Note that most ice loss has happened along the margins of the glaciers.

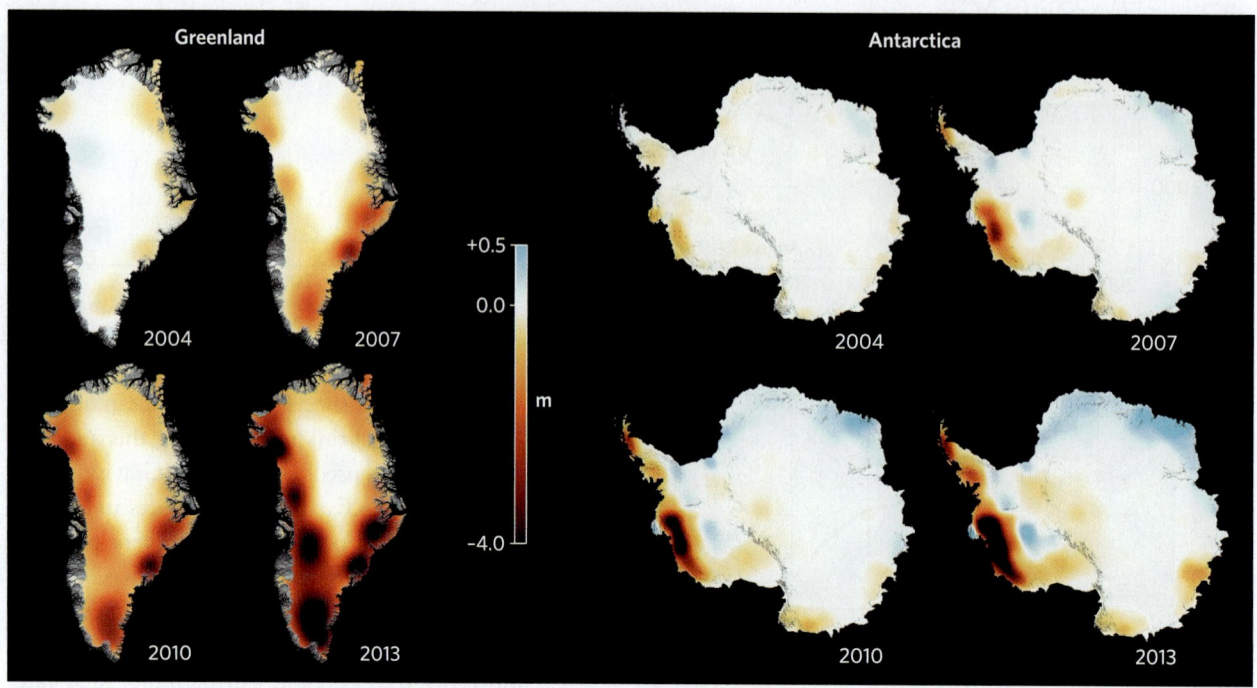

indicates that globally, ice volume has decreased significantly in the last half century (Fig. 20.24b).

These measurements of ice loss have been confirmed by direct observations. Researchers have found that large lakes of meltwater accumulate on the surface of the Greenland ice sheet in the summer; some of these lakes suddenly drain through crevasses down to the base of the glacier, then flow out to sea through tunnels melted into the base of the glacier or collect in lakes under the ice (Fig. 20.25). Researchers have also found that the number of days during the year when melting takes place in Greenland has increased (Fig. 20.26). The width of the annual melt zone along the edges of the ice sheet has also increased. Greenland has experienced most of its ice loss on its perimeter. The Antarctic ice sheet, meanwhile, has experienced both ice loss and ice gain on its perimeter in different locations. The gain is due to heavier snowfall. Melting trends also affect periglacial environments. For example, the upper layer of Arctic permafrost has started to melt earlier in the season, and over a broader area, than it did in the past (Fig. 20.27).

Sea-ice coverage in the Arctic Ocean has also been declining, so much so that in late summer, cruise ships can travel along the Northwest Passage, along the Arctic coast of Canada, from the Atlantic to the Pacific (Fig. 20.28a). A graph of September sea-ice area from 1979 to 2018 shows many ups and downs lasting a few years (Fig. 20.28b), but if the overall trend continues, it may be possible to sail to the North Pole within the next few decades. In Antarctica, large ice shelves have been breaking up rapidly. For example, a large portion of the Larsen B Ice Shelf on the coast of the Antarctic Peninsula disintegrated in 2002 (Fig. 20.29). In 2017, a slab of the Larsen C ice shelf (the next shelf to the south of Larsen B) measuring 160 km (100 mi) long by up to 50 km (30 mi) wide broke off to become an iceberg the size of the state of Delaware.

FIGURE 20.27 Lakes and puddles form when permafrost melts along the coast of the Arctic Ocean.

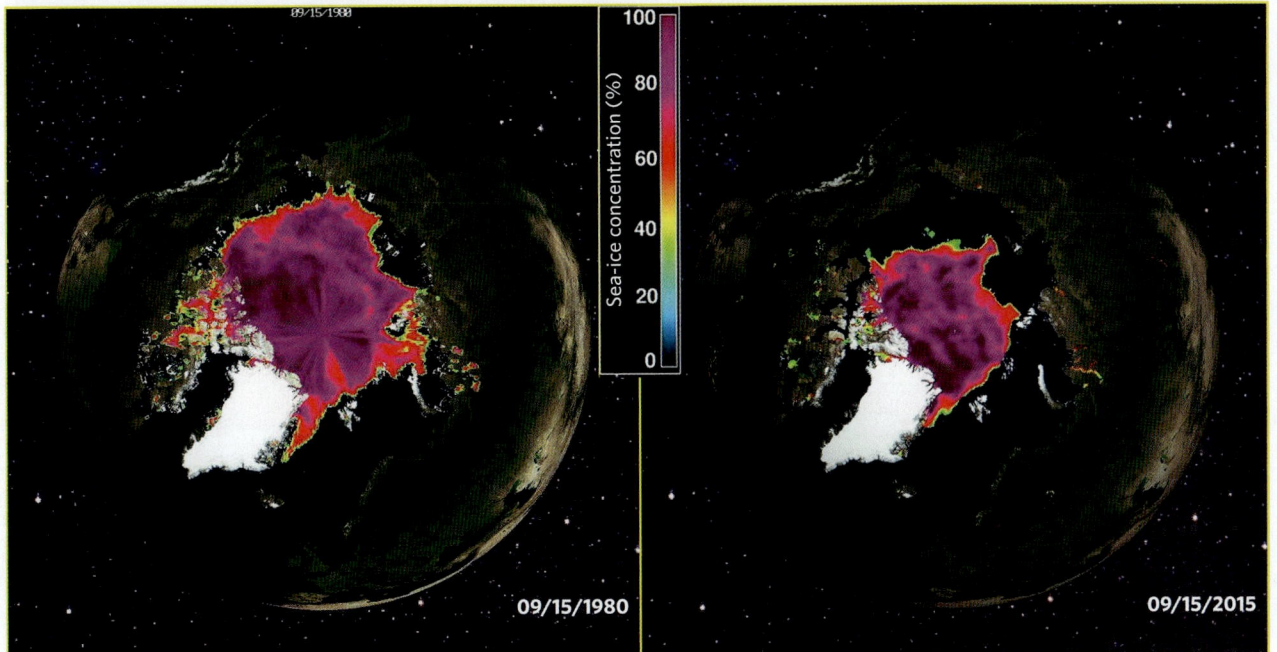

FIGURE 20.28 Sea-ice coverage of the Arctic Ocean has decreased.

(a) Satellite images of Arctic Ocean sea ice in 1980 and in 2015.

DISTRIBUTION OF ORGANISMS. Biological indicators also point toward global warming. For example, in the northern hemisphere, the growing season for plants has lengthened considerably. The time at which sap in maple trees starts to flow comes earlier in the year, and the time when leaves turn color comes later in the year. Comparing a plant hardiness zone map from 2012 with one from 20 years earlier emphasizes the change in the growing season **(Fig. 20.30)**. The ranges of birds and insects on continents have also changed noticeably in recent decades.

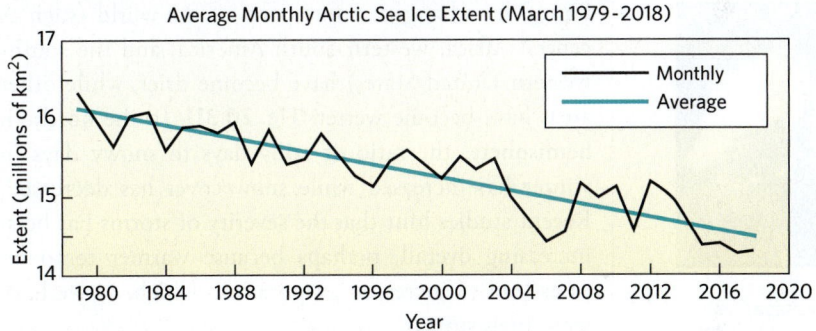

(b) A graph of sea-ice coverage relative to the average for 1979–2013 shows fluctuations, but also shows an overall downward trend.

FIGURE 20.29 A large portion of the Larsen B Ice Shelf on the coast of the Antarctic Peninsula disintegrated in 2002.

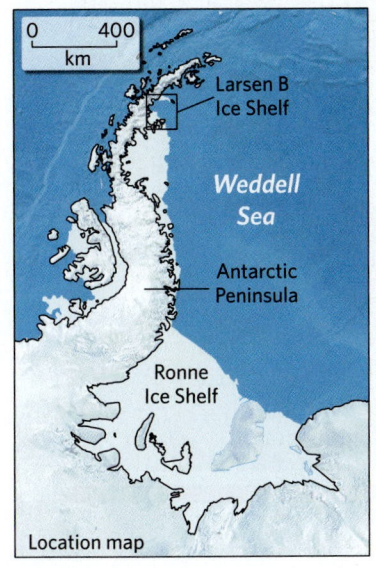

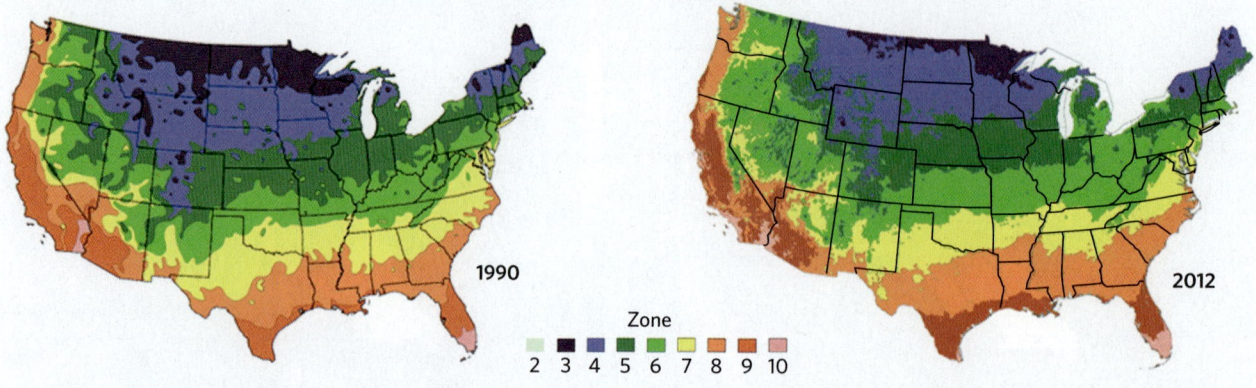

1990

2012

Zone
2 3 4 5 6 7 8 9 10

CHANGES IN GLOBAL WEATHER. Researchers have begun to detect changes in weather patterns over time. For example, during the past century, overall precipitation over land areas has increased by about 2% worldwide, and the distribution and character of this precipitation has changed. Some parts of the world (such as central Africa, western South America, and the southwestern United States) have become drier, while other areas have become wetter (Fig. 20.31). In the northern hemisphere, the ratio of rainy days to snowy days in winter has increased while snow cover has decreased. Recent studies hint that the severity of storms has been increasing overall, perhaps because warmer temperatures lead to increased evaporation, which, as we have seen, fuels storms.

Interpreting the Causes of Recent Climate Change

Which of the many causes of climate change that we discussed earlier in this chapter might be responsible for the observed global warming and associated sea-level rise of the past few centuries? The Keeling curve, as well as studies that track CO_2 and CH_4 concentrations over longer time frames (Fig. 20.32), have led nearly all climate scientists to conclude that an increase in the concentrations of these two greenhouse gases in the atmosphere serves as the main driver of global warming since the start of the industrial age. Does the change in greenhouse gas concentrations observed during the past century reflect natural causes, or does it reflect inputs into the Earth System

See for **yourself**

Northwest Passage, Arctic Ocean

Latitude: 74°22′33.45″ N
Longitude: 101°51′31.42″ W

Look down on this scene from an altitude of 5,246 km (3,260 mi) and you will see the Northwest Passage, the famed path between the Atlantic and the Pacific sought by ships over the centuries to avoid sailing all the way around the tip of South America. The passage has been blocked by sea ice for centuries. That is changing as the Arctic has warmed and Arctic sea ice coverage has diminished. In some recent years, open water has been found in late summer stretching along the coasts from Greenland to Alaska's northern coastline.

FIGURE 20.31 Trends in measured annual precipitation (% change) between 1900 and 2000. Yellow, orange, and red shading represent increases; green and blue shading represent decreases.

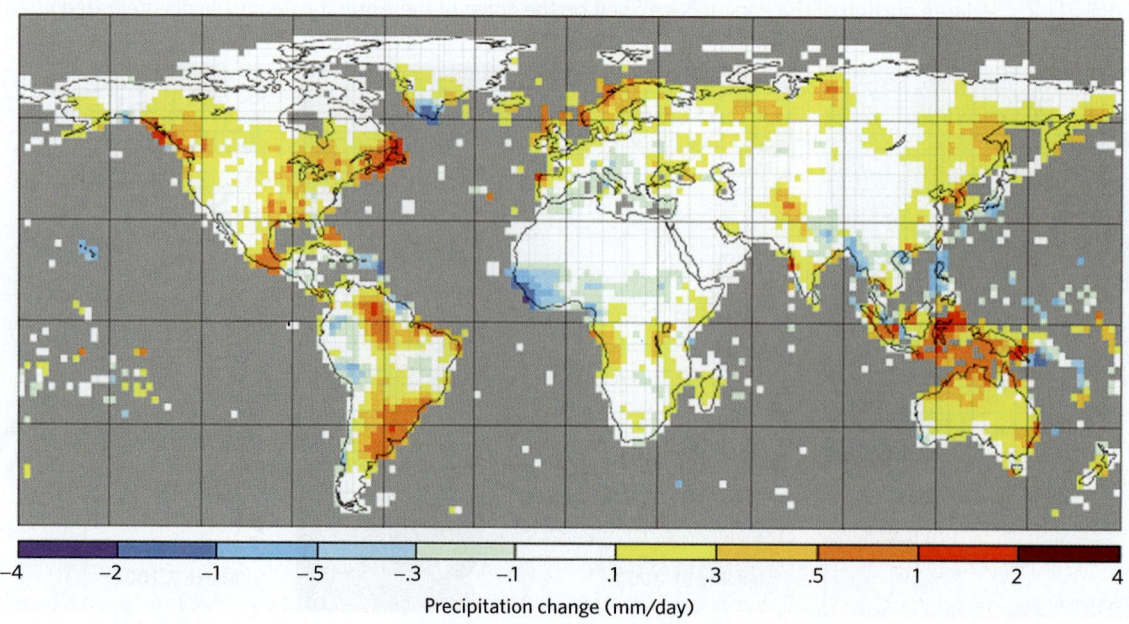

−4   −2   −1   −.5   −.3   −.1   .1   .3   .5   1   2   4
Precipitation change (mm/day)

FIGURE 20.32 Changes in atmospheric carbon dioxide and methane concentrations over time.

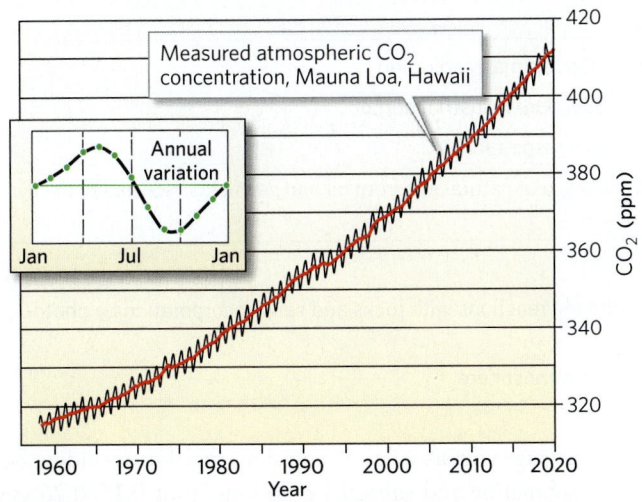

(a) The Keeling curve showing the change in CO_2 concentration.

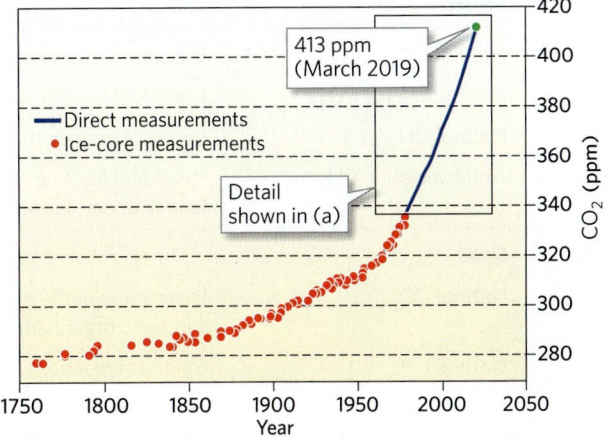

(b) Since the industrial revolution, the atmospheric CO_2 concentration has steadily increased.

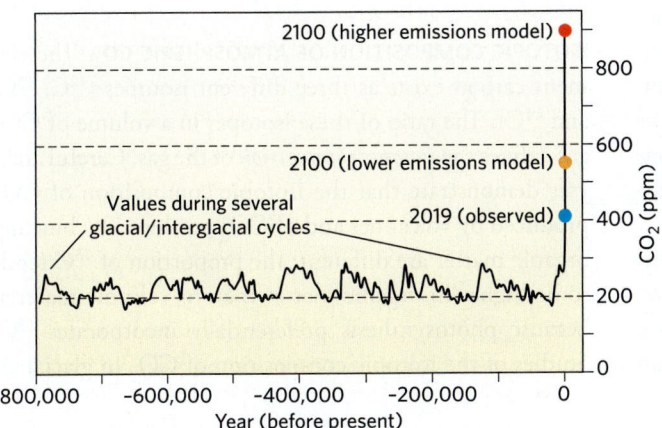

(c) Studies of ice cores from glaciers show that the CO_2 concentration varied between 180 and 300 ppm over the past 800,000 years. Now it is over 400 ppm.

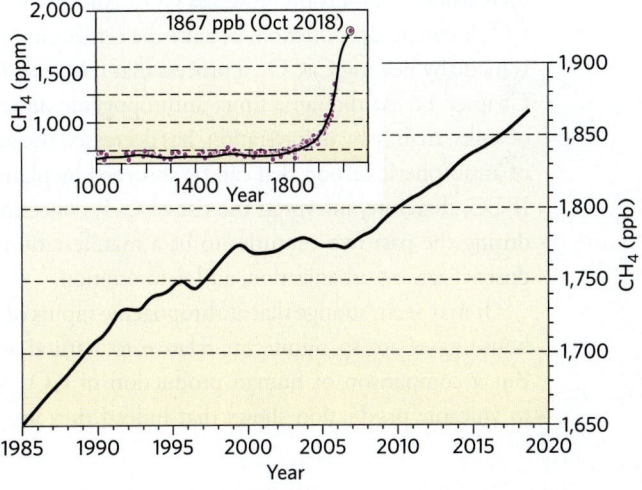

(d) The atmospheric concentration of CH_4, another greenhouse gas, has also been increasing.

from human activities? In other words, is the change due to natural causes or to anthropogenic causes? Insight into this question comes from several sources.

AIR-BUBBLE COMPOSITION IN GLACIAL ICE. Measurements of the composition of air bubbles in Antarctic ice cores provide a record of atmospheric CO_2 concentrations back through almost 800,000 years. This record indicates that during the alternating glaciations and interglacials of the Late Pleistocene, CO_2 concentrations varied between 180 and 300 ppm (see Fig. 20.32c). Therefore, the observed increases since the mid-19th century are beyond the range of the natural variation that has occurred during the last 800,000 years, suggesting that the change is anthropogenic.

THE CARBON BUDGET. As we discussed in Box 20.2, carbon passes through many components of the Earth System during the carbon cycle. If carbon transfers into the atmosphere in the form of CO_2 and CH_4 at a rate faster than

it transfers out of the atmosphere into the hydrosphere, biosphere, or geosphere, then the concentration of these greenhouse gases in the atmosphere will rise. Understanding whether greenhouse gases are being added at a rate faster than they are being removed requires that researchers characterize the Earth's **carbon budget** by quantifying *atmospheric carbon sources* that add carbon to the atmosphere and *atmospheric carbon sinks* that remove carbon from the atmosphere (Table 20.2). The difference between these two numbers indicates the amount of carbon that remains in the atmosphere in the form of CO_2 and CH_4.

After calculating the amounts of CO_2 and CH_4 being produced by their sources and comparing those quantities with the amounts being absorbed by sinks, researchers have concluded that significantly more greenhouse gases are being transferred from geologic reservoirs into the atmosphere than are being removed. Causes include dissolving in the ocean, being absorbed by biomass, or reacting with rock

> Earth's climate system is an ornery beast which overreacts even to small nudges.
>
> —WALLACE BROECKER
> (AMERICAN GEOCHEMIST,
> 1931–2019)

TABLE 20.2 Sources and Sinks of Carbon-Containing Greenhouse Gases

Sources	Causes
Natural CO_2	Volcanoes; decay of organic matter; animal respiration
Anthropogenic CO_2	Fossil fuel burning; concrete production; industrial output
Natural CH_4	Decay of organic matter; microbial respiration
Anthropogenic CH_4	Fossil fuel burning; venting or leakage of natural gas from oil and gas fields; rice paddy decay; landfill decay; livestock flatulence
Sinks	
Natural CO_2	Dissolution in the ocean; weathering reactions with rocks and soils; incorporation by photosynthetic organisms
Natural CH_4	Chemical reactions in soil and the atmosphere

(Fig. 20.33). Results summarized by the IPCC indicate that most of the CO_2 being released comes from burning fossil fuels whose combustion produces CO_2. Another source of CO_2 is concrete production, because the cement in concrete is made by heating $CaCO_3$, a process that releases CO_2; see Chapter 11. At the same time, anthropogenic destruction of sinks, mainly by deforestation, has decreased the amount of atmospheric carbon that can be absorbed by plants. The IPCC, therefore, interprets the rise of CO_2 concentrations during the past two centuries to be a manifestation of industrialization, urbanization, and deforestation.

It may seem strange that anthropogenic inputs of greenhouse gases are so significant relative to natural sources. But a comparison of human production of CO_2 relative to volcanic production shows that indeed they are. In an average year, all volcanic eruptions together, including both submarine and subaerial eruptions, emit 0.15–0.26 gigatons of CO_2. By comparison, human society emits about 35 gigatons of CO_2 each year (about 135 times as much).

ISOTOPIC COMPOSITION OF ATMOSPHERIC CO_2. The element carbon exists as three different isotopes (^{12}C, ^{13}C, and ^{14}C). The ratio of these isotopes in a volume of CO_2 gas defines the *isotopic composition* of the gas. Careful analyses demonstrate that the isotopic composition of CO_2 produced by volcanoes and of CO_2 produced by burning organic matter are different: the proportion of ^{13}C tends to be higher in organic matter than in volcanic material because photosynthesis preferentially incorporates ^{13}C. Studies of the isotopic composition of CO_2 in glacial air

FIGURE 20.33 Annual anthropogenic carbon emissions since 1800.

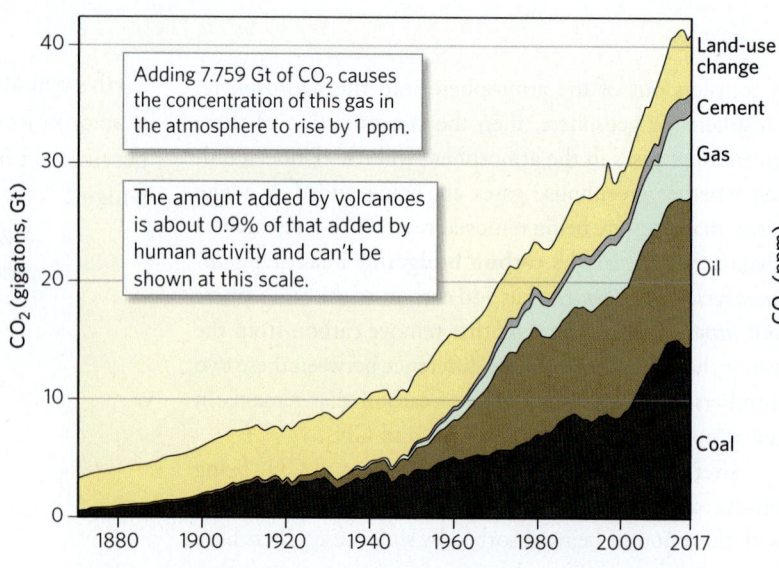

(a) The quantity of carbon emissions, overall, has increased dramatically since the start of the industrial age. Beginning in the mid-20th century, hydrocarbons dominated energy consumption.

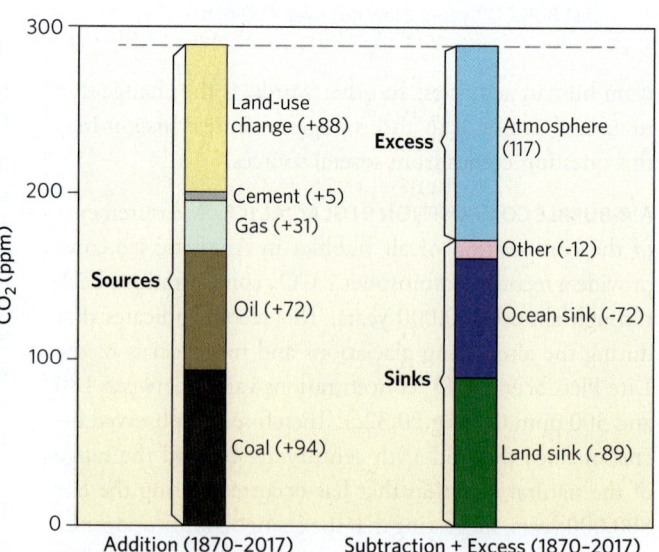

(b) Sources and sinks of atmospheric carbon since the start of the industrial age. The left column shows the relative proportions of different sources (in ppm). The right column shows where the added carbon goes. The "excess" is the amount remaining in the atmosphere, where it can contribute to global warming.

bubbles show that the proportion of ^{13}C in atmospheric CO_2 has increased since the industrial revolution. This change suggests that much of the new CO_2 entering the atmosphere must come from the burning of organic matter (biomass and fossil fuels), not from volcanoes.

Take-home message . . .

Global temperature and sea level have been rising since the mid-19th century. Many observed changes in the Earth System, such as changes in the volume of glacial ice and in the distribution of vegetation belts, support the conclusion that global warming is taking place. This warming corresponds to the largest increase in atmospheric CO_2 that has happened during the last 800,000 years. Researchers attribute this temperature increase to anthropogenic CO_2 production beginning with the industrial revolution.

Quick Question
How can global warming cause sea level to rise?

20.7 Climate in the Future

Global Climate Models

The development of high-speed supercomputers has allowed climate scientists to write programs, called **global climate models (GCMs)**, that simulate the behavior of climate over time scales of a century or more. These models, which use equations describing the physical and chemical interactions among the atmosphere, ocean, and land, are similar to models used for weather forecasting, but cover

much longer time spans. GCMs allow researchers, for example, to compare climate behavior in an imagined situation in which greenhouse gas concentrations remain at pre-industrial levels with the way the climate has actually behaved to date. These simulations show that if greenhouse gas concentrations had remained constant, the Earth would have experienced a slight cooling trend over the last century. The cooling trend would have been caused by decreases in incoming solar energy at high latitudes associated with the Milankovitch cycles. When observed greenhouse gas concentrations are added, however, the models predict the warming trend that we have actually experienced.

If global climate models have helped us interpret the past, can they help us predict the future? By adjusting the variables programmed into a GCM, researchers can provide a high emissions model (in which greenhouse gas emissions increase above current rates), a medium emissions model (in which the rate of greenhouse gas emissions stays about the same as today), and a low emissions model (in which the rate of greenhouse gas emissions decreases below current rates) **(Fig. 20.34a)**. The medium emissions model predicts that atmospheric temperatures will increase by 2°C–4°C (3°F–7°F) by the end of the 21st century, with the greatest warming taking place at high latitudes during the winter because the loss of sea ice will allow ocean water to moderate temperatures there **(Fig. 20.34b)**. The exact consequences of global warming remain uncertain, but according to researchers, the following events are possible:

- *A shift in climate belts and season length:* Temperate and desert climate belts will probably move to higher

FIGURE 20.34 Global climate model (GCM) predictions of future global warming.

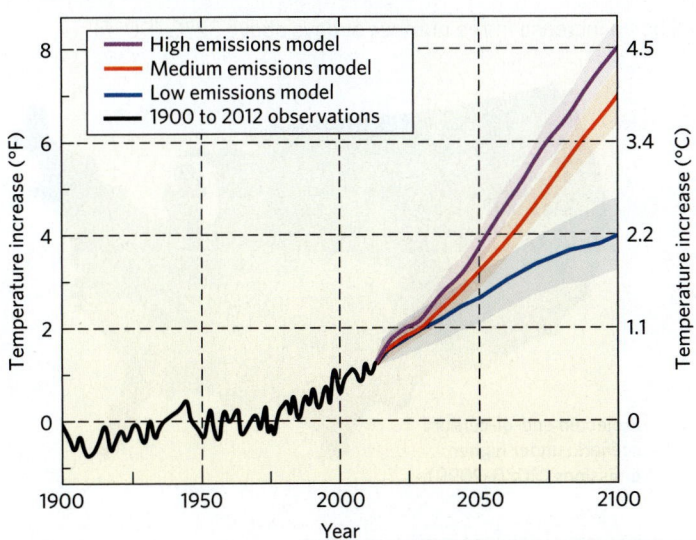

(a) GCM calculations for different rates of anthropogenic CO_2 emission. All suggest a significant global temperature increase by 2100.

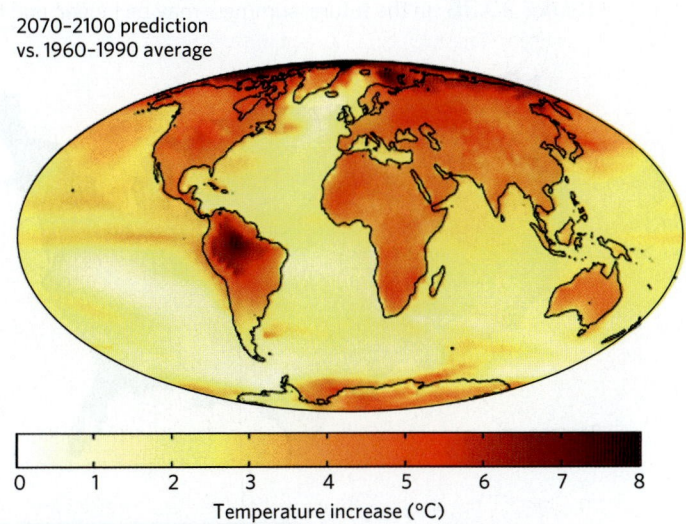

2070–2100 prediction vs. 1960–1990 average

(b) Global warming does not mean that all locations will warm by the same amount. This map shoes a model of possible variations for 2070–2100. Temperatures shown are relative to 1960–1990 averages.

FIGURE 20.35 In the future, temperate and desert climate belts will probably move to higher latitudes.

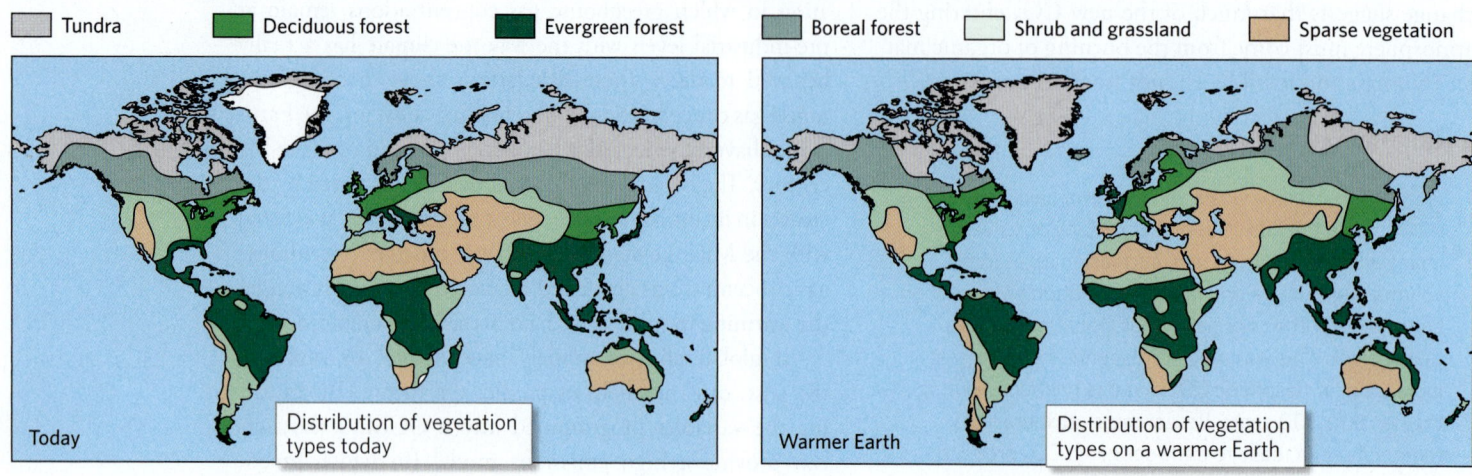

Legend: Tundra | Deciduous forest | Evergreen forest | Boreal forest | Shrub and grassland | Sparse vegetation

Today — Distribution of vegetation types today

Warmer Earth — Distribution of vegetation types on a warmer Earth

latitudes, so some present-day agricultural lands may become too hot to support the crops that they currently host, and climates appropriate for those crops will lie farther north **(Fig. 20.35)**. This change will negatively impact global agricultural productivity, for fields farther to the north have thinner soils and receive less sunlight than do those farther south.

- *More heat waves:* Because of warming, summers may lengthen by 4 to 6 weeks **(Fig. 20.36)**, and deadly heat waves may occur more often.

- *A change in precipitation patterns:* During the winter, more precipitation will take place in mid-latitudes, especially in western North America, where storms encounter mountain ranges **(Fig. 20.37a)**. During the summer, precipitation will probably increase at higher latitudes

and over the ocean, but will decrease over the southern United States and western Eurasia, so these regions will experience more frequent droughts **(Fig. 20.37b)**.

- *Ice retreat and snow-line rise:* Warming will drive further retreats of glaciers, and some will disappear entirely. Researchers are concerned that the ice sheet of West Antarctica may collapse (begin to melt and flow rapidly) if the ice shelves surrounding the ice sheet disintegrate. In nonpolar latitudes, the snow line in mountains will rise to higher elevations. A decrease in the volume of winter snowpack will cause a decrease in the volume of meltwater, an important component of the water supply in some regions.

- *Melting permafrost and methane:* Warming climates at high latitudes will cause areas of permafrost to

FIGURE 20.36 In the future, summers may be longer, and there will be an increase in the number of days above 32°C (90°F).

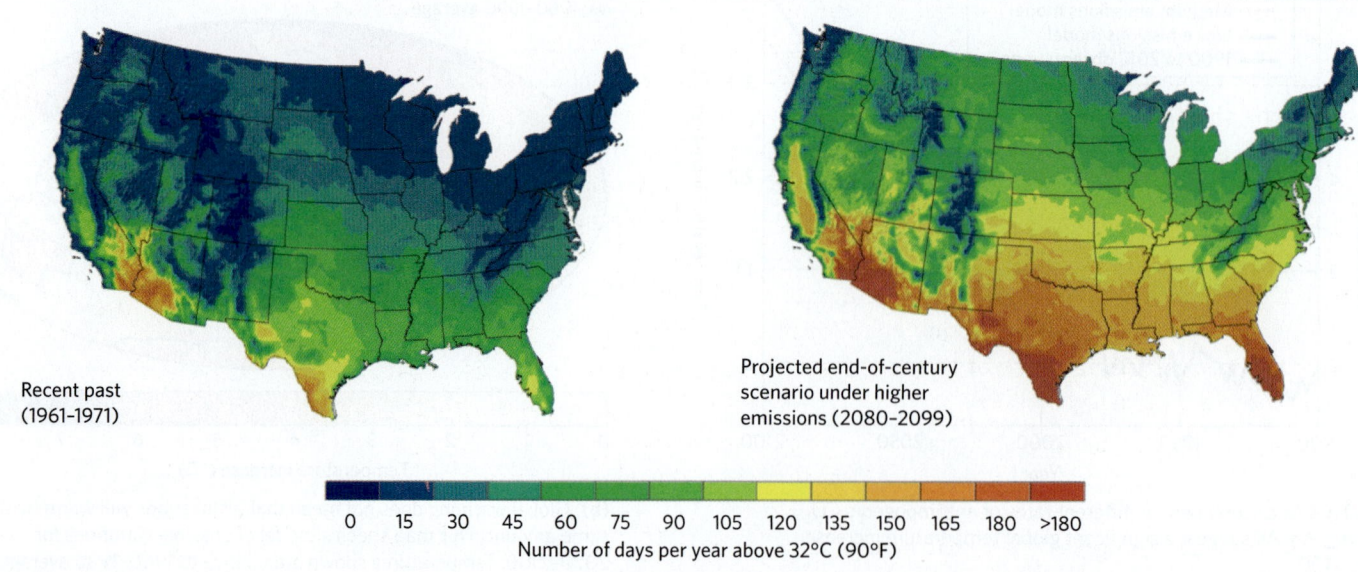

Recent past (1961–1971)

Projected end-of-century scenario under higher emissions (2080–2099)

0 15 30 45 60 75 90 105 120 135 150 165 180 >180
Number of days per year above 32°C (90°F)

thaw during the summer. When it defrosts, the organic matter in the former permafrost will decompose and release methane. Gas hydrate (methane) in seafloor sediments may also start to melt and release methane (see Fig. Bx20.2).

- *Additional rise in sea level:* Measurements suggest that sea level is currently rising by about 1.8 mm/yr (about an inch per decade). This rise has already caused flooding of coastal areas and has submerged some oceanic islands. Continued sea-level rise will only make such problems more widespread and may require that people abandon some communities (**Earth Science at a Glance**, pp. 660–661). Models suggest that by 2100, sea level may rise by an additional 20–60 cm (8–24 in) **(Fig. 20.38a)**. A sea-level rise of a meter or two could inundate regions of the world where 20% of the human population lives **(Fig. 20.38b)**. In some neighborhoods of Miami, Florida, particularly high tides already cause flooding **(Fig. 20.38c)**.

- *An increase in wildfires:* Warmer temperatures and drier conditions cause plants to dry out and therefore burn more easily. This means that global warming will lead to an increase in the frequency of wildfires. Given the number of immense wildfires in the past few years, this pattern has already started.

- *An interruption of thermohaline circulation:* Ocean currents play a major role in transferring heat across latitudes. According to some models, if global warming melts enough glacial ice, the resulting freshwater would dilute ocean surface water at high latitudes. This lower-density water would not sink, so the thermohaline circulation could slow or cease, preventing the transfer of heat to high latitudes.

- *Stronger storms:* An increase in average sea-surface and atmospheric temperatures will lead to increased evaporation from the sea. The additional moisture might nourish stronger storms that would drop greater amounts of precipitation, leading to increased amounts of flooding. A warmer climate could also shift jet streams to higher latitudes, causing the locations of storm tracks over the North Atlantic and North Pacific Oceans to shift.

FIGURE 20.37 Projected amounts of precipitation by 2070–2090, relative to those in 1980–2000, in the northern hemisphere under a medium emissions scenario.

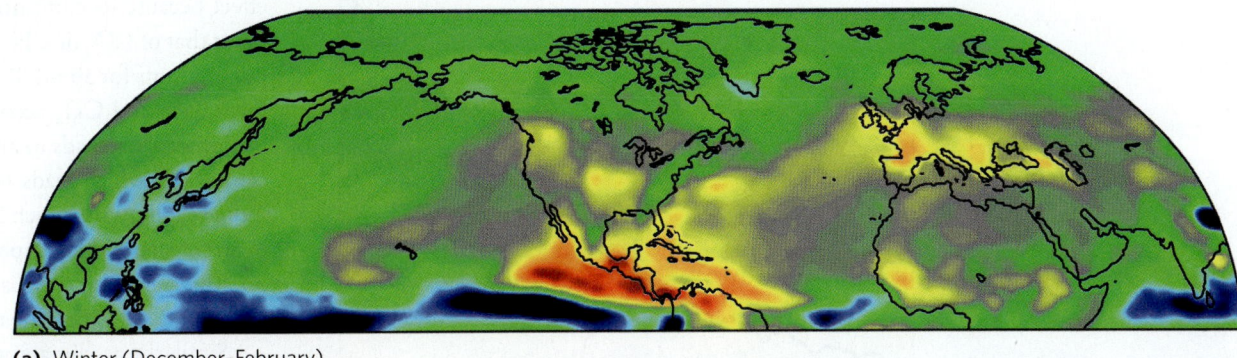

(a) Winter (December–February).

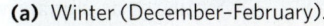

Seasonal precipitation change (cm)

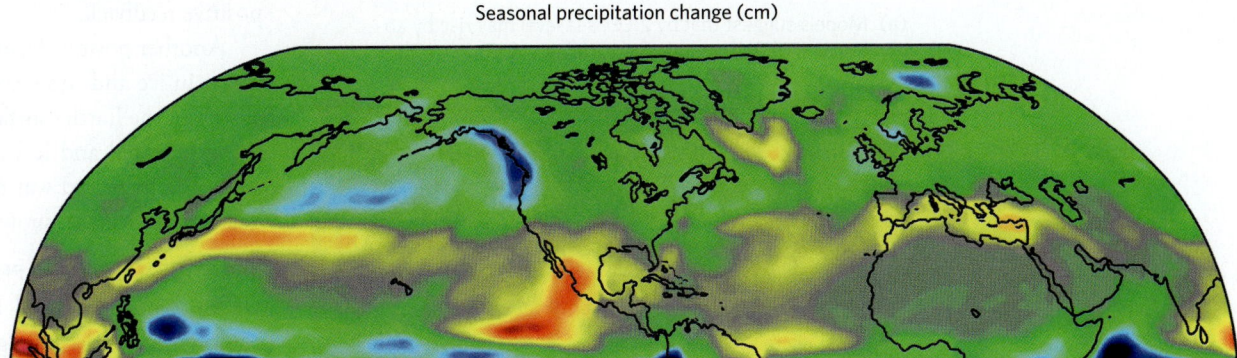

(b) Summer (June–August).

Complexities in Climate Predictions

The possible consequences of climate change over the next century that we describe in this book represent the consensus of most researchers. Complexities remain, however, and not all researchers agree on the quality of the data used in analyses, on the statistical approaches used to analyze these data, or on the importance of **feedback processes**, meaning phenomena that are themselves triggered by a change, and then either amplify and accelerate the change or subdue and slow down the change. Because of the importance of feedback processes, let's look at them more closely.

POSITIVE FEEDBACK PROCESSES. A *positive feedback process* amplifies the consequence of the initial change and may cause the rate of change to accelerate. The most important example of a positive feedback process during global warming involves the effect of warming on the concentration of water vapor in the atmosphere. Even though

FIGURE 20.38 Predictions of sea-level change.

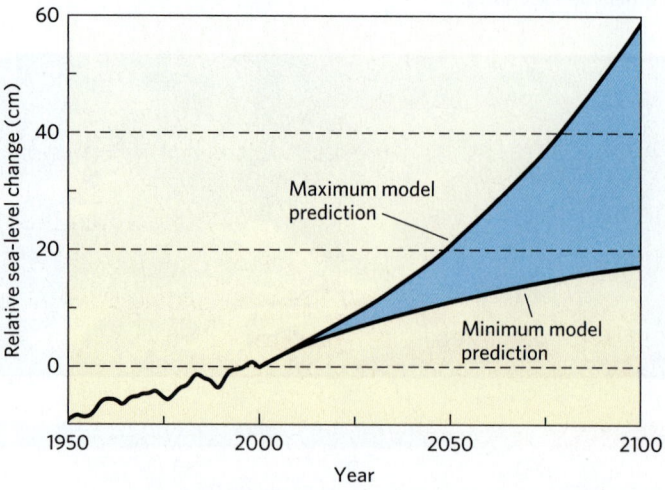

(a) Models suggest that by 2100, sea level may rise by an additional 20 to 60 cm, relative to today's level.

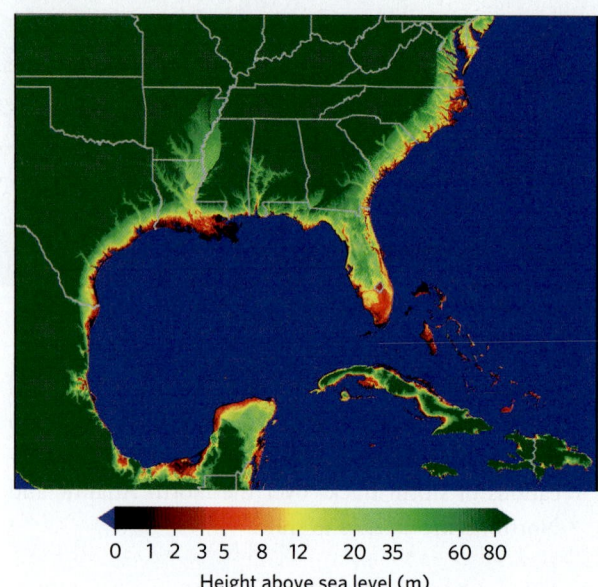

Height above sea level (m)

(b) If sea level rises by more than a meter, large areas of coastal lowlands, with large populations, may be flooded.

(c) Due to ongoing sea-level rise, when a particularly high tide takes place, parts of Miami Beach, Florida, get flooded with saltwater.

individual molecules of CO_2 or CH_4 are more efficient than a molecule of H_2O in trapping heat, water vapor overall causes most of the Earth's atmospheric greenhouse effect because its concentration in the atmosphere far exceeds that of CO_2 or CH_4. On a hot, humid day, water vapor accounts for about 2.3% of air, whereas CO_2 accounts for 0.04% and CH_4 accounts for 0.0002%. Warming of the atmosphere leads to an increase in its water vapor concentration, which leads to more warming and therefore, more evaporation, which in turn leads to a further increase in atmospheric water vapor—and this process goes on and on. This repeating cycle is a positive feedback process that probably accelerates global warming.

Similarly, as climate warms, permafrost and methane ice melts. This melting releases CH_4 into the atmosphere, which causes more warming, so this process also acts as a positive feedback.

Another positive feedback process results from the decrease in ice and snow cover, which in turn decreases the albedo of the Earth's surface, leading to less solar radiation being reflected and less infrared radiation being re-radiated. The increased warming that results leads to further decreases in ice and snow cover.

NEGATIVE FEEDBACK PROCESSES. A *negative feedback process*, when triggered by an initial change, dampens or slows down the change. For example, release of CO_2 into the atmosphere will warm the atmosphere in such a way that plants and trees may be able to grow at high latitudes where currently only tundra exists. These new plants, in turn, would absorb CO_2, reducing the amount of CO_2 in the atmosphere, and therefore would slow the rate of warming. Because plants need CO_2 to carry out photosynthesis, an increase in CO_2 will increase plant growth rates, allowing plants to remove more CO_2.

UNCERTAIN FEEDBACK PROCESSES. The effect of clouds on global temperature remains uncertain. Global warming increases the amount of water vapor in the air and thus leads to more clouds. Clouds reflect solar radiation, reducing the amount reaching the Earth's surface and cooling the Earth, so an increase in clouds could be a negative feedback. But clouds also trap infrared radiation from the Earth's surface and warm the Earth further, so they could also act as a positive feedback. It is unclear which of these effects will dominate in the future, particularly since so many types of clouds occur at different levels within the atmosphere (see Chapter 17).

Climate Policy

The predictions described above, along with estimates of the large economic cost of global warming, explain why the issue of climate change is a subject of international discussion. But what can be done? The nations that signed the *Kyoto*

BOX 20.3 > Putting Earth Science to Use

Reducing your personal carbon footprint

Combatting climate change requires a global effort. So what can we do individually to contribute to a reduction in greenhouse gas emissions? More than you might think!

In our homes, we can implement some easy changes that make our energy use more efficient. Changing light bulbs to LED lights, regularly replacing filters in furnaces, using programmable thermostats that reduce heating and cooling in our houses when nobody is home, unplugging electronic devices when not in use, washing clothes in cold water, adding insulation to our homes, and adding solar panels to our roofs all reduce energy use. When shopping, choose efficient appliances with Energy Star labels, and bring your own bags to the grocery store. Recycle everything that can be recycled, because recycling takes less energy than it takes to extract new resources from the ground.

Raising cattle and sheep produces vast amounts of methane, a powerful greenhouse gas. Reducing consumption of meat can reduce carbon. Using (and reusing) your own water bottle rather than purchasing plastic bottles of water also reduces carbon; plastic production requires energy and creates plastic pollution. (This is also a good way to save money!) Planting trees could help. Researchers calculate that 1 trillion more trees (about 130 trees per person) could remove 25% of atmospheric CO_2.

We all have to get from here to there. When possible, consider walking or bicycling instead of driving, or taking public transit if it's available. If you do drive, make sure your tires are properly inflated to increase fuel economy. Turn off your engine when you are stuck in stopped traffic. Carpool with friends. Other changes could include switching to an electric or hybrid car, or having more online meetings rather than traveling by plane.

And don't underestimate the impact you can have in your community. You can support local efforts to create green spaces, and to clean up and green-up highways and waterways. Advocate for better building codes, energy efficiency, regulations that require lowering emissions, and public transportation. And most important, let your representatives in government know your positions. Vote for those representatives who best represent your beliefs concerning stewardship of our world and its future.

Protocol at a 1997 summit meeting held in Japan proposed that the first step would be to slow the input of greenhouse gases into the atmosphere by decreasing the burning of fossil fuels. This approach was reinforced by the *Paris Agreement*, signed in April 2016, which committed 195 countries to efforts that will keep the global average temperature from exceeding 2°C above the pre-industrial temperatures.

Worldwide, individuals, private companies, nongovernmental organizations, and some governments are working to find ways to decrease greenhouse gas emissions. Approaches to this problem include decreasing the waste of energy, developing more efficient vehicles and appliances, and switching to alternative energy sources with a smaller **carbon footprint** (the total amount of greenhouse gases produced by a process or phenomenon) **(Box 20.3)**. Another, more controversial, approach involves collecting CO_2 produced at large sources, such as power plants, so that it can be condensed and then injected into deep wells to be stored in rock underground; this overall process is called *carbon capture and sequestration*.

Regardless of these efforts, the warming trends already happening will require society to adapt. Climate change has taken place throughout Earth history, and in fact, the Earth has seen higher atmospheric CO_2 concentrations in its past: 50 million years ago, during the Eocene, they reached 1,000 ppm. But the rapid rate of change now under way is unprecedented. The coming century will pose unique challenges that will demand innovative solutions to problems of energy production, water and food security, and coastal preservation.

Take-home message . . .

Predictions of future climate change based on global climate models remain imperfect because of the complexity of possible feedback processes. Nevertheless, the consensus of researchers is that significant changes will occur in the distribution of climate belts and in sea level in the coming decades, suggesting that society needs to decrease its carbon footprint.

Quick Question

How might global warming affect global agricultural productivity?

Consequences of Sea-Level Change

Antarctica

If Antarctica suddenly became ice free, most of West Antarctica would be underwater.

A large volume of water resides in the glaciers of Antarctica and Greenland. If all this ice were to melt suddenly, these two landmasses would become mostly dry land. (Parts would remain submerged until glacial rebound takes place, because subsidence due to the weight of ice has pushed the crust's surface below sea level.) Transfer of water from the glacial reservoir back to the oceanic reservoir would cause a sea-level rise of about 65–70 m (215–230 ft). Even if only 10% of this rise were to happen, coastal cities would flood.

San Francisco (after 7-m rise)

Greenland

A suddenly ice-free Greenland would host a central lake.

New Orleans
(after 7-m rise)

Measured sea-level rise varies with location. The current rate, 3 cm per decade, is expected to increase so that sea level rise will be 1 m (3 ft) over pre-industrial levels by the end of this century. At this rate, the 7-m rise depicted here in the city images will not happen for several centuries, but even a 1-m rise will lead to a costly increase in nuisance flooding and in storm-surge damage.

New York
(after 7-m rise)

A sea-level rise of 70 m would flood large areas of coastal land worldwide (areas shown in purple). In fact, most of Florida would be underwater, and New York City buildings would be submerged up to about the 20th floor. Even a 7-m (20-ft) rise would turn streets in many major cities into canals, as shown in the insets.

⟲20 CHAPTER REVIEW

Chapter Summary

- *Weather* refers to atmospheric conditions at a given location and time. *Climate* refers to the average and range of variation of weather conditions in a region over the course of decades or more. Climate change is a shift in average climate conditions.

- Two variables, temperature and precipitation, largely determine the climate a region will experience.

- Daily and seasonal changes in temperature and precipitation depend on latitude and on proximity to the ocean.

- Researchers classify the Earth's climate zones, using the Köppen-Geiger climate classification, into five climate groups—tropical, arid, temperate, cold, and polar—and several subcategories.

- Over time, the amount of solar energy arriving at the Earth from the Sun must balance the amount of infrared energy radiated away from the Earth. The atmosphere traps some of the radiated energy, keeping the Earth's surface warmer than it might otherwise be.

- Because of the tilt of the Earth's axis of rotation, the amount of solar radiation arriving at a particular latitude changes over the course of the year; this change causes seasons.

- Atmospheric circulations, ocean currents, distribution of land and oceans, elevation, and position relative to a mountain range also affect climate in a region.

- Various paleoclimate indicators allow researchers to characterize the climates of the past. Paleoclimate studies indicate that there has been both long-term and short-term climate change over the course of Earth history.

- The stratigraphic record, which permits reconstruction of long-term climate change, indicates that icehouse and hothouse periods alternate over time, but that the durations of these periods vary.

- The record of short-term climate change comes from studies of isotope ratios and pollen in sediment cores. Evidence also comes from studies of isotope ratios and air bubbles in ice cores from glaciers, as well as from the study of growth rings in trees and other organisms.

- Since the most recent glaciation, there have been significant episodes of global warming and global cooling, including the Medieval Warm Period and the Little Ice Age.

- Long-term climate change may be a consequence of changes in atmospheric concentrations of greenhouse gases, the positions of continents and oceans, the flow of ocean currents, and the uplift of mountain ranges.

- The primary controls on short-term climate change appear to be the Milankovitch cycles, volcanic aerosols in the atmosphere, modifications of surface albedo, and abrupt changes in greenhouse gas emission or absorption.

- Since the industrial revolution, average global temperature appears to have increased at rates much greater than in the previous millennia. Observations of glacial and sea-ice retreat, sea-level rise, and shifting of plant hardiness zones support this conclusion.

- The atmospheric CO_2 concentrations observed during the past two centuries are much greater than at any other time during the past 800,000 years. Global climate models suggest that, in the absence of anthropogenic CO_2 emissions, the climate might be experiencing a weak cooling trend rather than the observed warming.

- Global warming could significantly affect habitability of coastal areas, agricultural productivity, and even the frequency of wildfires.

- Positive feedback processes (increasing evaporation, releases of CH_4, decreasing albedo) may be accelerating warming. Negative feedbacks, such as expansion of regions where plants can grow, may be slowing warming.

- Because of the likely role that anthropogenic CO_2 plays in driving global warming, nearly all, governments are negotiating polices to decrease emissions of greenhouse gases worldwide.

Key Terms

albedo (p. 645)
carbon budget (p. 653)
carbon cycle (p. 644)
carbon footprint (p. 659)
climate (p. 628)
climate change (p. 628)
critical zone (p. 628)

energy balance (p. 631)
feedback process (p. 657)
global climate model (GCM) (p. 655)
global cooling (p. 638)
global warming (p. 638)
growth ring (p. 640)

hothouse period (p. 638)
icehouse period (p. 638)
Intergovernmental Panel on Climate Change (IPCC) (p. 647)
Keeling curve (p. 627)

Köppen-Geiger climate classification (KGCC) (p. 631)
paleoclimate (p. 638)
season (p. 635)
weather (p. 628)

Letters in parentheses correspond to the chapter's learning objectives.

1. Distinguish between *weather* and *climate*. What factors control a region's climate? **(A)**

2. What characteristic did Köppen and Geiger use to map out climate zones? **(B)**

3. What factors control a region's rainfall? What factors control a region's temperature and temperature variation? **(C)**

4. Explain the cause of the atmospheric greenhouse effect and how it affects the Earth's energy balance. **(C)**

5. What is the cause of the Earth's seasons? How does the figure illustrate why the Earth has seasons? **(C)**

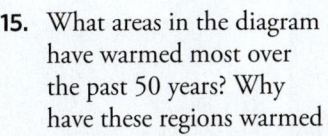

6. Why does the distribution of landmasses and mountain ranges affect climate in a region? **(B, C)**

7. Explain the difference between long-term and short-term climate change. **(E)**

8. How does a large volcanic eruption affect global temperature in the short term? **(E)**

9. What are the Milankovitch cycles, and what role do they play in controlling climate? **(E)**

10. What techniques can researchers use to determine how climate varied in the past? **(D)**

11. Describe the key factors that play a role in causing long-term climate change and those involved in short-term climate change. **(E)**

12. What is a global climate model? **(G)**

13. Based on paleoclimate records, how has climate changed over the course of the last few millennia? **(F)**

14. What evidence led researchers to the conclusion that the Earth's climate has warmed during the past few centuries? **(G)**

15. What areas in the diagram have warmed most over the past 50 years? Why have these regions warmed more dramatically than others? **(F, G)**

16. What kinds of feedbacks may amplify the process of global warming, and why? **(G)**

17. What efforts may help to diminish human society's carbon footprint? **(G)**

18. If the CO_2 concentration in the atmosphere doubled during the 21st century, which of the values for the Earth's energy balance in the diagram would increase, and which would decrease? **(C, E)**

19. Mosquitoes carry diseases (malaria, dengue fever, Zika) that are dangerous to humanity. The species of mosquitoes that carry such diseases prefer warmer climates. If current observed trends in global temperature change continue, will the number of people at risk of contracting these diseases increase or decrease during the next century? Explain your answer. **(G)**

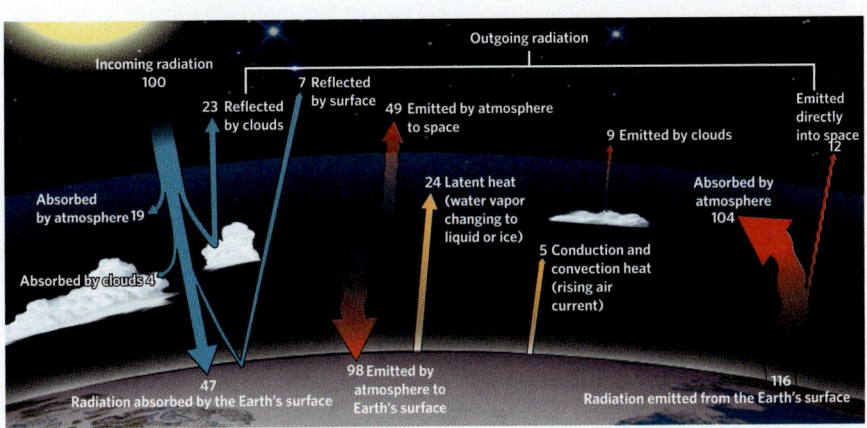

21 INTRODUCING ASTRONOMY
Looking Beyond the Earth

By the end of the chapter you should be able to . . .

A. describe how our ideas about the organization and movement of celestial objects evolved from ancient to modern times.

B. relate the observed movements of celestial objects to the Earth's rotation on its axis and to its orbit around the Sun.

C. define a constellation and the zodiac.

D. explain the paths of the planets and Moon across the sky, why the Moon has phases, and why eclipses take place.

E. discuss how astronomers use electromagnetic radiation to determine a star's temperature, composition, velocity, and distance.

F. list the units that astronomers use to describe vast distances between celestial objects.

G. explain how various kinds of telescopes function and why we use space telescopes.

21.1 Introduction

Centuries before Columbus, Polynesian sailors guided sailing canoes across vast reaches of the Pacific Ocean and colonized over 100 island groups **(Fig. 21.1)**. When a crew journeyed 4,200 km (2,600 mi) from Tahiti to Hawaii, they had to find a target only 120 km (75 mi) across—a feat equivalent to hitting a small coin from a distance of 35 km (22 mi)—without using GPS satellites, compasses, clocks, or charts. How did they do it? To read their surroundings as we read a map today, Polynesian navigators merged knowledge of where distinct groups of stars rose above the horizon with knowledge of how ocean swells move, birds migrate, and cloud characteristics vary with latitude. They then embedded navigational information in songs and myths that they could pass on to the next generation.

You can see the same lights that led Polynesian sailors across the ocean by gazing upward on a dark night. Most of the lights—the *stars*—look like twinkling points that stay fixed in position relative to their neighbors even as they make their nightly journey across the sky **(Fig. 21.2a)**. A few of the lights—the *planets*—have a steadier glow and move relative to the backdrop of stars and relative to one another, so their name, from the Greek word for wanderers, seems apt **(Fig. 21.2b)**. If it's a particularly clear and dark night, you'll also see a glowing haze—the *Milky Way*—forming a

<< Astronomical observatories at Kitt Peak, Arizona, provide a window to the sky.

FIGURE 21.1 Polynesian sailing canoes, similar to these modern replicas, were used for transpacific voyages.

band that spans the entire arc of the heavens **(Fig. 21.2c)**. On many nights, however, you'll be able to see only the brighter stars and planets, for the glow of the Moon masks the fainter lights. During the day, of course, the Sun's intense light completely obscures all planets and stars.

Humans have wondered about the nature of all these **celestial objects** since conscious thought began. In historic time, this wonder evolved into the science of **astronomy**, the systematic study of the **Universe**

FIGURE 21.2 Stars, planets, and galaxies.

(a) Look into the sky on a dark night, and you'll see a myriad of stars.

(b) Four planets rise over the Andes just before dawn: from top to bottom, Venus, Mercury, Jupiter, and Mars.

(c) On a particularly clear, moonless night, in a dark location, you can see the Milky Way.

(or *cosmos*), meaning all of space and all the celestial objects within it. In the final part of this book, we offer a brief survey of astronomy to provide a context within which you can understand the origin and evolution of our home, the Earth. Our study of astronomy will help you complete your personal image of your natural surroundings by explaining what you see when you look up and beyond the Earth. This chapter provides an introduction to the study of astronomy by tracing the path of ideas that early Western philosophers and, later, scientists took as they tried to fathom humanity's place in the cosmos. We then introduce the modern concept of how celestial objects move in the sky, knowledge that reflects scientific discoveries over several centuries. This concept relies on the study of electromagnetic radiation that comes from space, so we next explore the basic properties of this radiation and describe the tools astronomers use to measure it in ways that allow us to peer at unfathomably distant places and see events that happened in the distant past.

21.2 Building a Foundation for the Study of Space

The oldest hint of humanity's interest in the heavens dates to 35,000 B.C.E., as indicated by petroglyphs depicting celestial objects that cave dwellers chiseled into rock surfaces. By 14,000 B.C.E., cave paintings found at many localities portrayed distinctive clusters of dots that resemble star clusters **(Fig. 21.3)**. As civilization dawned, observers knew that the Sun, Moon, and stars rose at specific points on the horizon at times of seasonal change, so the study of the sky became a basis for calendars that people used to plan the sowing or reaping of crops.

As ancient peoples watched the motion of the Sun, Moon, and stars across the sky, they came to believe that the Earth lay at the center of the Universe, so that celestial objects followed orbits around the Earth, and all stars lay on the surface of a distant sphere outside it. (Note that an **orbit**, in a general sense, is a closed path in space that one object takes around another; the complete orbit of one object around another is called a *revolution*.) This idea, a **geocentric model** of the Universe, was promoted by Ptolemy (90–186 C.E.), a mathematician living in Egypt. The concept dominated European thinking about astronomy through the Middle Ages.

The first seismic shift in the way humanity viewed the Universe came during the Renaissance with the work of Nicolaus Copernicus (1473–1543). Copernicus re-examined planetary orbits and concluded that the geocentric model could not be real. He wrote a book, *De Revolutionibus Orbium Coelestium* (*On the Revolutions of the Heavenly Spheres*), in which he demonstrated that a **heliocentric model** provided a better explanation of celestial motions. This idea, which came to be known also as the *Copernican model*, did not gain widespread acceptance for many decades. It took the work of Galileo Galilei (1564–1642), the great Italian scientist, to provide the observations that ultimately led to acceptance of the Copernican model. Galileo, the first astronomer to examine the skies with a telescope, saw that Jupiter has moons, and that the moons cross the face of Jupiter and disappear behind the planet. He reasoned, correctly, that if moons orbit Jupiter, then not all celestial objects orbit the Earth. In 1632, he published a book in which he presented the case for the Copernican model, and launched the modern era of astronomy.

Kepler and His Laws of Planetary Motion

Johannes Kepler (1571–1630), a German astronomer, made the next great astronomical discovery. Kepler developed a set of simple mathematical relationships that completely explained the motions of planets in the context of the Copernican model. In these relationships, now known as **Kepler's laws** of planetary motion, Kepler for the first time portrayed planetary orbits as ellipses instead of as circles. Because of the importance of ellipses to picturing planetary orbits, let's quickly review their geometry. An **ellipse** resembles a flattened circle or the outline of an egg. It has two axes, which intersect at the center of the ellipse. The *major axis* stretches across the widest part of the ellipse, and the *minor axis* stretches across the narrowest part of the ellipse. A line running from the center to the rim along the major axis is called the *semi-major axis*, and one running from the center to the rim along the minor axis is the *semi-minor axis*. An ellipse has two *foci*, positioned along the major axis.

FIGURE 21.3 Earth observations of celestial objects include cave paintings thought to depict groups of stars.

Painted star cluster

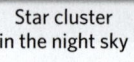

Star cluster in the night sky

FIGURE 21.4 Kepler's laws of planetary motion. (The eccentricity of the ellipses shown is exaggerated.)

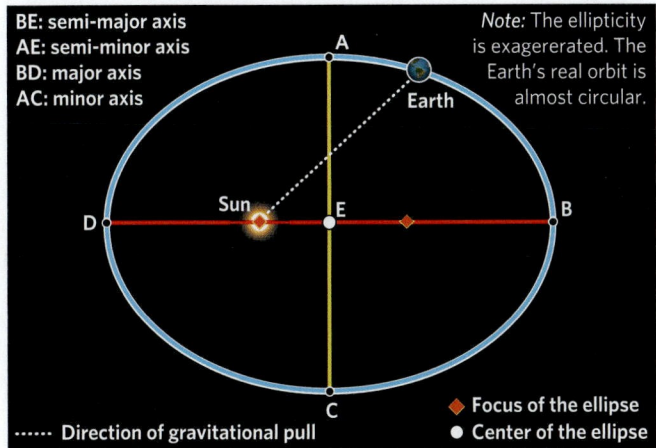

BE: semi-major axis
AE: semi-minor axis
BD: major axis
AC: minor axis

Note: The ellipticity is exaggererated. The Earth's real orbit is almost circular.

••••• Direction of gravitational pull

◆ Focus of the ellipse
○ Center of the ellipse

(a) Kepler's first law: The orbit of every planet is an ellipse, with the Sun at one of the ellipse's two foci.

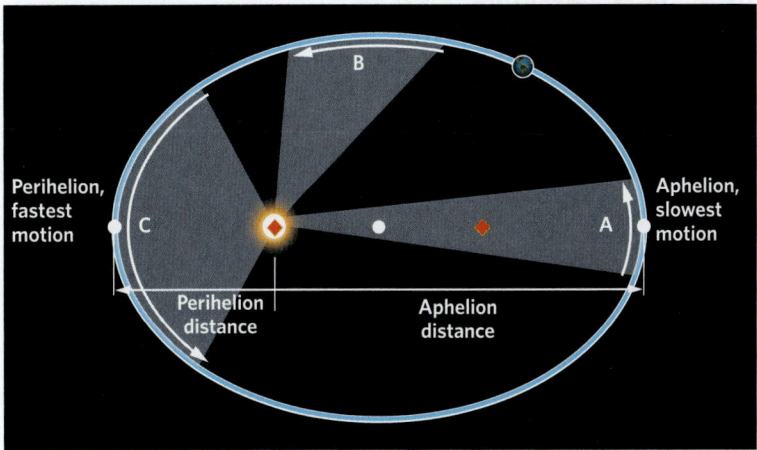

Perihelion, fastest motion

Aphelion, slowest motion

Perihelion distance

Aphelion distance

(b) Kepler's second law: As a planet orbits the Sun, an imaginary line joining that planet and the Sun sweeps out equal areas (labeled A, B, and C) during equal intervals of time. As a result, planets move fastest at perihelion, when they are close to the Sun, and slowest at aphelion, when they are farthest away.

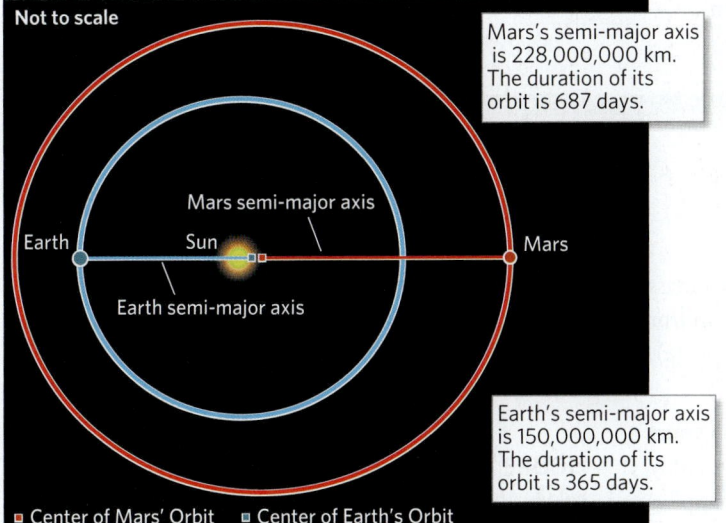

Not to scale

Mars's semi-major axis is 228,000,000 km. The duration of its orbit is 687 days.

Mars semi-major axis

Earth

Sun

Earth semi-major axis

Mars

Earth's semi-major axis is 150,000,000 km. The duration of its orbit is 365 days.

■ Center of Mars' Orbit   ■ Center of Earth's Orbit

(c) Kepler's third law: The larger the semi-major axis of a planet's orbit (a measure of the planet's distance from the Sun), the longer the planet's orbital period, the time it takes the planet to orbit the Sun.

3. The farther a planet is from the Sun, the longer its *orbital period* (the time it takes for the planet to go around the Sun). Put simply, this law states that we can calculate a planet's orbital period if we know the dimensions of its orbit **(Fig. 21.4c)**. In detail, the square of a planet's orbital period is directly proportional to the cube of the semi-major axis of the orbit.

Newton and His Laws of Physics

Isaac Newton (1642–1726) was born in England in the year of Galileo's death. To say that Newton was a genius would be an understatement, for Newton contributed more to the establishment of modern science than any other person. Not only was he one of the inventors of calculus, but he also discovered that white light can be broken into a spectrum of colors, and he invented the type of telescope favored by research astronomers today. Newton's most important contributions appeared in *Philosophiae Naturalis Principia Mathematica*. In this book, generally known simply as *Principia*, Newton proposed the three *laws of motion* as well as the *law of gravity*. Using these laws, he mathematically derived Kepler's laws, and by doing so he showed why planets have the motions and orbits that they do **(Box 21.1)**.

- *First law of motion:* Every object has *inertia*, meaning that an object will remain in a state of rest, or in a state of uniform motion in a straight line, unless compelled to change by the application of a force.

The distance between the foci determines the ellipse's *eccentricity*. An ellipse with high eccentricity looks flattened and elongate, and one with low eccentricity looks more circular. A circle, in effect, is a special ellipse—it's one whose major and minor axes have the same length, and whose foci both lie at the circle's center.

Kepler's laws can be stated as follows:

1. The orbit of every planet is an ellipse, not a circle, with the Sun at one of the ellipse's two foci **(Fig. 21.4a)**.

2. An imaginary line extending from the Sun to a planet sweeps out equal areas during equal intervals of time as a planet moves along its orbit. Therefore, a planet's speed along its orbit changes over the course of a year **(Fig. 21.4b)**.

BOX 21.1 ▶ A Deeper Look

Launching a satellite into orbit

What does it take to put an object into orbit around the Earth **(Fig. Bx21.1)**? Imagine that we place a gun, anchored on a post, high above the Earth, and fire projectiles horizontally out of it. All projectiles are pulled back toward the Earth. A slow projectile follows a curved path and lands not far from the post. The faster the projectile, the farther it lands from the post. If a projectile has orbital velocity, it keeps falling, but never gets back to the surface.

So how can a satellite be placed into orbit around the planet? The answer is speed. If a rocket propels a satellite into space fast enough, the arc of the satellite's fall due to the pull of gravity no longer brings it back to the Earth's surface, but rather takes it entirely around the planet. Gravity still pulls the satellite downward, but the Earth curves away from the satellite just as quickly, so the satellite stays at the same altitude above the Earth's surface. For example, to stay in orbit around the Earth, NASA's space shuttles traveled 8 km/s (18,000 mph) and orbited the Earth every 90 minutes!

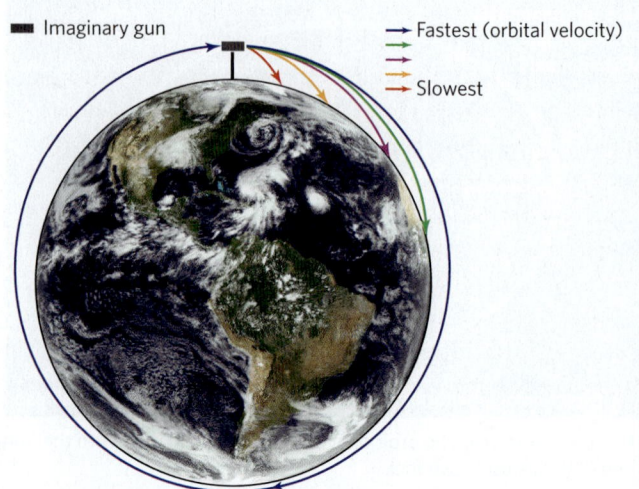

FIGURE Bx21.1 Paths of projectiles launched at different speeds.

- *Second law of motion:* When a force (F) acts on a mass (m), it causes the mass to change its speed and/or direction, meaning that it causes an acceleration (a). This law can be written as $F = ma$.

- *Third law of motion:* For every action, there is an equal and opposite reaction. So, for example, when we apply a force to a ball by hitting it with a bat, the ball applies the same force to the bat in the opposite direction.

- *Law of gravity:* Every object in the Universe attracts every other object. Objects with larger masses produce a greater gravitational force than objects with smaller masses, and objects closer to each other attract each other more strongly than do those that are farther apart.

Astronomy Enters the Modern Era

In the 18th and 19th centuries, astronomers developed new and more powerful telescopes that allowed them to see celestial objects more clearly and to see deeper into space. Such work allowed astronomers to understand details of the structure of our **Solar System** (the Sun and all the objects that orbit around it) and provided new information about the physical nature of celestial objects. As the 20th century dawned, astronomers learned how to measure distances to celestial objects. With this knowledge, they came to realize that our Sun lies in the Milky Way, and that the Milky Way is a **galaxy**, a collection of hundreds of billions of stars

distributed in a spiral or disk that slowly spins around a *galactic center* **(Fig. 21.5)**. Astronomers then discovered that some of the features that earlier observers knew only as puzzling hazy patches of light were actually **nebulae**, vast

FIGURE 21.5 The M-81 Galaxy looks much like our galaxy, the Milky Way. Our Sun resides on one of our galaxy's spiral arms.

clouds of dust and gas, and that others were independent galaxies located far beyond the edge of the Milky Way. Eventually, astronomers learned that the Universe contains over a trillion galaxies, and that the distant galaxies are all moving away from one another, meaning that the Universe is expanding. This realization set the stage for the development of the *Big Bang theory*, a scientific model of the origin of the Universe (see Chapter 1). Meanwhile, physicists developed new theories to explain the behavior of matter and energy, which provided astronomers with ways of using measurements of electromagnetic radiation emitted by celestial objects to characterize those objects.

We now know that our Solar System lies on an outer arm of the Milky Way. Together with neighboring stars and their planets, the Sun orbits the galactic center about once every 230 million years. This means that the Sun zips through space at about 250 km/s (155 mi/s) relative to an observer outside our galaxy, and that it has orbited the center of the Milky Way 24 times since it first formed. The Milky Way is one of many galaxies, all moving relative to one another. A photo of the Earth and Moon as viewed from Mercury emphasizes that we are all riding on a small speck as it journeys through the vast reaches of space **(Fig. 21.6)**. But because we live here, it is a very important speck indeed.

Take-home message . . .

Early astronomers thought the Earth lay at the center of the Solar System, with planets and the Sun following circular orbits around it. This idea had to be discarded in the wake of work by Copernicus, Galileo, Kepler, and Newton. Astronomers can now demonstrate that planets follow elliptical orbits around the Sun, and those orbits can be explained by Newton's laws. Modern discoveries reveal that the Sun is one of hundreds of billions of stars within one of more than a trillion galaxies.

Quick Question

Does the velocity of a planet along its orbit remain the same all year?

21.3 Motions of the Stars and Planets

The Celestial Sphere

To portray the locations of celestial objects relative to one another, as we see them in the night sky, astronomers mark points representing these objects on the **celestial sphere**, an imaginary surface in space that surrounds the Earth. We can picture the celestial sphere as a giant glass shell **(Fig. 21.7)**. To determine the position of an object

FIGURE 21.6 The two tiny specks close to each other in this image are the Earth and the Moon, as viewed by the *Messenger* space probe orbiting Mercury. The Earth is but a speck in the vastness of space.

on the celestial sphere, astronomers draw a line from the center of the Earth to the object. The point where the line pierces the celestial sphere represents the object's position. Note that specifying this position does not provide any information about the distance of the object from the Earth. Astronomers orient the celestial sphere so that its axis aligns with the Earth's **axis of rotation**, the imaginary line passing through the center of the Earth around which our planet spins. The axis of rotation intersects the

FIGURE 21.7 The concept of the celestial sphere.

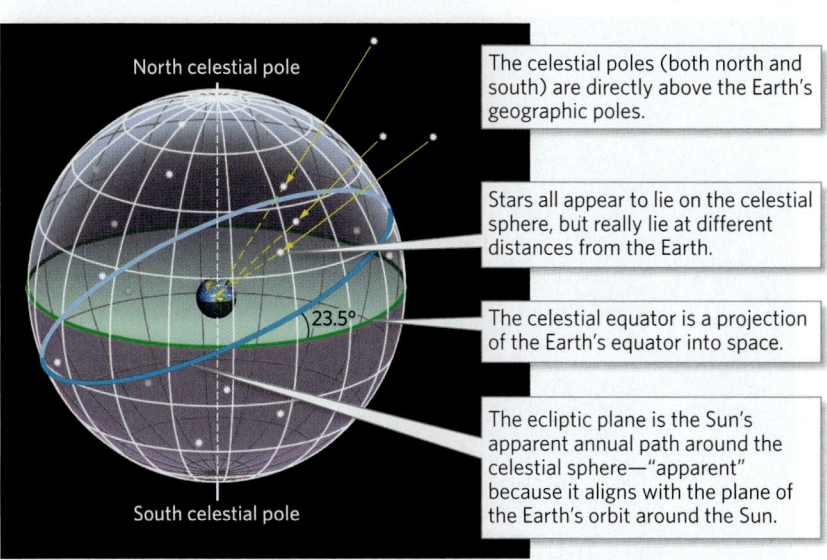

North celestial pole

The celestial poles (both north and south) are directly above the Earth's geographic poles.

Stars all appear to lie on the celestial sphere, but really lie at different distances from the Earth.

23.5°

The celestial equator is a projection of the Earth's equator into space.

The ecliptic plane is the Sun's apparent annual path around the celestial sphere—"apparent" because it aligns with the plane of the Earth's orbit around the Sun.

South celestial pole

FIGURE 21.8 The path of the Sun across the sky.

The path of the Sun, as seen from a high northern latitude, during the winter solstice.

(a) If you watch the Sun during the course of the day, you see that it rises at dawn, arcs across the sky, reaching its highest point at noon, and sets at dusk.

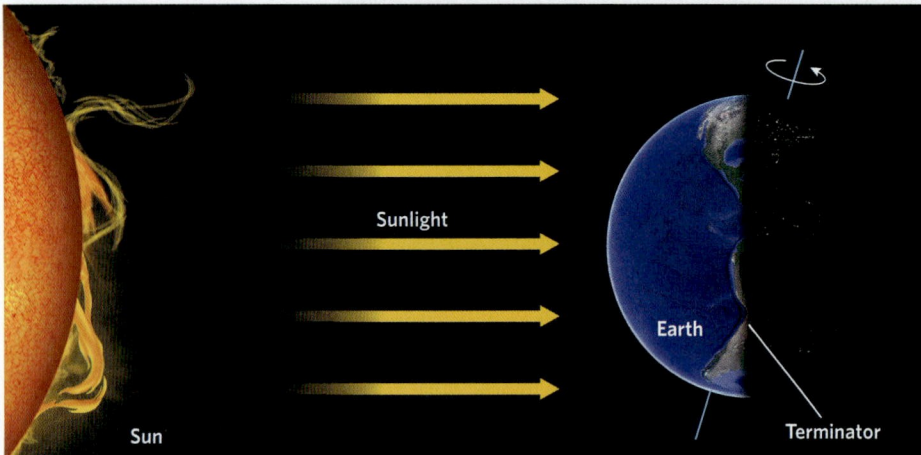

Sunlight

Earth

Sun

Terminator

(b) The terminator is the boundary between the lit and dark sides of the Earth. As the Earth rotates, the terminator moves from east to west.

FIGURE 21.9 The eccentricity of the Earth's real orbit is so small (0.017) that it is actually almost circular.

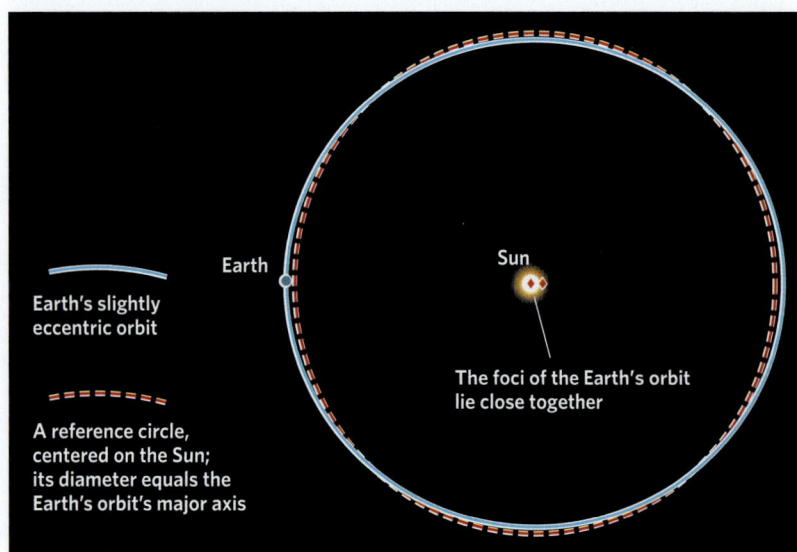

Earth

Sun

Earth's slightly eccentric orbit

A reference circle, centered on the Sun; its diameter equals the Earth's orbit's major axis

The foci of the Earth's orbit lie close together

surface of the Earth at our planet's **geographic poles**. On the celestial sphere, we position the *north celestial pole* as the point directly above the north geographic pole of the Earth, and the *south celestial pole* as the point directly above the Earth's south geographic pole.

The Journey of Helios: A Consequence of the Earth's Rotation

In ancient times, people thought that the Sun moved along the celestial sphere while the Earth stood still, and they concocted various myths to explain this movement. The Greeks, for example, attributed it to the god Helios, who drove the chariot of the Sun across the sky once a day. Now, of course, we understand that the Sun's apparent movement actually occurs because our planet rotates on its axis **(Fig. 21.8a)**. Because the Earth rotates counterclockwise, as viewed looking down on the North Pole, the *terminator*, meaning the boundary between night and day, moves from east to west **(Fig. 21.8b)**. When the terminator moves past a point on the Earth's surface, the Sun rises or sets at that point. Sunset and sunrise do not take place instantaneously because the atmosphere scatters sunlight.

How long does the Earth take to make one **rotation**, one complete turn on its axis? If we use the Sun as a reference and measure the amount of time it takes for the Sun to appear again at the same point in the sky, the Earth rotates, on average, once every 24 hours. Astronomers call this measure of time a **solar day**. A **sidereal day** (pronounced side-e-re-al), in contrast, represents the time it takes for a distant star to appear at the same point in the sky on successive days. A sidereal day has 4 fewer minutes than a solar day because the Earth changes its position relative to the Sun over the course of a day as it orbits the Sun. A sidereal day provides the true measure of one full rotation of the Earth.

The Earth's Orbit and the Ecliptic Plane

Our planet orbits the Sun once each year, traveling a distance of 940 million km (584 million mi) over a period of 365.256 solar days. So, when you think you're sitting still, you—and the ground beneath you—actually zip along an orbit at an average speed of about 107,000 km/h (66,000 mph) relative to the Sun. To accommodate the extra quarter day, the *Gregorian calendar*—the standard calendar of the Western world since its introduction by Pope Gregory in 1582—includes a *leap year* every 4 years, when we add an extra day to February. At the turn of each century, except those centuries divisible by 400, the calendar skips adding the leap day to take into account the extra 0.006 day.

As Kepler recognized, the Earth follows an elliptical orbit, and the Sun lies at one focus—not the center—of the ellipse (see Fig. 21.4a). The distance between the Earth and the Sun, therefore, changes over the course of a year, ranging from 147.1 million km (91.4 million mi) at *perihelion*, when the Earth is closest to the Sun, to 152.1 million km (94.5 million mi) at *aphelion*, when the Earth is most distant from the Sun (see Fig. 21.4b). Note that these distances differ by only about 4%, so the Earth's orbit is actually fairly close to being circular **(Fig. 21.9)**.

As viewed from space, the Earth's orbit around the Sun defines the Earth's **ecliptic plane (Fig. 21.10)**. Because the Earth's axis of rotation tilts at 23.5° relative to the plane of the Earth's orbit around the Sun (see Fig. 20.7), the ecliptic plane tilts at an angle of 23.5° relative to the celestial equator (see Fig. 21.7). We can picture the ecliptic plane as an imaginary surface that contains the ellipse of the Earth's orbit.

Because the tilt of the Earth's axis of rotation does not change as the Earth orbits the Sun, the arc that the Sun takes across the sky, from the perspective of an observer on the Earth, varies with both latitude and time of year. Therefore, the duration of daylight also varies with latitude and time of year. To understand this concept, consider how the Sun progresses across the sky from the perspectives of three observers at different latitudes—one standing on the equator, one at mid-latitude in the northern hemisphere, and one at the Arctic Circle—at three different times of year **(Fig. 21.11a)**.

Note that during half of the year, the northern hemisphere inclines toward the Sun, and during the other half of the year, it inclines away from the Sun **(Fig. 21.11b)**. This tilt, as we saw in Chapter 20, gives the Earth its seasons (see Fig. 20.7). In the northern hemisphere, the time when the Earth's axis transitions from tilting toward the Sun to tilting away defines the *autumnal equinox* (September 22), and the time when it transitions from tilting away from the Sun to tilting toward it defines the *vernal equinox* (March 20). The northern hemisphere's summer solstice occurs on June 21, when the northern hemisphere reaches its maximum tilt toward the Sun. The winter solstice occurs on December 21, when the northern hemisphere reaches its maximum tilt away from the Sun.

On the scale of a human lifetime, the Earth's orbit doesn't change noticeably. But on the scale of millennia, it does. Specifically, the orientation of the orbit's major axis changes relative to distant stars over the course of several millennia, and the eccentricity of the orbit changes over the course of millennia. These changes cause the Milankovitch cycles that we described in Chapter 20.

Navigating Darkness: The Night Sky and Its Constellations

Imagine yourself on a boat with no lights, floating in the middle of the ocean on a moonless, cloudless night. How many stars can you see? Astronomers estimate that there

FIGURE 21.10 The Earth's ecliptic plane contains the ellipse of Earth's orbit.

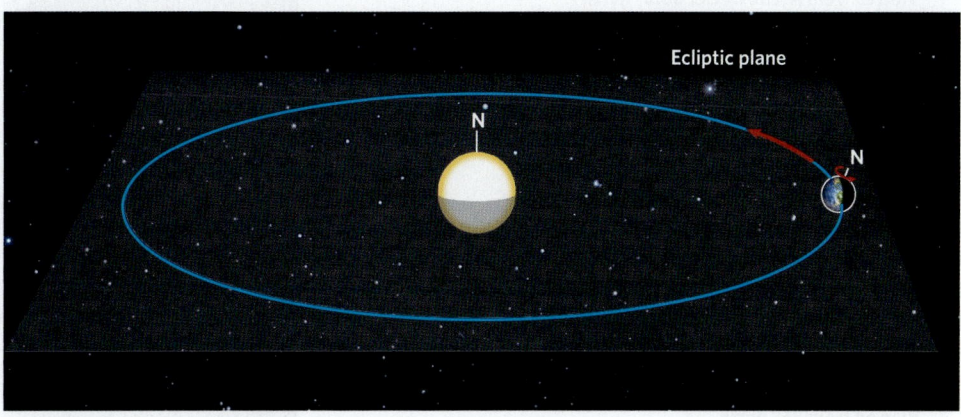

are 6,000 visible stars in the entire night sky. At any given moment on any given night, about half of these will be visible.

Now imagine that you're standing at the North Pole, so that the top of your head aligns with the north celestial pole. If you were to aim a camera straight up and set it to record starlight on a single image all night long as the Earth spins, the image would show circular *star streaks*. In the northern hemisphere, the star Polaris, positioned directly above the North Pole, would lie closest to the center of the star streak circles, so we refer to Polaris as the *North Star* **(Fig. 21.12)**. If you were to move to a lower latitude, you would find that the North Star does not lie directly overhead, but rather sits lower in the northern sky. Furthermore, you would record star streaks that have larger diameters and are partial circles that run into the horizon. If you were to go farther south, you would see the position of the North Star progressively lower in the sky, and when you reached the equator, you'd see the North Star right at the northern horizon. You wouldn't see the North Star at all from the southern hemisphere.

The identity of the star that serves as the North Star changes over millennia because the Earth's axis undergoes **precession**. This means that it wobbles like a top, so that the axis of rotation points to different stars on the celestial sphere **(Fig. 21.13)**. It takes about 23,000 years for the Earth's axis to make a complete circle. Because of precession, about 12,000 years from now, the North Star will be Vega, not Polaris. This wobble contributes to the Milankovitch cycles discussed in Chapter 20.

Ancient observers in many cultures found that if they wished to identify specific locations in the night sky, it helped to group stars into recognizable patterns or arrangements, which came to be known as **constellations (Fig. 21.14a)**. These patterns were typically named after

FIGURE 21.11 The path of the Sun across the sky varies with latitude and season.

Viewed from Arctic Circle, 66.5° north

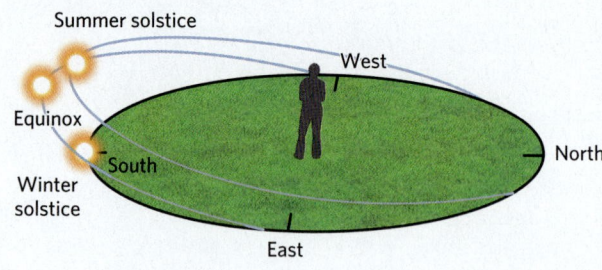

Viewed from 35° north

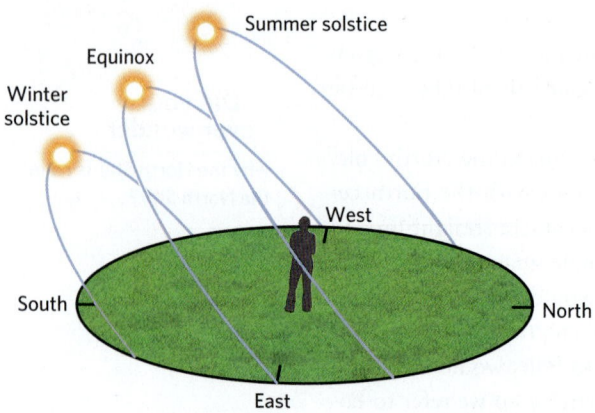

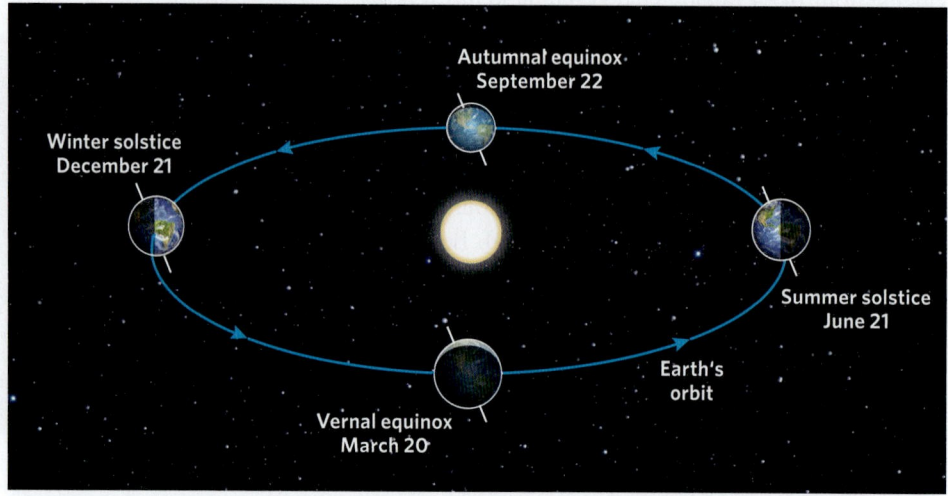

(b) The Earth revolves around the Sun once a year. The Earth's axis tilts 23.5° with respect to the Earth's orbital plane, so that the Earth's polar regions are in darkness at the winter solstice and in perpetual light at the summer solstice.

animals, warriors, or mythical beings. Each culture came up with its own set of constellations—Greek constellations are not the same as Chinese constellations—emphasizing that constellations are entirely a human invention.

The times and locations at which specific constellations appear at the horizon after sunset provide a useful basis for navigation and calendars. The International

Viewed from equator

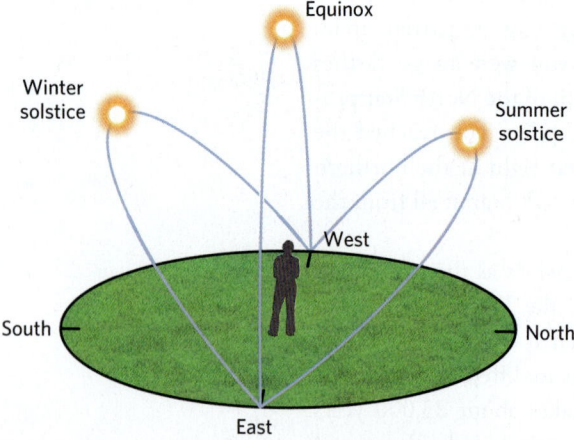

(a) The paths the Sun takes across the daytime sky at three times of year, each as viewed from a point on the equator, at about 35° N, and at the Arctic Circle. At the Arctic Circle, the Sun is up for 24 hours at the summer solstice, and it only briefly peeks above the south horizon at noon at the winter solstice before disappearing for the remainder of the day.

FIGURE 21.12 In the northern hemisphere, star streaks form circles around the North Star, Polaris. Some streaks intersect the horizon. The distance of Polaris above the horizon depends on latitude.

FIGURE 21.13 Precession of the Earth's axis. The Earth's axis of rotation wobbles, much like a top. This movement is called precession.

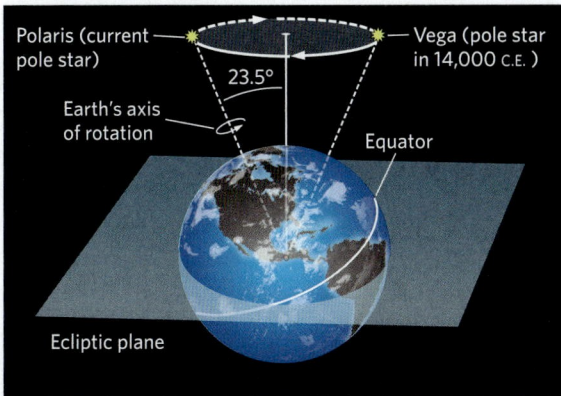

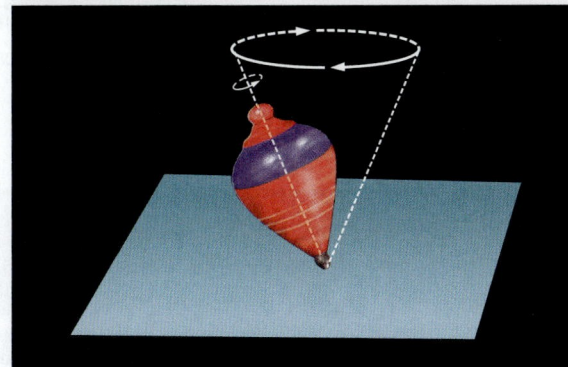

Astronomical Union recognizes 88 constellations, which stargazers use like ZIP codes to direct attention not just to the stars in the constellation, but to a given sector of the celestial sphere. On successive months throughout the year, different constellations make their appearance in the night sky because our planet has moved to a different position along its orbit.

The 12 constellations that lie along the ecliptic plane constitute the **zodiac (Fig. 21.14b)**. Each constellation lies within a 30°-wide arc of the ecliptic plane. Star charts and, more recently, apps on smartphones can be used to locate constellations.

Movements of the Planets as Seen from the Earth

Early observers of the heavens knew that aside from the Sun and Moon, all but five of the celestial objects in the sky remained fixed in position relative to one another as they moved smoothly across the night sky, and that night after night, each of these objects retained the same brightness. The five exceptions, however, are not so well behaved. Although their overall paths remain close to the ecliptic plane, these objects seem to vary in brightness and to "wander," changing position relative to the stars and to one another as the months pass. As we've noted, these objects are not stars at all. They are the nearer planets of our Solar System: Mercury, Venus, Mars, Jupiter, and Saturn. Two more distant planets, Uranus and

FIGURE 21.14 Constellations.

(a) A starry sky and the constellation of Orion above a birch forest in winter.

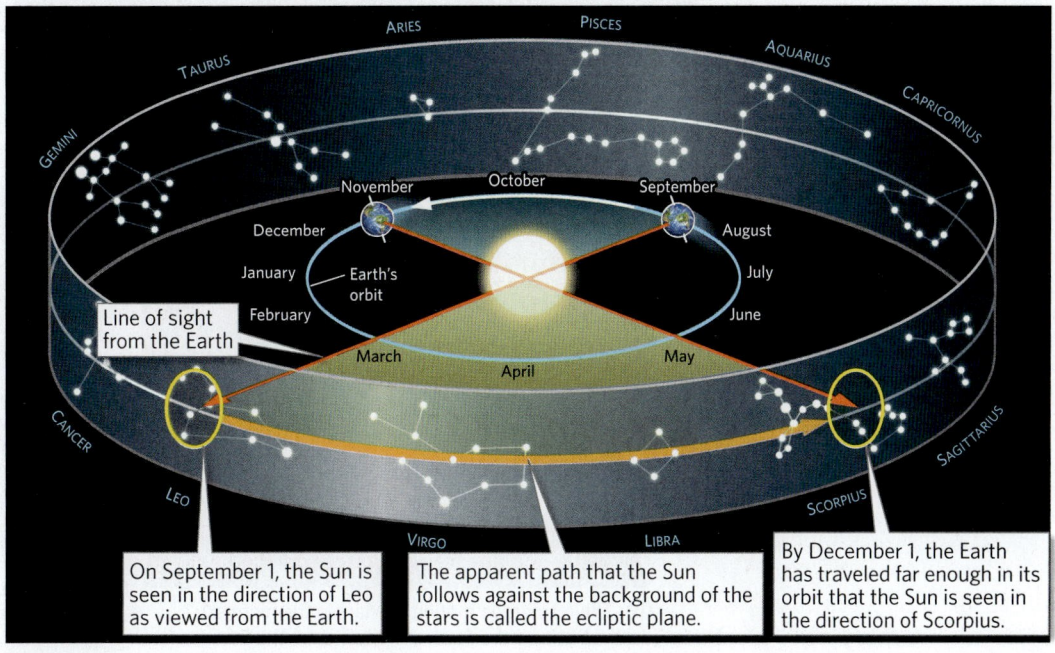

On September 1, the Sun is seen in the direction of Leo as viewed from the Earth.

The apparent path that the Sun follows against the background of the stars is called the ecliptic plane.

By December 1, the Earth has traveled far enough in its orbit that the Sun is seen in the direction of Scorpius.

(b) The zodiac is the set of 12 constellations that lie along the ecliptic plane.

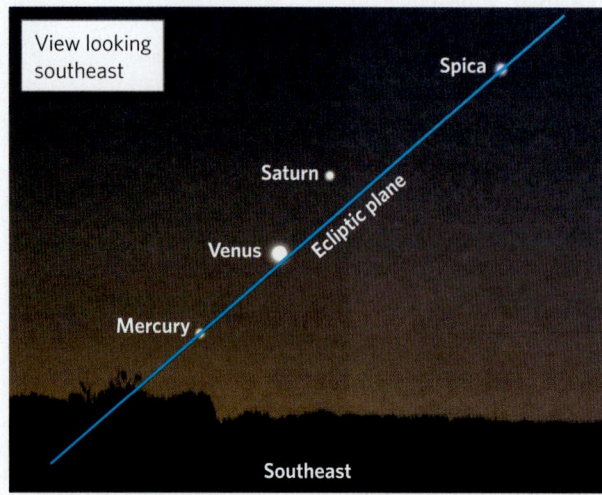

View looking southeast

Spica

Saturn

Venus

Ecliptic plane

Mercury

Southeast

Neptune, which can be seen only with telescopes, follow the same pattern of motion as the other planets.

Because their orbital planes roughly parallel the Earth's ecliptic plane, you can use the positions of planets to find the ecliptic plane. To do this, go outside right after sunset or right before sunrise and find one or more planets. Today, you can easily download a smartphone app to help you find them. Because they are brighter than most stars, they are among the first celestial objects (not counting

the Moon) to be visible at night and the last to be visible in the morning, so you can often identify them at dusk or dawn. Once you've picked out a planet, draw an imaginary line connecting it with the position of the setting or rising Sun. This line shows the trace of the ecliptic plane across the celestial sphere **(Fig. 21.15)**.

All of the planets orbit the Sun in a counterclockwise direction, as viewed looking down on the Sun's north pole. From an earthbound observer's viewpoint, therefore, the planets usually display **prograde motion**, meaning that they traverse the celestial sphere from west to east. Occasionally, however, a planet displays **retrograde motion** (east to west) on the celestial sphere **(Fig. 21.16a)**. Why does retrograde motion take place? You can visualize the answer by keeping in mind that planets orbit the Sun at different distances and therefore at different velocities. As a result, the line of sight from the Earth to another planet continually changes **(Fig. 21.16b, c)**.

FIGURE 21.16 Why do planets display retrograde motion?

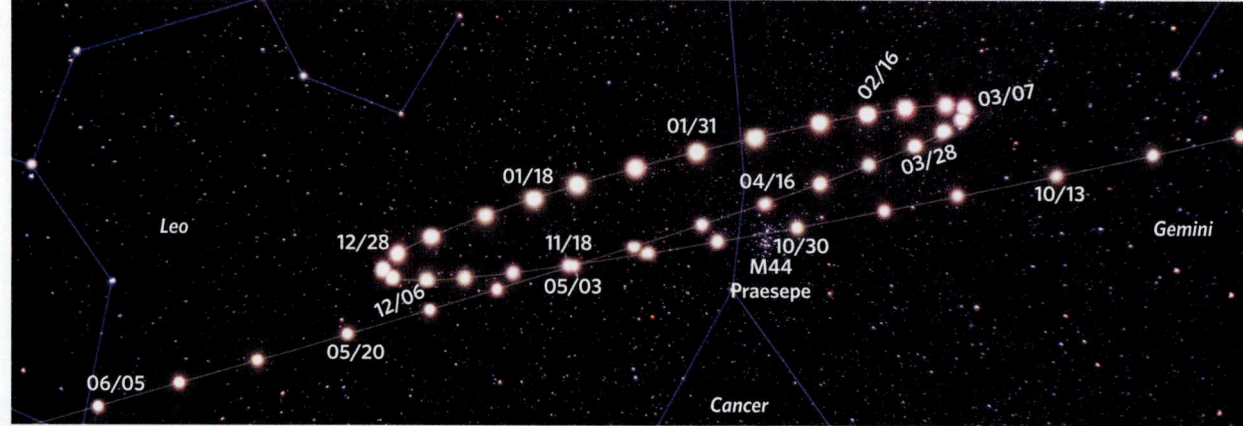

(a) Observed apparent retrograde motion of Mars.

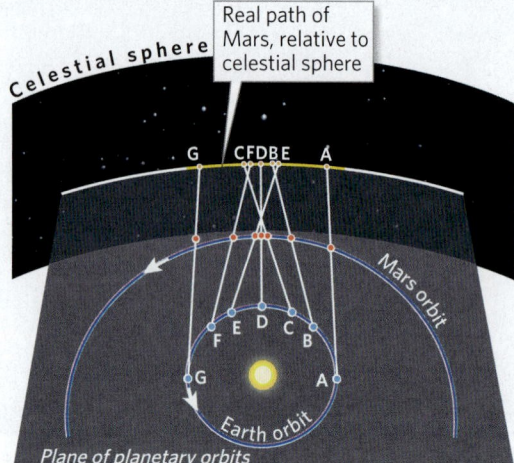

(b) The retrograde motion of Mars in the night sky occurs because the Earth and Mars orbit the Sun at different speeds, so the view of Mars from the Earth changes.

(c) Apparent path of Mars across the sky, relative to the Earth's horizon. Its path appears to go up and down because, due to the Earth's axial tilt, the trace of its orbital plane above the horizon moves up and down over the course of a year.

Why does the brightness of planets, as seen from the Earth, vary over time? Unlike stars, planets do not produce their own light. The light that we see when looking at them comes from reflected sunlight. The brightness of a planet as viewed from the Earth depends on several factors: its distance from the Sun, which determines how much sunlight it receives; its size, which determines the dimensions of its reflecting surface; its *albedo*, meaning the reflectivity of its surface; and its position relative to the Earth at a particular time, which determines its distance from the Earth and how much of its lit side faces the Earth. As the distance to the planet from the Earth increases, or as the planet reaches a position that allows us to see light reflected from less of its Sun-facing surface, the planet's light appears fainter to an Earth-based observer.

Take-home message . . .

To represent the positions of celestial objects as seen from the Earth, astronomers plot them on the celestial sphere. We see the movement of these objects in the night sky because the Earth rotates on its axis, which is tilted with respect to its orbital plane. The Earth revolves around the Sun once a year, causing an observer on the Earth to see the constellations move across the sky in predictable paths.

Quick Question
Why do planets sometimes display retrograde motion?

21.4 The Earth's Traveling Companion

The Moon, our planet's closest neighbor, has a fascinating surface displaying both smooth, dark *maria* (singular: *mare*, Latin for sea), and rugged, light-colored highlands **(Fig. 21.17a)**. If you watch the Moon night after night, you'll notice that these features are always visible. In other words, from the Earth, we always see the same face of the Moon. That's because, as the Moon moves along its orbit around the Earth, the Moon spins on its axis at just the right rate to keep the same side visible **(Fig. 21.17b)**. This synchrony developed over geologic time because the tug of the Earth's gravity generates a small tidal bulge on the Moon's surface. Friction doesn't allow this bulge to move as fast as the Moon spins, so the Earth pulls on one side slightly more than on the other. The pull slowed the Moon's spin until it locked into a rate that keeps one side facing the Earth—the synchrony of this tidal locking isn't perfect, though, so we actually see 59% of the Moon's surface over the course of a month. Because of tidal locking, humans didn't see the *far side* of the Moon until 1959, when the Soviet *Luna 3* space probe sent back pictures.

FIGURE 21.17 The Moon and its orbit.

(a) Distinct features of the Moon's surface are visible from the Earth with a telescope.

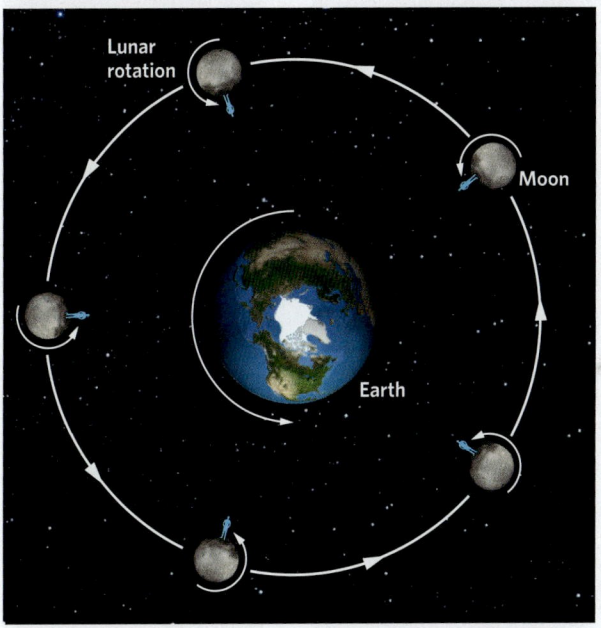

(b) The synchrony between the Moon's rotation and its orbit around the Earth causes the same side of the Moon to face the Earth continuously.

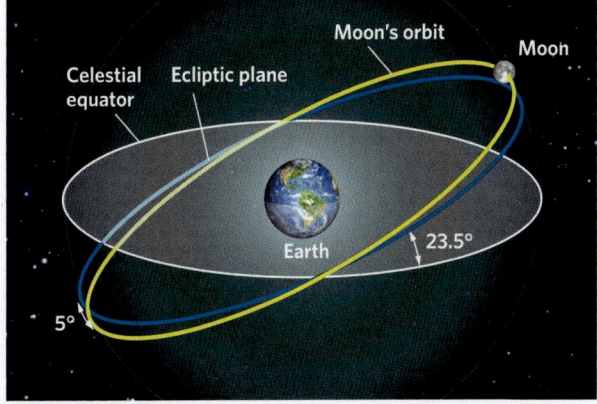

(c) The Moon's orbit is 5.2° off the ecliptic plane as seen on the celestial sphere.

FIGURE 21.18 The phases of the Moon.

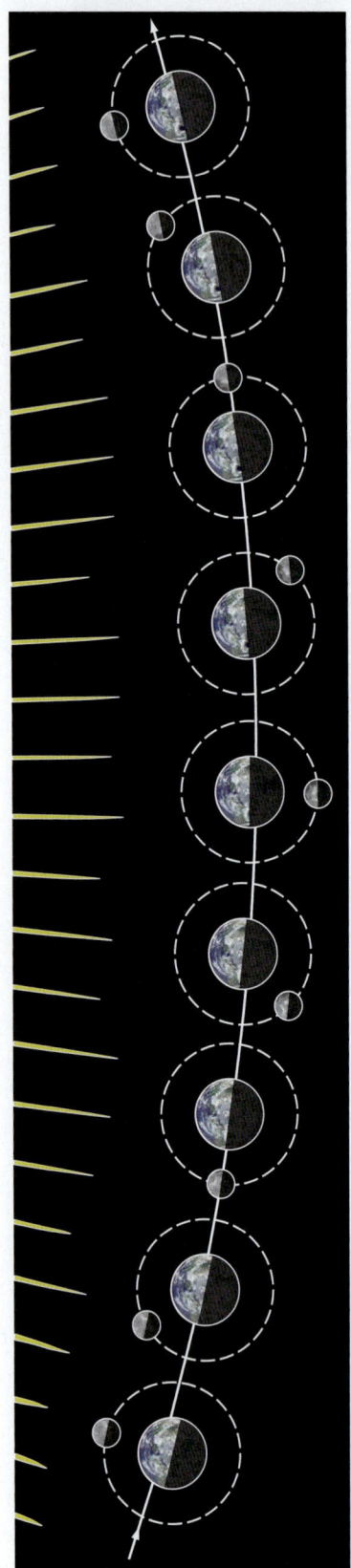

New
Moon

Waning
Crescent

Last
Quarter

Waning
Gibbous

Full
Moon

Waxing
Gibbous

First
Quarter

Waxing
Crescent

New
Moon

(a) The Moon's position relative to the Sun and the Earth determines how much of the illuminated part of the Moon can be seen from the Earth.

(b) The Moon completes its phases in 29.5 days.

Ancient Greeks used geometric methods to determine that the Moon lies about 30 times the Earth's diameter away from the Earth. Modern measurements indicate that, on average, it's 384,400 km (238,900 mi) away. The Apollo 11 astronauts traversed this distance to land on the moon in 1969. The Moon is the only celestial object outside of the Earth that humans have visited.

Because of its orbit around the Earth, the Moon's position in the sky arcs from east to west along a path defining an orbital plane tilted at about 5.2° to the Earth's ecliptic plane **(Fig. 21.17c)**. Relative to a fixed point on the Earth, the Moon speeds along its orbit at 1 km/s (3,600 km/h, or 2,200 mph).

Phases of the Moon

Even casual watchers of the sky know that if you look at the Moon night after night, you'll see the shape of the Moon's illuminated face change in a predictable cycle. Each successive stage of this cycle is known as a **phase** of the Moon. A single cycle between identical phases takes 29.5 days, a period called a **lunar month**. Lunar months were once used as the basis for calendars, but they don't work well over multiple years because they get out of synchrony with the seasons.

Why do phases of the Moon happen? Except during times when the Moon passes into the Earth's shadow—an event called a *lunar eclipse*, which we will discuss shortly—sunlight always shines on the Moon, lighting up half of its surface. This light, which reflects from the Moon's surface, is what we see when looking at the Moon **(Fig. 21.18a)**. When the Moon and the Sun lie on opposite sides of the sky, so that the Sun sets as the Moon rises, we see the Moon's entire lit face. As the Moon orbits the Earth, we see a progressively smaller portion of the lit face until the Moon lies on the same side of the Earth as the Sun, and the side of the Moon facing us receives no direct sunlight.

When a *new Moon* occurs, the Moon rises at approximately the same time as the Sun, so the Sun is behind the Moon, and the side of the Moon facing us is shaded. Every night after the new Moon, the visible part of the Moon *waxes*, or grows, as sunlight illuminates a progressively larger part of the Moon's surface **(Fig. 21.18b)**. The Moon passes sequentially through phases known as the *crescent Moon*, *quarter Moon*, and *gibbous Moon*. Halfway through a lunar month, the Moon rises as the Sun sets, and the entire face of the Moon that we see from the Earth is lit, producing a *full Moon*. On subsequent days, the Moon rises progressively later than the Sun sets, so the lit portion seen from the Earth *wanes*, or shrinks, through gibbous, quarter, and crescent phases. The cycle comes to completion when the Moon is once again in its new phase.

The Nature of Eclipses

The Moon and the Earth occasionally pass through each other's shadows. When they do, an *eclipse* takes place. If the Moon blocks the light of the Sun from reaching the Earth, we see a **solar eclipse**, whereas if the Earth blocks the light of the Sun from reaching the Moon, we see a **lunar eclipse**. Because such shadows can be cast only when the Earth, Sun, and Moon line up, and because the Moon's orbit inclines at 5.2° to the ecliptic plane, eclipses do not happen frequently. They take place only when a line drawn from the Sun to the Earth crosses the orbital plane of the Moon and when the Moon is either full or new (Fig. 21.19).

Our Moon, with a radius of 1,737 km (1,079 mi), is clearly much, much smaller than the Sun, which has a radius of 695,800 km (432,450 mi). Because the distance to the Moon from the Earth is much less than the distance to the Sun, when viewed from the Earth, both objects seem almost identical in size (Box 21.2). As a result, the Moon can occasionally block all the light of the Sun. In a general sense, the shadow that forms when an object obscures light coming from a nonpoint source (meaning a source that has a significant width) consists of two parts: an *umbra* of complete darkness and a *penumbra* of partial

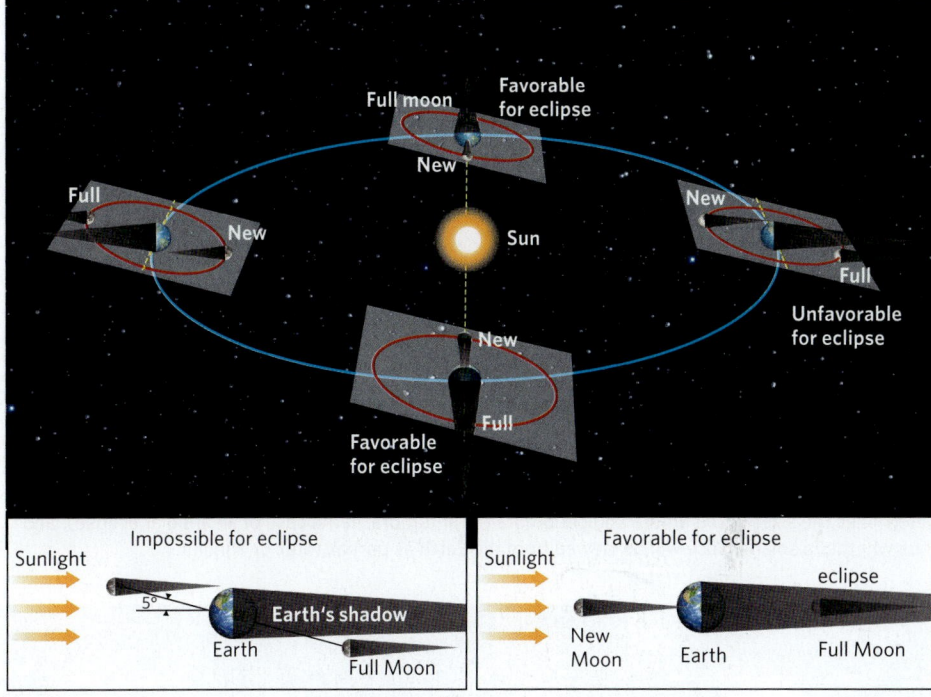

FIGURE 21.19 The causes of eclipses. Because the Moon's orbital plane is oriented at an angle to the ecliptic plane, there are only two times in a year when the Moon's orbit and the ecliptic plane intersect and an eclipse is possible. The Moon must be new (for a solar eclipse) or full (for a lunar eclipse) during these brief times.

BOX 21.2 ▶ # How can I explain . . .

The relative sizes of the Sun and the Moon

What are we learning?
How the apparent sizes of celestial objects in the sky depend on their distance from the Earth.

What you need:
- A basketball
- A measuring device that uses millimeters
- A piece of paper containing a typewritten period
- A calculator

Instructions:
- Use the basketball as your scale model for the Sun. Measure the diameter of the basketball and convert it to millimeters (if you don't have a basketball, look up its diameter on the Internet).
- To calculate the scale-model size of the Moon, we need to calculate proportions:

$$\frac{\text{true diameter of Sun (km)}}{\text{true diameter of Moon (km)}} = \frac{\text{measured diameter of basketball (mm)}}{\text{scaled diameter of Moon (mm)}}$$

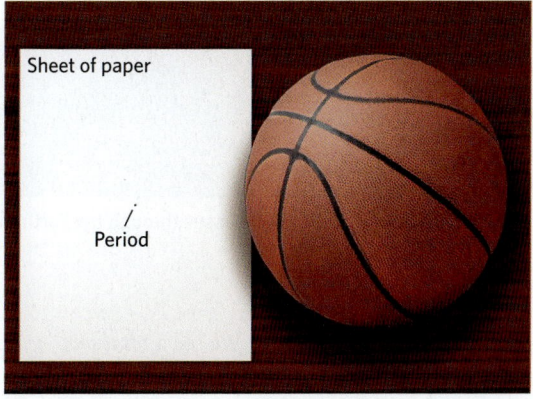

The unknown in this equation is the scaled diameter of the Moon. Now calculate what the diameter of the Moon would be if the Sun were the size of a basketball.

What did we see?
If you did the calculation correctly, you arrived at a very small number (0.5 mm). That is the diameter of a period at the end of a sentence, so if the Sun were the size of a basketball, the Moon would be the size of a period. Yet they appear the same size in the sky because the Moon lies much closer to the Earth.

FIGURE 21.20 The nature of eclipses.

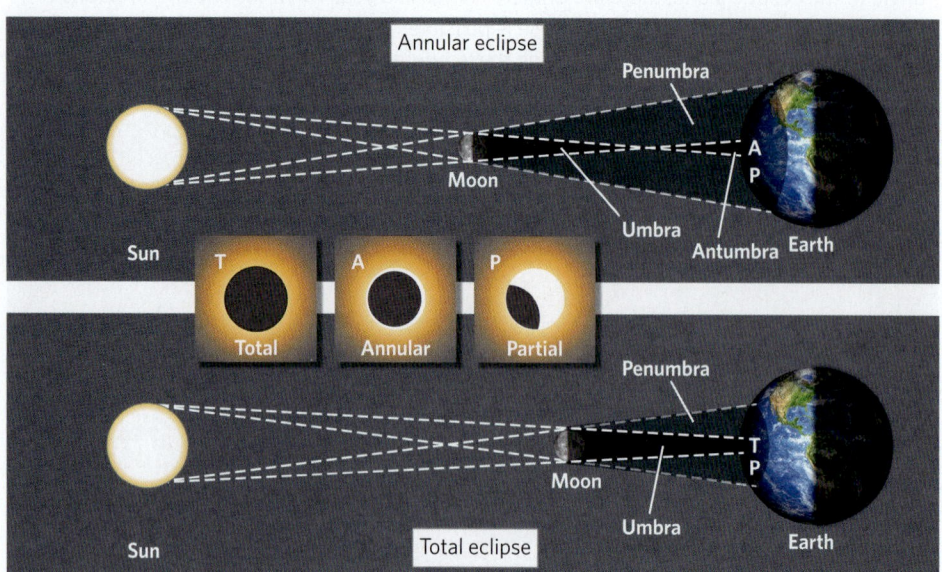

(a) We see a solar eclipse when the Moon blocks the Sun. The distance between the Earth and Moon determines the sizes of the umbra and penumbra (or antumbra, in the case of an annular eclipse), and thus whether a solar eclipse will be viewed from the Earth as partial, total, or annular.

(b) During the total eclipse of the Sun in 2017, visible across the US, observers could see gases streaming from the Sun into space. Millions looked up at the amazing sight when the sky went dark.

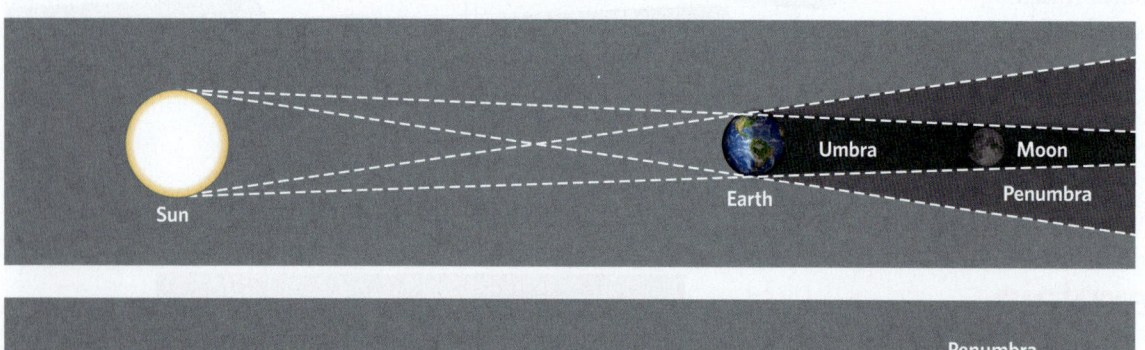

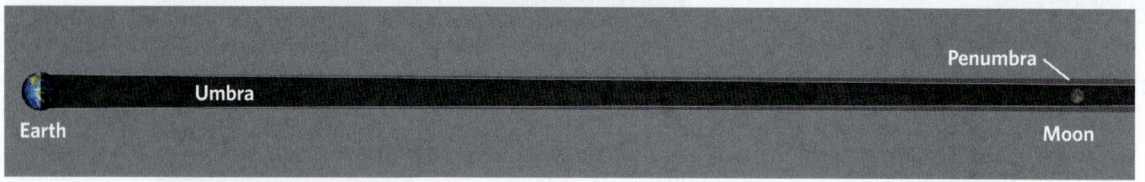

(c) We see a total lunar eclipse when the Moon passes through the Earth's umbra and a partial lunar eclipse when it passes through the penumbra.

(d) The Moon during a total lunar eclipse appears red because light from the Sun passes through the Earth's atmosphere and is bent toward the Moon.

darkness. We see a *total solar eclipse* when the Earth passes through the Moon's umbra and the Sun becomes completely obscured **(Fig. 21.20a)**. The path of the umbra on the Earth's surface is quite small, so relatively few people on the Earth see a particular total solar eclipse, and umbra paths differ for different eclipses. The movement of the Moon and the Earth along their orbits prevents a total solar eclipse from lasting more than 7.5 minutes. For observers in the penumbra, the Moon only partially obscures the Sun, producing a *partial solar eclipse*. When the Moon lies in a position far from the Earth, it appears smaller in the sky, so that an outer ring of the Sun is still visible around the shadow, which is called an *antumbra*. This phenomenon is called an *annular eclipse*. Except

during *complete totality*, when the viewer is in the umbra, enough radiation reaches the Earth during a solar eclipse that viewing it directly can severely damage your eyes. Never look at a solar eclipse without taking precautions **(Fig. 21.20b)**!

A lunar eclipse takes place when the Moon passes through the Earth's shadow. The Earth has a large shadow relative to the Moon, so total lunar eclipses happen more frequently than do total solar eclipses, and they can usually be seen from anywhere during nighttime on the Earth **(Fig. 21.20c)**. During a lunar eclipse, the Moon doesn't go black, but rather turns a ruddy red **(Fig. 21.20d)**. Why? Sunlight that passes through the Earth's atmosphere undergoes scattering and bending, and some of this light, after leaving the atmosphere, follows a path to the Moon. Because the atmosphere preferentially scatters blue light (see Chapter 17), the light that remains after passing through the atmosphere and reaching the Moon is red.

Take-home message . . .

The Moon revolves around its axis at the same rate that it orbits the Earth, so we always see the same side of the Moon. We see phases of the Moon because the side that we do see is not always fully lit by the Sun. Occasionally, the Moon casts a shadow on the Earth, so we see a solar eclipse, and occasionally the Earth casts a shadow on the Moon, so we see a lunar eclipse.

Quick Question
What is the difference between a total and partial eclipse?

21.5 Light from the Cosmos

Our ancestors knew that celestial objects existed because they could see the visible light that travels from these objects to the Earth. **Visible light** is a type of electromagnetic radiation that can be detected by the nerves in our eyes. In Chapter 17, we introduced electromagnetic radiation to explain how energy from the Sun heats the Earth's atmosphere. Modern astronomers study electromagnetic radiation—not just visible light, but other types as well—because examining this radiation helps astronomers to understand the composition, size, temperature, and movement of celestial objects. Therefore, to provide a foundation for discussing these objects, let's review the characteristics of electromagnetic radiation and consider how that radiation interacts with the Earth's atmosphere.

The Electromagnetic Spectrum, Revisited

Electromagnetic radiation travels across space in the form of **electromagnetic waves**, which, unlike sound or water waves, can move through a vacuum. We distinguish among different types of electromagnetic waves by specifying their **wavelength** (the distance between wave crests) or their **frequency** (the number of wave crests passing a point in space in 1 second). In a vacuum, all wavelengths of electromagnetic radiation travel at the same speed, the speed of light, so frequency (f) can be related to wavelength (w) by the equation $f = c \div w$. (In this equation, c stands for the speed of light.) Electromagnetic radiation occurs in a vast range of wavelengths, the entire range

of which constitutes the **electromagnetic spectrum (Fig. 21.21a)**. Visible light constitutes only a small part of this spectrum. Radio waves, microwaves, and infrared radiation have longer wavelengths than visible light, whereas ultraviolet (UV) light, X-rays, and gamma rays have shorter wavelengths.

The amount of energy carried by electromagnetic waves depends on the frequency of the waves: higher-frequency (shorter) waves carry more energy than do lower-frequency (longer) waves. In the early 20th century, researchers realized that to explain the way electromagnetic radiation interacts with atoms, radiation must be pictured not only as a succession of waves, but also as a stream of particles. These particles, known as **photons**, have no mass. Each photon contains a very specific amount of energy determined by the wavelength of the radiation.

Looking through Atmospheric Windows

If all wavelengths of electromagnetic radiation could pass through the atmosphere, we would not need to send up satellites to study the heavens. Unfortunately for ground-based astronomers, gas molecules in the atmosphere absorb most wavelengths of electromagnetic radiation. In fact, there are only two *atmospheric windows*, meaning ranges of wavelengths that can pass through air without being absorbed **(Fig. 21.21b)**. The **optical window** is *transparent* to visible light and *translucent* to infrared energy. This means that nearly all visible light waves penetrate the atmosphere, but only some infrared radiation does. The **radio window** is transparent to radio waves with wavelengths between 1 cm (0.4 in) and 10 m (33 ft).

We refer to the study of visible wavelengths as *optical astronomy* and the study of radio wavelengths as *radio astronomy*. *Optical telescopes* capture visible light and *radio telescopes* record radio signals. As we'll see, *space telescopes* can detect other wavelengths of radiation—such as X-rays and gamma rays—that can't pass through the atmospheric windows.

Using Spectroscopy to Answer Questions about Celestial Objects

WHAT'S IT MADE OF? If you've ever played with a prism, you've seen how it spreads a beam of white light into a spectrum of different colors **(Fig. 21.22a)**. That's because different wavelengths of light *refract* (bend) by slightly different amounts as they pass through the prism. Astronomers use **spectrometers**, devices that spread incoming electromagnetic radiation into a very broad spectrum, to analyze the radiation from stars.

If you look closely at an image produced by a spectrometer, you'll see that the light from celestial objects

FIGURE 21.21 The electromagnetic spectrum.

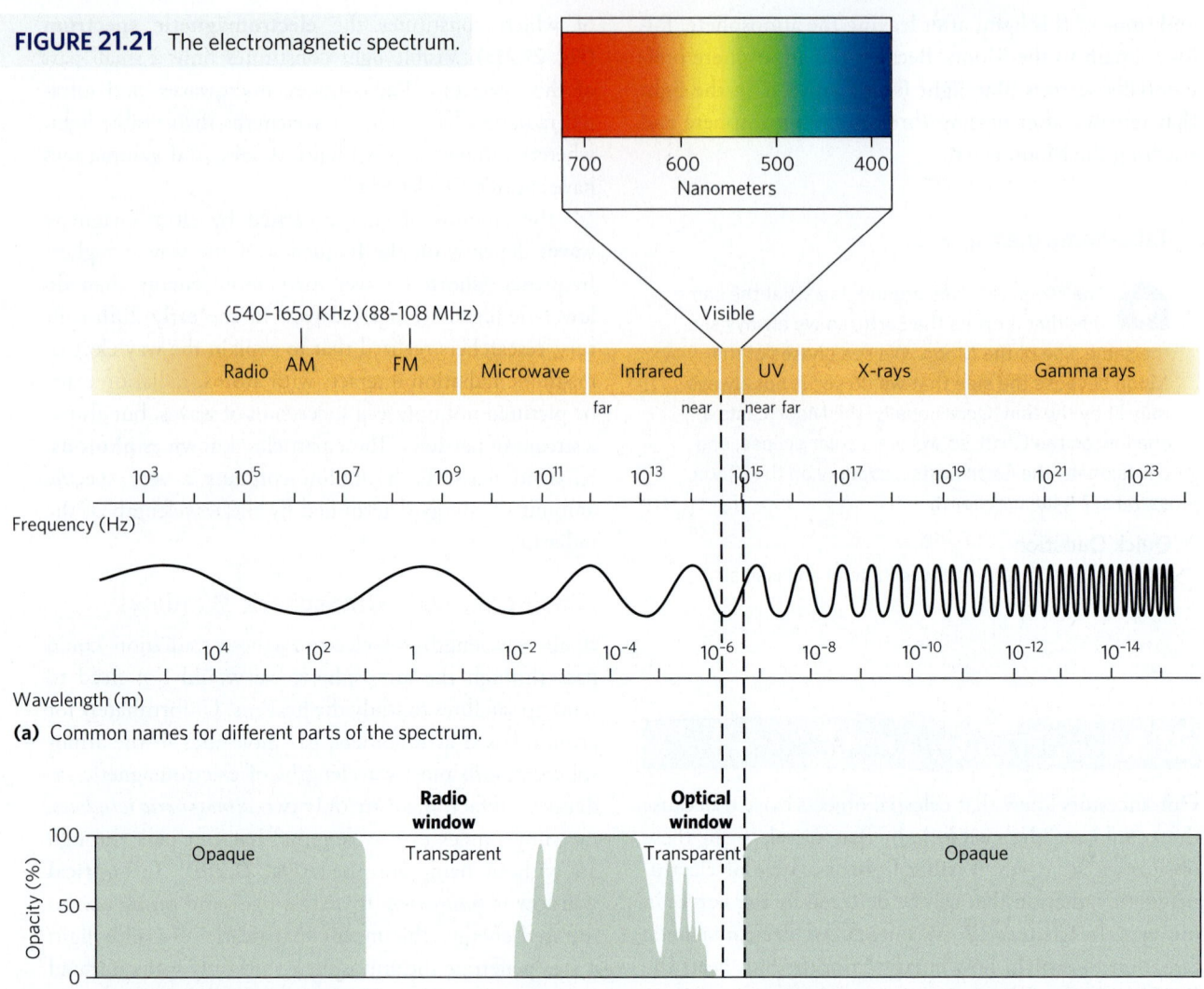

(a) Common names for different parts of the spectrum.

(b) Absorption of specific wavelengths by the atmosphere. Electromagnetic radiation from space can reach the Earth only through the radio and optical windows.

does not contain all possible wavelengths, but rather displays distinct **spectral lines**, bright or dark bands at specific wavelengths **(Fig. 21.22b)**. Spectral lines can serve as a fingerprint to characterize the chemical composition of the radiation source. *Emission lines* show up as narrow bright bands in the spectrum. Each emission line corresponds to the wavelength of the energy emitted by a specific element within the radiation source. If the radiation emitted by a source gets absorbed by a material between the source and the Earth, distinct wavelengths of the radiation do not reach the Earth. When this happens, the spectrum displays dark bands, called *absorption lines*, at specific wavelengths.

Astronomers use different kinds of spectral lines to study different kinds of celestial objects. For example, to learn about the chemical composition of glowing nebulae, astronomers use emission lines, for the specific wavelengths of radiation sent into space depend on the identity of gas atoms or molecules producing energy. In

contrast, clues to the chemical composition of stars comes mostly from examining the absorption lines in their spectra. That's because energy produced by a star comes from its very hot interior, and this energy must pass through the star's cooler outer layer before heading into space (see Chapter 23). The specific wavelengths of radiation absorbed by the outer layer identify the gases in the layer.

HOW FAST IS IT MOVING? Listen to the sound of a whistling train as it approaches you, passes you, and then moves away **(Fig. 21.23a)**. As it approaches, the sound gets louder, but its *pitch*, the note on the musical scale, remains constant. The instant that the train passes, however, the pitch of the sound changes to a lower note and stays at this note even as the train rumbles away and the sound gets softer. What's happening?

The pitch of sound depends on the frequency of the sound waves: higher-pitched sound has a higher frequency, and therefore a shorter wavelength, than does lower-pitched sound. As the train approaches, the sound waves "compact" into shorter wavelengths because both the source and the waves are moving toward the observer. Each successive wave has a shorter distance to travel, so the intervals between the arrival times of successive waves decrease. An observer hears the higher-frequency waves as a steady high pitch that gets louder as the train approaches. When the train passes and starts moving away, sound waves from its whistle still move toward the observer, but they "stretch" because the whistle and the waves are moving in opposite directions. Each successive wave has a longer distance to travel, so arrival times spread farther apart. Therefore, the frequency of the arriving waves decreases,

FIGURE 21.22 Spectral lines.

(a) A prism can be used to break white light into a spectrum of colors.

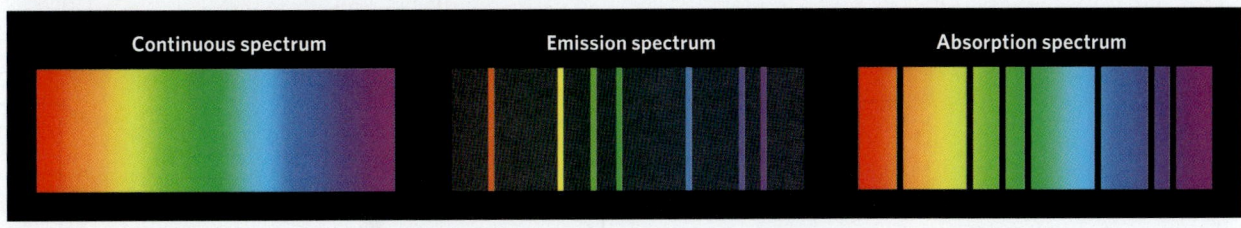

(b) The spectral lines of an element compared with the continuous visible spectrum. Bright lines are emission lines; dark lines are absorption lines. Note that emission and absorption lines occur at the same wavelengths for a given element.

and the observer hears a steady lower pitch that becomes quieter as the train moves away. The faster the train, the greater the frequency shift that the observer hears. Scientists refer to a shift in the frequency of waves caused by the movement of their source as the **Doppler effect**, named for the Austrian physicist Christian Doppler (1803–1853), who first explained it.

Just as the pitch of sound depends on the frequency of sound waves, the color of light depends on the frequency of electromagnetic waves. Electromagnetic waves can undergo a Doppler shift when emitted by objects moving toward or away from an observer. The faster the object travels, the greater the Doppler shift. A shift toward a higher frequency (shorter wavelength) happens when the object moves toward the observer, whereas a shift toward a lower frequency (longer wavelength) happens when the object moves away from the observer **(Fig. 21.23b)**. The wavelength of blue light is shorter than that of red light. Therefore, a light source moving toward an observer exhibits a **blue shift**, whereas a light source moving away from an observer exhibits a **red shift**. In Box 17.2, we learned that meteorologists use the Doppler shift of radar beams to identify possible tornadoes. Astronomers use the Doppler shift to determine the movements of celestial objects.

To detect blue or red shifts, astronomers need to compare the spectrum received from a celestial object with the spectrum produced by a stationary source of light. They can use the spectrum of light from the Sun **(Fig. 21.23c)**, or more sophisticated instrumentation, such as a spectrum tube. (A *spectrum tube* is a glass tube containing gas of a known composition; when an electric current flows through a filament at the base of the tube, the gas glows.) For example, when we say that a distant galaxy exhibits a red shift, we mean that its spectral lines have shifted toward the red end of the spectrum, relative to equivalent spectral lines displayed on the spectrum of light from the Sun.

How do astronomers use Doppler shifts to determine the velocities of celestial objects? Analysis of the Doppler effect for an individual star provides a basis for characterizing the star's rotation speed because the side of the star moving toward us exhibits a blue shift, while the side moving away from us exhibits a red shift, as the star spins on its axis. Doppler shifts can also indicate how fast a star is moving toward or away from the Earth. The same type of analysis can allow astronomers to determine how fast a galaxy is rotating and how fast it is moving toward or away from the Earth. When astronomers studied spectra from distant galaxies, they discovered that all display a red

FIGURE 21.23 The Doppler effect.

The Doppler effect for sound

Stationary whistle

Moving whistle

(a) The wavelength of the sound waves emitted by a stationary train whistle is the same in all directions. The sound waves behind a moving train whistle have longer wavelengths than those in front of it.

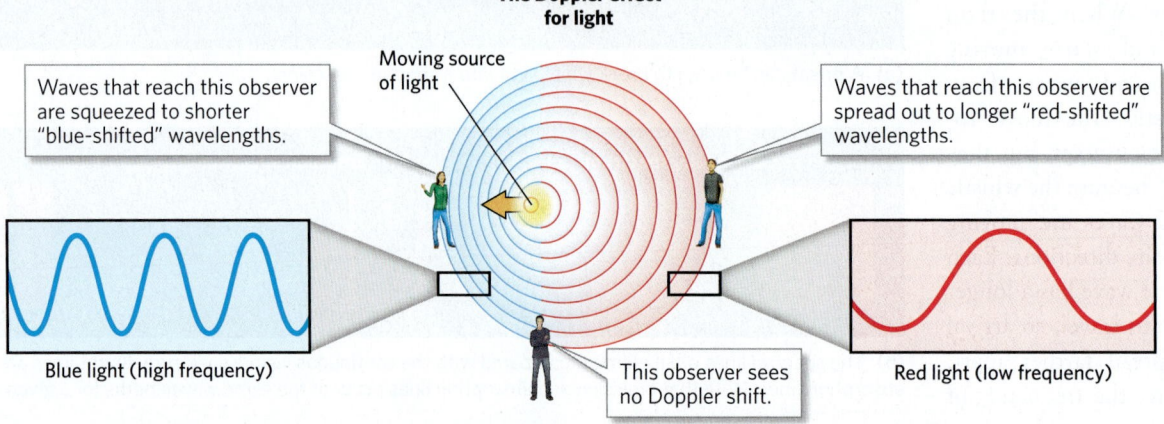

The Doppler effect for light

Moving source of light

Waves that reach this observer are squeezed to shorter "blue-shifted" wavelengths.

Waves that reach this observer are spread out to longer "red-shifted" wavelengths.

Blue light (high frequency)

This observer sees no Doppler shift.

Red light (low frequency)

(b) If a light source moves toward an observer, the radiation detected by the observer will be shifted toward shorter wavelengths (blue-shifted). If the light source moves away, the radiation will be shifted to longer wavelengths (red-shifted).

Sun

Distant galaxy

(c) The shift in wavelengths between radiation from a distant source, such as a galaxy or a star, and from the Sun can be used to determine the velocity of the source's movement toward or away from the Earth.

shift, indicating that all distant galaxies are moving away from the Earth. They also learned that galaxies farther from the Earth are moving away faster. This realization led Edwin Hubble to propose the theory of the expanding Universe (see Chapter 1).

WHAT'S ITS TEMPERATURE? According to the laws of *blackbody radiation* that we discussed in Chapter 17, the wavelength at which a glowing object emits the greatest amount of energy depends on the surface temperature of the object. Hotter objects emit the most energy at shorter wavelengths of light (toward the blue end of the spectrum), while cooler objects emit the most energy at longer wavelengths of light (toward the red end of the spectrum).

By using this concept, astronomers have determined that blue stars have a surface temperature of 20,000°C (36,000°F), yellow stars have a surface temperature of 6,000°C (10,800°F), and red stars have a surface temperature of 3,000°C (5,400°F). (We'll take a closer look at these spectral classes of stars in Chapter 23.)

WHAT'S ITS SIZE? Modern telescopes are sensitive enough to allow direct measurement of the sizes of a few nearby stars, using special techniques discussed in more advanced books. Astronomers using these telescopes have found a huge range of star sizes. For distant stars, direct measurement of a radius isn't possible because currently available instruments only see the star as a point of light. Instead, astronomers calculate the dimensions of these stars in-directly using equations that relate the temperature and brightness of a star to its surface area. From the surface area, simple geometric formulas yield the radius.

Take-home message . . .

Energy from stars passes through space in the form of electromagnetic radiation, but only part of this radiation makes it through the Earth's atmosphere. Astronomers use the wavelengths of radiation emitted by stars to determine their compositions, how fast they are moving toward or away from the Earth, how fast they are spinning, and what their surface temperatures are.

Quick Question

What does the Doppler shift tell us about motion of a star or galaxy relative to the Earth?

21.6 A Sense of Scale: The Vast Distances of Space

The Units That Astronomers Use

How big is the Earth? No one knew until a clever Greek astronomer, Eratosthenes (ca. 276–194 B.C.E.), figured it out. He read that at noon on the first day of summer, sunlight lit the bottom of a deep well in the town of Syene in southern Egypt. Eratosthenes found that at noon, the Sun's rays strike the ground at an angle 7.2° off of vertical in Alexandria, 800 km to the north. Knowing the distance between Syene and Alexandria (5,000 stadia, a unit of measurement in ancient Egypt), and knowing that a circle contains 360°, Eratosthenes calculated the Earth's circumference using a simple geometric equation and came up with a number within 2% of the modern accepted value (40,008 km, or 24,865 mi) **(Fig. 21.24)**.

To describe the mind-boggling distances between celestial objects, units such as kilometers and miles aren't big enough to be practical. So instead, astronomers use three units that are much bigger than the familiar ones we use in daily life. The first, called an **astronomical unit** (AU), is the average distance from the center of the Earth to the center of the Sun, equals about 150 million kilometers (93 million miles). The second, called a **light-year**, became available when researchers determined that light travels at a constant speed of about 300,000 km/s (186,000 mi/s) in a vacuum. Because light moves at a constant speed, it travels a specific distance in a given time. A light-year, the distance light travels through the vacuum of space in 1 year, equals about 9.5 trillion km (5.9 trillion mi). Nontechnical discussions of astronomy generally describe distances using light-years. Professional astronomers, however, use

FIGURE 21.24 The geometric calculation that Eratosthenes used to determine the circumference of the Earth.

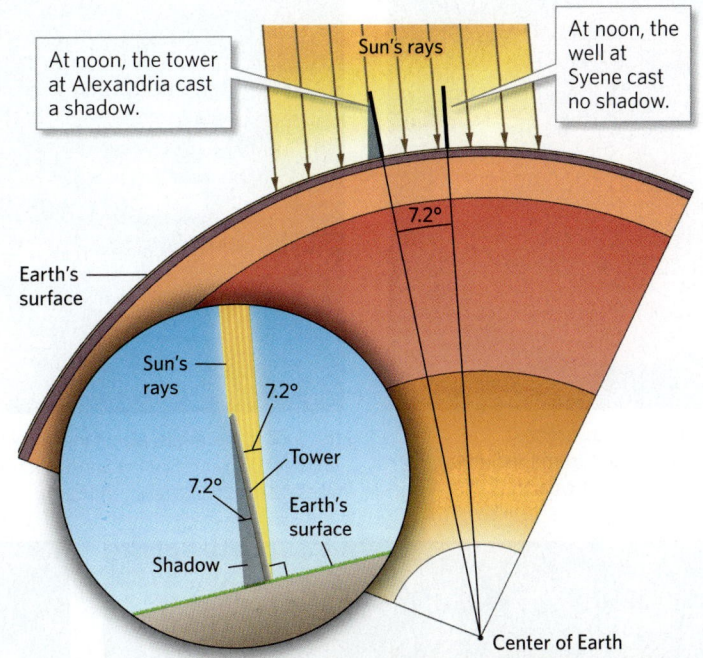

Eratosthenes's calculation:

$$\frac{360°}{x} = \frac{7.2°}{5,000 \text{ stadia}}$$

$$x = \frac{360° \times 5,000 \text{ stadia}}{7.2°}$$

$$x = 250,000 \text{ stadia}$$

250,000 stadia × 0.1572 km/stadium = 39,300 km (or, 24,421 miles)

an even larger unit, the *parsec*—roughly 3.3 light-years—for measuring distances. We'll discuss the basis for defining a parsec later in this section.

Because light takes time to travel, we're looking into the past when we look into space **(Fig. 21.25)**. The light we see when looking at the Moon left the Moon 1.3 seconds ago, light arriving from the Sun left the Sun 8 minutes ago, and light arriving from Alpha and Beta Centauri left those stars 4.2 years ago. The farther we peer into deep space, the further we peer back into deep time. Light arriving from the most distant visible object in the Universe left that object billions of years before the Earth even existed.

How Can We Measure Distances in Space?

Modern astronomers can measure distances from the Earth to the planets, to the stars, and even to galaxies that are millions of light-years from the Earth. How do they do it? Astronomers today rely on five basic tools for measuring the great distances of space.

Did you ever wonder . . .

how astronomers can tell how far a star is from the Earth?

FIGURE 21.25 Looking into deep space and back in time.

(a) The Sun, our nearest star, is about 150,000,000 km away. Light from the Sun takes about 8 minutes to arrive at the Earth.

(b) Jupiter, the largest planet in the Solar System, is 778,000,000 km from the Sun. Light from Jupiter takes 42 minutes to reach the Earth.

(c) By contrast, Alpha and Beta Centauri, the nearest stars to our Sun, are 39,740,000,000,000 km (4.2 light-years) from the Earth. Light from Alpha and Beta Centauri takes 4.2 years to reach the Earth.

(d) Our home galaxy, the Milky Way, is 100,000 light-years wide. Light arriving at the Earth from the edge of the Milky Way left that location 100,000 years ago.

(e) Magnification of a tiny area in a Hubble Space Telescope image of distant galaxies reveals the farthest object yet found in the known Universe. The light took 13.3 billion years to reach the Earth.

RADAR. Astronomers can use radar to determine the distances to objects in the Solar System. To make a radar measurement, they use a transmitter to send a pulse of radio waves out from an antenna. When the pulse bounces off an object, it returns to the antenna, where it is detected by a receiver. Because a radar pulse travels at the speed of light, the distance to the object can be calculated by precisely measuring the time between the transmission and return of the pulse.

GEOMETRIC PARALLAX. If you hold your thumb out at arm's length, and first close one eye and then the other, you'll see an example of **geometric parallax**, the apparent displacement of a foreground object relative to background objects when an observer's position, and therefore their line of sight, changes **(Fig. 21.26a)**. The movement of the Earth from one side of its orbit around the Sun to the other side provides enough distance so that astronomers can detect the slight geometric parallax that closer stars display relative to more distant stars **(Fig. 21.26b)**. By precisely measuring angles, astronomers can calculate the distance to relatively nearby stars, meaning those less than 500 light-years away.

The concept of geometric parallax provides the basis for defining a parsec. A **parsec** (short for parallax second) represents the distance that an imaginary star would have to be from the Earth for its angular shift to be exactly one arc-second (1/3,600 of a degree) if an observer were to move from a point at the Sun's center to a point on the Earth's surface, 1 AU away.

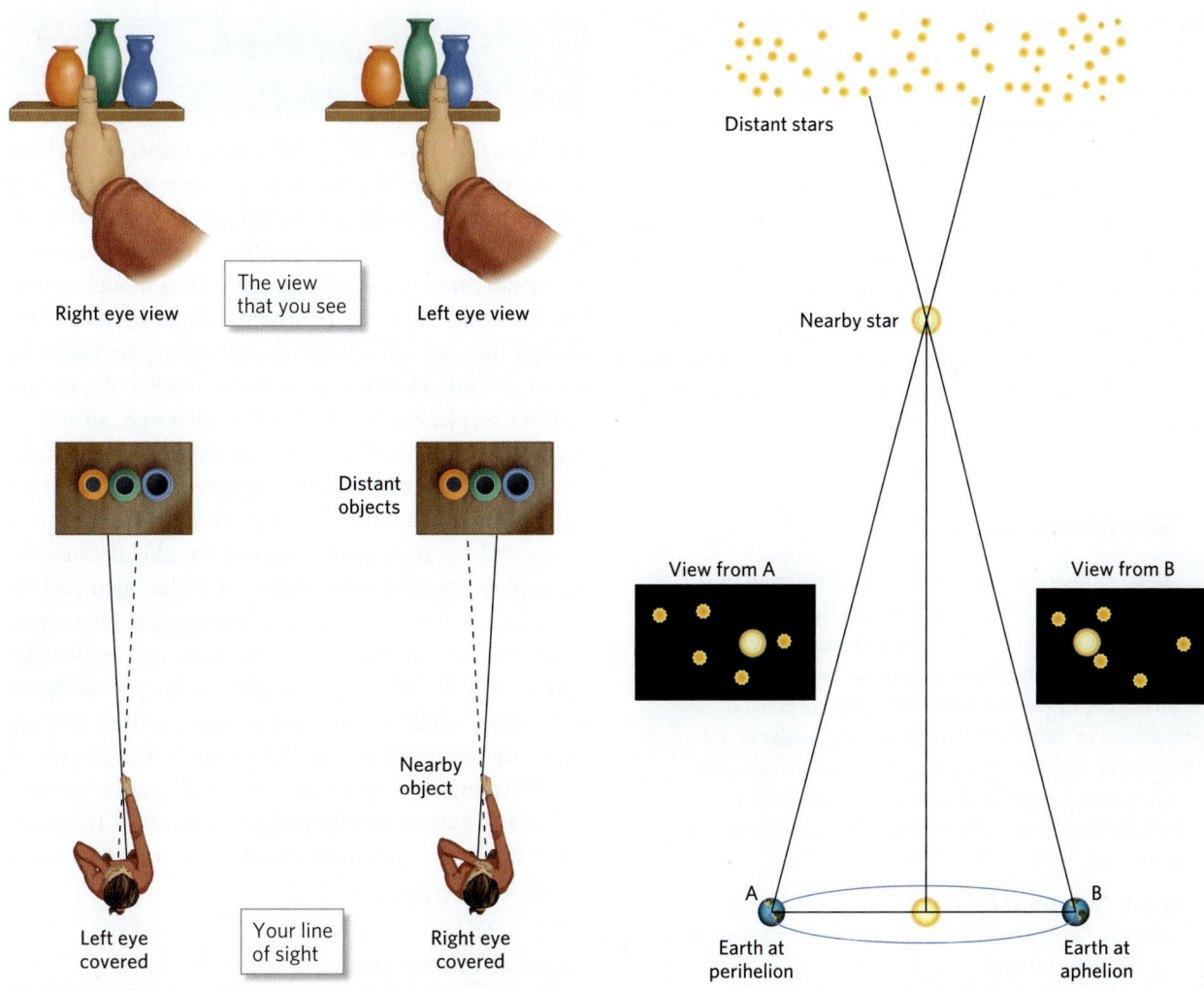

FIGURE 21.26 Geometric parallax.

Right eye view

The view that you see

Left eye view

Distant objects

Nearby object

Left eye covered

Your line of sight

Right eye covered

Distant stars

Nearby star

View from A

View from B

A
Earth at perihelion

B
Earth at aphelion

(a) You can observe geometric parallax by holding your thumb out at arm's length. When you close one eye and then the other, your thumb covers different distant objects.

(b) A nearby star shifts its position relative to the background stars when viewed from opposite sides of the Earth's orbit.

STELLAR BRIGHTNESS. The geometric parallax method can't be used to determine how far away a distant star lies because the displacement of the star against its background is too small to detect. For distant stars, astronomers rely on the simple relationship between apparent brightness and distance. The *apparent brightness* of a light source (the brightness you see) decreases by the square of its distance from an observer because its light energy spreads over an increasing area, even if its *true brightness* (the amount of energy emitted) remains unchanged. We'll discuss these measurements of brightness in more detail in Chapter 23. If astronomers can estimate the true brightness of a star and measure its apparent brightness in the Earth's sky, they can calculate the star's distance from the Earth.

CEPHEID VARIABLES. Measurements of brightness can determine stellar distances up to about 10,000 parsecs. Measurements of even greater distances come from the study of special kinds of stars that pulsate, meaning that they expand and contract periodically. When a pulsating

star expands, it has a larger surface area, but its temperature doesn't change, so its apparent brightness increases. Henrietta Leavitt (1868–1921) discovered that one class of pulsating stars, known as *Cepheid variables*, display a predictable relationship between true brightness and the periodicity of their pulsations. So, if astronomers know the period of a Cepheid variable star's pulsations (usually between 1 and 100 days), they also know its true brightness. Then, by comparing this true brightness with the apparent brightness of the star, astronomers can calculate the star's distance. If a Cepheid variable star is within a distant galaxy, the distance from the Earth to the galaxy can be measured by examining the changing brightness of that star.

DOPPLER SHIFTS. For very distant galaxies, none of the measuring techniques we have described so far will work. A breakthrough came when astronomers measured red shifts associated with closer galaxies using Cepheid variable stars. They determined that with greater distance between a galaxy and the Earth, the galaxy displays a greater red shift, and therefore, the velocity of its movement away from the Earth is greater. This relationship can be expressed in the form of an equation: $v = Hd$. (In this equation, v is velocity, H is a constant, and d is distance.) Edwin Hubble studied these phenomena and determined the value of H. To recognize the importance of Hubble's work, astronomers now refer to H as *Hubble's constant*, and to the equation relating velocity to distance as **Hubble's law**. Once Hubble's law had been calibrated using Cepheid variables, it could be used to measure the distance to the most remote objects in space.

Take-home message . . .

Astronomers use three units of measure to describe the vast distances in the Universe: astronomical units, light-years, and parsecs. Astronomers measure the distance from the Earth to closer objects using radar or geometric parallax and the distance to more remote objects using stellar brightness. For very distant objects, they use Hubble's law, which states that the velocity of a distant galaxy, as indicated by its red shift, depends on its distance from the Earth.

Quick Question

Why are Cepheid variable stars important for determining stellar distances?

FIGURE 21.27 How a refracting telescope works.

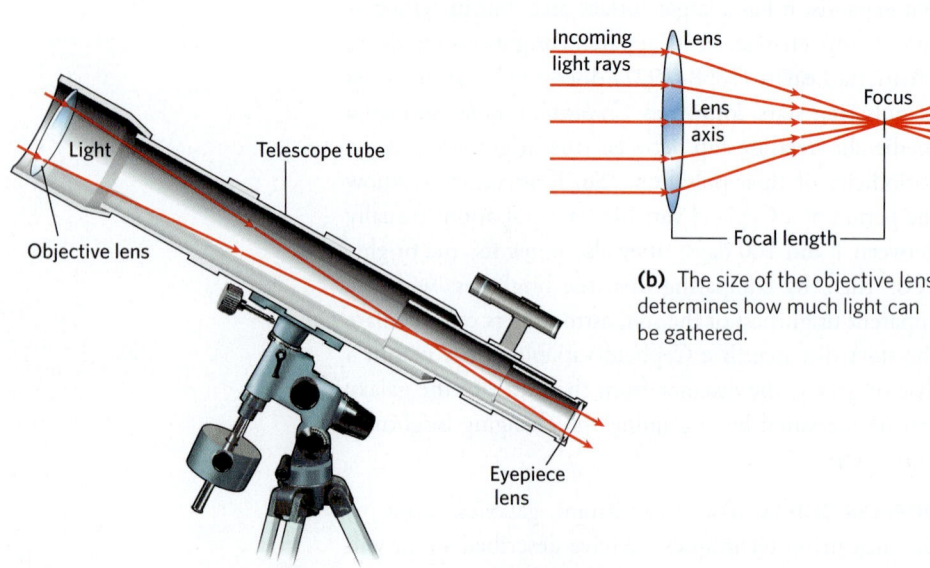

(b) The size of the objective lens determines how much light can be gathered.

(a) A modern amateur refracting telescope uses an objective lens to gather light and direct it to a focus. The eyepiece magnifies the light.

21.7 Our Window to the Universe: The Telescope

The human retina—the light-sensing tissue in the back of our eyes—doesn't have enough nerves to detect very faint light or to *resolve* (see the details of) objects that are far away. Because of this limitation, an earthbound observer can see only about 3,000 stars, even though countless stars exist, and can't see the volcanoes of Mars, even though they are vastly larger than those on the Earth. To detect distant celestial objects and to resolve the surface features of planets, we must use a **telescope**, an instrument designed to collect and focus electromagnetic radiation in order to make objects appear brighter and larger. A land-based **observatory** is a place where astronomers set up and use telescopes to observe the sky. Because the atmosphere permits transmission of visible light and radio waves, astronomers can use both **optical telescopes**, which collect and magnify visible light, and **radio telescopes**, which collect and amplify radio waves. **Space telescopes**, which observe space from satellites orbiting above the atmosphere, can detect other wavelengths of the electromagnetic spectrum, such as X-rays and gamma rays (**Earth Science at a Glance**, pp. 688–689). To understand how telescopes work, let's look at the designs of a few different types.

Optical Telescopes

TYPES OF OPTICAL TELESCOPES. Optical telescopes work by gathering all the light striking a broad area and concentrating it into a small area. Telescopes can be configured to collect light in two different ways. *Refracting* telescopes were developed first (in 1608), followed by *reflecting* telescopes (in 1668).

When Galileo spotted the moons of Jupiter for the first time, he used a **refracting telescope**. In this type of telescope, an *objective lens*—a disk of glass that has been ground and polished so that it has curved surfaces on its top and/or bottom—refracts incoming light and concentrates it at a *focus* behind the lens (**Fig. 21.27**). A second lens, the *eyepiece*, magnifies the concentrated light so that objects appear larger. Most hand-held telescopes are refracting telescopes. The quality of a refracting telescope depends on how clear and how perfectly curved its lenses are, and its light-gathering ability depends on the size of the objective lens.

When Newton began studying optics, he developed the **reflecting telescope**, in which a concave mirror reflects light to a focus in front of the mirror (**Fig. 21.28**). The larger the area of the mirror, the more light the telescope can gather. A smaller mirror at the focus sends the concentrated light to an eyepiece at the back or at the side of the telescope.

FIGURE 21.28 How a reflecting telescope works.

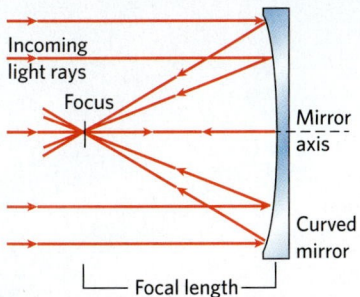

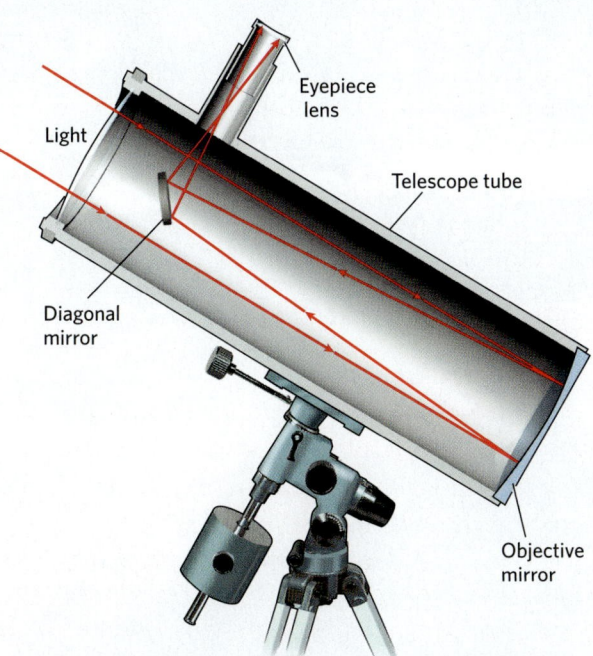

(b) The size of the concave mirror determines how much light can be gathered.

Incoming light rays
Focus
Mirror axis
Curved mirror
Focal length

(a) A reflecting telescope uses a concave mirror to gather light and direct it to a focus.

Eyepiece lens
Light
Telescope tube
Diagonal mirror
Objective mirror

Astronomers using a modern research telescope don't stare through the eyepiece directly, but rather examine photographs of the light coming through the eyepiece. Large research telescopes **(Fig. 21.29)** can be moved by motors so that they stay aimed at exactly the same spot in space for a long time, even as the Earth spins. Therefore, the camera can accumulate all the light collected for many minutes or even hours. Prior to the electronic age, cameras used film to record light coming through the eyepiece. Today, film has been replaced with an electronic screen containing millions of light-sensitive pixels. Such a device can be configured to produce a digital visual image or to serve as a photon counter that can accurately determine the brightness of an object.

CONSIDERATIONS IN TELESCOPE DESIGN AND LOCATION. The ability of a ground-based optical telescope to collect light depends on the size of the objective lens (in a refracting telescope) or mirror (in a reflecting telescope). Because it's difficult and expensive to grind lenses, and because lenses are very heavy and can warp due to their own weight, the largest objective lens ever made for a refracting telescope has a diameter of only 100 cm (40 in). It's easier to design and manufacture large curved mirrors,

FIGURE 21.29 Large research telescopes.

(a) The Hale Telescope at the Palomar Observatory has a mirror that is 5.1 m across.

(b) At night, the dome opens at the McDonald Observatory in Texas.

Observatories of the World

Visible

Hubble Space Telescope

James Webb Space Telescope

Infrared

Spitzer Space Telescope

Radio

Ground-based radio telescopes

Astronomical observatories on the Earth and in space observe the Universe using a wide array of telescopes that employ different wavelengths of electromagnetic radiation.

The wavelengths arriving from space that are captured by ground-based telescopes are those to which the atmosphere is transparent. Two types of ground-based telescopes are radio telescopes and the more traditional optical telescopes.

Scientists use radio telescopes, which employ large antennas, to study the center of our galaxy, distant objects beyond it, and other strong radio-wave sources. Radio telescopes can be used in the daytime as well as at night. They allow us to study astronomical radio-wave sources, such as stars, nebulae, and galaxies, in the distant reaches of the Universe. Optical telescopes are used to image a wide array of objects that emit light in wavelengths that human eyes can detect. Optical telescopes have a long history of opening windows to the Universe, from the first observations by Galileo to the sophisticated mountaintop observatories of today.

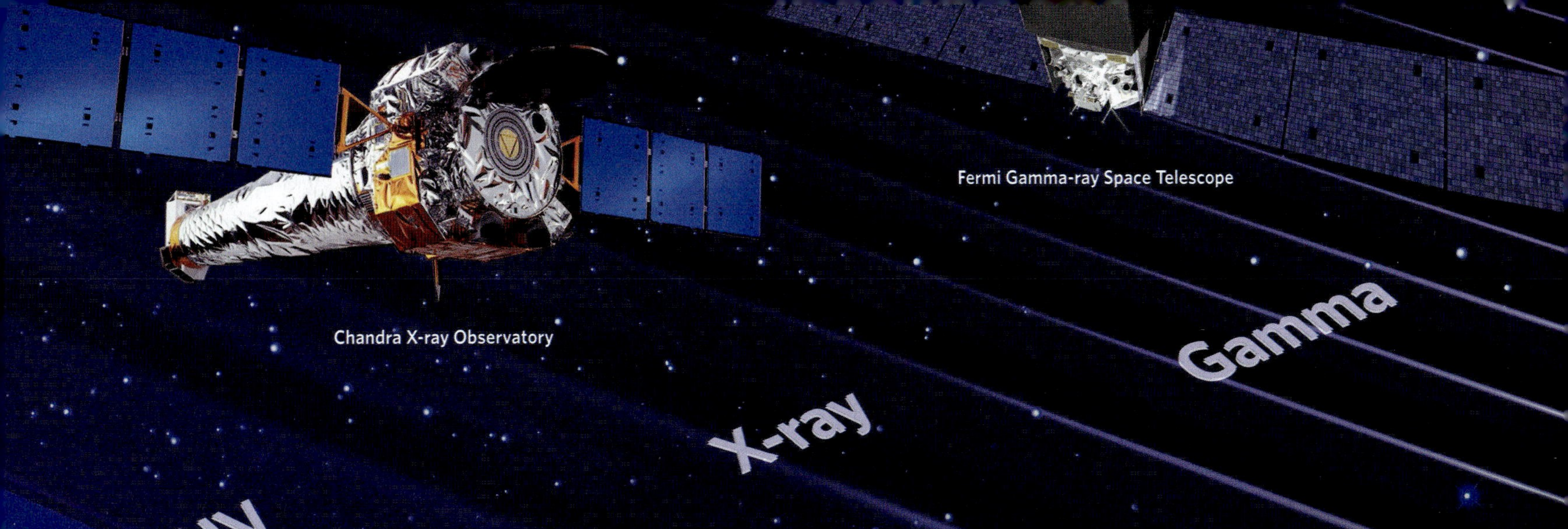

Chandra X-ray Observatory

Fermi Gamma-ray Space Telescope

Gamma

X-ray

UV

Space telescopes, on the other hand, have no limitations with regard to wavelength. They typically sample small ranges of wavelengths, with each telescope focusing on a different region of the electromagnetic spectrum.

The Spitzer Space Telescope has used the infrared spectrum to reveal the existence of exoplanets and has allowed scientists to study forming stars, while the new James Webb Space Telescope will be used to study the history of the Universe, from the first luminous glow after the Big Bang to the formation of solar systems capable of supporting life. The Hubble Space Telescope (which uses the visible spectrum) has astounded the world with fantastic views of space, from star nurseries shrouded with dust to clusters of galaxies at the far reaches of the Universe. The Chandra X-ray Observatory allows scientists to study very hot regions of the Universe, such as exploded stars, clusters of galaxies, and matter around black holes. The Fermi Gamma-ray Space Telescope targets strange phenomena of the Universe, such as supermassive black holes, neutron stars, and streams of hot gas moving close to the speed of light.

Five satellite-based telescopes have been launched so far, and some remain in operation today. At the time of this writing, the expected launch date for the James Webb Space Telescope is March 2021. Space telescopes have not only provided insight into the architecture of galaxies and other celestial objects, but have also yielded thousands of mind-boggling images of distant galaxies and nebulae. With every new image released, people the world over gasp at the astounding beauty of the Universe.

Ground-based optical telescopes

FIGURE 21.30 Atmospheric seeing. The path of light passing between space and an observer at the ground is altered by atmospheric turbulence, causing images to go out of focus.

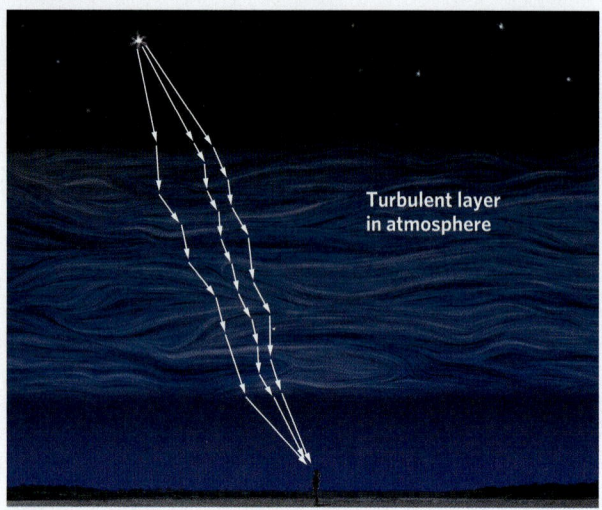

Turbulent layer in atmosphere

Did you ever wonder . . .

why astronomers like to locate observatories on isolated mountaintops?

and such mirrors can be supported from behind to prevent warping. As a result, reflecting telescopes can be much bigger than refracting telescopes, and today's observatories rely on reflecting telescopes for their research. The largest reflecting telescope mirror has a diameter of 10.4 m (409 in).

Even if a telescope has great light-gathering power, the telescope's **resolving power**—its capacity to distinguish between two faraway objects that lie close together—limits an observer's ability to discern details. The resolving power of a telescope can be affected by several phenomena, including *light diffraction*, which happens when straight waves bend into curves due to interaction with an obstacle. The best resolution possible in a modern telescope permits an observer to distinguish two coins placed side by side at a distance of 100 km (62 mi).

Even with the best telescopes, ground-based astronomers must contend with **atmospheric seeing**, the blurring or fuzzing of an object's image that happens because light from the object had to pass through the atmosphere **(Fig. 21.30)**. Random variations in air density associated with turbulence cause incoming light to refract in an irregular way. Atmospheric seeing can be reduced by using *adaptive optics*, a technology that uses computers to warp mirrors slightly as needed to compensate for distorted light. This technology has vastly improved the quality of images obtained by ground-based telescopes.

Any atmospheric phenomenon that decreases atmospheric clarity, telescope stability, or the darkness of the sky limits the image quality provided by ground-based telescopes. Snow, rain, clouds, and wind, for example, obscure views and blur images. In recent decades, **light pollution** (the glow of the sky due to reflection and scattering of light from sources such as streetlights, windows, and billboards) has become a significant problem for observatories. This glow can wash out faint starlight coming from space **(Fig. 21.31)**.

FIGURE 21.31 Light pollution.

(a) A photo of the United States from space at night shows the extent of urban areas with bright lights.

(b) From the ground, in a well-lit city, only the brightest celestial objects are visible, even on a clear night.

Paranal Observatory, Chile

Mauna Kea Observatories, Hawaii

Roque de los Muchachos Observatory, Canary Islands, Spain

For all these reasons, when locating optical observatories, astronomers must take into account elevation, climate, and proximity to urban areas. The largest observatories built in the last century perch on isolated mountaintops in arid climates, where there is less air between their optical telescopes and space. These locations offer less atmospheric seeing, less bad weather, and less light pollution **(Fig. 21.32)**. Examples include the Paranal Observatory of Chile, the Mauna Kea Observatories of Hawaii, the Kitt Peak National Observatory of Arizona, and the Roque de los Muchachos Observatory on the Canary Islands. All of these observatories lie at elevations between 2 and 5 km (1.2 – 3.1 mi) above sea level.

Radio Telescopes

In the late 1920s, Karl Jansky, a scientist at Bell Labs, began to investigate sources of *static* (unwanted electronic noise) that interfered with ground-based radio transmissions. Jansky built a dish-shaped antenna—shaped much like the mirror of a reflecting telescope—capable of collecting very weak radio signals, and pointed it skyward. He discovered that some of the static came from

thunderstorms. But he also recorded a faint, mysterious background hiss. After measuring this signal over a period of several days, he realized that a particularly strong hiss repeated every 24 hours, and that this hiss appeared when the telescope faced a particular location in space. Jansky concluded that the hiss must be coming from space to the Earth, and he eventually determined that it originated in the galactic center of the Milky Way. In effect, Jansky's dish-like antenna was detecting electromagnetic radiation that had come from space through the radio window of the atmosphere, so the antenna can be considered to be the first radio telescope.

Modern radio telescopes work much like reflecting telescopes. They consist of a dish that reflects incoming radio-wave energy to a focus, where instruments collect and analyze that energy. Because radio-wave energy coming from space is weak and its wavelengths are long, radio telescopes need to be very large **(Fig. 21.33a)**. In recent decades, astronomers have built **radio interferometers**, which superimpose the waves collected by numerous radio telescopes arranged in an array **(Fig. 21.33b)**. Radio interferometers effectively act like a single huge radio telescope

whose diameter equals the distance from one side of the array to the other.

Even with huge antennas, the resolution of radio telescopes can't come close to that of optical telescopes. Nevertheless, radio telescopes have advantages. The Sun does not emit significant radio-wave energy, so radio telescopes can be used both day and night. Also, since radio waves pass through clouds and precipitation, radio telescopes work in all types of weather. Finally, radio telescopes can detect some objects that cannot be seen with optical telescopes. For example, nebular dust absorbs visible light coming from the center of the Milky Way, so we can't see that light optically. Radio waves pass through this dust, so radio telescopes can detect energy sources at the galactic center.

Space Telescopes

A *space telescope*, one placed on a satellite or spacecraft that orbits above the Earth's energy-absorbing atmosphere, has the distinct advantage of being able to detect any part of the electromagnetic spectrum **(Box 21.3)**. Astronomers use space telescopes, therefore, to detect particularly short wavelengths (X-rays and gamma rays) and particularly long wavelengths (infrared radiation) that can't pass through the optical or radio window

to the ground. As a result, space telescopes can detect particularly hot objects (which emit short-wavelength energy), as well as particularly cold ones (which emit long-wavelength energy). Furthermore, because space telescopes do not have to contend with atmospheric seeing, they can produce amazingly sharp images. Space telescopes are, however, expensive and difficult to operate.

Take-home message . . .

A telescope can gather much more electromagnetic radiation than a human eye can, allowing us to detect and resolve objects deep in space. Optical telescopes collect visible light and provide visual images. Radio telescopes collect radio waves and can detect some objects that cannot be seen with optical telescopes. Space-based telescopes have allowed astronomers to view all parts of the electromagnetic spectrum and have led to revolutionary discoveries about the nature of the Universe.

Quick Question

Why can we detect energy from the center of the Milky Way with a radio telescope, but not with an optical telescope?

FIGURE 21.33 Radio telescopes and radio interferometers.

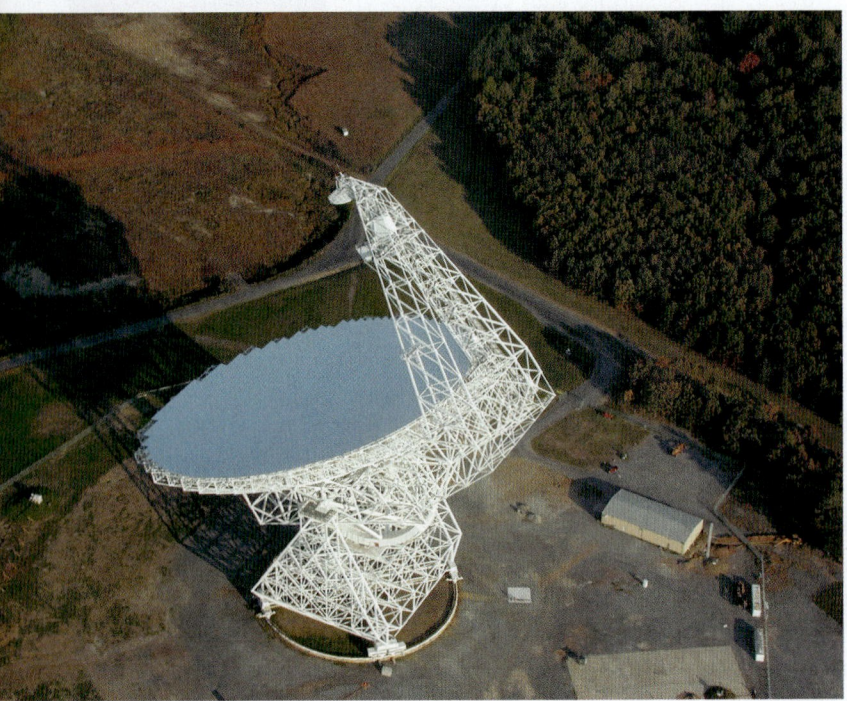

(a) The telescope at the National Radio Astronomy Laboratory in West Virginia has a dish 105 m (16 ft) in diameter.

(b) The Very Large Array, a radio interferometer in New Mexico.

Putting Earth Science to Use

Awesome views of our Universe from NASA's space telescopes

Astronomers working with the US National Aeronautics and Space Administration (NASA) used the space shuttle fleet to place five scientific space telescopes in orbit. Each was named in honor of a famous astronomer or physicist. With their capacity to detect all parts of the electromagnetic spectrum, these telescopes have revolutionized our understanding of the Universe.

The first and most famous space-based observatory, the Hubble Space Telescope, was launched in 1990 **(Fig. Bx21.3a)**. It's the only space telescope to be serviced in space by astronauts. The original mirror of Hubble was flawed and could not provide sharp images. Engineers designed a series of small mirrors that corrected the light coming from Hubble's primary mirror before it reached the telescope's detectors. Once the new equipment was installed, the telescope worked spectacularly. Hubble has given us amazingly sharp visual images of celestial objects **(Fig. Bx21.3b)**, as well as improved measurements of the distances to stars and a better estimate of the rate at which the Universe is expanding. When Hubble imaged patches of the night sky once thought to be empty, it found countless new galaxies, billions of light-years away.

Engineers have designed each space telescope for a distinct mission. For example, the Compton Gamma Ray Observatory, launched in 1991, detected gamma rays, the most powerful form of radiation, generated by gases with temperatures exceeding 10 billion degrees centigrade (18 billion degrees Fahrenheit). Its observations yielded insight into new classes of very energetic objects that researchers are still struggling to understand. The observatory operated until 2000, when its operators intentionally sent it plummeting into the Pacific Ocean. The Chandra X-ray Observatory, placed in orbit by the space shuttle *Columbia* in 1999, also detects radiation at the high-energy end of the spectrum and has provided new insight into the nature of exploding stars and their products. Launched in 2003, the Spitzer Space Telescope detects infrared radiation, the cooler end of the spectrum, so it has the capability of studying the nebulae in which new stars form. The Kepler Space Telescope has made headlines since its 2009 launch, for its instruments have detected thousands of *exoplanets*, planets orbiting other stars. On the basis of observations made by Kepler, astronomers now estimate that as many as 40 billion Earth-sized planets exist in the Milky Way.

FIGURE Bx21.4 The Hubble Space Telescope.

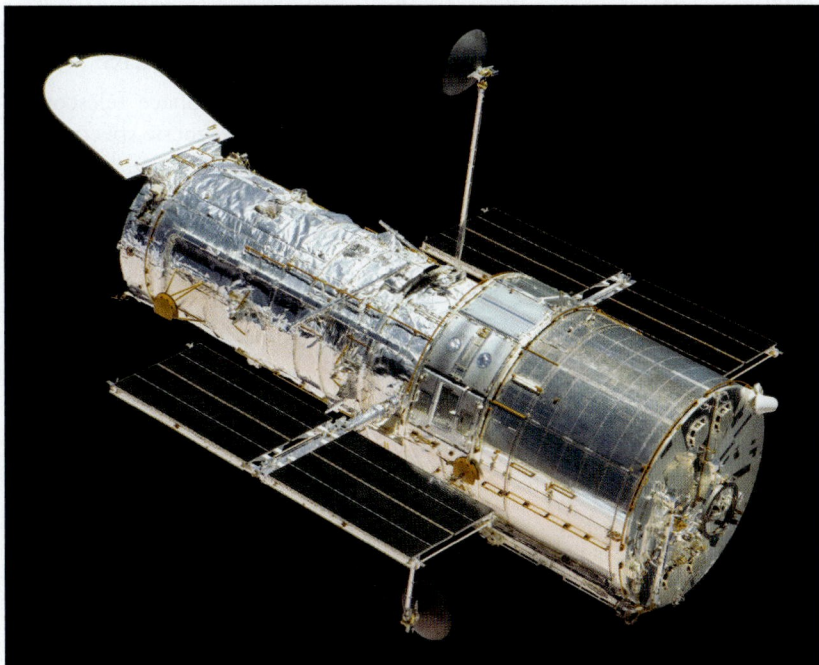

(a) The telescope.

(b) A spectacular Hubble image of the Pillars of Creation, a portion of a nebula where stars are born.

◉21 CHAPTER REVIEW

Letters in parentheses correspond to the chapter's learning objectives.

1. Contrast the geocentric model of the Solar System with the heliocentric model. **(A)**

2. Describe the contributions of Copernicus, Galileo, and Kepler to modern astronomy. **(A)**

3. What is the celestial sphere? What is the ecliptic plane, and how can you determine its location on the celestial sphere? **(B)**

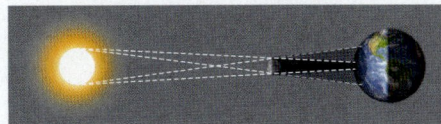

4. Identify the perihelion and aphelion locations in the diagram. Does the Earth move faster in its orbit when it is at perihelion or aphelion? **(B)**

5. What are constellations? What determines which constellations constitute the zodiac? **(C)**

6. What is a solar eclipse, and what is the difference between a total eclipse and a partial eclipse? Identify the areas where a total and a partial eclipse would be seen in the diagram. **(D)**

7. What is a lunar eclipse, and why can most people see a total lunar eclipse? **(D)**

8. What is meant by an atmospheric window in the electromagnetic spectrum? What atmospheric windows do earthbound astronomers use to peer into space? **(E)**

9. What is a spectrometer? What are the spectral lines visible in the spectrum produced by a spectrometer related to, and what do those spectral lines tell us about the composition of stars? **(E)**

10. What information can astronomers obtain about stars and galaxies using the Doppler effect? **(E)**

11. How do astronomers determine the surface temperature of a star? **(E)**

12. What is the difference between an astronomical unit and a light-year? Which is best for measuring the distance between objects in our Solar System? **(F)**

13. How does the figure relate to the way astronomers measure distances to nearby stars? **(F)**

14. Why do large astronomical observatories all use reflecting telescopes? **(G)**

15. How does a radio telescope work? Why must radio telescopes be so large? **(G)**

16. Give two reasons why the world's great astronomical observatories are all located on isolated mountaintops. **(G)**

17. What advantages do space telescopes have over earthbound telescopes? **(G)**

18. Imagine you are an astronomer who has just used a spectrometer to measure the emission lines of radiation from a distant galaxy. When you look at the emission lines of hydrogen, you note that all of the lines you measured are shifted from their expected positions toward the red end of the spectrum. Is the galaxy moving toward you or away from you? **(E)**

19. Why do you think cell phones and weather radar systems transmit signals in the radio and microwave regions of the electromagnetic spectrum, rather than the infrared or ultraviolet regions? **(E)**

ANOTHER VIEW Earthrise—from the vantage point of the Moon, our home planet appears as a blue marble within a dark and distant Universe.

⊙22 OUR NEIGHBORHOOD IN SPACE
The Solar System

By the end of the chapter you should be able to . . .

A. define a planet and understand the difference between true planets and dwarf planets.

B. explain the overall structure of the Solar System.

C. interpret the surface features of our nearest neighbor, the Moon.

D. distinguish the terrestrial planets from one another.

E. identify the key characteristics of the gas-giant and ice-giant planets.

F. describe the distinctive characteristics of the larger moons in our Solar System.

G. describe the origins and characteristics of smaller objects in the Solar System.

H. visualize where comets and meteoroids come from, and evaluate the threat they pose.

22.1 Introduction

On August 24, 2006, the International Astronomical Union (IAU), the world's principal organization of professional astronomers, held a fateful vote. When the ballots were counted, Pluto had been impeached. Henceforth, it would no longer be a planet. For three generations, students had memorized the names of the nine planets of the Solar System in order of their distance from the Sun: Mercury, Venus, Earth, Mars, Jupiter, Saturn, Uranus, Neptune . . . and Pluto. The next generation of students would stop at eight **(Fig. 22.1)**.

Clyde Tombaugh, a 23-year-old amateur astronomer, discovered Pluto in 1930. Headlines blared that "Planet X" had been found, and a schoolgirl, thinking about how lonely it must be so far from the Sun, suggested naming it after the god of the underworld. Pluto might still be the ninth planet were it not for the discovery in the 1990s that Pluto is, in fact, only one of many icy objects in a ring known as the *Kuiper Belt*, that lies beyond the orbit of Neptune. If Pluto is a planet, then are all other large Kuiper Belt objects planets, too? This was the dilemma facing the IAU, so their vote on Pluto was really a decision on the fundamental question, "What is a planet?"

After much deliberation, the IAU decided that to be a **planet**, an object must pass the following tests: (1) it must orbit a star directly, so **moons**, natural objects that orbit a planet, are not planets; (2) it must be spherical, which implies that it's large enough for internal gravitation to smooth out bumps and dimples on its surface; and (3) it must have cleared its orbit of other objects, either by incorporating them through collision or by trapping them as moons. Pluto passed the first two tests, but not the third—it has not cleared its neighborhood of

other Kuiper Belt objects. Astronomers now refer to objects that pass only the first two tests as **dwarf planets**. The demotion of Pluto didn't dim its mystery, however, so in 2006, NASA launched a space probe, *New Horizons*, to study it. In July 2015, humanity received its first close-up photos of this dwarf planet **(Fig. 22.2)**.

This chapter introduces all components of the Solar System except the Sun, which we'll describe in Chapter 23. We begin by characterizing the overall structure of the Solar System. With this background, we analyze distinctive features of the planets, starting with those closest to the Sun. We finish the chapter by discussing the smaller objects of the Solar System, such as the asteroids that lie between Mars and Jupiter. As you wander our Solar System, keep in mind the following themes: (1) There's no place like home! Each object in our Solar System is unique, and none shares the features of the Earth that make our planet livable. (2) Without data returned to us

<< Curiosity, a car-sized rover, has explored Gale Crater on the planet Mars since landing on August 6, 2012. It was still roving the planet at the end of 2019.

FIGURE 22.1 The eight planets of the Solar System.

Mercury
Venus
Earth
Mars
Jupiter
Saturn
Uranus

0   20,000
Km

Neptune

(a) Images of the eight planets emphasize that each is unique.

(b) All eight planets are much smaller than the Sun.

Sun

FIGURE 22.2 A composite photo in true color of Pluto, as seen by the *New Horizons* space probe in 2015.

22.2 Structure of the Solar System

The inventory of objects that make up the Solar System continues to grow as astronomers make new discoveries. In fact, in 2016, scientists who had performed computer simulations of the orbits of small Kuiper Belt objects announced that a large (Neptune-sized), yet-unseen planet may be lurking in the far reaches of the Kuiper Belt. Astronomers now classify many different types of objects by their size, composition, and orbital characteristics, as summarized in **Table 22.1**.

How are the objects of the Solar System arranged? The planets follow slightly elliptical orbits, and as we saw in Chapter 21, they display *prograde motion*, meaning that they move counterclockwise around the Sun, as viewed from above the Sun's north pole. Nearly all have orbital planes that lie within 3° of the plane of the Earth's orbit (the *ecliptic plane*) **(Fig. 22.3)**. In fact, a side-on view emphasizes that only Mercury's orbit deviates noticeably (7°) from the ecliptic plane. Mercury's orbit is even more tilted than that of the Moon, whose orbit is slightly greater than 5° from the ecliptic plane.

When astronomers defined the orbits of asteroids and Kuiper Belt objects, they realized that the majority of these objects also display prograde motion, but that some follow highly elliptical paths, called **eccentric orbits**. Also, while some of these objects have orbital planes close to the ecliptic plane, others follow paths that incline significantly to the ecliptic plane.

from space probes, we would know very little about our neighbors. (3) While many discoveries have been complete surprises, established scientific principles provide a basis for understanding newly discovered features of the Solar System.

TABLE 22.1 Objects in the Solar System

Objects	Description
• 1 star	Our Sun, a nuclear inferno emitting intense energy
• 8 planets	Relatively large spherical objects that orbit the Sun and have cleared their orbits of other objects
• 5 dwarf planets	Spherical objects that orbit the Sun, but have not cleared their orbits of other matter; there could be as many as 100 more to be discovered
• 181 moons	Objects orbiting planets or dwarf planets; some moons are spherical and some are not
• Asteroids	Rocky and/or metallic objects in the asteroid belt, which lies mostly between Mars and Jupiter; 3 have diameters over 500 km (300 mi) and are spherical; millions more are small and irregularly shaped
• Kuiper Belt	A band of icy objects in orbit around the Sun beyond the orbit of Neptune; about 1,000 have been seen, of which 100 have diameters exceeding 300 km (180 mi); 100,000 more may exist with diameters over 100 km (60 mi), and countless smaller ones may also exist
• Oort Cloud	A hypothetical spherical cloud of icy objects, held by the Sun's gravity but extending a quarter of the distance to the nearest star
• Comets	Icy and dusty objects, with diameters from tens of meters to a few kilometers, that revolve around the Sun in highly elliptical orbits; comets emit a tail of glowing gas when they approach the Sun; over 5,000 are known
• Meteoroids	Objects less than 1 m (3 ft) across that orbit the Sun

FIGURE 22.3 The arrangement of the planets and the orientations of their orbital planes.

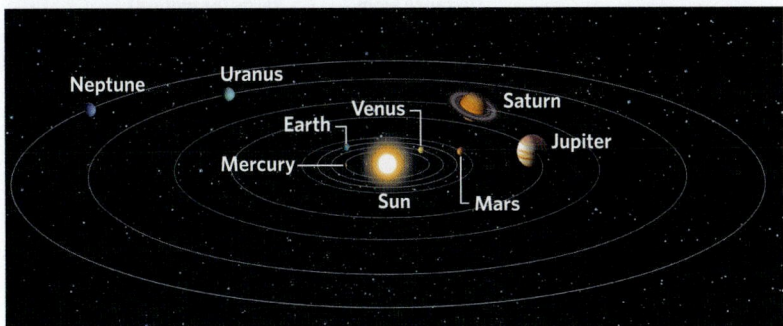

(a) Oblique view of all eight planetary orbits.

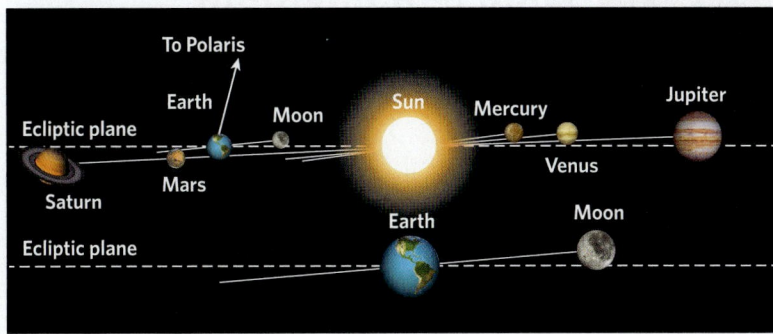

(b) The orientation of the planetary orbits (top) and of the Moon's orbit (bottom) relative to the ecliptic plane.

Table 22.2 reveals that the planets differ from one another in many ways. As they orbit the Sun, all the planets rotate counterclockwise on their axes—in the prograde direction—except for Venus, which rotates in the *retrograde direction*, meaning clockwise as viewed from above the Sun's north pole. Astronomers describe the orientation of a planet's axis of rotation by specifying its *axial tilt*: an axis perpendicular to the planet's orbital plane has a 0° tilt, and an axis parallel to the orbital plane has a 90° tilt. If we compare the planets, we see that they display a range of tilts **(Fig. 22.4)**. The tilt determines whether or not a planet has seasons, because the axis stays in the same orientation as the planet orbits the Sun. When a planet's axis has a tilt, the *insolation* (the amount of incoming solar radiation) striking a location on the planet varies over a year.

TABLE 22.2 Comparing the Planets

Name	Duration of Day*	Axial Tilt	Rotation Direction[†]	Duration of Year[‡]	Known Moons	Mass (× Earth)	Radius (km)	Radius (× Earth)	Distance to Sun (AU)	Density (g/cm³)
Mercury	59 d	0.1°	CCW	88 d	0	0.06	2,440	0.38	0.39	5.4
Venus	243 d	177°	CW	224 d	0	0.8	6,052	0.95	0.72	5.2
Earth	1 d	23°	CCW	365 d	1	1.0	6,371	1.00	1.00	5.5
Mars	1 d	25°	CCW	687 d	2	0.1	3,390	0.53	1.52	3.9
Jupiter	10 h	3°	CCW	12 yr	67	317.8	69,911	10.97	5.20	1.3
Saturn	10 h	27°	CCW	29 yr	62	95.2	58,232	9.14	9.54	0.7
Uranus	17 h	98°	CCW	84 yr	27	14.5	25,362	3.98	19.18	1.2
Neptune	16 h	28°	CCW	165 yr	14	17.2	24,622	3.86	30.06	1.7

*A day is the time it takes for a planet to rotate once on its axis, specified in Earth days or hours (h = hours, d = days).
[†]CW = clockwise; CCW = counterclockwise.
[‡]A year is the time it takes a planet to complete one orbit around the Sun, specified in Earth days or years (d = days; yr = years).

FIGURE 22.4 Each planet's axis of rotation tilts differently with respect to the ecliptic plane. Uranus's axis is almost parallel to the ecliptic plane, and Venus's axis is nearly upside down.

| Mercury 0.1° | Venus 177° | Earth 23° | Mars 25° | Jupiter 3° | Saturn 27° | Uranus 98° | Neptune 28° |

Red arrow = counterclockwise rotation
Green arrow = clockwise rotation

Different planets take different amounts of time to make a complete rotation. The Earth, as we know, takes one day (24 hours) to spin on its axis. In comparison, Venus is a slowpoke, taking 243 Earth days to rotate. Jupiter, Saturn, Uranus, and Neptune whirl completely around in less than 17 hours. Different planets also take different amounts of time to orbit the Sun because their distances from the Sun range so widely **(Box 22.1)**. As Kepler's third law states, the farther a planet lies from the Sun, the longer the planet takes to orbit the Sun (see Chapter 21). For example, Mercury makes its journey in just 88 Earth days, but Neptune takes 165 Earth years.

When we consider planetary size and density, we see that the planets fall into two categories, as Table 22.2 makes clear. The *inner planets*, those closer to the Sun (Mercury, Venus, Earth, and Mars), are relatively small (Earth-sized or smaller) and have relatively high average densities (greater than the density of basalt). A high density indicates that the mass of a planet consists mostly of rock or metal. Because of the similarities of the other inner planets to the Earth, the four inner planets are known collectively as the **terrestrial planets**.

The *outer planets*, those farther from the Sun (Jupiter, Saturn, Uranus, and Neptune) are all much larger than the Earth. In addition, they all have much lower average densities, indicating that they are composed primarily of gases or ices. In this context, *ice* refers not only to frozen water, but also to the solid versions of other compounds (such as frozen carbon dioxide, methane, or ammonia). Because of the similarity of the other outer planets to Jupiter, the four outer planets have been traditionally known as the **Jovian planets**.

Finally, all planets except Mercury and Venus have moons, but no two planets have the same number of moons. In addition, all the Jovian planets have **rings**, thin bands of tiny objects in orbit along each planet's equator.

The Solar System contains many other objects in addition to the planets (see Table 22.1). All display prograde motion, but some have very eccentric orbits, some of which are inclined to the ecliptic plane. Despite their great numbers, all the objects of the Solar System, other than the Sun, together account for only a tiny fraction (0.15%) of the Solar System's mass; 99.85% of the mass lies within the Sun itself.

FIGURE 22.5 The Apollo landings.

(a) The 35-story-high *Saturn V* rocket that launched the Apollo missions.

(b) The lunar module and the lunar rover on the Moon in 1972.

22.3 Our Nearest Neighbor: The Moon

History of Lunar Exploration

Long before the invention of the telescope, people realized that the Moon orbits the Earth, that we see it because it reflects sunlight, and that its surface has dark and light patches. They were also able to measure its dimensions, so they knew that the Moon has a radius of 1,737 km (1,079 mi), about one-fourth of the Earth's radius. But beyond that, early astronomers knew very little about our nearest neighbor.

Scientific understanding of the Moon took a great leap forward in the 1960s, when unmanned space probes and, later, Apollo spacecraft orbited the Moon **(Fig. 22.5a)**. In 1966, an unmanned Soviet space probe made the first successful landing on the Moon. Then, between 1969 and 1972, the United States landed six manned spacecraft on the lunar surface **(Fig. 22.5b)**. Altogether, the astronauts brought back 381 kg (840 lb) of Moon rocks. By analyzing these rocks, researchers were able to determine the chemical composition and the numerical age of the Moon. Modern orbiting space probes have produced high-resolution digital topographic maps of the entire lunar surface, and seismometers placed on the Moon have even detected moonquakes. Let's look at the lessons learned from all this work.

General Characteristics of the Moon

Unlike the Earth, the Moon has no active volcanoes or plate tectonics, for most of its mantle has become too cool for the conditions leading to melting to occur. Without volcanism to supply gases, the Moon has no atmosphere, and therefore, no oceans, rivers, glaciers, or life. And without an atmosphere or ocean to hold and redistribute heat, ground temperatures on the Moon undergo violent swings, ranging from over 120°C (248°F) on the sunlit lunar equator to −175°C (−283°F) on the Moon's dark side.

The lack of an atmosphere, and of the light scattering that it causes, also means that we can see the lunar surface clearly. This surface displays varied landscapes with topography ranging from a high of 10.8 km (6.7 mi) above mean (average) elevation to 9.1 km (5.7 mi) below. Lower-elevation regions of the Moon, called **maria** (singular: **mare**, pronounced mar-ay), have fairly smooth, dark surfaces **(Fig. 22.6a)**. Their name comes from the Latin word for sea because they were originally thought to be flooded with water. Based on the study of Moon rocks, we know that the maria consist of plains underlain by basalt lava flows, which have a low *albedo* (reflectivity). Notably, maria lie only on the near side of the Moon, so the near and far sides of the Moon look very different from each other **(Fig. 22.6b)**. The regions at higher elevation, the **lunar highlands**, are relatively rugged and have a high albedo, so they display a whitish glow when viewed from the Earth. They are underlain by a light-colored rock, called *anorthosite*, that consists largely of the mineral plagioclase.

Because so much of the Moon consists of anorthosite, a relatively low-density rock, the Moon has a lower average density than the Earth (3.4 g/cm³ for the Moon vs. 5.5 g/cm³ for the Earth). This characteristic, along with its smaller size, means that the Moon has a mass only about 1/80 of the Earth's mass. Less mass means weaker gravity, so a 68-kg (150-lb) astronaut weighs only 11 kg (25 lb) on the Moon.

That's one small step for a man, one giant leap for mankind.

—NEIL ARMSTRONG (AMERICAN ASTRONAUT, ON TAKING HIS FIRST LUNAR STEP, 1969)

FIGURE 22.6 The near and far sides of the Moon.

(a) The near side of the Moon hosts the maria, which lie at relatively low elevations.

(b) The far side of the Moon consists entirely of heavily cratered highlands.

Lunar Craters and Regolith

When Galileo turned his telescope toward the Moon in 1609, he discovered that bowl-shaped indentations, or **craters**, pockmark the surface. For centuries after Galileo, researchers assumed that the craters formed during volcanic eruptions. In 1892, however, an American geologist, G. K. Gilbert, argued that the craters were due to meteorite impacts and, to prove his point, made laboratory models of craters by dropping clay balls onto wet sand. Lunar exploration has confirmed that the lunar craters formed as a result of meteorite impacts. There are over 100,000 craters on the Moon with a diameter of 1 km (0.6 mi) or more.

Why do we see so many more craters on the Moon than on the Earth? First, fewer objects strike the Earth than the Moon because many of those that reach the Earth burn up or explode in its atmosphere. Second, plate tectonics and erosion destroy craters over time on the Earth, but these processes don't happen on the Moon, so a lunar crater, once formed, can survive for billions of years, and we can see craters that have accumulated over long periods. Though cratering has happened throughout lunar history, and continues today, the most intense cratering occurred during the **late heavy bombardment**, a time between 4.0 and 3.9 Ga when the gravitational pull of the outer planets sent huge swarms of objects careening across the orbits of other planets.

Not all lunar craters look the same. Their character depends on the size and speed of the impacting object. Smaller *simple craters* tend to be smooth-floored bowls,

whereas larger *complex craters* have a central uplift and may be surrounded by concentric ridges **(Fig. 22.7a)**. The central uplift forms when the impact disrupts the crust deeply enough to cause it to spring upward. Impacts also send out debris, which forms an *ejecta blanket* around the crater. In some cases, the debris forms a pattern of spoke-like streaks.

As the first astronauts to walk on the Moon discovered, a layer of **lunar regolith**, consisting of dust and debris, covers the Moon's surface **(Fig. 22.7b)**. This regolith consists of fragmented rock blasted out of craters, glass droplets frozen from rock melted by impacts, and accumulations of *micrometeorites* (meteorites less than 2 mm, or 0.08 in, in diameter). The top part of the regolith tends to be very fine, probably due to the pulverization of surface material by micrometeorite impacts and by absorption of *cosmic rays* (high-energy atomic nuclei from space). Researchers refer to such breakdown of material as **space weathering**. Because of space weathering, older craters tend to be blunter and somewhat smoothed, while newer ones have sharper features.

Age and Origin of the Moon

Based on analyses of Moon rocks, together with other data, most researchers favor a model in which the Moon formed from debris ejected into orbit around the Earth following a catastrophic collision of the newborn Earth with a protoplanet named Theia. This event happened at about 4.53 Ga, shortly after the interior of the Earth had undergone differentiation. The debris, mostly from the

FIGURE 22.7 The lunar surface.

(a) Lunar craters, as seen from an orbiting spacecraft. Larger complex craters have a central uplift and several ridges. Smaller simple craters are smooth bowls.

(b) Astronaut footprint in lunar regolith.

mantles of the colliding object and the Earth, first formed a ring around the Earth, then accumulated to become a sphere, the Moon **(Fig. 22.8)**.

When the Moon first formed, its surface was probably a magma sea. As the Moon cooled, denser mafic minerals formed and sank, while less dense felsic minerals, such as plagioclase, rose. The outer layer of less dense rock solidified into an anorthosite crust, exposed in the lunar highlands today. Geologists use standard isotopic dating methods to determine the ages of Moon rocks. The dates obtained so far fall between 4.36 Ga and 3.16 Ga. How the maria formed remains a subject of research. According to one model, they began as immense basins carved by meteorite impacts between

4.3 and 3.8 Ga. Melt from the Moon's mantle eventually broke through the thin crust on the floors of these basins in a succession of volcanic eruptions. These eruptions began about 100 million years after the basins had formed and continued on and off for the next several hundred million years. The crust of the Moon's near side is thinner and weaker than that of its far side, so while the crust could crack and permit volcanic eruptions on the near side, it could not on the far side, explaining the difference between the two sides. No one knows for sure why the melt formed. Some melt may have remained from the Moon's formation, but some may have formed due to decompression melting or frictional heating associated with tidal forces.

FIGURE 22.8 The formation of the Moon may have begun with an immense impact that blasted debris into orbit.

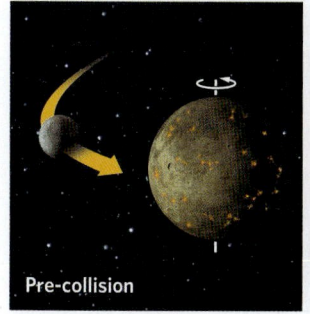

Pre-collision

Collision

Debris sprays into orbit

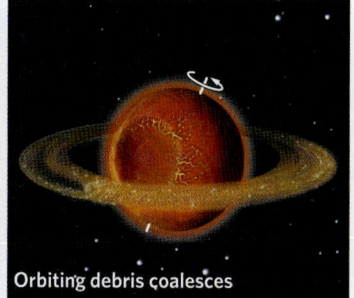

Orbiting debris coalesces

The Moon becomes a sphere

Time

Take-home message . . .

The Moon has no atmosphere, ocean, or active plate tectonics. Its surface displays dark, low-lying maria and light-colored, elevated lunar highlands. Craters pockmark the Moon's surface, and a layer of regolith covers most regions. The Moon formed around 4.53 Ga, most likely from debris blasted into space when a protoplanet struck the Earth.

Quick Question ——————————

Why are there so many more craters on the Moon than on the Earth?

22.4 The Terrestrial Planets

The inner solar system is the realm of four planets, including the Earth. Aside from their rocky consistency, however, the other three *terrestrial planets*, Mercury, Venus, and Mars, barely resemble our home planet. In this section, we explore these three planets. We'll see that our neighbors have surfaces that range from moonlike to volcanic, and temperatures that range from icy cold to a blistering inferno **(Fig. 22.9)**.

Mercury: A Moon-Like Landscape

THE FAST PLANET. The English name for Mercury came from the Roman messenger god known for his speed, because ancient astronomers saw that the planet moves across the sky faster than any other planet. It takes Mercury only 88 days to go around the Sun—orbiting rapidly because it is the planet closest to the Sun. Mercury's orbit has greater eccentricity than that of any other planet. At its closest, it lies 70 million kilometers (43.5 million miles) from the Sun, and at its farthest, it lies 74 million kilometers (46 million miles) away. Though it revolves around the Sun quickly, it rotates on its axis relatively slowly; one day on Mercury equals 59 days on the Earth.

Because Mercury orbits so close to the Sun, and because its surface consists of dark rock that absorbs radiation, surface temperatures on the planet can rise far above the boiling points of volatile materials. These characteristics, along with the lack of volcanic activity to produce new gas, mean that the planet has no ocean and virtually no atmosphere. The trace of gas surrounding Mercury consists of atoms released by space weathering, and the strong solar wind eventually blows these atoms into space. Without an ocean or atmosphere to transport heat around the planet, the temperature at a location depends only on the angle of insolation. Mercury's axis lies almost

FIGURE 22.9 Mercury, Venus, the Earth, the Earth's Moon, and Mars, with representations of their internal structures.

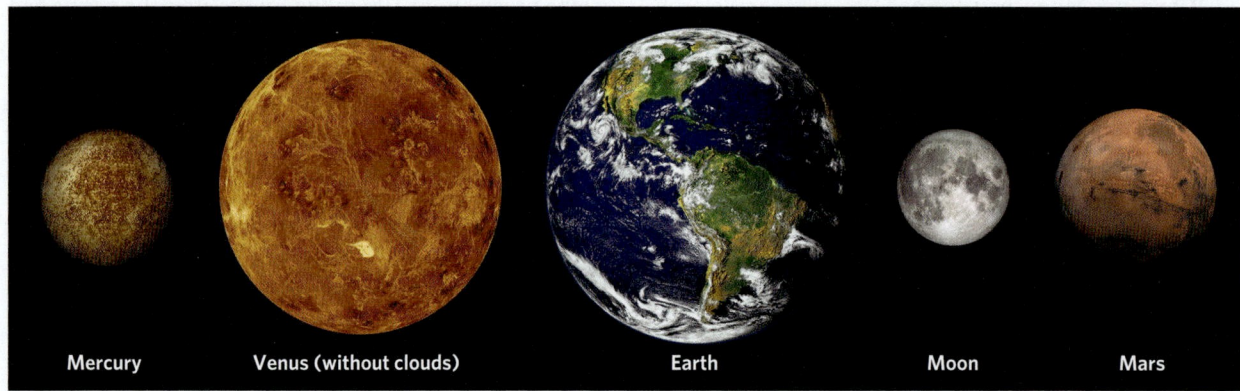

Mercury   Venus (without clouds)   Earth   Moon   Mars

(a) The surfaces and diameters of the terrestrial planets differ markedly from one another.

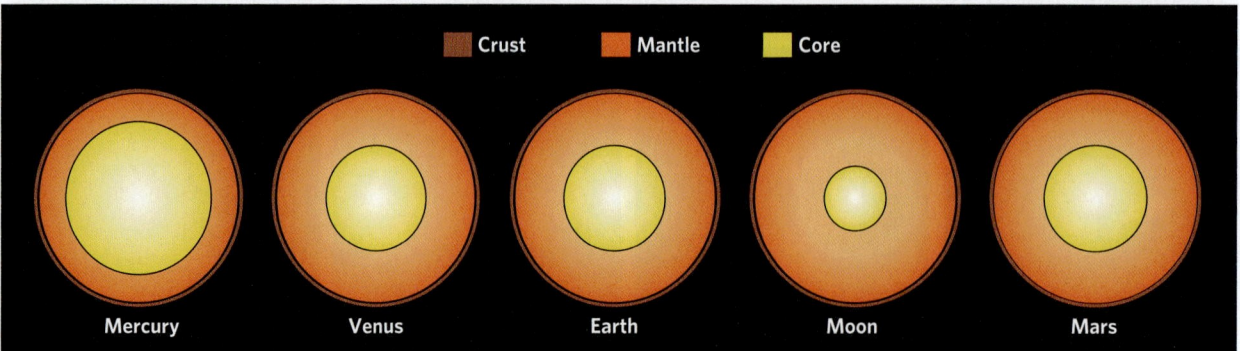

Crust   Mantle   Core

Mercury   Venus   Earth   Moon   Mars

(b) The interiors of the terrestrial planets differ from one another, too. In this figure, all the planets are resized to the same diameter, so as to emphasize differences in the proportions of crust, mantle, and core.

exactly perpendicular to its orbital plane, so at its poles, the temperature remains a constant −93°C (−136°F) all year, whereas at its equator, the temperature ranges from 427°C (800°F) at noon to −173° (−280°F) at midnight; this difference represents the largest temperature range of any planet in the Solar System.

Mercury's mass is about 5% that of the Earth, so a 150-pound astronaut would weigh about 57 pounds on Mercury. Researchers suggest that Mercury's metallic core accounts for a large portion of its interior (see Fig. 22.9b), perhaps because the accretion disk from which the Solar System formed contained a higher proportion of refractory material at the orbit of Mercury than existed farther from the Sun. Mercury is the only terrestrial planet besides the Earth to possess a magnetic field, perhaps because of its metallic core.

SURFICIAL SIMILARITIES TO THE MOON. Mercury resembles the Earth's Moon in many ways. The two bodies are similar in size (Mercury's diameter is 1.4 times that of the Moon's), both have heavily cratered surfaces, both lack active volcanoes and plate tectonics, both have a surface layer of regolith, and neither has an atmosphere or ocean. The *Messenger* space probe, which began to orbit Mercury in 2011, has mapped the planet's topography in exquisite detail **(Fig. 22.10)**, revealing that Mercury hosts wide plains that look like lunar maria. Mercury's plains, however, may be covered with pyroclastic debris instead of lava flows. Close-up photos indicate that large faults have offset craters on Mercury **(Fig. 22.11)**.

Venus: Veiled by Clouds

AN UPSIDE-DOWN PLANET? Venus, the second planet from the Sun **(Fig. 22.12a)**, shines brighter than any star in the Earth's night sky. Named for the Roman goddess of beauty, Venus shines brightly because it has a high albedo and lies fairly close to the Earth. Also, because it is relatively close to the Sun, it receives lots of solar radiation. The planet resembles the Earth in size, but it rotates much more slowly on its axis, taking 243 Earth days to spin once. Because Venus takes 224 days to orbit the Sun, its rotation rate exceeds its orbital period, so a day on Venus takes longer than a year on Venus. Venus and the Earth have about the same average density, suggesting that, like the Earth, Venus has an iron core. In addition, as we noted in Section 22.2, Venus differs from all the other planets in that it rotates clockwise around its axis (see Fig. 22.4). Astronomers speculate that a collision with a protoplanet early in the history of the Solar System flipped the planet over. In other words, Venus might have a retrograde rotation because the planet is upside down.

A CAUSTIC ATMOSPHERE. Venus hosts an atmosphere consisting almost entirely of carbon dioxide (96.5% by

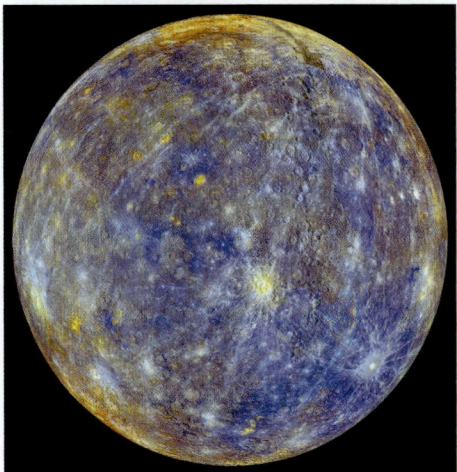
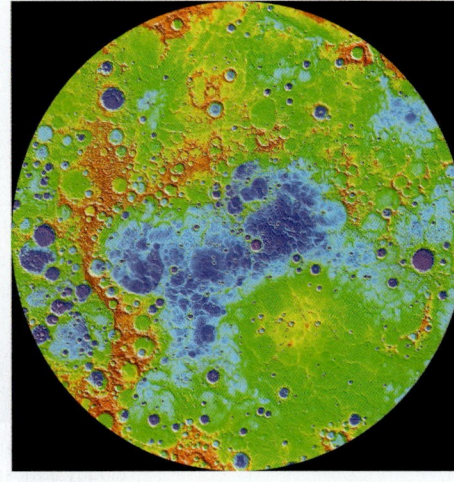

(a) A photograph, in false color, of the cratered surface of Mercury.

(b) A digital topographic map of Mercury's surface. Blue is lowest; red is highest.

volume), with the remainder made up of nitrogen molecules (3.5%) and trace amounts of water vapor, sulfur dioxide, and carbon monoxide **(Fig. 22.12b)**. This atmosphere has a much greater density than that of the Earth—it's so dense that if you were to stand on Venus's surface, the air pressure (93 atm) would equal the pressure you would feel at a depth of 800 m (0.5 mi) beneath the sea on the Earth. Venus's air pressure exceeds the Earth's both because Venus's atmosphere contains more gas and extends farther into space than does the Earth's atmosphere, and because the weight of CO_2 molecules exceeds that of N_2 or O_2 molecules. The presence of sulfuric acid clouds makes Venus's atmosphere so caustic that it would burn through metal.

FIGURE 22.11 A large fault dropped part of Mercury's surface to a lower elevation.

FIGURE 22.12 Venus has the densest atmosphere in the Solar System.

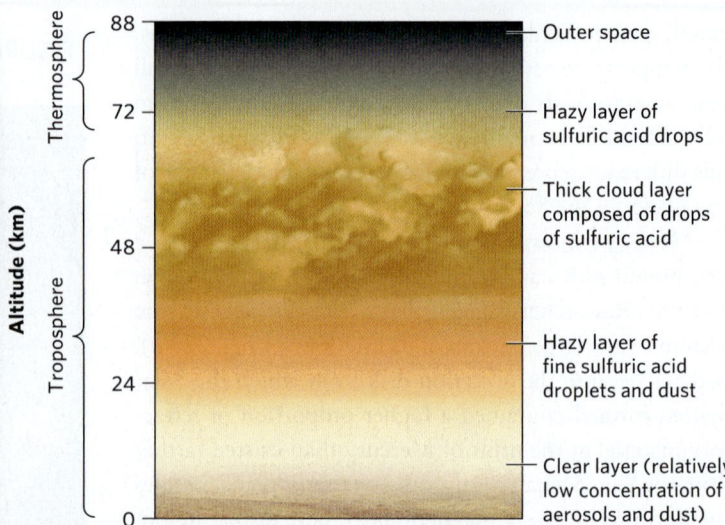

88 — Outer space

72 — Hazy layer of sulfuric acid drops

— Thick cloud layer composed of drops of sulfuric acid

48 — Hazy layer of fine sulfuric acid droplets and dust

24 — Clear layer (relatively low concentration of aerosols and dust)

0

Altitude (km)

Thermosphere / Troposphere

(a) A *Mariner 10* image of Venus shows a thick, cloudy atmosphere.

(b) A cross section of Venus's atmosphere. The atmosphere consists mostly of carbon dioxide. Its clarity depends on the concentration of sulfuric acid aerosols and dust.

The intense greenhouse effect caused by the abundance of CO_2 in Venus's atmosphere keeps the planet's surface temperature at about 450°C (840°F), hot enough to melt lead. Venus may once have had milder conditions, but at some point, the greenhouse effect warmed the planet enough for all surface water to evaporate, and the additional water vapor in its atmosphere amplified the greenhouse effect. The result—a *runaway greenhouse effect*—made the planet's atmosphere so hot that water molecules in the air broke apart; the hydrogen atoms drifted into space, and the oxygen atoms were absorbed by weathering reactions with surface rocks, so Venus's water disappeared for good. Winds blow so strongly on Venus that its atmosphere circulates very rapidly. This atmospheric movement ensures that hardly any temperature difference exists between Venus's poles and equator or between its night and day sides.

A WORLD OF VOLCANOES. We can't see the surface of Venus from the Earth, or even from orbit around Venus, because of the planet's dense, cloudy atmosphere. In fact, it has been photographed only twice, by Soviet space probes that landed on the Venusian surface **(Fig. 22.13a)**. Each probe, before rapidly succumbing to the extreme conditions on the planet, returned images of flat, broken rocks. To map the surface, researchers use radar, which can penetrate the veil of gas. Radar mapping reveals lava domes, shield volcanoes, and volcanic craters, but few impact craters **(Fig. 22.13b, c)**. The lack of craters on Venus suggests that the planet has been resurfaced by lava flows since the late heavy bombardment—that is, like a layer of asphalt filling in potholes on a highway, lava filled in the craters. Mapping has not revealed any plate boundaries, so it appears that hot-spot activity, rather than plate tectonic processes, caused Venus's volcanism.

Did you ever wonder . . .

if you could breathe the air on Venus?

Mars: Could Life Exist on the Red Planet?

Mars has never hosted life in the form of animals or plants. However, researchers wonder whether microbial life may have existed there in the past. Because life as we know it requires water, much of the effort to search for life on Mars has focused on finding evidence that liquid water existed there. Interest in the possibility of life on Mars, on moons of Jovian planets, and elsewhere in the galaxy, has led to the development of a field of study called *exobiology*.

Growing evidence suggests that long ago, Mars indeed had liquid water, as well as a thicker atmosphere. That environment has long since vanished, however, and the Mars of today is a dry, cold, desolate place, covered with rock and dust that contains oxidized iron. This "rust" led to the name Mars, after the Roman god of bloodshed, and to the planet's nickname, the *Red Planet*.

EXPLORING MARS. Because of its proximity to the Earth and its mostly transparent atmosphere, we can see fascinating rocks and structures on Mars **(Fig. 22.14)**. Several space probes have landed successfully on the planet, and several have surveyed Mars from orbit. The United States has also deployed four Martian rovers (*Sojourner* landed in 1997, *Spirit* and *Opportunity* landed in 2004, and *Curiosity* in 2012), vehicles that can travel across the Martian surface (see the chapter opening photo). These rovers have discovered minerals and sedimentary structures that may have formed underwater in the distant past. Since 2006, the *Mars Reconnaissance Orbiter* has sent back high-resolution images of landscape features on Mars, and in 2008, *Phoenix*, a stationary lander, touched down in Mars's north polar region and drilled into the surface, confirming the presence of subsurface water ice.

FIGURE 22.13 Images of the surface of Venus.

(a) The surface of Venus, as seen by the USSR's *Venera 13* space probe.

(b) An oblique radar image of Maat Mons, a volcano on Venus.

(c) A radar image of the planet from the *Magellan* space probe.

in the history of the planet when volcanoes were active, but most of the gas formed then either froze, reacted with the surface, or escaped into space. The thin air does move, and researchers have clocked winds in the range of 30–100 km/h (20–60 mph). These winds can churn up fine dust, producing haze that can last for months, but because the air has such low density, the winds cannot move larger objects.

FIGURE 22.14 Mars, the most studied planet beyond the Earth.

A THIN ATMOSPHERE. Air on Mars is much less dense than the Earth's air, so the air pressure on Mars's surface reaches only about 0.6% of that on the Earth. Compositionally, the Martian atmosphere resembles that of Venus—it contains 95.3% carbon dioxide, 2.5% nitrogen, 1.6% argon, and just traces of oxygen, carbon monoxide, and water vapor. Even though CO_2 makes up most of the Martian air, there's not enough to cause a significant greenhouse effect, so the surface of Mars remains frigid, with an average surface temperature of –55°C (–67°F). The air might have been denser earlier

See for yourself

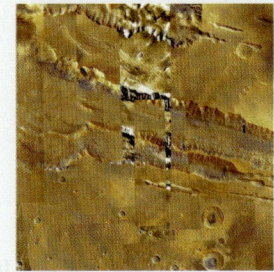

Valles Marineris: The Grand Canyon of Mars

You should visit this location using Google Earth, and then switch Google Earth to Google Mars by selecting the "Saturn" button on the button bar and selecting Mars.

Latitude: 14°35'57.61" S
Longitude: 62°31'11.52" W

From an elevation of 800 km (500 mi), you will see Valles Marineris, the Grand Canyon of Mars, lying below you. This canyon dwarfs the Grand Canyon of Arizona. It is 4,000 km (2,500 mi) long, 600 km (373 mi) wide, and reaches depths of 8 km (5 mi). The canyon covers almost 20% of the circumference of the planet.

FIGURE 22.15 The moons of Mars.

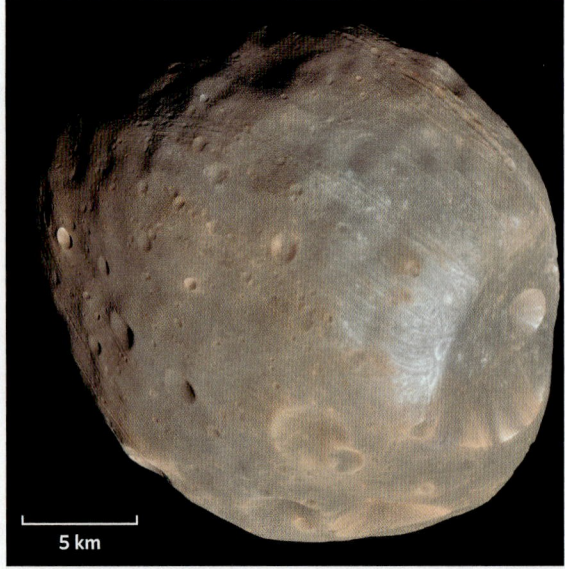

(a) Phobos.

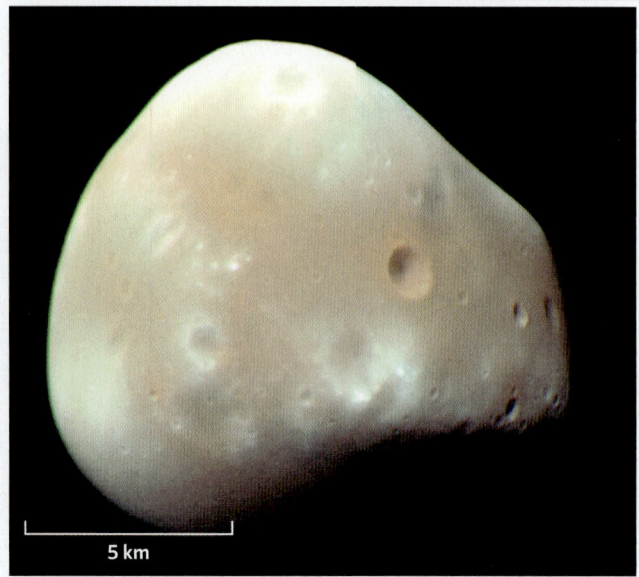

(b) Deimos.

GENERAL CHARACTERISTICS OF MARS. A day on Mars (24.6 hours) approximately equals a day on the Earth, and because Mars has an axial tilt of 25°, it, like the Earth, has seasons. But seasons on Mars are more extreme than the Earth's, for the orbit of Mars is more eccentric. Summer in the Martian southern hemisphere occurs when the planet lies closest to the Sun, so the southern hemisphere summer is warmer than the northern hemisphere summer.

The radius of Mars is only about half that of the Earth. Like the Earth, Mars has an iron core, and the core's radius is about one-half that of the entire planet. Mars has only about 10% of the Earth's mass, so a 150-pound astronaut would weigh just 57 pounds on Mars.

Mars has two moons, Phobos (Fear) and Deimos (Panic), only 22 km (13.7 mi) and 13 km (8 mi) in diameter, respectively **(Fig. 22.15)**. Because of their small size, they have large bumps and dimples that gravity didn't smooth out. Numerous craters speckle their surfaces. The moons, which are probably asteroids that were captured by Mars's gravity, now orbit Mars at such low altitudes (9,400 km and 23,500 km, or 5,841 and 14,600 mi, respectively) that a human standing on the Martian surface

FIGURE 22.16 Digital topographic maps of both sides of Mars. The dashed line shows the border of the Tharsis Bulge.

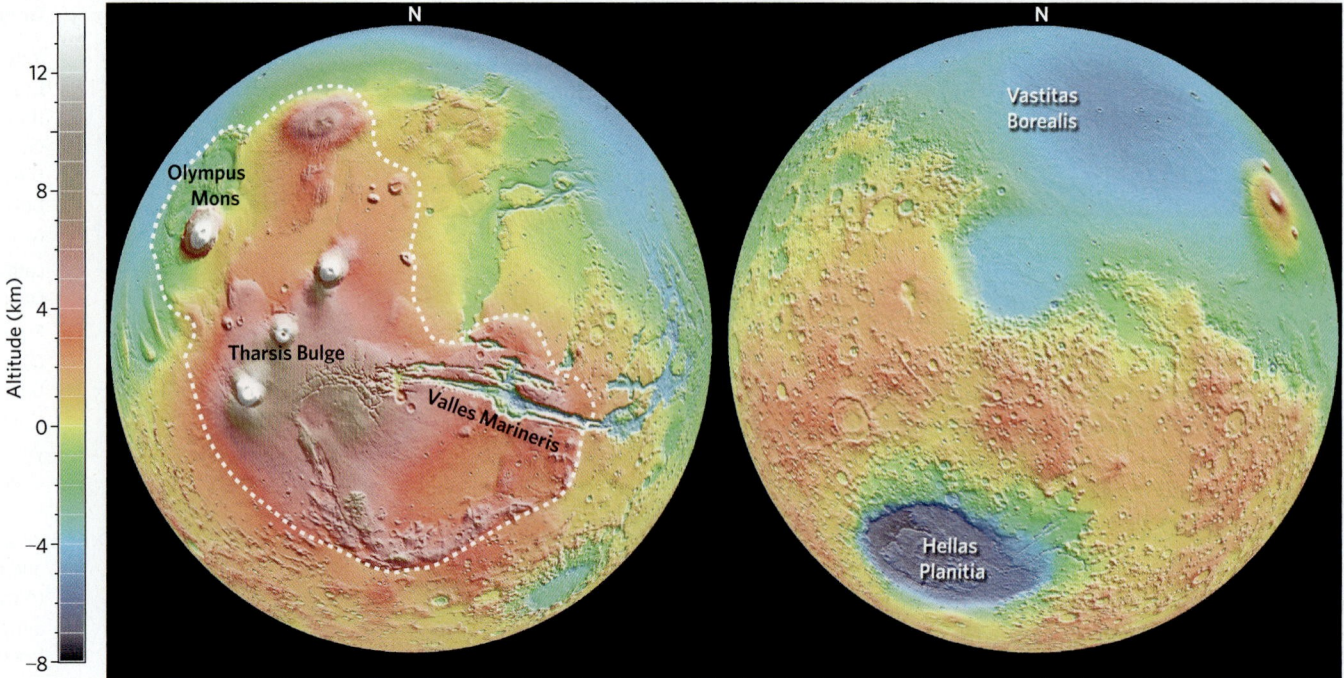

could easily see them with the naked eye, even though they're so small.

THE SURFACE OF MARS. The surface of Mars is marked by distinct topographic features **(Fig. 22.16)**. Researchers divide the planet into two topographic provinces. The northern lowlands province, Vastitas Borealis, a gigantic lava-covered plain encompassing 40% of the planet's surface, lies at a low elevation. It has fewer craters than the rest of the planet, suggesting that basaltic lava covered it after the late heavy bombardment. The rest of the Martian surface consists of cratered highlands, with the exception of two huge impact basins, one of which—Hellas Planitia—includes the lowest point on Mars. Near the equator, the Tharsis Bulge, a plateau as big as North America, rises above the other highlands. Relatively few craters exist on the plateau's surface, suggesting that it's younger than the rest of the highlands. An immense canyon, Valles Marineris, defines part of the northern edge of the Tharsis Bulge **(Fig. 22.17a)**. This structure, which probably formed as a tectonic rift, is 4,000 km (2,500 mi) long, 600 km (373 mi) across, and up to 8 km (5 mi) deep, so it dwarfs the Earth's Grand Canyon, which is only 250 km (155 mi) long, 10–15 km (6.2–9.3 mi) wide, and up to 1.6 km (1 mi) deep.

Olympus Mons, the largest volcano in the Solar System, rises from the plains to the west of the Tharsis Bulge **(Fig. 22.17b)**. With a base diameter of 700 km (435 mi) and an elevation of 25 km (15.5 mi), this volcano is 2.5 times the size of the Earth's largest volcano, Hawaii's Mauna Loa. Like the volcanoes of Hawaii, Olympus Mons formed as a shield volcano that erupted relatively low-viscosity basalt. Notably, some of the flows on the flanks of Olympus Mons have so few craters that they may have spilled out of the volcano within the last 100 million years. An 80-km (50-mi)-wide caldera has collapsed at the crest of the volcano.

Mars has an ice cap at each pole. *Martian ice caps* consist of seasonal ice—frozen CO_2 (dry ice) that accumulates when temperatures at the poles drop in the winter—on top of permanent ice, which lasts all year. The permanent ice of the north pole consists of a 1–3-km (0.6–1.9-mi)-thick sheet of water ice, while that of the south pole consists of frozen CO_2.

Data collected by probes suggest that large amounts of liquid water existed on Mars in the past. For example, surface soils contain clay, calcite, and gypsum, minerals that form only in the presence of water. Images reveal landscape features that look like river-carved channels, drainage networks, flow-eroded islands, and even deltas **(Fig.22.18)**. The presence of deltas and of terraces that look like remnants of beaches along the edge of Vastitas Borealis suggest that the plain was once an ocean. Does any liquid water exist on Mars today? Maybe. In 2015, the *Mars Reconnaissance*

FIGURE 22.17 Dramatic landscape features on Mars.

(a) Valles Marineris, the largest canyon in the Solar System.

(b) Oblique view of Olympus Mons, with a caldera visible at its summit.

Orbiter obtained images of streaks on hillslopes that become darker, as if damper, during part of the year.

Take-home message . . .

The terrestrial planets, although similar in their overall composition, differ in many ways. Mercury resembles the Earth's Moon, but is hotter on its sunlit side. A very dense, CO_2-rich atmosphere cloaks Venus. Mars is now a dry, cold planet with a very thin, CO_2-rich atmosphere. All but Venus contain abundant craters. Venus and Mars also host large volcanoes.

Quick Question
Do plate tectonic processes happen on all terrestrial planets?

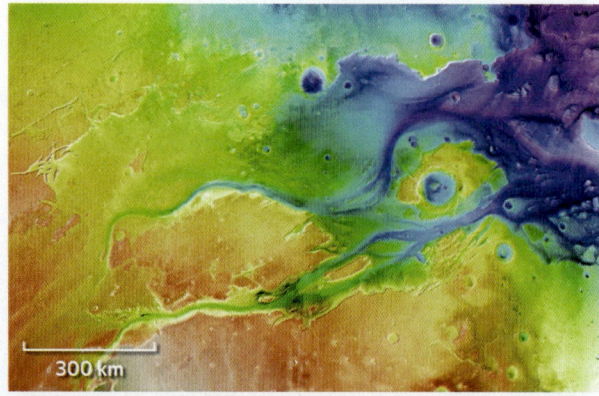

(a) Kasei Valles, showing what appears to be an outflow channel that, at times, contained flowing water.

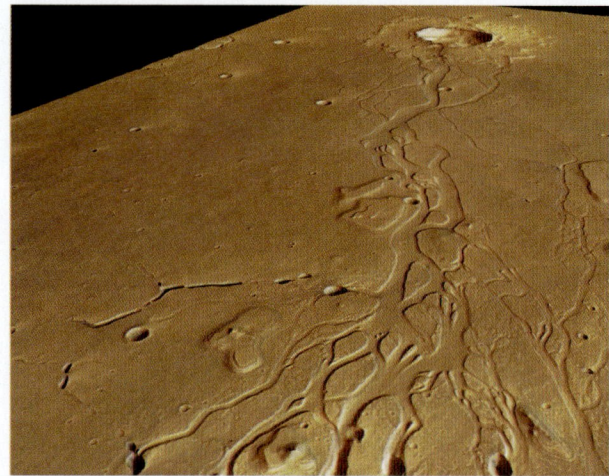

(b) Close-up of stream-like channels, as viewed obliquely from a space probe.

(c) The thin laminations in this rock look like cross beds deposited by currents.

22.5 The Jovian Planets and Their Moons

Each of the Jovian planets beyond Mars is a giant, compared with the terrestrial planets, and each holds on to many moons. Because of the state of the material within the Jovian planets, astronomers refer to the two nearer to the Sun (Jupiter and Saturn) as the **gas giants** and the farther two (Uranus and Neptune) as the **ice giants** **(Fig. 22.19)**. Although Jupiter and Saturn can be seen in the night sky, we knew little about these distant orbs until space probes sent back stunning images and other data. Some observations raise the possibility that liquid water may lie beneath the icy surfaces of some of the moons circling the Jovian planets . . . and if there's water, could there be life? This mystery remains to be solved.

Jupiter: The Giant among Giants

Jupiter, the Roman king of the gods, serves as an apt namesake for the fifth planet from the Sun, for Jupiter contains more matter than all the other planets combined. In fact, 70% of our Solar System's total mass, not including the

Sun, lies within Jupiter. This huge planet rotates much faster than any other planet and completes a day in only 10 hours. The centrifugal force associated with this fast rotation makes the planet's equator bulge, so its equatorial radius (11 times the Earth's) exceeds its polar radius by 7%.

Jupiter produces its own internal heat, radiating twice as much energy as it receives from the Sun. This heat does not come from nuclear fusion inside Jupiter, for the planet would have to be a hundred times more massive for fusion to occur. Rather, it's left over from the compression of gas during Jupiter's formation.

THE MOST COLORFUL ATMOSPHERE OF ALL. Even an amateur astronomer looking through a telescope at Jupiter from the Earth can see that the giant planet's atmosphere differs markedly from the Earth's. Orange and tan bands, oriented parallel to the equator, surround the planet **(Fig. 22.20a)**. The composition of these bands remained uncertain until 1994, when *Galileo* entered orbit around Jupiter and dropped a probe into the atmosphere. The probe sent back data for 78 minutes as it descended for a few hundred kilometers, until its instruments succumbed to the intense temperature and pressure of its surroundings.

The *Galileo* orbiter and its probe revealed that Jupiter's outer atmosphere consists primarily of hydrogen (86.1%) and helium (13.8%), but also contains traces of methane (CH_4), ammonia (NH_3), ammonium hydrosulfide (NH_4SH), and water vapor (H_2O). Different components concentrate in distinct layers. Jupiter has no solid surface, but the gases become denser and denser toward the center of the planet. Researchers arbitrarily define the base of the atmosphere—and, therefore, the "surface" of the planet—as the elevation at which the atmosphere has a pressure of 10 atm (10 times that of the Earth). Temperatures at this elevation reach about 300°C (570°F).

The colors of the visible atmospheric bands come from trace elements, such as sulfur and phosphorus, in the

(a) The diameters and colors of the giant planets differ markedly from one another. All of the giant planets are much larger than the Earth.

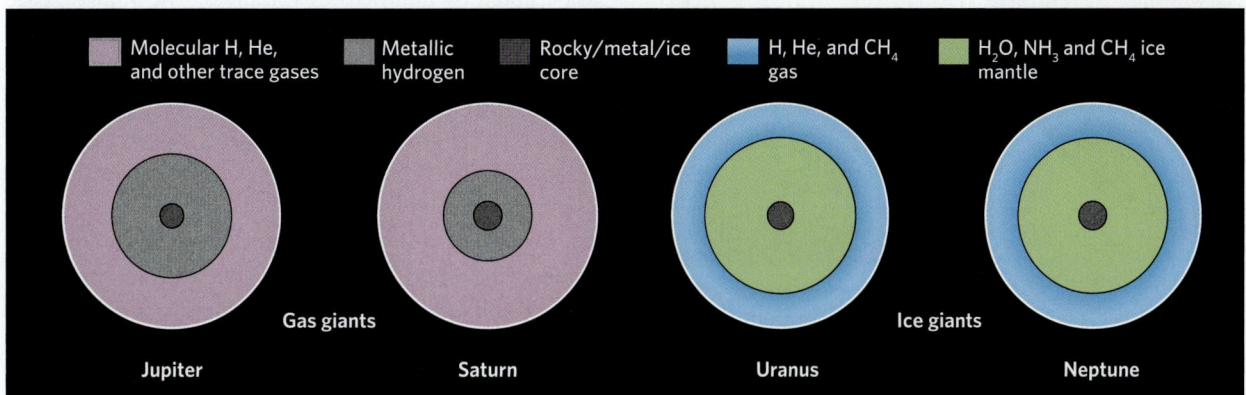

(b) The interiors of the giant planets differ from one another too. In this figure, all of the planets are resized to the same diameter, so as to emphasize the differences of their interior components.

FIGURE 22.19 Characteristics of the Jovian planets.

cloud layers. Researchers suggest that the bands indicate the presence of alternating upwelling (brighter colors) and downwelling (darker colors) regions of the atmosphere. Due to the Coriolis force, these bands align parallel to the equator. The winds within these bands flow at different velocities and even in different directions, so swirling eddies form at the boundaries between bands (Fig. 22.20b).

A giant ellipse of reddish gas, the **Great Red Spot**, dominates images of Jupiter's southern hemisphere (Fig. 22.20c). Detailed cloud patterns in and around the spot indicate that it's a huge, counterclockwise-rotating storm, big enough to contain two to three Earths. The Great Red Spot has existed for centuries and may be permanent. Its origin and longevity remain a mystery. Other huge storms form on Jupiter, but they exist for a shorter time before dissipating or combining with other storms.

WHAT'S INSIDE JUPITER? While the volume of 1,320 Earths could fit inside the volume of Jupiter, the giant planet's mass is only 318 times that of the Earth (see Table 22.2). That's because Jupiter has an average density that's only about one-fourth that of the Earth, for Jupiter contains mostly hydrogen and helium while the Earth consists mostly of rock and metal.

What would we see if we could slice through Jupiter? At the base of the atmosphere, the planet consists of compressed hydrogen and helium gas. Perhaps 100 km (60 mi) farther down, the pressure becomes great enough that the gas turns into liquid. This liquid hydrogen still consists of H_2 molecules, but the molecules can get close enough to one another to interact and bond.

The liquid hydrogen layer extends downward for perhaps 20,000 km (12,400 mi). Below this depth, where pressures reach 3 million atmospheres and temperatures reach 11,000°C (20,000°F), hydrogen undergoes a transition into a strange material, called **metallic hydrogen**, that normally doesn't exist on the Earth. Metallic hydrogen was created in a laboratory for the first time in 2017. It flows like a liquid, but the atoms have been squeezed together so tightly that they can't hold their electrons in orbit, so the electrons flow freely, as in a metal. The flow of electrons in Jupiter's metallic hydrogen layer generates a very strong magnetic field around the planet and causes brilliant aurorae near its poles. The magnetic field, in turn, traps charged particles to produce intense radiation belts thousands of times stronger than the Earth's Van Allen belts. Because of these radiation belts, space probes exploring Jupiter must have strong shielding to prevent their electronics from frying. The metallic hydrogen layer

FIGURE 22.20 Color bands, winds, and storms on Jupiter.

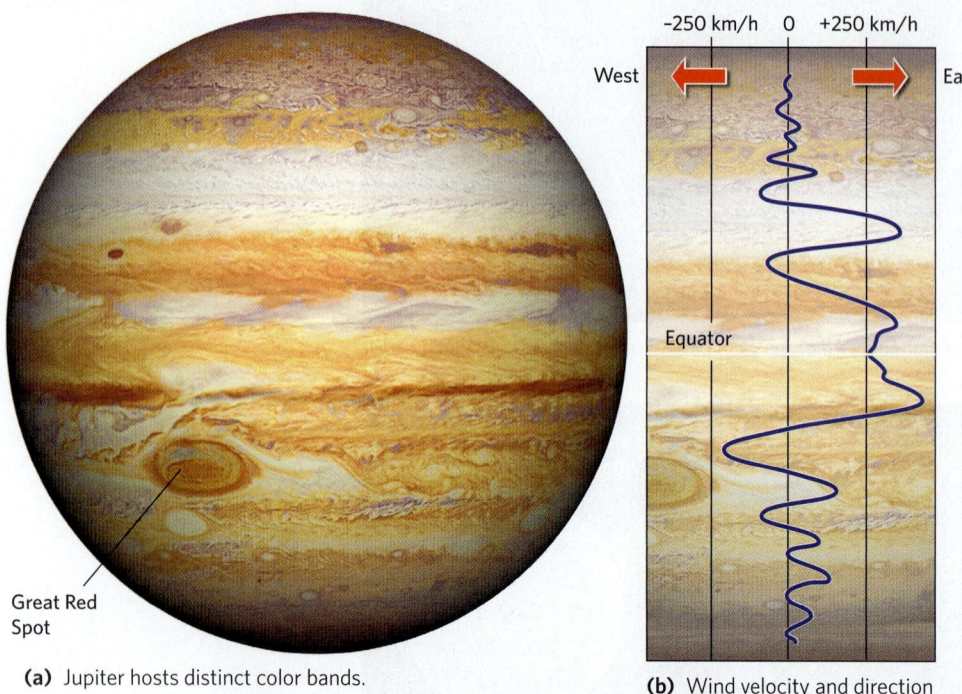

(a) Jupiter hosts distinct color bands.

Great Red
Spot

(b) Wind velocity and direction on Jupiter vary with latitude.

West   −250 km/h   0   +250 km/h   East

Equator

(c) A close-up of the Great Red Spot, a giant storm that rotates counterclockwise. The spot is bigger than the Earth.

continues down to a depth of 60,000 km (37,300 mi). Below that depth, researchers speculate that Jupiter has a core consisting of refractory material (rock and metal), which may contain about 12 to 45 times the mass of the Earth, though it is only a small part of Jupiter's mass.

AMAZING MOONS. Jupiter has 79 confirmed moons (**Earth Science at a Glance**, pp. 716–717). The four largest,

Io, Europa, Ganymede, and Callisto—known as the *Galilean moons* because Galileo observed them in 1610—reach planetary dimensions (**Fig. 22.21**). Ganymede, the largest moon of Jupiter, has a diameter 8% larger than that of Mercury and 150% larger than that of the Earth's Moon, which makes it the largest moon in the Solar System. The other Galilean moons are also comparable to the Earth's Moon in size, but all four differ markedly from our Moon and from one another in several ways.

Io, the closest Galilean moon to Jupiter, looks like a weirdly colored pizza, with splotches and spots of rich reds, golds, yellows, browns, and blacks (see Fig. 22.21a). The colors come from sulfur and other elements in the ash and lava erupted by hundreds of volcanoes. The *Voyager* space probe caught eight of these volcanoes in the act of erupting. What provides the heat that drives this volcanism? Io experiences internal heating because its orbit is synchronized with those of the other Galilean moons. Europa pulls on Io twice during Io's orbit, bringing it closer to Jupiter, and also resulting in changing tidal pulls that generate heat within the moon. The volcanoes erupt explosively, constantly resurfacing the moon, so it retains no impact craters. Beneath the colorful surface, Io probably resembles the terrestrial planets in having a rocky mantle and an iron core.

The remaining three Galilean moons, Europa, Ganymede, and Callisto (see Fig. 22.21b–d), have water-ice surfaces surrounding rocky interiors. Measurements made by the *Galileo* space probe suggest that both Ganymede and Europa have a solid, brittle outer crust of water ice on top of a warmer, softer ice layer (**Fig. 22.22**). Europa may host a vast subsurface ocean of liquid water, up to 100 km (62 mi) deep, beneath an ice layer. On both moons, the water layers overlie an internal shell of rock that surrounds an iron core. The existence of an ocean beneath Europa's surface has scientists wondering whether primitive life inhabits this moon. Callisto differs from other large moons, and from any terrestrial planet, in an important way: it has not differentiated into a core and mantle, which suggests that this moon accumulated from debris orbiting Jupiter later in Solar System history, when the accreting material was too cool to melt.

Jupiter is encircled by rings of debris, but they can't be seen from the Earth. They were detected for the first time by the *Voyager* space probe in 1979. These rings circle the planet in the plane of its equator. They are a few thousand kilometers wide from the innermost edge to their outer edge. A few small moons lie close to the rings and may have been the source of the debris within the rings.

Saturn: The Ringed Planet

Saturn, named for the Roman god of agriculture, has such a high albedo and is so big that you can see it without the aid of a telescope, even though it's so far away. Saturn

(a) Io has a diameter of 3,636 km (2,259 mi).

FIGURE 22.21 Jupiter's large (Galilean) moons.

(b) Ganymede has a diameter of 5,268 km (3,273 mi).

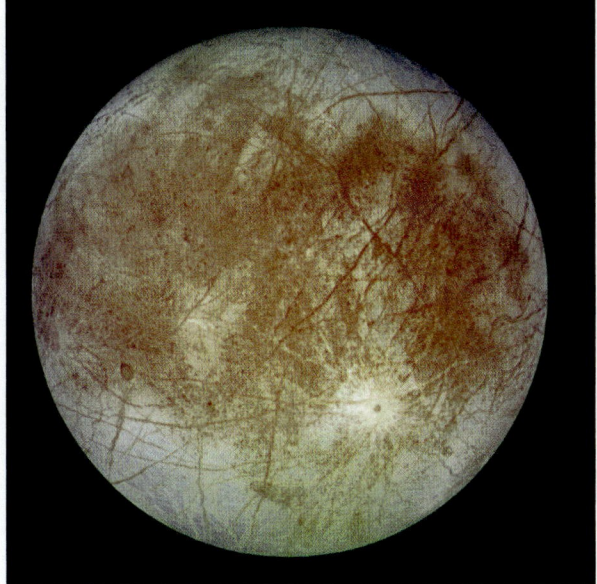

(c) Europa has a diameter of 3,100 km (1,930 mi).

(d) Callisto has a diameter of 4,821 km (2,995 mi).

resembles its bigger brother Jupiter in many ways, but Saturn's rings, in contrast to those of Jupiter, are very large and bright enough to be seen from the Earth.

SATURN'S ATMOSPHERE AND INTERIOR. Like Jupiter, Saturn has a low average density because it consists mostly of hydrogen (96% by volume). In fact, Saturn has the lowest average density of all the planets (0.7 g/cm³, less than that of water). We know its density is less than that of Jupiter because it doesn't exert as much gravitational pull, so the hydrogen within Saturn doesn't undergo as much compression as that in Jupiter.

Like Jupiter, Saturn lacks a solid surface, so researchers arbitrarily define the base of its atmosphere as the elevation where pressure is about 10 atm. And like Jupiter, Saturn rotates so fast—a day on Saturn lasts 10.75 hours—that the planet bulges at its middle, and its equatorial radius exceeds its polar radius by 11%. Saturn's thick atmosphere, composed mostly of hydrogen (92.4%) and helium (7.4%), contains cloud layers that resemble Jupiter's in composition, but Saturn's atmosphere doesn't convect as vigorously as Jupiter's, so Saturn has duller stripes and an overall tannish hue **(Fig. 22.23)**. Nevertheless, Saturn has intense winds that race along at up to 1,500 km/h (930 mph).

FIGURE 22.22 Speculative models of the possible watery interiors of Ganymede and Europa.

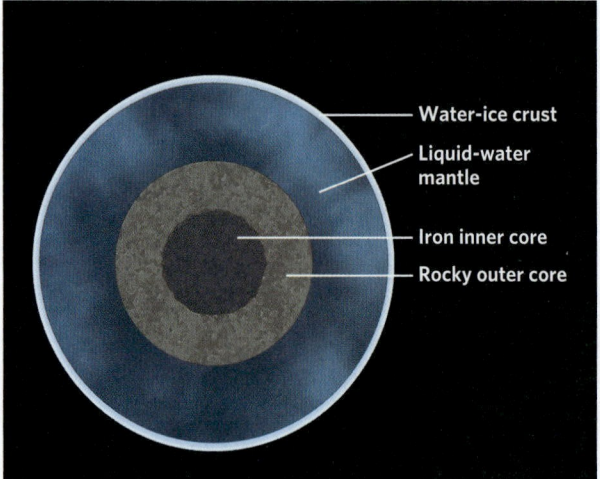

- Water-ice crust
- Liquid-water mantle
- Iron inner core
- Rocky outer core

(a) Ganymede may have a liquid-water mantle.

Ice crust

Liquid ocean under ice

- Metallic core
- Water layer
- Rocky interior

(b) An ocean may exist beneath an ice crust on Europa.

Model of Europa's outer layers

FIGURE 22.23 *Cassini* views of the rings of Saturn.

(a) Picture of the rings of Saturn during an eclipse (the Sun is behind Saturn). Note the fainter, more distant rings.

(b) False colors represent different particle sizes. Purple means that particles larger than 5 cm (2 in) are common; green means that particles smaller than 5 cm are common. White indicates dense rings.

Did you ever wonder . . .

why Saturn has spectacular rings?

The core of Saturn, which probably consists of rock and metal, has a mass about 10 to 20 times that of the Earth and a temperature of about 11,700°C (21,000°F). Outside this refractory core, Saturn consists of a layer of dense ice (composed mostly of ammonia and water), a layer of metallic hydrogen, a layer of helium (which probably developed when drops of liquid helium formed and fell like rain toward the interior), and a layer of liquid molecular hydrogen. Because Saturn has less internal pressure than Jupiter does, Saturn's metallic hydrogen layer is thinner than Jupiter's, and Saturn has a weaker magnetic field. Saturn generates about 2.5 times as much energy as it receives from the Sun. While some of this energy comes from compression of the planet's gases, astronomers speculate that some comes from the friction generated by the downward rain of helium.

AMAZING RINGS. When Galileo first looked at Saturn, he saw ear-like protrusions to either side of the planet, but had no idea what they could be. A half century later, an astronomer using a more powerful telescope recognized the protrusions as rings of matter that surrounded the planet and reflected light. In 1675, Giovanni Cassini showed that Saturn has several distinct rings extending over 100,000 km (62,000 mi) out into space. No other planet displays rings as bold and beautiful as Saturn's.

Our understanding of the rings improved greatly when the *Cassini* space probe sent back high-resolution photographs revealing fainter rings, which extend beyond the main rings, and showing that each main ring consists of hundreds of thin ringlets (10–15 m, or 32–50 ft, thick) with tiny gaps in between **(Fig. 22.23)**. The rings consist of countless particles of water ice and rocky dust, ranging in size from less than a millimeter to a few meters wide.

Why do Saturn and the other Jovian planets have rings? Astronomers suggest that the rings were formed by tidal forces acting on moons orbiting the planets. A planet's gravitational tug stretches the moons toward the planet, and moons closer to the planet experience a stronger pull. When close moons were subjected to strong tidal forces, their internal gravity couldn't hold them together, and eventually, each moon shattered into fragments, which dispersed into a ring in what had been the orbital plane of the moon. Some gaps between rings exist because they have been cleared by small moonlets, bodies 10–20 km (6–12 mi) in diameter that gravitationally vacuum in debris. Other gaps have formed because Saturn's moons act as shepherds that gravitationally "herd" particles into rings and out of gaps.

THE MANY MOONS OF SATURN. Saturn hosts 82 known moons, of which only 13 have diameters larger than 50 km (30 mi). Most of the smaller moons don't spin on an axis, but rather tumble through space as they follow their orbit. They probably represent captured planetesimals or fragments from colliding planetesimals.

Each of the seven largest moons of Saturn—those that became big enough to become roughly spherical—has unique characteristics **(Fig. 22.24a)**. Titan, the largest by far, has a diameter larger than Mercury's and an atmosphere 10 times denser than the Earth's. Unlike the atmosphere of Saturn itself, Titan's atmosphere consists of nitrogen (98%), methane (2%), and other trace gases, including a variety of hydrocarbons. Images made by the *Huygens* space probe, which parachuted to the surface of Titan in 2005, suggest that methane clouds hover in Titan's lower atmosphere and that liquid methane rains from these clouds, collecting in hydrocarbon lakes and rivers on the surface. Radar mapping reveals volcanic vents that emit gases, as well as huge dunes, presumably composed of ice and frozen methane grains.

FIGURE 22.24 The moons of Saturn.

Mimas (363 km)

Titan (5,152 km)

Enceladus (505 km)

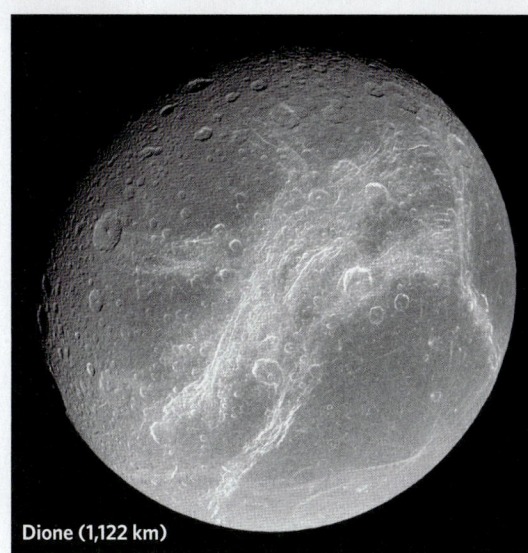
Dione (1,122 km)

(a) Four of Saturn's 62 known moons.

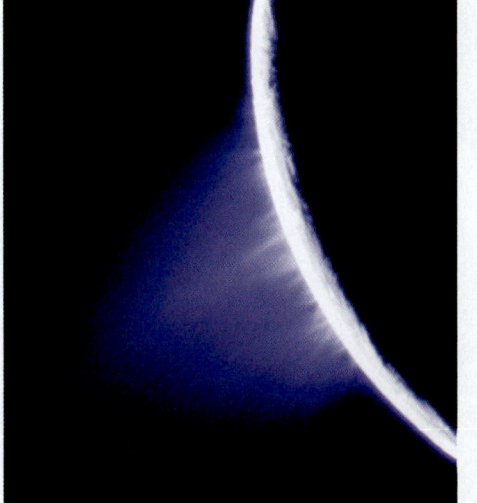

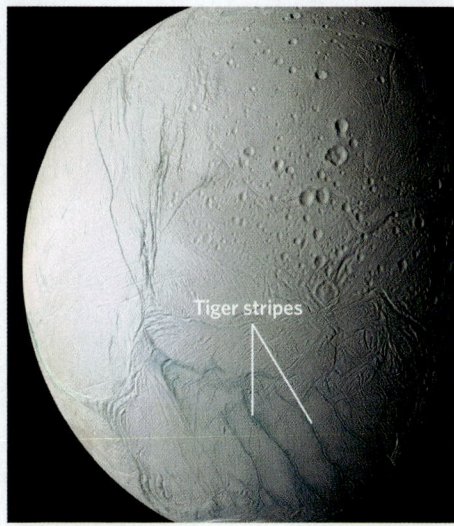
Tiger stripes

(b) Jets of volatiles erupt from the "tiger stripes" of Enceladus's south pole.

A Menagerie of Moons

The moons of the solar system vary in size, shape, composition, and surface characteristics. Some, like Io, have active volcanoes. Others, like Enceladus, are covered with water ice, and may harbor an ocean beneath the icy surface. Some are cratered, while others have smooth surfaces. One, Titan, has an atmosphere. Many are spherical, but the smallest, such as Deimos, are not large enough for gravity to compress them into spherical shapes. The objects within the curving band are drawn to scale.

Deimos,
Mars's small moon

Mimas,
Saturn's cratered moon

Titan

Mercury

Callisto

Io

Moon

Europa

Iapetus,
Saturn's icy and cratered moon

Triton

Pluto

Titania

Oberon

Tethys

Enceladus

Enceladeus,
Saturn's icy moon

Mimas

Ganymede

Mars

Venus

Earth

Titan,
Saturn's moon
with an atmosphere

Io,
Jupiter's erupting moon

Europa,
Jupiter's water- and
ice-covered moon

FIGURE 22.25 The ice giants.

(a) The atmosphere of Uranus appears featureless. Although the planet has rings, they are not visible without enhancing the image.

(b) Neptune's atmosphere has bands and storms. Neptune has rings, but they're even fainter than those of Uranus.

The next six large moons are much smaller than Titan. None have atmospheres, and most appear to be similar to one another in composition, with rocky interiors beneath a water-ice shell. Regions on the surfaces of these moons have high crater densities, suggesting that these areas have been tectonically inactive since the early days of the Solar System. But some of the moons display cracks or ridges indicative of tectonic activity in the past, and one, Enceladus, has been substantially resurfaced by younger material. Photographs of Enceladus reveal geysers of water vapor and ice crystals erupting from a set of beautiful blue fissures, informally called tiger stripes, near the moon's south pole **(Fig. 22.24b)**.

Uranus and Neptune: The Ice Giants

DISCOVERY AND BASIC CHARACTERISTICS. Uranus and Neptune lie so far from the Earth that they cannot be seen with the naked eye, even on the darkest night. Astronomers using telescopes first detected Uranus in 1781 and named it after the Greek god of the sky. Irregularities in Uranus's orbit, which could only have been caused by the pull of yet another planet farther out, led astronomers to find Neptune in 1846. Uranus and Neptune are both much larger than the Earth—their radii are 4.0 times and 3.9 times the Earth's radius, respectively—but they are only about one-third the size of Jupiter.

Uranus's axis of rotation has a tilt of 98° (see Fig. 22.4), so it lies almost parallel to the planet's orbital plane! In effect, Uranus lies on its side. As a result, Uranus has the most extreme seasons of any planet in the Solar System. During its orbit, which takes 84 Earth years, its polar regions remain in complete darkness for 42 years and then in constant sunlight for 42 years. Neptune's axis also tilts, but at an angle of 28°, just slightly greater than that of the Earth's axis. Both planets rotate faster than the Earth: a day lasts 17.2 hours on Uranus and 16.1 hours on Neptune.

INTERIORS AND ATMOSPHERES OF THE ICE GIANTS. On Uranus and Neptune, interior pressures don't become great enough for metallic hydrogen to form, as on Jupiter and Saturn. The rocky and metallic cores of Uranus and Neptune appear, instead, to be surrounded by a thick layer of slush composed of water, ammonia, and methane ice. A layer of molecular hydrogen, helium, and methane surrounds the slush (see Fig. 22.19). The presence of icy slush inside Uranus and Neptune led to their designation as ice giants.

From space, all we can see when looking at Uranus is the thick, frigid (–215°C, or –355°F) haze that forms the top of its atmosphere, so Uranus appears as a featureless teal-colored globe without banding or storm spots **(Fig. 22.25a)**. Neptune, in contrast, has a deep blue atmosphere with identifiable features, including bands and white streaks caused by clouds **(Fig. 22.25b)**. Uranus's and Neptune's atmospheres consist primarily of hydrogen (84%) and helium (14%). The concentration of methane

FIGURE 22.26 Rings of the ice giants.

(a) The rings of Uranus.

(b) The rings of Neptune. The planet itself has been blacked out in this image, so that its brightness doesn't mask the rings.

determines the richness of the blue color of these planets, so Neptune (with 3% methane) appears deep blue, while Uranus (with 2% methane) appears paler.

STILL MORE MOONS AND RINGS. At present, astronomers have identified 27 moons orbiting Uranus and 13 orbiting Neptune, as well as thin rings around each planet **(Fig. 22.26)**. The five largest moons of Uranus are all smaller than the Earth's Moon. They appear to be composed of ice and rock and are heavily cratered **(Fig. 22.27a)**. Neptune has one large moon, Triton, which has a diameter of about 2,700 km (1,680 mi) **(Fig. 22.27b)** and is the only large moon in the Solar System to orbit in a retrograde direction. The frigid (−236°C; −393°F) surface of Triton consists of water ice with polar caps of nitrogen ice.

FIGURE 22.27 Moons of the ice giants.

(a) Titania, a moon of Uranus.

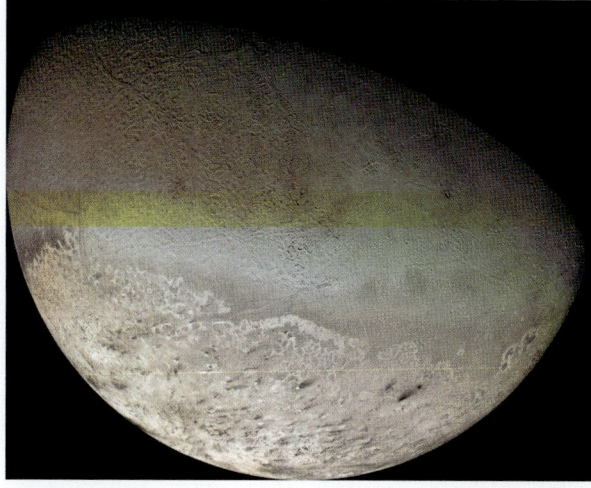

(b) Triton, a moon of Neptune.

Take-home message . . .

The Jovian planets all consist primarily of hydrogen and helium surrounding rocky cores. In Jupiter and Saturn, the larger of the Jovian planets, internal pressures transform hydrogen into a metallic form. The composition of clouds in the atmospheres of the Jovian planets determines their colors as seen from space. All have a large number of moons, and all have rings, with Saturn's the most prominent by far. Some of the moons may contain liquid water. The bluish color of Uranus and Neptune comes from the presence of methane.

Quick Question

Why do we refer to Jupiter and Saturn as the gas giants and to Uranus and Neptune as the ice giants?

FIGURE 22.28 Asteroids.

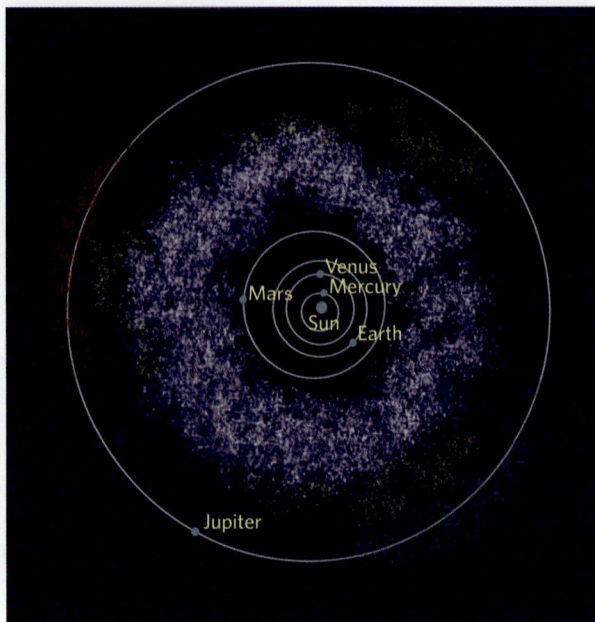

(a) Asteroids orbit the Sun in a broad ring, called the asteroid belt, that lies between the orbits of Mars and Jupiter.

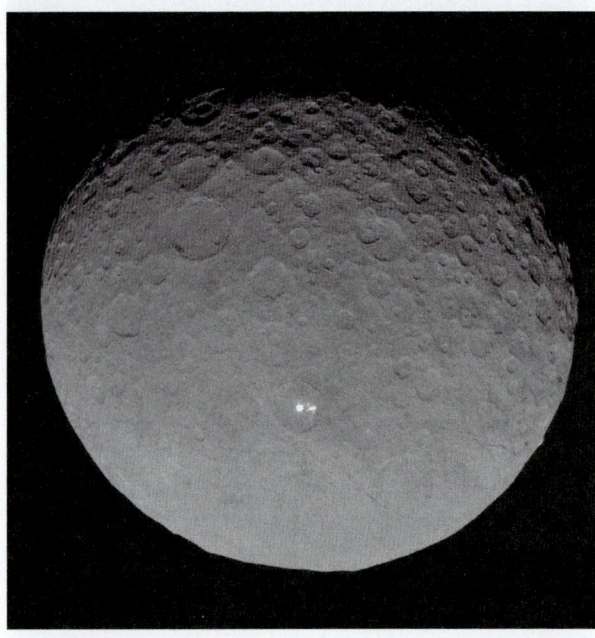

(b) The largest asteroid, Ceres, is now called a dwarf planet. The bright spot seen by the *Dawn* spacecraft in 2015 may be an ice eruption.

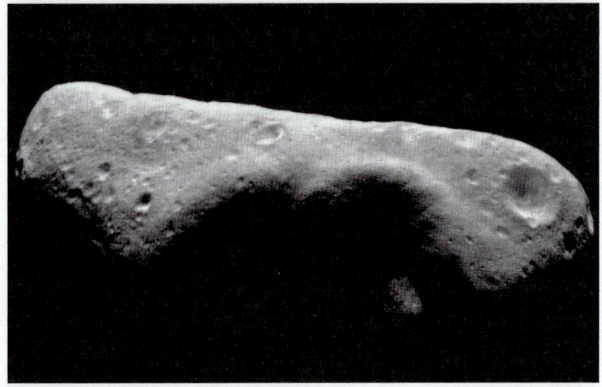

(c) An example of a small asteroid.

22.6 The Other "Stuff" of the Solar System

The eight planets and their moons represent the vast majority of the non-solar mass in the Solar System, but only a tiny fraction of the total number of objects. Researchers estimate that billions of other objects—nearly all too tiny or too far away to observe—move with the Sun as it orbits the Milky Way's galactic center. We can classify these objects into five primary groups—asteroids, Kuiper Belt objects, Oort Cloud objects, comets, and meteoroids—that differ from one another in terms of size, orbital characteristics, and location relative to the planets. (Astronomers now use the term *dwarf planet* for larger, spherical asteroids or Kuiper Belt objects.) An understanding of these objects not only enhances our understanding of how the Solar System developed, but also allows astronomers to evaluate the small, but real, risk that a large object may someday collide with our planet.

Asteroids

By convention, astronomers consider rocky or metallic objects larger than 100 m (330 ft) across to be **asteroids**. Astronomers estimate that there may be about 200 asteroids with diameters larger than 100 km (60 mi), 750,000 with diameters larger than 1 km (0.6 mi), and millions of smaller ones. Nearly all asteroids orbit the Sun within the **asteroid belt**, a band between the orbits of Jupiter and Mars **(Fig. 22.28a)**.

Where did asteroids come from? Researchers suggest that they include primitive planetesimals that never coalesced into a larger body as well as remnants of large planetesimals that had differentiated into a core and mantle before colliding and breaking up. The gravitational pull of Jupiter may have disturbed the orbits of the asteroids so much that they could never re-amalgamate into a planet.

Typical asteroids have irregular shapes, and may be elongate in one direction, because they are too small to have been remolded into spheres by gravity. Three asteroids have become spheres, but only Ceres, with a diameter of 940 km (584 mi), qualifies as a dwarf planet **(Fig. 22.28b)**. Ceres probably has a rocky interior, surrounded by a shell of water ice, which in turn is coated with a surface layer of micrometeorite dust. The *Dawn* space probe went into orbit around Ceres in 2015 and sent back images of its intensely cratered surface, on which a highly reflective mountain stand out. This mountain may represent an eruption of ice. Space probes have also taken close-up images of other asteroids **(Fig. 22.28c)**. In 2001, one landed on Eros, and in 2010, one landed on Itokawa, grabbed a sample, and returned it to the Earth.

FIGURE 22.29 *New Horizons* images of Pluto and Charon.

(a) Close-ups of Pluto's surface reveal icy mountains and fractured plains.

(b) Charon has few impact craters, suggesting that the moon has been resurfaced.

Pluto, Other Kuiper Belt Objects, and the Oort Cloud

Even before the dramatic vote demoting Pluto from planet to dwarf planet, astronomers realized that Pluto was an oddball among the other planets. Most notably, Pluto moves differently. Its orbit not only lies at an angle of 17° relative to the ecliptic plane, but is also highly eccentric: at its aphelion, Pluto lies 1.7 times farther from the Sun than it does at its perihelion. In fact, for part of a Pluto year, Pluto lies closer to the Sun than Neptune does, although the orbits never cross.

Pluto's surface characteristics remained unknown until 2015, when the *New Horizons* probe sent back amazingly detailed photos **(Fig. 22.29a)**. The photos reveal that this small world, which has a diameter about a fifth of the Earth's, has a surface with plains that display complex textures resembling snakeskin and mountains that rise to elevations of 3.5 km (2 mi) (see Fig. 22.2). Pluto has a low density (1.9 g/cm³), which gives it a mass that's about 0.2% of the Earth's, so it probably consists mostly of water ice and frozen nitrogen. Five moons orbit Pluto, of which the largest, Charon, has a diameter of 1,300 km (800 mi) **(Fig. 22.29b)**. All the moons appear to consist of water ice.

Pluto is one of several objects with diameters in the range of 1,000–2,500 km (600–1,500 mi) that lie outside the orbit of Neptune. One of these, Eris, may contain more mass than Pluto. Pluto and Eris are the largest known objects of the **Kuiper Belt**, named for the astronomer Gerard Kuiper (1905–1973). These objects, millions of which are tiny, but perhaps over 100,000 of which have a diameter larger than 100 km (60 mi), are icy bodies that circle the Sun in a region that extends from the orbit of Neptune (30 AU) out to a distance of about 50 AU from the Sun. Overall, many objects in the Kuiper Belt orbit near the ecliptic plane, but many other individual objects within it, such as Pluto, have orbits that are inclined to the ecliptic plane.

In 1950, a Dutch astronomer, Jan Oort (1900–1992), proposed that even more icy objects lie beyond the Kuiper Belt but remain gravitationally attached to the Sun. Researchers envision these objects as occupying a somewhat spherical shell, now called the **Oort Cloud**. Its inner edge lies at a distance of about 2,000 AU from the Sun, and its outer edge extends to a distance of 100,000 AU (1.5 light-years), more than a quarter of the distance to the nearest star. The Oort Cloud probably consists of debris that condensed during Solar System formation but never fell into the accretion disk.

Comets

Imagine how baffled ancient observers must have been when they looked skyward at an area that had previously contained only stars and saw a bright object with a long, glowing tail **(Fig. 22.30a)**. As they watched the object over successive nights, it moved across the backdrop of stars, dimmed, and then disappeared. It's no wonder that, for millennia, such objects, called **comets**, were thought to be omens of historical events. It wasn't until 1705 that the English astronomer Edmond Halley (1656–1742), by applying Newton's laws, calculated the path of a comet (now called Halley's comet) and showed that it orbits the Sun. Other astronomers in the 18th and 19th centuries proposed that comets were solid objects that emitted gas when they came close to the Sun, but it wasn't until 1950 that the current image of a comet—as a dirty snowball—came into favor. Modern space probes have provided spectacular images of comets. In 2005, the *Deep Impact* probe blasted debris off a comet that could be analyzed remotely, and in 2014, the *Rosetta* probe went into orbit around a comet and successfully placed a lander on its surface.

FIGURE 22.30 Comets.

(a) Hale-Bopp, a particularly dramatic comet, crossed the night sky in 1997.

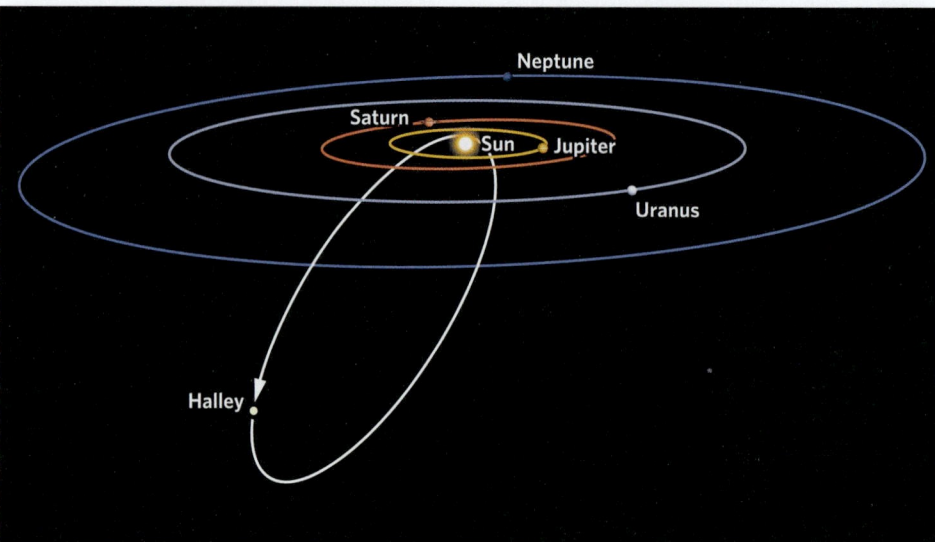

(b) Comets have very eccentric orbits.

Did you ever wonder...

why a comet has a tail?

We now understand that comets follow highly elliptical orbits with the Sun at one focus **(Fig. 22.30b)**. At the head of a comet lies a solid mass, the comet's *nucleus*. All well-studied comets have an irregularly shaped nucleus with a rubbly-looking surface **(Fig. 22.31)**. These nuclei range from 0.5–50 km (0.3–30 mi) long and 0.1–10 km (0.06–60 mi) wide, and they seem to consist of loosely bound aggregates of rock and ice (water ice, frozen methane, dry ice, and frozen ammonia), along with traces of more complex organic chemicals. As a comet gets to within about 3–4 AU of the Sun, its dark surface absorbs enough solar radiation that the comet's interior starts vaporizing. Volatiles vent into space through cracks on the comet's surface, producing geyser-like spouts of ionized (electrically charged) gas that carry refractory dust particles along with them. Some of the gas surrounds the nucleus to form a glowing atmosphere, or *coma*. The rest of the gas, along with all the dust, streams into space as a *comet tail*. The gas forms an *ion tail* whose orientation lies parallel to the direction of the solar wind, so it always points away from the Sun. The dust forms a *dust tail* that tends to curve, for the dust particles in it are partly following the comet's orbit, but they are also affected progressively by the solar wind **(Fig. 22.32)**. Some tails can be more than 1 AU long. The closer a comet comes to the Sun, the longer and brighter its tails become.

FIGURE 22.31 The nucleus of a comet has an irregular shape.

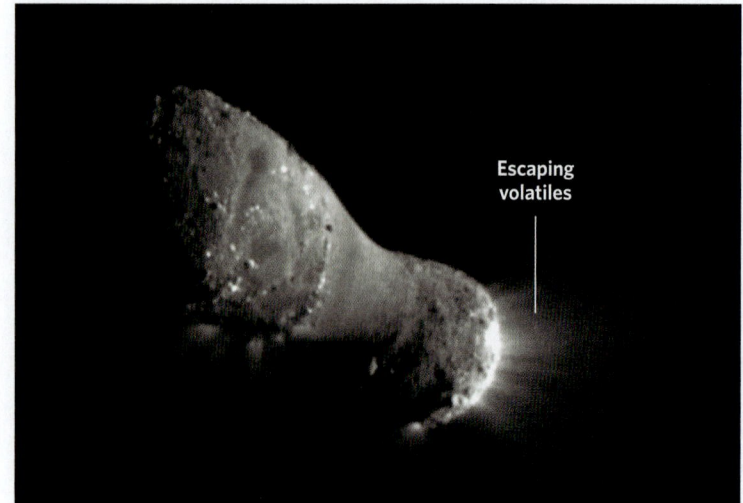

Escaping volatiles

(a) Comet Hartley 2, releasing volatiles. It is 2.0 km (1.2 mi) long.

(b) Comet 67P/Churyumov-Gerasimenko. It is 6.7 km (4.2 mi) long.

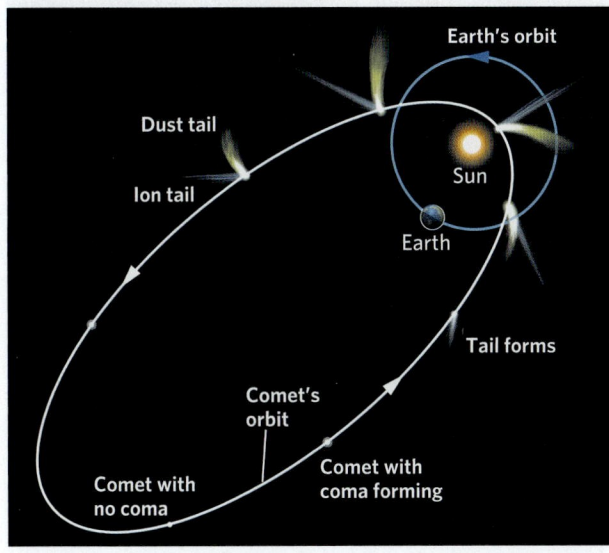

The formation of tails causes a comet to lose material each time it approaches the Sun. Small comets probably have little chance of survival beyond a few orbits, and even a large comet probably disintegrates after a few thousand orbits. Therefore, there must be a source that, over time, sends new icy bodies into the inner Solar System. Researchers conclude that comets come from both the Kuiper Belt and the Oort Cloud.

Researchers classify comets by the duration of their orbit. *Short-period comets* have less eccentric orbits and complete an orbit in less than 200 years, whereas *long-period comets* have more eccentric orbits that take more than 200 years—thousands of years in some cases—to complete. Researchers suggest that short-period comets originate as Kuiper Belt objects and that long-period comets come from the Oort Cloud. These objects become comets when gravitational tugs or collisions with other bodies send them careering toward the inner Solar System.

Meteoroids, Meteors, and Meteorites

Astronomers refer to small objects traveling through space as **meteoroids**. Larger objects are classified as small asteroids, small comets, or fragments of planets, moons, or asteroids ejected into space when struck by another object.

Meteoroids whose paths cross the Earth's orbit may collide with the Earth. A meteoroid that enters the Earth's atmosphere arrives at speeds of around 20 km/s (50,000 mph)—100 times the speed of a jet plane. As the speeding object enters the atmosphere, it compresses the air in its path intensely. Compression of a gas heats it up, so a meteoroid entering the air generates immense heat,

enough to cause some or all of the object to vaporize. Glowing gases from the meteoroid vapor, along with glowing superheated air molecules, produce a blazing light streak behind the object **(Fig. 22.33a)**. The light streak is called a **meteor** (or, misleadingly, a shooting star). Note that by this definition, a meteor is an atmospheric phenomenon. To produce a meteor, an incoming meteoroid must be larger than a sand grain. Note that frictional heating is not what produces the light of a meteor; air isn't dense enough in the upper atmosphere for friction to have much effect.

During a **meteor shower**, an observer can see from 10 to over 100 meteors per hour, all of which appear to emerge from a common point in the sky (see Fig. 22.33a). These showers, which last as long as 1 to 3 days, happen when the Earth passes through the orbit of a comet and intersects its dust tail. The Earth passes specific cometary orbits at the same time every year. Occasionally, meteor showers become *meteor storms*, during which over 1,000 meteors streak across the sky every hour. Not all meteoroids arrive in showers or storms, however; astronomers refer to those that arrive independently as *lone meteoroids*.

When a larger meteoroid enters the atmosphere, it becomes a very bright *fireball* that leaves a visible smoke-like trail across the sky. If it explodes, a fireball becomes a *bolide*. During the last century and a half, two bolides have exploded over Siberia. The first, known as the Tunguska bolide, released about 10 megatons of energy—1,000 times more than the Hiroshima atomic bomb—and flattened about 2,100 km² (830 mi²) of Russian forest—an

FIGURE 22.33 Meteors and meteor craters.

(a) The Perseids meteor shower.

(b) The bolide that created a huge shock wave over Russia on February 15, 2013.

2 cm

(c) An example of a meteorite.

(d) Meteor Crater in Arizona.

area almost twice the size of New York City. The bolide was probably about 120 m (400 ft) across and blew up at an elevation of 5–10 km (3–6 mi). A similar but smaller (0.5-megaton) bolide explosion took place over Russia in 2013 **(Fig. 22.33b)**. Dashboard video cameras recorded the streak of the fireball's tail, the blinding light of its explosion, and the chaos caused when its shock wave blew out windows and knocked down walls.

If a meteoroid makes it through the atmosphere without completely burning up, it strikes the Earth. The meteoroid then becomes a **meteorite (Fig. 22.33c)**. Smaller, slower meteoroids can excavate a small indentation when striking the ground, and the rare ones that have landed in populated areas have punched holes through roofs and dented cars. Larger impacts blast out a crater, whose size

depends on the size, speed, and density of the meteoroid. The 1.2-km (0.75-mi)-wide Meteor Crater of Arizona was formed about 50,000 years ago from the impact of a dense meteorite 50 m (160 ft) in diameter **(Fig. 22.33d)**. During the course of Earth history, some small comets and asteroids (1–15 km, or 0.6–9 mi, across) have collided with the Earth **(Box 22.2)**.

The Edge of the Solar System

Between the objects of our Solar System lies the vacuum of **interplanetary space**. This vacuum contains between 5,000 and 100,000 atoms or molecules per liter. (By comparison, air at sea level contains about 2.5×10^{22} molecules per liter.) Is there a boundary that defines the edge of the Solar System? Yes, and surprisingly, it lies closer to the Sun

Early detection of near-Earth objects

If devastating impacts with meteoroids, asteroids, and comets happened on the Earth in the past, then they will happen in the future. This concern has led astronomers to look for such objects that might be heading our way. So far, they have identified over 6,500 known objects that have Earth-crossing orbits. These **near-Earth objects** have the potential to collide with the Earth and cause devastation.

Statistical studies suggest that impacts with the potential to produce catastrophes have a recurrence interval of about 100,000 years, and those with extreme consequences, such as the impact that caused the K-Pg mass extinction (see Box 10.2), have a recurrence interval of about 10 million years. (Remember that the recurrence interval refers to the average time, not a defined period, between comparable events.) But even small impacts, if over or near a heavily populated area, could cause a catastrophe greater than any earthquake or hurricane. Imagine if the Tunguska bolide had exploded over a large city! Because of this concern, NASA has established the *Near-Earth Object Monitoring Program* to catalog the orbits of known objects that have the potential to cross the Earth's orbit within the next 100 years.

---

than the inner boundary of the Oort Cloud. Astronomers consider a bubble-like invisible surface, called the **heliosphere**, that lies 200 AU from the Sun to be the edge of the Solar System **(Fig. 22.34)**. Within the heliosphere, most of the atoms came as solar wind from the Sun. Outside, most particles, known as *cosmic rays*, came from distant stars or supernovae. Only two human-made objects until now, the *Voyager 1* and *2* space probes, launched in 1977, have crossed the heliosphere. At its current speed, it will traverse the distance to the nearest star in about 40,000 years. In the next chapter, we focus our attention on stars, galaxies, and other celestial objects out to the edge of the Universe.

Take-home message . . .

The Solar System encompasses an enormous number of small objects in orbit around the Sun. They occur primarily in three locations: the asteroid belt, between Jupiter and Mars; the Kuiper Belt, beyond Neptune; and the Oort Cloud, which lies beyond the outer fringes of the Solar System but remains tied to the Sun by gravity.

Quick Question —————————
What defines the edge of the Solar System?

FIGURE 22.34 The heliosphere encompasses all the planets and the Kuiper Belt; it lies at the center of the much larger Oort Cloud.

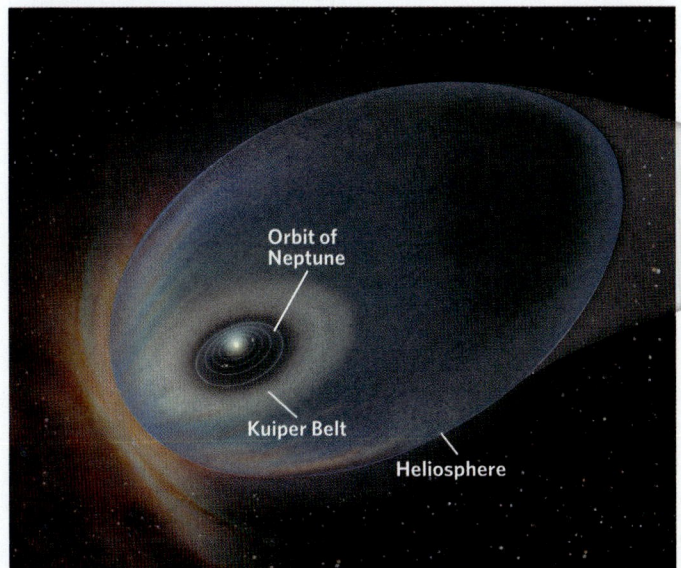

◉22 CHAPTER REVIEW

- A planet is a celestial body in orbit around the Sun that has a spherical shape and has cleared the neighborhood around its orbit. A dwarf planet is large enough to be spherical, but has not cleared the neighborhood around its orbit.

- A moon is a large object in orbit around a planet. All planets except Mercury and Venus have moons.

- The Earth's Moon rotates once each time it orbits the Earth, so that the same side always faces the Earth. Its surface includes both light-colored lunar highlands and dark-colored plain-like maria underlain by basalt. The Moon's entire surface is covered with regolith.

- Venus has a dense atmosphere, Mars a thin one, and Mercury has hardly any. Volcanoes rise from the surfaces of Venus and Mars.

- Mars is a cold planet with a very thin atmosphere composed of carbon dioxide. Except for water in its ice cap, and possibly local intermittent flows of liquid water, Mars is now dry.

- Mars has a relatively smooth basin in its northern hemisphere, highlands and volcanoes along its equatorial region, and highlands in the south. It hosts a canyon much bigger than the Earth's Grand Canyon.

- The Jovian planets have rocky and metallic cores, but consist mostly of hydrogen and helium. Jupiter and Saturn, the gas giants, contain gases in various states. Uranus and Neptune, the ice giants, contain icy slush. All have dense, colorful atmospheres.

- All the Jovian planets have rings and a variety of moons. The asteroid belt, between the orbits of Mars and Jupiter, contains millions of rocky and metallic chunks, fragments of planetesimals that never coalesced into a planet.

- The Kuiper Belt includes millions of icy objects, including the dwarf planet Pluto. The Oort Cloud lies farther out.

- A comet consists of rock, dust, and ice. It follows a highly eccentric orbit that takes it close enough to the Sun to start vaporizing and develop a coma and tail.

- A meteoroid is an object that is smaller than a comet or an asteroid. When a meteoroid enters the Earth's atmosphere, it generates a streak of light known as a meteor. A meteorite is a meteoroid that strikes the planet's surface.

- NASA monitors near-Earth objects that have the potential to cross the Earth's orbit and cause a collision within the next 100 years.

- Astronomers define the edge of the Solar System as an invisible surface called the heliosphere.

Key Terms

asteroid (p. 720)
asteroid belt (p. 720)
comet (p. 721)
crater (p. 702)
dwarf planet (p. 697)
eccentric orbit (p. 698)
gas giant (p. 710)
Great Red Spot (p. 711)

heliosphere (p. 725)
ice giant (p. 710)
interplanetary space (p. 724)
Jovian planet (p. 700)
Kuiper Belt (p. 721)
late heavy bombardment (p. 702)
lunar highland (p. 701)
lunar regolith (p. 702)

mare (plural: maria) (p. 701)
metallic hydrogen (p. 711)
meteor (p. 723)
meteorite (p. 724)
meteoroid (p. 723)
meteor shower (p. 723)
moon (p. 697)
near-Earth object (p. 725)

Oort Cloud (p. 721)
planet (p. 697)
ring (p. 700)
space weathering (p. 702)
terrestrial planet (p. 700)

Letters in parentheses correspond to the chapter's learning objectives.

1. List the succession of celestial objects in the Solar System, starting from the Sun. **(B)**

2. Are the axes of all the planets parallel to one another? If not, how do they differ? **(D, E)**

3. How do the orientations of the orbital planes of other planets compare with the Earth's ecliptic plane? **(B)**

4. What characteristics must an object have to be formally considered a planet? How does a dwarf planet differ from a planet? **(A)**

5. The figure depicts a massive object striking our planet. Sketch how this impact could have led to the formation of the Moon. **(C)**

6. What do the different landscapes of the Moon consist of, and how did they form? How can we determine which lunar landscapes are older than others? **(C)**

7. What characteristics do Mercury and the Earth's Moon share, and in what ways are they different? **(D)**

8. Why do the Earth and Venus have vastly different atmospheres and surface temperatures? **(D)**

9. What is the evidence for water on Mars in the present and in the planet's ancient past? **(D)**

10. Why does Jupiter have distinct rings, and what is its Great Red Spot? **(E)**

11. Summarize the distinctive characteristics of the largest moons of the Jovian planets. Which host active volcanoes? **(F)**

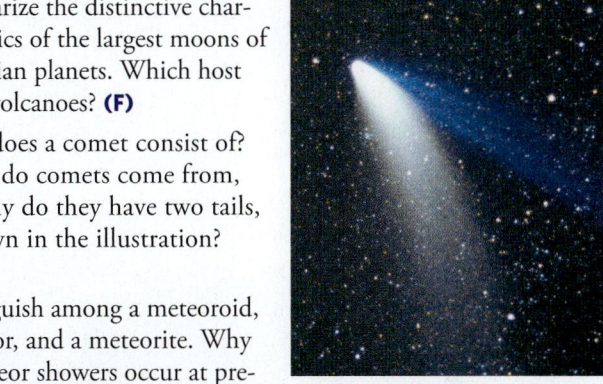

12. What does a comet consist of? Where do comets come from, and why do they have two tails, as shown in the illustration? **(H)**

13. Distinguish among a meteoroid, a meteor, and a meteorite. Why do meteor showers occur at predictable times? **(H)**

14. What feature defines the edge of the Solar System? Do Oort Cloud objects lie within or outside this boundary? **(G)**

On Further Thought

15. Name three unique characteristics of each planet that make it different from the other seven. **(D, E)**

16. Of the terrestrial planets, why is abundant water and oxygen found only on the Earth? **(D)**

17. Mars and Venus are different sizes. Yet a 150-lb person would weigh the same on either planet. Why? **(D)**

ANOTHER VIEW Jupiter's amazing clouds.

23 THE SUN, THE STARS, AND DEEP SPACE

By the end of the chapter you should be able to . . .

A. describe what's inside the Sun and other stars, and explain how they produce energy.

B. explain what is meant by apparent magnitude and absolute magnitude when describing a star's brightness.

C. demonstrate how astronomers classify stars using an H-R diagram.

D. discuss how stars form and evolve, explain what happens when they run out of fuel, and describe remnants of dead stars such as white dwarfs, neutron stars, and stellar black holes.

E. describe the overall structure of a galaxy, the differences among galaxies, and the relationship of a supermassive black hole to a galaxy.

F. visualize current ideas concerning the overall structure of the Universe.

23.1 Introduction

Gazing at a moonless black sky, awash with bright stars, makes for a truly memorable night. Such a view may be hard to come by for an urban dweller, given the light pollution and haze of cities, but look up from a desert mountain peak on a summer night and prepare to be amazed. In addition to stars and planets, you'll see the central part of the Milky Way Galaxy slicing across the sky from horizon to horizon. The Sun, and every individual star that you can see from the Earth, lies within the Milky Way. However, not all the points of light in the night sky are single stars, nor do they all lie within the Milky Way. Some are nebulae, some are clusters of stars, some are galaxies, and some are unusual objects that astronomers still struggle to interpret. Observations made with the Hubble Space Telescope emphasize that even a tiny part of the sky that looks dark to the naked eye contains billions upon billions of celestial objects **(Fig. 23.1)**. Most are so far away that the light you see has traveled for millions or even billions of years since it began its journey across *deep space*—the Universe far beyond the limits of our Solar System—to your eyes.

In this chapter, we complete our introduction to astronomy by exploring the vast variety of objects, in addition to the planets and moons of our Solar System, that constitute the Universe. We begin with our own Sun because it provides a basis for understanding all stars. Next, we turn our attention to other stars, to galaxies, and finally to a host of objects with strange-sounding names—including nebulae, dwarf stars, giant stars, neutron stars, quasars, novae, supernovae, and black holes—that no one had even dreamed of before the 20th century. This chapter, and this book, ends by pondering the possible fate of the Universe in a time long after the Earth will have ceased to exist.

23.2 Lessons from the Sun

Our Sun is a fairly ordinary star—not too big or too small, and not too hot or too cool. Because of its proximity to our planet **(Fig. 23.2)**, we can observe our Sun in fine detail, and our observations have led to an understanding of its energy generation, its composition, and its internal structure. This knowledge provides a basis for interpreting distant stars.

<< The central arc of the Milky Way galaxy graces a sky filled with stars.

FIGURE 23.1 The immensity of space as revealed by the Hubble Space Telescope.

(a) A speck of the sky, 5% of the area covered by the Moon, looks black when viewed from the Earth.

(b) When Hubble magnifies this speck, it reveals hundreds of galaxies, each containing hundreds of billions of stars. The point of light with spikes is a relatively nearby star in our own galaxy.

Did you ever wonder...

why the Sun shines?

FIGURE 23.2 The Sun, as seen by astronauts orbiting the Earth.

Composition of the Sun

In 1925, astronomer Cecilia Payne-Gaposchkin, by studying the spectra of sunlight and starlight (see Chapter 21), determined that the Sun, and stars like it, consist primarily of hydrogen (71.0% by mass) and helium (27.1%). Because hydrogen atoms weigh less than helium atoms, 91.2% of the individual atoms in the Sun are hydrogen, and only 8.7% are helium. The next eight most common elements, in order of their contribution to the Sun's mass, are oxygen, carbon, nitrogen, silicon, magnesium, neon, iron, and sulfur. The Sun contains traces of all the other naturally occurring elements as well.

Hydrogen and helium in the Sun do not exist in the same form in which we find these elements on the Earth, for at the temperatures found in the Sun, atoms lose some or all of their electrons and attain an electric charge. A material consisting almost entirely of such *ionized atoms*, circulating along with *free electrons* (which move independently of atoms), is called **plasma**. Physicists consider plasma to be the fourth state of matter.

The Source of the Sun's Energy

Discoveries made during the first half of the 20th century first provided an explanation for the Sun's energy generation. The key to solving this mystery came from the work of Albert Einstein (1879–1955). According to his iconic equation, $E = mc^2$, annihilation of a tiny amount of mass produces an enormous amount of energy. In the

1920s, physicists discovered that when temperatures are high enough, two atomic nuclei can collide with enough force to fuse together to form a single nucleus, and that during this process, called **nuclear fusion**, a tiny amount of matter converts into a large amount of energy. During the 1930s, further work suggested that nuclear fusion takes place in the Sun. By 1939, researchers had worked out the series of fusion reactions that take place inside the Sun and could, therefore, explain its energy production.

Almost all of the nuclear fusion in the Sun involves the bonding of protons (the nuclei of hydrogen atoms) to form helium nuclei. This reaction, known as the *proton-proton chain reaction*, involves several steps **(Fig. 23.3)**. Simplistically, the reaction begins when two protons fuse to form deuterium (^{2}H), an isotope of hydrogen whose nucleus contains one proton and one neutron; during the reaction, one of the colliding protons transforms into a neutron. Next, a deuterium nucleus collides with another proton to form ^{3}He, an isotope of helium containing two protons and only one neutron. In the final step, two ^{3}He isotopes collide to form a normal helium nucleus (^{4}He), releasing two excess protons. The complete chain reaction releases energy.

Because of nuclear fusion, the Sun produces an inconceivable 4×10^{26} watts of power. (A *watt* is a unit of power; it represents the expenditure of 1 joule of energy in 1 second.) By comparison, burning all the hydrocarbon reserves on the Earth in 1 second would yield 10^{21} watts, and a typical nuclear power plant produces 5×10^8 watts. If all of the Sun's energy were focused on the Earth, the planet would evaporate. Fortunately for us, the Earth receives only a tiny fraction of the Sun's energy, while most of the rest spreads out into space.

FIGURE 23.3 A nuclear fusion reaction in the Sun produces helium (He) nuclei from hydrogen (H) nuclei. The reaction, which takes place in stages, releases energy in the form of gamma rays, and releases subatomic particles (positrons and neutrinos).

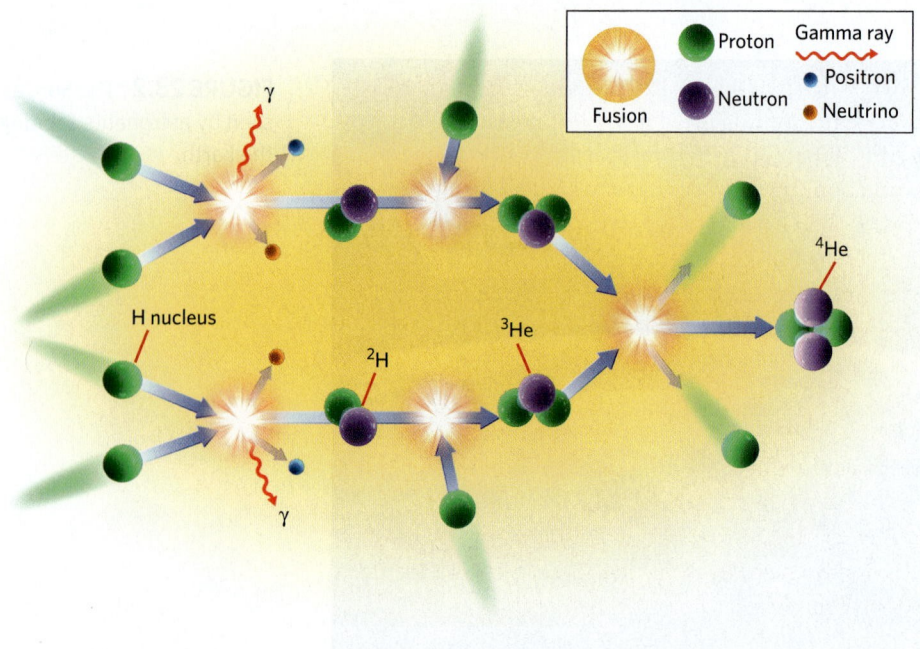

Internal Structure of the Sun

The radius of the Sun, defined as the distance from its center to its visible surface, is 696,000 km (432,000 mi), 110 times that of the Earth. Researchers distinguish three concentric layers within the Sun: the *solar core* at the center, the *radiative zone*, and the *convective zone* **(Fig. 23.4)**. Let's examine the characteristics of these layers, starting from the center.

THE SOLAR CORE. The **solar core** extends from the center outward for 160,000 km (100,000 mi), and therefore accounts for about 23% of the Sun's radius. Temperatures reach 15 million °C (27 million °F) at the center of this region and diminish to about 8 million °C (14.4 million °F) at its top. Because of the inward gravitational pull of the Sun's immense mass, the pressure at the center of the solar core may be over 10,000 times that found at the center of the Earth. About 99% of the Sun's energy

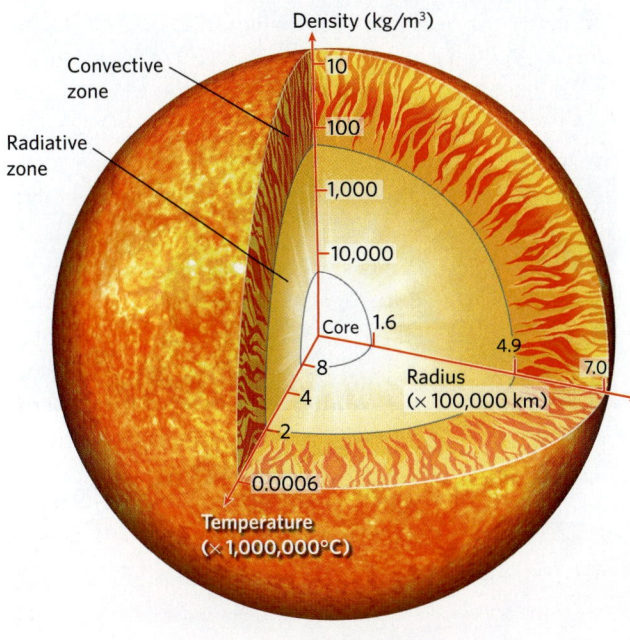

FIGURE 23.4 The internal structure of the Sun. The axes show how physical characteristics of the Sun change from the core to the top of the convective zone.

layer of plasma to about 2 million °C (3.6 million °F). This heat decreases the density of plasma, making it relatively buoyant. The buoyant plasma rises upward and carries thermal energy with it. As it rises toward the Sun's surface, the plasma cools, and it eventually sinks to replace other rising buoyant plasma. Astronomers refer to the region of the Sun in which this convective transport of energy takes place as the **convective zone**. This zone, which surrounds the radiative zone, has a thickness of about 200,000 km (125,000 mi).

The upwelling and downwelling of plasma in the convective zone organizes into *convective columns*, which are very tall convective cells. Plasma rises at the center of each convective column and sinks back down along the sides. The columns range from 500 – 2,000 km (300–1,200 mi) across—about a tenth to a third the width of North America. The column center, where hot plasma rises, is hotter, and therefore brighter, than the sides, where cooler plasma sinks. This pattern of light and dark gives the Sun's surface a grainy appearance, known as **solar granulation (Fig. 23.5)**. At any given time, about 4 million granules (the visible tops of convective columns) exist, but the pattern constantly evolves, with each granule lasting for only a few minutes before dissipating as new granules form.

The Solar Atmosphere

The Sun doesn't have a solid crust like that of the Earth, so it's a bit tricky to decide how to define the boundary between the Sun's interior and its atmosphere. Astronomers place the boundary at the depth where the density of material in the Sun becomes great enough that it is opaque to visible light. The material above this boundary is the **solar atmosphere**, which includes the distinct layers we describe next.

THE PHOTOSPHERE. All the light that we see from the Sun comes from glowing gases within the **photosphere**, a 50–500-km (30–310-mi)-thick layer between the top of the convective zone and the rest of the solar atmosphere. Transparency decreases gradually from the top to the base of the photosphere. At the base, the plasma of the Sun is too dense for light to escape, so the base of the photosphere serves as the visually sharp boundary of the Sun that you see at sunset or sunrise **(Fig. 23.6a)**. The range of temperatures from the top (4,300°C; 7,800°F) to the base (5,700°C; 10,300°F) of the photosphere resembles the range of temperatures found at the center of the Earth, but is vastly cooler than the inferno at the center of the Sun. Electromagnetic energy emitted by the photosphere heads off into space, reaching the Earth about 8 minutes later.

production comes from nuclear fusion reactions in the core. These reactions convert 600 million tons of hydrogen into helium every second. The remaining 1% of the Sun's energy comes from the base of the radiative zone. Temperatures elsewhere in the Sun are too cool for fusion to be possible.

THE RADIATIVE ZONE. Researchers refer to the thick layer surrounding the solar core as the **radiative zone** because energy passes through this zone in the form of electromagnetic radiation. The radiative zone extends from the core out to about 490,000 km (300,000 mi) from the center of the Sun, a distance of 70% of the radius, so it has a thickness of about 330,000 km (205,000 mi). It accounts for about 43% of the Sun's radius and about 48% of the Sun's mass. In the radiative zone, photons of energy emitted by fusion reactions in the core travel only a short distance before they interact with other particles of matter and are redirected in a new, random direction. These photons then travel a short distance before they, too, encounter another particle and are redirected in a new direction. Energy, in effect, wanders nearly randomly as it travels through the radiative zone, so a given packet of energy may take about 170,000 years to reach the zone's outer surface.

THE CONVECTIVE ZONE. When energy reaches the top of the radiative zone, it heats the base of the overlying

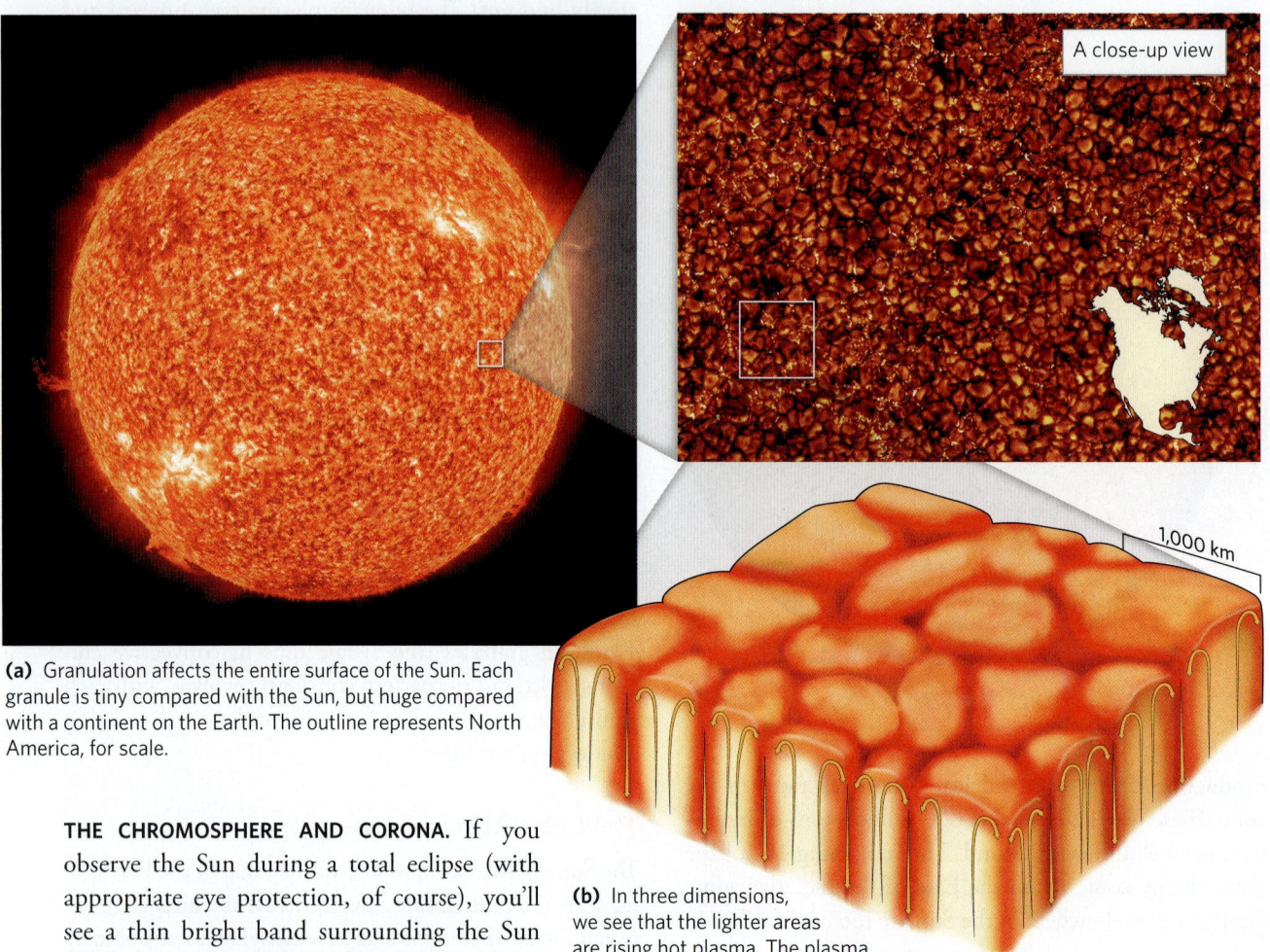

(a) Granulation affects the entire surface of the Sun. Each granule is tiny compared with the Sun, but huge compared with a continent on the Earth. The outline represents North America, for scale.

A close-up view

1,000 km

(b) In three dimensions, we see that the lighter areas are rising hot plasma. The plasma cools and sinks in the darker areas.

THE CHROMOSPHERE AND CORONA. If you observe the Sun during a total eclipse (with appropriate eye protection, of course), you'll see a thin bright band surrounding the Sun **(Fig. 23.6b)**. This band, the **chromosphere** (sphere of color), is so named because when viewed during an eclipse, it displays flashing tints of brilliant red. The chromosphere is 3,000–5,000 km (1,800–3,000 mi) thick. It has a density of only about 0.00000001 that of the Earth's atmosphere at sea level.

At the top of the chromosphere, across a thin (100-km, or 60-mi) *transition zone*, the temperature increases dramatically. In the overlying **corona**, the outer layer of the solar atmosphere, temperatures rise to 1 million °C (1.8 million °F) at a distance of 10,000 km (6,200 mi) above the photosphere. We can see the inner part of the corona during an eclipse, when it appears as a wispy glowing cloud **(Fig. 23.6c)**. When viewed in ultraviolet light, the outer part of the corona appears as bright streaks radiating far out from the Sun's surface **(Fig. 23.6d)**.

The Solar Wind

At high temperatures, particles move very fast, so it's no surprise that some of the particles in the extremely hot corona achieve escape velocity, break free of the Sun's gravitational pull, and head off into space. The high-speed stream of particles—mostly protons and electrons, along with some helium nuclei—flowing permanently away from the Sun makes up the **solar wind**. Note that the solar wind consists of moving matter, not just radiation, so it really is a "wind" like the winds of the Earth's atmosphere, except that solar wind particles move at enormous speeds and are ionized, whereas molecules in the Earth's wind are neutral and travel relatively slowly. Because of the solar wind, the Sun loses about 1.8 billion kilograms (2 million tons) of matter every second. But even at this seemingly huge rate, only about 0.1% of the Sun has been lost to space since its nuclear furnace first ignited. Solar wind particles travel at about 500 km/s (300 mi/s), much slower than the speed of light, so they take a few days to reach the Earth. Because they have an electric charge, the Earth's magnetosphere deflects most of them. But some reach the atmosphere and stream toward the poles to produce aurorae (see Chapter 17).

FIGURE 23.6 The solar atmosphere.

(a) At sunset, the edge of the Sun that you see is the base of the photosphere.

(b) The chromosphere, the thin red and white ring close to the Sun's surface, as seen in visible light during an eclipse. To make this layer visible, the light of the corona has been blocked.

(c) This photo shows the same image as part (b), but here the gases of the inner part of the corona are visible.

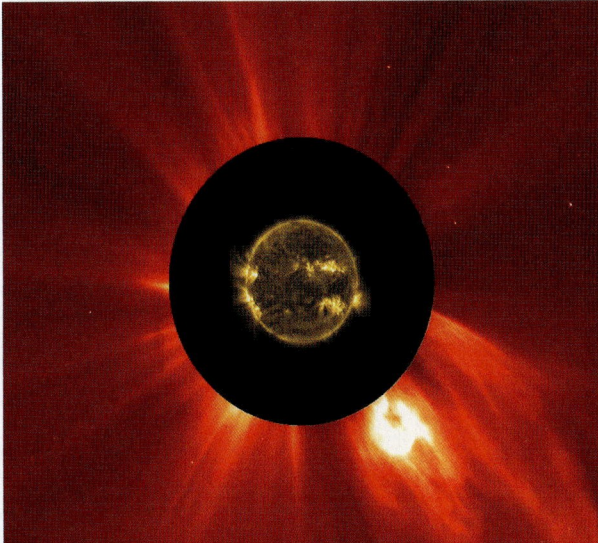

(d) In this photo, taken at a different time, the light of the inner corona has been blocked. The outer corona, as viewed in ultraviolet light, streams far into space.

Take-home message . . .

The Sun is a star, and like other stars, it produces huge amounts of energy by nuclear fusion reactions. These reactions can take place because it's so hot inside the Sun that nuclei can collide with enough force to fuse together. The interior of the Sun can be divided into layers. Most nuclear fusion takes place in the solar core. The resulting energy passes through the radiative zone as electromagnetic energy and then heats the base of the convective layer, producing convective columns that carry energy to the surface of the Sun. In the photosphere, the energy is emitted into space as electromagnetic radiation. Beyond the photosphere, the solar atmosphere consists of the chromosphere and the corona. Particles escaping from the corona head into space as solar wind.

Quick Question
What do the Sun's atmosphere and solar wind consist of?

FIGURE 23.7 The Sun's magnetic field and the formation of sunspots.

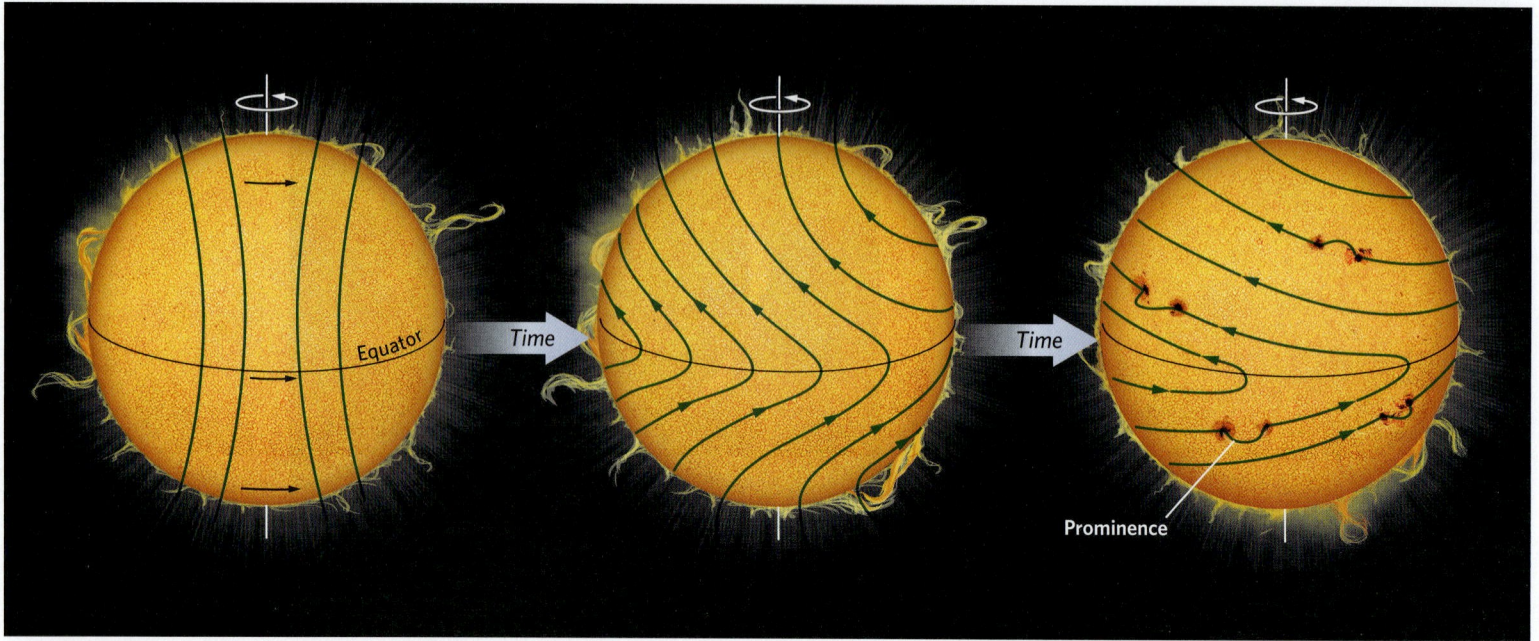

(a) The rotation rate of the Sun is faster at the equator than at the poles, so over time, magnetic field lines bend and become almost parallel to the equator.

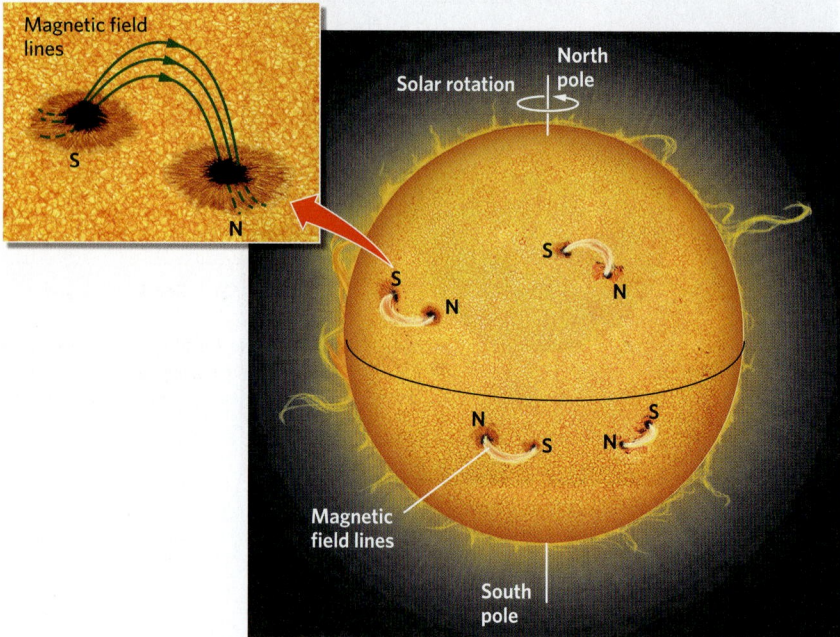

(b) The magnetic field can become so intense that field lines arc out from the solar surface, creating sunspots. The lines rise from one sunspot and re-enter at another.

23.3 The Sun's Magnetic Field and Solar Storms

The Sun possesses an intense magnetic field, much stronger than that of the Earth. The complex behavior of this field yields *sunspots* (dark patches) on the Sun's surface and, at times, causes particularly strong emissions of matter to blast from the surface. These *solar storms* send out such strong solar winds that they can have adverse effects on the Earth. In this section, we examine the Sun's magnetic field and sunspots, as well as their relationship to solar storms.

The Sun's strong magnetic field develops in response to the rapid circulation of plasma within the convective zone, because plasma is an electrical conductor. Unlike the magnetic field lines of the Earth, which arc from pole to pole, those of the Sun trend at a small angle to the solar equator. They point in one direction in the northern hemisphere and in the opposite direction in the southern hemisphere. This unusual orientation of the Sun's magnetic field lines reflects the nature of its rotation. The Sun is a fluid whose rotation rate varies with latitude: plasma at the equator rotates faster around the Sun's axis (once every 25 days) than does plasma near the poles (once every 38 days). This differential motion effectively wraps magnetic field lines around the Sun **(Fig. 23.7a)**.

Recall from Chapter 2 that on the Earth, the polarity of the magnetic field reverses (flips) at intervals of thousands to millions of years. On the Sun, reversals happen about once every 11 years. If, before a reversal, magnetic field lines in the Sun's northern hemisphere point eastward and those in the southern hemisphere point westward, then after a reversal, those in the northern hemisphere point westward and those in the southern hemisphere point eastward.

Sunspots

Over the course of several years, the wrapping of magnetic field lines around the Sun squeezes the lines together, making the magnetic field locally so intense that the field lines arc into space **(Fig. 23.7b)**. At the entrance and exit points of these magnetic fountains, the magnetic field can be strong enough to inhibit the upwelling of hot plasma in the underlying convective zone. Therefore, patches on the surface of the photosphere may become 1,300°C–2,700°C (800°F–1,700°F) cooler than brighter regions of the photosphere. These patches are known as **sunspots** because their lower temperatures make them look darker than the surrounding regions **(Fig. 23.8a, b)**. Individual sunspots move with the Sun's rotation and can survive for days. But during this time, they constantly change in shape and dimensions as the local magnetic field evolves. Sunspots always occur in pairs, one located at the point where the magnetic field lines arc upward and the other at the point where they arc downward (see Fig. 23.7b).

The number of sunspots on the Sun's surface varies over time as a consequence of the reversals in the Sun's magnetic field **(Fig. 23.8c)**. Over the course of 5.5 years, the field strengthens, and the number of sunspots increases to a maximum of about 50 to 120 per month. Then the polarity of the field reverses, after which, over the next 5.5 years, the number of sunspots decreases to nearly zero. During the next 11 years, the pattern repeats, but the sunspots have the opposite magnetic polarities. The 11-year cycle is known as the *sunspot cycle*, or **solar cycle**, and the 22-year cycle is called the *magnetic cycle*.

Solar Storms

Occasionally, local intensification or disruption of the Sun's magnetic field triggers the ejection of particularly large amounts of high-energy particles into space. These events, called **solar storms**, seem to take place in association with sunspot maximums in the solar cycle.

Astronomers distinguish among several different types of solar storms. During a **solar prominence**, outbursts of glowing gas and plasma emerge from one sunspot on the Sun's surface and then follow magnetic field lines back to the other sunspot of the pair (see Fig. 23.7). Solar prominences look like particularly huge arcs of plasma, mostly within the chromosphere, and they can last for hours, days, or weeks **(Fig. 23.9a)**. Occasionally, a **solar flare**, an even brighter eruption of plasma, shoots out of the Sun's surface **(Fig. 23.9b)**. A solar flare releases a huge number of particles as well as intense energy in the form of X-rays and UV rays, all in a matter of minutes, so a flare

FIGURE 23.8 Sunspots appear as dark patches on the Sun's surface.

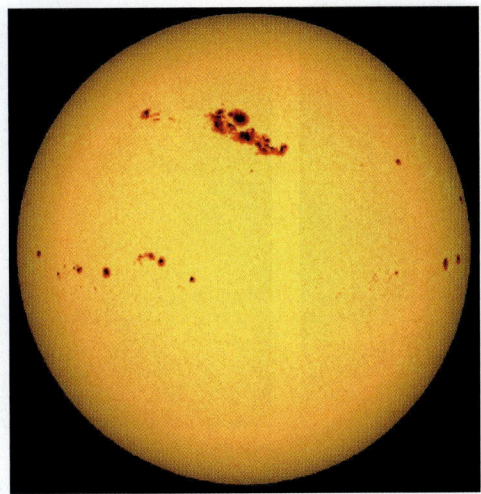
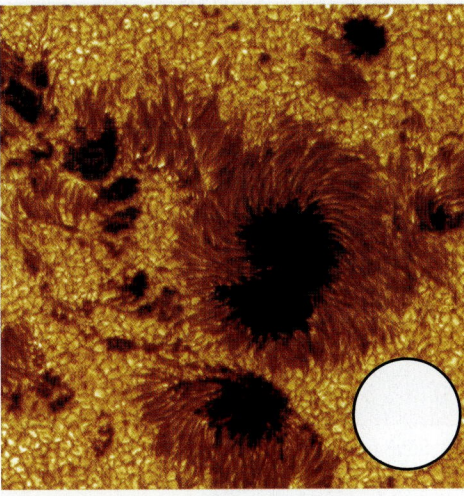

(a) At any given time, sunspots cover 0–0.4% of the area of the Sun.

(b) A close-up shows that a sunspot has a cooler, darker central area and a warmer outer area. The circle represents the Earth, for scale.

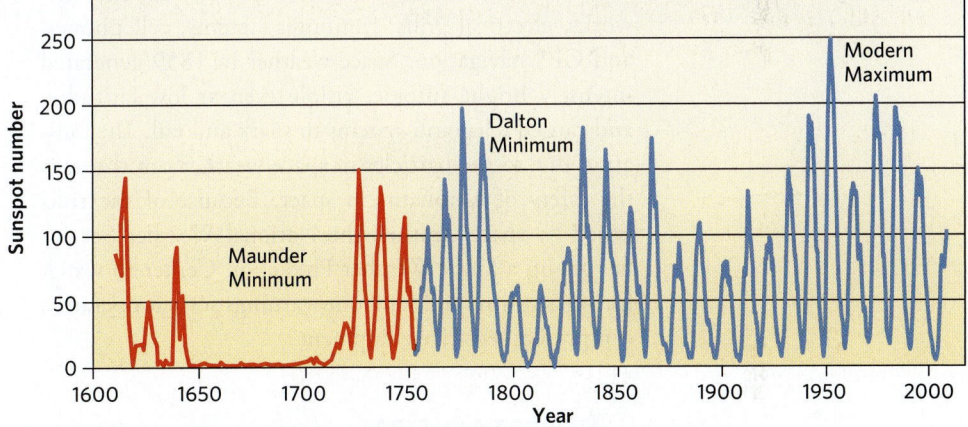

(c) The number of sunspots changes over time, showing that the solar cycle is about 11 years long. Note that the number of sunspots is not the same during each cycle. Astronomers have assigned names to time intervals with an anomalous number of sunspots. Note that from about 1650 to 1710, there were hardly any sunspots.

effectively represents an explosion at the Sun's surface. In fact, one flare emits 15% of the total average energy output of the Sun over the same period. The plasma in a flare can be extremely hot, reaching 200,000°C–1,500,000°C (360,000°F–2,700,000°F). Prominences and flares that occur together produce particularly intense solar winds **(Fig. 23.9c)**.

Space weather refers to the presence of charged particles and electromagnetic radiation from solar wind in the region of space around the Earth. Mild space weather happens when typical solar winds sweep toward the Earth. More intense space weather develops a few days after the eruption of a large solar prominence. Solar flares unleash the most severe space weather. Severe space weather can cause real problems for human society because the influx of ionized particles can disrupt electronics, particularly in satellites. In fact, severe space

FIGURE 23.9 Examples of solar storms.

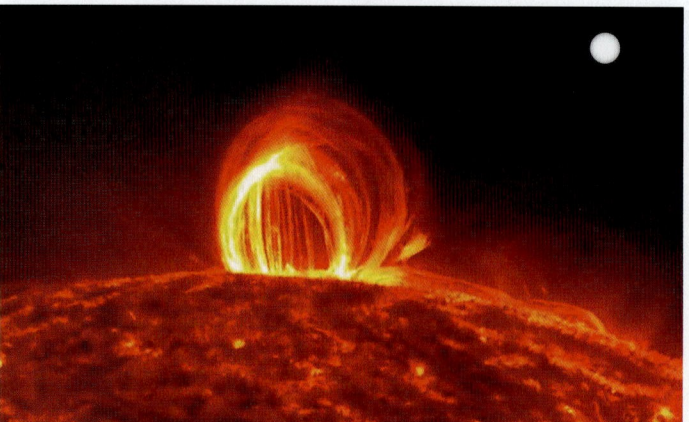

(a) Solar prominences loop into the corona, following curving magnetic field lines. The circle represents the Earth, for scale.

(b) Solar flares are like giant explosions that release energy quickly into space.

(c) Prominences and flares together eject large amounts of plasma into space, yielding a strong pulse of solar wind.

weather can potentially hobble communications networks, electrical grids, computer systems, cell phones, and GPS navigation. Space weather in 1859 generated intensely bright aurorae, visible even at low latitudes, and caused telegraph systems to spark and fail. The radiation due to the particles in space weather can threaten the safety of astronauts in space. Because of the risks posed by space weather, the National Weather Service has set up a Space Weather Prediction Center to watch for solar storms and send out warnings so that operators can protect sensitive equipment.

Take-home message . . .

The Sun generates an intense magnetic field. Local intensification of this field causes magnetic field lines to arc into space, forming sunspots. Solar prominences are arcs of plasma that flow between sunspots. Particularly intense disturbances yield solar flares, which send highly energetic plasma into space.

Quick Question
What is space weather, and how can it affect human society?

23.4 A Diversity of Stars

Did you ever wonder . . .
whether all stars are the same?

From the Earth, all stars look like points of light, but some definitely appear brighter than others. Are the differences in brightness that we see due to variations in the amount of energy that a star emits, in the distance of the star from the Earth, or in the size of the star? Research demonstrates that all three factors play a role. Stars do

indeed lie at different distances from the Earth, but they also vary in diameter, mass, and temperature, and all these variables affect the amount of light that a star emits. However, it is only the temperature of a star's photosphere that determines the wavelengths of light that the star emits, so hotter stars look blue or white, while cooler stars appear yellow or red.

Variations in Stellar Brightness

Astronomers refer to the amount of electromagnetic energy emitted by a star in a specified amount of time as the star's **luminosity**, or true brightness. You can think of luminosity as a measure of power: a more luminous star produces more power than a less luminous star does, just as a more powerful light bulb produces more light than a less powerful light bulb does. Luminosity doesn't depend on the distance between an observer and the star, or on the direction from which an observer views the star.

The brightness that an earthbound observer sees when looking at a star, however, depends not only on the star's luminosity, but also on the star's distance from the Earth. That's because as energy radiates from a source, it spreads out. You see the same phenomenon when you look toward a streetlight, or a light bulb, at night. At a distance, the light looks small and dim, but close up it appears large and bright **(Fig. 23.10a)**. The size and power of the light haven't changed, but your eye detects a smaller amount of the energy from a distant light than from a nearby light.

The amount of energy from a star that passes through a square meter at the surface of the Earth in 1 second is the star's **apparent brightness**. The objects we see in space have a huge range of apparent brightness. When astronomers describe the apparent brightness of stars, they use a scheme that dates back to the ancient Greek astronomer Hipparchus (ca. 190–120 B.C.E.). Hipparchus's original scheme classifies stars visible to the naked eye using a scale divided into **apparent magnitudes**, with 1 being the brightest and 6 being the faintest **(Fig. 23.10b)**. Modern astronomers have extended the scale in both directions to include fainter objects that Hipparchus couldn't see, and much brighter celestial objects, such as the Sun and the Moon. Catalogs of stars used by amateur and professional astronomers today provide the apparent magnitudes of stars and other objects in the sky. Smaller numbers on the apparent magnitude scale imply brighter objects. Therefore, objects such as the Sun, that are brighter than stars with an apparent magnitude of 1, have a negative apparent magnitude. Specifically, the Sun's apparent magnitude is –26.7, the full Moon's is –12.5, and Venus's is –4.4. Stars that you can't see with the naked eye, but which can be detected with telescopes, have apparent magnitudes greater than 6. The Hubble Space Telescope can detect objects with an apparent magnitude of 30.

To distinguish among stars based on luminosity, astronomers use two different scales. The first, the **absolute magnitude** scale, represents a star's apparent magnitude if the star were magically placed at a distance of exactly 32.6 light-years (10 parsecs) from the Earth **(Fig. 23.10c, d)**. The second scale, the one preferred by astronomers, expresses a star's luminosity in solar units. A **solar unit** is equal to the luminosity of the Sun. A star's luminosity, expressed in solar units, is that star's true luminosity divided by the Sun's luminosity. Rigel, the brightest star in the constellation Orion, has a luminosity of 10,000 solar units, meaning that it generates 10,000 times as much energy per second as the Sun. Our nearest neighbor, Proxima Centauri, has a luminosity of less than 0.0001 solar units, so it generates only one ten-thousandth as much energy per second as the Sun.

Variations in Stellar Surface Temperature and Color

Different stars emit visible light of different colors. As we discussed in Chapter 17, the wavelengths of radiation that an object emits depend on the temperature of the object (according to the rules of blackbody radiation), so the color of starlight tells us the temperature of the star's photosphere. The coolest stars are red, warmer ones (like our Sun) are yellow, hot stars are white, and the hottest stars are blue. There are gradations between these color categories, so each category includes stars in a range of temperatures. Astronomers divide stars into **spectral classes** based on their surface temperatures;

FIGURE 23.10 The concept of stellar magnitude.

(a) The same light bulb looks dimmer to you as its distance from you increases.

(b) Sirius looks particularly bright relative to the other stars around it in the night sky, so its apparent magnitude is greater than that of its neighbors.

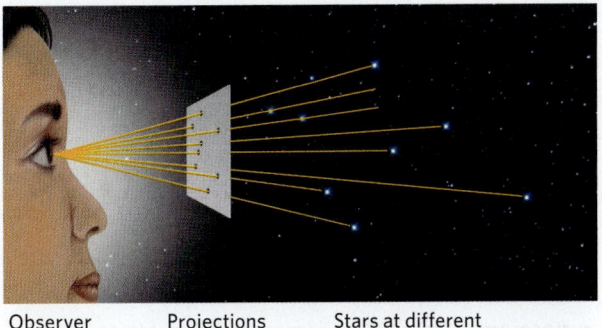

Observer on Earth Projections of stars Stars at different distances

(c) The absolute magnitude is the brightness a star would have if positioned at a specific distance (32.6 light-years) from the Earth.

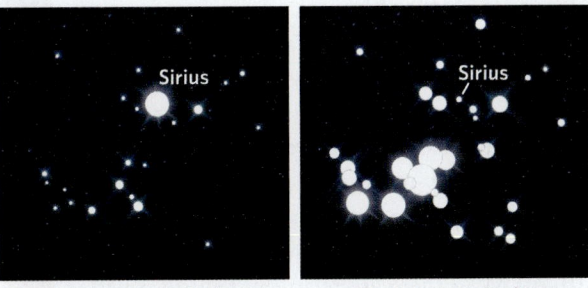

Apparent magnitudes Absolute magnitudes

(d) A comparison of apparent magnitude and absolute magnitude for the stars around Sirius.

TABLE 23.1 Spectral Classification of Stars

Spectral Class	Approximate Surface Temperature	Color	Example
O	30,000°C (54,000°F)	Electric blue	Mintaka (O9)
B	20,000°C (36,000°F)	Blue	Rigel (B8)
A	10,000°C (18,000°F)	White	Vega (A0)
F	7,000°C (12,600°F)	Yellow-white	Canopus (F0)
G	6,000°C (10,800°F)	Yellow	Sun (G2)
K	4,000°C (7,200°F)	Orange	Arcturus (K2)
M	3,000°C (5,400°F)	Red	Betelgeuse (M2)

these categories, named from hottest to coldest, are O, B, A, F, G, K, and M **(Table 23.1)**. To add greater specificity, each letter category can be divided into 10 subcategories, numbered 0 through 9 (with 0 being hotter than 9). According to this classification scheme, our Sun, a G2 star, is cooler than an A0 star and hotter than an M2 star.

Variations in Stellar Dimensions

Stars come in a tremendous range of diameters and masses **(Fig. 23.11)**. When classifying a star, astronomers usually indicate its size as well as its spectral class because color doesn't necessarily correlate with size. Our Sun, for example, is classified as a *yellow dwarf*. Astronomers have found numerous examples of very small stars (with diameters only slightly larger than that of the Earth) that glow white because of their high surface temperature; these stars are known as *white dwarfs*. Astronomers also recognize red dwarfs and brown dwarfs, which are in the same size range but are much cooler. **Giant stars** have diameters up to 50 times that of the Sun, and **supergiant stars** have diameters from 50 to over several hundred times that of the Sun. If the red supergiant Betelgeuse were placed at the position of our Sun, its surface would extend close to the orbit of Jupiter.

Stars also come in a huge range of masses. Astronomers describe the masses of stars by comparing them with our Sun: one **solar mass** (M_S) represents the mass of our Sun. Low-mass stars contain less than 0.5 M_S, intermediate-mass stars contain 0.5–10 M_S, and high-mass stars contain 10–40 M_S. (There are relatively few *supermassive* stars, with masses greater than 40 M_S; the largest of these stars has a mass of 150 M_S.)

As we'll see later in this chapter, stellar size does not necessarily correlate with stellar mass, and stellar mass controls the life history of a star.

The H-R Diagram

In the early 20th century, two astronomers, Ejnar Hertzsprung of Denmark and Henry Norris Russell of the United States, independently examined the relationship between stellar luminosity and surface temperature. The diagram they developed, now known as the *Hertzsprung-Russell diagram* (commonly abbreviated as **H-R diagram**), has become one of the most important tools for understanding stellar life cycles **(Fig. 23.12)**. An H-R diagram is a graph with surface temperature (which corresponds directly to spectral class) on the horizontal axis and luminosity on the vertical axis. Plotting observations of luminosity and temperature for thousands of real stars on the diagram shows a remarkable order to the properties of stars, as indicated by the relationship between luminosity and spectral class. Most stars, including our Sun, lie in a narrow band, called the **main sequence**, which stretches diagonally across the diagram from the upper left to the lower right. A star that falls on or below the main sequence, and has a mass less than about 20 M_S, is classified as a **dwarf star**. This classification includes our Sun, which, in its current stage of life, is a yellow dwarf star.

While the main sequence of an H-R diagram includes a great variety of stars that differ from one another in terms of mass, radius, temperature, and luminosity, not all stars fall on the main sequence.

FIGURE 23.11 The diameters of stars vary significantly, as we can see by comparing several stars that astronomers have named. Betelgeuse has a diameter 1,000 times that of our Sun.

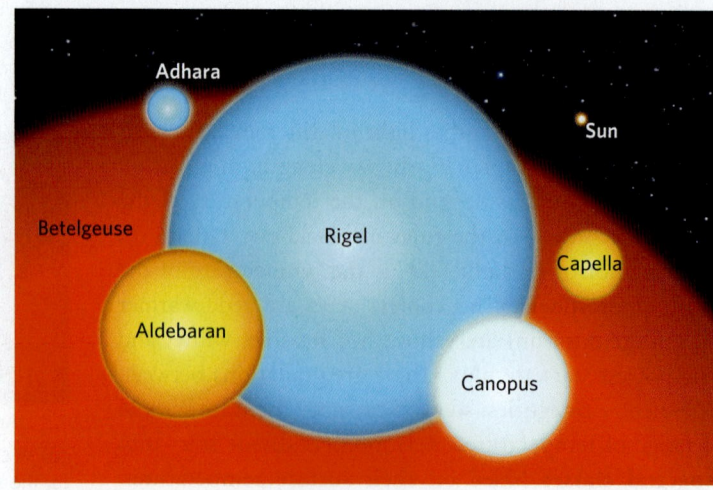

Smaller numbers of stars fall into three other areas. *Red giants* populate an area above and to the right of the main sequence. These stars have radii of 5 to over 100 times that of the Sun, but have surface temperatures of only 3,000°C–5,000°C (5,400°F–9,000°F). Above the red giants on the diagram lie the even rarer *bright giants* and *supergiants*. White dwarfs fall below the main sequence.

FIGURE 23.12 An H-R diagram shows the relationship between luminosity and temperature (represented by spectral class). Most stars, including the Sun, fall into the main sequence.

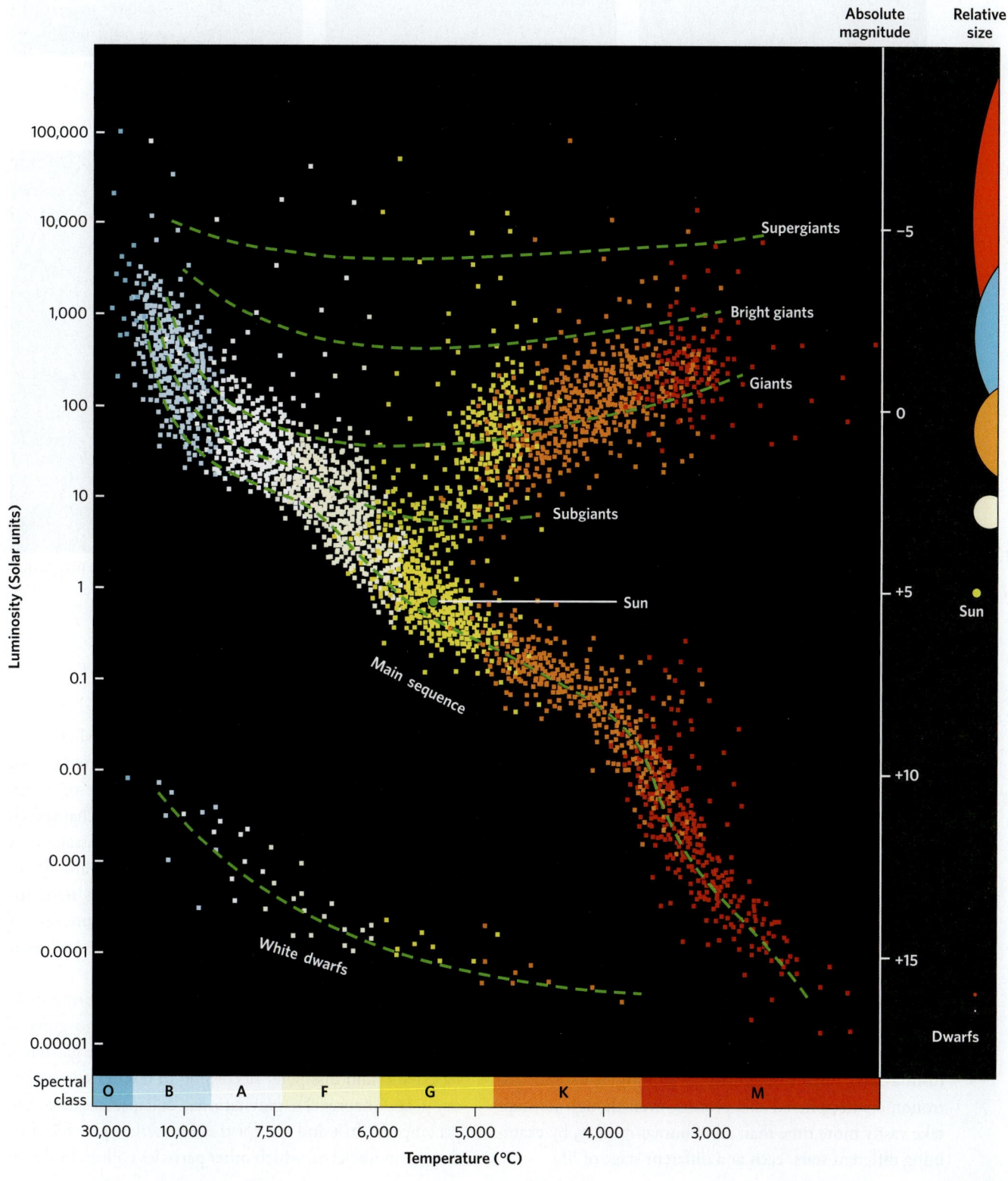

FIGURE 23.13 The evolution of a nebula into a star nursery may begin when shock waves from a supernova pass through the nebula.

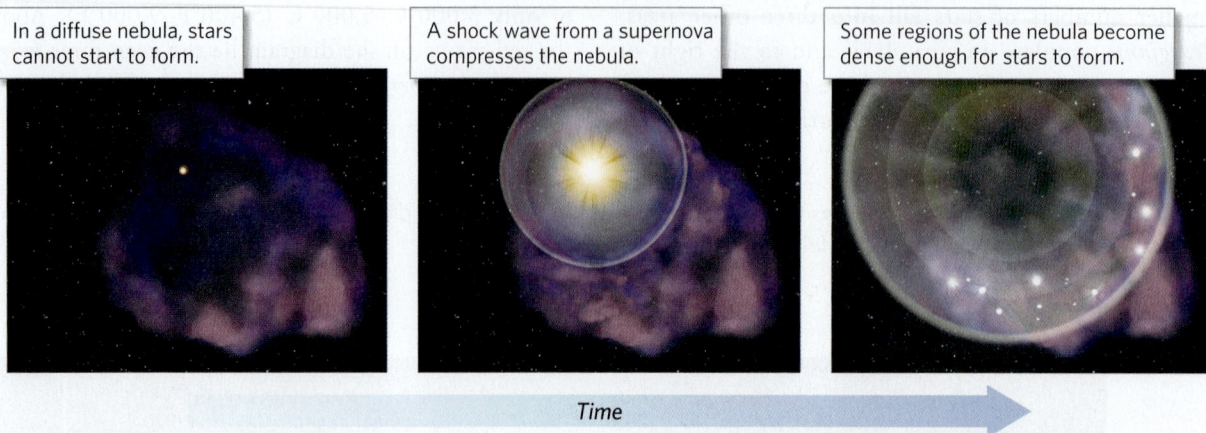

In a diffuse nebula, stars cannot start to form.

A shock wave from a supernova compresses the nebula.

Some regions of the nebula become dense enough for stars to form.

Time

(a) Compression due to a shock wave causes pockets of matter to become dense enough for gravitational collapse to begin.

Take-home message ...

The apparent brightness (apparent magnitude) of a star that we see from the Earth depends not only on the star's luminosity (the amount of energy it emits), but also on its distance from the Earth. Absolute magnitude—the brightness we would see if the star were exactly 32.6 light-years away—provides a better characterization of a star's luminosity. Astronomers usually indicate luminosity using solar units, a comparison with the luminosity of the Sun. Stars vary greatly in their luminosity, temperature, size, and mass. We can represent variations in luminosity and temperature on an H-R diagram. Most stars fall on the diagram's main sequence.

Quick Question

What characteristic defines a star's spectral class?

(b) The Henize 206 Nebula, 163,000 light-years from the Earth, contains stars that are less than 10 million years old.

23.5 The Life of a Star

Our Sun looks the same every day, except for the patterns of sunspots that speckle its surface, and the other stars look the same every night, except for their movement across the sky as the Earth rotates. A single glance at the sky provides only an instantaneous snapshot of the Sun and stars. If you could observe stars over the course of billions of years, and in some cases, just millions of years, you would see significant changes in their character, manifested primarily by changes in their luminosity, radius, and mass. In other words, stars evolve over time. The rate of change, however, is not uniform, for stellar evolution happens rapidly during the birth and infancy of a star, slowly during most of the star's life, and then very rapidly at the end of the star's life. In this section, we consider stellar evolution from a star's birth through the beginning of its end. Astronomers deciphered this process, even though its steps take vastly more time than all of human history, by examining different stars, each at a different stage of life.

The Evolution of a Star from Birth through Middle Age

The process of star formation probably begins when a disturbance causes the *interstellar medium*, the material (gas, dust, and ice) found between stars, to become relatively denser in a localized region of a nebula (see Chapter 1). Such disturbances might be caused by the passage of a **shock wave**—a pulse of pressure—sent outward by an immense supernova explosion, which squeezes mass together **(Fig. 23.13a)**. Typically, many distinct pockets of higher density develop within a given nebula, and each pocket becomes a separate star **(Fig. 23.13b)**.

Even when a shock wave causes local compression in a nebula, the compressed region won't undergo gravitational collapse if it consists only of gas, because gas tends to expand and dissipate. According to condensation theory (see Chapter 1), gravitational collapse requires the presence of dust and ice, for these materials provide condensation nuclei on which other particles collect, building

▶ **Animation**

Stellar Evolution: Birth of Stars

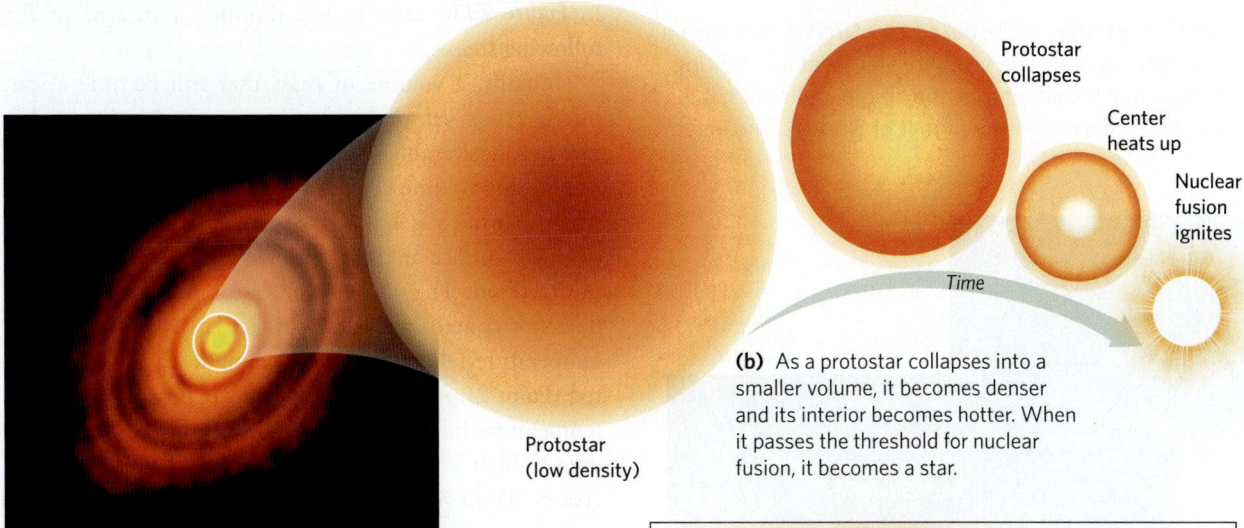

FIGURE 23.14 Formation of a star from a protostar.

Protostar collapses

Center heats up

Nuclear fusion ignites

Time

Protostar (low density)

(b) As a protostar collapses into a smaller volume, it becomes denser and its interior becomes hotter. When it passes the threshold for nuclear fusion, it becomes a star.

(a) A Hubble photo of HL Tauri, an accretion disk about 450 light-years from the Earth. Astronomers estimate that this disk is only 100,000 years old.

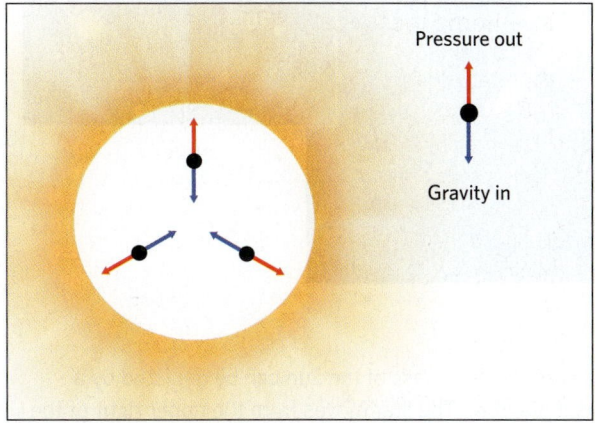

Pressure out

Gravity in

(c) The concept of equilibrium pressure in a star.

into masses that become large enough that their gravity can pull other matter, including gas, inward. As gravity pulls matter inward, the collapsing pocket of the nebula starts spinning. Gravity, together with this spin, flattens the collapsing material into an *accretion disk* **(Fig. 23.14a)**. At the center of this disk, where rotation is slower, a nearly spherical ball—a *protostar*—grows.

As we have seen, compression of a gas increases its temperature. At the outer edges of the protostar, heat generated through compression escapes into space, so the temperature remains relatively low. But in the opaque interior of the protostar, radiation can't escape, so as gravitational collapse progresses and the radius of the protostar shrinks, the temperature in its core progressively rises. A protostar that contains an amount of mass comparable to that of our Sun reaches a milestone about 20 million years after gravitational collapse begins: its core temperature reaches the threshold for nuclear fusion (10 million °C, or 18 million °F). When this happens, the protostar becomes a true star and begins to emit bright light into space **(Fig. 23.14b)**. Over the next 30 million years or so, a Sun-sized star contracts further, its core temperature increases to about 15 million °C (27 million °F), and its photosphere temperature reaches about 6,000°C (10,800°F). At this point, the star reaches an *equilibrium condition*, in that its temperature and radius stabilize and remain much the same for a long time (billions of years for a star such as our Sun). The radius of the star at this time is called the *equilibrium radius*, and the luminosity represented by its surface temperature is called the *equilibrium luminosity*. The equilibrium radius of a star is determined by a balance between two forces:

the inward pull of gravity, and an outward-directed *thermal pressure* caused by the tendency of hot material, the plasma in a star's interior, to try to expand **(Fig. 23.14c)**.

The temperature of a star at equilibrium depends on its mass. A more massive star has a denser core and a hotter core temperature; therefore, fusion happens faster in its core. Mass affects equilibrium luminosity as well: specifically, a star with more mass produces more energy than one with less mass.

Significantly, not all accretion disks produce a single star. In some cases, an object forming from rings of matter in the accretion disk becomes large enough that nuclear fusion begins within it. This situation yields a **binary star**, a pair of stars orbiting a common center of mass **(Fig. 23.15)**. More than half of all stars are binary stars.

Depicting the Evolution of a Star on an H-R Diagram

The position of a given star on an H-R diagram reflects the changes that take place as the star forms and evolves. Here, we summarize these changes for a star the size of our Sun. The numbered points on an H-R diagram

Did you ever wonder . . .

what will happen to the Earth when the Sun's hydrogen fuel is exhausted?

FIGURE 23.15 An example of a binary star. To the naked eye, Sirius A looks like a single star. A Hubble telescope image reveals two separate stars that are orbiting each other. The second, Sirius B, is much smaller than the first, but although it is smaller, it is a white dwarf star with a mass that is half that of Sirius A. The inset shows the orbit.

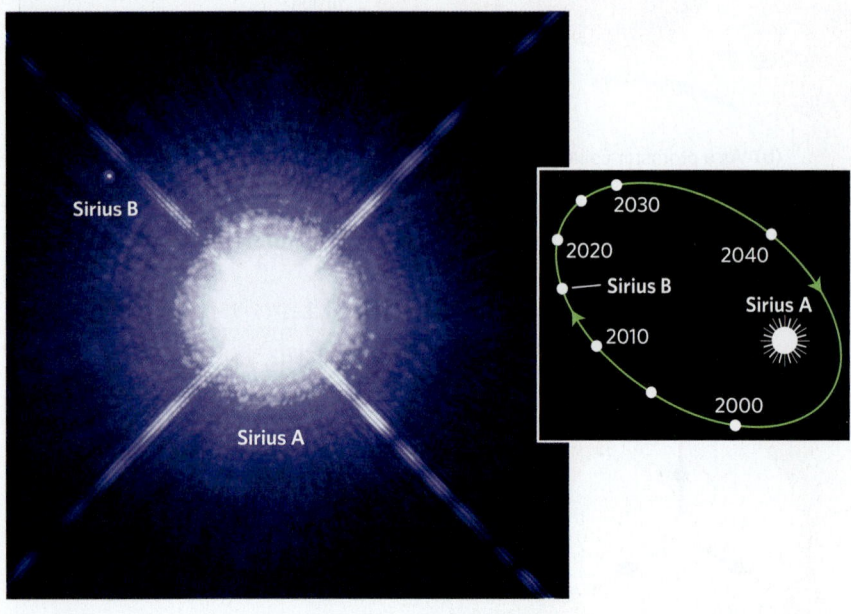

FIGURE 23.16 The life of a star with the mass of the Sun can be depicted by a succession of points on an H-R diagram. The star moves from the upper right of the diagram, down to the main sequence.

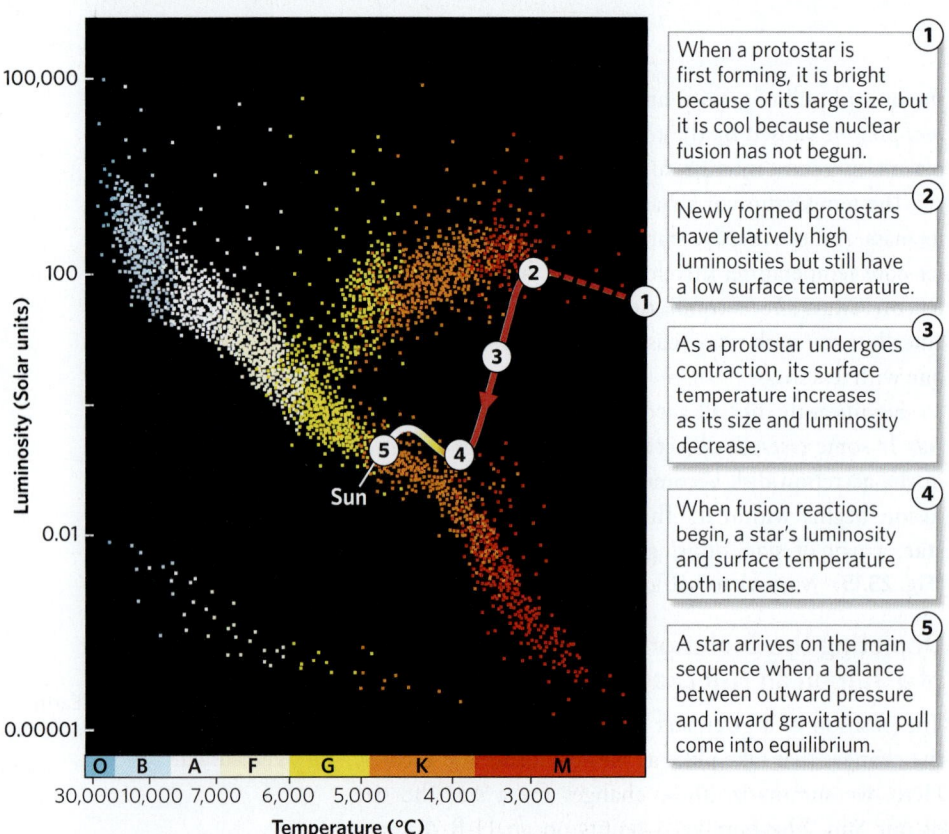

1. When a protostar is first forming, it is bright because of its large size, but it is cool because nuclear fusion has not begun.

2. Newly formed protostars have relatively high luminosities but still have a low surface temperature.

3. As a protostar undergoes contraction, its surface temperature increases as its size and luminosity decrease.

4. When fusion reactions begin, a star's luminosity and surface temperature both increase.

5. A star arrives on the main sequence when a balance between outward pressure and inward gravitational pull come into equilibrium.

in **Figure 23.16** refer to the numbers indicated in the following steps.

Initially, a volume of mass that will become a star, but has not yet undergone gravitational collapse, has low luminosity and a low surface temperature. As time passes, this volume collapses inward and becomes a protostar. A protostar has a large radius, and because it has so much surface area, it can have a relatively high luminosity (Point 1). As it continues to contract, because it does not contain a nuclear furnace, it has a low surface temperature (about 2,700°C, or 4,900°F), so it appears red (Point 2). As a protostar contracts further, its surface temperature increases and its radius decreases. Therefore, even though its temperature is greater, its luminosity decreases (Point 3). Once fusion reactions begin, the star's surface temperature and luminosity both increase, and its position on an H-R diagram begins to move toward the main sequence (Point 4). Eventually, the star achieves an equilibrium size and luminosity. At this time, the star arrives at a position on the main sequence of an H-R diagram (Point 5). Subsequently, its position doesn't change much until it starts to die. In other words, the main sequence represents stars that have achieved equilibrium. Stars stay on the main sequence for most of their lifetime, as long as they remain in equilibrium.

The specific spot where a given star falls on the main sequence of an H-R diagram depends on the mass of the star. Stars with a mass similar to the Sun's follow tracks similar to that shown in Figure 23.16. Those with a mass greater than the Sun's have a greater equilibrium size and luminosity, and they end up at the upper left of the main sequence. Stars with a mass less than the Sun's have a smaller equilibrium size and luminosity and end up at the lower right of the main sequence.

How Long Does a Star Survive?

A star remains in equilibrium until the hydrogen fuel in its core has been used up. How long this equilibrium lasts depends on the star's total mass. The more massive the star, the higher the temperature in the core of the star, and the faster the rate at which hydrogen fuses to form helium. A Type-G star with a mass of 1 M_S can remain on the main sequence for about 10 billion years. In contrast, a Type-B star with a mass of 5 M_S resides on the main sequence for only about 100 million years, and a Type-O star with a mass of 10 M_S exhausts its hydrogen fuel in just 20 million years. Thus, lower-mass stars have longer lives. In fact, stars with masses between 0.08 M_S and 0.25 M_S can survive for hundreds of billions of years. Such low-mass stars are called **red dwarfs**. Given the current estimated age of the Universe (13.8 billion years), no star with a mass this low has yet consumed its hydrogen and left the main sequence.

23.6 The Deaths of Stars and the Unusual Objects Left Behind

In this section, we explore the strange and cataclysmic events that take place when a star finally dies. We'll see that the products of stellar death, which can be quite spectacular, depend on the mass of the star.

Death of an Intermediate-Mass Star and White-Dwarf Formation

Though it seems huge to us, astronomers consider our Sun to be a dwarf star with intermediate mass. As we have just seen, low-mass stars burn so slowly that none have yet exhausted their hydrogen fuel, and they remain red dwarfs. An intermediate-mass star, in contrast, eventually exhausts its hydrogen fuel so that only helium remains in its core. It then begins to undergo a succession of major changes in luminosity and temperature, as indicated by the numbered points on an H-R diagram in **Figure 23.17**.

Just before it begins to die, a Sun-like intermediate-mass star lies on the main sequence of an H-R diagram (Point 5 in Fig. 23.17). Without hydrogen-to-helium fusion occurring in the core, the outward-directed thermal pressure decreases, so the inward gravitational force causes the core to contract. The resulting compression raises temperatures high enough for fusion to take place in a layer of hydrogen surrounding the helium core (Point 6).

Once ignited, fusion reactions in this hydrogen shell continue at a furious rate, even as the helium core continues to contract. The thermal pressure generated by this burning shell causes the outer layers of the star to expand. Therefore, the diameter of the star increases, and its luminosity increases, but the surface temperature of the expanding star actually decreases. Over the next 100 million years, the hydrogen shell burns outward, like a spreading forest fire, and the outer layers of the star continue to expand. Eventually,

the star grows to be 100 times its former size, and its photosphere cools to about 3,000° C. As time passes, the helium core also collapses inward, and its temperature rises. When the temperature reaches 100 million °C (180 million °F), fusion reactions that bind helium nuclei together to produce carbon commence rapidly. As a carbon core forms and increases in size, the outer edges of the burning shells of helium and hydrogen progress outward, and the star expands further—when our Sun passes through this stage, some 5 billion years from now, its photosphere will extend close to the orbit of Mars, and the Earth will have been incinerated. Such a dying star has grown to become a **red giant** (Point 7).

Over time, the star's luminosity decreases as its temperature increases (Point 8). At the end of the red-giant stage, the burning helium and hydrogen shells become unstable and start to alternately expand and contract (Point 9). This process spews matter from the star's outer layers into space to form a nebula. Finally, with its nuclear fuel exhausted, the carbon core cools and thermal pressure decreases, so the star contracts. Over tens of thousands of years, the nuclear fires of this residual carbon core die,

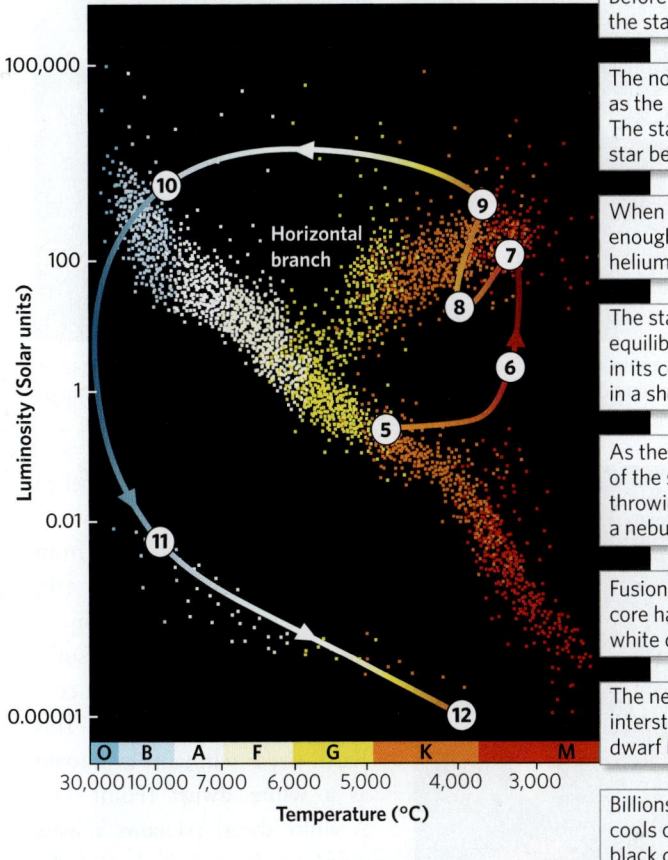

FIGURE 23.17 The death of a star with the mass of the Sun can be depicted by a succession of points on an H-R diagram. The star leaves the main sequence and follows a large loop around the diagram.

5 Before it exhausts its hydrogen fuel, the star resides on the main sequence.

6 The nonburning helium core contracts as the hydrogen shell around it burns. The star's outer layer expands, and the star becomes a red giant.

7 When the core temperature gets hot enough, the entire core starts fusing helium to carbon in a "flash."

8 The star reaches another state of equilibrium. Helium fuses to carbon in its core. Hydrogen fuses to helium in a shell surrounding the core.

9 As the carbon core grows, outer layers of the star expand and contract, throwing material into space to form a nebula.

10 Fusion reactions have ceased, and the core has collapsed into a superdense white dwarf, surrounded by a nebula.

11 The nebula has disappeared into interstellar space, leaving a white dwarf behind.

12 Billions of years later, the white dwarf cools completely and becomes a black dwarf.

 Animation

Stellar Evolution: Intermediate Mass Stars

FIGURE 23.18 Development of a nova from a binary star.

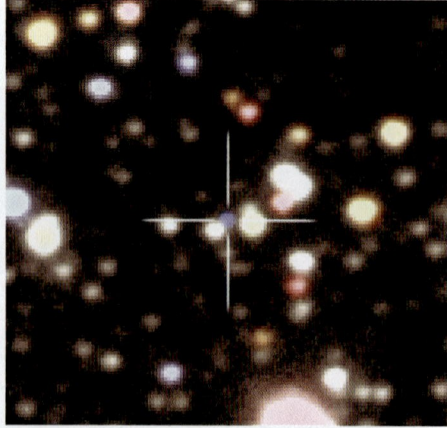

(a) This nova, seen here through a telescope, suddenly appeared in May 2009 and maintained intense brightness for about a week. It continued to fade over the next 7 years. The left image shows a sector of space during the nova, whereas the right image shows the same sector after the nova.

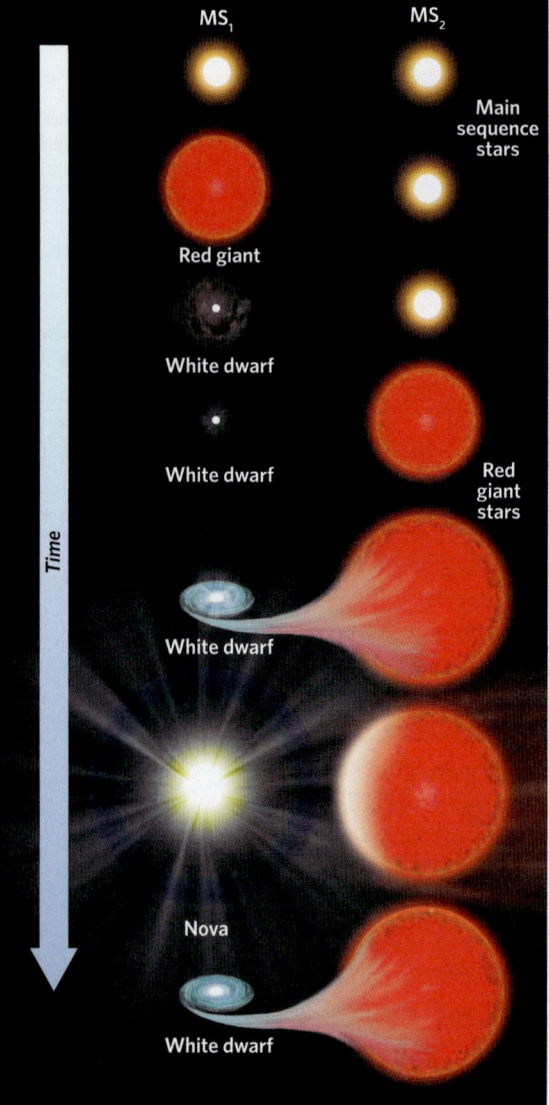

(b) Stages during the development of a nova, which forms when one star in a binary star gravitationally draws in the mass of its partner and eventually explodes.

but the outer nebula of expelled gas prevents the gravitationally collapsed white-hot core from becoming visible (Point 10). Eventually, the gases dissipate into space and become sufficiently transparent that the core of the star becomes visible. This Earth-sized remnant is known as a **white dwarf** (Point 11). A white dwarf contains a mass equal to about half that of the Sun packed into a volume only about 1% of the Sun's, which gives it an immense density: 1 cm^3 (0.06 in^3) of a white dwarf contains about 1,000 kg (1.1 tons) of matter. Ultraviolet radiation emitted by the white dwarf causes the gas of the nebula surrounding it to glow.

As the energy of a white dwarf radiates into space, the star cools and dims, for nuclear reactions are no longer taking place within it. Trillions of years later, the white dwarf will darken until it disappears from the night sky, becoming a cold cinder known as a **black dwarf** (Point 12). No black dwarfs currently exist because the Universe, with an age of 13.8 billion years, simply isn't old enough for any to have evolved.

Death of a Binary Star and Nova Formation

The term *nova*, from the Latin word for new, refers to a bright star that suddenly appears where there had been no bright star before. Specifically, a **nova** is a star that was once too dim to see, but for a period ranging from 20–80 days, it may attain a luminosity of 50,000–100,000 solar units **(Fig. 23.18a)**.

Modern research indicates that novae develop when mass passes from one star to another in a dying binary star in which one of the individual stars has become a white dwarf. The exchange of mass typically begins when the remaining star enters its red-giant stage and swells in size **(Fig. 23.18b)**. When the surface of the red giant gets close enough to the white dwarf, the white dwarf's gravity starts pulling matter away from the red giant. The material from the red giant (most of which is hydrogen) accretes onto the surface of the white dwarf, where it keeps building up and becoming denser. The high temperature of the white dwarf causes the dense, accreted material to suddenly ignite so that fusion reactions take place throughout it all at once. The sudden onset of fusion reactions in the captured hydrogen results in a violent hydrogen-bomb-like explosion on a stellar scale, causing the star's luminosity to increase quickly and dramatically, and much of the remaining accreted mass gets blasted back into space. It's this explosion that we see from the Earth as a nova.

The nuclear fuel that brightens a nova burns up quickly. Once the hydrogen that was captured by the white dwarf has been converted to helium, both the dwarf and its red-giant companion return to their original states. Later on, the white dwarf again draws in mass from its companion red giant until another nova occurs. White dwarfs can undergo several novae during the lifetime of a binary star.

Death of a High-Mass Star and Supernova Formation

The evolution of a high-mass star differs dramatically from that of an intermediate-mass star because so much gravitational compression takes place in the high-mass star that its core becomes much hotter than that of a lower-mass star. As a high-mass star's core temperature rises, the rate at which hydrogen fusion takes place in its core increases. In fact, as we've seen, fusion takes place so rapidly in a large star that the star can exhaust its hydrogen in just a few million years.

The early stages in the death of a high-mass star resemble those for an intermediate-mass star: when the core's hydrogen fuel has been used up, gravity compresses the helium that remains until helium fusion ignites and produces carbon. Meanwhile, the outer layers of the star expand. Because the star contains so much mass, however, a *red supergiant*, instead of a red giant, forms. Because of its immense mass, the late stages of a supergiant's evolution differ from those of a giant. Gravity causes much more inward compression within the core of a supergiant than within the core of a giant. This intense compression, in turn, drives the temperature of the supergiant's core so high that carbon nuclei can collide with enough energy to fuse. Such reactions don't take place in the somewhat cooler core of a giant because carbon nuclei don't collide at a high enough speed. Carbon fusion reactions produce nuclei of even heavier elements, such as neon, silicon, magnesium, and oxygen. When carbon fuel runs out, gravity again overcomes outward pressure, and the core of the red supergiant collapses even more and becomes even hotter. As a result, the products of carbon fusion themselves fuse to produce still heavier elements. In fact, a succession of fusion reactions, each producing progressively heavier nuclei, continues until iron nuclei have been produced. At each stage, fusion reactions involving successively lighter nuclei take place in shells around the star's core. So, for example, a shell of oxygen fusion surrounds a shell of neon fusion, and a shell of carbon fusion surrounds a shell of oxygen fusion. In fact, while very heavy nuclei are fusing in the core of a red supergiant, hydrogen fusion (to form helium) is still taking place in the outer shells of the star **(Fig. 23.19)**.

Once the core of a high-mass star contains only iron, its fusion reactions end. Without the outward thermal pressure produced by the energy of fusion reactions, nothing can counter the relentless inward pull of gravity, and the star suddenly collapses inward. During this *implosion*, matter in the core undergoes such intense compression that its temperature rises to 10 billion °C (18 billion °F). At these incredible temperatures, it takes only about 1 second for all the core's atomic nuclei to split apart into free protons and neutrons. Further compression of the core causes protons to combine with electrons to form neutrons until the core consists entirely of neutrons packed tightly together in an inconceivably dense mass.

But gravity hasn't finished with the star yet. The neutrons rush inward into an ever-smaller space until they compact into a point with a density approaching 1 trillion kg/cm³ (67 million tons/in³). This mass suddenly becomes catastrophically unstable and rebounds outward, generating a violent shock wave that destroys the star and blows all the mass lying outside the core into space. This explosion generates a brilliant but short-lived light in the night sky called a **supernova (Fig. 23.20a, b)**. Recorded supernovae have had apparent magnitudes of up to −7.5. Some have been so bright that they can be observed in the daytime sky.

The matter blasted into space by a supernova forms an expanding nebula **(Fig. 23.20c)**. For example, the spectacular Crab Nebula, so named because the radiating streaks of glowing gas within it look like the legs of a crab **(Fig. 23.20d)**, came from a supernova observed in 1054 C.E. During supernova explosions, elementary particles such as protons and neutrons move so fast that when they collide, the resulting fusion reactions can produce all the natural elements, including heavy elements such as uranium. These newly forming elements blast into the expanding nebula and can then be incorporated into the next generation of stars—stars like our Sun. Indeed, the elements that compose your body today originated in supernovae that occurred early in our Universe's evolution. Astronomers estimate the recurrence interval for supernovae in the Milky Way at about 100 years. Hundreds of supernovae have been observed in other galaxies. Notably, some of the supernovae in other galaxies have actually outshone their parent galaxies.

Corpses of High-Mass Stars: Neutron Stars and Black Holes

What's left after a supernova blast? We've seen that most of the star's matter spreads out into space as a nebula. The

Animation
Stellar Evolution: Massive Stars

FIGURE 23.19 Production of elements in shells within the core of a high-mass red-supergiant star just before a supernova.

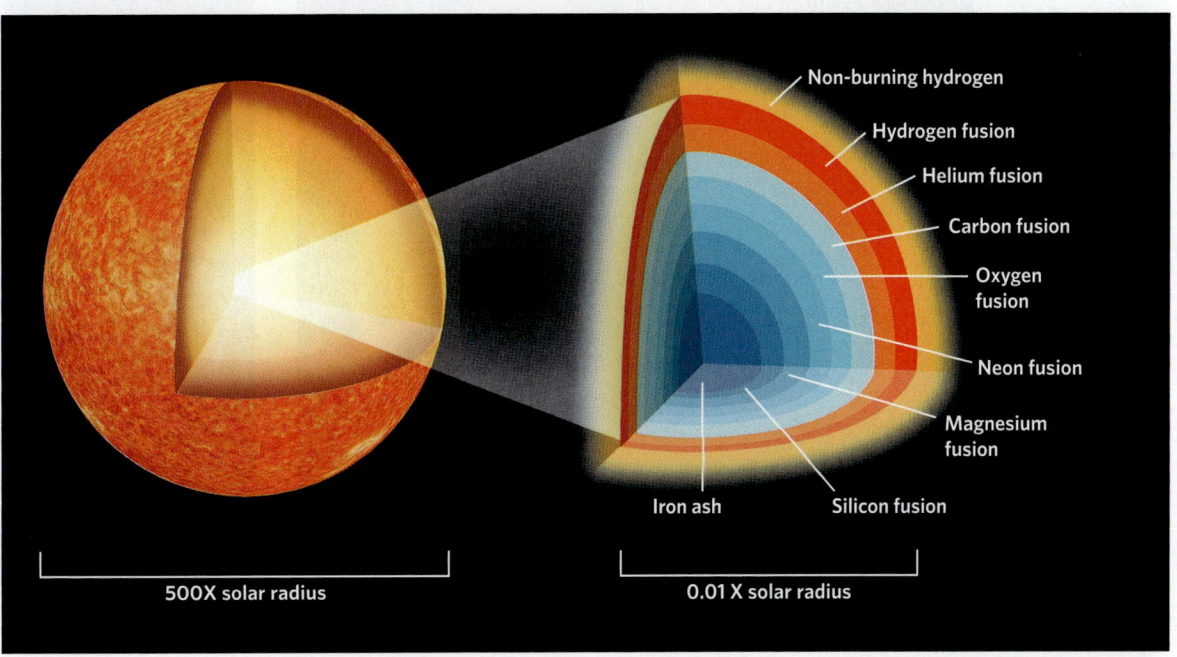

Non-burning hydrogen

Hydrogen fusion

Helium fusion

Carbon fusion

Oxygen fusion

Neon fusion

Magnesium fusion

Iron ash

Silicon fusion

500X solar radius

0.01 X solar radius

**Did you
ever wonder . . .**

what's black about a black
hole?

▶ **Animation**

Stellar Evolution:
Supermassive Stars

innermost core of the star, however, composed of highly compressed neutrons, survives. This remnant object, known as a **neutron star**, is tiny by stellar standards—only about 10–20 km (6–12 mi) in diameter—but its mass can be 1.1–3 M_S. A cubic centimeter of neutron star weighs about 100 billion kilograms (100 million tons)!

Astronomers can detect neutron stars, despite their small size, because neutron stars generate very strong magnetic fields and rotate rapidly. As a result, they emit radio waves at regular intervals. The pulses of energy from rotating neutron stars arrive on the Earth with clockwork accuracy, so researchers also refer to these objects as **pulsars**. There are well over 1,000 pulsars known in the Milky Way, including one at the center of the Crab Nebula. A pulsar's rotation rate slows and its magnetic field weakens over millions of years, so a pulsar eventually stops sending out its beacon-like radiation.

What happens when a very high-mass star, one with a mass greater than about 25 M_S, enters its death throes? After such a star undergoes a supernova explosion, the remnant neutron star can contain the mass of four Suns or more, crushed into a sphere with a diameter of 15–20 km (9–12 mi), the width of New York City. The overwhelming inward pull of gravity in this mass catastrophically crushes the neutrons into an even smaller volume. Eventually, the gravitational force of this mass becomes so great that within a region whose boundary is called the **event horizon**, nothing can escape, not even electromagnetic

FIGURE 23.20 A supernova produces an expanding nebula.

(a) The M-82 Galaxy (11 million light-years from the Earth) before and after a supernova. The supernova was visible in 2014.

(b) An artist's rendering of an immense supernova explosion.

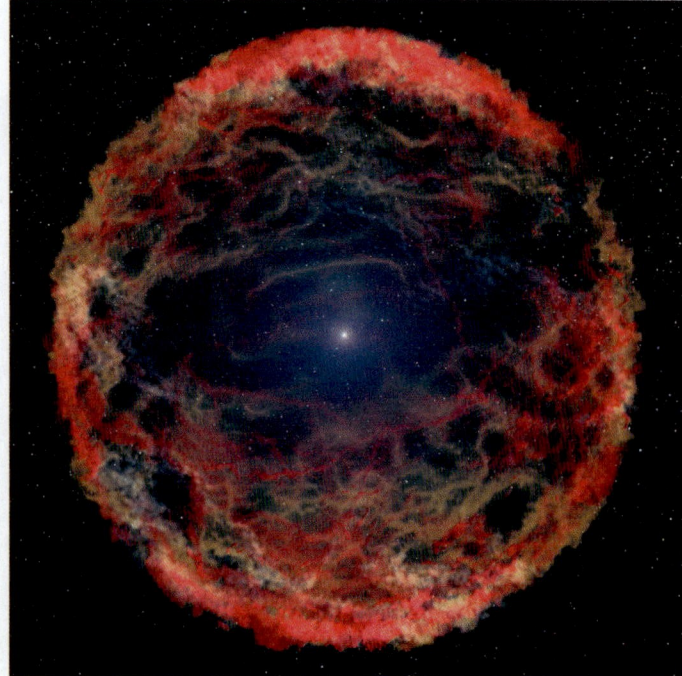 

(c) Gas and dust from the explosion expand outward to form a nebula. A tiny, but supermassive, bit of the star remains.

(d) The Crab Nebula, shown here, was formed by a supernova explosion in 1054 C.E.

radiation or high-speed atomic particles. Astronomers refer to the region within the event horizon as a **black hole** **(Fig. 23.21)**. Because no light comes out of a black hole, it would look completely dark to an observer.

Black holes are not all the same size. *Stellar black holes*, formed from individual stars, typically contain 10–20 M_S. Currently, the smallest known black hole contains 3.8 M_S. The largest black holes, which we examine in Section 23.7, are called *supermassive black holes* and are a product of galaxy formation. They can contain the mass of millions of stars within the volume of a single star or less.

Since black holes have mass, they are real physical objects. The formation of a black hole destroys matter as we know it. Matter returns to its fundamental, elementary particle form and then to an unknown state that scientists currently have no mathematics to describe. A black hole can also be considered a hole in space because all communication to or from the stuff within it is lost. Because black holes emit no light or particles, the only way they can be detected is by identifying their gravitational influence on other bodies in their vicinity. A lone black hole can be completely undetectable, but if a black hole has a binary companion, the behavior of the companion can be used to determine the black hole's mass and location. Science-fiction stories sometimes depict black holes as deadly galactic scavengers, lurking in space and gobbling hapless planets that pass near their location. In reality, black holes act like all other objects in space. Beyond the event horizon, objects orbit black holes just as they do other celestial bodies.

Clearly, the way in which a star's demise takes place depends on the star's mass. Smaller stars end up as white dwarfs, larger ones as neutron stars, and the largest as black holes **(Fig. 23.22)**.

Take-home message . . .

🏠 No low-mass stars (red dwarfs) have yet exhausted their nuclear fuel. Intermediate-mass stars die by expanding to become red giants and, eventually, collapsing to become dense white dwarfs surrounded by a nebula consisting of expelled gas. A high-mass star passes through a red-supergiant stage before eventually dying in a violent explosion called a supernova, which creates a nebula relatively rich in heavy elements. After the supernova, an extremely dense object, known as a neutron star or pulsar, remains. If this object is sufficiently massive, it collapses into a tiny object of incredible density called a black hole, from which not even light can escape.

Quick Question ——————————————
What's the difference between a nova and a supernova?

FIGURE 23.21 An artist's rendition of a black hole. No radiation can escape past the event horizon. The inset shows the first photo of a real black hole, captured by the Event Horizon Telescope in 2019.

Event horizon

23.7 Galaxies

Our Sun is one of hundreds of billions of stars in the Milky Way Galaxy. Beyond the Milky Way lie over a trillion other galaxies in the vastness of space **(Box 23.1)**. Each **galaxy** contains hundreds of billions of stars, as well as all of their planets and moons, and collections of interstellar gas, dust, and ice. Gravity binds all of this matter together, so it travels through the Universe together. In this section, we look at the character of galaxies and the variation among them.

The Milky Way Galaxy: A Closer Look

The Milky Way is our home galaxy, so when we look at it from the Earth, we can't see its overall form. But, by

FIGURE 23.22 Life cycles of stars with different masses.

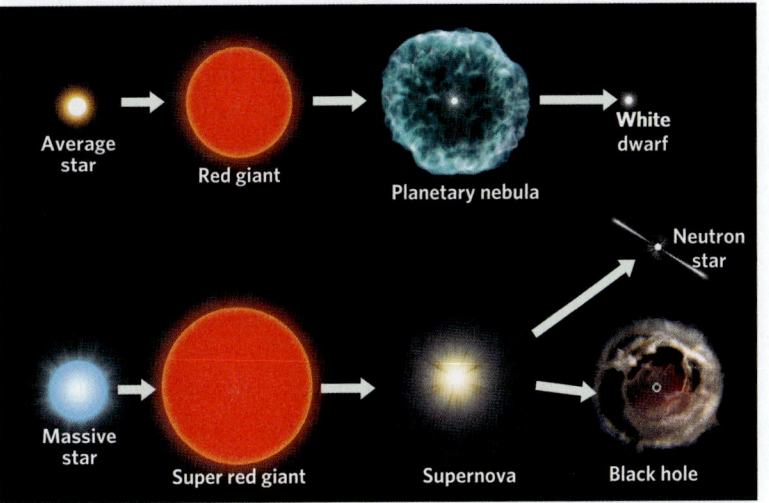

Average star → Red giant → Planetary nebula → White dwarf

Massive star → Super red giant → Supernova → Black hole / Neutron star

BOX 23.1

How can I explain . . .

The vastness of space

What are we learning?
How to think about the vast distances of interstellar space.

What you need:
- A small ball to represent the Sun
- A tiny ball to represent the Earth
- A small ball to represent the Sun's nearest neighbor, Proxima Centauri
- A map of your local area that extends at least 25 km (15 mi) beyond your location

Instructions:
- Place the ball representing the Sun on the floor.

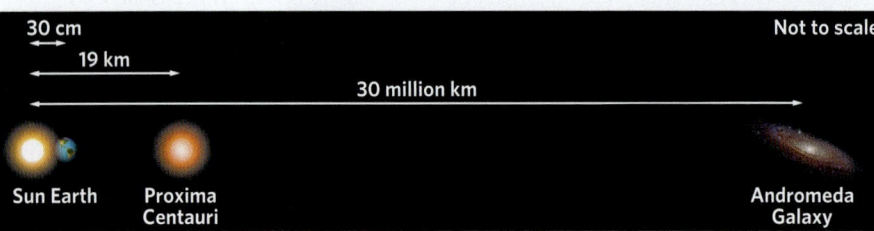

- Now carefully place the ball representing the Earth 30 cm (1 ft) away from the first ball. Note that that this 30-cm distance represents 148 million km (92 million mi).
- So where would the nearest star be located? That star is Proxima Centauri, and it is 4.3 light-years, or 40 trillion km (25 trillion mi), away from the Earth. Scaled to the two balls, Proxima Centauri would be 19 km (12 mi) distant. Find a familiar spot on the map that is 19 km (12 mi) away from where you are. That is where the next ball should be placed!
- Now, using the same scale (30 cm = 148 million km), determine where the nearest galaxy, Andromeda, would be located. (Andromeda is 2.5 million light-years from the Sun.)

What did we see?
- The goal of this exercise is to get some sense of the vastness of space. Picturing the distance you would have to travel from your scale-model Earth to your scale-model Proxima Centauri provides a dramatic example of the vastness of our Universe.
- Amazingly, if 30 cm represents 148 million km, the nearest large galaxy, Andromeda, would still be about 48 million km (30 million mi) away from the ball representing the Sun!

---

analyzing images and by comparing it with other galaxies, astronomers have determined that the Milky Way is a *barred spiral galaxy* **(Fig. 23.23a)**. The term *barred* refers to a concentration of stars that form a straight bar-like shape along the center of the galaxy, and the term *spiral* refers to the existence of concentrations of stars in bands, or arms, that curve into the center of the galaxy. The spiral arms of the Milky Way lie within the **galactic disk**, which flattens toward its rim **(Fig. 23.23b)**. Overall, the Milky Way has a diameter of 100,000 light-years and a thickness, on average, of 1,000 light-years, and it contains between 200 and 400 billion stars. The Sun resides along the outer edge of one of the spiral arms, about 27,000 light-years from the galactic center. The Milky Way slowly rotates, so our Sun makes one orbit around the galactic center about once every 230 million years, a period known as a *galactic year*.

At its center, the galactic disk of the Milky Way thickens into a **galactic bulge** containing a relatively dense concentration of stars. Starlight in the galactic bulge is so bright that night on a planet orbiting a star within it would be as bright as day. The galactic bulge contains an abundance of interstellar gas and dust and serves as a nursery for new stars. It's also a violent place, where stars collide and supernovae explode. At the center of the bulge, at the galactic center, lies a **supermassive black hole**, an object with the mass of 4 million Suns. Its event horizon has been estimated to have a diameter of 60 million km. The size of the object within the event horizon is unknown. Needless to say, the density of this supermassive black hole is beyond comprehension. Its gravity is so strong that it binds the Milky Way galaxy together.

A **galactic halo** surrounds the Milky Way. This halo contains hardly any dust or gas, but includes over 150 **globular clusters**, spherical accumulations of hundreds of thousands of stars that may be as close as 1 light-year apart **(Fig. 23.24)**. Most stars in globular clusters are small reddish stars that have probably been burning since the formation of the galaxy. New stars do not form in these clusters, and all the larger stars that once existed there have long since died.

FIGURE 23.23 The Milky Way Galaxy.

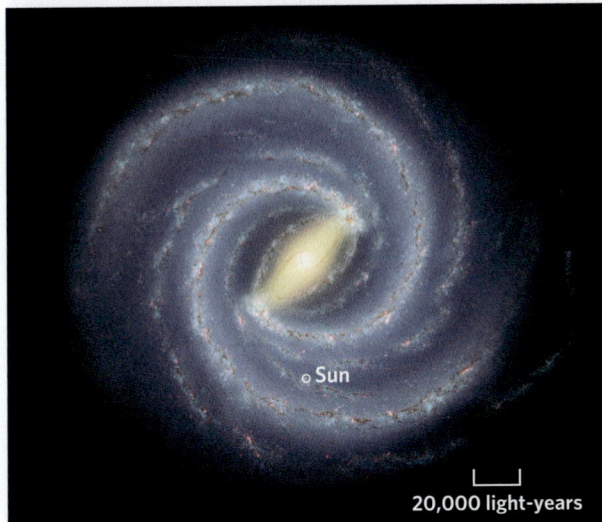

(a) Computer graphic of the Milky Way, as viewed from the top, with components labeled. Note the position of the Sun.

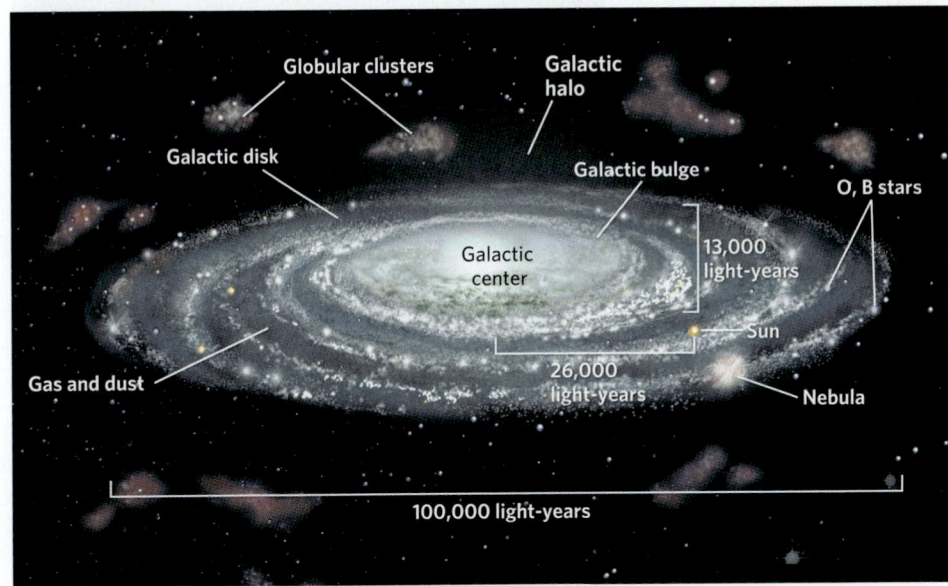

(b) Artist's rendering of the Milky Way as viewed from the side. The galactic bulge lies at the center.

Galaxies and Quasars

With the spectacular images of galaxies that we now have, it's hard to imagine that less than a century ago, astronomers didn't even realize that galaxies outside the Milky Way existed. They had seen objects they called "spiral nebulae," but thought that these objects lay within the Milky Way. It was only in the 1920s, when astronomers determined that spiral nebulae were actually spiral galaxies outside the Milky Way, that they began to catalog distant galaxies. Edwin Hubble classified galaxies into four basic types based on their shapes: *spiral, barred spiral, elliptical,* and *irregular* **(Fig. 23.25)**.

In addition to distinguishing among galaxies by shape, astronomers now categorize them by the nature of the energy they emit. About 60% of galaxies, our own Milky Way included, are **normal galaxies** in that they emit their greatest amounts of radiation at or near visible wavelengths. Normal galaxies range in size from *dwarf galaxies*, with luminosities starting at about 1 million solar units, to *giant galaxies*, with luminosities of up to 1 trillion solar units. **Active galaxies**, unlike normal galaxies, emit most of their energy in the X-ray and radio regions of the spectrum, so their energy cannot simply be the sum of all the light produced by stars in the galaxy. Furthermore, most of the energy of an active galaxy comes from the galactic center.

Astronomers refer to a subset of active galaxies as **quasars** (an abbreviation of *quasi-stellar radio source*). Quasars differ from other active galaxies in that they display an extreme red shift, meaning that they are the most distant objects detectable within the Universe **(Fig. 23.26)**. In fact, the energy we receive from quasars today started its journey at a very early stage in the Universe's history,

long before our Sun formed (see Chapter 21). Quasars emit enormous amounts of radiation, each having a luminosity approaching that of 1,000 Milky Ways.

As we've seen, a supermassive black hole occupies the center of the Milky Way. Recent research indicates that similar supermassive black holes occur in most, and perhaps all, galaxies. The discovery of supermassive black

FIGURE 23.24 This globular cluster, known as M-13, is 25,000 light-years from the Earth and lies within the Milky Way. It has a diameter of about 145 light-years and contains about 300,000 stars.

(a) Spiral galaxy NGC 1365, known as the Pinwheel Galaxy.

(b) Barred spiral galaxy M-86.

(c) Irregular galaxy NGC 1569.

(d) Elliptical galaxy ESO 325-G004.

holes provided a basis for explaining the difference between normal and active galaxies. In a normal galaxy, not much matter surrounds the central supermassive black hole, whereas in an active galaxy, the space around the central black hole contains abundant matter. Therefore, in a normal galaxy, the region around the black hole is fairly quiet, whereas in an active galaxy, it's a violent place where extreme tidal forces rip apart stars. The resulting stellar matter collects into an accretion disk that initially surrounds the event horizon of the black hole and eventually spirals into it **(Fig. 23.27)**. Friction among particles in this material, as well as compression of the material, generates the immense amount of energy that active galaxies

emit. But active galaxies don't stay active forever. Once all the stellar material in the vicinity of the supermassive black hole has been drawn into it, energy production decreases, and the active galaxy becomes a normal galaxy.

What conditions lead to the formation of an active galaxy? In the case of quasars, because they are so old, we are probably seeing the earliest galaxies in their formation stage, a time during which their central supermassive black holes were still acquiring their mass. In the case of nearby active galaxies, we are probably seeing the consequence of a collision between two galaxies **(Fig. 23.28)**. As galaxies pass through each other's space, gravitational forces can deflect stars dangerously close to the central supermassive black hole, where gravity traps them. Once near the black hole, the material of these stars swirls into an accretion disk and then eventually into the black hole itself.

Take-home message . . .

A galaxy is an assemblage of stars, their planets and moons, and interstellar gas, dust, and ice, all bound together by gravity. Evidence suggests that most, if not all, galaxies have a supermassive black hole at the center, whose effects determine whether a galaxy is normal or active.

Quick Question ———————————
What is a quasar?

FIGURE 23.26 A quasar 10 billion light-years from the Earth, as seen in an X-ray photograph by the Chandra X-ray Observatory.

23.8 The Structure and Evolution of the Universe

When astronomers mapped the distribution of galaxies, they discovered that galaxies are not randomly scattered across the Universe and that they do not move independently from one another, but rather, they feel the gravitational pull of their neighbors. For example, the Milky Way has links to 14 smaller *satellite galaxies*, including the Large and Small Magellanic Clouds. Our galaxy and its satellites, together with a few other large galaxies, including the Andromeda Galaxy **(Fig. 23.29)**, and their satellites, constitute the **Local Group**, which is about 10 million light-years across **(Fig. 23.30a)**. All together, the Local Group includes 54 galaxies. The Andromeda Galaxy, our nearest large neighbor in the Local Group, lies about 2.5 million light-years away.

Within a group, galaxies may be moving toward or away from one another. Andromeda and the Milky Way, for example, are moving toward each other at about 400,000 km/h. At this rate, the two galaxies will collide in about 4 billion years. Stars in a galaxy are so far apart, however, that during a collision, individual stars generally don't bash into one another. Rather, gravitational forces may merge the two galaxies into a single elliptical galaxy.

Groups, in general, contain fewer than 100 galaxies. Larger associations of galaxies, which may contain hundreds to thousands of galaxies, are known as *clusters*, and even larger associations, containing on the order of

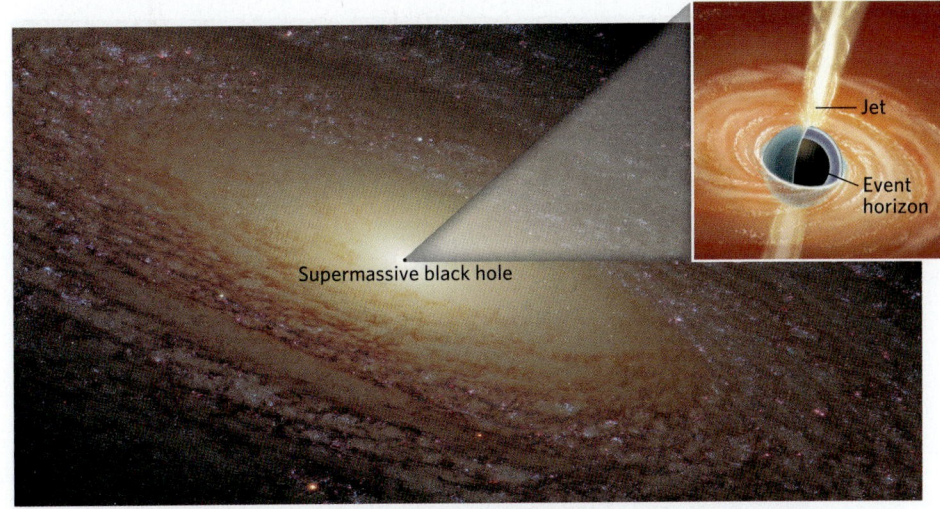

FIGURE 23.27 An artist's depiction of the accretion disk around a supermassive black hole in an active galaxy.

Jet

Event horizon

Supermassive black hole

100,000 galaxies, are called *superclusters* **(Fig. 23.30b)**. The Milky Way is part of the Virgo Supercluster, which in turn is part of the even larger Laniakea Supercluster, a region of galaxies that is over 500 million light-years across and contains about 100,000 times the mass of the Milky Way (**Earth Science at a Glance**, pp. 754–755).

At a still larger scale, the configuration of galaxies defines a three-dimensional mesh composed of string-like *filaments* and screen-like *walls*, whose long dimensions are approximately 30 to 300 million light-years **(Fig. 23.30c)**. Walls and filaments surround *voids*, bubble-like regions of space on the order of 300 million light-years across,

FIGURE 23.28 Examples of colliding galaxies.

(a) These colliding galaxies (NGC 4676B and NGC 4676A, 300 light-years from Earth) are called The Mice because of their long tails of stars and gas.

(b) Two spiral galaxies (NGC 2207 and IC 2163, 80 million light-years away) interacting.

FIGURE 23.29 Andromeda, the nearest large galaxy to the Milky Way, lies 2.5 million light-years from the Earth.

within these regions. The composition and character of this undetectable dark matter remains unknown. If it exists, then each galaxy, and the Universe as a whole, contains vastly more mass than telescopes can detect.

As we noted in Chapter 21, astronomers have observed that all distant galaxies display a red shift and are therefore moving away from the Earth at a high rate of speed. If all galaxies are moving away from one another, then earlier in the history of the Universe, galaxies must have been closer together. Mathematical simulations led astronomers to conclude that about 13.8 billion years ago, all the matter and energy in the Universe was located in a single point in space with an indescribable temperature and density—a **singularity**. For reasons that we have yet to understand, the Universe came into being when this singularity exploded and began to expand. As we discussed in Chapter 1, this event, the beginning of the Universe, is called the **Big Bang**. Since that time, space itself has enlarged as the Universe has expanded, and galaxies outside of groups and clusters have moved farther apart.

Will the Universe continue to expand forever, or will it, at some point in the future, reach its maximum extent and start to contract, eventually collapsing back into a singularity? This question lies at the heart of cosmologic

in which galaxies do not occur. Astronomers remain puzzled as to how gravity can extend across such large areas to hold together filaments and walls. Some speculate that **dark matter**, invisible material that has mass, exists

BOX 23.2 ▶ Putting Earth Science to Use

Exploring the Universe

For nearly all of history, the mysteries of space beyond the Earth lay far beyond our reach, except within the imagination. Human exploration of the Universe was limited to going outdoors on dark evenings and staring at the heavens. Today, in the age of technology and information, the Universe at last sits at our fingertips. With a few clicks on your smartphone or taps on your computer keyboard, crystal clear images of the Pillars of Creation, the amazing Hubble far-field image of thousands of galaxies, or the Sombrero Galaxy, appear right before your eyes **(Fig. Bx23.2)**. You can watch the clouds of Jupiter spin about the Great Red Spot, marvel at Saturn's rings, or be amazed at the strange heart-shaped landscape of Pluto. Planetariums around the world project these glorious images on their domes. Large-screen theaters can put you on the launch pad of the Saturn-V rocket and take you to the Moon with the astronauts. Hobbyists now purchase amateur telescopes with sufficient power and steering capability to create clear images of galaxies over one hundred light-years from the Earth. People

FIGURE Bx23.2 The Sombrero Galaxy.

today have an open door to our vast, incredible universe—walk through it, explore, and learn about the strange and fascinating objects, such as black holes, quasars, and pulsars that lurk out there in a real cosmos beyond the realm of science fiction!

FIGURE 23.30 Galaxies are organized in space at different scales.

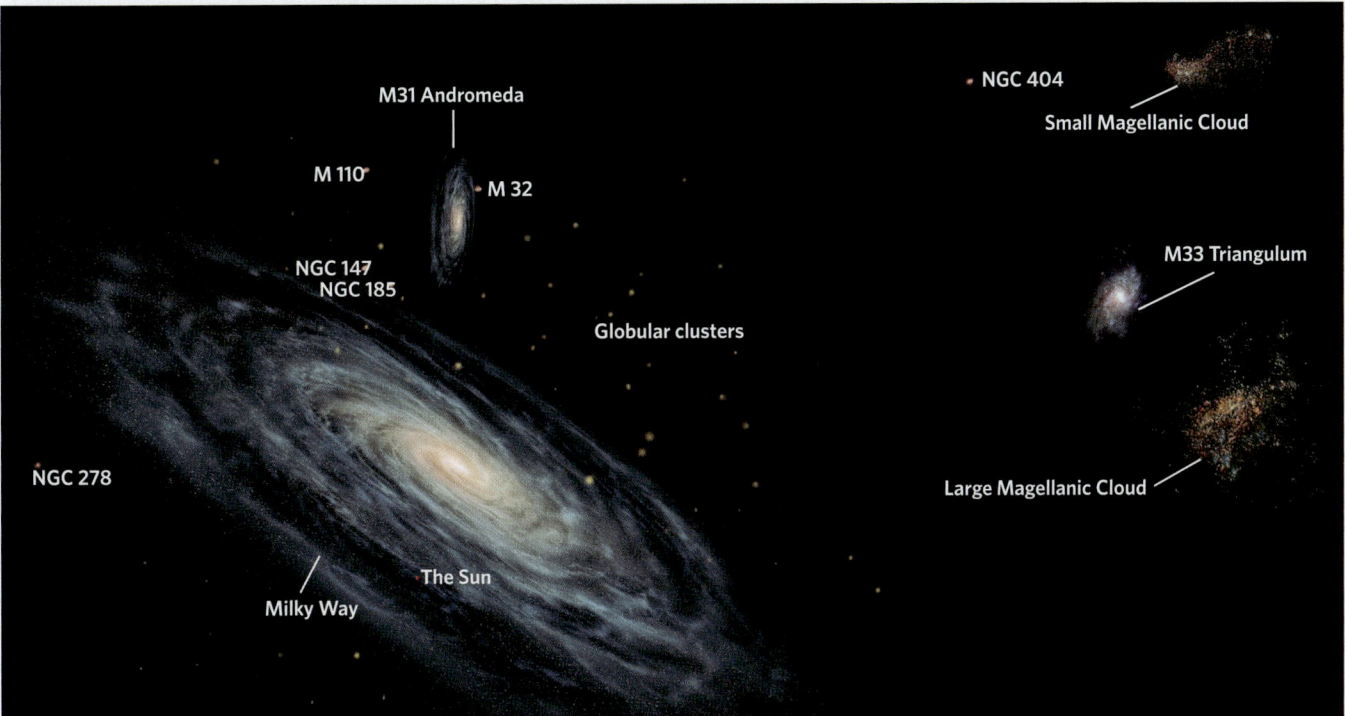

(a) The Local Group of galaxies in the vicinity of the Milky Way.

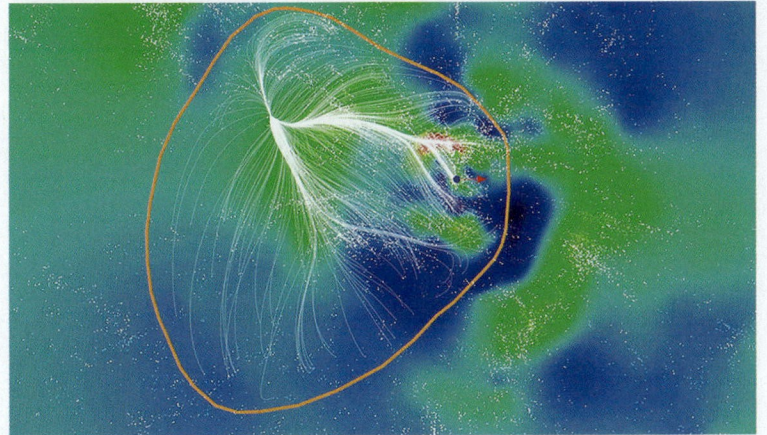

(b) A simulation of the distribution of galaxies in the Laniakea Supercluster, which is within the orange line and includes the Local Group. The Milky Way is located at the blue dot. The supercluster is over 500 million light-years across.

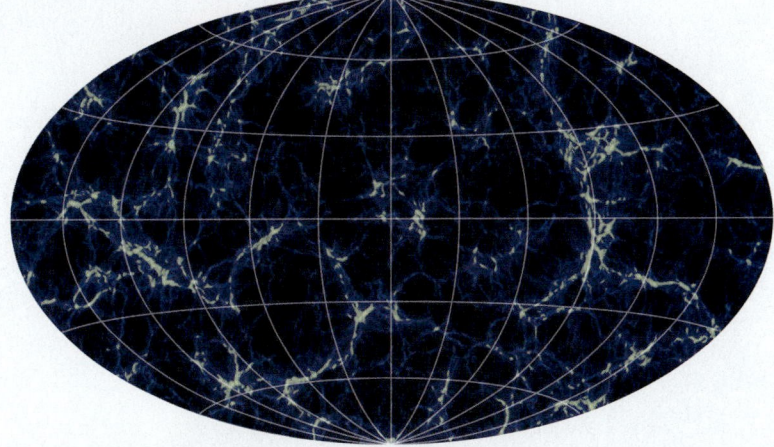

(c) At the scale of the entire Universe, the distribution of superclusters displays a web-like configuration, in which filaments and walls surround voids.

research today. The answer depends on whether gravity is strong enough to slow and eventually reverse the expansion of the Universe. The fate of our Universe is still an open question, one that scientists continue to unravel as they ponder the future of everything. But as that process of discovery continues (perhaps with your participation!), we hope that the explorations of this chapter, this book, and this course have left you with a new understanding of our beautiful planet and its place in the Universe **(Box 23.2)**.

Take-home message . . .

Galaxies are arranged in clusters throughout the Universe. Galaxies are moving away from one another at a constant rate as a consequence of the Big Bang. Researchers continue to gather data to determine whether this expansion will continue forever or reverse in the future.

Quick Question

What is the largest galaxy other than the Milky Way in our Local Group?

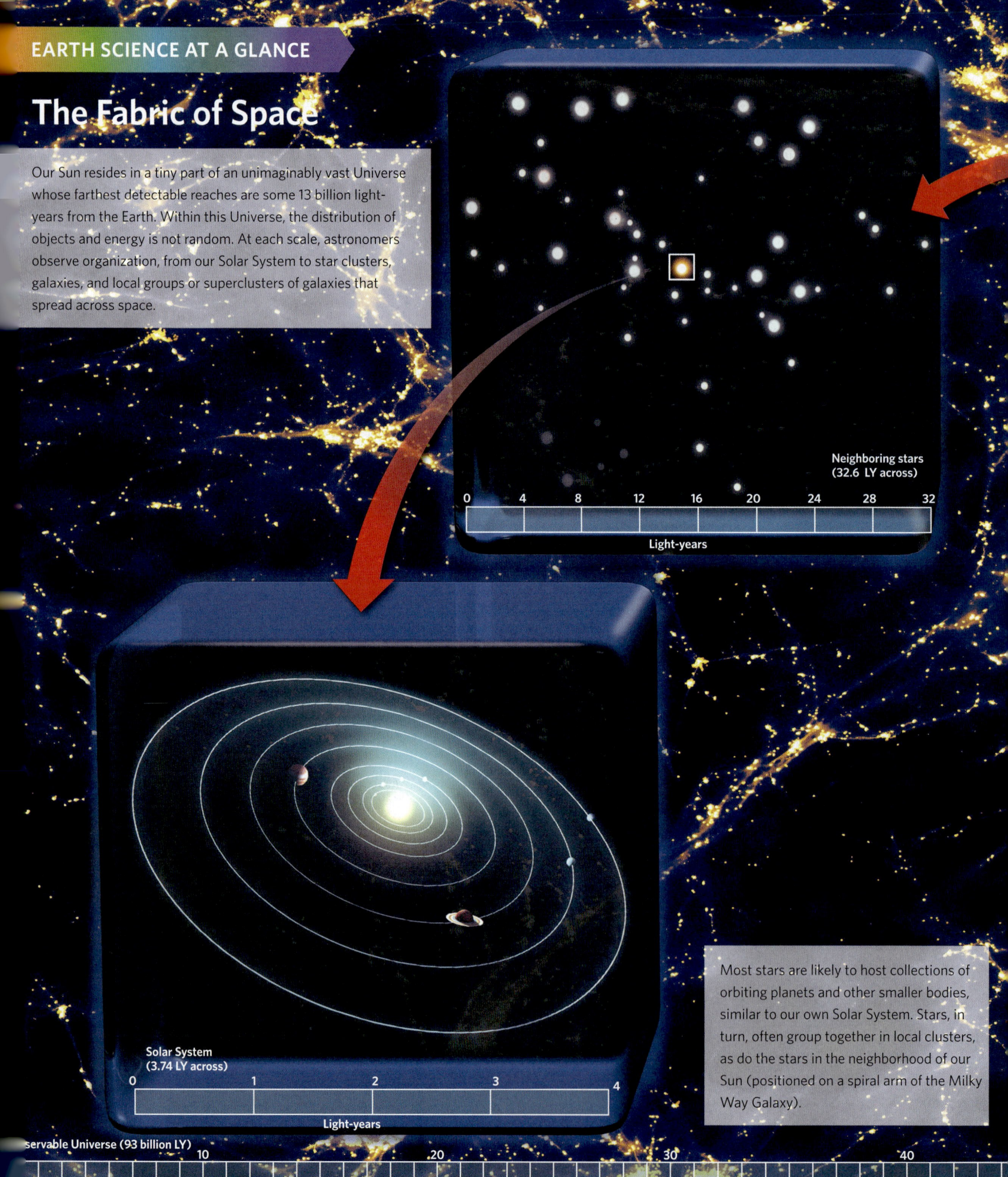

The Fabric of Space

Our Sun resides in a tiny part of an unimaginably vast Universe whose farthest detectable reaches are some 13 billion light-years from the Earth. Within this Universe, the distribution of objects and energy is not random. At each scale, astronomers observe organization, from our Solar System to star clusters, galaxies, and local groups or superclusters of galaxies that spread across space.

Neighboring stars
(32.6 LY across)

0   4   8   12   16   20   24   28   32

Light-years

Most stars are likely to host collections of orbiting planets and other smaller bodies, similar to our own Solar System. Stars, in turn, often group together in local clusters, as do the stars in the neighborhood of our Sun (positioned on a spiral arm of the Milky Way Galaxy).

Solar System
(3.74 LY across)

0   1   2   3   4

Light-years

servable Universe (93 billion LY)

10   20   30   40

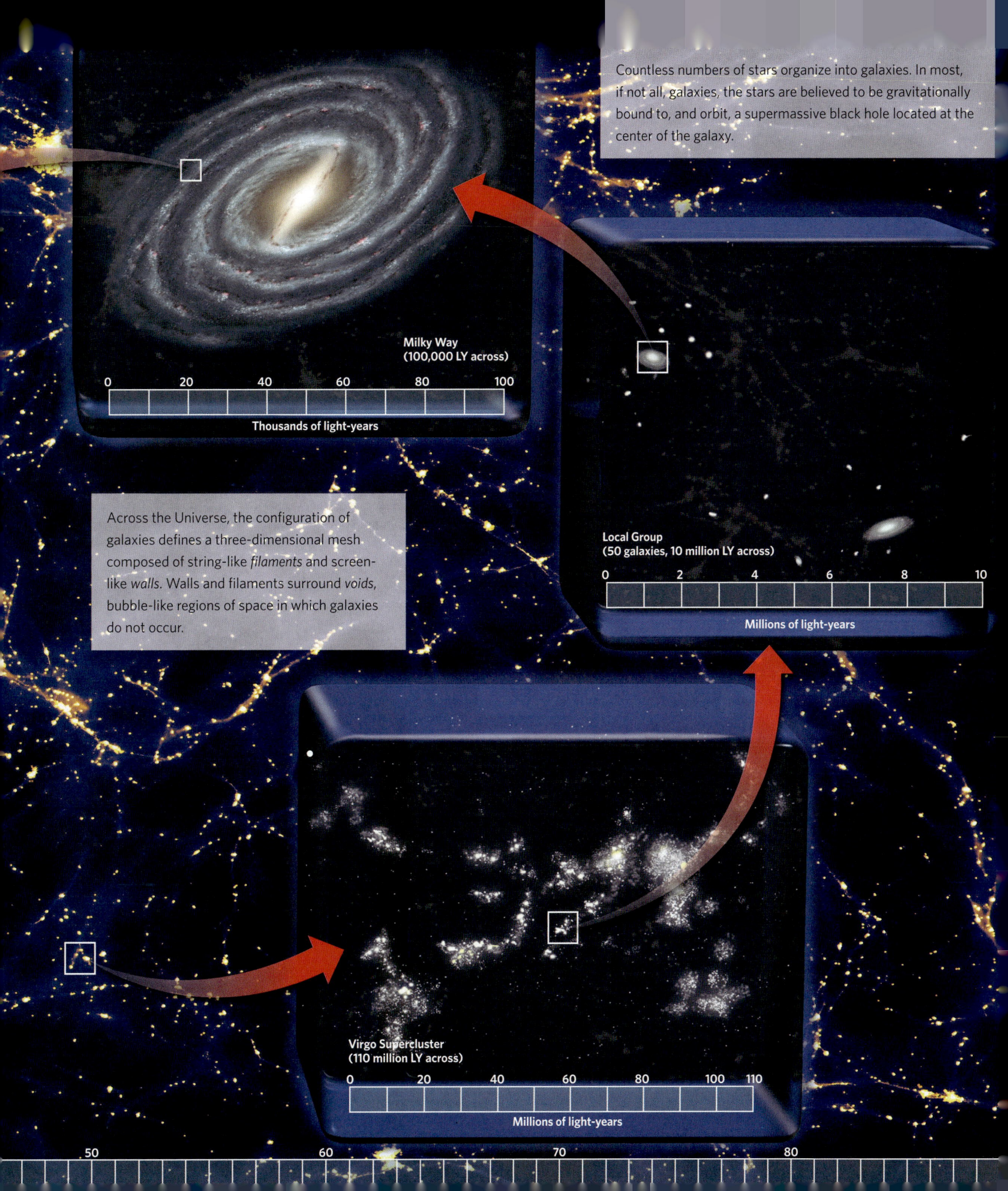

Countless numbers of stars organize into galaxies. In most, if not all, galaxies, the stars are believed to be gravitationally bound to, and orbit, a supermassive black hole located at the center of the galaxy.

Milky Way
(100,000 LY across)

0   20   40   60   80   100

Thousands of light-years

Across the Universe, the configuration of galaxies defines a three-dimensional mesh composed of string-like *filaments* and screen-like *walls*. Walls and filaments surround *voids*, bubble-like regions of space in which galaxies do not occur.

Local Group
(50 galaxies, 10 million LY across)

0   2   4   6   8   10

Millions of light-years

Virgo Supercluster
(110 million LY across)

0   20   40   60   80   100   110

Millions of light-years

⊙23 CHAPTER REVIEW

Chapter Summary

- The Sun consists primarily of hydrogen and helium. Other elements make up only about 1% of its mass.

- The Sun has a core in which nuclear fusion of hydrogen to form helium generates energy. Surrounding this solar core is the radiative zone, through which radiation from the core passes. In the outer layer, the convective zone, convection transports energy outward.

- The Sun has an atmosphere consisting of the photosphere, the chromosphere, and the corona. The light of the Sun radiates from the photosphere. The corona is an extremely hot, low-density gas layer that can be seen only during an eclipse.

- A large number of high-energy particles flow outward from the Sun at high speeds. This flow is called the solar wind.

- The Sun has a very strong and complicated magnetic field. Magnetic field lines sometimes arc upward from the photosphere. Where they do, sunspots, solar prominences, and solar flares occur. These phenomena affect space weather on the Earth.

- Stars vary in luminosity, a measure of the energy they emit, and in surface temperature. These two quantities are closely related to a star's size and mass.

- Astronomers use several scales to classify a star's brightness. Apparent magnitude indicates how bright a star appears to be from the Earth; absolute magnitude indicates how bright a star would appear to be if it were located 32.6 light-years from the Earth. Astronomers often compare the absolute magnitude of a star to the Sun, using solar units.

- An H-R diagram, which plots the luminosities of stars against their spectral classes, provides a tool for understanding the life cycles of stars.

- Stars form from the gravitational collapse of gas and dust within nebulae, which is triggered by disturbances, such as shock waves, that pile up mass locally.

- As gravity compresses a protostar's mass into a smaller and smaller volume, its core temperature increases. A protostar becomes a star when its core reaches a temperature of 10 million °C, at which nuclear fusion can occur.

- The lifetime of a star depends on its mass. Less massive stars reside much longer on the main sequence than do more massive stars because fusion reactions happen faster in more massive stars.

- After a star consumes the hydrogen in its core, the core collapses and ignites nuclear reactions of heavier elements, and the outer layers of the star swell. An intermediate-mass star becomes a red giant, whereas a high-mass star becomes a red supergiant.

- A star's life ends when it burns through its nuclear fuel. An intermediate-mass star's core collapses into an extremely dense white dwarf, while its outer layers disperse into space, creating a nebula.

- When one member of a binary star is a white dwarf, it may capture matter from its companion. When this matter ignites, a nuclear explosion called a nova takes place.

- The core of a high-mass star collapses to become a neutron star or black hole, while most of the star's mass is blasted into space in a titanic explosion called a supernova. White dwarfs, neutron stars, and black holes contain superdense matter.

- A black hole's gravity is so strong that no matter or energy can escape from it. The sphere surrounding a black hole within which nothing escapes is called the event horizon.

- A galaxy is a collection of stars, planets and moons, black holes, and interstellar gas, dust, and ice bound together by gravity. Over a trillion galaxies exist. Our home galaxy, the Milky Way, consists of a galactic disk with spiral arms surrounding a central star-filled galactic bulge anchored by a supermassive black hole at the galactic center.

- Distant galaxies are moving away from one another, implying that the Universe is expanding.

- The expansion of the Universe implies that about 13.8 billion years ago, all of the Universe's mass and energy emerged from a single point during the Big Bang.

- The future of the Universe remains uncertain and is the subject of cosmologic research.

Key Terms

absolute magnitude (p. 737)
active galaxy (p. 749)
apparent brightness (p. 736)
apparent magnitude (p. 737)
Big Bang (p. 752)
binary star (p. 741)
black dwarf (p. 744)
black hole (p. 747)

chromosphere (p. 732)
convective zone (p. 731)
corona (p. 732)
dark matter (p. 752)
dwarf star (p. 738)
event horizon (p. 746)
galactic bulge (p. 748)
galactic disk (p. 748)

galactic halo (p. 748)
galaxy (p. 747)
giant star (p. 738)
globular cluster (p. 748)
H-R diagram (p. 738)
Local Group (p. 751)
luminosity (p. 736)
main sequence (p. 738)

neutron star (p. 746)
normal galaxy (p. 749)
nova (p. 744)
nuclear fusion (p. 730)
photosphere (p. 731)
plasma (p. 730)
pulsar (p. 746)
quasar (p. 749)

Review Questions

Letters in parentheses correspond to the chapter's learning objectives.

1. What are the Sun's three internal layers? In which layer is energy produced, and by what mechanism? How does energy move across the other layers? **(A)**

2. What is a solar flare? How are solar flares, solar prominences, and sunspots related? **(A)**

3. Distinguish among luminosity, apparent magnitude, and absolute magnitude. **(B)**

4. What information does an H-R diagram give us about a star? What stars appear at the bottom of the graph shown here? What stars appear at the top right? **(C)**

5. The Sun currently falls on the main sequence of an H-R diagram. Explain how it will evolve as it nears the end of its life as a visible star. **(C, D)**

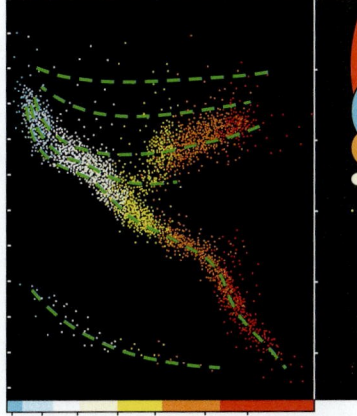

6. What is a nebula? What role(s) do nebulae play in the life cycle of stars? **(D)**

7. What property of a star is most important in determining whether it will end its life as a white dwarf, a neutron star, or a black hole? **(D)**

8. What are the differences among black-, white-, and red-dwarf stars in terms of their size, mass, temperature, formation mechanism, and life cycle stage? **(D)**

9. What is the difference between a quasar and a pulsar? **(D, E)**

10. What provides the energy of a nova? What provides the energy of a supernova? **(D)**

11. Describe the characteristics of a black hole. Can a planet orbit a black hole? Identify the event horizon in the figure. **(D)**

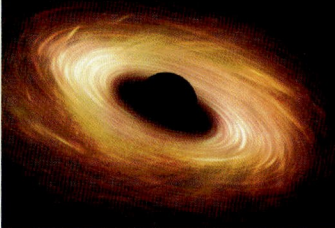

12. What are the key structural features of the Milky Way? How do galaxies differ from one another? **(E)**

13. Describe the current model of the overall structure of the Universe in terms of how galaxies are arranged over vast distances. **(F)**

On Further Thought

14. When astronomers plot the positions of thousands of stars on an H-R diagram, most of those stars fall along the main sequence. Why are so many stars found along this line rather than in other parts of the diagram? Why does the mass of a star determine its residence time on the main sequence? **(C)**

15. What objects in your pocket provide clear evidence that the Sun is not a first-generation star; that is, that it was not one of the first stars that formed after the Universe came into existence? Explain. **(D)**

ADDITIONAL CHARTS

Metric Conversion Chart

Length

1 kilometer (km) = 0.6214 mile (mi)

1 meter (m) = 1.094 yards = 3.281 feet

1 centimeter (cm) = 0.3937 inch

1 millimeter (mm) = 0.0394 inch

1 mile (mi) = 1.609 kilometers (km)

1 yard = 0.9144 meter (m)

1 foot = 0.3048 meter (m)

1 inch = 2.54 centimeters (cm)

Area

1 square kilometer (km^2) = 0.386 square mile (mi^2)

1 square meter (m^2) = 1.196 square yards (yd^2)

= 10.764 square feet (ft^2)

1 square centimeter (cm^2) = 0.155 square inch (in^2)

1 square mile (mi^2) = 2.59 square kilometers (km^2)

1 square yard (yd^2) = 0.836 square meter (m^2)

1 square foot (ft^2) = 0.0929 square meter (m^2)

1 square inch (in^2) = 6.4516 square centimeters (cm^2)

Volume

1 cubic kilometer (km^3) = 0.24 cubic mile (mi^3)

1 cubic meter (m^3) = 264.2 gallons

= 35.314 cubic feet (ft^3)

1 liter (l) = 1.057 quarts

= 33.815 fluid ounces

1 cubic centimeter (cm^3) = 0.0610 cubic inch (in^3)

1 cubic mile (mi^3) = 4.168 cubic kilometers (km^3)

1 cubic yard (yd^3) = 0.7646 cubic meter (m^3)

1 cubic foot (ft^3) = 0.0283 cubic meter (m^3)

1 cubic inch (in^3) = 16.39 cubic centimeters (cm^3)

Mass

1 metric ton = 2,205 pounds

1 kilogram (kg) = 2.205 pounds

1 gram (g) = 0.03527 ounce

1 pound (lb) = 0.4536 kilogram (kg)

1 ounce (oz) = 28.35 grams (g)

Pressure

1 kilogram per
square centimeter (kg/cm^2)* = 0.96784 atmosphere (atm)

= 0.98066 bar

= 9.8067 × 10^4 pascals (Pa)

1 bar = 0.1 megapascals (Mpa)

= 1.0 × 10^5 pascals (Pa)

= 29.53 inches of mercury (in a barometer)

= 0.98692 atmosphere (atm)

= 1.02 kilograms per square centimeter (kg/cm^2)

1 pascal (Pa) = 1 kg/m/s^2

1 pound per square inch = 0.06895 bars

= 6.895 × 10^3 pascals (Pa)

= 0.0703 kilogram per square centimeter

Temperature

To change from Fahrenheit (F) to Celsius (C):

$$°C = \frac{(°F - 32°)}{1.8}$$

To change from Celsius (C) to Fahrenheit (F):

°F = (°C × 1.8) + 32°

To change from Celsius (C) to Kelvin (K):

K = °C + 273.15

To change from Fahrenheit (F) to Kelvin (K):

$$K = \frac{(°F - 32°)}{1.8} + 273.15$$

*Note: Because kilograms are a measure of mass whereas pounds are a unit of weight, pressure units incorporating kilograms assume a given gravitational constant (g) for the Earth. In reality, the gravitational for the Earth varies slightly with location.

Periodic Table of the Elements

Symbol He 2 **Atomic number**
Helium
4.002 **Name**
Atomic weight

Alkali metals

Inert gases

Nonmetals

Transition elements (metals)

H	1
Hydrogen	1.007

Li	3	Be	4
Lithium	6.941	Beryllium	9.0121

Na	11	Mg	12
Sodium	22.989	Magnesium	24.305

| K | 19 | Ca | 20 | Sc | 21 | Ti | 22 | V | 23 | Cr | 24 | Mn | 25 | Fe | 26 | Co | 27 | Ni | 28 | Cu | 29 | Zn | 30 | Ga | 31 | Ge | 32 | As | 33 | Se | 34 | Br | 35 | Kr | 36 |

- Potassium 39.098
- Calcium 40.078
- Scandium 44.955
- Titanium 47.88
- Vanadium 50.941
- Chromium 51.996
- Manganese 54.938
- Iron 55.847
- Cobalt 58.933
- Nickel 58.693
- Copper 63.546
- Zinc 65.39
- Gallium 69.723
- Germanium 72.61
- Arsenic 74.921
- Selenium 78.96
- Bromine 79.904
- Krypton 83.80

| Rb | 37 | Sr | 38 | Y | 39 | Zr | 40 | Nb | 41 | Mo | 42 | Tc | 43 | Ru | 44 | Rh | 45 | Pd | 46 | Ag | 47 | Cd | 48 | In | 49 | Sn | 50 | Sb | 51 | Te | 52 | I | 53 | Xe | 54 |

- Rubidium 85.467
- Strontium 87.62
- Yttrium 88.905
- Zirconium 91.224
- Niobium 92.906
- Molybdenum 95.94
- Technetium 98.907
- Ruthenium 101.07
- Rhodium 102.905
- Palladium 106.42
- Silver 107.868
- Cadmium 112.411
- Indium 114.82
- Tin 118.710
- Antimony 121.757
- Tellurium 127.60
- Iodine 126.904
- Xenon 131.29

| Cs | 55 | Ba | 56 | La | 57 | Hf | 72 | Ta | 73 | W | 74 | Re | 75 | Os | 76 | Ir | 77 | Pt | 78 | Au | 79 | Hg | 80 | Tl | 81 | Pb | 82 | Bi | 83 | Po | 84 | At | 85 | Rn | 86 |

- Cesium 132.905
- Barium 137.327
- Lanthanum 138.905
- Hafnium 178.49
- Tantalum 180.947
- Tungsten 183.85
- Rhenium 186.207
- Osmium 190.2
- Iridium 192.22
- Platinum 195.08
- Gold 196.966
- Mercury 200.59
- Thallium 204.383
- Lead 207.2
- Bismuth 208.980
- Polonium 208.982
- Astatine 209.987
- Radon 222.017

Fr	87	Ra	88	Ac	89
Francium	223.019	Radium	226.025	Actinium	227.027

Nonmetal / Metalloid block:

B	5	C	6	N	7	O	8	F	9	He	2
Boron 10.811		Carbon 12.011		Nitrogen 14.006		Oxygen 15.999		Fluorine 18.998		Helium 4.002	

Al	13	Si	14	P	15	S	16	Cl	17	Ne	10
Aluminum 26.981		Silicon 28.085		Phosphorus 30.973		Sulfur 32.066		Chlorine 35.452		Neon 20.179	

Argon 18, 39.948

Lanthanide series:

| Ce | 58 | Pr | 59 | Nd | 60 | Pm | 61 | Sm | 62 | Eu | 63 | Gd | 64 | Tb | 65 | Dy | 66 | Ho | 67 | Er | 68 | Tm | 69 | Yb | 70 | Lu | 71 |

- Cerium 140.115
- Praseodymium 140.907
- Neodymium 144.24
- Promethium 144.912
- Samarium 150.36
- Europium 151.965
- Gadolinium 157.25
- Terbium 158.925
- Dysprosium 162.50
- Holmium 164.930
- Erbium 167.26
- Thulium 168.934
- Ytterbium 173.04
- Lutetium 174.967

Actinide series:

| Th | 90 | Pa | 91 | U | 92 | Np | 93 | Pu | 94 | Am | 95 | Cm | 96 | Bk | 97 | Cf | 98 | Es | 99 | Fm | 100 | Md | 101 | No | 102 | Lr | 103 |

- Thorium 232.038
- Protactinium 231.035
- Uranium 238.028
- Neptunium 237.048
- Plutonium 244.064
- Americium 243.061
- Curium 247.070
- Berkelium 247.070
- Californium 251.079
- Einsteinium 252.083
- Fermium 257.095
- Mendelevium 258.10
- Nobelium 259.100
- Lawrencium 262.11

The modern periodic table of the elements. Each column groups elements with related properties. For example, inert gases are listed in the column on the right. Metals are found in the central and left parts of the chart.

GLOSSARY

a'a' A lava flow with a rubbly surface.

ablation The removal of ice at the toe of a glacier by melting, sublimation, and/or calving.

absolute magnitude A measure of the luminosity of a star; specifically, the apparent magnitude the star would have if it were located at a standard distance of exactly 32.6 light-years (10 parsecs) from the Earth.

absolute plate velocity The movement of a plate relative to a fixed reference point in the mantle.

abyssal plain A broad, relatively flat region of the seafloor that lies at least 4–5 km (1.2–1.5 mi) below sea level.

accretion The addition of material. It can pertain to the addition of crust to the margin of a continent, sediment to the accretionary prism of a subduction zone, sediment to a coastline, or matter to an accretion disk.

accretion disk A flat, rotating disk of gas and dust surrounding an object, such as a protostar, a protoplanet, a collapsed star in a binary system, or a black hole.

accretionary coast A coastline that's widening because it receives more sediment than erodes away.

accretionary prism A wedge-shaped mass of sediment and rock scraped off the top of the downgoing plate at a convergent boundary.

acid mine runoff A dilute solution of sulfuric acid, produced when sulfur-bearing minerals in a mine react with rainwater or groundwater and flow out of the mine.

acid rain Rain formed from atmospheric water that reacted with pollutant aerosols to become a weak solution of sulfuric acid.

active galaxy A galaxy that emits most of its energy in the X-ray and radio regions of the electromagnetic spectrum, with most of the energy coming from the galactic center.

active margin A continental margin that is also a plate boundary.

active volcano A volcano that has erupted within the past few centuries or millennia and has the potential to erupt again.

adiabatic expansion An expansion of a volume of air during which no mass or energy exchanges with the environmental air surrounding the volume.

aerosols Microscopic solid particles or liquid droplets small enough to remain suspended in the atmosphere.

aftershock One of a series of smaller earthquakes that follow a major earthquake.

air The mixture of gases that make up the Earth's atmosphere.

air mass A large body of air (typically several kilometers thick and hundreds to thousands of kilometers wide) within which temperature and humidity are relatively uniform.

air temperature A measure of the average kinetic energy of air molecules, as measured by a thermometer or thermistor.

albedo The reflectivity of a surface.

alloy A metal containing more than one metallic element.

alluvial fan A gently sloping apron of sediment dropped by an ephemeral stream at the base of a mountain in arid or semiarid regions.

alluvium Sorted sediment deposited by a stream.

Alpine-Himalayan chain The collisional orogen that formed where continental fragments and India collided with the southern margin of Eurasia during the Cenozoic. It includes the Alps and the Himalayas.

amplitude Half of the wave height, where wave height is the vertical distance from crest to trough.

angle of repose The angle of the steepest slope that a pile of uncemented material can attain without collapsing from the pull of gravity.

angular unconformity An unconformity at which strata below the unconformity have a different orientation than the strata above have; such unconformities form when the strata below were tilted or folded before the strata above were deposited.

annual probability The likelihood of an event, such as a flood or an earthquake, happening in a given year, as specified by a percentage equal to 1 divided by the recurrence interval.

Anthropocene An informal name for the time during which human activity has had a major influence on climate, the environment, landscapes, and biogeochemical cycles.

anthropogenic change Change in the Earth System that is caused by human activity.

anticline A fold with an arch-like shape, so that the limbs dip away from the hinge.

anvil cloud The upper part of a large cumulonimbus cloud that spreads laterally at the tropopause to form a broad, flat top.

aphotic zone The deeper part of the ocean into which sunlight does not penetrate.

apparent brightness The amount of energy from a star that passes through a square meter at the surface of the Earth in 1 second.

apparent magnitude A measure of apparent brightness of a celestial object, such as a star, as seen by a person on the Earth. On the scale for defining apparent brightness of a star, a magnitude 6 star is dimmer than a magnitude 1 star, and very bright objects can have negative magnitudes.

apparent polar-wander path A path on the globe along which a magnetic pole appears to have wandered over time; in fact, the continents drift, while the magnetic pole stays fairly fixed.

aquifer Sediment or rock that transmits water easily.

aquitard Sediment or rock that does not transmit water easily and therefore slows the motion of groundwater.

Archean Eon The middle eon of the Precambrian. It spans the time from 4.0 Ga–2.5 Ga.

arête A knife-edged ridge of rock that separates two adjacent cirques in glacially carved mountains.

arrival time The time at which a specific seismic wave, such as a P-wave or an S-wave, arrives at a given seismometer.

artesian well A well in which water rises without pumping, due to pressure within a confined aquifer.

ash Very fine particles of glass or pulverized rock erupted by a volcano.

assimilation The incorporation by magma of chemicals dissolved from the wall rock through which it rises, or from blocks of wall rock that fall into the magma.

asteroid Rocky or metallic objects, larger than 100 m (about 330 ft) across, that lie in the asteroid belt; they include both primitive planetesimals that never became incorporated in larger bodies, as well as remnants of larger planetesimals that had differentiated into a core and mantle before breaking up.

asteroid belt The zone between Mars and Jupiter where most asteroids reside.

asthenosphere The layer of the mantle that lies directly below the lithosphere, in which the mantle can flow plastically and, therefore, undergo convection.

astronomical unit (AU) The average distance from the Sun to the Earth. It is approximately 150 million km (93 million mi).

astronomy The scientific study of celestial objects (planets, stars, galaxies, nebula, etc.), and of the Universe as a whole.

atmosphere A layer of gases that surrounds a planet.

atmospheric instability The condition in the atmosphere in which a parcel of air, after undergoing initial lifting, can remain buoyant and continue to rise.

atmospheric pressure The force applied by air on a unit area of a surface; equivalent to the weight of a column of air above the unit area.

atmospheric science The study of the air layer that surrounds the Earth.

atmospheric seeing The blurring or fuzzing of a celestial object's image in a telescope, caused by turbulence in the atmosphere.

atom The smallest piece of an element that has the properties of the element; it consists of a nucleus surrounded by an electron cloud.

atomic mass The amount of matter in an atom of an element; it is approximately equal to the total number of protons plus neutrons in the atom's nucleus. Different isotopes of an element have different atomic mass.

atomic number The number of protons in an element's nucleus; all isotopes of an element have the same atomic number.

aurora A translucent curtain or streak of varicolored light that appears across the night sky when charged particles from the Sun interact with ions in the ionosphere.

axis of rotation An imaginary line around which an object spins.

backwash The gravity-driven flow of water back down the slope of a beach after a wave has washed up the beach.

baked contact A boundary between wall rock and an igneous intrusion, in which the wall rock has been altered by heat emitted from an igneous intrusion.

banded iron formation (BIF) Iron-rich sedimentary layers consisting of alternating beds of gray iron oxide minerals and reddish iron-rich chert.

bar (1) A sheet or elongate lens or mound of sediment deposited by flowing water. (2) A unit of atmospheric pressure approximately equal to 1 atm.

barometer An instrument that measures atmospheric pressure.

barrier island An offshore sand bar that rises above the mean high-water level, forming an island.

basalt A fine-grained, dark-colored igneous rock with a mafic composition.

base level The lowest elevation down to which a stream's water can flow at a given locality.

basin In the context of geologic structures, a fold or depression in strata whose shape resembles a right-side-up bowl.

Basin and Range Province A broad Cenozoic continental rift that has affected a portion of the western United States in Nevada, Utah, and Arizona; in this province, tilted fault blocks form ranges, and alluvium-filled valleys are basins.

batholith A vast composite, intrusive, igneous rock body up to several hundred kilometers long and tens of kilometers wide, formed by the intrusion of numerous plutons.

bathymetric map A map illustrating the shape of the ocean floor or of a lake's floor.

bathymetry Variation in depth of a water body (lake or ocean).

beach A band of sediment, parallel to the shoreline, that undergoes sorting and shifting by the swash and backwash of waves.

beach nourishment The process of adding sediment, generally pumped or dredged from offshore, or trucked in from elsewhere, to replace beach sediment that has been eroded away.

beach profile The variation in elevation of a beach, as measured along a line perpendicular to the shoreline.

bearing The compass heading of a linear structure with respect to the horizontal. Bearing and plunge together uniquely define the orientation of a linear structure.

bedding Layering or stratification in sedimentary rocks.

bedrock Rock still attached to the Earth's crust.

benthic zone The region encompassing the seafloor and the region just below the seafloor.

benthos Organisms that live on or just beneath the seafloor (that is, in the benthic zone).

Big Bang The cataclysmic explosion of a singularity that, according to theory, represents the formation of the Universe.

Big Bang nucleosynthesis The formation of low-mass atomic nuclei (mainly H and He) during the first few minutes after the Big Bang.

binary star A system in which two stars are close enough to each other to be bound to each other gravitationally, and to orbit around a common center of mass.

biochemical sedimentary rock Sedimentary rock formed from material produced by living organisms, such as shells.

biodiesel A type of liquid hydrocarbon fuel manufactured from plant or animal oils.

biodiversity The total number of different species of organisms that exist on Earth at a given time, as an indication of the variety of life that our planet hosts.

biofuel Gas or liquid fuel made from plant material (biomass). Examples include alcohol, made from fermented sugar, and biodiesel, made from vegetable oil or wood.

biogeochemical cycle The exchange of chemicals between living and nonliving reservoirs in the Earth System.

biomass The total mass of biological material (living organisms) in a given volume

biosphere The portion of the Earth System inhabited by life; it includes the region from a few kilometers below the Earth's surface to a few kilometers above it.

blackbody An object that absorbs all radiation incident upon it and re-radiates energy according to the three laws of blackbody radiation.

black dwarf The last stage of life for a low-mass star, in which its dense core has lost so much energy that it no longer emits detectable amounts of radiation.

black hole An extremely massive object so dense that its gravitational force does not allow even light to escape; some form from the collapse of very massive stars, and some from the collapse of matter into the central area of a galaxy.

black smoker A vent along a mid-ocean ridge that spews hot, mineral-rich water; its dissolved sulfide components instantly precipitate when the water mixes with seawater and cools.

blizzard A snowstorm in which winds exceed 56 km/h (35 mph) for at least 3 hours, with reduced visibility due to falling or blowing snow.

block In the context of volcanoes, a large, angular chunk of pyroclastic debris.

blowout An event during which hydrocarbons spew from a well. It spills hydrocarbons onto the land or into the sea, and can vent them into the air, or where they can catch fire if sparks or flames are present. Gushers are a type of blowout where oil fountains out of a well.

blue shift The phenomenon in which a source of light moving toward an observer appears to have a higher frequency.

body wave In the context of discussing earthquakes, a seismic wave that passes through the interior of the Earth.

boundary layer The layer of the atmosphere adjacent to the Earth's surface in which friction with the surface significantly affects air movement.

Bowen's reaction series The sequence in which different silicate minerals crystallize during the progressive cooling of a magma or lava.

braided stream A stream in which the water slows and deposits sediment in numerous elongate gravel and sand bars, thereby choking the channel and forcing the stream to divide into strands that weave back and forth among the bars.

breaker A water wave in which water at the top of the wave curves over its base.

brittle deformation The cracking and fracturing of a material subjected to stress.

caldera A large circular or elliptical depression with steep walls and a fairly flat floor, formed after a volcanic eruption causes the central area of a volcano to collapse into the drained magma chamber below; major calderas form during large explosive eruptions.

Cambrian explosion The rapid diversification of life, as indicated by the fossil record, that occurred at the beginning of the Cambrian Period.

capacity The total quantity of sediment a stream can carry; the capacity of a stream depends on both its competence and its discharge.

carbonate rock Rock containing calcite and/or dolomite.

carbon budget The balance between additions and subtractions of carbon within the different reservoirs of the Earth System.

carbon capture and sequestration A process of decreasing the input of carbon dioxide into the atmosphere by trapping the CO_2 produced by large sources such as power plants, liquefying it, and pumping it into reservoir rocks deep underground.

carbon cycle The exchange of carbon among the Earth System's carbon reservoirs.

carbon footprint The total amount of greenhouse gases produced by an anthropogenic process or activity.

celestial object An object or feature in the night sky.

celestial sphere An imaginary sphere surrounding the Earth, oriented so that its axis aligns with the

Earth's axis of rotation; it has no physical existence, but is a convenient reference frame for specifying the positions of celestial objects.

cement In the context of sedimentary rocks, mineral material that precipitates from water and fills the spaces between sediment grains, holding the grains together.

cementation In the context of sedimentary rocks, the aspect of lithification during which cement partially or completely fills the spaces between sediment grains and attaches grains to their neighbors.

centigrade scale A temperature scale in which the melting point of ice is set at 0 degrees and the boiling point of water at sea level is set at 100 degrees.

chain reaction A succession of fission reactions that takes place when particles released from one reaction trigger further reactions.

channel A trough dug into the ground surface by flowing water.

chemical sedimentary rock Sedimentary rock made up of minerals that precipitate directly from water solutions.

chemical weathering The process in which chemical reactions alter or destroy minerals when rock comes into contact with water solutions and air.

chert Any sedimentary rock composed of quartz grains that are too small to be seen without extreme magnification.

chromosphere The region of the Sun's atmosphere located between the photosphere and the corona.

cinder cone A subaerial volcano consisting of a cone-shaped pile of lapilli whose slope approaches the angle of repose.

cirque A bowl-shaped depression carved by a glacier on the side of a mountain.

cirrus cloud A wispy, elongate cloud that develops very high in the troposphere (above 6 km, or 20,000 ft).

clast A single particle or grain of sediment.

clastic sedimentary rock Sedimentary rock consisting of clasts derived from the weathering of pre-existing rock that have been packed together and cemented to one another.

clastic texture The texture of a rock made of grains held together by cement.

cleavage The tendency of a mineral specimen to break along preferred planes whose orientation is defined by the internal crystal structure of the specimen. Cleavage planes are not the same as crystal faces; a specimen can display many cleavage faces in the same orientation.

cliff retreat Change in the position of a cliff face, over time, due to erosion.

climate The average, range, and seasonality of weather conditions in a given region over the course of decades to millennia.

climate change A long-term shift in the average of one or more climate conditions.

climate science The study of a region's climate and how it changes over time.

closure temperature The temperature below which parent and daughter atoms are locked into a mineral grain.

cloud A visible mass of condensed water vapor, consisting of tiny water droplets, ice crystals, or both, floating in the atmosphere, typically high above the ground.

cloud cover A measure of the percentage of the sky covered by cloud.

cloud-to-cloud lightning Lightning that occurs between clouds and does not come into contact with the ground.

cloud-to-ground lightning Any lightning stroke that extends from a cloud to the ground.

coal A black organic sedimentary rock, consisting of more than 50% carbon, formed from the lithified remains of plant material.

coal rank A measure of the carbon content of coal; higher-rank coal forms at higher temperatures.

coal seam A layer or bed of coal.

coast The region of land and sea adjacent to the shoreline.

coastal plain A low-relief region of land adjacent to a coast; commonly the term applies to a narrow band less than a few kilometers wide, but in the eastern and southern United States it applies to a broad band over 200 km wide.

coastal wetland A near-shore area that becomes submerged by shallow water for all or part of the day; such wetlands host salt-resistant vegetation and many other organisms.

cold front A boundary between a cold air mass and a warm air mass, at which the cold air advances forward and lifts the warmer air mass.

collision In the context of plate tectonics theory, the process during which two buoyant pieces of lithosphere push together, generally building a mountain range. Collision happens when these pieces converge at a subduction zone.

collisional orogen A mountain belt formed when two relatively buoyant blocks of lithosphere push together in response to a collision after the ocean floor between them has been entirely subducted.

collision-coalescence The process by which different-sized cloud droplets come into contact and merge with one another to form raindrops.

columnar jointing A pattern of cracks, produced when igneous rock shrinks as it cools, that yields roughly hexagonal columns of rock.

comet An object composed of ice and dust that revolves around the Sun in a highly elliptical orbit; it produces a tail as it nears the Sun.

compaction The phase of lithification during which the weight of overburden squeezes out water and air that was trapped between clasts of sediment, thereby pressing the clasts more tightly together.

competence In the context of describing streams, the maximum particle size of sediment that a stream can carry.

compression A stress that pushes on or squeezes a material.

compressional wave A seismic wave that causes particles of a material to move back and forth, parallel to the direction in which the wave itself moves.

conchoidal fracture The smoothly curving, clam-shell-shaped surface that may form when a mineral with no cleavage planes breaks.

condensation theory The theory describing how a nebula containing dust and ice contracts to form a central star and planetary system, such as our Solar System.

cone of depression The downward-pointing, cone-shaped surface of the water table in a location where the water table has experienced drawdown due to pumping at a well.

constellation An imaginary image in the sky formed by an apparent pattern of stars. The 88 constellations defined on the celestial sphere, in western cultures, are used by astronomers to locate celestial objects.

consumer An organism in the food chain or food web that eats other organisms.

contact The boundary surface between two rock bodies, such as between two stratigraphic formations, between an igneous intrusion and wall rock, between two igneous rock bodies, or between rocks juxtaposed by a fault.

contact metamorphism Change in the mineral assemblage in wall rock due to heating by an igneous intrusion; also called *thermal metamorphism*.

contaminant plume In the context of discussing groundwater, a subsurface cloud of pollution mixed with or dissolved in groundwater that has moved with groundwater flow from the pollutant's source.

continent A very large land area underlain by continental lithosphere.

continental arc A volcanic arc formed where an oceanic plate subducts beneath a continent.

continental drift The idea that continents have moved, and are still moving, slowly across the Earth's surface. The "drift" of continents happens because they are part of moving plates.

continental glacier A vast sheet of ice that spreads over thousands of square kilometers of continental crust; also called a *continental ice sheet*.

continental margin The boundary between an ocean basin and a continent.

continental shelf A broad region of shallow sea at a continental margin. The widest continental shelves occur over passive margins.

contour line A line on a map along which a parameter has a constant value; for example, all points along one contour line on a topographic map are at the same elevation.

convective cell A circulation pattern caused by convective flow.

convective flow A process of circulation and heat transfer that happens when a deeper layer of material warms up and, therefore, expands and becomes less dense than the overlying layer of cooler material. The deeper material becomes buoyant and rises, while the cooler material sinks to take its place, if the material is weak enough to flow.

convective zone The layer of the Sun in which convective transport of energy between the radiative zone and the photosphere takes place.

conventional reserve A hydrocarbon reserve that can be accessed simply by drilling into it and pumping the hydrocarbons from the reservoir rock.

convergence In the context of atmospheric flow, a net inflow of air molecules into a region of the atmosphere, caused by slowing wind speed or changes in wind direction, which increases atmospheric pressure at the Earth's surface.

convergent boundary A plate boundary, associated with a deep-sea trench, where one plate (the downgoing plate) sinks and slips beneath the other (the overriding plate).

convergent-boundary orogen A mountain belt, such as the Andes, formed on the overriding plate next to a convergent plate boundary.

Copernican model A model of the Solar System, proposed by Copernicus, which states that the Sun lies at the center and other objects, including the Earth, orbit the Sun.

coral reef A realm of shallow water underlain by a mound of coral and coral debris and associated organisms.

core In the context of discussing planets, the innermost layer of a planet's concentric layers; it includes the planet's center. The Earth's core consists of an iron alloy.

core-mantle boundary In discussing the Earth, the interface at a depth of 2,900 km (1,802 mi) below the Earth's surface that separates the metallic core and the rocky mantle.

Coriolis effect The deflection of a moving object or material relative to the surface of a rotating object beneath it.

Coriolis force The apparent force that causes the Coriolis effect.

corona The hottest, outermost layer of the Sun's atmosphere.

correlation The process of defining age relationships between the strata at one locality and the strata at another.

cosmologic time Time intervals related to events in the history of the Universe. The term can also be used to specify the time since the origin of the Universe.

crater (1) A bowl-shaped depression at the summit of a volcano. (2) A bowl-shaped depression formed by the impact of a meteorite.

craton A region of continental lithosphere consisting of rock that has not been affected by orogenies for at least the last 1 billion years. Cratons are commonly located in the interior of a continent.

cratonic platform A portion of a craton in which Phanerozoic strata bury most of the underlying Precambrian rock.

creep The gradual downslope movement of regolith.

crevasse A large crack that develops by brittle deformation in the top 60 m (197 ft) of a glacier.

critical zone The surface and near-surface realm of the Earth System that contains the resources and conditions that sustain life.

cross beds A sedimentary structure represented by thin layers, inclined at an angle to the top of a thicker sediment layer, which represent preserved slip faces of dunes or ripple marks.

cross-cutting relations Relationships between geologic features in which one cuts across another; the feature that has been cut is older than the one that cuts it.

crust The rock that makes up the outermost layer of the Earth.

cryosphere The frozen component of the hydrosphere on the Earth.

crystal A single, continuous piece of a mineral inside which atoms, molecules, or ions are fixed in an orderly arrangement.

crystal face Flat surfaces on a crystal that developed naturally as the crystal grew; crystal faces intersect at sharp edges.

crystal habit The general shape of a crystal or cluster of crystals that grew unimpeded.

crystalline igneous rock A rock, formed by solidification of a melt, with a texture characterized by interlocking crystals that grew together.

crystalline texture The texture of rock whose grains grew to interlock with one another.

crystalline solid A material made of atoms or molecules locked into an orderly arrangement.

crystal structure The orderly arrangement of atoms inside a crystal.

cumulus cloud A vertically growing cloud with cauliflower-like lobes.

current A band of flowing water that moves distinctly faster than adjacent water.

cycle A series of interrelated events or steps that occur in succession and can be repeated, perhaps indefinitely.

cyclone (1) A large low-pressure system with a surface air flow that moves counterclockwise in the northern hemisphere and clockwise in the southern hemisphere. Hurricanes and typhoons are types of cyclones. (2) The name used specifically for a hurricane formed over the Indian Ocean or affecting Australia. See *mid-latitude cyclone, tropical cyclone.*

dark matter Hypothetical matter in the Universe that does not emit or absorb electromagnetic radiation, so it cannot be observed directly. Dark matter is thought to constitute most of the mass of the Universe, but its character remains unknown.

data A set of observations, measurements, or calculations.

daughter isotope An atom that is the product of radioactive decay.

dead zone A region of the sea near the mouth of a river where nutrients brought in by the river cause algal blooms; the subsequent death and decomposition of the algae deplete the water of oxygen, so it cannot sustain oxygen-breathing marine organisms.

debris flow A downslope movement of mud mixed with larger rock fragments.

decompression melting Melting that occurs when hot mantle rock rises to shallow depths in the Earth, where pressure decreases but temperature remains unchanged.

deep-sea current A current that flows along the ocean bottom; commonly a result of the thermohaline circulation.

deep-sea trench An elongate trough, where the ocean floor may reach depths of 7–11 km (4–7 mi); these delineate the trace of a convergent boundary.

deformation A change in the shape, position, or orientation of a material by bending, breaking, or flowing.

delta A wedge of sediment formed at the mouth of a stream. When a stream enters standing water, its current slows, so that the stream loses competence, and sediment settles out.

density Mass per unit volume.

deposition The process by which sediment settles out of a transporting medium.

depositional environment A setting in which sediments accumulate; its character reflects local conditions such as the presence or absence of water, the velocity of the fluid carrying the sediment, and whether the setting hosts living organisms.

depositional landform A landform resulting from the deposition of sediment.

desalinization The process of removing salt from seawater to make freshwater.

desert A region that receives average rainfall (or snowfall equivalent) of less than 25 cm (10 in) per year; deserts are so dry (arid) that they generally don't contain permanent streams and support vegetation on more than 15% of their surface.

desertification The conversion of a semiarid region into a desert by natural drought or human activities.

desert pavement A mosaic-like surface of rock fragments that covers the ground in a desert.

desert varnish A dark, rusty-brown coating of iron oxide and magnesium oxide that forms on the surface of rock in deserts.

dew point The temperature at which air will become saturated if it is cooled at constant pressure.

differential stress A condition in which the push or pull in one direction differs from that in another direction, or in which shearing is taking place.

differentiation In the context of planetary evolution, the process by which denser materials sink toward the center of a molten or fluid planetary interior, resulting in internal layering, defined by a distinct core and mantle.

digital elevation model (DEM) A topographic map produced by a computer from a set of data in which each location is represented by three coordinates: latitude, longitude, and elevation.

dike A vertical, wall-shaped tabular intrusion of rock that cuts across a layer of wall rock.

dimension stone Rock that is intended to be cut for architectural purposes.

dip The angle of a planar structure's slope with respect to the horizontal, as measured in a vertical plane. Strike and dip together uniquely define the orientation of a planar structure.

dipole A magnetic field with two poles, which can be represented by an imaginary arrow that points from the north end to the south end of a magnet.

dip-slip fault A fault in which displacement occurs up or down the slope (dip) of the fault.

directional drilling Drilling using technology that allows the trajectory of the drill bit to be other than vertical, so that the drill hole can be angled or even horizontal.

discharge In the context of discussing stream flow, the volume of water that passes through a cross section of a stream, drawn perpendicular to the stream, in a given time (usually 1 s).

discharge area location where groundwater flows back up to the land surface, where it may emerge at springs.

disconformity An unconformity separating two sedimentary units, both of which have the same bedding orientation.

displacement The amount of movement or slip across a fault; also called *offset*.

dissolution A process during which materials dissolve in water.

divergence A net outflow of air molecules from a region of the atmosphere, caused by increasing wind speed or changes in wind direction, which decreases atmospheric pressure at the Earth's surface.

divergent boundary A plate boundary where two oceanic plates move apart by the process of sea-floor spreading; also called a *spreading boundary*.

doldrums A belt along the equator where little horizontal air movement occurs, so there is barely any wind.

dolostone A carbonate rock that contains a substantial proportion of the mineral dolomite.

dome A warping of strata or a fold with the shape of an overturned bowl.

Doppler effect The phenomenon in which the frequency of observed wave energy changes as its source moves toward or away from an observer.

Doppler radar A radar system that detects precipitation and wind by transmitting microwave energy and measuring shifts in the frequency of the returned energy.

dormant volcano A volcano that does not exhibit current activity and has not erupted for hundreds to thousands of years, but does have the potential to erupt again in the future.

downcutting The process by which water flowing along a stream channel cuts into the substrate and deepens the channel relative to its surroundings.

downdraft A downward-moving air flow.

downslope windstorm A strong flow of wind down the leeward side of a mountain; North American examples include the chinook and Santa Ana winds.

downwelling The sinking of near-surface water to greater depth in the ocean.

drainage basin A region that provides the water to a drainage network; within a drainage basin, all the tributaries feed into the same trunk stream.

drainage divide A highland or ridge that separates one drainage basin from another.

drainage network An array of interconnecting tributary streams that all flow into the same trunk stream.

drumlin A streamlined, elongate hill formed when a glacier overrides and molds glacial till; the long axis of a drumlin is parallel to the flow direction of the glacier.

dry wash The channel of an ephemeral stream when it contains no water.

dune A relatively large pile of sand deposited by a current of wind or water; its shape depends on current direction and velocity.

dust In the context of space science, it consists of particles made of rocky or metallic solids, known to chemists as refractory materials because they melt only at high temperatures; in discussing the atmosphere, it consists of very fine particles that can remain in suspension, or slowly settle in air.

dust storm An event in which strong winds strip fine-grained sediment from unvegetated soil and send it skyward to form rolling dark clouds that block out the Sun.

dwarf planet A spherical object that orbits a star directly, but has not cleared smaller bodies from the neighboring regions around its orbit; examples include Pluto and Ceres.

dwarf star A star of relatively small size and low luminosity. The majority of stars on the main sequence of the H-R diagram, including the Sun, are dwarf stars.

dynamothermal metamorphism Metamorphism that occurs over a large region as a consequence of heating, pressure, and differential stress, generally in association with mountain-building events; also called *regional metamorphism*.

earthquake An episode of shaking caused by the sudden breaking or frictional sliding of rock in the Earth.

earthquake engineering The design of buildings that can withstand ground shaking.

earthquake zoning Restrictions on the location and construction of buildings based on the susceptibility of a locality to earthquake damage.

Earth Science The study of the nature, origin, and evolution of all our natural surroundings through disciplines that include geology, oceanography, atmospheric sciences, and astronomy.

Earth System The geosphere, hydrosphere, atmosphere, and biosphere, along with the intricate ways in which those realms interact with one another over time.

eccentric orbit A highly elliptical path taken by one object rotating around another.

ecliptic plane The plane defined by the Earth's orbit about the sun.

eddy An isolated swirl of current in water or air.

Ediacaran fauna An assemblage of shell-less invertebrates that lived during the last 100 million years of the Proterozoic.

effusive eruption A volcanic eruption dominated by low-viscosity mafic lava.

Ekman transport The overall movement of a mass of water in a direction at a 90° angle to the wind direction; Ekman transport reflects the Ekman spiral, which characterizes the change in motion of water with depth.

elastic deformation A change in the shape of a solid that can be reversed by removing the stress that caused the change.

elastic-rebound theory The concept that elastic deformation builds up in rock adjacent to faults, prior to an earthquake, and the rock then rebounds when the fault slips, generating seismic waves.

electromagnetic radiation Energy that travels in the form of electric and magnetic fields and can pass through the vacuum of space.

electromagnetic spectrum The full range of possible frequencies or wavelengths of electromagnetic radiation, from gamma rays through radio waves, including the visible radiation our eyes can detect.

electromagnetic wave A wave consisting of oscillations in the electric- and magnetic-field strength.

electromagnetism A field force produced by magnetic objects or by electric currents.

electron A negatively charged subatomic particle that orbits the nucleus of an atom; an electron is about 0.0005 times the size of a proton.

element A material consisting entirely of one kind of atom; elements cannot be subdivided, and cannot be changed into other elements by chemical reactions.

El Niño A flow of warm water from west to east in the equatorial Pacific Ocean that reverses the upwelling of cold water along the western coast of South America and causes significant global changes in weather patterns.

embayment A location along a coast where the shoreline curves inland.

emergent coast A coast where the land is rising relative to sea level or sea level is falling relative to the land.

energy The capacity to do work. *Energy* can exist in forms such as electrical, mechanical, chemical, thermal, or nuclear, and can be transformed from one form to another.

energy balance In the context of the Earth System, the relationship between the amount of solar energy arriving at the Earth and the amount of infrared energy radiated by the Earth back to space, averaged over time.

Enhanced Fujita (EF) scale A scale for classifying the intensity of a tornado based on the damage it causes, using correlations of observed damage with wind velocities.

enrichment The process of increasing the concentration of the most radioactive isotope of uranium in a quantity of uranium.

ENSO (El Niño/Southern Oscillation) The overall seesaw pattern of shifts in sea-level atmospheric pressure across the Pacific Ocean, in association with the Walker circulation, that causes alternations between El Niño and La Niña conditions.

environmental lapse rate The rate at which environmental temperature changes with height in the atmosphere.

eon The largest subdivision of geologic time.

ephemeral stream A stream that flows during only part of a year.

epicenter The point on the surface of the Earth directly above the focus of an earthquake.

epicontinental sea A shallow sea overlying a continent.

epoch An interval of geologic time representing the largest subdivision of a period.

equant Having similar dimensions in all directions.

equilibrium line On a glacier, the boundary between the zone of accumulation and the zone of ablation.

era An interval of geologic time representing the largest subdivision of the Phanerozoic Eon.

erosion The grinding away and removal of the Earth's surface materials by moving water, air, or ice.

erosional coast A coastline where sediment is not accumulating and wave action grinds away at the shore.

erosional landform A landform that results from the breakdown and removal of rock or sediment.

erratic A boulder or cobble picked up by a glacier and deposited far from the outcrop from which it detached.

eruptive style The character of a particular volcanic eruption; geologists name eruptive styles based on typical examples (such as Hawaiian, Strombolian).

esker A ridge of sorted sand and gravel that snakes across a ground moraine; deposited by meltwater that tunneled through a glacier,

estuary An inlet in which seawater and river water mix; created when a coastal valley is flooded because of either rising sea level or land subsidence.

eutrophication The transformation of a well-oxygenated body of water into a poorly oxygenated one by an influx of nutrients that causes an algal bloom.

evaporite A sedimentary deposit formed by accumulation of salts, such as halite and gypsum, that happens when salty water evaporates.

event horizon The effective "surface" of a black hole. Nothing inside this surface—not even light—can escape from the black hole.

exhumation The process, involving uplift and erosion, that brings deeply buried rocks to the Earth's surface.

expanding Universe theory A theory, based on the observation that galaxies are moving away from one another, that the volume of the Universe is increasing over time.

explosive eruption A violent volcanic eruption that produces clouds and avalanches of pyroclastic debris.

external energy In the context of the Earth System, energy that comes to the Earth in the form of electromagnetic radiation from the Sun.

extinction The dying off of the last representative of a species of organism.

extinct volcano A volcano that was active in the past, but will not erupt in the future because the geologic conditions that could cause an eruption no longer exist.

extrusive igneous rock Rock that forms by the freezing of lava above ground, after it flows or explodes (extrudes) onto the Earth's surface and comes into contact with the atmosphere or ocean.

eye In the context of cyclone structure, the relatively calm, circular, clear region in the center of a hurricane or typhoon.

eye wall The cylinder of clouds that surrounds the eye of a hurricane.

facet A smooth, shiny face on a gem that forms sharp angles with their neighbors; created by gem cutters by grinding and polishing the stone, rather than by natural cleavage planes or natural crystal faces.

Fahrenheit scale An English-system measure of temperature in which the difference between the freezing point (32°F) and the boiling point (212°F) of water is divided into 180 units.

failure surface The plane on which a mass of material moves downslope during mass wasting.

fault A fracture on which one body of rock slides, or slips, past another.

faulting The process of forming or reactivating faults.

fault scarp A step on the ground surface where one side of a fault has moved vertically with respect to the other.

fetch The distance across a body of water along which a wind blows.

firn Compacted granular ice, derived from snow, that forms where snow is deeply buried and partially recrystallizes; if buried more deeply, firn turns into glacial ice.

fissure eruption Activity of a volcano during which lava comes out of an elongate crack, or from several aligned vents, rather than from a single circular vent.

fjord A deep, glacially carved, U-shaped valley flooded by the sea to form a narrow, elongate bay, or filled with freshwater to form a lake.

flash flood A flood during which the discharge of a stream increases very fast, as may occur during unusually intense rainfall or as the result of a dam collapse.

flood An event during which the volume of water in a stream becomes so great that it covers areas outside the stream's normal channel.

flood basalt Vast sheets of mafic lava that spread from a volcanic vent over the land; flood basalt may form where a rift develops above a continental hot spot.

floodplain The broad, flat area on either side of a stream that becomes covered with water during a flood.

flood stage The level at which a rising river begins to extend beyond its normal channel, and has the potential to damage property.

flux melting The transformation to produce a liquid melt that occurs when volatile material, such as H_2O or CO_2, is added to a hot solid.

focus In the context of seismology, the location where a fault slips during an earthquake; also called a *hypocenter*.

fog A cloud that forms at ground level.

fold A bend or wrinkle in a rock layer that forms as a consequence of deformation.

fold-thrust belt A region containing an assemblage of thrust faults and related folds; such belts typically form during mountain building.

foliation See *metamorphic foliation*.

food web A network of organisms defined by their feeding relationships.

force A push, pull, or shear that can cause an object to move, change shape, or rotate.

foreshock A smaller earthquake that precedes a major earthquake.

fossil A remnant, or trace, of a living organism that has been preserved in rock or sediment.

fossil assemblage A group of fossil species found in a specific sequence of sedimentary rock.

fossil fuel An energy resource, such as oil or coal, that comes from organisms that lived long ago; it stores solar energy that reached the Earth at the time the organisms lived.

fossilization The process of fossil formation.

fossil succession The principle that fossils found higher in a sequence of strata are younger than those found lower in the sequence; fossil succession is a consequence of life's evolution.

fractional crystallization The process by which a magma becomes progressively more felsic as it cools, because mafic components become preferentially incorporated in early-forming minerals that settle out.

fracture zone A narrow band of vertical fractures in the ocean floor, lying roughly at right angles to a mid-ocean ridge. The actively slipping part of a fracture zone is a *transform fault*.

fragmental igneous rock An igneous rock formed from pyroclastic debris that is either cemented or welded together.

frequency The number of waves that pass a point in a given time interval.

freshwater Water that contains very little, if any, salt.

frictional drag The force applied by a moving material to another material across an interface due to the roughness of the interface.

front In the context of atmospheric science, the boundary between air masses that differ in temperature and humidity.

fuel A substance that burns or undergoes a reaction that produces heat.

gabbro A coarse-grained mafic igneous rock.

galactic bulge The central region of a spiral galaxy where stars cluster in a bulge that extends above and below the galactic disk.

galactic disk The plane in which most stars and spiral arms of a galaxy reside.

galactic halo An extended, roughly spherical component of a galaxy that extends beyond the main, visible features.

galaxy An immense accumulation of hundreds of billions of stars that revolves around a supermassive black hole; the Milky Way is a galaxy.

gas A state of matter in which atoms and molecules move about freely. A gas can flow, and can expand or contract when the size of its container changes.

gas giant A giant planet formed mostly of hydrogen and helium. In our Solar System, Jupiter and Saturn are gas giants.

gem A mineral specimen that has been cut and polished and is particularly beautiful or valuable.

general circulation The global-scale pattern of air flow, characterized by semi-permanent features such as the subtropical highs, the subpolar lows, the trade winds, and the jet streams.

geocentric model An old, incorrect idea suggesting that the Earth sits motionless in the center of the Universe, while the stars, other planets, and Sun orbit around it.

geochronology The science of assigning numerical ages to geologic events, usually based on isotopic (radiometric) dating.

geode A cavity in rock lined with euhedral crystals that precipitate out of water solutions passing through the rock.

geographic poles The locations (north and south) where the Earth's axis of rotation intersects the planet's surface.

geologic column A composite stratigraphic chart that represents the entirety of the Earth's history.

geologic cross section A graphical representation of a vertical slice through the Earth that displays the configuration of rock units and geologic structures underground.

geologic history A description of the sequence of geologic events that has taken place in a region.

geologic map A two-dimensional representation showing the distribution of rock units and geologic structures across a region as they would appear if projected onto a horizontal surface.

geologic structure A feature that has formed as a consequence of deformation; examples include faults, folds, joints, and foliation.

geologic time The span of time since the formation of the Earth.

geologic time scale A scale that delineates intervals of geologic time and includes numerical ages assigned to eons, epochs, and periods.

geology The study of the Earth, including our planet's composition, behavior, and history.

geometric parallax The apparent displacement of a foreground object relative to a background object when an observer's line of sight changes.

geosphere The solid Earth component of the Earth System, from its surface to center.

geostrophic balance The balance that exists when the pressure gradient force and the Coriolis force are equal and opposite.

geostrophic current A roughly circular ocean current that results from geostrophic balance.

geostrophic wind A wind that flows parallel to isobars when air in the atmosphere is in geostrophic balance.

geotherm The change in temperature with depth in the Earth.

geothermal energy Heat and electricity produced by using the internal heat of the Earth.

geothermal gradient The rate of change in temperature with depth in the Earth.

geyser A fountain of steam and hot water that erupts periodically from a vent in the ground in a geothermal region.

giant star A star with a substantially larger diameter and luminosity than a main sequence star (on the H-R diagram) with the same surface temperature.

glacial advance The forward movement of a glacier's toe when the rate of accumulation exceeds the rate of ablation.

glacial outwash Sand and gravel deposited on a low-relief landscape by meltwater streams coming from a glacier.

glacial rebound The rising of the land surface after a large overlying ice sheet melts away and the weight of the ice is removed.

glacial retreat A change in position of a glacier's toe back toward the glacier's origin when the rate of ablation exceeds the rate of accumulation.

glacial striation A scratch or groove on rock produced by clasts embedded in flowing glacial ice.

glacial subsidence The sinking of the land surface due to the weight of a large ice sheet.

glacial till Unsorted sediment transported by flowing ice and deposited beneath a glacier or at its toe.

glaciation In the context of discussing ice ages, an interval of time during which glaciers grew and covered substantial areas of the continents.

glacier A sheet or stream of ice that slowly flows across the land surface and lasts all year.

glass An inorganic solid in which atoms are not arranged in an orderly pattern.

glassy igneous rock Igneous rock consisting entirely of glass, or of tiny crystals surrounded by a glass matrix.

global change Transformations or modifications of the Earth System over time.

global climate model (GCM) A numerical model consisting of mathematical equations that simulate changes in global climate conditions over time.

global cooling A fall in the average global atmospheric temperature over time.

global positioning system (GPS) A satellite network that allows determination of the location of a receiver with great accuracy; people can use it to navigate their cars; geologists can use it to measure rates of movement of portions of the Earth's crust relative to one another.

global warming A rise in the average global atmospheric temperature over time.

globular cluster A spherical accumulation of hundreds of thousands of stars.

gneiss A high-grade metamorphic rock composed of alternating bands of dark- and light-colored minerals.

gneissic banding Metamorphic layering in a gneiss, as manifested by alternating lighter and darker bands, also called *compositional banding*.

Gondwana A supercontinent of the Paleozoic Era that consisted of today's South America, Africa, Antarctica, India, and Australia; also called Gondwanaland.

grade See *metamorphic grade*.

graded bed A sedimentary layer in which grain size changes progressively from coarser at the base to finer at the top. Graded beds are formed by turbidity currents.

grain In a geological context, it is a relatively small fragment of a mineral or rock.

granite A coarse-grained, felsic intrusive igneous rock.

graupel A small (<3–4 mm), soft ball of ice that results when an ice crystal collects supercooled water droplets that freeze on its surface.

gravitational energy The potential energy stored in a mass if the mass lies within a gravitational field.

gravity The attractive force that one object exerts on another; its magnitude depends on the amount of matter in the objects, and the distance between the objects.

great oxygenation event The dramatic increase in atmospheric oxygen between about 2.4 and 1.8 Ga, when the Earth's surface rocks and water could no longer absorb or dissolve all the oxygen produced by photosynthetic organisms, so oxygen accumulated in the air.

Great Red Spot The giant oval, brick-red rotating storm seen in the atmosphere of Jupiter's southern hemisphere.

greenhouse effect The trapping of heat in the Earth's atmosphere by carbon dioxide and other greenhouse gases, which absorb infrared radiation.

groundwater Water that resides under the surface of the Earth, mostly in pores or cracks in rock or sediment.

growth rings A pattern of distinct rings that develops in trees and shelly organisms whose growth rates vary with the seasons.

gust front The leading edge of the outflow of a thunderstorm's rain-cooled downdraft; its arrival is marked by a sudden increase of wind speed.

guyot A seamount whose peak has been flattened by erosion before submergence, or by the growth of a coral reef over it while its top was still in shallow water.

gyre A large ocean surface current with a circular path that follows the margins of the basin in which it has formed.

Hadean Eon The oldest eon of the Precambrian; roughly, the time between the Earth's origin and the formation of the first rocks that have been preserved.

Hadley cells Low-latitude convection cells in the troposphere that extend from the equator to a latitude of about 25° in each hemisphere.

hail Frozen precipitation particles with diameters ranging from 3 mm to 20 cm, resulting from the attachment of supercooled liquid droplets to graupel in the strong updrafts of thunderstorms.

half-life The time it takes for half of the atoms of a radioactive isotope to decay.

halocline The boundary in the ocean between surface-water and deep-water salinities.

hand specimen A roughly fist-sized piece of rock that can be examined with a hand lens (magnifying glass).

hanging valley A glacially carved tributary valley whose floor lies at a higher elevation than the floor of the trunk valley with which it intersects.

hardness A measure of the relative ability of a mineral to resist scratching; it represents the resistance of bonds in the mineral to being broken.

haze Murky humid air that exists when the air contains a relatively high concentration of aerosols; the presence of haze decrease visibility.

headland Coastal land that protrudes into the sea.

headward erosion The lengthening of a stream channel by erosion at the stream's origin.

heat The total thermal energy contained in a material, meaning all the energy due to vibrations and movements of all atoms or molecules in a volume of the material.

heat flow The rate at which heat transfers from one location to another, such as from the Earth's interior to its surface.

heat-transfer melting Melting that results when heat from a hotter magma flows into a cooler rock and causes melting of the rock.

heliocentric model A model of the Solar System in which all of its objects, including the Earth, orbit the Sun.

heliosphere A bubble-like invisible surface in space within which most particles (not counting Oort Cloud objects) come from our Sun's solar wind; outside it, most particles come from other stars.

heterosphere A term for the upper portion of the atmosphere, in which gases separate into distinct layers on the basis of composition.

high-pressure system The circulation surrounding a high-pressure center; it is generally associated with sinking air and clear skies.

hinge In the context of geologic structures, the portion of a fold where its curvature is greatest.

homosphere The lower part of the atmosphere, in which the gases have been stirred into a homogeneous mixture by air circulation.

horn A pointed mountain peak on which glaciers have carved cirques on at least three sides.

hornfels A fine- to medium-grained nonfoliated metamorphic rock formed by thermal metamorphism; in such rocks, inequant grains are randomly oriented.

hothouse period An interval of geologic time during which the Earth's atmosphere is warmer than average and ice sheets do not cover polar regions.

hot spot A location where igneous activity occurs independently of plate interactions; they possibly develop where a mantle plume causes melting at the base of the lithosphere, and can exist in plate interiors or along plate boundaries.

hot-spot track A chain of extinct volcanoes transported off a hot spot by the movement of a lithosphere plate.

hot spring A source at which very warm to boiling groundwater spills out of the ground.

H-R diagram The Hertzsprung-Russell diagram, which shows the relationship of a star's luminosity to its surface temperature. Stars organize into distinct groups on the diagram.

Hubble's law A statement that the speed at which a galaxy is moving away from the Earth is proportional to the distance of that galaxy from the Earth.

hurricane A strong tropical cyclone over the North Atlantic or eastern Pacific Ocean in which sustained wind speeds reach 119 km/h (74 mph) or higher.

hurricane track The path that a hurricane follows.

hydrocarbon A chain-like or ring-like molecule made of hydrogen and carbon atoms; examples are petroleum and natural gas.

hydrocarbon reserve A significant quantity of potentially extractable oil or gas underground.

hydrofracturing The process of injecting water and other chemicals into a drill hole under high pressure to generate cracks in the surrounding rock that provide permeability pathways for hydrocarbons; informally known as fracking.

hydrologic cycle The flow of water from reservoir to reservoir in the Earth System.

hydrolysis The process by which water chemically reacts with minerals and produces new minerals; the transformation of feldspar into clay during weathering is an example.

hydrosphere The Earth's solid and liquid water, including surface water (lakes, rivers, and oceans), groundwater, glaciers, and permafrost.

hydrothermal fluid A hot water or steam solution that circulates through rock underground.

hypothesis A possible explanation for a set of data.

ice In the context of space science, the solid state of a volatile material; in everyday English, it refers to the solid-state version of water.

ice age An interval of geologic time in which the climate was colder than it is today, glaciers occasionally advanced to cover large areas of the continents, and mountain glaciers grow; an ice age can include many glaciations and interglacials.

iceberg A large block of ice floating in the sea; typically, they form by calving from a tidewater glacier or an ice shelf.

ice giant A giant planet formed mostly of solid volatile substances (ices). In our Solar System, Uranus and Neptune are ice giants.

icehouse period An interval of geologic time when the Earth's atmosphere is cooler than average and ice sheets cover polar regions.

ice shelf A broad, flat region of ice along the edge of a continent formed where a continental glacier flows into the sea.

ice storm A winter storm that produces an accumulation of ice at the ground surface, as well as on trees, power lines, and buildings.

igneous activity The overall process during which rock deep underground melts, producing magma that rises up into the crust where it can fill magma chambers or intrude existing rock; in some cases,

magma rises to the Earth's surface where it erupts as lava or pyroclastic debris.

igneous intrusion A body of rock that solidifies underground.

igneous rock Rock that forms when magma or lava cools and freezes solid.

inclusion In a general sense, a fragment or droplet of a relatively older material contained within a relatively younger rock.

inequant Having different dimensions in different directions; for example, the length and width in an inequant object are not the same.

infiltration The migration of a liquid through a porous material.

inner core The central portion of the Earth, extending from 5,155 km (3,203 mi) deep, to the Earth's center at 6,371 km (3,159 mi); it consists of solid iron alloy, in contrast to the outer core, which consists of liquid iron alloy.

intensity In the context of discussing earthquakes, the degree of ground shaking caused by an earthquake at a locality.

interglacial An interval of time between two glaciations.

Intergovernmental Panel on Climate Change (IPCC) An international group of scientists charged with evaluating and summarizing research pertaining to how the Earth's climate is changing and how humans are influencing those changes.

internal energy In the context of geology, the thermal energy that rises from the interior of the Earth.

interplanetary space The space between planets.

intertidal zone The area of coastal land across which the tide rises and falls.

intertropical convergence zone (ITCZ) The zone where the trade winds (the surface flow of the Hadley cells) of the northern and southern hemispheres converge in the vicinity of the equator.

intraplate earthquake An earthquake that occurs away from plate boundaries.

intrusive igneous rock Rock formed by the freezing of magma underground.

inversion A layer in the atmosphere in which the temperature increases with altitude.

ionosphere The region of the upper atmosphere, above 60–100 km (37–62 mi), that has a high concentration of ions.

island arc A chain of volcanic islands that builds up on the seafloor where one oceanic plate subducts beneath another.

isobar A contour line on a map along which all points have the same atmospheric pressure at a given elevation.

isostasy The condition that exists when the buoyancy force pushing lithosphere up equals the gravitational force pulling lithosphere down.

isotope A version of a given element that has the same atomic number as other versions, but a different atomic mass.

isotopic dating A tool used by geologists to provide a rock's numerical age, meaning the age of the rock in years; also called radiometric dating.

joint A naturally formed crack in rock; no shear displacement has happened on a joint.

Jovian planet One of the four outer planets of the Solar System—Jupiter, Saturn, Uranus, and Neptune—all of which are much larger and less dense than any of the *terrestrial planets* and lack a solid surface.

karst landscape A region underlain by caves formed in limestone bedrock, some of which have collapsed, so that the land surface has sinkholes separated by limestone ridges or spires.

katabatic wind Downslope near-surface wind caused by drainage of very cold, dense air from a location where the Earth's surface is higher to where it is lower; such winds are common along the edges of ice sheets and large glaciers.

Keeling curve A graph, begun by Charles Keeling in 1958, showing changes in the concentration of atmospheric carbon dioxide over time; it reveals both seasonal changes and the overall increase that takes place from year to year.

Kepler's laws The three rules of planetary motion inferred by Johannes Kepler from data collected by Tycho Brahe.

kerogen The waxy molecules into which the organic material in shale transforms on reaching 50°C–90°C. At higher temperatures, kerogen transforms into oil.

kettle hole A circular depression in a glacial moraine made where a block of ice buried by sediment at the toe of a glacier later melts.

kinetic energy The energy of motion.

Köppen-Geiger climate classification (KGCC) A characterization of the Earth's climate zones based on temperature and precipitation, developed by Wladimir Köppen and Rudolf Geiger.

K-Pg extinction (formerly known as the K-T extinction) A sudden mass die-off of all species of dinosaurs, and many other species, at the end of the Cretaceous Period (66 Ma). It has been attributed to an immense meteorite impact.

Kuiper Belt A diffuse ring of icy objects that orbit the Sun outside the orbit of Neptune.

lag deposit The coarse sediment left behind in a desert after wind erosion removes the finer sediment.

lagoon A body of shallow seawater separated from the open ocean by a barrier island.

lahar A flowing slurry composed of volcanic ash and debris mixed with water, formed when rain or melting snow and ice provides abundant water on the flank of a volcano during or after an ash-rich volcanic eruption.

lake-effect storm A snowstorm over and immediately downwind of a large lake, triggered by the flow of very cold air over relatively warm lake water.

land An area of the Earth's surface not covered by water.

landform A distinctive natural feature of the land surface.

landscape The character and shape of the land surface in a region.

landslide A general term for a mass-wasting event during which rock and/or regolith moves downslope.

landslide-potential map A map, based on hazard-assessment studies, on which regions are ranked according to the likelihood that mass wasting will occur there.

La Niña A strengthening of the Walker circulation and the trade winds that causes a flow of water from east to west in the equatorial Pacific Ocean and, therefore, increases the upwelling of cold water along the western coast of South America; the phase of the Southern Oscillation opposite to an El Niño event.

lapilli Pyroclastic particles 2–64 mm in diameter (marble- to golf-ball-sized) consisting of frozen lava clots, fragments of pre-existing rock, or ash clumps.

large igneous province (LIP) A region where huge volumes of flood basalts erupted over a relatively short interval of geologic time.

late heavy bombardment The battering of the inner planets of the Solar System by meteorites during the time interval between about 4.0–3.9 Ga; it may have pulverized and remelted the Earth's surface.

latent heat The energy required for, or released by, a transition between two states (solid, liquid, gaseous) of a substance. Condensation releases latent heat to air, while evaporation absorbs it from air.

Laurentia A continent that existed during the Paleozoic Era and consisted of today's North America and Greenland.

lava Molten rock that has flowed out onto the Earth's surface.

lava dome A dome-like mass of rhyolitic lava built over a volcanic vent.

lava flow A sheet or mound of lava that flows onto the ground surface or seafloor in molten form and then solidifies.

lava fountain An eruption of lava that is under pressure, during which lava spurts into the sky.

lava tube The empty tunnel-like space left when lava flowing under the already solidified crust of a lava flow drains away.

lifting mechanism A process that causes air near the ground surface to rise to higher elevation. Examples include the ascent of air at a frontal boundary, air flow over mountains, and convergent winds near the Earth's surface.

lightning An electrostatic discharge in the atmosphere that occurs when negatively and positively charged particles become separated within a cloud; the discharge can take place between different parts of the cloud, or between the cloud and the ground.

light pollution Brightening of the atmosphere produced by city lights that can obscure stars in the night sky so that people can't see them from the Earth's surface.

light scattering The absorption of light, and its immediate re-emission in random directions, by molecules and particles in the atmosphere or ocean.

light-year The distance that light travels in one Earth year (about 9.5 trillion kilometers, or 6 trillion miles).

limb In the context of geologic structures, a side of a fold, with less curvature than a hinge.

limestone Sedimentary rock composed mostly of calcite.

liquid A state of matter in which atoms or molecules can move relative to one another, but remain in contact. A liquid can flow and conform to the shape of its container.

lithification The transformation of loose sediment into solid rock through compaction and cementation.

lithosphere The relatively rigid outer layer of the Earth, consisting of the crust and the top part of the mantle; it is about 100–150 km (62–93 mi) thick.

lithospheric mantle The part of the Earth's mantle that lies within the lithosphere.

littoral zone The nearshore area of a body of water. In the ocean, it includes the intertidal zone but may also include adjacent shallow waters.

Local Group The cluster of relatively nearby galaxies that includes the Milky Way and Andromeda.

loess Layers of fine-grained sediments (clay and silt) deposited by the wind.

longshore current A flow of water parallel to the shore just off a coast that develops when waves move toward the shore obliquely.

longshore drift The net transport of sediment laterally along a beach that occurs when waves wash up a beach obliquely.

lower mantle The deepest section of the Earth's mantle, extending from 670 km (416 mi) down to the core-mantle boundary.

low-pressure system The circulation surrounding a center of low pressure; it is generally associated with rising air, clouds, and precipitation.

low-velocity zone (LVZ) A region of the mantle just below the oceanic lithosphere in which seismic waves travel more slowly than they do in the lithospheric mantle above or in mantle deeper down; it is thought to occur due to slight partial melting of the mantle rock.

luminosity The amount of electromagnetic energy emitted by a star in a specified amount of time.

lunar eclipse A darkening of the Moon's surface, as seen from the Earth, that occurs when the Moon is partially or entirely in the Earth's shadow.

lunar highland High-elevation regions on the Moon that appear light-colored from the Earth relative to the darker maria.

lunar month The interval of time between successive new Moons (roughly 29½ days).

lunar regolith The loose surface material found on the Moon, including micrometeorites, pulverized rock scattered from impacts, and dust produced by space weathering.

luster The way a mineral surface reflects light.

magma Molten rock beneath the Earth's surface.

magma chamber A space below ground filled with magma or a mush of magma and crystals.

magma source The rock from which a magma was extracted.

magnetic anomaly The difference between the expected strength of the Earth's magnetic field at a certain location and the actual measured strength of the field at that location.

magnetic field The region affected by the force emanating from a magnet.

magnetic pole The end of a magnetic dipole; all magnetic dipoles, including the Earth, have a north magnetic pole and a south magnetic pole.

magnetic reversal A change in the Earth's magnetic polarity; when a reversal occurs, the field flips from normal to reversed polarity, or vice versa.

magnetic-reversal chronology The history of magnetic reversals and their durations through geologic time.

magnitude (1) A number, on a logarithmic scale, representing the amount of energy released by an earthquake. (2) The luminosity of a star; see *absolute magnitude, apparent magnitude*).

main sequence The band on the H-R diagram where most stars appear. Main-sequence stars are fusing hydrogen nuclei into helium nuclei in their cores.

mainshock The main or largest earthquake of a sequence; it may be preceded by foreshocks and is always followed by aftershocks.

mantle The thick layer of almost entirely solid rock below the Earth's crust and above the core.

mantle plume A column of very hot rock rising through the mantle.

marble A metamorphic rock composed of calcite and transformed from a protolith of limestone.

mare (plural: maria) A broad, dark area on the Moon's surface, consisting of flood basalts that erupted over 3 billion years ago and spread out across a lunar lowland.

marine geology The study of the seafloor and of oceanic coasts.

marine magnetic anomaly One of a series of alternating positive and negative magnetic anomalies on the seafloor, which together define a pattern of alternating bands parallel to mid-ocean ridges.

mass The amount of matter in an object; mass differs from weight in that its value does not depend on the strength of gravity.

mass-extinction event A relatively brief interval of geologic time when vast numbers of species abruptly vanish.

mass wasting The gravitationally caused downslope transport of rock, regolith, snow, or ice.

matter The material substance of the Universe, which consists of atoms and has mass.

maturity The degree to which a sediment has changed between the place where it originated and the place where it was deposited.

meander A snake-like curve in a stream channel.

meandering stream A stream that contains meanders.

melt In the context of igneous rocks, the molten (liquid) state of rock.

mesocyclone A strong, long-lasting, rotating updraft within a supercell thunderstorm, typically 5–10 km (3–6 mi) in diameter; tornadoes may form in the mesocyclone.

mesosphere The cooler layer of atmosphere overlying the stratosphere.

metal A solid composed almost entirely of atoms of metallic elements; it is generally opaque, shiny, smooth, malleable, and can conduct electricity.

metallic hydrogen A form of hydrogen that occurs only under the tremendous pressures found near the core of a large planet such as Jupiter.

metamorphic aureole The region around an igneous intrusion in which heat transferred into the wall rock has metamorphosed the wall rock.

metamorphic conditions The temperature and pressure at which metamorphic minerals and textures form.

metamorphic facies A set of metamorphic mineral assemblages indicative of metamorphism under a specific range of temperatures and pressures.

metamorphic foliation Parallel surfaces or layers that develop in a rock as a result of metamorphism; schistosity and gneissic banding are examples.

metamorphic grade An informal indication of the intensity of metamorphism to which a rock has been subjected.

metamorphic mineral A mineral, not present in the protolith, that grows during metamorphism.

metamorphic reaction A process during which atoms separate from pre-existing minerals, diffuse through the solid mineral, then bond in a different configuration to form mineral grains that were not present in the protolith.

metamorphic rock Rock that forms when pre-existing rock is transformed into new rock, without first becoming a melt or a sediment, in response to changes in temperature, pressure, and/or interaction with chemically active fluids.

metamorphic texture A distinctive arrangement or orientation of mineral grains in metamorphic rock, not found in the protolith.

metamorphism The process by which one kind of rock is transformed into a different kind of rock by an increase in temperature and/or pressure, or by shearing under elevated temperatures, or by reaction with chemically active fluids.

metasomatism The process by which a rock's overall chemical composition changes during metamorphism because of reactions with hydrothermal fluids that bring in or remove elements.

meteor The incandescent trail produced by a small piece of interplanetary debris as it travels through the atmosphere at very high speeds.

meteorite A piece of rock or metal alloy that fell from space and landed on the Earth.

meteoroid A small fragment of planetary debris (less than 1 m, or 3 ft, across) that enters the Earth's atmosphere.

meteorologist A person who studies and forecasts weather and its consequences.

meteorology The study of weather, as well as of the movement of air and its consequences.

meteor shower A larger-than-normal display of meteors that occurs when the Earth passes through the orbit of a comet and intersects its dusty tail.

microcontinent A block of continental crust too small to be considered a continent; geologists use the term in reference to blocks of continental crust that collide with a larger continent.

mid-latitude cyclone A large, comma-shaped low-pressure system that forms along the polar-front jet stream between about 30° and 60° latitude, and produces many types of weather, some of which are hazardous; these systems are also known as extratropical cyclones.

mid-ocean ridge A submarine belt of elevated seafloor that forms along a divergent plate boundary.

Milankovitch cycle Any of the three cycles of change in the Earth's orbit shape, axial tilt, and wobble that occur over tens to hundreds of thousands of years and influence climate on the Earth.

Milky Way galaxy The galaxy in which our Solar System resides.

mineral A naturally occurring, homogeneous, crystalline solid that has a definable chemical composition and, in most cases, is inorganic.

mirage An optical illusion caused by specific atmospheric conditions that cause light rays to bend.

Modified Mercalli Intensity scale A scale for assessing an earthquake's intensity by the damage it causes and by people's perception of the shaking.

Moho The seismic-velocity discontinuity that defines the boundary between the Earth's crust and mantle.

Mohs hardness scale A list of 10 minerals, in a sequence of relative hardness, with which other minerals can be compared.

molecule A particle that consists of two or more atoms attached by chemical bonds.

moment magnitude scale A logarithmic scale for assessing the size of an earthquake using measurements of the amplitudes of seismic waves and the dimensions of the slipped area and the displacement on the fault; it has replaced the Richter scale in modern descriptions of earthquake size.

monocline A fold whose shape resembles that of a carpet draped over a stair step.

monsoon A seasonally changing air circulation in tropical of the world in which summer winds blow from the ocean toward the land, bringing heavy rain, and winter winds blow from the land toward the ocean.

moon A sizable solid object that orbits a planet. The word is usually capitalized when referring to the Earth's Moon.

moraine A pile of till deposited by a glacier.

mountain belt An elongate region of uplifted crust and rugged topography.

mountain building The set of events that produces a mountain belt.

mountain glacier A glacier that exists in or adjacent to a mountainous region; also called an alpine glacier.

mudcrack A sedimentary structure formed when mud dries, shrinks, and breaks into roughly hexagonal plates separated by open gashes.

mudflow A downslope movement of mud at slow to moderate speed.

natural arch A bridge-like span of rock that forms when erosion along joints leaves a narrow wall of rock, and the lower-middle part of the wall then erodes while the upper part remains intact.

natural change Change in the Earth System due to geologic, meteorological, or astronomical events that would happen whether or not people inhabited the Earth.

natural gas A hydrocarbon, such as methane or propane, that can exist in gaseous form at the Earth's surface.

natural levee Ridges that develop on either side of a stream as a result of the accumulation of sediment deposited naturally during flooding.

natural variation The range of changes in a system caused by the variability of natural phenomena affecting the system.

near-Earth object An asteroid, comet, or large meteoroid whose orbit intersects or nearly intersects the Earth's orbit.

nebula A cloud of gas, ice and dust in space, often a location of star formation.

nebular theory The theory that stars form within patches of nebulae that collapse inward in response to gravity.

negative feedback A process that leads to a change that reduces the rate at which the original process occurs.

nekton Organisms, such as fish, that actively swim in open water and can move against a current.

neritic zone The region from the edge of the continental shelf to the shore, where the sea is less than 200 m (660 ft) deep and sunlight can penetrate to the seafloor.

neutron A neutral subatomic particle in the nucleus of an atom.

neutron star The superdense core of a high-mass star left behind by a supernova.

nonconformity An unconformity in which sedimentary strata are in contact with underlying igneous or metamorphic rocks.

nonmetallic mineral resource Mineral resources that do not contain metals; examples include building stone, gravel, sand, gypsum, phosphate, and salt.

nonrenewable resource A resource that nature will take a long time (hundreds to millions of years) to replenish.

normal fault A dip-slip fault on which the hanging-wall block moves down the slope of the fault.

normal galaxy A galaxy that emits most of its radiation at or near visible wavelengths of light.

normal polarity An orientation of the Earth's magnetic dipole in which the north magnetic pole is near the north geographic pole, as it is today.

nova (plural: novae) A stellar explosion that results from runaway nuclear fusion in a layer of accreted material on the surface of a white dwarf in a binary star system.

nuclear fission A reaction in which a nucleus splits into two smaller nuclei.

nuclear fusion A reaction in which the nuclei of atoms fuse together, yielding new, larger atoms.

nuclear reactor The part of a nuclear power plant where nuclear fission occurs.

nuclear waste Radioactive byproducts or contaminated materials that could pose a danger to people.

nucleus (plural: nuclei) The cluster of protons and neutrons (except for hydrogen, whose nucleus contains only a proton) that occurs at the center of an atom.

numerical age The age of a geologic material or feature specified in years; also known as absolute age.

nutrient A chemical essential for the survival of living organisms.

observatory A place where astronomers use telescopes to observe objects and features in space.

obsidian An igneous rock consisting of a solid mass of volcanic glass.

occluded front A boundary between air masses that forms when a cold front overtakes a warm front and lifts the cold air behind it, so that the warm front no longer intersects the ground surface, and the two cold air masses come into contact.

ocean A large body of saltwater between continents.

ocean basin A broad, low area of the Earth's surface floored by oceanic crust and filled with saltwater.

oceanic island A volcanic peak that has built up above the surrounding seafloor and rises above sea level.

oceanic zone The deep, open-water portion of the ocean that does not interact with the coast.

oceanography The study of the water and life in the oceans, as well as the way in which ocean water moves and interacts with land and air.

oil A substance composed of hydrocarbons that exists in liquid form at room temperature.

Oil Age A period of human history, extending to the present, during which the economy has depended on oil.

oil seep A site where hydrocarbons naturally seep out onto the surface of the Earth from underground.

oil shale Shale containing kerogen.

oil window The narrow range of temperatures under which oil can form in a source rock.

Oort Cloud A cloud of icy objects that orbit the Sun in a region outside of the heliosphere.

open-pit mine A large excavation used to access energy or mineral resources that are relatively close to the Earth's surface.

optical telescope A type of telescope that employs visible light to observe objects in space.

optical window The range of visible and near-infrared wavelengths to which the atmosphere is transparent.

orbit The path taken by one object revolving around another object under the influence of their mutual gravitational or electrical attraction.

ordinary thunderstorm A thunderstorm that does not rotate, meaning that it does not contain a mesocyclone.

ore Rock containing native metals or an accumulation of ore minerals in sufficient concentration that the rock might be worth mining.

ore deposit A region where geologic processes have concentrated ore minerals.

ore mineral A mineral that contains metal in high concentrations and in a form that can be easily extracted.

organic chemical A carbon-containing compound that occurs in living organisms or resembles compounds produced in living organisms, in which the carbon atoms are often arranged in chains or rings and bonded to hydrogen, nitrogen, or oxygen atoms.

organic coast A coast along which living organisms, such as mangroves or corals, control landforms along the shore.

organic sedimentary rock Sedimentary rock (such as coal) formed from plant debris.

original horizontality The geologic principle stating that sedimentary beds, when originally deposited, are nearly horizontal.

orogen A linear belt in which mountain building produced deformation, metamorphism, igneous rock, and elevated land, that can still be visible even when the topography has been eroded away.

orogeny The process of producing an orogen.

orographic lifting The forced ascent of air on the windward side of a mountain range.

outcrop An exposure of bedrock at the Earth's surface.

outer core The section of the Earth's core that lies between the mantle and the inner core, extending from 2,900-5,150 km (1,802- 3,200 mi) deep; it consists of liquid iron alloy.

oxbow lake A meander that has been cut off, yet remains filled with water.

ozone hole An area of diminished ozone concentration in the stratosphere over the Antarctic polar region.

pahoehoe A lava flow with a surface texture of smooth, glassy, rope-like ridges.

paleoclimate The past climate of the Earth.

paleogeographic map A portrayal of the distribution of land and sea at times in the past.

paleomagnetism A record of the orientation of the Earth's magnetic field preserved in rock.

paleontology The study of fossils.

paleopole The supposed position of the Earth's magnetic pole in the past with respect to a particular continent.

Pangaea A supercontinent that assembled at the end of the Paleozoic Era and included nearly all land.

Pannotia A supercontinent that may have existed at some time between 800 Ma and 600 Ma.

parent isotope A radioactive isotope that undergoes decay.

parsec Short for *parallax second*; the distance to a star with a geometric parallax of 1 arc-second (1/3,600 of a degree) using a base of 1 astronomical unit (AU). One parsec is approximately 3.3 light-years.

partial melting The melting in a rock during which only a small percentage of the rock actually becomes liquid; partial melts tend to be more felsic than the rock from which they were derived.

passive margin A continental margin that is not a plate boundary.

passive-margin basin A region along a tectonically inactive continental margin that had been stretched during rifting, continued to subside due to cooling after rifting ceased, and has filled with a very thick accumulation of sediment.

patterned ground A polar landscape in which the ground splits into pentagonal or hexagonal shapes.

peat Compacted and partially decayed vegetation that accumulates beneath a swamp.

pelagic zone The region of the ocean above the deep seafloor.

perched water table The top surface of a lens of groundwater that lies above a layer of impermeable rock above the regional water table.

peridotite A coarse-grained ultramafic rock.

period The time it takes for a regularly repetitive process to complete one cycle.

permafrost Permanently frozen ground.

permanent stream A stream that flows year-round.

permeability The degree to which a material allows fluids to pass through it via an interconnected network of pores and cracks.

Phanerozoic Eon The most recent eon, an interval of time from 541 Ma to the present.

phase One of the various shapes of the sunlit surface of the Moon, which change over the 29½-day cycle caused by change in the location of the Earth relative to both the Sun and the Moon. Also, the state of a material (solid, liquid or gas).

phase change (1) The transformation of a mineral into another mineral that has the same chemical composition but a different crystal structure. (2) The change of state of a substance, such as water to ice, or water to vapor.

photic zone The upper layer of the ocean that sunlight penetrates.

photomicrograph A photograph taken through the lens of a microscope.

photon A discrete unit or particle of electromagnetic radiation.

photosphere The visible surface of the Sun, from which the light that we see radiates.

photosynthesis The process by which plants, algae, and cyanobacteria convert energy from sunlight into chemical energy.

photovoltaic cell A wafer of materials that together can transform solar radiation into electricity.

phyllite A fine-grained metamorphic rock with foliation defined by the preferred orientation of very fine-grained mica.

physical weathering The process by which intact rock breaks into smaller grains or chunks.

phytoplankton Plankton that can carry out photosynthesis.

pillow basalt Glass-encrusted basalt blobs that form when mafic magma extrudes under water and cools very quickly.

planet A large spherical mass that directly orbits a star and has incorporated all the matter that lies in or near its orbit.

planetesimal A solid body with a diameter greater than 1 km that forms in the protoplanetary disk of a collapsing nebula.

plankton Organisms that float in or on a water body, but are not capable of moving against a current.

plasma A material consisting almost entirely of *ionized atoms* and free electrons; considered by physicists to be the fourth state of matter.

plastic deformation A process by which materials subjected to stress change shape without cracking or breaking.

plate In the context of plate tectonics, one of about 20 distinct pieces of the Earth's lithosphere.

plate boundary The border between two adjacent lithosphere plates, defined by seismicity and other geologic features.

plate interior A region away from plate boundaries, which consequently experiences few earthquakes.

plate tectonics The theory that the outer layer of the Earth (the lithosphere) consists of separate plates that move with respect to one another.

playa A flat, typically salty lake bed that remains when all the water evaporates from a desert lake.

Pleistocene Ice Age The interval of time, from about 2.6 Ma to 14,000 years ago, during which the Earth experienced an ice age.

plunge The angle of a linear structure with respect to the horizontal, as measured in a vertical plane; the angle at which a linear structure tilts.

pluton An irregular or blob-shaped igneous intrusion; plutons can range in size from tens of meters to tens of kilometers across.

pluvial lake A lake formed at a distance from a continental glacier as a result of enhanced rainfall during an ice age.

point bar A wedge-shaped deposit of sediment on the inside bank of a meander.

polar front The boundary between the cold air mass located over the polar and subpolar regions and the warm air mass located over the tropics and subtropics; it encircles the globe in each hemisphere.

polar-front jet stream A band of strong winds, found at the top of the troposphere above the polar front, that encircles each hemisphere at middle or high latitudes. Its trace defines a wavelike pattern in map view; it is most prominent during the winter months, although it exists during all seasons.

polarity chron A time interval between reversals in the polarity of the Earth's magnetic field.

pore A small, open space within sediment or rock.

porosity The total volume of empty space (pore space) in a material, usually expressed as a percentage.

positive feedback A process that leads to a change that increases the rate at which the original process occurs.

potential energy Energy stored in a material that can be released later.

pothole A bowl-shaped depression carved into a streambed by a long-lived whirlpool carrying sand or gravel.

Precambrian The interval of geologic time between the Earth's formation, about 4.57 Ga, and the beginning of the Phanerozoic Eon, about 541 Ma.

precession The conical path traced out by the Earth's axis of rotation; simply put, it is the "wobble" of the axis.

precipitation (1) The process by which atoms dissolved in a solution come together and form a solid. (2) Atmospheric water that condenses to a liquid or solid state and falls from the sky.

preferred orientation The metamorphic texture that exists where platy grains lie parallel to one another and/or elongate grains align in the same direction.

preservation potential The likelihood that an organism will be buried and transformed into a fossil.

pressure Force per unit area, or the push acting on a material in cases where the push is the same in all directions.

pressure gradient The rate of pressure change over a given horizontal distance.

pressure-gradient force The force applied to a small region of air or water due to the variation of pressure over a small distance around it.

primary producer An organism in the food web that does not survive by eating other organisms, but rather by photosynthesis or by extracting chemical energy from minerals.

Proterozoic Eon The latest eon of the Precambrian.

protolith The original rock from which a metamorphic rock formed.

proton A positively charged subatomic particle in the nucleus of an atom.

protoplanet A body that is growing in the protoplanetary disk of a collapsing nebula by the accumulation of planetesimals, but has not yet become big enough to be called a planet.

protoplanetary disk The outer part of the accretion disk around a young star, from which a planetary system may form.

protostar A dense ball of gas, at the center of an accretion disk, that is collapsing inward because of gravitational forces and has begun to emit radiant energy.

pulsar A rapidly rotating neutron star that emits radio waves at regular intervals.

pumice A glassy igneous rock that forms from frothy felsic lava and contains abundant small vesicles.

P-wave shadow zone A band between 103° and 143°, as measured around the Earth from either side of an earthquake's epicenter, where P-waves do not arrive at seismometer stations because they are refracted by the core-mantle boundary.

pycnocline A boundary between layers of water of different densities in an ocean or lake.

pyroclastic debris Material ejected by an erupting volcano that lands on the ground or seafloor in solid form. It can include both newly solidified lava and debris that had already been part of the volcanic edifice.

pyroclastic flow A fast-moving, extremely-hot avalanche that occurs when volcanic ash and debris mix with air and flow down the side of an erupting volcano.

pyroclastic rock Rock made from fragments that are ejected by a volcano, then either are welded together or undergo compaction and cementation.

quartzite A metamorphic rock composed of interlocking quartz grains, formed by recrystallization of a quartz-sandstone protolith.

quasar Short for *quasi-stellar radio source*. Extremely bright centers of active galaxies; they have a luminosity approaching that of a thousand Milky Ways.

radiation Energy that travels away from its source through a medium or a vacuum.

radiation window A range of radiation wavelengths that can pass between the Earth and space without being absorbed by the atmosphere; also called an *atmospheric window*.

radiative zone A thick layer surrounding the solar core through which energy passes in the form of electromagnetic radiation.

radioactive decay The process by which an atom of a radioactive element releases subatomic particles. As a consequence, the atomic number of the atom changes, so the atom becomes a different element.

radioactive element An element whose atoms spontaneously decay by releasing subatomic particles from the nucleus.

radio interferometer An instrument that superimposes the waves collected by numerous radio telescopes arranged in an array, effectively acting like a single huge radio telescope whose diameter equals the distance from one side of the array to the other.

radio telescope An instrument for detecting and measuring radio-frequency emissions from celestial sources.

radio window The range of radio wavelengths to which the atmosphere is transparent.

rainbow An arc in the sky displaying all colors of visible light, formed by the refraction and reflection of light through water droplets in the atmosphere.

raindrop A small, near-spherical ball of liquid water that is heavy enough to fall through the atmosphere; surface tension smooths the surface of a raindrop.

rapid A reach of a stream in which the water surface is particularly turbulent. Mixing water with air in a rapid produces whitewater.

recharge area A location where water enters the ground, infiltrates downward to the water table, and adds to the groundwater supply at a locality.

recrystallization A metamorphic process that changes the shape and size of mineral grains in a rock, and in some cases the way in which the grains are held together, without changing the identity of the minerals involved.

recurrence interval The average time between successive geologic events. The term is often used in the context of characterizing the frequency of damaging geologic events, such as floods and earthquakes.

redbed A nonmarine clastic sedimentary layer that has a reddish color due to the presence of oxidized iron minerals, which were formed by reaction of iron with oxygen.

red dwarf A small, relatively cool star that resides on the main sequence of the H-R diagram.

red giant A huge, relatively cool star formed when a Sun-sized star starts to die and expands; it resides in an area above and to the right of the main sequence of the H-R diagram.

red shift A phenomenon in which light emitted by a source moving away from an observer appears to shift to a lower frequency.

reef bleaching Loss of color and death of a coral reef resulting from the loss or death of the corals' symbiotic algae.

reflecting telescope A telescope that uses mirrors for collecting and focusing incoming electromagnetic radiation to form an image.

refracting telescope A telescope that uses objective lenses for collecting and focusing incoming electromagnetic radiation to form an image.

refractory material A substance that has a relatively high melting point and tends to exist in solid form at the Earth's surface.

regional metamorphism Metamorphism over a broad region, generally due to deep burial and deformation of rock during mountain building; also called *dynamothermal metamorphism*.

regolith A general term for unconsolidated material at the Earth's surface; it includes soil, uncemented sediment, and weathered and broken-up rock.

regression The seaward migration of a shoreline caused by a lowering of sea level.

relative age The age of one geologic feature with respect to another.

relative humidity (RH) The ratio of the amount of water vapor in the atmosphere to the atmosphere's capacity for holding water vapor at a given temperature; expressed as the vapor pressure divided by the saturation vapor pressure.

relative plate velocity The rate and direction of movement of a lithosphere plate with respect to another lithosphere plate.

relief The difference in elevation between adjacent high and low regions on the land surface.

renewable resource A resource that can be replenished in a time span of years to decades.

reservoir In the context of the Earth System, a realm that can contain a quantity of material.

reservoir rock Rock with high porosity and permeability that can contain an abundant amount of easily accessible oil and gas.

residence time The average length of time that a material stays in a particular reservoir.

resolving power The ability of a telescope or film to separate or distinguish small or nearby objects.

resource In a geological context, an Earth material used by society for any of a variety of purposes, such as construction, energy production, tool making, and fertilizer.

reversed polarity An orientation of the Earth's magnetic dipole in which the north magnetic pole is near the south geographic pole.

reverse fault A dip-slip fault on which the hanging-wall block undergoes displacement up the slope of the fault.

Richter scale A logarithmic scale that defines earthquakes based on the amplitude of the largest ground motion that would be recorded by a seismometer at a specified distance from the earthquake.

ridge An elongate area of high atmospheric pressure. In the vicinity of the polar-front jet stream, it is located where the jetstream bows toward the poles.

ridge-push force The outward-directed push in a plate caused by the gravitational potential energy of the elevated lithosphere at a mid-ocean ridge.

rift In a geological context, a distinct linear belt in which rifting is taking place or has occurred.

rifting The process of stretching and breaking a continent apart.

ring In the context of space science, a thin band of tiny objects in orbit around a planet's equator; all the Jovian planets have rings, though only Saturn's can be easily seen from the Earth.

ripple A very small wave formed by the frictional drag of the wind on the surface of a water body.

ripple marks Small wave-like ridges and troughs on the surface of a sedimentary layer, formed by flowing air or water and preserved in rock.

roche moutonnée A glacially eroded hill that is elongate in the direction of glacial flow and asymmetrical in a cross section drawn parallel to its length. Glacial rasping smooths the upstream part of the hill into a gentle slope, while glacial plucking erodes the downstream part into a steep slope.

rock A coherent, naturally occurring solid, consisting of an aggregate of minerals or a mass of glass.

rock cycle The progressive transformations that result in the passage of atoms through different rock types over geologic time.

rockfall Mass wasting that involves the sudden drop of a mass of rock from a vertical cliff or overhang, so that for part of its downward journey, the rock free-falls through air.

rockslide Mass wasting that occurs when a mass of rock detaches from its substrate on a failure surface and moves down a non-vertical slope.

rocky coast An area of coast where bedrock rises directly from the sea and beaches are absent.

Rodinia A supercontinent that may have existed around 1 Ga.

rogue wave An isolated wave that rises more than twice as high as most large waves passing a locality during a specified time interval.

rotary drill A machine for producing a hole in bedrock, consisting of a metal tube tipped by a drill bit composed of material hard enough to grind into rock, which cuts into the rock as it rotates.

rotation The process of turning around an axis.

runoff Water that flows downslope over the land surface as sheetwash or in stream channels.

salinity The concentration of salt in water.

saltation The movement of a sediment in which grains bounce along their substrate, knocking other grains into the air in the process.

salt lake A standing body of water that contains a high concentration of salt. They typically form when a lake has no outlet, so that evaporation of water concentrates salt in the water that remains.

salt wedging A physical weathering process in which salt dissolved in groundwater crystallizes and grows in open pore spaces in rocks and pushes apart the surrounding grains.

sand dune A relatively large ridge of sand formed by wind deposition, within which cross bedding is typically found.

sand spit A ridge of sand, parallel to the shore, that stretches out into open water, typically across part of the mouth of a lagoon or bay.

saturation vapor pressure The vapor pressure at which air is saturated at a given temperature.

schist A medium- to coarse-grained metamorphic rock with schistosity, a type of foliation defined by the preferred orientation of visible crystals of mica (generally muscovite or biotite).

science The systematic analysis of natural phenomena based on observation, experiment, and calculation.

scientific law A concise statement that completely describes a specific relationship or phenomenon and applies without exception for a defined range of conditions.

scientific method A sequence of steps for systematically analyzing scientific problems in a way that leads to verifiable results.

scientific research The process of seeking to understand natural phenomena through the scientific method.

scientist A person who searches for ideas that can explain the way natural phenomena on Earth, or elsewhere in the Universe, operate.

scoria A glassy, mafic igneous rock containing large, abundant air-filled holes.

sea breeze A circulation that develops along shorelines during daytime as warm air over land rises and moves seaward aloft, while cool air offshore descends and moves onshore to replace the warm air.

seafloor spreading The gradual widening of an ocean basin as new oceanic lithosphere forms at a mid-ocean ridge axis and then moves away from the axis.

sea ice Ice formed by the freezing of the sea surface.

sea level The elevation that the sea surface would have if there were no waves. Sea level on a map usually refers to mean sea level, which is the average level over time, lying roughly halfway between high tide and low tide.

seamount A volcano edifice whose peak lies below sea level.

season Each of the four divisions of the year (spring, summer, autumn, and winter) characterized by different weather.

sea stack A chimney-shaped column of rock produced by wave erosion along a rocky coast.

seaweed A non-microscopic multicellular alga that grows in the sea.

sediment An accumulation of loose mineral grains of any size that are not cemented together.

sedimentary basin A depression in the Earth's surface that fills with sediment.

sedimentary rock Rock that forms at or near the Earth's surface by the compaction and cementation of rock fragments, or shells and shell fragments, by the accumulation and alteration of organic matter, or by the precipitation of mineral crystals in the shells of organisms or directly from water solutions.

sedimentary structure A shape, texture, or layering that develops in sedimentary rock as a consequence of depositional processes; examples include bedding, ripple marks, cross beds, mudcracks, and graded beds.

sediment budget The difference between input of sediment into a region and the removal of sediment from a region.

sediment liquefaction A phenomenon that happens when seismic shaking causes wet sediment that has been consolidated, but not cemented, to disaggregate and form a slurry of sediment and water that cannot support weight.

sediment load The total volume of sediment carried by a stream.

seismic belt A linear band of the Earth's crust in which earthquakes occur relatively frequently.

seismicity The occurrence of earthquakes.

seismic-reflection profile A cross-sectional image of subsurface layering made by measuring the time it takes for artificial seismic waves to be reflected by boundaries between different layers of rock.

seismic-velocity discontinuity A boundary within the Earth at which seismic-wave velocity changes abruptly.

seismic waves Vibrations generated by an earthquake that pass through the Earth or along its surface.

seismogram The record of an earthquake produced by a seismometer.

seismometer An instrument that can record the ground motion from an earthquake.

semi-permanent high A high-pressure system in the atmosphere that develops as a result of cooling of the air over a broad region; the system lasts for a relatively long time, and gets regenerated in roughly the same place when the air mass containing it moves away.

severe thunderstorm A thunderstorm with the potential to produce hail with diameters of 2.5 cm (1 in) or larger, winds that exceed 50 knots (92 km/h, or 56 mph), or a tornado.

shaded-relief map A map that emphasizes the topography of an area by simulating shadows the would be produced by the Sun when located low in the sky.

shale gas Hydrocarbons trapped in organic shale in gaseous form.

shear The movement of one part of a material sideways past another part, or the stress that causes such movement.

shear wave A seismic wave that causes particles of a material to move back and forth perpendicular to the direction in which the wave itself moves.

shield A portion of a craton in which Precambrian rock is exposed over a broad area.

shield volcano A subaerial volcano with a broad, gentle dome. The term generally implies that the edifice was constructed from many mafic lava flows; it is also used in reference to broad volcanic edifices formed from felsic tuff.

shock wave A vibration caused by a sudden push against an elastic, or somewhat elastic, material; for example, a meteorite impact produces shock waves in the Earth's crust.

shoreline The boundary between the water and the land.

sidereal day The Earth's period of rotation, measured by the time it takes for a distant star to return to the same point in the sky (about 23 h 56 min), and thus the true measure of one full rotation of the Earth on its axis; it differs from the *solar day* because of the Earth's motion around the Sun.

silicate mineral A mineral composed principally of silicon-oxygen tetrahedra linked in various arrangements.

silicon-oxygen tetrahedron The fundamental building block of silicate minerals, which consists of one silicon atom surrounded by four oxygen atoms, producing a pyramid-like shape with four triangular faces; it is informally known as a silica tetrahedron.

sill A tabletop-shaped tabular intrusion that forms where magma intrudes between layers of sedimentary wall rock; if the wall rock is not sedimentary, the term generally implies that the intrusion is roughly horizontal.

singularity In the context of cosmology, the point in space containing an extremely small, infinitely dense, and infinitely hot volume that exploded during the Big Bang.

sinkhole A circular depression in the land surface that forms when an underground cavern collapses.

slab-pull force A plate-driving force caused by the sinking of a downgoing plate, which pulls the rest of the plate behind it; it exists because the downgoing slab is denser than the surrounding asthenosphere, so gravity pulls it downwards.

slate Fine-grained, low-grade metamorphic rock formed by the metamorphism of shale; it splits into thin sheets

slaty cleavage A type of foliation in low-grade metamorphic rock that forms partly due to rotation of pre-existing clay flakes and partly due to the growth of new flakes in the plane of cleavage.

sleet Precipitation consisting of frozen raindrops.

slickenside A polished fault surface produced by the slip of a fault.

slip In the context of rock deformation, the movement on a fault.

slope failure The downslope movement of material on an unstable slope.

slow-onset flood Flooding that develops over several days and take days to weeks to subside, as can happen when snow melts or rainfall occurs over a long time in the watershed.

slump A type of mass wasting that involves the displacement of rock or regolith above a spoon-shaped failure surface. Slumps take place at slow to moderate rates.

smelting The heating of an ore to a temperature at which the ore decomposes to yield metal plus a nonmetallic residue called slag.

snow avalanche The downslope movement of a mass of snow, either as a turbulent cloud of powder or as a slurry of snow and water.

snowball Earth A condition thought to have happened in the late Proterozoic when nearly all of the Earth's land surface was covered by glaciers and nearly all of the sea surface was covered by sea ice.

snowflake A composite of hexagonal ice crystals that forms in the atmosphere.

soil Sediment that has undergone changes at the surface of the Earth due to reactions with air, water, and life, and has mixed with organic material.

soil erosion The removal of soil by water or wind.

soil horizon One of several distinct zones within a soil, distinguished from one another by factors such as chemical composition and organic content.

soil profile A vertical sequence of soil horizons.

solar atmosphere The photosphere, chromosphere, and corona of the Sun.

solar core The central part of the Sun where nuclear fusion occurs.

solar cycle The 11-year and 22-year cycles of variations in sunspot frequency that occur on the Sun's surface in association with reversals in the Sun's magnetic field.

solar day The Earth's period of rotation, measured by the time it takes for the Sun to return to the same point in the sky (24 h); what people commonly mean when they use the word day.

solar eclipse An eclipse that occurs when the Sun is partially or entirely blocked by the Moon.

solar flare An explosion of plasma on the Sun's surface, sending an intense burst of radiation into space; a type of solar storm.

solar granulation The grainy appearance of the Sun's surface caused by temperature variations within convective cells of its convective zone.

solar mass ($M_\odot$) The mass of the Sun.

solar prominence A large, bright arc of plasma extends outward from the Sun's surface between a pair of sunspots; a type of solar storm.

solar storm An event in which the Sun ejects particularly large amounts of high-energy particles into space, this can result in a disturbance of the Earth's magnetic field.

Solar System The Sun and all the materials that orbit it, including planets, moons, asteroids, meteors, Kuiper Belt objects, and Oort Cloud objects.

solar unit A unit of luminosity equal to the luminosity of the Sun.

solar wind A stream of charged particles with enough energy to escape from the Sun's gravity and flow outward into space.

solid A state of matter in which atoms or molecules remain fixed in position. A solid retains its shape regardless of changes in the size of its container.

solifluction A type of creep characteristic of tundra regions that occurs during summer when the uppermost layer of permafrost melts and the resulting soggy, weak layer of ground flows slowly downslope in overlapping sheets.

sorting (1) The range of clast sizes in a collection of sediment. (2) The degree to which sediment has been separated by flowing currents into different-sized fractions.

source rock Shale containing the organic raw materials from which hydrocarbons form.

Southern Oscillation The shift in sea-level atmospheric pressure back and forth across the Pacific Ocean in association with the Walker circulation.

space telescope A telescope mounted on a satellite in orbit above the Earth's atmosphere.

space weather Conditions in the region of space surrounding the Earth that result from interactions with the solar wind; the intensity of space weather can be affected by solar flares.

space weathering The breakdown of surface material by micrometeorite impacts and by absorption of *cosmic rays* (high-energy atomic nuclei from space).

specific gravity A number representing the density of a mineral, as specified by the ratio between the weight of a volume of the mineral and the weight of an equal volume of water.

spectral class Any of the categories in a classification system for stars that is based on their surface temperatures, as determined by the presence and relative strength of spectral absorption lines in a spectrometer image.

spectral line A dark or bright line representing a specific wavelength of light in an image produced by a spectrometer.

spectrometer A device that spreads incoming electromagnetic radiation into a very broad spectrum in order to characterize the chemical composition of the radiation source.

speleothem A formation that grows in a limestone cave by the accumulation of travertine precipitated from a water solution that drips from the ceiling or flows down the wall of the cave.

spring A natural outlet from which groundwater flows to the ground surface.

squall-line thunderstorm One thunderstorm in a distinct line of several thunderstorms.

stalactite A downward-pointing speleothem that forms when a water solution drips from the ceiling of a cave.

stalagmite An upward-pointing speleothem that forms when a water solution drips from the ceiling of a cave and hits the floor.

star A celestial object in which fusion reactions occur continuously, producing vast amounts of energy; our Sun is a star.

state of matter One of four different forms in which matter can exist; the forms are *solid*, *liquid*, *gas*, and *plasma*.

stationary front A boundary between air masses where the colder air mass is neither advancing nor retreating at the Earth's surface.

steady-state condition The situation that occurs in cycles when the proportions of a material in different reservoirs remain fairly constant, even though flow of the material continues among reservoirs.

stellar nucleosynthesis A process in which smaller nuclei fuse together within a star to form larger nuclei; the largest nuclei formed in a star are those of the element iron.

stick-slip behavior Stop-start movement along a fault caused by friction, which prevents movement until stress builds up sufficiently to overcome it.

storm surge A rise in local sea level, above normal tide level, that develops in response to a storm, when strong winds push the sea into a mound. It occurs to a lesser extent because low air pressure allows the sea surface to rise; when storm surge reaches the shore, it can contribute to coastal flooding; storm surges are particularly high during hurricanes.

straight-line wind A strong wind produced by a downdraft that blows in a single direction at the ground surface, rather than in a rotational pattern.

strain A change in the shape of a material in response to the application of a stress.

strata A succession of sedimentary beds.

stratigraphic column A chart representing the succession of stratigraphic formations in a region, and their relative thicknesses; the formations appear in order from oldest at the bottom to youngest at the top.

stratigraphic formation A recognizable layer consisting of sedimentary beds of a specific rock type or set of rock types, deposited during a certain time interval, that can be traced over a broad region.

stratosphere The layer of the Earth's atmosphere, in which the temperature increases with elevation, that lies directly above the troposphere.

stratovolcano A large, cone-shaped, subaerial volcano consisting of alternating layers of lava and pyroclastic debris.

stratus cloud A sheet-like, widespread cloud below 3 km (10,000 ft) in altitude.

streak The color of the powder produced by pulverizing a mineral on an unglazed ceramic plate.

stream A body of flowing water; streams flowing on land are confined to a channel, except during floods; a large stream is commonly known as a river.

stream gradient The slope of a stream's channel in the downstream direction.

stream rejuvenation The renewed downcutting of a stream into the land surface caused by a rise in the surface of the land or a drop of the base level.

stress The push, pull, or shear that a material experiences when subjected to a force; formally, the force applied per unit area.

strike The compass orientation of a horizontal line on a plane. Strike and dip together uniquely define the orientation of a planar structure.

strike-slip fault A fault in which one block slides horizontally past another (that is, in a direction parallel to the strike line of the fault surface) so that there is no relative vertical displacement.

subduction The process by which one oceanic plate slides beneath the edge of another plate and sinks into the asthenosphere beneath.

subduction zone The region along a convergent boundary where one plate slides beneath another.

submarine canyon A canyon cut into a continental shelf and slope, entirely beneath sea level.

submarine debris flow Mass wasting of a slurry of pebbles, cobbles, and boulders suspended in a mud matrix down a submarine slope.

submarine fan A wedge-shaped accumulation of sediment at the base of a submarine slope, usually at the mouth of a submarine canyon.

submarine slump Mass wasting of a semicoherent block of sediment down a submarine slope.

submergent coast A coast where the land is sinking relative to sea level or sea level is rising relative to the land.

subsidence The slow vertical sinking of the Earth's surface in a region.

subsoil A layer of soil (the B-horizon) that lies beneath the topsoil and represents the zone of accumulation.

substrate A general term for material at and just below a specified feature.

subtropical jet stream A band of strong (>160 km/h, or 100 mph) winds at the top of the troposphere, typically found between 20° and 35° latitude.

Sun The star at the center of our Solar System.

sunspot A relatively cool, transitory region on the solar surface produced when loops of the Sun's magnetic field break through the surface of the Sun.

supercell thunderstorm A rotating thunderstorm; such storms often become severe and can produce hail, strong straight-line winds, and tornadoes.

supergiant star A star that is thousands of times brighter than the Sun and has a relatively short lifespan of only about 10–50 million years.

supermassive black hole A huge black hole, containing the mass equivalent of 1,000 Suns or more, that resides in the center of a galaxy.

supernova A short-lived, very bright object in space that results from the cataclysmic explosion marking the death of a high-mass star; it ejects large quantities of matter into space to form new nebulae.

supernova nucleosynthesis The fusion of smaller nuclei to form larger ones during a supernova explosion; it can produce heavy elements, such as uranium.

superposition The geologic principle stating that, in a succession of sedimentary rock, younger rocks were deposited over older rocks.

supervolcano A volcano that emits more than 1,000 km³ (240 mi³) of volcanic material during a single explosive eruption; none have erupted during recorded human history.

surface current An ocean current in the top 100–400 m (330–1,300 ft) of water.

surface load Sediment transported by rolling or saltation along a surface by wind or flowing water.

surface water The portion of the hydrosphere that occurs in oceans, lakes, streams, rivers, swamps, snow, and glaciers on the Earth's surface.

surface wave A seismic wave that travels along the Earth's surface.

suspended load Fine-grained sediment carried by water or wind without settling out of the current.

sustainability The capacity to maintain or improve society's standard of living without running out of resources or ruining the environment.

suture The boundary formed during a collisional orogen between two formerly separate, relatively buoyant crustal blocks that have collided.

swash The upward surge of water on the surface of a beach.

S-wave shadow zone A band between 103° and 180°, as measured along the Earth's surface from an earthquake's epicenter, where S-waves do not arrive at seismometer stations because they cannot pass through the Earth's liquid outer core.

swell Long-wavelength, periodic ocean waves that have traveled a long distance from their source and are not being driven by the local wind.

symmetry The condition in which the shape of one part of an object is a mirror image of the other part.

syncline A trough-shaped fold whose limbs dip toward the hinge.

talus Rock fragments that have fallen from a cliff or steep slope.

talus pile An accumulation of fallen rock fragments along the base of a cliff, typically forming an apron whose surface is at the angle of repose.

tar A substance composed of hydrocarbons that exist in solid form at room temperature.

tar sand Sandstone reservoir rock in which small hydrocarbon molecules have either escaped or been consumed by microbes, so that only very viscous, nonflowing oil remains.

taxonomy The study and classification of the relationships among organisms.

telescope The basic tool of astronomers, designed to collect and focus electromagnetic radiation in order to make celestial objects appear brighter and larger.

temperature A measure of the average speed of movement of molecules in a fluid, or of the vibration of molecules in a solid; commonly, the hotness or coldness of a substance as measured with a thermometer.

tension In the context of discussing geologic deformation, a stress that pulls on a material and could lead to stretching.

terminal moraine The end moraine at the farthest limit of glaciation.

terrestrial planet One of the four inner planets of the Solar System—Mercury, Venus, Earth, and Mars—all of which are made of rock and metal and have a solid surface.

texture The arrangement of grains in a rock, as manifested by the way grains connect to one another and the degree to which inequant grains align parallel to one another.

theory A scientific idea, supported by an abundance of evidence, that has passed many tests and has failed none.

theory of evolution The idea that new species of life appear as others go extinct, so that the assemblage of species on the Earth changes over time, by the process known as "natural selection by the survival of the fittest."

thermal energy The total kinetic energy in a material due to the vibration and movement of atoms in that material.

thermal metamorphism Metamorphism caused by the heat of an igneous intrusion unaccompanied by change in pressure or differential stress; also called *contact metamorphism*.

thermal pressure An outward push generated when heat causes a gas to expand.

thermocline A boundary between layers of water with differing temperatures.

thermohaline circulation A global oceanic circulation that involves both surface and deep-sea currents and the upwelling and sinking of ocean water, driven by contrasts in water density, which are due in turn to differences in temperature and salinity.

thermometer An instrument used to measure temperature, often constructed by partially filling a sealed glass tube with a liquid such as alcohol.

thermosphere The hot outermost layer of the atmosphere containing very few gas molecules.

thin section An 0.03-mm-thick slice of rock, thin enough for light to pass through, that can be examined with a petrographic microscope.

thrust fault A gently dipping reverse fault on which the hanging-wall block moves up the slope of the fault.

thunderstorm A large cumulonimbus cloud, produced where there is atmospheric instability, that produces lightning and thunder.

tidal bore A visible wave of water produced when a flood (rising) tide arrives in the mouth of an estuary, where it moves against a river current.

tidal energy Electricity produced by using tidal flow to drive generators.

tidal flat A broad, nearly horizontal plain of mud and silt that is exposed or nearly exposed at low tide but totally submerged at high tide.

tidal reach The difference in sea level between high tide and low tide at a given location.

tide The rising or falling of sea level that generally occurs twice daily.

tide-generating force The force that generates tides, caused in part by the gravitational attraction of the Moon and Sun, and in part by the centrifugal force created by the revolution of the Earth-Moon system.

tidewater glacier A glacier that has entered the sea along a coast.

tight oil Oil that occurs within the source rock, an organic shale, that has passed through the oil window so that the organic matter has transformed into oil and gas. It is an unconventional hydrocarbon that can be extracted only by using hydrofracturing; also known as shale oil.

tillite A rock formed from lithified deposits of glacial till, consisting of cobbles and boulders distributed through a matrix of sandstone and mudstone.

topographic map A map that uses contour lines to represent variations in elevation.

topographic profile A trace of the ground surface as it would appear on a vertical plane that sliced into the ground; it represents elevations along a traverse in a given direction.

topography Variation in the elevation of the land surface.

topsoil The top layer of soil (the O- and A-horizons, or in some cases the E-horizon), which lies within the zone of leaching and may be dark and nutrient-rich.

tornado A nearly vertical, funnel-shaped cloud, in which air rotates violently around the axis of the funnel; they commonly form in association with supercell thunderstorms, and can cause extreme damage.

tornado track The path that a tornado takes across the ground.

trade winds Tropical surface winds that blow from northeast to southwest between 30° N and the equator in the northern hemisphere and southeast to northwest between 30° S and the equator in southern hemisphere; in both hemispheres, the winds curve to blow nearly from east to west near the equator.

transform boundary A plate boundary where one plate slips sideways relative to the other along a transform fault, but no new lithosphere forms, and no old lithosphere subducts. It consists of a vertical strike-slip fault or a belt of related strike-slip faults.

transform fault A strike-slip fault marking a transform boundary. Most transform faults are the actively slipping segment of a fracture zone between two segments of a mid-ocean ridge, but a few cut through continental crust.

transgression The inland migration of a shoreline caused by a rise in sea level.

transition zone The lower part of the upper mantle, extending from 410 down to 660 km (255–410 mi) deep, in which there are several seismic-velocity discontinuities.

trap A configuration of impermeable seal rock lying above reservoir rock that confines hydrocarbons in a restricted area underground.

travel time In the context of seismology, the interval of time that it takes for a seismic wave to travel from the focus of an earthquake to a given seismometer.

tributary A smaller stream that flows into a larger stream.

triple junction A point where three plate boundaries intersect.

tropical cyclone A large, spiral-shaped low-pressure system that originates over warm tropical ocean waters and has an organized rotation around its center. *Hurricanes* are very strong tropical cyclones.

tropical storm A weak tropical cyclone with sustained winds between 63 and 118 km/h (39–73 mph).

tropics Latitudes of the Earth between 23.5° N and 23.5° S, the Tropic of Cancer and the Tropic of Capricorn, respectively, where cold weather is rare.

troposphere The layer of the Earth's atmosphere that extends from the Earth's surface to the tropopause and contains all the Earth's weather.

trough In the context of atmospheric science, an elongate area of low atmospheric pressure. At jetstream levels, it is where the polar-front jet stream bows toward the equator.

tsunami A wave produced by displacement of the seafloor, or more rarely, a meteor impact; tsunamis differ from storm waves in that they have a much greater wavelength. They can build into large waves when they reach the shore, and because of their wide wavelength, they can flood areas inshore of the beach.

tuff A fine-grained pyroclastic rock composed mainly of volcanic ash.

turbidity current A submarine avalanche of sediment and water that flows down a submarine slope.

turbulence The chaotic twisting, swirling motion in a flowing fluid.

typhoon A strong tropical cyclone over the western Pacific Ocean in which sustained wind speeds reach 119 km/h (74 mph) or higher. It is the same kind of storm as an Atlantic hurricane.

unconformity A boundary between two strata representing an interval of time during which new strata were not deposited and/or old strata were eroded away. It can appear in an outcrop as a contact between two stratigraphic formations, or between an igneous body and strata, or between metamorphic basement and strata. The strata above the unconformity are younger than the rocks below.

unconventional reserve A hydrocarbon reserve that can be accessed only by using expensive technology such as directional drilling and hydrofracturing.

underground mine A system of tunnels and shafts used to access energy or mineral resources underground (generally 100 m or more below the Earth's surface).

uniformitarianism The geologic principle stating that the same physical processes that we observe today happened in the past at roughly the same rates; simply put, the present is the key to the past.

Universe All of space, and everything within it.

updraft An upward-moving air flow.

uplift The upward vertical movement of the Earth's surface in a region.

upper mantle The uppermost section of the mantle, extending from the Moho down to a depth of 660 km (410 mi).

upwelling The rising of deep, cold ocean water to the ocean surface.

U-shaped valley A valley carved by glacial erosion; it has very steep sides and, on a profile drawn perpendicular to the valley, has the form of a U.

vacuum A volume that contains very little matter. Interplanetary and interstellar space is a vacuum.

vapor pressure That part of the total atmospheric pressure exerted by water vapor molecules.

vesicle An open hole in igneous rock formed by the preservation of a gas bubble in lava as the lava cools into solid rock.

viscosity The resistance of material to flow.

visibility The maximum distance from which a person with normal vision can distinguish objects when looking through the atmosphere.

visible light Electromagnetic radiation that can be detected by the human eye.

volatile material An element or compound, such as H_2O or CO_2, that can melt at relatively low temperatures and can exist in gaseous forms at the Earth's surface.

volcanic arc A curving chain of active volcanoes formed on the overriding plate adjacent to a convergent boundary.

volcanic bomb A large fragment of pyroclastic debris that is still soft when erupted, so that the fragment becomes stream lined as it flies through the air.

volcanic breccia A pyroclastic rock composed of volcanic blocks.

volcanic eruption An event during which lava and/or pyroclastic debris is expelled from a vent in a volcano.

volcanic hazard assessment map A map that delineates those areas near a volcano that lie in the path of potential lava flows, lahars, or pyroclastic flows.

volcano (1) A vent from which melt from inside the Earth spews out onto the planet's surface. (2) A mountain formed by the accumulation of extrusive volcanic rock.

Walker circulation An atmospheric circulation in the equatorial regions of the Pacific Ocean characterized by rising air over the western Pacific, eastward air flow near the tropopause, sinking air over the eastern Pacific, and westward air flow near the Earth's surface.

wall cloud A region of rotating clouds that extends below the rain-free base of a supercell thunderstorm within the region of the updraft. The formation of a wall cloud often precedes tornado formation.

wall rock Pre-existing rock adjacent to the surface of an igneous intrusion.

warm front A boundary between air masses where the colder air mass is retreating and the warmer air mass is advancing at the Earth's surface.

waterfall A feature that forms where the gradient of a stream becomes so steep that some or all of the water free-falls above the streambed.

water mass In the context of oceanography, a large volume of water in the ocean with relatively uniform density characteristics, due to its relatively uniform temperature and/or salinity.

water table The underground boundary, approximately parallel to the Earth's surface, that separates the unsaturated zone (nearer the ground surface) in which air partially fills pores, from the underlying saturated zone in which groundwater completely fills the pores.

water vapor Water in its gaseous form.

wave In a general sense, the periodic motion of a material in an ocean, lake or the atmosphere during the transmission of energy. In a water body, for example, it is the periodic up-and-down motion of the water's surface to produce distinct troughs and ridges. Seismic energy, sound, and electromagnetic energy are also transmitted by waves. See also *electromagnetic wave, seismic wave*.

wave base The depth in a body of water, approximately equal in distance to half a wavelength, below which there is no wave movement.

wave-cut platform A shelf of rock, cut by wave erosion, at the low-tide line that was left behind a retreating cliff.

wave erosion The grinding or breaking away of the coastline by the action of water waves.

wavelength The horizontal difference between two adjacent wave troughs or two adjacent wave crests.

wave refraction The bending of water waves as they approach a shore so that their crests make no more than about a 5° angle with the shoreline as they wash up on a beach.

weather Local-scale atmospheric conditions as defined by temperature, atmospheric pressure, relative humidity, wind speed, and precipitation.

weathering The processes that gradually modify and weaken rock exposed to air and water, eventually transforming it into sediment.

well A hole in the ground dug or drilled in order to obtain water or oil.

white dwarf A stellar object formed when an intermediate-mass star has exhausted its hydrogen fuel, undergone gravitational collapse to develop a dense core, and lost its outer layers.

wind The horizontal movement of air.

wind-chill temperature The temperature that human skin feels due to heat loss caused by the combined effects of both cold and wind.

wind direction The direction from which the wind blows. For example, a westerly wind results in the movement of air from west to east.

wind-driven wave A wave formed by the interaction of moving air with the surface of a water body.

wind speed The speed at which air moves horizontally across the Earth's surface, as measured by an anemometer.

zodiac The array of 12 constellations, defined in western culture, that lie along the ecliptic plane.

zone of accumulation The layer of soil in which new minerals precipitate out of downward-percolating water, and fine clay leached from above that is left behind.

zone of leaching The top layer of soil, where downward-percolating water picks up ions and clay and transports them to the layer below.

zooplankton Microscopic organisms floating in water, that survive by ingesting other organisms, living or dead.

CREDITS

p. 162: Stephen Marshak; (graded beds): Marli Miller/Visuals Unlimited; p. 163 (top a-b): Stephen Marshak; (right): Google Earth; (bottom a): Emma Marshak; (bottom b): Stephen Marshak; (bottom c): Marli Miller/Visuals Unlimited; p. 164: The Natural History Museum/Alamy Stock Photo; p. 165: Google Earth; p. 166 (left): Stephen Marshak; (right): David Wall/Alamy Stock Photo; p. 169: Stephen Marshak.

Chapter 6
Page 170: Stephen Marshak; p. 171 (caterpillar): Scott Camazine/Science Source; (butterfly): John Serrao/Science Source; (shale): Science Stock Photography/Science Source; (schist): Science Stock Photography/Science Source; p. 176 (a): Stephen Marshak; (center b): Emma Marshak; (right b): Stefano Clemente/Alamy Stock Photo; (bottom both): Stephen Marshak; p. 177 (both): Stephen Marshak; p. 178 (a): lillisphotography/Getty Images; (b): Google Earth; (c-d): Stephen Marshak; p. 180 (both a): Stephen Marshak; (left inset a): Science Photo Library/Science Source; (right a): Stephen Marshak; (right inset a): Dirk Wiersma/Science Source; (inset left b): Bernardo Cesare/Visuals Unlimited; (both b): Stephen Marshak; (inset right b): Kurt Freihauf; (bottom c): Stephen Marshak; p. 183 (top): Stephen Marshak; (bottom): Google Earth; p. 185 (top): Google Earth; (a): Michael Stewart, University of Illinois; (b): Stephen Marshak; p. 186: Google Earth; p. 187 (a-c): Stephen Marshak; (right): Google Earth; p. 191: Stephen Marshak.

Chapter 7
Page 192: Stephen Marshak; p. 194: NOAA/ETOPO1382; p. 196 (all): Stephen Marshak; p. 197 (both): Stephen Marshak; p. 199 (a): Aurora Photos/Alamy Stock Photo; (b): Stephen Marshak; (bottom): Google Earth; p. 201: Stephen Marshak; p. 202 (a): Stephen Marshak; (b): USGS; (c): Lloyd Cluff/Corbis Getty Images; p. 203 (all): Stephen Marshak; p. 204: Google Earth; p. 205 (a-d): Stephen Marshak; (e): Landsat/USGS; p. 207(top): Stephen Marshak; (bottom): Google Earth; p. 208 (both): Google Earth; p. 209: Stephen Marshak; p. 210 (bottom left): Stephen Marshak; (bottom center): Google Earth; (bottom right): Stephen Marshak; p. 211: Kurt Burmeister; p. 216 (both): Stephen Marshak.

Chapter 8
Page 220: The Asahi Shimbun via Getty Images; p. 222 (a): AFP/Getty Images; (b): JIJI Press/AFP/Getty Images; (c): AP Photo/Kyodo News; (bottom): Stephen Marshak; p. 223: NOAA/NGDC, USGS; p. 224: Google Earth; p. 234 (b): National Archives; (c): AP Photo/Paul Sakuma, file; (right): Omar Havana/Getty Images; p. 235 (left): Anna Kompanek/CIPE; p. 239: (a): AP Photo/Str; (b): Pacific Press Service/Alamy Stock Photo; (c): M. Celebi, U.S. Geographical Survey; (d): Javier Casella/ AFP/Getty Images; p. 240 (top): Joseph Sohm/Shutterstock; (a): NOAA/National Geophysical Data Center (NGDC); (b): AP Photo/New Zealand Herald, Geoff Sloan; (c): Adek Berry/AFP/Getty Images; p. 241 (a): National Geophysical Data Center (NGDC); p. 242: Google Earth; p. 244 (a): AFP/Getty Images; (b): Photo by David Rydevik. 2004. Wikimedia, public domain; (both c): Jessica Wilson/NASA/Science Source; (bottom): AP Photo/Kyodo News; p. 245: (c): Brendan Hoffman/Alamy Stock Photo; (d): Stocktrek Images, Inc./Alamy Stock Photo; p. 247: NOAA/NOA Center for Tsunami Research; (inset): Stephen Marshak.

Chapter 9
Page 256: Stephen Marshak; p. 258 (a-b): Stephen Marshak; p. 259 (a-b): Stephen Marshak; p. 261 (a-b): Stephen Marshak; (c): Reynolds Sumayku/Alamy Stock Photo; p. 263 (a): Sovfoto/UIG via Getty Images; (b): Dirk Wiersma/Science Source; (c,e,g,h): Stephen Marshak; (d): Naturfoto Honal/Getty Images; (f): Kevin Schafer/Corbis/Getty Image; (i): Biophoto Associates/Science Source; p. 265: Illustration by Karen Carr and Karen Carr Studio, Inc. © Smithsonian Institution; p. 267 (a): Stephen Marshak; (bottom): Google Earth; p. 268 (a-b): Stephen Marshak; p. 272: Google Earth; p. 274: USGS; p. 278: Stephen Marshak; p. 281: Stephen Marshak; p. 283: Stephen Marshak.

Chapter 10
Page 284: Stephen Marshak; p. 286 (a): © William K. Hartmann; (d): Richard Bizley/Science Source; (e): (artwork c) Don Dixon/cosmographica.com; (bottom): Science Photo Library/Science Source; p. 288 (a): Courtesy of Dr. J. William Schopf/UCLA; (c): Stephen Marshak; (d): Bill Bachman/Alamy Stock Photo; p. 290: USGS; p. 291 (a): John Sibbick; (b): Stephen Marshak; p. 292: Courtesy of Dr. Paul Hoffman, Harvard University; p. 294 (all): Deep Time Maps; p. 295 (all): Deep Time Maps; p. 297 (top): Tom McHugh/Science Source; (b): Deep Time Maps; p. 298 (a): Chase Studio/Science Source; (b): Corbin17/Alamy Stock Photo; (bottom b): Deep Time Maps; p. 299: Mackenzie, J. 2012. Hillshaded Digital Elevation Model of the Continental US. http://www.udel.

edu/johnmack/data_library/usa_dem.png; p. 300 (a): Science History Images/Alamy Stock Photo; (b): Stephen J. Krasemann/Science Source; (center b): Stephen Marshak; (right b): Deep Time Maps; (bottom): Richard Bizley; p. 302: Deep Time Maps; p. 303 (a): Mohamad Haghani/Stocktrek Images/Science Source; (b): De Agostini Picture Library/Getty Images; (bottom): Google Earth; p. 304 (inset): Deep Time Maps; (a): NASA, JPL; (b): Google Earth; p. 305: Deep Time Maps; p. 306: Stephen Marshak; p. 307: Google Earth; p. 308 (a): Mauricio Anton/Science Source; (b): P. Plailly/E. Daynes/Science Source; (bottom both): Deep Time Maps; p. 309 (left): Daniel Eskridge/Alamy Stock Photo; (right): Photo by Felix Andrews (Floybix); p. 311: Stephen Marshak; p. 313: Deep Time Maps.

Chapter 11
Page 314: Luis Gutierrez/NortePhoto.com/Alamy Stock Photo; p. 315 (both): Stephen Marshak; p. 317: choja/Getty Images; p. 320: Data courtesy of Fugro. Credit: Virtual Seismic Atlas, http://www.seismicatlas.org; p. 321: Kanok Sulaiman/Getty Images; p. 322 (a-b): Stephen Marshak; (c): Calvin Larsen/Science Source; (left): Google Earth; p. 323: Andrew Harrer/Bloomberg via Getty Images; p. 325 (left): dan_prat/Getty Images; (a): Field Museum Library/Getty Images; (c): Department of Natural Resources, Alaska; p. 327 (a): Stephen Marshak; (c): Cultura Creative/Alamy Stock Photo; p. 328 (a): RGB Ventures/SuperStock/Alamy Stock Photo; (b): Jonathan Plant/Alamy Stock Photo; p. 329: Science Source; p. 330: G. R. Roberts © Natural Sciences Image Library; p. 332 (all): Stephen Marshak; p. 333 (top inset): Layne Kennedy/Getty Images; (center): Stephen Marshak; (bottom a): Richard P. Jacobs/JLM Visuals; (bottom b): Stephen Marshak; p. 336 (a-b): Stephen Marshak; (bottom): Google Earth; p. 337 (a): Richard P. Jacobs/JLM Visuals; (b): Stephen Marshak; (bottom): Stephen Marshak; (left): Doug Sokell/Visuals Unlimited.

Chapter 12
Page 342: Stephen Marshak; p. 343 (a-f): Stephen Marshak; p. 345: NOAA; p. 346 (a): G. R. Roberts © Natural Sciences Image Library; (b): Julie Dermansky; p. 347 (all): Stephen Marshak; p. 349 (a): Lloyd Cluff Consulting Earthquake Geologist. San Francisco, California lloydcluff@gmail.com; p. 349: Lloyd Cluff/Getty Images; p. 352 (inset): Stephen Marshak; (d): Marli Miller/Visuals Unlimited, Inc; p. 353 (a): Tom Myers/Science Source; (c-d): Stephen Marshak; p. 354 (a): BrazilPhotos.com/Alamy Stock Photo; (b): Cascades Volcano Observatory/USGS; (c): Stephen Marshak; (d): Xinhua/Alamy Stock Photo; (bottom): Google Earth; p. 355 (top left): Mike Eliason/Santa Barbara County Fire Department via AP, File; (top right): Google Earth; (a): Stephen Marshak; (b): Yann Arthus-Bertrand/Getty Images; p. 356 (all): Stephen Marshak; p. 357 (a): Stephen Marshak; (bottom): Google Earth; p. 358 (a): Jacques Lange/Getty Images; (b): Alaska Stock/Alamy Stock Photo; p. 359 (a): Jerome Neufeld and Stephen Morris (c): 2002 Nonlinear Physics, University of Toronto; (b): USGS/ Barry W. Eakins; (c): Google Earth; p. 360: Stephen Marshak; p. 362: Breck P. Kent/JLM Visuals; p. 363: AP Photo/Ted S. Warren; p. 364 (a): Google Earth; (b): Bob Schuster/USGS; p. 365: Google Earth; p. 366: Ralph A. Haugerud/USGS; p. 369: Stephen Marshak.

Chapter 13
Page 370: Stephen Marshak; p. 372 (a): Google Earth; (b): Stephen Marshak; p. 373: Google Earth; p. 374: Stephen Marshak; p. 375: NASA; p. 376 (both): Stephen Marshak; p. 377 (all): Stephen Marshak; p. 378 (a-b): Stephen Marshak; p. 379: Google Earth; p. 380 (both): Stephen Marshak; p. 381 (all): Stephen Marshak; p. 382 (a, e): Stephen Marshak; (d): P.A. Lawrence, LLC./Alamy Stock Photo; p. 383 (a): Marti Miller/Visuals Unlimited; (b): Christina Neal/Alaska Volcano Observatory/USGS; p. 384 (inset): NASA/GSFC/meti/ersdac/jaros, and US/Japan ASTER Science Team; p. 385: nimor72/Shutterstock; p. 386 (both): NASA images created by Jesse Allen, Earth Observatory, using data provided courtesy of the Landsat Project Science Office-copyright 2008; p. 387 (a): Mamunur Rashid/NurPhoto/Sipa via AP Images; (b): Randy Pench/The Sacramento Bee via AP; p. 388 (a): Reuters/Yoray Cohen/Eilat Rescue Unit; (b): Reuters/Danish Siddiqui/Newscom; (c): Stephen Marshak; (d): USGS; p. 389 (a-b): Stephen Marshak; (d): AP Photo/The News-Star, Margaret Croft; p. 391: Stephen Marshak; p. 395: Google Earth; p. 396: Stephen Marshak; p. 397: Stephen Marshak; p. 398 (b): Allan Tuchman; (c): Thom Foley; (d-e): Stephen Marshak; p. 399: Stephen Marshak; p. 400 (limestone): Stephen Marshak; (sinkholes): G. R. 'Dick' Roberts/NSIL/Getty Images; (Mexico): SCPhotos/Alamy Stock Photo; p. 401 (bridge): Design Pics Inc/Alamy Stock Photo; (spelunker): Ashley Cooper/Corbis/Getty Images; (stalactites): Stephen Marshak; p. 402 (a): USGS; (left and c): Google Earth; (bottom a-b): Stephen Marshak; p. 403 (a): George Steinmetz/Corbis/Getty Images; (b): Stephen Marshak; (bottom): Google Earth; p. 404 (a): Chinch Gryniewicz/UIG/Science Source; (b): Stephen Marshak; p. 405 (b): Photo courtesy of the Bureau of Reclamation.

Chapter 14

Page 410 (both): Stephen Marshak; p. 412: Stephen Marshak; p. 413: Professor Andre Danderfer; p. 414 (all): Stephen Marshak; p. 415 (a-c): Stephen Marshak; (bottom): Google Earth; p. 416 (a): Image courtesy Jacques Descloitres MODIS Rapid Response Team/NASA; (b): Daniel J. Bryant/Getty Images; p. 417 (all): Stephen Marshak; p. 418 (top all): Stephen Marshak; (bottom): Google Earth; p. 419 (c): Stephen Marshak; (right): Google Earth; p. 420 (top, b): Whit Richardson/Alamy Stock Photo; (bottom b): Stephen Marshak; p. 421 (all): Stephen Marshak; p. 424 (top): Eye Ubiquitous/Newscom; (bottom): Google Earth; p. 425 (from left to right): Stocktrek Images, Inc./Alamy Stock Photo; (2): USGS; (bottom): BDR/Alamy Stock Photo; p. 426 (a): Shutterstock; (b both): Stephen Marshak; (c top): Emma Marshak; (c, bottom): Ted Spiegel/National Geographic Creative; p. 427 (b-c): Stephen Marshak; (d): SRTM Team NASA/JPL/NIMA; p. 428 (top): Stephen Marshak; (center): Stephen Marshak; (bottom): National Geophysical Data Center/NOAA; p. 432 (b): Ralph A. Clevenger/Corbis/Getty Images; (c both): ESA; (d): Stephen Marshak; p. 433 (all): Stephen Marshak; p.434 (left): Google Earth; (a): Shutterstock; (b): Stephen Marshak; (c): 1986 Keith S. Walklet/Quietworks; p. 435 (top b): Marli Miller/Visuals Unlimited, Inc; (bottom a): Google Earth; (bottom b): Wolfgang Meier/Corbis/Getty Images; p. 436: Google Earth; p. 437 (all): Stephen Marshak; p. 439 (all but noted): Stephen Marshak; (right e): Kevin Schafer/Alamy Stock Photo; p. 440: Stephen Marshak; p. 441 (b): Stephen Marshak; (d): All Canada Photos/Alamy Stock Photo; p. 442 (b): National Geographic Image Collection/Alamy Stock Photo; p. 443 (b): Google Earth; (bottom): Lynda Dredge/Geological Survey of Canada; p. 444: Detail of mural by Charles R. Knight, American Museum of Natural History, #4950(5), Photo by Denis Finnin; p. 451: Stephen Marshak.

Chapter 15

Page 452: Stephen Marshak; p. 453: Photo by Alexander Kaiser; p. 454 (top): NASA's Earth Observatory; (a): Library of Congress; (b): Stephen Marshak; p. 455 (background): Stephen Marshak; (a): TopFoto/The Image Works; (b): Woods Hole Oceanographic Institution; (c): National Geographic Image Collection/Alamy Stock Photo; (d): Ingo Wagner/picture-alliance/dpa/AP Images; (e): Adam Soule, WHOI, WHOI-MISO Facility, Nadir submersible Pilots, Dalio Explore Fund, Woods Hole Oceanographic Institution; (f): NASA; p. 457: Deep Time Maps; p. 458 (all): Stephen Marshak; p. 460: Juan José Pascual/age fotostock; p. 462: NASA/SVS; p. 463: NOAA; p. 470 (top): ESA/European Space Agency; (bottom): Los Alamos National Laboratory; p. 471 (left): Los Alamos National Laboratory; (c): NOAA; p. 472 (both): Stephen Marshak; p. 473: Blend Images/Alamy Stock Photo; p. 474: Stephen Marshak; p. 475: NOAA; p. 477 (a): Emma Marshak; (b): Stephen Marshak; p. 479: Google Earth; p. 480 (a): Andrew Syred/Science Source; (b): D.P. Wilson/ FLPA/Science Source; (c): Steve Gschmeissner/Science Source; (d): Seaphotoart/Alamy Stock Photo; (bottom): Google Earth; p. 481 (top a-b): WaterFrame/Alamy Stock Photo; (c): Martin Habluetzel/Alamy Stock Photo; (a): Lorna Roberts/Alamy Stock Photo; (b): Andrew J. Martinez/Science Source; p. 482 (a): Reinhard Dirscherl/ullstein bild via Getty Images; (b): Mark Conlin/Alamy Stock Photo; (c): Stephen Marshak; (d): SeaWiFS Project and NASA/Goddard Space Flight Center; p. 485 (squid): WaterFrame/Alamy Stock Photo; (boats): Stephen Marshak.

Chapter 16

Page 486: Stephen Marshak; p. 487: (a): Google Earth; (center, b): Chris Kelley/HURL, data from Schimdt Ocean Institute's ship Falkor; (bottom a): Mountains in the Sea Research Team; the IFE Crew; and 0NOAA/OAR/OER; (bottom b): Emory Kristof And Alvin Chandler/National Geographic; p. 488 (a, c): William Crawford, Integrated Ocean Drilling Program/TAMU; (b): International Ocean Discovery Program (IODP); p. 491: Google Earth; p. 492 (a): Google Earth; (bottom): NOAA; p. 493 (top a): Ryan, W. B. F., S.M. Carbotte, J. Coplan, S. O'Hara, A. Melkonian, R. Arko, R.A. Weissel, V. Ferrini, A. Goodwillie, F. Nitsche, J. Bonczkowski, and R. Zemsky (2009), Global Multi-Resolution Topography (GMRT) synthesis data set, Geochem. Geophys. Geosyst., 10, Q03014, doi:10.1029/2008GC002332; (top b): Google Earth; (bottom): NOAA Office of Ocean Exploration and Research; p. 494 (a): © Jose Fuste Raga/age fotostock; (b): LDEO: Marine Geoscience Data System. Reference: Ryan, W.B.F., S.M. Carbotte, J.O. Coplan, S. O'Hara, A. Melkonian, R. Arko, R.A. Weissel, V. Ferrini, A. Goodwillie, F. Nitsche, J. Bonczkowski, R. Zemsky (2009), Global Multi-Resolution Topography synthesis, Geochem. Geophys. Geosyst., 10, Q03014, 10.1029/2008GC002332; (c): Image created using GeoMapApp and shows elevation data from GMRT, the Global Multi-Resolution Topography synthesis. http://www.geomapapp.org Ryan, W.B.F., S.M. Carbotte, J.O. Coplan, S. O'Hara, A. Melkonian, R.Arko, R.A. Weissel, V. Ferrini, A. Goodwillie, F. Nitsche, J. Bonczkowski, and R. Zemsky (2009), Global Multi-Resolution Topography synthesis, Geochem. Geophys. Geosyst., 10, Q03014, doi:10.1029/2008GC002332; p. 495 (a-b): www.michaelmarten.com; (c-d): Stephen

Marshak; p. 496 (all): Stephen Marshak; p. 500: R. Ray, TOPEX/Poseidon: Revealing Hidden Tidal Energy, GSFC, NASA; p. 501 (a-c): Stephen Marshak; (d): Manfred Gottschalk/Alamy Stock Photo; p. 502 (b-c): Stephen Marshak; (bottom): Education Images/UIG via Getty Images; 503 (top): Mark Benham/Alamy Stock Photo; (bottom): NOAA; p. 506 (all): Stephen Marshak; p. 508 (b): Panther Media GmbH/Alamy Stock Photo; (c): David Wall/Alamy Stock Photo; (d): NASA; (left): Google Earth; p. 509: Wikimedia, public domain; p. 510: (top a-b): Stephen Marshak; (bottom, a): G. R. Roberts © Natural Sciences Image Library; (bottom b): Stephen Marshak; p. 511 (all but noted): Stephen Marshak; (bottom right): Emma Marshak; p. 512 (top right): Cody Duncan/Alamy Stock Photo; (b): The National Trust Photolibrary/Alamy Stock Photo; p. 513 (top): Google Earth; (a-b): Stephen Marshak; p. 514 (top a): Stephen Marshak; (top b): Reinhard Dirscherl/ullstein bild via Getty Images; (c): Steve Bloom Images/Alamy Stock Photo; (bottom both): Stephen Marshak; p. 515 (top): NASA / GSFC/LaRC/JPL, MISR Team; (a): DeAgostini/Getty Images; (b): Marcello Bertinetti/Science Source; (c): Caroline von Tuempling/Getty Images; p. 516: Google Earth; p. 517 (both): USGS; p. 518 (both): Stephen Marshak; p. 519 (both): Stephen Marshak; p. 520: (top): imageBROKER/Alamy Stock Photo; (others): Stephen Marshak; p. 521: Stephen Marshak; p. 523: Stephen Marshak.

Chapter 17

Page 524: Stephen Marshak; p. 525: (a): Sean Heavey; (b): Desintegrator/Alamy Stock Photo; (c): Zuzana Dolezalova/Alamy Stock Photo; (d): Ai Shieu/Snapwire/Alamy Stock Photo; p. 526: (a-b): Stephen Marshak; p. 526 (inset b): Nigel Cattlin/Alamy Stock Photo; (c): Jeffrey Frame; (inset c): Ted Kinsman/Science Source; p. 527: (all a): James Anderson, Arizona State University; (b): Luis Sinco/Los Angeles Times via Getty Images; (c): NASA image courtesy Jeff Schmaltz, LANCE/EOSDIS MODIS Rapid Response Team at NASA GSFC; (d): Travelart/Alamy Stock Photo; (e): Stephen Marshak; p. 530: Google Earth; p. 534: (foggy): Dan Lloyd/Alamy Stock Photo; (all but noted): Stephen Marshak; p. 535 (a): Roger Dauriac/Science Source; (b): Catherine Hoggins/Alamy Stock Photo; p. 536 (a): Mark Boulton/Alamy Stock Photo; (c): Cameron Beccario; (bottom): NASA; p. 538 (a): © 2016 UCAR, photo by Scott Ellis; (b): Robert Rauber; (all c): SSEC/University of Wisconsin-Madison; p. 539 (a): Larry Oolman; (b): NOAA; p. 541 (a): robertharding/Alamy Stock Photo; (inset): AP Photo/James MacPherson; p. 542: NASA; p. 543 (top): Suranga Weeratuna/Alamy Stock Photo; (bottom): Google Earth; p. 546 (cirrus): John Spragens, Jr./Science Source; (cirrostratus): Mark Schneider/Visuals Unlimited; (cirrocumulus): Pekka Parviainen/Science Source; (altostratus): Lesley Pardoe/Alamy Stock Photo; (altocumulus): Doug Allan/Science Source; (stratocumulus): Bildarchiv Okapia/Science Source; (stratus): Robert Rauber; (cumulonimbus): Howard Bluestein/Science Source; (cumulus): Jim Corwin/Science Source; (cumulus congestus): Dr. Jeffrey Frame; p. 547 (lenticular): Ron Sanford/Science Source; (a): William Brooks/Alamy Stock Photo; (b): Sylvia Schug/Getty Images; (c): Rob Crandall/Alamy Stock Photo; p. 554 (a): Mike Hill/Alamy Stock Photo; (c): Joanne Weston/Alamy Stock Photo; (bottom a): Kent Wood/Science Source; (bottom, b): Jack Stephens/Alamy Stock Photo; p. 555: Science Source; p. 556: Stephen Marshak; p. 559: (cumulonimbus): Howard Bluestein/Science Source.

Chapter 18

Page 560: Juergen Schonnop/Alamy Stock Photo; p. 561 (top): Pictorial Press Ltd/Alamy Stock Photo; (bottom): NASA; pp. 562-563 (4 wind maps): Cameron Beccario; p. 565 (a): Greg Vaughn/Alamy Stock Photo; (b): JeffG/Alamy Stock Photo; p. 567 (b): Sergio Mendoza Hochmann/Getty Images; (right b): blickwinkel/Alamy Stock Photo; p. 571 (desert): Frans Lemmens/Alamy Stock Photo; (grasslands): FLPA/Alamy Stock Photo; (forest): Hemis/Alamy Stock Photo; p. 572 (b): Image Courtesy GOES Project Science Office/NASA; p. 574 (both): NOAA; p. 576 (b): STR/AFP/Getty Images; (c): Sam Panthaky/AFP/Getty Images; (left): Google Earth; p. 579: (b): NOAA; (right): Google Earth; p. 580 (top): Tiffany Marie Green/Dreamstime; (bottom): Peter Carey/Alamy Stock Photo; p. 581 (a): NASA/NOAA image; (b): John G. Walter/Alamy Stock Photo; (c): NOAA; p. 583: Greg Dimijian/Science Source; p. 585: Cooperative Institute for Research in the Atmosphere (CIRA).

Chapter 19

Page 586: Jason Weingart/Barcroft Images/Barcroft Media via Getty Images; p. 587 (both): NOAA; p. 590 (top): Jeffrey Frame, University of Illinois at Urbana-Champaign; (bottom): NASA image by Marit Jentoft-Nilsen, based on data provided by the Global Hydrology and Climate Center Lightning Team; p. 591: Iowa Environmental Mesonet, Iowa State University; p. 593 (top): Wild Horizons/UIG via Getty Images; (bottom): Stephen Marshak; p. 596 (top): Andrew Fox/Alamy Stock Photo; (top inset): Judith Collins/Alamy Stock Photo; (bottom): Ryan McGinnis/Alamy Stock Photo; (bottom inset): Domenic Di Girolamo; **p. 597:** NOAA; **p. 598:** Iowa Environmental Mesonet, Iowa State University; **p. 599 (a):** NOAA; **(b):** NASA; **(c):** Mike Hollingshead/Alamy Stock Photo; **p. 601:** Mike Hollingshead/Science Source; **p. 602 (a):** NOAA; **(b):** Tribune Content Agency LLC/Alamy Stock Photo; **p. 603 (a):** Roger Hill/Science Source; **(b):** Dan Ross/Shutterstock; **p. 604:** Terra satellite, part of NASA's Earth Observing Satellite system (NASA); **p. 607 (a):** Christine Prichard/ZUMA Press, Inc./Alamy Stock Photo; **(b):** AP Photo/Gerald Herbert; **p. 608 (a):** NASA; **(b):** NOAA; **p. 609:** From "Understanding basic tornadic radar signatures" by Kathryn Prociv, 2/14/2013; **pp. 610-611** (background spread): Google Earth; **p. 610 (EF-0):** Brednan Fitterer/Zuma Press/Newscom; **(EF-1):** Gary Coronado/ZUMA Press/Newscom; **(EF-2):** Benjamin Simeneta/Dreamstime.com; **p. 611 (EF-3):** NOAA; **(EF-4):** Us Department of Labor/ZUMA Press/Newscom; **(EF-5):** National Weather Service, Birmingham, AL; **p. 614 (c):** Terry Livingstone/Alamy Stock Photo; **(d):** © Dennis MacDonald/age fotostock; **p. 614 (a):** Cameron Beccario; **p. 615** (top): Mandel Ngan/AFP/Getty Images; **(b):** Ron Austing/Science Source; **p. 617 (a):** U.S. Air National Guard photo by Staff Sgt. Daniel J. Martinez; **(b):** Erika P. Rodriguez/The New York Times/Redux; **(c):** AP Photo/Carlos Giusti; **p. 619 (b):** NOAA; **p. 621 (left):** U.S. Air Force photo/Staff Sgt. James L. Harper Jr., public domain; **(right):** REUTERS / Smiley N. Pool; **p. 622 (a):** Nilfanion & NASA; **p. 623:** Scott Olson/Getty Images.

Chapter 20

Page 626: David J Slater/Alamy Stock Photo; **p. 627:** NOAA; **p. 628:** Stephen Marshak; **p. 630 (both):** University of Washington Joint Institute for the Study of the Atmosphere and Ocean; **p. 630:** University of Washington Joint Institute for the Study of the Atmosphere and Ocean; **p. 632 (top):** Peel MC, Finlayson BL & McMahon TA (2007), Updated world map of the Köppen-Geiger climate classification, Hydrol. Earth Syst. Sci., 11, 1633-1644; **p. 633 (all):** Stephen Marshak; **p. 636 (both):** Eric Snodgrass; **636:** Eric Snodgrass; **p. 637 (overview):** Google Earth; **(stream):** Stephen Marshak; **(desert):** Shutterstock; **p. 640 (a):** Nick Krug/Lawrence Journal-World; **(top b):** Bob Sacha/Getty Images; **(bottom b):** Diomedia/ISM UCBL; **640 (right b):** NOAA; **p. 641 (a):** Philippe Desmazes/AFP/Getty Images; **(b):** Karim Agabi/Science Source; **(inset b):** NOAA; **(inset d):** NOAA; **(bottom left):** Stephen Marshak; **(bottom inset):** Provided courtesy of JRTC & Fort Polk. Photography by Bruce Martin, Natural Resources Management Branch, ENRMD; **p. 642 (top c):** akg-images; **p. 642 (bottom):** Stephen Marshak; **(a):** NASA; **(inset a):** Wikimedia, public domain; **p. 643 (Venus only):** NASA; **(right both):** Lunar and Planetary Institute; **p. 644 (all):** Stephen Marshak; **p. 646 (c):** J.D. Griggs/USGS; **p. 647:** NASA; **p. 648 (top c):** Stephen Marshak; **(bottom left):** USGS; **(bottom right):** USGS photograph by Bruce Molnia; **p. 649 (top b):** Stephen Marshak; **(bottom):** Nature Picture Library/Alamy Stock Photo; **(bottom inset):** Stephen Marshak; **p. 650 (top left):** National Geophysical Data Center/NOAA; **(top right):** Jacques Descloitres, MODIS Rapid Response Team, NASA/GSFC; **(top):** Steven Kazlowski/SuperStock/Alamy Stock Photo; **p. 651 (top a both):** The Cryosphere Today, University of Illinois, Urbana-Champaign; **(bottom both):** NASA/Goddard Space Flight Center Scientific Visualization Studio; **p. 652 (left):** Google Earth; **(bottom):** NASA; **p. 658 (c):** Joe Raedle/Getty Images; **pp. 660-661:** Google Earth.

Chapter 21

Page 664: Bryan Allen/Corbis/Getty Images; **p. 665 (top):** Douglas Peebles Photography/Alamy Stock Photo; **p. 665 (a):** Sally Stevens/EyeEm/Getty Images; **(b):** Alan Dyer/Visuals Unlimited/Getty Images; **(c):** John A Davis/Shutterstock; **p. 666:** Photononstop/Superstock; **(inset):** angelinast/Shutterstock; **p. 668:** NASA/JPL-Caltech/ESA/Harvard-Smithsonian CFA; **p. 669:** NASA/Johns Hopkins University Applied Physics Laboratory/Carnegie Institution of Washington; **p. 670:** Pekka Parviainen/Science Source; **p. 672 (left):** Pekka Parviainen/Science Source; **(right):** Dave Nunuk; **p. 673:** Erkki Makkonen/Alamy Stock Photo; **p. 674 (top):** Bob King; **p. 674 (a):** Tunc Tezel (TWAN); **p. 675:** NASA/GSFC/Arizona State University; **p. 678 (both b):** Stephen Marshak; **(d):** Primož Cigler/Shutterstock; **p. 681:** artpartner-images/Getty Images; **p. 684 (a):** xfox01/Shutterstock; **(b):** NASA; **(c):** Alan Dyer/VWPics/Science Source; **(d):** Maya Karkalicheva/Getty Images; **(e):** NASA, ESA, and M. Postman and D. Coe (Space Telescope Science Institute), and the CLASH team: **p. 687 (right):** Richard Wainscoat/Alamy Stock Photo; **(a):** Ian Dagnall/Alamy Stock Photo; **(b):** Larry Landolfi/Getty Images; **p. 688 (Spitzer):** NASA/JPL; **(James Webb):** NASA; **(Hubble):** STSI/NASA; **p. 689 (Chandra):** NASA; **(Fermi):** NASA; **pp. 688-689 (ground telescopes):** Joseph Sohm/Shutterstock; **p. 690 (a):** NASA Earth Observatory image by Robert Simmon, using Suomi NPP VIIRS data provided courtesy of Chris Elvidge (NOAA National Geophysical Data Center). Suomi NPP is the result of a partnership between NASA, NOAA, and the Department of Defense; **(b):** Iceink/Shutterstock; **p. 691 (top):** ESO; **(center left):** paranyu pithayarungsarit/Getty Images; **(center right):** ESO/R. Hook; **(bottom right):** EUROPEAN SOUTHERN OBSERVATORY/G. HUDEPOHL, ATACAMAPHOTO.COM/Science Source; **p. 692 (a):** NRAO/AUI; **(b):** Dave Finley; NRAO/AUI/NSF; **p. 693 (a):** NASA/Hubble Heritage Team (STSI/AURA); **(b):** NASA, ESA, and the Hubble Heritage Team (STScI/AURA); **695:** NASA, Apollo 8, Bill Anders, Processing: Jim Weigang.

Chapter 22

Page 696: NASA/JPL-Caltech/MSSS; **p. 697 (a):** International Astronomical Union; **(b):** NASA/SDO; **p. 698:** NASA/Johns Hopkins University Applied Physics Laboratory/ Southwest Research Institute; **p. 701 (both):** NASA; **p. 702 (a):** NASA/GSFC/Arizona State University; **(b):** NASA/Goddard/Arizona State University; **p. 703 (both):** NASA; **p. 704 (all): Mercury:** NASA/Johns Hopkins University Applied Physics Laboratory/ Carnegie Institution of Washington; **Venus:** NASA-JPL PIA00104; **Earth:** NASA-Johnson Space Center(AS17-148-22727); **Mars:** NASA, ESA, and The Hubble Heritage Team (STScI/AURA); **p. 705: (a-b):** NASA/Johns Hopkins University Applied Physics Laboratory/Carnegie Institution of Washington; **(bottom):** NASA/JHUAPL/CIW; **p. 706 (a):** Raw data: Mariner 10 - NASA/JPL. Digital processing: Mattias Malmer; **p. 707 (a):** © Don P. Mitchell; **(b):** NASA/JPL; **(c):** NASA/JPL; **(bottom center):** NASA; **(bottom right):** NASA/USGS; **(right):** Google Mars; **p. 708 (a-b):** NASA/JPL-Caltech/University of Arizona; **p. 708 (bottom):** NASA/JPL; **p. 709 (a):** NASA/JPL/Arizona State University; **(b):** ©02/2004 by Wolfgang Wieser; **p. 710 (a):** SA/DLR/FU Berlin (G.Neukum) and NASA/JPL/MSSS; **(b):** ESA/DLR/FU Berlin (G. Neukum); **(c):** NASA/JPL-Caltech/ MSSS; **p. 711:** NASA/JPL; **p. 712 (a):** NASA/JPL/University of Arizona; **(b):** NASA's Goddard Space Flight Center. NASA/JPL; **(c):** NASA; **p. 713 (a):** NASA /JPL/ University of Arizona; **(b):** NASA/JPL; **(c):** NASA/JPL/DLR; **(d):** NASA; **p. 714 (a):** NASA/JPL-Caltech/SSI; **(b):** NASA/JPL; **p. 715 (mimas):** NASA/JPL/SSI; **(titan):** NASA/JPL/ SSI and NASA/JPL/University of Arizona; **(dione):** NASA/JPL/Space Science Institute; **(enceladus):** NASA/JPL/Space Science Institute; **(jets of volatiles):** NASA/JPL/Space Science Institute; **(tiger striped):** NASA/JPL/Space Science Institute; **pp. 716-717 (all):** NASA; **p. 718 (a):** NASA/JPL-Caltech; **(b):** NASA/JPL; **p. 719 (top a-b):** NASA; **(bottom) (a):** NASA/JPL; **(b):** NASA/JPL/USGS; **p. 720 (a):** Photo Researchers, Inc/ Alamy Stock Photo; **(b):** NASA/JPL-Caltech/ UCLA/MPS/DLR/IDA; **(c):** NASA/ JHUAPL; **p. 721 (all):** NASA/JHUAPL/SwRI; **p. 722 (top a):** Frank Zullo/Science Source/Getty Image; **(bottom a):** NASA/JPL-Caltech/UMD; **(bottom b):** ESA / Rosetta / MPS for OSIRIS Team; MPS/UPD/LAM/IAA/SSO/INTA/UPM/DASP/IDA; **p. 723 (right):** Dr. Fred Espenak/Science Source; **p. 724 (a):** Alexandros Maragos/Getty Images; **(b):** AP Photo/AP Video; **(c):** Stephen Marshak; **(d):** USGS/D. Roddy; **p. 727 (top right):** Dr. Fred Espenak/Science Source; **(bottom):** NASA/JPL-Caltech/SwRI/JunoCam.

Chapter 23

Page 728: VW Pics/UIG via Getty Images; **p. 729 (a):** Deyan Georgiev Creative collection/Alamy Stock Photo; **(b):** Hubble Deep Field North. Hubblesite.org; **(bottom):** NASA; **p. 732:** NASA/SDO (AIA); **(inset):** Daniel Müller et al., Kiepenheuer Institute for Solar Physics/Swedish Vacuum Solar Telescope (2001); **p. 733 (a):** Stephen Marshak; **(b):** Space Frontiers/Hulton Archive/Getty Images; **(c):** Luc Viatour/www.Lucnix.be; **(d):** NASA/SDO; **p. 735 (a):** Courtesy of the SOHO-MDI consortium. SOHO is a project of international cooperation between ESA and NASA; **(b):** Observed with the Swedish 1-m Solar Telescope (SST) by the Institute for Solar Physics; **p. 736 (a):** NASA/ SDO; **(b):** NASA/SDO/STEREO/Duberstein; **(c):** NASA/SDO/STEREO/Duberstein; **p. 737 (b):** Babak Tafreshi/National Geographic Collection; **p. 740 (b):** NASA/JPL-Caltech/V. Gorjian(JPL); **p. 741:** ALMA (ESO/NAOJ/NRAO); **p. 742:** NASA, ESA, H. Bond (STScI), and M. Barstow (University of Leicester); **p. 744 (top both):** Jan Skowron/ Warsaw University Observatory; **p. 746 (a: before & after):** Courtesy E. Guido, N. Howes, and M. Nicolini; **(b):** tose/Shutterstock; **(c):** NASA, ESA, and G. Bacon (STScI); **(d):** NASA, ESA, J. Hester and A. Loli (Arizona State University); **p. 747 (top right):** NASA, ESA, J. Hester and A. Loli (Arizona State University); **(left):** Hallowedland/ Shutterstock; **(left inset):** EHT Collaboration/ESO; **p. 749 (a):** NASA/JPL-Caltech; **(bottom):** Stocktrek Images, Inc./Alamy Stock Photo; **p. 750 (a):** NASA, ESA, K. Kuntz (JHU), F. Bresolin (University of Hawaii), J. Trauger (Jet Propulsion Lab), J. Mould (NOAO), Y.-H. Chu (University of Illinois, Urbana), and STScI; **(b):** ©SSRO-South (R. Gilbert, D. Goldman, J. Harvey, D. Verschatse)-PROMPT (D. Reichart); **(c):** NASA / A. Aloisi (STScI/ESA) et al.; **(d):** NASA, ESA, and the Hubble Heritage Team (STScI/ AURA). Acknowledgment: J. Blakeslee (Washington State University); **(bottom):** NASA/ CXC/A.Siemiginowska(CfA)/J.Bechtold(U.Arizona); **p. 751 (a):** NASA (http://www. nasa.gov/); H. Ford (JHU), G. Illingworth (UCSC/LO), M. Clampin (STScI (http:// www.stsci.edu/)), G. Hartig (STScI (http://www.stsci.edu/)), the ACS Science Team, and ESA (http://hubble.esa.int/)); **p. 751 (b):** Debra Meloy Elmegreen (Vassar College) et al., & the Hubble Heritage Team (AURA/STScI/NASA); **p. 752:** PlanilAstro/Shutterstock; **p. 753 (a):** Ron Miller/Stock Trek Images/Getty Images; **(b):** R. Brent Tully (U. Hawaii) et al., SDvision, DP, CEA/Saclay; **(bottom):** NASA and the Hubble Heritage Team (STScI/AURA); **p. 757:** Hallowedland/Shutterstock.

INDEX

marine organisms, 476, *478*, 480–481, *480–481*
mass-extinction events, **271**, *272*, 282, 296, 299, 300
Mesozoic era, *309*
preservation potential, **261**
soil formation and, 151, *152*
taxonomy, **264**
in tidal flats, 509, 511
in wetlands, 513
See also life
anion, 87
annual probability
of earthquakes, 246
of flooding, **390**
annular eclipse, 678
anorthosite, 701, 703
Antarctic bottom water, 461
Antarctic Circumpolar Current, 467, *471*
Antarctic ice sheet, *428*, 639, *649*, 649–650, 656
Antarctica, *626*
glaciers and sea ice, *3*, 3–4, *51*, 431, *432*
mirages in, 555
ozone hole over, *542*
paleoclimate, 640
as part of Pangaea, *51–52*
sea-level change and, *648–649, 660*
temperature at, 531
anthracite, 326
Anthropocene Period, 16, 310, **311**
anthropogenic change, *16, 17*, 19, 41–42, *43*, 662
anticline trap, *320*
anticlines, **203**, 204, *204, 205*, 218
anticyclones, 608
Antler orogeny, 297
anvil clouds, 546, **595**, *596*, 599, *599*
aphanitic igneous rocks, 120
aphelion, *667, 671*
aphotic food webs, 481–482
aphotic zone, **476**, *476, 479*, 484
Apollo spacecraft, 701
Appalachian Fold Belt, *204*
Appalachian fold-thrust belts, 298–299, *299*
Appalachian Mountains, 70, 187, *210*, 216, *284, 298, 299*
apparent brightness, 684–685, **736**, *737, 740*
apparent magnitude, **737**, *737*, 740, 756
apparent polar-wander path, *75*, **75**–76, 78, 82
aquifers, **389**, 392, *393*
confined, 389
unconfined, 389
aquitards, **389**, 391, *393*
Arabian Desert, 412
Arabian Plate, *80*
Aral Sea (Asia), 424, *425*
Archaea, 264, 271
Archaeopteryx, 300
Archean Eon
about, *14, 271, 271, 280*
Earth history in, 287–289, *294–295*, 312
evolution of life, *287*, 288–289, *310*
geological considerations, **287**
global climate during, *639*

arches, 418, *420*
Arches National Park (Utah), *199, 420*
Archimedes' principle, 215, 457
arcs, 555, *556*
Arctic, 555
arctic air mass, 588, *588*
Arctic Ocean
Northwest Passage, *652*
sea-ice coverage, *432, 432*, 650, *651*
arete, *433*, **434**, *447*
argon, in air, 38
arid climate, 631, *632*, 633
aridisol, 152, *152*
aridity, 411
Arizona, *125, 272, 273, 405, 414, 419, 421, 422*
Kitt Peak, *664*
See also Grand Canyon (Arizona)
arkose, 155, *155*
Armenia earthquake (1988), *239*
Armero (Colombia), 133
arrival time (seismic wave), **228**
arroyos, 415
arsenic poisoning, 92
arsenopyrite, 92
artesian well, **396**, *397*
artificial levees, 387, *389*
artificial satellites. *See* satellites
asbestos, 92, *92*
ash, 107, **117**, *119*, 131, 133, 142
Asia
collision of India with, 69, *69, 208*, 212–213, *305, 312*
continental drift and, *51*
monsoons in, *575–576*, **575**–576
assay, 334–335
assimilation, **111**
asteroid belt, 33, **720**, *720*, 726
asteroids, **33**, *33*, 698, **720**, *720*
asthenosphere, 45, 137
plastic behavior and flow, **56**, 60, *64*, 82
rifting, 70
subduction, 62, *63*, 66
astronomers, *5*, 10, 669
astronomical unit (AU), **683**, 694
astronomy, *6*, 664–695
astronomical unit (AU), **683**, 694
celestial sphere, **669**, *669*, 694
constellations, 672–673, *673*, 692
defined, **665**
distance
measuring in space, 683–686
units of, 683
Doppler shift, 681–682
eclipses, 677–679, 694
lunar eclipses, **677**, *678*, 678–679, 694
solar eclipse, **677**–678, *679*, 694
electromagnetic radiation used in, 679–683
geocentric model, 23
history of, 670, 694
ancient interest in the heavens, *666, 666*
Copernican revolution, 666
geocentric model, **666**

heliocentric model, **666**
Kepler's laws of planetary motion, **666**–667, *667*
modern era, 668–669
Newton's laws of physics, 667–668
Ptolemy's geocentric model, 666
light-year, **27**, 46, **683**, 694
observatories, **686**, *688–689*, 691
optical astronomy, 679
parsec, 683, 694
Pillars of Creation (image), *693*, 752
radio astronomy, 679
radio interferometers, **691**, *692*
spectroscopy, 679–683
telescopes, **686**–693
design and location of, 687, *687*, 690–691, *691*
ground-based telescopes, 692
large research telescopes, 687, 694
optical telescopes, **686**–687, 690–691, 692, 694
radio telescopes, **686**, 691–692, *692*, 694
reflecting telescopes, *686*, **690**, 694
refracting telescopes, **686**–687, *687*, 694
resolving power, **690**
space observatories, 692
space telescopes, 679, **686**, 692, 693, *693*, 726
theory of expanding Universe, **28**, *28*, 29, 669, 682, 751–753, 756
units used, 683
use by Polynesian navigators, 665, *665*
zodiac, **673**, *673*, 694
See also planets; Solar System; stars; Sun; Universe
Atacama Desert, 412, *413*
Atlantic Ocean
air masses over, 588, *588*
formation of, 312
size of, 14
water masses and current, *478*
See also North Atlantic Ocean
atmosphere, *12–13*, 16, 525–526
about, *12–13*, **38**, *38*, 525–526
aerosols, 133, **527**, *527*, 528, 544, 550, 645, 662
albedo of, 449, **645**, 662
aurora, *37*, 38, **543**, *557*, 736
boundary layer, **566**, *567*, 624
carbon dioxide in (*See* atmospheric carbon dioxide)
change in, 16, *16*
characteristics of, *38*, 38–39
clouds, 16, *16*, 39, *39*, *524*, 526, **527**
composition of, 38, 42, 292, 448–449, *526*, 526–528, 529, 643–644
defined, **11**, 525
energy balance of Earth and, **631**–633, *634*
evolution over time, 528–530, *529*
first atmosphere, 528
great oxygenation event, **291**, **529**, *529*
second atmosphere, 528
third atmosphere, 528–530
formation of in early Earth, *35*, 286, 289
general circulation, **561**, *561*
great oxygenation event, **291**, **529**, *529*
greenhouse effect, 541, **551**, 552–553, *553*, 632
greenhouse gases, 553, 627–628, 632, 638, 643, 644, *654*, 662